PRINCIPLES, PRACTICE AND DESIGN OF HIGHWAY ENGINEERING

(Including Airport Pavements)

For B.E./B.Tech. & M.E./M.Tech. students of Civil Engineering and Practicing Engineers and Designers

Dr. S.K. Sharma

B.Tech. (Hons.) (Civil Engg.)
M.Sc. (Highway Engg.) (Distinction)
Ph.D. (Civil Engg.) F.I.E.

Former Professor and Head of Civil Engineering,
PEC University of Technology,
Chandigarh

S Chand And Company Limited

(ISO 9001 Certified Company)

S. CHAND
PUBLISHING

S Chand And Company Limited

(ISO 9001 Certified Company)

Head Office: D-92, Sector–2, Noida – 201301, U.P. (India), Ph. 91-120-4682700

Registered Office: A-27, 2nd Floor, Mohan Co-operative Industrial Estate, New Delhi – 110 044, Phone: 011-49731800

www.schandpublishing.com; e-mail: **info@schandpublishing.com**

Marketing Offices:

Chennai : Ph: 23632120; chennai@schandpublishing.com
Guwahati : Ph: 2738811, 2735640; guwahati@schandpublishing.com
Hyderabad : Ph: 40186018; hyderabad@schandpublishing.com
Jalandhar : Ph: 4645630; jalandhar@schandpublishing.com
Kolkata : Ph: 23357458, 23353914; kolkata@schandpublishing.com
Lucknow : Ph: 4003633; lucknow@schandpublishing.com
Mumbai : Ph: 25000297; mumbai@schandpublishing.com
Patna : Ph: 2260011; patna@schandpublishing.com

© S Chand And Company Limited, 1985

All rights reserved. No part of this publication may be reproduced or copied in any material form (including photocopying or storing it in any medium in form of graphics, electronic or mechanical means and whether or not transient or incidental to some other use of this publication) without written permission of the copyright owner. Any breach of this will entail legal action and prosecution without further notice.

Jurisdiction: *All disputes with respect to this publication shall be subject to the jurisdiction of the Courts, Tribunals and Forums of New Delhi, India only.*

First Edition 1985
Second Revised Edition 2012; Reprint 2014
Third Revised Edition 2015; Reprints 2016, 2017, 2019, 2022

Reprint 2023

ISBN: 978-81-219-0131-4 **Product Code:** H3HWY46ARCH10ENAC15R

PRINTED IN INDIA

By Vikas Publishing House Private Limited, Plot 20/4, Site-IV, Industrial Area Sahibabad, Ghaziabad – 201 010 and Published by S Chand And Company Limited, A-27, 2nd Floor, Mohan Co-operative Industrial Estate, New Delhi – 110 044

PREFACE TO THE THIRD REVISED EDITION

Highways in a country have been rightly compared to the arteries of a human being and their importance in the social and economic upliftment of India cannot be over emphasized.

There has been a tremendous growth of both passenger and goods traffic in the country in recent years. To cope with this traffic, we need a net-work of good roads and airfield pavements.

It was felt by the author that there is hardly any book worth its name on Highway Engineering in the market and therefore was encouraged to undertake the task of bringing out a comprehensive text on the subject incorporating latest information and design. Though written primarily to meet the needs of students both at the undergraduate and post graduate levels, its treatment of important recent developments and design methods for highway and airport pavements will be of interest to the practicing engineers also.

It has been the endeavour of the author to present a well organized account of all the phases of highway engineering, including airfield pavements. The various topics have been dealt with in a manner in which they most commonly occur in the development of a highway system. Not only the basic principles underlying each topic have been developed, but their applications in the design, construction and maintenance of highways have been covered. To clearly make the students understand the theory, a large number of problems have been worked out, wherever feasible.

The book has been divided into six parts. **Part I** covers *introduction, highway plans, economics, administration, finance and traffic engineering. Highway location including hill roads, project preparation, drainage and geometric design* are dealt with in **Part II**. The geometric standards laid down by the Indian Roads Congress have been mainly incorporated. **Part III** *includes highway soil classification and earthwork. Pavement types and their material requirements and construction methods* are covered in **Part IV**. The *subject of soil-stabilized roads* have been given a comprehensive treatment to encourage the use of local materials for the low-cost roads. Equipment needed for construction has been mentioned. **Part V** deals with *pavement design*. General considerations and factors involved in the design of both flexible and rigid pavements are discussed in chapter 19. Latest design methods for flexible and rigid pavements, including airfield pavements, are treated in chapters 20 and 21. **Part VI** covers *highway distress, pavement evaluation and rehabilitation (strengthening) and maintenance of roads*. The information will be of great importance to the designers and the practicing engineers.

Every effort has been made to give credit for all original contributions in the field. The latest publications of Indian Roads Congress on each topic have been referred to. The author is grateful to Dr. Umesh Sharma, Associate Professor in Civil Engineering, PEC University of Technology, Chandigarh for providing latest literature on the subject. I will be failing in my duty if I do not mention the name of my dear wife, Sudesh Sharma for her inspiration and perseverance. For all omissions of credit, where due, the author expresses his regrets.

Although every effort has been made to check minutely the proofs and the subject matter, still there may be some errors and omissions. The author shall feel obliged for the critical and constructive suggestions for the improvement of the text.

Dr. S.K. Sharma

Disclaimer : While the author of this book has made every effort to avoid any mistake or omission and has used his skill, expertise and knowledge to the best of their capacity to provide accurate and updated information. The author and the publisher do not give any representation or warranty with respect to the accuracy or completeness of the contents of this publication and are selling this publication on the condition and understanding that they shall not be made liable in any manner whatsoever. The publisher and the author expressly disclaim all and any liability/responsibility to any person, whether a purchaser or reader of this publication or not, in respect of anything and everything forming part of the contents of this publication. The publisher shall not be responsible for any errors, omissions or damages arising out of the use of the information contained in this publication.
Further, the appearance of the personal name, location, place and incidence, if any; in the illustrations used herein is purely coincidental and work of imagination. Thus the same should in no manner be termed as defamatory to any individual.

CONTENTS

DETAILED CONTENTS

PART – I

PART – III

PART – IV

PART – VI

PART – I

1. Introduction
2. Highway Planning and Administration
3. Highway Economics and Financing
4. Traffic Engineering

Chapter

INTRODUCTION

THE IMPORTANCE OF HIGHWAY TRANSPORTATION

In the modern world, speed has become the measure of economic development of a nation and the indication of speed can be judged only from the means of transport available in different parts of the country. In a village, life moves at a slow pace, but in a city, every thing appears to be dynamic. The bigger the city, the greater the rush. Aeroplanes, railway trains, automobiles, boats and steamers, which are symbols of activity and hence of speed, indicate that people are being engaged in creating wealth.

An effective system of transport is, therefore, essential for economic prosperity and industrial development. In fact, transport facilities in a country can be rightly compared to arteries in human body. Just as arteries maintain man's health by enabling circulation of blood, means of transport increase nation's wealth by keeping people and goods moving. The two important means of transport are the railways and the highways. The railways, have done much to increase the prosperity of India and their role in shaping the economic life of the country must inevitably remain dominant. But, however efficient a railway system might be, it could not provide more than the arterial means of transport. Without depreciating the values of the railways, it is apparent that they must be supplemented by roads— good roads capable of carrying motor vehicles of all sorts. Behind the railways must lie a network of roads by which the rail heads can be reached. Unless the roads are there, the railways can serve no more than a fractional portion of the territory in which they run.

1.1. ADVANTAGES OF GOOD ROADS

It is clear, therefore, that if India is to progress towards greater prosperity, the first essential is the provision of an adequate network of roads. The obvious advantages of good roads are:

1. No civilized society can survive without roads. The businessman dashing about his business in his car, the doctor rushing to his patient, the official going on a scooter to his office, the teacher cycling to his school and the workman proceeding to his place of work require roads, more roads and better roads.

2. The main function of roads is to serve the traffic. The total economic cost or value of a road system is the cost of roads themselves plus the cost of operating vehicles on them. If all vehicles using the roads in India today were able to use better roads, the annual saving in transport costs would exceed the total road expenditure.

3. Agriculture depends for its very existence, and certainly for its efficient distribution on adequate transport facilities being available. In regard to the food problem, the roads have a very vital part to play. A good and balanced network of roads will enable the villagers to transport their yield to the market speedily and easily. Today, the cultivator produces mainly such things which are unperishable and can be easily sold. Moreover, a sizeable portion of the perishable produce of India runs to waste on account of delay in reaching the market, consequent upon limitation of transport. Improved roads which mean faster and larger transport facilities could convert this waste into wealth. In addition, the advent of better roads, would encourage more production of perishable commodities like fruits, vegetables, dairy products which fetch good profits to the villagers. Increased profit can be well utilized towards the purchase of better seeds, improved implements and manure to effect improvements in the land.

4. Roads have a great cultural value also. With the development of roads, mobile libraries, newspapers and magazines will reach every door step of rural inhabitant, thus making him an active and alert member of the community to learn about the other parts of the country, and strengthening the national unity.
5. Roads enhance land value greatly. Fuel or service stations, hotels and such high revenue buildings are necessary adjuncts to any road system, and these pay well for profitable sites. Industry and commerce gather around good transport facilities. Land along a good road is a very profitable investment.
6. Roads have a most important role to play in the defence of India. We have a border extending to four thousand and five hundred kilometres and it is practically impossible to station troops all along this border throughout the year. Modern army move on wheels and are greatly dependent upon the highways of the country. These require a network of roads to move on and to concentrate the army on the threatening point of the border.
7. Roads are the symbol of a country's progress. At this juncture, India is busy with the task of moral and material reconstruction. Plans are underway to harness rivers, modernize and expand industry and so on. In any comprehensive scheme of development, roads must have a first priority.
8. Road development and road transport have the highest employment potential of all our economic activities. Even today, more than one crore men and women find employment in road construction and maintenance activity, and the road transport operation.

HISTORY OF HIGHWAY DEVELOPMENT

1.2. EARLY ROADS

In tracing the early history of roads, we have to refer back to the origin of civilization and culture; for the desire of movement has been the fundamental feature of any form of civilization. The history of transport and communication, it is aptly remarked, is the history of civilization. The pioneers of society are distinguished from their contemporaries by the attention which they paid to the building of good roads. The road makers are the torch bearers of enlightened progress.

The early concept of a road was a path or track on which a foot-passenger could travel. Trees were blazed or marked to show the direction. Men travelled by compass or by stars or by watching their own shadows or by noting the direction of the wind. Successive travellers, following the same route, tred down a forest path which was the first step towards road making. On such a road, rivers were crossed by swimming or by rafts. Felled trees were used for crossing narrow streams, while ranges of hills were passed by following the beds of mountain torrents.

As social intercourse increased, various animals were used as beasts of burden. Pack horses were employed in England down to a very late period; camels, horse mules, asses and bullocks, were used all over the East, while even sheep were used to carry tea and salt over the Tibet passes. Employment of any of these animals necessitated the improvement of tracks. So the foot-paths were widened, jungles were cleared, crude bridges of logs of wood were constructed.

As animal power is more economically employed in drawing than in carriage, carts were soon made for conveying goods and passengers. Hence attention was required towards gradients. The roads must be raised clear of inundation from rain or river. Permanent bridges were provided over the water courses, and finally, the surface metalled to diminish friction, loss of power of animals and wear and tear of vehicles.

Road have been put to use from a very early times. The first hard surfaces were constructed in Mesopotamia soon after the discovery of the wheel about 3500 B.C. A stone surface road was traced as early as 1500 B.C. on the island of Crete in the Mediterranean sea. Reference has been made in the Bible about a road constructed between Babylon and Egypt after 539 B.C. With the fall of the Roman Empire, the art of road building was lost for a considerable time. It was again in the 18th century that a good system of roads was developed in France. In England, the development of roads followed soon after that in France. Mc Adam (1756–1836) used crushed aggregate as a road surfacing material in a manner which is still widely applied with some modification. The road development in the U.S.A. was also at a very slow rate till the end of the 19th century. A few roads were constructed by companies organized to gain profit through toll collection.

The first two decades of the 20th century saw the improvement of the motor vehicle from a "rich man's toy" to fairly dependable method for transporting persons and goods in the U.S.A. It was during this period that road improvement was recognized as a matter of Federal and State concern rather than of purely local interest. Organizations were established and small amounts of money appropriated by the congress and the state governments to deal with road problems.

1.3. EARLY INDIAN ROADS

The history of Indian roads can be traced back to the remote past. Mention has been made in the Rig Veda (Part I, para 5) about roads as a means of communication. The excavations of cities like Mohenjo-daro and Harappa, which existed between 3500 B.C. and 2500 B.C., have revealed the existence of broad streets in India. Coming to the Aryan period, the ancient scriptures refer to the existence of a highway; a road built by King Bimbisara in 6th century B.C. which has survived the ravages of times to this day in Rajgir in Patna. Written in the 6th century B.C. *Artha Sasthra*, the well-known treatise on administration, gives a good deal of information on the subject of roads in the Aryan period. Even the width of the roads for different purposes and traffic have been laid down in this treatise. It is noteworthy to mention that Chandra Gupta Maurya had a department of communications to look after the public roads. Pillars, which served as sign posts and milestones, were erected at regular intervals. A grand trunk road was built which connected north western frontier to the capital Pataliputra (Patna). Emperor Ashoka also paid special attention to roads. On the roadsides, banyan trees were planted, mango gardens cultivated at each *kos*, wells dug and rest houses built for the comfort of the travellers. During Sher Shah's time, road maintenance was continued. Road development received good attention of the Mughal rulers also. Twenty-four roads formed a network of those times, thirteen of which have been traced. Some of the main highways of the Mughal period received great admiration from foreigners who visited India in those days. The road system of India in those days could be claimed to be one of the best systems in the world.

Then followed a century of retarded programme during the British rule up to 1919. Railways came on the Indian scene and the roads gradually lost the interest of the government. Only roads of military importance were of some concern to the central government and the maintenance of the remaining roads was shoved on the provincial governments, which in turn placed the greater road lengths in charge of local bodies.

1.4. MODERN HIGHWAY DEVELOPMENT IN INDIA

After the World War I, cheap supply of surplus military vehicles was made available to Indian public which created revolution in India's transport system. Under the continued effect of high speed motor transport, cart roads were wearing and the local bodies could not cope with the maintenance work. In the year 1927, Jayakar Committee was set up to investigate the causes of bad roads. The main finding of this committee was that the road development was passing beyond the capacity of local bodies and the provincial governments and it was becoming a matter of national importance. The Committee suggested that further development of the system was desirable for the general welfare of the country as a whole; and in particular:

(*a*) for the better marketing of agricultural products

(*b*) for the social and political progress of the rural population; and

(*c*) as a complement to railway development.

It is surprising that even after more than 80 years, the above objectives still seem extremely relavent.

On the recommendations of the Jayakar committee the central government began to realize the importance of roads and took steps to promote the creation of a semi-government body known as Indian Roads Congress. Thus Indian Roads Congress came into existence in the year 1934. The main objective of this body was to pool up information and knowledge with regard to the roads and circulate the same among the members.

The World War II broke out in the year 1939 and at that time, the provinces were solely responsible for our roads. Due to the depression of pre-war period the provincial governments lacked funds and could not meet the requirements of the road development schemes. The Japanese offensive proved that the road system was very unsatisfactory. Thus the entire outlook underwent a change. It was realized that the country must have an effective system of roads and this system could be kept only if the centre took over the development and maintenance of important roads. Thus a conference of the chief engineers was called in 1943 at Nagpur which made very useful recommendations. This is commonly known as the Nagpur plan. The construction of important roads needed from the strategic point of view for the defence of India, was made the responsibility of the central government.

The Nagpur plan was the starting point of road development in India on a national scale. Previous efforts were mostly local and dispersed and had no overall perspective.

With the initiation of five year plans in 1951, plans were prepared for road development. Table 1.1 gives an idea about the trend in the increase of road kilometreage during the various periods. In 1958, the chief engineers were again given the assignment of preparing a 20 year road development plan for the period 1961–1981. This is known as the Chief Engineers Plan or Bombay Plan. This plan aimed at raising the density of a road kilometreage from 16 kilometres to 32 kilometres per 100 square kilometres during the plan period (1961–81) with an estimated cost of ₹ 5200 crores.

During the early stages of planned road development, the country did not have a single trunk road which allowed through uninterrupted traffic. The cities of Mumbai, Kolkata, Delhi and Chennai were not linked by through routes. A large number of missing links and missing bridges remained as plague spots. Therefore, the first task of the planners was to forge a continuous link on arterial routes. The country has now a dependable through trunk road system connecting the main cities. The length of the trunk routes which are known as National Highways has also increased considerably.

The vast land of India is criss-crossed by innumerable rivers, big and small, and a major effort was directed to the construction of bridges allowing uninterrupted traffic in all types of weather. Over a span of 25 years from 1951 to 1976, a total number of 266 major bridges were constructed on National Highways alone.

Apart from increasing the length of road network, many other improvements were carried out to meet the needs of increasing traffic. The pavements were improved by adding extra crust. Single lane pavements were widened and strengthened to cope with the increasing traffic and to make traffic movement quick and safe. Thus apart from quantitative increase in kilometreage, there has been a marked qualitative improvement in the highways all over the country.

India has one of the largest road networks in the world. In spite of so much increase in the road kilometreage, our system of road communication has been all along too meagre and defective to meet

Table 1.1. Road length in India during the years 1951–2021

Road category	*Road length in kilometres as on 31st March*				
	1951 (Beginning of First Five Years Plan)	*1981 (End of Bombay Plan)*	*2001 (End of Lucknow Plan)*	*2011 (End of 11th Five Years Plan)*	*2021 (Vision Plan)*
1. National Highways	19,811	31737	57700	76818	80,000
2. State Highways	42,809	95491	124300	166129	160,000
3. Major District Roads	89,802	1,53,000		3,33,188	320,000
4. Other District Roads	81,272	912684 (4 and 5)	2994000 (3 to 6)	3264765 (4 and 5)	No target set
5. Village Roads	1,66,249				
6. Unclassified, Urban, Project Roads	—	3,09,785			
Total	3,99,943	1502697	3176000	38,40,900	
1. Surfaced	1,57,019	6,58,119			
2. Unsurfaced	2,42,924	8,43,578			

even our reasonable requirements. India has approximately 1.168 kilometre of road per square kilometre. The corresponding values for the United Kingdom and the U.S.A. are comparatively more than we have in India. Thus more attention should be paid to our roads by the government. It has been recognized that the rate of growth of motor transport in India is very slow. The reasons for this are:

1. Absence of sufficient length of good roads.
2. Low standard of living.
3. Expensiveness of motor vehicles and their high cost of operation.
4. Inadequate repair facilities.
5. The predominantly agricultural economy of the country.
6. Preference to railways than roads.
7. Lack of proper organization of the motor transport industry.
8. Policies pursued by government including the high level of taxation, mode of control and fear of nationalization.

The present condition of road transport facilities in India is not very satisfactory. In terms of capacity, out of a total of 38,40,900 km of roads, National Highways, State Highways and Major District Roads from only 15 per cent of roads. Major portion—about three-fourth of these roads is still single lane. Our commercial vehicles are able to cover only 300–400 km a day compared to 600–700 km in developed countries. Further high cost of operation, excessive fuel consumption, slow speeds, congestions and high accident rates characterize the present state of affairs in the Road Transport sector. Moreover the overall connectivity of villages is still poor. More than 40 per cent of villages are yet to be provided with all-weather roads.

Our country is large and the population exceedingly high. The communications in the cities could perhaps be considered tolerable but if one moves away even a few kilometres from the main urban areas, one would feel that one had gone to a different land altogether. Except for a few important roads, which also are interspersed with gaps, even the metalled roads of local boards are in poor condition. And then, more than half of our roads are *kutcha,* which means undulating dusty pathways passing through narrow lanes in the villages and an arduous journey throughout. Truely speaking, the condition of roads in the rural areas is beyond description. And then, we have large hilly tracts where roads can be the only means of transport. The cost of construction in such tracts is relatively much higher, but the fact remains that we must do a lot more for opening up these areas by building interconnected roads.

We are dedicating ourselves to the formidable and challenging task of raising the standard of living in the country by increasing production in field and factory and ensuring a more equitable distribution of the national income and a fair deal for the common man. We are engaged at the present time not only to make ourselves self-sufficient in food and industry but also to export to the developing countries. In case this great task of national uplift is to be fulfilled and further if the Socialistic Pattern of Society, which is the declared objective, is to have any real meaning, the villagers who comprise 70 % population—must be given an equal chance to share the new ways of life and rising levels of prosperity. How can all the social services that are now planned or are already in operation, and all our schemes for imparting free education to every citizen, hospitals and regulated markets with their godowns for storing grains and other primary produce ever operate successfully? How can all our much publicized plans for rural uplift ever come to anything tangible and permanent without an adequate and well planned network of roads? A bold policy on road development is of immediate necessity whereby more investment on roads could reduce the cost of transport and in consequence, prices of transported commodities, conserve fuel and improve the quality of life.

The following suggestions are made for further improvements in the system:

1. Regulator for Roads. There is no independent regulatory authority for India's Roads and Highways sector. Current arrangement both at Centre and States (MORTH, NHAI, MPRDC, PWDs and so on) results in a potential conflict as the rule making body is also the implementing body and there is no independent assessment of its performance across various parameters. There is, therefore, a need for a regulator whose key functions should include tariff setting, regulation of service quality, assessment of concessionaire claims, collection and dissemination of sector information, service-level benchmarks and monitoring compliance of concession agreements.

2. Many policy and implementation deficiencies have to be redressed. These include: thin spreading of resources; delay in pre-construction activities due to delay in land acquisition; delay ii environmental clearance and shifting/removal of utilities; weak management by contractors due to improper deployment of human resources and equipment; and poor implementation capacities of the state Public Works Departments (PWDs).

1.5. HIGHWAY CLASSIFICATION

Roads may be broadly classified into two main categories, namely:

1. Non-Urban Roads; and
2. Urban Roads.

1.5.1. Non-Urban Roads

There are six categories of non-urban roads in the country:

1. ***Expressways*:** These cater to motor traffic moving at high speeds. No slow moving traffic, parking, loading and unloading of goods and passenger traffic is permitted on these roads.
2. ***National Highways:*** Connecting State Capitals, Ports, large industrial and tourist centers and neighbouring countries.
3. ***State Highway:*** Connecting district headquarters and important cities within the state.
4. ***Major District Roads:*** are roads within the district serving markets, industrial towns and connecting with the main highways.
5. ***Other District Roads:*** These serve rural areas of production, tehsil and block development headquarters, etc.
6. ***Village Roads:*** Connect villages and to the nearest other roads.

1.5.2. Urban Roads

There are five categories of urban roads; as shown in Fig. 1.1.

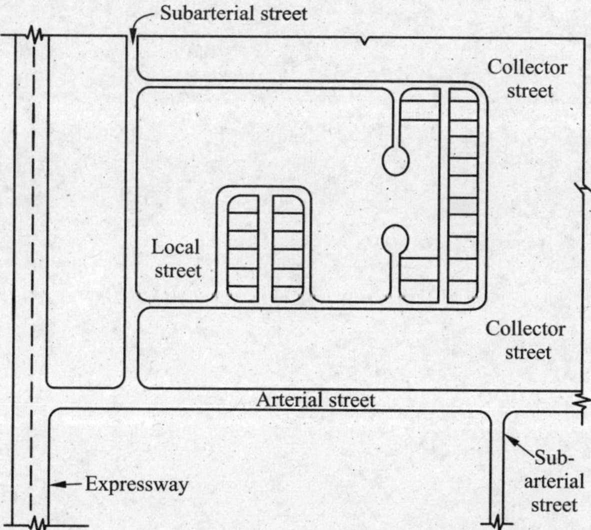

Fig. 1.1. Categories of urban roads.

1. **Expressway.** These serve the same function as non-urban expressways.
2. **Arterial Streets:** This system of streets, along with expressways where they exist, serves as the principal network for through traffic flows. Significant intra-urban travel, such as, between central business district and outlying residential areas or between major sub-urban centres takes place on this system. These streets may generally be spaced at less than 1.5 km in highly developed central business areas and at 8 km or more in sparsely developed urban fringes. The arterial streets are generally divided highways with full or partial access. Parking, loading and unloading activities are usually restricted and regulated. Pedestrians are allowed to cross only at intersections.
3. **Sub-arterial Streets:** These are functionally similar to arterial streets but with somewhat lower level of travel mobility. Their spacing may vary from about 0.5 km in the central business district to 3-5 km in the sub-urban fringes.
4. **Collector Streets:** The function of collector streets is to collect traffic from local streets and feed it to the arterial and sub-arterial streets or vice versa. These may be located in residential neighbourhoods, business areas and industrial areas. Normally, full access is allowed on these streets from abutting properties. There are few parking restrictions except during the peak hours.
5. **Local Streets:** These are intended primary to provide access to abutting property and normally do not carry large volumes of traffic. Majority of trips in urban areas originate from or terminate on these streets. Local streets may be residential, commercial or industrial, depending on the predominant use of the adjoining land. They allow unrestricted parking and pedestrian movements.

❏❏❏

Chapter 2 HIGHWAY PLANNING AND ADMINISTRATION

HIGHWAY PLANNING

2.1. HIGHWAY PLANNING

Before undertaking any development programme, planning is of paramount importance since it is the basic requirement for any new project or the expansion of an existing facility. Planning assumes greater significance when the funds available are limited and the requirement is much higher, particularly in all developing countries like India. The aim should be to utilise the available resources in the best possible way and in a systematic manner.

Highway planning is equally important for its development. The objectives of highway planning can be summarized as follows:

(*i*) To establish an integrated highway network capable of accommodating all highway travel in an orderly, safe, efficient and economical manner. The planner should account for not only the initial cost of construction, but maintenance and renewal costs and the vehicle operation costs should also be taken into consideration.

(*ii*) To forecast the future requirements of road kilometreage needed in different parts of the country.

(*iii*) To fix up priorities and schedule for construction of roads and bridges in accordance with the available resources and the utility of the project or in other words phasing of the road development programme.

(*iv*) To work out financing system.

2.1.1. Need for highway planning

The need for highway planning is the result of four separate demands; namely

(*i*) *Last ground demand i.e.,* the backlog of construction which has been postponed during the past twenty years. The backlog may be due to lack of resources, materials, manpower and equipment. In the Five year plan emphasis has to be laid to complete the missing links.

(*ii*) *Replacements.* Old facilities are constantly wearing out and need replacements. These must be replaced according to a regular schedule.

(*iii*) *Expansion.* This is the construction necessary to accommodate an increasing population and an increasing use of highway travel.

(*iv*) *Increasing quality of service.* The motorists demand higher standards of construction and maintenance so that they can travel faster, farther, easier and more comfortably.

The urgency of these demands is so great that a careful planning is a necessity.

2.1.2. Advantages of highway planning

The obvious advantages of highway planning are:

1. It co-ordinates the activities of various agencies connected with highway construction. In addition to the central government, the states and the municipalities are responsible for the construction of roads. Unless co-ordination is done, construction plans may come in coflict with each other.

2. Highway planning eliminates waste, duplication and extravagance in the use of public money. With advanced planning, purchase of sites for bridges and rights of way can be done at an economical rate, before the property is developed and becomes very costly.
3. Lower interest rates are possible, if funds are collected by an infrastructure bond issue.
4. Careful planning can minimize costly mistakes in design and construction.
5. Planning also provides a basis for proper allocation of funds to the states and permits a wiser programme of construction.
6. The quantity of work to be done can be estimated.
7. Projects can be initiated with fore-thought and careful investigation.
8. Future demands of highway needs can be anticipated.

2.2. SCOPE OF HIGHWAY PLANNING

Good highway planning should be long range, comprehensive and co-ordinated.

2.2.1. Long-Range Planning

This means to fore-cast the needs as for ahead as the life of the facilities which are to be built initially. Long-term plans for highways in India have been made for 20 years. Long term planning can produce several tangible results such as (*i*) effective use of physical resources, (*ii*) gaining support for the plan, (*iii*) proper development of the area, and (*iv*) obtaining sufficient finances for the plan.

2.2.2. Comprehensive Planning

A good plan should have breadth in addition to being long-range. Transportation plan should be only a part of an overall plan—the master plan of the area. The master plan considers all the needs and desires of the area and determines the facilities and services which are necessary. It determines the best plan for future growth and places the highway plan in its proper perspective. Highway plans may be a part of a national plan, a state plan, or metropolitan plan in an urban area.

2.2.3. Co-ordinated Planning

Co-ordinated planning means that the plans of overlapping government agencies such as centre and state, the state and city are considered and steps taken to arrive at agreements which are mutually satisfactory. The development of the plan requires the co-operation of the centre and the states to ensure that the various facilities in the plan are constructed efficiently and for the benefit of the entire nation.

Long term, comprehensive and co-ordinated planning will give a new highway the benefits of both planning and engineering.

HIGHWAY PLANS

2.3. THE NAGPUR PLAN

A conference of chief engineers was called by the government of India at Nagpur in December 1943 to consider the planning of road development. The outcome of the conference is commonly known as the Nagpur plan.

2.3.1. Classification of Roads

According to the Nagpur plan, the network of roads in India was divided into four categories, a brief description of each of them follows:

(i) ***National highways*** were defined as the main highways running through the length and breadth of India, connecting ports, foreign highways, capitals of provinces and including roads needed for strategic movements for the defence of India. For underdeveloped areas, at first, it was proposed to construct National traits so that later on, they could be converted into roads to suit the traffic requirements. It was also agreed by the delegates that these National highways should, form the framework of roads on which should be based the entire road communication of India. They should be selected in such a manner that they afford uninterrupted road communication throughout the country. Thus it was not necessary that they should all be of the same type and dimensions.

(ii) ***Provincial (or State) highways*** were defined as all other main trunk or arterial roads of a province, connecting up with the National highways, or highways of adjacent provinces, district headquarters, and important cities within the province. They should serve as the main arteries of traffic to and from District roads. The commerce by road within a province should be taken by such roads. It might be possible at certain places that they carry heavier traffic than even the National highways but this should not be taken as a basis for changing their designation or function.

(iii) ***District roads*** were defined as roads traversing each district, serving areas of production and markets and connecting these with each other or with highways and railways. These roads should be capable of taking traffic into the heart of rural areas throughout the year without much interruption. It was further decided that every village in a highly populated area should be within about 3.2 kilometres of such a road and within 8 kilometres or so in other areas. These roads should be main branches from the National or State highways and should be capable of taking traffic into the interior of each district or similar unit of the area.

District roads were further divided into two classes keeping in view the expected traffic and the type of finish provided. This sub-division is as follows:

(a) ***Major district roads*** to be roughly of the same type as the Provincial highways.

(b) ***Other district roads*** to be of a somewhat lower specifications than the above type. Many of these roads remained closed to traffic for long periods, particularly during the monsoons. It was aimed at in the Nagpur plan, that these roads should be constructed and drained properly so that they could sustain traffic throughout the year.

(iv) ***Village roads*** were defined as roads connecting villages and group of villages with each other and to the nearest district road, main highway, railway or river ghat. These roads should in essence, form tracks to carry village produce to the market. They should not be neglected, but must be planned well so as to form a regulated system.

2.3.2. Road Kilometreage

The plan was based upon the estimated requirements of the country for the first twenty years of the post-war-period. The crux of the plan was to have a balanced development of all classes of roads so that every village in a well developed agricultural area was brought within reasonable reach of a main road.

A network of main roads was provided to connect towns and large villages. Other district and village roads were to be developed on a corresponding scale. The plan included both the improvement of the existing roads and the construction of new roads. The road kilometreage for the National and Provincial highways and Major district roads was calculated in every province by the following formula:

$$\left.\begin{array}{l}\text{Kilometreage of National, Provincial}\\ \text{Highways and Major district roads}\end{array}\right] = \frac{A}{8} + \frac{B}{32} + 1.6\,N + 8T + D - R$$

where A = agricultural area of the province concerned in km^2

B = non-agricultural area in km^2

N = number of towns and villages having a population of 2001–5000.

T = number of towns and villages having a population over 5000
D = an allowance for agricultural and industrial development during the next 20 years; and
R = length of railway track (kilometres) in the area under consideration.

On the basis of this formula, it was estimated that further 24,000 kilometres of these roads are required in India at the completion of Nagpur plan; against 126,400 kilometres existing at that time.

The formula for the calculation of road kilometreage for other district and village roads is given by:

$$\left.\begin{array}{l}\text{Kilometreage of other district}\\ \text{and village roads}\end{array}\right\} = 0.32V + 0.8Q + 1.6R + 3.2S + D$$

where V = number of villages with population of 500 or less
Q = number of villages with population of 501–1000
R = number of villages with population of 1001–2000
S = number of villages with population of 2001–5000
D = an allowance for agricultural and industrial development during the next 20 years.

It was found out that a total of 404,800 kilometres of such roads were deemed essential against 260,800 kilometres existing in the year 1943.

Problem 2.1 *Estimate the additional road kilometreage required for*

(*i*) *National Highways (N.H.), State Highways (S.H.) and Major District Roads (M.D.R) and*
(*ii*) *Other District Roads (O.D.R.) and Village Roads (V.R.) from the following data, as per Nagpur plan.*

(*a*) *Agricultural area* = *5,700 sq km*
(*b*) *Non-agricultural area* = *8,400 sq km*
(*c*) *Railway kilometreage in the area* = *340 km*
(*d*) *Assume the rate of agricultural and industrial development in the area to be 20 per cent*
(*e*)

Population	*< 500*	*501–1000*	*1001–2000*	*2001–5000*	*> 5000*
Number of villages and towns	*640*	*370*	*120*	*50*	*12*

(*f*) *Existing length of N.H., S.H. and M.D.R.* = *560 km*
(*g*) *Existing length of O.D.R. and V.R.* = *732 km*

Solution:

Kilometreage of N.H., S.H. and M.D.R. required as per Nagpur plan is given by:

$$\text{Road kilometreage} = \frac{A}{8} + \frac{B}{32} + 1.6N + 8T + D - R$$

Here

$$A = 5{,}700 \text{ sq km},\ B = 8{,}400 \text{ sq km},\ N = 50,\ T = 12,\ D = 20\ \%,\ R = 340 \text{ km}$$

N.H., S.H. and M.D.R. kilometreage as per Nagpur plan

$$= \left\{\frac{5700}{8} + \frac{8400}{32} + 1.6 \times 50 + 8 \times 12\right\} + 20\% \text{ increase} - 340$$

$$= \{712.5 + 262.5 + 80 + 96\} + 20\% \text{ increase} - 340$$

$$= 1151 + \frac{20 \times 1151}{100} - 340$$

$$= 1381.2 - 340 = 1041.2 \text{ km}$$

Existing length of N.H., S.H. and M.D.R. = 560 km

$\therefore$ Additional road kilometreage required = 1041.2 − 560 = 481.2 km

Kilometreage of Other District Roads and Village Roads is given by

$$\overset{K}{O}.D.R. + V.R. = 0.32V + 0.8Q + 1.6R + 3.2S + D$$

Here $V = 640$, $Q = 370$, $R = 120$, $S = 50$, $D = 20\%$

$$\overset{K}{O}.D.R. \text{ and } V.R. = \{0.32\times640 + 0.8\times370 + 1.6\times120 + 3.2\times50\} + 20\%$$

$$= (204.8 + 296.0 + 192.0 + 160.0) + 20\%$$

$$= 852.8 + 20\%$$

$$= \frac{852.8\times120}{100} = 1023.36 \text{ km}$$

Existing length of O.D.R. and V.R. = 732 km

Additional road kilometreage required = 1023.36 − 732.00 = 291.36 km

2.4. BOMBAY PLAN

By the end of the Second-Five-Year plan, *i.e.*, 1961, the targets of the Nagpur plan had been more or less achieved with the exception of the National and State highways. Due to the great political, economic and social changes that had taken place since the formulation of the Nagpur plan, it was considered necessary to review the road kilometreage position in India again. This subject was taken up by the Chief Engineers of the State Public Works Departments in May 1957, at Shillong and they appointed a committee comprising the Consulting Engineer (Road Development) and the Chief Engineers of the five states representing each of the five zones in the country to draw a long term plan for the 20-year period commencing with the Third Five-Year plan. The plan prepared by this committee was approved by the general meeting of the Chief Engineers held in January 1959 and is commonly known as Bombay plan (1961-1981).

The plan takes into account the special requirement of developed and under-developed areas. The broad order of priorities envisaged in the 20-year Bombay plan were that:

1. On all arterial routes, missing bridges should be provided and the road surfaces improved at least to one-lane black topped specification.

2. the main roads in the vicinity of large towns should be widened to two lanes or more.

3. major arterial routes should have at least two-lane carriageway; and

4. rural roads should be improved to fair-weather roads.

The objective was to bring every village:

(*i*) in an agricultural area within about 6.4 kilometres of a metalled road and 2.4 kilometres of any road;

(*ii*) in a semi-agricultural area within about 12.8 kilometres of a metalled road and 8 kilometres of any road;

(*iii*) in a non-agricultural area within 19.2 kilometres of a metalled road and 8 kilometres of any road.

This will give a network of 52 kilometres per 100 sq. kilometres.

Table 2.1 compares the targets of the Nagpur plan, the road kilometreage by 1960–61, and the road kilometreage in the 20-year Bombay plan.

Table 2.1 *Road kilometreage under different plans*

	Road kilometreage under Nagpur plan	*Road kilometreage 1960–61*	*Road kilometreage under 1961–81 Bombay plan*
MAIN ROADS			
National Highways	32,000	22,080	31,737
State Highways	84,800	56,000	95,491
Major District Roads	80,000	152,320	1,53,000
Total	196,800	230,400	280,228
OTHER ROADS			
Other District Roads	112,000	125,280	912,684
Village Roads	220,800	250,720	
Total	332,800	376,000	
Grand Total	532,960	606,400	11,92,912

The road kilometreage as envisaged for the above plan, have been worked out from the following modified 'grid and star' formulae:

1. National Highways (km)

$$= \frac{A}{64} + \frac{B}{30} + \frac{C}{96} + (32K + 8M) + D$$

2. Natinoal Highways and State Highways (km)

$$= \frac{A}{20} + \frac{B}{24} + \frac{C}{32} + (48K + 24M + 11.2N + 1.6P) + D$$

3. Natioanl Highways, State Highways and Major District Roads (km)

$$= \frac{A}{8} + \frac{B}{16} + \frac{C}{24} + (48K + 24M + 11.2N + 9.6P + 6.4Q + 2.4R) + D$$

4. National Highways, State Highways, Major District Roads and Other District Roads (km)

$$= \frac{3A}{16} + \frac{3B}{32} + \frac{C}{16} + (48K + 24M + 11.2N + 9.6P + 12.8Q + 4R + .8S + 32T)$$

5. National Highways, State HIghways, Major District Roads, Minor District Roads and Village Roads (km)

$$= \frac{A}{4} + \frac{B}{8} + \frac{C}{12} + (48K + 24M + 11.2N + 9.6P + 12.8Q + 5.9R + 1.6S + 0.64T + 0.2V) + D$$

where A = developed and agricultural area in square km
B = semi-developed area in square km
C = under-developed and uncultivated area in square km
K = number of towns with population over 1,00,000
M = number of towns with population between 1,00,000 and 50,000
N = number of towns with population between 50,000 and 20,000
O = number of towns with population between 20,000 and 10,000
Q = number of towns with population between 10,000 and 5,000
R = number of towns with population between 5,000 and 2,000
S = number of towns with population between 2,000 and 1,000
T = number of towns with population between 1,000 and 500
V = number of towns with population below 500; and
D = allowance (assumed as 5 per cent) for future development and other unforeseen factors.

Problem 2.2 *Estimate the road kilometreage for National highways, State highways, Major District Roads, Other District Roads and Village Roads in accordance with the Bombay plan (1961–81) from the following data:*

(*i*) *Total area = 34,000 sq km*
(*ii*) *Developed and agricultural area = 11,000 sq km*
(*iii*) *Semi-developed area = 6,000 sq km*
(*iv*) *Number of towns and villages for different populations are as given below:*

Population range	*Number of towns and villages*
500	450
500–1,000	580
1,000–2,000	800
2,000–5,000	350
5,000–10,000	90
10,000–20,000	38
20,000–50,000	15
50,000–1,00,000	7
Over 1,00,000	4

Solution:

Undeveloped area, G = Total area − (developed and agricultural area, A + semi-developed area, B)

$$= 24000 - (11000 + 6000)$$

$$= 7000 \text{ sq km}$$

(*i*) Kilometreage (National Highways)

$$\overset{K}{NH} = \frac{A}{64} + \frac{B}{80} + \frac{C}{96} + (32K + 8M) + D$$

Here $A = 11{,}000\ \text{km}^2;\ B = 6{,}000\ \text{km}^2;\ C = 7{,}000\ \text{km}^2$

$K = 4;\ M = 7;\ D = 5\%$

$$\overset{K}{NH} = \left\{\frac{11000}{64} + \frac{6000}{80} + \frac{7000}{96} + (32 \times 4 + 8 + 7)\right\} + 5\%$$

$$= \{171.88 + 75.00 + 72.92 + (128 + 56)\} + 5\%$$

$$= 503.8 \times \frac{105}{100}$$

$$= 528.99 \text{ km}$$

(*ii*) $$\overset{K}{NH} + SH = \frac{A}{20} + \frac{B}{24} + \frac{C}{32} + (48K + 24M + 11.2N + 1.6P) + D$$

Here $A = 11{,}000\ \text{km}^2;\ B = 6{,}000\ \text{km}^2;\ C = 7{,}000\ \text{km}^2$

$K = 4,\ M = 7,\ N = 15,\ P = 38,\ D = 5\%$

$$\overset{K}{NH+SH}=\left\{\frac{11000}{20}+\frac{6000}{24}+\frac{7000}{32}+(48\times4+24\times7+11.2\times15+1.6\times38)\right\}\times\frac{105}{100}$$

$$=\{550+250+218.75+(192+168+168+60.8)\}\times1.05$$

$$=1607.55\times1.05$$

$$=1687.93\text{ km}$$

∴ Kilometreage (State highways)

$$=1687.93-528.99$$

$$=1158.94\text{ km}$$

(*iii*) $$\overset{K}{NH+SH+MDR}=\frac{A}{8}+\frac{B}{16}+\frac{C}{24}+(48K+24M+11.2N+9.6P+6.4Q+2.4R)+D$$

Here $Q=90,\ R=250$

$$\overset{K}{NH+SH+MDR}=\left\{\frac{11000}{8}+\frac{6000}{16}+\frac{7000}{24}+(48\times4+24\times7+11.2\times15+9.6\times38+6.4\times90+2.4\times350)\right\}+5\%$$

$$=(1375+375+291.67+192+192+168+364.8+576+840)\,1.05$$

$$=4374.47\times1.05$$

$$=4593.19\text{ km}$$

∴ $$\overset{K}{MDR}=4593.19-1687.93$$

$$=2905.26\text{ km}$$

(*iv*) $$\overset{K}{NH+SH+MDR+ODR}=\left\{\frac{3A}{16}+\frac{3B}{32}+\frac{C}{16}+(48K+24M+11.2N+9.6P+12.8Q+4R+0.8S+0.32T)\right\}+D$$

Here $S=800,\quad T=580$

$$\overset{K}{NH+SH+MDR+ODR}=\left\{\frac{3\times11000}{16}+\frac{3\times6000}{32}+\frac{7000}{16}(+48\times4+24\times7+11.2\times15+9.6\times38+12.8\times90+4\times350+0.8\times800+0.32\times580)\right\}\frac{105}{100}$$

$$=(2062.5+562.5+437.5+192+168+168+364.8+1152+1400+640+185.6)\times1.05$$

$$=7332.9\times1.05$$

$$=7699.55\text{ km}$$

∴ $$\overset{K}{ODR}=7699.55-4593.19$$

$$=3106.36\text{ km}$$

(v) $$\overset{K}{NH + SH + MDR + ODR + VR} = \left\{ \begin{array}{l} \frac{A}{4} + \frac{B}{8} + \frac{C}{12} + (48K + 24M + 11.2N + 9.6P + 12.8Q \\ \quad + 5.9R + 1.6S + 0.64T + 0.2V) \end{array} \right\} + D$$

Here $$V = 460$$

$$\overset{K}{NH + SH + MDR + ODR + VR} = \left\{ \begin{array}{l} \frac{11000}{4} + \frac{6000}{8} + \frac{7000}{12} + (48\times4 + 24\times7 + 11.2\times15 \\ +9.6\times38 - 12.8\times90 + 5.9\times350 + 1.6\times800 + 0.64\times580 \\ +0.2\times460) \end{array} \right\} \frac{105}{100}$$

$$= \left(\begin{array}{r} 2750 + 750 + 583.33 + 192 + 168 + 168 + 364.8 + 1152 + 2065 \\ +1280 + 371.2 + 92 \end{array} \right) \times 1.05$$

$$= 9936.33 \times 1.05$$

$$= 10433.15 \text{ km}$$

$$\therefore \quad \overset{K}{VR} = 10433.55 - 7699.55$$

$$= 2733.60 \text{ km}$$

2.5. ROAD DEVELOPMENT PLAN (1981–2001)

The country had done well, within the constraints of resources, during the last four decades of highway planning. The Nagpur Plan and the Bomaby Plan served as beacon lights to the highway engineers. Great strides have been registered in this period in highway engineering, bridge engineering, highway planning, transport planning and transport economics.

At the same time, new challenges and fresh issues had surfaced, which demanded major changes in highway planning strategies for the future. Constraints of funds remained as serious as ever. Energy, environment and roads safety were causing serious concerns. Traffic on the roads had increased beyond expectations and the high axle loads had exposed the serious weaknesses in the pavement structure. The aspirations of the people for better roads had been aroused due to better education, keener awareness and increased spending capacity. Clamour for connectivity for villages so far left out of the transport network had grown. Road transport, being a vital infrastructure needed for all development activities, occupied a pivotal role in the future development programmes of the country. It was against this background that the Road Development Plan for 1981-2001 had been conceived.

2.5.1. Technology Trends

Considerable changes were expected to occur in road transport by 2001. Some of the possible changes were:

1. Multi-axled vehicles will be operating on roads.
2. More fuel efficient vehicles will be pressed into service.
3. The use of light weight materials such as aluminium, fibre glass and plastics will result in a reduction of unladen weight of vehicles, thus leading to increased payloads.
4. New fuels such as LPG, gassified coal and hydrogen will be making their entry.
5. Solar powered batteries will be a source of motive power for light vehicles such as two-wheelers and cars.

6. Expressways will be constructed connecting major traffic centres.
7. Bullock-carts will still be present but with improved designs and pneumatic wheels.
8. Bicycles will continue to be popular.
9. Cement concrete surfacings will be used on roads in view of bitumen shortages.
10. Recycling of bituminous pavements will be commonly adopted. Binders such as lime-flyash will be used for base courses.

2.5.2. Goals and Policies

The goals and policies for the road plan 1981-2001 are given below:

1. Road network should preserve the rural oriented economy. All villages with a population of 500 and above should be connected by an all-weather road. For villages of population less than 500, an all-weather road should be available at a distance of less than 3 km in plain areas and 5 km in hilly areas.
2. National Highways carry a substantial proportion of road traffic. They carry roughly 45 per cent of total freight movement by road and one-tenth of the bus passenger movement. The National Highways network should be expanded so as to form a square grid of 100 km side. This will ensure that no part of the country will be more than 50 km away from a National Highway.
3. Expressways should be constructed on major traffic corridors to provide speedy travel.
4. Energy conservation by improving roads (widening, strengthening, improving the riding quality and geometrics) should be given a high priority.
5. The environmental quality of roads and the areas through which they pass must be maintained and improved.
6. Road safety measures must be undertaken to contain and even bring down the level of road accidents.
7. State Highways should be extended to serve district headquarters, sub-divisional headquarters, major industrial centres, places of commercial interest, places of tourist attraction, major agricultural market centres and ports.
8. Major District Roads should serve and connect all towns and villages with a population of 1500 and above which are not directly connected by NH or SH. They should also provide inter-connections between towns.
9. Other District Roads should serve and connect villages with a population of 1000-1500.
10. The maintenance of roads already constructed should receive priority attention.

2.5.3. Expressways

Certain through routes have a very high volume of traffic at present and represent the obvious choice for a superior highway facility. It is desirable to develop such routes as a separate class, to be termed *Expressways*. These are superior type facilities with a divided carriageway, controlled access, grade separation at cross-roads and fencing. They permit only fast moving vehicles. The Expressways may be owned by the Central Government or State Government depending upon whether the route is a National Highway or a State highway. It is proposed to construct 1000 kms of expressways by the end of Twelth Five Years Plan. One such expressway is already in operation from Delhi to Agra.

2.5.4. Modified Classification of the Road System

It is necessary to introduce the concept of primary, secondary and tertiary road systems for purposes of transport planning, functional identification, earmarking administrative jurisdictions and assigning priorities on a road network.

The following classification was suggested:

(*a*) Primary System
   1. Expressways
   2. National Highways

(*b*) Secondary System
   1. State Highways
   2. Major District Roads

(*c*) Tertiary System (Rural Roads)
   1. Other District Roads
   2. Village Roads.

Urban roads will form a separate entity to be taken care of by the respective urban authority.

2.5.5. Estimated Lengths of Roads by 2001

(1) *Total length of National Highways*

Total area of the country = 3287782 sq. km

Taking a grid of 100 km. square,

$$\text{Length of N.H.} = \frac{3287782}{50} = 65,756 \text{ km}$$

Say 66000 km.

(2) *Total Length of State Highways*

Total number of towns = 3364
(1981 Census)

Assuming that the N.H. and S.H. must pass through each of these towns

$$\text{Size of sq. grid} = \sqrt{\frac{3287782}{3364}} = 31.26$$

Say 31.25 km

$$\text{Thus length of N.H. and S.H.} = 2\times31.25\times3364 = 210250 \text{ km}$$

$$\therefore \quad \text{Length of S.H.} = 210250-66000 = 144250 \text{ km}$$

Say 145000 km.

Note: Length of S.H. in each State can be worked out from either of the following formulae

$$(1)\ \text{S.H. length in km} = \frac{\text{State Area in Sq. km}}{25}$$

$$(2)\ \text{S.H. length in km} = 62.5\times\text{No. of towns} - \frac{\text{Area in Sq. km}}{50}$$

(3) *Length of Major District Roads*

The length of M.D.R. in each State can be obtained from either of the following formulae:

$$(1)\ \text{MDR length in km} = \frac{\text{Area in Sq. km}}{12.5}$$

$$(2)\ \text{MDR length in km} = 90\times\text{No. of towns}$$

On this basis, the total length of MDR's in the country works out to be 300,000 km.

(4) *Total Length of Roads in the Country*

The present density of roads in the Country is 0.46 times the area. Taking the density of roads in the year 2001 to be 0.8, the total length of roads in the country

$= 0.8 \times 3287782 \ = 2630225$ km

Say 2700000 km.

(5) *Length of Tertiary Roads i.e. Other D.R. and V.R.*

= Total length − Length of (N.H.+S.H).− Length of MDR

$= 27,00,000 - 210250 - 300,000 \ = 2189750$ km

Say 22,00,000 km

2.6. ROAD DEVELOPMENT PLAN VISION 2021

IRC published the Road Development Plan Vision 2021 in the year 2001. The major thrust areas covered in this vision document relate to mobility in respect of main roads and accessibility in respect of rural roads to connect all the villages in the country in a time bound programme.

By the end of Lucknow Plan (1981-2001), the National Highway network stands at 57737 km against a target of 66,000. For the next 20 years (2001-2021), the attempt should be to consolidate the existing NH network and go in for expansion upto 80,000 km, keeping in view the following factors:

1. Connecting major ports, industrial complexes, important growth nodes, pilgrimage and tourist centres.
2. Filling up the grid of 100 km square side, in pockets of various regions without a NH.
3. Providing linkages with adjoining countries.
4. Connecting capitals of the states being carved out now or likely in future.

It was further proposed that the State Highways be increased from 124300 km by 2001 to 160,000 km by 2021; keeping in view the following factors:

1. Providing linkage with minor ports, industrial towns, pilgrimage and tourist centres.
2. Connecting the remaining towns with population 5000 and above.
3. Connecting the Capitals of newly carved out States with the district headquarters.

As a vision, a length of 320,000 km of Major District Roads (MDR's) *i.e.* twice the anticipated length of SH's by 2021 was considered to be the reasonable target.

By the end of 2000, only 54 per cent of villages in the country were connected by roads. The goal set for village roads was to connect all villages with a population below 500 by the end of 2010. The work of further improvement of village roads and additional links, where necessary, was proposed to be taken up in the subsequent decade (2011-2021).

2.7. PRESENT STATUS OF ROADS IN INDIA

Indian roads are one of the largest road networks in the world. The total length of roads in the country is estimated to be around 38.41 lakh kms.

The length of National Highways, also termed primary roads, is 76818 km at the end of 11th Five Years' Plan (Marth 31, 2011). This constitutes about 2 per cent of the total length of roads but caters to 40 per cent of the total road traffic. Out of the total of 76818 kms of NHs, 30537 kms were constructed by the National Highway Authority of India (NHAI) under the National Highway Development Programme (NHDP), 42483 kms constructed by the states' PWDs and the remaining 3798 kms by the Border Roads Organisation (BRO). At present, out of 76818 kms of NHs, about 23 per cent length is of 4 lanes and above standard, 54 per cent length is of 2 lane standard and 23 per cent length is of single and intermediate standard. Despite the progress in NHs, only 23 per cent of their total length is wider than two lanes, leading to heavy congestions. Shortfalls in the construction of bypasses, inadequate capacity, insufficient pavement thickness and weak, narrow and distressed bridges/culverts are some of the other deficiencies.

The secondary roads comprise State Highways and Majaor District Roads and their total length is

4,99,317 kms, which constitutes about 13 per cent of the total roads in the country. These roads also cater to about 40 per cent of the total road traffic. Out of the total of 4,99,317 kms, 1,66,129 kms are the State Highways (SHs) and the balance 3,33,188 kms are the Major District Roads (MDRs). Despite the efforts of the government to increase the road network in the country and to strengthem them, the road network remains grossly inadequate in various respects. These roads suffer from low investment, inadequate width of carriageway to meet traffic demand, weak pavement and bridges, congested stretches passing through cities/towns, poor safety features and road geometrics, and inadequate formation width in hilly and mountainous regions, missing links and bridges and several railway level crossings requiring urgent replacements with road over bridge (ROB)/road under bridge (RUB) to improve safety and faster traffic movement. A broad assessment slows that over 50 per cent of SHs and MDRs network have poor riding quality and the annual losses due to poor condition of these roads is around ₹ 6000 crores.

At the tertiary level are the Other District Road (ODRs) and the Village Roads (VRs). These are also termed Rural Roads (RRs). Their length is about 32.65 lakh kms and are mostly of single lane or intermediate type and of low quality (*Katcha roads*). These once adequately developed and maintained, hold the potential to provide rural connectivity vital for generating higher agricultural incomes and productive employment opportunities besides promoting access to economic and social services. The Government of India has also identified 'rural roads' as one of six components of "Bharat Nirman" with a goal to provide connectivity to all habitations with a population of 1000 persons and above in plain areas and 500 persons and above in hilly or tribal areas with an all weather road.

It is recognised that rural connectivity is a key component of rural development and poverty alleviation in India. The main mechanism for enhancing rural connectivity is through a centrally sponsored scheme, known as Pradhan Mantri Gram Sadak Yojana (PMGSY), launched on 25th December, 2000. The programme seeks to connect all habitations with a population of 500 persons and above in plain areas and 250 persons and above in Hill States, Tribal areas, the Desert areas and in the 82 selected and Tribal Backward Districts with an all weather road. It is estimated that 158891 habitations are eligible under this scheme, for the achievement of which 3,67,673 kms of additional roads and upgradation of 3,748,44 kms of roads would be required. By the end of March, 2012, 53 per cent of the target has been achieved (84414 habitations have been connected) by the construction/improvement of 2,09,500 kms of rural roads through the National Rural Roads Development Agencies (NRRDA).

ROAD ADMINISTRATION

2.8. CENTRAL ADMINISTRATION

To deal with the problems of roads, Roads Wing, under the Ministry of Transport and Communication was set up by the Government of India on the recommendations of the Nagpur conference. It was proposed by the delegates of the conference that the centre should take full responsibility for the development and maintenance of National highways and should assist in co-ordination and control standards and specifications, setting priorities and general administration and inspection of National highways. For this purpose, an effective administrative and technical organization was considered essential at the centre. The outcome was the creation of the Roads Wing. The Roads Wing was also entrusted with the responsibility of administrating the existing Central Road Fund and other funds allocated by the centre for the development and maintenance of National highways and other State roads and co-ordinate the road policies of the centre and the states. This organization was also to function as a central design office and to distribute technical information on roads and bridges.

From 1950 onward, the Roads Wing was also given the charge of developing and maintaining roads in the centrally administrated areas (now Union Territories). The organization was also entrusted with the responsibility for the development of certain selected roads, *viz.* roads under the charge of the central P.W.D. in Sikkim, roads in the North-East Frontier Agency, the West Coast Road in Tamil

Nadu and Maharashtra States, the Passi-Baddarpur Road in Assam and the Dhan-Udhampur Road in Jammu and Kashmir State.

The subjects dealt with by the Roads Wing are listed below:

1. National highways.
2. Roads other than National Highways, and bridges in the Union Territories.
3. Roads of inter-state or economic importance for which the Government of India had agreed to give grants from a special provision approved by the Planning Commission.
4. Central road fund.
5. Road and bridge standards, road research and co-ordination of road development.
6. Administrative control over the execution of road works financed wholly or partly by the central government.
7. Statistics relating to highways.

Later on, an independent Ministry of Road Transport and Highways (MORT and H) was formed at the centre which is responsible for policy and planning of development and maintenance of expressways and national highways. This also serves as an apex institution in technical excellence.

The Ministry of Road Transport and Highways was intended to serve the following functions:

1. To administer funds sanctioned by the Central Govt. for the development of National Highways.
2. To prepare development and maintenance plans for NH in consultation with State P.W.D.'s.
3. To recommend the projects for financial approval, prepared by State P.W.D.'s and found technically sound.
4. To supervise the works executed by the State P.W.D.'s on agency basis.
5. To administer the Central Road Fund.
6. To administer the works carried out in the Union Territories.
7. To review the problems concerning the development of road projects with the State authorities either in the IRC annual meeting or meeting of the council which is held during the mid-year.
8. To administer road projects other than N.H., such as Inter-State roads, Border roads, roads of economic importance, etc.
9. To co-ordinate research work carried out by research organizations such as C.R.R.I, H.R.B., State research laboratories.
10. To administer Indian Roads Construction Corporation, a public sector undertaking.
11. To evolve standards and specifications for roads and bridges.
12. To disseminate technical information and knowledge on road and bridge works.

This Ministry helps Indian Roads Congress (IRC) in evolving standards, guidelines and manuals on various facets of highways. The IRC have an apex body, the Highway Research Board (HRB) (Constituted in 1973) to guide the profession on research and development (R&D) needs. The HRB identifies new research projects, monitors those which are under implementation and disseminates the findings.

A Central Organisation, National Highway Authority of India (NHAI) was set up in 1988 under an act of Parliament. However, this became operational in 1995. This authority has been made responsible for execution of the externally aided projects by the World Bank, Asian Development Bank, etc. NHAI is mandated to implement the National Highway Development Project (NHDP) comprising the following:

Golden Quadrilateral : Delhi-Mumbai-Kolkata-Delhi

North-South Corridor : Srinagar to Kanyakumari with Spur from Salem to Cochin

East-West Corridor : Silchar to Porbander

NHAI has a lean and thin organizational structure and outsources most of the work through consultants and contractors.

Apart from the above the Ministry of Road Transport and Highways has a Liaison and Inspectorate Organization consisting of Engineering Liaison Officers attached to the Chief Engineers of the various states. The Engineer Liaison Officers are expected to make frequent inspections of road works in states financed by the centre.

In additional to the MORT and H, there are other central ministries, which are looking up the road construction works e.g. roads in the military areas are constructed, maintained and financed by the Ministry of Defence through the Military Engineering Service (M.E.S). Cantonment Boards are responsible for the construction and maintenance of roads in the cantonment areas serving civil needs, these projects being financed by the Ministry of Defence.

The Border Roads Organization (BRO) under the Ministry of Defence was set up in the year 1960 to accelerate the development of the Northern, North Eastern and Western borders, with the responsibility of construction and maintenance of roads in the border areas. This organization has expanded considerably and is self-sufficient in man-power, machinery and equipment.

The Border Roads Organization (BRO) is playing a vital role in the construction and maintenance of roads in difficult terrain.

The Ministry of Railways undertake the construction, maintenance and financing of roads adjoining to railway lines.

The Ministry of Rural Development is responsible for the upkeep of village roads programmes. The Prime Minister has announced a nation-wide Rural Roads programme, popularly known as *Pradhan Mantri Gram Sadak Yojana* (PMGSY) which aims to connect every village with a population of over 500 habitants in plain areas and 250 in hilly areas through a good all weather road. It is also proposed to upgrade about 7.5 lakh km of existing rural roads. For this purpose, a National Rural Road Development Agency (NRRDA) has been formed under the Ministry of Rural Development.

2.8.1. Central Road Research Institute

Central Road Research Institute (CRRI), New Delhi was set up in the year 1950 under the Council of Scientific and Industrial Research (CSIR), as one of the national laboratory to carry out the following functions:

1. To undertake basic and applied research for investigations, design, construction and maintenance of different types of roads and runways.
2. To undertake research for the economical use of locally available material for the construction and maintenance of roads and runways.
3. To undertake research on road traffic and transportation, including traffic economics and safety.
4. To devise ways and means for the construction of low cost all-weather roads.
5. To adopt such technologies which can achieve self sufficiency through development of appropriate tools, machinery, equipments and instruments.
6. To render technical advice and consultancy services to different organizations.
7. To disseminate results of research through display centres, publication, library and documentation centres.
8. To conduct training and refresher courses for the personnel of agencies connected with road work.

2.9. STATES ADMINISTRATION

The states are responsible to play a critical role in the execution of road projects. Works on NH are also implemented on ground by the states except the segments of National Highways entrusted to Border Roads Organization and to the National Highway Authority of India. In some of the States, District Roads and Village Roads are under the charge of the local bodies, but they are not in a position to handle these works effectively due to scanty resources available. The Public Works Departments in the States are basically a strong institution. With the stress on private financing of roads and entry of Build, Operate and Transfer (BOT) entrepreneurs, the role of the states PWDs becomes all the more critical. Volume and size of works to be handled is also increasing rapidly.

Highways, in the states, in general, are under the State Public Works Department with a Chief Engineer as the head. In some states such as Tamil Nadu and Andhra Pradesh, there exists a separate highway department which is responsible for the development and maintenance of highways. Under the Chief Engineer, are Superintending Engineers who are in charge of their circles (definite region). Each circle is further split into divisions under the care of Divisional or Executive Engineers. Each division is again divided into Sub-Divisions under the charge of Assistant Engineers or Sub-Divisional Officers.

The administrative set-up of roads in one of the states of India is shown below:

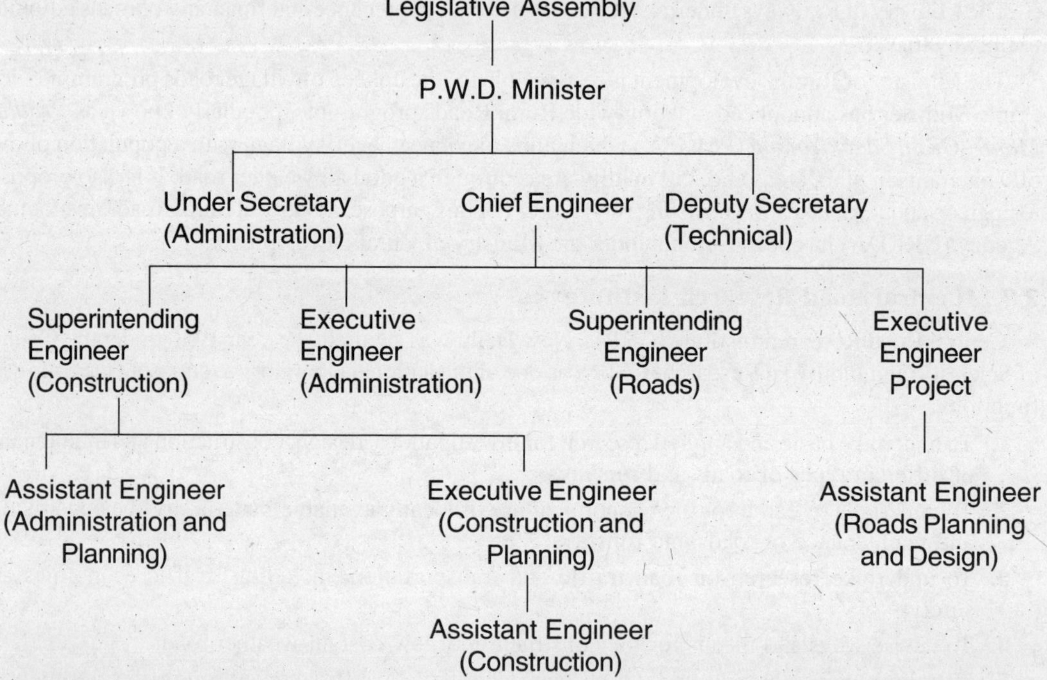

The problem of highway improvement cannot be successfully solved so long as every state does not have a separate highway department. Also there is a great need of inter-state relations and of an all-India administration. There is clearly a need for bettter co-ordination and cooperative action. The provision of maximum highway service today, requires that the several systems of roads and streets which make up the complete highway network in a state, be integrated, adequately improved and efficiently managed. The successful results of this great task calls for the development of new and extensive co-operative relationship at all levels.

❑❑❑

3 HIGHWAY ECONOMICS AND FINANCING

Chapter

GENERAL

Highways are economic ventures. Economic analysis are therefore, becoming of increased importance for the design engineer to reach at the most economical and serviceable design and still more to decide whether a highway facility should be built at all or not. In situation where construction of certain proposed highway facility is obviously desirable economically and otherwise, it requires careful economic analysis to determine the comparative advantages of various proposals or features of location, design and construction programme priority.

It is, therefore, necessary that each proposed highway improvement should be examined to ascertain if it is economically or otherwise justified and to examine the alternates of location and design features to ensure the overall soundness of the investment. To determine the economy of proposed highway facilities (*i*) the annual cost of owning and operating the highway facility, (*ii*) the annual cost of motor-vehicle operation over that facility, and (*iii*) the value of other economic benefits, have to be carefully examined. These considerations form the basis for the study of cost and economy of the project.

The objective of economic analysis is to answer the question: would the monetary benefits accruing from the investment on the proposed highway be greater than the cost of construction? Answering this question would mean deciding whether public resources should be committed to a long term investment in the specific highway facility under consideration, a different highway facility, water resources project, hospitals, education facility or a personal or private venture.

Thus the analysis inherently involves taking decisions on the alternate commitments for the nation, land, labour and natural resources. The objective is not only to justify a proposed highway facility but to select only those proposals which have the highest, most certain and most sound justification keeping in view the alternate decisions of (*i*) not to construct anything, (*ii*) adopt another highway facility, and (*iii*) invest the money in some other venture.

In highway economic analysis, one has to consider the total transportation cost; comprising of:

1. Cost of construction of the facility initially.
2. Periodic cost of maintaining the facility over its design life; and
3. Road user cost.

These three cost components are interdependent. The designer has to select that alternative in which the sum total of the three is minimum.

Highway cost is the sum total of the first two components, *i.e.* cost of construction initially plus the periodic maintenance cost.

The cost of construction of the facility includes:

(*i*) Survey, investigation and design costs

(*ii*) Land acquisition costs

(*iii*) Construction costs

(*iv*) Physical contingencies (unforeseen items and unforeseen increase in cost not attributable to esclation and unforeseen increase in quantities.

(*v*) Supervision, quality control and administration charges.

The cost of maintenance of the facility includes:

(*i*) Ordinary repairs, such as patch repairs, pot-hole filling, dressing earthwork, etc.

(*ii*) Periodic repairs, such as renewals and resurfacing.

(*iii*) Operational expenses, such as traffic signals, traffic aid-post, policing etc.

(*iv*) Supervision and administration charges.

The road user cost is composed of the following components

(*i*) Vehicle operating cost

(*ii*) Time cost

(*iii*) Accident cost

The following components affect the road user costs:

(1) ***Roadway factors***

(*i*) Pavement width

(*ii*) Pavement surface type and its riding quality as measured by its roughness

(*iii*) Vertical profile

(*iv*) Horizontal geometry

(*v*) Number of junctions per kilometer (in case of accidents cost)

(2) ***Vehicle factors***

(*i*) Type

(*ii*) Age

(*iii*) Make

(*iv*) Engine horse power

(*v*) Power-weight ratio

(3) ***Traffic factors***

(*i*) Traffic volume

(*ii*) Traffic composition

(*iii*) Speed

(*iv*) Congestion

The ultimate aim is to achieve the lowest cost of transportation on the highway when all costs are included. Thus the purpose of design and operation of the highway is to minimize the highway cost, rendering at the same time adequate and safe service. It is here that highway engineers and administrators have an occasion to effect certain economics in vehicle operation through control of highway design, though they cannot do much about the cost of operating the motor vehicle as a machine.

3.1. DEFINITIONS

The following terms have been defined to convey their proper meanings in highway economics:

(*a*) ***Road user's services*** are the facilities, advantages, conveniences and privileges provided by the completed highway including the provision for safe, fast, certain, comfort sale and economical transportation by the most direct reasonable route; accruing to the owner or driver of the vehicle.

(*b*) ***Road user's benefits*** are the gains, opportunities and realizations tangible and intangible—that are realized by the owner or driver of the vehicle as a result of highway improvement. These benefits are measures of improvements upon prior conditions, facility or situation. Benefits may occur in the form of savings in motor-vehicle operating costs, value of time of travel saved and intangible benefits such as safety, comfort, convenience and satisfaction enjoyed.

(*c*) ***Annual cost*** is the actual or estimated cost comprising total maintenance and operating costs and the capital costs of depreciation and interest for one year of operation.

(*d*) ***Capital cost*** is the investment in the property having long life, *i.e.*, a service life greater than one accounting period (one year). Capital cost comprises the money cost incurred on the highway, other property or equipment kept for service. This includes the cost for planning, construction and overhead cost of administration and supervision.

(*e*) ***Maintenance cost*** is the cost of maintaining the property in a serviceable condition. This includes costs of repair, parts, fuel and labour. Cost incurred in excess of the cost necessary for the maintenance of the highway in a serviceable condition, which results in restoring the highway to a condition of excellence or service superior to its condition when originally constructed are called betterment costs and are regarded as additional capital investment. These are not included in the maintenance cost.

(*f*) ***Highway operating cost.*** Costs of highway petrol and general services provided to highway users not directly related to maintenance of the physical components of the highway are regarded as highway operating costs. Signs, markings, signals, snow control and removal are also sometimes classed under operating cost rather than under maintenance cost.

(*g*) ***Salvage value*** is the value of the property for further use in some secondary capacity or as scrap when it is removed from initial service. While estimating the overall depreciation cost chargeable to the period of initial service, the salvage value realized by the disposal of the salvage is deducted from the original capital cost.

(*h*) ***Depreciation*** is the overall value or total cost of the property less its salvage value at any particular time. The depreciation, in the counting sense, is an element of cost of production. It is chargeable to the end-product, produced with the service of the property. For example, in estimating the cost of operating a truck the expenditure for fuel, tyres and repairs are as much the elements of cost as the cost of the new truck itself less its trade-in allowance (salvage). Allowance has to be made for the estimated depreciation of the truck while estimating the cost of its operation. Until the property is retired, depreciation cost is an estimated figure that allocates a fraction of the estimated depreciation of the property to each accounting period or unit of service rendered, usually applied on some systematic basis.

(*i*) ***Interest*** is an element of cost paid to the owner of the capital for the capital investment, expressed as a percentage rate per annum. Interest cost always exists regardless of how the property is financed, who owns and operates the property and by whom the capital is furnished.

(*j*) ***Capital recovery*** is the combined sum of return received for (*i*) the depreciation of capital and (*ii*) ownership cost of the capital investment (interest), recovered annually, for each year of operation spread over the service life (or estimated probable life). The full depreciation cost together with interest on the undepreciated cost may be realized at the beginning of each year.

(*k*) ***Service life*** is the period of time extending from the date of beginning of service for the facility to the date when it is removed from service. Each component of the highway— grading, bridge, pavement and right of way, or any item of physical property, such as motor, truck, sign post, office building, air compressor—has its service life. Except for such components of property which have been retired from service, the life is always an estimated figure and therefore probable service life.

(*l*) ***Analysis life*** is the period of life of the facility chosen for purposes of economic analysis. It is usually much less than the estimated life of the facility. Physically, service life of a highway may be expected to be 50 to 100 years, but estimates extending far into the future are highly uncertain and any analysis involving economic advantages of constructing a particular highway facility based upon such extended future period is unrealistic. In order to arrive at realistic conclusions, a comparatively short period of about 30 years should be taken.

(*m*) *Motor vehicle running costs* are the costs incurred in operating motor vehicles. These costs comprise items such as fuel, tyres, lubrication, repair, a part of depreciation costs and other costs necessary in order to run the vehicle. Costs of licence fee, driver licence, garage, parking, insurance, interest, a part of depreciation costs which are not dependent upon kilometres driven are called ownership costs and these are not included in the running costs. The running costs and ownership costs when added together give the total motor-vehicle operating cost.

(*n*) *Value of occupants' time* is an estimation of the value of clock-time in terms of money to the owner of the vehicle, the driver or other occupants. The element of time is an important factor of cost to highway users. The saving in time occasioned by highway improvement or otherwise, may be taken into account by converting the clock-time shortened or lengthened to its equivalent money cost at fixed unit value in the process of economic analysis.

(*o*) *Overall travel time* is the period of time travelled plus stop and delays, except those off the travelled way.

(*p*) *Running time* is the period of time during which the vehicle remains in motion, from the beginning till end of the journey (route) under study.

(*q*) *Running speed* over a specified section of the highway, is the distance of the highway divided by the running time. The average running speed for all traffic or component there of is the summation of distance travelled divided by the summation of the running time.

(*r*) *Nominal speed* is the safe running speed. In case there is no traffic interference, normal speed is just the running speed at which drivers operate on a specified section of the highway. In the case of test runs drivers are instructed to attempt to operate at a speed at which running is safe. This is called the nominal speed.

(*s*) *Benefit-ratio* is a measure of the desirability of constructing a specific improvement. It is obtained by dividing the benefits acquiring to the road users per year expressed in terms of money by the annual highway cost.

VEHICLE OPERATION COST (VOC)

What is the cost of operating a motor vehicle on the road? Although there is no single answer to this question, yet it must be found out to evolve an economical design of the highway facility.

The cost of motor-vehicle operation comprises of the following:

1. Those which are directly proportional to the kilometreage driven, *i.e.*, running costs. In other words, their cost per vehicle kilometre is relatively constant. Under this category are included costs of such items as petrol, mobile oil, tyres, maintenance and repairs, and a part of the depreciation cost resulting from wearing out.
2. Those which depend upon the speed of travel, *e.g.*, value of travel of the passengers and the driver.
3. Those which increase with time but are constant for a given period (one year). If expressed in terms of costs per vehicle kilometre, they are inversely proportional to the number of kilometres driven. Registration fee, driver's licence, garage rent, insurance and portion of depreciation attributable to inadequacy or being out of date—costs of all these fall under this class. These costs are constant for all vehicles and need not be considered in economic studies.
4. Accident costs.
5. Other costs.

These costs are often affected by roadway improvement and should be carefully examined, as highway improvement can be justified only if savings in operating costs offset the expenditure required. These are discussed below:

3.2. AVERAGE RUNNING COSTS

(a) *Vehicle fuel costs.* The consumption of fuel by vehicle for every kilometre travelled varies with :

(*i*) Running speed attained
(*ii*) Condition and type of road surface
(*iii*) Grade or slope of the road
(*iv*) Curvature and superelevation
(*v*) Degree of congestion on the road
(*vi*) Number and duration of stops
(*vii*) Weight and size of the vehicle
(*viii*) Adjustment and efficiency of the engine
(*ix*) Skill of the driver
(*x*) Temperature and
(*xi*) Elevation above sea-level.

(b) *Mobile oil consumption.* The influence of highway design upon mobile oil consumption is difficult to isolate. It has been observed that mobile oil consumption does not vary substantially with increased speed. Mobile oil consumption is increased as the roadway changes from pavement to loose gravel to unsurfaced. Little information is available on other variables, such as frequency of oil change and use of filters.

(c) *Tyre costs.* The wear of tyre varies with the type of surface, grade, speed, curvature and the type of operation, *i.e.*, free, normal or restricted. Due attention in respect of inflation pressure, wheel balance, tyre rotation, overload control, *etc.*, must be paid. Tyre wear and tyre costs are much increased at higher speeds, *e.g.*, wear is increased by three times when the speed of travel is changed from 30 km/hr to 50 km/hr. Similarly grade curves increase wear sharply. Wear is almost double on 7 per cent to 9 per cent grades than on a level stretch. On curves, which are not properly superelevated, side friction is called upon to keep the vehicle in its curved path. Wear is related to the coefficient of side friction. Tyre costs also vary with slowing or stopping or speeding up of vehicles.

(d) *Maintenance and repair of vehicles.* These costs depend upon the type of road surface. On concrete and other high type surfaces, the maintenance and repair costs are the least and they are more with surfaces such as earth roads, gravel roads, and unsurfaced roads. Maintenance costs inlcude engine, chasis and body servicing and repairs and lubrication.

(e) *Depreciation.* Depreciation may be defined as **"the inevitable march to the junk heap"**. A part of the depreciation cost, say 50 per cent, attributable to the wearing out is fixed on kilometreage basis and rest is attributable to obsolescence or being inadequate or out of date and is assigned on time basis (one year). In economy studies, only that proportion of the total depreciation cost that is assigned to be attributable to kilometreage use of the vehicle is considered. The other proportion which depends upon the age or time factor is usually omitted from the calculation as this remains constant for all conditions. If studies are made on alternate routes of equal length, depreciation charges on kilometreage basis will be the same for each alternative and will not affect result at all. These have to be only considered when routes of different lengths are being compared.

(f) *Slowing, stopping and standing costs.* When ever a motor vehicle slows or comes to a stop and then resumes speed, extra fuel and oil are consumed, breaks and tyres have greater wear and maintenance cost increases. Costs per stop vary with speed, vehicle type and characteristics and the driver's manouvre. These costs are more for heavy trucks in comparison to motor cars. These costs go on mounting if a vehicle has to wait for a longer period at a traffic signal.

(*g*) ***Traffic volume costs.*** As the traffic volume increases, there is large interference to free movement from other vehicles plying on the road. This causes not only the reduction in the average speed but also repeated slow downs and speed ups. The traffic congestion will increase motor-vehicle operation cost.

(*h*) ***Costs due to restricted sight distances.*** Because of restricted sight distances, over-taking becomes impracticable and thus the overtaking vehicle has to slow down to the speed of the vehicle being overtaken and maintain this speed until it gets opportunity to pass. This causes a reduction both in the speed and running cost.

(*i*) ***Costs relative to grades.*** Since greater energy is needed to travel uphill, costs increase uniformly with steepness. It has been found that the increase in cost is about 20 per cent more for a six per cent grade as compared to a level road. The reduction in operation cost is more or less the same as the increase on upgrades for four per cent slopes or flatter. There is, however, little decrease in cost for down slopes greater than 4 per cent, because breaks have to be applied.

(*j*) ***Costs associated with curves and corners.*** Wear of tyres takes place on curves and at corners such as intersections and at diamond crossings.

Travelling at high speeds on sharp curves or around corners can considerably increase running costs *e.g.* costs are increased by almost 100 per cent as compared to those on level roads, if a car travels around a 6° (287 m radius) curve at 100 km/hr. However the speeds around curves and hence increase in running costs are limited by superelevation and side friction.

(*k*) ***Costs related to roadway surface.*** The running costs, discussed so far, are for high type surfacing such as bituminous roads. The running cost will increase substantially with lower type pavements such as water macadam roads, stabilised roads, earth roads, *etc.*

3.3. VALUE OF TIME OF TRAVEL

Most of the proposed highway improvements effect an increase in the speed of travel, thereby resulting in appreciable time saving for the driver and the passengers over travel on existing routes. For passengers cars-used for business purposes by salesman and others—buses, trucks and other commercial vehicles, reduced time of travel will effect savings in wages of drivers and cleaners. There has been a controversy whether the money value of this time-saving should be included in the economic studies or not. There is a great difficulty in fixing the proper money value to it. Further, there are many drivers and different types of trips for which saving in time is of no avail. There are management labour contracts and conditions of operations where time saving does not reduce the cost of driver wages.

Moreover, a major part of the passenger-car use is devoted to necessity travel like trips to office, for business and for family affairs. In case there occurs a saving in the time of travel, the passengers would not mind paying more if this time could be made available for other purposes such as business, pleasure or rest. However, there is no yardstick to measure its worth. Some scholars consider this time saving as intangible benefits and would not include them in tangible benefits. There are others who determine its value in terms of the costs of driving extra distance in order to save time.

A conservative basis is desirable in the economic justification. Estimates of time saved and value assigned to it should, therefore, be conservative and considered in only those proposals where such time saving has a realistic value. In all economic studies, analysis should be made separately for the time factor so that economic justification may be based completely on realised savings in motor-vehicle operating cost, on the value of time saved or known weighting of these two factors.

3.4. ACCIDENT COST

Accident not only causes suffering and misery to the persons involved in it, but also wastes nation's resources. Motor-vehicle accidents cost a lot in advanced countries like the U.K. and the U.S.A. and are nearly half to three-fourth of the expenditure incurred on construction, maintenance and operation of a highway system. Costs of accidents are classified as follows:

Table 3.1

Additional cost of passenger car stop over that of continued operation at a given speed

(Time value neglected)

Additional cost per stop in Rupee

Approach speed km/hr	*Fuel*	*Tyre and brakes*	*Other operating costs*	*Total operating costs*
16	0.47	0.10	0.19	0.76
32	1.12	0.29	0.43	1.84
48	1.69	0.62	0.76	3.07
64	2.25	1.27	1.19	4.71
80	3.66	2.28	2.09	8.03
96	4.50	3.10	2.66	10.26

(*i*) Property damage, usually to motor-vehicles only

(*ii*) Medical expenses for personal injuries

(*iii*) Deaths-value of human life

(*iv*) Overhead cost of insurance

(*v*) Projected net future losses in earnings of persons killed or injured

(*vi*) Costs associated with disruptions or delays to economic activities because the services of specialised personal are lost or important goods are destroyed or damaged or their delivery delayed.

Property damage and medical aid can be easily determined with accuracy from accident reports and records. These are the realistic monetary costs to highway users involved in accidents. In the case of fatalities, costs associated with death and burial can be worked out but there is considerable dispute on the subject of monetary value of human life. This value may be judged either by taking the present worth of the future earnings of persons or their future worth as members of society. It is usually recommended not to take into account the value of time lost by injured persons, although, in some cases, time lost from gainful employment is a real loss.

For economy study, accidents are classified into three classes namely:

(*i*) Those involving fatalities

(*ii*) Those non-fatal accidents involving injuries, and

(*iii*) Those resulting in only property damage.

Only realistic and reliable values of tangible cost of motor-vehicle accidents, which can be predicted from existing highway conditions, should be included in the economic analysis. Accident records from the existing highways should be utilized rather than general average results. For a proposed new highway, accident cost should be found out from available records of existing similar facility in similar use. However, benefits accruing from accident reduction, when an existing highway is replaced with modern facility are difficult to evaluate. There is no doubt that this saving is large but it is not always so. There are instances where due to excessive speed and drivers rashness, accident costs could not be reduced even on an improved facility. Thus adequate knowledge of driver's behaviour and characteristics and careful, uniform and complete reports of accidents are prerequisites for the evaluation and assignment of accident costs.

3.5. OTHER COSTS

Other costs which are doubtful factors in the analysis of highway economics are:

(*i*) Convenience and comfort to drivers and passengers. Some drivers prefer to take a long through route than a short congested route, although the cost of operation in the former case is more. They are prepared even to pay heavy toll taxes. Thus there is a substantial proof that drivers place a money value on the comfort and convenience provided by improved highway facility.

(*ii*) Mental strain and physical strain to drivers and passengers.

(*iii*) Roughness and vibration.

(*iv*) Confusion and congestion (poor sign boards, inconsistent design standards, *etc*).

(*v*) Noise level.

Economists agree that these factors bring an economic benefits but they fail to agree on how these should be treated in economy studies. There is no basis available by which these factors can be converted into money value. So long as there are opportunities to invest nation's wealth in highway improvement, which will produce positive savings in direct motor-vehicle operating costs, there appears to be little justification in taking into consideration these indeterminable factors which depend mainly on mental and physical comfort.

3.6. BENEFITS FROM HIGHWAY IMPROVEMENTS

Benefits from highway improvements can be broadly classified under two categories:

1. *Road user benefits*

(*i*) Vehicle operating cost savings

(*ii*) Value of travel time savings

(*iii*) Value of savings in accidents cost

(*iv*) Savings in maintenance cost

2. *Social benefits*

(*i*) Improvements in administration, law and order and defence

(*ii*) Improvements in health and education

(*iii*) Improvements in agriculture, industry, trade and mining

(*iv*) Improvements in environmental levels.

(*v*) Appreciation in value of land adjacent to roads.

(*vi*) Increased values of natural resources

(*vii*) Development of recreational values such as sports, sight-seing, tourists, *etc*.

To be on the conservative side, these social benefits are not used in economic analysis to justify highway improvements.

ANNUAL HIGHWAY COST

The annual cost of ownership and operation of a highway comprises the following items:

1. Administration (portion not chargeable to construction or maintenance)

(*a*) Personal services.

(*b*) Equipment and building operation, including capital cost.

(*c*) Office requirements and services.

(*d*) Insurance and other overhead costs.

2. Highway operation

(*a*) Personal services of highway fuel.

(*b*) Equipment and building operation, including capital cost.

(*c*) Office requirements and services.

(*d*) Insurance and other overhead costs.

(*e*) Vehicle operation, including capital costs.

3. Highway maintenance

(*a*) Personal services.

(*b*) Equipment and building operation, including capital cost.

(*c*) Office requirement and services.

(*d*) Insurance and overhead costs.

(*e*) Material, *etc*.

(*f*) Equipment operation including capital cost.

4. Highway capital costs

(*a*) Depreciation cost on the expenditure in the physical highway and other structures.

(*b*) Interest on the expenditure.

3.7. ANNUAL COST FORMULAE

The annual highway cost of ownership and operation for a specific facility, project or road system may be expressed by the following relation:

$$C_a = A + O + M + C_D \qquad \text{... (3.1)}$$

Where C_a = average annual cost of ownership and operation

A = average annual cost of administration (portion not included elsewhere)

O = average annual highway operation cost

M = average annual highway maintenance cost

C_D = average annual highway capital costs of depreciation plus interest charges on investment; capital recovery/return on capital.

Costs of administration include those general overhead, supervisory, administrative, legal and other central office costs which cannot be allotted to construction and maintenance expenditures. These are usually independent of the amount and character of traffic.

The operation costs are those which are incurred on highway patrol or other such organizations. These depend primarily on the character and volume of traffic. In other words these are the costs of furnishing services to traffic and may include signs, signals and other traffic control devices.

Expenditure on labour, materials, equipment operation and depreciation and other investments required to maintain the roadway and all other allied structure in a serviceable condition are included in the maintenance costs. Costs of snow removal and prevention also come under this class although they are also sometimes put under operation costs.

The capital cost includes depreciation on the investment in the highway facility plus an interest or return on investment charge for the use of that capital. Depreciation cost is an appropriate item to be included in the estimation of the annual cost, because the investment was made through payments for construction work, which are apportionable to future years or operation. Depreciation cost is included on such a basis that this is in the form of a prepaid expense, paid after the performance of the facility, but applicable to the future period of use of that facility. Interest is also included in the annual cost because there must be some charge representing rent for the use of the funds invested in the highway facility. Interest charge is always considered in the economic analysis irrespective of the method adopted for financing the highway construction. This factor takes into account the differences in the time value of money.

Average annual capital cost C_D may be found by the use of any one of the following formulae:

(*i*) Straight line formula:

$$C_D = \frac{(C - S_n)}{n} + i\frac{(C - S_n)}{2}\frac{(n+1)}{n} + iS_n$$

(*ii*) Sinking fund formula:

$$C_D = (C - S_n)\left(\frac{i}{(1+i)^n - 1}\right) + iC$$

(*iii*) Capital recovery with return formula:

$$C_D = (C - S_n)\left(\frac{i\,(1+i)^n}{(1+i)^n - 1}\right) + iS_n$$

Where C = Construction cost or total investment

S_n = Salvage value at the end of n years

i = Rate of interest applicable

n = Expected life period of the project.

Out of the above three formulae, the capital recovery with return method is recommended for use in economic studies of highways. This is very simple to apply when the salvage value is zero. Also it gives identical results with the sinking fund formula. The straight line formula is not reliable as it under-estimates the annual capital costs, particularly at higher values of n and i.

Other formulae which are useful in highway cost and economy studies are given below:

Accumulation of a sum, C at interest rate i for n years

$$S = C\,(1+i)^n$$

Present worth of a sum S receivable n years hence

$$C = S\left[\frac{1}{(1+i)^n}\right]$$

Year and deposit in a sinking fund to accumulate to S in n years (year and annuity deposit)

$$D = S\left(\frac{i}{(1+i)^n - 1}\right)$$

Capital recovery with return on capital

$$C_D = C\left(\frac{i\,(1+i)^n}{(1+i)^n - 1}\right)$$

Accumulation of an annual year-and annuity payment D in n years

$$A = D\left[\frac{(1+i)^n - 1}{i}\right]$$

Present worth of an annual year-and annuity payment D for n years

$$C = D\left[\frac{(1+i)^n - 1}{i\,(1+i)^n}\right]$$

where C = Principal sum

S = Compound amount; Principal plus interest

A = Accumulation of an annuity

D = Sinking fund deposit.

3.8. CALCULATION OF THE ANNUAL COST OF A HIGHWAY FACILITY

Each component of the annual cost is to be estimated. It is therefore, essential that planning and design for the proposed highway facility must have reached such a state that reasonable estimates for each factor are possible. Materials to be used for construction, type of surface, geometric design elements, etc. should also have been decided. Such estimates are prerequisite for the economic comparison of possible alternatives.

While comparing the economy of different proposals for a given facility, use is made of equation (3.1) to determine the annual cost for existing condition and as well as for each proposal for improvement. The choice of each factor should not be done on the general average value but by keeping in view the specific characteristics of the existing highway and of each proposed alternative. Annual costs for A, D and M should be taken at what is forecasted to prevail for the next year or route section under analysis or average for the period of analysis for the specific route.

3.8.1. Administrative Costs

Choice of location, type and design of a highway facility has very little effect on the administrative costs. As such, they are usually neglected in economic studies. Administrative costs should be taken into consideration only when there is a sufficient reason to believe that they are an important factor influencing the economic comparison.

3.8.2. Operation Costs

These are the costs of providing services to the traffic as contrasted to the costs of maintaining the roadway. If any highway department includes the costs of maintaining signs, signals and markings in the maintenance costs, then operation costs comprise only highway patrol services costs. When the operation costs cannot be determined to vary in between alternatives included in the economic analysis, which is usually true, this factor may be neglected in the calculation.

3.8.3. Maintenance Costs

Maintenance costs can be found out from the Public Works Department for any specific existing section of a highway. If these records are not available, estimates may be made by reference to the maintenance costs of identical routes. For any proposed facility, the maintenance cost can be assumed on the basis of experience with the existing highway having similar traffic, physical and geographic characteristics.

3.8.4. Construction Costs

Construction costs or invested capital are determined by estimating the cost of the components of the proposed highway facility. To obtain uniform and accurate results in economic analysis, the costs of the following components of the highway are considered:

(*i*) Right of way land

(*ii*) Right of way damages, demolition, removal of structures, *etc.* which are not for specific permanent property

(*iii*) Earthwork, including road side development and land-scaping
(*iv*) Drainage, culverts and small structures
(*v*) Pavement
(*vi*) Bridges
(*vii*) Traffic signs, signals and markings.

Construction cost factor C may be taken as zero for any existing highway when it has reached the end of its main usefulness. When a part or whole of it is to be utilised in another proposal, suitable values may be assigned for C, based on its salvage value for use in reconstruction as compared with the salvage value one or two years hence. The annual capital cost woule be the amount which the existing highway would decrease in value the next year plus one year's interest on its present salvage value. Naturally, if the existing highway is to be abandoned, its value is zero and hence the factor C_D becomes zero.

3.8.5. Salvage Value

In the annual cost formula, the salvage to be used is the cash-sale value, the money value for continued use in place, the value for continued use in some other place or the value for some other purpose of the facility at the end of n years. Thus, salvage value is an estimate value of a property at some future date. The price forecasted to prevail when the salvage is to be realized is used in estimating salvage value. In the cost analysis, only net salvage value is used which is the value of the property minus the cost of disposal at the time of reconstruction or abandonment. Salvage value also varies with the chosen useful life of a facility. Its determination becomes tedious when property is used in place such as when an old concrete pavement is being used as a base for a new bituminous surface. The salvage value in such cases is assigned on the basis of what it would cost to construct a new base of the same quality at the time of salvaging minus appropriate allowances for difference in the quality of service and estimated service lives. In some cases, the existing construction is used again in place because of the high cost of its disposal. Salvage value should be estimated on the basis of the value to the proposed facility without giving any regard to the original cost of the property being salvaged.

For pavements and other components, which cannot be used in future reconstruction, salvage should be kept as low as possible. Salvage value is expressed as a percentage of the capital cost C and should be consistent with the estimate of n used in the analysis. For large values of n, S_n should be low. For many highway facilities S_n is taken as zero in economic analysis.

3.8.6. Interest Rate

There are divergent views on the rate of interest to be used in highway economy studies. Generally, three viewpoints are advocated:

(*i*) No interest should be charged at all, except where the facility is constructed by borrowing money.
(*ii*) Interest should be charged at the current rates at which the government can borrow money on bond issue. The charge is included even though the facility is constructed from current revenues.
(*iii*) Interest should be charged at a rate which represents a fair return of earning on the invested capital. This is somewhat higher than the cost of borrowing money.

AASHO and State governments in most of the countries adopt the second view-point. The third view-point comes closer to the one found in private sector. In this case interest is regarded as the cost of an opportunity foregone. This opportunity may be to invest money in a project or business other than highway or making use of it for some immediate satisfaction. Money is to be provided by the public either in the form of taxes or bond issues. It is, therefore, reasonable to consider that the capital invested in highway project should pay a rate of return commensurate with the interest rate which the public could earn in other investments.

Moreover, the rate of interest to be used in economic analysis is independent of the method of financing the highway project, the construction of which is going to last for a number of years. A suitable rate of interest should be fixed on the basis of the estimated benefits accruing from the project for the period of years used in the analysis. An interest rate varying from five to eight per cent is suggested for use in economic studies of highway improvements.

3.9. PROBABLE LIFE AND ANALYSIS LIFE

In economic analysis and highway annual cost estimates, probable life of various components of the highway facility should be selected with a considerable care. Sufficient records are not available on which probable life can be suitably estimated. Since it is the life in the future, it can only be judged.

Probable lives are of three types, namely:

(*i*) Actual physical life

(*ii*) Service life

(*iii*) Economic life.

The actual physical life ends with the physical deterioration of the material. The service life is the period for which the original facility can be used without major rebuilding. When the existing highway is abandoned or rebuilt, its service life comes to an end. Economic life is the life which is ended when the services rendered by the existing facility could be achieved at a lower cost by rebuilding it or by providing an alternate new facility. Economic life is not necessarily the same period as is physical life or service life. A facility may be put to use beyond its economic life because of the lack of availability of funds for its improvement.

When the physical life, service life and economic life are unduly large, it is desirable to use, for n in the formula for capital cost, a term of years which is less than these lives. This is called analysis life. This life is normally taken as 30 to 40 years. A highway facility is designed in such a manner that the invested capital cost will be returned with interest during the analysis life. A large useful period such as 60, 80 or 100 years is not taken because there is every likelihood that a more desirable investment may arise in the mean time.

ECONOMIC ANALYSIS

The basis of economic analysis is to ascertain from the monetary point of view whether it is profitable to make the improvement or which alternative is more economical. Economic study is made to answer the questions "why do it at all? Why do it this way?" In other words it has to be decided whether it is economical to adopt a proposed highway facility or the money could be invested in some other profitable enterprise. Of course, the final decision of the government or the local authorities will be based on all the factors involved, and not only on the findings of the economic analysis.

Thus it is required to determine whether the money value of the benefits accruing from highway improvements are greater than the invested capital or not. It is, therefore, essential that all costs and benefits be reduced to a common basis so that a direct comparison could be made. The usual common basis is to reduce costs and benefits to rupees or dollars or pounds per year.

Economic studies are simple in the case of business enterprises where benefits are the net profits from the business. The problem is somewhat complicated in highway studies. Here the benefits are not the profits realized from sales but are the profits measured in terms of the reduction in road-user costs and money value of other benefits arising from the improvement of the highway facility. Thus in the economy study of highway improvements, the invested capital should be returned to the government or the public with a reasonable return on it. Also it is required to make the choice of that proposal which produces the highest rate of return.

Table 3.2 lists the various elements of costs and benefits, which are the direct effects of a highway construction and have been discussed earlier.

Economic studies, connected with highways, are generally classified into three categories:

1. Proposed highway improvements

(*a*) Undertaking a new facility so that the existing facilities are of no consequence in the analysis.

(*b*) Replacement of existing facility with new facility.

(*c*) Rebuilding and improving the existing facility.

Table 3.2 Factors of costs and benefits of a highway

	Factor	*Description*	*Units*	*Time period (years)*
A—Market values	(Quantifiable)			
1.	Cost of highway (*a*) Planning (*b*) Right of way (*c*) Construction (*d*) Maintenance (*e*) Operation	Capital cost and annual cost of planning, construction maintaining and operating the highway	₹	Not applicable 50–100 20 Annual Annual
2.	Costs (benefits) to highway users			
	(*a*) Vehicle operating cost (including congestion costs)	Net increase or decrease in costs of vehicle operation per year	₹	Annual
	(*b*) Travel time saving	Net increase or decrease in travel time × ₹ value of travel time	₹	Annual
	(*c*) Motorist safety Economic cost of accident	Net change in expected number of accidents × average cost per accident	₹	Annual
B—Non-market values	(Quantifiable)			
3.	Costs (benefits) to highway users			
	Traveltime savings (non-commercial	Minutes saved per vehicle trip	Minutes or hours	Annual
	Non-market values (Quantifiable)			
4.	Cost (benefits) to highway user			
	(*a*) Motorist safety	Accident costs of pain, suffering and deprivation	₹	Annual

(*b*) Comfort and convenience	Discomfort, inconvenience and strain of driving	₹	Annual
(*c*) Aesthetics from driver view-point	Benefits of pleasing views and scenery from the road	₹	Annual

2. Priority schedule of improvements

In a phased facility, determination of priority for specific projects which will afford greater benefits.

3. Selection of design features

(*a*) Location—distance, gradient, alignment features

(*b*) Construction materials—earth, gravel, bitumen, concrete pavements, *etc*.

(*c*) Type of structure—bridge or tunnel, open cut or tunnel, truss or girder bridge, earth fill or bridge, etc.

(*d*) Improve existing facility or construct a new one.

(*e*) Size of facility—number of lanes to be built, bridge or culvert capacity.

The economic analysis is the comparison of various alternatives. A proposal which is economical in the long run and also provides adequate safety, service and aesthetics should be adopted.

Selection of Economic Elements

Discretion has to be exercised in the analysis of economic justification of a proposed facility, specially in respect of factors on which depends:

(*i*) annual highway cost.

(*ii*) unit cost per vehicle kilometre.

(*iii*) average daily traffic volume for every alternative and each class of vehicle.

Factors such as cost, production and consequences should be thoroughly examined while comparing alternative design features which are independent of traffic volume *e.g.* size of culvert or bridge. Acceptance or rejection of a particular proposal should be based on the specific factors which apply to that proposal. Factors such as construction costs, maintenance costs, average daily traffic, proportion of passenger cars to commercial vehicles and trucks, curves, grades, steps, unit operative costs, etc. must be adequately scrutinized for the prevalent conditions. The analysis life n and interest rate i should be selected by careful judgement.

3.10. METHODS OF ECONOMIC STUDIES

Usually four methods are adopted for economy studies. These are:

(*i*) Annual cost method

(*ii*) Net present value method

(*iii*) Rate of return method

(*iv*) Benefit-cost ratio method.

3.10.1. Annual Cost Method

In this case, the annual cost of ownership and operation of different alternatives is determined. Other things being equal, an alternative with the smallest total annual cost of owning and operating the facility will be the best choice. It has already been shown in formula 3.1 that annual cost is given by

$$C_a = A + D + M + C_D$$

Table 3.3

Capital recovery factors for various lives (n) and interest rates (i)

n	0%	2%	4%	6%	8%	10%	12%	15%	20%	25%
1	1.00000	1.02000	1.04000	1.06000	1.08000	1.10000	1.12000	1.15000	1.20000	1.25000
2	0.50000	0.51505	0.53020	0.54544	0.56077	0.57619	0.59170	0.61512	0.65455	0.69444
3	0.33333	0.34675	0.36035	0.37411	0.38803	0.40211	0.41635	0.43798	0.47473	0.51230
4	0.25000	0.26262	0.27549	0.28859	0.30192	0.31547	0.32923	0.35027	0.38629	0.42344
5	0.20000	0.21216	0.22463	0.23740	0.25046	0.26380	0.27741	0.29832	0.33438	0.37184
6	0.16667	0.17853	0.19076	0.20336	0.21632	0.22961	0.24323	0.26424	0.30071	0.33882
7	0.14286	0.15451	0.16661	0.17914	0.19207	0.20541	0.21912	0.24036	0.27742	0.31634
8	0.12500	0.13651	0.14853	0.16104	0.17401	0.18744	0.20130	0.22285	0.26061	0.30040
9	0.11111	0.12252	0.13449	0.14702	0.16008	0.17364	0.18768	0.20957	0.24808	0.28876
10	0.10000	0.11133	0.12320	0.13587	0.14903	0.16275	0.17698	0.19925	0.23852	0.28007
11	0.09091	0.10218	0.11415	0.12679	0.14008	0.15396	0.16842	0.19107	0.23110	0.27349
12	0.08333	0.09456	0.10655	0.11928	0.13270	0.14676	0.16144	0.18448	0.22526	0.26845
13	0.07692	0.08812	0.10014	0.11296	0.12652	0.14078	0.15568	0.17911	0.22062	0.26454

n	0%	2%	4%	6%	8%	10%	12%	15%	20%	25%
14	0.07143	0.08260	0.09467	0.10758	0.12130	0.13575	0.15087	0.17469	0.21689	0.26150
15	0.06667	0.07783	0.08994	0.10296	0.11683	0.13147	0.14682	0.17102	0.21388	0.25912
16	0.06250	0.07365	0.08582	0.09895	0.11298	0.12782	0.14339	0.16795	0.21144	0.25724
17	0.05882	0.06997	0.08220	0.09544	0.10963	0.12466	0.14046	0.16537	0.20944	0.25576
18	0.05556	0.06670	0.07899	0.09236	0.10670	0.12193	0.13794	0.16319	0.20781	0.25459
19	0.05263	0.06378	0.07614	0.08962	0.10413	0.11955	0.13576	0.16134	0.20646	0.25366
20	0.05000	0.06116	0.07358	0.08718	0.10185	0.11746	0.13388	0.15976	0.20536	0.25292
25	0.04000	0.05122	0.06401	0.07823	0.09368	0.11017	0.12750	0.15470	0.20212	0.25095
30	0.03333	0.04465	0.05783	0.07265	0.08883	0.10608	0.12414	0.15230	0.20085	0.25031
40	0.02500	0.03656	0.05052	0.06646	0.08386	0.10226	0.12130	0.15056	0.20014	0.25003
50	0.02000	0.03182	0.04655	0.06344	0.08174	0.10086	0.12042	0.15014	0.20002	0.25000
100	0.01000	0.02320	0.04081	0.06018	0.08004	0.10001	0.12000	0.15000	0.20000	0.25000

The administrative, operation and maintenance cost can be easily calculated on a yearly basis. But the value of C_D is to be estimated from the cost of construction. This is found out from the following relation:

$$C_D = R \times C$$

where R = Capital recovery factor

$$= \frac{i(1+i)^n}{(1+i)^n - 1}; \text{ and}$$

C = Construction cost of proposed improvement

Thus from known values of i, n and C, the value of R and hence C_D can be calculated. Usually tables have been computed for the values of recovery factor R for different values of analysis life n and rate of interest i. These values are given in Table 3.3.

3.10.2. Net Present Value Method (NPV)

In this method the present worth of benefits minus costs are compared for each alternative. The benefits/costs associated with the project over the life time of the project are calculated and discounted at a selected discount rate to give the present value. Benefits are treated as positive and costs as negative and the summation gives the net present value (NPV). The study should cover the same time period.

The most attractive alternative is one, which gives the highest positive difference between the present worth of benefits and the present worth of costs.

The NPV is algebraically expressed as:

$$NPV_0 = (B_0 - C_0) + \frac{B_1 - C_1}{(1+i)} + \frac{B_2 - C_2}{(1+i)^2} + \ldots + \frac{B_n - C_n}{(1+i)^n} \quad \ldots (3.2)$$

Where

NPV_0 = Net present value in the year 0.

B_1, B_2 = Value of benefits which occur in the years 1, 2...

C_1, C_2 = Value of costs which occurs in the years 1, 2...

i = discount rate per annum in decimals

n = number of years taken for analysis.

Problem 3.1 *An existing highway, 30 km long, was improved at a cost of ₹. 4.5 lacs per km. Road users costs, accidents costs and maintenance costs, with and without improvements are given in Table 3.4 for a period of 10 years. Using NPV method, determine whether the project is economically viable or not? The discount rate may be taken as 10 per cent.*

Solution.

Table 3.4

All amounts in ₹ Lacs

Year	Road user costs		Accident costs		Maintenance costs		Benefits		
x	With improvement	Without improvement	With improvement	Without improvement	With improvement	Without improvement	costs (3+5+7) –(2+4+6)	$B_x - C_x$	$\frac{B_x - C_x}{\left(1+\frac{10}{100}\right)^x}$
(1)	(2)	(3)	(4)	(5)	(6)	(7)	(8)	(9)	(10)
0	–	–	–	–	–	–	–	– 135.0	–135.0
1	206.4	227.4	11.2	13.2	8.6	6.6	21.0	21.0	+19.01
2	211.2	233.1	11.2	13.2	8.6	6.6	21.9	21.9	+ 18.01
3	216.7	239.8	11.3	13.6	8.6	6.6	23.4	23.4	+ 17.58
4	222.5	246.7	11.3	13.8	8.6	6.6	24.7	24.7	+ 16.87
5	228.5	254.0	11.4	13.9	8.6	6.6	26.0	26.0	+ 16.14
6	235.0	262.0	11.4	14.1	8.6	6.6	27.7	27.7	+ 15.63
7	241.6	269.8	11.5	14.3	8.6	6.6	29.0	29.0	+ 14.88
8	248.7	278.0	11.6	14.5	8.6	6.6	30.2	30.2	+ 14.09
9	256.0	287.1	11.7	14.8	8.6	6.6	32.2	32.2	+ 13.66
10	263.8	296.1	11.7	15.0	8.6	6.6	33.6	33.6	+ 12.95
									158.82
									– 135.00
									+ 23.82

Cost of improvement = 30 × 4.5 = ₹ 135 lacs

NPV = 158.82 – 135 = 23.82 lacs

Since NPV is positive, hence the project is economically viable.

3.10.3. Rate of Return Method

This method involves the determination of the rate of interest at which two or more alternatives will have equal annual costs. In this approach as alternative which would promise the highest rate of return on the investment is to be preferred.

To justify any proposed facility, the total annual net benefits should be equal to or greater than the average annual cost of ownership and operation. Therefore in the limiting case:

$$B = A + O + M + C_D$$

where B = Annual net benefits

But $C_D = R \times C$

$\therefore$ $B = A + O + M + R \times C$

or $R = \dfrac{B - A - O - M}{C}$

In case two facilities are to be compared, the above relation can be written as:

$$R = \frac{(U_1 - U_2) - (A_1 + O_1 + M_1) - (A_2 + O_2 + M_2)}{C}$$

where R = Capital recovery factor

U_1 = Road user costs on existing facility

U_2 = Road user costs on proposed facility

A_1, O_1, M_1 = Annual administration, operation and maintenance costs on existing facility

A_2, O_2, M_2 = Annual administration, operation and maintenance costs on proposed facility

Thus the value of R can be easily found out. Knowing the values of R and n the value of i is determined from Table 3.4. Similarly other alternatives can be compared and the one which gives the maximum rate of return should be selected.

Alternatively, in this method, the discount rate in determined which makes the stream of cash flows to zero. Equation (3.1) can be rearranged as follows (B_0 being zero)

$$C_0 = \frac{B_1 - C_1}{1+i} + \frac{B_2 - C_2}{(1+i)^2} + \ldots + \frac{B_n - C_n}{(1+i)^n}$$

$$= \sum_{x=1}^{n} \frac{B_x - C_x}{(1+i)^x}$$

The solution to above equation can be done by trial and error method, a rather tedious process. The work can be simplified with a computer programme. If the rate of return calculated from the above formula is greater than the interest rate obtainable by investing in the open market the scheme is considered economically justified.

The advantages of the rate of return method are:

1. It lends itself to the 'capital rationing' concept. The funds available for highway improvements are limited. The economic desirability of projects can be determined by listing them in the descending rates of return.
2. A cut off point at which all the funds are exhausted can be established. This will indicate the minimum attractive rate of return. Thus a standard can be established to judge all design and administrative decisions that have economy aspects.
3. Rate of return is clearly understood by a layman. Thus explanations in the parliament and legislative assemblies and to the public will be simpler.
4. Disagreement over appropriate interest rates are avoided.

The disadvantages of the method are:

1. The method is based on trial and error, whereas the solution is direct in other methods.
2. The solution may have multiple roots for problems where the stream of annual consequences swing between benefits and costs.

3.10.4. Benefit-Cost (B/C) Ratio Method

Benefit-cost method compares the value of alternatives by the ratio of annual benefits to annual costs. The benefit-cost ratio is expressed by :

$$\text{Benefit-cost ratio} = \frac{\text{benefits from improvement}}{\text{cost of improvement}} = \frac{B}{C}$$

where B = Discounted benefits

C = Cost of proposed facility.

Obviously, the benefit-cost ratio should be greater than one to justify the proposed improvement or alternative. The benefit-cost ratio method has limited use. This cannot be applied to drainage and resurfacing projects. Moreover, the choice of rate of interest seriously affects the results of benefit cost solutions.

Problem 3.2 *It is proposed to widen 25 km of an existing single lane road to two lanes. The cost of widening is ₹12 lakhs per km. The vehicle operating costs, accident costs and maintenance costs, for a period of 10 years, with and without widening is given in Table 3.5.*

The discount rate is 12 per cent. Determine, by the benefit-cost ratio method, whether the widening is economically justified or not?

Solution

Table 3.5 *All costs in ₹ Lacs*

Year	*Road User Costs*		*Accident Costs*		*Maintenance Costs*		*Benefits*	*Discounted*
t	*With widening*	*Without widening*	*With widening*	*Without widening*	*With widening*	*Without widening*	*costs* B_t [3+5+7] – [2+4+6]	*benefits* $\frac{B_x}{(1+i)^x} = \frac{B_x}{(1+0.12)^x}$
(1)	(2)	(3)	(4)	(5)	(6)	(7)	(8)	(9)
1	111.6	170.8	7.5	8.6	15.0	12.5	57.8	51.6
2	115.7	178.3	7.6	8.7	15.0	12.5	61.2	48.8
3	120.3	186.4	7.7	8.8	15.0	12.5	64.7	48.1
4	126.3	195.3	7.8	8.9	15.0	12.5	67.6	43.0
5	132.4	200.1	7.9	9.0	15.0	12.5	66.3	37.6
6	138.5	209.6	7.9	9.0	15.0	12.5	69.7	35.3
7	145.7	220.1	8.0	9.1	15.0	12.5	73.0	33.0
8	153.3	229.6	8.1	9.2	15.0	12.5	74.9	30.3
9	159.2	238.3	8.2	9.3	15.0	12.5	77.7	28.0
10	164.7	250.2	8.2	9.3	15.0	12.5	84.1	27.1
								382.8

Cost of widening = 25 × 12 = 300 lacs

Discounted benefits = 382.8 lacs

$$\therefore \quad \frac{B}{C} \text{ ratio} = \frac{382.8}{300} = 1.27 > 1$$

Hence the widening is economically justified.

3.11. ECONOMIC STUDY IN DEVELOPING COUNTRIES

The methods of economic analysis, discussed so far, are applicable to projects where there is already an access by motor vehicle and comparisons are made to the effects on the costs due to improvements or alternatives.

In developing countries, like India, economic analysis are intended to evaluate the effects of providing access for the first time in conjunction with agricultural or other resource developments. In such cases, comparison of costs before and after the provision of the facility are meaningless. Infact, the analysis must evaluate the entire investment, with transportation as only one of the elements.

Some roads may be constructed to extend medical, education and other services to isolated people or villagers. The aim is to bring such people into contact with the rest of the world and to bring higher living standards to them. Money value for such facilities cannot be specified.

An appropriate method for comparing projects will be the ratio between the number of people who are provided access and the cost of providing an access.

Problem 3.3 *A section of length 18 km of an existing road follows a circuitous route, has inadequate width of roadway and poor surface condition. There are three alternatives for improvement, namely:*

(A) To follow the existing alignment by bringing the road to fairly high standards by grading, widening of carriage-way and re-surfacing.

(B) To abandon the existing route and realigning the road along new alignment of length 13.5 km.

(C) To follow another route of length 13 km.

Estimated useful life and the costs of the three alternatives are given below:

		Cost in thousand rupees		
Element	*Estimated useful life (years)*	*Alternative A*	*Alternative B*	*Alternative C*
Right of way	100	0	280	430
Grading	50	350	740	910
Structures	50	210	280	300
Surface	15	280	2380	1960

The present average annual daily traffic is 600 cars with negligible commercial vehicles. The traffic is expected to increase by 50 per cent in 15 years. The average speed of vehicles on alternative A will be 60 km.p.h. and on the new alignments B and C 80 km. p.h. Effects of congestion can be neglected. Accident hazard is equal on all routes. Maintenance cost for alternative A is ₹8000 per km and for other alternatives ₹ 6000 per km.

Assuming the rate of interest as 6 per cent, work out the economics of the various proposals.

Solution.

(1) *Annual cost method (Interest at 6%)*

Annual cost of capital recovery

	Alternative A		Alternative B		Alternative C	
Element	*First cost (th ₹)*	*Annual cost (th ₹)*	*First cost (th ₹)*	*Annual cost (th ₹)*	*First cost (th ₹)*	*Annual cost (th ₹)*
From Table for $n = 100$ and $i = 6\%$, Right of way $R = 0.06018$	0	0	280	16.85	430	25.88
Grading R ($n = 50$) $= 0.06344$	350	22.20	740	46.95	910	57.43
Structures R ($n = 50$) $= 0.06344$	210	13.32	280	17.76	300	19.03
Surface R ($n = 15$) $= 0.10296$	280	28.83	2380	245.04	1960	210.80
Total annual cost of capital recovery		64.35		326.60		304.33

Assuming the road user cost for the various alternatives as follows:

For alternative A = ₹ 1/km

For alternatives B and C = ₹ 1.05/km.

The road user cost for alternatives B and C has been assumed to be higher since the average speed on these alternatives is more and the cost increases with increase in speed.

Road user costs (₹/km/year)

Vehicle type	*Vehicle km/year*	*Road user cost (thousand ₹)*	
		Alternative A (@ ₹ 1/km)	*Alternative B and C (@ ₹ 1.05/km)*
Cars	600 × 1.5 × 365 = 328500	328.5	344.925

Summary of annual costs

Annual cost of capital recovery

Cost item	*Alternative A* length	*Alternative A* Annual cost (th ₹)	*Alternative B* length	*Alternative B* Annual cost (th ₹)	*Alternative C* length	*Alternative C* Annual cost (th ₹)
Capital recovery		64.35		326.60		304.33
Maintenance	18 km @ ₹ 800	144.00	13.5 km @ ₹ 600	81.00	13 km @ ₹ 600	78.00
Road user cost	@ ₹ 328500	5913.00	@ ₹ 344925	4656.50	@ ₹ 344925	4484.03
Total annual cost		6121.35		5064.10		4866.36

Alternative C is somewhat better than other alternatives from economy study point of view.

In addition, alternative C will save some vehicle-hour of passenger car time annually, which is an added advantage. This alternative has worked out to be cheaper because of the assumption of lower rate of interest of 6%. If the rate of interest is enhanced, alternative B may work out to be slightly cheaper than alternative C. At some higher interest rate, alternative A will become the most attractive. The results are very sensitive to changes in assumptions in traffic volume and composition, running speeds and time costs. However the findings will be of great assistance in making a rational decision.

(2) ***Solution by present worth method*** **(Interest rate 6%)**

Converting the above result to present worth form.

Present worth of 40 year service.

Annual cost (th ₹)		Multiplier	Present worth (th ₹)	Advantage of alternative C
Alternative A	6121.35		73222	15012
Alternative B	5064.10	$\frac{1}{0.0836}$	60575	2365
Alternative C	4866.36		58210	0

(3) *Solution by benefit-cost-method*

Benefit-cost-ratio (*B* vs *A*)

$$= \frac{5913.0 - 4656.5}{(326.6 + 81) - (64.35 + 144.0)}$$

$$= \frac{1256.5}{199.25}$$

$$= 6.3$$

Benefit-cost-ratio (*C* vs *A*)

$$= \frac{5913.0 - 4484.03}{(304.33 + 78) - (144 + 64.35)}$$

$$= \frac{1428.97}{173.98}$$

$$= 8.2$$

Benefit-cost-ratio (*C* vs *B*)

$$= \frac{4656.5 - 4483.03}{(326.6 + 81) - (304.33 + 78)}$$

$$= \frac{173.47}{25.27}$$

$$= 6.86$$

Thus, if 6 per cent is the minimum attractive return, alternative *C* represents a higher benefit-cost-ratio and is economically a better proposition.

FINANCING OF HIGHWAYS

There are two general methods of financing highways, one commonly known as "*pay-as-you-go*" method, involves paying for all highway improvements and costs of maintaining and operating the highway system from current revenues. This is the method which is currently in use by many government agencies.

The second method is that of borrowing money to pay the costs of highway improvements, the borrowed sum plus interest being repaid over a large period of time from future income. This method is generally termed "*credit financing*".

3.12. PAY-AS-YOU-GO METHOD

In this case funds have to be collected for undertaking highway projects. Since highways are constructed by the government, it becomes the responsibility of the government to impose taxes upon or to extract charges from the public. The government is immediately faced with two problems:

(*i*) Who should pay for highways?

(*ii*) What should be the most appropriate method of raising funds?

The first problem is of vital importance. It must be decided as to who should pay for highway improvements-whether the general public or only a particular section *i.e.*, highway users. In the words of Owen: "*Is it desirable to include transportation facilities in the same category with general government services such as education and defence or whether transportation should rather be looked upon as similar to the supplying of food and clothing, of which it is a part, and therefore financed by the user*". There are very divergent views on this point. On one extreme are those who advocate "*every body benefits, so every body should pay*". They are of the view that highway facilities are needed for national defence, fire and police protection, sanitation and health, delivery of the mails, school buses and transit lines, telephone lines, pedestrians, etc. Roads increase the value of land both in urban and rural areas. Thus according to them highways should be financed from the general revenue funds. At the other extreme are those who want the road users to bear the entire cost of highway projects. The case for charging highway users for highway costs rests on the contention that while there are general benefits from improved highways, most of these benefits are the by-products of direct benefits to highway users from improvements. The direct saving to highway users made possible by improved highways, are more than sufficient to justify charging highway users for the costs of the highway projects. They compare highways to other productions which have subsequent benefits to others than the customers of the products produced. They argue that many other services, not financed by the government, also help to enhance the country's capacity for defence and promote the development of land and other resources. Since their cost is borne by the users, similarly highway facilities should be financed by road users. Also highway users are required to pay for their vehicles, fuels and other necessities. There appears little reason why they should not also pay for the highway services which constitutes only a small fraction of total highway transportation costs. Further, charging highway users for road improvements is necessary to avoid competition and discrimination between highway travel and other means of transportation. This is to fulfil the fundamental of tax neutrality among transportation agencies. From economic considerations, each transportation agency should bear full economic costs so that traffic may be allocated among them in relation to the economy and fitness of each. The assessment of user charges against highway carriers is a direct method of charging against them, and hence against their customer, the cost of supplying highway service.

There are still others who take a mid-way position. There is wide acceptance of the view that since highway users derive benefits from road improvements, they should pay a part of the cost of highway programmes, at least for the major traffic facilities. But at the same time, others should also contribute their share as highways enhance the value of their property, *etc*. Of course, the next issue of dispute is the extent to which others may be called upon to share the burden through special levies of one kind or the other. Highway costs between highway users and others may be based on the strength of benefits to them. There are two approaches to this problem. One known as the "*relative use*" approach assigns a part of costs of every highway to the general tax payers, this being low for the main roads and high for the approach roads to houses or property. In the other approach, called the "*predominant use*" approach, all costs of through highways is assigned to the road users and all costs of approach or land access road are attributed to the general tax payers with the reasoning that the benefits to users on approach roads and those to others on the through roads would tend to balance each other.

There are certain highway facilities which do not bring much or proportionate benefits to highway users. Under this category come approach roads to property, roads required for the development of natural resources, expensive tunnels and bridges *etc*. On such facilities, it is proposed that, if practicable, toll tax may be imposed to meet costs in excess of users'earnings, where toll is not a practical solution since these roads are required for the general interest, such as defence and education, the excess cost may be met out from the general revenue funds. If such roads provide access to land or natural resources, special taxes may be collected from the owners of the property to pay for the excessive costs.

Whatever, may be the theory, the general trend has been that those who derive benefits from highways, should pay for the road improvements. Thus it appears that the most appropriate method of raising funds is by taxing the beneficiaries of highway improvements. On this basis, the revenues for the "*pay-as-you-go*" method are usually collected by the following methods:

1. Highway users' taxes

2. Property taxes

3. General revenue funds

4. Miscellaneous sources, *e.g.*, tolls.

3.12.1. Highway Users' Taxes

Highway users' taxes are defined as those taxes which motor-vehicle operators are required to pay for highways over and above their obligations for the support of the general governments. This taxation was primarily developed in response to the demands for better roads. The objective of these taxes is to raise money with convenience and certainty in order to finance highway programmes. Their purpose is to recover for government costs of supplying highway service through direct charges on those utilizing the facilities.

In levy of this tax, it must fulfil fundamental tax equity among tax payers. Whatever may be the users' aggregate share, a question still remains to be solved : how to divide the share, among road users? If all motor vehicles were alike and make use of roads to the same extent every year, there would be no problem. This cannot happen and both passenger cars and commercial vehicles like trucks, lorries, which vary considerably in their weight, have to use the road to a varying degree. Moreover, they impose different requirements on the highway design. The problem is, therefore, very complex. In spite of the fact that considerable thought has been given by the research scholars to this problem, they have not been able to devise a single method for the equitable distribution of the users' burden. Four basic theories have been advanced for solving this problem. These are:

(*i*) Increment cost theory

(*ii*) Relative use theory

(*iii*) Value of service theory

(*iv*) Price rationing theory.

(*i*) *Increment cost theory.* According to this theory, highways have to be constructed for joint use of passenger cars and trucks. The cost of making a basic road suitable for passenger cars only should be realized by dividing the costs equally among all classes of users. The heavier vehicles should pay all costs in excess to the above costs. In others words each class of users should contribute in proportion to the costs occasioned by each. The taxes would be levied on the following basis:

(*a*) Costs common to all classes is to be charged among the various weight classes on some reasonable basis. The costs of right of way and administration would be allocated in proportion to the number of vehicles in each group. Traffic control costs and part of maintenance and construction expenses which are not affected by the weight of the vehicle would be divided on the basis of vehicle kilometres or axle kilometres or tonnes kilometres.

(*b*) Costs of greater base and pavement thickness, extra maintenance, etc. would be charged on a cumulative basis from heavier vehicles. Thus all classes would be required to bear the expenses in the basic increment applicable to passenger cars. Second weight group vehicles would share the added costs required to provide additional thickness for them. This continues till the highest weight group of vehicles keep paying the highest assignment of increment costs.

Tha heavier vehicles have caused extensive damage to highways is quite apparent. Therefore design of highways have to be considerably improved on account of them. But the theory of incremental costs is not so simple as it appears to be, because it has been found to be extremely difficult and sometimes impossible to separate the costs which are solely responsible for given classes of vehicles. This is particularly true when highway project is taken as a whole rather than a particular highway. Further, the assignment of costs which are common to more than one class of vehicles is a perplexing problem. This theory involves many arbitrary engineering judgements and is therefore not preferred.

(ii) Relative use theory. This is also termed as "*tonne-kilometre*" theory. According to this theory tonne-kilometre is considered to be an appropriate measure of highway use and thus can be used as a basic unit for distributing highway users' costs. Every vehicle moving on the road would be required to pay a fraction of costs in proportion to the tonne-kilometres operated. Thus a truck with a gross-weight of 10 tonnes and travelling one kilometre would be assigned ten times the highway costs than a passenger car weighing one tonne and travelling the same distance. Thus simplicity is one of the virtues of this theory. The theory is becoming popular because the major factors which affect highway costs are vehicle weights and distance travelled.

But both above mentioned theories, *i.e.*, increment costs theory and tonne kilometre theory, do not attempt to distinguish the value of highway services to individual users of particular classes of vehicles. These do not involve the "*value-of-service*" concept of dividing the cost.

(iii) Value of service theory. Theory involves the value of service concept of pricing. It attempts to distinguish the value of highway services to individual users of given class of vehicles. Value of service pricing is a departure from the method of pricing on the basis of fully associated costs. It may be discriminatory, since some users may pay more and others less than full costs. Value pricing helps to achieve optimum utilization of the highway facility. The project could be regarded a constant facility in the short run, and therefore, value pricing might be designed to secure larger volume of traffic when there is under-utilization of the plant with the result that contribution to the fixed costs which are not variable with use would be larger.

In fixing the user charges, two considerations weigh against value pricing theory. At any given time while some highways may be congested, many others are likely to be underutilized, but since user charges will be uniform, price fixed to reduce traffic on congested segments would also tend to reduce traffic on segments which are underutilized and similarly an attempt designed to increase traffic on underutilized segments would also increase traffic on the segments which are already congested. Secondly, in the long run there is much variability in cost than is supposed. The project can be adjusted according to the need of traffic.

However, in special circumstances the value of service considerations may be adopted with advantage. Take for example toll facilities. Once constructed a particular toll facility will have a high proportion of fixed costs. Now, when fixed costs are fully allocated, if a particular class of vehicle, say heavy trucks not using the facility, though such vehicles could be accommodated without congestion, are extended the facility, it would be beneficial to all users because of the additional contribution by the heavy vehicles to the constant costs, which otherwise would have been borne by the other users alone.

(iv) Price rationing theory. Another way of introducing the value pricing concept suggested by a number of economists is price rationing. It is observed that over-congestion of highway facilities creates substantial social costs by way of losses of time and direct money costs. If congestion could be eliminated, the social costs would be reduced. The method of price rationing attempts at achieving this objective by setting the user charges high enough so that such users of the facility which are less important would be reduced. The objective is to minimize economic and social costs whether or not highway expenditure are met.

Considered in the context of dynamic system, this theory is not acceptable as long as, meeting their effective demand, the users are willing to meet the full economic costs of highway. Also the expansion of the plant will, over a time, reduce if not eliminate entirely the costs of congestion. Additionally successful manipulation at a state-wise or area-wise user charge system for the achievement of short run objective sought would be difficult in actual application through rationing of space through the adept manipulation of tolls might be used on particular facilities.

Finally, the theory implies the elastic nature of the demand for highway service though in actual practice there is avoidance of a rather considerable in-elasticity and for that matter rather extreme changes in charges would be required in order to make a significant impact on traffic volumes.

The following taxes are imposed on the highway user:

(*i*) Fuel taxes
(*ii*) Registration taxes
(*iii*) Special taxes on commercial vehicles.

(i) Fuel taxes. Taxation of motor fuel, mainly petrol and diesel is the major element of the users' tax structure. This tax has a basic appeal that it is directly proportional to the use to which highways are put. It has further advantage in that the fuel consumption increases with an increase in the vehicle sizes and weights. The tax not only results in collecting large amount of revenue, but also makes the collection easy and simple thus requiring very little administration. The tax is charged from the producers and distributors of fuel who in turn are expected to collect from the consumers who find it convenient to pay at the time of purchasing the fuel.

This system has a loop-hole. Refunds are sometimes made for petrol and diesel not used on highways. Thus there is likelihood that producers and distributors may make false statements to evade the tax on fuel consumed for some highway uses. Moreover, the use of diesel fuel by heavy motor vehicles has complicated the matter still further. Since most of the diesel fuel is used for non-highway purposes, it will not be feasible to adopt the same procedure as is done in the case of petrol. In this case, it is suggested that the retail distributor may collect the tax on diesel fuel when he fuels motor vehicles. The problem is further complicated because of the difference in rates of consumption of petrol and diesel fuel. It may be recalled that the purpose of the tax is to metre highway use on the basis of litres of fuel consumed. If one type of fuel yields more kilometres per litre than another for the same vehicle, identical tax rates will result in different charges per kilometre of road use. It is therefore, suggested that higher rates than those of petrol should be charged on diesel fuel because diesel vehicles consume less fuel per kilometre than petrol powered vehicles.

(ii) Registration taxes. The procedure of registration was first adopted as a regulatory device but now it has become a source of revenue to supplement the fuel tax. It is usually charged one time or annually from the time of registration of the vehicle. A portion of the cost is utilized for meeting the expenses associated with the process of registration. The main object of this tax is to recover, especially from heavy vehicles, assigned highway costs not met by fuel taxation. Registration taxes are collected according to the weight and size of vehicles. The most common measure being adopted is the maximum gross weight of the vehicle and loaded weight as declared by the owner at the time of registration although empty weight and rated capacity are also widely used.

Registration tax has also its drawbacks. Firstly it does not account for the kilometreage travelled by each vehicle group. Same amount of taxes are levied on vehicles of equal weight and size irrespective of the kilometres driven. Another problem posed by registration taxes is the method of its collection for vehicles which operate in more than one state. Should each state charge full annual tax from these vehicles? This will amount to hardship to the operators and the inter-state commerce may receive a great set back. Thus the notion of tax reciprocity was developed. One state would not tax vehicles registered in another, if that state did not tax its registered vehicles. This would work well so long as there is a balance of vehicle movement between the two states.

(iii) Special taxes on commercial vehicles. The practice of assessing special fees against motor vehicle used for commercial purposes including both passengers and freight carriers has developed gradually. Special fees are levied against this class of motor vehicles in proportion to their use of highway system and roughly in proportion to their effect upon the physical design of the highway.

These imposts have generally taken the form of graduated fees. The special nature of these vehicles is also taken into account by the graduated registration fee. It is an admitted fact that certain classes of commercial vehicles, *e.g.*, heavy trucks, tractor-trailer combinations are the main sources for whom the design thickness of pavement has to be increased, and thereby increasing the cost of highway. The thickness of pavement, which is required to carry heavy trucks may be considerably greater than that required for adequate support of passenger automobiles. Similarly a direct effect may be seen in providing extra lanes for heavy trucks on long routes and high grades. The increased cost of highway should be charged from these commercial vehicles by levying additional taxes.

3.12.2. Property Taxes

These are of two types:

(*a*) Valorem property tax.

(*b*) Special assessments against real estate lying contiguous or adjacent to a highway improvement.

Valorem property tax is extensively used in financing highway improvements by local bodies. The common concept of the use of general property taxes for highway improvements is justified if improvements which are effected are of general benefit to the community. Money derived from this source goes into the general revenue fund and is then budgeted, generally on an annual basis for highways, education, *etc*.

Special assessments are those which are made directly upon the owners of properties which lie on or adjacent to a highway or street in proportion to the benefits which are expected to accrue to the owners of such properties. Special assessment is made on the basis of frontage rather than on the basis of property value.

3.12.3. General Revenue Fund

Highways are of economic importance to the society. Thus it is proper that these should be financed by general revenues upto a certain extent. There are roads which are vitally important for movement of troops for the defence of the country, especially in the border areas and these roads should be financed partly from the general revenue. The thickness of pavement and other structures have to be designed according to types of military vehicles which will be run on the road. Increased thickness or pavement will naturally increase the cost of the highway. The increased cost of construction and maintenance of such highways should be financed from general revenue.

3.12.4. Toll Roads (Miscellaneous Sources)

Sometimes important bridges and roads are constructed by the centre, state governments or local bodies and then toll is collected from the traffic. Collection is made in varying amounts depending upon the type of vehicle and the extent of damage it might cause to the roadway. It is least for passenger cars and motor cycles and heaviest for trucks laden with load. In the U.S.A. some of the earlier free-ways and express-ways were constructed by local bodies by borrowing money, and then levying toll. These highways could attract through traffic from congested roads. This method has certain drawbacks which may be summarized as follows:

(*a*) These roads do not offer a nation-wide solution to the urgent need of highways.

(*b*) These roads are difficult to be fitted in an overall highway system.

(*c*) Toll collection is difficult.

(*d*) Many highway users prefer to travel on other route than pay tolls and also it is inconvenient to vehicles making short trips.

3.13. CREDIT FINANCING

If the earlier use of money is more valuable than the interest which must be paid, then as a general rule, it is not objectionable for the government to borrow money. Thus credit financing of highways is a time honoured practice of governments.

In this case it is essential to consider the advantages that will be derived from the credit financing of highway improvements and the conditions under which these advantages are likely to be realised to the fullest degree.

The advantages resulting from highway improvements include (*i*) reduction of vehicles operating cost, (*ii*) reduction of accident, and (*iii*) lessening of interruptions to steady flow of traffic. Since all these benefits will be realized at an earlier date under an accelerated bond issue programme, their accumulated values over a given time will be much greater than under a long term current revenue programme. The issue of bonds will be justified if the excess benefits due to acceleration are greater than the interest charges on the loan.

The principle of acceleration is the key-note to the credit financing of highways. The most favourable condition for credit financing is that of a truly accelerated programme which contemplates a short period of abnormally high capital outlay expenditures, during which the highway plan will progress rapidly toward a condition of adequacy. A subsequent lull in construction activity during which the need for replacement will accumulate very slowly will provide an opportunity for the retirement of bonds.

3.13.1. Credit Financing versus Pay-as-you-go Method

Highway financing is a continuing operation, especially when higher user taxes are employed. Revenues are generated by the entire highway plan that can be utilized from the improvement of segments. Now improvements too when put into operation, generate revenues which can again be used elsewhere. The situation may be compared to the undistributed corporate profits that might be utilized for expansion of other industries.

The pay-as-you-go method has a few advantages over the credit financing method. Forecasting future needs and rates of obsolescence are at best the estimates which may or may not prove to be accurate. There are numerous instances that state governments and local bodies have been found paying for capital facilities even after they have outlived their utility. In the process of growth, every estimate of future needs record an increase upon the previous one. Would it not be appropriate for the people, therefore, to doubt the wisdom of passing on the future a part of the burden which will compound problems by requiring future users not only to meet burdens of ever greater magnitude in providing for their own needs but also to pay for past improvements?

It is also argued that interest has to be paid on borrowed funds which leads to accumulation of liabilities for future users, though on grounds of sound economic reasoning it is difficult to support the argument.

Whatsoever the arguments against the use of credit financing we have yet to attain that stage of accuracy in preparing estimates in highway financing where we can confidently finance on a budget basis, using credit for major improvements, providing for maintenance and operations, depreciation and obsolescence interest and possibly tax equivalents, all out of current revenues.

Use of credit in the financing of capital facilities of long life has many advantages. The burden of highway could be supported over a time, by means of credit financing. The costs of the project could be equitably spread out. It promotes neutrality of public policy for competing carriers. The users of a publicly provided facility are required to meet only the economic costs. They are not subjected to the burden of a particular expenditure programme which may be more or less than costs in any particular year. For financing a credit programme, the tax structure would be closer to actual economic costs than the tax structure for a pay-as-you-go programme. The use of credit permits a larger programme than would be possible with current taxes. This helps to overcome accumulated deficiencies and to avail of the savings associated with highway improvements at an earlier time.

In actual practice, the question of adopting pay-as-you-go or credit method has to be considered in the broad context of general economic policy including the important issue of economic stabilization. Highway finance is still closely tied, and rightly so, to general fiscal considerations. The decisions for the use of credit to highway are taken by the state governments and local bodies in the light of their general financial position. Constitutional and statutory debt limits have to be kept in view. While taking decisions, projects other than highway; for which ready sources of current revenue will be required, equally draw the attention of the authorities.

During the period when underemployment prevails, the credit is likely to be helpful to increase aggregate purchasing power and thereby to stimulate employment when pay-as-you-go programme will simply transfer resources from individuals to the government. On the other hand, when there is full employment in the country it may not be advisable for the central government to resort to credit, for it will most likely promote inflation. In effect, during the period when full employment prevails and when inflationary pressure appear to be strong, a strict pay-as-you-go approach may be accepted whereas in the face of an economic recession and substantial unemployment temporary deficit financing may be resorted to in order to accelerate the programme.

FINANCING ROADS IN INDIA

3.14. TAXES LEVIED BY THE CENTRE AND STATES

The various taxes levied by the Centre and States are :

Centre

1. Import or excise duty on motor vehicles, motor parts and accessories.
2. Import or excise duty on motor tyres and tubes
3. Import or excise duty on petrol, diesel, motor-oil and other fuel oils
4. Toll at selected bridges, tunnels and roads on National Highways

States

1. Sales tax on motor-vehicles, motor parts and accessories
2. Sales tax on petrol and diesel
3. Motor vehicle tax
4. Taxes on passengers and goods
5. Driver's licence free
6. Permit fees
7. Toll levied by the state
8. Wheel tax

In India roads are financed by the central or state government or even local bodies, depending upon the class of road. The different agencies of road financing are summarized below:

3.15. FINANCING OF ROADS

3.15.1. Roads Financed Wholly by Central Government

Certain highways have been classified as national highways by the National Highways Act, 1956. All national highways including their construction, maintenance, improvements and construction of minor and major bridges thereon, are financed by the central government. The total length of national highways in India is about 76818 kilometres. Another 10000 kms are expected to be added in the next

10 years. Financing of NHs under the National Highway Development Plan (NHDP) is done by the following methods:

1. Gross Budgetary Support (GBS) and Additional Budgetary Support (ABS).
2. Dedicated accurals under the Central Road Fund. Present rate of cess is ₹ 2.00 per litre on both petrol and diesel. A part of this cess is allocated to NHAI to fund the NHDP.
3. External Assistance through World Bank, ADB, JBIC, and so on.
4. Ploughing back of toll revenue including toll collection, negative grant, premium and revenue share deposited by NHAI into Consolidated Fund of India and equivalent amount to be released to NHAI for ploughing back in its projects.
5. Private Sector Investment under Public Private Partnership (PPP) frameworks that os-BOT-(Toll) BOT (Annuity). Special Purpose Vehicle (SPV) with Equity participation by NHAI.
6. Market Borrowing by NHAI as authorised by GOI to bridge the gap between the available resources and funds requirement.

3.15.2. Roads financed by the state and centre

The Central Government helps the state governments in the construction and improvements of roads of secondary importance by giving grants in the following manner:

1. By allocating funds to the states from the central road fund which is non-lapsable.
2. By giving special grants approved by the Planning Commission for construction and improvements of roads of Inter-state or economic importance.
3. By giving funds for improving geometric design standards of roads and bridges.

3.15.3. Roads financed by state governments

State government finance the construction and maintenance of all roads other than national highways from the state revenues. The funds are allotted year by year and are lapsable.

3.15.4. Roads financed by local bodies

Local bodies including district boards or panchayats, municipalities and corporation finance all roads under their jurisdiction.

3.15.5. Private sector investment

Under Public Private Partnership (PPP) frameworks that is BOT (Build, Operate and Transfer) scheme.

3.16. TAXES LEVIED

The roads are financed by collecting funds from the levy of taxes by the centre and state governments. The details are given below:

3.16.1. Taxes levied by the centre

The existing sources of financing by the central government are :

1. Government budget including sovereign borrowings
2. Fee/toll on roads and bridges
3. Central road fund
4. Market committee fee
5. Other levies/surcharge

3.16.1.1. Government budget

Funds for road development are provided in the budget, Central Government for National Highways and States for other roads. Presently funds to the tune of ₹ 1500000 crores are made available for all categories of roads. This also includes loan assistance given by World Bank, Asia Development Bank

(ADB), Japan Bank for International Cooperation (JBIC). National Bank for Agriculture and Rural Development (NABARD) etc.

3.16.1.2. Fee/toll on roads and bridges

Fees/tolls are levied by both the Central/State Government on bridges/roads on N.H. and S.H. These funds are spent on upgradation/improvement of roads.

3.16.1.3. Central road fund

Funds to the tune of ₹ 50 crores come from Central Road Fund through levying of 3.5 paise per litre of petrol and diesel. Grants are allotted to states on non-lapsable basis.

The division is as follows:

(*i*) 20 per cent of the annual revenue is retained as a central reserve for giving grants by G.O.I. for meeting expenditure on the administration of funds, R&D, and suitable roads and bridges in the states.

(*ii*) The balance 80 per cent is allocated to the states on the basis of the actual petrol consumption.

A revised resolution was adopted in the Parliament in 1988 which provided for setting aside an amount not less than 5 per cent of the basic price out of the duty of customs and excise levied on petrol and diesel. The G.O.I again decided to annul this resolution and create a dedicated funds for roads. The major source of this fund is the additional excise duty of ₹ 2.00 per litre on petrol levied since 2.6.1998 and on high speed diesel since 1.3.1999. This generates a fund of over 35000 crores per year and is likely to grow with increase in the consumption of fuel. The allocation of Dedicated Road Fund is an under:

(*i*) 50 per cent of the proceeds from additional excise duty on petrol and diesel is set aside for rural roads (being essential for rural development).

(*ii*) The balance 50 per cent to be distributed as follows:

57.5 per cent : National Highways
27.0 per cent : States Roads
3.0 per cent : Roads of interstate and economic importance
12.5 per cent : Railways for safety works *e.g.* rail-road over bridges, manning of level crossings

3.16.1.4. Market committee fee

Some states e.g. Punjab, Haryana, Rajasthan, Uttar Pradesh levy cess on food grains through their market committees and part of fund thus collected, is utilized for the construction of link roads and their maintenance.

3.16.1.5. Other levies/surcharge

Funds for roads are made available from other schemes such as Jawahar Rozgar Yojana, Command Area Development Agencies and from budgets of other sectors such as tourism, mining, power plants *etc.*

3.16.2. State taxes

State vehicle tax is levied under state motor vehicle taxation acts. These acts were enacted in various states between 1924 and 1952, and are revenue measures. The rates of taxes vary from state to state. This consists of :

(*a*) Registration fee for registration of vehicles
(*b*) Issue and renewal of driving licenses, issue of duplicate licenses
(*c*) Levy of permit fees on transport vehicles
(*d*) A tax on passengers or goods carried by motor vehicles.
(*e*) Sales tax on motor vehicles, parts and accessories including tyres, tubes and on motor spirit.

These are levied at varying rates and under varying conditions in different states.

3.16.3. Local Bodies Taxes

The main tax on motor vehicles levied by local bodies is the wheel tax which exists only in a few states. The rate also varies in municipalities. In addition, some local bodies levy tolls and other similar taxes. Details of these taxes are not available, but generally taxes are small in magnitude and neglected.

3.17. RECOMMENDATIONS FOR MOBILISATION OF ADDITIONAL FUNDS

No one would dispute that India needs more and better roads which in fact, are the main arteries through which the life blood of economic activity flows. But then we would require very large sums of money, much beyond the limits of available resources. For example in India, if the targets set in Road Development Plan Vision 2021 are to be achieved, we will require in the next 10 years (2011-21) funds to the ture of ₹ 140,000 crores for National Highways and ₹ 300,000 crores for State Highways and Major District Roads. In addition, huge funds will be required for Other District roads and Village roads.

Some of the recommendations for raising more funds are given below:

(1) ***Independent road funds.*** The taxes collected by centre or state governments from road users, *e.g.*, import and excise duty on motor fuel, tax on petrol, motor-vehicle registration, *etc*. should form a separate revenue head. There should be no diversion of highway user taxes to purposes other than road financing. As regards the revenue derived from other sources such as import duty on motor vehicles, tyres and spare parts, sales tax on motor vehicles, tyres petrol, *etc*. a certain percentage of the total income should be earmarked for road works and should be pooled into the road fund mentioned above.

(2) ***Additional taxes on commercial vehicles.*** The design of road is considerably affected if highways are to accommodate trucks etc. Such heavier vehicles have caused extensive damage to the existing highways. These trucks should bear the increased cost of highways. Additional taxes should be levied on the basis of incremental cost theory.

The bullock cart and other iron tyred vehicles cause a lot of damage to pavements. According to the theory of price rationing, the traffic which requires extra maintenance, *etc*. should be taxed for financing highways. At present there is no tax on these vehicles. It is not fair to other vehicle operators to tax for other's benefit. A nominal tax of even ₹ 1000 per year per bullock cart will do a lot in financing. Moreover, these iron tyred vehicles damage the road most. We should make a legislation incorporating that these vehicles should have a minimum width of wheel, at least 15 cm. The pneumatic tyred bullock cart should be encouraged by taxing less.

(3) ***Bond issue.*** India still requires a network of highways. Every village is not connected with major highways. Most of our roads are not metalled. For all these improvements and new construction, we require an accelerated highway programme. But the present difficulty is of finance.

The government should issue 20 to 30 year bonds at a certain rate of interest free of tax. These bonds can bring money from public and from private sector. The advantages resulting from highway improvements will be a cause for reduction of vehicle operating cost, reduction of accidents, etc. In addition there will be developmental advantages derived from highway improvement by rural and urban communities. Since all of these benefits will be realized at an early date and under an accelerated bond issue programme, their accumulated values over a given time, will be much greater than under a long term current revenue programme.

These bonds can be retired by periodic payments upon the principal plus periodic payments of interest. In their most common forms constant payments are made upon the principal each 5 year, with interest payments which decreases as time passes. Bonds are completely retired at the end of their term of issue.

(4) ***Public Private Partnership (PPP).*** We know that railways in India were built by private companies to start with and government entered into a contract with these companies, according to which, these companies earned profits and were owners of the railways for a fixed number of years. After this, the railways were government property when the contract period expired.

On the same basis, government may enter into contract with private companies for the construction, operation and maintenance of certain important highways. The companies should be authorized to collect tolls from the people for a fixed number of years and when this period expired, these highways should be transfreed to the government to become national property. By adopting this method, we can give the public the benefits, *etc.* and company can charge a certain toll fixed by the government.

The NHAI have identified several segments of National Highways where 4-laning/6-laning is proposed to be taken up through private sector on Build-Operate-Transfer (BOT) basis. Three models are being tried. These relate to actual toll from the user, annuity payments and shadow tolls. Success has been achieved by collecting actual toll from the users, in respect of small and medium sized projects such as bridges, bypasses, railway over bridges, flyovers and four laning projects in the state sector. Such schemes should be thoroughly examined taking into consideration several factors such as:

(*i*) Minimum spacing of toll plazas so that users are not made to stop too frequently. Minimum spacing of 80 km is considered reasonal on National Highways.

(*ii*) Integrating the on going spot projects where tolls are already being collected with new upgrading projects through a formal agreement between the parties.

(*iii*) Considering the need of local traffic including agricultural tractors, *etc.* to use such facility either free or on nominal fee basis, provision of service roads in both urban and rural areas will take care in addressing this concern. This might involve some additional expenditure on additional acquisition of land for service lanes and providing grade-separated facilities at some suitable locations. This will, however, allow much higher capacity of the facility and level of service to the users paying tolls.

(5) *Road levy.* A road levy on beneficiaries, such as agriculturists, business concerns, *etc.* should be imposed.

(6) *Increased tax on diesel oil.* In view of the fact that more and more vehicles are being run now-a-days on diesel fuel and as a litre of diesel gives more kilometreage than a litre of petrol, highway tax levied on diesel oil used on road transport should be increased suitably and this money should also go to the separate road fund known as '*Dedicated Road Fund*'.

(7) *Levy on community.* Betterment levy on the communities served by new roads and owners of adjacent property should be charged.

As the construction of new roads will greatly reduce the transport cost and also offer manifold benefits to the community and the owners of properties adjacent to roads, they may be taxed to contribute towards the cost of construction and maintenance of new roads. It is also suggested that in the case of taxes to be levied on communities, payment may be accepted in the form of labour or materials or transportation of materials, which may be equated in terms of money value.

(8) *Borrowings from domestic and outside agencies.* NABARD in India provides loan assistance for construction of rural roads in several states. Other agencies like HUDCO, IDFC *etc.* also provide finances for road projects on commercialisation principles.

Similarly World Bank, ADB and JBIC are providing loan assistance for highway projects to the Centre and States.

All such sources should be tapped in future also.

3.18. OUTLAY FOR THE 12th FIVE YEAR PLAN (2012-2017)

The Twelth Plan budgetary support for central sector roads is ₹ 144769 crores. In addition, the sector is expected to generate IEBR amounting to ₹ 64834 crore and private sector investment of ₹ 214186 crores during the period.

The 12th plan budgetary support for Rural Roads (PMGSY) is ₹ 126491 crores.

❑❑❑

TRAFFIC ENGINEERING

Traffic engineering is defined as that phase of highway engineering which deals with the planning and design of highways and abutting lands, and with the traffic operations thereon for the safe, convenient and economic transportation of persons and goods.

HISTORICAL REVIEW

It was in the twenties of the last century that the role of a traffic engineer was recognized in any city. However, the installation of the first traffic signal goes back to 1912. In those days, the scope was very much restricted, being mainly the management of the traffic signals. No traffic studies, such as the volume studies, speed studies, parking studies, origin-destination studies, road life studies, etc. were made to form the basis for the establishment of traffic regulations for planning streets and highways.

No national standards for uniform design and application were made for traffic control devices and the traffic signs usually followed local design.

Today most of the big cities employ traffic engineers. In India traffic problems are fast coming up in big cities such as New Delhi, Mumbai, Kolkata, Chennai, *etc.* and the role of traffic engineering is becoming increasingly important.

With the increasing traffic development and safety requirements, it is absolutely essential that the Highway Departments in the States should, have a suitable cell to deal with traffic engineering, road standards, sizes and weights of vehicles and provision of amenities for road users.

4.1. SCOPE OR FUNCTIONS OF TRAFFIC ENGINEERING

The scope of traffic engineering can be covered by the following five categories:

1. *Traffic characteristics:*

(*a*) Vehicular limitations, *e.g.,* the weight, size and power of the vehicles (to accelerate and decelerate)

(*b*) Human limitations

(*i*) Physical limitations, *e.g.,* vision, hearing, fatigue, *etc.*

(*ii*) Mental limitations, *e.g.,* intelligence, skill, experience of drivers, *etc.*

(*iii*) Emotional limitations, *e.g.,* attentiveness, impatience, ability to follow regulations, *etc.*

2. *Traffic operations*, which constitute the traffic regulations, traffic control devices like the traffic signs, signals, markings, channelization, *etc.*

3. *Traffic planning,* *e.g.,* programme of construction, planning of major streets and freeways, terminals, off-street parking, *etc.*

4. *Traffic geometric design,* involving the design of expressways, streets, interchanges, intersections, parking lots, garages, shopping centres, *etc.*

5. *Traffic administration,* involving what are popularly known as the 3 Es, *viz.,* Engineering, Enforcement and Education.

Table 4.1 indicates the principles of traffic engineering and their functional relationship.

Table 4.1 Function of traffic engineering

Traffic investigation and study	*Planning*	*Design*	*Operation*
Transportation surveys Parking surveys	National and regional development of land and facilities	New road design, including intersections	Capacity characteristics
Traffic and pedestrian characteristics Road utilisation	Integration of transportation road, rail, air and water	Redesign of intersections and existing street systems standards	Signs, markings and signal systems street lighting
Road safety and accidents	Environmental planning	Service areas parking	Regulation and control of traffic
Environmental surveys	Utilisation of resources	Pedestriation and terminals	Traffic segregation
The study and improvement of highway transportation	Community satisfactions and individual needs	Design for safety and economy	Efficient operation of the road system

Education	*Enforcement*	*Engineering*
Schools	Legislation for:	Vehicle engineering
Adult education	Driver and vehicle	Highways
Public press	Licensing	Traffic engineering
Radio-Television	Police and courts	Pollution control
Re-education of offenders	Regulatory policy	Information, Engineering

TRAFFIC STUDIES

A detailed knowledge of the operating characteristics of the traffic is essential to form a basis for the establishment of traffic control or for design of streets and highways. The results of data collected are used in (*i*) traffic planning (*ii*) traffic management (*iii*) economic studies (*iv*) traffic and environmental control; and (*v*) monitoring trends, both for the establishment and updating of design standards.

Traffic studies can broadly be classified into two categories, namely:

(1) *Those concerned with the characteristics of traffic in transit, e.g.,*

(*a*) Volume studics and characteristics

(*b*) Speed and delay studies

(*c*) Origin-destination studies

(*d*) Road-life studies

(*e*) Motor-vehicle use studies

(2) *Those related to land use movements, e.g.,*

(*a*) Parking studies; and

(*b*) Accident studies.

These studies are discussed, in detail in the following articles.

4.2. VOLUME STUDIES AND CHARACTERISTICS

Volume of traffic is a very important variable and is essentially the quantity of movement per unit of time at a specified location. The quantity of movement may be either of single traffic unit-pedestrians, cars, buses or goods vehicles or of composite groups such as tractor-trailor, *etc*. Time period will depend on the purpose of study. The required level of accuracy will determine the frequency, duration and sub-division of the particular flow.

Volume studies are basically useful to establish:

(*i*) the relative importance of any route

(*ii*) the fluctuations in flows

(*iii*) the distribution of traffic on the road system; and

(*iv*) the trends in the road use.

In addition to the above, traffic volume studies have many other purposes. Table 4.2 lists the various uses of volume studies.

Table 4.2 Uses of traffic volume studies

General	*Traffic design*	*Traffic studies*
Highway classification and budget allocation	Capacity requirements Geometric design	Daily, weekly or monthly distribution of volume on road net-work
Highway user trends	Accident rates	Classified volume for design of roads and terminals
Planning	Economic computations	Structural design
Checking traffic	Traffic and environmental management	Noise and pollution
Calculations and scaling surveys	Traffic control	Parking

Specific methods of the various traffic volume studies are given below:

(*a*) *Manual counting*. Trained persons are posted at each leg of an intersection, each being asked to count and record the number of trucks or buses, cars, bullock carts, *etc*. The observers' work can be simplified by the use of mechanical or electrical tally counters.

(*b*) *Automatic recorders or counters*. These are mainly of two types:

(*i*) Permanent counting recorders which may be of any of the following three types (permanently fixed up on the road.)

(*a*) Magnetic detectors

(*b*) Pressure sensitive detectors

(*c*) Electronic detectors.

(*ii*) Portable recorders which are smaller in size and actuated by air switches, as shown in Fig. 4.1.

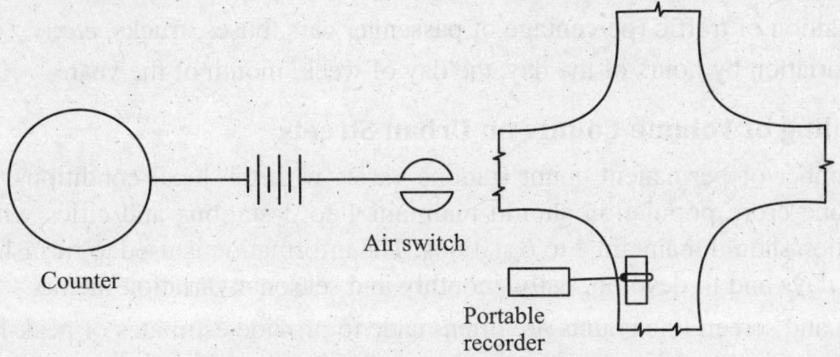

Fig. 4.1 Portable recorder.

(c) *Moving car method of counting*. (Wardrop Charlesworth method, developed in England). This method consists in counting the number of cars met, number of cars overtaken, and the time taken to travel by the observer, once moving against the traffic and once moving along with the traffic.

If q = Traffic flow in one direction (vehicles per minute)

x = Number of cars met, when moving against the traffic

y = Number of overtaking minus number of cars overtaken by the observer, when travelling along with the traffic

t_a = Time taken by observer to traverse the route against the traffic

t_w = Time taken by observer to traverse the route with the traffic

T = Average time for the traffic to travel the route.

Then the traffic volume in one direction for each section of the route for each class of vehicle is given by:

$$q = \frac{x + y}{t_a + t_w}$$

The average journey in minutes of the particular class of vehicle in the stream can be obtained from the relation:

$$t = t_w - \frac{y}{q}$$

This information is of use in speed and delay studies.

4.2.1. Scheduling of Volume Counts on Rural Highways

Indian Road Congress in 1972 has recommended a repetitive 7 day count taken twice every year, once during the peak season of harvesting and marketing and the other during the lean period at pre-selected stations. These counts are intended mainly for N.H., S.H and M.D.R. No fixed criteria is laid down for the selection of number of stations.

Later on, a continuous count of traffic at a number of stations in the country was done for 9, 10, 11, 12 and 13 days a year; with the objective to get an accurate data for hourly, daily, weekly variation of traffic. In case the study is carried out continuously, it can also give an idea of annual rate of growth of traffic, along with the monthly and seasonal variations.

The classification counts to be taken at needed control stations should take into consideration the following factors:

(*i*) Classification of highway (*e.g.,* National, State, District or Village Road). In case of urban areas Arterial streets, Sub-arterial streets, Collector streets, local streets *etc.*)

(*ii*) Whether inter-state or inter-city.

(*iii*) Composition of traffic (percentage of passenger cars, buses, trucks, *etc*.)

(*iv*) Time variation by hours of the day, the day of week, month of the year.

4.2.2. Scheduling of Volume Counts on Urban Streets

(*i*) The number of permanent count stations varies with the local conditions. However, cities under one crore population should maintain 1 to 3 stations and cities with 1 to 5 crores population should maintain 4 to 6 stations. The information is used to furnish volume data on special days and to develop, daily, monthly and seasonal variation factors.

(*ii*) Cordon and screen line counts are often made to provide estimates of peak-hour volume.

(*iii*) Evening peak hour counts at intersections of two major streets, once a month or so.

(*iv*) Quarterly counts on representative streets and different parts of the city to provide appropriately expansion factors.

(*v*) Coverage counts to provide current data.

4.2.3. Types of Volume Counts

(*a*) *ADT or average daily traffic volume* is the average of 365 days (24 hours a day). It can be used for the following purposes:

(*i*) Planning major streets.

(*ii*) Improvement, construction or reconstruction of roads.

(*iii*) Computing accident rates.

(*iv*) Computing highway user revenue.

(*b*) *Classified volumes* involving the composition of traffic, *i.e.*, trucks, cars, rickshaws, bullock carts, *etc*. These volumes can be used for geometric and structural design of roads, computing capacities of highways user revenues.

(*c*) *Hourly volumes* (as shown in Fig. 4.2) usually required within cities) are used for:

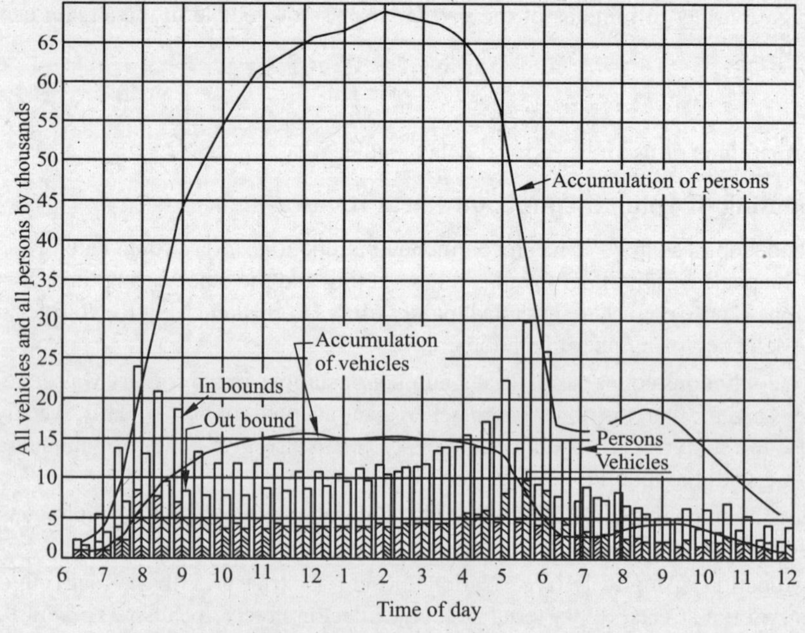

Fig. 4.2 Traffic variations by hours of the day.

(*i*) Determining deficiency in capacities, geometric designs, *etc*.

(*ii*) Determining number and width of lanes

(*iii*) Parking demands

(*iv*) Planning traffic control (traffic signs, signals, timing of signals, *etc*.)

(*v*) Location of interchanges.

(*d*) *Directional distribution.* Out of the four lanes leading to a central business district, we may have three lanes IN and one lane OUT in the morning when most of the vehicles will be proceeding towards the central business district and follow the reverse arrangement in the evening. It is also used for parking prohibitions, capacity analysis, *etc*.

(*e*) *Cordon count volumes.* For planning parking facilities, modes of transportation, *etc*. the cordon count volumes are used to obtain accumulation of vehicles inside the cordon area.

(*f*) *Pedestrian volumes.* These are utilized in planning the cross walks and signals for pedestrians.

(*g*) *Turning movements counts.* These are used in the design of intersections and interchanges, planning of signal timings and turn prohibitions, channelization, *etc*.

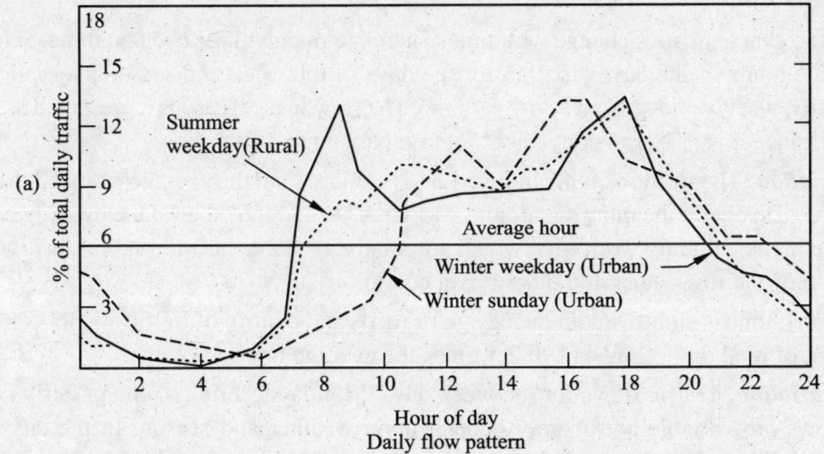

Daily flow pattern

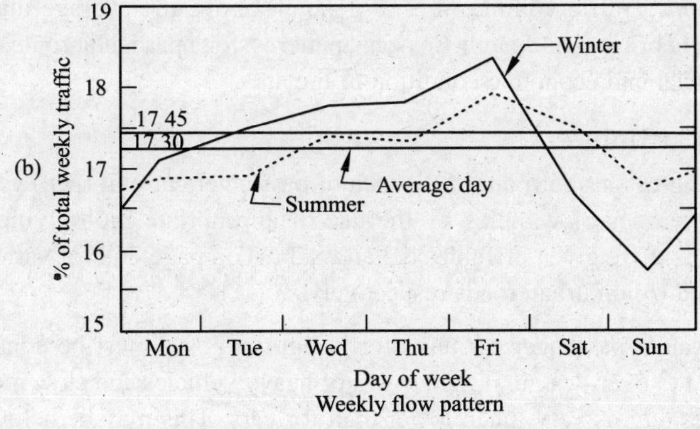

Weekly flow pattern

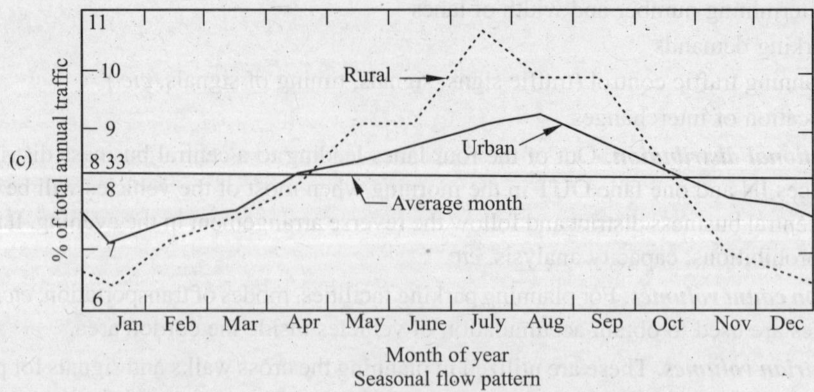

Fig. 4.3 Daily, weekly and seasonal flow patterns.

4.2.4. Volume Characteristics

The volume characteristics change with time. There are mainly three cyclical trends, *i.e.*, variation throughout the hours of the day, variation for the days of the week and seasonal variation. Fig. 4.3 shows the daily, weekly and seasonal flow patterns. The flow in each case is represented as percentage of total flow, since it is easy and convenient for comparison and study.

Daily variation. The traffic volume in Fig. 4.3 (*a*) is more than the average from 7 A.M. There are two distinct peaks, one in the morning around 9 to 10 A.M. and the other in the evening at about 4 to 6 P.M. These include mainly work trips which are relatively stable in time and insensitive to change from day-to-day and to weather and other travel conditions.

Peak hour volume is significant in the design of roads and control of traffic and this ranges from 8 to 10 per cent of total daily flow or 2 to 2.5 times the average hourly volume.

Weekly variation. Traffic flows for the week days. Monday to Friday, remain fairly constant but weekend flows are variable and mainly depend upon weather and season. In the city centre, the weekend flows will be less; but there will be an increase in flow towards the picnic and recreational spots, particularly if the season is mild and is neither too hot nor too cold.

Seasonal variation. Traffic volumes are, in general, above the average during the months of September, October, February and March. Seasonal patterns, for a particular route, are fairly consistent and represent the social and economic condition of the area.

4.2.5. Passenger Car Units

Indian Roads Congress has introduced the idea of passenger car unit (pcu) which allows relative effects of different classes of vehicles by the use of appropriate multiplying factors known as Equivalency factors and are given in Tables 8.3 and 8.3 (*a*) on page 268 for various type of vehicles (*a*) on rural roads and (*b*) on urban roads respectively.

The use of equivalent passenger car units for design purposes must be done with considerable caution as it can lead to over-design, if the growth of heavy vehicles and slow moving bullock carts, horse drawn vehicles, *etc.* is very much lower than for cars. This may be helpful only for a broad comparison of the importance of different routes.

4.2.6. Peak Flow

Peak volume is of vital importance as it is indicative of design requirements. The usual method to determine peak flow in U.K. is to take maximum hour's flow during a 7-day August traffic census, or as suggested by the Road Research Laboratory, U.K. to take the average flow in a selected peak hour of each day during the highest 13-week period in June, July and August. The average is taken as the peak flow hour and may be used for design purposes after making suitable allowance for future growth.

The practice to determine peak flow is different in the United States of America. There the 30th highest hour flow after expansion is taken as the peak flow for design purposes. The 30th hour flow is correctly expanded, say for 20 year design period, then 29 hour over-loading of the highway would occur a year at the end of that period.

In India, it will be sufficient if 60th to 80th hour peak flow is taken for design purposes for highways, after correct expansion for the design period. This will affect considerable saving as the number of lanes required will be less and at the same time there will not be much congestion.

4.2.7. Volume Trends

Fig. 4.4 shows the yearly increase of travel. The increase rate is not uniform, probably due to slower rate of economic development. Traffic trend studies are very important for the prediction of future traffic loads. Tanner has suggested the following relation for predicting the increase in number of vehicles.

$$y_n = \frac{y_s y_o}{y_o + (y_s - y_o)\, e^{-\beta}}$$

where, y_n = Vehicles per head in the *n*th year

y_s = Vehicles per head in the base year

y_s = Saturation or maximum vehicles per head likely to be reached: and

$\beta = \dfrac{y_s x_n}{(y_s - y_s)}$; where 100 x = annual percentage growth during base year.

This curve varies from $y = O$ at $n = -\infty$ to $y = y_s$ at $n = +\infty$ and is S-shaped. Fig. 4.5 shows the logistic growth curve for car ownership.

The assumptions made in this method are:

(*i*) Increase in vehicle ownership follows a logistic curve

(*ii*) Separate predicted growth of population

(*iii*) The average annual kilometres per vehicle remains constant.

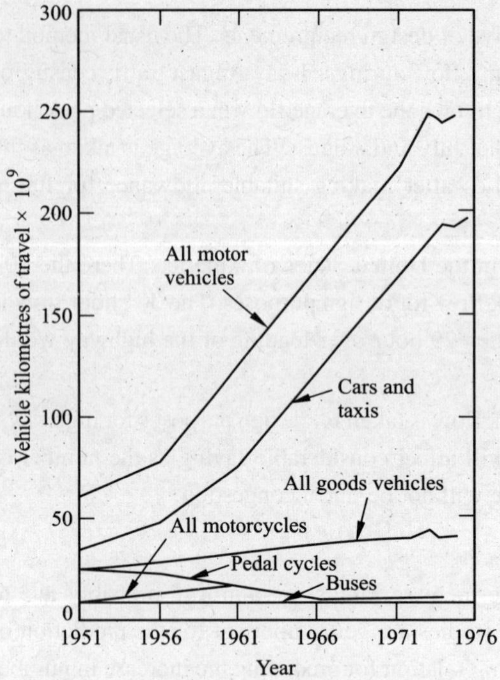

Fig. 4.4 Vehicle kilometres of travel by vehicle-categories. (1951 to 1976)

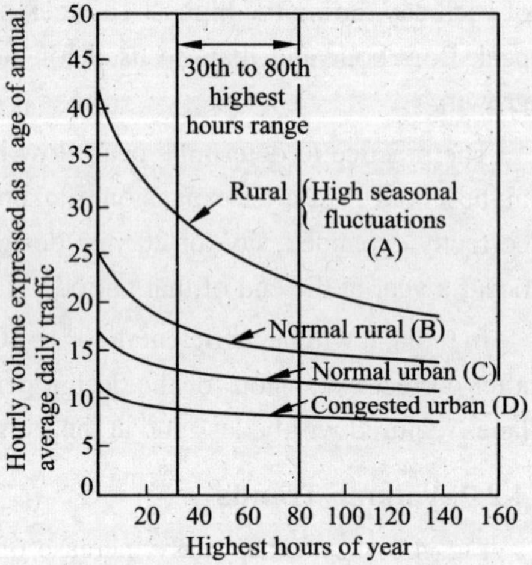

Fig. 4.5 Highest flow curves

4.2.8. Volume Flow Maps

Flow maps are prepared to show the distribution of volume by location. Such maps show the width of the route in proportion to the volume for a specified period. On a long route, annual average daily volume and in cities, where congestion occurs, average peak hour volumes will be more informative. Fig. 4.6 shows a flow map.

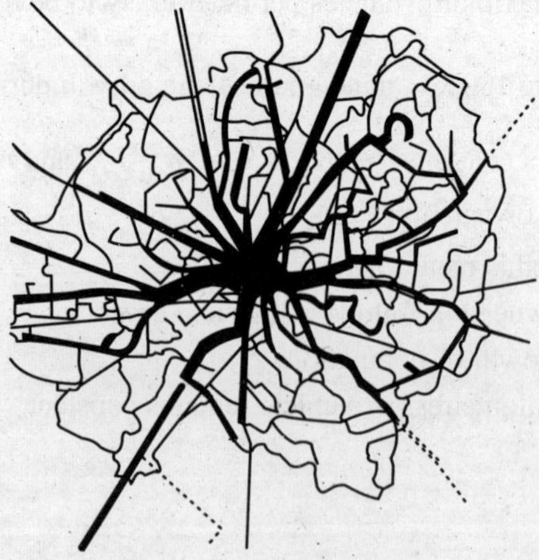

Fig. 4.6 Flow map

At intersections, where the number of turning vehicles is important, flow maps may be prepared in a similar manner *i.e.,* the width of the band showing the volume. As an alternative, volume may be represented by lines with figures indicating the number or percentage of vehicles undertaking the various manoeuvers. Fig. 4.7 depicts the flow map at an intersection in Chandigarh.

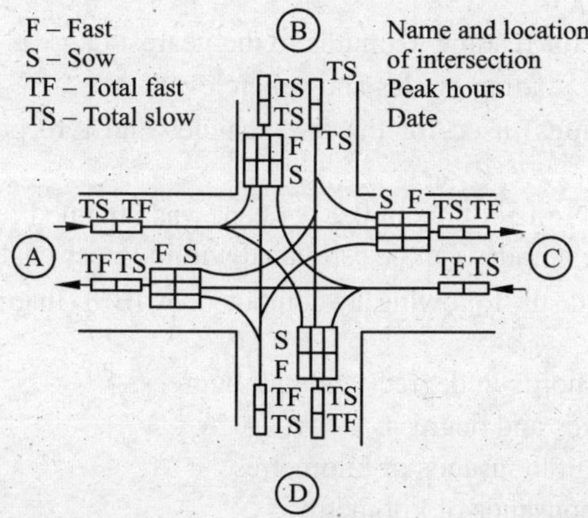

Fig. 4.7 Peak flow traffic flow diagram for design of intersection.

4.2.9. Road Inventory

The purpose is to collect information for all highway planning surveys. In this study, a physical check is made with provisions for revision, improvement, reconstruction and maintenance.

The data is collected to permit preparation of large-scale maps showing all public roads outside the cities, incorporated towns and villages, location of important structures on the road, principal highway connections to the various communities, *etc.*

Items to be recorded in road inventory are:

(*i*) Section designation of roadway.

(*ii*) Road width, pavement width, method of dividing traffic lanes, surface type and condition.

(*iii*) Side-walks and kerbs adjacent to the roads.

(*iv*) Alignment and gradient of the road.

(*v*) Cultural things lying on each side of the road-land usage.

(*vi*) Safety devices and signs.

(*vii*) Reservoirs, lakes, rivers, *etc.*

(*viii*) Structures of over 6 m length—bridges.

(*ix*) Function of the structure—type and materials used, number of spans, length of longest span, overall length, width, side walks.

(*x*) Ferries—their capacity, motive power, number of scheduled trips, hours of service.

(*xi*) Estimated width of street traversed—whether free or toll facilities.

(*xii*) Railroad crossings and their exact location, number of tracks, *etc.*

This data is then processed and final maps produced. Maps may be of the following types:

(*i*) General highway maps showing culture served by roads and has surface type of roads marked by symbols in the road plans.

(*ii*) Traffic map also shows culture and has the ADT's marked by symbols in the road plans.

(*iii*) Country road system map showing all the country roads, names and numbers, *etc.* State highways and other roads are also marked in the road plan.

The following equipment is used for road inventory:

(*i*) Odometer reading to $\frac{1}{625}$ km accuracy (close to 1.5 m).

(*ii*) Horizontal gyroscope for reading azimuths to the nearest degree.

(*iiii*) Vertical gyroscope for reading grades and superelevations.

(*iv*) Test vehicles (no springs) used for the test vehicles and tyre pressures are checked quite frequently.

Using the odometer, an accuracy of about 0.2 per cent was obtained for a study conducted in the U.S.A. for length of 4500 km. Grades can be estimated within ½ per cent accuracy.

Using the analysis of the data, the following are computed, by IBM (International Business Machine) method:

(*i*) Azimuths of curves, chords in degrees and half degrees.

(*ii*) Central angles of curves and degrees.

(*iii*) Intervening distances in thousands of kilometres.

(*iv*) Length of chords in thousands of kilometres.

(*v*) Radii of curves in metres.

(*vi*) Safe driving speed at curves in kilometres per hour.

(*vii*) Progressive latitudes in thousands of kilometres.

(*viii*) Progressive departures in thousands of kilometres.

(*ix*) Grade profile in tenths of metre.

Uses of Road Inventory

The various users are:

(*i*) Basis of more extensive surveys like the road use studies, the road live studies, traffic studies, sufficiency rating, *etc*.

(*ii*) Special studies can be done quickly as all the data is available on IBM cards.

(*iii*) Maps used by general public to establish new business, sales territories, service routes, *etc*. can be prepared.

(*iv*) Used for regional planning by state government agencies.

(*v*) Used for national planning by central government agencies.

(*vi*) Used for marking speed zones.

(*vii*) Used to issue and renew special trip permits for over-weight and over sized vehicles.

4.3. SPEED AND DELAY STUDIES

Speed is the rate of travel expressed in kilometres per hour (km/hr). Speed is of three main types:

(*a*) Spot speed

(*b*) Running speed

(*c*) Journey speed

Spot speed is the instantaneous speed of a vehicle at any specified point. Running speed is the average speed over a particular course while the vehicle is moving and equals the length of the course divided by the time the vehicle is in motion. Journey speed is the effective speed of the vehicle on a journey between two points and is found by dividing the distance between the two points by the total time taken for the vehicle to complete the journey, including stoppage time due to traffic delays and traffic control devices. In a normal journey, where stopped delays are likely to occur, journey speed must be less than the running speed and spot speed will vary from zero to a maximum, in excess of the running speed. High journey speeds are desirable on highways to effect saving in journey time but in urban areas, this is not very significant.

4.3.1. Purposes of spot speed studies

These studies are conducted mainly for the following purposes:

1. Planning traffic control-establishing speed zones, traffic signals, location of warning, regulatory and information signs, non-passing zones, etc.

2. Determining the speed trends.

3. For the accident studies:

(*a*) Relation of accident to speed.

(*b*) Analysis of high accident location.

(*c*) Effectiveness of remedial measures put in.

4. Used in capacity studies.

5. Used for geometric design.

4.3.2. Location and Time of Spot Speed Study

Location

(*a*) At all major highways.

(*b*) At all high accident frequency points.

(*c*) At all points where installation of traffic signals and stop signs are contemplated.

(*d*) At other representative locations for collecting basic data for future planning.

Time of study

(*a*) One hour between 9 to 12 A.M.

(*b*) One hour between 3 P.M. to 6 P.M.

(*c*) One hour between 6 P.M. to 10 P.M.

The studies are usually made in good and normal weather.

4.3.3. Methods for Spot Speed Studies

(*i*) *Mannual method* as used by the Central Road Research Institute, Delhi consists in using stop watches and noting down the time. The distance between two fixed points is divided by this time.

(*ii*) *Electrical motor method.* In this method, a condenser with an electric charge begins discharging as the first impulse is recorded. The change gives the speed directly by calibration.

(*iii*) *Radar meter method.* It is operated on the basis of Dopller effect—an electromagnetic wave, which is deflected by the vehicle on the highway. The speed of the vehicle in relation to the meter, changes the wavelength and returns to the receiving unit of the motor. This change is calibrated to indicate speeds directly in the meter. A graphic recorder can be connected to the meter, so that a permanent record of the speeds observed can be kept. The accuracy is 1.6 km/hr (shielded car is not recorded).

(*iv*) *Enoscope method.* In this method an L-shaped box, open at both ends, with mirrors set inside at 45° angle is used. The time between the successive reflections is noted by a stop watch.

(*v*) *Spaced photographic method.* Pictures of traffic streams are taken with a motion picture camera operated by a constant speed-motor. From this movie, the speed can be studied, the speed of motor in the camera being constant.

4.3.4. Analysis of Spot Speed Measurement

At a given location, the spot speed measurements may be averaged to obtain arithmetic mean of the speeds, but this information is of little use to the traffic engineer. He is more concerned with the distribution, the range and the dispersion of the speeds. Standard statistical method can be employed for their determination. These are briefly given below:

Distribution

Table 4.3 shows frequency distribution for spot speed of cars on a two-lane state highway. Speeds are grouped in class intervals (column 1). The average speed in the interval is given in column 3, *e.g.*, out of 152 cars, 52 were travelling at speeds between 60 and 70 km/hr.

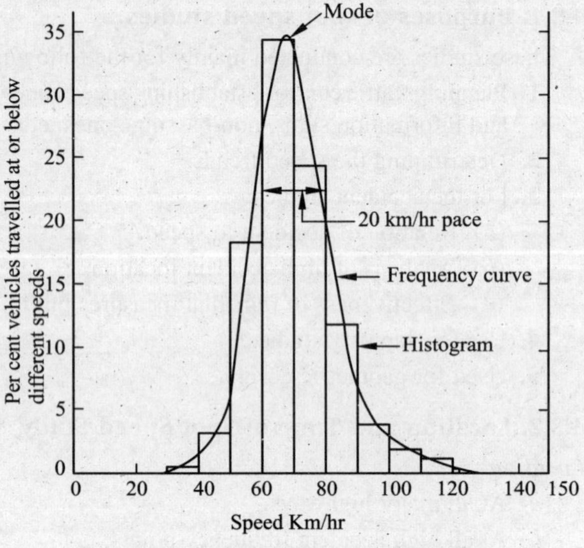

Fig. 4.8 Histogram and frequency curve.

Arithmetic Mean

The arithmetic mean or average speed is a measure of central tendency. It is determined from the relation:

$$\bar{x} = \frac{\Sigma f \cdot x}{\Sigma f}$$

where, $\bar{x}$ = mean spot speed in km/hr

f = frequency in each class

x = Average speed in each class in km/hr

In this particular example

$$\bar{x} = \frac{10,600}{152} = 69.7 \text{ km/hr}$$

The data and results of table 4.3 can be best represented by histogram and frequency curves shown in Fig. 4.8. Historgram is a graphical plot between speed interval (column 1) and frequency per cent (column 4). Frequency curve is found by rounding off the histogram so that the area of the frequency curve is equal to the area of the histogram. Peak of the curve gives the modal speed, *i.e.*, the speed occurring most frequently. The curve is also useful to find the pace of the vehicles, which is the speed range for some nominal increment of speed, usually 20 km/hr which contains the most vehicles. From Fig. 4.9, it may be seen that the mode is 67 km/hr and the pace is 60 to 80 km/hr.

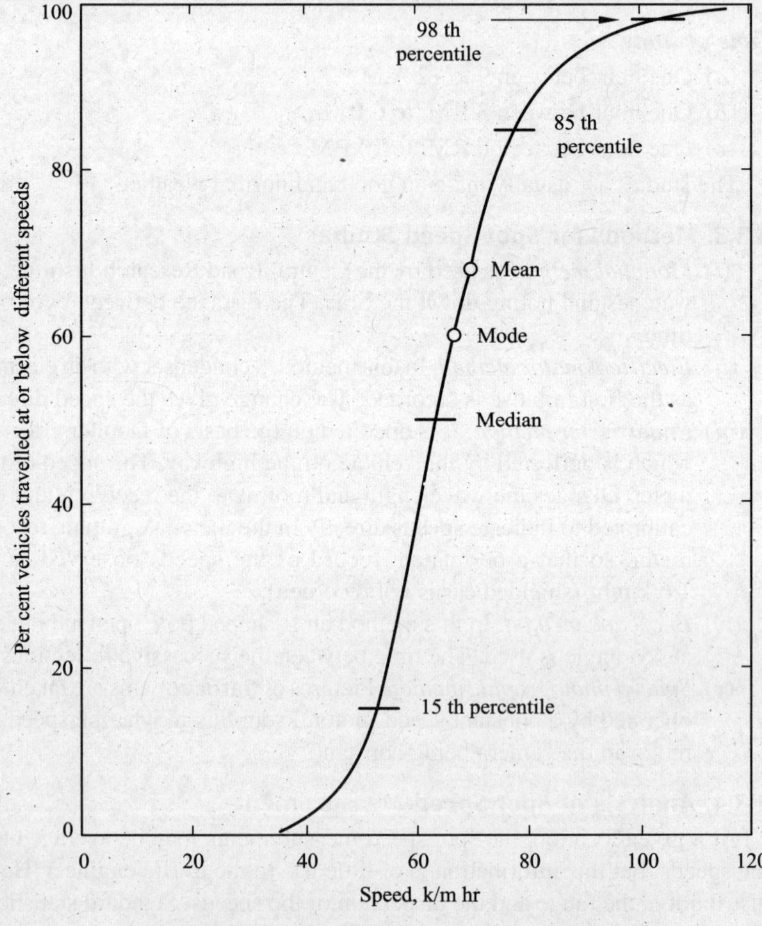

Fig. 4.9 Cumulative frequency curve-spot-speeds.

Table 4.3 Spot speed distribution

Speed interval km/hr	*Average speed km/hr*	*Frequency (f)*	*Frequency %*	*Cumulative frequency; %*	*f.x*	*Deviation from assumed mean ; d_1*	fd_1	fd_1^2
1	2	3	4	5	6	7	8	9
30–40	35	1	0.7	0.7	35	– 4	– 4	16
40–50	45	5	3.3	4.0	225	–3	– 15	45
50–60	55	28	18.4	22.4	1540	–2	– 56	112
60–70	65	52	34.2	56.6	3380	– 1	– 52	52
70–80	75	37	24.3	80.9	2775	0	0	0
80–90	85	18	11.8	92.7	1530	1	18	18
90–100	95	6	4.0	96.7	570	2	12	24
100–110	105	3	2.0	98.7	315	3	9	27
110–120	115	2	1.3	100.0	230	4	8	32
Total		152	100.0	—	10600	—	– 80	326

Cumulative frequency

To determine the number of vehicles travelling above or below a given speed, cumulative frequency curve is plotted between the average speed (column 2) and percentage cumulative frequency (column 5). The median speed which is another measure of central tendency, is the speed below which 50 per cent of the vehicles are moving. For this road, the median speed is 63 km/hr.

Sometimes percentile speeds, which are the speeds below which a specified percentage of vehicles are travelling, are determined. Of particular interest are the 98th percentile (used for design speed in geometric layout), 85th percentile (used in overtaking distances or speed limit imposition) and 15th percentile (showing slow vehicles causing interference within the traffic stream). Their values, in this case, are 105 km/hr, 79 km/hr and 53 km/hr respectively.

Measurement of dispersion

The important measurements of dispersion are the standard deviation and the coefficient of variation.

The standard deviation of a grouped frequency distribution is given by

$$s = \sqrt{\frac{\text{summation of [frequency} \times \text{(deviation)}^2]}{\text{sum of frequencies}}}$$

or

$$s = \sqrt{\frac{\Sigma\,(fd^2)}{\Sigma f}}$$

where s = standard deviation

d = deviation of average from arithmetic mean

$= x - \bar{x}$

However, it is convenient to compute 's' by using an assumed mean and applying the following relation

$$s = \sqrt{\frac{\Sigma f d_1^{\,2}}{\Sigma f} - \left(\frac{\Sigma f d_1}{\Sigma f}\right)^2} \times i$$

where d_1 = deviation from an assumed mean, expressed as a multiple of class interval and

i = class interval

In this example

$$\text{Standard deviation, } s = \sqrt{\frac{326}{152} - \left(\frac{-80}{152}\right)^2} \times 10$$

$$= \sqrt{2.15 - 0.28} \times 10$$

$$= 13.7 \text{ km/hr}$$

To obtain a measure of dispersion, related to the mean, coefficient of variation is determined. This is given by

$$V = \frac{100\,s}{\bar{x}}$$

A higher value indicates wider scatter about the mean and *vice versa*

$$\therefore \quad V = \frac{100 \times 13.7}{69.7}$$

$$= 19.7\ \%$$

4.3.5. Running speed, Journey speed and Delay studies

These are related to each other. The difference in running speed and journey speed is due to delay in the traffic.

These studies are made mainly for the following purposes:

1. The speed and delay data provides facts on the amount, location and delay. This data is used in remedy of particular situation of congestion.
2. Sufficiency rating—being based on travel time information.
3. Traffic assignment for a new facility is based upon relative travel time.
4. Economic studies utilise travel time and delay data.
5. Travel time being a good indication of efficiency of roadway.
6. Before and after studies—made to check effectiveness of changes in traffic control, likely parking prohibition, signal timing revisions, new one-way street, *etc.*
7. Delay represents non-productive time and when converted to monetary values, show the cost to the community of an inadequate road.

This enables determination of economic return for improvements and permits to fix priorities.

Delay can be categorized into two types:

(i) Fixed delay. It is the delay to which traffic is subjected to regardless of the amount of traffic volumes and interferences present on the highway. This is not due to the characteristics of traffic streams. This includes traffic signals, stop signals, railroad crossings, *etc*. This delay can occur even with only one vehicle on the highway.

(ii) Operational delay. This is also known as congestion delay. This is the delay caused by interference with other components of traffic. The difference between travel time over a route during an extremely low and during very high traffic volume indicates the amount of operational delay. This can be of two types:

(*a*) Caused by other traffic movements that interfere with the stream flow, is by parking or imparking vehicles, turning vehicles, pedestrians, *etc.*

(*b*) Caused due to internal function of stream flow like congestion due to high volume, lack of capacity, and waiting for a gap to cross street traffic.

4.3.6. Methods for determining running speed, journey speed and delay

(1) *In-vehicle method.* The moving car method, described earlier enables the determination of running speed, journey speed and delay. The mean journey time can be calculated from the relation:

$$i = t_w - \frac{y}{q}$$

The mean speed of the vehicles or journey speed along a section of length 1, can be found out by applying the formula:

$$V_s = \frac{1}{t}$$

where V_s = the space mean speed.

This is different from the time-mean speed, V_t. The two are related to each other, by the formula

$$\overline{V_t} = \overline{V_s} + \frac{S_s^2}{V_s}$$

where $\overline{V_t}$ = Time mean speed = $\frac{\Sigma V}{n}$

$$\overline{V_s} = \text{Space mean speed} = \frac{1}{\Sigma t / n} = \frac{1}{t}$$

S_s = Standard deviation of the speed distribution of the speeds.

The mean running speed can be determined by dividing the length of the section 1, by the difference in mean journey time t and stopped delays recorded by the moving car observer.

Similar data, for selected routes, can be obtained by the trailing observer method.

(2) *Registration number method.* Two observers are stationed, one at each end of the section. With the help of synchronised watches, time and registration number of each vehicle passing is recorded. Then by matching the registration numbers, the journey time of vehicles and hence the journey speed is estimated. This does not give the running speed and the cause, location or duration of delays.

(3) *Vantage point method.* In urban areas, where the length of the section is short and the view is unobstructed, course of randomly selected vehicles is traced and the time of entering the section, the duration and nature of delay and the time of leaving are recorded. This enables the determination of delays, journey speed and running speed. Closed circuit television over a network of streets covered by cameras or serial photography can also be used to get details on traffic flows, congestion points, speeds, delays and parkings.

4.4. ORIGIN-DESTINATION SURVEY

The major purpose of this study is to collect factual information about travel desire to provide the most efficient transportation system for the movement of persons and goods.

The information needed is as follows:

(*i*) Where the people want to go-their origins and destinations, irrespective of the present routes of travel.

(*ii*) How they travel-by automobile, bus, truck, on foot, *etc*.

(*iii*) When they travel-time, direction, *etc*.

(*iv*) Why they travel-purpose of the trip.

(*v*) Where they stop-to determine the accumulation of vehicles in a certain area for estimating parking demands.

For collecting this information, the following surveys are made:

(1) *External survey.* It determines the origins and destinations of persons entering or leaving the area under study. The survey collects information concerning through trips, *i.e.*, trips having both origin and destination outside the survey area. The data can be collected also for external-internal trips, *i.e.,* the trips having the origin or destination, but not both, inside the survey area.

The study is based on the assumption that the whole population of motorists, day in and day out, will behave like the ones interviewed, *i.e.*, the total demand can be determined by expanding the data collected and using appropriate factors.

The details of the external survey are:

(*a*) A cordon line is established around the survey area which is generally the most highly populated section of the city.

(*b*) Interview stations are located at points or at all major points where highways cross the cordon line. Stations are located in such a manner that at least 95 per cent of traffic entering or leaving the survey area can be inter-surveyed. It is further recommended that 80 per cent of peak hour traffic should be interviewed.

(*c*) Operating schedule is usually kept for 16 hrs *i.e.,* from 6 A.M. to 10 P.M. When the volume between 10 P.M. and 6 A.M. exceeds 500 vehicles, the station is operated for 24 hrs.

Classification counts are taken in addition to interviews for checking and expansion purposes. Mechanical counts are taken for a sufficient period to determine the ADT in order to convert the interview data to an average per day.

(*d*) Motorists are stopped at the interview stations and the following information is obtained for each vehicle:

(*i*) Station number, date, hour, direction of travel.

(*ii*) Vehicle type, *i.e.*, passenger car, bus, truck, *etc.*

(*iii*) Origin and destination of the trip.

(*iv*) Purpose of trip.

(*v*) Any intermediate stop for through passengers or through vehicles.

(2) ***Internal survey*****.** This collects information concerning the trips made within the survey area. In this case the procedure is as follows:

(*i*) A sample of houses is selected in a systematic manner from the census block statistics, telephone directories or actual field investigation. The size of the sample is as follows:

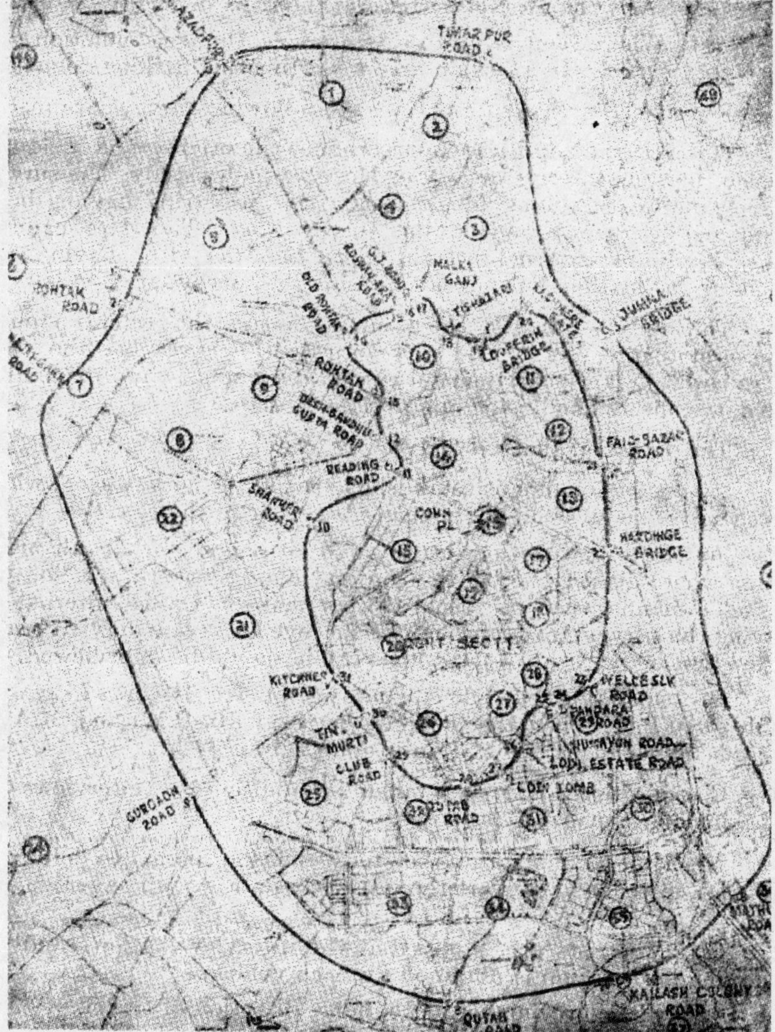

Fig. 4.10 Map showing the 48 zones, the principal areas and cordon lines for O.D. Survey of Delhi.

(*a*) For a population under 50,000, size is 20 per cent.
(*b*) For a population between 50,000 to 300,000, size is 10 per cent.
(*c*) For a population between 300,000 to 1,000,000, size is 5 per cent.

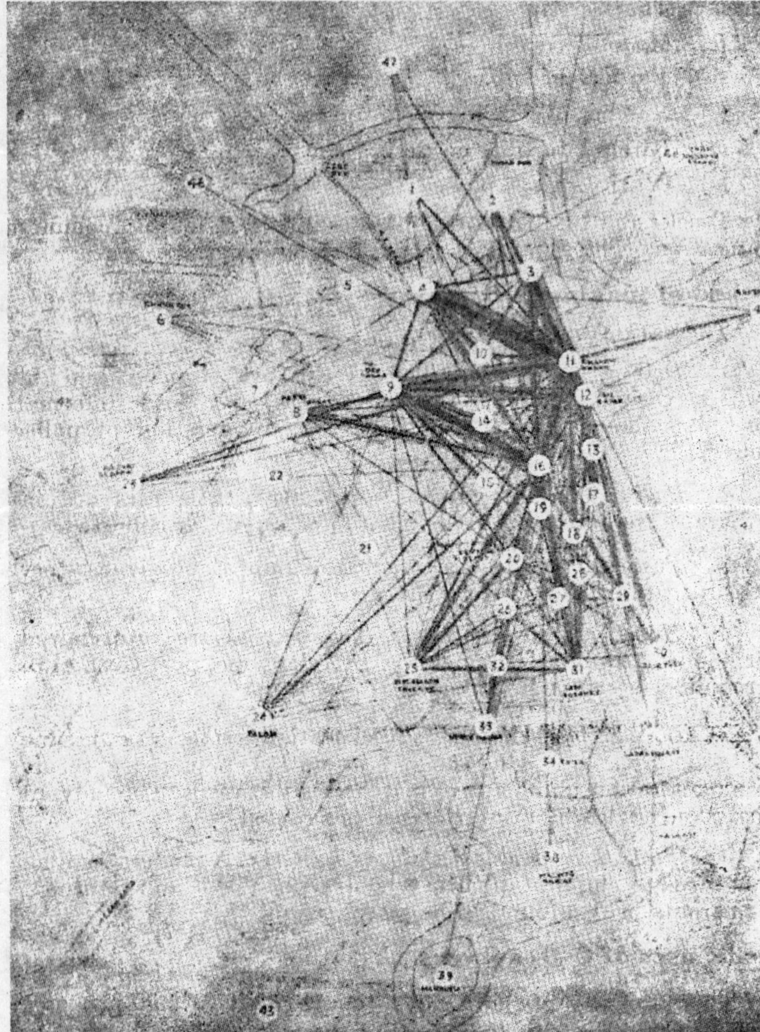

Fig. 4.11 Desire lines of fast moving vehicles.

(*ii*) The data on the number of residents, occupation, number of vehicles, *etc*. in addition to the data on all trips by all modes of travel made by all occupants on the previous day is gethered. For each trip, the following information is collected:

(*a*) Origin and destination.
(*b*) Time of trip.
(*c*) Mode of travel.
(*d*) Purpose of trip.
(*e*) Number of passengers in the car.
(*f*) Kind of parking needed.
(*g*) Control points passed.

Trucks and taxi trips are obtained by interviewing commercial organizations. The size varies from 20 to 50 per cent.

4.4.1. Method of making O-D survey

The various methods are:

1. *Drivers' interview.* The drivers are stopped and interviewed at various points throughout the city and information collected as above. In this method, the co-operation of police is required.

2. *Parker survey.* Persons parking their cars are interviewed and the information, as mentioned above, is collected.

3. *Post-card survey.* Drivers are stopped and given interview post-cards to fill-in and return by mail.

4. *Home-interview method.* The drivers are interviewed in their homes, arranging the meeting by telephone, post-card, *etc.* and the required information collected.

5. *Licence plate Method.* Registration number and time of vehicles entering or leaving the area at station points are recorded. These registration numbers are matched afterwards. The advantage is that slowing down or interference is avoided.

6. *Cards through school children method.* A modern method of giving cards to children in their schools to be filled and returned by their parents is also found to be satisfactory.

4.4.2. Specific uses of O-D surveys

The information gathered from O-D surveys is put to the following uses:

1. Collected data is used by both the engineers and city planners in the design of major facilities, including major street system, freeway locations, location of interchanges, *etc.*
2. Improving the present facilities if adequacy of the existing street system can be found out by determining its capacity and comparing it with the actual volume of traffic. The data is also used for other improvements such as designation of one-way streets, limited parking, traffic signals, *etc.*
3. ***Planning of future facilities.*** Future needs are determined by expanding the data from O-D survey and the facilities provided are designed to meet the future needs rather than any need existing at present.

4.5. ROAD LIFE STUDIES

The service life is defined as the period of time elapsing between the construction of the roadway and its eventual removal from the highway system by retirement. The specific purposes of road life studies are:

(*i*) To obtain data concerning the service lives of the various types of highway pavements.

(*ii*) To investigate construction cost, maintenance cost, traffic services and salvage value of retired roadway to determine the total revenue required for roadway.

(*iii*) To plan roadways which incorporate elements of correlated service lives.

(*iv*) For this study, the various organizations follow different classifications. For example in California in U.S.A. the classification is as follows:

1. *Soil surfaced roads.* Of natural soil, the surface of which has been improved by the addition of some borrowed material.

2. *Gravel roads.* The surface of gravelly material with or without a dust palliative.

3. *Bituminous surface treated roads.* Soil surfaced or gravel roads to which a bituminous surface of less than 2.5 cm thickness has been added.

4. *Mixed bituminous roads.* On which the surface greater than 2.5 cm of aggregate mixed with bituminous material has been placed.

5. *Bituminous penetration road.* Similar to mixed bituminous roads except that mixing is accomplished by soaking instead of physical handling.

6. ***Bituminous concrete.*** Bituminous surface greater than 2.5 cm in thickness and the mix prepared under rigid control, usually in the central plant.

7. ***Portland cement concrete roads.***

A highway can be retired due to any one of the following reasons:

(*i*) Structural deterioration—failure of top surface, base or sub-base.

(*ii*) It may be functionally obsolute, which may be due to poor alignment which reduces the speed too much and also reduces sight distances or due to obsolete construction and standards.

(*iii*) Need for construction of highway improvements, *e.g.*, grade separations, intersections, traffic rotaries, *etc*. and other miscellaneous factors like flooding by reservoirs, slides or other such-like disasters may occur.

Surface life of pavement. A pavement can be retired due to any one of the following reasons: (*i*) re-surfacing; (*ii*) re-construction; and (*iii*) abandonment of the pavement due to changes in alignment, *etc*.

***Survivor curves*.** These are curves drawn for percentage of the various types of surfaces *vs* life in years.

Data for 25 states in the U.S.A. showed the most probable values:

Soil surface	2 years
General roads	5 years
Bituminous surface	10 years
Mixed bituminous surface	12 years
Bituminous penetration surface	$14\frac{1}{2}$ years
Bituminous concrete surface	$18\frac{1}{2}$ years
Portland cement concrete surface	28 years

Various factors which will determine the life of pavements are: (*a*) oil conditions; (*b*) climatic conditions-amount of precipitation, freezing and thawing, snowing, *etc*. and (*c*) construction and control methods.

The fiscal studies provide the following information: (*i*) Suggest long range plan, (*ii*) Give an idea regarding the cost, (*iii*) Suggest as to how to finance it-considering the budget of the state, road user revenues, money collected and money spent on road, *etc*.

4.6. MOTOR VEHICLE USE STUDIES

These studies permit an estimate to be made of the total traffic carried by the road and streets of the state, the relative amount carried by several systems and the relative amounts of travel performed by the residents of various geographical and political divisions on the various highway systems.

The data can thus be used in determining where highway user revenues should be spent to benefit the greatest number of motorists and to provide for the most essential needs on all classes of roads and streets. The specific purposes of the study are:

(*i*) To find as to how the ownership of automobiles and trucks is distributed, geographically among rural and urban areas.

(*ii*) To determine the characteristics of the use made of private and commercial vehicles by the various groups including estimates of annual travel of vehicles, road systems used, trip length and purpose of trip.

(*iii*) To determine the age distribution of the automobiles and the age and weight of commercial vehicles owned by different groups.

(*iv*) To know something about the rates of the fuel consumption classified according to residence of owner, vehicle type and age group.

(*v*) To collect information about the transportation media used by gainfully employed workers in getting to and from work.

No traffic pattern of travel is found; instead, the total travel on the systems by owners residing in the various political sub-divisions is determined.

The method of collecting the data are:

1. *Home-interview*. This is the most generally used method where the sample is first determined by selecting 2 to 3 per cent sample houses according to the various occupational groups.

Collection of this data is done in the different seasons of the year. Then at the actual interview, the following data are to be collected:

(*a*) Listing of residents of the dwelling unit indicating sex, approximate age, and whether or not those of legal driving age possess a driving licence.

(*b*) Occupational information about these residents who are gainfully employed-distance to place of work and means of getting to work.

(*c*) Inventory of automobiles and trucks, regularly driven by the residents of their dwelling unit.

(*d*) Estimated total annual travel of vehicles and the fuel consumed by them.

(*e*) Description of motor-vehicle trips made by residents on a certain day. The information includes the origin and destination of such trips, purpose, route adopted, trip length and number of persons carried in the vehicle.

In case of some truck trips the general description and weight of commodities is also collected.

2. Other methods like road side interview or giving questionnare to the school-children to be filled up by their parents are also used.

After the data are collected, these are analysed by the following methods:

(*a*) Percentage of motor-vehicle owners residing in the various political sub-divisions.

(*b*) Type of trip; whether it is business, like going to work, shopping or to rail or bus station, to market, *etc.* or whether it is a pleasure trip or social trip *i.e.*, like going to temple, theatre, club, *etc.*

(*c*) Use of highway systems, whether it is a national highway, state highway; major or minor district road or village road.

These data can be put to the following uses:

1. In the development of long-range plans for construction and reconstruction of the several highway systems in the state.
2. In determining on equitable basis the allocation of motor-vehicle user revenues for highway purposes.
3. In times of national emergency, the data helps in determining the allocation of critical materials and equipment for the construction and maintenance of highways.
4. The data is also used for allocation of critical materials such as steel and rubber to industry for the manufacture of parts, tyres, *etc.* to keep the vehicle in service.
5. The data is used by the industry in planning and developing their production and market.

4.7. PARKING STUDIES

Terminal facilities form an integral part of any transportation system. Traffic usually travels towards a destination and the vehicle must be parked while some business *e.g.*, private, public, recreational or servicing, is transacted. Failure to provide suitable parking facility can result in congestion and frustration, ultimately leading to the decline in the importance and value of areas considered at present to be most desirable for the day-to-day business of a city by its inhabitants. As a general rule, increase in vehicle ownership results in increased parking demand.

4.7.1. Purposes of parking studies

The purposes of parking studies are:

(1) To determine the congestion in the city or town areas.

(2) To assess the suppressed parking demand.

(3) The capacity of the existing parking facility.

(4) To estimate the desires and demands of the public for parking facility.

(5) To decide the capacity, location and type of future parking facilities.

4.7.2. Parking surveys

A parking survey can be made by the following methods:

4.7.2.1. Inventory studies

(*a*) *Inventories* consist of counting all available facilities, and tabulating block by block for the area. The total figure is obtained by collecting data on the following: (*i*) Kerb spaces, (*ii*) Parking lots, (iii) Parking garages (*iv*) Spaces available but not used for parking and (*v*) Time restriction: duration and restrictions by time of day; free and metered and rates charged give the accumulation of vehicles in the area.

(*b*) *Cordon count.* This consists of a count of all cars and other vehicles entering and leaving the area hourly or in shorter periods of time. The difference between the vehicles entering the area and those which are leaving will give the accumulation of vehicles in the area. This accumulation represents the number of vehicles parked and on the move in the area; and is a measure of the required parking facility. A typical accumulation curve is shown in Fig. 4.12.

(*c*) *Direct interview.* Individuals parking in an area are interviewed about their origin and destination and the purpose for parking. This information alongwith the time the vehicle is parked, enables the determination of major parking characteristics.

The following information is collected by the interviewer for each vehicle:

(1) Registration number : for identification

(2) Vehicle classification : car, scooter, taxi, truck, bus *etc*.

(3) Nature of parking : kerbside, off-street, garage, *etc*.

(4) Time at which vehicle stopped.

(5) Time at which vehicle started.

(6) Last point at which driver made an essential trip.

(7) Drivers' destination after he has left the vehicle.

(8) The purpose for the stop: work, shopping, business, loading, off-loading, *etc*.

(*d*) *Patrol survey.* The area is divided into small sections which can be toured once in a suitable interval *e.g.*, half hourly, hourly, etc. On each patrol, the number of vehicles parked in the area is counted, thus revealing the parking accumulation for the study period. Sometimes the registration number is also recorded to know the parking duration of the vehicle. The section may be patrolled on foot or in a car. The trip interval depends upon the amount of time, money and man-power available for the survey and upon the characteristics of the parked vehicles.

4.7.2.2. Land use method

(*a*) Demand is determined by the studies of the correlation between the sales and parking area required for monetary unit of sale. This depends on the type of business, city and cost of living and other facilities.

(*b*) *Floor area and store area.* A correlation has been made of the floor area and store area of a commercial establishment.

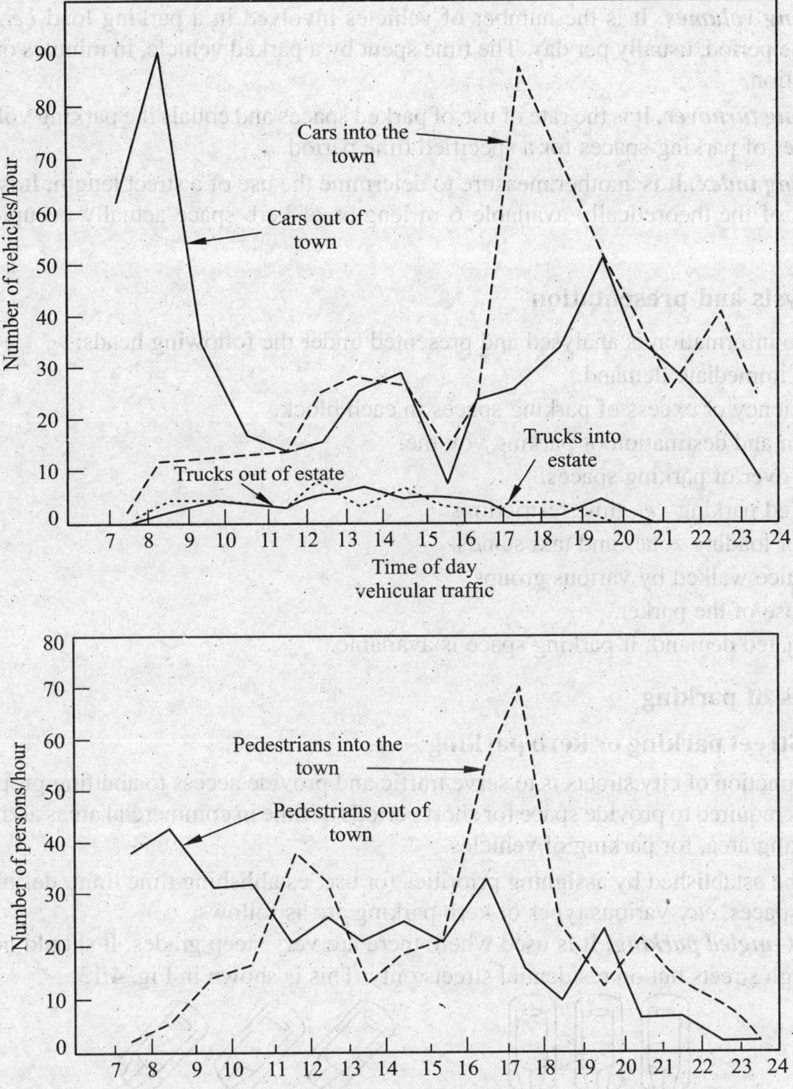

Fig. 4.12 Total movements into and out of estate by cordon survey.

4.7.2.3. Customer interview method

The customers are interviewed as they enter the commercial stores, restaurants, hotels, cinemas, offices and other places of work. The suppressed demand for parking can be established from the difference between the desired and present users. It will give us an idea as to how many people will like to bring their vehicle, if parking facility is available.

4.7.3. Parking Measurements

In parking surveys, the following are the main items of measurement:

(1) ***Parking accumulation***: It is the number of vehicles parked in an area at any specified moment. It can be divided into journey purpose categories *e.g.,* shopping, work, other purpose, etc. Parking load in vehicles hours per specified time period can be found by determining the area under the accumulation curve or by integration of the accumulation curve over a specified period.

(2) ***Parking volumes.*** It is the number of vehicles involved in a parking load *i.e.,* vehicles per specified time period, usually per day. The time spent by a parked vehicle, in minutes or hours, is the parking duration.

(3) ***Parking turnover.*** It is the rate of use of parked spaces and equals the parking volume, divided by the number of parking spaces for a specified time period.

(4) ***Parking index.*** It is another measure to determine the use of a street length. It is expressed as a percentage of the theoretically available 6 m lengths of kerb space actually occupied by parked vehicle.

4.7.4. Analysis and presentation

The above information is analysed and presented under the following heads:

1. Total immediate demand.
2. Deficiency or excess of parking spaces in each block.
3. Origin and destination of parking volume.
4. Turn-over of parking spaces.
5. Limited parking *i.e.*, time restrictions.
6. Use of loading zones and taxi stands.
7. Distance walked by various groups.
8. Purpose of the parker.
9. Estimated demand, if parking space is available.

4.7.5. Types of parking

4.7.5.1. Street parking or kerb parking

Primary function of city streets is to serve traffic and provide access to abutting properties. Along with this, it is required to provide space for short periods of time in commercial areas and for unlimited time in outlying area, for parking of vehicles.

This can be established by assigning priorities for use, establishing time limit, definitely marking the parking spaces, *etc*. Various types of kerb parking are as follows:

1. ***Right-angled parking.*** It is used where there are very steep grades. It should not be used on through streets but on residential streets only. This is shown in Fig. 4.13.

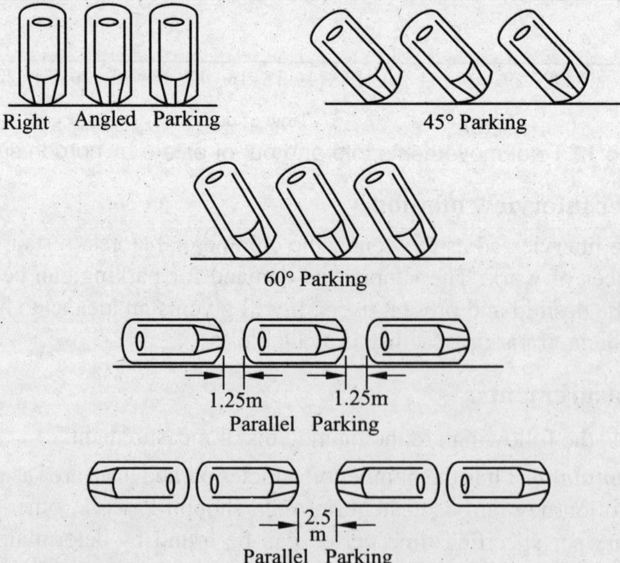

Fig. 4.13 Right angled, angled and parallel parking.

2. ***Angle parking***. This requires more spacing than the parallel parking.

   It provides quicker and easier parking, but it is more difficult and hazardous to impark. Since this takes greater width of street, it should be used only when need for parking takes priority over the need for traffic of vehicles.

3. ***Parallel parking***. Parking the vehicles parallel to the centre line of the road, with spaces in between two vehicles as shown in Fig. 4.13. Parallel parking has been found to be satisfactory from the safety point of view and the use of street-width.

4.7.5.2. Off-street parking

In big cities like Mumbai, Kolkata, Delhi, Chennai etc. street parking is limited. It is, therefore, necessary to augment the parking capacity by providing off street parking facilities. These are:

(1) Surface parks
(2) Multi-storey garages
(3) Under-ground garages
(4) Composite development
(5) Mechanical garages; and
(6) Drive-in facilities.

The design standards for surface parks and parking structures are based on the dimensions of the vehicles in use. In India, most of the off-street parking is done in surface parks. The construction costs of surface car parks is minimum but less efficient use of land is made. The various lay-outs of surface car parking are shown in Fig. 4.14.

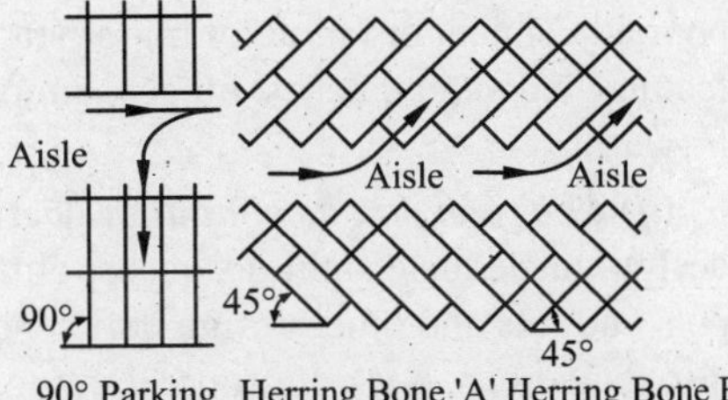

Fig. 4.14. Layout of surface car parking.

4.7.6. Prohibition of parking

Parking should be prohibited on the following locations:

(*i*) Side-walks
(*ii*) In front of drive-way entrance
(*iii*) Within a cross-section
(*iv*) Within 5 m of fire hydrants
(*v*) In a cross-walk
(*vi*) In a pedestrian safety zone
(*vii*) On a highway bridge or tunnel.

It is further recommended that parking should be prohibited within 5 m of cross-walk and within 10 m of stop sign signal or other warning signs. No parking on road of width less than 7.5 m should be permitted. Parking is allowed on one side only if the roadway width is within 7.5 to 9 m.

Prohibit parking when volume in peak hours exceeds the practical capacity of street or intersection.

4.8. TRAFFIC ACCIDENTS

Road accidents have become a major problem especially in big cities like Mumbai, Delhi, Kolkata, Chennai, etc. Responsibility for road safety lies initially with authority but finally rests on the individual. An effective working partnership between the police, engineers, planners, legislators, journalists and doctors and in the end with the individual citizen, is necessary for fuller realisation of the problem, of the many factors which contribute to road accidents, and the effective measures to reduce the occurrence of accidents. The traffic engineer can make an important contribution to the better design and control of the road system.

The annual road deaths, caused by accidents, can be approximately calculated by Smeeds' relation, according to which:

$$D = 0.0003\ (nP^2)^{1/3}$$

where, D = Annual road deaths

n = Number of registered motor vehicles ; and

P = Population

4.8.1. Causes of accidents

Basically, four factors are involved in road accidents:

(*i*) The vehicle

(*ii*) The road and its condition

(*iii*) The road user; and

(*iv*) The environment.

These causes are discussed below:

(i) The vehicle. There may be inherent design limitations, defects due to lack of maintenance, poor adjustment or failure of important components such as brakes, tyres, lights, etc. External vehicle features *e.g.* door catches, hinges, lamps, ornaments etc. also can entangle clothing and drag a victim forward.

(ii) The road and its condition. Deficiency in roads *e.g.* poor alignment, steep grades, poor curve design, inadequate sight distances, blind intersection, inadequate shoulders, holes in the pavements, pots and ruts and other design deficiencies can cause accidents. Slippery or skidding road surface is also dangerous. Lack of proper lighting and changing of signal timings without any prior information can also cause accidents.

(iii) The road user. The road user involved in the accident may be driver of one or more vehicles involved, the passengers, the pedestrians or stray animals on the road. Violation of traffic rules and faulty action of the road user *e.g.* excessive speed and rash driving, driving on the wrong side of the road, right of way violation, poor reaction under the influence of alcohol and other intoxicating drugs, failure to look, misperception and panic reaction, etc. are the factors responsible for causing accidents. Lack of experience and emotional maturity and exhibitionism among young drivers cause high accident rates. It is observed that drivers in the age group of 30 to 50 years make less mistakes.

Some accidents also take place due to drivers' using lighting inappropriate to the circumstances *e.g.* parking without lighting, throw glare on coming drivers, *etc.*

Passengers alighting or getting into moving vehicles and pedestrians violating regulations and using the carriage way meant for vehicular traffic are involved in accidents.

(iv) Environmental condition. Unfavorable weather conditions such as mist, fog, rain, dust, snow, etc. reduce visibility. People are more prone to accidents during night hours due to darkness and glare.

Besides the above causes, increase in population and wealth, enabling more people a greater amount of individual travel and man's physical and emotional limitations to live safely in a mechanised environment are responsible for the growth in road accidents.

4.8.2. Accident reporting and records

Accident reporting

In India, the police authorities are responsible for reporting accidents and the compilation of records. Data on accidents may be required for the following purposes:

(*i*) For law enforcement and the distribution of man-power by the police.

(*ii*) To determine the need for road improvement and to initiate programmes for propaganda and educational purposes by the government ; and

(*iii*) Accident investigation by research organizations.

Standard forms for reporting the accidents are used. These should contain some or all the information given in Table 4.4.

Table 4.4 Information required for accident reports

No.	*Element*	*Information required*
1.	General	Time, date (say, month and year). Locality of event and weather conditions. Highway classification.
2.	Road users	*Personal*: Age, sex, marital status, occupation and physical disabilities. Travel mode and journey purpose, and previous accident record. If a driver, experience. *General*: Position of fatalities and injured. Type of injuries and property damage. If in vehicle—driver or passenger and number of passengers. Impairments —drinks, drug or illness.
3.	Vehicles	Type, make, year of manufacture. External and internal features — ornaments, etc. Condition of tyres, brakes, suspension. Equipment check and functioning — lights and indicators. Damage sustained and position of vehicles. Seating capacity. Vehicle use at that time and loading condition. Type of movement. Ancillary equipment—safety belts and crash helmets.
4.	Road environment	*Traffic Control*: Signs (type) and other controls (one way, speed, parking, loading, bus stop, etc). pedestrian crossings. Road markings. *Traffic*: Volume, speeds and traffic composition, Public service vehicles. *Road design features*: Grade, alignment, width and cross-sectional elements, intersection layout, bends, crossfall, kerbs and barrier, rails. Sight distances. Street furniture. *Road surface*: State and type of surface, skid resistance values, defects, drainage and lighting conditions. *Adjacent land use*: Special building—schools, factories, etc. position of accesses. *Special consideration*: Movement of vehicles and pedestrians. Animals involved.

Accident records

It is essential to compile, analyse and present accident data in a convenient and legible form. Such records are useful for the following purposes:

(*i*) To establish national safety standards and formation of legislation for education and safety programmes.

(*ii*) To improve police patrolling of the road net-work.

(*iii*) To improve the design and operation of motor vehicle.

(*iv*) For the design of highway facility, including geometric layout, road drainage and materials of construction.

(*v*) For the provision of emergency services, medical facilities, etc.

(*vi*) To decide compensation for injury.

(*vii*) To launch education programmes for improving public consciousness to road safety.

Accident records are maintained, giving full information of the accident, location and other details. The various methods of recording are:

1. *Location files.* These are maintained by each police station for its jurisdiction. These are useful to identify points of high accident rates and to keep a check on the location of accidents.

2. *Card index system.* A detailed and index filing system can be employed, giving details of each accident with street and intersection locations. Different coloured cards may be used to differentiate between different types of accidents. A site sketch showing the layout and main features is also necessary.

3. *Computer analysis.* Routine programme is prepared to enable records on individual location to be analysed on the computer. If a graphical plotter is available, information can be displayed by a diagram also.

In addition to the above systems, it is essential that accident maps are also maintained with a cross-reference to the field data. Maps quickly indicate danger points on the whole road system and should be read directly with flow, speed, lighting, traffic control and construction details of road surfacing and drainage.

The accident records can be represented diagrammatically in the following ways:

1. *Spot maps.* Spot maps show accident location by spots, pins or symbols on the diagram. A map is drawn to a convenient scale *e.g.* 1 cm = 40 to 60 m on urban roads. Fig. 4.15 shows a spot map.

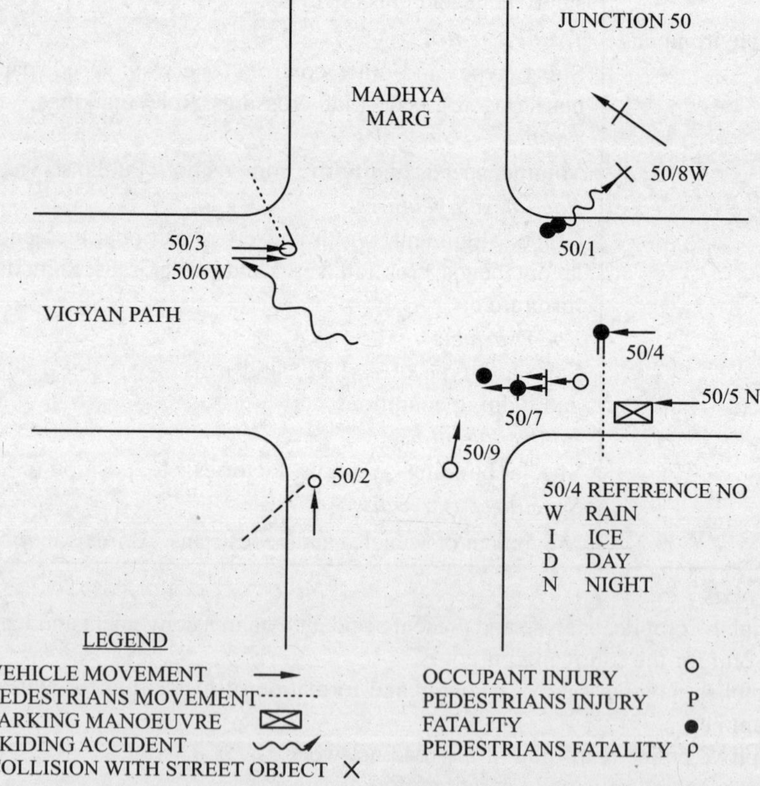

Fig. 4.15 Spot-map of accident records.

2. *Collision diagram.* It is a sketch showing, by means of arrows, the approximate paths of vehicles and pedestrians involved in accidents. The diagram (See Fig. 4.17) is not to scale and shows only the approximate paths. The various symbols used in the collision diagram are shown in Fig. 4.16.

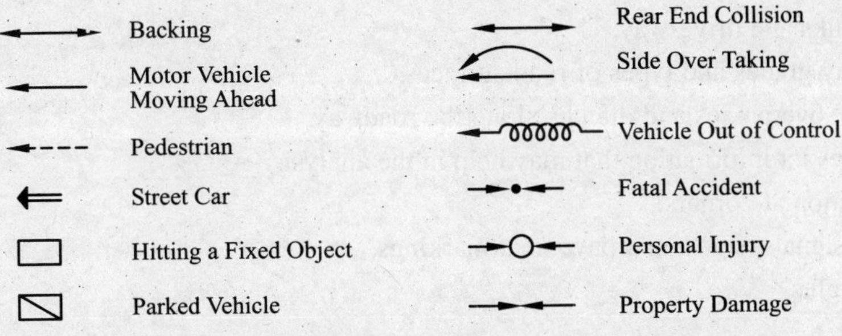

Fig. 4.16 Symbols for collision diagram.

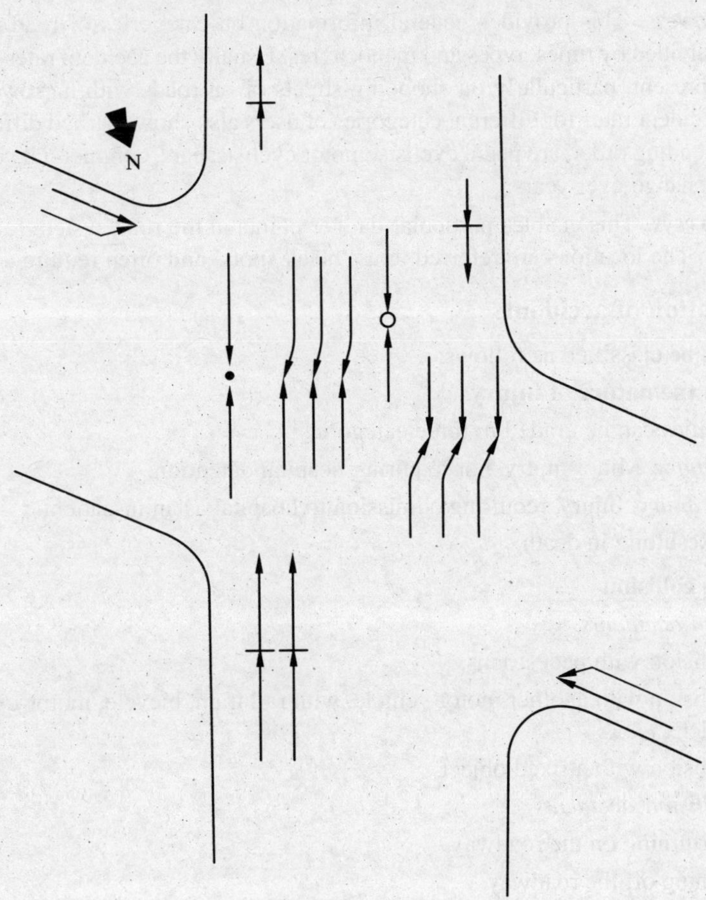

Fig. 4.17 Collision diagram.

3. *Condition diagram.* It is drawn to scale and gives all the physical conditions at a location to be studied as an aid in the interpretation of accidents.

This usually shows the following:

1. Physical obstruction in the roadways.
2. The curve length and roadway limits.
3. Property lines.
4. Side-walks and drive-way.
5. Roadway grades and types of road surfaces.
6. Bridges, overpasses and ditches along the roadway.
7. Any relevant information that may help in the analysis.
8. Obstruction at corners.
9. Traffic signals, signs, and pavement markings.
10. Street lights.

4.8.3. Accident surveys

Accident surveys are of two broad types:

(a) Macro-surveys. This provides general information on categories of road user with vehicle and location subdivided by times, types and manoeuvres. Usually the accident rates are highest where mixed traffic is present, particularly on shopping streets or on roads with narrow rights of way. A comparison of accident rates for different categories of users also show marked differences. The risks involved, in descending order, are pedal cyclists, motor cyclists, cars, commercial vehicles and buses. However trends change over years.

(b) Micro-surveys. This enables particular danger points in the road system to be identified and causes evaluated. The locations are referred to as 'black spots' and often require a detailed study.

4.8.4. Classification of accidents

Accidents can be classified as follows:

(1) Based on the nature of injury

Traffic casualities can be graded in three categories:

(*a*) *Slight injury*. Minor injury, not requiring hospital detention.

(*b*) *Serious injury*. Injury, requiring admission to hospital as an in-patient.

(*c*) *Fatal*. Resulting in death.

(2) Based on collision

(*a*) *Collision accidents*

(*i*) Collision with pedestrians.

(*ii*) Collision with another motor vehicle, with rail train, bicycle, motor-cycle, any body on vehicle, *etc.*

(*iii*) Collision with a fixed object.

(*b*) *Non-collision accidents*

(*i*) Overturning on the roadway.

(*ii*) Running off the roadway.

(*iii*) Any other non-collision.

(3) Damage-only accidents

Physical damage to vehicles or property but not to a person.

4.8.5. Accident rates

For comparison purposes and to determine the degree of exposure to risk, accident rates are determined in proportion to population, vehicle registration and vehicle travel. Further break down may be required into urban and rural accidents, by the time of day, day of week and season. Sometimes accident rates are calculated by a breakdown into road user categories, ages and sexes. The most useful form of accident rate relates traffic volume and length of road for a specified time period. This can be expressed by the following two indices:

(*a*) Personal injury accidents per 10 lacs vehicle km per year

$$= \frac{\text{Number of personal injury accidents per year} \times 10^6}{\text{Length of road (km)} \times \text{Traffic flow per year}}$$

(*b*) Fatalities per 10 crores vehicle km per year

$$= \frac{\text{Number of fatalities per year} \times 10^8}{\text{Length of road (km)} \times \text{Traffic flow per year}}$$

Another measure which is useful in comparing the sites of accidents, is the accident severity rate, calculated for selected sites and time periods, as follows:

$$\frac{\text{Fatal accidents}}{\text{Fatal and serious accidents}} \quad \text{or} \quad \frac{\text{Fatal and serious accidents}}{\text{All injury accidents}}$$

4.8.6. Accident and highway design

Effect of volume. The accident rate increases in two-lane roads up to an average daily volume of 900 vehicles per day and then decreases where volumes exceeds 900 vehicles per day.

This drop is believed to be the effect of congestion—the slower speeds of congested traffic resulting in fewer reportable accidents.

Effect of pavement width. It directly effects the accident rate as is clear from Table 4.5.

Table 4.5 Number of accidents for different pavement widths

Pavement width	*Accidents per 10 lacs vehicle km.*
Less than 5.4 m	5.2
5.4 m to 5.9 m	3.8
6 m to 6.9 m	3.5
7 m and over	3.4

Other studies have shown that factors such as type and width of shoulders and lateral clearance to walls, poles and similar roadside features must be considered jointly with pavement width in the evaluation of highway design.

Effect of curves. Studies indicate that accident rate on curves is not much different from that for straight sections, but there is a considerable variation in the rate with curve sharpness and frequency. Results indicate that where curves are infrequent, they are more hazardous than where they form a part of a continuous winding alignment. At isolated curves, the rate shows a marked increase with curve sharpness.

4.8.7. Pedestrian Accidents

Pedestrian accidents in cities and towns are much more than non-pedestrian accidents. The death rate amongst the pedestrian is highest for children and aged persons.

Also the death rate has been found to be much higher in autumn and winter than in any other season. Most of the pedestrians are killed at night. In a study it was found that 75 per cent were killed in cross-walks where both pedestrian and motorists had the maximum opportunity to see each other. In rural areas, pedestrian accidents are more than one-sixth of the total accidents and are caused where pedestrians cross the streets.

The following factors lead to pedestrian accidents:

(*i*) Excessive speed.

(*ii*) Intoxicating effect of alcohol on drivers.

(*iii*) Physical conditions of the roadway.

(*iv*) Unsafe vehicles.

(*v*) Vision obstruction of the roadway.

(*vi*) Weather conditions.

(*vii*) Driver's age. Effect of age of drivers on accidents is shown in Table 4.6.

(*viii*) Traffic signal installation.

Table 4.6 Effect of driver's age on accidents

Drivers age in years	*Accidents*
Less than 24	89 per cent higher than average
24–64	24 per cent higher than average
65 and above	58 per cent higher than average

During the age group of 20 to 50 years, it was found that there is a self improvement among the drivers.

4.8.8. Measures for preventing accidents

Effective measures for improving safety and preventing accidents can be divided into three main groups, viz.

(*i*) Engineering

(*ii*) Enforcement ; and

(*iii*) Education.

These are sometimes called "3 Es'" and are discussed below:

4.8.8.1. Engineering Measures

(1) Road design

Efficient and economic road transport is largely dependent on road layout. The general alignment must be compatible with traffic conditions. The following considerations in road design should be paid due attention:

(*a*) *Segregation of traffic.* A traffic engineering principle of over-riding importance is the segregation of traffic categories by function and in space and time *e.g.* the segregation of vehicle and pedestrian of different classes of vehicles and of different types and purposes of vehicular movement. Raised foot paths with vertical kerbs should be provided for pedestrians. Guard railing should be installed at hazard crossings.

(*b*) *Lighting system.* Visual observation is markedly poorer at night and the driver requires additional reaction and perception time even under good street lighting. Considerable reduction in accident rates can be achieved after the installation of effective street lighting; particularly at intersections, bridges and at other places where there are restrictions to traffic movements.

Whether a street light should be installed or not can be found out from the following relation:

$$I_c \geq CNA_c$$

where I_c = The annual installation cost

C = The fractional reduction in accidents expected after the installation of lighting (0.25 to 0.45)

N = The average number of injury accidents, before the installation of lights

A_c = Average cost of the injury accident.

Similarly, a volume warrant for the lighting cost (I_c)

$$= 365\ D_v\ V_n\ R\ C\ A_c\ S$$

where D_v = Annual average daily volume

V_n = The fraction of the annual average daily volume carried at night

R = The general night injury accident rate

S = The length of the installation.

On well lit roads, the vehicles can use dim dipper lighting system. Reflectorised and illuminated signals and street lighting should be well maintained, if they are to be effective.

(*c*) ***Speed.*** A general expression relating accident rate and speed is given by

$$r = 3.3 - 0.027\ S$$

where, r = All injury accidents 10 lacs vehicle km on rural roads ; and

S = Mean speed over the length of road in km/hr.

(*d*) ***Double white lines.*** The prohibition to overtake at bends and summits can be indicated by double white line marking. These may be marked either on the criterion for safe stopping distance or for the time required to complete an overtaking manoeuvre at various speeds. The required visibility distance, D, can be computed from the relation;

$$D = 1.23\,\overline{V} + 0.0173\,\overline{V}^2$$

where $\overline{V}$ = Mean speed of vehicles in km/hr.

(*e*) ***Alignment.*** Accident rates have a marked relationship with road width, curvature and sight distance. They have a psychological effect on the driver. Lane width, horizontal and vertical curves and sight distances should all be satisfactory for the design speed, traffic volumes and vehicle types.

(*f*) ***Road surface.*** Selection of correct road materials to suit traffic needs and avoid skidding accidents is very important. Surfaces with high skidding resistance should be provided where braking and turning is frequent *e.g.* at curves, intersections, round abouts, on gradients and at approaches to bus stops and pedestrian crossings. Calcined bauxite aggregates have resistance to polishing and are recommended for use at such locations.

(*g*) ***Margins.*** All main roads, especially those carrying mixed traffic and pedestrians, should have clear margins free from parked vehicles and vertical obstructions such as bridge parapets, piers, wall abutments and street furniture such as telegraph posts, street lighting columns. These should be set back from the pavement edges by at least 3 m and where it is not feasible, special precautions in the form of deflector rails, etc. should be taken. Trees must be planted well clear of carriageway edges. Obstructions, close to the road margins, not only reduces passing clearance distances and capacity, but also cause greater accident potential and severity.

(*h*) *Warning and control devices.* Warning and control devices are sometimes set into the road surface at the approach to or hazard areas to alert the drivers. Examples are rumble areas (alternating areas of rough and smooth road surfaces) and giggle bars (sequence of raised strips). Jiggle bars may be 10 to 15 mm in height and 50 to 200 mm in width.

Speed breakers or so called 'sleeping police men', usually about 100 mm high and designed to be crossed safely and without discomfort at 10 to 20 km/hr are also constructed as control devices. These increase discomfort at higher speeds and if impacts are high, suspension and tyres can be damaged and lead to subsequent failure. Adequate warning signs should be installed to indicate their existence ahead.

Warning horns, bells and light can also be used at dangerous sites, but their use is not very common.

(*i*) *Drainage*: Surface drainage must be adequate for prevailing weather conditions. The surface should be regular to prevent water ponding.

(2) *Design and maintenance of vehicles.* The vehicles must be designed in such a manner that there are least failures and minimum number of projected parts. The following measures, if adopted, will be very helpful in reducing the accident rates.

(*a*) Comprehensive testing of vehicles, to create general awareness among the vehicle owners to the need for adequate maintenance and up-keep of the vehicles should be done at regular intervals through inspection of brakes, steering, suspension, chassis and other auxiliary equipment.

(*b*) High penetration laminated glass should be used in vehicles. This will reduce the incidence of head injury, provide visibility without impact fragmentation and because of its plastic interleaving, decelerate restraint when impacted.

(*c*) All round lighting system for heavy vehicles should be made.

(*d*) The lower beam setting system should be used for a clear view ahead, consistent with speed and freedom from glare.

(*e*) Improved methods for cleansing and heating glazed surfaces should be used.

(*f*) The mirror system for rear view should be replaced by a rearward closed circuit television with in-cab monitor.

4.8.8.2. Enforcement measures

(1) *Speed control.* Tachometers should invariably be installed in the trucks and buses to check the drivers to accelerate their vehicles beyond a certain speed. Surprise checks for spot speed should also be made of all fast moving vehicles. Fine should be imposed and legal action instituted against those who violate speed limits.

(2) *Traffic control devices.* Traffic control devices *e.g.* signs, markings, channelizing islands and signals, pedestrian control at turns, parking, pedestrian cross walks, etc. should be provided for proper segregation and direction of traffic.

(3) *Training and supervision.* Strict tests should be conducted for the issue of licences to the drivers, especially in the case of taxi and heavy commercial vehicle drivers. Drivers correction centres for accident repeaters should be set up.

(4) *Medical check.* Drivers should be tested for vision and reaction time, at least once in every three years.

(5) Use of crash helmets and seat belts should be made obligatory.

4.8.8.3. Education

It is very important that all citizens are trained and educated to be good road users. The education may be imparted either at home or in the school or in the field. The rules of the road should be taught *e.g.* pedestrians should know the correct timing and place of crossing the road, the drivers of cars, motor cycles, buses, *etc.* should be conversant with the signal system *etc.* Posters may be exhibited showing the serious results of careless driving.

Traffic safety week may be organized by the transport authorities and traffic police to impress upon the road users "*Do's and do nots*". Documentary films may be shown to the public.

TRAFFIC SIGNS AND MARKINGS

4.9. TRAFFIC SIGNS

A traffic sign is a device mounted on a fixed or portable support whereby a specific message is conveyed by means of words or symbols for the purposes of regulating, warning or guiding traffic. They are usually installed where special regulations apply or unusual conditions or hazards exist, which are not self evident, or to furnish directional information.

Indian Motor Vehicles Act has classified the signs into three categories. Complete details of the sign are given in IRC-67 'Code of Practice for Road Signs'.

1. ***Mandatory/Regulatory signs.*** These signs are used to inform road users of certain laws and regulations to provide safety and free flow to traffic. These include all signs which give notice of special obligations, prohibitions or restrictions with which the road users must comply. The violation of these signs is a legal offence.

2. ***Cautionary/Warning signs.*** These are used to warn road users of hazardous conditions either on or adjacent to the road. Cautionary signs indicate the approach to a place where caution is required.

3. ***Informatory signs.*** These are intended to guide the road user and to give him such other information as may be of interest or use to him.

Traffic signs included in the Act are:

4.9.1. Mandatory/Regulatory signs

1. Stop and giveway signs

(*i*) Stop

(*ii*) Giveway

2. Prohibitory signs

(*i*) Straight prohibited/No entry
(*ii*) One way
(*iii*) Vehicles prohibited in both directions
(*iv*) All motor vehicles prohibited
(*v*) Truck prohibited
(*vi*) Bullock cart and hand cart prohibited
(*vii*) Bullock cart prohibited
(*viii*) Tonga prohibited
(*ix*) Hand cart prohibited
(*x*) Cycle prohibited
(*xi*) Pedestrian prohibited
(*xii*) Right/left turn prohibited
(*xiii*) U-turn prohibited
(*xiv*) Overtaking prohibited
(*xv*) Horn prohibited

All dimensions in mm.

Fig. 4.18 Some of the mandatory/regulatory signs.

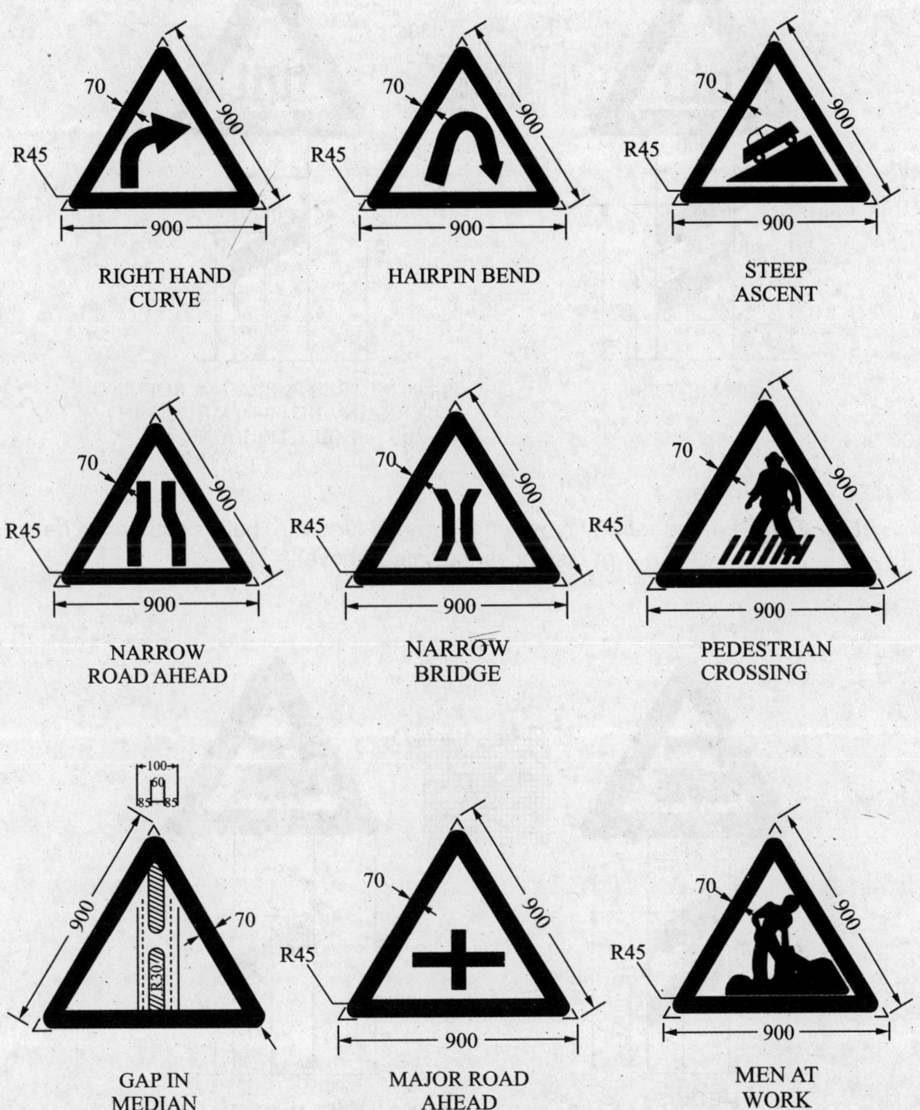

All dimensions in mm.

Fig. 4.19 Some of the cautionary signs.

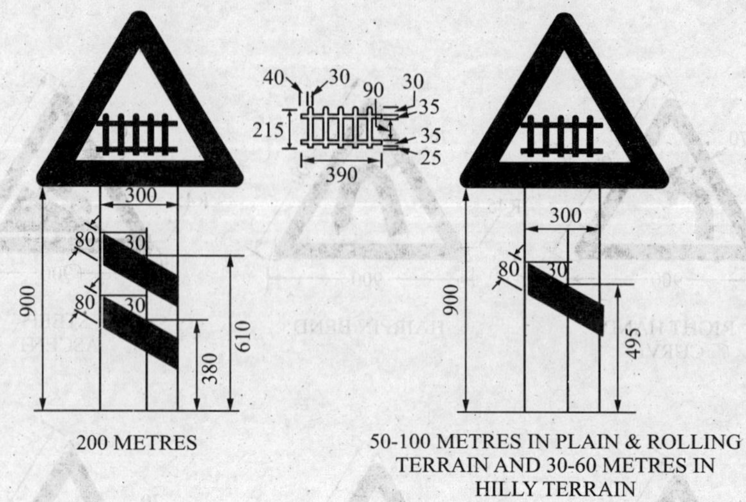

Fig. 4.20 Guarded railway crossing (For each crossing, both signs are to be at distances indicated above).

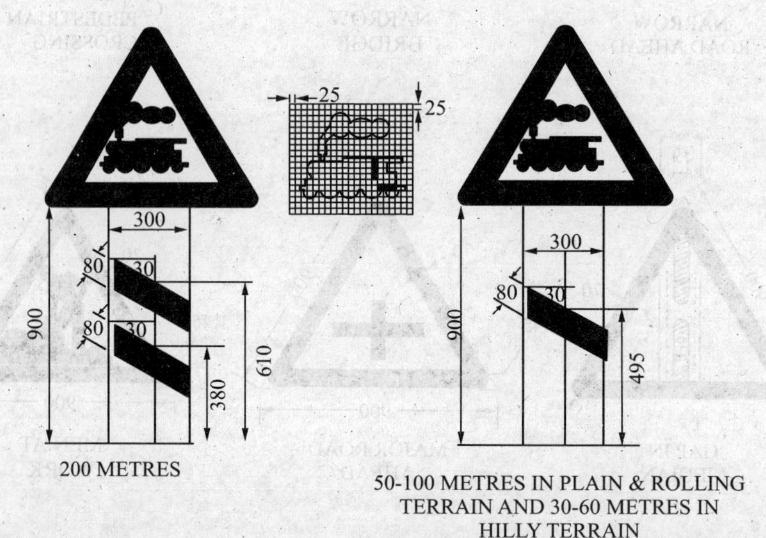

All dimensions in mm.

Fig. 4.21 Unguarded railway crossing (For each crossing, both signs are to be at distances indicated above).

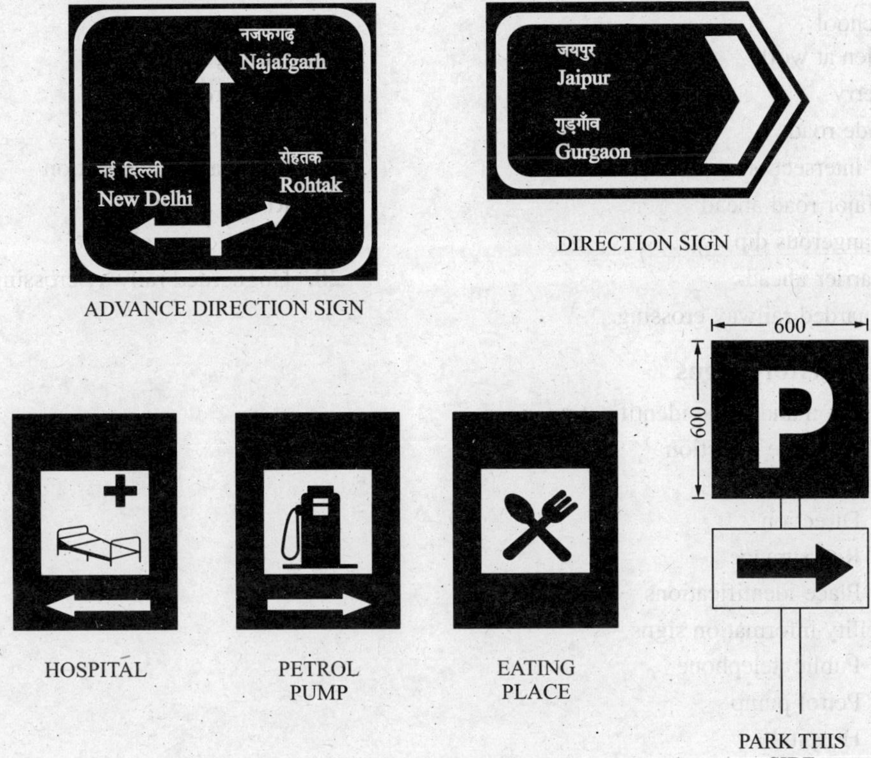

Fig. 4.22 Some of the informatory signs.

3. No parking and no stopping signs

(*i*) No parking (*ii*) No stopping/standing

4. Speed limit and vehicle control signs

(*i*) Speed limit (*ii*) Width limit

(*iii*) Height limit (*iv*) Length limit

(*v*) Load limit (*vi*) Axle load limit

5. Restriction ends signs

6. Compulsory direction control and other signs

(*i*) Compulsory turn left/right (*ii*) Compulsory ahead only

(*iii*) Compulsory turn right/left ahead (*iv*) Compulsory ahead or turn right

(*v*) Compulsory ahead or turn left (*vi*) Compulsory keep left

(*vii*) Compulsory cycle truck (*viii*) Compulsory sound horn

4.9.2. Cautionary/Warning signs

1. Right hand/left hand curve **2.** Right/left hairpin bend

3. Right/left reverse bend **4.** Steep ascent/descent

5. Narrow bridge **6.** Narrow road ahead

7. Road widens ahead **8.** Gap in median

9. Slippery road **10.** Loose gravel

11. Cycle crossing **12.** Pedestrian crossing

13. School
14. Cattle
15. Men at work
16. Falling rocks
17. Ferry
18. Cross roads
19. Side road
20. T-intersection
21. Y-intersection
22. Straggared intersection
23. Major road ahead
24. Round about
25. Dangerous dip
26. Hump or rough road
27. Barrier ahead
28. Unguarded railway crossing
29. Guarded railway crossing.

4.9.3. Informatory Signs

1. Direction and place identification signs
(*i*) Advance direction
(*ii*) Destination
(*iii*) Direction
(*iv*) Reassurance
(*v*) Place identifications

2. Facility information signs
(*i*) Public telephone
(*ii*) Petrol pump
(*iii*) Hospital
(*iv*) First aid post
(*v*) Eating place
(*vi*) Light refreshment
(*vii*) Resting place

3. Other useful information signs
(*i*) No through road
(*ii*) No through side road

4. Parking signs

5. Flood gauge

4.9.4. Location of Signs

Mandatory signs, indicating weight limit, total prohibition, direction, no parking, overtaking prohibitions, and use of sound signal prohibition, should be located in the immediate vicinity of the point, where a driver or vehicle is required to observe the mandate or prohibition shown by the sign, and if necessary, at further points where regulation continues. Similarly, cautionary signs showing school and narrow bridge should be put up in their immediate vicinity. Informatory sign depicting flood gauge should be placed on the upstream edge of the bridge at the lowest point and at suitable distances all along it, keeping in view the longitudinal section affected by flood water side of the lowest point. The marking on all the flood gauges on a causeway should indicate the depth of flow at its lowest point.

All other signs should be fixed at the following distances in advance of the hazards warned against:

1. ***In rural locations***. As per Table 4.7.

2. ***In urban locations***. In cities warning signs should usually be placed nearer the points of hazard than on rural roads.

Table 4.7 Distances at which signs should be fixed on rural roads

	In plain and rolling country	*In hilly or mountainous country*
On National and State Highways	120 metres	60 metres
On Major District Roads	90 metres	50 metres
On Other District Roads	60 metres	40 metres
On Village Roads	40 metres	30 metres

4.9.5. Colour of signs

The signs should be painted in colours as given in IRC 67. The reverse side of all sign plates, discs, or triangles should be painted in an unobtrusive grey colour.

The sign post should be painted in 25 cm bands alternately in black and white. The lowest band (next to the ground) should be in black.

4.9.6. Material of signs

The traffic sign plates, discs, or triangles may be made of stove-enamelled metal plate, or other suitable local materials like plywood, timber planks with or without metal sheet lining, or reinforced concrete.

The posts for traffic signs may consist of suitable mild steel sections, galvanized iron pipes (5 cm dia), reinforced concrete (10 cm × 10 cm) or timber scantlings (10 cm × 10 cm).

When pipes are used they should be anchored to prevent turning.

4.9.7. Design of signs

The effectiveness of sign depends upon:

(*i*) Uniformity of design, application and installation.

(*ii*) Attention or target value depends upon the size, shape, colour, colour contrast, and level of illumination.

(*iii*) Priority value, *i.e.,* the characteristics which determine the order in which the signs are read. It will depend primarily on the placement and size of message compared with all others in a group.

(*iv*) Legibility which may be either pure legibility or glance legibility depending upon the conditions. Pure legibility is the distance at which the sign can be read in unlimited time. Glance legibility is the distance at which sign can be read under normal traffic conditions and speed. Both are dependent upon the letter size, shapes, level of illumination, etc. and familiarity of the motorist with the particular design.

From research, the following findings have been arrived at:

1. A maximum of three words can be read at a glance.
2. Familiar symbols are more easily recognized and understood at a glance than equivalent word messages.
3. Rounded letters are more easily read than block letters.
4. Letter spacing considerably affects the ability to understand or read the sign. Too narrow spacing results in letters running together and too wide spacing results in difficulty in recognizing the entire word at a glance.
5. Capital and lower case letters are better than all capitals on large destination signs.
6. White letters on dark background are better for large destination signs than the reverse combination.

Recommendations for the design of signs

The following recommendations are made for the design of signs:

1. Use large signs on high-speed roads.
2. Wider spacing between letters, with optically equal spacing, depending upon the type of adjacent strokes, increases the legibility.
3. Use a maximum of three words.
4. Reflectize or illuminate the signs to be read at night.
5. Location of the signs will depend on the speed of vehicles and the size of letters on the sign.
6. Keept uniformity in (*a*) design, *i.e.,* shape, colour, (*b*) size of sign (*c*) symbols, (*d*) word messages, (*e*) illumination (*f*) lettering.
7. Distraction or advertisement signs and other unnecessary signs should be eliminated whenever possible.
8. It is recommended that two signs for different purposes should not be placed on the same sign post, but should be separated by at least 30 m, if possible.
9. Location of the signs with respect to the carriageway:

On an unkerbed road, the sign should be erected at a distance of 2 to 3 metres from the edge of carriageway depending on local circumstances, but in no case should any part of the sign come in the way of vehicular traffic.

On a kerbed road, the sign post should not be less than 60 cm away from the edge of kerb; nor should any part of the sign be less than 30 cm away from the edge of the kerb.

Traffic sign should be fixed on the left side of the road, at right angles to the pavement and facing approaching traffic. This rule, however, does not apply to flood gauge and parking signs.

The signs should be so erected that the centre of the definition plate, or device plate or of the main sign when there is no definition or device plate, is 140 cm above the crown of the carriageway. However, the lowest edge of the sign should in no case be less than 75 cm above the crown of the carriageway.

In location, where they will obstruct pedestrians, the signs should be mounted at a height of not less than 215 cm to the lowest edge of the sign.

4.10. TRAFFIC MARKINGS

These can be broadly classified as: (*i*) pavement markings; and (*ii*) object markings.

4.10.1. Pavement markings

Pavement markings are done on the carriageway itself whereas object markings are applied on objects such as kerbs, piers, traffic islands, abutments, *etc.*

Pavement markings include:

(*i*) Centre lines (*ii*) Lane-lines (*iii*) Edge lines (*iv*) Stop-lines (*v*) No overtaking or passing zone marking (*vi*) Pedestrian crossings (*vii*) Cyclist crossing (*viii*) Carriageway width restriction markings (*ix*) Route direction arrows (*x*) Parking space limits (*xi*) Intersection approach markings, *etc.*

These should be used whenever the following conditions exist:

(*a*) Two-lane roadway with a minimum ADT of 2,000 vehicles.

(*b*) Two-lane roadway less than 6 m wide with a minimum ADT of 1,000 vehicles.

(*c*) Two-lane roadways less than 5.4 m wide with a minimum ADT of 500 vehicles.

(*d*) All four or six-lane undivided highways.

(*e*) Wherever safety or other conditions warrant it.

Centre lines. Various patterns have been used for centre-line delineation. On two-lane rural highways, the most common design has 3 m dashes separated by 4.5 m spaces. For a multi lane highway, the centre line may be continuous or broken.

Lane-lines: These guide traffic and help in the proper utilization of the roadway. These lines comprise of 1.5 m dashes with a 4.5 m gap. They should be used as follows:

(*a*) On all rural highways with an odd number of lanes.

(*b*) On all multilane highways.

(*c*) On all congested locations where the roadway will accommodate additional lanes.

(*d*) On all important streets and highways where maximum usage is desired.

The centre line and lane lines are shown in Fig. 4.23.

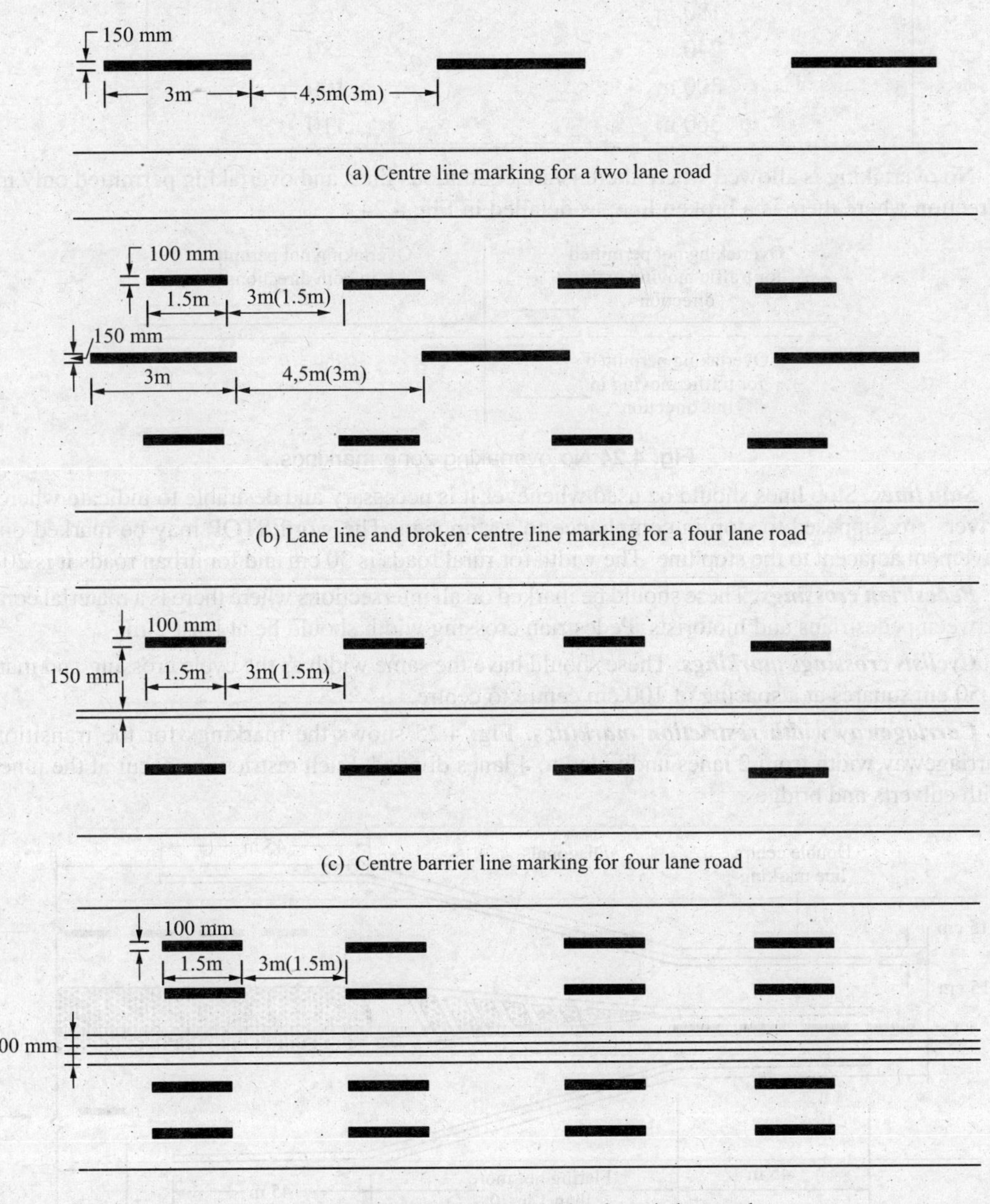

Fig. 4.23 Centre line and lane line marking for urban areas.

Edge lines: Those carriageways which have no kerbs, are provided edge lines to serve as a visual guidance for the drivers about the road edge. These are useful during bad weather conditions and poor visibility. Edge lines are continuous lines about 15 cm away from the edge.

No overtaking or passing zones. These are provided on summit curves and horizontal curves where the minimum sight distances are less than those specified for different speeds as follows:

Minimum sight distances	*Speed (km/hr)*
150 m	50
180 m	65
240 m	80
300 m	100
360 m	110

No overtaking is allowed where there is/are continuous lines and overtaking permitted only in the direction where there is a broken line; as detailed in Fig. 4.24.

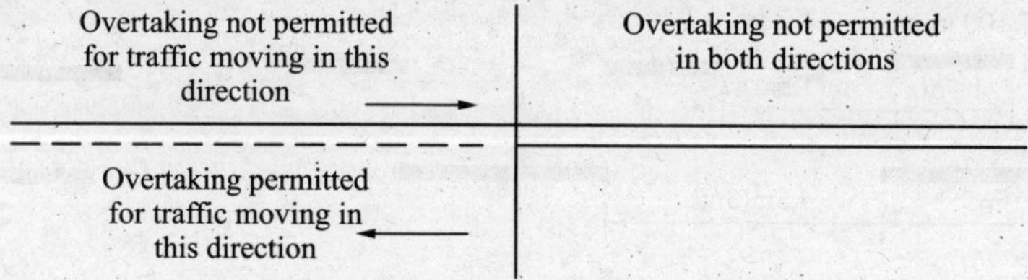

Fig. 4.24 No overtaking zone markings.

Stop lines. Stop lines should be used whenever it is necessary and desirable to indicate where the drivers are supposed to stop in compliance to a stop sign. The word STOP may be marked on the pavement adjacent to the stop line. The width for rural roads is 30 cm and for urban roads it is 20 cm.

Pedestrian crossings. These should be marked on all intersections where there is a material conflict between pedestrians and motorists. Pedestrian crossing width should be at least 2 m.

Cyclists crossings markings. These should have the same width as the cycle crossing and marked in 50 cm squares at a spacing of 100 cm centre to centre.

Carriageway width restriction markings. Fig. 4.25 shows the markings for the transition of carriageway width from 2 lanes undivided to 4 lanes divided. Such restrictions occur at the junction with culverts and bridges.

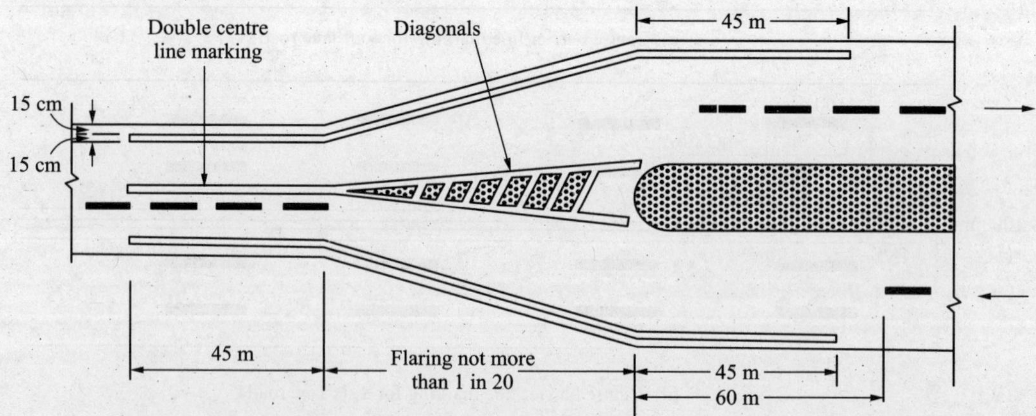

Fig. 4.25 Markings on transition from 2-Lane undivided to 4-Lane divided carriageway.

Route direction arrows. To guide the traffic in the correct direction and lane, route direction arrows are marked. These markings are elongated in the direction of the traffic to enable the drivers to see them properly and legibly.

Parking space limits. Parking space limits enables efficient use of parking space and prevents encroachment on bus stops, loading and unloading zones, fire-hydrants etc.

Intersection approach markings. Markings at the intersections are a combination of centre lines, lane lines, stop lines, route direction arrows to guide the traffic at the intersections.

4.10.2. Object Markings

These include objects within the roadway and objects adjacent to the roadway e.g. guard rails, trees, drums, *etc*. For this, alternate white and black strips are put up at least 10 cm wide at 45°. Object markings are usually either illuminated or reflectized.

4.10.3. Kerb Markings

These are put in to indicate parking restriction and loading zones, and are given with appropriate colours indicating the instruction to the drivers. They are as follows:

(*i*) Parking, standing and stopping strictly prohibited with the exception that buses may stop and load and may be indicated by red colour.

(*ii*) Commercial loading zones effective only during the business hours and may be indicated in yellow colour.

(*iii*) Passenger loading zones may be indicated in white and for very short parking time limits in green colour.

4.10.4. Reflector Markers

These are provided at kerbs, *etc*. These consist of single reflectizing button or clusters of buttons of small panels covered with reflecting coatings.

TRAFFIC SIGNALS

A traffic signal is defined as any power operated traffic control device, or a sign by which traffic is warned or directed to take some specific action. A traffic control signal is a signal, which through its indications, directs the traffic to stop and permits it to proceed alternatively. Typical details of a signal post and its foundation are depicted in Fig. 4.26.

Advantages of a traffic control signal

(*i*) It provides for an orderly movement of traffic.

(*ii*) It will reduce the frequency of certain type of accidents, in particular the right angled accidents.

(*iii*) It provides a means of interrupting heavy traffic to cross.

(*iv*) It promotes driver confidence by assuring the right of way.

(*v*) It may afford considerable economy over manual control at intersections where alternate assignment of the right of way is necessary.

Disadvantages

(*i*) It may increase certain type of accidents particularly the re-entrant collision.

(*ii*) When improperly located, it promotes disrespect for this type of control device, and when improperly tuned it will cause excessive delays.

(*iii*) It may encourage the drivers to prefer alternate routes thus dispersing traffic on minor streets rather than concentrating on a major street.

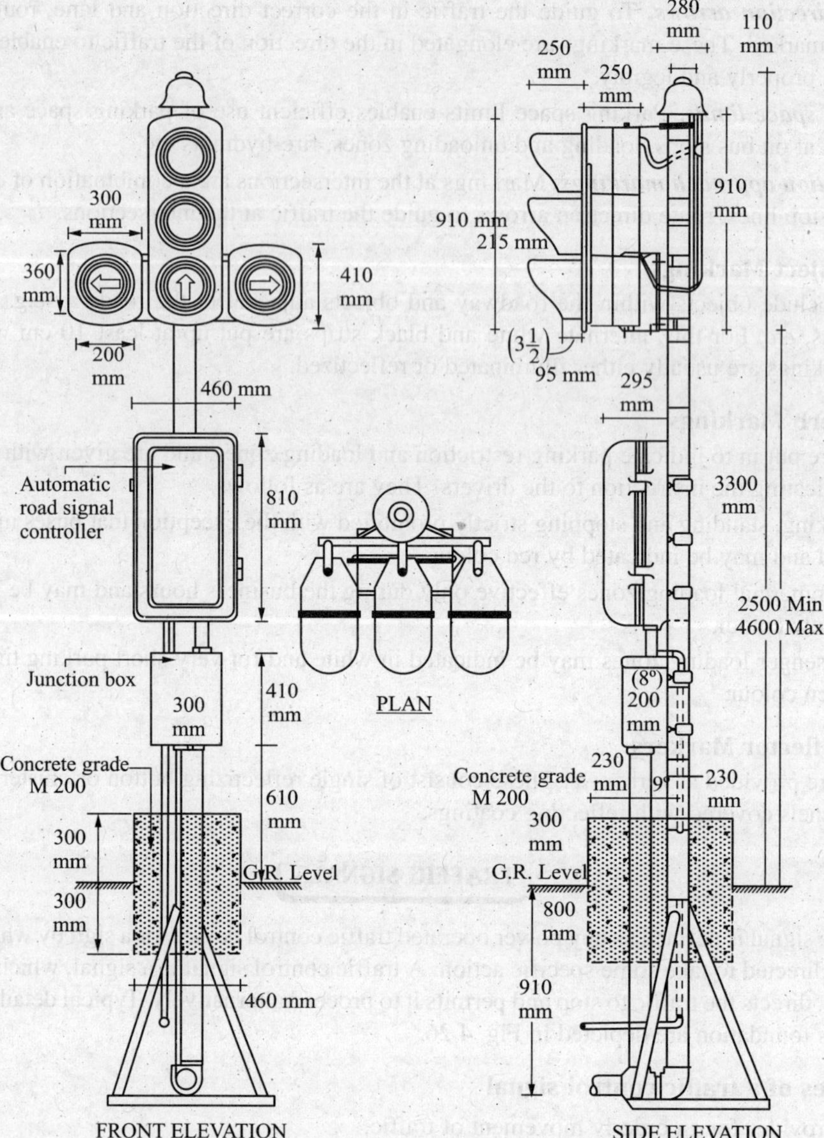

Fig. 4.26 Typical details of signal pole and its foundation.

4.11. CLASSIFICATION OF SIGNALS

Signals can be classified as follows:

1. Traffic control signals

(*a*) Fixed-time signal

(*b*) Traffic actuated signal

(*i*) Fully actuated signal

(*ii*) Semi-actuated signal

(*iii*) Speed control signal.

2. Pedestrian signals.

3. Special traffic signals.

More important of them are described below:

4.11.1. Fixed-time Signals

These signals are set to repeat regularly a cycle of red-yellow-green lights. These have the advantage of having the minimum delay for a group of vehicles travelling along a series of intersections, with interconnected signals. The disadvantage, however, is that it may hold vehicles from one direction when there is no traffic in the other, resulting in delay and sometimes in a decrease in capacity. This disadvantage is removed in the case of traffic-actuated signals, as explained in the following paragraph.

4.11.2. Traffic Actuated Signals

The main characteristic of the traffic-actuated signal is that it recognizes the demand of the flow of traffic and distributes the time cycle or the green time accordingly. Time cycle is the number of seconds required for one complete sequence of signal indications *i.e.,* green-yellow-red.

Its main advantages are the following:

(*i*) Usually reduces delay.

(*ii*) It is adaptable to short-term fluctuation in traffic.

(*iii*) Usually increases the capacity.

(*iv*) It provides continuous operation under low volume conditions and is specially effective at multiple phase intersection. Generally it is most efficient for isolated intersections.

Its main disadvantages are:

(*i*) It is uneconomical. The cost is about two to four times the cost of a fixed time installation.

(*ii*) The actuated controller (complete electronic mechanism for controlling the operation of traffic control signals, including timer and all necessary appliances mounted in a cabin) and detectors are much more complicated than fixed-time controllers, thus increasing the maintenance and inspection costs.

(*iii*) Detectors are very costly to install and present very difficult maintenance problems.

Traffic actuated signals are of two types:

Fully-actuated traffic signals. These have detectors located on each approach and assign the right of way to the various traffic movements on the basis of demand.

Semi-actuated traffic signals. Where traffic on heavy-volume of high speed arteries must be interrupted for relative light cross-traffic, semi-actuated traffic signals are often installed. For these signals the detectors are placed only on the minor street.

Detectors. Vehicle detectors are of various types:

1. Pressure sensitive detectors. These are of two types: (a) non-directional detectors, (b) directional detectors. These are installed underground.
2. Magnetic detectors which operate on the principle that all moving vehicles have a magnetic field which will induce certain voltages in the magnetic circuit of the detector. These are also installed underground.
3. Radar detectors which detect the vehicles on a leg of intersection by the detector mounted on a post, resembling in appearance to a high street lamp-post.
4. Sound detectors which detect the sound waves set up by the passage of vehicle.
5. Light sensitive detectors which detect the vehicle by the interruption of light.

4.12. SIGNAL SYSTEMS

The various signal systems are:

1. *Simultaneous system also known as synchronised system.* In this system all signals along the given street always show same indication at the same time. The division of the cycle is the same at all signalized intersection systems and only one controller is used to operate a series of intersections.

The disadvantages of a simultaneous signal are:

(*i*) It is not conducive to allow continuous movement of all the vehicles.

(*ii*) It encourages speeding of drivers between stops.
(*iii*) The overall speed is often reduced.
(*iv*) Because the division of the cycle time is the same at all the intersections, inefficieny is inevitable at some intersections.
(*v*) The simultaneous stoppage of a continuous line of traffic at all intersections often results in difficulty for the side street vehicles in turning into or crossing the main side street.

2. *Alternative system.* Here the alternate signals or group of signals along a given road show opposite indications at the same time. The system is operated with a single controller. This permits the vehicles to travel one block in half the cycle time.

Some of the disadvantages of this system are:
(*i*) The green times for both the main and side streets have to be substantially equal, resulting in inefficiency at most of the intersections.
(*ii*) In situations where the block lengths are unequal, the system is not well suited.
(*iii*) Adjustments are difficult for changing traffic conditions.

3. *Simple progressive system.* In this system the signals controlling a street give green indications according to a pre-determined schedule to permit continuous operation of groups of vehicles along the street at a planned rate of speed, which may vary in different parts of the system. A typical time-space diagram is shown in Fig. 4.27.

4. *Flexible progressive system.* In this system it is possible at each intersection automatically to vary the following:
(*i*) Cycle time and division at each signal depending upon the traffic.
(*ii*) The offset thus enabling two or more completely different plans.
(*iii*) It is possible to introduce flashing or shut down during off-peak hours.

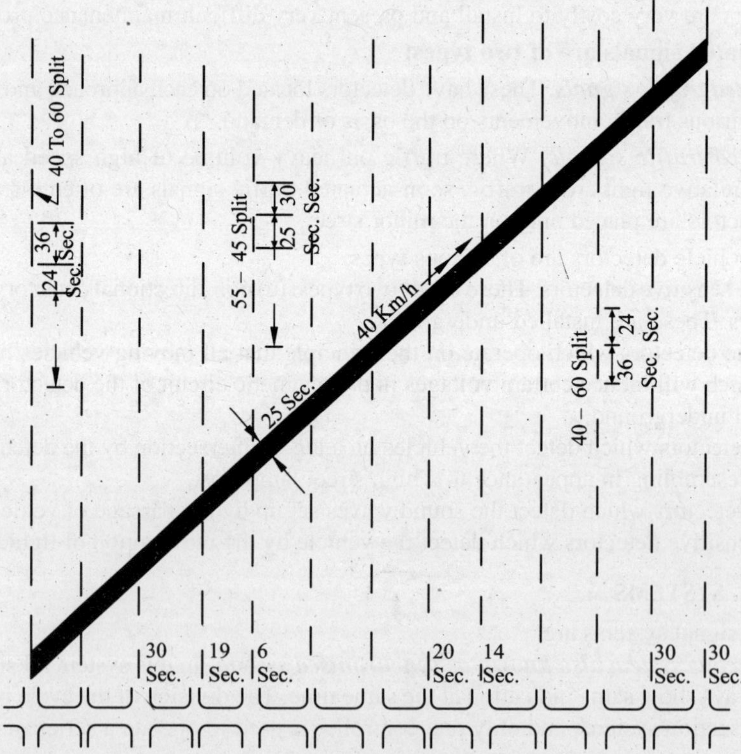

Fig. 4.27 Time-space diagram.

4.13. TIMING OF SIGNALS

4.13.1. Timing an isolated signal.

It should be determined on the following lines (given in steps) according to the traffic requirements. Cycle time should normally be from 40 to 60 sec (best timing would be the shortest possible under the traffic conditions). The maximum cycle time should be 120 seconds.

1. Determine pedestrian crossing time of all approaches based upon pedestrian walking speed generally taken as 1 m/s.
2. These values will be minimum green plus amber time for each phase.
3. Based on these minimum, compute green plus amber time in proportion to approach volumes per approach lane. No phase should be less than 15 sec.
4. Adjust cycle time (sum of all phases) to next higher 5 sec interval and recompute phase values.
5. Select amber periods based upon approach speeds. Table 4.8 is recommended as a guide for selecting appropriate amber periods.
6. Compute percentage value (to nearest per cent) for all phases (total being 100 per cent). It is necessary to use percentages since controller settings are in per cent of signal cycle.
7. Computed timing should be installed in the controller and the operation of the intersection observed, especially during peak conditions.

Field correction of the timing may be necessary to provide smooth flow.

Table 4.9 Amber periods for different speeds

Approach Speed (Km/hr)	*Amber period (second)*
0–50	3
50–65	4
65–80	5
80 or more	5 (plus all-red period)

4.13.2. Timing a co-ordinated system

1. The prerequisite of any co-ordinated system is that all signals in the system must operate on the same cycle length. The division of the cycle may vary with the individual intersection but the total cycle length must remain constant.
2. Usually the critical intersection is tuned according to the above system and the resulting cycle length used throughout the system.
3. Determination of offsets: An offset is the interval in seconds between a certain instant arbitrarily used as a tune-reference base and the start of the "go" indications at a given signal at that system.

Problem 4.1 *The intersection of Madhya Marg and Udhyan Path in Chandigarh is to be signalized. Madhya Marg is 13 m wide, having an approach volume of 600 vehicles per hr and 70–30 split during the peak hour. Approuch speed is 55 km/hr.*

Udhyan Path is 7 m wide having an approach volume of 450 vehicles/hour with a 80–20 split and approach speed of 40 km/hr. There is a lot of pedestrian and bicycle traffic at the intersection.

Determine the cycle time, green and amber times for each street.

Solution. *Step I.* Pedestrian consideration:

Time required to indicate the pedestrian to start = 5 sec

Madhya Marg: Time required to cross Madhya Marg = (green + amber) time for Madhya Marg

$$= \frac{7}{1} + 5 = 12\,\text{sec}.$$

Udhyan Path: Time required to cross Udhyan Path = (green + amber) time for Udhyan Path

$$= \frac{13}{1} + 5 = 18\text{ sec}.$$

Step II. Considering approach speed:

Minimum amber time for Madhya Marg

(speed = 55 km/hr) = 4 sec

Minimum amber time for Udhyan Path

(speed = 40 km/hr) = 3 sec

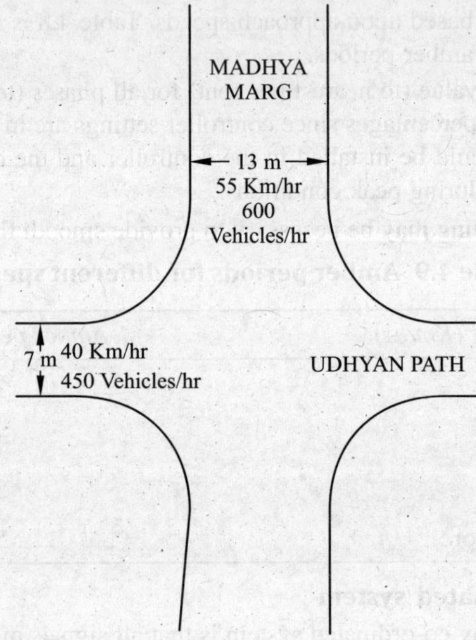

Step III. Clearance interval:

Stopping sight distance for vehicles at 55 km/hr on Madhya Marg = 75 m

Stopping sight distance for vehicles at 40 km/hr on Udhyan Path = 50 m

Total distance to be travelled to clear intersection for vehicles moving on Madhya Marg

= 75 + 7 = 82 m

Total distance to be travelled to clear intersection for vehicles moving on Udhyan Path

$= 50 + 13 = 63$ m

Time required by vehicles on Madhya Marg $= \dfrac{82 \times 3.6}{55}$

$= 5.7$ sec say 6 sec.

Time required by vehicles on Udhyan Path $= \dfrac{63}{40} \times 3.6$

$= 5.67$ sec. say 6 sec.

Assume the total cycle length of signals = 40 sec.

Amber time for Madhya Marg $A_M = 6$ sec.

Amber time for Udhyan Path $A_U = 6$ sec.

Green time for Madhya Marg and Udhyan Path $= 40-(6+6) = 28$ sec.

This time is to be divided in the ratio of approach volumes

$$\frac{G_U}{G_M} = \frac{V_U}{V_M}$$

Volume of traffic on Udhyan Path/lane $= \dfrac{450\times 80}{100}$

$= 360$ vehs/hr

Volume of traffic on Madhya Marg/lane $= \dfrac{600\times 70}{100}\times\dfrac{1}{2}$

$= 210$ vehs/hr.

$$\frac{G_U}{G_M} = \frac{360}{210}; G_U + G_M = 28 \text{ sec.}$$

$$\frac{28-G_M}{G_M} = \frac{12}{7} \text{ or } \frac{28}{G_M} = \frac{19}{7}$$

$$G_M = \frac{7}{19}\times 28 \simeq 10.32 \text{ sec. Take } G_M = 10 \text{ sec.}$$

$\therefore \quad G_U = 18$ sec.

For Madhya Marg : $G_M = 10$ sec.

For Udhyan Path : $G_U = 18$ sec.

Distribution of time-interval of the cycle in the two directions is as follows:

Cycle time $= 40$ sec	(100%)
$G_U = 18$ sec	(45%)
$A_U = 6$ sec	(15%)
$G_M = 10$ sec	(25%)
$A_M = 6$ sec	(15%)

Minimum phase of (green + amber) time for Madhya Marg = 10 + 6 = 16 sec.

These times also satisfy the pedestrian requirements.

Considering equal headway on both roads = 2.5 sec.

Madhya Marg

Volume entering the intersection/lane $= \dfrac{420}{2}$

$= 210$ vehs/hr

Volume entering the intersection during the cycle length $= \dfrac{210}{60\times 60}\times 40 = 3$ vehs

Time required to clear the intersection $= 3\times 2.5 = 7.5$ sec.

Actual time provided = 10 sec

$\therefore$ Safe

Udhyan Path

Volume entering intersection per lane = 360 vehs/hr

Volume entering intersection during cycle length $= \dfrac{360 \times 40}{60 \times 60} = 4$ vehicles

Time required to clear the intersection $= 4 \times 2.5 = 10$ sec

Actual time provided = 18 sec.

$\therefore$ Safe

Problem 4.2 *If there is no pedestrian and cycle traffic in problem 4.1, what will be the minimum green and amber time for each street, assuming:*

(*a*) *Equal headways on both sides.*

(*b*) *4 sec. headway on Madhya Marg and 3 sec. headway on Udhyan Path.*

Solution. (*a*) From the previous problem we note that the results were not affected by the pedestrians' considerations (clearance interval and approach volume)

Assuming equal headway, the same timings can be applied, *viz.*,

Total cycle = 40 sec (100%)

$G_U = 18$ sec (45%)

$A_U = 6$ sec (15%)

$G_M = 10$ sec (25%)

$A_M = 6$ sec (15%)

(*b*) Assuming total cycle length = 40 sec

From clearance consideration : $A_M = 6$ sec

$A_U = 6$ sec

Green time for Madhya Marg and Udhyan Path $= 40 - 12 = 28$ sec.

Madhya Marg

Volume entering the intersection per cycle time = 3 vehicles

Green time required to clear the intersection $= 3 \times 4 = 12$ sec

Udhyan Path

Volume entering the intersection per cycle time = 4 vehicles

Green time required to clear the intersection = 4 × 3 = 12 sec.

Allowing, $G_U = 16$ sec

Cycle time = 40 sec (100%)

$G_U = 16$ sec (40%)

$A_U = 6$ sec (15%)

$G_M = 12$ sec (30%)

$A_M = 6$ sec (15%)

4.14. DESIGN OF SIGNALS

The cycle time should be such that delay at all approaches of the intersec tion is minimum. This is given by the relation:

$$C_0 = \frac{1.5T + 5}{1 - X} \text{ seconds}$$

where C_0 = Optimum cycle time

T = Lost time due to starting time and termination time

$$X = x_1 + x_2 + x_3 \ldots + x_n$$

$$x_1, x_2, x_3 \ldots x_n = \frac{f_1}{S_1}, \frac{f_2}{S_2}, \frac{f_3}{S_3}, \ldots \frac{f_n}{S_n}$$

where $f_1, f_2, f_3 \ldots f_n$ represent the flow in legs 1, 2,...*n*

and, $S_1, S_2, S_3 \ldots S_n$ represent the saturation flow in legs 1, 2, 3...*n*

The effective green time is proportional to $x_1, x_2 \ldots x_n$ such that

$$G_1 : G_2 : G_3 \ldots G_n = x_1 : x_2 : x_3 \ldots : x_n$$

where $G_1, G_2, G_3 \ldots G_n$ ate the effective green time in the legs 1, 2, 3...*n*.

The amber time is assumed as 2 seconds in each leg.

The saturation flow is expressed as

$$S = 525 \text{ W PCU/hr}$$

where S = saturation flow

W = width of roadway

Problem 4.3 *A four-legged right angled intersection is to be signalised with a fixed time 2 phase signal.*

The design hour flow and saturation flow are as under:

	North (N)	*South (S)*	*East (E)*	*West (W)*
Design hour flow	*900*	*500*	*800*	*700*
Saturation flow	*2500*	*2000*	*3200*	*3000*

The lost time may be taken as 2 seconds per arm. Determine the optimum cycle time and apportion the green times in the two phases.

Solution. $\frac{f_N}{S_N} = \frac{900}{2500} = 0.36$

$$\frac{f_S}{S_S} = \frac{500}{2000} = 0.25$$

$\therefore$ Max. value of $\frac{f}{S}$ in the *N-S* direction = 0.36

$$\frac{f_E}{S_E} = \frac{800}{3200} = 0.25$$

$$\frac{f_W}{S_W} = \frac{700}{3000} = 0.33$$

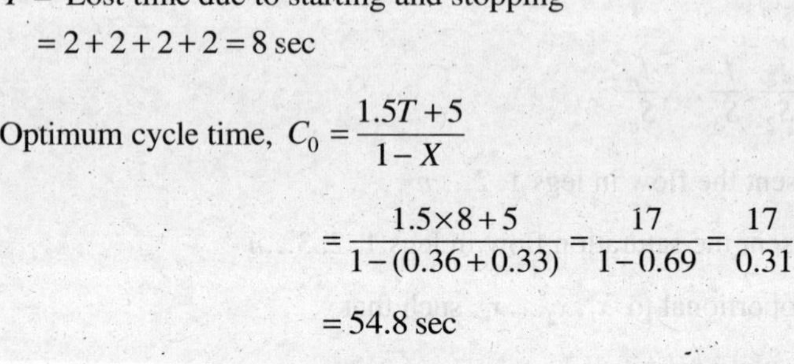

$\therefore$ Max. value of $\frac{f}{S}$ in E-W direction = 0.33

T = Lost time due to starting and stopping

$= 2 + 2 + 2 + 2 = 8$ sec

Optimum cycle time, $C_0 = \frac{1.5T + 5}{1 - X}$

$$= \frac{1.5 \times 8 + 5}{1 - (0.36 + 0.33)} = \frac{17}{1 - 0.69} = \frac{17}{0.31}$$

$= 54.8$ sec

say 55 sec

$\therefore$ Effective green time per cycle $= C_0 - T$

$= 55 - 8 = 47$ seconds

Effective green time per phase is given by

$$G_{NS} = \frac{X_{NS}}{X}(C_0 - T)$$

$$= \frac{0.36}{0.69} \times 47$$

$= 24.5$ sec

say 25 sec

$$\therefore \quad G_{EW} = \frac{X_{EW}}{X} \times 47 = \frac{0.33}{0.69} \times 47 = 22.48 \text{ sec}$$

say 22 sec

The cycle time diagrams are drawn below:

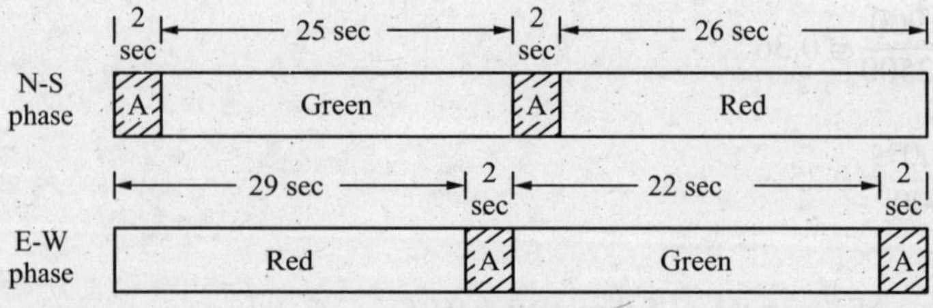

Cycle time diagram

CHANNELIZATION

Improvement to a plain intersection can be done by channelizing traffic. This is done by providing obstacles called islands and thus guiding vehicles at intersection into definite paths. The main function of channelization is to reduce conflicting movements as far as possible and afford one way movement.

4.15. ISLANDS

Island is defined as the area or certain areas within the roadway from which the vehicular traffic is diverted by pavement marking or other warning devices or by physical construction.

The various functions of the islands are:

(*i*) To segregate the pedestrians and vehicles.
(*ii*) To separate traffic into specified paths.
(*iii*) To reduce the conflict area in order to minimize hazards.
(*iv*) To divert the traffic from obstacles and expedite the traffic flow.
(*v*) To increase safety.

The various types of islands are:

(*i*) *Channelizing islands*. These are located in the roadway to confine specific movements of traffic to definite channels.

(*ii*) *Refuge islands* (Pedestrian islands). To raise safety zones; these are located in a cross-walk to provide refuge for the pedestrians.

(*iii*) *Loading islands*. They raise safety zones and are provided at regular bus or tram stops when it is near the middle of a street and give protection to the passengers.

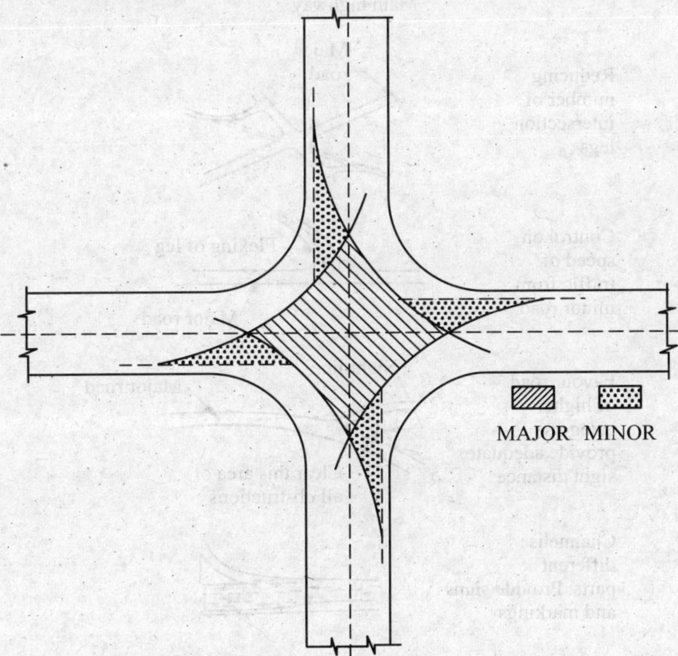

Fig. 4.28 Conflicting zones at an intersection.

4.16. CONDITIONS WARRANTING CHANNELIZATION

(*i*) Excessive areas or improper use of the area.
(*ii*) High accident rate.

(*iii*) Complex intersection.
(*iv*) Conflicting turning movements.
(*v*) Speeds too high for conditions.
(*vi*) Very high traffic volume.
(*vii*) Providing protection to the pedestrians due to traffic congestion and at times dangerous midblock turning movement.

4.17. DESIGN OF ISLANDS

The various factors to be considered in the design of traffic islands are as follows:

1. *Traffic factors* which should take into account the following points for the design:

(*i*) Earlier traffic capacity.
(*ii*) Capacity of the roadway.
(*iii*) Size and shape characteristics of vehicles.
(*iv*) Turning movements.
(*v*) Pedestrian volumes and movements.
(*vi*) Transit operation.
(*vii*) Accidents, *etc*.

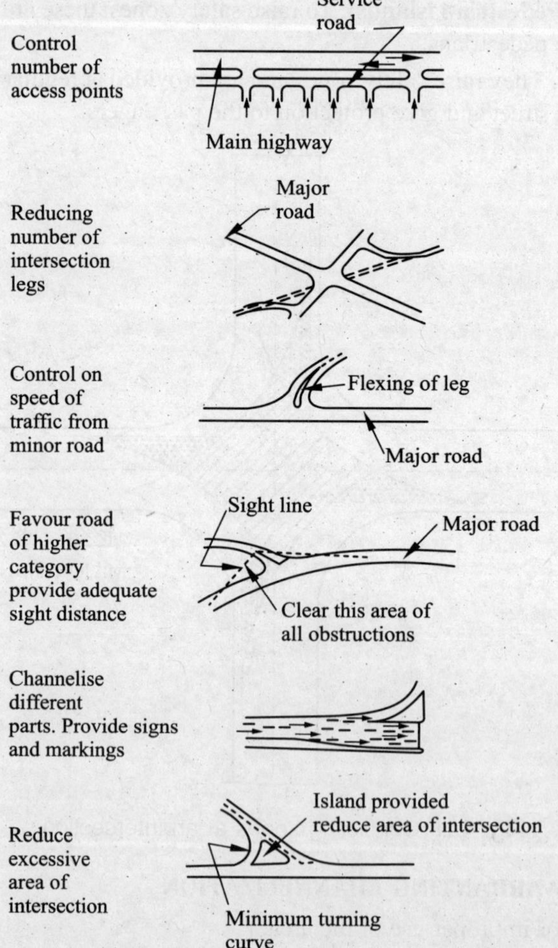

Fig. 4.29 Main considerations in the planning and design of at-grade intersection.

2. ***Physical Factors*** which should take into account the following points:

(*i*) Areas of conflict.
(*ii*) Angles of entering the streets.
(*iii*) Cross-sections
(*iv*) Steep grades.
(*v*) Sight distances.
(*vi*) Right of way lines (indicating how much area is to be at disposal).
(*vii*) Location of buildings and other limiting physical features.
(*viii*) Existing traffic control devices.
(*ix*) Type of pavement.
(*x*) Lighting conditions.

3. ***Human Factors*** which should take into account the following points:

(*i*) Intelligence and skill of the drivers.
(*ii*) Reaction time.
(*iii*) Driving habits.
(*iv*) Vision.
(*v*) Pedestrian walking speeds.

4. ***Economic Factors*** which should take into account the following points:

(*i*) Cost of improvement.
(*ii*) Reduction in the number of accidents.
(*iii*) Time saved.
(*iv*) Reduction in the number of stops (user benefits).

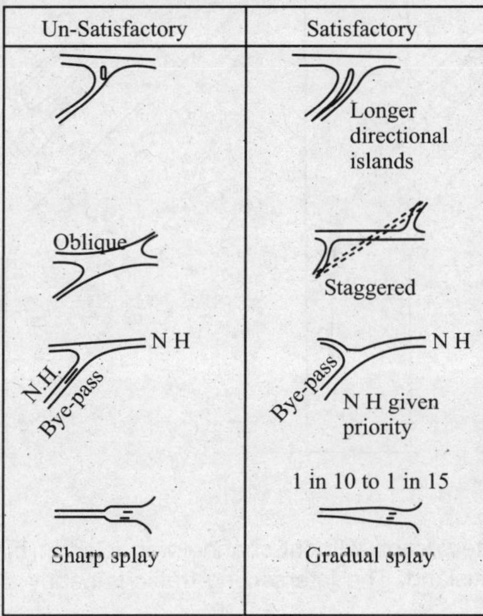

Fig. 4.30 Un-satisfactory and satisfactory design of at-grade intersections.

4.18. SPECIFIC EXAMPLES OF CHANNELIZATION

Main consideration in the planning and design of at-grade intersections are given in Fig. 4.29. Un-satisfactory and satisfactory designs of at-grade intersections are shown in Fig. 4.30.

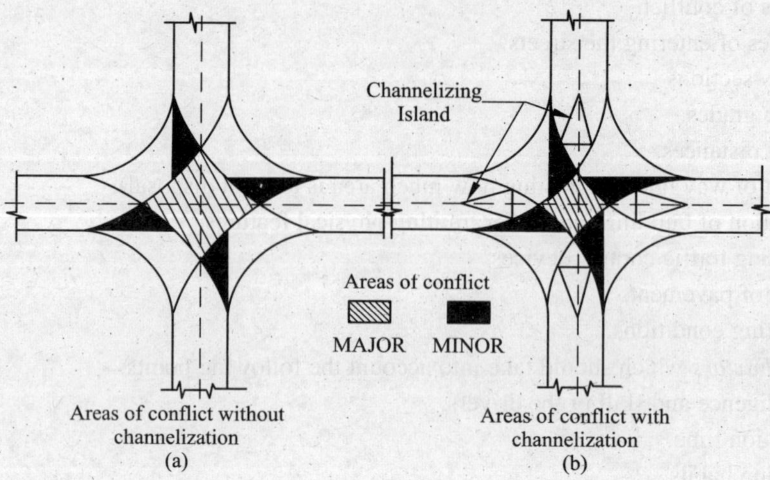

Fig. 4.31 Conflicting area (*a*) without and (*b*) with channelizing islands

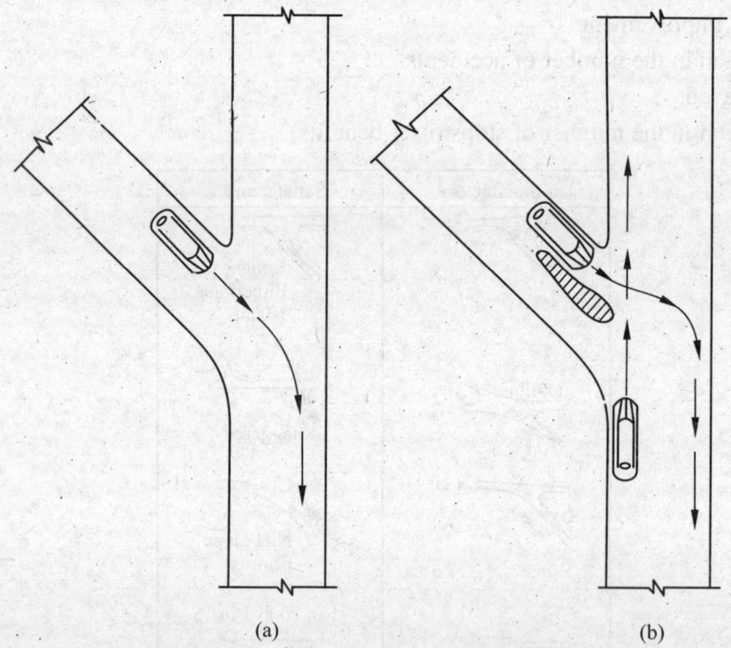

Fig. 4.32 (*a*) Intersection without channelizing island. (*b*) Intersection with channelizing island. The intersecting traffic is nearly at right angles.

(*i*) To reduce the area of conflict zone by having channelizing islands, as shown in Fig. 4.31.

(*ii*) To reduce the conflict zone of two intersecting traffic streams by making them cross at or nearly at right angles, as shown in Fig. 4.32.

(*iii*) To have the traffic streams merging at the smallest possible angle of about 10° to 15°. The idea is to prevent head-on-collision of vehicles by having them almost parallel to each other. This is illustrated in Fig. 4.33.

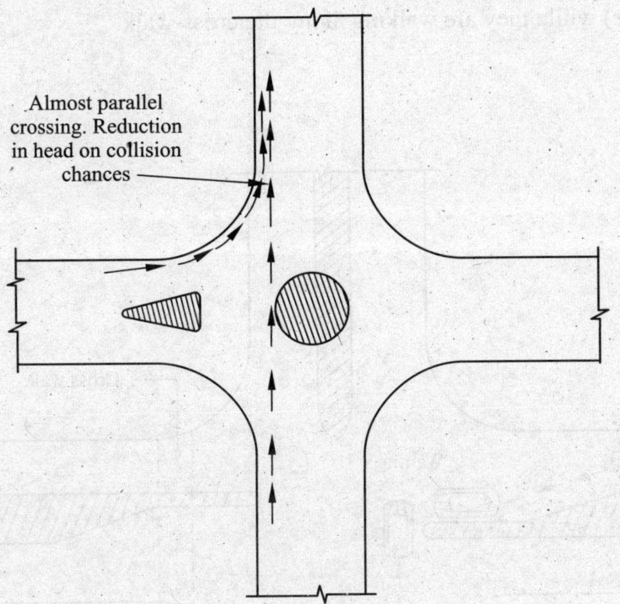

Fig. 4.33 Traffic streams merging.

(*iv*) The speed of the vehicles can be controlled by bending the traffic stream. It is to be noted. however, that the main stream is not bent. This is illustrated in Fig. 4.34.

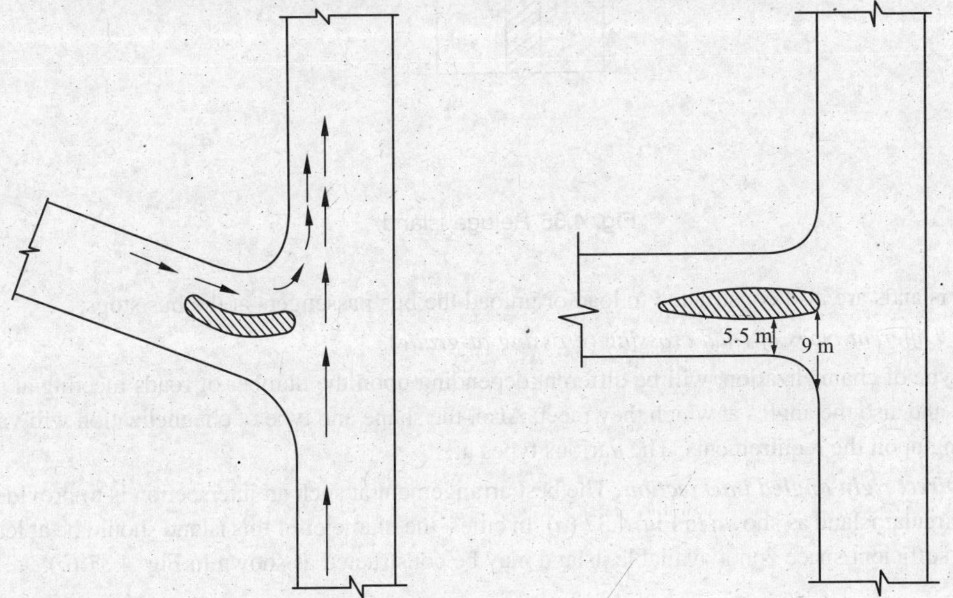

Fig. 4.34 Control by bending the traffic stream.

Fig. 4.35 Funnelling.

(*v*) Another way to reduce the speed is by funnelling *i.e.*, gradual reduction in the width of the road, as shown in Fig. 4.35.

(*vi*) The refuge islands are usually placed, as shown in Fig. 4.36, to enable the pedestrians to wait (to seek refuge) while they are walking along the cross-walk.

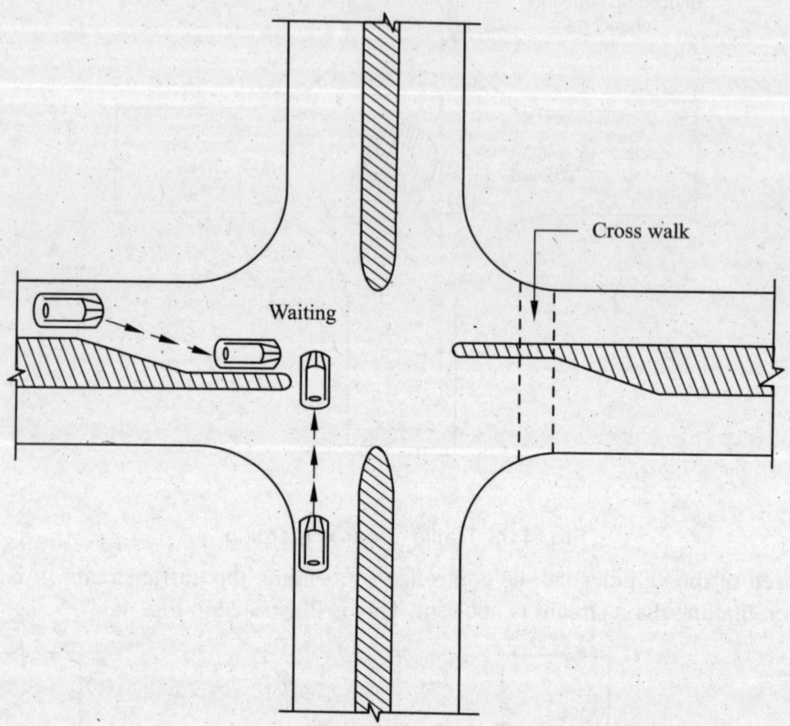

Fig. 4.36 Refuge island.

Such islands are also constructed to load or unload the bus passengers at the bus stops.

(*vii*) *Different types of road crossing (crossing at-grade).*

The type of channelization will be different depending upon the number of roads meeting at the crossing and also the angles at which they meet. Also, the shape and type of channelization will vary depending upon the requirements. The various types are:

(*a*) ***Direct right angled intersection.*** The best arrangement at such an intersection is to provide a central circular island as shown in Fig. 4.37 (*a*). In cities, the diameter of this island should be at least 20 m. If sufficient space is not available, island may be constructed as shown in Fig. 4.37 (*b*).

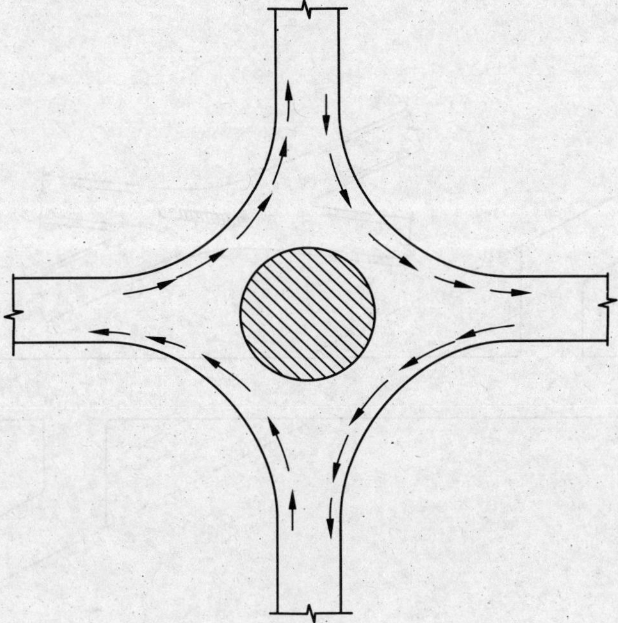

Fig. 4.37 (*a*) Central circular island.

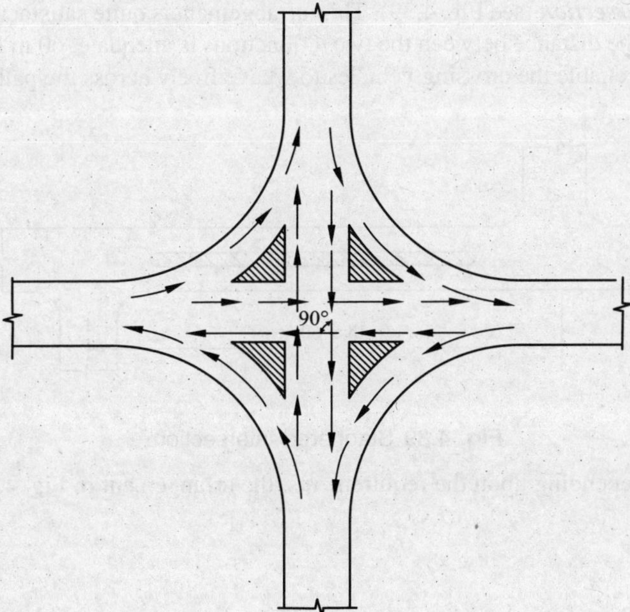

Fig. 4.37 (*b*) Right-angled intersections.

(b) Direct acute angled intersection. These should be avoided as the vehicles may directly come in conflict due to insufficient sight distances.

If a vehicle has to turn from the main road to a side street, direct head-on-conflict may result with the opposing line of traffic.

It is desirable to develop this type of intersection into a staggered intersection, as shown in Fig. 4.38 (b).

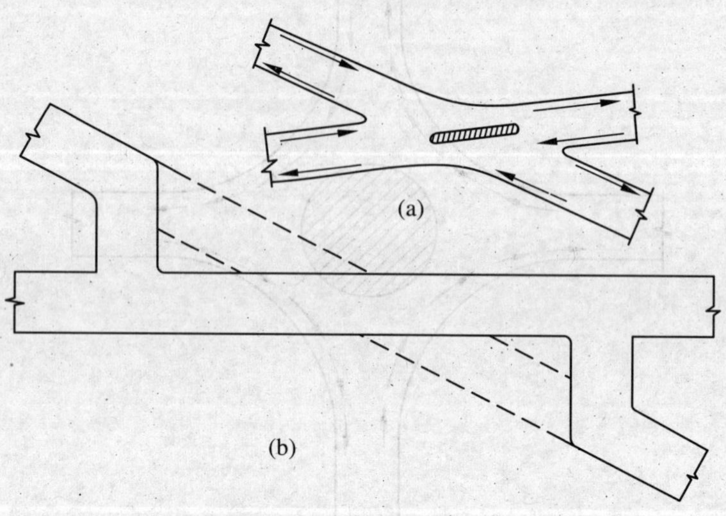

Fig. 4.38 Direct acute-angled intersection.

(*c*) ***Staggered intersection*** (see Fig. 4.39). This arrangement is quite satisfactory provided proper signs are put up and the distance between the two T-junctions is adequate. 60 m is considered to be a minimum distance to enable the crossing vehicles to weave freely across the path of opposite traffic.

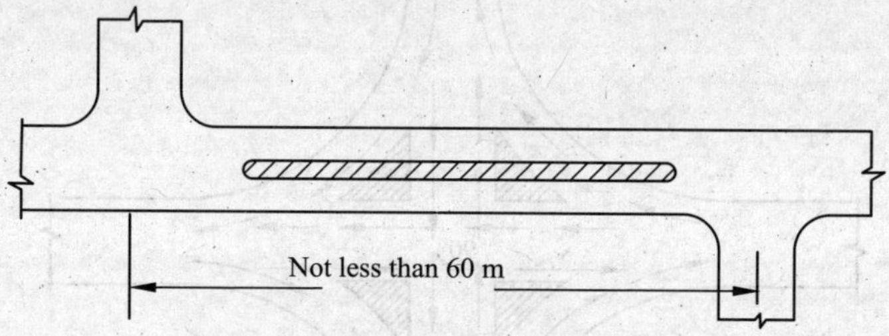

Fig. 4.39 Staggered intersection.

(*d*) ***T-junction.*** Depending upon the requirements, the arrangement in Fig. 4.40 may be chosen.

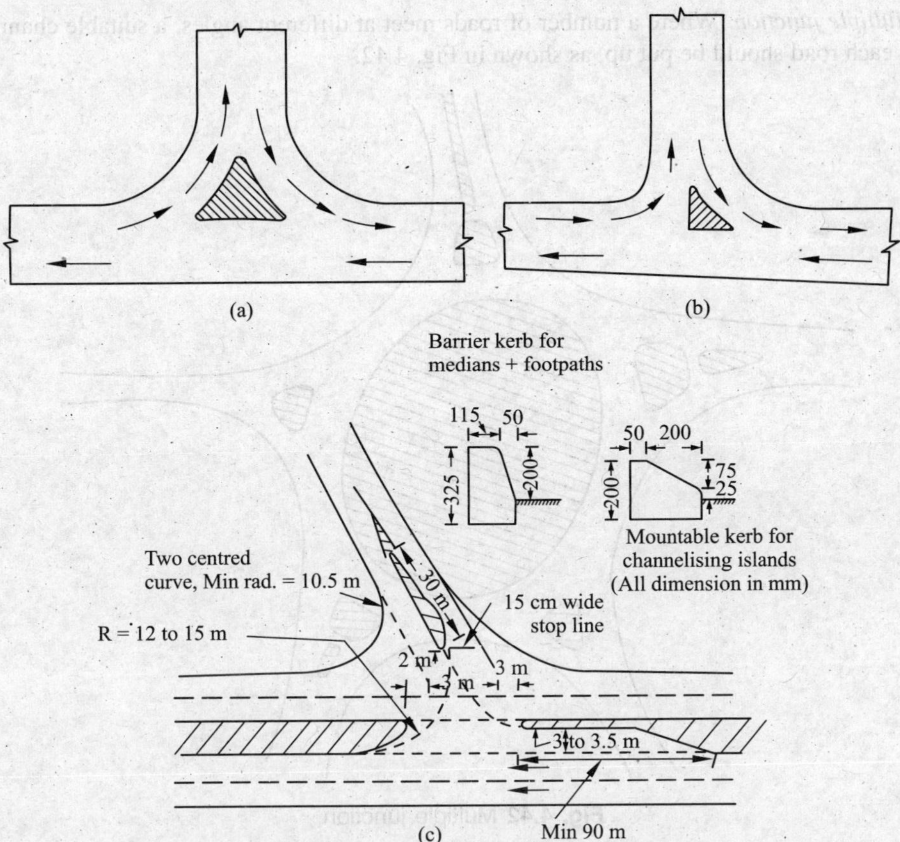

Fig. 4.40 Typical layouts of T-intersection.

(e) Y-junctions. The type, generally adopted is shown in Fig. 4.41. This type of junction should be avoided as far as possible as accident rate has been found to be very high at these junctions and are virtually a death trap.

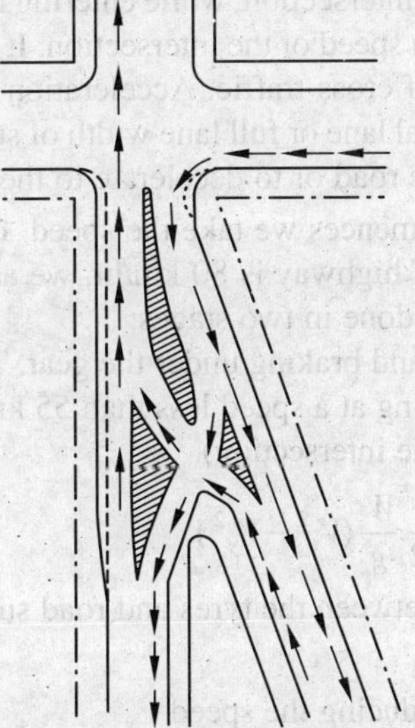

Fig. 4.41 Y-junction

(*f*) *Multiple junction.* Where a number of roads meet at different angles, a suitable channelizing island at each road should be put up, as shown in Fig. 4.42.

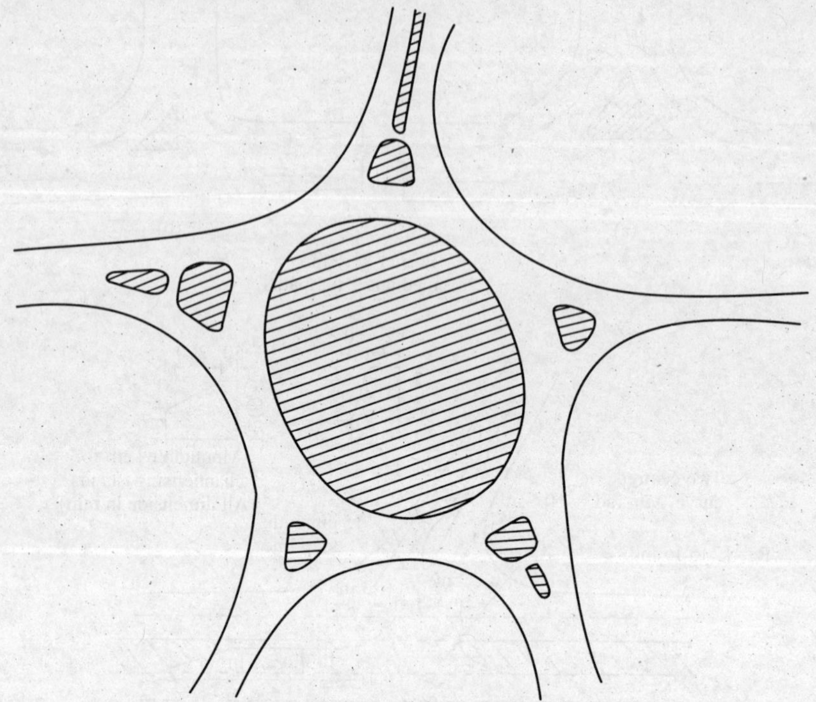

Fig. 4.42 Multiple junction.

ACCELERATION AND DECELERATION LANES

Especially on important roads, we should provide acceleration and deceleration lanes to enable the vehicle to accelerate to the design speed of the road, after emerging out of the intersection, or to decelerate to the design speed of the intersection, while entering it. The vehicle has to pull itself to left so that it can decelerate to the design speed of the intersection. It is not necessary, however to provide both, depending upon the amount of cross-traffic. Acceleration and deceleration lane need not be a full width lane. It only means a partial lane or full lane width of sufficient length to enable a vehicle to accelerate to the design speed of the road or to decelerate to the design speed of intersection.

Just before the deceleration commences we take the speed of the road as equal to 0.7 times the design speed. If the design speed of highway is 80 km/hr, we assume $V = 0.7 \times 80 = 56$ km/hr say 55 km/hr. The reduction in speed is done in two stages:

1. The foot off the accelerator and braking under the gear. This time is taken as 3 sec.
2. After 3 sec, it will be travelling at a speed less than 55 km/hr ($V = 0.7 \times 80$) and greater than 30 km/hr (design speed of the intersection).

Now
$$fWd = \frac{1}{2}\frac{W}{g}(V_x^2 - V_y^2)$$

where f = coefficient of friction between the tyres and road surface

W = weight of vehicle

d = distance travelled in reducing the speed

V_x = 80 km/hr (design speed of highway)

$V_y = 30$ km/hr (design speed of intersection)

Deceleration is the average instantaneous rate of deceleration with complete release of accelerator and is given in Table 4.9.

Table 4.9

Speed (km/hr)	*Speed change rate (km/hr/sec)*
50	2.0
65	2.5
90	3.0

Speed at beginning of deceleration lane = 55 km/hr. Speed after 3 sec will be

$$V = u - at$$
$$= 55 - 2.17 \times 3$$
$$= 48.5 \text{ km/hr}$$

where 'a' is the speed change rate, corresponding to 55 km/hr = $\left(2 + \frac{2.5-2}{15} \times 5\right)$

Distance (S) travelled in reducing the speed from 55 km/hr to 48.5 km/hr is given by

$$\frac{(55^2 - 48.5^2)}{(3.6)^2} = \frac{2 \times 2.17 \times S}{3.6}$$

$$\therefore \quad S = \frac{(3025 - 2352)}{4.34 \times 3.6}$$

$$= 43.1 \text{ m}$$

It has been found that a deceleration rate of 14 km/hr/sec is supposed to be a convenient rate of deceleration

$$fWd = \frac{1}{2}\frac{W}{g}(V_x^{\,2} - V_y^{\,2})$$

$$0.2 \times d = \frac{1}{2} \times \frac{1}{9.81}(48.5^2 - 30)^2 \times \frac{1}{(3.6)^2}$$

or $$d = 28.56 \text{ m}$$

as recommended $f = 0.2$ for lower speeds

$= 0.3$ for higher speeds

$$f = \frac{a}{g} \text{ where } a \text{ is the speed change rate}$$

$$= \frac{14}{9.81 \times 3.6} = 0.4 \text{ for 14 km/hr/sec}$$

$$= 0.3 \text{ for 10.5 km/hr/sec}$$

$$= 0.2 \text{ for 7 km/hr/sec}$$

The value of f recommended for calculating the length of acceleration lane is 0.2.

We can have a deceleration length = 43.1 + 28.56 = 71.66 m.

When the overtaking manoeuvre is being undertaken, the following maximum acceleration values will be used as recommended by *IRC*.

Table 4.10. Accelerations for different speeds

Speed in km/hr	*Normal acceleration km/hr/sec*
50	2.25
30	2.80
25	3.20

Taking design speed of vehicle before entering the intersection as $0.7 \times 80 \cong 55$ km/hr and design speed of intersection = 30 km/hr the time (t) to reduce the speed will be given by,

$$55 = 30 + 2.80\,t$$

$$t = \frac{25}{2.8}$$

$$= 8.93 \text{ sec}$$

$$\text{Distance travelled } = ut + \frac{1}{2}at^2$$

$$= \frac{30 \times 8.93}{3.6} + \frac{1}{2} \times \frac{2.80}{3.6} \times (8.93)^2$$

$$= 74.42 + 31.01 = 105.43 \text{ m}$$

which gives the distance required for the acceleration lane or the deceleration lane.

TRAFFIC ROTARY

A traffic rotary is a specialised form of '*at grade*' intersection where vehicles from the converging arms are forced to move round an island in one direction in an orderly and regimental manner and '*weave*' out of the rotary movement into their desired directions.

4.19. DEFINITIONS

1. **At-grade intersection:** An intersection where all roadways join or cross at the same level.
2. **Diverging** : The dividing of a single stream of traffic into separate streams.
3. **Intersection angle** : The angle between two intersection legs.
4. **Merging** : The converging of separate streams of traffic into a single stream.
5. **Rotary** : A road junction laid out for movement of traffic in one direction round a central island.
6. **Rotary island** : A traffic island located in the centre of an intersection to compel movement in a clock-wise direction and thus substitute weaving of traffic around the island instead of direct crossing of vehicle pathways.
7. **Weaving** : The combined movement of merging and diverging of traffic streams moving in the same general direction.
8. **Weaving length** : The length of a section of a rotary in which weaving occurs.

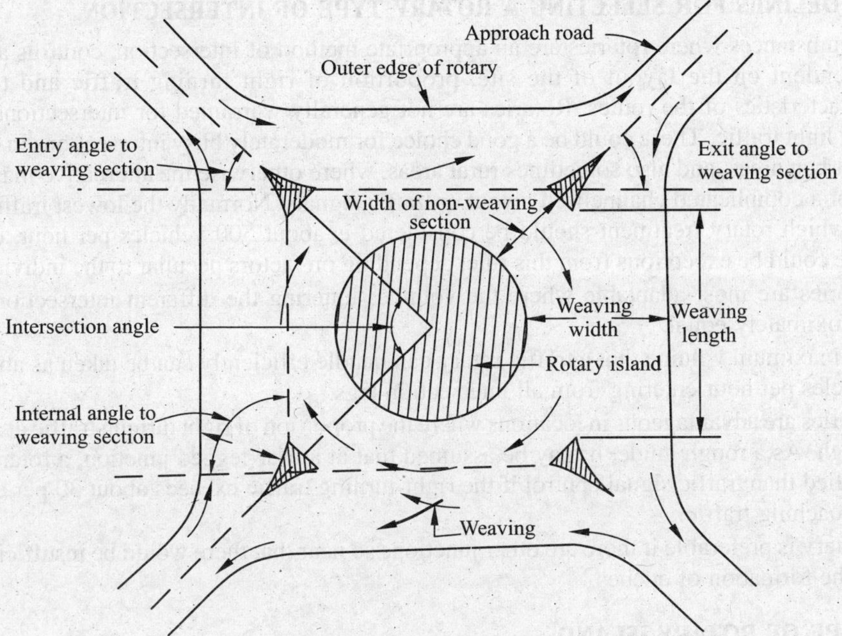

Fig. 4.43 Rotary elements.

4.20. ADVANTAGES OF ROTARY INTERSECTIONS

(*a*) An orderly and regimented traffic flow is provided. Individual traffic movements are subordinated, in favour of traffic as a whole.

(*b*) All traffic proceeds at a fairly uniform speed. Frequent stopping and starting are avoided.

(*c*) Weaving replaces the usual crossing movements at typical at-grade intersections. Direct conflict is eliminated, all traffic streams merging or diverging at small angles. Accidents occurring from such movements are usually of a minor nature.

(*d*) Rotaries are especially suited for intersections with five or more intersection legs, though these can also be adopted at intersections with 3 or 4 legs.

(*e*) For moderate traffic, rotaries are self-governing and need no control by police or traffic signals.

4.21. DISADVANTAGES OF ROTARY INTERSECTIONS

(*a*) As the flow increases and reaches the capacity, '*weaving*' generally gives way to a 'stop and go' motion, as vehicles force their way into the rotary being followed by vehicle waiting in the queue behind them. Under such conditions, vehicles, once having got into the rotary, may not be able to get out of it, because of vehicles across their path and the rotary may 'lock-up'. Once the rotary has 'lock up', the movement of vehicles completely stops and the traffic will have to be ultimately sorted out by the police.

(*b*) A rotary requires a comparatively larger area and may not be feasible in many built-up locations.

(*c*) Where pedestrian traffic is large, a rotary by itself is not sufficient to control traffic and has to be supplemented by traffic police.

(*d*) Where the angle of intersection between two roads is too acute, it becomes difficult to provide adequate weaving length.

(*e*) The provision of rotaries at close intervals makes travel troublesome.

(*f*) Traffic turning right has to travel a little extra distance.

4.22. GUIDELINES FOR SELECTING A ROTARY TYPE OF INTERSECTION

(*a*) Circumstances where rotaries are an appropriate method of intersection, controls are largely dependent on the layout of the site, proportion of right turning traffic and the traffic characteristics of the routes. Rotaries are not generally warranted for intersections carrying very light traffic. These could be a good choice for moderately busy intersections in urban and suburban areas, and also sometimes rural areas, where otherwise the alternative may be to go in for a complicated channelised layout or traffic signals. Normally the lowest traffic volume for which rotary treatment should be considered is about 500 vehicles per hour, of course, there could be exceptions from this rule, depending on factors peculiar to the individual sites.

(*b*) Rotaries are most adaptable where the volumes entering the different intersection legs are approximately equal.

(*c*) The maximum volume that a traffic rotary can handle efficiently can be taken as about 3,000 vehicles per hour entering from all intersection legs.

(*d*) Rotaries are advantageous in locations where the proportion of right turning traffic at a junction is high. As a rough guide, it may be assumed that at a four-legged junction, a rotary is more justified than traffic signal control if the right-turning traffic exceeds about 30 per cent of all approaching traffic.

(*e*) A rotary is preferable if there are other junctions so near that there would be insufficient space for the formation of queues.

4.23. SHAPE OF ROTARY ISLAND

The shape and disposition of the rotary island depends upon various factors, such as the number and disposition of the intersecting roads and the traffic flow pattern. The design of the rotary is developed by connecting the one-way entrance and exit roads to form a closed figure, with at least the minimum weaving lengths interposed between two intersecting legs and then adjusting for the minimum radius of the rotary corresponding to the design speed. In doing so, it may be necessary to try out a number of alternatives, before selecting the best. While finalising the shape of the rotary island, traffic streams within the rotary should be given dominance over the streams of traffic entering from different roads. Asymmetric shapes, either wholly curved or with a combinations of straight and curves may often provide the only satisfactory solution. The possibility of realigning one or more of the intersecting legs could also be considered to achieve the minimum weaving lengths and the desired intersection angles. Some of the more common shapes and disposition of the rotary islands are discussed below.

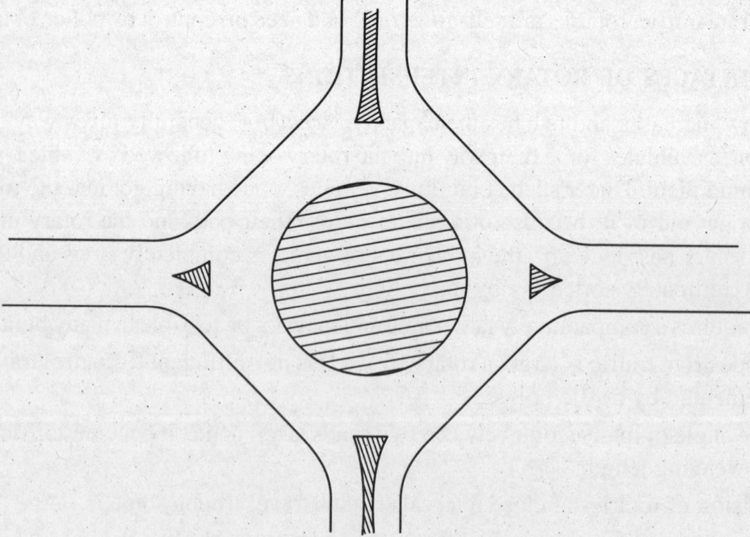

Fig. 4.44 Circular shaped rotary.

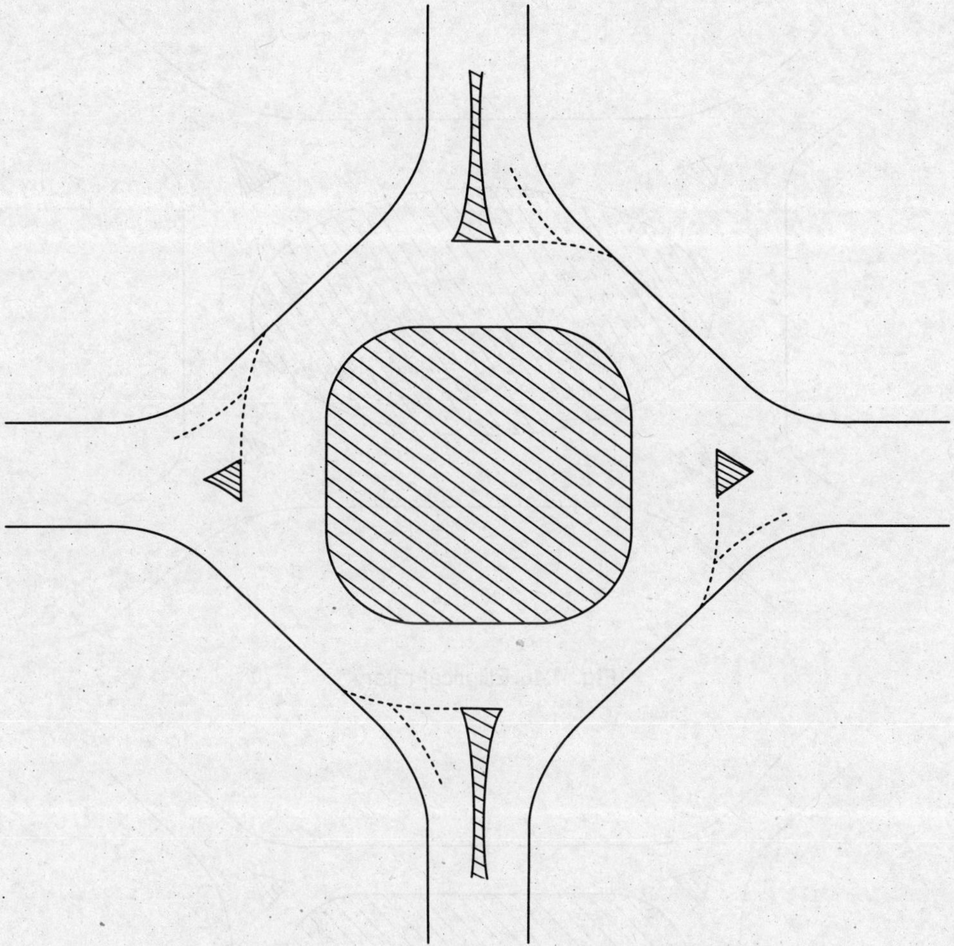

Fig. 4.45 Squarish rotary with rounded edge.

(A) Circular

A circular shape is suited where roads of equal importance intersect at nearly equal angles and carry nearly equal volume of traffic, as shown in Fig. 4.44. Under these conditions, with a circular shape, constant and regular flow is achieved.

(B) Squarish with rounded edge

This is a modification of the circular shape and is composed of four straights or four large radii curves roughly forming four sides of a square, and four small radii curves as corners (Fig. 4.45). The advantage of this layout is that it is suitable for predominantly straight ahead flows.

(C) Elliptical, elongated oval or rectangular shapes

The above shapes are provided to favour through traffic, to suit the geometry of the intersecting legs, or to provide a longer weaving length. Figs. 4.46 and 4.47 are illustrative.

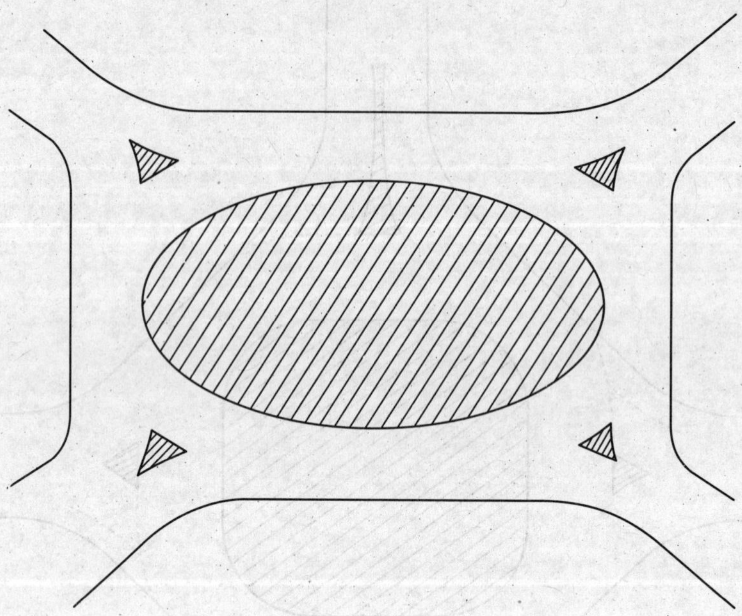

Fig. 4.46 Elliptical rotary

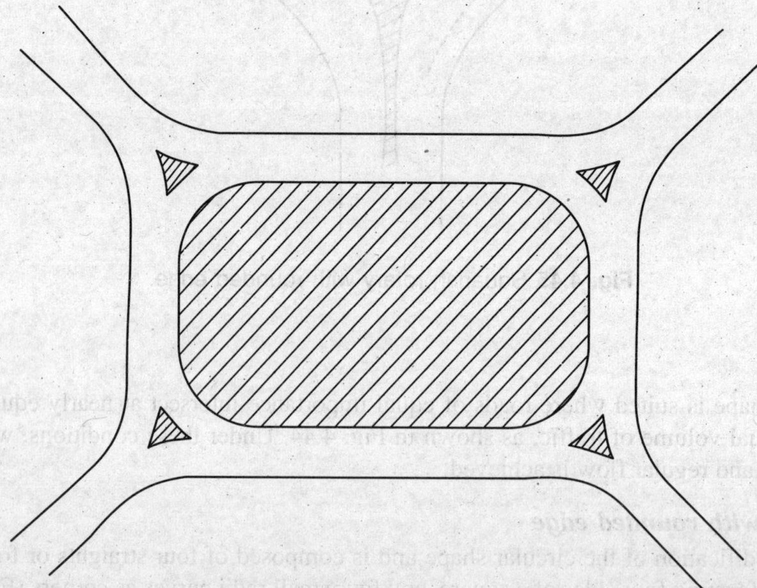

Fig. 4.47 Rectangular shaped rotary.

(*D*) *Complex intersection with many approaches*

Fig. 4.48 gives a layout of a complex intersection, whose shape is directed by the existence of a large number of approaches.

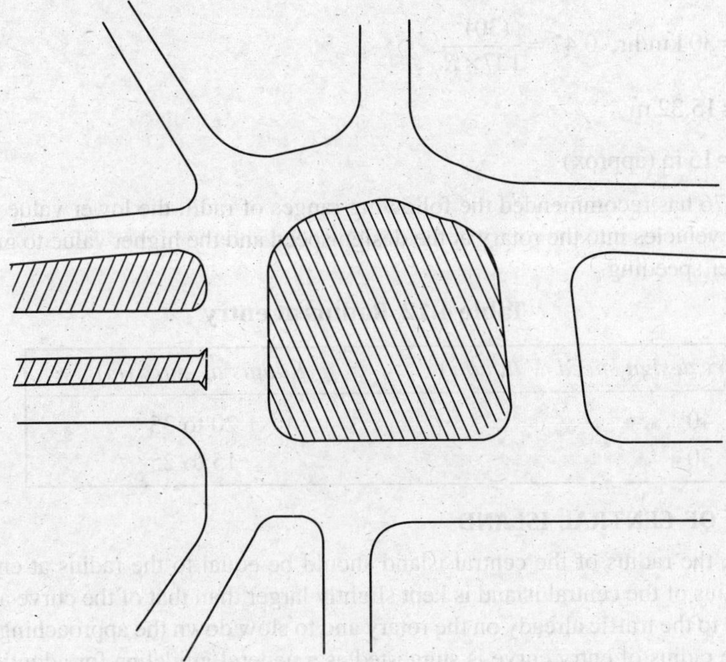

Fig. 4.48 Complex rotary intersection, with many approaches.

4.24. DESIGN SPEED OF ROTARY

The Indian Roads Congress has suggested a design speed of 40 km/hr for rotaries in rural areas and 30 km/hr for rotaries in urban areas and other restricted locations.

4.25. RADII OF CURVES

At entry the following method is usually adopted to find out the radius of the curve.

Let R = radius of rotary in m

V = design speed of rotary in km/hr

f = friction factor

Now,
$$e + f = \frac{V^2}{127R}$$

If we neglect the super-elevation, *i.e.*, taking $e = 0$,

$$f = \frac{V^2}{127 \cdot R}$$

The values of f recommended are given in Table 4.11.

Table 4.11 Values of *f*

f	*Design speed V in km/hr*
0.47	30
0.43	40

Thus for $V = 30$ km/hr, $0.47 = \dfrac{(30)^2}{127 \times R}$

$$R = 15.32 \text{ m}$$

$$= 15 \text{ m (approx)}$$

IRC: 65—1976 has recommended the following ranges of radii, the lower value meant to ensure easy entrance of vehicles into the rotary at the design speed and the higher value to guard against any tendency for over speeding.

Table 4.12 Radius at entry

Rotary design speed, V (km/hr)	*Radius at entry (m)*
40	20 to 35
30	15 to 25

4.26. RADIUS OF CENTRAL ISLAND

Theoretically, the radius of the central island should be equal to the radius at entry. In practice, however, the radius of the central island is kept slightly larger than that of the curve at entry to give a slight preference to the traffic already on the rotary and to slow down the approaching traffic. A value of 1.33 times the radius of entry curve is suggested as a general guideline for adoption.

4.27. RADIUS AT EXIT

The radii of curves at exist should be larger than that of the central island and at entry so as to encourage the drivers to pick up speed and clear away from the rotary expeditiously. The radius of the exit curve is kept 1.5 to 2.0 times the radius of the entry curves. If, however, there is a large pedestrian traffic across the exit road, radii similar to those at entrances should be provided to keep the exit speeds reasonably low.

Centrifugal Ratio $\dfrac{P}{W}$ should always be greater than 0.2 and less than 0.4.

4.28. ENTRY AND EXIT ANGLES

Entry angle should be larger than exit angle and it is desirable that the entry angles should be 60° if possible. The exist angles should be small, even tangential.

4.29. WEAVING ANGLE

It is the angle between the path of a vehicle entering the rotary and that of another vehicle leaving the rotary, and therefore crossing the path of the former.

In Fig. 4.49, *GH* gives the direction of a vehicle as it emerges into the flow and *EF* the direction as the vehicle emerges out of the flow into the adjacent road. The angle between these two lines is the weaving angle.

Theoretically the lines *GH* and *EF* are the common tangents of curves parallel to the central island and the curves at the entry and exit at a distance of 1.5 m from the kerbs. This 1.5 m is the assumed distance of the centre line of the motion of vehicle from the kerbs.

The weaving angle is different from the angle of intersection of two radial roads.

If the weaving angle is too small, the outgoing traffic will merge with the incoming traffic, which is not desirable. Also small angle means large diameter of island. Taking all these factors into consideration, it is recommended that the weaving angles should not be less than 15°.

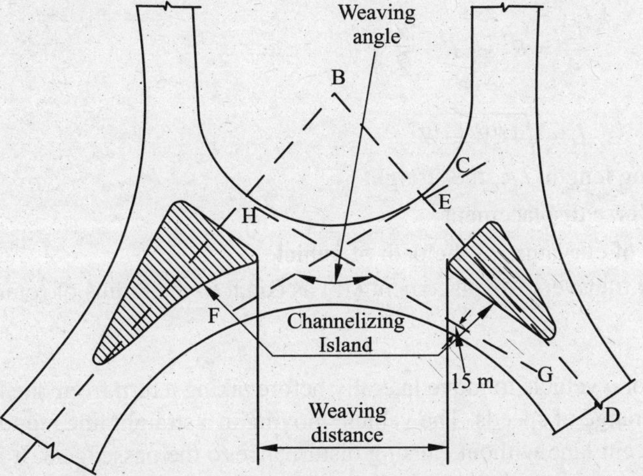

Fig. 4.49 Weaving angle and weaving distance.

If, however, too large a weaving angle is adopted, it may need a vehicle to come to a stop for allowing others to pass.

4.30. WEAVING LENGTH

It is measured along the centre of the one-way road between the noses of the channelizing islands, though in actual case, the length in which the vehicles may weave in or weave out may be shorter than this length.

It has been found that the freedom of movement in a rotary for a given speed depends upon the size of the weaving area and the effectiveness of this area is more dependent upon the available length than on the available width, if the width does not fall below a certain minimum.

If we have unused space, the vehicles will take their own routes unless guided into proper lanes.

Unfortunately there is no precise scientific manner to determine the weaving length. Thus, to estimate the weaving length, the following approximate method is adopted.

A weaving movement is considered as the change in lateral displacement. For practical purposes, we assume it to be a reverse circular curve of radius, R (Fig. 4.50).

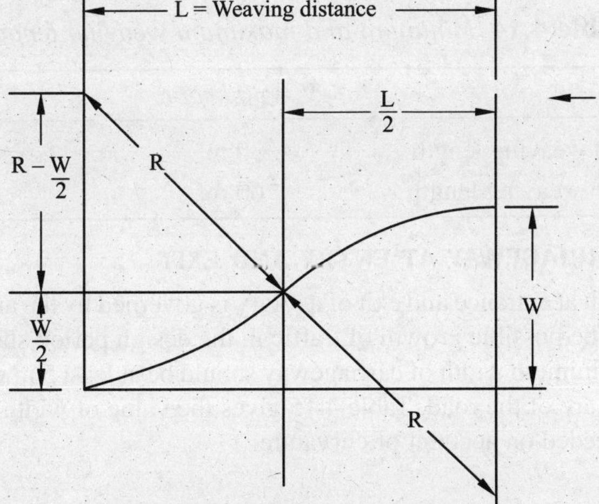

Fig. 4.50 A weaving movement.

$$\frac{L^2}{4} = R^2 - \left(R - \frac{W}{2}\right)^2$$

or
$$L = \sqrt{4RW - W^2}$$

where L = weaving length, *i.e.*, the straight

W = transverse displacement

R = radius of curvature of the path of vehicle

Putting the transverse displacement (W) as equal to the radius of rotary (R),

$$L = \sqrt{3}\,R$$

The time taken for a vehicle to move laterally before taking a turn from one lane to another lane is constant for a large range of speeds. The vehicle moving in a straight line requires 3 to 4 sec to move laterally to the adjacent lane without causing disturbance to the passengers. It is also found that one more second is required for one additional lane-width. Usually a time of 4 sec is taken.

For V = 30 km/hr (Design speed of intersection) weaving distance = 0.278 × 30 × 4

= 33.36 m, (Say 35 m)

For, V = 40 km/hr, weaving distance = 0.278 × 40 × 4

= 44.48 ≃ 45 m

We have thus the values as in Table 4.13.

Table 4.13

	V = 30 km/hr	*V = 40 km/hr*
R	15 m	20 m
$L = \sqrt{3}R$	26 m	39 m
$L = 0.278 \times V \times t$	35 m	45 m

Minimum and maximum weaving lengths. Increasing the weaving length means increasing the overall length and hence incurring more cost. If the length is too much, it will encourage speeding up.

The maximum and minimum weaving lengths, as recommended by IRC are given in Table 4.14.

Table 4.14 *Minimum and maximum weaving lengths*

	V = 30 km/hr	*V = 40 km/hr*
Minimum weaving length	30 m	45 m
Maximum weaving length	60 m	90 m

4.31. WIDTH OF CARRIAGEWAY AT ENTRY AND EXIT

The carriageway width at entrance and exit of a rotary is governed by the amount of traffic entering and leaving the rotary. The possible growth of traffic in the design period should be considered. It is recommended that the minimum width of carriageway should be at least 5 m with necessary widening to account for the curvature of the road. Table 4.15 gives the value of width of carriageway at entry inclusive of widening needed on account of curvature.

Table 4.15 Width of carriageway at entry and exit

Carriageway width of the approach road	*Radius of entry (m)*	*Width of carriageway at entry and exit (m)*
7 m (2 lanes)	25 – 35	6.5
10.5 m (3 lanes)		7.0
14 m (4 lanes)		8.0
21 m (6 lanes)		13.0
7 m (2 lanes)	15 – 25	7.0
10.5 m (3 lanes)		7.5
14 m (4 lanes)		10.0
21 m (6 lanes)		15.0

4.32. WIDTH OF ROTARY CARRIAGEWAY

1. *Width of non-weaving section*

The width of non-weaving section of the rotary should be equal to the widest single entry into the rotary and should generally be less than the width of the weaving section.

2. *Width of weaving section*

The width of the weaving section of the rotary should be one traffic lane (3.5 m) wider than the mean entry width. Referring to Fig. 4.51.

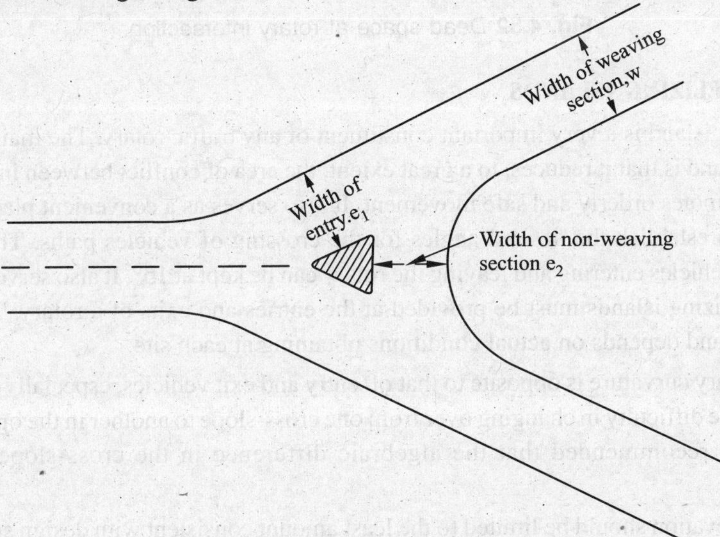

Fig. 4.51 Width of rotary carriageway.

$$w = \frac{e_1 + e_2}{2} + 3.5$$

where w = width of weaving section in m

e_1 = width of entry in m

e_2 = width of non-weaving section in m

4.33. AT OTHER CURB LINE

Provision of the reverse curve on the external curb line of weaving section should always be avoided. It has given rise to dead space (See Fig. 4.52). It must always be made straight or a combination of flat curves.

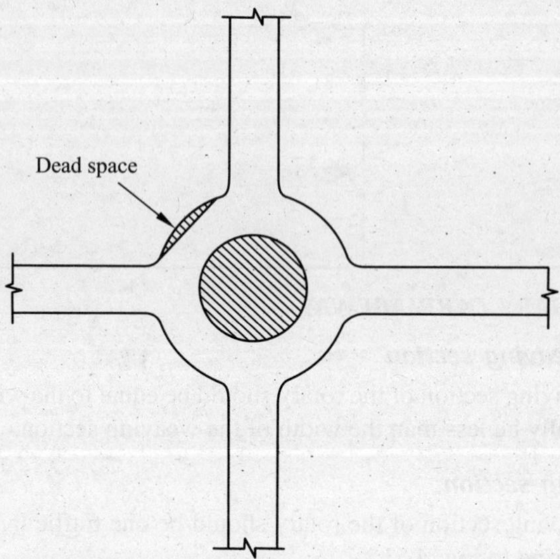

Fig. 4.52 Dead space at rotary intersection.

4.34. CHANNELIZING ISLANDS

Channelizing island is a very important constituent of any traffic rotary. The main function of the channelizing island is that it reduces, to a great extent, the area of conflict between intersecting traffic streams and promotes orderly and safe movement. It also serves as a convenient place for lights, etc. Also, it helps to establish the desired angles for the crossing of vehicles paths. Thus, the angle of intersection of vehicles entering and leaving the rotary can be kept at 15°. It also serves as a pedestrian refuge. Channelizing islands must be provided at the entries and exits of a rotary. The shape of the channelizing island depends on actual conditions obtaining at each site.

Since the rotary curvature is opposite to that of entry and exit vehicles, especially heavy buses and trucks experience difficulty in changing over from one cross-slope to another in the opposite direction. It is, therefore, recommended that the algebraic difference in the cross-slopes be limited to about 0.07.

The super-elevation should be limited to the least amount consistent with design speed. The crown line, which is the line of meeting of opposite cross-slopes, should as far as possible, be located such that vehicles cross it while travelling along the common tangent to the reverse curve. Channelizing islands should be situated on the peak with the road surfaces sloping away from them to all sides. Whenever possible, the cross-slope at an entrance should be carried around on the outer edge of the rotary to the adjacent exit, altering the slope slightly to suit the curvature in the rotary and the exit. A typical disposition of cross-slopes in a rotary is shown in Fig. 4.53.

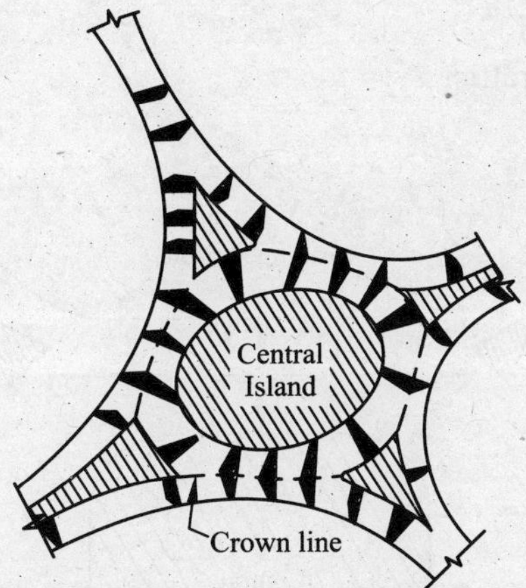

Fig. 4.53 Camber and super-elevation at a rotary.

4.35. GRADES

A rotary should preferably be located on level ground. The maximum grade should not exceed 2 per cent. It is, however, not desirable that a rotary be located in two planes having different inclinations to the horizontal.

A rotary may, with advantages be located on a summit. Such locations assist deceleration while approaching and acceleration while leaving the rotary. But it is essential that sufficient sight distance is available.

Rotaries in valleys always provide a full view to the approaching vehicles, but are likely to induce greater approaching speeds and have drainage problems.

4.36. SIGHT DISTANCE

On approaches to the rotary, the sight distance available should enable a driver to discern the channelizing and rotary islands clearly. A stopping sight distance appropriate to the approach speed should be ensured.

On the rotary itself, the sight distance should be adequate for vehicles first entering a rotary to see vehicles to their right at a safe distance. Similarly, once a vehicle is on a rotary in the middle of the weaving section, it should be possible for it to see another vehicle ahead of it in the next weaving section at a safe distance. As a general guideline, the sight distance for the 30 to 40 km/hr speed, should range between 30 to 45 m.

4.37. DRAINAGE

Adequate attention should be paid to drainage within the area of the rotary junction. Particularly the water likely to accumulate at the edges of the rotary island should be drained by means of curb and gutter section having an outlet to underground pipes through appropriately placed gulley traps.

4.38. CAPACITY OF THE ROTARY

It is important that the geometric design evolved for the rotary should be able to deal with the traffic flow at the end of the design period for the rotary. The practical capacity of a rotary is synonymous with the capacity of the weaving section which can accommodate the least traffic.

Capacity of the individual weaving sections depend on factors such as:

(*i*) Width of the weaving section

(*ii*) Average width of entry into the rotary

(*iii*) The weaving length; and
(*iv*) Proportion of weaving traffic.

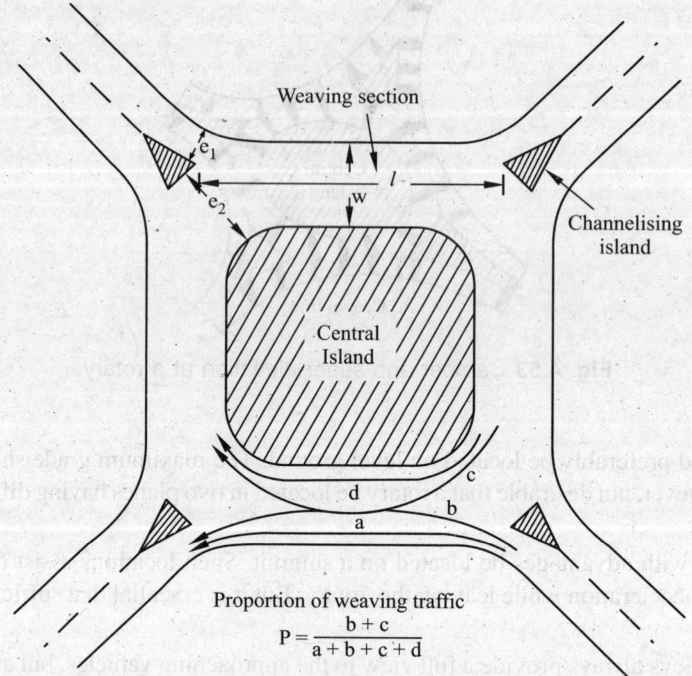

Fig. 4.54 Relevant dimension of weaving section, and properties of weaving traffic for use in capacity formula for rotaries.

The capacity of the weaving section can be calculated from the following formula

$$Q_p = \frac{280\, w \left(1 + \frac{e}{w}\right)\left(1 - \frac{p}{3}\right)}{1 + \frac{w}{l}}$$

where Q_p = Practical capacity of the weaving section of the rotary in passenger car units (*PCU*) per hour.

w = width of weaving section (see Fig. 4.54) in *m* (range 6 to 18 *m*)

e = average entry width in *m* (average of e_1 and e_2), e/w to be within a range of 0.4 to 1.00.

l = length in *m* of the weaving section between the ends of channelising islands w/l to be within the range of 0.12 to 0.4.

p = proportion of weaving traffic *i.e.*, ratio of sum of crossing streams to the total traffic on the weaving section $\left(p = \frac{b+c}{a+b+c+d}\right)$, range of p being 0.4 to 1.0

IRC has suggested the following adjustments in the capacity calculated from the above formula:

(*i*) Where the entry angle is between 0° and 15°, deduct 5 per cent from the capacity of the weaving section.

(*ii*) Where the entry angle is between 15° and 30°, deduct 2.5 per cent from the capacity of the weaving section.

(*iii*) Where exit angle is between 60° and 75°, deduct 2.5 per cent from the capacity of the weaving section.

(*iv*) Where the exit angle is greater than 75°, deduct 5 per cent from the capacity of the using section.

(*v*) Where the internal angle is greater than 95°, deduct 5 per cent from the capacity of the weaving section.

Care should be exercised that weaving sections are adequate for the required capacity so that merging and diverging manoeuvres take place smoothly. As a major disadvantage with rotaries is the reduction in speed, the weaving section should preferably be kept slightly longer than just necessary for capacity, say 33 to 50 per cent more.

TRAFFIC SEGREGATION

Traffic segregation means separation of traffic types: (*i*) in relation to destination; (*ii*) in relation to speed; (*iii*) in relation to direction; (*iv*) in relation to grade; and (*v*) in relation to time.

The types (*i*) to (*iv*) come under "*place segregation*" and (*v*) under "*time segregation*".

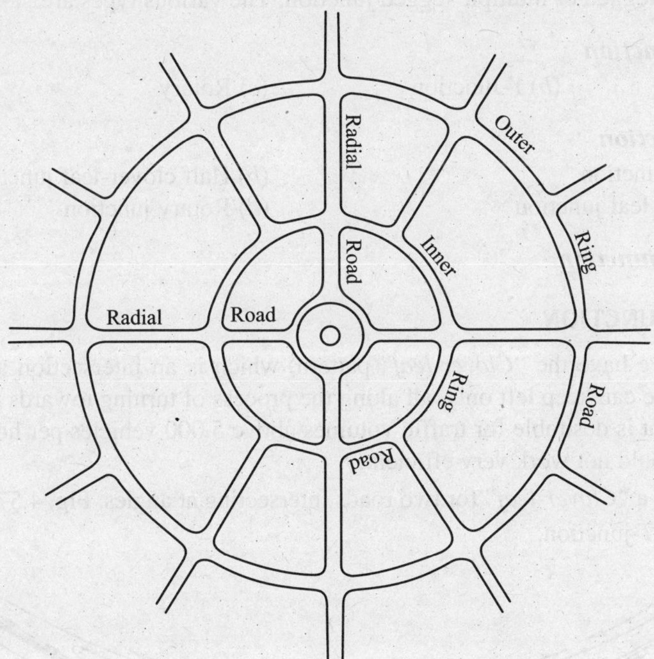

Fig. 4.55 Outer and inner ring roads.

(*i*) Segregation in relation to destination is done by means of a ring road or by means of a bypass. These are the means by which we do not mix local and through traffic (see Fig. 4.55).

The purpose of the inner ring road is to see that the traffic there is kept moving, though at a slower speed. The bus centres provided within inner ring road at the heart of the city should not be more than 100 m from the shopping centre or theatres, *etc*.

It is better if the outer ring road is provided at the edge of town, preferably at the future periphery of the town. The ring road is to provide facilities for traffic to go from one part of a city to another, without going into congested and commercial block. One can drive at a higher speed on an outer ring road. The junction of radial road to the ring road is a traffic rotary. In some countries fly-overs (intersections at different levels) are provided depending upon the traffic and importance. The access to the ring road should be controlled.

(*ii*) Segregation in relation to speed means segregating the fast moving traffic from the slow moving traffic. Out of a 10.5 m road the outer 3.5 m lane is usually used by the slow-moving vehicles, *e.g.*, scooters, bullock carts. Depending upon the traffic conditions prevailing, separate tracks for cyclist, etc. are made *e.g.* in Delhi, Chandigarh, *etc*.

(*iii*) Segregation in relation to direction, *e.g.*, "*one-way traffic restrictions*". For example, if the volume of traffic on a six lane road in the morning is mainly of the office-goers in the direction to the office, four out of the six lanes may be set apart for the traffic going in the direction towards the office. The arrangement may be reversed in the evening when the main traffic is from the office.

(*iv*) Segregation in relation to grade, *e.g.*, the construction of fly-overs (intersection at different levels), underpasses or over-bridges.

These grade separated intersections are without an interchange because the traffic moves separately at different levels and there is no provision for an interchange. When two or more roads have a facility or provision for interchange, the junction is known as the grade separated intersection with interchange.

The number of legs served by an intersection defines the form of grade-separated junction *e.g.* three legged, four legged or multiple legged junction. The various types are:

Three legged Junction

(*a*) *T*-Junction (*b*) *Y*-Junction (*c*) Rotary

Four legged Junction

(*a*) Diamond junction (*b*) Half clover-leaf junction
(*c*) Full clover-leaf junction (*d*) Rotary junction

Multiple legged junction

4.39. ROTARY JUNCTION

For example, we have the "*Clover-leaf*" pattern, which is an intersection arranged at different levels that a vehicle can keep left only, all along the process of turning towards the road on its right. Such a arrangement is desirable for traffic volumes above 5,000 vehicles per hour, when a rotary at the intersection would not work very efficiently.

Fig. 4.56 shows a "*Clover-leaf*" for two roads intersecting at angles. Fig. 4.57 shows a "*Trumpet*" type fly-over for a *T*-junction.

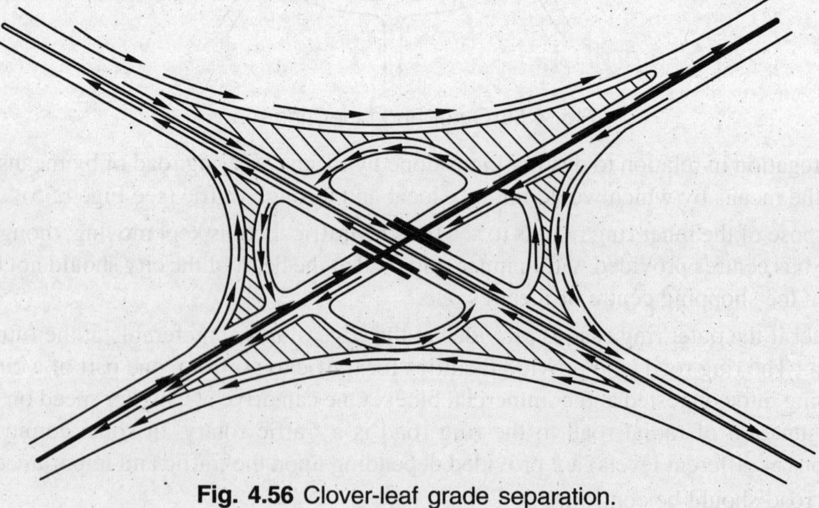

Fig. 4.56 Clover-leaf grade separation.

The arrows indicate the paths which the vehicles on a particular lane would take in order to get into the desired lane, always keeping to their left.

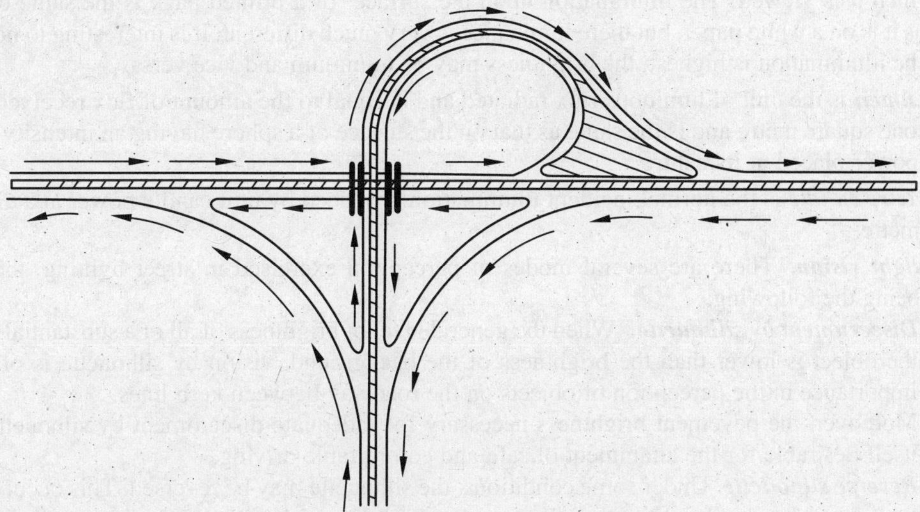

Fig. 4.57 Trumpet or *T* type fly-over.

(*v*) Time segregation is done by traffic light signals. In any urban road one or more of these methods have to be employed depending on the different requirements. In urban areas, we attain a compromise between the different conflicting units, though at a slow speed giving first consideration to safety.

HIGHWAY STREET LIGHTING

In the olden days when there was not much of fast moving traffic on the street in towns and cities, the ordinary oil lamps on the roadside were more than sufficient for the pedestrians moving at 3.5 km/hr and the slow moving traffic such as the horse-drawn vehicles. The relative position of the moving traffic was not much of importance at that time. The purpose of putting these oil lamps in the streets was more for the psychological benefit of the users rather than the actual showing of path to avoid accidents.

With the fast moving traffic on the streets and highways these days, not only the dusk and darkness increased the hazards of travel for both pedestrians and automobiles owing to impaired vision, but studies have shown that the fatalities attributable to this impaired visibility total to about 10,000 per year in the normal times over and above those that would be suffered if visibility at night were as good as in day time. Though no similar statistics is available in India, yet it can be safety presumed that here too, accidents must be happening quite often because of unsatisfactory lighting on our highways.

In the modern times, it is felt that suitable improvements in the lighting of streets and highways generally are necessary as it will contribute likewise to the reduction of crime and to the expedition of traffic which are principal purposes of such lighting.

4.40. DEFINITIONS

1. *Illumination* is the measure of the amount of light flux which falls on a surface, it is independent of the directions from which the light comes, or of the size or kind of light sources, or their positions, or of the kind of surface on which it falls. Illumination is same if surface is white or black.

2. *Brightness* (*Luminance*) is a measure of the amount and concentration of light flux which leaves some surfaces; it is by this light that the object can be seen. It is this brightness which controls the intensity of sensation which the eye receives from the object. The luminance of a surface does, in general, depends upon some or all of the above quantities of which the illumination is independent, such as the property of the surface, the direction from which light reaches the surface and the direction from which it is viewed. The illumination upon the surface of a printed page is the same on dark letters as it is on a white paper, but there luminance is very much different. It is interesting to note that where the illumination is highest, the brightness may be minimum and vice versa.

3. ***Lumen*** is the unit of luminous flux radiated and is equal to the amount of flux received on an area of one square metre and is the same as that on the surface of a sphere having an intensity of one candle power placed at its centre.

4. ***Metre candle*** is the normal incident illumination produced by unit candle power at a distance of one metre.

5. ***Night vision.*** There are several modes of perception exercised in street lighting, the main modes being the following:

(*a*) ***Discernment by silhouette.*** When the general level of brightness of all or a substantial part of the object is lower than the brightness of the background, vision by silhouette is of major importance in the perception of objects on the roadway between kerb lines.
Moreover, the pavement brightness necessary for adequate discernment by silhouette is in itself desirable for the attainment of safe and comfortable driving.

(*b*) ***Reverse silhouette.*** Under some conditions, the silhouette may be reversed. This occurs when the general level of brightness of all or a substantial part of the object is higher than brightness of its background.

(*c*) ***Discernment due to direct illumination.*** The intensity of illumination on the object is sufficient to permit discrimination of surface details without regard to its general contrast with its background. Where the pavements are largely obscured by the shifting pattern of vehicles, such as on streets carrying a large volume of vehicular and pedestrian traffic in business districts, discernment by surface detail become the predominant method to achieve good visibility. For this, a high order of illumination is necessary than the one for silhouette.

(*d*) ***Discernment by shadow.*** The presence of an object is inferred from its shadow.

(*e*) ***Discernment by glint.*** The object or details of the object being perceived through specular reflection of light from some of its polished surface.

4.41. FACTORS AFFECTING STREET LIGHTING

Efficiency of any street lighting system depends upon:

1. Types and power of sources
2. Siting of posts
3. Spacing of lamps
4. Mounting height
5. Overhang
6. Distribution of light
7. Avoidance of glare.

These are discussed below:

4.41.1. Types and Power of Sources

Light sources for street and highway lighting are usually of the filament, sodium vapour, mercury vapour or fluorescent type. All of these produce about the same visibility for the same level of illumination, since colour differences do not materially affect vision. Of the four types, the filament lamp has its greatest usage. It produces a pleasing colour, is inexpensive and is available in a variety of sizes from 1,000 to 25,000 lumens. Sodium vapour lamps produce a distinctive yellow orange colour. They are particularly adapted for hazardous locations, since drivers are familiar with yellow orange colour as an indication of caution. They have high efficiency and long life. Mercury vapour lamps are being employed increasingly because of their light efficiency. Their light is bluish white with little of red component. Fluorescent lamps are employed generally overhead in tunnels and under passes. They offer a "warm" light, less direct glare and wider bands of direct glare and wider bands of direct reflection from the pavement surface to driver's eye and thus are being used in highway and street lighting. For purposes of illumination, the classification made is given in Table 4.16.

Table 4.16 Traffic volume for different illuminations

Classification type	*Volume of vehicular traffic (max night hour; both directions)*
Very light	Under 150
Light	150–500
Medium	500–1,200
Heavy	1,200–2,400
Very heavy	2,400–4,000

It is further suggested that traffic be further classified as to light, medium and heavy pedestrian traffic. Average metre-candle values for these classifications are given in Table 4.17.

Table 4.17 Average metre-candle values

Pedestrian traffic	*Average metre-candle values for vehicular traffic*		
	Light	*Medium*	*Heavy*
Heavy	0.8	1.0	1.2
Medium	0.6	0.8	1.0
Light	0.4	0.6	0.8

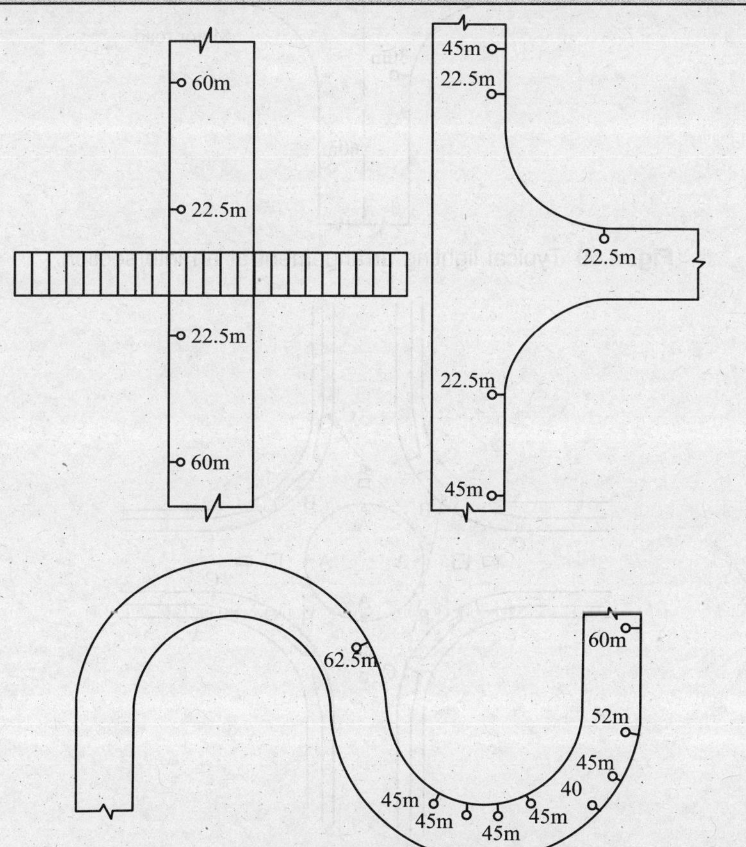

Fig. 4.58 Typical lighting arrangements.

The above values are minimum and can always be increased at intersections, dangerous locations and places near some temple or school, *etc*.

4.41.2. Siting of Posts

There are three systems of location of lights:

(*a*) All the lamps on one side of the road.

(*b*) Centrally hung system of lights.

(*c*) The staggered light system on two sides.

Systems (*b*) and (*c*) are used in case of broad thorough fares as these are pretty costly due to the extra cost of poles and other fittings and accessories.

Lighting arrangements at typical places are shown in Figs. 4.58 to 4.60.

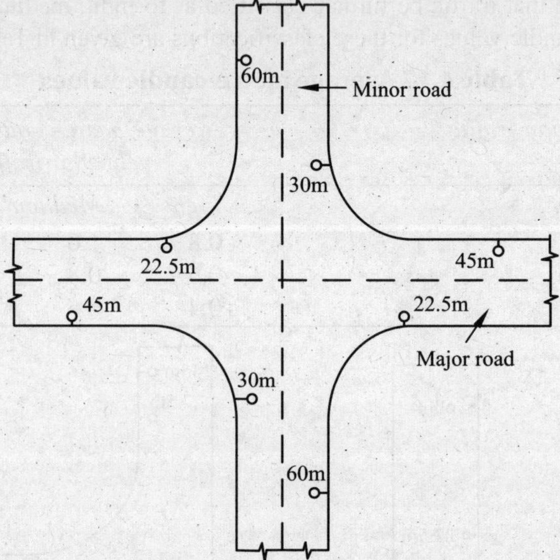

Fig. 4.59 Typical lighting arrangement at an intersection.

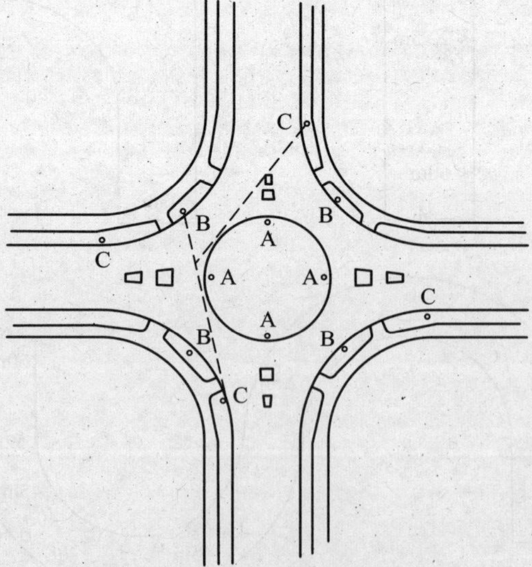

Fig. 4.60 Illumination of a rotary.

4.41.3. Spacing of Lamps

It is always better to have closer spacing of lights since it gives better brightness. Actually the spacing of light shall depend upon the importance of the street.

Experiments have shown that usual spacing ranges from 35 m to 55 m for good brightness for all types of roads.

The spacing between lamps is determined by the following relation:

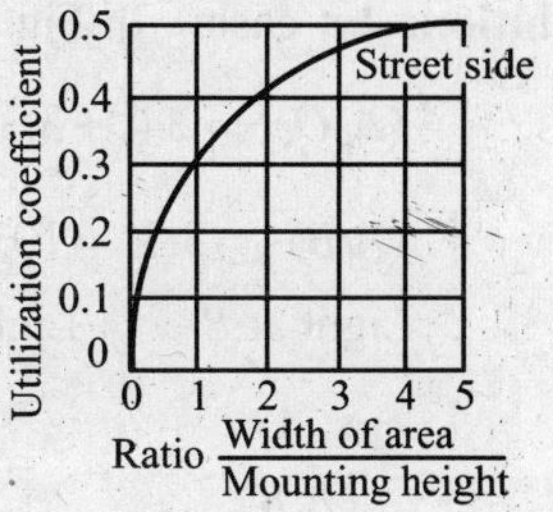

Fig. 4.61 Utilization coefficients.

$$\text{Spacing} = \frac{\text{Lamp lumen} \times \text{coefficient of utilisation} \times \text{maintenance factor}}{\text{Average flux} \times \text{width of road}}$$

The maintenance factor is generally, assumed to be 0.8. Coefficient of utilisation is determined from utilisation coefficient charts. One such typical chart is shown in Fig. 4.61.

Problem 4.4 *Design a highway lighting system with the following data:*

Street width = 15 *m*
Mounting height = 9 *m*
Lamp size = 6000 *lumens*
Flux = 5

Solution. Ratio of $\frac{\text{street width}}{\text{mounting height}} = \frac{15}{9} = 1.67$

From Fig. 4.61, coefficient of utilisation = 0.36

Assume maintenance factor = 0.8

Spacing between lamps

$$= \frac{\text{Lamp lumen} \times \text{coeff. of utilisation} \times \text{maintenance factor}}{\text{Average flux} \times \text{width of road}}$$

$$= \frac{6000 \times 0.36 \times 0.8}{5 \times 15}$$

$$= 23.04 \text{ m}$$

Adopt a spacing of 20 m

4.41.4. Mounting Height

It is important from the following two points of view:

(*a*) The distribution of light point of view.

(*b*) The glare point of view.

From the law of "*inverse squares*" the intensity of illumination produced by light from a small source varies inversely as the square of the distance from the source.

Problem 4.5 *Given, height of lamp at* A_1 = *3.3 m*

Spacing of lamps = *48 m*
Height of lamp at A_2 = *6.6 m*
Spacing of lamps = *48 m*
Determine in which case the distribution of light is better.

Solution. 1st case: In Fig. 4.62

$$(A_1O)^2 = 24^2 + 3.3^2 = 576 + 10.9 = 586.9$$

$$(A_1B)^2 = 3.3^2 = 10.9$$

Light at B : Light at O

$$= \frac{1}{(A_1B)^2} : = \frac{1}{(A_1O)^2}$$

or $(A_1O)^2$: $(A_1B)^2$

or 586.9 : 10.9

or 53.8 : 1

$\therefore$ Total light at B : Total light at O : : 26.9 : 1

(Since O receives light from A_1 and also the same amount from A_1').

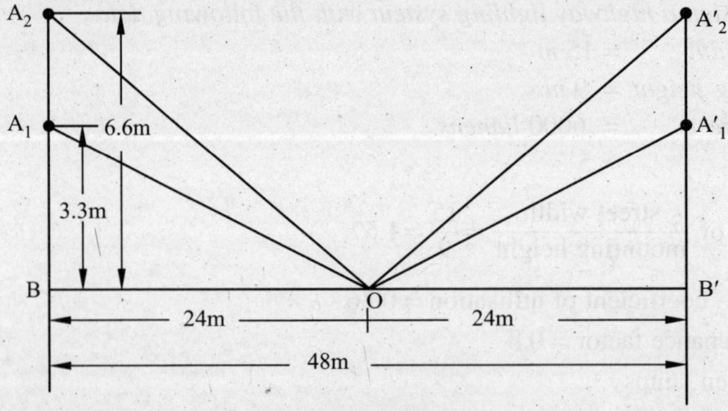

Fig. 4.62

2nd case: $(A_2O)^2 = 24^2 + 6.6^2 = 576 + 43.56 = 619.56$

$$A_2B^2 = 6.6^2 = 43.56$$

Light at B : Light at O

$$= \frac{1}{(A_2B)^2} : = \frac{1}{(A_2O)^2}$$

or $(A_2O)^2$: $(A_2B)^2$

or 619.56 : 43.56

or 14.22 : 1

$\therefore$ Total light at B : Total light at O : : 7.11:1.

(Since O receives the same amount of light from the other source at A_2' also)

Hence the distribution of light is much better in the second case as there is much curtailment of light at B as compared to the light at O and there is no concentration of light at one place.

Thus, greater mounting heights give a much better distribution though at the cost of less quantity of actual light falling on the surface.

Secondly if the light is mounted low it might directly reach the eyes of the driver as he approaches it and may lead to accidents.

4.41.5. Overhang

If the lamp is provided just above the kerb *i.e.*, at *A* (Fig. 4.63), the light will be most efficient near the kerb and will give perhaps long shadows of the moving traffic. The trouble is more predominant in the case of roads with overhanging lights only on one side as it produces light and shade successively in case of a moving vehicle which causes glare in the eyes of the road users which might lead to accidents. Proper overhang is thus essential.

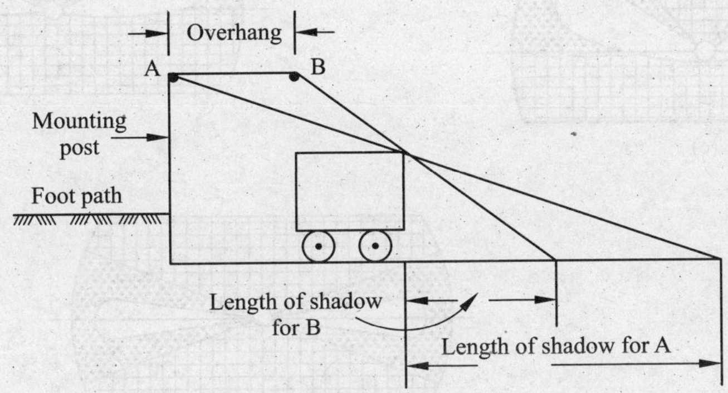

Fig. 4.63 Overhang.

4.41.6. Distribution of Light

It is very difficult to have a uniform distribution of light but at the same time, a fair distribution of light can always be obtained. Uneven and patchy distribution of light is harmful. Actually the shape of light made by a lamp depends upon the nature of the road surface and the direction of light from the lamp. The lanterns should be so arranged that the patches of light from each source should touch each other. Also it is better to use "cut-off" system so that oblique rays of light which are the biggest trouble are eliminated although this may involve larger spacing of lights. The various distributions of light are shown in Fig. 4.64.

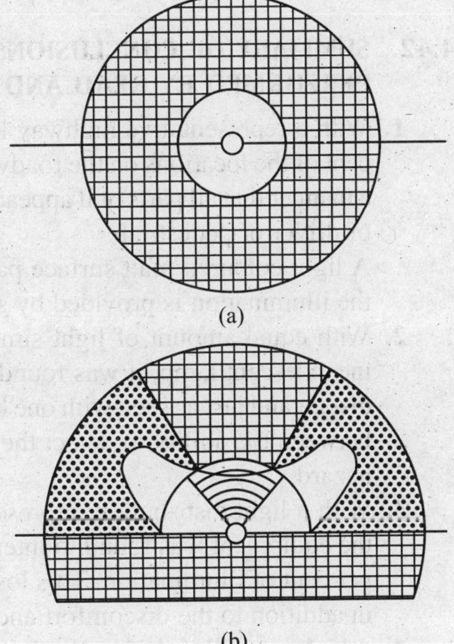

Fig. 4.64 (*a*, *b*) Distribution of light.

4.41.7. Avoidance of Glare

Glare causes much waste of illumination. Anything which reduces the contrast between the glaring light source and the roadway and surrounding area or which increases the illuminated area within view or which separates bright light source from the road surface reduces the glare effect.

The blinding effect of direct glare is greatly reduced by high mounting, *e.g.* for the same candlepower, a lamp mounted 3 m above the pavement produces 13 times the glare than one at 9.0 m.

As a result of laboratory and field tests, the following facts have been established:

(*a*) All the effects of glare are diminished as the luminaries are removed from the centre of field of vision.
(*b*) Increasing the brightness of surroundings reduces contrast and improves visibility.
(*c*) The discomforting effect of glare is reduced by decreasing luminaire brightness by increasing its area.

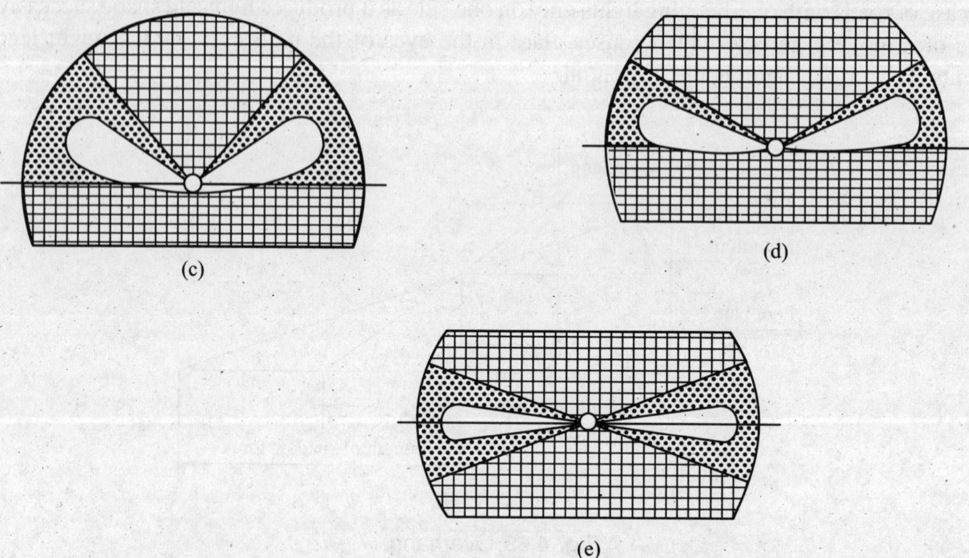

Fig. 4.64 (contd.) Distribution of light (*a*) Symmetrical (*b*) Wide asymmetric lateral (*c*) Medium width asymmetric (*d*) Narrow asymmetric lateral (two-way or four-way) (*e*) Two way or four-way light).

4.42. SUMMARY OF CONCLUSIONS (DRAWN AFTER A NUMBER OF EXPERIMENTS PERFORMED BY READ AND CHANON)

1. With a representative highway lighting system, objects are seen as silhouette in over 80 per cent of the locations on the roadway. This would make it desirable to light the highway in such a manner that all parts of it appear substantially uniformly bright when viewed from the position of driver or pedestrian.
   A light coloured matt surface pavement offers the best visibility and safety at night whether the illumination is provided by street lamps or automobile head lights.
2. With equal amount of light similarly distributed from sodium vapour, mercury vapour and incandescent lamp, it was found that one sees objects from the stand point of safety equally distant and just as seen with one kind of light as with others. There appeared to be no difference between the ability to detect the presence of an obstacle and ability to discriminate between hazard.
3. With a light distribution, representative of current highway lighting practice, the glare from the equipment was found to interfere with seeing to such an extent as to waste as much as 50 per cent of illumination. This loss in seeing value is much greater than generally recognized, in addition to the discomfort and distracting influence of glare.
4. A highway with a uniform horizontal illumination throughout its length presents a large variation in brightness when viewed by drivers or pedestrians.
   For uniform brightness, the light must be distributed to take account of specular reflection of light at small angles from the surface.

5. Using the standard pole spacing of 37.5 m, it is feasible to obtain this high order of uniform brightness economically and at the same time to shield the sources sufficiently to obviate to a very large extent the seeing loss from glare characteristics of all past systems.

4.43. VARIOUS OTHER RECOMMENDATIONS

A few other agencies suggest the following points:

1. Where traffic volume and pedestrian numbers are small, an average surface brightness of about 2 horizontal metre candle is sufficient, Intensities should be increased for more important highways and at intersections.

2. The maximum recommended brightness is for business streets and is 12 horizontal metre-candles.

3. For highways with controlled access, the level recommended by AASHO is 6 metre candles. All the above values are for pavements having average reflectance.
Where reflectance is poor, illumination should be raised as much as 50 per cent but where it is excellent, vision may be decreased by about 25 per cent.

4. A light coloured pavement or a seal coat that reflects light back to the driver is highly desirable.

5. Surfaces that become mirror-like, when wet, are particularly to be avoided.

William E Barrows of University of Mainse, U.S.A. suggests: "*If a particular street is wide, its brightness depends upon the location of the lamps with respect to the street as well as upon the amount of light delivered upon the street*".

If the lamps are mounted over the kerbs of wide streets, the streaks of light will be along the sides of the streets and the middle will appear dark. Better results will be secured by placing the lamps on brackets extending out over the street. Then the far lights will produce streaks of light near the middle of the road and the whole street surface will appear fairly bright. Two rows of lamps are better than one to produce silhouette effect. A third row *e.g.* over the centre is likely to improve conditions by increasing the number of streaks on the roadways.

If the street is long and straight, the distant lamps are particularly effective in providing streaks of light well out towards the middle of the street.

Lamps mounted over the kerb are not so effective in these respects as are lamps mounted somewhat out over the driveway. With the latter mounting and with the lamps on both sides of the street, the illumination is likely to be excellent.

Streaks are increased in size by the crudity of the street surface. They will also be longer and wider as the size of the light source is increased, hence giving a large area of brightness of relatively greater effectiveness thereby rendering the street surface bright.

PART – II

5. Highway Location and Project Preparation
6. Hill Roads
7. Drainage and Drainage Structures
8. Geometric Design of Highways

HIGHWAY LOCATION AND PROJECT PREPARATION

LOCATION

By the term *"location"* of a highway is meant the position occupied by the central line of the road surface. Location comprises two components; one horizontal known as the *"line"* and the other vertical called *"grade"*. Both these components play a vital role in determining the alignment of a road. Except in flat areas, due consideration to grade has also to be given before selecting the best line for a road. Once the location is fixed, it becomes very difficult, sometimes impossible, to change it later on because of the exorbitant cost of the abutting land which had developed along the new road. The best location of a highway should be made keeping in view the design standards.

5.1. FACTORS GOVERNING HIGHWAY LOCATION

Under ideal conditions, a highway location should conform to the following points:

(*i*) Minimum length or directness
(*ii*) Low grades
(*iii*) Easy curves
(*iv*) Good sight distances
(*v*) Strong foundation
(*vi*) Adequate drainage
(*vii*) Least earthwork
(*viii*) Wide right of way
(*ix*) Suitable sites for crossing rivers, railway lines and other highways.
(*x*) Road materials required for its construction should be available at reasonable distance; and
(*xi*) Freedom from slides, *etc*.
(*xii*) Minimum interference to agriculture and industry.

A direct route between two terminal points is the cheapest one as it gives the shortest length. It should therefore be always a factor to be considered. However, deviations may be required in the following cases:

(*i*) To avoid marshy places and low lying land.
(*ii*) To avoid bad river crossing and thus to secure good foundation for bridges and to avoid costly approach banks.
(*iii*) To avoid forest, built-up areas and to avoid destruction to property.
(*iv*) To avoid crossing high hills or high ridges to achieve permissible grades
(*v*) To get as near as possible to materials required for construction such as gravel, metal and even water.

(*vi*) To avoid costly land acquisition.

(*vii*) To serve the largest number of villages lying between the terminal points.

(*viii*) To avoid foreign country.

Easy gradients are even more important than the directness of the line. Gradients affect haulage power of animal drawn vehicles and operation cost of motor vehicles. At the same time, ineffective rise and fall is to be avoided. Curves should have large radii of curvature so that these could be negotiated without a decrease in the design speed. Moreover a curve much sharper than others on the same road will invariably be a place of accident hazards. Sight distances are also essential for safety. No road surface is better than its foundation. Thus a highway engineer must look for better foundation and should avoid areas of poor soil and poor drainage. Construction cost depends much on the type and amount of material to be excavated. Boulders and rocks are difficult to excavate and should, therefore, be avoided as far as possible. Right of way cost is much increased if the line passes through an urban area. Good sites for crossing rivers and suitable locations for intersecting railway lines and other highways should be found. The cost of construction will also depend upon the nearness or otherwise of the materials required for the road. To have public safety and less maintenance costs the line must be free from slides. Finally since roads are constructed to serve traffic, the location should facilitate the development of the area traversed.

Location on hills. This is discussed in detail in the next chapter.

Location in deserts. The location of roads in desert areas should preferably be not transverse to dune and valley lines, but should follow the normal "*grain*" of the country. A clear and unobstructed land width of 100 m should be provided in order to ensure the free carriage of sands by winds, thus preventing its accumulation on the road. If possible, a forest belt of 100 to 150 m should be developed for the protection of roads from sand drifts.

Any one location will rarely afford all the desirable features and thus a compromise has to be made.

Control points. In locating a highway, there are certain points which exert a controlling influence. These are either obligatory points, that must be touched by the road or other points which must be avoided.

The obligatory points are:

(*i*) The two terminal points of the road

(*ii*) River crossings

(*iii*) Railway crossings

(*iv*) Irrigation and navigation canals

(*v*) Saddles and passes in the case of hill roads

(*vi*) Towns to be served by the road

(*vii*) Spectacular view of a canyon for a scenic park highway

(*viii*) Particularly good area of agricultural land for a farm road.

Swamps, grave yards, religious places, buildings and valuable agricultural land, etc. are points, which should be avoided as far as possible.

The identification of control points is one of the very early and important steps in highway location. When the two terminal points of a highway are separated by a considerable distance and it is proposed to construct the road in stages, location of the entire road should be decided in the beginning. It will not suffice to work from one control point to another. Surveys should be conducted beyond the temporary terminals to make sure that the later sections of the overall project will fit properly with the earlier units.

5.1.1. General Guidance on Route Selection and Highway Location

I.R.C. has given the following guide lines for highway location and route selection:

Prefer alignment which

- is as direct as possible between points to be linked.
- is clear of obstruction and avoids interference with industry/agriculture, places of worship, etc.
- fully integrates with the surrounding landscape.
- passes through better soil area and has good natural drainage.
- runs close to sources of embankment and pavement materials.

Keep in view

- obligatory and control points from physical, administrative, strategic and other considerations. the need for adopting a uniform design speed, and easy grades and curvature.
- the needs of major river crossings.
- better aesthetics of the road.

Avoid

- frequent crossing/re-crossing of railway lines and water courses.
- areas which are unstable subject to flooding/water-logging/seepage flows, etc.
- erosion/landslide prone areas are unsuitable for embankment construction purpose. Clay having liquid limit and plasticity index more than 70 and 45 respectively is also unsuitable for embankment construction.
- *Others:* Subsoil water level, floodability of the area, and any special geological features.

In general, approaches to railway overbridges or high level bridges will warrant special time-consuming investigations. It is necessary that the investigation, design and construction phases of such embankments are coordinated with the construction of the bridge structure so that the bridge and the approaches are completed side by side without the need for one to wait for the other.

HIGHWAY LOCATION SURVEYS

The surveys to be carried out for locating a highway are:

1. Reconnaissance
2. Preliminary survey
3. Location survey

5.2. RECONNAISSANCE

The main objective is to get an idea of the general character of the area. Reconnaissance is the survey without instruments. This is carried out for:

(*i*) Ascertaining the possibility of building a road through a country

(*ii*) Choosing the most feasible route to follow and recording it on a map

(*iii*) Drafting a report and estimating probable cost.

Reconnaissance is usually carried out in two stages, namely area reconnaissance and route reconnaissance. Area reconnaissance is done from the existing topsheets and aerial photographs prepared by the Surveyor General of India, Dehra Dun. Top sheets are available to the scale of 1:2,50,000, 1:50,000 and 1:25,000. On these maps, all the possible routes are marked and, if possible, selection made. Control points are also indicated on these maps. In case aerial photographs are available a mosaic is prepared from them, studied under a stereoscope and possible routes identified.

If neither maps nor aerial photographs are available, then area reconnaissance and route reconnaissance merge together. The only way then left is to traverse the country between two terminals by foot, horse, camel, car or even by a plane and investigate different specific routes. By this the practicability of each of these suggested routes is checked on the ground. In this way, more details are picked up as aerial photographs do not reveal subsurface conditions, condition and value of a building, *etc*.

I.R.C:S.P:19-2001 and IRC:SP:48-1998 have given a list of points on which data may be collected during ground reconnaissance in rural areas and in hilly tracts respectively. This is as follows:

1. Details of route *vis-à-vis* topography of the area, whether plain, rolling, mountainous or steep.

2. Length of the road along various alternatives.

3. Bridging requirements, number, length.

4. Geometrics:

(*a*) Gradients that are feasible, specifying the extent of deviation called for.

(*b*) Curves and hair-pin bends, *etc*.

5. Existing means of communication—mule path, jeep track, earthen cart tracks, *etc*.

6. Right-of-way available, bringing out constraints on account of built-up area, monuments and other structures.

7. Terrain and soil conditions:

(*a*) Geology of the area

(*b*) Nature of the soil, drainage conditions and nature of hill slopes

(*c*) Road length passing through:

(*i*) mountainous terrain

(*ii*) steep terrain

(*iii*) rocky stretches with indication of the length in loose rock stretches

(*iv*) areas subject to avalanches and snow drifts

(*v*) areas subjected to inundation and flooding

(*vi*) areas subjected to sand dunes including location of dunes

(*vii*) areas of poor soil and drainage conditions.

(*d*) Cliffs and gorges

(*e*) Drainage characteristics of the area including susceptibility to flooding

(*f*) General elevation of the road indicating maximum and minimum heights negotiated by main ascents and descents in hill sections

(*g*) Total number of ascents and descents in hill sections

(*h*) Disposition and location of sand dunes

(*i*) Vegetation—extent and type

8. Climatic conditions:

(*a*) Temperature—monthly maximum and minimum readings

(*b*) Rainfall data—average annual, peak intensity, monthly distribution (to the extent available)

(*c*) Snowfall data—average annual, peak intensities, monthly distribution (to the extent available)

(*d*) Wind direction and velocities

(*e*) Fog conditions

(*f*) Exposure to sun

(*g*) Water table and its variation between maximum and minimum.

9. Facilities/Resources

(*a*) Landing ground in case of hilly stretches

(*b*) Dropping zones in case of hilly stretches

(*c*) Foodstuffs

(*d*) Labour—local availability and need for import

(*e*) Construction materials (timber, bamboo, sand, stones, shingle, *etc*.) with extent of their availability, leads involved, ease of access etc.

(*f*) Availability of water, especially in arid zones

(*g*) Availability of local contractors.

10. Value of land—agriculture land, irrigated land, built-up land, forest land, *etc*.

11. Approximate construction cost of various alternatives.

12. Access points indicating possibility of induction of equipment.

13. Period required for construction.

14. Strategic considerations.

15. Recreational potential.

16. Important villages, towns and marketing centres connected.

17. Economic factors:

(*i*) population served by the alignment

(*ii*) agricultural and economic potential of the area

(*iii*) marketing centres.

18. Other major developmental projects being taken up in the area, e.g. hydro-electric projects, dams, reservoirs, *etc*.

19. Crossings with railway lines and other existing highways.

20. Location of existing or proposed utilities along the alignment.

21. Necessity of bypasses for towns and villages.

22. Position of ancient monuments, burial grounds, cremation grounds, religious structures, hospitals and schools.

23. Ecology or environment factors.

24. Aspects needing coordination with other administrative authorities.

The information collected by reconnaissance is presented in the form of a report. This is very helpful as it gives details of the considerations kept in view for a particular choice and recommendations of the survey form the basis for the preliminary survey.

The report should, as far as possible, include the following items:

(*i*) Name of the project.

(*ii*) Terminal points and approximate length of the proposed highway.

(*iii*) Traffic conditions and design standards as assumed by the surveyor.

(*iv*) Maps, aerial photographs or mosaics, *etc*. showing all the possible routes, selected route being indicated specifically.

(*v*) Control points to be touched and to be avoided.

(*vi*) Nature, extent and date of map and field studies and by whom made.

(*vii*) General character of the country, soil type, drainage, etc.

(*viii*) Existing features and the use to which land is being put.

(*ix*) Transportation facilities available.

(*x*) Local materials available.

(*xi*) Recommended type of road surface.

(*xii*) Drainage and traffic facilities required to be provided.

(*xiii*) Reasons for recommending a particular location.

(*xiv*) Discussion of alternate routes and suggestions for further investigation.

(*xv*) Approximate cost per km for the type of surface and design standards recommended.

(*xvi*) Officer responsible for report.

Reconnaissance is comparatively easier in the plains than in the hills.

5.3. PRELIMINARY SURVEY

Preliminary surveys are conducted for the following purposes:

1. To collect general information about traffic, soil, drainage, etc.
2. To establish reference bench marks
3. For transverse survey
4. To determine fly levels and draw cross-sections
5. For map preparation.

The main objective of the preliminary survey is to determine the location of horizontal line of the proposed highway, of course giving due consideration to the grade also. Before a final selection of the route is made, an engineer must equip himself on the three W's where, what and when.

1. *Where* is the road to begin and end? Which are the places to be touched and which to be avoided? Some places may also have to be avoided on political and strategical considerations and also in the interest of the railways.

2. *What* is the type of road to be built?

(*a*) Normal width

(*b*) What traffic to be catered because this effects radius of curvature, gradients, sight distances, road section, etc. The aim should not be to provide the minimum standards but efforts should be made to secure as better standards as are feasible.

(*c*) Extent to which the streams are to be bridged. If a stream is not to be bridged, a wide detour is necessary.

(*d*) Whether the road is to be metalled or not.

(*e*) Information on building, camping grounds, water supplies, etc.

(*f*) Position of sand and gravel for bridge and culvert construction.

(*g*) Length and height of probable earth retaining structure.

(*h*) Position and quantity of water sources for consolidation.

3. *When* are the report and the road required? If the report is not required for several months, it may be advisable to wait till snow conditions in mountains and flood conditions in plains are known. If the road is very urgently required for military reasons, the shortest length and ease of construction would be more important although this road may not be the best for peace time use.

For collecting this information, surveys of the area have to be done. The preliminary surveys are made by transit and chaining methods. Use may also be made of the aerial photographs supplemented by information collected in the field.

Ground surveys. The base line of the preliminary survey, commonly known as "*P*" line, should be run as close to the expected final line as is feasible. It is not essential to include run-in curves. For approximating long turns, short courses avoiding deflection angles greater than 30 degrees may be used. The topographic features of the land are noted as the traverse is carried forward. The width covered on each side of the base line varies with topography and land use, but usually ranges from 30 to 250 m. It should cover the entire right of way of the proposed road and must make adequate allowances for possible shifting of the centre line away from the traverse when the final surveys are completed. It is, therefore, not essential that the topography should cover a band of uniform width on either side of the base line and should be widened at places of uncertainty. Note should be made about the rock ledges and important existing buildings and structures as they will affect the location of the proposed road.

Bench marks are established at 250 to 500 m intervals and they are tied to the *GTS* bench marks or existing monuments, if the permanent bench marks are not within a reasonable reach. Elevations are then taken with a level at 30 m intervals and at breaks in the ground. Thus the profile of the ground can be plotted from this. On the longitudinal profile, the semi-final location line (*L*-line) and tentative grades are laid out, if the profile is fairly representative of the surrounding. Cross-sections may also be required where relative altitudes appear necessary for the location of the central line. Intervals for cross-sections should be:

Plain terrain	:	50–100 m
Rolling terrain	:	50–75 m
Built-up area	:	50 m
Hilly terrain	:	20 m

In some cases where neither profile nor cross sections show the ground properly, contours at 1 m intervals are plotted.

Aerial maps and supplementary ground investigation. Aerial maps cannot be used for making a topographic map if the area is covered with forest or vegetation or snow. Thus their use is limited only to uncovered land. Aerial mapping is cheap and take considerably less time than ground surveys. If aerial photographs have been procured for reconnaissance, they can be used for the preparation of topographic maps also. If the aerial map is to be used its accuracy is checked in the field and missing data plotted. The missing data includes the property lines, the names of the owners, type and conditions of buildings, underground structures, etc.

Plotting topographic map. The information collected by the preliminary survey is plotted to get a topographic map. Topographic maps are drawn as the survey progresses. The scale to which it is drawn varies depending upon the topography of the area covered. I.R.C. has recommended the following scales:

(*i*) Built up areas and stretches in hilly terrain — 1 : 1000 for horizontal scale and 1:100 for vertical scale

(*ii*) Plain and rolling terrain — 1 : 2500 for horizontal scale and 1 : 250 for vertical scale.

For difficult locations such as steep terrain, hairpin bends, etc. larger scales may be adopted.

A topographic map should show the following details:

(*i*) Buildings—residential, commercial, industrial, sheds, *etc*.
(*ii*) Elevations to the extent needed.
(*iii*) Boundary walls, fences, etc. along with the names of the property owners.
(*iv*) Wells, hydrants, man-holes, underground sewers, water lines *etc*.
(*v*) Existing roads, their types.
(*vi*) Railway lines.
(*vii*) Existing drainage structures.
(*viii*) Valuable cultivated land and trees.
(*ix*) Natural water courses.
(*x*) Ponds, swamps and other places to be avoided.
(*xi*) Position of rock strata.
(*xii*) Any other feature affecting the final location of the road.

Plotting the preliminary central line. When the topographic map (longitudinal profile, cross-sections and contour maps wherever required) has been prepared, the next step is to locate the position of the semi-final central line of the road. This is often called the "Paper" location. In case of ground survey, the proposed central line should be marked by the engineer who conducted the surveys as he would have gathered an adequate knowledge of the road during the survey. Base line should serve as a guide in locating the central line. The principles governing the location of the base line should be kept in mind. The site may be re-examined if any doubt arises in the mind of the locator. The following points should guide the selection of the tentative central line:

1. Preserve and save, where possible, all roadside shade trees located outside the shoulder of a highway.

2. Use a minimum number of curves.

3. Use existing adequate bridges and large culverts.

4. Secure, as nearly as possible, right angle crossings of intersecting roads, railroads and streams.
5. Locate the central line in stable soils, if possible.
6. Co-ordinate horizontal curves with vertical curvature for good appearance.
7. Leave local and drainage in as good condition as it is found.
8. Consider the safety of children at all schools.
9. Leave all religious places, grave yards, monumental buildings undisturbed.
10. Hold all right of way, building and property damage costs to a minimum.
11. Avoid successive horizontal curves (flat backs) in the same direction. Make longer single curve, compound, if necessary.
12. Avoid short reversed (dog leg) alignment.
13. When reverse curves are necessary, distance between curves should be 100 m minimum, and length of curves 200 m minimum.

Field review of preliminary central line. After establishing a preliminary central line location on paper, review is made of the line in the field. The entire location is walked over and the proposed central line compared with the ground. Careful examination should be made of all such locations where minimum design standards have to be provided. Such areas which might affect the design and construction of the road such as water bearing soil, unstable ground, etc. should also be carefully looked into. Right of way acquisition problems should be reduced to a minimum. The topographic map is then revised to take into account the additional information gathered in the field and adjustments in the location of the preliminary central line made accordingly. This location of the central line is then recommended and the preliminary survey is considered complete.

5.4. LOCATION SURVEY

The location survey comprises two distinct stages, namely:

1. Staking the central line of the proposed road on the ground; and
2. Collecting other information required for:
   (*i*) The design of the highway.
   (*ii*) The preparation of construction drawings.
   (*iii*) The calculations of earthwork quantities.
   (*iv*) Evolving suitable specifications.
   (*v*) Working out the cost of the project.
   (*vi*) Procuring right of way.

Staking the central line. Once the central line location is finalized by the preliminary survey and marked accurately on the topographic map, steps are taken next to stake it on the ground. The staking out of the central line is carried out by means of a continuous transit survey. All angles are measured with a transit. Double reversal method should be adopted at all horizontal intersection points (H.I.P.) and intermediate points of transit (P.O.T.) on long tangents. Methods of referencing H.I.P's and P.O.T's is illustrated in Fig. 5.1. As the staking is in progress, other details of the topography are taken from

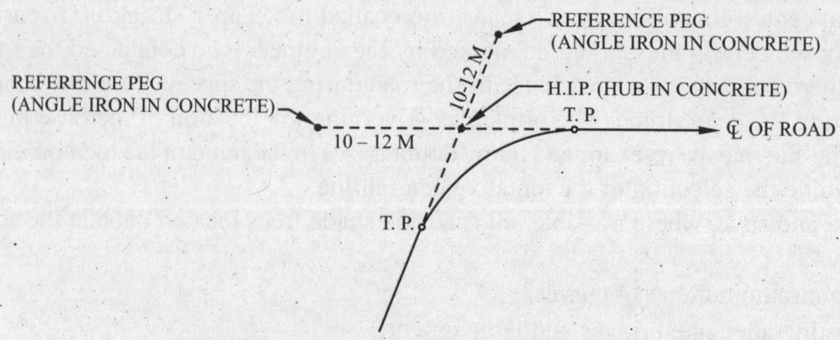

(*a*) Reference pegs at horizontal intersection point (H.I.P.)

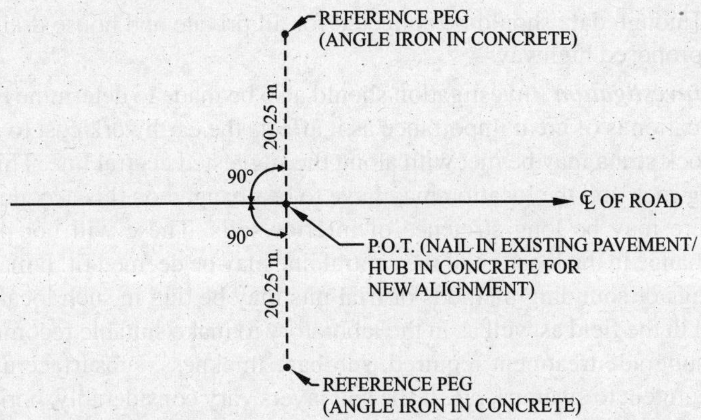

(*b*) Reference pegs at point of transit (P.O.T.)

Fig. 5.1 Method of referencing horizontal intersection point (H.I.P.) and point of transit (P.O.T.)

the staked central line. Every minute details should be noted. For example, in the case of a building, notice should be made of the steps, and open courts in front and back. If a part or whole of the building comes in the right of way, all dimensions to a hundred of metre should be taken. Intersection with other highways and railway lines should be shown giving due regard to the right of way, type, dimensions and conditions of existing railway crossing and other protective works required to be provided. Sites for the location of culverts and bridges, grade separations, etc. should also be decided. The location of the central line on the field should be checked with respect to the topographic details already taken in the preliminary survey. Staking out of the central line on the field offers opportunity for minor corrections such as small shifts of the line, adjustments in grades, and positioning of structures, channels, *etc*.

Longitudinal profile and cross-sections. The reduced levels at 30 m intervals have already been found out in the preliminary survey. From this, longitudinal profile can be drawn.

After staking out the central line, cross-sections are taken at 50-100 m intervals in plain terrain and 50-75 m in rolling terrain. In hilly areas the interval should be 20 m for a considerable distance on both sides of the central line. Cross-sections are also taken at points where there is a considerable change in the general profile, *i.e.*, at breaks. These cross-sections should be of adequate width to cover anticipated construction and should be taken with respect to the bench mark established in the preliminary survey at every 250 m (exceptionally 500 m). The central profile for each existing road should be determined for a considerable distance to determine their existing grades and to indicate any change in grade due to intersection. At railway crossings, elevations of the top of rails should be found out. The elevations of all important drainage structures, trees, *etc*. in the vicinity should be recorded.

Plotting the location survey. The survey party should plot all the additional information gathered in the field on the maps. The longitudinal profile is then plotted. This is done to the same horizontal scale as used for the topographic map. The vertical scale is usually ten times the horizontal scale. Then proposed grade lines are also put on the *L*-section. The cross-sections are also plotted on tracing sheets and these are attached to the final set of construction drawing. In rolling country, the vertical and horizontal scales are 1 m to a cm while in plains, they are drawn to a scale of 0.5 m to a cm. Cross-sections are started from the bottom of the sheet and are one above the other.

Surface drainage investigation. Notice should be made of all defined natural courses of water which either cross the central line of the proposed road or lie adjacent to it. Information should be gathered whether they carry water continuously or only in rainy seasons, drainage basin, nature and extent of run-off, high flood level etc. to enable the designer to provide adequate drainage structures. In case there exist culverts and bridges, their adequacy or inadequacy should be ascertained from the neighbouring villagers. Enough data should be collected for all private and house drains which may affect the design of the proposed highway.

Soil and subsurface investigation. Investigation should also be made to determine the type of soil in that area. Soil identification is of great importance as it affects the earthwork cost to a great extent. It may be possible that rock strata may be met with along the suggested central line. This will involve enormous cost in cutting rock and the location may have to be changed on this account. Similarly, it is also possible that there may be long stretches of inferior soils. These will not provide strong foundations and thus a change in the location of the central line may be deemed fit. It may be desirable to resolve to auger borings or sounding methods or trial pits may be dug in such localities. The soil samples should be tested in the field as well as in the laboratory to make suitable recommendations in respect of safe slopes, subgrade treatment required, sub-base thickness, subsurface drainage under slopes and shoulders, treatment to swamps, *etc*. If the soil layers vary considerably, borings should be made at regular interval, along the entire central line of the proposed locations.

Road materials investigation. The suitability or otherwise of the excavated material for use on embankments is also known from the soil investigation. In case the excavated material is adequate in quality and quantity, borrow-pits will be required. If borrow-pits are needed, investigation should be made to find out whether material can be borrowed within the right of way or it has to be procured from outside. Location and accessibility of the borrow-pits should also be determined to work out the cost of the earthwork required for embankments. Borrow-pits should be so located that the haulage is as minimum as possible.

Transportation facilities. For the construction of roads, machinery and material may have to be carried from great distances. Survey should be made of the available transportation facilities such as railways or roads around the proposed project. If temporary roads have to be made their feasibility and cost should also be looked into.

Right of way investigation. Right of way may not involve much problems in rural areas but considerable care has to be exercised in fixing the right of way in urban localities. It is not sufficient to tell the designer only about the shapes, sizes and horizontal and vertical locations in and near the right of way. But the location survey should also provide adequate information regarding the existing use to which the abutting land is put, condtion and cost of the buildings and their boundary lines. This will help the designer to fix the elevation of the highway with respect to abutting houses, width of the right of way to be taken and other design features as he will be in a position to find out the extent of damage that might occur by dismantling the abutting buildings.

Bridge location investigation. Since the purpose of bridges, culverts and grade separations as well as of highways is to carry traffic, considerable thought should be given to their location and positioning after the general alignment and grade have been fixed. At such places, greater details of the topography and needed cross-sections should be taken at every 3 m intervals along the central line of the proposed bridge and also at breaks in the stream bed. These should extend to 15 m beyond the right of way both upstream and downstream. These sections should be taken not only for the length of the bridge but also for approaches at each end (usually 30 m). Cross-sectional area should be adequate to include all structures such as abutments, wing walls, *etc*. Profile of the central line of stream bed should be taken at every 7.5 m interval for at least 150 m both upstream and downstream. Care should be taken to note holes in the stream bed or other places of scouring. If the stream width is large or consists of more than one channel, additional stream profiles should be taken. The angle between the central line of the proposed highway and the profile of the stream should be carefully found out. Area of the catchment basin, high flood level, *etc*. should be ascertained to calculate the waterway required. Borings and soundings may be made to determine the soil type and suitable foundations. If there exists a bridge on or near the central line of the road, information should be gathered regarding the number of spans and their length, width of roadway, size of waterway and its adequacy or otherwise, high flood level, velocity of water in the channel, type of bridge, *etc*.

Similarly, surveys should be conducted at grade separations between a highway and a railway. Profile of the tracks for 300 m on either side of the highway central line should be drawn. The angle of intersection, *etc*. should also be ascertained.

5.5. SURVEY REPORT

It is considered desirable to prepare a report about the surveys conducted for the proposed highway project. This will briefly summarize all the factors that helped in the location of the central line. It will also serve as a check for the engineer to find out if he has missed any important information. A suggested out-line for such a report is given below:

5.5.1. Identification and history of survey

(*i*) Name or other identification of the project.

(*ii*) Terminal points—reasons for their selection, length of the road, relation with any existing system.

(*iii*) Dates on which different surveys were started and completed, interruptions and delays, if any.

(*iv*) Proceedings of the meetings of officials.

5.5.2. Traffic

A brief summary of the findings of traffic surveys and geometric requirements is supplied to the surveyor. It should include the following information:

(*i*) Design speed.

(*ii*) Type and width of pavement, number of lanes, median width, if provided.

(*iii*) Width and character of shoulder.

(*iv*) Maximum allowable grade.

(*v*) Minimum radius of curvature on curves.

(*vi*) Sight distances required, frequency of overtaking.

(*vii*) Intersections—their type

(*viii*) Extent of any control of access.

(*ix*) Any other imposed restrictions.

5.5.3. Topography, existing land use and utilities

(*i*) General description of topography of area showing locations of swamps, rocks, and other obstacles.

(*ii*) Present use to which the land along the central line is being put, locations of grave-yards, monumental buildings etc.

(*iii*) Location, size, type, profile and owner's name of underground utilities like sewer, gas and water mains.

5.5.4. Proposed alignment and grades

(*i*) Relationship to existing road or roads.

(*ii*) Control points, their influence on location.

(*iii*) Different locations possible, reasons or adopting a particular one, and disadvantages of others.

(*iv*) Critical points governing the choice of the final grade line, minimum and maximum gradients provided, reasons for the selection of grade line, possibilities of adjusting grades, *etc*.

(*v*) Any weak spots in the proposed locations, reason for keeping them and discussion why they could not be dispensed with.

5.5.5. Foundations conditions

(*i*) Soil classification, either from mosaics or by investigation methods.

(*ii*) Locations of rock occurrences as revealed from borings, side slopes recommended in cuts and fills of earth and rock, any further sub-surface investigation still required.

(*iii*) Locations of swamps and other weak soils, their depth to hard starta underneath.
(*iv*) Possible locations where slides can occur.
(*v*) Recommendations for sub-base and treatment to be given in areas of weak soils.

5.5.6. Drainage

(*i*) Rainfall in cm, intensity, *etc.* in that area.
(*ii*) General review of the drainage scheme most suitable for that locality.
(*iii*) A list of existing drainage structures such as culverts, bridges, *etc.* giving details of their location, size, length, condition and adequacy. Recommendations in regard to the reconstruction of existing structures, possibilities of making new ones of parallel roads or proposed project.
(*iv*) A list of proposed culverts and bridges with location, size, length, skew, end walls, *etc.*
(*v*) Review of existing sewer systems, ditches, gutters and other drains.
(*vi*) List of proposed drains, ditches, gutters, etc.
(*vii*) Location of any area requiring high grade line to avoid flooding of road.

5.5.7. Construction materials

(*i*) Description of available local materials—their location, quality and quantity.
(*ii*) Possibilities of borrow-pits within right of way.
(*iii*) Recommendations for borrow-pits outside right of way; their location, cost of hauling and procuring land.
(*iv*) Location of quarries with details of materials available, their quality distance from quarry to the place of work.
(*v*) Availability of water.

5.5.8. Right of way

(*i*) Description of right of way to be acquired, condition and cost of buildings along the central line for the entire route.
(*ii*) Damages to be incurred in dismantling buildings, trees, *etc.*
(*iii*) Right of way difficulties involved in setting grade.

5.5.9. Maintenance of traffic (when required)

(*i*) Proposals for handling traffic, whether through project or diversion.
(*ii*) Proposed diversions, their length, type, condition and cost.

5.5.10. Other considerations

(*i*) Cost of removing trees along the proposed route, right of way with magnitude of work involved.
(*ii*) A list of approaches to be constructed.
(*iii*) Proposed treatment to intersections with highways and railway lines.
(*iv*) Retaining walls to be built.
(*v*) A list of walls, springs, *etc.* with proposed treatment.
(*vi*) Type and condition of existing property line, fences, posts, guard rails, *etc.*
(*vii*) List of survey data-maps, profiles, sections, *etc.* should accompany the report.
(*viii*) Designation of engineer responsible for the report with the list of personnel involved on surveys.
(*ix*) Date of report.

I.R.C has suggested the following check-list of items for a highway project report:

A. PROJECT REPORT

1. Preliminary

(*i*) Name of work and its scope
(*ii*) Authority and plan provisions
(*iii*) History, geography, climate, *etc*.
(*iv*) Necessity
(*v*) Details of previous improvement work carried out.

2. Road Features

(*i*) Route selection
(*ii*) Alignment
(*iii*) Environmental factors
(*iv*) Cross-sectional elements
(*v*) Traffic

3. Road Design Specification

(*i*) Road design
(*ii*) Pavement design
(*iii*) Masonry works
(*iv*) Specifications

4. Drainage Facilities including Cross-Drainage Structures

(*i*) General drainage conditions, HFL, water-table, seepage flows
(*ii*) Surface drainage, catch water drains, longitudinal side drains
(*iii*) Sub-surface drainage, blanket courses, sub-drains
(*iv*) Cross-drainage structures

5. Material, Labour and Equipment

(*i*) Sources of construction materials, transport arrangements
(*ii*) Labour, availability, amenities
(*iii*) Equipment.

6. Rates

(*i*) Schedule of rates. Mention year and district to which applicable.
(*ii*) Rate justification

7. Construction Programming

(*i*) Working season
(*ii*) Schedule of completing the work

8. Miscellaneous

(*i*) Rest houses, temporary quarters and other site amenities
(*ii*) Diversions and haul roads. Traffic control devices, Temporary diversion of traffic during construction
(*iii*) Wayside amenities

(*iv*) Roadside plantations, turfing, landscaping

(*v*) Road safety measures. Comprehensive improvement of accident prone sections. Safety in construction zone.

B. ESTIMATE

1. General abstract of cost
2. Detailed estimates for each major head
   (*i*) Abstract of cost
   (*ii*) Estimates of quantities
   (*iii*) Analysis of rates
   (*iv*) Quarry/material source charts

C. PROJECT DRAWINGS

(*i*) Locality map-cum-site plan
(*ii*) Strip plan showing the location of utilities, right of way, trees, junctions, etc.
(*iii*) Land acquisition plans
(*iv*) Plan and longitudinal section
(*v*) Typical cross-section sheet
(*vi*) Detailed cross-sections
(*vii*) Drawings for cross-drainage structures
(*viii*) Road junction drawings
(*ix*) Drawings for retaining walls and other structures.

LOCATION OF ROADS IN URBAN AREAS

The following additional points should be kept in mind while locating roads in urban areas:

1. Arterial traffic should be deflected from built-up areas as far as possible.
2. Where heavy arterial traffic cannot be so diverted, consider practicability of constructing an elevated or sub-surface road.
3. On important traffic routes mechanically propelled, pedal cyclist and pedestrian traffic should be segregated on separate tracks.
4. The reduction frontage access and reduction of side road constructions should be related to the function of the road and the volume and speed of traffic using it.
5. Service roads if properly designed is one method of restricting frontage access and of limiting side road connections, but outside the central area of the city, buildings should be grouped as far as possible on subsiding development roads.
6. Business premises directly fronting busy streets should be provided with means of loading and unloading at the rear of and within the compounds of buildings.
7. Buildings attracting large number of vehicles should have space for parking other than on highway.
8. Where traffic is heavy and continuous, intersection should be planned so as to enable cross traffic to pass at different levels.
9. Where considerable streams of traffic cross at the same level, eliminate tendency for direct cuts or right hand turns by providing traffic rotary or control by traffic light signal.
10. Ample sight distances should be provided at intersections and vehicles from minor roads should be led to enter main roads at reduced speeds.

11. Where vehicular and pedestrian traffic are both heavy, subways or overbridges with ramped approaches should be constructed for pedestrians and if necessary for cyclists. Otherwise provide pedestrians crossings with refuge islands at suitable points and erect barriers to make pedestrians use the crossing.
12. Footpaths should be of adequate width and well lighted.
13. Schools, playgrounds etc. should be located away from the main road and should be provided with access roads.
14. Public conveniences should not be located on central islands of traffic rotaries, unless access to them is provided through subways.
15. Where justified, provide specific parking areas.
16. Bus and public service vehicles should be provided with properly designed stops.
17. Road signs and street name plates should be conspicuous, of standard design and sited on uniform principles.
18. Adequate lighting *i.e.,* to enable drivers to drive without headlights should be provided.
19. Houses, shops and other buildings should be systematically numbered so that they can be easily identified without causing obstruction.
20. Roadside advertisement that tend to disturb attention of drivers or obstruct their line of sight should be controlled.

5.6. PUBLIC TRANSPORT SERVICES

Public service vehicles are indispensable in any major community. Considered as traffic units, they are more obstructive than private cars by reason of their size and frequency of stops. But from considerations of road space occupied per passenger they have marked advantage.

Before considering road development schemes in towns and cities, the transport authorities should be consulted on (*a*) frequency of services, (*b*) siting of stops, and bus and tram depots and stations. Motor buses and trolley buses have advantage of flexibility in routing and ability to pick up and set down passengers at the kerbside.

Trams are most rigid as far as their alignment is concerned. Trolley buses are limited by overhead lines. Turning radius of trolley buses and tramlines has to be much larger. When there are double decker buses, do not increase camber beyond 1 in 40.

The following provisions are required to be made in the case of public transport services:

Waiting bays. Waiting bay 3 m wide and of length equal to number of buses occupying the bay at one time is to be provided. Thirty metres ends have to be given smooth approaches from road. These bays are to be exclusively used by buses. In shopping streets, waiting bays need not be provided separately, one additional lane would be required for buses and parked vehicles.

Queue barriers and shelters. These should be provided for orderly arrangement of waiting passengers and for their shelter from weather. Double line queue barriers 1.25 m wide and 45 cm removed from kerb edge should be provided. The footpath may have to be set back at these places. Hence this is an important element requiring careful consideration.

Bus stations. No general rule can be laid down. These should be near railway stations, markets, important places of public resort. They must be set away from the main road.

Provision in housing estates. Due consideration should be given for facilities for mass transportation and road width determined to enable public transport vehicles to traverse the whole length through middle and perimeter. Due regard should be given to bus stops. Tramways are a danger especially if track is badly maintained. Where tracks are not segregated, tram cars by themselves are a greater potential cause of danger than other vehicles. When tracks are in the middle of a road, passengers getting down or boarding are often exposed to risk of accident. Tram track introduces serious complication in the layout of road intersections. They should, as soon as possible, be replaced by vehicles of greater flexibility except on through routes of heavy traffic. If possible, locate them in their own fenced location outside highway limits.

Problems concerning goods vehicles. Goods vehicles have greater overall length, width of body and load. Larger turning circles is required for them. At badly designed corners, these have a tendency to pull out to take the turn. In central areas, at peak traffic flow hours, the turning of large vehicles should be prohibited. Special turning circles should be provided.

Use of bypasses and their limitations. The function of the bypass is to provide an attractive alternate route for traffic which has no valid reasons to pass through the busy streets of built-up area, thereby eliminating conditions producing congestion, delay and danger.

From a traffic survey it was found that in towns of up to:

10,000 population	80% of entering traffic is bypassable
10,000 to 50,000	65% of entering traffic is bypassable
50,000 to 1,00000	45% of entering traffic is bypassable
1,00,000 to 250,000	20% of entering traffic is bypassable
250,000 to 500,000	15% of entering traffic is bypassable.
500,000 to 1,000, 000	less than 10% of entering traffic is bypassable

So the general proposition that larger the town, the smaller the bypassable traffic has been confirmed by these observations. Special cases of tourist centres are exceptions.

A bypass should not be too long; should not have difficult grades; should be free from ribbon development; and no cross road inter-sections should be provided.

Radial roads. The points of convergence of radial roads in most towns is the crux of the traffic problem. Main radials serve primarily the needs of through traffic and minor radials those of local through traffic. Since early development has been on the basis of radial lines, these roads are less amenable to improved treatment than new roads, and their improvement can only be a slow and costly process. It may more often be possible to consider an existing parallel road for development for drawing away the through traffic.

Outer ring roads. In most towns, the primary function of an outer ring road is to serve the traffic of the town itself by linking up the outer communities and acting as a distributor between the radials. So it should be located not outside the limits of the town but within the outer fringe of present potential urban development.

The location of an outer ring road should be based on the following principles:

(*a*) Its predominate function is to serve local through traffic.

(*b*) It should be located within the outer fringe of present and potential development.

(*c*) It should be designed to compensate for the detour involved providing facilities for higher speed, added convenience, safety and amenity.

(*d*) Provision should be made on it for all classes of traffic, *i.e.*, vehicular, pedal cyclist, pedestrian.

(*e*) Frontages should not be permitted.

Ring roads should be of dual carriageways and cross road should be provided with fly-overs, traffic rotaries or the intersections should be controlled by traffic signals. The number and character of ring road necessary for traffic requirements depend mainly on the population and spread of urban development. Towns with a population of say 2 lacs may need only a single ring road whereas a town with 10 lacs or so may require both an inner and an outer ring road.

The objections to frontage development can be minimized to some extent by means of service roads, preferably without direct connection with the ring road.

Outer ring roads can be developed as parkways. Connections with radial roads and access from subsidiary roads should be effected by means of fly overs, round abouts or intersections controlled by traffic signals.

Intermediate ring road. These serve primarily the needs of traffic whether from a distance or of local origin to reach points between the outer and inner ring roads. They will normally be located in areas already particularly developed and to larger extent than the outer ring roads. It should be the aim in locating the intermediate ring road to limit as far as possible the connections with subsidiary roads and also frontage access. Where frontage development exists or is inevitable, it should wherever possible, be provided for by means of parallel service roads.

Connections with minor or main radials are made by use of round abouts.

Inner ring roads. These should not be regarded as a route for through traffic other than that having its points of origin or destination in the central area. Its traffic function is to serve the needs of local through or local traffic and its purpose is to promote the convenient use and amenity of the central area by deflecting from it all vehicles which have no need to traverse it, whilst affording convenient means of entry to those for which access is essential.

Generally, the central area will contain the public buildings and principal shopping centre which has grown up around the hub of radial roads. As such, public service vehicles will form substantial proportion of vehicular traffic reaching the centre. If design requirements necessitate that public service vehicles are to be wholly excluded from the inner ring road, the inner ring road should be so located that the passengers do not have to walk more than 300 m to reach the principal shopping centre. The area enclosed by the inner ring road should not have a diameter greater than 600 m to ensure this aspect.

Safety and free flow require that frontage access to the inner ring road should be restricted. Places of public resort should not derive their means of access from the inner ring road, and as far as practicable all premises fronting it should derive their means of vehicular access from the rear. As far as practicable the inner ring roads should not be developed as a shopping centre. The treatment of the intersection of the inner ring road should be determined based on the following main principles:

1. Discouragement of passage through the central area of traffic having no business there.
2. Maintenance of free flow on the ring.
3. Provision for safe and convenient movement of vehicular traffic and pedestrians between the central area and the rest of the town.

5.7. LOCATION OF SHOPPING STREETS

The area enclosed by an inner ring road can best be considered suitable for a shopping centre, places of amusement, civic buildings, banks, hotels etc. Desirable characteristics of ideal shopping streets are:

(*a*) Facilities by which the pedestrians may, in comfort and safety:
  - (*i*) move along the street
  - (*ii*) cross the street; or and
  - (*iii*) view the goods displaced in the shop windows.

(*b*) Absence of vehicular traffic other than serving the shops.

(*c*) Easy accessibility to public service vehicle routes.

(*d*) Space in the street in which vehicles can be parked for a limited period while occupants do their shopping.

(*e*) Adjacency of car parks.

(*f*) Provision of separate loading and unloading facilities.

(*g*) Continuity of shopping frontage on both sides of the street.

(*h*) Limitation of carriageway width to the minimum required for the service of the shops.

5.7.1. Ample footway width.

It is seen that the requirements of through or local traffic are incompatible with the conditions which should obtain in shopping streets where convenience and safe crossing of streets is essential. If provision for means of approach for vehicles were not necessary and in particular for waiting vehicles, the consideration of safety of shoppers would suggest that shops could best be located in arcades or other purely pedestrian ways. The minimum width of carriageway is one lane 3.8 m wide for traffic. 3 m for parking, 1.25 m central refuge, and footpaths not less than 4.5 m in width. Retail shopkeepers do not favour wide streets as they detract from the intimacy with purchasers which is a psychological factor. Where the shopping street should also take through traffic, prohibit waiting vehicles. But stops should not be too far away. Rear access roads should be used only for loading and unloading of goods and not be developed as subsidiary shopping streets. Arcading over footways in street and projecting balconies over footways are not desirable. The recessing of shop fronts with projecting canopies, however, are desirable to protect shoppers from the weather. Supporting columns or piers on the footways should be avoided.

PROJECT PREPARATION

The preparation of a project for the proposed highway can be done under the following heads:

(*i*) Preparation of key map
(*ii*) Preparation of index map
(*iii*) Review of the survey data and plans
(*iv*) Preparation of construction plans
(*v*) Field review of the above plans
(*vi*) Preparation of final construction drawings
(*vii*) Drafting specifications
(*viii*) Preparation of cost estimates
(*ix*) Preparation of right of way plans and their cost estimates, if required.

Key map. A key map should indicate the location of the road, important towns and industrial areas in the vicinity, modes of existing communications such as roads, railways, waterways, *etc*. The key map is to give an overall picture of the project in relation to the whole system of communications in that area. The proposed road should be shown in red colour on the key map. North direction should also be marked.

Index map. This map should show the topographic features of the area such as hills, rolling or plain terrain, drainage system, existing industrial centres, main prevailing communication systems. The index map should also show average rainfall in the area, north line, latitudes and longitudes where needed. The map should also indicate the alignment and phasing of the project.

Survey data and plans. The topographic maps prepared by the survey party are reviewed and checked. They show the following survey data:

(1) Important topographic features including all property lines and names of the landowners.
(2) Diversions, curves and village boundaries.
(3) Detailed designs for masonry works.
(4) Drawings for dak bungalows, rest houses, *etc*.
(5) Land plans for land to be acquired for quarries, *etc*.
(6) Other details:

(a) Detailed plan and longitudinal section. The ground plan should show the following details:

(*i*) The proposed alignment of the road clearly indicating the right of way boundaries. If an existing road is to be improved, the existing and the proposed width of the roads should be shown.

(*ii*) Contours at an interval of 1 m in plains or rolling country. These may be drawn at 3 m intervals in hilly areas.

(*iii*) All existing structures, *e.g.*, religious places, graveyards, buildings, residential, industrial and commercial, wells, level crossings drainage courses, tanks, ponds, *etc*.

(*iv*) Nature and description of the soil, position of quarries.

(*v*) In the case of bends: deviation angles, tangent points, radius of curve, nature of transition, length of curve, *etc*. Separate particulars with regard to the curves should accompany the estimate and in complicated cases, separate drawings may be drawn.

(*vi*) The location of the drainage crossings. These should be numbered serially giving information about the nature of bridging proposed. The names of rivers, streams, major drainage courses and nallahs should also be given.

(*vii*) The location of kilometre stones.

(*viii*) The reduced distance of cross-sections. They should also be serially numbered.

(*ix*) Bench-marks clearly indicating their position and reduced levels.

(*x*) A north line.

(*xi*) Key chart at the bottom of plan showing the bearing of the sheet to other connected sheets.

(*xii*) Reference to the field book from which the survey data has been taken.

(*xiii*) The central line of proposed road showing cutting by dotted red lines and embankment by firm red lines.

The detailed plan can be drawn from the topographic map and the survey report prepared by the survey party.

The longitudinal section, if already drawn by the survey party can be used. In case this is not done, longitudinal section may be plotted from the reduced levels taken at every 30 m interval and also at breaks. This section should show:

(*i*) Datum

(*ii*) Ground or existing road profile

(*iii*) Formation levels

(*iv*) Height of embankment or depth of cutting

(*v*) Type of soil on the surface

(*vi*) Gradients

(*vii*) Reduced distances and position of km stone

(*viii*) Position of cross roads

(*ix*) Vertical curves

(*x*) Drainage crossings with number and name corresponding to the plan and giving rough indication of the high flood level and the nature of crossing proposed.

(*xi*) Reference to the field book from which the longitudinal section has been plotted.

(*xii*) Positions of cross-sections with numbers

(*xiii*) Formation line in dotted red in cutting and firm red in embankment.

(b) Detailed cross-sections. It should be aimed to draw detailed cross-sections on separate standard sheets for one plan and longitudinal section sheet. If the country is fairly level and there are not many undulations, only typical cross-sections will be enough and if possible, these may be drawn on the longitudinal profile sheet. Thus, separate detailed cross-section sheet will not be required.

Care must be taken to plot cross-section at all points where there is an abrupt break in the ground profile. In plain country, cross-section at 100 m will serve the purpose all right but in hilly areas, cross-sections should be drawn at closer intervals depending upon the area and the judgment of the engineer incharge. Each cross-section should show:

(*i*) Its number as marked on the longitudinal section.

(*ii*) Reduced distance.

(*iii*) Area of cutting and/or filling.

(*iv*) Reference of the field book from which the data has been taken.

The cross-section should be taken over a width equal to at least the width of the right of way proposed at that section. If an improvement to the existing road has to be done, positions of lines of avenue trees should be indicated. The cross-sections should also show side drains, catch drains, etc.

(c) *Detailed drawings for masonry.* Detailed drawings should also be drawn for masonry works to be constructed along the road such as bridges, etc. Ground plan, longitudinal and cross-sections and elevations should be accommodated in one sheet as for as possible. Cross-sections of streams should be shown to a scale of at least 1 m to 10 m.

(d) *Land plan for quarries.* In case quarries for materials required for the construction purpose have to be acquired, separate land plans and estimates should be prepared. These should be attached to the project estimates.

In order to maintain uniformity, all the above drawings should be prepared to definite sizes. The sizes and scales recommended by the IRC are given in Table 5.1.

Table 5.1 Standards for sizes of maps and scales for road works

Name of drawing	*Size of map*	*Scales recommended*
1. Key map	32.5 cm by 20 cm	Depending on the extent of area to be covered 1.5 cm to 1 km, 0.75 cm to 1 km, 0.4 cm to a km
2. Index map	32.5 cm by 20 cm in successive sheets	1.5 cm to 1 km where the project extends over several km, index plans for each section to a scale of 1.5 cm to 1 km should be provided.
3. Preliminary survey or location plans	82.5 cm × 45 cm 67.5 cm × 50 cm 50 cm × 37.5 cm 40 cm ×32.5 cm 32.5 cm ×20 cm depending on the extent of area to be covered	12 cm to a km or 24 cm to a km
4. Detailed plan and longitudinal section	82.5 cm × 35 cm	(*i*) Rolling open country, 1 cm to 25 m for ground plan and horizontal distances of the longitudinal section (*ii*) Close country, 1 cm to 12.5 m ground and horizontal distances of longitudinal section (*iii*) Vertical scale for longitudinal section, 1 cm to 1 m
5. Cross-sections	82.5 cm × 45 cm	1 cm to 2 m (natural)
6. Land plans for road quarries, rest houses etc.		
7. Masonary works, Dak bungalows, Rest houses	One of the following: 100 cm × 67.5 cm 82.5 cm × 45 cm 67.5 cm × 50 cm 40 cm × 37.5 cm 32.5 cm × 20 cm	Working drawings 1 cm to 1 m. Details 2 cm or 4 cm or 8 cm to *a* m. Enlarged drawings 1/8th or 1/4th or or 1/2 full size.

The drawings should be drawn in particular colours also. These are recommended by IRC and are as given in Table 5.2.

Table 5.2 Recommended colours for drawings and maps

Name of drawing	*Original drawing*	*Tracing (in block)*
Plans		
Preliminary survey	Blue line	Thin line
Contours (normal)	Thin brown line	Thin chain-dotted line
Contours (every fifth)	Thick brown line	Thick chain dotted line
Grade contour	Green line	Dotted line
Located centre line	Thick black line	Thick line
Other details	Thin black line	Thin line
Sections		
Ground line	Thick black line	Thick line
Grade line	Read line dotted in cutting and firm in embankment	Thin line dotted in cutting and firm in embankment.

(e) Other details. Keeping in view the geometric design standards as discussed in Chapter 8, decision should be made regarding: (*i*) number and width of lanes (*ii*) width of shoulder medium strips (*iii*) other elements of the road cross-section (*iv*) gradients (*v*) horizontal and vertical curves (*vi*) location of guard rails, signs, *etc*.

Also drainage structures, to be provided should be decided and details worked out. In case a bridge is required to be provided, bridge approaches are designed and bridge details also decided.

Field review of construction drawings. After all the details have been worked out regarding the construction proposed to be undertaken for the project, a thorough field inspection is carried out to check up any defects. Moreover, this being done at a different date and different season than when the surveys were conducted, will enable the area to be inspected under different conditions and thus any change in the soil condition or sub-surface water can be noted. The inspection should take into account any factors regarding adequacy of the road to anticipated traffic, construction cost, use of land and maintenance of roadway and roadside, correctness of the survey map, suitability of cross-section, grades, curves, *etc*.

Preparations of final construction drawings. On the basis of the findings of the inspection party, the construction drawings are corrected according to the suggestions made in the inspection note and final drawings prepared.

Drafting specifications. General specifications should then be prepared for the execution of the project. Specifications which are in vogue in that particular area such as PWD specifications should serve as a guide for drafting specifications for a particular project. These specifications should include materials to be used, the tests which they must satisfy, construction methods. The method of measurement of each item and the basis of payment, where labour rates or conditions of employment, *etc*. are prescribed, these should also form a part of the specifications or special provisions made for them.

Preparation of cost estimates. The cost estimates of the project should be prepared on the estimate sheet and should cover each item systematically. The quantities of each item should be worked out and then the cost estimated keeping in view the schedule of rates in vogue in that locality. To the total cost of items so obtained, a percentage, often ten per cent, is added to meet the contingencies and supervision charges. Thus total cost of the project should be calculated. Due consideration should be given to the quantities involved, freight charges to the site, length of haul, type of excavation and facilities available, depth of swamps, time allowed for the execution of work, etc.

APPLICATION OF THE CRITICAL PATH METHOD (CPM) TO HIGHWAY PROJECTS

The use of modern management techniques such as Programme Evaluation and Review Technique (PERT) and Critical Path Method (CPM) have been successfully employed to achieve economy and efficiency. These techniques help management in efficient and economic use of resources in the accomplishment of programme objectives. They also help in effective planning scheduling, evaluating progress and controlling of projects and programmes. Though PERT and CPM techniques are basically similar, the CPM technique is more widely used in highway projects.

The CPM is a schematic representation of a project by means of a diagram or 'newtork' depicting the sequence and interplay of the numerous component events that go to form the project and the utilisation of the data contained in the network for determining the most suitable programme for the implementation of the project. The final programme is so selected as to result in the lowest cost consistent with the time factor. Its use in the highway departments has been widely accepted in areas such as planning, design, construction, research programme and maintenance. In India, where there are a large number of constraints due to shortage of resources such as steel and cement, non-availability of skilled labour, seasonal availability of unskilled labour and climatic variation (monsoon period, the season of floods in streams, *etc*.), the planning and execution of highway projects could be done in a more efficient manner by the application of CPM.

Advantages of the application of CPM to highway projects

The application of CPM to highway projects results in several advantages over the conventional reporting system of bar chart. These are:

(*a*) The pre-requisite of CPM analysis requires a thorough and detailed examination of the project.

(*b*) It enables the planner to chalk out a logical programme with inter-dependence of the various activities and restraints.

(*c*) It provides a useful method of scheduling the resources to the best of advantage.

(*d*) It indicates and emphasizes the likely activities which may be the cause of the trouble and delay in the project.

(*e*) It provides a basis for reporting progress.

(*f*) It indicates, in case some activities are delayed, where extra effort has to be applied to restore the progress and effect timely completion.

(*g*) It facilitates any change in programme when the situation warrants.

(*h*) It provides for an easy and clear method of communicating the engineer's plan to the others.

(*i*) It can be applied to various fields such as planning, design, construction and maintenance, which a highway department is called upon to deal with.

(*j*) Considerable saving in time and money is possible with the application of CPM.

Limitations of CPM

Though CPM has a very wide range of application in highway engineering, yet it has its own limitations and problems. The drawbacks of this system are:

(1) Any basic error in feeding the data to the network cannot be checked and this would result in completely erroneous results.

(2) CPM costs more to implement than other planning methods since it necessitates a very detailed study of the project.

(3) Expert advice is needed on factors like duration of activity, restraints, cycle of activities, etc.

5.8. DEFINITIONS

Before the procedure of constructing the network is discussed, it will be necessary to be conversant with the terms commonly used. The definitions of these terms with suitable examples are as follows:

5.8.1. Network of flow diagram

It is diagrammatic representation of the entire project, where in the order in which the various items of works must be performed is shown. A typical CPM network is depicted in Fig. 5.2. Generally, in a network the general flow of events is from the left to the right.

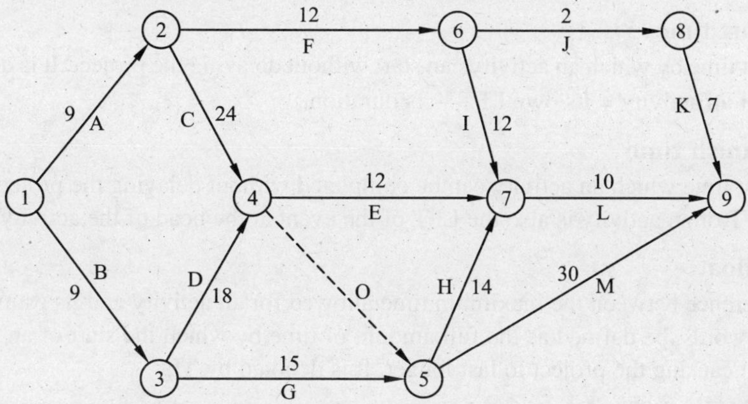

Fig. 5.2 Typical CPM network.

5.8.2. Activity

An activity is a part of the project denoted by an arrow on the network, the tail representing the start of the activity and the head the end of the activity. One and only one arrow can be used to represent one activity or operation. Each activity has a given time. However, length and direction of the arrow have no significance. For example in Fig. 5.2, an activity can be represented as 1-2 or by 'A'.

5.8.3. Event

An event is a stage or a point where all previous jobs are complete and the succeeding ones are to start. The event is represented on the network by a circle (node) at the junction of arrows. An event cannot occur until all the activities leading to it are completed. Similarly, no activities emanating from an event can start until the event has occurred. These events are numbered in their sequential order on the network.

5.8.4. Durations

Duration is the estimated time required to complete an activity and is denoted on the activity in the network. For example the duration for activity 1-2 (A) is 9 days in the network shown in Fig. 5.2, the duration can be expressed in any convenient unit such as hours, days, week, months, *etc*.

5.8.5. Dummy activity

It is a fictitious activity with zero duration and no cost and is used to maintain the sequential order of the activities in the network. It is generally represented by dotted lines on the network. For example in Fig. 5.2, activity 4-5 is a dummy activity and denotes that all activities commencing from event 5 cannot start before completion of event 4.

5.8.6. Restraint

It is a restriction very similar to a dummy activity but has a duration which can be either negative or positive. It is frequently used to fix intermediate dates within the network and thereby fix the relative start or finish of the parallel activities, when these activities are not coincident.

5.8.7. Early start time

It is the earliest possible time an activity can start without changing the sequence of activities in the network. It is denoted as EST. EST of an activity is also the Early Finish Time of the event proceeding it, *i.e.*, the event at the tail of the activity arrow.

5.8.8. Early finish time

It is the earliest time by which an activity can be completed. It is denoted by EFT.

EFT of an activity = Its EST ÷ duration of activity.

5.8.9. Late start time

It is the latest time by which an activity can start without delaying the project. It is denoted as LST.

LST of an activity = Its own LFT—its duration.

5.8.10. Late finish time

It is the latest time which an activity can be completed without delaying the project. It is denoted as LFT. The LFT of an activity is also the LFT of the event at the head of the activity arrow.

5.8.11. Total float

It is the difference between the maximum time allowed for an activity and its estimated duration. It can, in other words, be defined as the full amount of time by which the start of an activity can be delayed without causing the project to last longer. It is denoted by TF.

TF = LFT – (EST + Duration)

= LFT – EFT

or TF = LST – EST

5.8.12. Free float

The free float of an activity is the amount of time by which the activity can be delayed without interfering with the start of succeeding activities. It is denoted as FF.

FF = EST of its following activity — its own EFT

5.8.13. Interfering float

It is the difference between the total float and the free float. Consumption of any of the interfering float time in a delayed start of an activity will necessarily retard some of the following activities, but will not delay the overall project time. Interfering float is denoted by IF.

IF = TF – EF

5.8.14. Critical path

The events which have no float are the critical events, *i.e.,* they have the same EFT and LFT or in other words no leeway. These events must be completed on schedule if the project is to be completed in the minimum total time. The path joining such critical events is called the critical path of the network.

5.8.15. Critical activities

The activities lying on the critical path are called critical activities. Critical activity has a zero float.

5.9. CPM TECHNIQUE

The critical path method involves the following steps:

(1) Network construction.

(2) Assignment of duration for activities.

(3) Determination of project schedule and critical path.

(4) Cost-time balancing.

(5) Resources (manpower, equipment) scheduling.

(6) Budgeting and actual cost.

A brief description of these follows:

Network construction. The basic step in analysing a project by the CPM techniques is to draw up the network, which schematically represents the entire project after it is split up into a number of activities. Care must be taken to see that none of the activities involved are left over, as an error at this stage would be a basic one. The number of activities into which a project is to be split will depend entirely upon the extent of detailed control and information feedback desired. The next step is to establish a logical sequence of the activities and identify the necessary constraints.

Assignment of duration for activities. When the network has been completed, it is necessary to assign duration to each activity. Duration is the time required to complete the activity. No regard is given to the effect and delays caused from the adjoining activities. The unit of time used in the network for assigning duration could be hours, shifts or days. The duration is written above or below the arrow line representing the activity.

Determination of project schedule and critical path. The next objective is the preparation of the project schedule and the critical path. For this it is necessary to determine Early Finish Time (EFT) and Late Finish Time (LFT) of each activity. The node at the start of the net work is assigned a value of zero. As one progresses, the duration of each succeeding activity is added to the EFT of an event to determine the EFT of the next event. In doing so, it will be seen that, the EFT of each event is one which is arrived at by the longest path. For example, in Fig. 5.3, the EFT of event (4) works out to be 33 hours. All the EFT thus calculated are marked on the left half of the 'oval time space' on the network. After proceeding in this manner, EFT of the last event is known *i.e.,* the earliest possible time by which the project can be completed. This equals the sum of the duration of the longest time path through the network. This is entered in the right half of the 'oval time space' of the final event. From the LFT of the final event, one works backwards to find the LFT of each event. The LFT is controlled by all the activities starting from the event concerned and is the minimum figure so obtained. The LFT of each event is worked out in this manner and is shown in the right half of the oval time space.

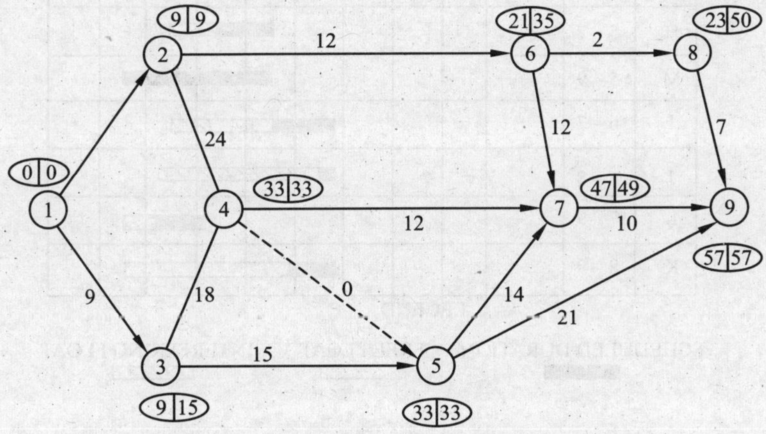

Fig. 5.3 Project schedule and critical path determination.

The events in which EFT and LFT are the same or in other words there is no leeway, fall on the critical path. The critical path is shown in heavy lines on the network. The critical path will be a continuous line beginning from the start of the project and ending at its completion. All the activities on the critical path must be completed on schedule and cannot be prolonged. On the other hand, the activities outside the critical path, can be prolonged within the limits indicated by their floats without prolonging the total project time.

5.9.1. Activity times, floats and bar graphs

After deriving EFT and LFT for all the events in the project, the next step is to find the activity times and the floats. Having done this, the essential time relationships between the various activities can be analysed and decisions taken concerning the scheduled timing of the activities in the construction works. To appreciate the scheduling of time better, often the CPM bar graphs are drawn.

The CPM bar graph for the activity times and the floats, given in Table 5.3 is shown in Fig. 5.4. Though this chart is identical to the conventional construction programme, but in addition it shows the critical activities and the float times. Hence one can see at a glance as to which are the activities which cannot be delayed and also in which activities delay can be tolerated and by how much time. Thus the repercussions of any delay in the entire project are exactly known and remedial measures, if necessary, can be undertaken in time.

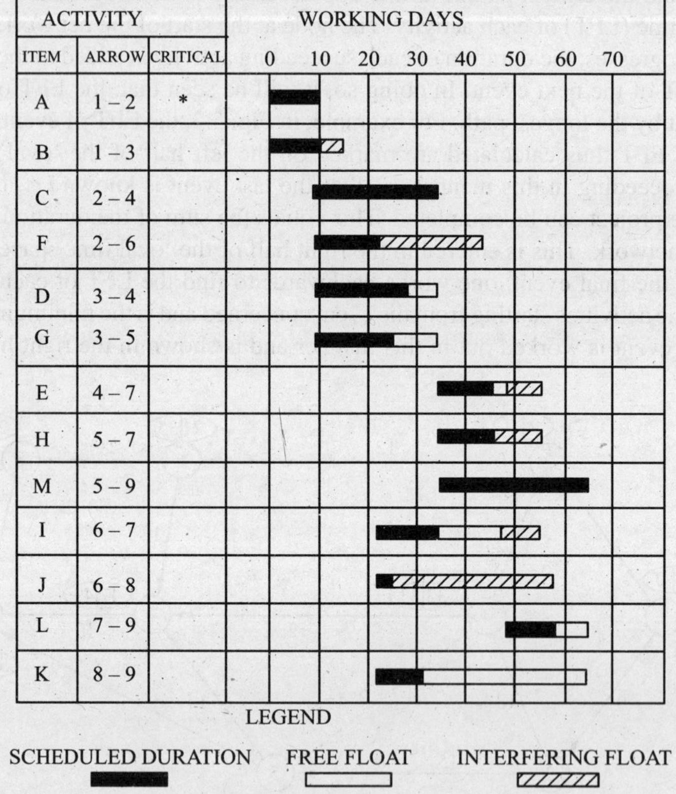

Fig. 5.4 Bar graph.

Table 5.3 EST, LST, EFT, LFT and floats for different activities

Activity Item	Activity Arrow	Duration in days	EST	LST	EFT	LFT	TP	FF	IF	Remarks
A	1–2	9	0	0	9	9	0	0	0	Critical
B	1–3	9	0	6	9	15	6	0	6	—
C	2–4	24	9	9	33	33	0	0	0	Critical
F	2–6	12	9	29	21	41	20	0	20	—
D	3–4	18	9	15	27	33	6	6	0	—
G	3–4	15	9	18	24	33	9	9	0	—
Dummy	4–5	0	33	33	33	33	0	0	0	Critical
E	4–7	12	33	41	45	53	8	2	6	—
H	5–7	14	33	39	47	53	6	0	6	—
M	5–9	30	33	33	63	63	0	0	0	Critical
I	6–7	12	21	41	33	53	20	14	6	—
J	6–8	2	21	54	23	56	33	0	33	—
L	7–9	10	47	53	57	63	6	6	0	—
K	8–9	7	23	56	30	63	33	33	0	—

Cost-time balancing. There is a marked effect of altering the duration of a particular activity on the project, as a whole. For example, if in the network of Fig. 5.3, the activity 5–9 can be carried out in 21 days, instead of 30 days, the network gets modified as shown in Fig. 5.5.

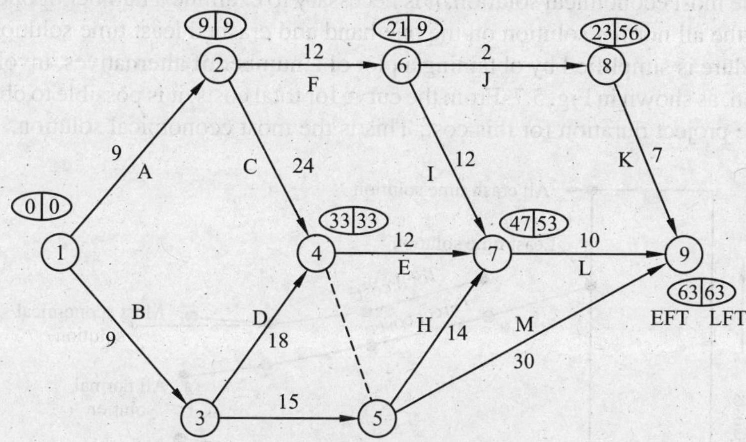

Fig. 5.5 Cost time balancing.

In order to evaluate the monetary effect of reducing or increasing the duration of each activity, it is necessary to have activity time versus cost relations. These are established only on the basis of past records of project and cost and time. A typical time-cost curve for an activity is illustrated in Fig. 5.6. The lowest cost (least cost) is also called normal cost and the time corresponding to the cost is called the normal time. As the time is reduced, the cost tends to go up curvilinearly, till a point of least time is reached. Further reduction in time is not possible and the cost increases with no reduction in time. The least time is called '*crash time*' and the speeding up of the project at the crash time is called the '*crash cost for least time*'. For approximation, the curvilinear curve can be treated as a linear curve, shown as dotted line in Fig. 5.6. The slope of the straight line gives an indication of the sensitivity of the activity so far as its time-cost relation is concerned.

In a network, if the project is to be completed in a shorter period, it is necessary to '*crash*' a number of activities. Only those activities which are critical need be crashed. The solution contained by crashing the activities along the critical path is called the "*Least Time Solution*".

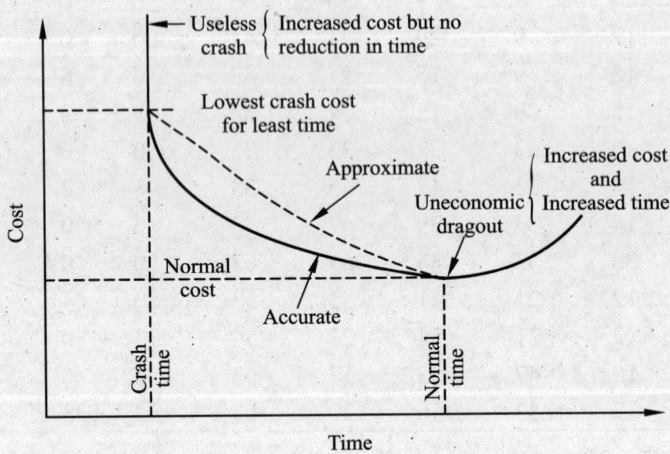

Fig. 5.6 Activity cost-time relationship.

To arrive at the most economical solution, it is necessary to examine a number of optimal solutions that lie between the all normal solution on the one hand and crashed least time solution on the other hand. The procedure is simplified by obtaining a plot of a number of alternatives, involving different costs and duration, as shown in Fig. 5.7. From the curve for total costs, it is possible to obtain minimum total cost and the project duration for this cost. This is the most economical solution.

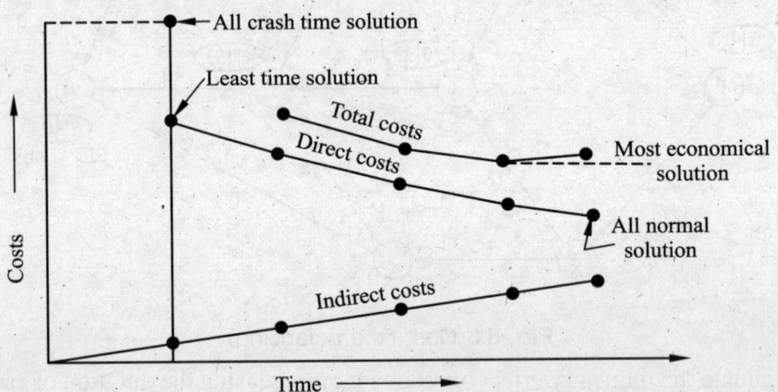

Fig. 5.7 Cost time curve.

***Resources (manpower, equipment) scheduling*.** When the initial *CMP* network is drawn up the duration of each activity is fixed on the assumption that all resources are available and can be applied for the completion of each activity. In actual practice this ideal condition is hardly met with, as there will always be a limit to the availability of resources, such as labour, plant, equipment, etc. It, therefore, often becomes necessary to schedule the requirements of the resources on a time base. Obviously the activities which are critical cannot be re-scheduled. Only those activities that have a float can be adjusted. So rescheduling of the non-critical activities within the leeway provided by their float is done, so as to make use of resources within the constraints of resources availability.

5.9.2. Budgeting and Actual costs

In most projects, the planner is time as well as cost bound. He has to co-ordinate his resources so as to meet the actual requirements and at the same time keep up the time schedule. It is desirable to draw graph between resources and time. Normally manpower constitutes the most common resource and a graph of number of man hours versus time can be drawn to indicate to the planner the actual manpower requirements during a particular period and thus give him sufficient warning to arrange the same timely.

Similarly from the data available from *CPM* network, it is possible to draw a graph of total expenditure *Vs* time. In this very graph, expenditure actually incurred can also be shown so as to indicate to the planner, whether the actual expenditure is likely to exceed the budgetary provision and if so what precautionary steps have to be taken.

PROCEDURE FOR LAND ACQUISITION

Highway departments, along with many other public bodies are vested with the power of compulsory land acquisition. This is the right of the government to take private property for public use, with or without the owner's consent.

Whenever land in any locality is needed for constructing a road or for any public purpose, a notification under section 4 of the Land Acquisition Act is published in the official gazette and the Collector concerned also gives public notice of such notification at convenient places in the said locality. Any person interested in any land which has been notified under section 4 can file objection in writing to the Collector, within 30 days after the issue of notification. The Collector gives the owners an opportunity to hear their objections. After this notification, the government servant can enter this land for preliminary survey work, etc.

Once the Government is satisfied, after hearing all objections, that the particular land is needed for public use, notification under section 6 of the Act is made in the official gazette giving the onwer's name and *Khasra number* of the land. Road land plans showing the land to be acquired are also prepared for the purpose.

Suitable compensation for the private property thus acquired is paid by the collector to the land owners after which the ownership of the particular land vests with the government. The compensation to be paid is usually worked out on the basis of average cost of land transactions in the vicinity during the last 5 years, as verified from revenue records. Fifteen per cent extra compensation is paid over and above the average value of property due to compulsory land acquisition.

When due to certain emergency it is not possible to notify separately under section 4 and section 6 and the land is required immediately for public use, the notification is made under section 17 of the Act which combines notification under section 4 and 6 and thus the property is acquired immediately after the notification.

Chapter

HILL ROADS

A hill road is considered as one passing through *'mountainous'* or *'steep'* terrain. For arriving at a few possible alternative alignments, the investigation should start only from obligatory summit points and proceed downwards. The alignment finally selected, linking the obligatory and control points, should fit in well with the landscape. It should satisfy the requirements of geometric *vis-a-vis* the needs of traffic, as also the terrain and climatic conditions. Optimum alignment will be one which yields the least overall transportation cost, taking into account the costs of construction and maintenance of the road as well as the recurring cost of vehicle operation, and at the same time have least adverse impact on the environment and ecological balance.

The route should avoid the introduction of hair-pin bends as far as possible. However if such a provision becomes inevitable, the number of hair-pin bends should be reduced to absolute minimum. Further the bends should be located on stable and flat hill slopes, and their location in valleys avoided. Also, a series of hair-pin bends on the same face of the hill should be avoided.

In hill roads, additional care has to be taken for ecological considerations, such as:

(*i*) Stability against geological disturbances

(*ii*) Land degradation and soil erosion

(*iii*) Destruction and denudation of forest

(*iv*) Interruption and disturbance to drainage system

(*v*) Aesthetic considerations

(*vi*) Siltation of water reservoirs

(*vii*) Geological disturbances

6.1. PROCEDURE FOR FIXING THE ALIGNMENT

The alignment of a hill road is fixed and translated on the ground in several operations:

(*a*) Reconnaissance

(*b*) Preliminary survey

(*c*) Determination of final centre line

(*d*) Final location survey.

During reconnaissance, a general route for the alignment is selected. A trace is cut thereafter corresponding to this so as to provide an access for the subsequent surveys. The final alignment to desired geometrics is marked on the ground in the last phase.

It is imperative that the personnel in charge of survey and design should keep in view the salient principles of route selection pertinent to hill roads during the various phases of alignment finalization, particularly at the time of initial reconnaissance and the preliminary surveys following the trace cut. It will be a good practice to associate the Forest and Geology Departments in all stages of alignment selection. Broad principles applicable to hill roads location are spelt out below:

6.2. GUIDING PRINCIPLES OF ROUTE SELECTION AND LOCATION APPLICABLE TO HILL ROADS

6.2.1. General

The alignment should be as direct as possible between the obligatory and control points to be linked. A direct highway link results in economy in construction, maintenance and operation.

The alignment should be along the side of the hill, which is stable because a common problem in hill roads is that of landslides.

The location should result in minimum interference to agriculture and industry. It should steer clear of obstructions such as cemeteries, burning ghats, places of worship, archaeological and historical monuments, and public facilities like hospitals, schools, playgrounds, etc.

Where the proposed location interferes with utility services like overhead transmission lines, water supply lines, etc. decision between changing the highway alignment or shifting the utility services should be based on a study of relative economics and feasibility.

As far as possible, frequent crossing and recrossing of railways, canals, water courses, ridges, etc. should be avoided.

The alignment should avoid large scale cutting and filling, and follow the line of the land as far as possible. The cutting and filling of earth for the construction of hill road on hill side causes steeping up of existing slopes which affects its stability. Use of tunnel to avoid deep cuts should be considered where feasible and economical. If the road has to be in cutting, the location and the grade line should permit the adoption of half-cut and half-fill type of cross-section which involves least disturbance to the natural ground, subject however to considerations of economy and road stability being satisfied.

6.2.2. Obligatory points

The obligatory points to be connected from administrative, strategic or other considerations should be ascertained and taken into account while finalizing the highway alignment. Similarly, control points like mountain passes, saddles, river crossings, etc. should be kept in view when deciding the alignment.

When crossing mountain ranges, the highway should preferably cross the ridges at their lowest elevation. In certain cases it may be more expedient to negotiate high mountain ranges through tunnels. This decision should be taken after considering the relative economics or the strategic requirements.

6.2.3. Grade and curvature

The route should enable ruling gradient to be attained in most of its length.

As far as possible, the alignment should permit adoption of a uniform design speed and easy curvature in the entire length.

The route should avoid the introduction of hair-pin bends as far as possible and their location in valleys avoided. The bends should be located on stable and flat hill slopes. Also, a series of hair-pin bends on the same face of the hill should be avoided.

Needless rise and fall must be avoided where the general purpose of the route is to gain elevation from lower to a higher point. Also, deep cuts involving destabilisation of natural hill slopes should be avoided.

6.2.4. River crossings

It is preferable that crossings of major rivers (waterway exceeding 100 m) should be at right angles to the river flow with highway alignment subordinated to considerations of the bridge siting. Crossings of medium/minor streams may also sometimes govern the choice of alignment in the case of hill roads due to foundation problems, though their position will be determined generally by requirements of the highway proper, and the crossings could be even skew or on curve if necessary.

As far as possible, efforts should be made to locate bridges where:

(*i*) the river is straight both on the upstream and downstream side
(*ii*) the location is sufficiently away from confluence of tributaries
(*iii*) the channel is well-defined and narrow; and
(*iv*) the banks are high, rocky/firm and well defined above the HFL.

6.2.5. Areas to be avoided

As far as possible, attempt should be made to avoid the following areas:

(*i*) unstable hill features and areas having perennial landslide or settlement problems
(*ii*) areas subjected to seepage flow, springs, hydel channels, subterrenean channels, etc.
(*iii*) steep hill sides
(*iv*) areas subject to flooding or waterlogging
(*v*) areas liable to snow drift or avalanches; and
(*vi*) locations involving unnecessary and expensive destruction of wooden areas.

6.2.6. Miscellaneous

Location along a river valley has the inherent advantage of comparatively gentle gradients, proximity of inhibited villages, and easy supply of water for construction purposes. But this solution is set with disadvantages such as the need for a large number of cross-drainage structures and protective works against erosion. These pros and cons should be kept in view while making initial selection of the alignment.

The location should be such that the highway is fully integrated with the surrounding landscape of the area. It would be desirable to study the environmental impact of the highway and ensure that adverse effects are kept to a minimum.

An alignment likely to receive plenty of sunlight should receive preference over the one which will be in shade.

HILL ROAD SURVEYS

6.3. RECONNAISSANCE

General. The reconnaissance survey may be conducted in the following sequence:

(*a*) Study of topographical survey sheets, geological and meterological maps, and aerial photographs where available.
(*b*) Aerial reconnaissance (where necessary and feasible).
(*c*) Ground reconnaissance.
(*d*) Final reconnaissance of inaccessible and difficult stretches.

Study of survey sheets, maps, etc. Reconnaissance begins with a study of all the available maps. In India, topographical sheets are available to the scale of 1 : 50,000. After study of the topographical features on the maps, a number of alignments feasible in a general way are selected keeping in view the guiding principles set forth above.

If photographs of the area are not available, but their need is considered imperative, aerial photography may be arranged for further study in the interest of overall economy. These may be to a scale of 1 : 20,000 to 1 : 50,000 to supplement the information from topographic maps. If stereoscopic techniques are applied, aerial photographs can yield quantitative data, and if studies by a skilled photo-interpreter is made, also significant soil and sub-soil information can be gathered.

Aerial reconnaissance. Aerial reconnaissance will provide a bird's eyeview of the alignments under consideration alongwith the surrounding area. It will help to identify factors which call for rejection or modification of any of the alignments. Final decision about the alignments to be studied in detail on the ground could be taken on the basis of the aerial reconnaissance.

Ground reconnaissance. The various alternative routes found feasible as a result of map and aerial photograph study and aerial reconnaissance are further examined in the field by ground reconnaissance. As such, this part of the survey is an important link in the chain of activities leading to selection of the final route. If possible, a geotechnical engineer should be associated with this phase of the survey work.

Ground reconnaissance consists of general examination of the ground by walking or riding along the probable routes and collecting all available information necessary for evaluating the same.

It will be advantageous to start reconnaissance from an obligatory point situated at the highest level. If an area is inaccessible for ground reconnaissance, recourse may be had to aerial reconnaissance to clear the doubts.

While reconnaissance on the ground, it is advisable to leave reference pegs to facilitate further survey operations.

6.3.1. Points on which Data may be Collected during Ground Reconnaissance

Points on which data may be collected during ground reconnaissance are listed below:

1. Details of route *vis-à-vis* topography of the area.

2. Length of the road.

3. Bridging requirements—number, length.

4. Geometrics.

(*a*) Gradients

(*b*) Curves, hair-pin bends, *etc*.

5. Existing means of communication—mule path, jeep track, *etc*.

6. Right-of-way bringing out constraints on account of built-up area, monuments, and other structures.

7. Terrain and soil conditions.

(*a*) Geology of the area.

(*b*) Nature of the soil.

(*c*) Road length passing through

(*i*) mountainous terrain,

(*ii*) steep terrain,

(*iii*) rocky stretches with induction of the length in loose rock stretches,

(*iv*) areas subject to avalanches and snow drifts, and

(*v*) slip-prone areas,

(*d*) Cliffs and gorges.

(*e*) Drainage characteristics of the area including susceptibility to flooding.

(*f*) General elevation of the road indicating maximum and minimum heights negotiated by main ascents and descents.

(*g*) Total number of ascents and descents.

(*h*) Vegetation-extent and type.

8. Climatic conditions:

(*a*) Temperature—monthly maximum and minimum readings.

(*b*) Rainfall data—average annual, peak intensities, monthly distribution (to the extent available).

(*c*) Snowfall data—average annual, peak intensities monthly distribution (to the extent available).

(*d*) Wind direction and velocities.

(*e*) Fog conditions.

(*f*) Exposure to sun.

(*g*) Unusual weather conditions like cloud-bursts, *etc*.

9. Facilities/Resources:

(*a*) Landing ground.

(*b*) Dropping zones.

(*c*) Foodstuffs.

(*d*) Labour—local availability and need for import.

(*e*) Construction materials (timber, bamboo, sand, stones, shingle, *etc*.)-extent of their availability and loads involved.

10. Value of land—agriculture land, irrigated land, built-up land, forest land, *etc*.

11. Approximate construction cost.

12. Access points indicating possibility of induction of equipment.

13. Period required for construction.

14. Strategic considerations.

15. Recreational potential.

16. Important villages, towns and marketing centres to be connected.

17. Economic factors:

(*i*) Population served by the alignment.

(*ii*) Agricultural and economic potential of the area.

18. Other major developmental projects being taken up in the area *e.g*. hydro-electric projects.

19. Miscellaneous such as camping sites, law and order problems, royalty charges, availability of contractors for collection and carriage of construction materials, working period available for construction work, *etc*.

Final reconnaissance of inaccessible and difficult stretches. Ground reconnaissance may disclose certain difficult stretches which call for detailed examination. A trace cut might be specially made in such sections for inspection. Apart from this, before the alignment is accorded final approval by the competent authority, it may be desirable to have one last round of aerial reconnaissance to resolve the remaining doubts.

Reconnaissance report. Based on information collected during the reconnaissance survey, a report should be prepared. It should include all relevant information collected during the survey, a plan to the scale of 1:50,000 showing the alternative alignments studied alongwith their general profile and rough cost estimates. It should also discuss the merits and demerits including the erosion potential and the expected adverse impact on the environment of the different alternatives to help the selection of one or more alignments for detailed survey and investigation.

6.4. PRELIMINARY SURVEY

General. The preliminary survey consists of pegging the route previously selected on the basis of the reconnaissance survey, cutting a trace 1.0 to 1.2 m wide and running an accurate traverse line along it for the purpose of taking longitudinal and cross-sections and establishing bench marks. The data collected at this stage forms the basis for the determination of the final centre line of the road. For this reason it is essential that every precaution should be taken to maintain a high degree of accuracy. Besides the above, general information concerning traffic, soil conditions, construction material, drainage, *etc*. which may be relevant for fixing the design features is also collected during this phase.

Pegging and trace cut. The line and grade of the selected alternative is pegged and the trace is cut along the pegged route. The gradients to be followed at this stage should be easier than those proposed to be achieved for the road by a margin of 10 to 20 per cent *i.e.*, if 5 per cent gradient is proposed to be achieved on the road, the gradient of the pegged alignment should be 4 to 4.5 per cent. The indication about the grade should be provided at conspicuous locations so as to be easily visible from a distance. This can be done by debarking a portion of any nearby tree and writing there on the number, direction, distance and relative elevation of the peg with paint. Where no trees are available for this purpose, about 2 m high poles, with a cross piece attached thereto giving this information, may be firmly fixed in the ground.

Trace cut consists of 1.0 to 1.2 metre wide track constructed along the selected alignment to facilitate access to the area for inspection and survey. It may not be possible to cut a trace where the pegged route traverses precipices and other areas affected by major landslides. These stretches may, therefore, be detoured by cutting the trace either along the top or bottom periphery of these areas. Further, in rocky areas where cutting of trace may be a difficult task, platforms made out of local timber or bamboos (***mechans***) supported over a '*bally*' framework resting on ledges can be provided as shown in fig. 6.1. These mechans can also be supported by cane or ropes hung from trees above. The mechans can also be constructed in a very dense jungle where trace cut may not be feasible. It is desirable that a senior engineer should walk over the trace cut before further survey work is undertaken to derive benefit from his experience for selection of the best possible route.

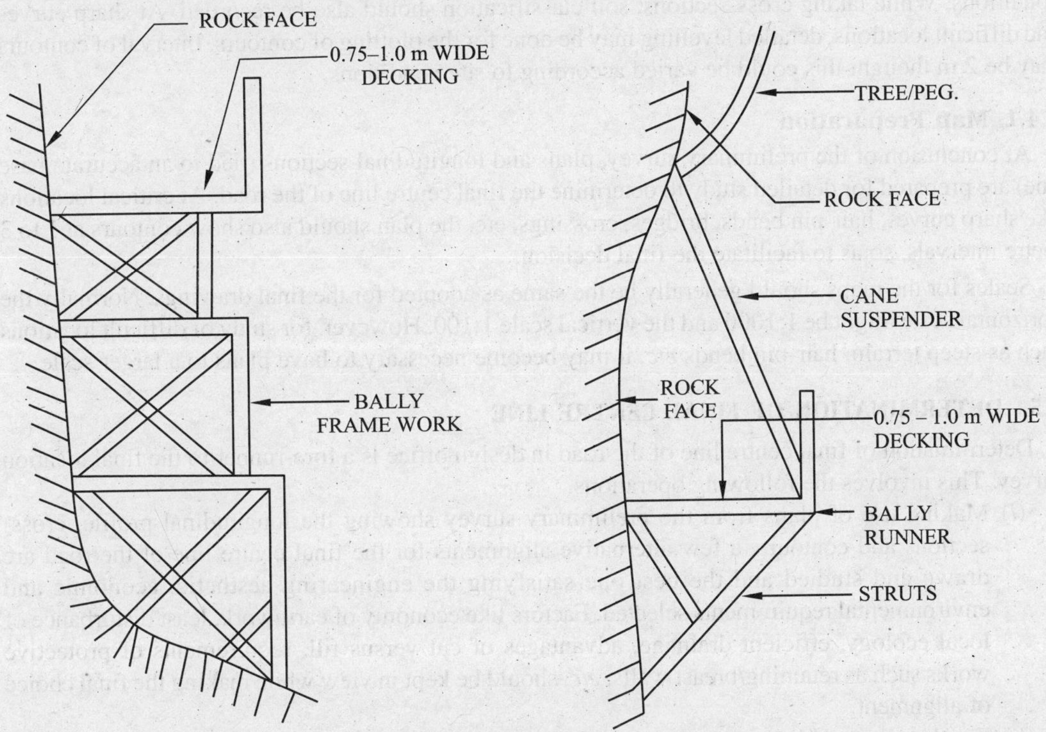

Fig. 6.1 Machans.

Survey procedure. The survey should cover a strip of sufficient width taking into account the degree and extent of cut/fill, with some allowance for possible shift in the centre line of the alignment at the time of final design. In the normal course, a strip width of about 30 m in straight or slightly curving reaches (*i.e.,* 15 m on either side of centre line) and 60 m at sharp curves and hair-pin bends (*i.e.,* 30 m on either side of centre line) should meet the requirement. Strip width could be somewhat less in the case of lower category roads.

Traverse along the trace should be run with a theodolite and all angles measured by double reversal method. Alternatively, this can be done by prismatic compass, particularly for less important roads. Distances along the traverse line should be measured with metallic tape or chain.

No hard and fast rule can be laid down as regards distance between two consecutive transit stations. In practice, the interval will be dictated by directional changes in the alignment, terrain conditions and visibility. The transit stations should be marked by means of stakes and numbered in sequence. These should be protected and preserved till the final location survey.

Physical features such as buildings, monuments, burial grounds, burning places, places of worship, pipelines, power/telephone lines, existing roads and railway lines, stream/river/canal crossings, cross-drainage structures, etc. that are likely to affect the project proposals should be located by means of offsets measured from the traverse line.

Levelling work includes taking ground levels along the trace out at intervals of 10 m and at abrupt changes in slope and also establishing bench marks at intervals of 250 metres, (exceptionally 500 metres) by running check levels on a closed traverse basis independently. While levelling along the centre line, readings of bench marks should also be taken so as to have a cross check in regard to accuracy of field work. It is particularly important that single datum preferably GTS datum should be used to tie up all levels.

Cross-sections should be taken at intervals of 20 m and at points of appreciable change in soil conditions. While taking cross-sections, soil classification should also be recorded. At sharp curves and difficult locations, detailed levelling may be done for the plotting of contours. Interval of contours may be 2 m thought this could be varied according to site conditions.

6.4.1. Map Preparation

At conclusion of the preliminary survey, plans and longitudinal sections (tied to an accurate base line) are prepared for detailed study to determine the final centre line of the road. At critical locations like sharp curves, hair-pin bends, bridges, crossings, etc. the plan should also show contours at 1 to 3 metre ıntervals, so as to facilitate the final decision.

Scales for the maps should generally be the same as adopted for the final drawings. Normally the horizontal scale might be 1:1000 and the vertical scale 1:100. However, for study of difficult locations such as steep terrain, hair-pin bends *etc*. it may become necessary to have plans to a larger scale.

6.5. DETERMINATION OF FINAL CENTRE LINE

Determination of final centre line of the road in design office is a fore-runner to the final location survey. This involves the following operations.

(*i*) Making use of plans from the preliminary survey showing the longitudinal profile, cross-sections and contours, a few alternative alignments for the final centre line of the road are drawn and studied and the best one satisfying the engineering aesthetic, economic and environmental requirements selected. Factors like economy of earthwork, least disturbance of local ecology, efficient drainage, advantages of cut versus fill, requirements of protective works such as retaining/breast walls, *etc*. should be kept in view while making the final choice of alignment.

(*ii*) For the selected alignment, a trial grade line is drawn taking into account the controls which are established by mountain passes, intersections with other roads, railway river crossings, unstable areas, *etc*. In the case of improvements to an existing road, the existing levels are also kept in view.

(*iii*) For the alignment finally chosen, a study of the horizontal alignment in conjunction with the profile is carried out and adjustments made in both if necessary for achieving proper condition.

(*iv*) Horizontal curves including spiral transitions are designed and the final centre line marked on the map.

(*v*) The vertical curves are designed and the profile shown on the longitudinal section.

Where warranted, the alignment determined in the design office may also be cross-checked in the field.

6.6. FINAL LOCATION SURVEY

General. The purpose of the final location survey is to lay out the final centre line of the road in the field based on the alignment selected in the design office and to collect necessary data for the preparation of working drawings. The completeness and accuracy of the project drawing and estimates of quantities depends to a great deal on the precision with which this survey is carried out.

The two main operations involved in the survey are: staking out the final centre line of the road by means of a continuous transit (theodolite) survey; and detailed levelling.

Transit Survey. The centre line of the road, as determined in the design office, is translated on the ground by means of a continuous transit survey and pegging of the centre line as the survey proceeds. All angles should be measured with a transit theodolite. It will be necessary to fix reference marks for this purpose. These marks should be generally 20 m apart in straight reaches and 10 m apart in curved reaches. To fix the centre line, reference pillars/control *burjis* should be firmly embedded in the ground. These should be located beyond the expected edge of the cutting on the hill side. The maximum spacing of reference pillars may be 100 m.

The following information should be put down on the reference pillars:

(*a*) Reduced distance

(*b*) Horizontal distance from the centre line of the road.

(*c*) Reduced level at the top of the reference pillar.

(*d*) Formation level of the road.

The reference pillars should be so located that these will not be disturbed during construction. Description and location of the reference pillars should be noted for reproduction on the final alignment plans. Distance of the reference pillars should be measured along the slope, the slope angle determined with theodolite, and the actual horizontal projection calculated.

The final centre line of the road should be suitably staked, stakes being fixed at 20 metre intervals. The stakes are intended only for short period for taking levels of the ground along the centre line and cross-sections with reference thereto. In the case of existing roads, paint marks may be used instead of stakes.

Distance measurements along the final centre line should be continuous, following the horizontal curves where these occur.

At the road crossings, the angles which the intersecting roads make with the final centre line should be measured with the help of a transit. Similar measurements should be made at railway level crossings.

Bench marks. To establish firm vertical control for location, design and construction, bench marks established during the preliminary survey should be re-checked and where likely to be disturbed during construction re-established at intervals of 250 metres (exceptionally 500 metres), and at or near all drainage crossings.

Longitudinal sections and cross-sections. Levels along the final centre line should be taken at all staked stations and at all breaks in the ground.

Cross sections should be taken at 20 m intervals. In addition cross-sections should be taken at points of beginning and end of spiral transition curves, at the beginning, middle and end of circular curves, and at other critical locations. All cross-sections should be with reference to the final centre line, extend normally upto the right-of way limits and show levels at every 2 to 5 metre intervals and all breaks in the profile.

Centre line profile should normally be continued at least 200 metres beyond the limits of the project. This is intended to ensure proper connecting grade at both ends. With the same objective, profile along all intersecting roads should be measured upto a distance of about 150 metres. Further, at railway level crossings, the level of the rails, and in the case of subways the level of the roof, should be noted. On existing roads, levels should be taken at all points of intersection in order to help the final fixation of profile.

Proper protection of points of reference. The final location survey is considered complete when all necessary information is available and ready for the designer to be able to plot the final road profile and prepare the project drawings. Among other things, field notes should give a clear description and location of all the bench marks and reference points. This information should be transferred to the plan drawings, so that at the time of construction, the centre line and the bench marks could be located in the field without any difficulty.

At the time of execution, all construction lines will be set out and checked with reference to the final centre line established during the final location survey. It is important, therefore, that not only all the points referencing the centre line should be protected and preserved but these are so fixed at site that there is little possibility of their being disturbed or removed till the construction is completed.

In the last stage of alignment survey, hydrological and soil investigations for the route should be carried out. These will enable details of drainage and protective works to be decided.

GEOMETRIC DESIGN

A uniform application of design standards is most desirable from the view-point of road safety and smooth flow of traffic. Optimum design standards should be aimed to reduce the possibility of early obsolescence of the facilities.

Elements of a Roadway in hills and Road-land widths are depicted in Figs. 6.2 and 6.3.

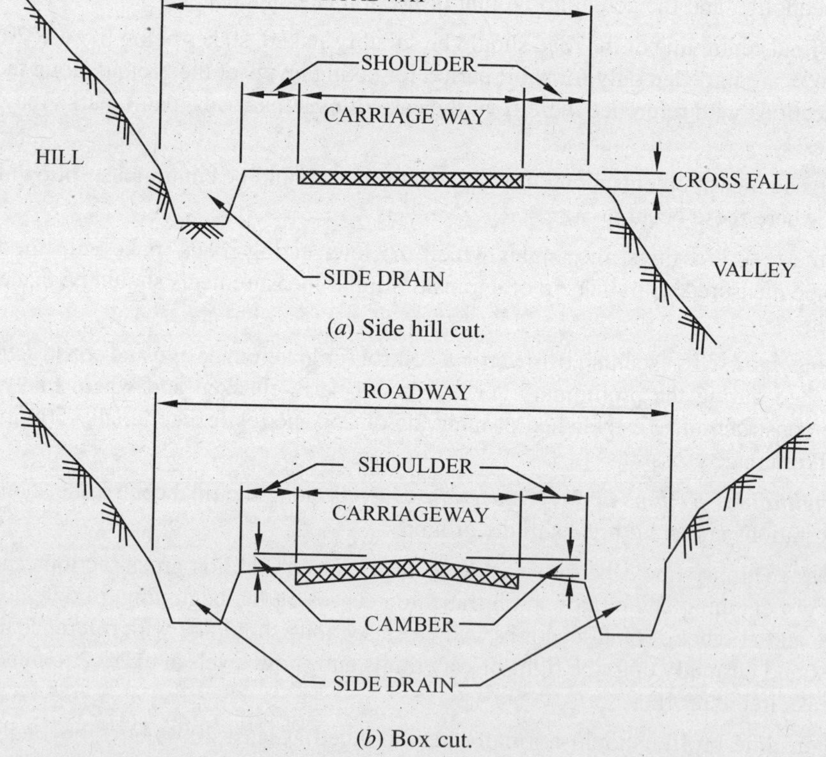

(*a*) Side hill cut.

(*b*) Box cut.

Fig. 6.2 Elements of a roadway.

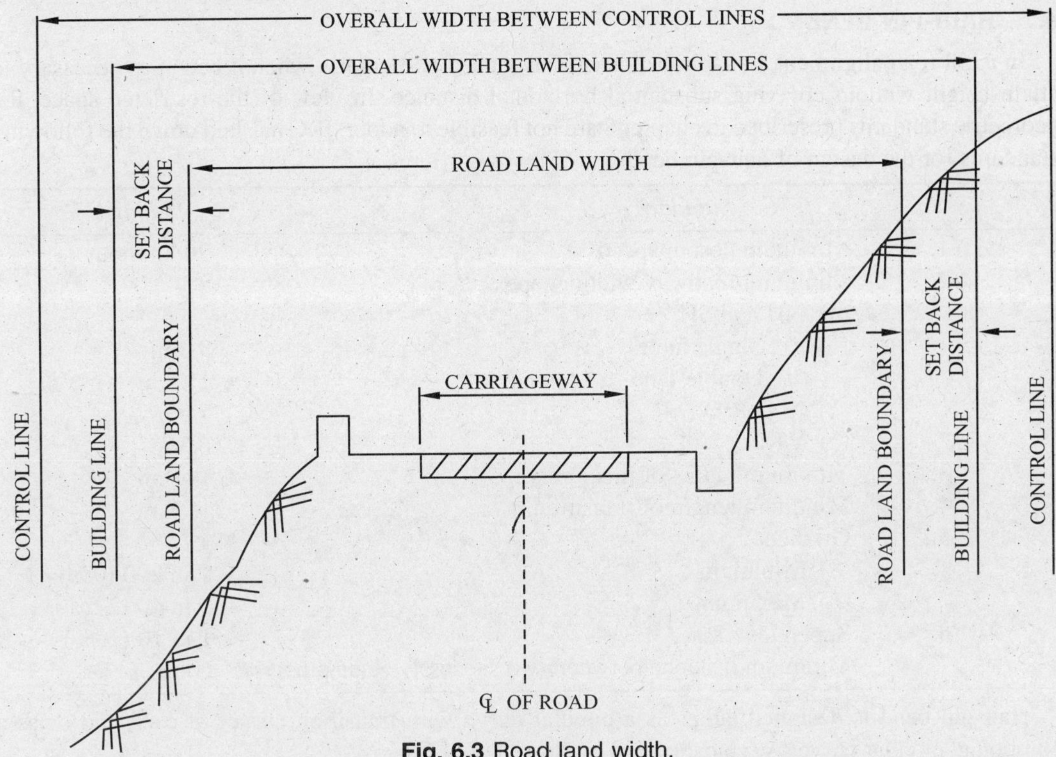

Fig. 6.3 Road land width.

As a general rule, geometric features of a highway except cross-sectional elements do not lend to stage construction. Particularly in the case of hill roads, improvement of features like grade and curvature at a later stage can be very expensive and may sometimes be impossible. Thus ultimate geometric requirements of hill roads should be kept in view right in the beginning.

Development of cross-section in stages should be decided only after very careful consideration, since hill roads need a lot of permanent works like retaining walls, breast walls, catch water drains etc. which may have to be altogether rebuilt.

The design standards recommended in chapter 8 are absolute minimum. However the minimum values should be applied only where serious restrictions are placed on technical or economic considerations. General effort should be to exceed the minimum values as far as possible. Where even the minimum design standards cannot be adopted for inescapable reasons, proper signs should be put up sufficiently in advance to inform the road users of the reduction in design speed.

SPECIAL FEATURES CONNECTED WITH HILL ROADS

In addition to the design standards discussed in chapter 8, there are certain special features connected with hill roads, which need elaboration. The important features are:

(1) Hair-pin bends
(2) Tunnels
(3) Hill road drainage
(4) Protective works:
  (*a*) Breast walls
  (*b*) Retaining walls
  (*c*) Parapet walls
  (*d*) Check walls
  (*e*) Gabion walls
(5) Land slides
(6) Maintenance including snow removal.

6.7. HAIR-PIN BENDS

In a hill road alignment, hair-pin bends are introduced particularly when it becomes necessary to attain height without covering substantial horizontal distance. In view of the restricted space, the geometric standards prescribed in chapter 8 are not feasible to adopt. IRC has laid down the following standards for the design of hair-pin bends.

S.No.	*Standard*	*Limit*
1.	Minimum design speed	20 km/hour
2.	Minimum roadway width at apex	
	(*a*) NH and SH	
	(*i*) Single lane	9.0 m
	(*ii*) Double lane	11.5 m
	(*b*) M.D.R/O.D.R	7.5 m
	(*c*) V.R.	6.5 m
3.	Minimum radius of inner curve	14.0 m
4.	Minimum length of transition	15.0 m
5.	Gradient	
	(*a*) Minimum	1 in 200 (0.5%)
	(*b*) Maximum	1 in 40 (2-5%)
6.	Superelevation	1 in 10 (10%)
7.	Minimum distance between two successive hair-pins	60 m

Hair-pin bend is designed either as a circular curve with transition curves at each end or as a compound circular curve. A typical design is illustrated in Fig. 6.4.

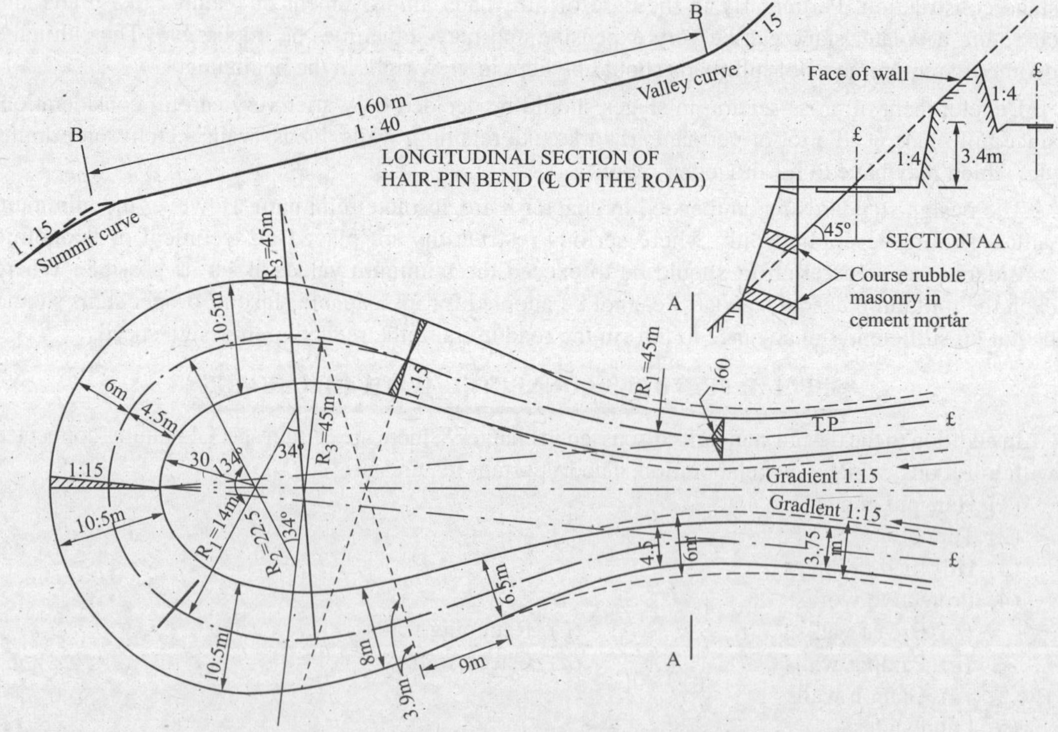

Fig. 6.4 Typical hair-pin bend.

6.8. TUNNELS

Provision of a tunnel becomes necessary under the following situations:

(1) The cost of an open cut through a very high and hard rocky stretch becomes exorbitantly high as compared to the cost of a tunnel.

(2) When a sudden high rocky cliff is encountered along a hill face.

(3) When the route becomes very circuitous and long due to steep climb and descent over a long stretch of road and it is economical to cut across the mountain range.

(4) The normal route encounters areas of avalanches, snow cover, reckfalls etc. and the provision of a tunnel at a lower altitude can avoid these locations.

(5) Tunnels are provided under rivers and channels.

The disadvantages of providing a tunnel are:

(1) Cost is very high

(2) There is a problem of maintaining lighting and ventilation.

(3) Seepage water, if any, will cause trouble for clearance.

6.9. TYPES OF TUNNELS

These are of two types:

(1) Half tunnels

(2) Full tunnels

6.9.1. Half Tunnels

In very hard rocky stretches, where formation width is to be provided by rock blasting, it is a normal practice to provide half tunnel, giving due regard to the minimum head clearance. The rock above the half tunnel is left as an overhang, as shown in Fig. 6.5.

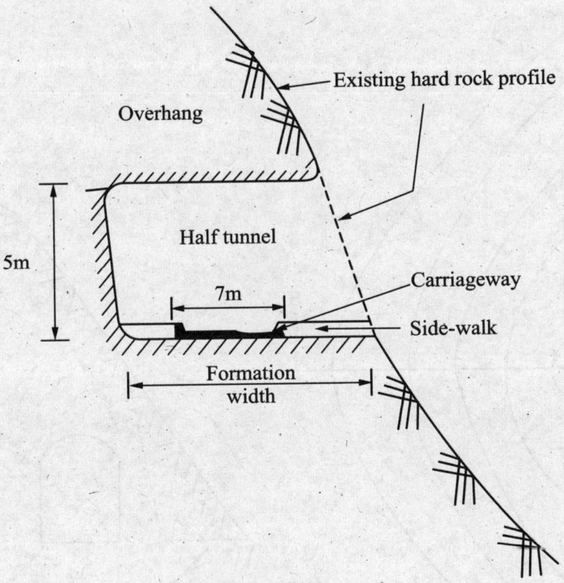

Fig. 6.5 Half tunnel.

6.9.2. Full Tunnel Design Considerations

(1) Align the tunnel along a straight section.

(2) If horizontal curves become absolutely essential provide a minimum radius of 200 m. In exceptional cases, this may be reduced to 100 m.

(3) For effective drainage, provide a longitudinal gradient of 1 in 50. This may be provided

continuously, if the length of the tunnel is less than 400 m, otherwise provide gradient in two stages. Maximum permissible gradient for long tunnels is 1 in 25.

(4) The minimum carriageway should be two lanes *i.e.* 7 m.

(5) Additional sidewalks of 75 cm should be provided on both sides.

(6) Minimum vertical clearance should be 5 m.

(7) Provision of ducts for inlet and exhaust ventilation be made in the circular portion above 5 m; especially for tunnels longer than 400 m. The fresh air input should be at least 0.3 m^3/m length of tunnel and velocity of air limited to 5 m/sec.

(8) The lighting system should be such that the driver entering / leaving the tunnel should be capable of adjusting with the artificial / natural light respectively.

A cross-section of the tunnel incorporating the above design standards is depicted in Fig. 6.6.

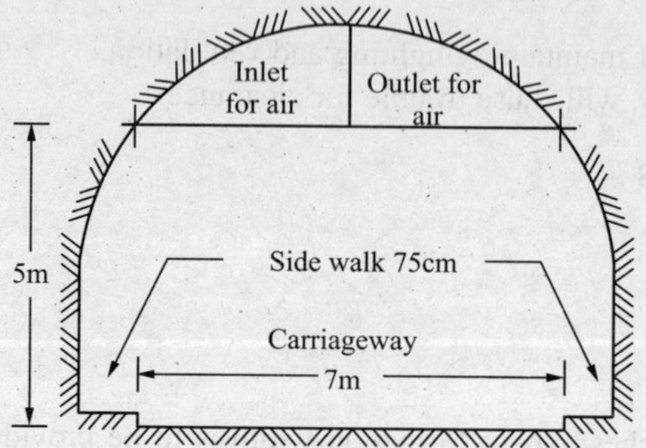

Fig. 6.6 Full tunnel cross-section.

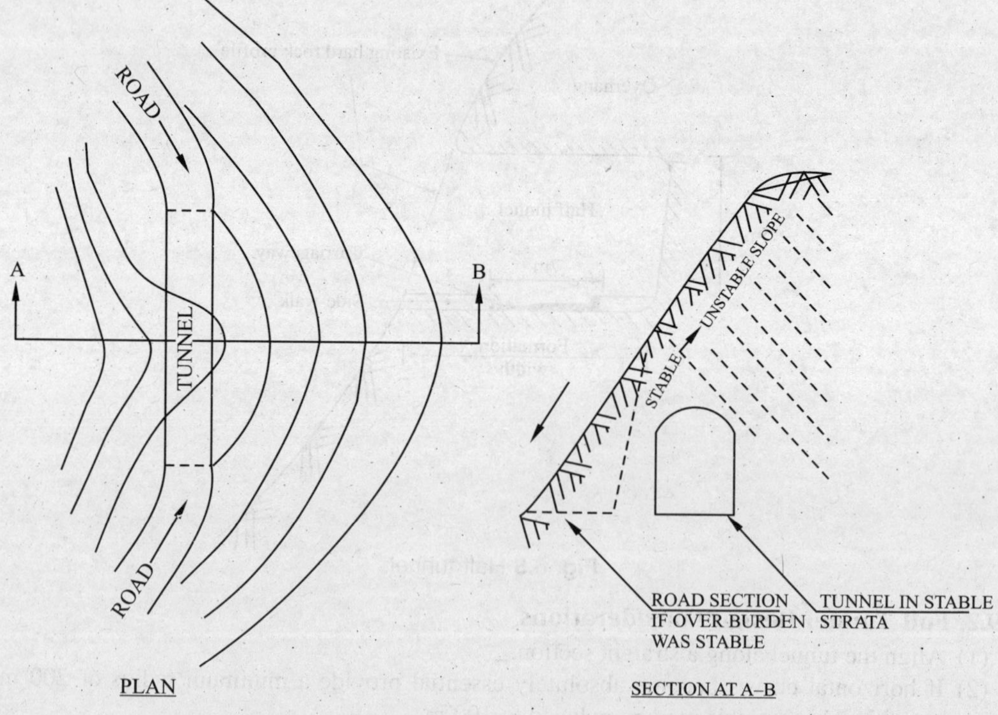

Fig. 6.7 Plan and section of a tunnel to avoid unstable slope.

6.10. HILL ROAD DRAINAGE

For proper maintenance of hill roads, an effective drainage system is very important. In the case of hill roads surface water is drained by the provision of side drains and catch water drains.

In the plains, side drains are provided on both sides of the road to drain off water. In the case of hills, drains can be provided on both sides of the roadway only where the road runs in a through cutting. This is not possible in the case of side cutting. So side drain is recommended only on one side. Moreover, if any drain is located by the side of the road for surface water drainage, it will reduce the effective width of the roadway. It is, therefore, considered essential that the drain should be of such a form that it could function both as a drain and also as a part of the road surface in emergency. Thus, if a vehicle is forced to move to the extreme edge of the road (*i.e.*, into the drain) to avoid accident, it should be able to come out easily. On this account, side drain should be either angle or saucer or kerb and channel type as shown in Fig. 6.8. The minimum depth should be 30 cm, but the section should be suitably enlarged to meet the actual requirement of the discharge. Where there are soft formations, the drains should be suitably lined.

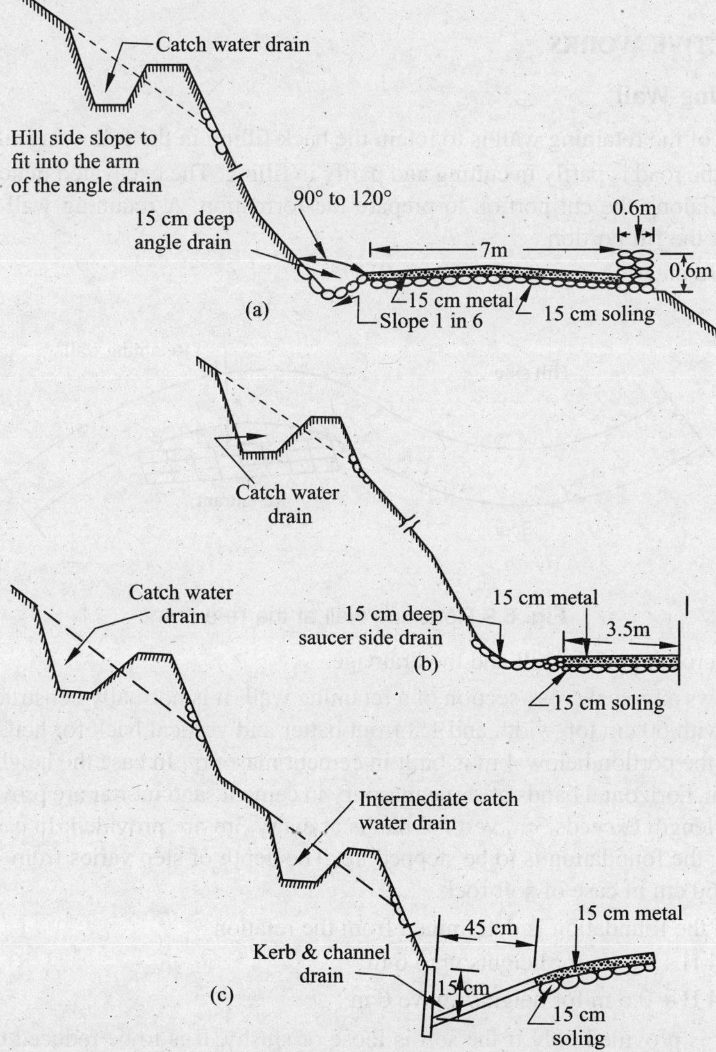

Fig. 6.8 Typical cross-section of hill-roads with (a) angle side-drain (b) saucer side-drain and (c) kerb and channel side-drain.

When the road goes in steep side long ground, the water from the upper slope would rush on the road and wash it away. This is prevented by interrupting the water at some distance away from the road edge by means of what are commonly known as catch drains. Thus where slips are likely to occur and where large quantities of the surface flow are likely to come on to the road, catch water drains should be provided. These drains are provided on the same side as the side drain, usually laid along the contours up the hill, to intercept the run-off from the hill slope and divert it to the nearest cross-drainage work such as a culvert or sopper or to a natural stream. A careful watch should be kept on the condition of these drains in soft strata so that no damage is done to the hillside by percolation of water from these drains. The lower side and bottom of a catch water drain should be paved with rough stone or brick if the soil is easily scourable. These should not be located closer than 4.5 m from the edge of the road to avoid land-slide due to water seeping through down the slope. These catch drains are 1 m by 1 m generally. When one catch water drain is not sufficient to take care of total run-off, two or more catch water drains may be provided. The other drains are called intermediate catch water drains. Where provided they should be connected at intervals to the roadside drains by means of counterfort drains.

6.11. PROTECTIVE WORKS

6.11.1. Retaining Wall

The function of the retaining wall is to retain the back filling in the following situations:

(*i*) When the road is partly in cutting and partly in filling. The excavated material is dumped or stacked along the cut portion to prepare the formation. A retaining wall is constructed to support the fill portion.

(*ii*) At the re-entering curves (See Fig. 6.9).

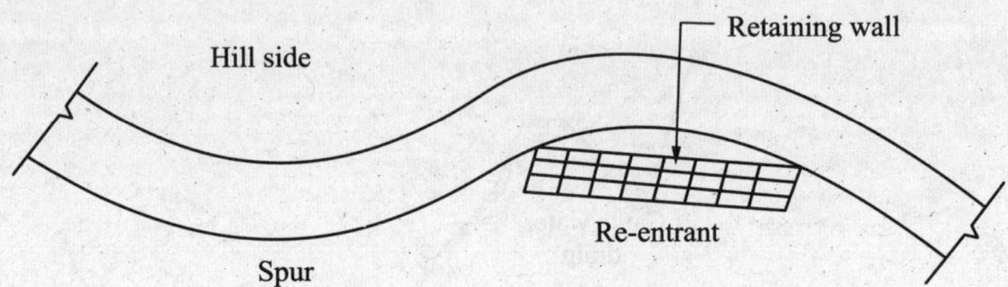

Fig. 6.9 Retaining wall at the re-entrant.

(*iii*) At the crossing of the hill and the drainage.

Fig. 6.10 shows a typical cross-section of a retaining wall. It is normally constructed in dry rubble stone masonry with 60 cm top width and 1:3 front batter and vertical back for heights upto 4 m. For greater heights, the portion below 4 m is built in cement masonry. In case the height of the retaining wall exceeds 4 m, horizontal bands of stone masonry in cement sand mortar are provided at every 3m. Similarly if the length exceeds 3m, vertical bands at every 3m are provided. In case the foundation consists of rock, the foundation is to be stepped up. The depth of step varies from 15 cm in the case of hard rock to 60 cm in case of soft rock.

The width of the foundation is determined from the relation

$B = 0.4\ H + 0.3$ m for heights upto 6 m

and $B = 0.4\ H + 0.6$ m for heights above 6 m

Full width B is provided only if the soil is loose or slushy. It is to be reduced to X-Y where the foundation soil is moorum, moorum mixed with boulders or stiff clay.

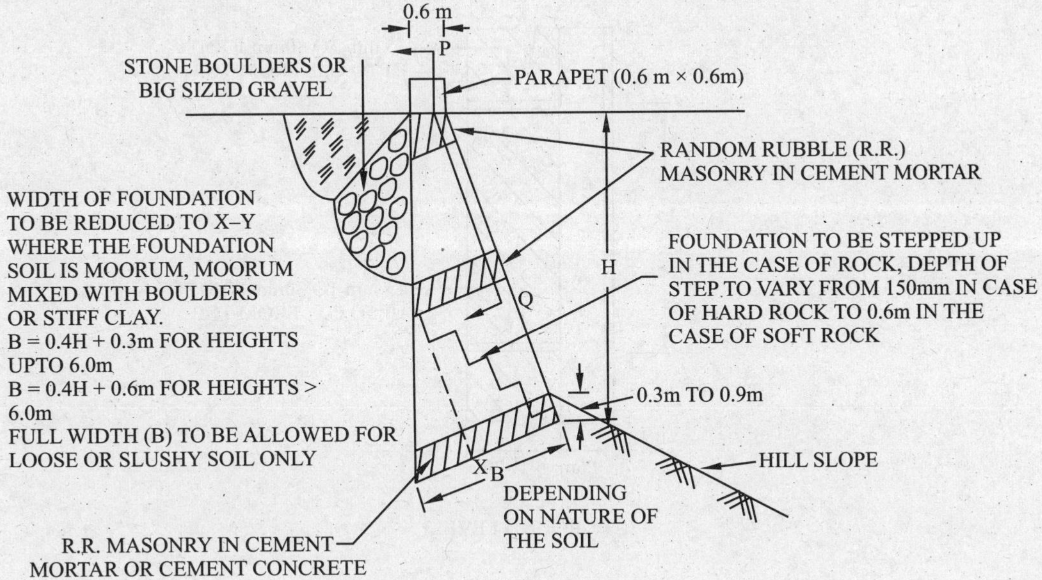

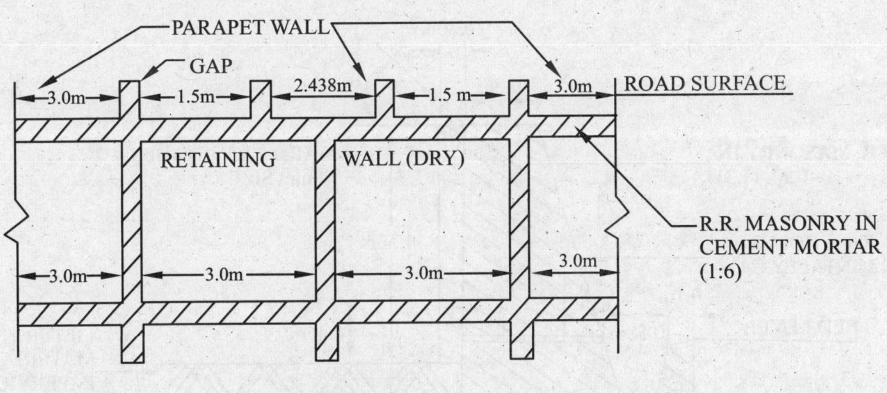

Fig. 6.10 Cross section and L-section of a retaining wall.

6.11.2. Breast Wall

The function of the breast wall is to support the uphill slope and prevent the cut-hill from sliding. This should be strong enough to withstand the earth pressure of the soil behind, along with the surcharge caused by the slope. Thus its cross-section depends upon its height and the angle of the slope.

To facilitate underground water drainage, weep holes are provided at 1 m vertical height and 1.2 m centre to centre horizontally. Sometimes vertical gutters connecting weep holes to side drains are provided.

In case the height of the breast wall is upto 2 m, it is entirely laid in rubble stone masonry. If the height exceeds 2m, only the top 2m is constructed in stone masonry and the lower portion of coursed masonry in 1 : 3 cement sand mortar.

Normally the front batter is 1 in 3 and the back vertical Fig. 6.11 depicts a typical breast wall at ground level and 1.5 m above ground level. The back batter may be provided in one straight batter 1 in 2 or by means of projections on the back, as shown in Fig. 6.12.

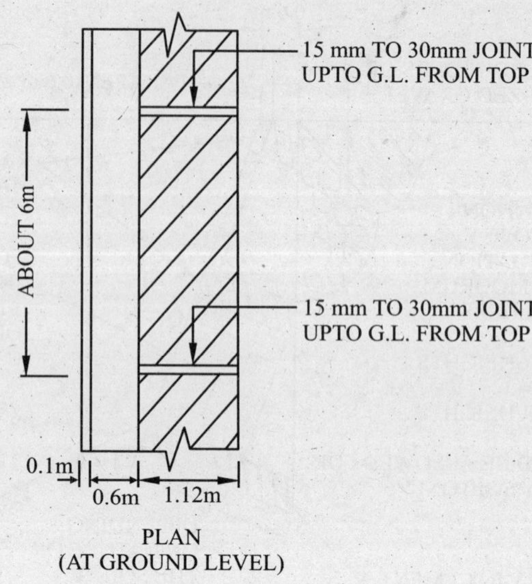

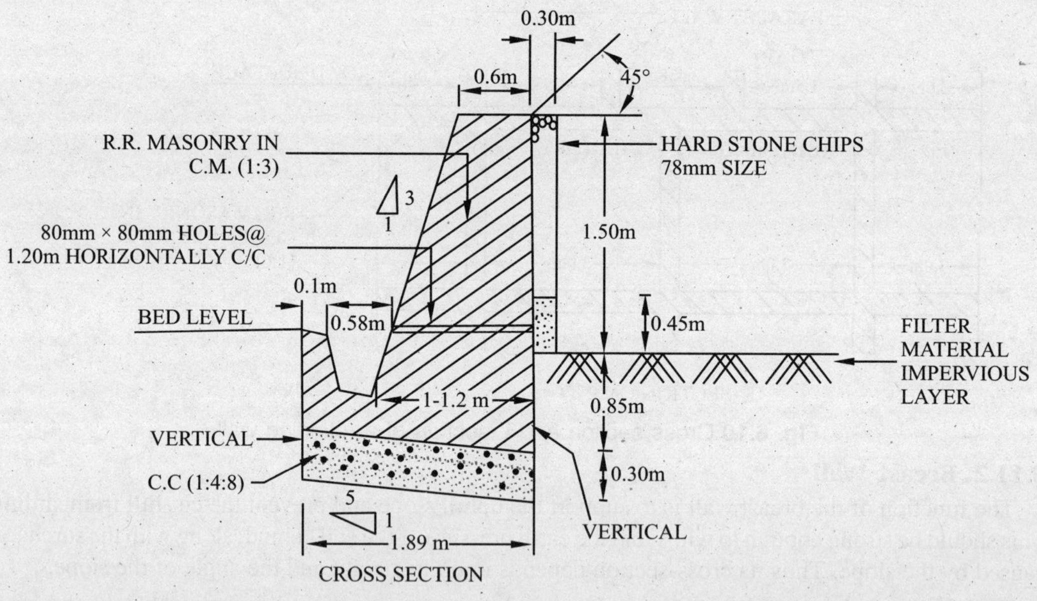

STANDARD 1.50m HIGH BREAST WALL

Fig. 6.11 Typical breast wall.

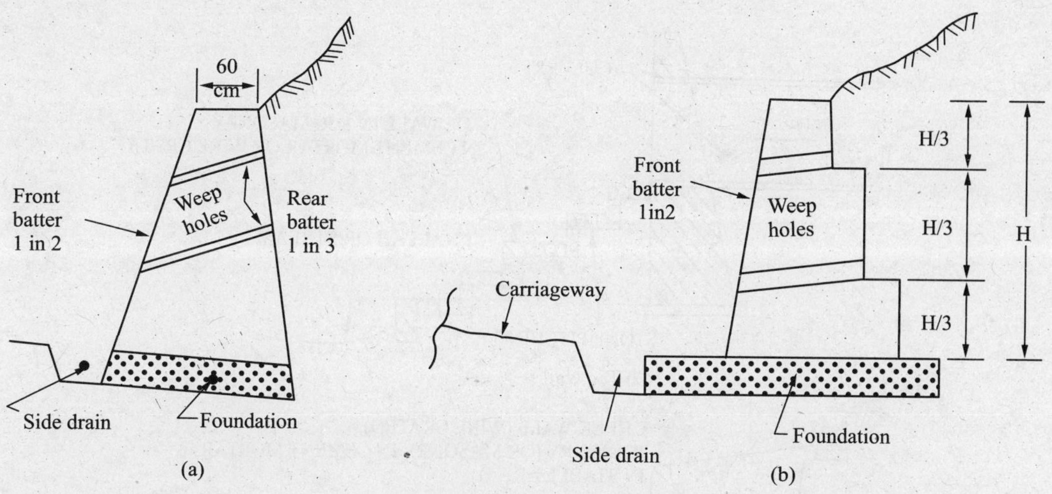

Fig. 6.12 Breast wall.

6.11.3. Parapet Walls

Parapet walls are needed to provide a sense of security psychologically to the drivers and the motorists while travelling on hill roads with steep valley slopes (*Khud*). These also prevent the wheels of the vehicles coming on the retaining wall.

There are mainly two types of parapets used in hill roads, namely:

(*i*) Wall type; and

(*ii*) R.C.C. post type

The wall type may be rectangular with a cross-section of 60 cm × 60 cm and length of 3.6 m, with a gap of 1.5 m. Trapezoidal shape is also provided with the slanting face towards the road. The slanting portion is sometimes filled with clay to serve as a flower bed in which turf or flowers are grown. The top portion of the parapet is given a weathered slope of 2.5 cm.

The gap of 1.5 m between the parapet walls serves the following purposes:

(*i*) To facilitate natural drainage, road maintenance and repairs including slip clearance

(*ii*) Reduces cost of construction, as these are provided at intervals.

R.C.C. post type are 1.45 m in height and 15 cm × 15 cm in cross-section. These are kept 1 m above the G.L. and spaced at 1 m interval.

6.11.4. Check Walls

These are the retaining structures along the hill slope at suitable intervals to check the slides or flow of avalanches (sliding snow mountains). They add to the overall stability of the hill face. The height is normally kept 1.5 m, width at top 60 cm and the bottom width 1 m. The top and bottom 15 cm are laid in 1:3 cement sand mortar and the rest in rubble stone masonry. Fig. 6.13 illustrates the use of check walls.

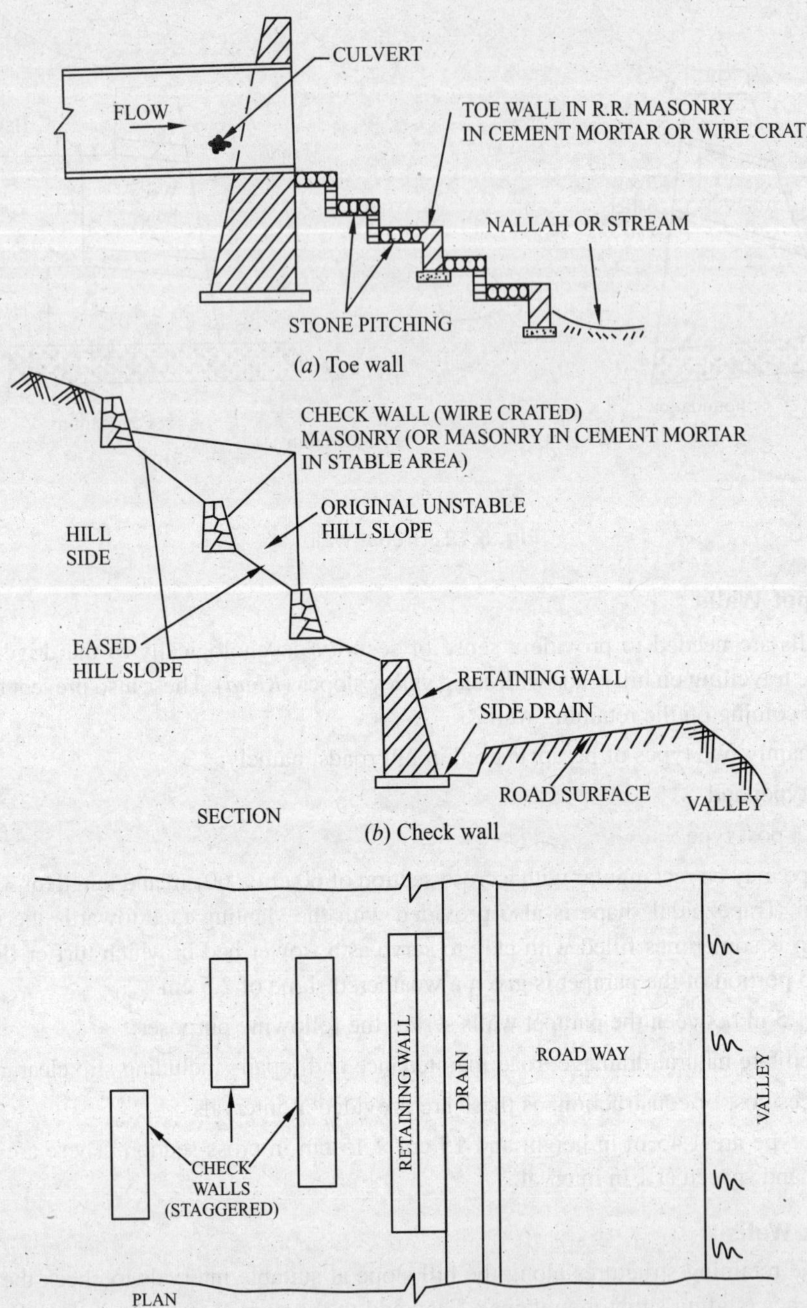

Fig. 6.13 Toe and check wall.

6.11.5. Gabion Walls

When the protective works such as retaining walls, breast walls, check walls are constructed in dry stone masonry, they are known as Gabion walls. These structures are flexible and can adjust to the settlement or disturbance that occur and thus do not get damaged. When the road runs parallel to the stream, gabion walls act as toe-protection walls.

6.12. LAND SLIDES

Land slides are a common feature in hill roads, particularly in the Himalayan ranges because these are geologically young formations and are subjected to settlements and disturbances. It has been found that the behaviour of the hill after excavation and formation of road is entirely different than what it was before excavation. This is due to the fact that the natural conditions get disturbed, the slopes become unstable and results in slips, subsidences and land slides. The stability of the slope depends upon the following factors:

1. Nature of slope whether barren or covered with vegetation.
2. Angle of slope whether less or greater than the angle of repose of the soil.
3. Geological conditions such as the composition of rock, the direction and angle of dip, grain size distribution of the soil.
4. Ground water conditions.

The unstable hill slope results in land-slides.

Factors resulting in land-slides can be grouped as:

(1) those due to increased shear stress or

(2) those due to decreased shear stress.

The factors resulting in increased shear stress are:

1. Erosion by streams, rivers and glaciers resulting in removal of underlying or lateral support.
2. Quarrying of stone which removes partly the lateral support.
3. Weight of rain, hail or snow, resulting in increased surcharge.
4. Earthquakes.
5. Weight of water stored in reservoirs, buildings *etc.* causing increased surcharge.
6. Increase in pore-water pressure.

Factors which results in the reduction of shear stress are:

(1) Faults, fissures, joints, cleavage planes.

(2) Increase in moisture content due to changes in the properties of the material.

(3) Progressive creep causing weakening of the soil structure.

(4) Weakening caused by burrowing animals and roots of trees.

(5) Increased percolation of water due to deforestation and removal of vegetable cover.

6.12.1 Preventive Measures

There are a number of ways by which land-slides can be prevented, the treatment depending upon the particular conditions prevailing at site. Fig. 6.14 shows the plan and section of some of the control measures for a slide. Some of the remedial measures are:

1. Provision of effective surface drainage system such as catch water drains, side drains and cross drainage works.
2. Provision of effective underground drainage system such as French drains, weep holes *etc.*
3. Provision of fencing to prevent grazing by animals.
4. Afforcstation.
5. Grouting of the weak rock.
6. Constructing breast walls, check walls, toe walls, *etc.*
7. Wire netting to prevent erosion.
8. By reducing angle of slope to less than the angle of repose.
9. Bitumen-mulch treatment.
10. Vegetation growth
11. Chemical treatment.

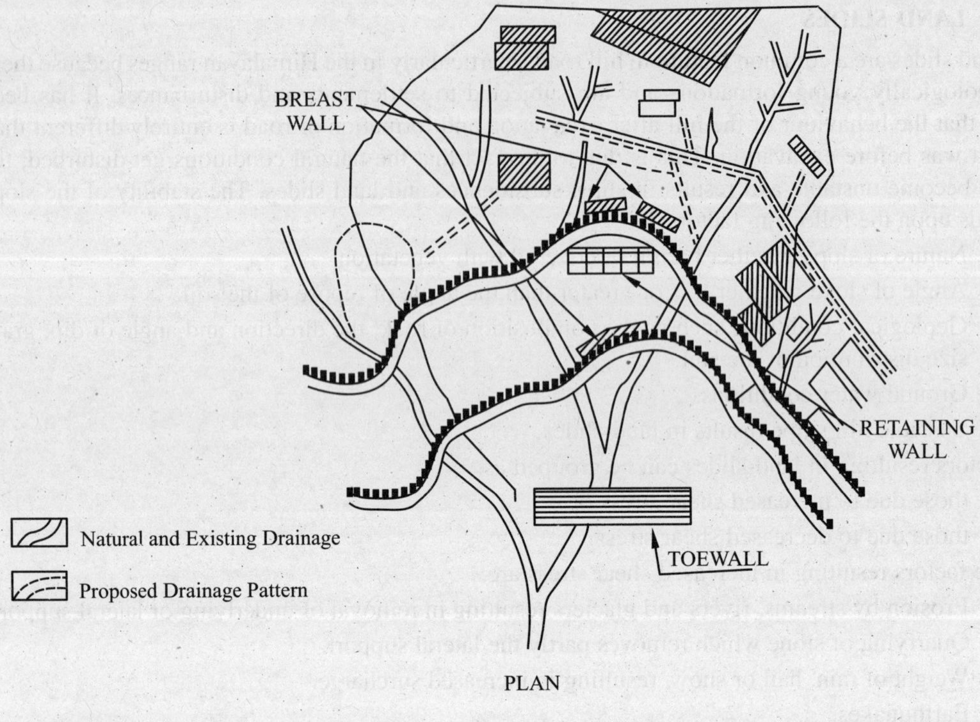

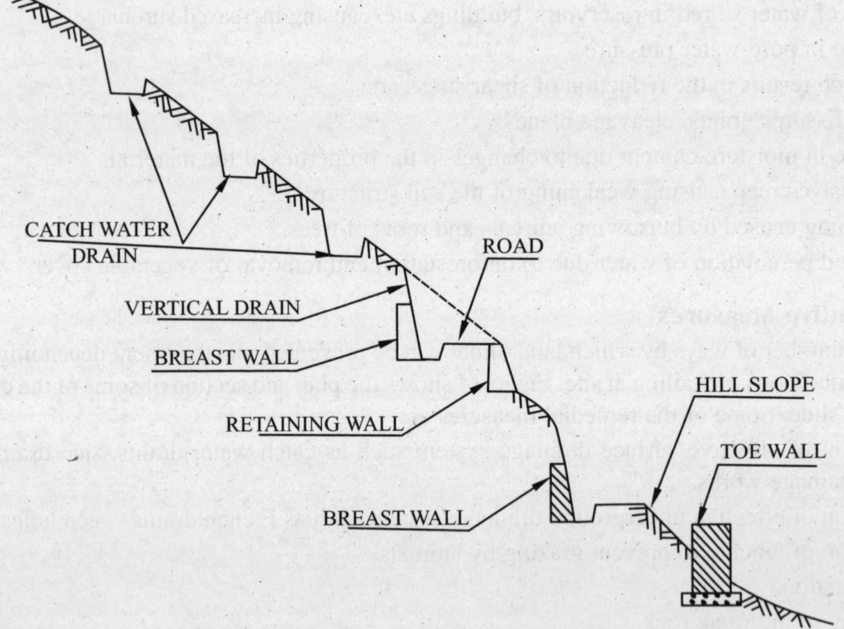

Fig. 6.14 Control measures for a slide-plan and section.

6.13. MAINTENANCE OF HILL ROADS INCLUDING SNOW REMOVAL

The maintenance procedure described in chapter 24 applies to hill roads also, but the hill roads maintenance has the following additional features:

(1) Slip clearance

(2) Special Repairs to Monsoon Damage (SRMD)

(3) Snow clearance

6.13.1. Slip Clearance

The soil on the uphill side slides occasionally during or after the rain and blocks the traffic. Sometimes these slips are very large. Effort should be made to clear the road for smooth flow of traffic either through the maintenance gang or additional muster roll labour or by auctioning the work and getting it done immediately to restore the road communication at the earliest.

6.13.2. Special Repairs to Monsoon Damages (SRMD)

These repairs are a unique feature of hill roads and are more extensive at early stages after construction. The damages caused by rains during the monsoon cannot be rectified within the maintenance grant. The usual practice followed in the hilly regions is to carry out the minimum repairs out of the maintenance grant and later on reappropriate these expenditures to the SRMD head after the sanction of the estimate. Major portions of the repairs are carried out after the sanction.

The SRMD estimate is initiated immediately after the rains along with the lowest tender so that the actual work can be undertaken when the working season in the hills starts sometimes in October. The proposals for widening or improvement to the road surface also, sometimes, is included in the estimate for SRMD.

6.13.3. Snow Clearance

The topic of snow and its clearance has been dealt in chapter 24.

❑❑❑

Chapter

DRAINAGE AND DRAINAGE STRUCTURES

GENERAL

One of the major causes of the failure of highways is water. It has been appreciated since roads were first built that their stability can only be maintained if the surface and foundation remain in a relatively dry condition. Water brings about the destruction of highways by:

(*i*) Softening the road surface when constructed of soil or sand-clay or gravel or water bound macadam.

(*ii*) Washing out unprotected areas of the top surface, erosion of side slopes forming gullies, erosion of side drains, *etc*.

(*iii*) Generally softening of the ground above and below the road where it goes along the high ground, giving rise to land slides or slips.

(*iv*) Softening the subgrade soil and decreasing its bearing power.

Preventive measures include changing or training of water courses, intercepting and leading off water, bank protection and soil treatment. Drainage works designed to protect the road from these effects may be grouped under the following heads:

(*i*) Interception and diversion of the surface water which would otherwise flow across the road or along it and cause erosion.

(*ii*) Interception and rapid removal of seepage or sub-surface water.

(*iii*) Interception and disposal of natural drainage water under the road surface.

These are termed as surface drainage works, sub-surface or subsoil drainage works and cross drainage works respectively and are discussed below:

SURFACE DRAINAGE WORKS

Surface drainage works provide for the removal of surface water flows such as from rainfall or other causes. Surface drainage comprises: (*i*) collecting water from the pavement; and (*ii*) disposing it off.

The collection of water falling on the road surface is made by allowing it to flow to the sides by providing a camber to the road, that is giving a cross-slope from the crown to the road edge. The road surface is also made impervious as far as possible to prevent the water entering sub-surface. At curves, super-elevation (raising of one edge of the road above the other) helps to allow the water to flow laterally or obliquely.

Further disposal of this water depends upon the topography and land use and will be considered under the following conditions:

(*a*) Drainage in rural plain area

(*b*) Drainage in urban plain area

(*c*) Drainage in hilly terrain

7.1. DRAINAGE IN RURAL PLAIN AREA

In rural plain area, the disposal of water depends on whether the road is in embankment or in cutting or on ground line.

When the roadway is in bank, it is a common practice to allow the flow to continue across the shoulder and down the side slopes to the natural ground, or side drain. Where the side slopes, are properly protected by turf or grass and flow from the road surface is distributed to give a fairly uniform sheet of water, erosion is very little. Where slopes are unprotected and sheet flow is not possible or pavement or shoulder irregularities concentrate the water into small streams, slopes are washed badly. In such cases, it is necessary to provide other means of protecting the shoulders and side slopes. One method is to retain the water at the outer edge of the shoulder as shown in Fig. 7.1 (*a*).

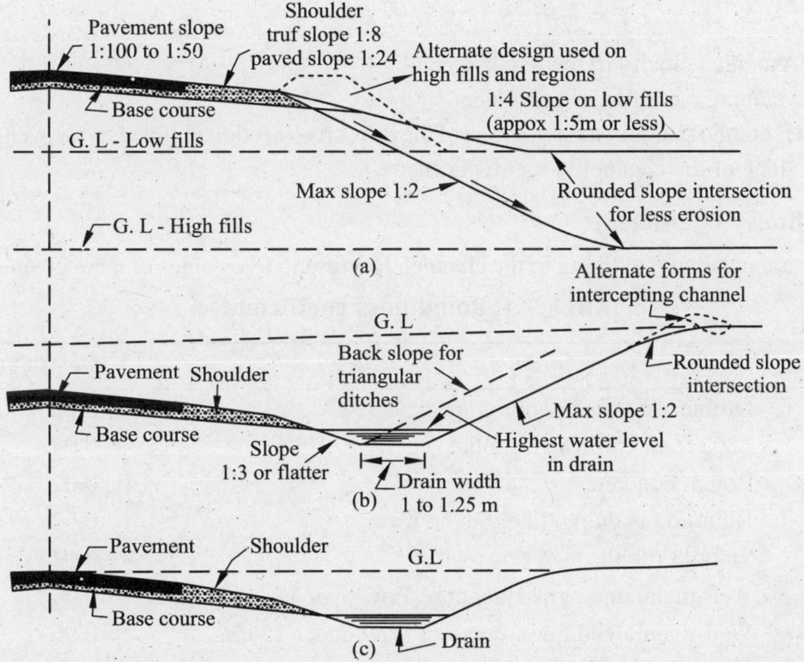

Fig. 7.1 Typical highway cross-sections, with good drainage features (a) in embankment, (b) in cutting and (c) on ground level.

In another method, a kerb is provided at the outer edge of the pavement. This carries the water, along the pavement to each basin or other collecting devices from which it is removed and led down the slopes to the natural ground or side drain. This design is not generally recommended since the water on the pavement edge creates some hazard to traffic, especially when frozen.

When the road is in cutting or on ground line, water is disposed of to the side drain, as shown in Fig. 7.1 (*b*) and (*c*) which should slope towards the nearest water course to flow away from the road. Side drains are required where the ground by the side of the road is at a higher level because if the ground is lower, the water will flow away by natural drainage. In the case of a long high embankment of poor soil, as in the case of approaches to bridges, the surface water can be collected through *V*-shaped drains near the edge of formation and then led down the slope by means of paved drains. Aprons should be provided at the foot of the slope. On low embankments the slopes can be protected by turfing or vegetation and the velocity also reduced by giving flat slopes corners. Convex corners at edges of embankment should be avoided. They should be rounded off. On side long ground, it is seldom necessary to have a drain at the top of the bank.

The water which enters the pavement or the berms is intercepted by providing a granular base below the surface, as shown in Fig. 7.1. Usually, this granular base is omitted in India and the berms are made up of ordinary soil which is turfed. Care should be taken that the depth of grass should not be deep otherwise it will obstruct the flow of water.

7.1.1. Design of Side Drain

Design for side drains and culverts partially full, is based on the principle of flow in open channels. Manning's formula is used for determining the velocity, assuming the flow to be steady and uniform. This formula assumes that the channel itself should be uniform and the slope of water surface constant and parallel to the slope of the bottom of the channel.

The formula is

$$V = \frac{1}{n} R^{2/3} S^{1/2}$$

where V = Average velocity in metre per second

n = Manning's roughness coefficient

R = Hydraulic radius in metres = area of the flow cross-section, divided by the wetted perimeter

S = Slope of the channel in metre per metre

7.1.2. Roughness Coefficient

It depends upon the type of lining to the channel. Representative values of n are given in Table 7.1.

TABLE 7.1 Roughness coefficient, n

Type of lining	*Value of n*
Ordinary earth, smooth graded	0.02
Jagged rock or rough rubble	0.04
Rough concrete	0.02
Bituminous lining, likely to be wavy	0.02
Smooth rubble	0.02
Well-maintained grass-depth of flow over 15 cm	0.04
Well-maintained grass-depth of flow under 15 cm	0.06
Heavy grass	0.10

Nomographs have also been prepared for the solution of Manning's formula; the one developed in India is shown in Fig. 7.2.

The velocity in the channel should be such that it will neither cause silting nor erosion.

Thus, the value of discharge can be found out as follows:

$$Q = VA = \left(\frac{1}{n} R^{2/3} S^{1/2}\right) A$$

where Q = discharge in cubic metre per second

A = area of flow cross-section in square metre.

The channel should be designed not only to accommodate the flow of water but also to provide sufficient waterway. To prevent erosion of the shoulder and undercutting of the toe of the slope, the cross-section of the channel should be selected in such a manner that the cost of construction and maintenance is low. Side slopes 2:1 or flatter are to be adopted except in rock or lined channels. In the case of unlined channels generally a channel which requires the least excavation is the best. Considerable care should be exercised in adopting a suitable value of n. Table 7.1 only serves as a guide but due consideration should be given to the channel condition, material of the bed and slopes, meandering of the channel and aquatic growth. When velocities are not excessive, earth or sodded side drains of triangular or trapezoidal cross-sections may be used. These have to be paved where ditch gradients are steep and are likely to be eroded. Rectangular, semicircular and other types of cross-sections may be adopted in the case of paved drains.

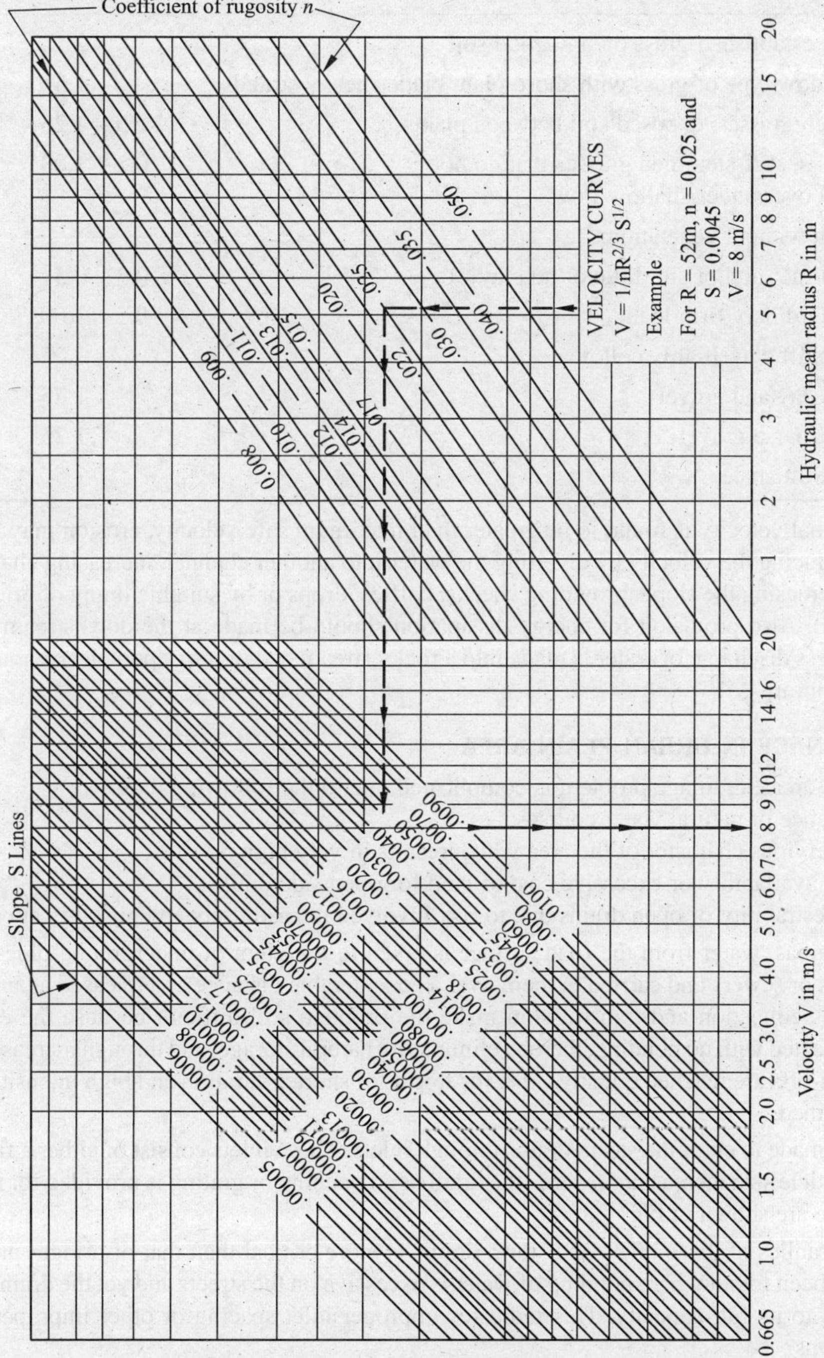

Fig. 7.2 Nomograph for the solution of Manning's equation.

Channel design is considered complete only when the possibility of erosion is eliminated. In this connection, it is necessary first to check the actual velocity against maximum safe velocities for unlined channels given in Table 7.2.

Table 7.2 Allowable velocities for different types of lining

S.No.	*Type of lining*	*Allowable velocity m/s*
1.	Well established grass on any good soil	2
2.	Meadow type of grass with short, plant blades heavy strands	1.5
3.	Bunch grasses, exposed soil between plants	0.5 – 1.25
4.	Grains, stiff-stemmed grasses that do not bend over under shallow flow	0.5 – 1.00
5	Earth without vegetation:	
	(*a*) Fine sand or silt, little or no clay	0.3 – 0.5
	(*b*) Ordinary firm loam	0.5 – 1.00
	(*c*) Stiff clay, highly colloidal	1.25
	(*d*) Clay and gravel	1.25
	(*e*) Coarse gravel	1.25
	(*f*) Soft shale	1.5

If the actual velocity is found to be greater than maximum safe velocity, erosion may be checked either by reducing the velocity by diverting the water into another channel, increasing channel length and thus decreasing the slope, providing checks, baffles, drops or by suitable lining of bricks, rubble, concrete, *etc*. Also provision for energy dissipation should be made at the downstream end of the channel unless discharge of water is made into a rocky stream bed or into a pool deep enough to force a hydraulic jump.

7.2. DRAINAGE IN URBAN PLAIN AREA

In Urban areas, drainage problem is complicated due to the following reasons:

(*i*) Absence of natural water courses.

(*ii*) Impervious character of the area which results in very high run off.

(*iii*) Non availability or excessive cost of land for open side drains.

(*iv*) Undesirability of open drains due to risk involved to road users and unsighty appearance.

In such areas, water from the road surface is taken to a system of underground pipes known as storm drains or sewers and carried in them over a considerable distance and released again as surface run-off. This collection and transmission make the problem still difficult because the entire water must be collected with no ponding, thereby eliminating natural storage, and through increased velocity, the peak run-offs are reached quickly. Also for storms of shorter duration and high intensity peaks are quickly attained.

Water is made to enter the storm drains through inlets. These inlets consist of either a filter system with the particle size increasing towards the drain or an opening or grating is provided for the entry of water. These are shown in Fig. 7.3.

The hydraulics of these inlets and storm drains is more critical than that of bridges and culverts. Cases have been found where ponding of water is excessive on the streets and yet the drainage system is not taxed to its full capacity. This indicates improper inlet spacing or other improper hydraulic considerations.

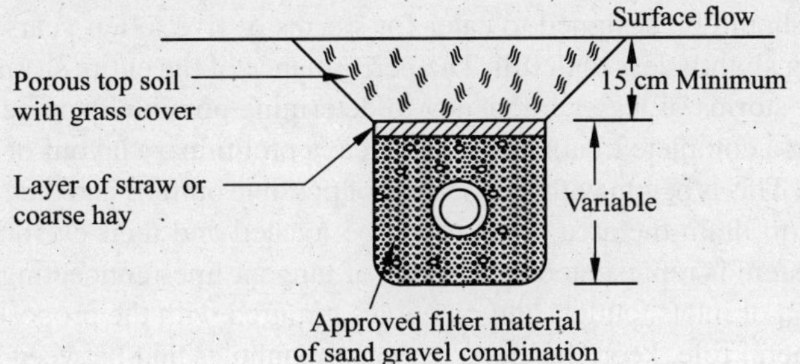

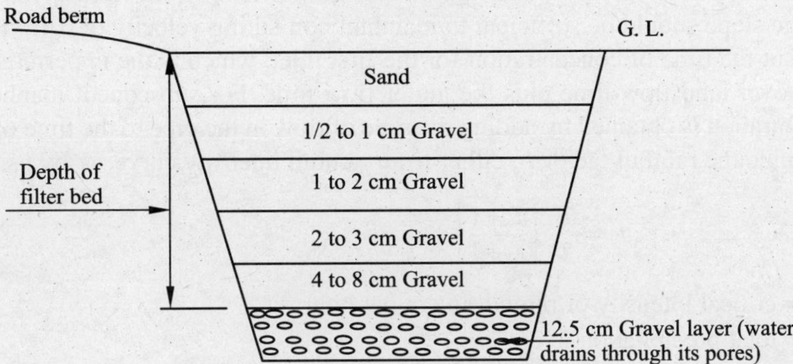

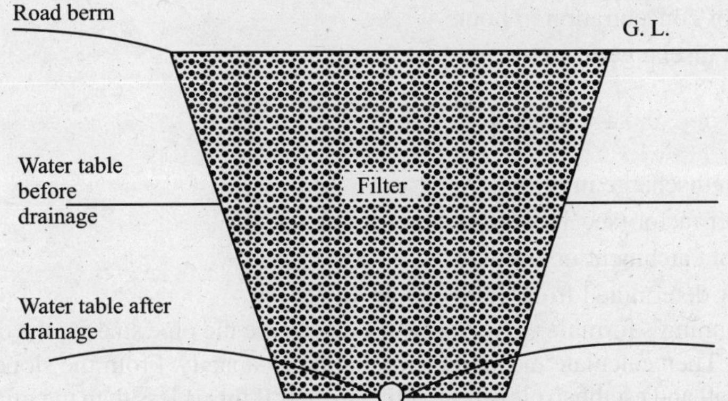

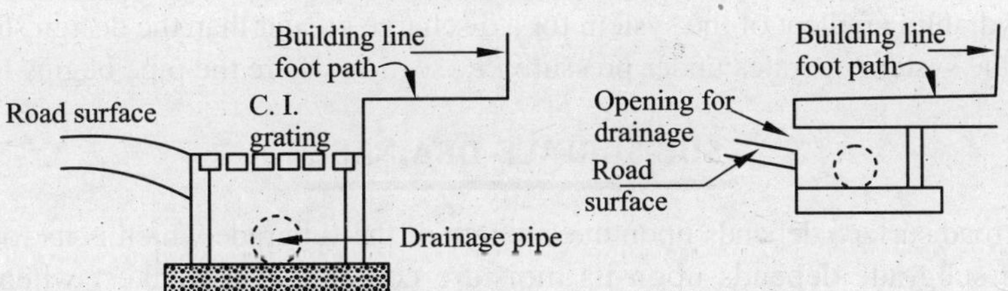

Fig. 7.3 Different types of storm-sewer inlets.

7.2.1. Design of Storm Sewers

A detailed design of storm sewers is beyond the scope of this book and reference may be made to any textbook on public health engineering or hydraulics. However, a brief outline of the design procedure is given below.

(*i*) Storm sewers should be designed to cater for storms of five to ten years' frequency with the conduit flowing slightly less than full. The performance of the entire storm sewer system must be checked for storms of lesser frequency to determine possible damage.

(*ii*) After preparing a complete treatment of the area, a preliminary layout of the drainage system should be done. This is begun with the location of possible outfalls and their flow line elevations. Inlets required to drain the area should then be located and their elevations calculated. The preliminary system is represented by a series of tangent lines connecting these points.

(*iii*) Plot the location of inlets, outlets and manholes required with the proposed sewer connecting them. As a general rule, keep the slope between manholes and between inlets and manholes constant. Avoid as far as possible abrupt changes of slopes from steep to mild and the overall average slope should be sufficient to maintain non-silting velocity of 0.75 m per sec.

(*iv*) Find out the time of concentration for the first inlet, which is the uppermost inlet. This will equal over land flow time plus the gutter flow time. For subsequent manholes, the time of concentration is obtained by adding the time of flow in the pipe to the time of the inlets. Then determine the rainfall factor I_c, either from rainfall intensity curve, or by using the formula:

$$I_c = \frac{F}{T}\left\{\frac{T+1}{t_c+1}\right\}$$

where I_c = critical intensity of rainfall in cm per hour
F = total precipitation in cm
T = duration of storm in hours
t_c = time of concentration in hours

Calculate the design discharge from the rational formula:

$$Q = \frac{1}{36} P A f I_c$$

where, Q = design discharge in m^3/s
P = run-off factor (see Table 7.3)
A = area of catchment in hectares
f = Factor determined from Table 7.5 or Fig. 7.15.

(*v*) By using Manning's formula or monographs, determine the pipe size for the desired slope and type of pipe. Then calculate mean velocity and pipe capacity. From the slope and the length, find out the fall and establish elevation of the inverts. If this is less than the minimum specified pipe size, adopt the minimum.

(*vi*) Repeat these calculations for the next manhole and so on for the entire drainage system.

Check the hydraulic gradient of the system for a discharge greater than the design storm discharge when a part of the system operates under pressure *i.e.*, section where the pipe begins to flow full.

SUB-SURFACE DRAINAGE

Stability of the road surface depends upon the strength of the subgrade which is its foundation. The strength of the subgrade depends upon its moisture content. It is weakest when saturated or oversaturated.

7.3. CAUSES OF MOISTURE VARIATION IN SUBGRADE

The various ways in which changes in the moisture content can occur in the road subgrade are:

1. By free water:

(*a*) Seepage of water from higher ground adjacent to the road.

(*b*) Penetration of water through the pavement.

(*c*) Transfer of moisture from the shoulders and pavement edges.

2. By ground water:

(*a*) Rise or fall of water table.

(*b*) Capillary rise from lower soil layers.

(*c*) Transfer of water vapour through soil.

These are illustrated in Fig. 7.4.

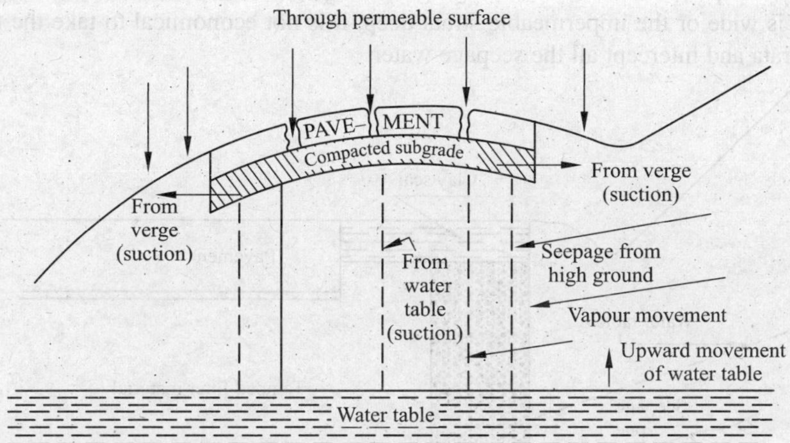

Fig. 7.4 Changes in moisture content in road sub-grade.

In the case of hill roads, water can enter the subgrade by seepage flow from adjoining slope. The top soil layer is permeable and underneath it is hard rock which is impermeable. The result is that the rain water seeps into the permeable layer and flows along the impermeable layer towards the road subgrade. The flow of water may approach the road from any direction and emerge as a spring.

When the pavement is pervious such as water bound macadam, earth, gravel or some bituminous surfaces, water finds its way through these porous surfaces. Water also percolates through joints and cracks in concrete roads.

Transfer of moisture from the shoulders and pavement edges occurs due to changes in moisture contents on account of seasonal variations. The shoulders and edges are usually dry in summer and wet in winter as compared to the subgrade. There is thus a tendency for the moisture to flow from shoulders and edges to the subgrade in winter and vice versa in summer.

Rise or fall in water table occurs due to seasonal variations. This is particularly so in flat low lying areas. If the water table is very close to the subgrade, water rises by capillary action. It has been found that for the maintenance of a stable subgrade, water table should not rise to a level less than 1.2 m below the formation.

If the soil layers beneath the subgrade are wetter than the subgrade itself at the time of construction, it will attract moisture from these layers by capillarity. This will stop when the equilibrium is reached.

Transfer of water vapours through the soil occurs due to difference in vapour pressure caused by the difference in moisture contents or temperatures between various parts of the soil.

7.4. SUBSOIL DRAINAGE SYSTEM

Under ideal conditions, there should occur no change in moisture content of the subgrade during the life time of a road. Thus the function of the subsoil drainage system is to prevent changes in subgrade moisture content. Before deciding the type of subsoil drainage system to be adopted, it is necessary to get the following information:

(*i*) Soil profiles, *i.e.*, type of soil and thickness of the various strata.
(*ii*) Water table position, if within 2 m from the formation level.
(*iii*) Position of seepage zones, if any.

The various controls of sub-surface flow are discussed in the following paragraphs.

7.5. CONTROL OF SEEPAGE FLOW

Two cases arise. In case the seepage zone exists within 1 m of the surface, install an intercepting drain just in the impermeable strata underlying the seepage zone as illustrated in Fig. 7.5. If the seepage zone is wide or the impermeable strata deep, it is not economical to take the trench to the impervious strata and intercept all the seepage water.

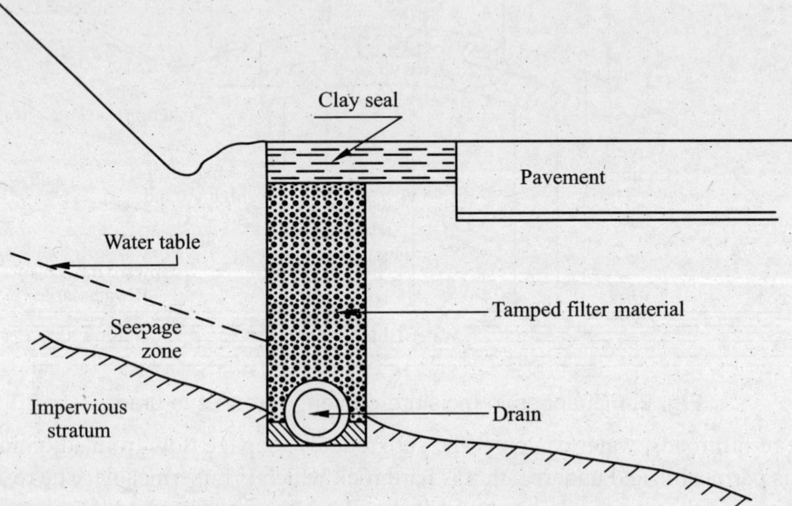

Fig. 7.5 Interception of shallow seepage zone.

A drain is provided to keep the water-table about 1.25 m below formation, as shown in Fig. 7.6.

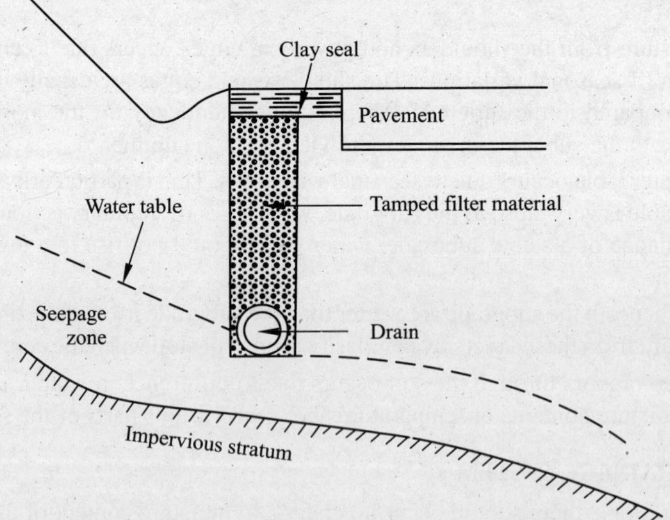

Fig. 7.6 Partial interception of deep seepage zone.

7.6. CONTROL OF HIGH WATER TABLE

Water table can be kept low by cutting deep side drains which would intercept the flow of water towards the road subgrade. In one case a ditch of 2m deep became necessary to intercept the spring.

Sometimes trenches are also cut along the centre of the road and filled with graded rough stone so as to collect the water which is led away by side drains. Metre drains at intervals are cut in the pattern of herring bone to lead away water to the side drains. These are also filled with rough graded stone.

A drainage system as shown in Fig. 7.7 may be installed to lower water table. In case a perched water table exists, vertical drains are installed to transfer the water to the main water table.

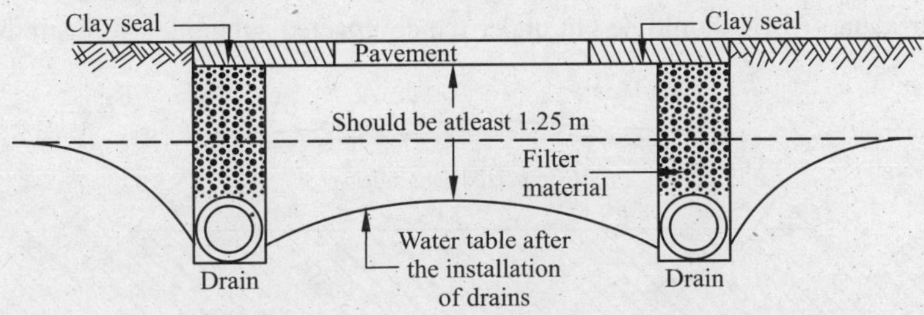

Fig. 7.7 Drains installed to lower water-table level.

7.7. CONTROL OF WATER ENTERING THE SUBGRADE THROUGH A PERVIOUS ROAD PAVEMENT

It is not practicable to construct a perfectly impervious road pavement. The entry of water into the subgrade through the pavement may be checked by the following methods:

(*a*) ***By porous sub-base.*** A porous sub-base 15 to 30 cm in thickness of sand, gravel, etc. is provided between the subgrade and the pavement, as shown in Fig. 7.8. The purpose of the sub-base is to intercept water percolating through the surface and leading it to the side trenches and drains. This should be given a proper camber and should be free from depressions.

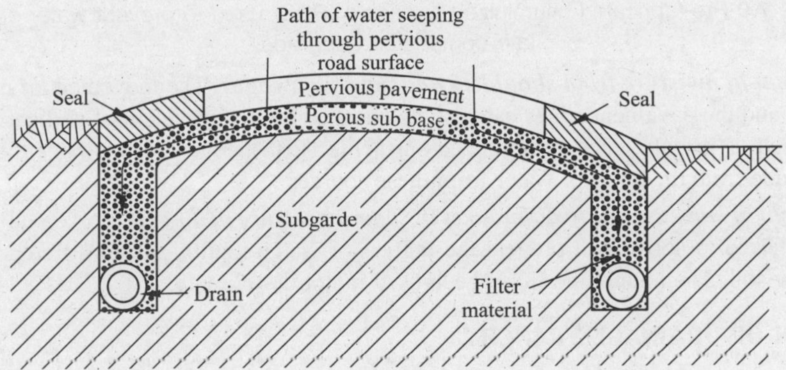

Fig. 7.8 Porous sub-base for minimizing sub-grade deterioration due to pervious road crust.

(*b*) ***Stabilized sub-base.*** Instead of providing a porous sub-base, a sub-base stabilized with cement, bitumen or other suitable material and of 7.5 to 15 cm thickness may be provided over the sub-grade.

(*c*) Surface dressing to the formation

7.8. CONTROL OF WATER ENTERING THE SUBGRADE DUE TO DIFFERENCE IN SOIL SUCTION AND VAPOUR PRESSURE

The moisture movements due to these reasons can also be controlled by the methods detailed above, but each case requires special treatment. The treatment to be given in each case is discussed below:

1. *Movement of moisture due to suction between the subgrade and the soil layers beneath.* A thin layer of coarse material like gravel is usually provided under the subgrade to prevent rise of moisture into the subgrade by capillary action through fine grained soils. This method has not proved very effective. It is recommended that a horizontal impermeable membrane *e.g.* prefabricated bituminized surfacing (PBS) should be put under the compacted subgrade as illustrated in Fig. 7.9.

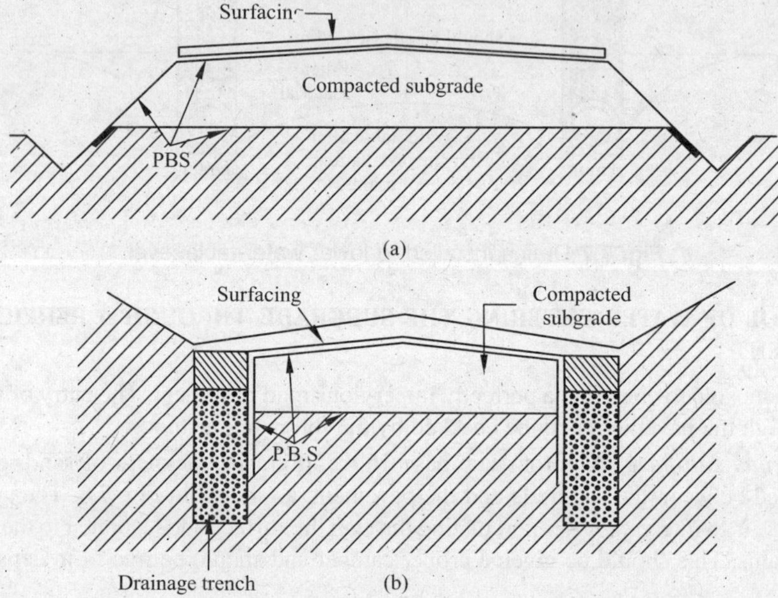

Fig. 7.9 Pre-fabricated bituminized surfacing (PBS) used to prevent water entry into compacted subgrade.

2. *Movement of moisture from shoulders and pavement edges.* The movement of moisture from the shoulders and the pavement edges can also be checked by the provision of vertical impermeable membrane such as prefabricated bituminized surfacing (PBS). Fig. 7.9 illustrates the methods of sealing subgrade from the entry of water from these sources.

3. *Movement of moisture due to differences in vapour pressure in the soil.* Horizontal impermeable membrane referred to above will prevent the ingress of water resulting from differences in vapour pressure in the soil. The membranes have been used in hot dry climate.

7.9. DESIGN OF DRAINAGE TRENCH

Proper design of the drainage trench is essential for an efficient working of the drainage system. Drainage action mostly depends upon the drainage trench as the drain pipe serves only to remove rapidly the drained water. The design of various components of the drainage trench are discussed below.

Backfill of drainage trench. The process previously adopted was to surround the drainage pipe with coarse rubble stone or shingle and refill the trench with the excavated material. This did not work well after some time as the trench used to get chocked due to silting up of the backfill. Also in silty or sandy soil, the silt was carried to the drain through rubble and large voids (internal erosion) occurred under the pavement which caused its failure.

Nowadays it is a common practice to recommend the use of specially selected filter material for backfilling the trench. The material is designed in such a manner that it offers practically no resistance to the flow of water to the drain and at the same time resists the movements of silt, thereby preventing clogging or internal erosion of the backfill. The design method has been developed by the U.S. Corps of Engineers (see Fig. 7.10) and the steps are listed below:

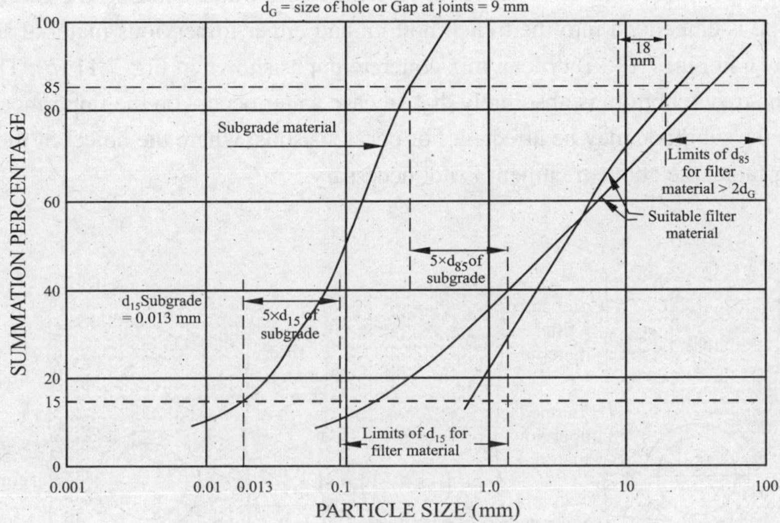

Fig. 7.10 Design method for filter material

(*i*) Obtain the particle size analysis of the subgrade soil where the drain is to be installed.

(*ii*) Plot the particle size distribution curve.

(*iii*) Base the limits of the particle size distribution of the filter material on the following considerations:

(*a*) To prevent silt being washed and entering into the drain, piping ratio

$$= \left(\frac{\text{15 per cent size of the filter material}}{\text{85 per cent size of the subgrade material}}\right) \text{ must be less than 5}$$

(*b*) For sufficient permeability of the filter material, permeability ratio

$$= \left(\frac{\text{15 per cent size of the filter material}}{\text{15 per cent size of the subgrade material}}\right) \text{ must be greater than 5}$$

(15 per cent size means the size of particles corresponding to the 15 per cent ordinate on the particle distribution graph)

(*iv*) From these requirements, the limits of the sizes for the 15 per cent size of the filter material should be between (see Fig. 7.11) 5 × 15 per cent size of the subgrade material and 5 × 85 per cent size of the subgrade material.

(*v*) Base the limits for the coarse particles of the filter material on the size of holes in the pipes in the case of perforated pipes or the gap at the joints for open jointed pipes. The 85 per cent size of the filter material should always be greater than twice the size of this gap. This requirement is not necessary for porous concrete pipes.

(*vi*) Based on above particle size distribution, use a filter material. It is desirable to use naturally occurring material. In case the particle size distribution of the subgrade soil and the gap between the drain pipes do not allow the use of one filter material, then use two filter materials. The coarse material should surround the pipe and the finer material used as the backfill.

Installation method of drain pipes in trench bottom. The methods to be used are illustrated in Fig. 7.11. For impervious soils it is desirable to remove all water entering the drain trench from above. The pipe is embedded into the trench bottom and either impervious material such as clay is tamped as shown in Fig. 7.11 (*a*) or lean mix concrete put as shown in Fig. 7.11 (*b*). Thus it must be ensured that the trench bottom is absolutely dry. In case water occurs on the impermeable soil at the trench bottom the subgrade may be affected. For pervious soils, where the object of the drain is only to lower water table, the above treatment is not necessary.

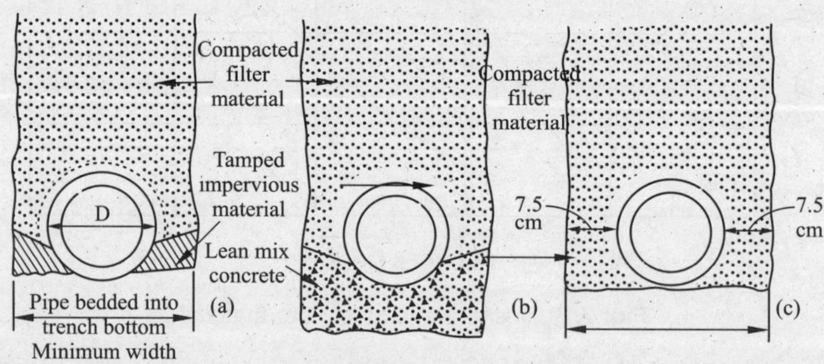

Fig. 7.11 Installation methods of drain pipes.

Sealing of the top of drainage trench. When a filter material is used, it is considered desirable to seal the top of the drainage trench with a layer of compacted clay (15 cm) or other similar soil. This will prevent silting up of the backfill from the top. The silt carried by the surface run-off may enter the trench at the top.

Grading requirements of filter material, as per IRC, are given in Table 7.3.

Table 7.3 Grading requirements of filter material

Sieve designation	*Per cent by weight passing the sieve*		
	Class I	*Class II*	*Class III*
50 mm	—	—	100
40 mm	—	—	95 – 100
25 mm	—	100	—
20 mm	—	90 – 100	50 – 100
10 mm	100	40 – 100	15 – 55

4.75 mm	90 – 100	25 – 40	0 – 25
2.36 mm	80 – 100	18 – 33	0 – 5
1.18 mm	50 – 95	—	—
600 micron	30 – 75	5 – 15	—
300 micron	10 – 30	0 – 7	—
150 micron	0 – 10	—	—
75 micron	0 –3	0 – 3	0 – 3

Notes:

(*i*) Where the soil met within the trench is of fine grained type (*e.g.* silt, clay or a mixture there of), the backfill material should conform to class I grading.

(*ii*) Where the soil met in the trench is of coarse silt to medium sand or sandy type, the backfill material should correspond to class II grading.

(*iii*) Where the soil met in the trench is gravelly sand, the backfill material should correspond to class III grading.

(*iv*) The thickness of backfill material around the pipe should be at least 150 mm allround in all the cases.

CROSS-DRAINAGE

The construction of a road interferes with the natural drainage of the area. Water which, after a rain, might have been flowing as a thin layer spread over a wide area would be held up by the road bank and concentrated at the lowest points. This water will have to be led across the road by some construction known as cross-drainage work. The usual methods of taking the water across the road are:

1. Culverts
2. Bridges and
3. Irish bridges or causeways.

Before dealing with them, it is essential to know the following details:

7.10. FLOOD DISCHARGE

The size of the flood depends upon the following major factors:

(1) Rainfall

(*a*) Intensity

(*b*) Distribution in time and space

(*c*) Duration

(2) Nature of the catchment

(*a*) Area

(*b*) Shape

(*c*) Slope

(*d*) Permeability of the soil and vegetable cover

(*e*) Initial state of wetness.

Various methods of finding the flood discharge are:

1. Empirical formulae
2. Rational method
3. From the conveyance factor and slope of the stream
4. From flood marks on existing structures
5. Unit hydrograph; and
6. Probability curves

7.10.1. Emperical methods

Various emperical formulae have been recommended for determining the flood flow. These have to be used with great care because they involve the use of constants, exact judgement of which depends upon the experience of the engineer.

Emperical formulae for flood discharge from a catchment have been proposed in the form

$$Q = CA^n$$

where Q = Maximum flood discharge in m^3/s

A = Catchment area in km^2

C = Constant depending on the nature and location of the catchment.

n = Constant

Some of the popular formulae are

(*i*) Dicken's formula.

$$Q = CA^{3/4}$$

where C = Constant which

= 11 to 14 where the annual rainfall varies from 60 to 120 cm

= 14 to 19 in Madhya Pradesh

= 32 in Western ghats

(*ii*) Ryves' formula (for Madras).

$$Q = CA^{2/3}$$

where C = 6.8 for flat tracts near areas within 25 km of the coast

= 8.5 for areas within 25 to 160 km of the coast

= 10.0 for limited areas near the hills.

(*iii*) Inglis' formula (for Bombay).

$$Q = \frac{125\,A}{\sqrt{A+10}}$$

These emperical formulae involve only one factor, *viz*, the area of the catchment and all the other factors that affect the run-off have to be taken care in selecting an appropriate value of the constant. This is an extreme simplification of the problem and cannot be expected to yield accurate results.

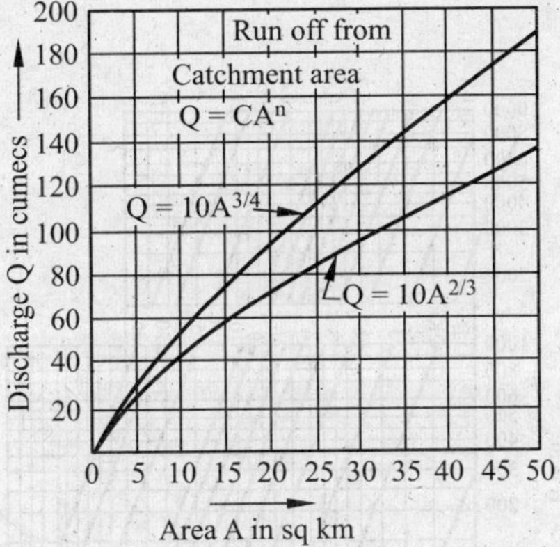

Fig. 7.12. Graphical solution of eqation $Q = 10\ A^n$.

To assist routine computations, Figs. 7.12 and 7.13 have been drawn to represent the equation $Q = 10A^n$. Two curves one for each of the common values of *n*, *viz.* $\frac{2}{3}$ and $\frac{3}{4}$ are plotted.

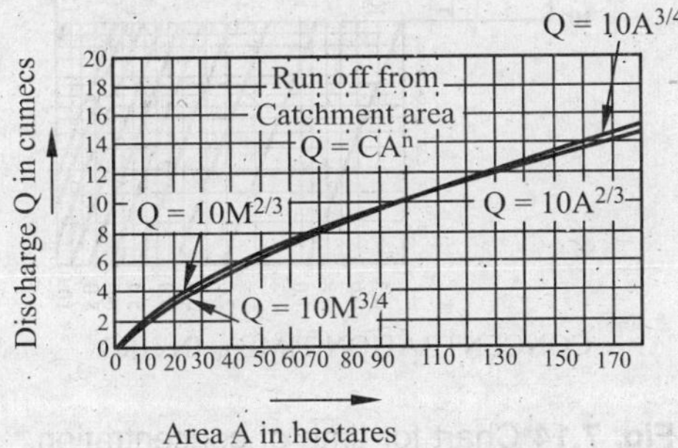

Fig. 7.13 Graphical solution of equation $Q = 10\ A^n$.

7.10.2 Rational method

This method is commonly used as it takes into account all the factors responsible for flood flow. According to this method

$$Q = \frac{1}{36}\ PAfI_c$$

where $I_c = \frac{F}{T}\left(\frac{T+I}{t_c+I}\right)$ (See page 212)

The value of t_c is found from the relation

$$t_c = \left(0.87 \times \frac{L^3}{H}\right)^{0.385}$$

where t_c = Concentration time in hours

L = Distance from the critical point to the culvert or bridge in km.

H = Fall in level from the critical point to the culvert in metre.

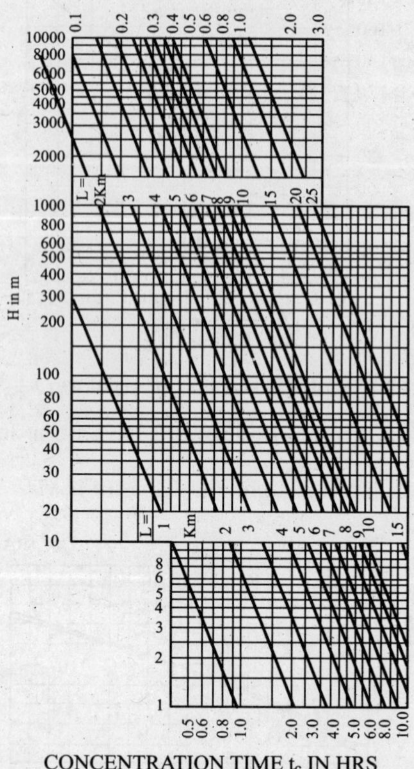

Fig. 7.14 Chart for time of concentration.

The value of t_c can also be taken from Fig. 7.14 for known values of *L* and *H*. The run off factor (*P*) is given in Table 7.4.

Table 7.4 Runoff factor '*P*'

Steep, bare rock, also city pavements	0.90
Rock, steep but wooded	0.80
Plateaus, lightly covered	0.70
Clayey soils, stiff and bare	0.60
Clayey soils, lightly covered	0.50
Loam, lightly cultivated or covered	0.40
Loam, largely cultivated	0.30
Sandy soil, light growth	0.20
Sandy soil, covered with heavy bush	0.10

f is a factor which takes into account the discrepancy due to the assumption that the spread of the storm is equal to the area of the catchment. This is taken from Fig. 7.15 or Table 7.5.

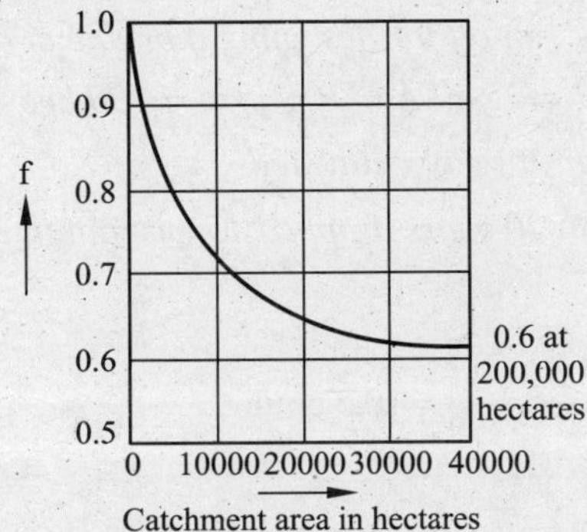

Fig. 7.15 Graph between *f* and spread of storm, in hectares

Table 7.5 Value of factor *f* in rational formula

Area (km²)	*f*	*Area (km²)*	*f*
0	1.000	80	0.760
10	0.950	90	0.745
20	0.900	100	0.730
30	0.875	150	0.675
40	0.845	200	0.645
50	0.820	300	0.625
60	0.800	400	0.620
70	0.775	2000	0.600

The run off for small catchments can be determined from the nomograph given in Fig. 7.16. This gives the discharge q in cubic metre per second for one hour rainfall (I_o) of one cm. I_o is calculated from the formula $I_o = \frac{F}{T}\left(\frac{T+1}{2}\right)$. The total discharge can be calculated by multiplying q with I_c. An example has been worked out in Fig. 7.16.

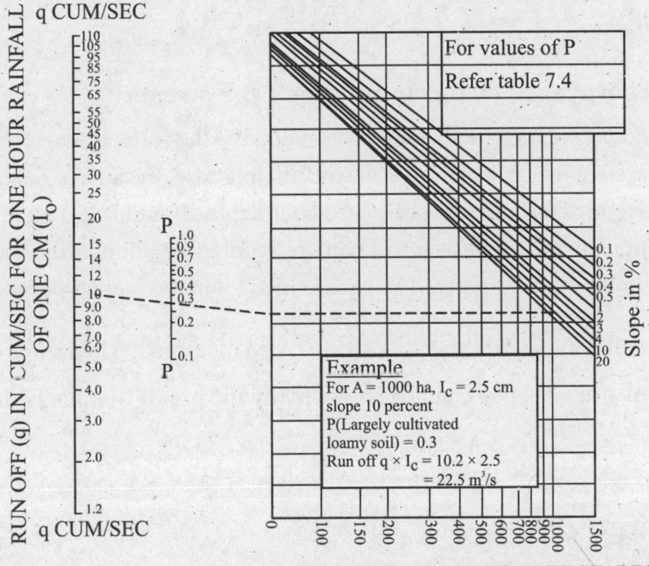

Fig. 7.16 Run-off chart for small catchments.

Problem 7.1 *Calculate the peak run off for designing a bridge across a stream, given*

Catchment L = 6 km, H = 25 m; A = 10 sq km = 100 hec

Loamy soil largely cultivated

Rainfall: The severest storm in 20 years dropped 15 cm rain in 2.5 hours.

Solution.

$$I_0 = \frac{F}{T}\left(\frac{T+1}{2}\right) = \frac{15}{2.5}\left(\frac{2.5+1}{2}\right) = 10.5 \text{ cm/hr}$$

$$t_c = \left(0.87\frac{L^3}{H}\right)^{0.385} = \left(\frac{0.87\times 6^3}{25}\right)^{0.385} = 2.17 \text{ hr}$$

$$I_c = I_0\left(\frac{2}{t_c+1}\right) = \frac{10.5\times 2}{2.17+1} = 6.62 \text{ cm/hr}$$

From Table 7.5,

$$f = 0.95$$

From Table 7.4 for loamy soil largely cultivated

$$P = 0.30$$

$$\therefore \quad Q = \frac{1}{36} Pf\, A\, I_C$$

$$= \frac{1}{36} 0.3\times 0.95\times 1000\times 6.62$$

$$= 52.4 \text{ m}^3\text{/s}$$

7.10.3. From the conveyance factor and slope of the stream

A stream with rigid boundaries (bed and banks) will have the same cross-section before and after flood. This will not be so in the case of alluvium soils. In this case, the size of the cross-section will be more during floods due to scouring action. So the scour depth should be ascertained from soundings and on the cross-section should be plotted the average scoured bed line. From the cross-section thus plotted, measure the cross-sectional area A in square metre and the wetted perimeter P in metres.

Calculate the hydraulic mean depth $R = \frac{A}{P}$ metre. Also measure the bed slope (S) from the plotted longitudinal section of stream. Then calculate velocity by the use of Mannings' formula:

$$V = \frac{1}{n} R^{2/3}\ S^{1/2}$$

The value of n is taken from Table 7.6.

Table 7.6 Values of n in the formula $V = \frac{1}{n} R^{2/3} S^{1/2}$

Surface	*Perfect*	*Good*	*Fair*	*Bad*
Natural streams				
1. Clean, straight bank full stage, no rifts or deep pools	0.025	0.0275	0.030	0.033
2. Same as (1), but some weeds and stones	0.030	0.033	0.035	0.040
3. Winding, some pools and shoals, clean	0.035	0.040	0.045	0.050
4. Same as (3), lower stages, more ineffective slope and sections	0.040	0.045	0.045	0.055
5. Same as (3) some weeds and stones	0.033	0.035	0.040	0.045
6. Same as (4) stony sections	0.045	0.050	0.055	0.060
7. Sluggish river reaches, rather weedy or with very deep pools	0.050	0.060	0.070	0.080
8. Very weedy reaches	0.075	0.100	0.125	0.150

$$Q = A \times V$$

$$= \frac{1}{n} AR^{2/3} S^{1/2}$$

$$= \lambda S^{1/2}$$

where $$\lambda = \frac{1}{n} AR^{2/3}$$

λ is called the conveyance factor and depends upon the size, shape, and roughness of the stream. Thus from the conveyance factor and the slope of the stream, the flood discharge can be estimated. Instead of calculating the velocity as above, it is better to measure velocity during floods and then calculate the discharge.

7.10.4. From flood marks on existing structures

The procedure adopted is as follows:

(*i*) From the distinct water marks on bridge piers and abutments, immediately following the flood, determine the head of water upstream. Let it be D_u.

(*ii*) Also measure the depth of water on the down-stream side. Let it be D_d. This can also be calculated from the conveyance factor and the slope of the tail race as it is independent of the existing bridge.

(*iii*) Find out $D_u - D_d$ *i.e.,* afflux, h

(*iv*) When h is greater than $\frac{1}{4} D_d$ *i.e.,* when a standing wave is formed, then apply the weir formula.

(*a*) The discharge just upstream of the bridge is given by

$$Q = u \times D_u \times W \quad \text{...(7.1)}$$

where Q = Discharge in cumecs

u = Velocity of approach up-stream in metre per second
W = Unobstructed width of the stream in metre

(*b*) According to the weir formula, discharge down-stream

$$Q = 1.705 \times C_w \times L\left(D_u + \frac{u^2}{2g}\right)^{3/2} \quad \ldots(7.2)$$

where C_w = A coefficient to account for the losses in friction and its values are given in Table 7.7.
L = Linear waterway of the bridge in metre

Table 7.7 Coefficient-C_w

1.	Narrow bridge opening with or without floors	0.94
2.	Wide bridge opening with floors	0.96
3	Wide bridge opening with no bed floors	0.98

(*c*) Equate equation 7.1 and 7.2 and find the value of u.

(*d*) Substitute the value of u in either equation 7.1 or 7.2 and find the value of Q.

(*v*) When h is less that $\frac{1}{4}D_d$, apply the orifice formula:

(*a*) Discharge just up-stream of the bridge is given by $Q = u \times D_u \times W$

(*b*) According to the orifice formula, discharge down-stream

$$Q = C_0\sqrt{2gL}\,D_d\left\{h + (1+e)\frac{u^2}{2g}\right\}^{1/2}$$

where C_0 = Constant to account for losses of head through the bridge and is found out from Fig. 7.17.
e = Constant, determined from Fig. 7.18.

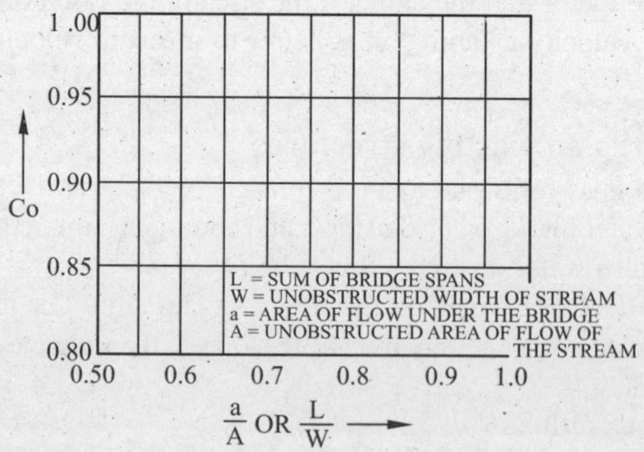

Fig. 7.17 Graph for C_0.

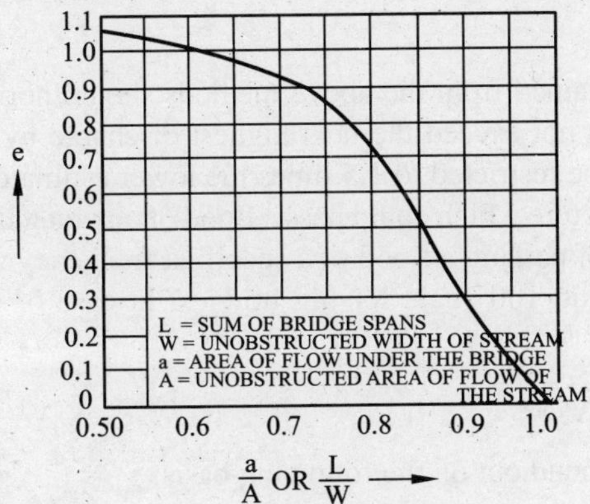

Fig. 7.18 Grpah for *e*.

(*c*) Calculate Q as before.

7.10.5. From unit hydrograph

K.L. Sherman has suggested the use of unit hydrograph for estimating flood discharge. A hydrograph is a graphical representation of discharge versus time. A unit hydrograph is a hydrograph representing one cm of run-off from a rainfall excess of some unit duration and specific areal distribution. The rainfall excess (rainfall—losses) is determined and the ordinates of the unit hydrograph multiplied with this value gives the discharge.

7.10.6. From probability curves

A very logical method to determine the flood discharge expected to occur is to base the prediction upon the records of the past. A graph is plotted showing the discharge as ordinate and per cent occurrence as abscissa on a logarithmic scale. Knowing the frequency of occurrence, the discharge can be read from the graphs. One such curve is shown in Fig. 7.19 which represents 153 floods of different intensities in 54 years. This method will yield good results provided sufficient records are available upon which to base the determination. If the discharge is required to be found for a flood expected to occur once in 1000 years the percentage occurrence $= \dfrac{100}{\dfrac{153}{54} \times 1000} = 0.035\ \%$. Against this, the value of Q from the curve is 154,00 cubic metre per sec. Based on actual observations, Kanwar Sain and Karpov have also produced graphs for Indian rivers.

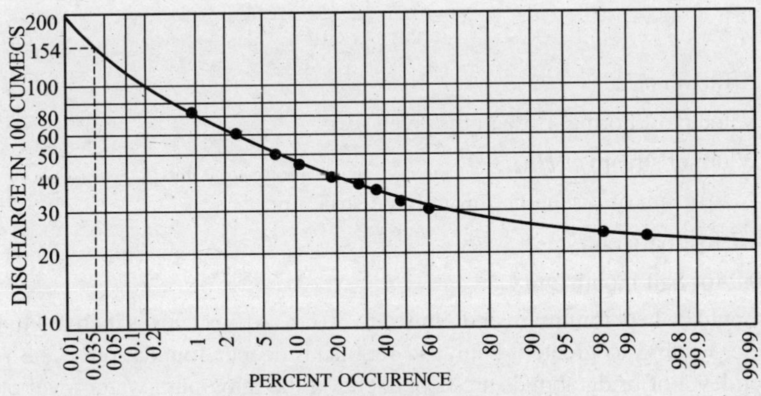

Fig. 7.19 Probability curve.

7.10.7. Design discharge

The highest discharge obtained from the above methods should normally be taken as the design discharge (Q) in case it does not exceed the next highest discharge by more than 50 per cent. If it exceeds this limit, it should be restricted to 1.5 times the lower estimate. It should not be the aim of the designer to provide a structure which could pass a flood of any magnitude during its life time. The structure should be capable of passing a flood of a specified frequency which is usually 20 years for culverts and small bridges and 100 years for big bridges. In case of an unusual flood, excessive damage should not be caused to the structure or to the road.

7.11. LINEAR WATERWAY

Linear waterway can be found out on the following basis:

(*a*) If the banks of the stream are rigid, well defined and high (rock or hard soil), but the bed is alluvial, linear waterway equal to the surface width of the stream, from one edge to the other edge of water along HFL, should be provided.

(*b*) In wholly rigid streams, the above rule holds good. In case it is possible to effect some reduction in the linear waterway, it should be done provided the velocity is not high.

(*c*) For large alluvial stream with undefined banks, the required linear water-way is determined using Lacey's formula,

$$W = C\sqrt{Q}$$

where W = The linear waterway in m

Q = Designed maximum discharge in m^3/s; and

C = A constant, usually taken as 4.75 for regime channels, but varies from 4.5 to 6.3 depending on local conditions.

It is not desirable to reduce the linear waterway below that for regime condition. If a reduction is unavoidable, due attention should be given to afflux and velocity of water under the structure. With reduced waterway, velocity increases and greater scour depth would be involved, requiring deeper foundations. Thus any attempt for possible saving from a reduced linear waterway would be offset by the extra expenditure on deeper foundations and protective works.

7.12. AFFLUX

It is the heading up of water over the flood level caused by constriction of waterway at the bridge site, which can be computed from the relation:

$$X = \frac{V^2}{2g}\left\{\frac{L^2}{C^2W^2} - 1\right\}$$

where X = Afflux in m

V = Velocity of normal flow in the river, m/s

L = Width of stream at HFL

C = Coefficient of discharge through the structure,

= 0.7 for sharp entry

= 0.9 for bell mouth entry.

The afflux should be kept minimum and limited to 30 cm. Afflux causes increase in velocity on the downstream side, leading to greater scour and requiring deeper foundations. The road formation level and the top level of bridge bunds are dependent on the maximum water level on the upstream side including afflux.

The increased velocity under the bridge should be kept below the allowable safe velocity for the bed material. Typical values of safe velocities are given in Table 7.8.

Table 7.8 Typical safe velocities for different materials

Sr. No.	*Bed material*	*Safe velocity (m/s)*
1.	Loose clay or fine sand	upto 0.5
2.	Coarse sand	0.5 – 1.0
3.	Fine gravel, sandy or stiff clay	1.0 – 1.5
4.	Coarse gravel, rocky soil	1.5 – 2.5
5.	Boulders, rock	2.5 – 5.0

7.13. NORMAL SCOUR DEPTH

In the case of semi-alluvial and alluvial streams, scour of the bed and sides take place, when they are in floods. In such cases, it is desirable to find out the normal depth of scour.

(*a*) In the case of alluvial streams, the normal or regime depth of scour is calculated from the Lacey's formula provided the linear waterway is not less than the regime width. According to Lacey's formula:

$$D = 0.473\left(\frac{Q}{f}\right)^{1/3}$$

where D = Normal scour depth in metre

Q = Design discharge is cumecs

F = Silt factor and is given in Table 7.9 and is equal to $1.76\sqrt{m_r}$

where m_r = Mean diameter of bed soil in mm.

Fig. 7.20 and 7.21 can be used to determine the normal or regime scour depth for known values of Q and f.

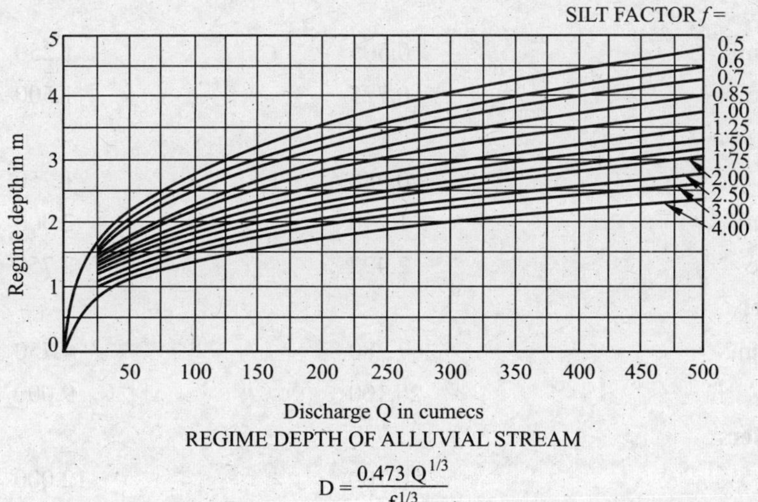

Fig. 7.20. Regime depth for discharge upto 500 cumecs.

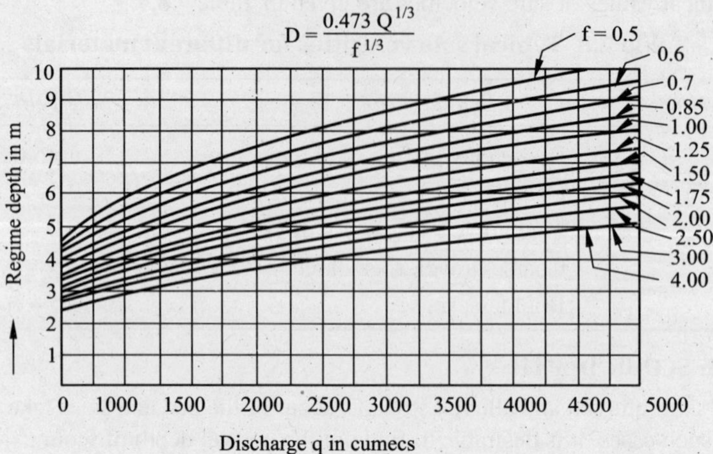

Fig. 7.21 Regime depth for discharge upto 5000 cumecs.

(*b*) When the stream has erodable bed but rigid banks (quasi-alluvial streams) provided the linear waterway is not less than natural unobstructed surface width of the stream, the normal scour depth D is found from the following formulae:

Table 7.9 Silt factor *f* in Lacey's equation

Soil type	*Size of grain in mm*	*Silt factor f*
Silt		
Fine	0.120	0.600
Medium	0.233	0.850
Standard	0.323	1.000
Sand		
Medium	0.505	1.250
Coarse	0.725	1.500
Bajri		
Fine	0.988	1.750
Medium	1.290	2.000
Coarse	2.422	2.750
Gravel		
Medium	7.280	4.750
Heavy	26.100	9.000
Boulders		
Small	50.100	12.000
Medium	72.500	15.000
Large	88.800	24.000

(*i*) $$D = \left(\frac{Qn}{WS^{1/2}}\right)^{3/5}$$

with usual notations

(*ii*) $$D = \frac{0.61\,Q^{0.63}}{f^{0.33} \times W^{0.60}}$$

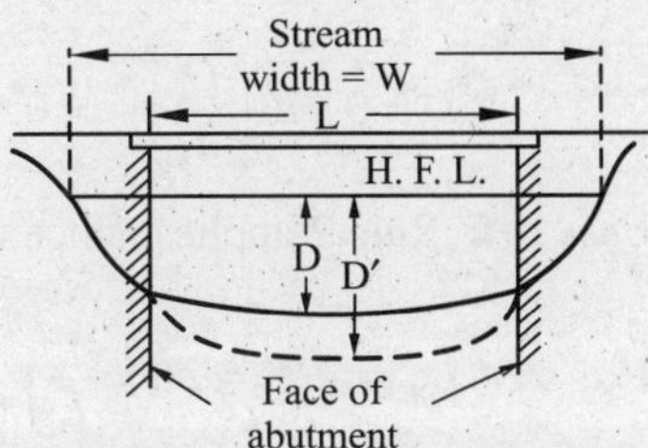

Fig. 7.22 Normal scour depth.

(*c*) When the bridge causes contraction, ($L < W$), as shown in Fig. 7.22, the normal scour depth (D') for both alluvial and quasi-alluvial streams is determined from

$$D' = D\left(\frac{W}{L}\right)^{0.61}$$

where L = Linear waterway of the bridge

W = Unobstructed natural width for quasi-alluvial streams and regime width for alluvial streams.

7.14. MAXIMUM SCOUR DEPTH

The rules for determining maximum scour depth from normal scour depth are:

(1) For a straight reach of the stream and when the bridge or the culvert is of a single span, maximum scour depth (D_m) is equal to 1.27 times the normal scour depth, modified for the effect of contraction where required.

(2) At curves or where diagonal currents prevail or when the structure is of a multiple span, maximum scour depth is equal to twice the normal scour depth, modified for the effect of contraction where required.

(3) For structures causing contraction, the maximum scour depth calculated by the above rules should be compared with that obtained from the following relation and the greater of the two values adopted:

$$D_m = D\left(\frac{W}{L}\right)^{1.56}$$

The maximum depth of scour D_m should be taken as follows:

(*i*) in a straight reach 1.27 D
(*ii*) at a moderate bend 1.50 D
(*iii*) at a severe bend 1.75 D
(*iv*) at a right angled bend 2.00 D
(*v*) at noses of piers 2.00 D
(*vi*) at upstream noses of guide banks 2.75 D.

Problem 7.2 *A bridge is proposed across an alluvial stream (f = 1.2) carrying a discharge of 500 m³/s. Calculate the depth of maximum scour when the bridge consists of :*

(a) 3 spans of 15 m, 20 m and 15 m.
(b) 5 spans of 15 m each.
(c) 4 spans of 26.5 m each.

Solution. Regime surface width of the stream

$$W = 4.75\,Q^{1/2} = 4.75 \times (500)^{1/2} = 106.2 \text{ m}$$

Regime depth

$$D = 0.473\left(\frac{Q}{f}\right)^{1/3} = 0.473\left(\frac{500}{1.2}\right)^{1/3} = 3.53 \text{ m}$$

Case (a): Rule 2 applies, since the bridge has piers

$$D_m = 2D' = 2D\left(\frac{W}{L}\right)^{0.61} = 2 \times 3.53 \times \left[\frac{106.2}{15+20+15}\right]^{0.61}$$

$$= 2\times3.53\times\left(\frac{106.2}{50}\right)^{0.61}$$

$$= 11.18 \text{ m}$$

For check, calculate D_m by equation

$$D_m = D\left(\frac{W}{L}\right)^{1.56}$$

$$= 3.53\left(\frac{106.2}{50}\right)^{1.56}$$

$$= 11.43 \text{ m}$$

Adopt 11.43 m say 11.5 m

Case (b): Rule 2 applies

$$D_m = 2D' = 2D\left(\frac{W}{L}\right)^{0.61}$$

$$= 2\times3.53\left(\frac{106.2}{75}\right)^{0.61}$$

$$= 8.73 \text{ m}$$

Check $D_m = D\left(\frac{W}{L}\right)^{1.56} = 3.53\left(\frac{106.2}{75}\right)^{1.56}$

$$= 6.07 \text{ m}$$

Adopt 8.73 m say 8.75 m

Case (c): Rule 2 applies

$$D_m = 2D = 2\times3.53 = 7.06$$

Equation 6 will not apply as $L = W$

Adopt 7.06 m say 7.10 m

7.15. DEPTH OF FOUNDATION

For small bridges and culverts, the depth of foundation is calculated from the following rules as suggested by *IRC*:

(*a*) For erodable beds, the foundations should be taken down to a depth below the maximum *HFL*, one-third greater than the calculated maximum scour depth subject to a minimum depth below the scour line of 2 m for arched bridges and 1.2 m for other bridges.

(*b*) For hard beds, when rock or other non-erodable material at maximum velocity is available, the foundation should be securely anchored into the hard stratum (about 0.3 m into rock and about 0.6 m into other material).

(*c*) The pressure on the foundation material should be well within its bearing capacity for all types of beds. This rule applies when no bed floor is provided under the structure and the stream is free to scour.

(*d*) In the case of culverts, where bed floor is usually provided, keep the top of the floor about 0.3 m below the bed level. Carry the foundations of the abutment 1.25 m deep below the top of the floor. Provide an up-stream curtain wall 1 to 1.5 m deep and a down-stream curtain wall 1.5 to 2.5 m deep from the top of floor. The exact depth of the curtain walls depends upon the velocity of flow through the structure.

7.16. LENGTH OF CLEAR SPAN AND NUMBER OF SPANS

7.16.1. Length of clear span

Most economical clear span is one that satisfies the relation:

Cost of superstructure = cost of sub-structure

In case of small structures, where abutments and piers can be constructed on open foundations, the following practices may be adopted:

(*i*) For masonry arch bridges

$S = 2H$

where S = Clear length of the span in metre

H = Total height of pier or abutment including the foundation. For arch bridges this is measured up to the intrados of the key-stone. This equals the depth of foundation (D_f) plus the vertical clearances to be provided as per Table 7.10.

Table 7.10 Vertical clearances for different discharges

Discharge in cumecs	*Vertical clearance in metre*
Below 0.3	0.15
0.3 – 3	0.45
3 – 30	0.60
30 – 300	0.90
300 – 3000	1.20
Above 3000	1.50

The vertical clearances are not normally provided in the case of culverts.

(*ii*) For *RCC* slab bridges

$S = 1.5\ H$

7.16.2. Number of spans

Effort should be made to provide as minimum number of spans as possible since greater the number of spans, greater the number of piers and hence more the obstruction to the flow under the structure. Especially in hilly terrain, where torrential velocities exist, it is desirable not to provide any pier and span from bank to bank. The number of spans are found out from the following rules;

(*a*) When the required linear waterway L is less than the economical span length S, it has to be provided, as it is in one single span.

(*b*) When L is greater than S, the number of spans are roughly found from the relation.

$$L = NS$$

where N is the number of spans. Since N must be a whole number and preferably odd, S is adjusted accordingly. There is no objection to providing varying span lengths in the structure.

Usually each pier causes contraction in the filaments of water amounting to its thickness, equally divided on both sides. Therefore length of clear spans should be such that their sum should be equal to linear waterway plus the sum of thickness of all piers. Thus in Fig. 7.23,

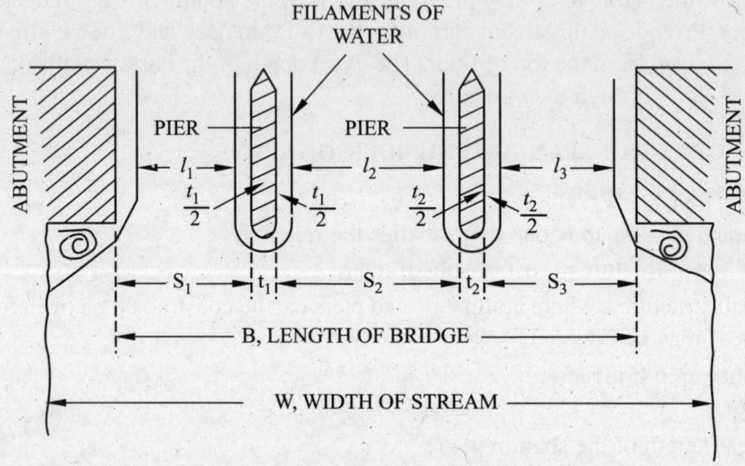

Fig. 7.23 Linear waterway.

$$\Sigma S = \Sigma l + \Sigma t \quad \text{and} \quad B = \Sigma l + 2\Sigma t$$

where Σl = effective linear waterway required

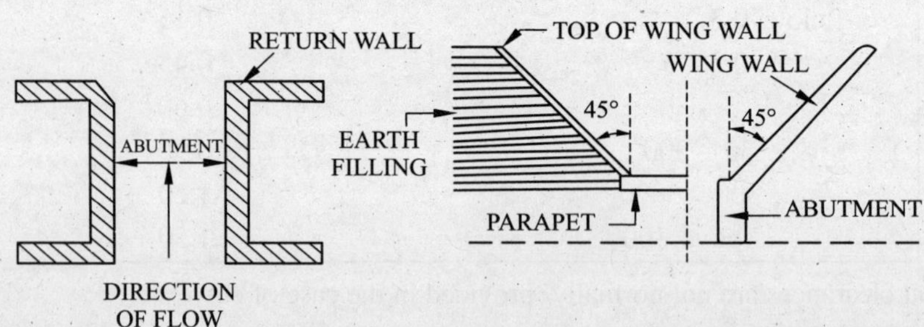

Fig. 7.24 Wing-walls and return-walls.

S = Sum of all clear spans

T = Sum of all pier thicknesses

B = Length of the bridge from face to face of abutments

7.17. ABUTMENTS, WING-WALLS AND RETURN-WALLS

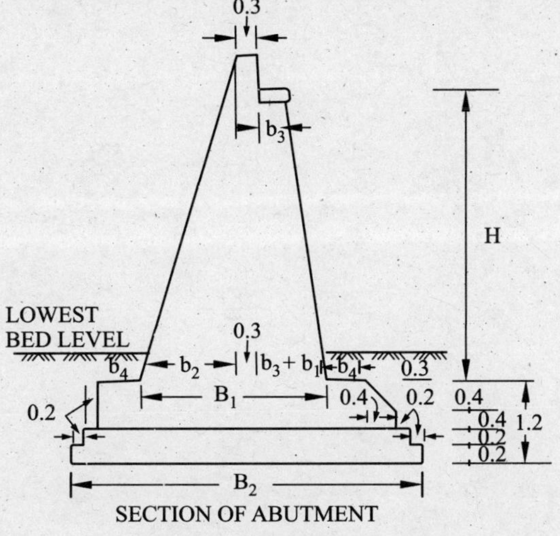

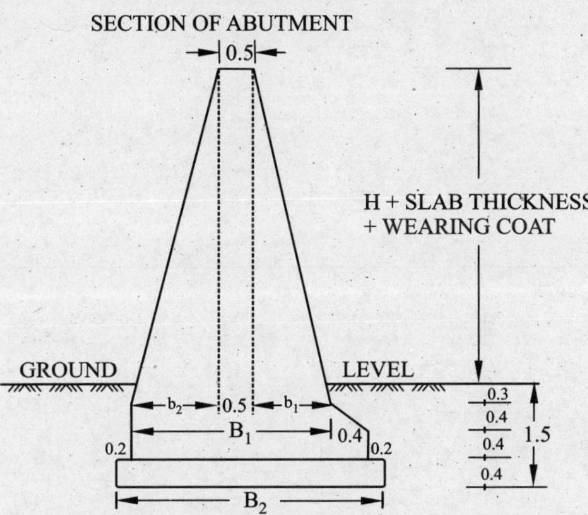

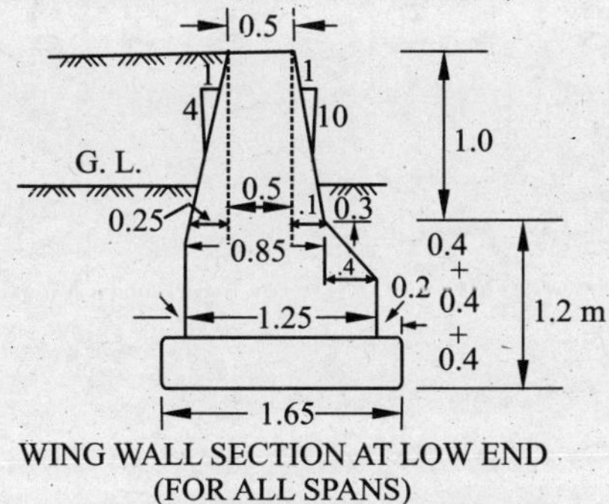

All dimensions in mm.

Fig. 7.25 Typical sections of abutments and wing-walls for culverts.

Table 7.11 Dimensions for abutment

Span	*6 m and 5 m*					*4 m, 3 m, 2 m, 1.5 m and 1 m*					
H	2.0 m	2.5 m	3.0 m	3.5 m	4.0 m	1.5 m	2.0 m	2.5 m	3.0 m	3.5 m	4.0 m
b_1	0.2	0.25	0.3	0.35	0.4	0.15	0.2	0.25	0.3	0.35	0.4
b_2	0.6	0.85	1.0	1.2	1.4	0.5	0.7	0.95	1.1	1.25	1.4
b_3	0.4	0.4	0.4	0.4	0.4	0.3	0.3	0.3	0.3	0.3	0.3
b_4	—	0.1	0.2	0.4	0.6	—	—	0.1	0.2	0.3	0.5
B_1	1.5	1.8	2.0	2.25	2.51	2.5	1.5	1.8	2.0	2.2	2.4
B_2	2.7	3.2	3.6	4.25	4.9	2.45	2.7	3.2	3.6	4.0	4.6

Table 7.12 Dimensions for wing-walls

Span	*Upto 2 metres*						*3 metres*						*4 metres*						*5 metres*					*6 metres*				
H	1.50	2.00	2.50	3.00	3.50	4.00	1.50	2.00	2.50	3.00	3.50	4.00	1.50	2.00	2.50	3.00	3.50	4.00	2.00	2.50	3.00	3.50	4.00	2.00	2.50	3.00	3.50	4.00
h_1	0.18	0.23	0.28	0.33	0.36	0.43	0.19	0.24	0.29	0.34	0.39	0.44	0.19	0.24	0.29	0.34	0.39	0.44	0.25	0.30	0.35	0.40	0.45	0.25	0.30	0.35	0.40	0.45
b_2	0.45	0.57	0.10	0.82	0.95	1.07	0.46	0.59	0.71	0.84	0.96	1.09	0.48	0.60	0.73	0.85	0.98	1.10	0.62	0.75	0.87	1.00	1.13	0.63	0.75	0.88	1.00	1.13
B_1	1.13	1.30	1.48	1.65	1.83	2.00	1.15	1.33	1.50	1.68	1.85	2.03	1.17	1.34	1.52	1.69	1.87	2.04	1.37	1.55	1.72	1.90	2.06	1.38	1.56	1.73	1.90	2.08
B_2	1.93	2.10	2.28	2.45	2.63	2.80	1.95	2.13	2.30	2.48	2.65	2.83	1.97	2.14	2.32	2.49	2.67	2.84	2.17	2.35	2.52	2.70	2.88	2.18	2.35	2.53	2.70	2.88

Notes.

1. Abutment and wing wall sections are applicable for a minimum bearing capacity of the soil of 16.5 T/M^2. For soils having lower bearing capacity the sections should be increased suitably.
2. Abutment and wing wall sections for intermediate heights to be adopted suitably.
3. The various dimensions to be suitably adjusted to suit the size of bricks where necessary.
4. The sections are applicable for culverts designed for IRC class 70 R or 2 lanes of class A loading whichever is more severe without provision of approach slabs.
5. The sections shall be in cement concrete 1 : 3 : 6, brick masonry in cement mortar 1 : 3 or coursed rubble masonry (IInd sort) in cement mortar 1 : 3. The foundation concrete shall be in cement concrete 1 : 3 : 6.

The base width of abutments should be adequate. It should be ensured that the compressive strength caused by them on the foundation soil should not exceed its safe bearing capacity. Thus the pressure at the toe of the abutment should be calculated to ensure that the soil is not overstressed.

Return walls and wing walls are also provided at the ends of the abutments. Return wall is at right angles to the abutment and supports earth-fill at its back. A wing wall is also provided at some angle, usually 30° to 45°, to the abutments as shown in Fig. 7.24. It serves double purpose, *i.e.*, retains earthfill at its back and as well as directs the flow into the culvert proper at the upstream and provides a transition from the culvert back to the stream at the downstream end. Typical sections of abutment and wing walls for culverts are shown in Fig. 7.25. Typical values of dimensions indicated in the figure for different spans are given in Table 7.11 and 7.12. To effect economy, foundations of the wing walls may be stepped up paying due regard to the soil profile.

7.18. WIDTH OF CARRIAGEWAY

The width of carriageway will depend on the intensity and volume of traffic. The width of carriageway is expressed in terms of traffic lanes.

Expect on very minor village roads all bridges must provide at least two-lane width. The minimum width of carriageway is 4.25 m for one-lane bridge and 7.5 m for two-lane bridge. For every additional lane, a minimum of 3.5 m must be allowed. Bridges should have a carriageway of two or four or multiple of two lanes. Odd lane bridges should not be constructed as they are conducive to the occurrence of accidents. In the case of a wide bridge, it is desirable to provide a central verge of at least 1.2 m width to segregate the traffic coming in the opposite directions.

From considerations of safety and effective utilization of carriageway, it is desirable to provide a foot path of atleast 1.5 m width on either side of the carriageway. In urban areas, it may be necessary to provide for separate cycle tracks besides the carriageway.

CULVERTS

A culvert is a cross-drainage structure for the purpose of carrying a road over a small streamlet or nallah. When the linear waterway required is 8 m or less, it is called a culvert. Culverts are in fact all closed conduits used for the drainage of highways with the exception of storm drains. The two main functions of the culverts are:

1. Collecting and leading the water across the road so as not to cause damage to raod bank or the stream bed by scour.
2. Allowing sufficient waterway to prevent heading up of water above the road surface.

Unsatisfactory and satisfactory location, alignment and profile for culverts is shown in Fig. 7.26.

The common types of culverts used are:

1. Pipe culverts
2. Slab culverts
3. Box culverts
4. Arch culverts.

7.19. PIPE CULVERTS

These are suitable when the stream carries a low discharge and where the road embankment is high. The water passes under the road through slightly inclined pipes. The pipe should have a minimum diameter of 60 cm to facilitate occasional cleaning and to avoid blocking of the vent by debris. It should also have a cover of at least 50 cm. Pipes may be made of stoneware, concrete, reinforced concrete, corrugated steel, etc. The length of a pipe culvert should be ascertained by the relative costs of the pipe and of the masonry for the head walls. The standard length of a concrete pipe is 2.5 m. A protective rubble apron should be provided in sandy or clayey beds. A concrete bedding of a suitable thickness should be provided below the pipe. The design of a pipe culvert involves the following:

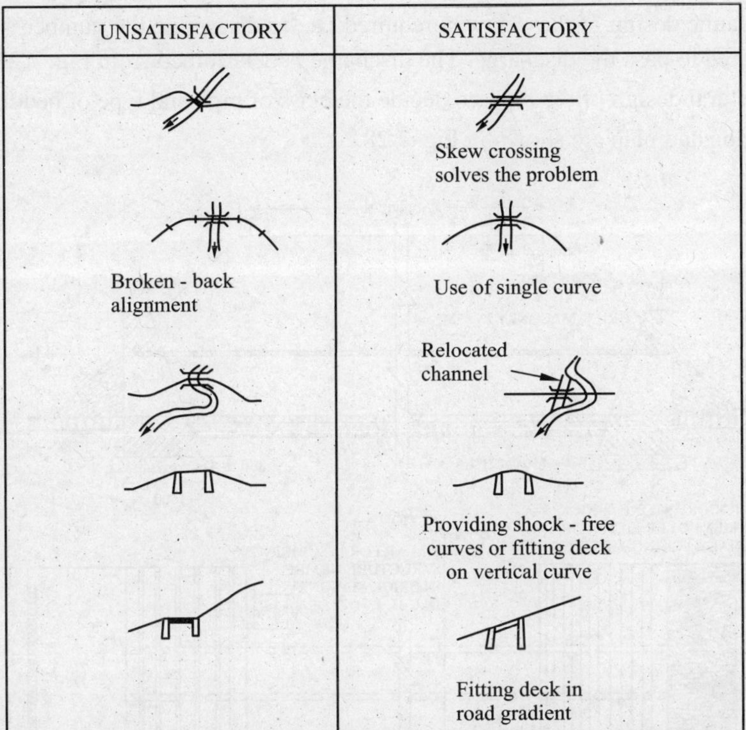

Fig. 7.26 Unsatisfactory and satisfactory location, alignment and profile for culverts.

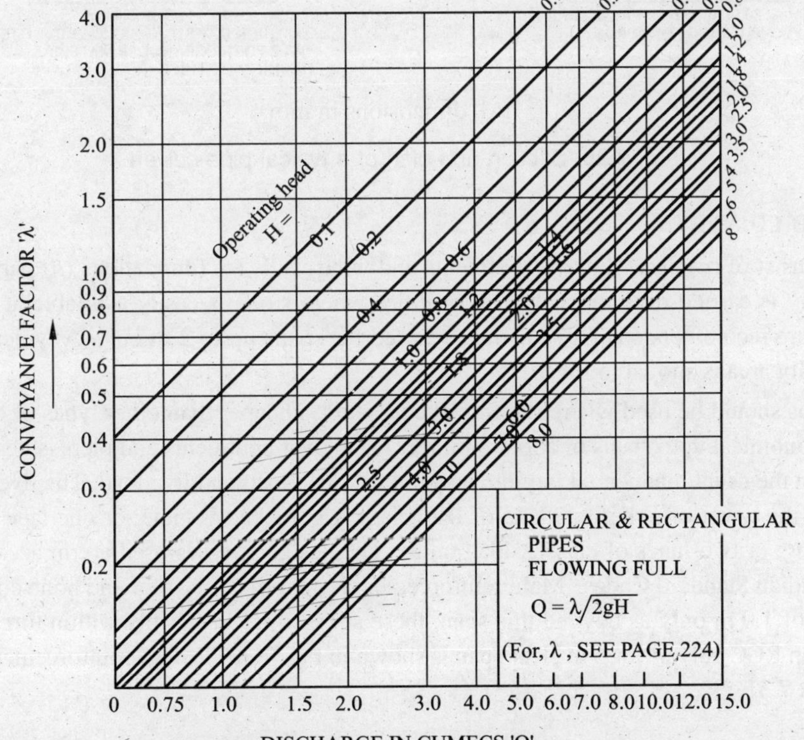

Fig. 7.27 Discharge for circular and rectangular pipes.

(*i*) Hydraulic design of the ventway required *i.e.*, to determine the number and size of pipes required to pass the discharge. The discharge is determined from Fig. 7.27.

and (*ii*) Structural design of the pipe, to decide the class of pipe and type of bedding.

Typical section and plan are shown in Fig. 7.28.

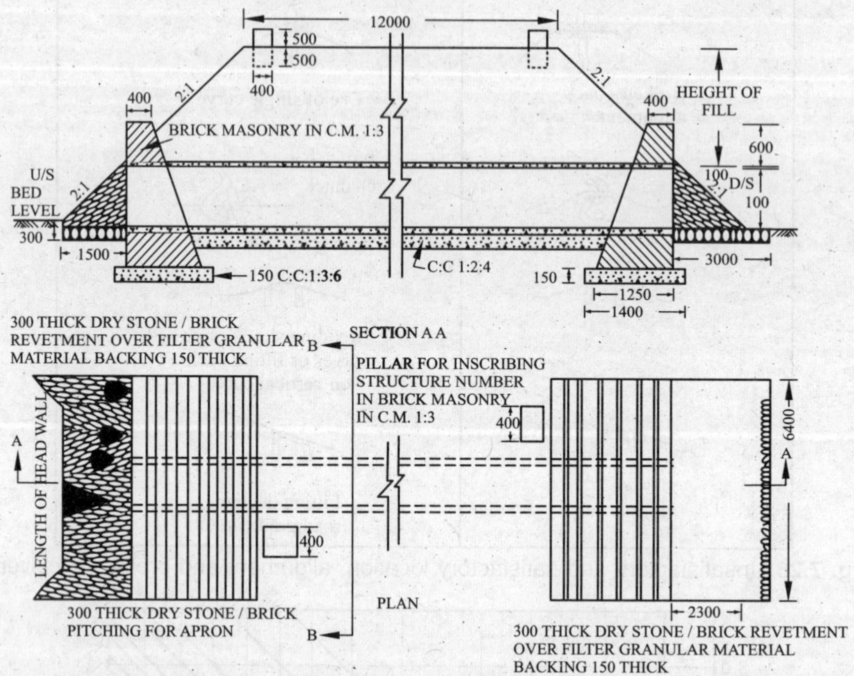

All dimensions in mm.

Fig. 7.28. Section and plan of a typical pipe-culvert.

7.20. SLAB CULVERTS

These consist of (*a*) deck slab (*b*) abutments and wing-walls (*c*) foundations (*d*) parapets, kerbs, etc. Roadway is carried over the slabs. In localities where stone is easily available at cheap rates, stone slabs are recommended. These should be used for spans up to 2 m only. A typical stone slab culvert in hilly area is shown in Fig. 7.29.

RCC slabs should be used where reinforced concrete is cheaper than other types of construction. They are economical upto spans of about 8 m. RCC slab over abutments and piers is designed freely supported, in the usual manner, taking into account the dead load and live load. The live load for the deck slab is the worst of either one lane of IRC class AA tracked vehicle; or one lane of class AA wheel loading; or two lanes of class A load trains. The design stresses used are in accordance with Bureau of Indian Standard Codes. Main reinforcement is governed by shear and bond considerations up to spans of 1.0 m only as beyond this span, these stresses are invariably within limits. A typical drawing of an RCC culvert for 6 m clear span is shown in Fig. 7.30. Typical reinforcement details are given in Fig. 7.31.

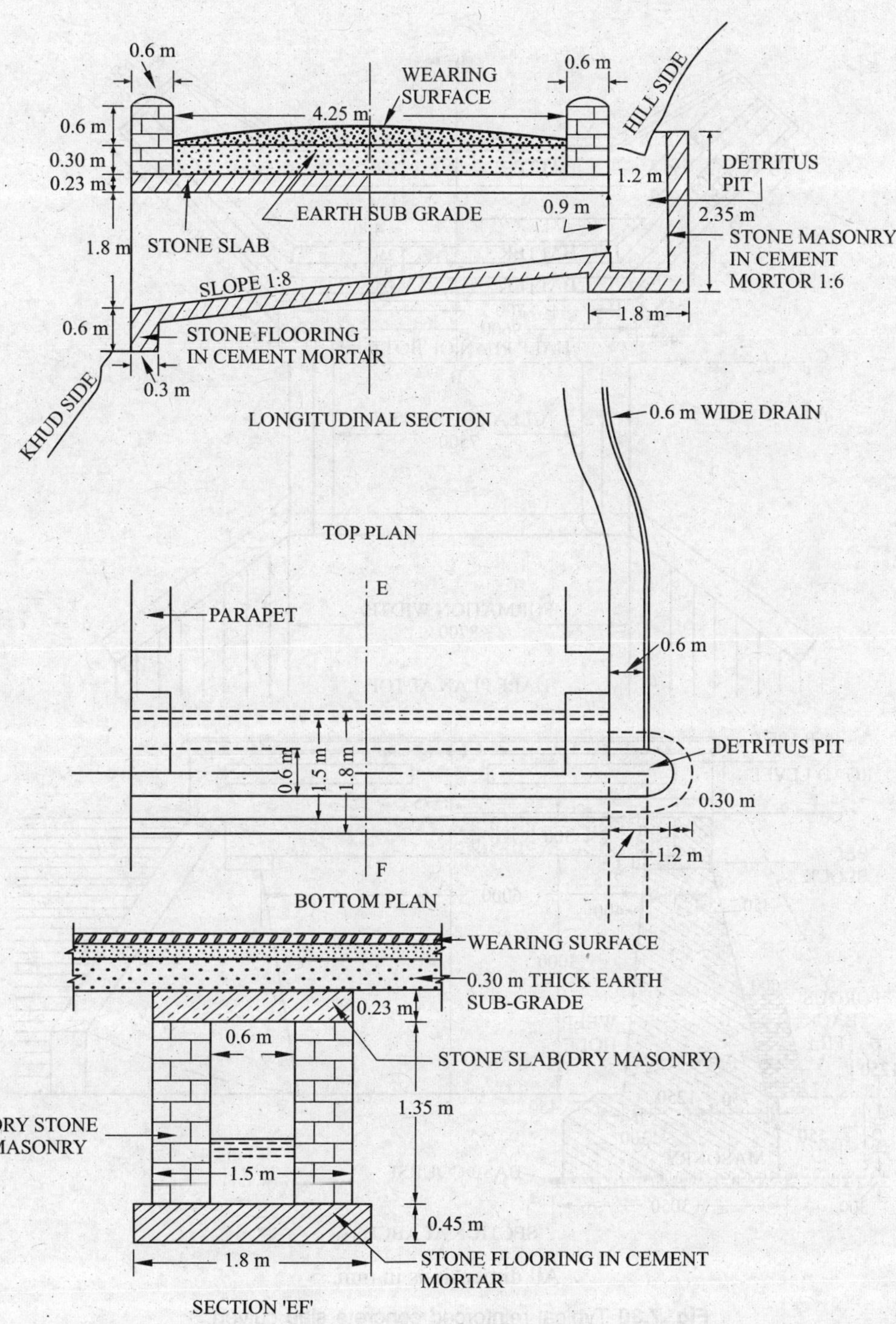

Fig. 7.29. Typical stone slab culvert in hilly area.

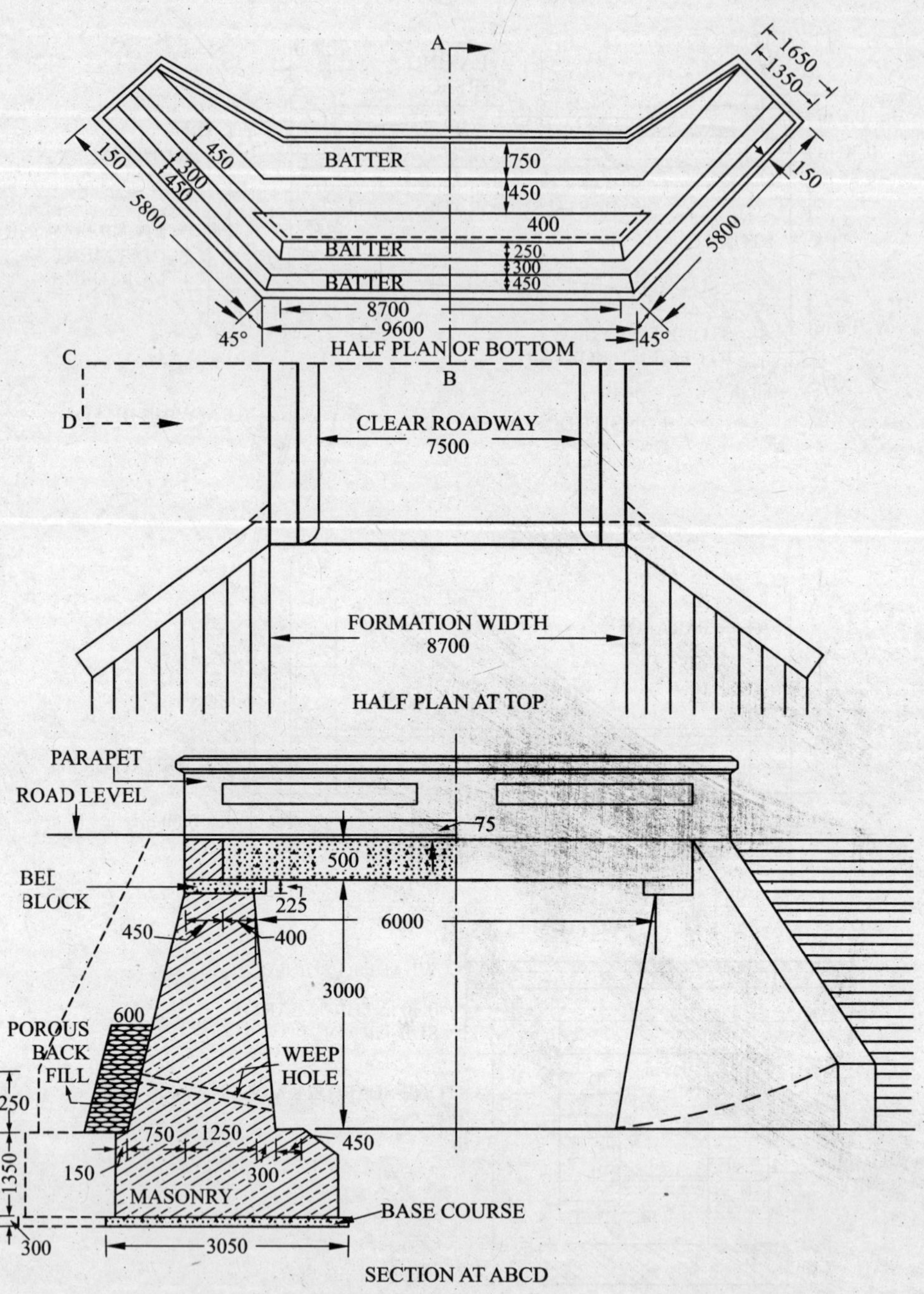

All dimensions in mm.

Fig. 7.30 Typical reinforced concrete slab culvert.

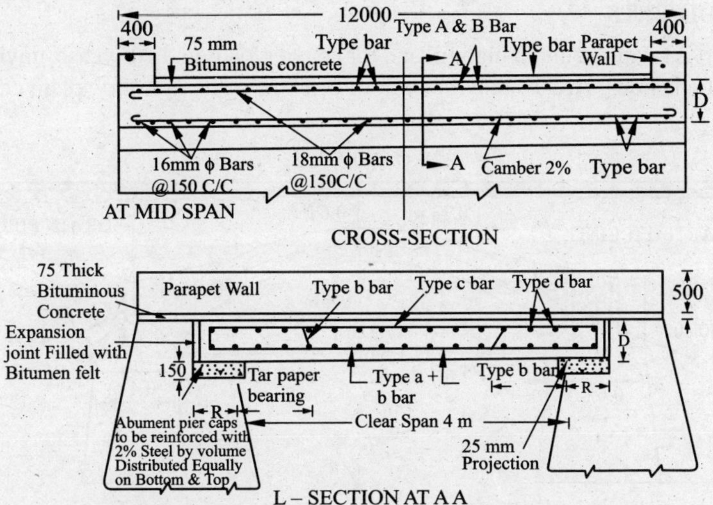

Fig. 7.31 Typical reinforcement details of RCC slab.

7.21. BOX CULVERTS

These are used where the nature of the soil below the foundation is not suitable for individual footings under piers and abutments. In such cases it is required to spread the load of the structure over a wide area. The minimum size of the opening is 0.6 m × 0.6 m. Single span box culverts vary in sizes from 0.6 m × 0.6 m to 1.35 m × 0.9 m. Rigid frame box culverts have sizes varying from 1.5 m × 1.5 m to 4 m × 3 m. Box culverts for short spans can be precast. The box culvert consists of the following components:

(*i*) Barrel of box section of sufficient length to accommodate the carriageway and the kerbs.

(*ii*) Wing walls splayed at 45° to retain the embankment and also to guide the flow of water into and out f the barrel.

Typical section for 3.6 m span is shown is Fig. 7.32.

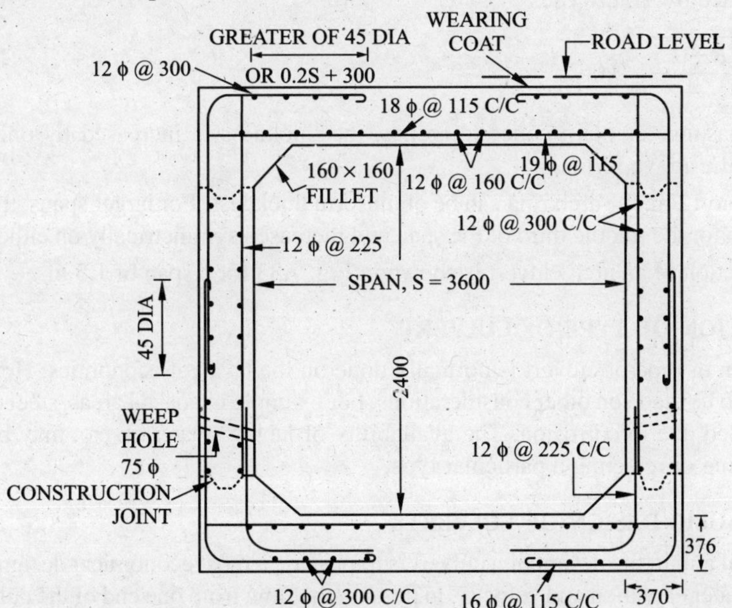

Fig. 7.32 Typical cross-section of box culvert for 3.6 m span.

7.22. ARCH CULVERTS

An arch culvert is favoured under high fill and for heavier loading. In this case, unyielding abutments are available on both sides. Arches may be built of brick or stone masonry, plain cement concrete or lime concrete.

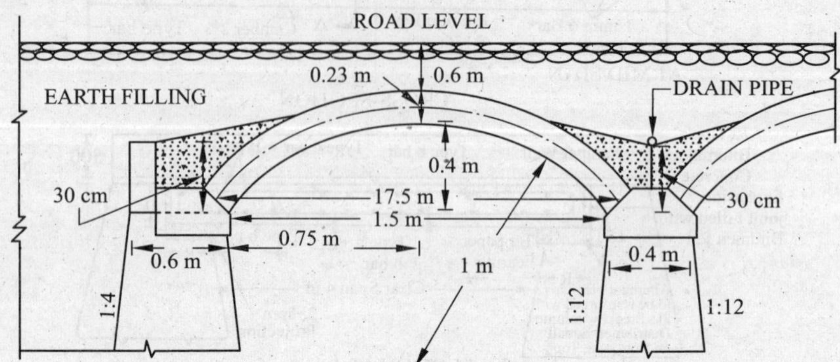

Fig. 7.33 Typical section of an arch culvert for 1.5 m span.

The dimensions of the arch ring are first assumed based on experience and then checked for adequacy. For economy in foundations, the rise of the arch should be at least one-fourth of the span, otherwise the horizontal thrust will be of high magnitude.

The radius of the segmental arch is computed from the relation:

$$R = \frac{l^2 + 4r^2}{8r}$$

where R = Radius of segmental arch in m

r = Rise of arch in m

l = Span of arch in m

The thickness 'd' of the arch ring in m for cut-stone masonry may be computed from the following formula advocated by Trautwyne.

$$d = \sqrt{\frac{R + 0.5}{4}} + 0.06$$

For brick masonry or plain concrete arches, the thickness is increased by one-third than that determined by the above formula.

For spans upto 12 m, the arch ring can be of uniform thickness. For larger spans, the thickness may be kept uniform for the middle third of the span and increased symmetrically on either side gradually.

A typical section of an arch culvert is shown in Fig. 7.33 for a span of 1.5 m.

7.23. SELECTION OF TYPE OF CULVERT

The selection of type of culvert is normally done on the basis of economics. However, selection may also have to be made on other considerations. For example in coastal areas, steel or metal culverts are not advocated due to corrosion. The availability of labour, material, etc. may be the governing factor in choosing or rejecting a particular type.

7.24. HYDRAULIC DESIGN OF CULVERTS

The principal aim in the design of a culvert is to provide a most economical design, within specific limits of headwater elevation and velocity, in passing the flow from one end of the culvert to the other. This depends upon a number of factors, hydraulic design being one of them.

In the hydraulic design of culverts we have to consider first of all the stream channel. A culvert put in a stream has little influence on the normal stream characteristics above the ponding area at the inlet and this effect is only upto a short distance below the outlet. Amount of turbulence or other disturbances caused by the presence of the structure govern this distance.

It is assumed that the quantity of water approaching the culvert is uniform and continuous during the peak run-off rates. Thus an equal amount of water will be leaving the culvert at the downstream end. There may be retardation of the flow for a short period till the storage capacity of the ponding area is reached, after which equilibrium will be attained between the rate of supply and down-stream run-off. Even if the opening in the culvert is insufficient, equilibrium will result through overflowing of the roadway.

In almost all cases, the primary control is the permissible elevation of water up-stream of the structure at the inlet, *i.e.*, head-water elevation. Thus it is essential to determine the relationship between headwater elevation and discharge. This depends upon the manner in which the culvert operates. Broadly there are two classifications:

1. Full flow at the outlet; and

2. Partially full flow at the outlet.

A culvert will flow full when laid on a grade less than that of the friction slope and if the outlet end is submerged by the tailwater elevation. This means that the normal depth of flow in the outfall channel is greater than the height of the culvert. If normal depth is less than the culvert depth, *i.e.*, the tailwater depth cannot submerge the structure and that the culvert is laid to a slope equal to or greater than that required to overcome friction, it will flow partially full.

Headwater elevation, for culverts flowing full, is computed by applying the Bernoulli equation

$$H = \left(1 + k_e + 2g\,\frac{n^2 L}{R^{4/3}}\right)\frac{V^2}{2g}$$

where H = Difference in elevation between headwater and tail water depths or between the headwater depth and crown of the culvert outlet.

$\dfrac{V^2}{2g}$ = Velocity head in metres

k_e = Coefficient of entrance loss and varies from 0.5 for squared entrances to 0.08 or less for bell mouthed entrances.

n = Mannings roughness coefficient

L = Length of the culvert in metre

R = Hydraulic radius in metre

The flow in culverts may be of several types. The methods of operation of standard culverts are classified as follows:

Class I. Culverts which flow with a free water surface through the control section and entrance not submerged (headwater depth is less than or equal to 1.2 × culvert height)

Class II. Culverts which flow with a submerged entrance (headwater depth greater than 1.2 × culvert height).

BRIDGES

When the waterway is more than 6 metre the cross-drainage structure provided to cross streams and other bodies of water is called a bridge. The cost of bridges is of a considerable amount in any highway project, although they are few in number. Special care must, therefore, be taken for their design. It is beyond the scope of this book to deal in detail about bridge design. However, important aspects are described below.

7.25. IDEAL BRIDGE

The characteristics of an ideal bridge are:

1. The line of the ideal bridge should not present any serious deviation from the line of the approach roads at either end. As far as possible the bridge should be in a straight line, but there is no objection to a bridge being on a curve provided the radius of the curve is not less than the general curvature of the road.
2. The bridge should be absolutely in level. If it has to be in gradient, it should conform to that of the roadway on both sides of the bridge. If the length of the bridge is large, camber may be provided throughout its length.
3. The width of the bridge should be adequate not only to cater for the present day traffic but also for the future anticipated traffic.
4. The bridge should be designed to carry *IRC* standard loading with a reasonable factor of safety.
5. The bridge crossing a stream should not produce undue obstruction. Thus, it should provide adequate waterway width, *i.e.,* at least equal to the width of the stream above the bridge.
6. The foundation should be of rock, if possible and should be capable of carrying the loads imposed on them. Foundations should be taken to a sufficient depth to avoid damage by floods.
7. For an un-navigable stream, head-room provided should allow a little clearance above the highest flood level recorded. In the case of a navigable stream, the head-room should be fixed on the basis of the height of the vessel likely to use the stream.
8. The bridge, as a whole, should fit into the surrounding landscape.
9. The ideal bridge should also provide for services of sewerage water, telephones, *etc.*
10. The road surface over the bridge should be similar to that of the roadway approaching the bridge at both ends.
11. Adequate provision should be made for the drainage of the road surface.
12. The bridge should be economical in cost and also in maintenance.

7.26. IDEAL LOCATION FOR A BRIDGE SITE

The characteristics of an ideal site for a bridge are:

1. A straight reach of the river. This is essential for locating piers parallel to the direction of flow and also for uniform depth of foundation.
2. Steady regime of the river and absence of serious whirls or cross-currents.
3. A narrow and well defined channel.
4. Rocky or other compact and non-erodable foundation close to the bed level.
5. Secure economical approaches which should not be very high, long or liable to flank attacks of the river and its spills during floods; nor should the approaches involve obstacles such as hills, frequent drainage crossings, sacred places, congested or built-up area requiring viaducts or troublesome land acquisition.
6. Reasonable proximity to a direct alignment of the road to be served, *i.e.*, avoidance of long detours.

7. Absence of sharp curves in the approaches.
8. If it is unavoidably necessary for the approaches of a bridge to cross the spill zone towards the river, face down-stream and not up-stream. Facing up-stream will cause heading up, pocket formation and danger to the approaches.
9. Absence of costly draining works and where such works are unavoidable, the possibility of executing them while the river is dry.
10. Avoidance of excessive construction work under water.

It is needless to say that an ideal site never exists in reality. At each site only a few favourable characteristics of ideal location are available and the site lacks in one or more of the ideal conditions. The main aim is to select a least objectionable site.

7.27. CHOICE OF BRIDGE TYPE

The choice of a particular type of bridge should be made so that it is most suitable to carry the desired traffic, adequately strong to support the design loads, economical and aesthetically pleasing. Some of the factors influencing the choice of the bridge type are:

1. Economy of the overall construction cost may require road-cum-rail bridge in two tiers across a very wide river *e.g.* Godavari bridge.
2. Large navigational clearances may dictate the use of particular types such as arches, cantilever bridges or suspension bridges *e.g.,* Howrah bridge.
3. A high level structure with un-interrupted traffic, as on a National Highway and the need to reduce the number of piers may necessitate a cantilever bridge or a cable stayed bridge or a series of simply supported trusses *e.g.,* Ganga bridge at Patna.
4. The climatic and environmental conditions unsuitable for certain types of bridges *e.g.* the corrosive atmosphere has precluded the use of a cable stayed steel bridge at Pamban near Rameshwaram and has dictated the use of cantilever construction with precast segments for the prestressed concrete navigation bridge.
5. Deck bridges are preferred to through bridges for highway traffic because of better scenic view.
6. The topographic and soil conditions may limit the choice *e.g.* a rocky ravine area is ideal for an arch bridge.
7. Weak subsoil conditions may limit the use of simply supported spans instead of continuous spans *e.g.* bridges in areas subjected to mining subsidence.
8. Availability of funds may necessitate the adoption of a submersible bridge instead of a high level bridge on a road with less traffic and this may in turn result in reinforced concrete slab decking.
9. The type of traffic may restrict the choice of bridge type. Suspension bridges are preferred for road traffic.
10. Personal preferences of the designer or specialization of the construction agency may influence the type of bridge adopted.

7.28. CLEARANCES FOR HIGHWAY VEHICLES ON BRIDGES

Recommendations of the IRC for horizontal and vertical clearances for highway vehicles are shown in Fig. 7.34. Unless otherwise stated, bridges for the use of road only or combined tramway and road traffic shall be constructed in such a manner that the minimum clearance for traffic is as shown. If the bridge happens to be on a horizontal curve with superelevated road surface, the horizontal clearance shall be increased on the side of the inner kerb by an amount equal to 5 times the super-elevation. The minimum vertical clearance shall be measured from the super-elevated level of the roadway. The maximum width and depth of a moving vehicle are assumed as 3.3 m and 4.5 m respectively.

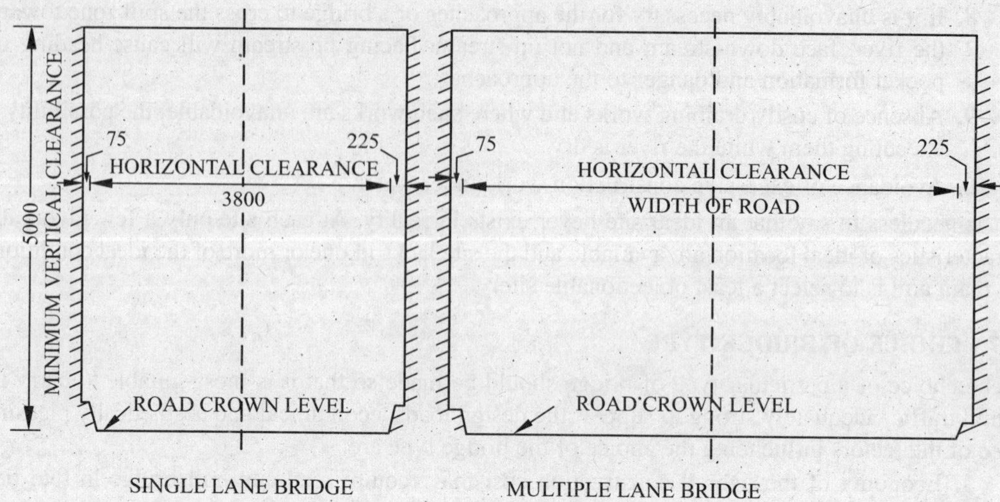

All dimensions in mm.

Fig. 7.34 Clearances diagram for highway traffic.

7.29. ABUTMENTS AND WING-WALLS

Typical sections of abutments and wing walls for slab bridges are shown in Fig. 7.35. Typical values of *A*, *B* and *C* as indicated in the figure are given in Table 7.13 for two values of allowable soil pressure *i.e.*, 20 and 40 tonnes/square metre.

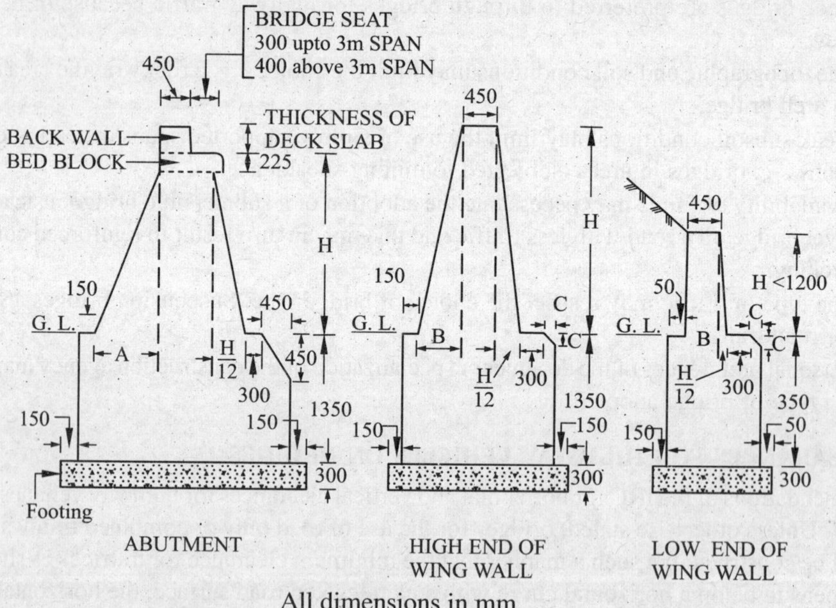

All dimensions in mm.

Fig. 7.35 Sections of abutments and wing-walls for slab-bridges.

Table 7.13 Dimensions of abutments and wing walls for slab bridges

Maximum pressure on soil t/m²	*Dimension*	*Value in metre for height H in metres* 1.5	2.0	2.5	3.0	3.5
20	*A*	0.75	0.90	1.20	1.65	2.60
	B	0.20	0.30	0.45	0.65	1.06
	C	0.45	0.45	0.45	0.45	0.45
40	*A*	0.30	0.45	0.55	0.75	0.95
	B	0.20	0.30	0.45	0.65	0.85
	C	0.15	0.15	0.15	0.15	0.15

7.30. BRIDGE LOADING AND STRESSES

Various loads, forces and stresses, to be taken into consideration, in the design of road bridges are:

(*i*) Dead load.

(*ii*) Live load.

(*iii*) Impact or dynamic effect of the live load.

(*iv*) Wind load.

(*v*) Horizontal forces due to water currents.

(*vi*) Longitudinal forces caused by the tractive effort of vehicles or by braking of vehicles and or those caused by restraint to movement of free bearings.

(*vii*) Centrifugal forces due to curvature.

(*viii*) Buoyancy.

(*ix*) Earth pressure.

(*x*) Deformation stresses due to frictional resistance of expansion bearings.

(*xi*) Temperature stresses.

(*xii*) Secondary stresses.

(*xiii*) Erection stresses; and

(*xiv*) Seismic force.

All members should be designed to carry safely the various loads, forces and stresses that can coexist. When any combination of forces (i) to (x) are considered, the net stresses should not exceed the permissible working stresses. With the temperature effect included, the permissible stresses may be exceeded by 15 per cent. When the effects due to forces at serial number (xi) to (xiii) are taken into account, increase in permissible stress may be allowed upto 25 per cent and upto 50 per cent when earthquake effects are also considered. Further it will be necessary to ensure that when steel members are used, the maximum stresses under any combination does not exceed the yield strength of the steel. It has been observed that wind loads and earthquake forces do not occur simultaneously.

Bridges are designed to carry standard wheeled or tracked vehicles or trains of vehicle according to their classification. Following types of loadings are recommended by IRC:

Class AA loading. This loading is to be adopted within certain municipal limits, existing or contemplated industrial areas, other specified areas, and along specified highways. Bridges designed for Class AA loading should be checked for Class A loading also, as under certain conditions, heavier stresses may be obtained under Class A loading.

Class A Loading. This loading is to be normally adopted on all roads on which permanent bridges and culverts are constructed.

Class B Loading. This loading is to be normally adopted for temporary structures and for bridges in specified areas. Structures with timber spans are to be regarded as temporary structures for the purpose of this class.

Details of the loadings are shown in Figs. 7.36 to 7.38.

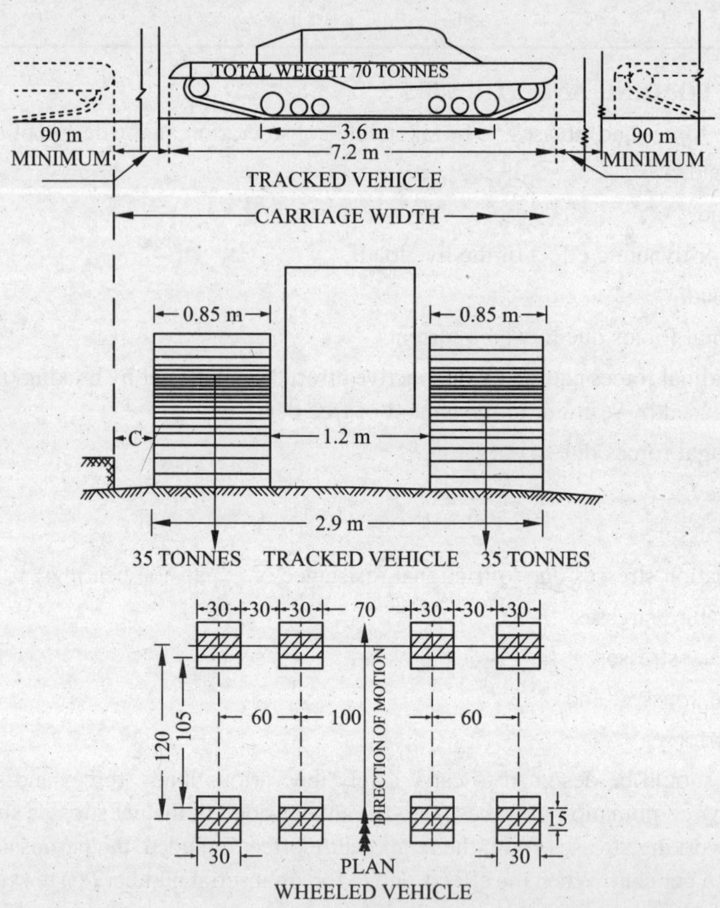

Fig. 7.36 IRC class 'AA' tracked and wheeled vehicle (all dimensions in cm).

The minimum clearance between the road face of the kerb and the outer edge of the wheel or track *C*, shall be as under for class *AA* tracked and wheeled vehicles:

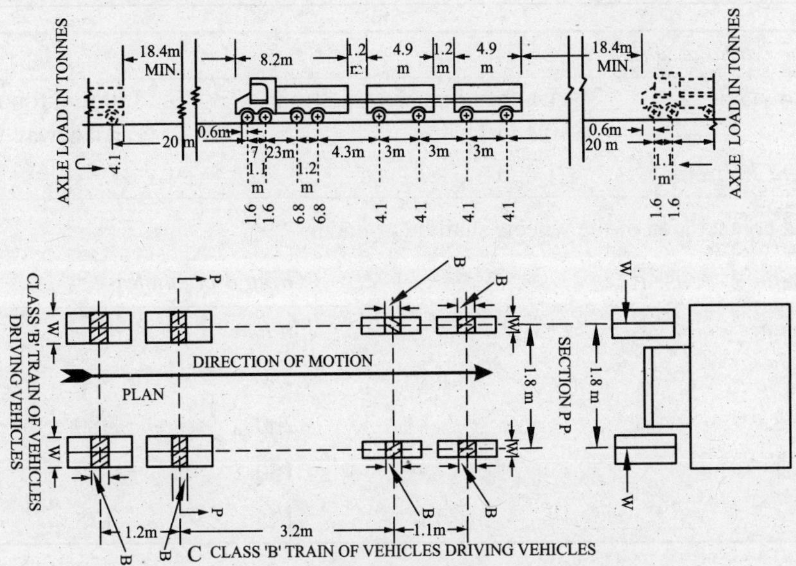

Fig. 7.37 Class 'A' train of vehicles, driving vehicles.

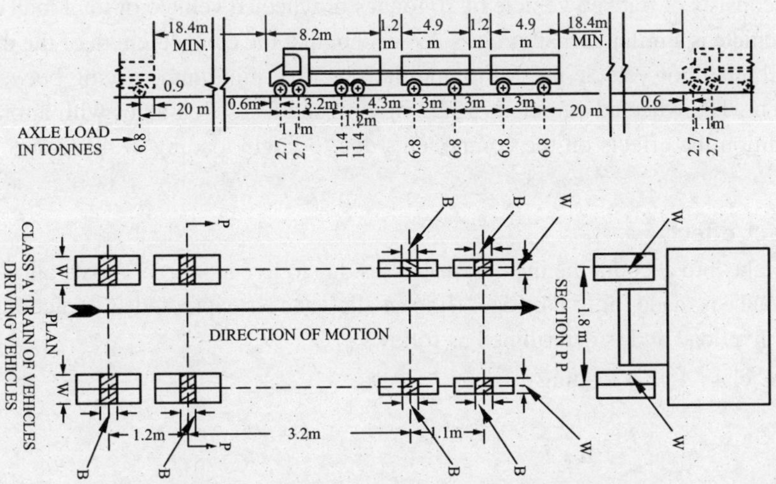

Fig. 7.38 Class 'B' train of vehicles, driving vehicles.

Carriageway width	*Minimum value of C*
Single lane bridges 3.8 m and above	0.3 m
Multiple lane bridges less than 5.5 m	0.6 m
5.5 m or above	1.2 m

The minimum clearance, *f*, between the outer edge of the wheel and the roadway face of the kerb, and the minimum clearance *g*, between the outer edges of passing or crossing vehicles on multilane bridges for both class *A* and class *B* train of vehicles shall be as given below:

Clear carriage width	*g*	*f*
5.5 to 7.5 m	Uniformly increasing from 0.4 to 1.2 m	150 mm for all carriageway width
Above 7.5 m	1.2 m	

The ground contact area of the wheels shall be as under:

Axle load in tonnes	*Ground contact area*	
	B in mm	*W in mm*
11.4	250	500
6.8	200	380
4.1	150	300
1.6	125	175

7.30.1 Class 70 R loading

An additional loading known as Class 70 *R* is sometimes specified for use instead of Class *AA* loading. This consists of tracked vehicle of 70 tonnes or wheeled vehicle of total load of 100 tonnes. The tracked vehicle is similar to that of class AA except that the contact length of the track is 4.57 m, the nose to tail length of vehicle is 7.92 m and the specified minimum spacing between successive vehicles is 30 m. The wheeled vehicle is 15.22 m long and has seven axles with a total load of 100 tonnes. In addition the effects on the components due to bogie loading of 40 tonnes are also to be checked.

7.30.2. Impact effect

In order to take into account the increased stresses due to live loads, an impact allowance is made. No such allowance is made for foot bridges. Impact allowance is expressed as a fraction or percentage of the applied live load and is determined as follows:

(*a*) For *IRC* class *A* or *B* loading

$$I = \frac{A}{B + L}$$

where I = Impact allowance in fraction

A = Constant

= 4.5 for *RCC* bridges and 9 for steel bridges

B = Constant

= 6 for *RCC* bridges and 13.5 for steel bridges

L = Span in metres

For spans less than 3 m, impact factor is 0.5 for *RCC* bridges and 0.545 for steel bridges. When the span exceeds 45 m, impact factor is 0.088 for *RCC* bridge and 0.154 for steel bridges. As an alternative impact factor is determined from Fig. 7.39.

(*b*) For *IRC* class *AA* or 70 *R* loading

(*i*) For spans less than 9 m

(a) For tracked vehicles–25% for spans upto 5 m–and linearly reducing to 10% for spans of 9 m

(b) For wheeled vehicles–25%.

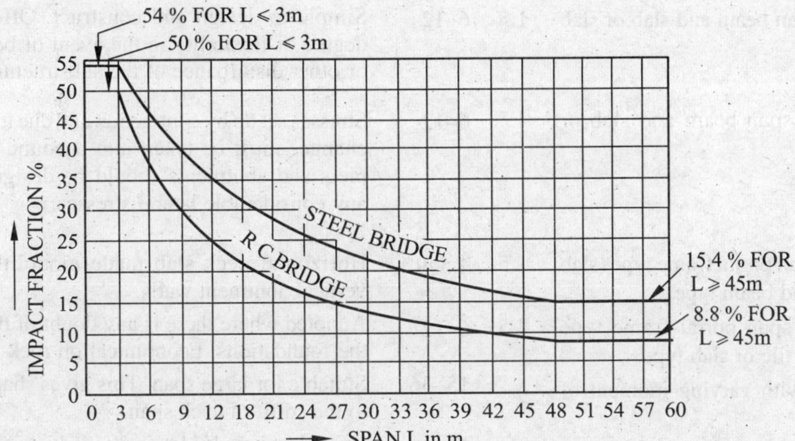

Fig. 7.39 Impact percentage curve for highway bridges for class 'A' and class 'B' loadings.

(ii) For spans of 9 m or more

(a) For tracked vehicles: For *RCC* bridges, 10% upto a spans of 40 m and in accordance with Fig. 7.39 for spans exceeding 40 m. For steel bridges–10% for all spans.

(b) For wheeled vehicles–For *RCC* bridges, 25% for spans upto 12 m and in accordance with Fig. 7.40 for spans exceeding 12 m.

For steel bridges, 25% for spans upto 23 m and in accordance with Fig. 7.39 for spans exceeding 23 m.

The span length to be considered above is determined as follows:

(i) Simply supported, continuous or arch spans—the effective span on which the load is placed.

(ii) Bridges having cantilever arm without suspended span: 0.75 of effective cantilever arm for loads on the cantilever arm and the effective span between supports for loads on the main span.

When there is a filling of not less than 0.6 m including the road crust as in spandrel filled arches, the impact allowance may be taken as half of that computed above.

Full impact allowance should be made for design of bearings. But for computing the pressures at different levels of the substructure, a reduced impact allowance is made by multiplying the appropriate impact fraction by the following factor:

(i) at the bottom of bed block : 0.5

(ii) for the top 3 m of the substructure decreasing below the bed block uniformly to zero : 0.5

(iii) for portion of substructure more than 3 m below the bed block : zero

7.31. BRIDGE TYPES

Various types of bridges used on highways are given below:

7.31.1. Concrete Bridges

S. No.	Bridge type	Span in m minimum	General use	Special points
1.	Single span beam and slab or slab type	1.5	6–12	Simple to design and construct. Offers a certain degree of flexibility in the event of bedding down or other disturbance of the substructure.
2.	Multiple span beam and slab or slab type	4.5	6–12	Stresses set up by contraction and due to temperature changes must be taken into account. Bearings on piers and abutments should be designed to avoid any considerable lateral stresses.
3.	Single span portal frame type (slab or slab and beam type)	1.5	4.5–9	Horizontal deck slab made monolithic with the vertical abutment walls.
4.	Multiple span portal frame type (slab and tile or slab type)	4.5	6–12	Adopted where there is any doubt of the rigidity of the foundations. Economical on rock foundations.
5.	Girders with varying moment of inertia	9	15–36	Suitable for large span. This gives shape of an arch to the soffit of each span.
6.	Double cantilever type with free centre span	12	18–40	Suitable where bridge is divided into several spans. Main feature is the construction of alternate spans with projecting cantilevers, the ends of which are used as supports for freely supported spans constructed between them. Particularly well suited to sites liable to some bedding down or settlement of the supporting piers.
7.	Double cantilever type	12	18–36	Consists of a single span between supporting piers but with projecting cantilevers on the landward side of the piers at both ends.
8.	Fixed barrel arch type single or multiple span	3	9–30	Arch fixed or made monolithic with the abutments or piers and provided with closed spandrels.
9.	Open spandrel ribbed arch type (fixed, 2 hinge or 3 hinge ribs)	21	30–60	Ribs monolithic with piers or abutments. Deck carried by column, beam and slab construction.
10.	Three-hinged barrel arch type (single or multiple span)	12	15–30	Suitable where there is a possibility of slight movement in the foundation. Also there is reduction in stresses due to contraction and temperature changes.
11.	Two-hinged barrel arch type (single or multiple span)	12	15–30	Not frequently used.
12.	Bowstring girder type	21	30–45	Consists of arch ribs constructed above the deck level of the bridge with horizontal ties connecting the springing and resisting the horizontal thrust of the arches. These tiers are in direct tension.
13.	Arch ribbed type with partially hung decking	36	36–75	Deck placed at a level above the apparent springing of the arch. Deck is suspended by ties from the arch rib for the major portion of its length, but towards the end is supported from the top of the arch ring by means of coloumns or cross walls.

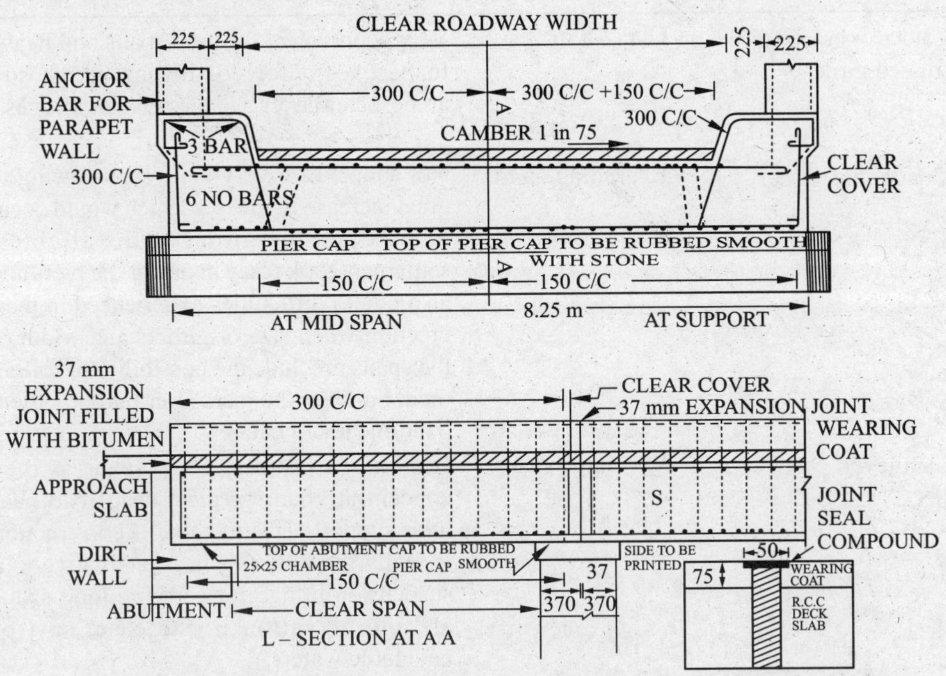

Fig. 7.40 Details of a *RCC* slab bridge without foot-paths and simply supported.

7.31.2 Iron and Steel Bridges

S.No.	*Bridge type*	*Span in m*	*Special points*
1.	Suspension	Up to 480 or more	Generally adopted on hill roads in India as it consists of very light members which are easily transported.
2.	Arch	Exceeding 105	Generally adopted over deep gorges or rivers in which pier foundations would be very expensive. Another advantage is that large metal arches can be built out from abutments when scaffolding is impossible in deep gorged or rapid stream.
3.	Bowstring girder	45 m	Not now met with to any considerable extent. Sometimes it may prove more economical than arch in respect of weight of steel used but more work is called for in the fabrication of the structure.
4.	Trough plate	Upto 4.5 m	Used for small road bridges.
5.	Rolled joist	Upto 6 m	Used for small road bridges.
6.	Plate girder	Between 6 m and 24 m	
7.	Tubular or box girder	Upto 135 m	Only varieties of plate girders. Practically it consists of two plate girders. They are now practically obsolete and are abandoned in favour of open web girders which are not only lighter but also offer much less surface to the wind.

8.	Lattice or braced girder or trussed girder	Over 18 m	Very economical for long spans and heavy loads except for conditions which suit arches, cantilever and suspension bridges.
9.	Continuous girder	Exceeding 45 m	Not adopted for very large spans owing to great increase in stresses which would occur in parts of the girder if the slightest settlement took place in one of the piers and also due to difficulties experienced in their erection when size of girders and width of the spans preclude the possibility of rolling them out over the piers after building them over one abutment.
10.	Cantilever	Very large spans	Best suited for very large spans as they concentrate their greatest dead load near their support. This method of construction is really the only practical alternative to a suspension bridge for carrying long spans at high elevation in the case, say, of navigable waters.
11.	Movable	Exceeding 30 m	Main types of movable bridge in use may be classified as swing bridges, Bascule bridges, either of the hinged or rolling lift class, vertical lift bridges, and transporter bridges. Most commonly met with is the swing bridge comprising of two or more girders or trusses supported somewhere about the centre so that they can be revolved upon a circular track or pivot resting upon the centre pier.

7.31.3. Prestressed concrete bridges

The application of the concept of prestressing to bridge engineering has resulted in a wide number of bridge types and has enlarged the span range possible with concrete.

Basically there are two types of prestressing.

(*i*) Pre-tensioning; and

(*ii*) Post-tensioning.

In the pre-tensioning method, the steel tendons (wires or strands) are tensioned before the concrete is placed in the moulds. The tendons are tensioned by hydraulic jacks bearing against strong abutments between which the moulds are placed. After the setting and hardening of the concrete, the tendons are released from the tensioning device and the forces in the tendons are transferred to concrete by bond.

Post-tensioning method basically requires the following steps:

(*i*) The prestressing tendon is assembled in a flexible metal sheath and anchor fittings are attached to its ends.

(*ii*) The tendon assembly is placed in the form and tied in place, along with other untensioned and auxiliary reinforcement;

(*iii*) Concrete is placed in the form and allowed to cure to the specified strength.

(*iv*) Tendons are stressed to computed extent and anchored.

(*v*) The space around the tendon within the sheath is grouted under pressure with cement grout, and

(*vi*) Anchor fittings are covered with a protective coating.

The prestressing tendon may be of the following types:

(*i*) Cable with a fixed number of parallel wires

(*a*) Freyssinet system…12 No. 5 mm ϕ

(*b*) Magnel—Blaton system….32 No. 7 mm ϕ

Number and sizes can be varied within certain ranges.

(*ii*) Seven wire strands e.g. Gifford-Udall strand system (13 mm nominal ϕ)

(*iii*) High tensile bars e.g. Lee Mcall bars—28 mm ϕ

The sheathing is generally of a flexible galvanized metal hose of interlocking construction.

The use of precast pre-tensioned members in bridge construction may follow one of the following methods:

(*a*) Members which are cast in the final shape and just assembled together at site. The solid and hollow precast slab segments shown in Fig. 7.41 (*i*) and (*ii*) and the beam members shown in Fig. 7.41. (*iii*) relate to this method. The openings shown in Fig. 7.41 (*ii*) may also be rectangular. This method is suitable for short spans.

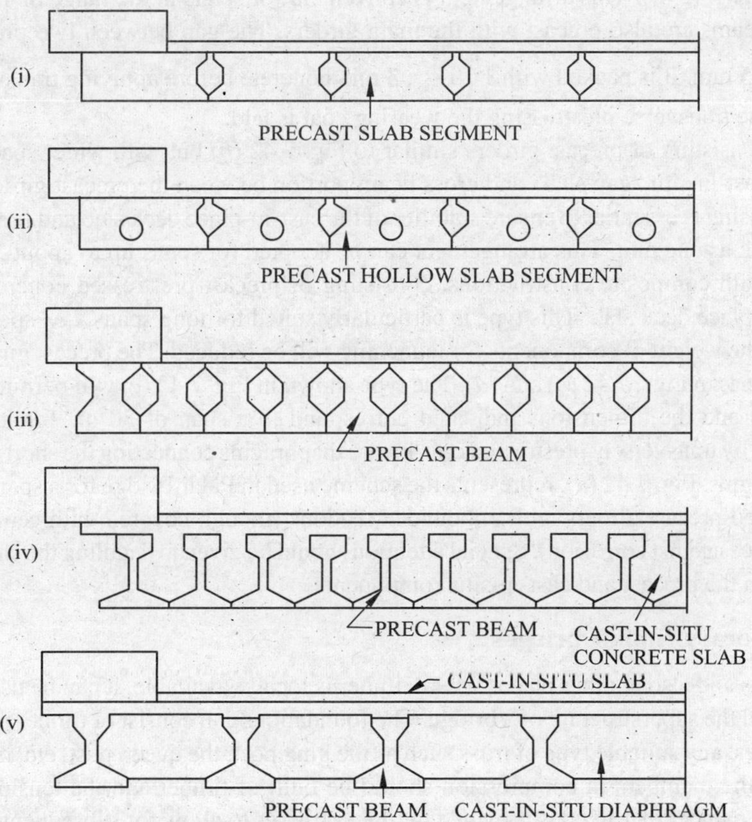

Fig. 7.41 Cross-section of bridges with pre-cast, pretensioned members.

(*b*) Beams of inverted T-shape placed side by side and the gaps filled with in-situ concrete to form an integral slab, as in Fig. 7.41 (*iv*). Transverse bond rods are provided just above the bottom flange passing through preformed holes in the precast beams. This method can be used upto a span of about 20 m. In view of the good transverse distribution of load possible in this type of construction, transverse prestressing can be eliminated. Other advantages include the elimination of propping and soffit shuttering, use of small cranes or hoisting devices for placing the beams in position, and the economy achieved by the use of relatively small prestressed members in conjution with large quantities of concrete cast in place. Sections have been standardized facilitating factory production.

(*c*) Composite deck consisting of precast girders placed at definite spacing and connected with cast-in place diaphragms and deck slab, as in Fig. 7.41 (*v*). These bridge girder sections are standardized. This procedure constitutes one of the most economical types of prestressed concrete bridges. Special attention should be given to the provision of adequate shear connectors.

Typical arrangements of bridge decks with post-tensioned girders suitable for simply supported constructions are shown in Fig. 7.42. Basically the arrangements may be one of the following types:

(*i*) Fully cast-in situ construction, as in Fig. 7.42 (*a*). This type is suitable when site conditions permit putting up the centering for the deck from the river bed and the river is dry for a major part of the year. It cannot be used if the river is perennial or if there is a considerable depth of water.

(*ii*) Bridge deck with precast prestressed girders, assembled together and transversely prestressed as in Fig. 7.42 (*b*). This arrangement is convenient for spans in the range of 15 to 20 m. The cross beams are also precast with the main girders. The gap between two precast girders is about 25 mm. It is packed with $1 : 1\frac{1}{2} : 3$ mix concrete before applying transverse prestress. After the transverse prestressing the wearing coat is laid.

(*iii*) Deck consisting of precast girders similar to Fig. 7.42 (*b*) but with wider spacing of girders and a cast-in-situ "gap slab" and cross beam portion between the precast girders. Transverse prestressing is essential to ensure stability of the cast-in-place deck slab and functioning of the full deck as one unit. This arrangement can be adopted for spans up to about 30 m.

(*iv*) Deck with composite constructions, consisting of precast prestressed concrete girders and cast-in-place deck slab. This type is particularly suited for long spans, *i.e.*, spans above 30 m in that the weight of components for launching will be reduced. The precast girder can be *T* or *U* shape as in Figs. 7.42 (*c*) and (*d*). The type shown in Fig. 7.42 (*c*) will permit use of bonded tendons and the dimensions indicated correspond to a span of 30 m. Torsional rigidity is secured by transversely prestressed cast-in-situ diaphragms connecting the short sloping precast diaphragms. Fig. 7.42 (*d*), represents the scheme used in Palar Bridge for a span of 27 m using U-shaped precast girders with unbounded tendons (though covered with cement mortar for resistance against corrosion). Special attention should be given to detailing the shear connection between the precast and cast-in-situ components.

7.31.4. Temporary timber bridges

In hilly areas and also at other places where timber is locally available, it can be used both for the foundations and the superstructure of a bridge. The foundations can consist of timber piles or trestles. For truss bridges, any suitable type of truss such as the king post, the queen post, etc. may be used. In these bridges, the members in compression should be built of timber and the tension members of steel. An initial camber of 1 in 60 is provided in timber trusses to allow for subsequent sagging under dead load and other causes. Sizes of the members should be suitably increased, if the local timber available is not of good quality.

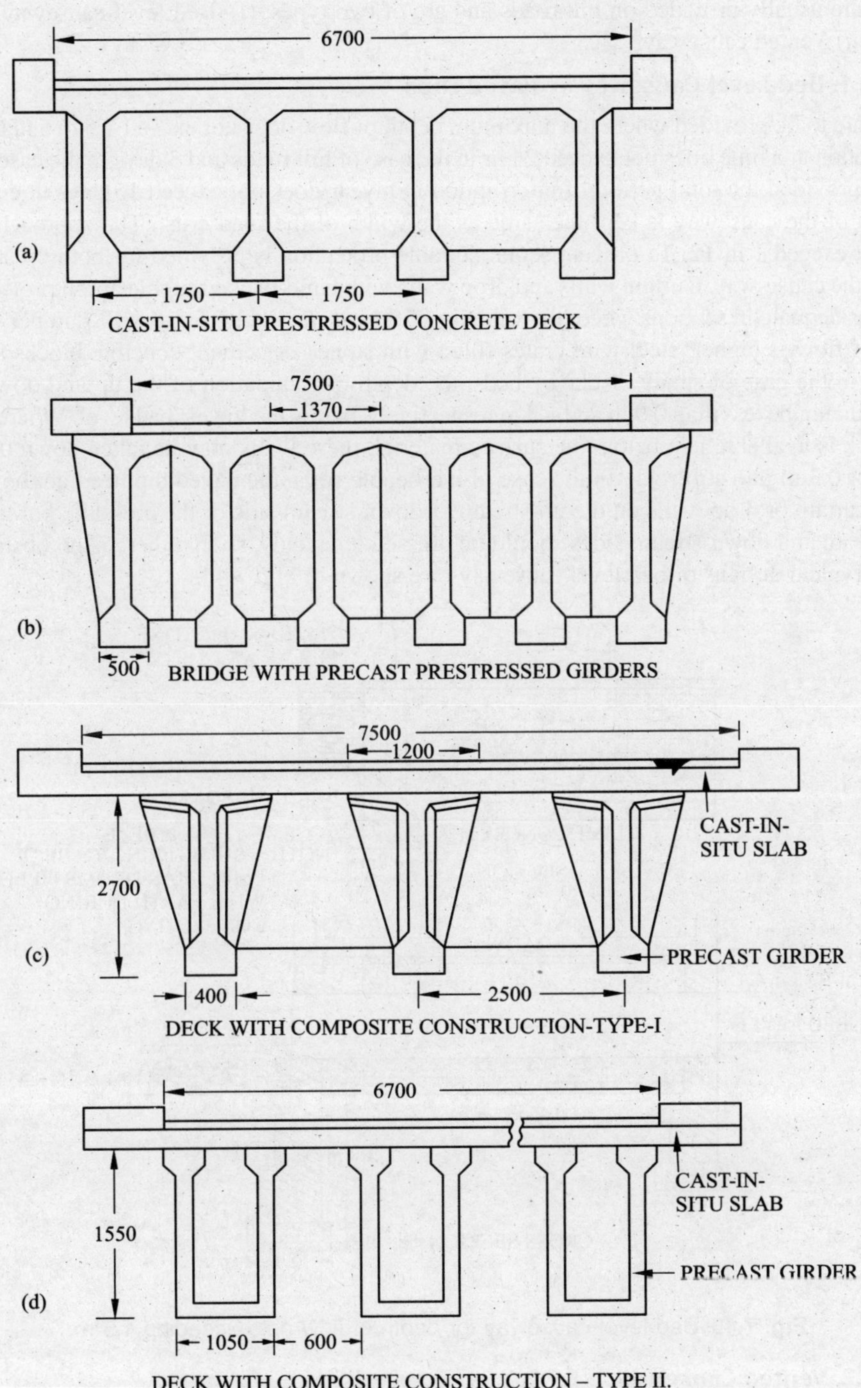

Fig. 7.42 Bridge decks with post-tensioned girders.

7.31.5. Irish bridges or Causeways

These are usually provided on hill roads and are of two types. (*i*) Bed level causeway or paved dips and (*ii*) Vented causeway.

7.31.5.1. Bed Level Causeway or Paved Dips

These are to be provided where the maximum depth of flow does not exceed 1.5 m and the period of interruption at a time does not exceed 24 hr in the case of hill roads and 3 days in the case of roads in plains, provided the total period of interruption in a year does not exceed 15 days in either case. The width of the causeway is kept the same as the width of the road crust. The approach gradient should not exceed 1 in 12. To prevent scour, suitable protection is provided for both up and down-stream of the cause-way. Curtain walls and drop walls with hand-packed rubble apron protection will usually be adequate in sections where the velocity of flow is not likely to exceed 1.8 m per sec. If the velocity of flow is higher, steel wire crates filled with stones or cement concrete blocks should be provided. In the case of sandy or clayey beds, the depth of foundation of the up and down-stream walls should not be less than 0.9 m and 1.8 m respectively, below the lowest bed level. Where moorum or soft rock is available just below the surface material, these walls may be taken down 0.9 m into moorum or 0.3 m into soft rock. In the case of inerodable soils, the paved dip need not be provided with any curtain or drop walls on the up-stream or down-stream side of the crossing, but the bed on the up-stream and down-stream sides should be dressed up evenly, so that there is no obstruction to the flow. Typical designs of bed level causeways are shown in Fig. 7.43.

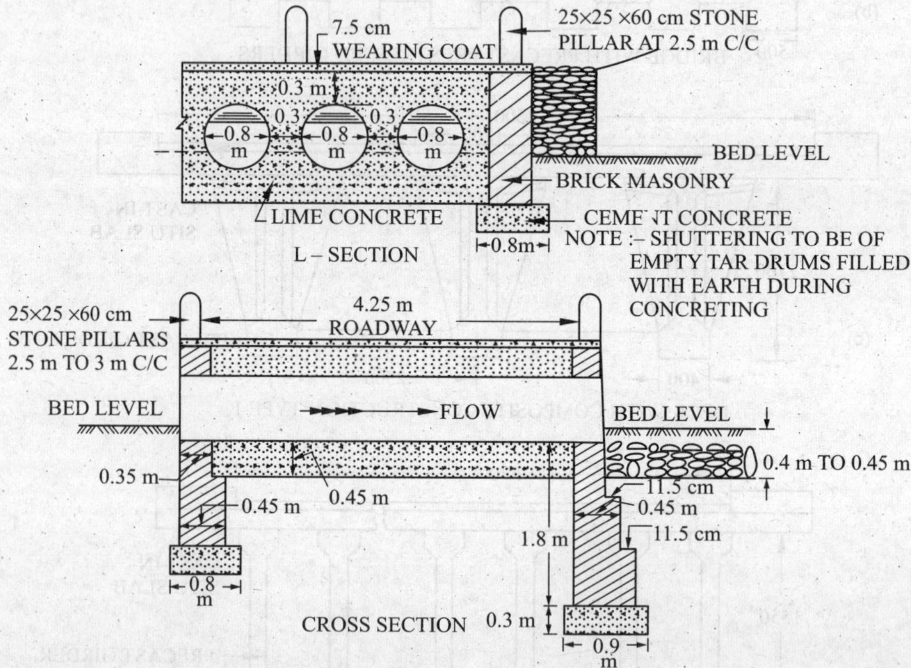

Fig. 7.43 Bed level causeway for depth of flow not exceeding 1.5 m.

7.31.5.2. Vented Causeway

In case a stream has a permanent dry weather flow-exceeding 0.3 m in depth, a suitable number of vents should be provided in a bed level causeway to carry the dry weather flow. The vents may consist of pipes, stone slabs, RCC slabs or arches depending upon the availability and cost of materials. Fig. 7.44 shows a typical vented causeway.

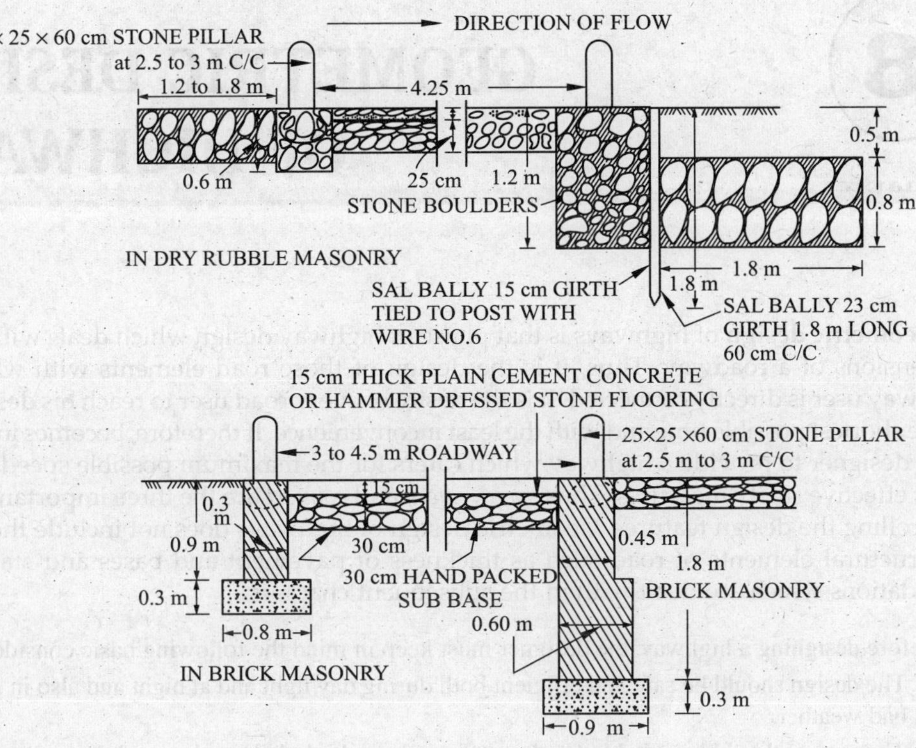

Fig. 7.44 Vented causeway with pipe opening.

❑❑❑

GEOMETRIC DESIGN OF HIGHWAYS

Geometric design of highways is that phase of highway design which deals with visible dimensions of a roadway. Thus, it is the design of those road elements with which the highway user is directly connected. It is the desire of every road user to reach his destination in the shortest possible time and with the least inconvenience. It therefore, becomes inevitable for a designer to provide a highway which caters for the maximum possible speed and the most effective safety measures. Thus *speed*, *safety* and *comfort* are the three important factors controlling the design features. Geometric design of highways does not include the design of structural elements of road such as thickness of pavement and bases and stability of foundations which are dealt with in the subsequent chapters.

Before designing a highway, the designer must keep in mind the following basic considerations:

1. The design should be safe and efficient both during day light and at night and also in good and bad weather.
2. The design should be suitable for the traffic volume both daily and at peak hours, and also for the future anticipated traffic.
3. The design should conform to the design speed, and characteristics of vehicles and their drivers using the road. It should take into account not only those vehicles that are at present using the roads but also those that may be expected to use it during its life time.
4. The design should be consistent, *i.e.*, there should not be abrupt changes so that the drivers are not confronted with difficult and serious situations.
5. The designer should have a full knowledge of the conditions under which a vehicle is going to operate. This has a bearing in the design of motor engines and gear box and the length of vehicle in the case of mountainous terrain, *etc*.
6. The design should provide an attractive and pleasing view to the road user and those who live along it.
7. The design must be complete. Traffic signs, signals, roadside treatment, *etc*. should be provided.
8. The design should be as simple as possible from the user's as well as from construction points of view. Too many changes in the cross-section or different types of surfaces will create difficulties in construction.
9. The maintenance cost should be as minimum as possible.

ELEMENTS OF GEOMETRIC DESIGN

The geometric design of highway will be considered under the following heads:

1. Dimensions and weights of road vehicles.
2. Terrain classification and design speed.

3. Design of the road cross-section. (*a*) width of carriageway; (*b*) width of roadway; (*c*) shoulders; (*d*) median strips; (*e*) side slopes (*f*) camber; (*g*) land width building lines and control lines; and (*h*) cross-section.
4. Super-elevation.
5. Curves-horizontal (circular and transition curves).
6. Sight distances.
7. Gradients.
8. Curves-vertical (summit and valley curves).
9. Alignment-horizontal, vertical and lateral and vertical clearances at under-passes.

DIMENSIONS AND WEIGHTS OF ROAD VEHICLES

The dimensions and weights of vehicles influence the road design to a great extent. The vehicle characteristics affecting the road design are listed below:

(*a*) The width of a vehicle affects (*i*) the width of traffic lanes (*ii*) the width of shoulders and (*iii*) parking space.

(*b*) The height of a vehicle affects the clearance to be provided to overhead structures such as under bridges, electrical service lines and to overhanging branches of road-side trees.

(*c*) The overall length (including trailer and semi-trailer combinations) affects (*i*) design of bridges where length of vehicles is a factor in the application of load; (*ii*) design of horizontal curves, regarding the radius, extra width, etc; (*iii*) design of valley curves; (*iv*) safety regulations governing passing and overtaking sight distances.

(*d*) Vehicle loads affect (*i*) design of bridges; (*ii*) design of pavement thickness; (*iii*) design of ruling gradients.

(*e*) Vehicle speeds affect (*i*) super elevation in curves; (*ii*) limiting radius; (*iii*) sight distances; (*iv*) width of pavement in straights and curves; (*v*) width of shoulder; (*vi*) grades; (*vii*) adequate provision of guard rails; (*viii*) increased protection at rail road crossing; (ix) special treatment of inter-sections; (*x*) capacity of a lane for traffic.

8.1. DIMENSIONS OF ROAD VEHICLES

Highways carry both passenger automobiles and trucks. The design standards should, therefore, meet the requirements of both. Usually the dimensions and weights of the trucks govern as they are more than those of passenger vehicles. The maximum dimensions of road vehicles have been recommended by the Indian Roads Congress. These are:

		Maximum dimensions
Width :		2.50 m
Height :	Single decked vehicle	3.80 - 4.20 m
	Double decked vehicle	4.75 m
Length :	Single unit with two axles	11.00 m
	Single unit with more than two axles	12.00 m
	Tractor semi-trailer combination	16.00 m
	Tractor and trailer combination	18.00 m

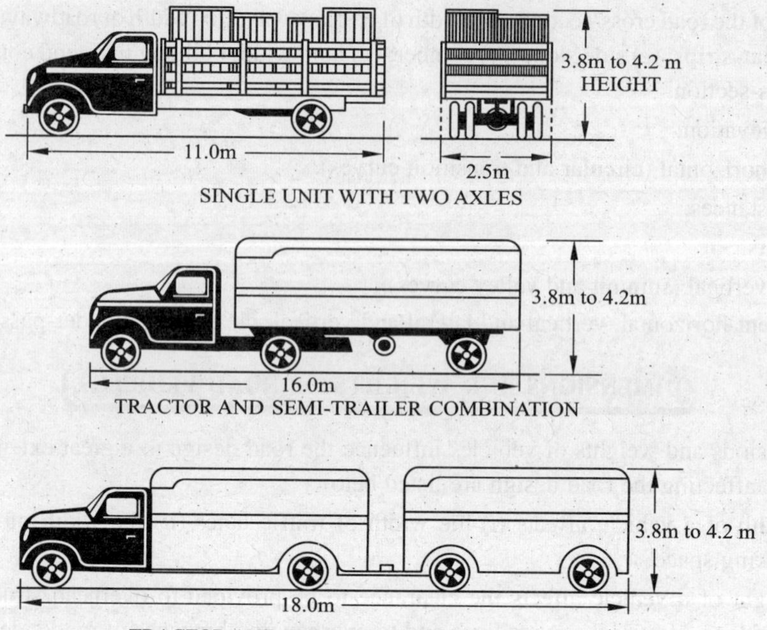

Fig. 8.1 Dimensions of road design vehicles.

No combination shall consist of more than two units and no such combination, laden or unladen, shall have an overall length exceeding 18.00 m.

These are illustrated in Fig. 8.1.

8.2. WEIGHTS OF VEHICLES

The weight of a vehicle is usually expressed in terms of its axle load which is the total load transmitted to the road by all wheels whose centres are included between two parallel vertical planes, one metre apart and transverse to the longitudinal axis of the vehicle. No single axle load, as suggested by IRC, should exceed 102 kN (10.2 tonnes) and for tandem axle 180 kN (18 tonnes). The gross load of any vehicle or combination of vehicles should not exceed the weight worked out by the following formula:

$$W = 1525\ (L + 7.3) - 14.7L^2$$

where W = The gross weight of the vehicle in kg.

L = The distance in metre between the extreme axles, measured parallel to the axis of the vehicle.

This relation holds good only when L is greater than 2.50 m. When its value is less than 2.50 m, the gross weight should not exceed 14515 kg.

TERRAIN CLASSIFICATION AND DESIGN SPEED

8.3. TERRAIN CLASSIFICATION

The geometric design of a highway is significantly influenced by terrain conditions. Economy dictates choice of different standards for different types of terrains. Terrain is classified by the general slope of the country across the highway alignment. The criteria for terrain classification, recommended by Indian Roads Congress is given in Table 8.1.

Table 8.1 Terrain classification

S. No.	*Terrain classification*	*Cross slope of the country*	
1.	Plain	0 – 10 per cent	More than 1 in 10
2.	Rolling	10 – 25 per cent	1 in 10 to 1 in 4
3.	Mountainous	25 – 60 per cent	1 in 4 to 1 in 1.67
4.	Steep	Greater than 60 per cent	Less than 1 in 1.67

This is shown in Fig. below.

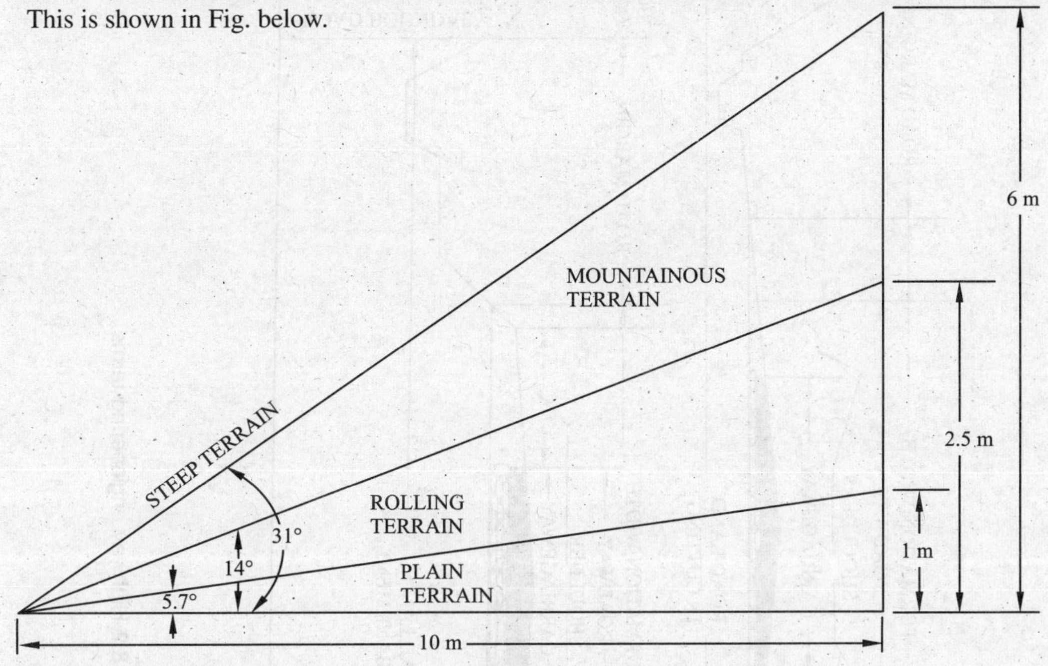

Classification of terrain.

8.4. DESIGN SPEED

It will not be economical to design all roads for very high speeds. A road has, therefore, to be designed for a specific speed known as "design speed". The design speed is defined as the maximum approximately uniform speed that will be adopted by the majority of the drivers. It is the basic parameter which determines all other geometric design features. Choice of design speed depends on the function of the road and the terrain conditions. Table 8.2 gives the design speed for various classes of roads.

Table 8.2 Design speeds for rural highways

S. No	*Road classification*	*Plain terrain*		*Rolling terrain*		*Mountainous terrain*		*Steep terrain*	
		Design speed (km/hr)							
		Ruling	*Minimum*	*Ruling*	*Minimum*	*Ruling*	*Minimum*	*Ruling*	*Minimum*
1.	National and State Highways	100	80	80	65	50	40	40	30
2.	Major District Roads	80	65	65	50	40	30	30	20
3.	Other District Roads	65	50	50	40	30	25	25	20
4.	Village Roads	50	40	40	35	25	20	25	20

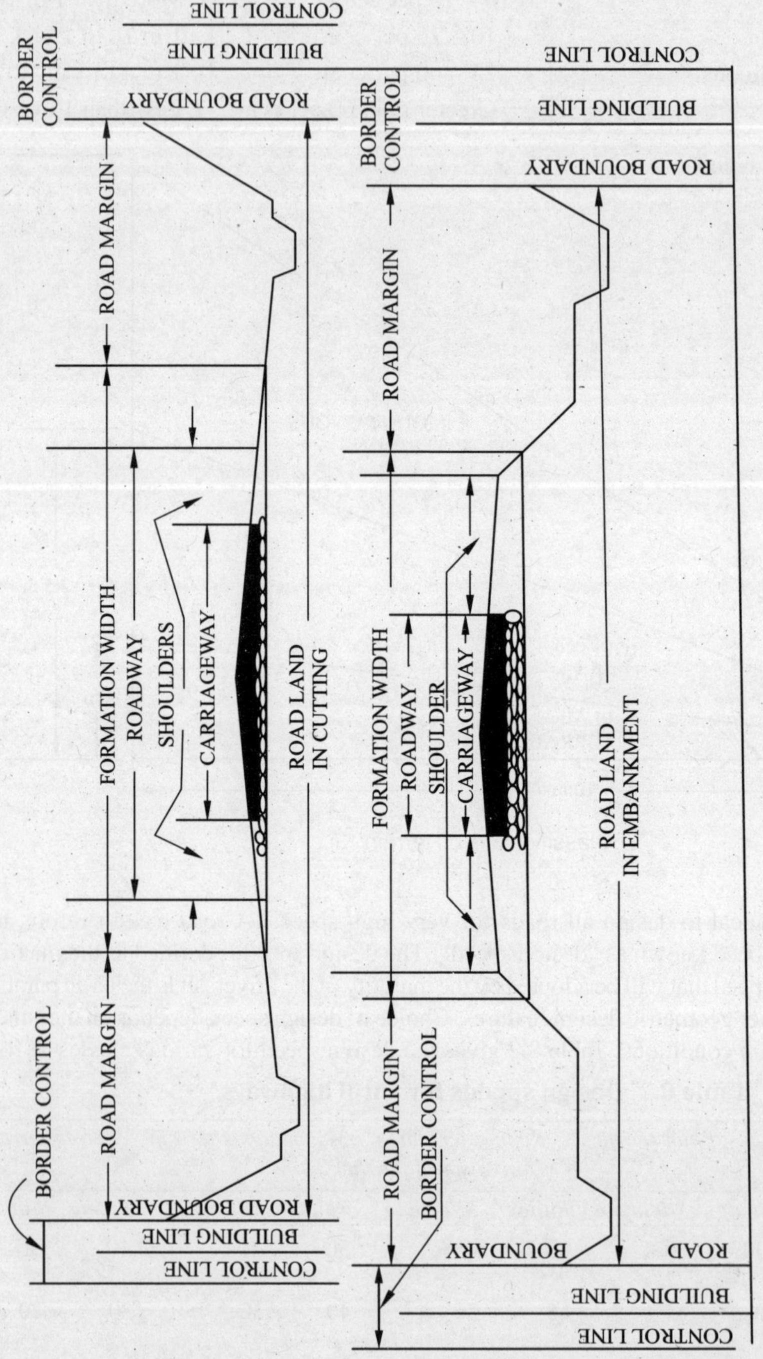

Fig. 8.2 Highway engineering terms.

The following design speeds are suggested for urban streets:

Arterial	80 km/hr
Sub-arterial	65 km/hr
Collector roads	55 km/hr
Local streets	30 km/hr

Normally ruling design speed should be the guiding criterion for correlating the various geometric design features. Minimum design speed may, however, be adopted in sections where site conditions, including costs, do not permit a design based on the ruling design speed.

The design speed should preferably be uniform along a given highway. But if changes in design speed become unavoidable due to variations in terrain, the design speed should be changed gradually by introducing successive sections of increasing/decreasing design speed so that the road users get conditioned to the change by degrees.

In some advanced countries even a minimum speed is specified which must be maintained by all drivers because a very low speed impedes or blocks the normal and reasonable movements of traffic. This condition arises only when the traffic reaches the saturation point. In India the problem may arise on Expressways or some National and State Highways.

CROSS-SECTION DESIGN

Fig. 8.2 represents some of the terms used in connection with the road cross-section The terms used to describe road cross-section for flexible and rigid pavements are given in Figs 8.3 and 8.4 respectively. Elements of the cross-section will be discussed in the following articles.

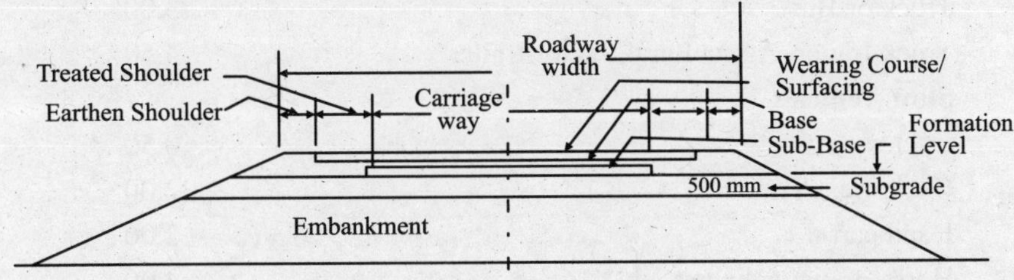

Fig. 8.3 Terms used to describe road cross-section elements in a flexible pavement.

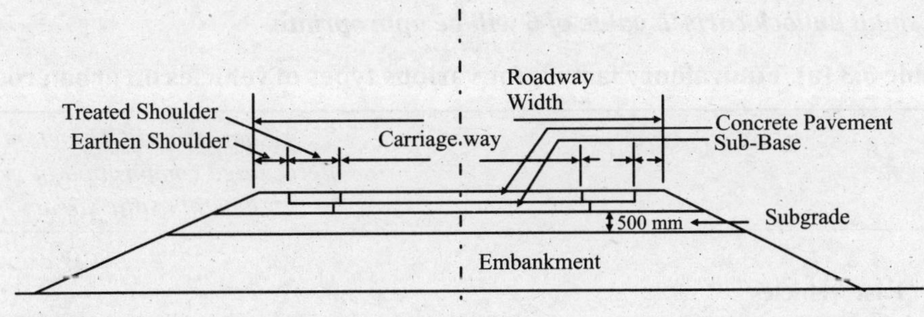

Fig. 8.4 Terms used to describe road cross-section elements in a concrete pavement

8.5. CARRIAGEWAY OR PAVEMENT WIDTH

8.5.1. Width of carriageway on straight highways

A carriage-way is that portion of a road which is constructed for vehicular traffic. Pavement width depends upon a number of factors, the most important are:

(*i*) Design traffic.
(*ii*) Traffic lane capacity.
(*iii*) Maximum overall width of vehicles using the road.
(*iv*) Clearance between the edge of the road and the body of the vehicle and between two vehicles or transverse placement of vehicles.

8.5.2. Design traffic

The width of carriageway should be sufficient for design traffic *i.e.,* traffic expected on the road in the design year. Design traffic depends upon the rate of growth of traffic, the design period, importance of road in the system, nature of roadside development, etc.

Under mixed traffic conditions, the different types of vehicles need to be converted to a common unit known as "*passenger car unit*" (PCU) by multiplying their number with relevant equivalency factors.

Tentative values of equivalency factors are given in Table 8.3 for rural and urban roads.

Table 8.3 Equivalency factors for various types of vehicles on rural roads

S. No.	*Vehicle type*	*Equivalence factor*
	Fast Vehicles	
1.	Motor-cycle or Scooter	0.50
2.	Passenger car, Pick-up van or Auto-rickshaw	1.00
3.	Agricultural tractor, light commercial vehicle	1.50
4.	Truck or Bus	3.00
5.	Truck-trailer, Agricultural tractor-trailer	4.50
	Slow Vehicles	
6.	Cycle	0.50
7.	Cycle-rickshaw	2.00
8.	Hand cart	3.00
9.	Horse-drawn vehicle	4.00
10.	Bullock cart*	8.00

**** For small bullock carts, a value of 6 will be appropriate.***

Table 8.3 (a) Equivalency factors for various types of vehicles on urban roads

S. No.	*Vehicle type*	*Equivalent PCU factors Percentage composition of vehicle type in traffic stream*	
		5%	10% and above
	Fast Vehicles		
1.	Two-wheelers - Motor Cycle or Scooter, etc	0.5	0.75
2.	Passenger car, Pick-up Van	1.0	1.0
3.	Auto-rickshaw	1.2	2.0
4.	Light commercial vehicle	1.4	2.0
5.	Truck or Bus	2.2	3.7
6.	Agricultural tractor - trailer	4.0	5.0

	Slow Vehicles		
7.	Cycle	0.4	0.5
8.	Cycle-rickshaw	0.5	2.0
9.	Horse-drawn vehicles	1.5	2.0
10.	Hand cart	2.0	3.0

8.5.4. Traffic lane capacity

The number of lanes to be provided depends upon the traffic lane capacity. This capacity will be maximum under ideal traffic conditions and is termed as *basic capacity*. Basic capacity of a traffic lane is the maximum number of vehicles which pass a definite point on the lane during one hour under the most ideal roadway and traffic conditions that can possibly be achieved. The following points are taken into consideration while calculating the basic capacity:

(*i*) Only one type of vehicles uses the road, *i.e.,* passenger cars.

(*ii*) All vehicles move at the same speed and maintain a minimum spacing between them.

(*iii*) There should be at least two lanes for the exclusive use of traffic moving in one direction *i.e.,* uni-directional traffic.

(*iv*) There should be no commercial vehicle using the road.

(*v*) Adequate widths of traffic lanes, shoulders and clearances should be provided.

(*vi*) There should be no restrictive sight distances, intersections, grades, improperly superelevated curves, *etc.*

(*vii*) There must be no interference by pedestrians.

This is calculated as follows:

Let L = The length of the passenger car in *m.*

T = Reaction time of the driver *i.e.,* the time taken to apply the brakes after those of the preceding car have been applied. This comprises the perception plus brake reaction time and is usually taken as one second.

S = Number of cars per hour per lane at saturation.

v = Speed of the vehicle in m/s = 0.28 V (approx.), where V is in km/hr

C = Clearance between two successive vehicles in m

f = Coefficient of friction between the vehicle and the road surface

$$H = \text{Headway in seconds} = \frac{C+L}{v} = \frac{C+L}{0.28\,V}$$

$$\therefore \quad S = \frac{60\times 60}{H} = \frac{3600}{H}$$

$$= \frac{3600}{\dfrac{C+L}{0.28V}}$$

$$\simeq \frac{1000V}{C+L}$$

Now C is the distance travelled by the car in reaction time, T plus the distance, d travelled by the vehicle to come to a stop after the application of the brakes.

$$\therefore \quad C = 0.28\,VT + d$$

$$= 0.28\,VT + \frac{V^2}{250f}$$

$$\therefore \quad S = \frac{1000\,V}{L + 0.28\,VT + \dfrac{V^2}{250f}}$$

or a general expression of the capacity would be:

$$C_p = \frac{1000\,V}{L + AV + BV^2} \quad \ldots(8.1)$$

Clayton has suggested the following formulae which have also been adopted by the Indian Roads Congress:

(*i*) Light vehicle capacity $= \dfrac{1000\,V}{4.5 + 0.28V + \dfrac{V^2}{250f}}$ ...(8.2)

(*ii*) Commercial vehicle capacity $= \dfrac{1000\,V}{6 + 0.28V + \dfrac{V^2}{170f}}$ ...(8.3)

The actual capacity of a traffic lane is less than the basic capacity due to various interferences. The various factors that reduce capacity are:

(*i*) *Mixed traffic.* It is seldom that only one type of vehicles uses the road. Some of the vehicles may be slow moving, while others fast moving. But if there is only one lane and even if one slow moving vehicle, such as bullock cart, comes on the road, the whole traffic behind it has to slow down as there is no provision of overtaking. Thus the capacity will be much reduced. Even if there are more than one lane, in the case of slow moving vehicles, the speed is less thereby causing a further reduction in capacity.

(*ii*) *Delays at the intersections.* It can so happen that vehicles may reach an intersection when traffic signal is at danger. Thus they will all come to a stop for a short while, thus reducing the capacity.

(*iii*) *Slipperiness of the road surface.* If rainfall occurs or water is poured on the surface, the road becomes slippery. The drivers will move at a slower speed to avoid accidents.

(*iv*) *Imperfect alignment.* Imperfect alignment and breaks in the profile reduces sight distances, thus reducing passing manoeuvres and hence the capacity.

(*v*) *Effect of grades.* On a down-hill grade, the spacing of the vehicles is more. Also, restricted sight distances may be provided on grades which decrease capacity.

(*vi*) *Insufficient lane width and edge clearances.* If sufficient lane width and edge clearances are not provided, a reduction in the speed will occur due to the fear of accidents. This will reduce capacity.

Keeping all these factors in view, it is customary to find out the possible capacity and the practical capacity. These are defined as below:

Possible capacity is the maximum number of vehicles which pass a definite point on a lane during one hour under the prevailing roadway conditions.

Practical capacity is the maximum number of vehicles which pass a definite point on a lane during one hour without causing any unreasonable delay hazard or restriction to the drivers overtaking under prevailing roadway and traffic conditions.

Thus to find the capacities of existing or contemplated roads, reduction in the basic capacity is made in view of the limitations cited above. Basic capacity is multiplied by the appropriate factors to get the practical capacity.

Problem 8.1 *Calculate the basic capacity for light weight and commercial vehicles for a lane on a State Highway in plain country. Assume the length of light vehicle as 4.5 m and commercial vehicle as 6.0 m.*

Solution. The traffic lane capacity for light weight vehicles is given by the relation 8.2.

$$C_l = \frac{1000\,V}{L + 0.28V + \dfrac{V^2}{250 f}}$$

For a State Highway in plain country

$$V = 100 \text{ km/hr}$$

For light weight vehicles, $L = 4.5$ m. Assume $f = 0.4$.

$$\therefore \quad C_l = \frac{1000 \times 100}{4.5 + 0.28 \times 100 + \dfrac{(100)^2}{250 \times 0.4}}$$

$$= \frac{1000 \times 100}{132.5}$$

$$= 755 \text{ vehicles per hour}$$

Traffic capacity for commercial vehicles is given by equation 8.3;

$$C_c = \frac{1000\,V}{6 + 0.28V + \dfrac{V^2}{170 f}}$$

$$= \frac{1000 \times 100}{6 + 0.28 \times 100 + \dfrac{(100)^2}{170 \times 0.4}}$$

$$= \frac{1000 \times 100}{6 + 28 + 147.1}$$

$$= \frac{1000 \times 100}{181.1}$$

$$= 552 \text{ vehicles per hour.}$$

Maximum overall width of vehicles. The lane width is influenced by the overall dimensions of the vehicles using the road. The legal overall width of commercial motor vehicle is 2.50 m. The lane width has to be increased suitably if military vehicles of larger dimensions have to use the road.

Transverse placement of vehicles. Based on extensive studies conducted in the U.S.A. on the transverse placement of vehicles while travelling free, passing vehicles in opposite directions and over-taking vehicles in the same direction, certain broad conclusions have been made. It is seen that trucks stay closer to the edge of pavements than passenger cars. The cars do not travel closer than 0.57 m while the trucks move at not less than 0.42 m from the edge the pavement. For safe overtaking of commercial vehicles, a body clearance of about 1.2 m is recommended.

For purposes of design, the capacity of different types of roads may be taken as given in Table 8.4.

Table 8.4 Capacity of different types of roads

Sr. No.	*Type of road*	*Capacity (passenger car units per day in both directions)*
1.	Single-lane roads having a 3.75 m wide carriageway with normal earthen shoulders	1,000
2.	Single-lane roads having a 3.75 m wide carriageway with adequately designed shoulders 1.0 m wide	2,500
3.	Two-lane roads having a 7 m wide carriageway with normal earthen shoulders	10,000
4.	Roads of intermediate width *i.e.,* having a carriageway of 5.5 m with normal earthen shoulders	5,000

Note: Capacity of highways having a dual carriageway will depend on factors like the directional split of traffic, degree of access control, composition of traffic, etc. Depending on the actual conditions capacity of a 4 lane divided highway could be upto 20,000 to 30,000. The standards in Table 8.4 are applicable where the visibility is unrestricted and there are no lateral obstructions within 1.75 m from the edge of pavement. These also presume that only a normal amount of animal drawn vehicles (say 5 to 10 per cent) are present in the traffic stream during the peak hour.

The Indian Roads Congress has suggested the minimum widths of carriageway for various classes of roads as given in Table 8.5.

Table 8.5 Width of carriageway for various classes of roads

A-Roads

Class of roads	*Width of carriageway in metres*			
	Single lane	*Two lanes without raised kerbs*	*Two lanes with raised kerbs*	*Multi-lane pavements per lane*
(*a*) National and state highways	3.75	7.0	7.5	3.5
(*b*) Major district roads	3.75			
(*c*) Other district roads	3.75			
(*d*) Village roads	3.75			

B-Bridges

	Clear width of carriageway on road bridges in metres
(*a*) Single lane bridges	3.75
(*b*) Two lane bridges	7.5
(*c*) Bridges on divided highways with four lane dual carriageway (2 lanes on either side of the central dividing verge)	7.5 for two lane carriageway

But on the basis of traffic classification, the carriageway width and number of traffic lanes, as recommended by IRC, are given in Table 8.5 (*a*).

Table 8.5 (a)

S. No.	*Traffic classification*	*Name of traffic*	*Average daily tonnage*	*Number of traffic lanes*	*Width of carriageway in m*
A.	Very light	Slow bullock cart traffic & occasional pleasure cars	Upto 100	1	3 (desirable 3.75)
B.	Light	Mainly slow traffic, mixed with occasional pleasure cars and not more than 45 vehicles exceeding 3 tonnes in weight in 24 hr	101 to 400	1	3.75
C.	Medium	Slow moving traffic mixed with pleasure cars including 46 to 250 vehicles exceeding 3 tonnes in 24 hr	401 to 1500	1	3.75
D.	Heavy	Mixed traffic including 251 to 450 vehicles exceeding 3 tonnes in weight in 24 hr	1501 to 4000	2 to 3	In case of total daily traffic exceeding 2500 tonnes per day in urban areas if the number of cyclists exceed 400 per hour in peak hour, one raised cycle track be provided on each side of the road.
E.	Extra heavy	Mixed traffic including 451 to 1500 vehicles exceeding 3 tonnes in weight in 24 hr.	4001 to 12000	4	Dual divided carriageway each 7.5 m wide with median (at least 5 m wide in rural areas & 1.5 m wide in urban areas). If the peak hour number of cyclists is 400 or more, one raised cycle track be provided on each side of the road.
F.	Extra heavy	Mixed traffic including 1501 to 4500 vehicles exceeding 3 tonnes in weight in 24 hr	12001 to 36000	6	Dual divided carriageway each 10.5 m with median (at least 5 m wide in rural areas and 1.5 m wide in urban areas). One raised cycle track and one 3 m wide footpath be provided on each side of the road.

8.6. WIDTH OF CARRIAGEWAY ON CURVES

On curves, a greater carriageway width has to be provided for the following reasons:

(*i*) On a curve a larger width is required by a vehicle since the following wheels do not tread the same path as the leading ones.

(*ii*) The drivers have a tendency to keep away from the edge of the carriageway while driving around curves.

(*iii*) The clearance between the vehicles while passing each other on a curve is more than that on a straight path.

It has been observed that, under normal speeds, the rear axle of a vehicle keeps itself in line with the radius of the curve. But the body of the vehicle is a rigid structure. It is therefore, essential that the leading wheels should twist at some angle to their axle to enable the vehicle to follow a route along the curve. Thus the leading wheels are required to follow a path of greater radius than their corresponding following wheels.

In Fig. 8.5, let

L = Wheel base length

R_1 = Radius of the outer following wheel

R_2 = Radius of the outer leading wheel

R = Radius of the curve

W_e = Extra width of pavement required for one lane

n = Number of traffic lanes

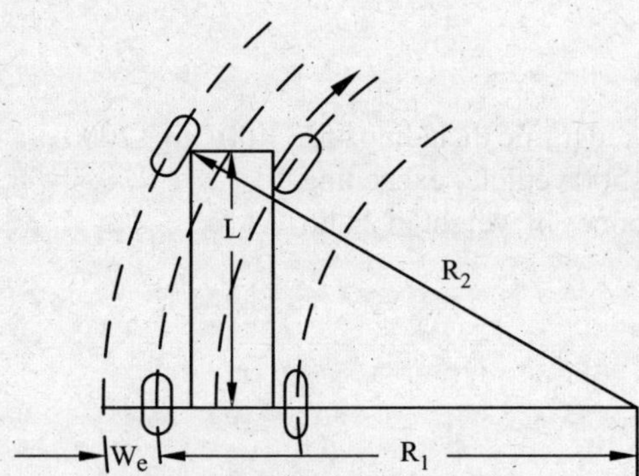

Fig. 8.5 Extra width on curves

Now, the extra width required per lane

$$W_e = R_2 - R_1$$

$$= R_2 - \sqrt{R_2^2 - L^2}$$

or $$R_2 - W_e = \sqrt{R_2^2 - L^2}$$

or $$(R_2 - W_e)^2 = R_2^2 - L^2$$

or $$R_2^2 + W_e^2 - 2R_2W_e = R_2^2 - L^2$$

Since the value of W_e is small as compared to R_2, W_e^2 may be neglected.

$\therefore \quad 2R_2W_e = L^2$

or $\quad W_e = \dfrac{L^2}{2R_2}$

$= \dfrac{L^2}{2R}$ (approx.)

where R is the radius of the curve.

If the number of traffic lanes are n, then total extra width (W_m) required for pavement is given by

$$W_m = \frac{nL^2}{2R}$$

This is called mechanical widening and takes into account only the first factor listed above.

Extra width has also to be provided for the other two factors as the judgement of the clearance by the driver becomes difficult on the curve. An emperical formula

$$W_p = \frac{0.105V}{\sqrt{R}}$$

is suggested for this, where V is the speed of the vehicle in km/hr, R is the radius of the curve in metres and W_p is the extra width required. This is termed as psychological widening.

$\therefore$ Total extra width required, W is given by

$$W = W_m + W_p = \frac{nL^2}{2R} + \frac{0.105V}{\sqrt{R}} \qquad \text{...(8.4)}$$

In most of the foreign countries, extra width for radius greater than 150 m is not recommended.

On two-lane or wider roads, it is necessary that both the above components *i.e.*, W_m and W_p should be fully catered for so that the lateral clearance between vehicles on curves is maintained equal to the clearance available on straights. Position of single-lane roads, is, however, somewhat different, since during crossing manoeuvres, outer wheels of the vehicles have in any case to use the shoulders whether on the straight or on the curve. It is therefore sufficient on single-lane roads if only the mechanical component of widening *i.e.*, W_m is taken into account.

Based on the above considerations, the Indian Roads Congress, has recommended the extra-width of carriageway to be provided at horizontal curves on single and two-lane roads as given in Table 8.6.

Table 8.6 Extra width of pavement at horizontal curves

Radius of curve (m)	Upto 20	21–40	41–60	61–100	101–300	above 300
Extra width (m)						
(*i*) Two lanes	1.5	1.5	1.2	0.9	0.6	Nil
(*ii*) Single lane	0.9	0.6	0.6	Nil	Nil	Nil

The widening should be effected by increasing the width at an approximately uniform rate along the transition curve. For transition curves, the widening will start at the beginning or tangent point and progressively increase at a uniform rate till the maximum designed widening is achieved at a point in the transition curve where the full designed super-elevation is reached. Further the maximum designed widening will be continued over the full length of the circular curve and to a point on the transition curve on the other side till the designed super-elevation starts reducing. This is shown in Fig. 8.6.

On curves having no transition *i.e.*, circular curves, widening should be achieved in the same way as the superelevation *i.e.*, two-third being attained on the straight section before start of the curve and one third one the curve.

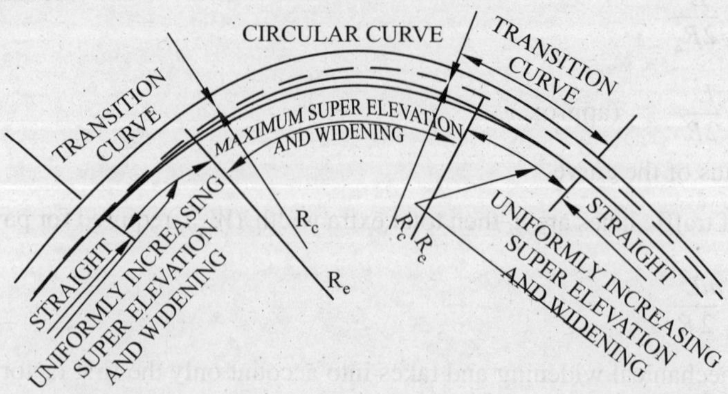

Fig. 8.6 Methods of effecting widening and superelevation.

Widening should be equally distributed on the inner and outer sides, except that on hill roads where it is preferred that the entire widening is done only on the inside. Similarly, the widening should be provided only on the inside when the curve is plain circular and has no transition.

Problem 8.2 *Determine the extra width required for a road of carriageway 7.5 m on a horizontal curve of radius 300 m. The longest wheel base of vehicle using the road may be taken as 6.1 m. Design speed is 80 km/hr.*

Solution. Extra width required, W, given by equation 8.4

$$= \frac{nL^2}{2R} + \frac{0.105V}{\sqrt{R}}$$

Here, $n = 2$ (carriageway of 7.5 m has two lanes)

$L = 6.1$ m

$R = 300$ m

$V = 80$ km/hr

$$\therefore \quad W = \frac{2 \times (6.1)^2}{2 \times 300} + \frac{0.105 \times 80}{\sqrt{300}}$$

$$= 0.124 + 0.485$$

$$= 0.609 \text{ m}$$

IRC has recommended a value of 0.6 m (Table 8.6).

Problem 8.3 *Calculate the total width of pavement on a horizontal curve for a National Highway in plain area with a ruling minimum radius. Assume data suitably.*

Solution. Adopt the following data:

(*i*) Design speed = 100 km/hr

(*ii*) Pavement width on straight stretch = 7.0 m

(*iii*) Number of lanes = 2

(*iv*) Wheel base of trucks = 6.1 m

(*v*) Ruling minimum radius = 360 m (From Table 8.13 or worked out from relation $R = \frac{V^2}{28}$)

Extra-width required on curves

$$= \frac{nL^2}{2R} + \frac{0.105V}{\sqrt{R}}$$

$$= \frac{2\times(6.1)^2}{2\times360} + \frac{0.105\times100}{\sqrt{360}}$$

$$= 0.10 + 0.55$$

$$= 0.65 \text{ m}$$

Total pavement width on horizontal curve

$$= 7.0 + 0.65$$

$$= 7.65 \text{ m}$$

8.7. WIDTH OF ROADWAY

A roadway consists of carriageway plus shoulders on either side or the bottom width of cutting. The Indian Roads Congress has recommended the minimum roadway widths for different classes of highway as follows:

Roadway width for single lane and two lane roads in plain and rolling terrain. The width of roadway for single and two lane roads in plains and rolling terrain is given in table 8.7.

Table 8.7 Width of roadway for single lane and two-lane roads in plain and rolling terrain

S. No.	*Road classification*	*Road width (m)*
*1	National Highways and State Highways (Single or two lanes)	12.0
2.	Major District Roads (Single or two lanes)	9.0
3.	Other District Roads	
	(*i*) Single lane	7.5
	(*ii*) Two lanes	9.0
4.	Village roads (Single lane)	7.5

Note : In case of State Highways having single lane pavement, the width of roadway may be reduced to 9 m if the possibility of widening the carriageway to two lanes is considered remote.

Width of roadway for single lane and two lane roads in mountainous and steep terrain. The width of roadway, exclusive of side drains and parapets, for single and two lane roads in mountainous and steep terrain is recommended in Table 8.8. In certain places, passing places may be required in addition.

Table 8.8 Width of roadway for single lane and two lane roads in mountainous and steep terrain

S. No.	*Road classification*	*Roadway width (m)*
1.	National Highways and State Highways	
	(*i*) Single lane	6.25
	(*ii*) Two lanes	8.8
2.	Major District Roads and other District Roads (Single lane)	4.75
3.	Village Roads (Single lane)	4.0

Notes: (1) The roadway widths given above are exclusive of parapets (usual width 0.6 m) and side drains (usual width 0.6 m).

(2) The roadway width for village roads are on the basis of a single lane carriageway of 3 m. If a higher pavement width is adopted, the roadway width should be increased correspondingly.

(3) In hard rock stretches, or unstable locations where excessive cutting might lead to slope failure, width of roadway may be reduced by 0.8 m on two-lane roads and 0.4 m in other cases. However, where such stretches occur in continuous long length, reduction in roadway width should not be effected unless requisite passing places are provided.

(4) On horizontal curves, the roadway width should be increased corresponding to the extra widening of carriageway for curvature (Table 8.6).

(5) On roads subject to heavy snowfall, where regular snow clearance is done over long periods to keep the road open to traffic, roadway width may be increased by 1.5 m for major district roads, other district roads and village roads.

8.7.1. Passing places for roads in mountainous and steep terrain

Passing places or lay-byes should be provided on single lane roads in mountainous and steep terrain to cater for the following requirements:

(*a*) To facilitate crossing of vehicles approaching from opposite direction; and

(*b*) To tow aside a disabled vehicle so that it does not obstruct the traffic.

Passing places are not necessary on two lane national and state highways. But on single lane sections having narrow roadway, it is desirable to provide some passing places depending on actual needs. On other roads, these should be provided, in general, at the rate of 2 to 3 places per km. Their exact location should be judiciously determined taking into consideration the available extra width on curves and visibility.

Normally the passing places/lay-byes should be 3.75 m wide, 30 m long on the inside edge *i.e.*, towards the carriageway side and 20 m long on the farther side.

Roadway width for multi-lane highways. For multi-lane highways, roadway width should be adequate for the requisite number of traffic lanes besides shoulders and central median. Width of shoulder, in general, is kept 2.5 m. The width of carriageway is decided in accordance with Table 8.5 and for width of median, refer to page 279.

8.8. ROADWAY WIDTHS AT CROSS-DRAINAGE STRUCTURES

Cross-drainage structures are difficult to widen at a later stage. As such, the roadway width for them should be decided very carefully at the planning stage itself. It is desirable to provide a higher roadway width at the cross-drainage structures right in the beginning for roads being built to lower standards initially or those which are expected to be upgraded/widened in the foreseable future. IRC has recommended the following values of minimum roadway widths at cross-drainage structures.

At culverts (upto 6 m span): In plain and rolling terrain, the overall width of culverts (measured from outside to outside of the parapet walls) should equal the normal roadway width given in Table 8.7. In mountainous or steep terrain, the clear roadway width available on the culverts (measured from inside to inside of parapet walls of kerbs) should be as below:

All roads other than 6 m span : As given in Table 8.8.

Village roads

(*i*) Minimum : As given in Table 8.8

(*ii*) Desirable : 4.5 m

At bridges (greater than 6 m span). At bridges, the clear width of roadway between kerbs should be as under:

Single-lane bridge : 4.25 m
Two-lane bridge : 7.5 m
Multi-lane bridge : 3.5 m per lane plus 0.5 m for each carriageway.

At causeways and submersible bridges, the minimum width of roadway (between kerbs) should be 7.5 m, unless the width is specially reduced.

Where a footpath is provided for the use of pedestrians, its width should not be less than 1.5 m.

8.9. SHOULDERS

It is that portion of the roadway which lies between the outer edge of the carriageway and the inner edge of the ditch, gutter, kerb or slope. Shoulders serve as a place for parking. In case shoulders are not provided or their width is small, roadway capacity is reduced and chances of accidents are increased. They also help in overtaking manoeuvres on a one-lane road.

Shoulders of very narrow widths ranging from 0.6 m to 1.2 m made up of earth, are usually provided. These shoulders could be used only in fair weather. Larger shoulder widths are now being recommended and 3 m widths are considered suitable for high type facilities. It is better to provide gravel or similar material to make the surface hard. It is common to pave the inside 0.5 m to 1 m with bituminous material or at least to apply a bituminous surface treatment. The following points are worth noting in regard to the provision of shoulders:

(*i*) Well maintained grass shoulders have the same effect on the transverse position of moving vehicles as well maintained gravel shoulders.

(*ii*) Bituminous treated shoulders 1.2 m or more in width adjacent to 7.0 m and 7.5 m pavement serve to increase the effective pavement width by about 0.6 m.

(*iii*) Shoulder width in excess of 1.2 m does not influence the effective pavement width for moving traffic, but it does not mean that shoulder widths greater than 1.2 m should not be provided. On major highways a width of 3 m is suggested, whereas in hilly areas, 1.2 m shoulder width is permitted.

(*iv*) Shoulder use increases rapidly with a decrease in pavement width below 7.0 m.

(*v*) Drivers do not reduce their speeds when meeting other vehicles on narrow pavements with unpaved shoulders but will squeeze down the centre clearances, thereby increasing accident hazards.

Shoulder width will be one-half the difference between the roadway width (Table 8.7) and carriageway width (Table 8.5).

8.10. MEDIAN STRIPS FOR DIVIDED HIGHWAYS

A road on which traffic in one direction of travel is separated from that in the opposite direction is called a divided highway and the dividing strip in the middle of the roadway is known as median strip. Positive segregation of traffic between opposing streams is essential for efficient and safe movement of vehicles in the two directions. Median strips are required on very busy roads, which have four or more lanes, especially in crowded cities like Mumbai, Delhi, Kolkata, Chennai, etc. If these strips are narrow, separation is provided by raised kerbs and where greater space is available, kerbs may or may not be used. Median strips from 3 to 9 m width are recommended. If the land is expensive, then narrow median strips may be provided. Medians are narrowed at grade separation to effect reduction in the length or width of structures. However, wide median strips are to be preferred for the following reasons:

(*i*) Chances of accidents, resulting in head-on collision, are reduced.

(*ii*) Head light glare at night from traffic in the opposite direction is less troublesome.

(*iii*) At intersections, these provide a refuge for the cross-traffic.

Indian Roads Congress has recommended minimum desirable width of medians on rural highways as 5 m, but this can be reduced to 3 m where land is restricted. On long bridges and viaducts, (road bridges over valleys in hilly areas) the width of median may be reduced to 1.5 m, but in any case this should not be less than 1.2 m.

As far as possible, the median should be of uniform width in a particular section of the highway. However, where changes are unavoidable, a transition of 1 in 15 to 1 in 20 must be provided.

8.11. SIDE SLOPES

Earth embankments of usual height will stand safely with side slope of 1 in $1\frac{1}{2}$ in cutting, a side slope of 1 : 1 will serve the purpose well. Rock cuts as steep as 1 in $\frac{1}{2}$ and occasionally 1 in $\frac{1}{4}$ are stable. These slopes were adopted in the past as they involved minimum amount of earth work. But the tendency in these days is to go in for flatter slopes as they result in safe operation of vehicles and the maintenance cost is also reduced. Steep slopes in embankments have many disadvantages. These are:

(*i*) Steep slopes result in accident hazards. If the driver loses control, due to a wheel of the vehicle going over the edge, it may result in overturning. In case slopes are flat, the vehicle can be continued safely down the slope or can also be directed back to the road easily.

(*ii*) Steep slopes on gutter ditches may also cause accidents.

(*iii*) The clearance between the edge of the slopes and the vehicle is more as the driver keeps away from the edge for fear of accidents. With flat slopes, the driver is closer to the edge and farther from opposing traffic.

(*iv*) Steep slopes erode badly. The maintenance cost is much increased.

(*v*) Tree plantation or grass growing is difficult, which further aggravates erosion trouble.

Indian Roads Congress has recommended the following side slopes in embankments.

(*a*) Over 0.6 m height : 1 in 4.

(*b*) Upto 0.6 m height: natural slope or 1 in 2 whichever is flatter.

In cutting, the side slopes should not be steeper than 1 in 2 except in solid rock.

8.12. CAMBER OR CROSS-FALL

It is the convexity provided to the cross-section of the surface of carriageway and is the difference of level between the highest point, known as the crown (which is usually the centre of the carriageway), and the edge. In the case of a curved profile, camber is expressed in terms of the slope of the chord joining the crown and the edge of the carriageway whereas in the case of a straight line profile, it is the slope of the actual surface, e.g. a camber of 1 in 48 in carriageway 7.0 m wide means a difference of 7.3 cm in level between the crown and the edge. It is designated by 1 in n which indicate that the cross fall is in the ratio of 1 vertical to n horizontal. It is expressed in percentage.

The reasons for providing camber on the road surface are:

(*i*) To drain surface water.

(*ii*) To separate the traffic in the two directions.

(*iii*) To improve the appearance of the road.

The first reason is very important and this will require a profile which can discharge water quickly to the gutters. The amount of camber to be provided for drainage will depend upon the intensity of rainfall in the terrain through which the road passes and its rate of run-off from the road surface. If the intensity is less severe and is spread over a longer period or the road is pervious *e.g.* earth roads or water bound macadam roads, a steep camber will be required. For heavy rainfall intensity and confined to a few months in a year or for an impervious road with a concrete surface, flat camber will be suitable.

A natural separation will be provided by the crown to demarcate the centre line. In the case of divided carriageways, the crown is replaced by the median strip or central reservation.

A road with camber provides a better appearance than a road without it. The minimum camber, as recommended by the Indian Roads Congress, for various types of roads is given in Table 8.9. In general a steep camber is to be preferred in areas of heavy rainfall and flat camber in regions of low rainfall.

Table 8.9 Camber/cross fall values for different road surface types

S. No.	*Surface type*	*Camber/Cross fall*
1.	High type bituminous surfacing or cement concrete	1.7 to 2.0 per cent (1 in 60 to 1 in 50)
2.	Thin bituminous surfacing	2.0 to 2.5 per cent (1 in 50 to 1 in 40)
3.	Water bound macadam, gravel	2.5 to 3.0 per cent (1 in 40 to 1 in 33)
4.	Earth	3.0 to 4.0 per cent (1 in 33 to 1 in 25)

Generally, undivided roads on straights should be provided with a crown in the middle surface on either side sloping towards the edge. However on hill roads, this may not be possible in every situation particularly in reaches with a winding alignment where straight sections are few and far between. In such cases, discretion may be exercised and instead of normal camber, the carriageway may be given a uni-directional cross fall towards the hill side having regard to factors such as the direction of superelevation at the flanking horizontal curves, ease of drainage, problem of erosion of the down-hill face, *etc.*

On divided roads *i.e.*, dual carriageways having a median, it is usual to have a uni-directional cross fall for each carriageway sloping towards the outer edge.

Cross fall for shoulders. The cross fall for earth shoulders should be at least 0.5 per cent steeper than the slope of the pavement subject to a minimum of 3 per cent. If the shoulders are paved, a cross fall appropriate to the type of surface should be selected from Table 8.9. In superelevated sections, the shoulder should normally have the same cross fall as the pavement.

8.12.1. Design of Road Camber

There are four methods by which road camber can be designed; as given below:

(*i*) A straight line camber [Fig. 8.7 (a)]

Let R = Rise of the crown

$W = \frac{1}{2}$ road width

Y = Ordinate at a distance X from the crown

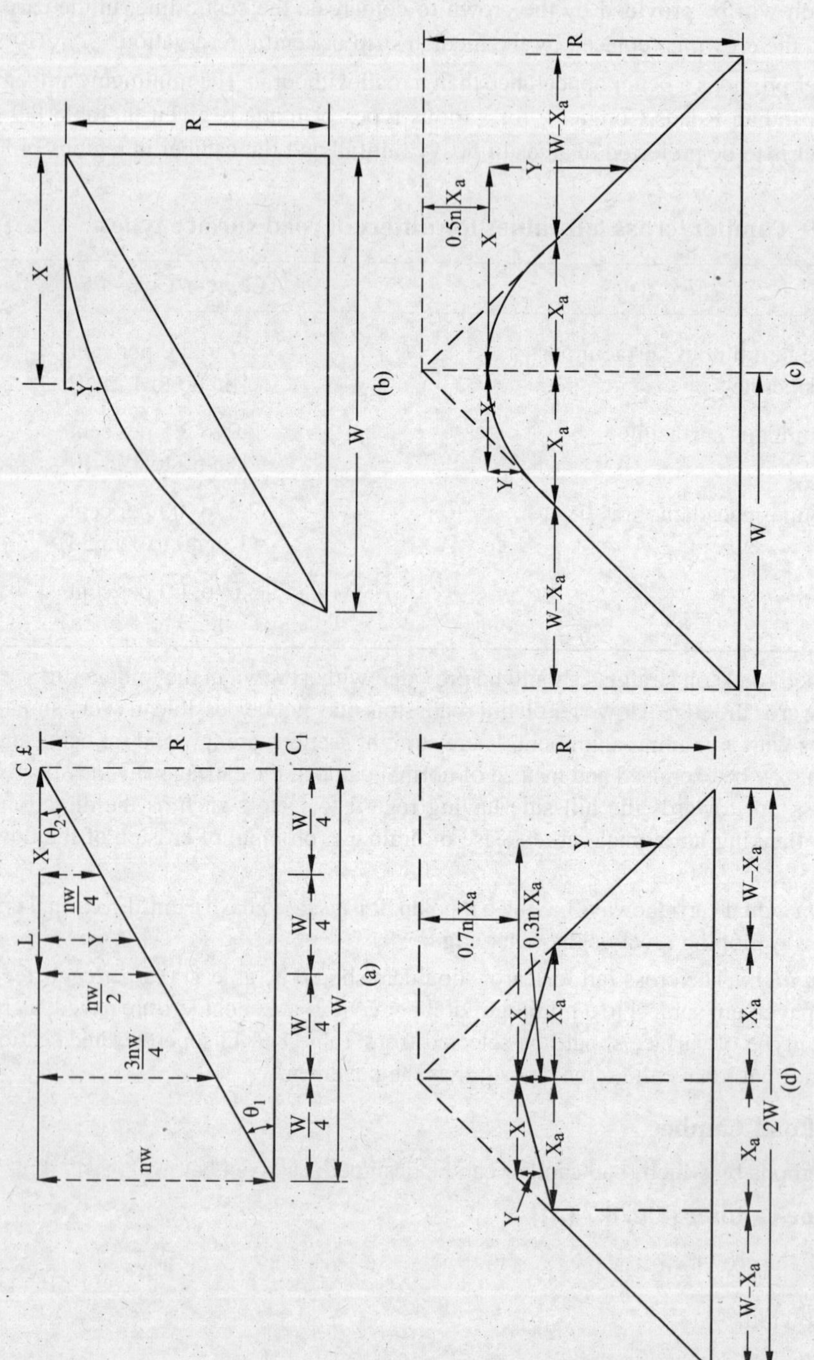

Fig. 8.7 Different types of cambers (a) straight line (b) parabolic (c) partly straight and partly parabolic ; and (d) a camber made up of two straight lines.

Road camber $n = \frac{R}{W}$

Then angles θ_1 and θ_2 being equal;

$$\frac{Y}{X} = \frac{R}{W} = n \qquad \text{...(8.5)}$$

$$\therefore \quad Y = nX$$

(ii) Parabolic camber [Fig. 8.7 (b)]

In this case, the ordinate varies as the square of the distance

$$\frac{Y}{X^2} = \frac{R}{W^2} = \frac{R}{W}\frac{1}{W} = \frac{n}{W}$$

$$\therefore \quad Y = \frac{nX^2}{W} \qquad \text{...(8.6)}$$

(iii) *Partly straight and partly parabolic [Fig. 8.7 (c)]*

For the portion X_a;

$$Y = \frac{n_1 X^2}{W} = \frac{nX^2}{2X_a}; \text{ since } W = X_a \text{ and camber} = \frac{1}{2}n. \qquad \text{...(8.7)}$$

For the remaining straight line portion $(W - X_a)$

$$Y = nX - \text{flattening of the crown}$$

$$= nX - 0.5n\,X_a$$

$$= n\,(X - 0.5\,X_a) \qquad \text{...(8.8)}$$

(iv) A camber made up of two straight lines [Fig. 8.7 (d)]

Here the camber is straight both in $W - X_a$ and X_a and the crown is flattened by $0.7nX_a$. The formulae are:

For X_a,

$$y = 0.3\,nX \qquad \text{...(8.9)}$$

(Since camber = 0.3 n)

for $W - X_a$,

$$Y = nX - \text{flattening of the crown}$$

$$= nX - 0.7\,nX_a = n\,(X - 0.7\,X_a) \qquad \text{...(8.10)}$$

This method is considered to be the best for Indian roads, as the contact area of the tyres with the road surface is claimed to be more than in other cases. Thus the damage to the road surface is less in this case.

At horizontal curves, the outer edge of the road is raised with respect to the inner edge for effecting drainage and providing desired superelevation. On straight roads the centre of the road is raised with *respect to the edges forming a crown at the centre.*

8.12.2. Disadvantages of Heavy Camber

(*i*) The road gets worn excessively at the central portion. This is due to the fact that the drivers have a tendency to keep their vehicles near the centre where the slope is flatter than the edges. In fact, the traffic needs little or no camber.

(*ii*) Very steep camber is not feasible on Indian roads on which bullock carts are predominant as their centre of gravity is very high.

(*iii*) This results in increase in accidents as the vehicles moving at high speed have a tendency to slip towards the edges.

(*iv*) The berms may get washed away due to increased velocity of drain water.

(*v*) The contact area in the case of iron-tyred vehicles will be less, with the result, there will be concentration of load on the road surface. Thus, there will be more damage to the road.

The only advantage in providing a heavy camber is that the water gets easily drained and there are little chances of water being collected in local sinkages on the road. This will avoid weakening of the foundations of the road. Moreover, water pools on the road will lend an ugly appearance.

8.12.3. Relation between Camber and Gradient

It may be concluded from above that flatter cambers are required for the purpose of smooth traffic flow and steeper cambers are necessary for the purpose of drainage. Both these objectives can be achieved on gradients.

When a road is in gradient, its effect will be to make the surface water travel a greater distance before reaching a channel. But it is much better if the rain water travels as minimum distance along the road as is possible. If camber is reduced, the rain water will tend to run down the hill nearly parallel to the central line. Thus camber in a gradient should be increased and not decreased. Water always travels along a path which is perpendicular to the contours of the road surface. For finding the relation between camber and gradient, it will be assumed that camber is a straight fall from the centre.

Let

C = Per cent grade of camber

G = Per cent longitudinal grade

AB = Half the width of carriageway

$= b/2$

In Fig. 8.8, let A be the point on the central line or crown of the road from which water is required to be drained. Let the longitudinal slope be from A towards F and E be the point on the road which has the same level as that of B, *i.e.,* EB is a contour line. Water will flow from A along the path AD so that AD is perpendicular to EB. BD equals b.

From triangles ABD and ABE;

$$\frac{DB}{AB} = \frac{AB}{EA}$$

or $$EA = \frac{AB^2}{DB} = \frac{\left(\frac{b}{2}\right)^2}{b} = \frac{b}{4}$$

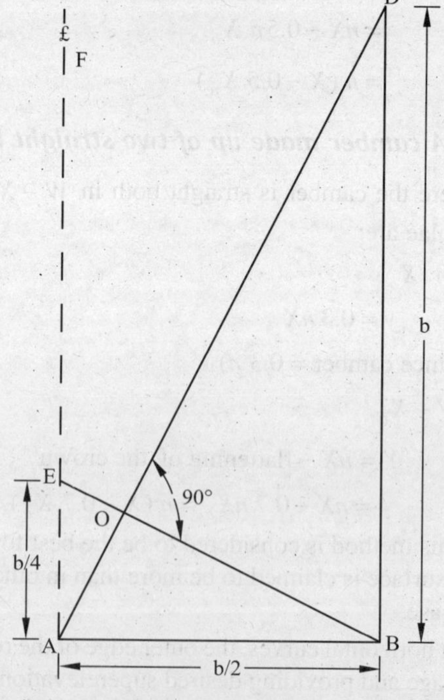

Fig. 8.8 Relation between camber and gradient

Also $AB \times n = AE \times g$

or $\frac{b}{2} \times n = \frac{b}{4} \times g$

or $n = \frac{g}{2}$ ...(8.11)

Thus camber should not be less than half the gradient. However, there will be difficulty in construction as the camber changes with gradients.

Problem 8.4 *Determine the height of the crown with respect to the edges of the road in the following cases:*

(*a*) *WBM road 3.8 m wide in areas of low rainfall.*

(*b*) *WBM road 7.0 m wide in areas of heavy rainfall.*

(*c*) *Bituminous road 7.5 m wide.*

(*d*) *Concrete raod 7.5 m wide.*

Solution. (*a*) Adopt a camber of 1 in 48 for WBM roads in areas of low rainfall.

Rise of crown w.r.t. edges

$$= \frac{1}{2} \times \text{width of road} \times \text{camber}$$

$$= \frac{1}{2} \times 3.8 \times \frac{1}{48}$$

$$= 0.0396 \text{ m}$$

$$= 3.96 \text{ cm}$$

(*b*) Adopt a camber of 1 in 36 for WBM roads in areas of heavy rainfall.

Rise of crown w.r.t. edges

$$= \frac{1}{2} \times 7.0 \times \frac{1}{36}$$

$$= 0.0972 \text{ m}$$

$$= 9.72 \text{ cm}$$

(*c*) Provide a camber of 1 in 60 for bituminous roads

Rise of crown w.r.t. edges

$$= \frac{1}{2} \times 7.5 \times \frac{1}{60}$$

$$= 0.0625 \text{ m}$$

$$= 6.25 \text{ cm}$$

(*d*) Select a camber of 1 in 72 for concrete roads.

Rise of crown w.r.t. edges

$$= \frac{1}{2} \times 7.5 \times \frac{1}{72}$$

$$= 0.0521 \text{ m}$$

$$= 5.21 \text{ cm}$$

Problem 8.5 *Prepare the camber boards for the following cases:*

(*a*) *Straight line camber of WBM village road in areas of light rainfall.*

(*b*) *Parabolic camber for WBM major district roads in areas of heavy rainfall.*

(*c*) *Partly straight and partly parabolic camber for bituminous State Highways in rural areas.*

(*d*) *Camber made up of two straight lines for concrete National Highways in built up areas.*

Solution. (*a*) ***Straight-line camber***

Adopt Camber = 1 in 48

Carriageway width = 3.8 m

∴ Rise of crown,

$$R = \frac{1}{2} \times 3.8 \times \frac{1}{48} \times 100$$

$$= 3.96 \text{ cm}$$

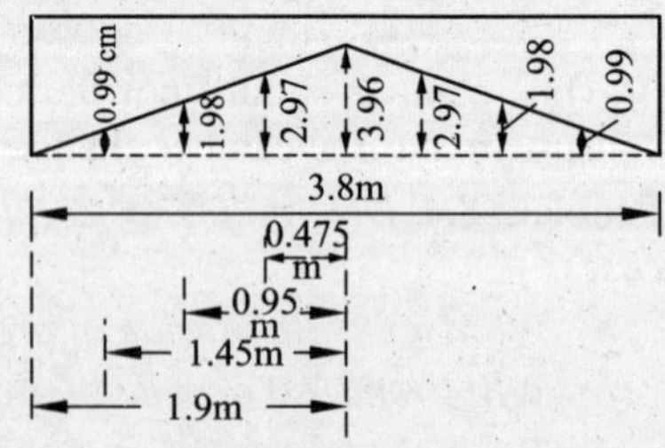

Fig. 8.9 For Problem 8.5 (*a*).

The ordinates Y at a distance X from the crown are given by the relation $Y = nX$. These are worked out below and shown in Fig. 8.9.

X (*m*)	0	$W/4 = 0.475$	$W/2 = 0.95$	$3W/4 = 1.425$	$W = 1.9$
Y (*cm*)	3.96	2.97	1.98	0.99	0.0

(*b*) ***Parabolic camber***

Adopt camber = 1 in 36

Carriageway width = 3.8 m

Rise of crown,

$$R = \frac{1}{2} \times 3.8 \times \frac{1}{36} \times 100$$

$$= 5.28 \text{ cm}$$

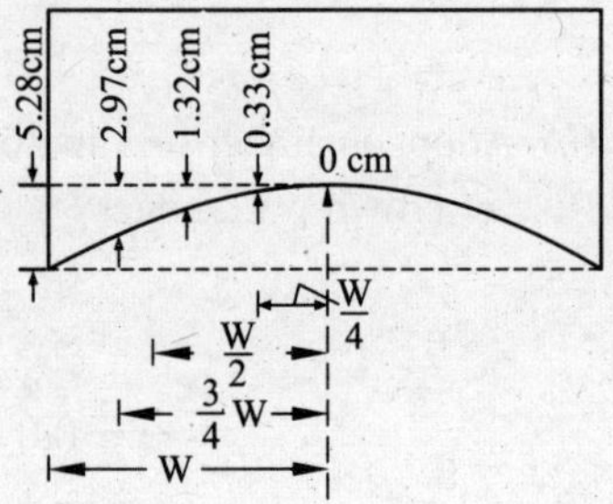

Fig. 8.10 For Problem 8.5 (*b*).

The ordinates Y at a distance X from the crown are given by the relation:

$$Y = \frac{nX^2}{W}$$

These are worked out below and shown in Fig. 8.10.

X (*m*)	0	$W/4 = 0.475$	$W/2 = 0.95$	$3W/4 = 1.425$	$W = 1.9$
Y (*cm*)	0.00	0.33	1.32	2.97	5.28

(*c*) ***Partly straight and partly parabolic camber***

Adopt straight line camber = 1 in 60

Take two-lane carriageway without raised kerbs

Width of carriageway = 7.0 m

Assume a parabolic camber for the central 2 m carriageway.

∴ $$\text{Rise of crown} = \left(\frac{nW}{2} - 0.5\, nX_a\right) \times 100$$

$$= \frac{1}{2} \times 7 \times \frac{1}{60} \times 100 - \frac{0.5}{00} \times 100$$

$$= 5.83 - 0.83 = 5.00 \text{ cm}$$

(*i*) For the parabolic portion

$$Y = \frac{nX^2}{2X_a}$$

X (*m*)	0	0.25	0.5	0.75	1
Y (*cm*)	0	0.052	0.21	0.47	0.83

(*ii*) For the straight line portion

$$Y = n\,(X - 0.5\, X_a)$$

X (*m*)	1.5	2.0	2.5	3.0	3.5
Y (*cm*)	1.67	2.50	3.33	4.17	5.00

This is shown in Fig. 8.11.

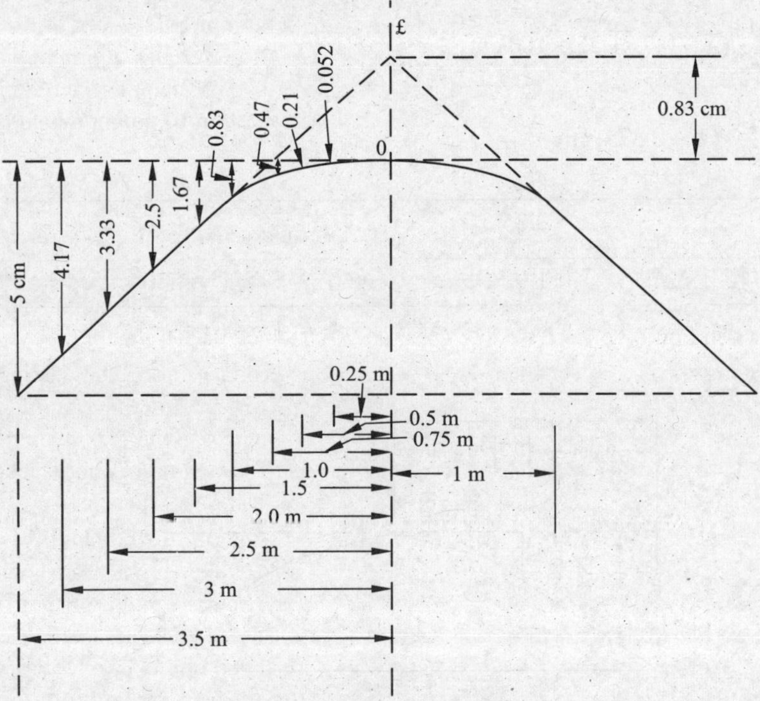

Fig. 8.11 For Problem 8.5 (*c*).

(*d*) *Camber made up of two straight lines*

Adopt straight line camber = 1 in 72

Take two-lane carriageway with raised kerbs

Width of carriageway = 7.5 m

Assume that the crown is flattened for the central 2.5 m portion

$$X_a = \frac{2.5}{2} = 1.25 \text{ m}$$

Rise of crown

$$= \left(\frac{nW}{2} - 0.7 \times nX_a\right) 100 \text{ cm}$$

$$= \left(\frac{1}{72} \times \frac{7.5}{2} - 0.7 \times \frac{1}{72} \times 1.25\right) 100$$

$$= 3.99 \text{ cm}$$

(*i*) For the flattened portion; $Y = 0.3\, nX$

X	0	0.25	0.5	0.75	1.00	1.25
Y	0	0.10	0.21	0.31	0.42	0.52

(*ii*) For the other portion

$$Y = n\,(X - 0.7\, X_a) \text{ metres}$$

$$= \frac{1}{72}(X - 0.7 \times 1.25) \times 100 \text{ cm}$$

$$= \frac{100}{72}(X - 0.875) \text{ cm}$$

X (m)	1.75	2.25	2.75	3.25	3.75
Y (cm)	1.22	1.91	2.60	3.30	3.99

This is shown in Fig. 8.12.

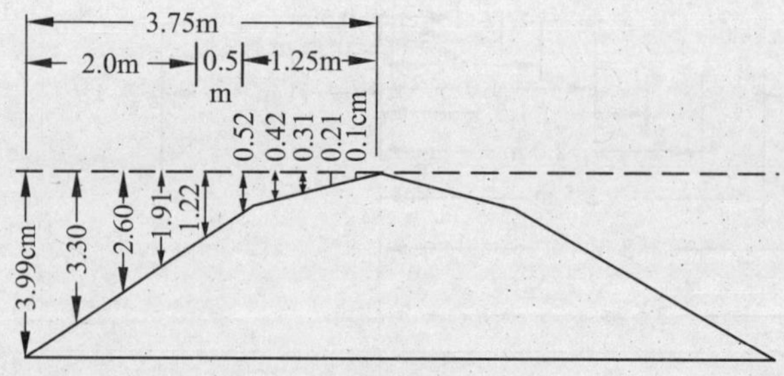

Fig. 8.12 For Problem 8.5 (*d*).

8.13. ROAD LAND WIDTH, BUILDING LINES AND CONTROL LINES

8.13.1. Road land width

Road land width is also termed right of way and is the land acquired for road purposes. Table 8.10 gives the desirable land width for different classes of roads.

Table 8.10 Recommended land width for different classes of roads

S. No.	*Road Classification*	*Plain and rolling terrain*				*Mountainous and steep terrain*	
		Open areas		*Built up areas*		*Open areas*	*Built up areas*
		normal	*range*	*normal*	*range*	*normal*	*normal*
1.	National and State Highways	45	30–60	30	30–60	24	20
2.	Major District Roads	25	25–30	20	15–25	18	15
3.	Other District Roads	15	15–25	15	15–20	15	12
4.	Village Roads	12	12–18	10	10–15	9	9

Land width should be acquired keeping in view the traffic requirements, topographical features, design needs and ultimate economy.

In high banks or deep cuts, the land width should be suitably increased. Similarly a higher value should be adopted in unstable or landslide prone areas. Near cities, towns, industrial areas and other important road intersections, acquisition of more width is desirable to provide for potential increase in traffic. If a road is expected to be upgraded to a higher classification in the foreseable future, the land width should correspond to the latter.

8.13.2. Building Lines and Control Lines

In order to prevent over crowding and preserve sufficient space for future road improvement, it is advisable to lay down restrictions on building activity along the roads. Building activity should not be allowed within a prescribed distance from the road, which is defined by a hypothetical line set back from the road boundary and called "building line".

In addition, it is desirable to exercise control on the nature of building activity for a further distance beyond the building line upto what are known as the "control lines". Building and control lines are illustrated in Fig. 8.13 with respect to the road centre line and road boundary.

Indian Roads Congress (IRC 73–1980) has recommended standards for building and control lines, which are given in Table 8.11.

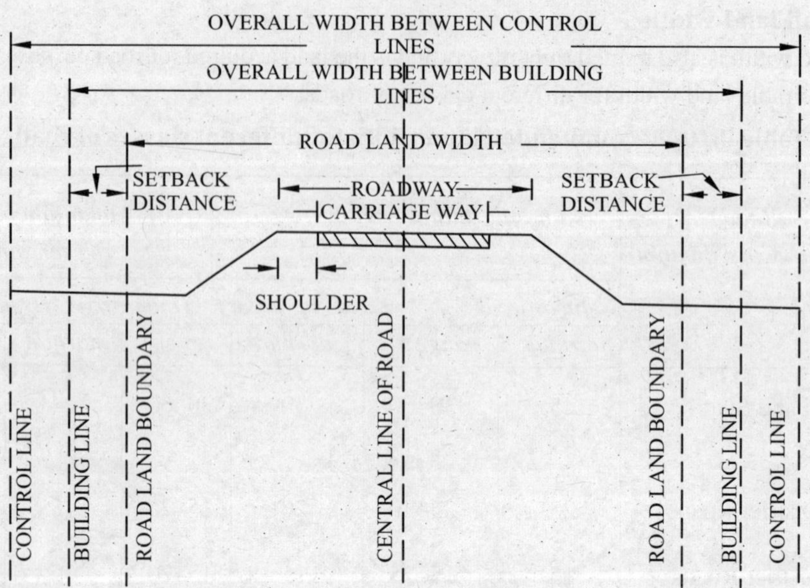

Fig. 8.13 Road land boundary, building lines and control lines.

Table 8.11 Recommended standards for building lines and control lines

S. No.	*Road Classification*	*Plain and rolling terrain*		*Mountainous and steep terrain*		
		Open areas	*Built up areas*	*Open areas*	*Built up areas*	
1.	*2*	*3*	*4*	*5*	*6*	*7*
1.	National and State Highways	80	150	3–6	3–5	3–5
2.	Major District Roads	50	100	3–5	3–5	3–5
3.	Other District Roads	25/30*	35	3–5	3–5	3–5
4.	Village Roads	25	30	3–5	3–5	3–5

Note: * If the land width is equal to the width between building lines indicated in this column, the building line should be set back 2.5 m from the road land boundary.

8.14. CROSS-SECTION

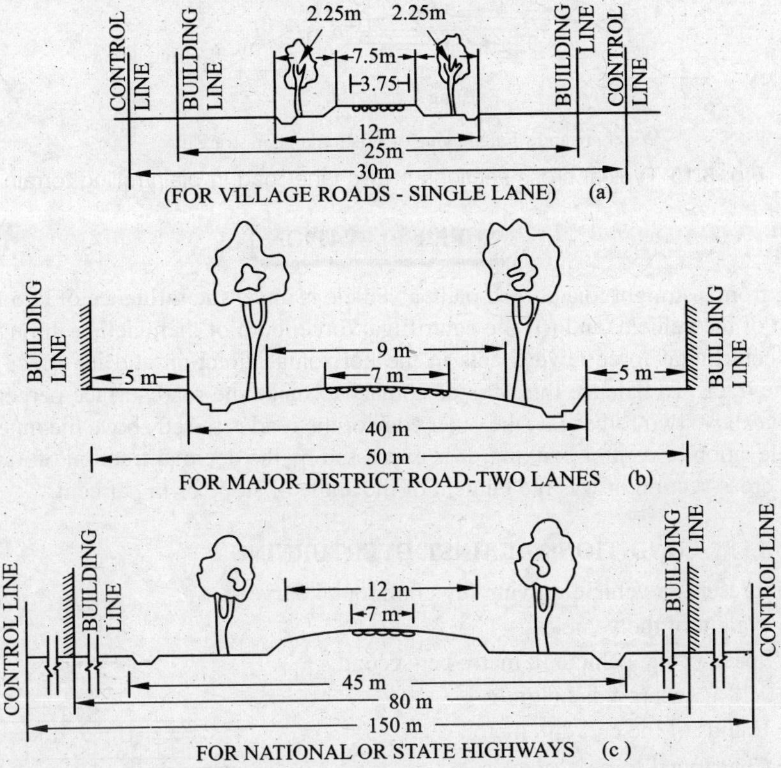

Fig. 8.14 Cross-section for various classes of highways in India (a) for village roads (single lane) (b) for major district roads (two lanes) (c) for national highways.

Cross-sections for various classes of highways in India are shown in Fig. 8.14. Each dimension is based on the recommendations made in the above articles.

Fig. 8.15 shows typical cross-sections of a two lane road in plain/rolling terrain for (a) a new road in straight (b) a new road in curve and (c) widening and strengthening of existing road in straight.

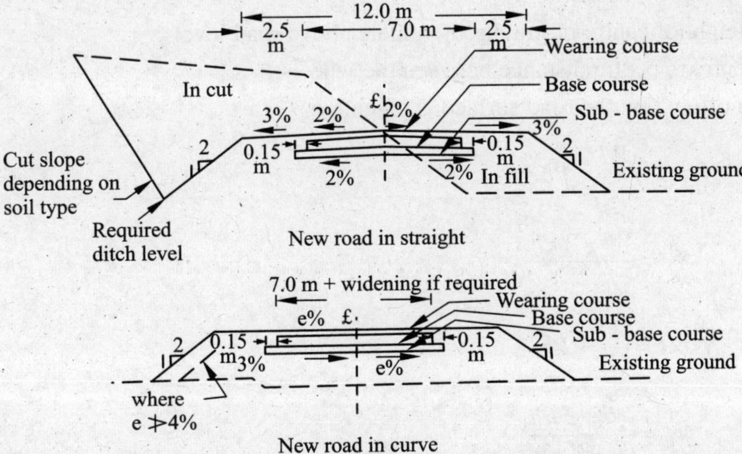

(Fig. 8.15 Continued on next page)

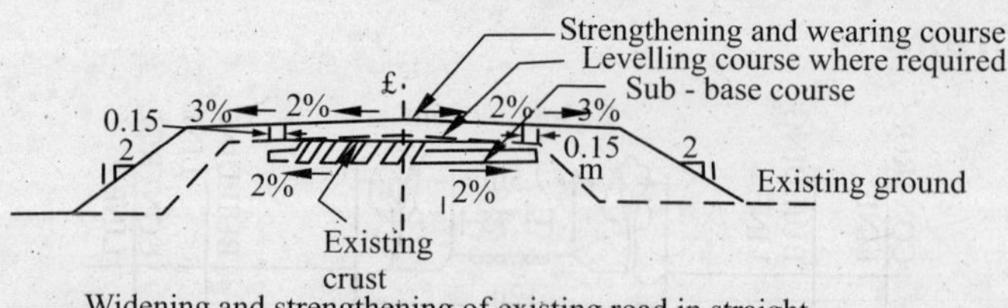

Fig. 8.15 Typical cross-sections of two-lane road in plain/rolling terrain.

SUPER ELEVATION

In passing from a straight to a curved path, a vehicle is under the influence of two forces, namely (*i*) the weight of the vehicle, and (*ii*) the centrifugal force; both of them acting through its centre of gravity. The centrifugal force always acts in the horizontal direction and its effect is to push the vehicle off the track. To balance this, it is customary to make the road surface perpendicular to the resultant of the above two forces *i.e.*, the outer edge of the road is raised above the inner edge. This is called superelevation or cant or banking. It is expressed by the decimal fraction of a *m* of rise per *m* of horizontal cross-section or it is the tangent of the angle of slope or in per cent.

8.15. STABILITY CONDITIONS AGAINST OVERTURNING

Fig. 8.16 represents a vehicle moving on a horizontal curve.

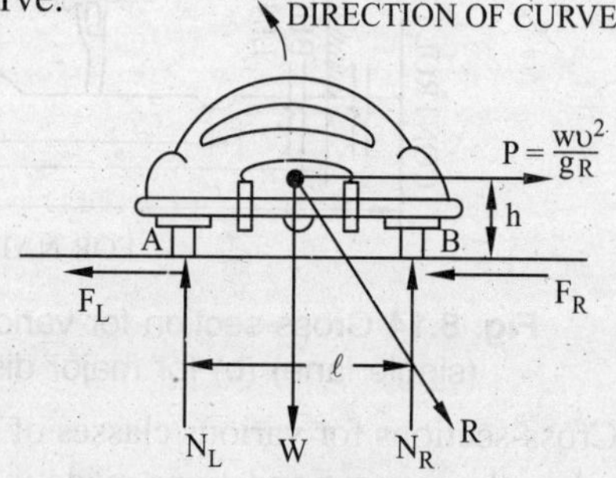

Fig. 8.16 Forces acting on a vehicle moving on a curve, with a level path.

Let W = Weight of the vehicle

v = Speed of the vehicle in metre per second

$= 0.28\ V$, where V is in km/hr

R = Radius of the curve in metre

P = Centrifugal force

$= \dfrac{Wv^2}{gR}$ in kg

N_L = Reaction of A normal to the surface

N_R = Reaction at B normal to the surface

F_L = Lateral friction at A

F_R = Lateral friction at B

h = Height of centre of gravity of vehicle above road level

l = Centre to centre distance between the wheels.

The reaction offered by the road surface is given by;

$$N_L \times l = \frac{Wl}{2} - Ph$$

or

$$N_L = \frac{Wl - 2Ph}{2l}$$

and

$$N_R \times l = \frac{Wl}{2} + Ph$$

or

$$N_R = \frac{Wl + 2Ph}{2l}$$

Also $N_l + N_R = W$

The vehicle will be on the verge of toppling over when the resultant of the centrifugal force and the weight of the vehicle passes through the centre of the right wheel.

Thus in the limiting condition;

$$Wl - 2\,Ph = 0$$

or centrifugal ratio $= \dfrac{P}{W} = \dfrac{l}{2h}$

$\therefore$ $\dfrac{P}{W}$ should always be less than $\dfrac{l}{2h}$

The centrifugal force P gives an overturning moment. Thus to avoid overturning, it is desirable that h should be as small as possible. It is mainly due to this reason that the modern passenger cars have low centre of gravity. But trucks have large centre of gravity so that large overturning moments are created.

The lateral outward thrust $P = \dfrac{Wv^2}{gR}$ is resisted by the lateral friction (F_L and F_R) between the tyres and the pavement. This lateral friction, f is dependent on the type of road surface and the type of tyre. If the lateral friction exceeds a certain limit, the vehicle will skid sideways. Thus another condition of stability is that centrifugal ratio $\dfrac{P}{W}$ should not be greater than the lateral friction between tyres and the pavement.

8.16. RELATION BETWEEN SUPERELEVATION, COEFFICIENT OF FRICTION AND CENTRIFUGAL RATIO

To counteract the effect of the centrifugal force, let the surface of the road be superelevated by an angle θ to the horizontal (as shown in Fig. 8.17), so that:

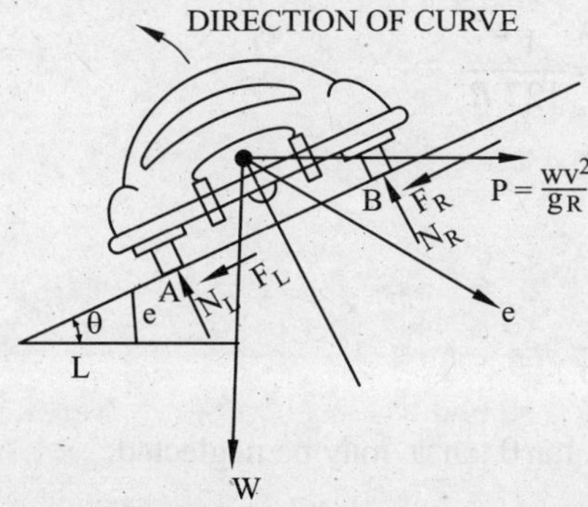

Fig. 8.17 Forces acting on a vehicle moving on a curve, with superelevated path.

$f = \tan\psi$ = Coefficient of lateral friction between the tyre and the road surface

$e = \tan\theta$ = Superelevation

R = Resultant of P and W

N_L = Reaction at A normal to the superelevated surface

N_R = Reaction at B normal to the superelevated surface

F_L = Friction at A along the superelevated surface

$= f\,N_L$

$$F_R = \text{Friction at } B \text{ along the superelevated surface}$$
$$= f\, N_R$$

Resolving horizontally and vertically

$$P = (N_L + N_R)\sin\theta + (F_L + F_R)\cos\theta$$
$$= (N_L + N_R)\sin\theta + (f\, N_L + f\, N_R)\cos\theta$$
$$= (N_L + N_R)(\sin\theta + f\cos\theta)$$

Similarly

$$W = (N_L + N_R)(\cos\theta + f\sin\theta)$$

$$\therefore \quad \frac{P}{W} = \frac{\sin\theta + f\cos\theta}{\cos\theta + f\sin\theta}$$

$$= \frac{\sin\theta + \dfrac{\sin\psi}{\cos\psi}\cos\theta}{\cos\theta + \dfrac{\sin\psi}{\cos\psi}\sin\theta}$$

$$= \frac{\sin\theta\cos\psi + \sin\psi\cos\theta}{\cos\theta\cos\psi + \sin\psi\sin\theta}$$

$$= \frac{\sin(\theta+\psi)}{\cos(\theta+\psi)}$$

$$= \tan(\theta+\psi)$$

Also $\quad \dfrac{P}{W} = \dfrac{v^2}{gR} = \dfrac{(0.28\,V)^2}{9.81\,R} = \dfrac{V^2}{127\,R}$

$$\therefore \quad \frac{V^2}{127\,R} = \tan(\theta+\psi)$$

$$= \frac{\tan\theta + \tan\psi}{1 - \tan\theta\tan\psi}$$

For small values of θ and ψ, $\tan\theta\tan\psi$ may be neglected.

$$\therefore \quad \frac{V^2}{127\,R} = \tan\theta + \tan\psi$$

$$= e + f \qquad \text{...(8.12)}$$

$$\therefore \quad \frac{P}{W} = C = e + f\text{; where } C \text{ is the centrifugal ratio} \qquad \text{...(8.13)}$$

This is a fundamental equation for the curve design. However, the following points may be noted:

(*i*) If the coefficient of friction is neglected (*i.e.*, $\psi = 0$);

$$\tan\theta = e = \frac{V^2}{127\,R}$$

This means that if a road surface is superelevated by $\tan^{-1}\dfrac{V^2}{127R}$ the frictional force will not be called upon to act and thus the pressure on both the wheels will be equal. The distribution of pressure in the case of a superelevated and level road is shown in Fig. 8.18.

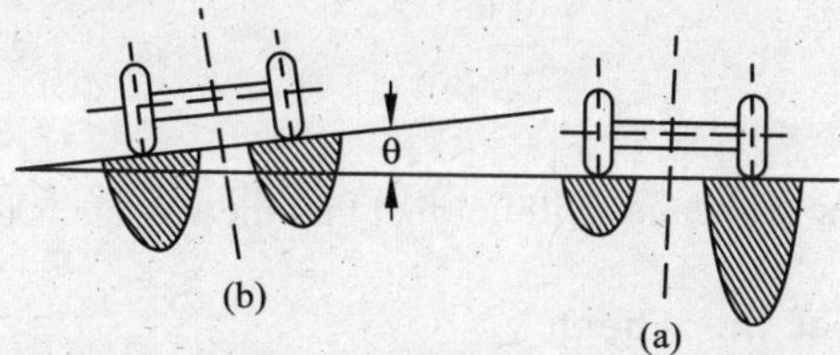

Fig. 8.18 Distribution of pressure on wheels on (*a*) level road (*b*) superelevated to $\tan^{-1}(V^2/127R)$.

(*ii*) If the superelevation provided is less than $\dfrac{V^2}{127R}$, lateral friction will come into play and will go on increasing as the super elevation gets reduced.

(*iii*) If no superelevation is provided (*i.e.*, $\theta = 0$)

$$\tan\psi = f = \frac{V^2}{127R}$$

(*iv*) If the coefficient of lateral friction between the tyre and the road surface is less than the required value, the vehicle will skid in the lateral direction.

8.17. EFFECT OF SUPERELEVATION ON PASSENGERS

So far, the effect of superelevation on the vehicle has been considered, where the lateral thrust is resisted by the lateral friction between the tyres and the road surface. But in the case of passengers, this lateral thrust may tilt the passengers in their seats.

Let W_1 = Weight of passenger

K = Lateral ratio

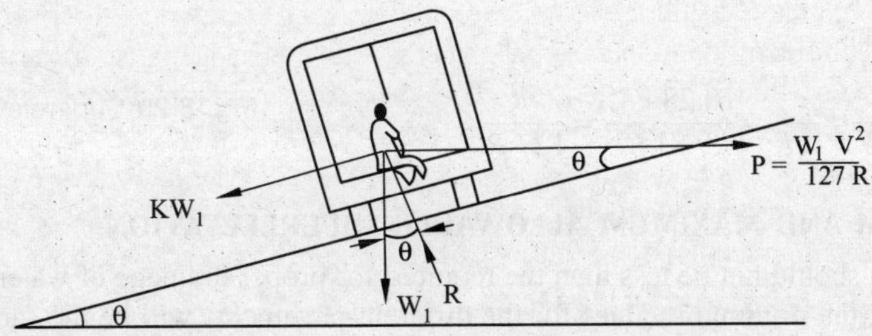

Fig. 8.19 Forces acting on a passenger, when the vehicle moves on a curve.

The forces acting on the passenger while the vehicle is moving on the curve are shown in Fig. 8.19.

Resolving the forces along the inclined plane, the net force KW_1, acting on the passenger is given by

$$KW_1 = P\cos\theta - W_1\sin\theta$$

$$= \frac{W_1 V^2}{127R}\cos\theta - W_1\sin\theta$$

or $$K = \frac{V^2}{127R}\cos\theta - \sin\theta$$

thus if $e = \tan\theta = \frac{V^2}{127R}$; then $K = \tan\theta\cos\theta - \sin\theta$

$$= \sin\theta - \sin\theta$$
$$= 0$$

$\therefore$ When $e = \frac{V^2}{127R}$ the passenger will not feel that the vehicle is taking a turn.

8.17.1. Drivers reaction at the wheel

If superelevation $= \frac{V^2}{127R}$ is provided on curves, the driver is not required to apply any force on the wheel. The speed at which this occurs is known as "hand off speed," at which

$$\frac{V^2}{127} = \sin\theta$$

or $$V = \sqrt{127\sin\theta}$$

With the increase in the value of KW_1, the force required to be applied on the wheel also increases. As the lateral ratio exerted on the driver is the same as that exerted on the vehicle, the driver automatically exerts force on his wheel to maintain the vehicle on track.

8.17.2. Other method of finding superelevation

For equilibrium (Fig. 8.20) the resultant of the weight W and the centrifugal force P must be equal and opposite to the reaction perpendicular to the road surface. Thus the inclination of the resultant R to the vertical is also θ.

$$\therefore \quad \frac{AB}{OB} = \frac{WN}{LW}$$

or $$\frac{h}{b} = \frac{P}{W}$$

or $$e = \frac{Wv^2}{gR.W} = \frac{v^2}{gR} = \frac{(0.28\,V)^2}{9.81\,R} = \frac{V^2}{127\,R}$$

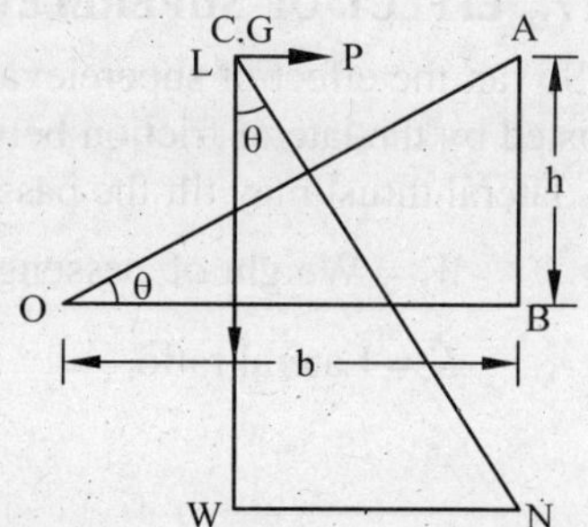

Fig. 8.20 Superelevation on a curve.

8.18. MINIMUM AND MAXIMUM ALLOWABLE SUPERELEVATION

Superelevation should not be less than the required for proper drainage of water from the surface of the road. Thus, the minimum values for the different pavements will be the same as allowed for cross fall drainage in flat country. Where the superelevation, as found out by the above relations, is less than the slope required for surface drainage, the normal cambered profile should be continued and no superelevation may be provided.

The maximum value of superelevation is governed in India by the bullock-cart traffic since these will be using the roads for a considerable time to come. The slope should not be too steep for these slow moving vehicles. The worst happens when the carts are transporting goods such as hay, cotton, etc. because the centre of gravity is pretty high in such cases. Keeping this in view, a limiting superelevation of 7 per cent is recommended by the Indian Roads Congress.

8.19. SUPERELEVATION FOR VARIOUS DESIGN SPEEDS

Super-elevation (e) is a direct function of the square of the speed of the vehicle $\left(e = \frac{V^2}{127R}\right)$. Therefore, higher the speed, the greater will be the value of superelevation. Thus on roads, which are to be used by the fast moving vehicles, the value of superelevation should be calculated on the basis of the design speed. As the speed of the vehicles moving on the road vary considerably, it has been suggested that the superelevation may be found out taking average speed as three-fourth of the design speed. This assumes that centrifugal force corresponding to three-fourth of the design is balanced by superelevation and rest counteracted by side friction.

$$\therefore \quad e = \frac{\left(\frac{3}{4}V\right)^2}{127\,R}$$

or $$e = \frac{V^2}{225\,R} \qquad \ldots(8.14)$$

Superelevation obtained from the above expression should however be kept limited to the following values:

(*a*) In plain and rolling terrain : 7 per cent
(*b*) In snow-bound areas : 7 per cent
(*c*) In hilly areas not bound by snow : 13 per cent.

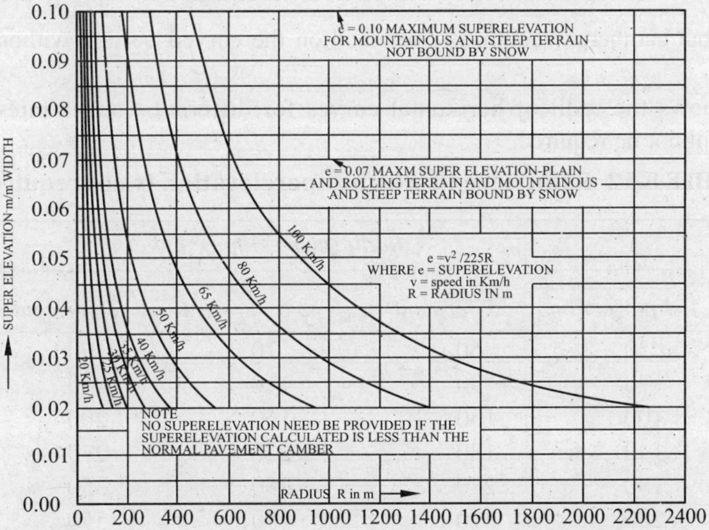

Fig. 8.21 Superelevation rates for various design speeds.

Fig. 8.21 indicates the superelevation for various design speeds on this basis.

8.20. LATERAL FRICTIONAL RESISTANCE

This depends upon the type and condition of the road surface, the presence of moisture, mud, snow, etc. on it and the condition of tread and air pressure of tyres. The value of f varies from 0.20 for muddy roads on a rainy day to 0.15 for newly constructed asphalt roads. Since the condition of Indian roads is not good, the Indian Roads Congress has recommended the least value of 0.20. Taking a factor of safety of $1\frac{1}{3}$, the value of safe coefficient of friction works out to be $0.2 \times \frac{3}{4} = 0.15$.

8.21. MAXIMUM VALUE OF CENTRIFUGAL RATIO

Since a limit has been imposed on the maximum value of superelevation and lateral frictional resistance, the centrifugal force cannot be allowed to develop beyond a certain value. Thus the maximum centrifugal ratio, that can be permitted, is determined from the limiting values of *e* and *f*. From above, it is clear that;

$$e_{max.} = 0.07$$

$$f_{max.} = 0.15$$

∴ Max. permissible centrifugal ratio $(C) = e_{max.} + f_{max.}$

$$= 0.07 + 0.15$$

$$= 0.22$$

But also $C = \dfrac{V^2}{127\,R}$

∴ $\dfrac{V^2}{127\,R} = 0.22$

or $R = \dfrac{V^2}{28}$ (approx.) ...(8.15)

Thus for any design speed, the minimum permissible radius of curvature is also fixed.

When the value of the super elevation obtained from the equation $e = \dfrac{V^2}{225\,R}$ is less than the road camber, the normal cambered section is continued on the curved portion without providing any superelevation.

Table 8.12 shows the radii of horizontal curves for different camber rates beyond which superelevation will not be required.

TABLE 8.12 Radii beyond which superelevation is not required

Design speed km/hr	*Radii (metres) for camber of*				
	4 per cent	*3 per cent*	*2.5 per cent*	*2 per cent*	*1.7 per cent*
20	50	60	70	90	100
25	70	90	110	140	150
30	100	130	160	200	240
35	140	180	220	270	320
40	180	240	280	350	420
50	280	370	450	550	650
65	470	620	750	950	1100
80	700	950	1100	1400	1700
100	1100	1500	1800	2200	2600

8.22. METHODS FOR BUILDING SUPERELEVATION

Roads on straight path are usually provided with a camber on both sides from the crown or the central line of a road. But on curves, this road has to be superelevated, *i.e.*, outer edge has to be raised with respect to the inner edge. Thus it is required to change the cambered section into a superelevated one. This is accomplished in two stages.

(*i*) Firstly the removal of adverse camber in outer half of the pavement, and

(*ii*) Secondly superelevation is gradually built up over the full width of the carriageway so that required superelevation is available at the beginning of the circular curve.

There are three different methods for attaining the superelevation:

(*i*) Revolving pavement about the centre line.

(*ii*) Revolving pavement about the inner edge, and

(*iii*) Revolving pavement about the outer edge.

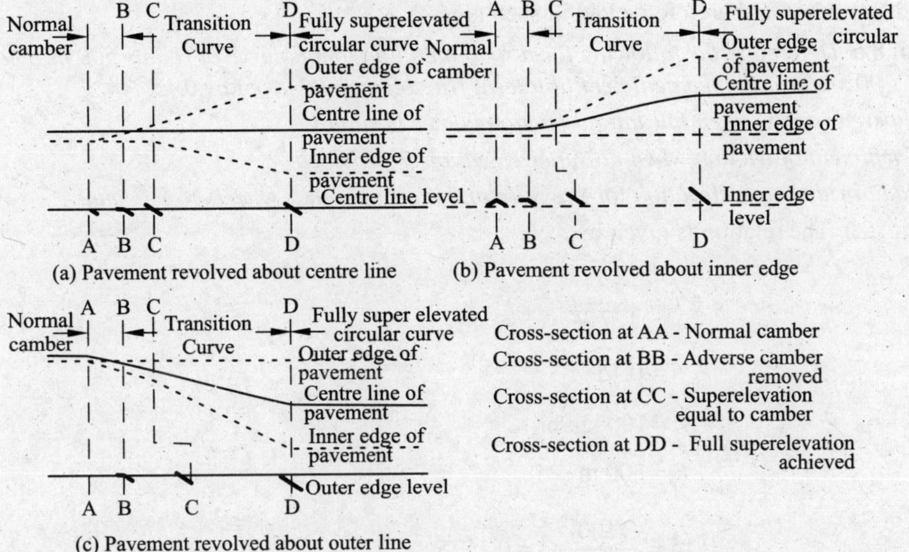

Fig. 8.22 Schematic diagrams showing different methods of attaining superelevation.

Fig. 8.22 illustrates these methods diagrammatically. The small cross-sections at the bottom of each diagram indicate the pavement cross slope condition at different points.

Each of the above methods is applicable under different conditions. The first method involves least distortion of the pavement and is found suitable in most of the situations where there are no physical controls and may be adopted in the normal course. This method is illustrated in Fig. 8.23.

The second method is preferable where the lower edge profile is a major control e.g. on account of drainage. Where overall appearance is the criterion, the last method is preferred since the outer edge profile which is most noticeable to drivers is not distorted.

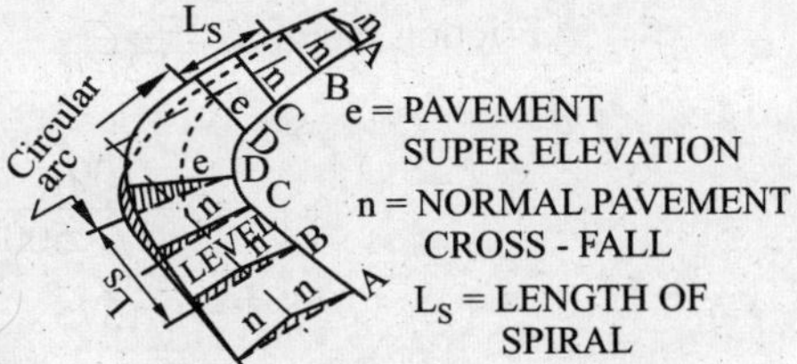

Fig. 8.23 Method of attaining superelevation, when pavement curves revolve about central line.

The superelevation should be attained gradually over the full length of the transition curve so that the design superelevation is available at the starting point of the circular portion. In cases where transition curve cannot be provided for some reason, two-third superelevation should be attained on the straight section before start of the circular curve and the balance one third on the curve.

In developing the required superelevation, it should be ensured that the longitudinal slope of the pavement edge compared to the centre-line (*i.e.*, the rate of superelevation) is not steeper than 1 in 150 for roads in plain and rolling terrain and 1 in 60 in mountainous and steep terrain.

When cross-drainage structures fall on a horizontal curve, their deck should be superelevated in the manner described above.

8.23. SUPERELEVATION IN BUILT-UP AREAS

The above discussion refers mainly to roads in open country. In the case of built-up areas, the amount of superelevation to be provided depends upon other considerations such as access to buildings, drainage, moving vehicles, parking facilities, intersections, etc. However, as the speed of the vehicle gets considerably reduced in built-up areas, greater amount of superelevation will not be required. No hard and fast rule can therefore be laid in such areas.

Problem 8.6 *Determine the following for a road on a horizontal curve of radius 500 m. The design speed is 100 km/hr and the coefficient of lateral friction may be taken as 0.15.*

(*a*) *Superelevation, when full lateral friction comes into play.*

(*b*) *Coefficient of friction when no super elevation is provided.*

(*c*) *Equilibrium superelevation for pressure on inner and outer wheels to be equal.*

Solution. (*a*) The relation is given by :

$$e + f = \frac{V^2}{127\,R}$$

Here

$$f = 0.15$$

$$V = 100 \text{ km/hr}$$

$$R = 500 \text{ m}$$

$$e + 0.15 = \frac{(100)^2}{127 \times 500} = 0.1575$$

$$\therefore \quad e = 0.1575 - 0.15$$

$$= 0.0075$$

i.e., superelevation rate of 0.75 per cent.

(*b*) When no superelevation is provided,

$$e = 0$$

$$\text{Friction factor } f = \frac{V^2}{127\,R}$$

$$= \frac{(100)^2}{127 \times 500}$$

$$= 0.1575$$

(*c*) When the pressure on the inner and outer wheels is equal, friction factor $f = 0$.

$$\therefore \quad \text{Equilibrium superelevation, } e = \frac{V^2}{127R} = \frac{(100)^2}{127 \times 500} = 0.1575$$

i.e., superelevation rate of 15.75 per cent.

However this rate of superelevation is greater than the maximum permissible *i.e.*, 7 per cent and hence cannot be provided.

Problem 8.7 *Calculate the superelevation required on a road curve of 240 m radius with mixed traffic conditions. The design speed is 80 km/hr. The coefficient of friction is 0.15. The road is passing through rolling terrain.*

Solution. For mixed traffic conditions, superelevation is given by equation 8.14.

$$e = \frac{V^2}{225 R}$$

Here $V = 80$ km/hr

and $R = 240$ m

$$\therefore \quad e = \frac{(80)^2}{225 \times 240}$$

$$= 0.1185$$

i.e., super elevation rate of 11.85 per cent.

Since the calculated value of superelevation rate is greater than the maximum permissible value of 7 per cent, the actual superelevation is restricted to 7 per cent or 0.07 m/m.

Check for coefficient of lateral friction

$$e + f = \frac{V^2}{127 R}$$

or

$$f = \frac{V^2}{127 R} - e$$

$$= \frac{(80)^2}{127 \times 240} - 0.07$$

$$= 0.210 - 0.07$$

$$= 0.140$$

Since the value is less than the maximum permissible value of 0.15, the design is safe for $e = 0.07$ or 7 per cent.

Problem 8.8 *How much should the outer edges of the pavement be raised with respect to the central line on a two-lane raod designed to cater for mixed traffic at a speed of 65 km/hr on a horizontal curve of radius 160 m, if*

(*a*) *the pavement is rotated w.r.t. the central line, and*

(*b*) *the pavement is rotated w.r.t. to the inner edge.*

Solution. For mixed traffic conditions, superelevation should counteract the centrifugal force for 75 per cent of design speed,

$$\therefore \quad e = \frac{(0.75V)^2}{127 R} = \frac{V^2}{225 R}$$

$$= \frac{(65)^2}{225 \times 160} = 0.1174$$

Since this value is greater than the maximum permissible value of 0.07, adopt $e = 0.07$ or 7 per cent.

(*a*) Raising of outer edge w.r.t. centre

Assume total width of 2 lane pavement = 7 m

Raising of outer edge w.r.t. centre $= \frac{7}{2} \times 0.07 = 0.245$ m

(*b*) Raising of outer edge w.r.t. inner edge $= 7 \times 0.07 = 0.49$ m

Problem 8.9 *Calculate the maximum permissible speed on a horizontal curve of radius 125 m of a highway designed for a speed of 65 km/hr to carry mixed traffic, if the super elevation is not to exceed 7 per cent.*

Solution. As per equation 8.14

$$e = \frac{V^2}{225\,R}$$

$$= \frac{(65)^2}{225 \times 125}$$

$$= 0.150 \text{ or } 15 \text{ per cent}.$$

Maximum permissible value of e is 0.07 or 7 per cent.

Check for lateral friction factor.

$$f = \frac{V^2}{127R} - 0.07$$

$$= \frac{(65)^2}{127 \times 125} - 0.07$$

$$= 0.266 - 0.07$$

$$= 0.196$$

Since this value is greater than the maximum permissible value of 0.15 and also the radius cannot be increased, the speed has to be restricted. The maximum permissible speed can be calculated by assuming:

$e = 0.07$ and

$f = 0.15$ *i.e.*, maximum allowable values

If V_1 is the maximum permissible speed, then

$$e + f = \frac{V_1^2}{127\,R}$$

or $$0.07 + 0.15 = \frac{V_1^2}{127 \times 125}$$

or $$V_1 = \sqrt{127 \times 125 \times 0.22}$$

$$= 59.1 \text{ km/hr}$$

Hence the speed has to be restricted to 59.1 km/hr on the curve.

CURVES

Curves are provided in highways in order that the change of direction at the intersection of straight alignments either in horizontal or vertical plane, shall be gradual. The necessity of providing curves arises due to the following reasons:

(*i*) Topography of the country.
(*ii*) To provide access to a particular locality.
(*iii*) Restrictions imposed by property.
(*iv*) Preservation of existing amenities.
(*v*) Avoidance of existing religious, monumental and other costly structures.
(*vi*) Making use of existing right of way.

Advantages

The obvious advantages of curves are:

(*i*) They provide comfort to the passengers. If there is an abrupt change in the direction or grade of a highway it will upset the passengers.
(*ii*) They help to avoid mental strain induced by the monotony of continuous journey along straight path.
(*iii*) The life of the vehicles is increased. An abrupt change will subject a vehicle passing over it to an impact that can be dangerous or injurious to the vehicles. In the case of sharp turns, brakes have to be applied more frequently. Thus the life of the tyres would be much reduced.
(*iv*) The drivers become alert due to the change in the direction of road.
(*v*) They help to keep the speed of the vehicles within limits. On a straight road, a driver is tempted to go at a much faster speed.

8.24. FACTORS AFFECTING THE DESIGN OF CURVES

The various factors which affect the design of curves are:

(*i*) Design speed of the vehicle.
(*ii*) Allowable friction.
(*iii*) Maximum permissible superelevation.
(*iv*) Permissible centrifugal ratio.

8.25. TYPE OF CURVES

Curves have been divided into two classes:

(*i*) ***Horizontal curve.*** A curve in plan to provide change in direction of the central line of a road.
(*ii*) ***Vertical curve.*** A curve in the longitudinal section of a roadway to provide for easy change of gradient.

The different types of curves used in highways are:

(*i*) ***Circular curves***
(*a*) Simple
(*b*) Compound
(*c*) Reverse

(*ii*) ***Transition curves***
(*a*) True spiral or clothoid
(*b*) Cubic spiral
(*c*) Cubic parabola
(*d*) Lemniscate

(*iii*) Parabolic curves

Circular and transition curves are used as horizontal curves and the parabolic curves as vertical curves. These are described below.

8.26. CIRCULAR CURVES

8.26.1. Simple curve

A simple curve consists of a single arc connecting two straights. In India, a curve is expressed in terms of degree of the curve which is the angle in degrees subtended at the centre, e.g. 1° curve, a 4° curve, etc.

The radius of the curve can be found from the following relation:

$$\frac{D}{2} \text{ radians} = \frac{T_1M}{T_1O} = \frac{l/2}{r}$$

or $$\frac{D}{2} \text{ radians} = \frac{l}{2r}$$

or $$\frac{D^\circ}{2} \times \frac{\pi}{180} = \frac{l}{2r}$$

or $$r = \frac{360 \times l}{2\pi \times D^\circ}$$

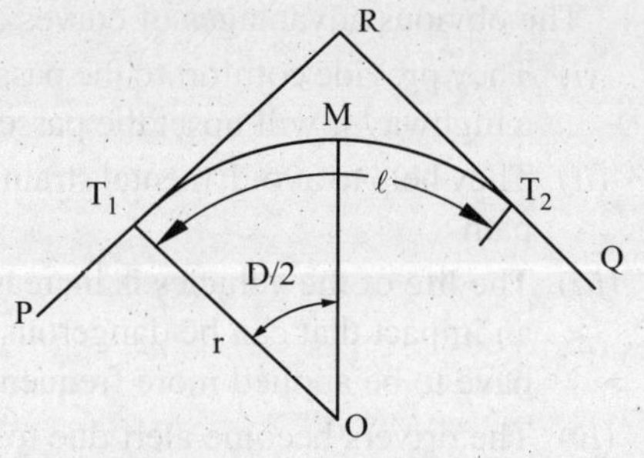

But $$l = 30 \text{ m}$$

$\therefore$ $$r = \frac{360 \times 30}{2\pi D^\circ}$$

or $$r = \frac{1720}{D^\circ} \text{ m} \quad \ldots(8.16)$$

$\therefore$ for 1° curve

$$r = 1720 \text{ m}$$

Fig. 8.24 Simple circular curve.

8.26.2. Radii of horizontal curves

On a horizontal curve, the centrifugal force is balanced by the combined effects of superelevation and slide friction. The basic equation for this condition of equilibrium is

$$\frac{v^2}{gR} = e + f$$

or $$R = \frac{V^2}{127\,(e+f)}$$

Based on this equation and the maximum permissible values of superelevation given on page 296 and maximum value of coefficient of friction of 0.15, radii for horizontal curves corresponding to ruling minimum and absolute minimum design speeds are worked out in Table 8.13.

On new roads, horizontal curves should be designed to have the largest practicable radius, generally more than the values corresponding to the ruling design speed. However, absolute minimum values based on minimum design speed may be resorted to if economics of construction or the site conditions so dictate. While improving existing roads, curves having radii corresponding to absolute minimum standards may not be flattened unless it is necessary to realign the road for some other reasons.

Table 8.13 Minimum radii of horizontal curves for different terrain conditions in metres

Classification of road	Plain terrain		Rolling terrain		Mountainous terrain				Steep terrain			
					Areas not affected by snow		*Snow bound areas*		*Areas not affected by snow*		*Snow bound areas*	
	Ruling minimum	*Absolute minimum*	*Ruling minimum*	*Absolute minimum*	*Ruling minimum*	*Absolute minimum*	*Ruling minimum*	*Absolute minimum*	*Ruling minimum*	*Absolute minimum*	*Ruling minimum*	*Absolute minimum*
1. National Highway, State Highway	360	230	230	155	80	50	90	60	50	30	60	33
2. Major District Roads	230	155	155	90	50	30	60	33	30	14	33	15
3. Other District Roads	155	90	90	60	30	20	33	23	20	14	23	15
4. Village Roads	90	60	60	45	20	14	23	15	20	14	23	15

Note: Absolute minimum and ruling minimum radii correspond to the minimum design speed and ruling design speed respectively vide Table 8.2.

8.26.3. Set-back distance at horizontal curves

Requisite sight distance, S (see Fig. 8.25) should be available across the inside of horizontal curves. Lack of visibility in the lateral direction may arise due to obstructions like walls, cut slopes, buildings, wooded areas, high farm crops, etc. Distance from the road centre line within which the obstruction should be cleared to ensure the needed visibility *i.e.,* set back distance is calculated from the equation:

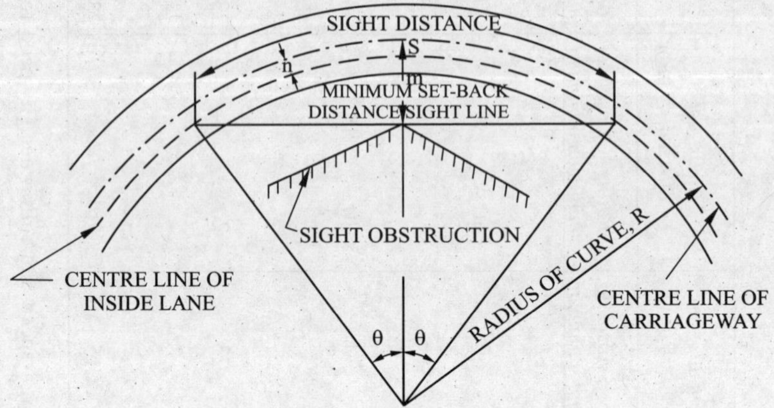

Fig. 8.25 Visibility at horizontal curves.

$$d = R - (R - n) \cos \theta$$

where $\theta = \dfrac{S}{2(R-n)}$ radians;

d = The minimum set back distance to sight obstruction in m (measured from the centre line of the road)

R = Radius at centre line of the road in m

n = Distance between the centre line of the road and the centre line of the inside lane in m; and

S = Sight distance in m.

In the above equation, sight distance is measured along the middle of inner lane. On single lane roads, sight distance is measured along centre line of the road and n is taken as zero.

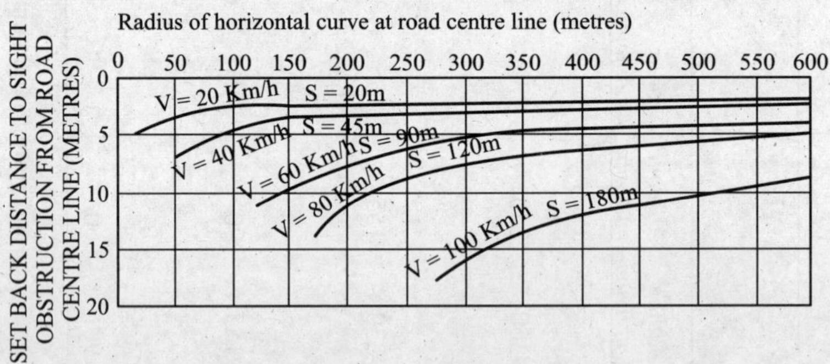

Fig. 8.26 Minimum set back distance required at horizontal curves for safe stopping sight distance.

Based on the above equation, design charts for set back distance corresponding to the safe stopping sight distance are given in Fig. 8.26. Set-back distance for overtaking and intermediate sight distance can be computed similarly but the clearance required is usually too large to be economically feasible except on very flat curves.

When there is a cut slope on the inside of the horizontal curve the average height of sight line can be used as an approximation for deciding the extent of clearance. For stopping sight distance which is the bare minimum requirement for design, the average height may be taken as 0.7 m. Cut slopes should be kept lower than this height at the line demarcating the set back distance envelope either by cutting back the slope or benching suitably. In the case of intermediate or overtaking sight distance, height of sight line above the ground should be taken as 1.2 m.

Where horizontal and summit vertical curves overlap, the design should provide for the required sight distance both in the vertical direction along the pavement and in the horizontal direction on the inside of the curve.

8.26.4. Compound Curve

A compound curve consists of a series of two or more simple curves that turn in the same direction and join at common tangent points. At each common tangent point, the adjacent curves have a common tangent and their centres are on the same side of the curve. In Fig. 8.27, T_2 is the common tangent point and QR is the common tangent.

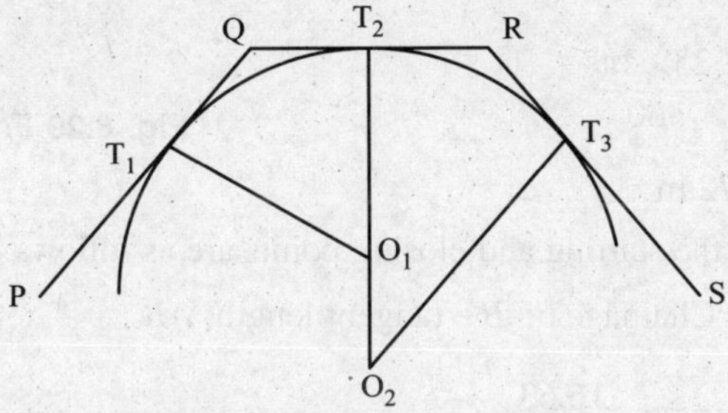

Fig. 8.27 Compound circular curve.

8.26.5. Reverse Curve

A reverse curve consists of two simple curves of opposite direction that join at the common tangent point called the point of reverse curve.

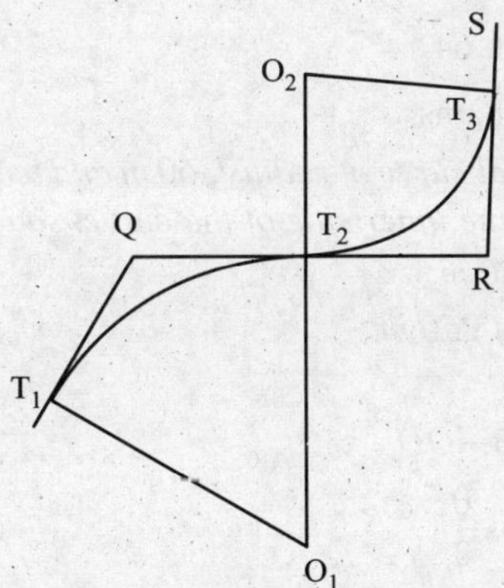

Fig. 8.28 Reverse circular curve.

Problem 8.10 *The straight lengths of a newly proposed road intersect at a point with a deflection angle of 25° to the right. A 75 m radius circular curve is proposed to be inserted in place of the straight intersections where the chainage at the intersection is 19.78 in 20 m units. Find out the length of the curve and the starting and closing chainages of the curve.*

Solution. In Fig. 8.29,

AB and *BC* are the two straight lengths of the newly proposed road, making a deflection angle $EBC = 25°$. *ADC* is the simple circular curve.

$$\angle AOC = 25°$$

$$\angle AOB = \frac{25}{2} = 12.5°$$

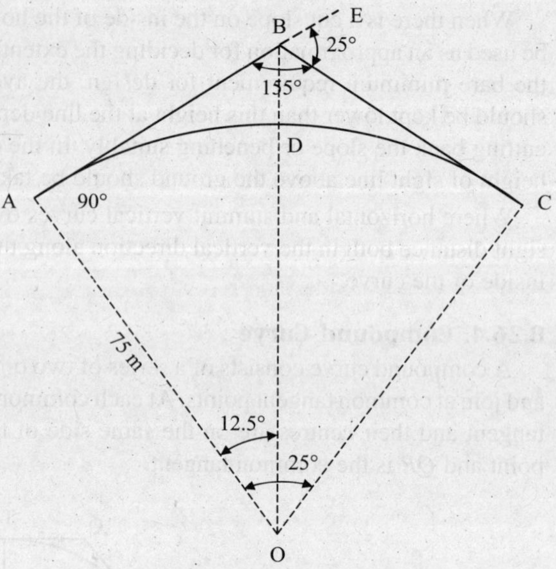

Fig. 8.29 For Problem 8.10.

Tangent length, $AB = AO \tan 12.5°$

$$= 75 \times 0.2217$$

$$= 16.63 \text{ m}$$

Length of the curve *ADC*

$$= R \times \theta$$

$$= 75 \times \frac{25 \times 2\pi}{360}$$

$$= 32.72 \text{ m}$$

∴ The chainages of the starting and closing points are as follows:

Chainage of A = Chainage of B − tangent length, AB

$$= 19.78 - \frac{16.63}{20}$$

$$= 18.95 \text{ chains}$$

Chainage of C = Change of A + curve length

$$= 18.95 + \frac{32.72}{20}$$

$$= 18.95 + 1.64$$

$$= 20.59 \text{ chains}$$

Problem 8.11 *There is a horizontal curve of radius 360 m and length 180 m. Calculate the clearance required from the central line on the inner side of the curve, so as to provide;*

(*a*) *Stopping sight distance of 80 m.*

(*b*) *Overtaking sight distance of 250 m.*

Solution. (*a*) Refer to Fig. 8.30

Clearance required $= BD = OB - OD$

$$= R - R\cos\frac{\theta}{2}$$

$$= R\left(1 - \cos\frac{\theta}{2}\right)$$

But $\dfrac{\theta}{2} = \dfrac{AB}{AO}$ radians

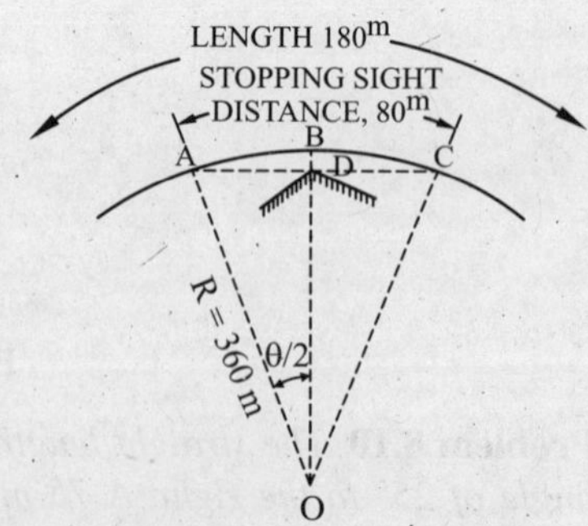

Fig. 8.30 For Problem 8.11.

$$= \frac{AB}{AO} \times \frac{360°}{2\pi} \text{ degrees}$$

$$= \frac{40}{360} \times \frac{360}{2\pi}$$

$$= 6.37°$$

$$\therefore \quad BD = R\left(1 - \cos\frac{\theta}{2}\right) = 360\,(1 - \cos 6.37°)$$

$$= 360\,(1 - 0.9939)$$

$$= 360 \times 0.0061$$

$$= 2.2 \text{ m}$$

Thus a clearance of 2.2 m is required to provide stopping sight distance of 80 m.

(*b*) The overtaking sight distance of 250 m is greater than the length of the circular curve, which is 180 m.

∴ Clearance required (see Fig. 8.31).

$$= CF + FG$$

$$\frac{\theta}{2} = \frac{180 \times 180}{2\pi \times 360}$$

$$= 14.32°$$

$$CF = R\left(1 - \cos\frac{\theta}{2}\right) = 360\,(1 - \cos 14.32°)$$

$$= 360\,(1 - 0.9689)$$

$$= 360 \times 0.0311$$

$$= 11.20 \text{ m}$$

$$FG = AB \sin\frac{\theta}{2}$$

$$= 35 \times \sin 14.32°$$

$$= 35 \times 0.2473$$

$$= 8.66 \text{ m}$$

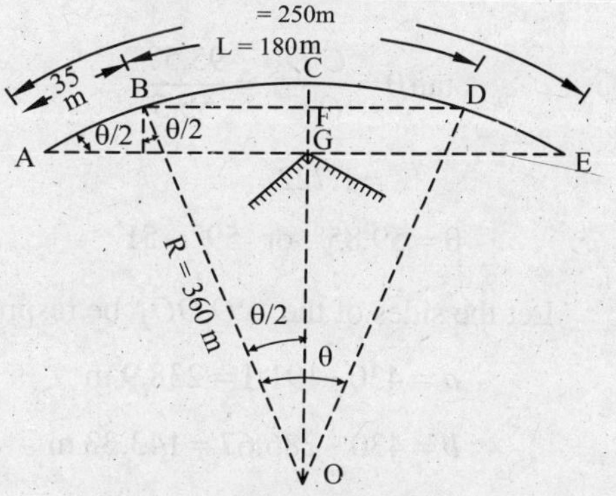

Fig. 8.31 For Problem 8.11.

∴ Total clearance required

$$= CF + FG$$

$$= 11.20 + 8.66$$

$$= 19.88 \text{ m}$$

Thus a clearance of 19.88 m is required to provide an overtaking sight distance of 250 m.

Problem 8.12 *A broken back curve consists of a 9° curve joined with a 6° curve with a straight length of 55.5 m. It is required to substitute a 4° curve in place of the straight. Calculate the necessary data for the compound curve.*

Solution. In Fig. 8.32, *AB* is a 9° curve and *CD* a 6° curve joined by a straight portion *BC* = 55.5 m. O_1 and O_2 are the centres of the two curves respectively. Let *O* be the centre of the 4° curve joining the two curves, so that $AB' C'D$ is the compound curve. Let R_1, R_2 and R be the radii of 9°, 6° and 4° curves respectively.

$$R_1 = \frac{1720}{D°} = \frac{1720}{9} = 191.1 \text{ m}$$

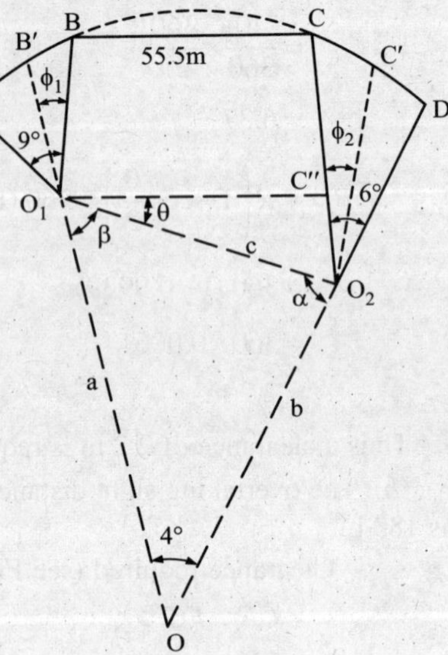

Fig. 8.32 For Problem 8.12.

$$R_2 = \frac{1720}{6} = 286.67 \text{ m}$$

$$R = \frac{1720}{4} = 430 \text{ m}$$

Join O_1 and O_2 and draw O_2C'' through O_1 and parallel to BC.

Let $\angle C'' O_1 O_2 = \theta$

$\angle OO_2O_1 = \alpha$, and

$\angle OO_1 O_2 = \beta$

Also $C''O_2 = CO_2 - CC''$

$= R_2 - R_1$

$= 286.67 - 191.1$

$= 95.57$ m

$$\tan\theta = \frac{C''O_2}{O_1C''} = \frac{95.57}{55.5}$$

$= 1.722$

∴ $\theta = 59.85°$ or $59° - 51'$

Let the sides of the ΔO_1OO_2 be respresented by a, b and c, such that

$a = 430 - 191.1 = 238.9$ m

$b = 430 - 286.67 = 143.33$ m

$c = \sqrt{95.57^2 + 55.5^2}$

$= \sqrt{9133.62 + 3080.25}$

$= \sqrt{12213.87}$

$= 110.52$ m

Let $2S$ = perimeter of the $\Delta O_1 OO_2$

$$\therefore \quad S = \frac{a+b+c}{2}$$

$$= \frac{238.9 + 143.33 + 110.52}{2}$$

$= 246.38$ m

$$\sin\frac{\alpha}{2} = \sqrt{\frac{(S-c)\,(S-b)}{bc}}$$

$$= \sqrt{\frac{(246.38-110.52)\,(246.38-143.33)}{143.33\times110.52}}$$

$$= \sqrt{\frac{135.86\times103.05}{143.33\times110.52}}$$

$$= 0.8838$$

$$\frac{\alpha}{2} = 62.1°$$

or $\alpha = 124.2°$ or $124° - 12'$

Also $\sin\frac{\beta}{2} = \sqrt{\frac{(S-a)\,(S-c)}{ac}}$

$$= \sqrt{\frac{(246.38-238.9)\,(246.38-110.52)}{238.9\times110.52}}$$

$$= \sqrt{\frac{7.48\times135.86}{238.9\times110.52}}$$

$$= 0.03849$$

$$= 2.21°$$

∴ $\beta = 2.21\times2 = 4.42°$

or $4° - 25'$

Let ϕ_1 and ϕ_2 are the angles which the radii of the 4° curve make with BO_1 and CO_2 respectively.

Then $\phi_1 = 90° - (\theta + \beta)$

$$= 90° - (59°51' + 4°25')$$

$$= 90° - 64°16'$$

$$= 25°44'$$

$$\phi_2 = 90° - (\alpha - \theta)$$

$$= 90° - (124°12' - 59°51')$$

$$= 90° - 54°21'$$

$$= 25°39'$$

Knowing the above data, the compound curve can be laid out.

TRANSITION CURVES

It is a common practice to introduce a curve of varying radius on highways. A transition or spiral or easement curve is one whose radius changes gradually from infinite to a finite value or vice versa for the purpose of giving easy change of direction of a path. This tends to counteract the swaying outwards of a vehicle when subjected to sudden application of a centrifugal force at the instant of its entering or leaving the curve. This may be provided either between a tangent and a circular curve or between two branches of a compound or reverse curve (Fig. 8.33).

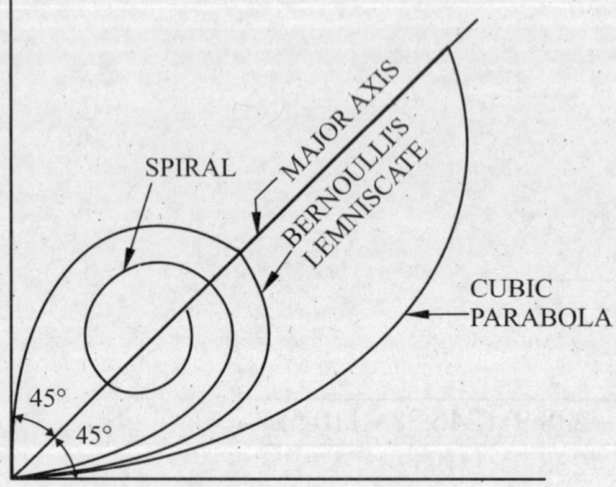

Fig. 8.33 Types of transition curves.

The *objectives* of providing a transition curve are:

(*i*) To obtain gradually the transition from the tangent to the circular curve and from the circular curve to the tangent.

(*ii*) To accomplish a gradual increase of curvature from a value of zero at the tangent point to that of the circular curve at their junction.

(*iii*) To have a gradual increase of superelevation from zero on the tangent point to a specific value on the main circular curve. The maximum superelevation is attained simultaneously with the curvature of the circular curve at the junction of the transition with circular curve.

The necessity of introducing a transition curve is to have a smooth change of direction of a vehicle from a straight to a curved path or *viceversa.* Jerks are avoided and there is no discomfort to the passengers. On a curve, a vehicle is acted upon by a centrifugal force (P) whose value varies directly to the square of the velocity and inversely to the radius $\left(P = \dfrac{WV^2}{127\,R}\right)$. When a vehicle enters from a tangent to a circular path, a sudden application of this force is encountered. Similarly when a vehicle passes from a circular to a tangent path, this causes jerks. To reduce these jerks either: (*a*) decrease the speed of the vehicle or (*b*) allow the vehicle to go off the track, *i.e.,* to leave its path or (*c*) combination of (*a*) and (*b*) or (*d*) to provide a transition curve, *i.e.,* a curve with varying radius. The radius changes from infinite at the tangent to a definite value at the junction with the circular curve or *vice versa*. This provides a gradual increase or decrease in the centrifugal force on the vehicle. Thus it enables a vehicle to maintain a constant speed and without causing any discomfort or jerk to the passengers. Therefore, the radial acceleration should change gradually.

A transition curve should satisfy the following conditions:

(*i*) It should meet the straight path tangentially.

(*ii*) It should meet the circular curve tangentially.

(*iii*) It should have the same radius as that of the circular curve at its junction with the circular curve.

(*iv*) The rate of increase of curvature and superelevation should be the same.

8.27. LENGTH OF TRANSITION CURVE

The length of a transition curve may be determined by the following methods and the larger value adopted for design.

(*i*) By the rate of change of radial acceleration.

(*ii*) By an arbitrary rate of change of superelevation.

(*iii*) By the time rate.

These are described below:

8.27.1. By the Rate of Change of Radial Acceleration

Radial acceleration at any point on a circular curve $= \frac{v^2}{R}$ m/sec^2 where v = velocity of the vehicle in m/sec = 0.278 V where V is in km/hr.

R = Radius of the curve at the point in *m*.

In order to have a gradual change of radial acceleration so as not to cause discomfort to the drivers, the curvature, $\left(\frac{1}{R}\right)$, must change at a definite rate from zero to a designed value. Thus if we fix a certain value of this rate of change of radial acceleration for a given minimum radius and given velocity, there will be a fixed length of the transition curve.

Let L_s = Length of the transition curve in *m*.

C = Rate of change of radial acceleration in m/sec^3

P = Centrifugal force in kg m

W = Weight of vehicle in kg

The time (t) taken by a vehicle to pass over the transition curve is given by:

$$t = \frac{L_s}{v} \text{ sec}$$

The maximum acceleration is to be achieved at the end of the transition curve.

∴ Rate of change of radial acceleration

$$= C = \frac{\text{max. radial acceleration}}{t}$$

$$= \frac{\frac{v^2}{R}}{t} = \frac{v^2}{Rt}$$

But $$t = \frac{L_s}{v}$$

∴ $$C = \frac{v^2}{R \times \frac{L_s}{v}} = \frac{v^3}{RL_s}$$

or $$L_s = \frac{v^3}{CR} = \frac{(0.278\,V)^3}{CR} = \frac{0.0215\,V^3}{CR} \quad \text{...(8.17)}$$

The value of C has to be so selected that there is least inconvenience to the passengers. In the U.K. and U.S.A., the value of C is recommended as 0.3 m/sec^3 and 0.66 m/sec^3 respectively. IRC has suggested a value given by

$$C = \frac{80}{75 + V} \quad \text{...(8.18)}$$

subject to a maximum of 0.8 and minimum of 0.5.

8.27.2. By an Arbitrary rate of change of Superelevation

The length can be such that the superelevation (e) is applied at a uniform rate of 1 in n.

Therefore $L_s = Wn\,e$

where L_s = Length of the transition curve in m

W = Total width of pavement in m

1 in n = Rate of superelevation or cant

e = Superelevation in m

In the case of a two lane highway the IRC has suggested that its outer edge with rspect to the central line of the road may be raised at a constant differential grade of 1 in 150 in the case of plain and rolling terrains, 1 in 100 in built-up and industrial areas and 1 in 60 in mountainous or steep topography. The formulae for minimum length of transition on this basis are:

For plain and rolling terrain

$$L_s = \frac{2.7\,V^2}{R}$$

For mountainous and steep terrain

$$L_s = \frac{V^2}{R}$$

Differential grade applies to the difference between the actual gradient to which superelevation has been added and the gradient of the road, had there been no superelevation. It must be ensured that the actual gradient does not exceed the ruling gradient. Thus for 7.5 m wide road and with maximum superelevation of 1 in 15, the length of the transition, in case road is level before the transition curve is given by;

$$L_s = \frac{\text{Total width of pavement} \times \text{Superelevation} \times \text{Rate of superelevation}}{2}$$

$$= \frac{7.5}{2 \times 15} \times 150 = 37.5 \text{ m}$$

This is when the pavement in rotated about the central line.

In case the pavement is rotated about the inner edge, then the length of transition is given by

$$L_s = \frac{7.5}{15} \times 150 = 75 \text{ m}$$

If widening of the road is done at the curve the total width of the pavement shuld be taken into account.

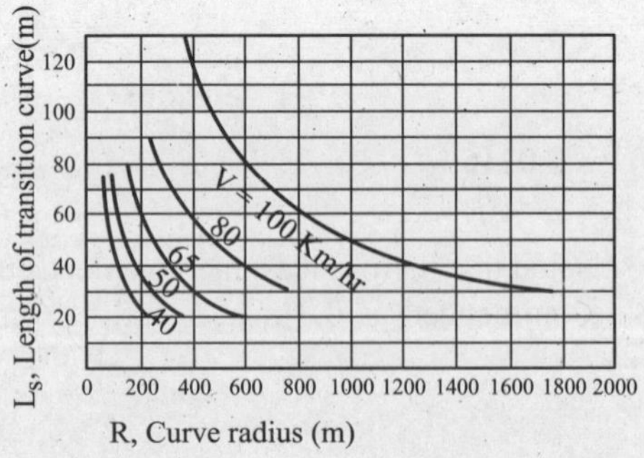

Fig. 8.34 Length of transition curve for roads in plain/rolling terrain.

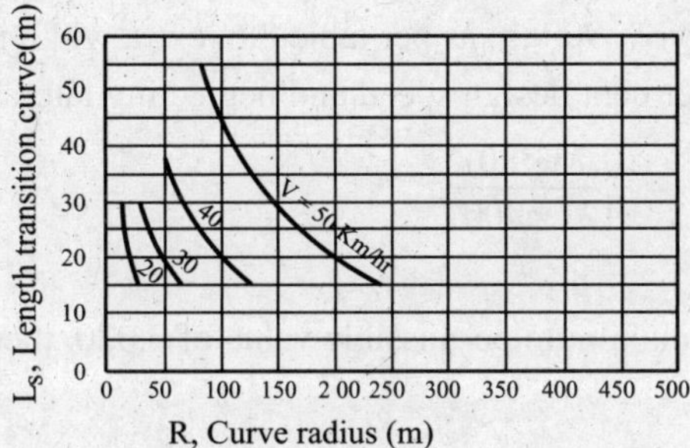

Fig. 8.35 Length of transition curve for roads in mountainous and steep terrains.

Graphs have also been plotted to determine the length of the transition curve L_s for different curve radii. Two such plots for roads in (*a*) plain rolling terrain and (*b*) mountainous and steep terrain are shown in Figs. 8.34 and 8.35.

8.27.3. By the Time Rate

The transition may be of such a length that the superelevation or cant is applied at an arbitrary time rate of P metre per second.

Let P = time rate

e = superelevation in m

Time taken by a vehicle in passing over the transition curve

$$t = \frac{L_s}{v}$$

Superelevation e attained during this period $= t \times P = \frac{L_s}{v} \times P$

$$\therefore \quad L_s = \frac{ev}{P} = \frac{0.28\, eV}{P}$$

Table 8.14 gives the minimum transition lengths recommended by IRC for different speeds and curve radii.

Problem 8.15 *Calculate the length of the transition curve from the following data:*

Design speed = 80 km/hr
Radius of circular curve = 900 m
Width of pavement on straight stretches = 7.0 m
The road is passing in plain terrain.

Solution. (*i*) *By the rate of change of radial acceleration.*

Rate of change of radial acceleration (Equation 8.18)

$$C = \frac{80}{75+V} = \frac{80}{75+80} = \frac{80}{155} = 0.516 \text{ m/sec}^3$$, which is within the limits prescribed by IRC. (0.5 to 0.8)

Length of transition curve

$$L_s = \frac{0.0215\, V^3}{CR} = \frac{0.0215\,(80)^3}{0.516 \times 900}$$

$$= 23.70 \text{ m}$$

(ii) By rate of change of superelevation

Since the radius of the curve = 900 m; as per table 8.6, extra width required on curve will be nil

Superelevation, e for 75 per cent design speed and neglecting lateral friction is given by:

$$= \frac{(0.75\,V)^2}{127\,R} = \frac{(0.75 \times 80)^2}{127 \times 900}$$

$$= 0.0315$$

As this value is less than the maximum permissible value of 0.070, therefore adopt $e = 0.0315$.

Check for lateral friction

$$f = \frac{V^2}{127\,R} - e$$

$$= \frac{(80)^2}{127 \times 900} - 0.0315$$

$$= 0.056 - 0.0315$$

$$= 0.0245$$

As this is lower than the maximum permissible value of 0.15, speed of 80 km/hr on the curve with superelevation $e = 0.0315$ is quite safe.

Assume rate of change of superelevation, n for plain terrain = 1 in 150

Also assume that the road is rotated about the central line

$\therefore$ Length of transition curve

$$= \frac{\text{Total width of pavement} \times \text{superelevation} \times \text{rate of superelevation}}{2}$$

$$= \frac{7.00 \times 0.0315 \times 150}{2}$$

$$= 16.5 \text{ m}$$

Adopting the higher value

$L_s = 23.70$ m

Say 24 m

Problem 8.16 *A two lane national highway without raised kerbs is passing through a plain terrain. Calculate the length of transition curve, if the radius of the circular curve is equal to the minimum ruling radius. Assume data as per IRC recommendations.*

Solution. As per IRC, assume the following data for a national highway passing through a plain terrain:

Design speed $V = 100$ km/hr

Maximum permissible superelevation = 0.07

Rate of superelevation, n = 1 in 150

Maximum coefficient of friction, $f = 0.15$

Wheel base of largest truck, $L = 6.1$ m

Width for pavement without raised kerbs on straight stretches,

$W = 7.0$ m

(*i*) Radius of curve = Ruling minimum radius

$$= \frac{V^2}{127\,(e+f)} = \frac{V^2}{127\,(0.07+0.15)} = \frac{V^2}{28}$$

$$= \frac{(100)^2}{28}$$

$$= 357.14 \text{ m}$$

Say 360 m (See Table 8.13)

(*ii*) Superelevation, for 75 per cent design speed, neglecting lateral friction

$$e = \frac{(0.75\,V)^2}{127\,R} = \frac{(0.75 \times 100)^2}{127 \times 360}$$

$$= 0.123 \text{ or } 12.3 \text{ per cent}$$

Since this value is greater than the maximum permissible value of $e = 0.07$, hence adopt $e = 0.07$

Check for coefficient of friction

$$f = \frac{V^2}{127\,R} - e$$

$$= \frac{(100)^2}{127 \times 360} - 0.07$$

$$= 0.22 - 0.07$$

$$= 0.15$$

Since the value of coefficient of lateral friction f equals the maximum permissible value of 0.15, the design speed of 100 km/hr with superelevation $e = 0.07$ is safe.

(*iii*) *Extra width*

Since the radius of the horizontal curve is greater than 300 m, no extra widening is required at the curves.

(*iv*) *Length of transition curve*

(*a*) By rate of change of centrifugal acceleration

$$C = \frac{80}{75+V} = \frac{80}{75+100} = 0.457 \text{ m/sec}^3$$

$$L_s = \frac{0.0215\,V^3}{CR} = \frac{0.0215 \times (100)^3}{0.457 \times 360}$$

$$= 130.68 \text{ m; Say } 130 \text{ m}$$

(*b*) By the rate of change of superelevation

Assuming the road to be rotated about the central line

$$L_s = \frac{Wne}{2} = \frac{7.0}{2} \times 0.07 \times 150$$

$$= 36.75 \text{ m}$$

Adopt transition length

$L_s = 130$ m (see Table 8.14)

Problem 8.15 *Calculate the length of transition curve required, for a carriageway of 7.5 m, width on straight portions, if the design speed is 65 km/hr and the road is passing through rolling terrain. The radius of the horizontal curve is 200 m.*

Solution. (*i*) By the rate of change of radial acceleration

Rate of change of radial acceleration, $C = \frac{80}{75+V}$

$$= \frac{80}{75+65} = 0.57 \text{ m/sec}^3$$

Length of transition curve

$$L_s = \frac{0.0215\, V^3}{CR}$$

$$= \frac{0.0215 \times (65)^3}{0.57 \times 200}$$

$$= 51.79 \text{ m}$$

(*ii*) By rate of change of superelevation

(*a*) Extra width required on curves $= \frac{nL^2}{2R} + \frac{0.105\, V}{\sqrt{R}}$

Here $n = 2$ (7.5 m pavement has two lanes)

Wheel base, $L = 6.1$ m

Extra width on curves

$$= \frac{2 \times (6.1)^2}{2 \times 200} + \frac{0.105 \times 65}{\sqrt{200}}$$

$$= 0.186 + 0.483$$

$$= 0.669 \text{ m}$$

Total width of pavement on curves

$$= 7.5 + 0.669 = 8.169 \text{ m}$$

(*b*) Superelevation for 75 per cent design speed, neglecting lateral friction

$$e = \frac{(0.75V)^2}{127\, R} = \frac{V^2}{225\, R} = \frac{(65)^2}{225 \times 200}$$

$$= 0.0938 \quad \text{or} \quad 9.38 \text{ per cent}$$

Since this value is greater than the maximum permissible value of $e = 0.07$, hence adopt $e = 0.07$

(*c*) Check for coefficient of friction; f

$$f = \frac{V^2}{127\, R} - e = \frac{(65)^2}{127 \times 200} - 0.07$$

$$= 0.166 - 0.07$$

$$= 0.096$$

Since the value of coefficient of lateral friction f is less than the maximum permissible value of 0.15, the design speed of 65 km/h with superelevation of 0.07 m/m is quite safe.

Table 8.14 Minimum transition lengths for different speeds and curve radii

Plain and rolling terrain							Mountainous and steep terrain					
Curve radius (metres)	*Design speed (km/hr)*						*Curve radius (metres)*	*Design speed (km/hr)*				
	100	*80*	*65*	*50*	*40*	*35*		*50*	*40*	*30*	*25*	*20*
	Transition length (metres)							Transition length (metres)				
45					NA	70	14				NA	30
60				NA	75	55	20				35	20
90				75	50	40	25			NA	25	20
100			NA	70	45	35	30			30	25	15
150			80	45	30	25	40		NA	25	20	15
170			70	40	25	20	50		40	20	15	15
200		NA	60	35	25	20	55		40	20	15	15
240		90	50	30	20	NR	70	NA	30	15	15	15
300	NA	75	40	25	NR		80	55	35	15	15	NA
360	130	60	35	20			90	45	25	15	15	
400	115	55	30	20			100	45	20	15	15	
500	95	45	25	NR			125	35	15	15	NR	
600	80	35	20				150	30	15	15		
700	70	35	20				170	25	15	NR		
800	60	30	NR				200	20	15			
900	55	30					250	15	15			
1000	50	30					300	15	NR			
1200	40	NR					400	15				
1500	35						500	NR				
1800	30											
2000	NR											

NA—Not applicable

NR—Transition not required.

(*d*) $$L_s = \frac{2.7\,V^2}{R}$$ (see page 314)

$$= \frac{2.7 \times 65^2}{200} = 57.00 \text{ m}$$

(*e*) Adopting a higher value (see table 8.14)

$$L_s = 57 \text{ m}, \quad \text{Say } 60 \text{ m}$$

8.28. EQUATION OF TRUE OR CLOTHOID SPIRAL CURVE

A clothoid or spiral is an ideal transition curve in which the radius of curvature at any point is inversely proportional to the distance of the point from the beginning of the curve. Thus the rate of change of acceleration is uniform. In Fig. 8.36.

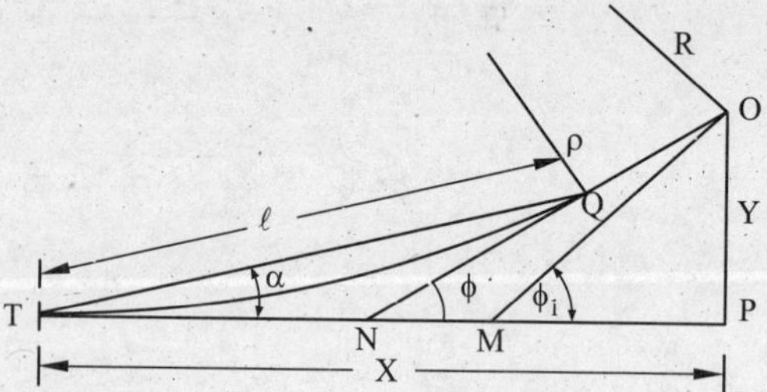

Fig. 8.36 True or clothoid spiral curve (Relation between *l* and ρ).

Let T = Starting point of the transition curve

TP = The initial tangent

O = The junction point of the transition curve and the circular curve

R = Radius of the circular curve

Q = Any point on the transition curve

ρ = Radius of the transition curve at Q

l = Length of the transition curve from the start to the point Q

L_s = Total length of the transition curve

ϕ = The inclination of the tangent at Q with the initial tangent TP

ϕ_1 = The inclination of the tangent TP and the tangent to the transition curve at the junction point O. This is called spiral angle.

From the definition of the spiral curve

$$\rho \propto \frac{1}{l} \quad \text{or} \quad \frac{1}{\rho} = c_1 l$$

where c_1 is a constant.

But for all curves, curvature $= \dfrac{d\phi}{dl} = \dfrac{1}{\rho}$

$$\therefore \qquad d\phi = \frac{1}{\rho}\, dl = c_1\, l\, dl$$

Integrating

$$\phi = \frac{c_1 l^2}{2} + c_2 \text{ (constant)}$$

But when $l = 0; \phi = 0$

$\therefore \quad c_2 = 0$

Hence, $\phi = \dfrac{c_1 l^2}{2}$ ...(8.19)

At the point O

$$l = L_s \text{ ; } \rho = R \text{ and } \phi = \phi_1$$

$$\therefore \quad \frac{1}{R} = c_1 L_s \text{ or } c_1 = \frac{1}{RL_s} \text{ and } \phi_1 = \frac{1}{RL_s}\frac{L_s^2}{2} = \frac{L_s}{2R}$$

Puttinng the value of c_1 in equation 8.19, we get

$$\phi = \frac{1}{RL_s}\frac{l^2}{2} = \frac{l^2}{2\,RL_s}$$

$$\therefore \quad l = \sqrt{2RL_s\phi}$$

or $\quad l = c\sqrt{\phi}$ ...(8.20)

where $\quad c = \sqrt{2RL_s}$

8.28.1. Equation in cartesian coordinates

If the curve has to be set out by the offsets from the point T, cartesian or rectangular coordinates have to be found out. TP is taken as X-axis and a line perpendicular to it as Y-axis.

In Fig. 8.37, let Q and Q_1 be two points on the curve at a distance of δl. Let the coordinates of Q and Q_1 be (x, y) and $(x + \delta x, y + \delta y)$ respectively and ϕ and $\phi + \delta\phi$ respectively be the angles between the initial tangent TP and the tangents at the points Q and Q_1. From the figure, in the limiting conditions:

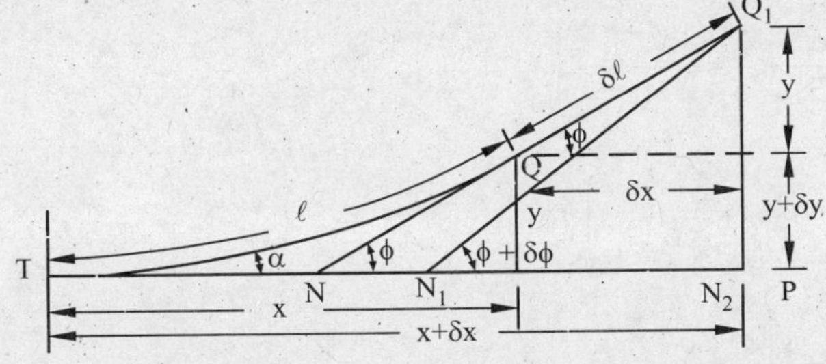

Fig. 8.37 Cartesian coordinates of the spiral.

$$\cos\phi = \frac{dx}{dl}$$

$$\therefore \quad dx = dl\cos\phi$$

$$= dl\left(1 - \frac{\phi^2}{2!} + \frac{\phi^4}{4!} - \frac{\phi^6}{6!} + \frac{\phi^8}{8!} - \ldots\right)$$

But $\quad l = c\sqrt{\phi}$

Differentiating we get, $dl = c \cdot \frac{1}{2} \cdot \phi^{-1/2} \cdot d\phi$

$$= \frac{cd\phi}{2\sqrt{\phi}}$$

$$\therefore \quad dx = \left\{ \frac{c}{2\sqrt{\phi}} 1 - \frac{\phi^2}{2!} + \frac{\phi^4}{4} - \frac{\phi^6}{6!} + \ldots \right\} d\phi$$

$$= \frac{c}{2} \left(\phi^{-1/2} - \frac{\phi^{3/2}}{2!} + \frac{\phi^{7/2}}{4!} - \frac{\phi^{11/2}}{6!} + \ldots \right) d\phi$$

Integrating we get

$$x = c \left(\phi^{1/2} - \frac{\phi^{5/2}}{10} + \frac{\phi^{9/2}}{216} - \ldots \right)$$

$$= c\sqrt{\phi} \left(1 - \frac{\phi^2}{10} + \frac{\phi^4}{216} - \ldots \right)$$

$$= l \left(1 - \frac{\phi^2}{10} + \frac{\phi^4}{216} - \ldots \right) \qquad \left(\because l = C\sqrt{\phi} \right)$$

$$l = \left(1 - \frac{l^4}{10c^4} + \frac{l^8}{216c^8} - \right) \qquad \left(\because \phi = \frac{i^2}{c^2} \right) \qquad \ldots(8.21)$$

Similarly

$$\frac{dy}{dl} = \sin \phi$$

$$\therefore \quad dy = dl \sin \phi$$

$$= dl \left(\phi - \frac{\phi^3}{3!} + \frac{\phi^5}{5!} - \frac{\phi^7}{7!} + \ldots \right)$$

$$= \frac{c}{2\sqrt{\phi}} \left(\phi - \frac{\phi^3}{3!} + \frac{\phi^5}{5!} - \ldots \right) d\phi \qquad \left(\because dl = \frac{c\,d\phi}{3\sqrt{\phi}} \right)$$

$$= \frac{c}{2} \left(\phi^{1/2} - \frac{\phi^{5/2}}{6} + \frac{\phi^{9/2}}{120} - \ldots \right) d\phi$$

Integrating we get

$$y = c \left(\frac{\phi^{3/2}}{3} - \frac{\phi^{7/2}}{42} + \frac{\phi^{11/2}}{1320} - \ldots \right)$$

$$= c\,\phi^{1/2}\,\frac{\phi}{3} \left(1 - \frac{\phi^2}{14} + \frac{\phi^4}{440} - \ldots \right)$$

$$= \frac{l^3}{3c^2}\left(1 - \frac{\phi^2}{14} + \frac{\phi^4}{440} - \ldots\right) \qquad \left(\because c\sqrt{\phi} = l \text{ and } \phi = \frac{l^2}{c^2}\right)$$

$$= \frac{l^2}{3c^2}\left(1 - \frac{l^4}{14c^4} + \frac{l^8}{440c^8} - \ldots\right)$$

$$= \frac{l^2}{6RL_s}\left(1 - \frac{l^4}{56R^2L_s^{\ 2}} + \frac{l^8}{7040\,R^4L_s^{\ 4}} - \ldots\right) \quad \left(\because\ c = \sqrt{2RL_s}\right) \qquad \ldots(8.22)$$

Taking first two terms of the equation 8.21 and 8.22, we get,

$$x = l\left(1 - \frac{l^4}{10c^4}\right) = l\left(1 - \frac{l^4}{40R^2L_s^{\ 2}}\right) \qquad \ldots(8.23)$$

$$y = \frac{l^3}{3c^2}\left(1 - \frac{l^4}{14l^4}\right) = \frac{l^3}{6RL}\left(1 - \frac{l^4}{56R^2L_s^{\ 2}}\right) \qquad \ldots(8.24)$$

From the equations 8.23 and 8.24, the coordinates of any point on the true or clothoid spiral curve can be calculated, the length l being measured along the curve.

8.28.2. Relation between deflection angle α and ϕ

Let $\alpha =$ The angle between the tangent at T and the line joining any point (Q) on the curve to T. This is known as a deflection angle.

Now from Fig. 8.37,

$$\tan\alpha = \frac{y}{x}$$

$$= \frac{\frac{l^3}{3c^2}\left(1 - \frac{\phi^2}{14} + \frac{\phi^4}{440} - \ldots\right)}{l\left(1 - \frac{\phi^2}{10} + \frac{\phi^4}{216} - \ldots\right)}$$

$$= \frac{l^2}{3c^2}\left(1 + \frac{\phi^2}{35} + \ldots\right)$$

$$= \frac{1}{3}\phi\left(1 + \frac{\phi^2}{35} + \ldots\right) \qquad \left(\because \frac{l^2}{c^2} = \phi\right)$$

$$= \frac{\phi}{3} + \frac{\phi^3}{105} + \ldots$$

Which is very nearly equal to the expansion of $\tan\frac{\phi}{3}$ *i.e.*,

$$\frac{\phi}{3} + \frac{\phi^3}{81} + \ldots$$

$\therefore$ $\tan \alpha$ is nearly equal to $\tan \frac{\phi}{3}$ or $\alpha = \frac{\phi}{3} - N_s$

where N_s is a constant.

A curve has been drawn between ϕ and N_s. The value of N_s for a definite value of ϕ can be determined from this curve.

For small values of ϕ

$$\alpha = \frac{\phi}{3} = \frac{l^2}{3\,c^2} = \frac{l^2}{6\,RL_s} \text{ radians} = \frac{1800 l^2}{\pi RL_s} \text{ minutes}$$

8.29. RELATIONSHIP BETWEEN TRANSITION AND MAIN CURVE

The effect of introducing transition curves at both the ends of a simple curve is to decrease the curvature. It appears essential that the curvature of the main curve should be increased in order to enable the whole curve to fit in the tangents. It is, however, preferable to preserve the radius proposed for the circular curve and accommodate the transitions by shifting the main curve to a position farther from the intersection point of the tangents.

In Fig. 8.38, let

TP = Original tangent

AP_1 = Shift tangent

S = Shift, the perpendicular distance between the two tangents

OD = Perpendicular from O to XC

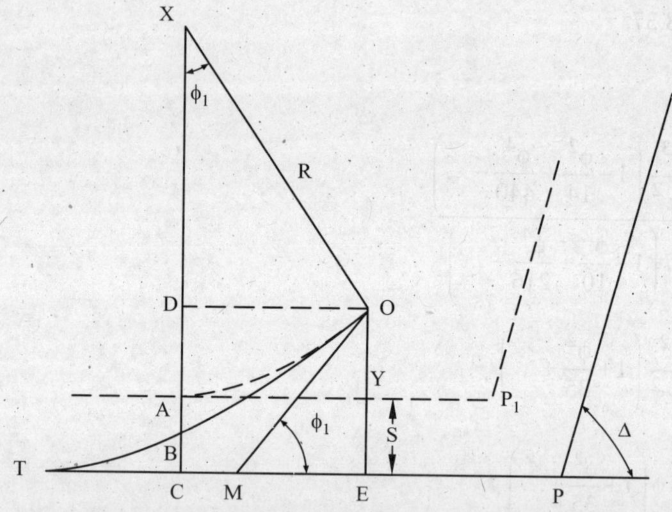

Fig. 8.38 Sketch showing characteristics of spiral curve.

Then

$$AC = S = DC - DA$$

$$= OE - DA$$

$$= Y - R\,(1 - \cos \phi_1)$$

$$= Y - 2R \sin^2 \frac{\phi_1}{2}$$

But $Y = \frac{L_s^3}{6\,RL_s} = \frac{L_s^2}{6\,R}$ and $\phi_1 = \frac{L_s}{2R}$

$\therefore \quad S = \frac{L_s^2}{6\,R} - 2\,R\sin^2\frac{\phi_1}{2}$

But for small values of ϕ_1, $\sin\frac{\phi_1}{2} = \frac{\phi_1}{2}$

$$\therefore \quad S = \frac{L_s^2}{6R} - 2R \times \frac{\phi_1^2}{4}$$

$$= \frac{L_s^2}{6R} - \frac{2R}{4} \times \left(\frac{L_s}{2R}\right)^2$$

$$= \frac{L_s^2}{6R} - \frac{L_s^2}{8R} = \frac{L_s^2}{24R} \qquad \ldots(8.25)$$

Also $\quad \phi_1 = \frac{L_s}{2R} = \frac{AO}{R}$

$\therefore \quad AO = \frac{L_s}{2}$

But in actual practice, there is very little divergence between AO and BO; thus AO can be replaced by BO.

$\therefore \quad BO = \frac{L_s}{2}$

or $\quad B$ is the mid-point of the transition curve.

Also, $\quad BC = \frac{(TB)^3}{6\,RL_s} = \frac{\left(\frac{L_s}{2}\right)^3}{6\,RL_s} = \frac{L_s^2}{48\,R} = \frac{1}{2}S \qquad \left[\because S = \frac{L_S^2}{24}\right]$

i.e., the transition curve bisects the shift.

The elements of a combined circular and transition curves are illustrated in Fig. 8.39. For deriving values of the individual elements like shift, tangent distance, apex distance, etc. and working out co-ordinates to lay the curves in the field, it is convenient to use curve tables. For this, reference may be made to IRC; 38 'Design tables for horizontal curves for highways'.

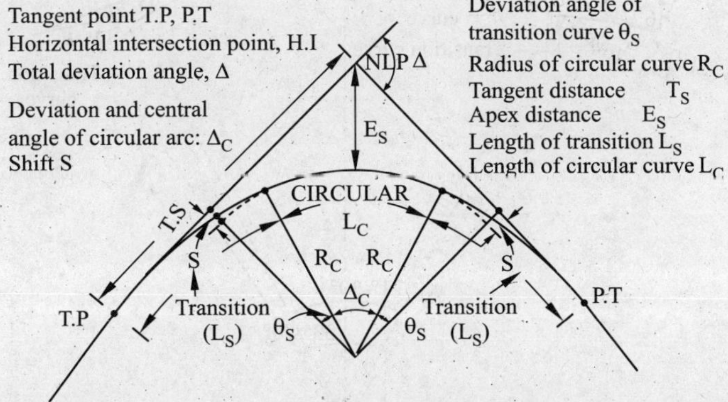

Fig. 8.39 Elements of a combined circular and transition curve.

Problem 8.16 *On a National Highway in rolling terrain, two straights intersect at chainage 82.50 with an angle of intersection of* $146°30'$. *It is proposed to put a circular arc of 15.5 chains radius with transition curves of 3.25 chain lengths (of 20 m) at each end. Determine the chainages at the beginning of the first transition length, circular curve and the second transition curve and show the computed data in a neat sketch.*

Solution. Length of the transition curve $= 3.25$ chains

$$= 3.25 \times 20 = 65 \text{ m}$$

Radius of circular curve $= 15.5 \times 20 = 310$ m

Shift, $S = \dfrac{L_s^{\,2}}{24\,R} = \dfrac{65 \times 65}{24 \times 310}$

$= 0.57$ m

Tangent Length of the curve

$$= (R + S)\tan\frac{\theta}{2} + \frac{L_s}{2}$$

$$= (310 + 0.57)\tan\frac{33° \cdot 30'}{2} + \frac{65}{2}$$

$= 310.57 \tan 16.75° + 32.5$

$= 310.57 \times 0.3 + 32.5$

$= 93.7 + 32.5$

$= 125.67$ m

$= 6.28$ chains

Angle ϕ, made by half length of the transition

$$= \frac{L_s}{2R} = \frac{65}{2 \times 310}$$

$$= 0.1048 \text{ radians} = \frac{0.1048 \times 360}{2\pi}$$

$= 6°$

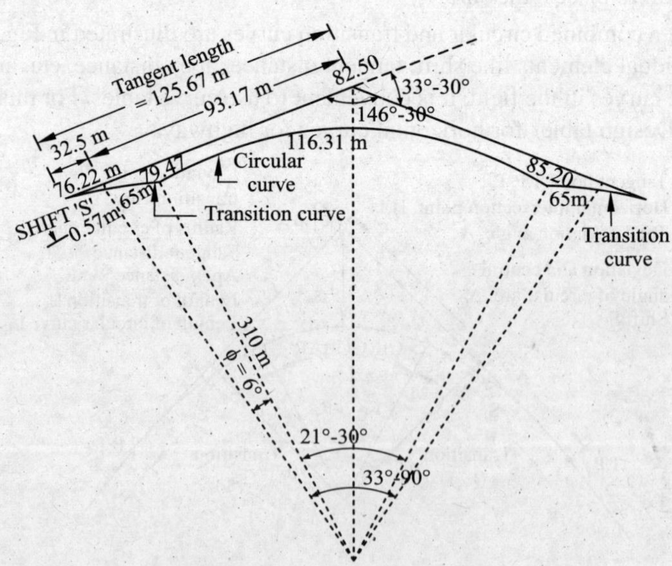

Fig. 8.40 For problem 8.16.

∴ The central angle of the circular curve

$= \theta - 2\phi = 33°.30' - 2 \times 6$

$= 21°.30'$

$= 0.3752$ radians

Length of circular curve $= R\,(\theta - 2\phi)$

$= 310 \times 0.3752$

$= 116.31$ m

$= 5.82$ chains

Chainages at the various points are given below:

Chainages at the point of intersection = 82.50

Chainage at beginning of first transition curve = 82.5 – 6.28

= 76.22 chains

Chainage at the beginning of circular curve = 76.22 + 3.25

= 79.47 chains

Chainage at the beginning of the second transition curve $= 79.47 + 5.82 = 85.29$ chains

8.30. ELEMENTS OF TRUE SPIRAL

The elements of the true spiral for the purpose of its layout are:

1. The coordinates of any point are

$$x = l\left(1 - \frac{\phi^2}{10}\right) = l\left(1 - \frac{l^4}{40\,R^2 L_s^{\,2}}\right)$$

$$y = \frac{l^3}{6RL_s}\left(1 - \frac{\phi^2}{14}\right) = \frac{l^3}{6\,RL_s}\left(1 - \frac{l^4}{56\,R^2 L_s^{\,2}}\right)$$

2. The coordinates of the end point are

$$X = L_s\left(1 - \frac{\phi_1^{\,2}}{10}\right) = L_s\left(1 - \frac{L_s^2}{40\,R^2}\right) = L_s\left(1 - \frac{3S}{5R}\right) \qquad \left[\because \frac{L_s^{\,2}}{R} = 24S\right]$$

$$Y = \frac{L_s}{6R}\left(1 - \frac{\phi_1^{\,2}}{14}\right) = \frac{L_s^{\,2}}{6R}\left(1 - \frac{L_s^{\,2}}{56\,R^2}\right)$$

where S is the shift.

3. The deflection angles are

$$\alpha = \frac{573 l^2}{RL_s} \text{ minutes} = \frac{Dl^2}{10\,L_s} \text{ minutes}$$

$$\alpha_n = \frac{573 L_s}{R} \text{ minutes} = \frac{DL_s}{10} \text{ minutes}$$

$$\phi_1 = \frac{3 \times 573\,L_s}{R} \text{ minutes} = \frac{DL_s}{10} \text{ minutes} = \frac{DL_s}{200} \text{ degrees}$$

4. Total tangent length is

$$= (R+S)\tan\frac{\Delta}{2} + x - R\sin\phi_1$$

$$= (R+S)\tan\frac{\Delta}{2} + \frac{L_s}{2}\left(1 - \frac{S}{5R}\right)$$

where Δ = Deflection angle between the two tangents.

8.31. CUBIC SPIRAL CURVE

For simplifying the calculations of the quantities required in setting out acute transition curve, various modified forms of this curve are in use. In the case of cubic spiral curve, in equation 8.22 all terms except the first one are rejected. Therefore, the equation of the cubic spiral is

$$y = \frac{l^3}{6RL_s}$$

The approximation made in this case is

$$\sin\phi = \phi \text{ or } \frac{dy}{dl} = \phi$$

But $\phi = \dfrac{l^2}{2\,RL_s}$

$\therefore$ $dy = \dfrac{l^2\,dl}{2RL_s}$

or $y = \dfrac{l^3}{6\,RL_s}$, the constant of integration being zero.

The elements of a cubic spiral are:

(*i*) $y = \dfrac{l^3}{6\,RL_s}$, l being measured along the curve

(*ii*) $\alpha = \dfrac{1800 l^2}{\pi RL_s}$ minutes $= \dfrac{573 l^2}{RL_s} = \dfrac{Dl^2}{10\,L_s}$ minutes

where D is the degree of curve.

$$\alpha_n = \frac{1800 L_s}{\pi R} = \frac{573 L_s}{R} = \frac{DL_s}{10} \text{ minutes}$$

(*iii*) $\phi_1 = 3\,\alpha_n = \dfrac{3\times 573 L_s}{R}$ minutes $= \dfrac{DL_s}{200}$ degrees

(*iv*) Total tangent length $= (R+S)\tan\dfrac{\Delta}{2} + \dfrac{L_s}{2}$

8.32. CUBIC PARABOLIC CURVE

In this case, it is desired to lay the curve by the cartesian coordinates and all terms except for the first one are rejected in both the equations 8.21 and 8.22.

$\therefore \qquad x = l$

and $\qquad y = \dfrac{l^3}{6\,RL_s} = \dfrac{x^3}{6\,RL_s}$

This is also known as Froude's transition or easement curve. Two approximations are made in this, *i.e.*, $\cos\phi = 1$ and $\sin\phi = \phi$. The error involved in the assumption that $x = l$ or $\cos\phi = 1$ is greater than that made in the approximation $\sin\phi = \phi$, since the cosine series is less rapidly converging than the sine series. The cubic parabola is, therefore, inferior to the cubic spiral but it has been extensively used owing to the ease with which it can be set out by rectangular coordinates. The elements of cubic parabola are:

(*i*) $y = \dfrac{x^3}{6RL_s}$ and $x = l$;

the length being measured along the x-axis

(*ii*) $X = L_s$ and $Y = \dfrac{L_s}{6R} = 4S$

(*iii*) The coordinates of the end points are

$$\alpha = \frac{573 l^2}{RL_s} \text{ minutes} = \frac{Dl^2}{10\,L_s} \text{ minutes}$$

$$\alpha_n = \frac{DL_s}{10} \text{ minutes}$$

$$\phi_1 = \frac{3\,DL_s}{10} \text{ minutes} = \frac{DL_s}{200} \text{ degrees}$$

(*iv*) Total tangent length $= (R + S)\tan\dfrac{\Delta}{2} + \dfrac{L_s}{2}$

8.33. BERNOULLI'S LEMNISCATE CURVE

This type of curve is mostly used in modern road construction as it is symmetrical and can be well adopted when the deflection angle between the tangents is large. It has got the following advantages over the spiral curve:

1. The radius of curvature decreases more gradually with length.
2. It is an autogenous curve, *i.e.*, follows a path which is actually traced by a vehicle when turning freely.
3. Towards the end of the curve, for deflection angles greater than 30°, the rate of increase of curvature decreases and this is not uniform.
4. This curve can be easily set by polar coordinates.

Figure 8.41 shows the shape of the curve in the first quadrant. The polar equation of the lemniscate is :

$$\rho = c\sqrt{\sin 2\alpha}$$

where ρ = Polar ray or radius

α = Polar deflection angle; and

c = Constant

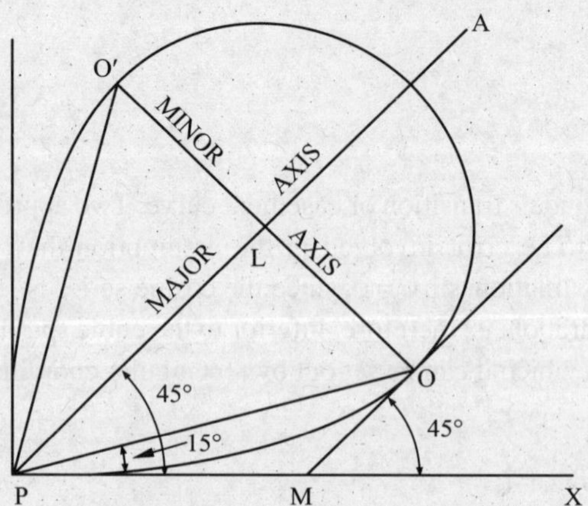

Fig. 8.41 Lemniscate curve.

The maximum value of $\rho = c$ when $\alpha = 45°$

while ρ is zero when $\alpha = 0°$ and $90°$.

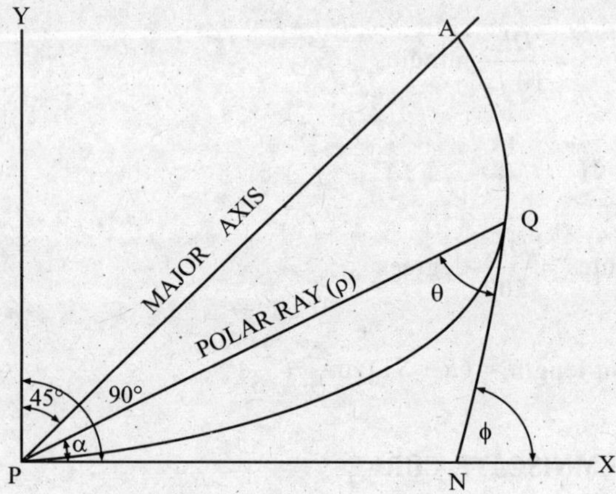

Fig. 8.42 Lemniscate curve.

In Fig. 8.42, let Q be any point on the curve, whose polar ray is ρ and polar deflection angle is α. Let *QN* be the tangent at *Q*.

From the properties of polar coordinates, we have

$$\tan\theta = \rho \cdot \frac{d\alpha}{d\rho}$$

But
$$\rho = c\sqrt{\sin 2\alpha}$$

$$\therefore \quad \frac{d\rho}{d\alpha} = c\frac{\cos 2\alpha}{\sqrt{\sin 2\alpha}}$$

$$\therefore \quad \tan\theta = c\sqrt{\sin 2\alpha} \cdot \frac{\sqrt{\sin 2\alpha}}{c\cos 2\alpha}$$

$$= \tan 2\alpha$$

$\therefore \qquad \theta = 2\alpha$

Now $\qquad \phi = \alpha + \theta = \alpha + 2\alpha = 3\alpha \quad$ (exactly)

Thus the main difference between the true spiral and the cubic spiral or the cubic parabola and the lamniscate is that in the former cases, $\alpha = \dfrac{\phi}{3}$ approx while in the case of lemniscate $\alpha = \dfrac{\phi}{3}$ exactly.

The radius of curvature can be found out by the usual formula for polar coordinates.

$$r = \frac{\left[\rho^2 + \left(\dfrac{d\rho}{d\alpha}\right)^2\right]^{3/2}}{\rho^2 + 2\left(\dfrac{d\rho}{d\alpha}\right)^2 - \rho\,\dfrac{d^2\rho}{d\alpha^2}}$$

where $\qquad \dfrac{d\rho}{d\alpha} = \dfrac{c\cos 2\alpha}{\sqrt{\sin 2\alpha}}$;

and $\qquad \dfrac{d^2\rho}{d\alpha^2} = -\dfrac{c}{(\sin 2\alpha)^{3/2}}\left[1 + \sin^2 2\alpha\right]$

$$\therefore \qquad r = \frac{c^2 \sin 2\alpha + \dfrac{c^2\cos^2 2\alpha}{\sin 2\alpha}}{c^2 \sin 2\alpha + \dfrac{2c^2\cos^2 2\alpha}{\sin 2\alpha} + \dfrac{c\sqrt{\sin 2\alpha}}{(\sin 2\alpha)^{3/2}} \cdot c\,(1 + \sin^2 2\alpha)}$$

or $\qquad r = \dfrac{\rho}{3\sqrt{\sin 2\alpha}}$

$\therefore \qquad c = \sqrt{3\rho r}$

Also $\qquad \dfrac{dl}{d\phi} = r = \dfrac{c}{3\sqrt{\sin 2\alpha}}$

Integrating we get,

$$l = \frac{c}{3}\left(\frac{1}{2}\sqrt{\tan\alpha} - \frac{1}{5}\sqrt{\tan 5\alpha} + \frac{1}{12}\sqrt{\tan 9\alpha} \ldots\right)$$

This is not a convenient series as it does not converge rapidly. Prof. FG Royal Dawson has, therefore suggested the following empirical formula:

$$l = \frac{\sqrt{2c\alpha}}{\sqrt{\sin\alpha}}\cos k\alpha = 6r\alpha - \sqrt{\cos\alpha}\cos k\alpha$$

where k is a coefficient which varies with α. Its values are given in the following table:

α	k	α	k	α	k
5°	0.190	20°	0.181	35°	0.168
10°	0.187	25°	0.177	40°	0.163
15°	0.184	30°	0.173	45°	0.159

Since the maximum value of α is 45°, $\cos\alpha \cos k\alpha$ is always less than unity.

$\therefore \quad l = 6r\ (\alpha - < 1)$

For small deflection angles,

$l = 6r\alpha$ where α is in radians

$$= \frac{6r\alpha^\circ}{57.3}$$

At the end of the curve

$$r = R;\ l = L_s;\ \phi = \phi_1 = 3\alpha$$

In Fig. 8.40, *PA*, the polar ray for $\alpha = 45°$, is the major axis. To determine the position of the minor axis, draw a polar ray *PO* making angle = 15° with *PX*. Then tangent at *O* makes angle $\phi = 45°$ with *PM*. Therefore *MO* is parallel to *PA*. Draw OLO' perpendicular to *PA* meeting the other side of the curve at O'. Then OO' is the minor axis. Triangle POO' being equilateral,

$$OO' = PO = c\sqrt{\sin 2\alpha} = c\sqrt{\sin 30^\circ} = \frac{k}{\sqrt{2}}$$

Now $\quad PA = c\sqrt{\sin 90^\circ} = c$

$$\therefore \quad OO' = \frac{PA}{\sqrt{2}}$$

or $\quad \dfrac{OO'}{PA} = \dfrac{1}{\sqrt{2}} = \dfrac{1}{1.4142}$

Radius of curvature at *A*

$$= \frac{c}{3\sqrt{\sin 2\alpha}} = \frac{c}{3\sqrt{\sin 90^\circ}} = \frac{c}{3} = \frac{PA}{3}$$

Thus the radius of curvature diminishes gradually from α at the origin (*P*) to a minimum value of $\frac{1}{3}PA$ at *A*.

Length of the curve $PQA = 1{\cdot}311115\ PA$

$$= 1.311115\ c$$

8.34. SETTING OUT LEMNISCATE CURVE

Two cases have to be considered, namely:

(*i*) When the curve between the tangents is entirely transitional, *i.e.,* with no circular curve.

(*ii*) Two transitional curves on either side of a circular curve

Case I: When the curve is transitional throughout, it may be set out by the following methods:

(*a*) When the apex distance (AT) and the deflection angle (Δ) are given. In Fig. 8.43:

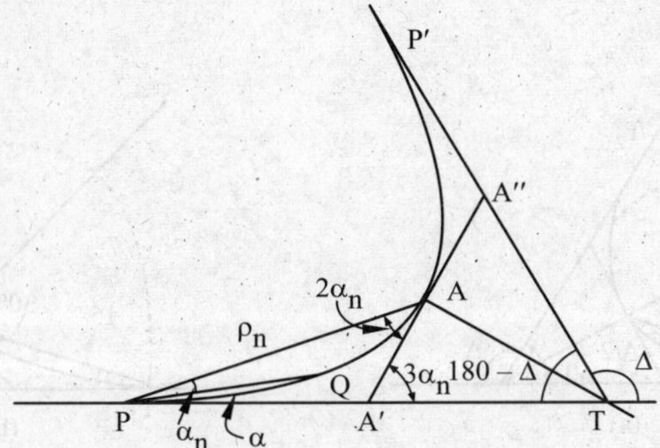

Fig. 8.43 Setting out lemniscate curve.

(*i*) Find out α_n by the equation

$$\alpha_n = \frac{\Delta}{6}$$

(*ii*) Find out the tangent length *PT* and *PA*, the polar ray of *A* by the sine formula

$$\frac{PT}{\sin PAT} = \frac{AT}{\sin APT}$$

or $$PT = AT \times \frac{\sin\left(90 - \frac{\Delta}{2}\right)}{\sin \alpha_n}$$

Similarly, $$PA = AT \frac{\sin\left(90 - \frac{\Delta}{2}\right)}{\sin \alpha_n}$$

(*iii*) Locate the tangent point *P*, knowing the tangent distance *PT*. Similarly P' can be located.

(*iv*) Find out the value of *c* from

$$c = \frac{PA}{\sqrt{\sin 2\alpha_n}}$$

(*v*) Determine different values of ρ by assuming succesive values of α by the formula

$$\rho = c\sqrt{\sin 2\alpha_n}$$

(*vi*) Then set out the lemniscate from *P* and P' by the method of deflection angles.

Fig. 8.44 shows the curves which are transitional throughout for deflection angles of 60°, 90° and 180°.

(*b*) When the radius of the curve (*R*) at the point *A* and deflection angle (Δ) are given:

(*i*) Find out α_n as above

(*ii*) Determine the value of *c* from

$$c = 3R\sqrt{\sin 2\alpha_n}$$

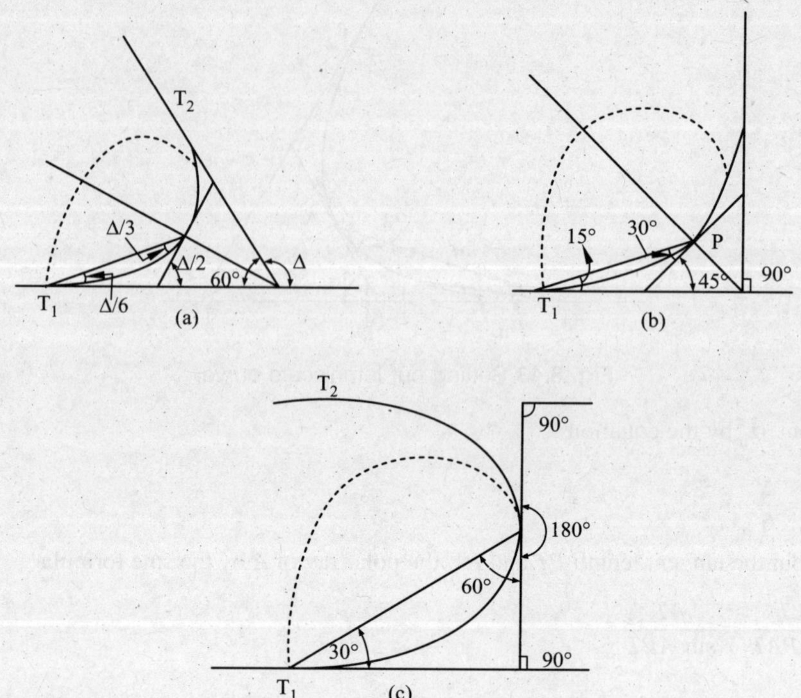

Fig. 8.44 Lemniscate curve (a) acute angle bend, 60° (b) right angle bend, 90° and (c) hair pin bend, 180°.

(*iii*) Find *PA*, the polar ray of *A* by the relation

$$\rho_n = c\sqrt{\sin 2\alpha_n}$$

(*iv*) Calculate the values of *PT* and *AT* by the sine formulae.

(*v*) Proceed further as above.

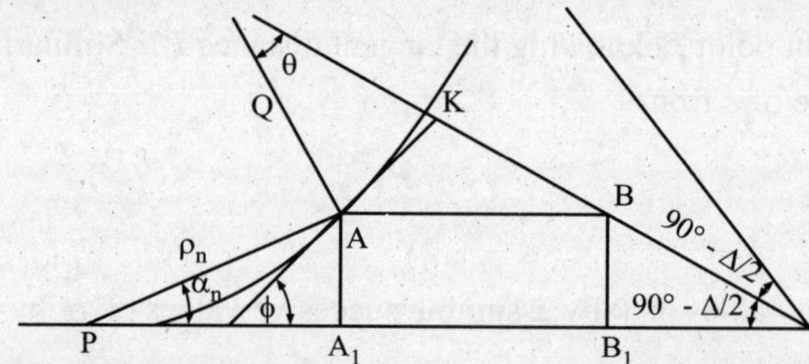

Fig. 8.45 Setting out of partly lemniscate and partly circular curve.

Case II: When value of α_n is less than $1/6\,\Delta$, it becomes essential to introduce a circular curve between the two lemniscates as shown in Fig. 8.45. The central angle 2θ of the circular curve is $\Delta - 2\phi_1 = \Delta - 6\alpha_n$. The tangent length is given by

$$PT = PA_1 + A_1B_1 + B_1T$$

$$= \rho_n \cos\alpha_n + R\left(\cos\phi_1 \cdot \tan\frac{\Delta}{2} - \sin\phi_1\right) + \rho_n \sin\alpha_n \tan\frac{\Delta}{2}$$

where *R* is the radius of the circular curve.

The curve is set out as in the previous case.

8.35. COMPARISON OF TRANSITION CURVES

For deflection angles up to 4°, there is no difference between spiral, cubic parabola and lemniscate curves.

True spiral or clothoid is the only curve which satisfies the relation Rl = constant at all points, *i.e.*, the radius of curvature is inversely proportional to the length traversed. The rate of change of acceleration is, therefore, uniform throughout its length. Thus this curve is the best from comfort point of view. However, the value of the deflection angle is very nearly equal to one third the angle which the tangent at that point makes with the same axis, *i.e.*, may be taken as equal to $\phi/3$ for values of ϕ up to 20°. Beyond this, α should be taken as equal to $\phi / 3 - N_s$ and value of N_s determined from the graph. This curve can be easily set from the tangent point by means of a theodolite or offset method. This has also to be preferred where the whole curve cannot be set out from the tangent point and the instrument is required to be shifted as the calculations are much easier in this case. Moreover the formula $L_s = 2\,R\phi$ in the case of spiral curve is very simple and as such the curve can be set out easily on rough ground.

In the case of cubic parabola, the radius of curvature varies inversely proportional to the length only for deflection angles up to 4°. From 4° to 9° deflection angles, there is a rapid fall in the value of the radius of curvature and beyond this, it again increases rapidly. Therefore this type of curve is suitable only for α up to 4°. On roads, this value is exceeded in most of the cases and as such cubic parabola is not suitable for roads. It has been used on railways where the degree of the curve is small.

In the case of lemniscate curve, the radius of curvature does not decrease rapidly with length. Only towards the end, *i.e.*, for ϕ greater than 30°, the rate of change of curvature decreases rapidly. For lemniscate curve, α equals $\frac{\phi}{3}$ exactly throughout the length of the curve and thus it can be easily calculated in the field. In the case of spiral α equals $\frac{\phi}{3}$ approximately for small deflection angles only. Therefore for greater values of α, spiral curves are difficult to calculate in the field.

8.36. HAIR-PIN BENDS

In hilly areas it may become difficult to avoid bends where direction of the road reverses. Such bends are commonly known as hair-pin bends.

A hair-pin bend may be designed as a circular curve with transition curves at each end. Alternatively, compound circular curves may be provided.

The following design criteria should be adopted normally for the design of hair-pin bends.

(*a*) Minimum design speed ... 20 km/hr

(*b*) Minimum roadway width at apex:

- (*i*) National/state highways ...11.5 m for double lane; 9.0 m for single lane
- (*ii*) Major district roads and other district roads ...7.5 m
- (*iii*) Village roads ... 6.5 m

(*c*) Minimum radius for the inner curve ...14.0 m

(*d*) Minimum length of the transition ...15.0 m

(*e*) Gradient

- Maximum 1 in 40 (2.5 per cent)
- Minimum 1 in 200 (0.5 per cent)

(*f*) Superelevation 1 in 10 (10 per cent)

Further the following points should also be given due consideration:

(*i*) Inner and outer edges of the roadways should be concentric with respect to central line of the pavement.

(*ii*) Where a number of hair-pin bends have to be introduced, minimum intervening length of 60 m should be provided between the successive bends to enable the driver to negotiate the alignment smoothly.

(*iii*) Widening of hair-pin bends at a later date is a difficult and costly process. Moreover gradients tend to become sharper, as widening can be achieved generally only by cutting the hill side.

(*iv*) At hair-pin bends, preferably the full roadway width should be surfaced.

These points should be kept in view at the planning stage, especially where a series of hair-pin bends are involved.

8.37. SELECTION OF HORIZONTAL CURVES

Simple circular curves should be provided where there is more of slow moving traffic, e.g. bullock carts and less of fast moving traffic. A transition curve with central circular curve is recommended when there is more fast moving traffic and the radius of the curve is less than 300 m. Transition curve is not necessary for radius of the curve greater than 300 m. Compound curves should be recommended only when the topography of the area compels for their adoption. In this case, the curve should begin from the tangent with the largest possible radius and may then decrease. Care should be taken that there is as least difference of radius between two adjacent curves as possible so that the centrifugal force may act uniformly on the vehicle. Reverse curves should not be provided on through routes as it is not possible to provide superelevation at their junction. These cannot be avoided on hilly roads. If possible a straight portion should be provided between the two components of a reverse curve to enable the provision of superelevation.

SIGHT DISTANCES

Visibility is an important requirement for the safety of travel on highways; particularly with the fast moving vehicles. For this, it is necessary that sight distance of adequate length should be available in different situations to permit drivers enough time and distances to control their vehicles so that there are no unwarranted accidents.

Three types of sight distances are relevant so far as the design of summit vertical curves and visibility at the horizontal curves is concerned.

Firstly the obstruction may be such that a vehicle is required to come to a stop before colliding with the object. In such a case, it must be ensured that the driver travelling at design speed has sufficient view ahead to enable him on seeing the object on the road, to apply brakes and come to a stop so as to avoid striking the object. Thus *stopping sight distance* has to be found out.

In the second case, obstruction can be due to vehicles, moving slowly in the same direction. In this case, the driver should be capable of overtaking slower vehicles without causing an obstruction or hazard to other traffic. *Overtaking sight distance* is, therefore, needed. Vehicles may also be moving in the opposite direction.

Thirdly *intermediate sight distances* afford reasonable opportunities to drivers to overtake with caution.

8.38. STOPPING SIGHT DISTANCE

It is the distance covered by a vehicle from the instance a danger is comprehended by the driver to the actual stop and depends upon the following factors:

(*i*) Total reaction time of the driver comprising perception time and brake reaction time.
(*ii*) Brake efficiency of the vehicle.
(*iii*) Frictional resistance or grip between the road and the tyre; and
(*iv*) Gradient or slope of the road.

Perception time is the time taken by the average motor vehicle driver to comprehend a danger ahead before trying to apply brakes or taking avoiding action. The obstruction is usually such that its presence would not normally be expected on the road. Perception time varies considerably and depends upon many factors such as distance to the object; its colour, condition, type and size; alertness and optical ability of the driver; atmospheric visibility; type and condition of the roadway, etc. After the danger has been perceived the driver may proceed to apply brakes. The brief interval or time lag between the perception of danger and the effective application of brakes is called the *brake reaction time*. These are closely interlinked and are considered, for all practical purposes together. It has been found that the perception plus reaction time equals 2.5 sec. This time interval is much increased when the driver is faced with complex situations. For example, at a right angled junction, the driver has also to look to both sides, decide a course of action and then carry out the muscular reaction. Thus, here, the procedure becomes perception—judgement and reaction. The eye can see clearly only in an arc covering about 5 degrees. Thus to look to all the sides, they must be focussed individually in all the directions, each one of which requires about 6 sec. The perception time will, therefore, be much increased. The judgement will require another one to three seconds and then brake reaction time will be needed. The distance travelled during the perception and brake reaction time is known as the *lag distance* and is equal to $d_1 = 0.278\, Vt$ in which V is the speed of the vehicle in km/hr and t is the perception plus reaction time in seconds.

PIEV theory has been advanced by traffic engineers to account for the total reaction time. According to this theory, the total reaction time has been split up into four parts, namely:

(*i*) Perception time
(*ii*) Intellection time
(*iii*) Emotion time; and
(*iv*) Volition time.

Perception time is the time taken for the sensations received by the ears or eye to be transmitted to the brain through the spinal chord and nervous system. Thus it is the time taken to perceive an object or obstruction. Intellection time is the time taken to compose the different thoughts, regrouping and registering new situations. Emotion time is the time spent during emotional disturbances and sensations such as anger, fear, etc. Volition time is the time required for the final action.

It is, sometimes, quite likely that the driver may apply brakes or take avoiding action by the reflex action.

These actions are shown in Fig. 8.46.

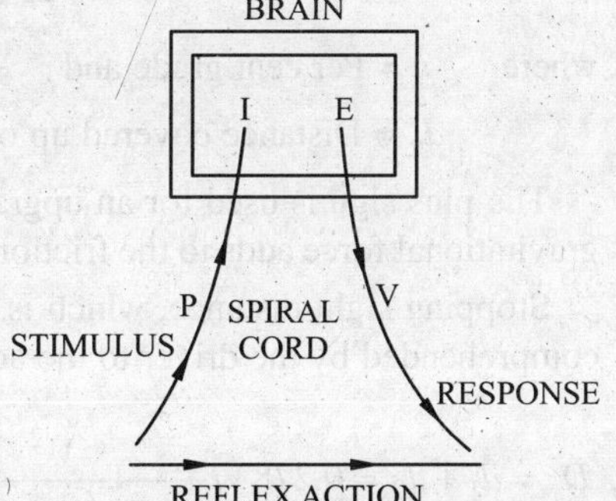

Fig. 8.46 PIEV process.

After this, the driver applies brakes. The motion of the vehicle gets retarded and after moving for some distance, it comes to a stop. The maximum effective braking effort is determined by the frictional resistance between the tyres and the road surface. The coefficient of friction between the road surface and the tyres of a vehicle depends upon:

(*i*) The type and condition of road surface.
(*ii*) The presence of moisture, snow, mud, etc. on the road surface.
(*iii*) The condition of treads of tyres.
(*iv*) Air pressure of tyres.

A value of 0.40 at 20 km/hr to 0.35 at 100 km/hr is recommended by the Indian Roads Congress for this frictional coefficient. The braking distance within which a moving vehicle can be brought to rest can be determined by equating the work done by friction to the difference in the kinetic energies of the vehicle at the time of applying brakes and at the time of stopping.

Let f = Coefficient of friction between the tyres and road surface
W = Weight of the vehicle
d_2 = Distance covered during retardation
v = Speed in m per sec at that instance
= 0.278 V where V is in km/hr
g = Acceleration due to gravitational force = 9.81 m per sec^2.

If the frictional force F is supposed to act at a uniform rate during retardation, then

Work done against friction = $Fd_2 = fWd_2$ to stop the vehicle.

The difference of kinetic energies of the vehicle at the time of applying brakes and at the time of stopping is given by

$$\text{K.E.} = \frac{1}{2} W \left(\frac{v^2}{g} - 0 \right)$$

$$= \frac{1}{2} \frac{Wv^2}{g} = \frac{Wv^2}{2 \times 9.81} = \frac{Wv^2}{19.62}$$

$$= \frac{W\,(0.278\,V)^2}{19.62}$$

$$= \frac{WV^2}{254}$$

$$\therefore \qquad fWd_2 = \frac{WV^2}{254}$$

or

$$d_2 = \frac{V^2}{254 f}$$

This relation holds good only on a level ground. When the vehicle moves on an incline, the formula is modified as;

$$d_2 = \frac{V^2}{254\left(f \pm \frac{n}{100} \right)}$$

where n = Per cent grade and
d_2 = Distance covered up or down the grade.

The plus sign is used for an upgrade and the minus sign for downgrade as the component of the gravitational force adds to the frictional resistance in the first case and neutralizes in the second case.

Stopping sight distance, which is the distance travelled by a vehicle from the time the danger is comprehended by the driver to the actual stop, is made up of lag distance plus braking distance *i.e.,*

$D_s = d_1 + d_2 = 0.278\,Vt + \dfrac{V^2}{254\left(f \pm \frac{n}{100} \right)}$. These distances for the various speeds, on a leval road are given in Table 8.15.

Table 8.15 Stopping sight distance for various speeds

Speed	*Perception and brake reaction distance*		*Braking distance*		*Safe stopping sight distance (metres)*	
V (Km/h)	*Time, t (sec)*	*Distance (metres)* $d_1 = 0.278\ Vt$	*Coefficient of longitudinal friction (f)*	*Distance (metres)* $d_2 = \frac{V^2}{254\ f}$	*Calculated values* $d_1 + d_2$	*Rounded off values for design*
20	2.5	14	0.40	4	18	20
25	2.5	18	0.40	6	24	25
30	2.5	21	0.40	9	30	30
40	2.5	28	0.38	17	45	45
50	2.5	35	0.37	27	62	60
60	2.5	42	0.36	39	81	80
65	2.5	45	0.36	46	91	90
80	2.5	56	0.35	62	118	120
100	2.5	70	0.35	112	182	180

Problem 8.17 *Calculate the stopping sight distance for a design speed of 65 km/hr assuming the coefficient of friction as 0.36 and total reaction time of drivers as 2.5 seconds.*

Solution. Design speed, $V = 65$ km/hr

Coefficient of friction, $f = 0.36$

Total reaction time, $t = 2.5$ seconds

Lag distance $= 0.278\ Vt$

$= 0.278 \times 65 \times 2.5$

$= 45.18$ m

Braking distance $= \frac{V^2}{254\ f}$

$= \frac{(65)^2}{254 \times 0.36}$

$= 46.21$ m

Stopping distance = Lag distance + braking distance

$= 45.18 + 46.21$

$= 91.39$ m

Say 90 m (see table 8.15)

Problem 8.18 *Calculate the stopping distance on a national highway in rolling terrain at an ascending gradient of 3 per cent. Assume relevant data as per IRC recommendations.*

Solution. As per Indian Roads Congress recommendations, for a national highway in rolling terrain:

Design speed, $V = 80$ km/hr

Coefficient of friction, $f = 0.35$

Total reaction time = 2.5 seconds

Stopping distance $= 0.278\,Vt + \dfrac{V^2}{254\left(f + \dfrac{n}{100}\right)}$

$$= 0.278 \times 80 \times 2.5 + \frac{(80)^2}{254\left(0.35 + \dfrac{3}{100}\right)}$$

$$= 55.6 + 66.31$$

$$= 121.91 \text{ m}$$

$$= 120 \text{ m} \quad \text{(say)}$$

8.39. MINIMUM SIGHT DISTANCE

It is the minimum distance required within which a vehicle moving at the design speed can be stopped. *IRC* has recommended minimum sight distance to be taken for a vertical curve as the length of the line that joins an object 15 cm high and an eye 1.2 m high above the curved road surface and touching the road curve at a point in between the object and the eye, as shown in Fig. 8.47. This is also sometimes referred as the non-passing sight distance or non-over taking sight distance. However the sight distance at every point of a highway should be as long as possible.

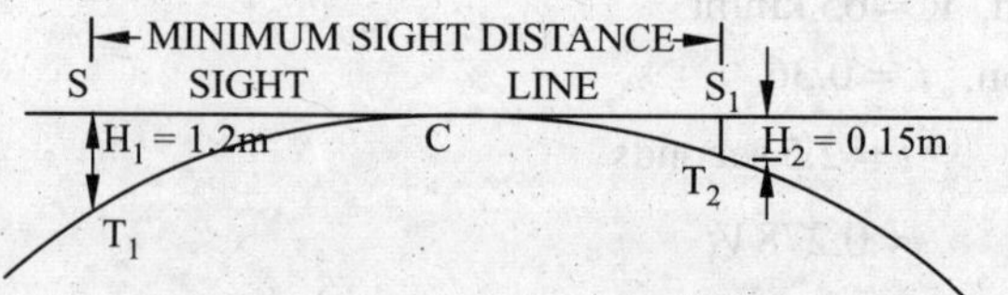

Fig. 8.47 Minimum sight distance.

Problem 8.19 *Determine the minimum sight distance for a two lane divided highway in hilly terrain on 4 per cent decending grade. Assume data as per IRC recommendations.*

Solution. Since the highway is a two lane divided highway, there exists no facility for overtaking. As such, the minimum sight distance to be provided is the stopping distance.

As per *IRC* recommendations:

Ruling design speed, $V = 60$ km/hr

Coefficient of friction, $f = 0.36$

Total reaction time, $t = 2.5$ sec (Table 8.15)

Stopping distance = Lag distance + Braking distance

$$= 0.278\,Vt + \frac{V^2}{254\left(f - \dfrac{n}{100}\right)}$$

$$= 0.278 \times 60 \times 2.5 + \frac{(60)^2}{254\left(0.36 - \dfrac{4}{100}\right)}$$

$$= 0.278 \times 60 \times 2.5 + \frac{(60)^2}{254 \times 0.32}$$

$$= 41.7 + 44.3$$

$$= 86 \text{ m}$$

$\therefore$ Adopt minimum sight distance $\simeq 85$ m

8.40. OVERTAKING SIGHT DISTANCE

Overtaking sight distance is the minimum sight distance that should be available to a driver on a two way road to enable him to overtake another vehicle safely. Optimum condition for design is one in which the overtaking driver can follow the vehicle ahead for a short time while he assesses his chances for overtaking, pulls out his vehicle, overtakes the other vehicle at design speed of the highway and returns to his own side of the road before meeting any on coming vehicle from the opposite direction travelling at the same speed. Thus the overtaking manoeuvre comprises three distinct operations as shown in Fig. 8.48.

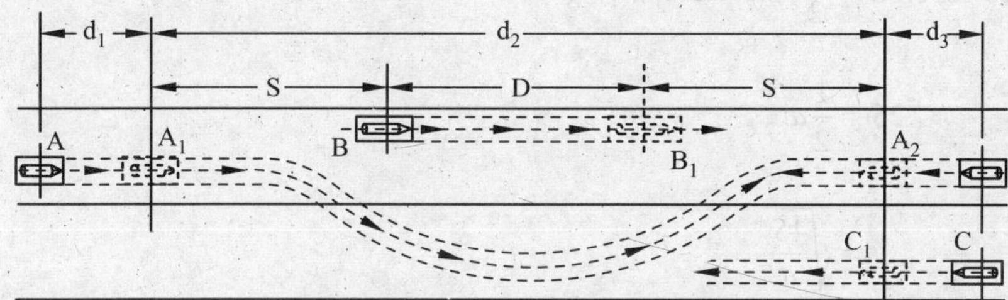

Fig. 8.48 Overtaking sight distance.

Let A be the overtaking vehicle travelling at the design speed of V km/hr and B, the overtaken vehicle moving at a speed of V_B so that the speed difference $m = V - V_B$. In travelling from A to A_1, the driver of the vehicle A sizes up the situation, moves at a reduced speed of $(V - m)$ and travels a distance d_1 in time internal t_1 which is usually taken as 2 sec.

$$\therefore \quad d_1 = 0.278\,(V - \text{m})\, t_1 = 0.278\,(V - m) \times 2$$

$$= 0.556\,(V - \text{m}) = 0.556\, V_B$$

At the point A_1, the driver of A is compelled to turn its vehicle out of its own path, accelerate, overtake B_1, return to his own path at A_2 and then move again at the design speed. The distance covered during this manoeuvre is given by

$$d_2 = 2S + b$$

where S = Safe spacing between vehicle in m

b = Distance travelled by B during the time interval t

From the observations made on the behaviour of traffic for safe spacing, the following relations have been suggested:

$$S = 0.69\,(v_B) + 6 \text{ if velocities are taken in m/sec}$$

and $$= 0.19\,(V - m) + 6 \text{ or } 0.19\, V_B + 6, \text{ if velocities are taken in km/hr}.$$

Also $b = 0.278\,(V - m)\,t = 0.278\,V_B\,t$

where m = Difference in speeds in km/hr; and

T = Time taken by A to gain 2 S metres on B.

To determine t, it is necessary to have an idea of the acceleration 'a' of vehicle A during the operation. The following values of 'a' have been suggested by IRC for various design speeds, assuming m = 16 km/hr.

Table 8.16 Overtaking acceleration for different speeds

Speed V km/hr	25	30	40	50	65	80	100
Acc. a m/sec^2	1.40	1.30	1.25	1.10	0.90	0.70	0.53
Acc. a in km/hr/sec	5.0	4.8	4.45	4.0	3.28	2.56	1.92

But distance covered $= \frac{1}{2} \times \text{acceleration} \times (\text{time})^2$

$$\therefore \qquad 2S = \frac{1}{2}at^2$$

or
$$t = \sqrt{\frac{4S}{a}}$$

Knowing S and a, t and hence b can be calculated. Thus it is possible to find the value of d_2. If the acceleration is taken in km/hr/sec then, time, t, is given by $t = \sqrt{=\frac{14.4S}{a}}$.

During this overtaking process, it is possible that another vehicle C travelling in the opposite direction at the design speed may approach vehicles A and B. This opposing traffic may come into view as the passing manoeuvre is being undertaken and cover a distance $d_3 = 0.278\,Vt$.

Overtaking sight distance:

$$d = d_1 + d_2 + d_3 = 0.556\,V_B + 2S + 0.278\,V_B\,t + 0.278\,Vt$$

or
$$d = 0.556\,(V - m) + 2S + 0.278\,(V - m)\,t + 0.278\,Vt \qquad \ldots(8.26)$$

The Indian Roads Congress has now simplified the procedure for the determination of overtaking sight distance. This is based on a time component of 9 to 14 seconds for the actual overtaking manoeuvre depending on design speed, increased by about two-third to take into account the distance travelled by a vehicle from the opposite direction during the same time. Design values for overtaking sight distances are given in Table 8.17.

Table 8.17 Overtaking sight distances for various speeds

Speed km/hr	Time components (seconds) For overtaking manoeuvre	For opposing Vehicle	Total time (t)	Safe overtaking sight distance (m) = 0.278 Vt (rounded to nearest five m)
40	9	6	15	165
50	10	7	17	235
60	10.8	7.2	18	300
65	11.5	7.5	19	340
80	12.5	8.5	21	470
100	14	9	23	640

8.41. INTERMEDIATE SIGHT DISTANCE

Intermediate sight distance is twice the safe stopping sight distance and affords opportunity to the drivers to overtake with caution. Design values of intermediate sight distance for different speeds are given in Table 8.18.

Table 8.18 Intermediate sight distances

Speed (km/h)	Intermediate sight distance (m)
20	40
25	50
30	60
35	80
40	90
50	120
60	160
65	180
80	240
100	360

8.42. APPLICATION OF SIGHT DISTANCE STANDARDS

8.42.1. Single/two lane roads

Normally an attempt should be made to provide overtaking sight distance in as much length of the road as possible. Where this is not feasible, intermediance sight distance, which affords reasonable opportunities for overtaking, should be adopted as the next best alternative. In no case, however, should the visibility correspond to less than the safe stopping distance which is the basic minimum for any road.

No hard and fast rule can be laid down for the application of overtaking sight distance since this will depend on site conditions, economics, etc. It will be good engineering practice however to use overtaking sight distance in the case of following situations:

(*i*) Straight sections of road with isolated over-bridges or summit vertical curves where the provision of overtaking sight distance would conveniently result in unobstructed visibility over a long length of the road; and (*ii*) Relatively easy sections of terrain adjacent to long reaches affording no opportunity for overtaking at all *e.g.* on either side of a winding road in hilly/rolling terrain.

8.42.2. Divided highways

On divided highways, *i.e.*, dual carriageways having a central median, the design should correspond at least to stopping sight distance given in Table 8.15. It will, however, be desirable for operational convenience and better appearance of the highway to design for somewhat more liberal values; say upto the intermediate sight distance.

8.42.3. Undivided four-lane highways

On undivided 4-lane highways, there are sufficient opportunities for overtaking within one-half of the carriageway and there should be no need to cross the centre line unless the capacity of the road is grossly deficient. Such roads may, therefore, be designed on the lines of divided highways.

8.42.4. Sight distance of valley curves

During day time, visibility is not a problem on valley curves. However for night travel, the design must ensure that the roadway ahead is illuminated by vehicle headlights to a sufficient length enabling the vehicle to brake to a stop, if necessary. Thus the visibility is reckoned in terms of the headlight sight distance which is the distance ahead of the vehicle illuminated by the headlights which is within the view of the driver. Headlight sight distance should at least equal the safe stopping sight distance.

In designing valley curves, the following criteria of measurement should be followed with regards to the headlight sight distance;

(*i*) Height of headlight above road surface is 0.75 m.

(*ii*) The useful beam of headlight is upto one degree upwards from the grade of the road; and

(*iii*) The height of object is nil.

SIGHT DISTANCE AT INTERSECTIONS

8.43. GENERAL

Visibility is an important requirement at intersection. To avoid collisions, it is essential that sufficient sight distance is available along the intersecting roads and their included corners, to enable the operators of vehicles simultaneously approaching the intersection to see each other in time.

At grade intersections can be broadly grouped under two headings:

(*i*) 'Uncontrolled intersection' where the intersecting roads are of more or less equal importance and there is no established priority; and

(*ii*) 'Priority intersection', like minor-major road intersections, where one road takes virtual precedence over the other. Traffic on the minor road may be controlled by STOP or GIVEWAY signs/road markings, making it clear that the other road has the priority.

8.43.1. Uncontrolled intersections

At these intersections, visibility should be provided on the principle that the drivers of vehicles on either highway are able to sight the intersection and the intersecting highway in good time to be able to halt their vehicles if that becomes essential. The area for clear visibility should be determined with respect to the stopping sight distance for each highway corresponding to the design speed.

Minimum sight triangles in the included corners of uncontrolled intersections, which must be kept free of all obstructions to sight, could be demarcated as in Fig. 8.49. If visibility conditions upto this standard are ensured, drivers of vehicles will be able to either stop or adjust their speed in the event of a dangerous situation ahead.

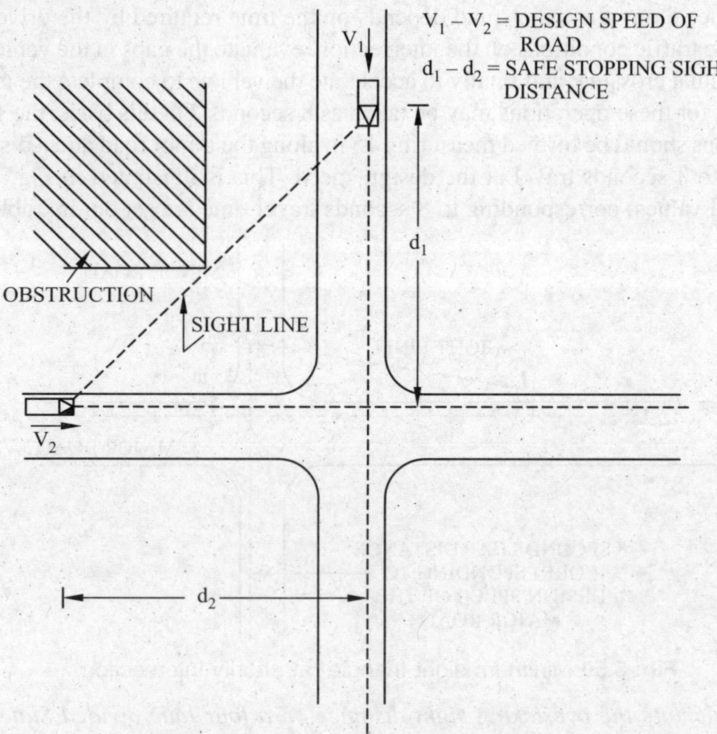

Fig. 8.49 Minimum sight triangle at uncontrolled intersection.

Occasionally, the size of the sight triangle available may be less than the desirable minimum, due to presence of an obstruction which cannot be removed except at prohibitive cost. In such circumstances, the vehicles must be appropriately warned to travel at speeds corresponding to the available sight distance and not at the design speed of the highway. One solution can be to permit vehicles on one of the roads to travel at the design speed and evaluate the corresponding critical speed for the other road which might be posted. Alternatively, the approach speed for both the road could be restricted in accordance with the sight triangle available by installing suitable speed limit signs.

8.43.2. Priority intersections

On priority intersections, the visibility provided should be such that drivers approaching from the minor road are able to see vehicles on the major road in adequate time and to judge whether the required gap is available in the main road traffic stream for a safe crossing so that the vehicle could be brought to a halt, if necessary. For this purpose, minimum visibility distance of 15 m along the minor road is recommended.

Table 8.19 Minimum visibility distance along major roads at priority intersections

Design speed of major road (km/h)	*Minimum visibility distance along major roads (metres)*
100	220
80	180
65	145
50	110

Visibility distance along the major road depends on the time required by the driver on the minor road to perceive the traffic conditions on the intersection, evaluate the gaps in the vehicle stream, take a decision about actual crossing, and finally to accelerate the vehicle to complete the manoeuvre. The total time required for these operations may be taken as 8 seconds. On this basis, the sight triangle at priority intersections should be formed measuring 15 m along the minor road and a distance along the major road equal to 8 seconds travel at the design speed. This is illustrated in Fig. 8.50. Visibility distances (rounded values) corresponding to 8 seconds travel time are set out in Table 8.19.

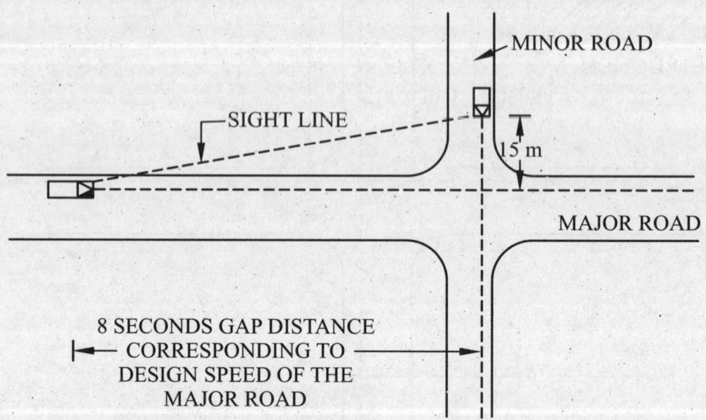

Fig. 8.50 Minimum sight triangle at priority intersection.

Problem 8.20 *Calculate the overtaking sight distance on a four lane divided State Highway in a plain terrain. Compare this with IRC recommendations.*

Solution. For a State Highway in plain terrain,

Ruling design speed = 100 km/hr (Table 8.2)

Since it is a four-lane divided highway, so there will be only overtaking and no vehicle in the opposite direction will be coming.

Overtaking sight distance = $d_1 + d_2$

Assume that the difference in speed between the fast moving vehicles and slow moving vehicles,

$$m = 16 \text{ km/hr}$$

$$V - m = 100 - 16 = 84 \text{ km/hr}$$

$$\text{Distance } S = 0.19\,(V - m) + 6$$
$$= 0.19 \times 84 + 6$$
$$= 15.96 + 6$$
$$= 21.96 \text{ m}$$

For $V = 100$ km/hr

Rate of acceleration, $a = 1.92$ km/hr per sec (Table 8.16)

$$\text{Time } t = \sqrt{\frac{14.4S}{a}}$$
$$= \sqrt{\frac{14.4 \times 21.96}{1.92}}$$
$$= 12.83 \text{ sec}$$

Distance $d_1 = 0.56\,(V - m) = 0.56 \times 84$

$= 47.04$ m

Distance $d_2 = 2S + 0.28\,(V - m)\ t$

$= 2 \times 21.96 + 0.28 \times 84 \times 12.83$

$= 43.92 + 301.76$

$= 345.68$ m

$\therefore$ Overtaking sight distance $= d_1 + d_2 = 47.04 + 345.68$

$= 392.72$ m say 390 m

As per IRC recommendations, t for overtaking manoeuvre = 14 secs

$\therefore$ Overtaking sight distance $= 0.278 \times V \times t$

$= 0.278 \times 100 \times 14 = 389.2$ m

$\simeq 390$ m

Thus the two values are approximately equal.

Problem 8.21 *A car travelling at* 22.22 *m/sec is overtaking another car moving at* 16.67 *m/sec on a two lane undivided highway. Assuming an acceleration of the overtaking car as* 0.7 *m/sec*^{2}*; Calculate (i) minimum overtaking sight distance; and (ii) minimum and desirable length of overtaking zones.*

Solution. Since it is a two-lane undivided highway, vehicles can overtake and there can be a vehicle coming in the opposite direction also.

Minimum overtaking sight distance $= d_1 + d_2 + d_3$

Assume design speed as the speed of overtaking vehicle

$v = 22.22$ m/sec

Also $(v - m) = v_B = 16.67$ m/sec

$a = 0.7$ m/sec^2

$d_1 = (v_B) \times t_1$

Taking $t_1 = 2$ sec.

$d_1 = 16.67 \times 2$

$= 33.34$ m

$S = 0.69\ v_B + 6$

$= 0.69 \times 16.67 + 6$

$= 11.50 + 6$

$= 17.50$ m

$$t = \sqrt{\frac{4 \times S}{a}}$$

$$= \sqrt{\frac{4\times17.5}{0.7}}$$

$$= 10 \text{ sec}$$

$$d_2 = 2S + v_B t = 2\times17.5 + 16.67\times10$$

$$= 35 + 166.7$$

$$= 201.7 \text{ m}$$

$$d_3 = v\times t$$

$$= 22.22\times10$$

$$= 222.2 \text{ m}$$

Minimum overtaking sight distance

$$= d_1 + d_2 + d_3$$

$$= 33.34 + 201.7 + 222.2$$

$$= 457.24 \text{ m}$$

Say 460 m

Minimum length of overtaking zone

= 3 × minimum overtaking sight distance

$$= 3\times460$$

$$= 1380 \text{ m}$$

Desirable length of overtaking zone

= 5 × minimum overtaking sight distance

$$= 5\times460$$

$$= 2300 \text{ m}$$

The overtaking zone is shown in Fig. 8.51.

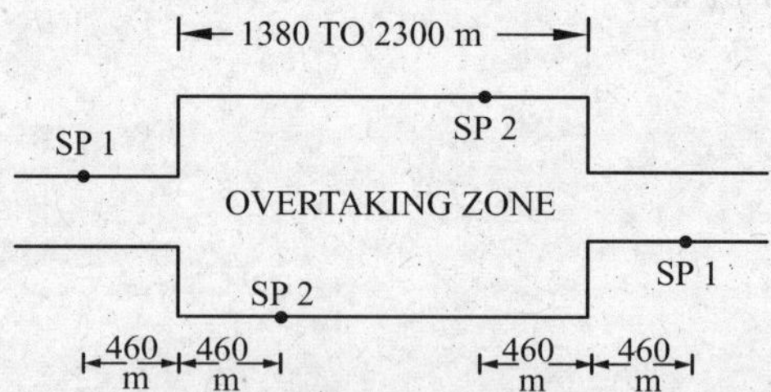

Fig. 8.51 For Problem 8.21.

GRADIENTS

Gradient is the rate of rise or fall of road level along its length. It is expressed either as rate of rise or fall to horizontal distance or as percentage rise or fall. Thus if there is an ascending or descending of the road profile by one metre for every twenty-five metres, the grade is said to be 1 in 25 or 4 per cent. As the gradient on the road is not very steep, the distance along the road is taken equal to the horizontal distance. Thus in expressing grades, the distance along the length of the road is assumed to be horizontal distance. Grade is limited by the hauling power of transport vehicles, engines or animals.

Longitudinal grades, once provided, constitute nearly the most permanent features of the highway. This is especially true in urban and built up areas. In order to evolve a balanced design, it is essential that a complete analysis of the factors governing this aspect of design be made before selecting alignment. The various factors on which longitudinal gradient depends are:

1. Nature of traffic expected on the road.

2. Physical factors associated with site:

(*a*) drainage (*b*) access to adjacent property

(*c*) appearance (*d*) safety.

3. Rail-road intersections, bridge approaches, *etc*.

4. Design speed.

8.44. TYPES OF GRADIENTS

The various types of gradients are:

(*a*) Exceptional gradient

(*b*) Maximum gradient or limiting gradient; and

(*c*) Ruling gradient.

Exceptional gradient. Such gradients are provided only in exceptional cases such as approaches to causeways, near hair-pin bends, etc. In a built up area, adherence to the ruling gradient may involve damage to expensive abutting property. Also in a hilly tract, short stretches of grades, steeper than the maximum, may become essential to gain elevation to reach an obligatory point on a fixed route. In such exceptional cases, gradients upto the exceptional value, may be permitted. They are allowed only for short stretches, not more than 100 m.

The bad effects of excessive grades are:

(*i*) More fuel consumption.

(*ii*) More friction losses.

(*iii*) Reduced engine efficiency.

(*iv*) Use of brakes to keep the speed of the vehicle to a safe limit on steep descending gradient.

(*v*) Early tiring of animals.

(*vi*) Increased wear of road surface and vehicle parts.

Thus exceptional gradients should be provided only in very difficult situations and where these cannot be avoided.

Maximum gradient. This is also called limiting gradient. It is the steepest gradient to be provided in any part of the road and should be exceeded only in exceptional cases. It is allowed to effect a large reduction in the cost of earth-work and when compelled by the topography of an area. Long stretches of such gradients should be separated by comparatively flatter gradient or level sections especially in hills, where limiting gradients have to be used rather frequently.

Ruling gradient. This is the desirable upper limit of a gradient on a road. Thus a designed road should aim to provide a ruling gradient keeping in view the type of traffic and the topography of the country. This gradient is such that animals and vehicles can overcome long stretches of this gradient without much fatigue or wear and within economical petrol consumption. Horses, bullocks, etc. can carry load on steep slopes; donkeys on less steep slope; rikshaws, tongas and carts on gentler slopes and automobiles on still flatter slopes. Thus the mode of transport in the locality will govern the ruling gradients to be adopted. This also depends upon the type of road surfacing.

Recommended gradients by IRC for different classes of terrain are given in Table 8.20.

Table 8.20 Gradient for roads in different terrains

Sr. No.	*Terrain*	*Gradient*		
		Ruling	*Limiting*	*Exceptional*
1.	Plain or rolling	3.3 per cent (1 in 30)	5 per cent (1 in 20)	6.7 per cent (1 in 15)
2.	Mountainous terrain and steep terrain having elevation upto 300 m above the mean sea level	5 per cent (1 in 20)	6 per cent (1 in 16.7)	7 per cent (1 in 14.3)
3.	Steep terrain having elevation more than 300 m above the mean sea level.	6 per cent (1 in 16.7)	7 per cent (1 in 14.3)	8 per cent (1 in 12.5)

IRC has made the following recommendations in regard to the adoption of gradient standards:

(*i*) Gradients upto the ruling gradient may be used as a matter of course in design. However in special situations such as isolated over-bridges in flat country or roads carrying a large volume of slow moving traffic, it will be desirable to adopt a flatter gradient of 2 per cent from the view point of aesthetics, traffic operations and safety.

(*ii*) The limiting gradients may be used where the topography of place compels this course or where the adoption of gentler gradients would add enormously to the cost. In such cases, the length of continuous grade steeper than the ruling gradient should be as short as possible.

(*iii*) Exceptional gradients are meant to be adopted only in very difficult situations and for short lengths not exceeding 100 m at a stretch. In mountainous and steep terrain, successive stretches of exceptional gradients must be separated by a minimum length of 100 m having gentler gradient *i.e.*, limiting gradient or flatter.

(*iv*) The rise in elevation over a length of 2 km shall not exceed 100 m in mountainous terrain and 120 m in steep terrain.

8.45. MINIMUM GRADIENTS FOR DRAINAGE

On unkerbed pavements in embankment, near-level grades are not objectionable when the pavement has sufficient camber to drain the storm water laterally. However in cut sections or where the pavement is provided with kerbs, it is necessary that the road should have some gradient for efficient drainage. Desirable minimum gradient for this purpose is 1.5 per cent if the side drains are lined and 1.0 per cent if these are unlined.

8.46. FACTORS CONTROLLING GRADIENT DESIGN

The factors which control the design of a gradient are:

(*a*) Tractive resistance

(*b*) Grade resistance; and

(*c*) Tractive effort.

(a) Tractive resistance. This is the resistance offered to the vehicle by a force whose line of action is parallel to the road surface and the longitudinal axis of the vehicle. It comprises two factors;

(*i*) Rolling resistance ; and

(*ii*) Air resistance.

Rolling resistance is offered by (*i*) the friction between the moving parts of the vehicle which is negligible for all types of vehicles except bullock carts, etc. and (*ii*) the movement of tyre on the road surface. Whenever a wheel moves over a road surface, distortion of the surface occurs which varies with the type of surface. For a yielding or plastic surface, *e.g.*, earth, mud, *etc*. a rut is formed. In case the surface is elastic, *e.g.*, concrete, bituminous, *etc.* it deflects slightly under the load, but subsequently resumes its original position as soon as the load is removed. Road surface resistance will be more in the case of plastic surface and very little for elastic surfaces. This is due to the reason that the wheel has to climb up each time the depression caused by the load. In the former case the depression will be more than the latter. Thus the wheel has to encounter considerably greater road resistance in the first case. The resistance to the movement of tyre of the vehicle depends upon a number of factors such as diameter, tread, width and pressure of tyre, number of tyres, weight of vehicle and the type of road surface. The tyre must be capable of affording proper grip on the road surface.

Air resistance also comprises a considerable part of the total tractive resistance. It depends upon the area of the vehicle exposed to the air and the speed of the vehicle relative to air. The air resistance is calculated from the following relation:

$$R_a = CAV^2$$

where R_a = Total air resistance in kg.

A = The projected area of cross-section of the vehicle in metres

V = Speed of the vehicle relative to air in km/hr

C = Experimental coefficient and from a number of experiments value has been estimated to vary from 0.0036 to 0.0048.

Total tractive resistance may be expressed as:

$$R = R_r + R_a$$

where R = Total tractive resistance in kg per tonne

R_r = Rolling resistance in kg per tonne

R_a = Air resistance in kg per tonne.

The value of R_r varies from 10 to 20 kg per tonne of vehicle for all types of roads in good condition and may reach a value of 45 kg to 70 kg per tonne for wet water bound macadam and earth roads.

(b) Grade resistance. The above discussion mainly refers to a level road. When the vehicle moves on incline, additional force of gravity equal to $+w\sin\theta$ for ascending gradient and $-w\sin\theta$ for descending gradient comes into play, where w is the weight of the vehicle in kg and θ is the inclination of the road surface to the horizontal. For small value of θ, $\sin\theta = \tan\theta$. Therefore grade resistance (R_g) is given by:

$$R_g = \pm w\tan\theta = \pm w \times \text{gradient}$$

$$= \pm \frac{1000W}{N}$$

where W = Weight of the vehicle in tonnes

N = Horizontal distance for one metre rise or fall

R_g = Grade resistance in kg per tonne.

(c) Tractive effort. This is the effort required by the vehicle to overcome both tractive resistance and grade resistance. It is expressed as:

$$F_T = \frac{P}{W}$$

where F_T = Tractive effort in kg per tonne

P = Total force required by the vehicle in kg

W = Total weight of the vehicle in tonnes

$\therefore$ Tractive effort $= R + R_g = R_r + R_a + R_g$

$$= R_r + CAV^2 + \frac{1000\,W}{N}$$

8.47. DESIGN OF RULING GRADIENT

8.47.1. Ascending or Plus Gradient

Most economical gradient for a vehicle to ascend a road is one which permits ascent in top gear at or near the design speed of the vehicle. A vehicle is designed in such a way that it has got available tractive effort to overcome both tractive resistance and grade resistance at that speed. When a vehicle moves on a level surface, its engine is not taxed to its full power as it has not to overcome grade resistance. Thus if from the tractive effort of the vehicle, tractive resistance on a level road is subtracted, the remainder power of the engine is available for overcoming grade resistance. Assuming that the engine is called upon to work at its maximum power and that the same speed as that on a level road is to be maintained on an incline, then the most economical gradient will be given by

$$\frac{1000}{N} = T - R \qquad \text{...(8.27)}$$

where $\frac{1}{N}$ = Maximum economical ascending gradient (1 in N)

T = Maximum tractive effort of the engine of the vehicle at its design speed in kg per tonne of weight of the vehicle and its load

R = Tractive resistance of the vehicle on the proposed road surface at that speed in kg per tonne of weight of the vehicle and its load.

It is evident from above that the economical gradients depend also upon the weight of the vehicle and its load. Thus steeper gradients are possible for passenger cars and flatter gradients are essential for commercial vehicles and trucks. From a series of experiments, it has been found that the slow down effect of adverse grades on passenger cars is not appreciable up to a gradient of 1 in 14 to 1 in 12, *i.e.*, 7 to 8 per cent, but the truck speed is reduced considerably even on a gradient of 1 in 50 or 2 per cent. On a gradient of 1 in 33, the truck speed is reduced to 25 km/hr and on 1 in 12 gradient, it may be only 1 to 10 km/hr.

At the same time the trucks are designed for tractive effort which is nearly equal to the tractive resistance on a level road. Thus they are taxed to their maximum capacity on a level stretch and do not have reserve power for grade resistance. Therefore, they depend more on the use of gears for moving on an upward incline. But driving at a lower gear is uneconomical due to greater fuel consumption on account of increased friction and decreased engine efficiency.

Thus in selecting a ruling gradient, care must be taken to provide a balanced economical gradient for all types of vehicles using the road.

Momentum effect. In climbing up an incline, a vehicle takes an advantage of the momentum gained on the level road. If there occurs a reduction in the speed of the vehicle, a part of the kinetic energy lost is utilized in doing work in going up the incline.

Let m = Mass of the vehicle and its load.

v_1 = Initial velocity in m per sec.

$= 0.28\,V_1$ where V_1 is in km/hr

v_2 = Reduced or final velocity in m per sec

$= 0.28\, V_2$ in km/hr

Loss in kinetic energy is given by

$$E_k = L_t + L_r = \frac{1}{2} m\,(v_1^2 - v_2^2) + L_r$$

where E_k = Loss in kinetic energy in kg

L_r = Loss due to corresponding change in kinetic energy of rotating parts.

L_r = Loss in kinetic energy of translation due to the change in speed.

The value of L_r is tedious to calculate. It has been determined from experiments that it may be taken as 5 per cent for passenger cars and 10 per cent for commercial vehicles of the kinetic energy of translation.

$$\therefore \quad E_k \text{ for passenger cars } = 1.05 \times \frac{m}{2}(v_1^2 - v_2^2)$$

$$= \frac{1.05}{2} \times m \times (0.28)^2 (V_1^2 - V_2^2)$$

$$= 0.041 \text{ m } (V_1^2 - V_2^2)$$

If L is the length of gradient in m in which the velocity has changed from V_1 to V_2, the average force in kg per tonne of vehicle load contributed by the change in kinetic energy;

$$= \frac{0.041 \times 1000}{9.81}\left(\frac{V_1^2 - V_2^2}{L}\right) = 4.18\left(\frac{V_1^2 - V^2}{L}\right)$$

Therefore equation 8.27 is modified as

$$\frac{1000}{N} = T - R + 4.18\left(\frac{V_1^2 - V_2^2}{L}\right)$$

or $$\frac{1}{N} = \frac{T}{1000} - \frac{R}{1000} + 0.00418\left(\frac{V_1^2 - V_2^2}{L}\right) \quad \ldots(8.28)$$

This relation will give the maximum economical gradient for passenger cars.

For commercial vehicles;

$$E_k = 1.1 \times \frac{m}{2}(v_1^2 - v_2^2)$$

$$= \frac{1.1}{2} \times m \times (0.28)^2 (V_1^2 - V_2^2) = 0.043\,(V_1^2 - V_2^2)$$

Reasoning as above, average force in kg per tonne of vehicle load contributed by the change in kinetic energy;

$$= 1.843 \times \frac{1000}{9.81}\left(\frac{V_1^2 - V_2^2}{L}\right)$$

$$= 4.38\left(\frac{V_1^{\ 2} - V_2^2}{L}\right)$$

$$\therefore \quad \frac{1000}{N} = T - R + 4.38\left(\frac{V_1^{\ 2} - V_2^2}{L}\right)$$

or $$\frac{1}{N} = \frac{T}{1000} - \frac{R}{1000} + 0.00438\left(\frac{V_1^{\ 2} - V_2^2}{L}\right) \quad \ldots(8.29)$$

8.47.2. Descending or Minus Gradient

When a vehicle is moving down an incline the power from the engine will be utilized to maintain the design speed of the vehicle if the rate of grade is below certain minimum. It will accelerate, without power, if the rate of grade is excessive. It will move at the design speed, without power, if this rate is just sufficient to maintain it. This will be the most economical grade and is known as the floating gradient.

Let V_2 = Velocity in km/hr at the beginning, *i.e.*, when it just starts moving a down grade.

V_1 = Velocity in km/hr at the end, *i.e.*, at the bottom of the down grade.

Then, if momentum is neglected:

$$\frac{1000}{N} = R, \text{ since } T = 0$$

$$\therefore \quad \frac{1}{N} = \frac{R}{1000}$$

If momentum is also considered; then

$$\frac{1000}{N} - 4.18\left(\frac{V_1^{\ 2} - V_2^2}{L}\right) = R$$

or for passenger cars $$\frac{1}{N} = \frac{R}{1000} + 0.00418\left(\frac{V_1^{\ 2} - V_2^2}{L}\right) \quad \ldots(8.30)$$

For commercial vehicles

$$\frac{1}{N} = \frac{R}{1000} + 0.00438\left(\frac{V_1^{\ 2} - V_2^2}{L}\right) \quad \ldots(8.31)$$

8.47.3. Grade compensation at curves on hill roads

When a vehicle moves on a grade, there always occurs a loss in the tractive effort. Similar effect is also noticed when the vehicle moves on a curve. Thus if in addition to the grade, road is also on a curve, the total effect of the grade and the curve should not exceed that of the limiting gradient specified. Very little information, based on experiments, is available to decide the amount of reduction in the grade to take into account the loss in the tractive effort on the curve. The following empirical relation, has been suggested by IRC.

$$\text{Compensation in per cent grade} = \frac{30 + R}{R}$$

subject to a maximum of $\frac{75}{R}$ where R is the radius of the curve in metres.

Curves which are greater than six degrees or of radius less than 300 m should always be compensated on grades of five per cent or over. Thus if on a grade of six per cent, a horizontal curve of 75 m radius is provided, the compensation in grade to be given is $\frac{75}{75} = 1$ per cent. The grade has, therefore, to be reduced to five per cent.

Grade compensation is not necessary for gradients flatter than 4 per cent.

VERTICAL CURVES

When two different or contrary gradients meet, they are connected by a curve in the vertical plane known as vertical curve. These are needed to secure a gradual change in grade so that abrupt change in grade at the apex is avoided. The disadvantages of not providing vertical curves are:

(*i*) ***Accident due to inadequate visibility.*** On a high and blind culvert over which the road profile has not been properly designed as a vertical curve, visibility is limited. If a driver is coming at a very high speed, there is every possibility of an accident at such spots. Thus there are much greater chances of accidents on a road which has been considerably widened without introducing correct vertical curves than on a narrow road which throughout its length sets a low limit on speed.

(*ii*) ***Discomfort and damage due to humps and troughs.*** If there exist a number of hogs and sags on a road, although these may not be a source of accidents, they make travelling very uncomfortable. Moreover these cause damage to both the vehicle and the surface. Stresses and oscillations in the springs, tyres and the framework of the vehicle are set up. These oscillations convey hammer blows to the road surface through its wheels and ruin the structure of the road.

(*iii*) ***It pays more to construct proper vertical curves.*** It is much cheaper and easier to construct good curves in the first instance than to set right road curves later on because it may amount to dismantling expensive road works such as culverts or acquiring costly land which has been developed along the road side.

Thus, vertical curves are provided to get;

(*i*) Safety and adequate visibility.

(*ii*) Comfort to the passengers.

There are two types of vertical curves:

(*i*) Summit curves; and

(*ii*) Valley curves

Curves should be provided at all grade changes exceeding those indicated in Table 8.21.

Table 8.21 Maximum grade change not requiring a vertical curve

S.No.	*Design speed (km/hr)*	*Maximum grade change (per cent) or deviation angle not requiring a vertical curve*
1	Up to 35	1.5
2	40	1.2
3	50	1.0
4	65	0.8
5	80	0.6
6	100	0.5

SUMMIT CURVES

A curve with convexity upward is called a summit curve. This occurs when an ascending gradient intersects a descending gradient or when an ascending gradient meets another ascending gradient or an ascending gradient, meeting a horizontal. These are shown in Fig. 8.52.

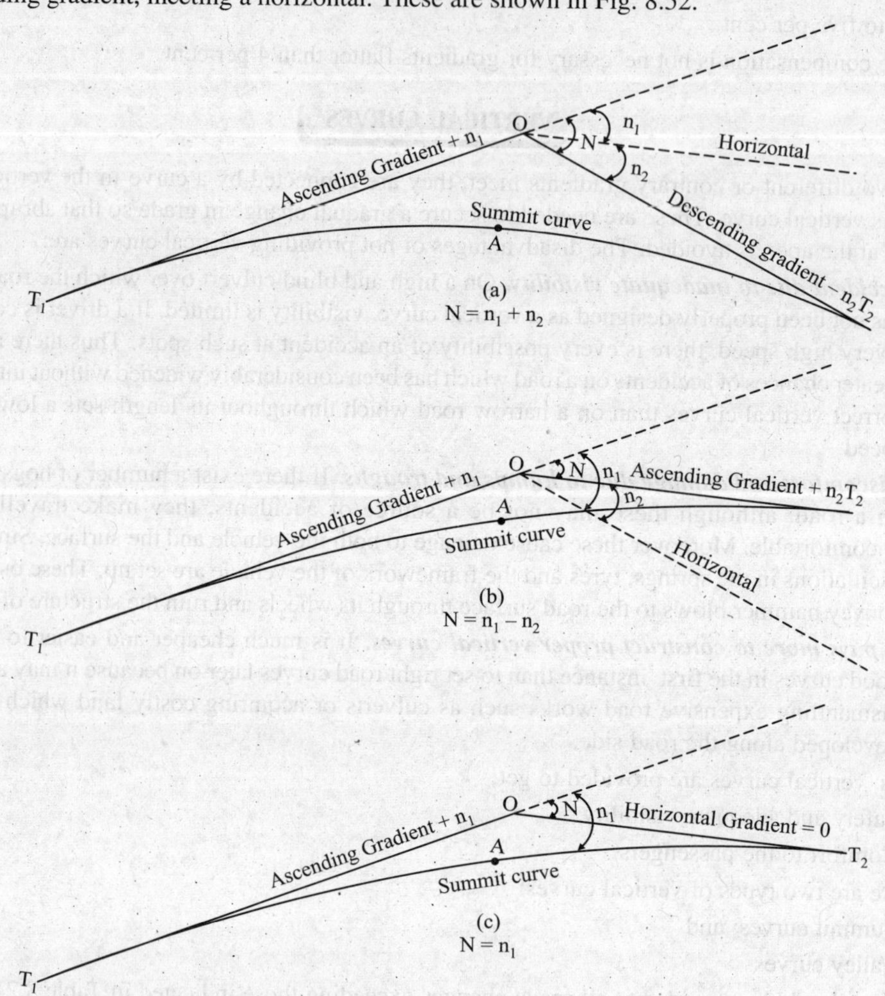

Fig. 8.52 Summit vertical curves, showing deflection angles.

8.48. DEVIATION ANGLE OF SUMMIT CURVES

Deviation angle of a curve is expressed by the algebraic difference of the grade angles. If n_1 and n_2 are the grade angles for the two curves, then assigning proper signs (*i.e.*, + for ascending and – for descending), the deviation angle N is given by:

In Fig. 8.52 (*a*)

$$N = \text{Angle EOT}_2 = (+n_1) - (-n_2) = n_1 + n_2$$

If $n_1 = +\frac{1}{25}$ or $+\frac{4}{100}$ or 4 per cent; and

$n_2 = -\frac{1}{20}$ or $-\frac{5}{100}$ or -5 per cent

Then $N = 4 - (-5) = 9$ per cent.

In Fig. 8.52 (*b*)

$N = \text{Angle EOT}_2 = (+n_1) - (+n_2) = n_1 - n_2 = 4 - 5 = -1$ per cent

In Fig. 8.52 (*c*)

$N = \text{Angle EOT}_2 = (+n_1) - O = n_1 = 4$ per cent

Thus deviation angle of a summit curve is the angle which measures the change of direction in the path of motion at the intersection of two grade lines.

8.49. TYPE OF SUMMIT CURVES

Sight distance is the governing factor in the design of a summit curve unless the summit is so small as not to interfere with visibility. The dynamics of movement over an ordinary summit curve is of less importance. The centrifugal force generated by a moving vehicle along the curve acts opposite to the force of gravity and is, therefore, helpful in relieving the pressure on the tyres and springs of the vehicle. Moreover, due to small vertical deviation angles and provision of long and easy summit curve on the consideration of sight distances, shock to the vehicle and the passengers is almost imperceptible.

Thus, transition curves need not be provided on the summit and simple circular arcs will serve well. A circular curve has a constant sight distance all along its length since its radius of curvature is uniform. This is an advantageous factor in the case of circular arcs. Transition curves cannot be provided because its radius of curvature decreases towards its apex and hence visibility is also least at the apex.

But, in actual practice, a simple parabolic curve is usually adopted for the following reasons:

(*i*) A simple parabola nearly resembles a circular arc between the same tangent points because on road work the vertical deviation angles are very small and lengths of curves are very large.

(*ii*) A parabola can be easily laid in the field.

Over small humps, a transition curve is better as it is a perfectly shock free curve. Thus at approaches to a culvert or over small humps which do not rise higher than the sight line, a transition curve is more appropriate.

8.50. DESIGN OF SUMMIT CURVES

Let T_1O and T_2O be the two intersecting grade lines having grade angles n_1 and n_2 respectively as shown in Fig. 8.53. Let T_1AT_2 be the simple parabolic curve between the tangent points T_1 and T_2. Let y be the intercept between the curve and the grade line T_1O at a horizontal distance x from T_1.

Then $\quad y = \dfrac{x^2}{C} \quad$ ...(8.32)

where C is a constant.

From Fig. 8.53,

$$PT_2 = PR + RT_2$$

$$= \frac{Ln_1}{2} + \frac{Ln_2}{2}$$

$$= \frac{L}{2}(n_1 + n_2)$$

$$= \frac{L}{2}N$$

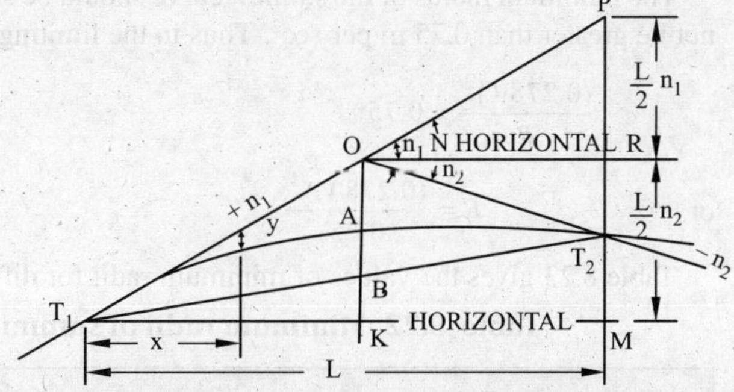

Fig. 8.53 Parabolic summit vertical curve.

At the end point T_2 of the curve

$$x = L \text{ and } y = PT_2 = \frac{L}{2} N$$

Substituting these values in equation 8.32, we get

$$\frac{LN}{2} = \frac{L^2}{C}$$

$$\therefore \quad C = \frac{2L}{N} \qquad \ldots(8.33)$$

8.51. RADIUS OF CURVATURE OF SUMMIT CURVE

Let (x, y) be the cartesian coordinates of any point on the curve and R, the radius of curvature at that point.

For a flat curve

$$\frac{1}{R} = \frac{d^2 y}{dx^2} \qquad \ldots(8.34)$$

The equation of a summit parabolic curve is

$$y = \frac{x^2}{C}$$

$$\therefore \quad \frac{dy}{dx} = \frac{2x}{C} \qquad \ldots(8.35)$$

and

$$\frac{d^2 y}{dx^2} = \frac{2}{C}$$

Equating 8.34 and 8.35,

$$\frac{1}{R} = \frac{2}{C}$$

or

$$R = \frac{C}{2}$$

which is constant. Thus parabolic curve is virtually a circular arc.

But

$$C = \frac{2L}{N}$$

$$\therefore \quad R = \frac{2L}{2N} = \frac{L}{N} \qquad \ldots(8.36)$$

The minimum radius of the summit curve should be such that the rate of radial deceleration should not be greater than 0.75 m per sec^2. Thus in the limiting case:

$$\frac{(0.278V)^2}{R} = 0.75$$

or

$$R = \frac{(0.278\,V)^2}{0.75}$$

Table 8.22 gives the values of minimum radii for different design speeds.

Table 8.22 Minimum radii of summit curves for various speeds

Speed km/hr	25	30	35	40	50	60	65	80	100
Minimum Radius (m)	65	95	130	170	250	380	440	670	1045

However, it may be mentioned that curves designed for the minimum sight distance prescribed shall have radii greater than the minimum radii given in Table 8.22.

8.52. LEGNTH OF THE SUMMIT CURVE

The length of a summit curve is a function of

(*i*) the deviation angle, *N*; (*ii*) the required sight distance.

Gradients and sight distances are fixed on the considerations mentioned earlier. From the gradient, deviation angle and the chainage of the intersecting point can be calculated.

In determining the length of the summit curve, two cases have to be considered;

8.52.1. For intermediate or overtaking sight distances

Case I. When the length of the curve exceeds the required sight distance *i.e.*, $L > S$.

In Fig. 8.54 (*b*), let T_1AT_2 be the parabolic curve having T_1 and T_2 as the tangent points.

$S_1S_2 = S$ = required sight distance = $(d_1 + d_2)$

H_1, H_2 = heights of the driver's eye above the road level.

N = deviation angle POT_2

L = horizontal projection T_1M of the curve which for all practical purposes is equal to the length of the curve $T_1\ A\ T_2$ as summit curves are long and flat.

From equation (8.32), $y = \dfrac{x^2}{C}$

and from eqn. (8.33), $C = \dfrac{2L}{N}$

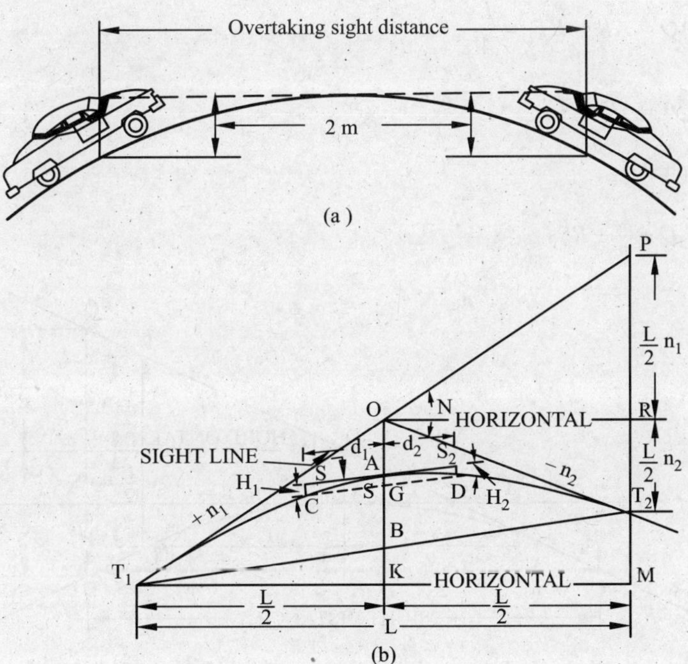

Fig. 8.54 Length of summit curve in relation to overtaking sight distance, when $L > S$.

Now $S_1S_2 = S_1A + AS_2$ ...(8.37)

$= d_1 + d_2$

Thus $H_1 = \frac{d_1^2}{C}$ and $H_2 = \frac{d_2^2}{C}$

or $d_1 = \sqrt{H_1 C}$ and $d_2 = \sqrt{H_2 C}$

Putting in eqn (8.37)

$$S_1S_2 = \left(\sqrt{H_1} + \sqrt{H_2}\right)\sqrt{C}$$

or $$S = \left(\sqrt{H_1} + \sqrt{H_2}\right)\sqrt{\frac{2L}{N}}$$

or $$S^2 = \left(\sqrt{H_1} + \sqrt{H_2}\right)^2 \frac{2L}{N}$$

or $$L = \frac{NS^2}{2\left(\sqrt{H_1} + \sqrt{H_2}\right)^2} \quad \text{...(8.38)}$$

Putting $H_1 = H_2 = 1.2$ m

$$L = \frac{NS^2}{9.6} \quad \text{...(8.39)}$$

Case II. When the length of the curve is less than the required sight distance, *i.e.*, $L < S$. Keeping the same notations as before

Clearly $QT_1 = AQ = AR = RT_2 = \frac{L}{4}$

or $QR = QA + AR = \frac{L}{2}$

Now $S = d_1 + d_2$

$$= S_1Q + \frac{L}{2} + RS_2 \quad \text{...(8.40)}$$

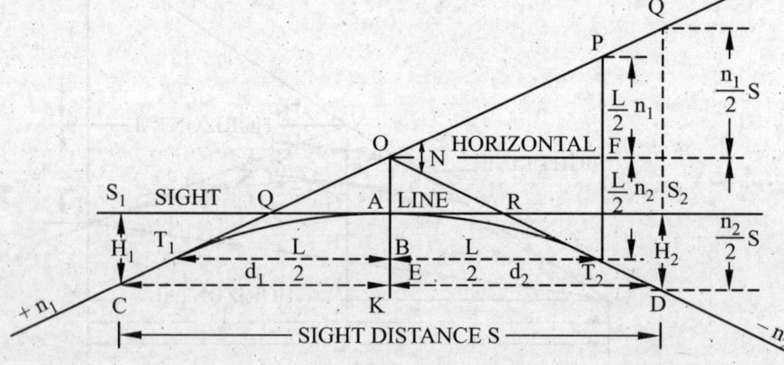

Fig. 8.55 Length of summit curve in relation to overtaking sight distance, when $L < S$.

Also $n_1 + n_2 = N$ = constant for a given curve

$$\therefore \quad \frac{dn_1}{dx} + \frac{dn_2}{dx} = 0 \quad \text{...(8.41)}$$

Also $S_1Q = \frac{H_1}{n_1}$ and $RS_2 = \frac{H_2}{n_2}$

Putting in equation 8.40

$$S = \frac{H_1}{n_1} + \frac{L}{2} + \frac{H_2}{n_2} \quad \text{...(8.42)}$$

Differentiating both sides w.r.t. x

$$O = \frac{H_1(-1)}{{n_1}^2}\frac{dn_1}{dx} + \frac{H_2(-1)}{{n_2}^2}\frac{dn_2}{dx}$$

$$\frac{H_1}{{n_1}^2}\frac{dn_1}{dx} = \frac{H_2(-1)}{{n_2}^2}\frac{dn_2}{dx} \quad \text{...(8.43)}$$

But from equation 8.41

$$\frac{dn_1}{dx} = \frac{dn_2}{dx}$$

Thus equation 8.43 becomes

$$\frac{H_1}{{n_1}^2} = -\frac{H_2}{{n_2}^2}$$

or $$n_2 = -n_1\sqrt{\frac{H_2}{H_1}}$$

$\therefore$ $$N = n_1 - n_2 = n_1\left\{\frac{\sqrt{H_1} + \sqrt{H_2}}{\sqrt{H_1}}\right\}$$

or $$n_1 = \frac{N\sqrt{H_1}}{\sqrt{H_1} + \sqrt{H_2}}; \text{ and}$$

$$n_2 = \frac{N\sqrt{H_2}}{\sqrt{H_1} + \sqrt{H_2}}$$

Putting the values in equation 8.42

$$S = \frac{\sqrt{H_1}\left(\sqrt{H_1} + \sqrt{H_2}\right)}{N} + \frac{\sqrt{H_2}\left(\sqrt{H_1} + \sqrt{H_2}\right)}{N} + \frac{L}{2}$$

or $$S = \frac{L}{2} + \frac{\left(\sqrt{H_1} + \sqrt{H_2}\right)^2}{N}$$

or $$L = 2S - \frac{2}{N}\left(\sqrt{H_1} + \sqrt{H_2}\right)^2 \quad \text{...(8.44)}$$

For $H_1 = H_2 = 1.2$ m.

$$L = 2S - \frac{9.6}{N} \qquad \ldots(8.45)$$

8.52.2. For safe stopping sight distances or non-overtaking sight distances

Case I. When the length of the curve exceeds the required sight distance, *i.e.*, $L > S$.

In this case, the driver views the top of an object lying beyond the apex of the curve and is shown in Fig. 8.56.

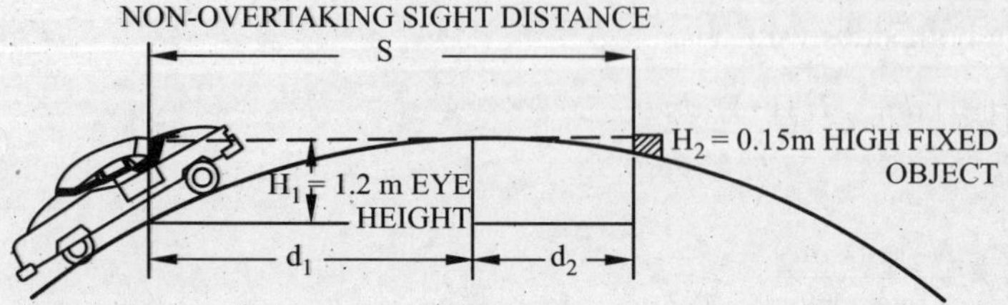

Fig. 8.56 Length of summit curve in relation to non-overtaking sight distance.

Let H_1 = Height of the driver's eye above the curve = 1.2 m

H_2 = Height of the object = 0.15 m

d_1 = Horizontal distance between the apex and the driver's eye.

d_2 = Horizontal distance between apex and the object.

$S = d_1 + d_2$

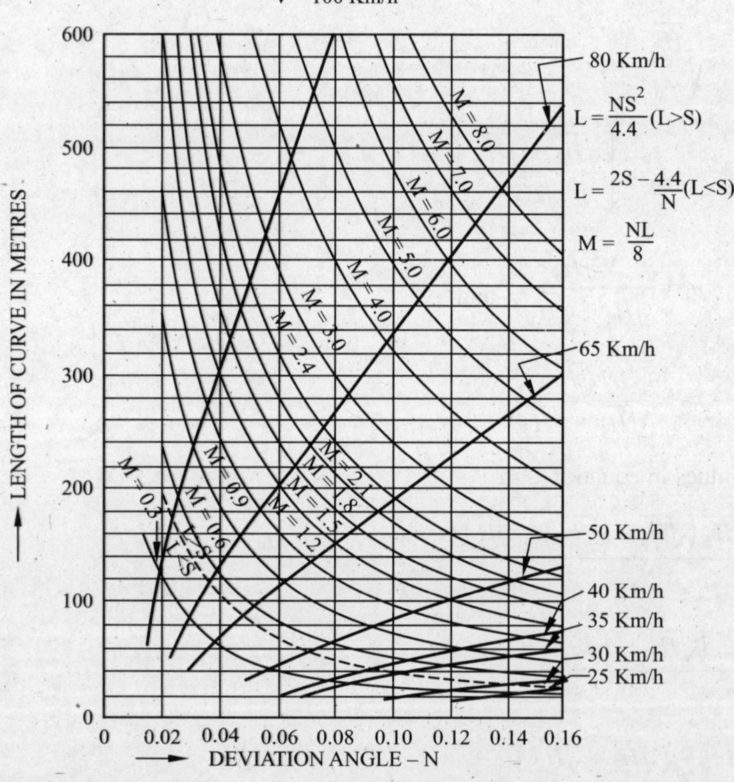

Fig. 8.57 Length of summit curve for stopping sight distance.

Then, from the equation 8.38 of the parabola

$$L = \frac{NS^2}{2(\sqrt{H_1} + \sqrt{H_2})^2}$$

Putting $H_1 = 1.2$ m and $H_2 = 0.15$ m

we get,

$$L = \frac{NS^2}{4.4} \qquad \ldots(8.46)$$

Case II. When the length of curve is less than the required sight distance *i.e.,* $L < S$,

Putting $H_1 = 1.2$ m and $H_2 = 0.15$ m in equation 8.44, we get

$$L = 2S - \frac{4.4}{N} \qquad \ldots(8.47)$$

The length of summit curve for various cases mentioned above can be read from Fig. 8.57 to 8.59. In these figures, value of the ordinate *M* to the curve from the intersection point of grade lines is also shown.

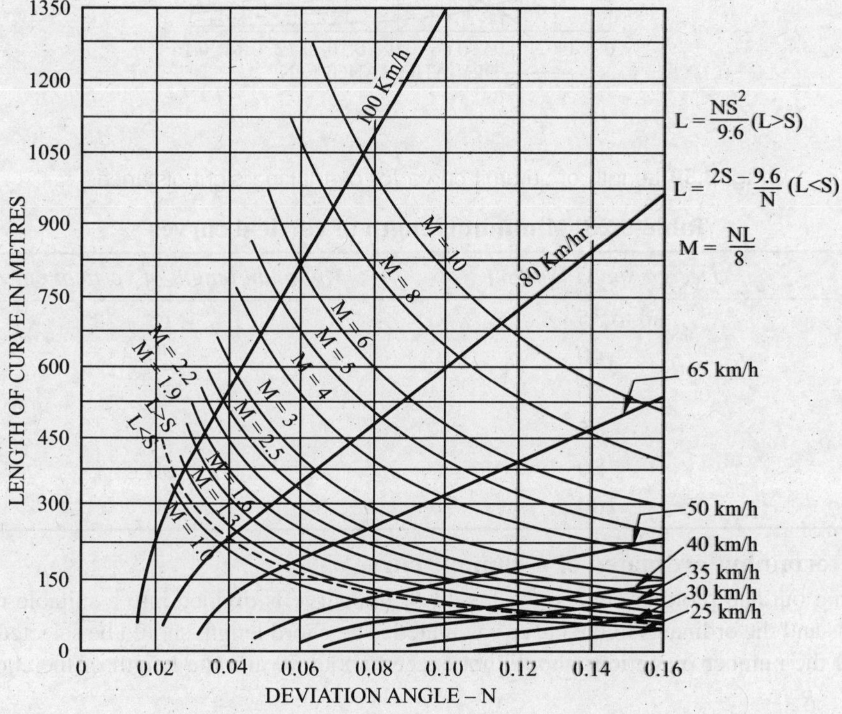

Fig. 8.58 Length of summit curve for intermediate sight distance.

8.52.3. Minimum length of summit curve

From the above formulae for the length of the curve, it can be seen that the length, *L* of a vertical curve decreases as *N* and or *S* decreases. In some cases on the basis of sight distances, very small length of the curve would be required. Moreover for very flat grades, no vertical curve is required for visibility. But for avoiding shock and for a comfortable ride vertical curves need to be provided.

For satisfactory appearance, the minimum length should be as given in Table 8.23.

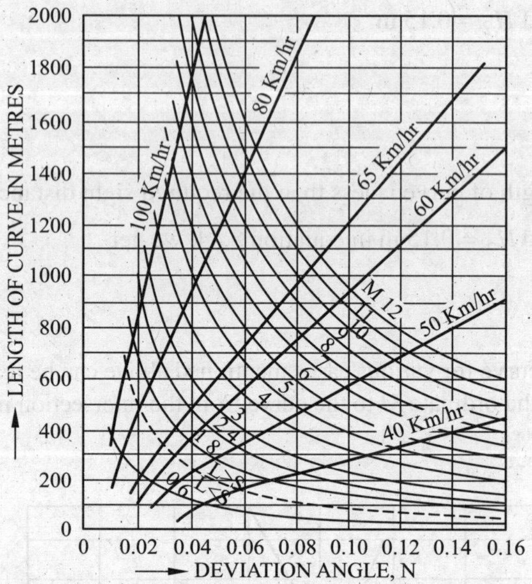

Fig. 8.59 Length of summit curve for overtaking sight distance.

Table 8.23 Minimum length of vertical curves

Sr. No.	*Design speed (km/hr)*	*Minimum length of vertical curve (m)*
1	up to 35	15
2	40	20
3	50	30
4	65	40
5	80	50
6	100	60

8.52.4. Determining ordinates of a summit curve

For setting out and plotting curves, the length of the curve is divided into a suitable number of equal chords and the ordinates to the curve calculated. The chord length should be selected in such a manner that the number of stations should not exceed about 50 and the length of the chord be not greater than $\frac{R}{60}$.

Ordinates $y_1, y_2, y_3 \ldots y_r$; as shown in Fig. 8.60, at stations 1, 2, 3...r are determined as follows:

Since $\quad y = \frac{x^2}{C}$

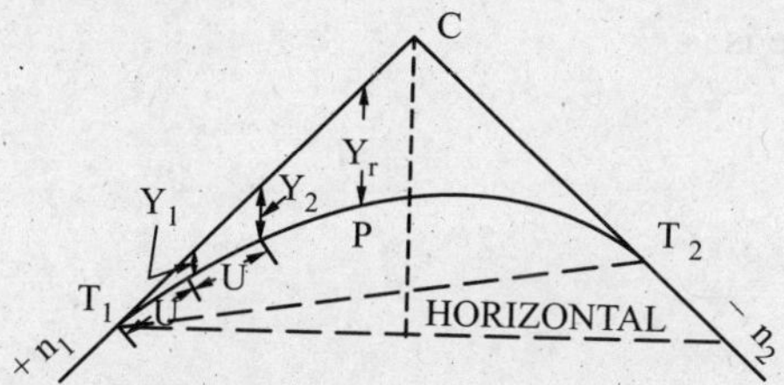

Fig. 8.60 Determining ordinates of summit curve.

$$y_1 = \frac{u^2}{C} \text{ (where } u \text{ is the chosen length of the chord)}$$

where $C = \dfrac{2L}{N}$ from equation 8.33

$$y_2 = \frac{(2u)^2}{C} \quad \text{or} \quad y_1 \times 2^2$$

$$y_3 = y_1 \times 3^2$$

$$y_r = y_1 \times (r)^2$$

If P is the point on the curve at the end of the rth sub-chord, Q the point on the grade line vertically above P and the reduced level (R.L.) of tangent point T_1 be 100.00.

Then R.L. of $Q = 100.00 + r\,(u \times n_1)$

R.L. of $P = \text{R.L. of } Q - y_r$

Similarly the reduced level of other points on the curve can be worked out.

Highest point on summit curve. It is sometimes important to know the position of the highest point of summit curve for the purpose of drainage and ascertaining bridge clearances. When the two gradients are equal, the curve is symmetrical about the bisector (vertical) of the intersecting angle. The highest point, in this case, also lies on this bisector. But, when the two gradients are unequal, the curve is tilted. The highest point does not fall vertically below the bisector of the intersecting angle but lies on the side of the flatter gradient.

Let H be the highest point which occurs at a horizontal distance x_0 from the tangent point T_1 (Fig. 8.61).

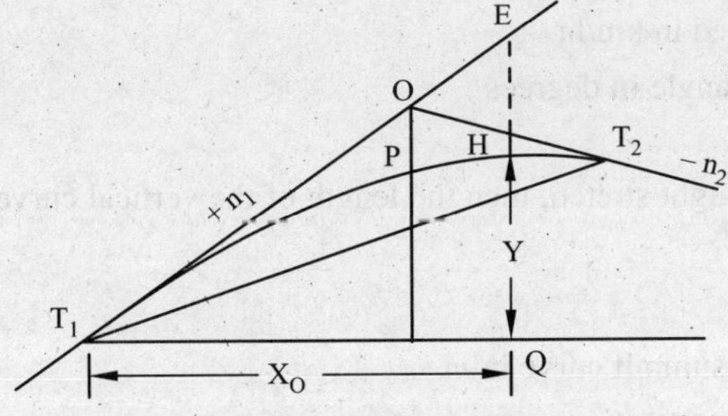

Fig. 8.61 Highest point on summit curve.

The equation of the curve is

$$y = \frac{x^2}{C}$$

$$EH = \frac{x_0^2}{C}$$

Also $\quad EQ = n_1 x_0$

$\therefore \quad HQ = y = EQ - EH$

$$= n_1 x_0 - \frac{x_0^2}{C}$$

H will be the highest point, when y is maximum

i.e., $\quad \frac{dy}{dx} = 0$

or $\quad n_1 - \frac{2x_0}{C} = 0$

or $\quad x_0 = \frac{C}{2} n_1$

But $\quad C = \frac{2L}{N} = \frac{2L}{n_1 + n_2}$

Hence $\quad x_0 = \frac{n_1 L}{n_1 + n_2}$

8.53. DESIGN OF HUMPS

Very often in the vertical alignment of a road, there occur humps over culverts, where the total height of the hump is less than the height of eye of the driver and the sight line is not obstructed. In such cases, in order to make the hump shock-free, it is necessary to employ transition curves on either side with a flat stretch at the centre. The length of the transition curves on either side is calculated from the empirical relation

$$l = 36\, V \tan \theta \quad \text{...(8.48)}$$

where $\quad l$ = Length of transition curve on either side in m *i.e.,* half the length of summit curve

V = Design speed in km/hr

θ = Deflection angle in degrees

or $\quad \tan \theta = N$

In case there is no straight stretch, then the length of the vertical curve will be equal to twice the length of transition curve,

i.e., $\quad L = 72\, V \tan \theta \quad \text{...(8.49)}$

where $\quad L$ = Length of summit curve in m.

The above relations are of little significance in the design of summit curves, since they do not yield adequate sight distances. However they are useful for finding transition lengths on humps. Properly designed humps should consist of four transitions with a level strip in between, as shown in Fig. 8.62.

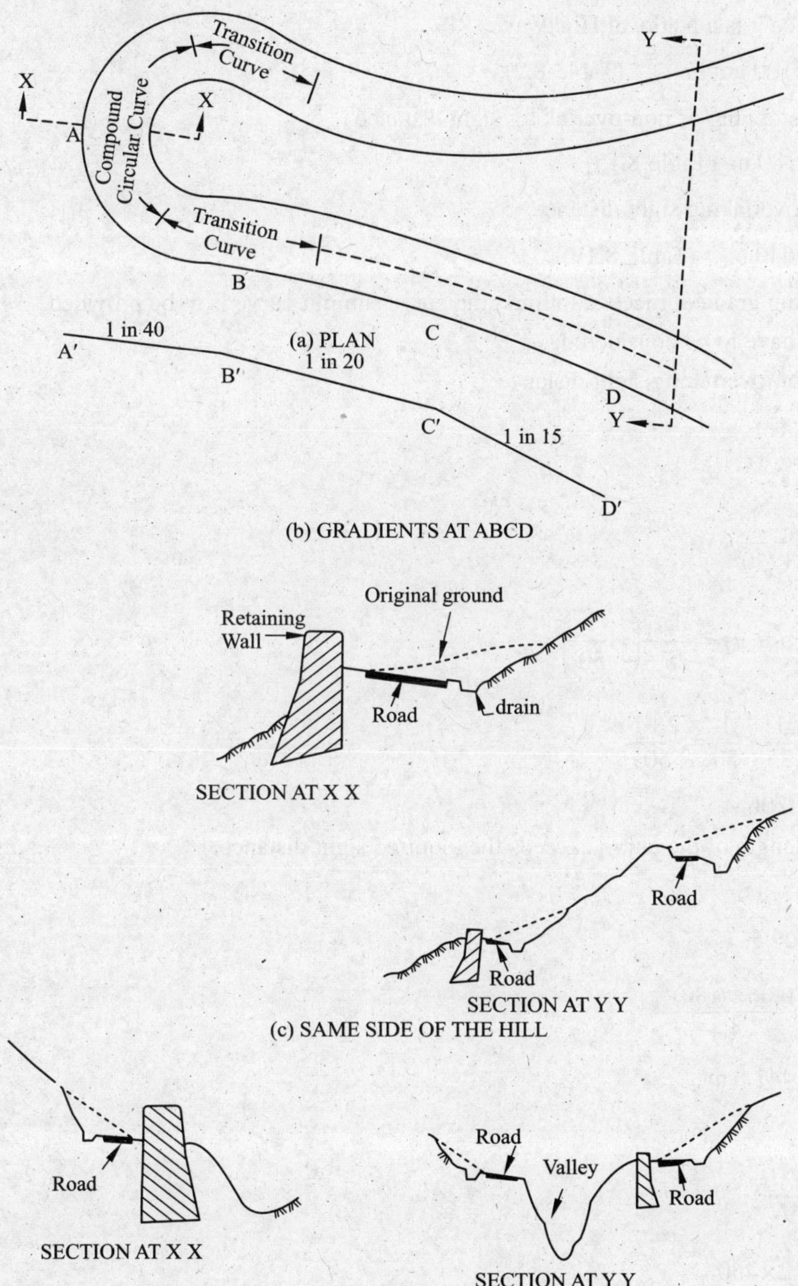

Fig. 8.62 Hump.

Problem 8.22 *A rising gradient of 1 in 25 meets a falling gradient of 1 in 50 on a National Highway. Design a vertical curve if the existing features near the locality permit the adoption of only minimum sight distance.*

Solution. Since it is a National Highway

$V = 100$ km/hr (Table 8.2)

Minimum stopping or non-overtaking sight distance

$= 180$ m (Table 8.15)

Minimum overtaking sight distance

$= 640$ m (Table 8.17)

Since a rising gradient meets a falling gradient, a summit curve is to be provided.

Two cases have to be considered:

Case I. Non-overtaking sight distance

$$n_1 = +\frac{1}{25}$$

$$n_2 = -\frac{1}{50}$$

$$N = n_1 - n_2 = \frac{1}{25} - \left(-\frac{1}{50}\right)$$

$$= \frac{1}{25} + \frac{1}{50} = \frac{2+1}{50}$$

$$= 0.06$$

When the length of the curve exceeds the required sight distance

$$L = \frac{NS^2}{4.4}$$

$$= \frac{0.06 \times (180)^2}{4.4}$$

$$= 441.8 \text{ m}$$

say 450 m

Then $C = \dfrac{2L}{N}$

$$= \frac{2 \times 450}{0.06} = 15{,}000$$

Taking ordinates at every 50 m interval

$$Y_1 = \frac{u^2}{C}$$

$$= \frac{(50)^2}{15000}$$

$$= 0.167$$

Table 8.24 Design for stopping sight distance of 180 m and speed 100 km/hr

First-half of the curve (1 in 25)					Second half of the curve (1 in 50)				
No. of stations from B.V.C.	*Chainage from B.V.C.*	*R.L. of points on first grade line (m)*	*Ordinate between curve & 1st grade line (m)*	*R.L. of station on curve (m)*	*No. of stations from B.V.C. (m)*	*Chainage from B.V.C. (m)*	*R.L. of points on second grade line (m)*	*Ordinates between curve & 2nd grade line (m)*	*R.L. of station on curve (m)*
Read downwards		Rise per 25 m = 1 m							
0 (B.V.C.)	0	100.00	0.000	100.00	10 (B.V.C.)	450	104.5	0.000	104.500
1	50	102.00	$y_1 = 0.167$	101.833	9	400	105.5	0.167	105.333
2	100	104.00	$2^2y_1 = 0.668$	103.332	8	350	106.5	0.668	105.832
3	150	106.00	$3^2y_1 = 1.503$	104.497	7	300	107.5	1.503	105.997 (highest point)
4	200	108.00	$4^2y_1 = 2.672$	105.328	6	250	108.5	2.672	105.828
5	225	109.00	$5^2y_1 = 3.382$	105.618 (nearest third decimal points)	5 Read upward	225	109.00 fall per 50 m = 1m	3.382	105.618

If the R.L. of the tangent point, *i.e.*, beginning of the vertical curve (B.V.C) T_1 is 100.00. Table 8.24 shows the design.

Highest point on the curve occurs at x_0 which is given by

$$x_0 = \frac{n_1}{n_1 + n_2} \times L$$

$$= \frac{0.04}{0.04 + 0.02} \times 450$$

$$= 300 \text{ m}$$

The R.L. of this point is found as follows:

$$Y_{max} = \left(\frac{300}{50}\right)^2 \times 0.167$$

$$= 36 \times 0.167$$

$$= 6.013 \text{ m}$$

R.L of the point along the $\frac{1}{25}$ gradient corresponding to the highest point on the curve.

$$= 100 + \frac{300}{100} \times 4.00$$

$$= 112.00$$

$\therefore$ R.L. of the highest point on the curve $= 112.00 - 6.013$

$$= 105.997$$

The solution is indicated in Fig. 8.63.

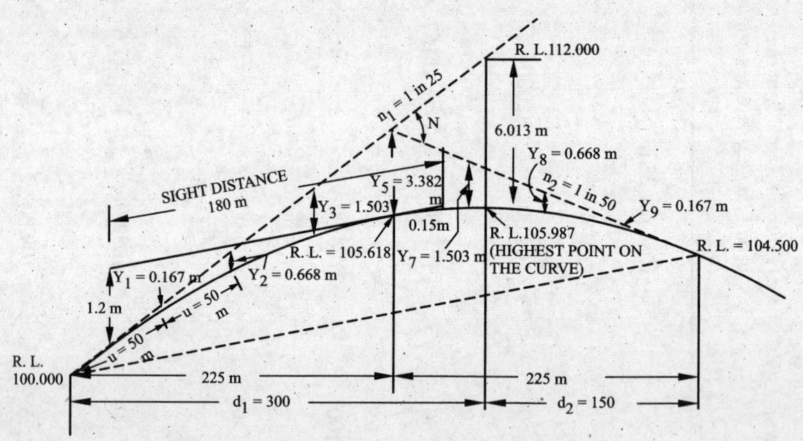

Fig. 8.63 Summit curve for Problem 8.22.

Case II. Overtaking sight distance

When the length of the curve exceeds the required sight distance,

$$L = \frac{NS^2}{9.6}$$

$$= \frac{0.06 \times 640^2}{9.6}$$

$$= 2560 \text{ m}$$

Say 2600 m

Divide the curve into twenty six stations of 100 m each

$$R = \frac{L}{N} = \frac{2600}{0.06} = 43333 \text{ m}$$

$$C = \frac{2L}{N}$$

$$= \frac{2 \times 2600}{0.06}$$

$$= 86667$$

$$Y_1 = \frac{u^2}{C}$$

$$= \frac{100 \times 100}{86667}$$

$$= 0.1154$$

Highest point of the curve occurs at x_0, which is given by

$$x_0 = \frac{n_1}{n_1 + n_2} L$$

$$= \frac{0.04}{0.04 + 0.02} \times 2600$$

$$= 1733 \text{ m}$$

$$Y_{max} = \left(\frac{1733}{100}\right)^2 \times 0.1154 = 34.66 \text{ m}$$

R.L. along the $\frac{1}{25}$ gradient $= 100.00 + 17.33 \times 4$

$$= 100 + 69.32$$

$$= 169.32 \text{ m}$$

R.L. of the highest point $= 169.32 - 34.66$

$$= 134.66 \text{ m}$$

The solution is shown in Fig. 8.64 and in the design Table 8.25.

Table 8.25 Design for an overtaking sight distance of 640 m; speed 100 km. p.h.

First-half of the curve (1 in 25)					*Second half of the curve (1 in 50)*				
No. of stations from B.V.C.	*Chainage from B.V.C.*	*R.L. of point on first grade line (m)*	*Ordinates between curve & first grade line (m)*	*R.L. of station on curve (m)*	*No. of stations from B.V.C. (m)*	*Chainage from B.V.C. (m)*	*R.L. of point on second grade line (m)*	*Ordinates between curve & second grade line (m)*	*R.L. of station on curve (m)*
Read downwards		Rise per station = 4 m							
0 (B.V.C.)	0	100	0.00	100.0000	26	2600	126	0.0000	126.0000
1	100	104	$y_1 = 0.1154$	103.8846	25	2500	128	0.1154	127.8846
2	200	108	$2^2y_1 = 0.4616$	107.5384	24	2400	130	0.4616	129.5384
3	300	112	$3^2y_1 = 1.0386$	110.9614	23	2300	132	1.0386	130.9614
4	400	116	$4^2y_1 = 1.8464$	114.1536	22	2200	134	1.8464	132.1536
5	500	120	$5^2y_1 = 2.8850$	117.1150	21	2100	136	2.8850	133.1150
6	600	124	$6^2y_1 = 4.1544$	119.8456	20	2000	138	4.1544	133.8456
7	700	128	$7^2y_1 = 5.6546$	122.3454	19	1900	140	5.6546	134.3454
8	800	132	$8^2y_1 = 7.3856$	124.6144	18	1800	142	7.3856	134.6144
9	900	136	$9^2y_1 = 9.3474$	126.6526	17	1700	144	9.3474	134.6526
10	1000	140	$10^2y_1 = 11.5400$	128.4600	16	1600	146	11.5400	134.4600
11	1100	144	$11^2y_1 = 13.9634$	130.0366	15	1500	148	13.9634	134.0366
12	1200	148	$12^2y_1 = 16.6176$	131.3824	14	1400	150	16.6176	133.3824
13	152	152	$13^2y_1 = 19.5026$	132.4974	13	1300	152	19.5026	132.4974
					Read upward		Fall/station = 2 m		

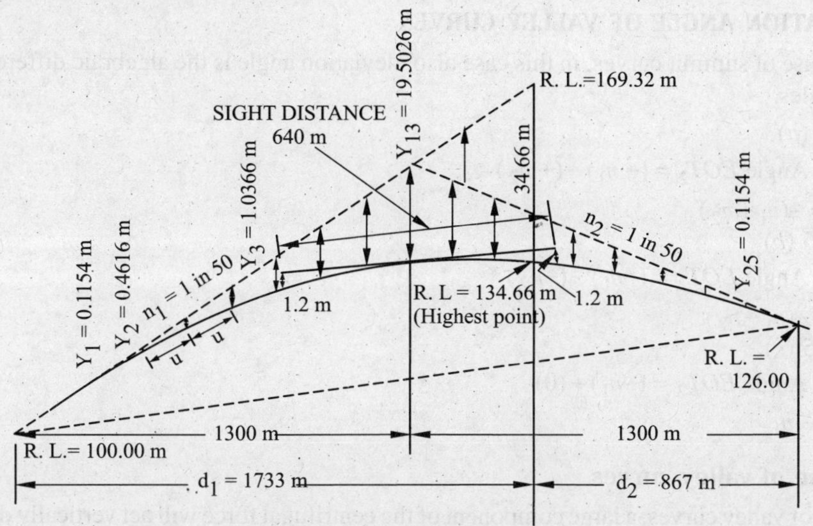

Fig. 8.64 Summit curve for Problem 8.22

VALLEY CURVES

A vertical curve, concave upwards, is called a valley curve. This is formed when a descending gradient intersects an ascending gradient or when a descending gradient meets another descending gradient or when a descending gradient joins a horizontal path, as shown in Fig. 8.65.

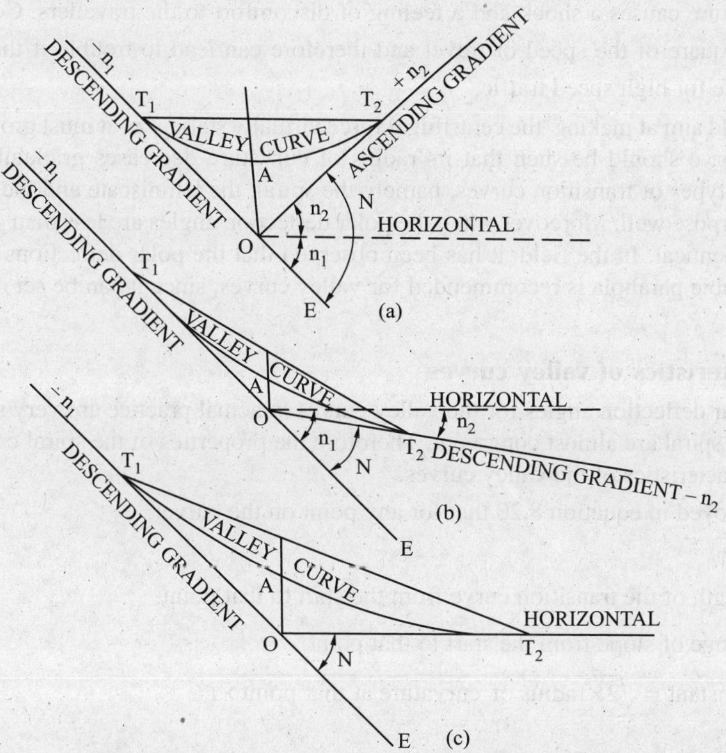

Fig. 8.65 Valley curves, showing total deflection angles.

8.54. DEVIATION ANGLE OF VALLEY CURVES

As in the case of summit curves, in this case also, deviation angle is the algebraic difference of the two grade angles.

In Fig. 8.6 (*a*)

$$N = \text{Angle } EOT_2 = (-n_1) - (+n_2)$$

$$= -(n_1 + n_2)$$

In Fig. 8.65 (*b*)

$$N = \text{Angle } EOT_2 = (-n_1) - (-n_2)$$

$$= -(n_1 - n_2)$$

In Fig. 8.65 (*c*)

$$N = \text{Angle } EOT_2 = (-n_1) - (0)$$

$$= -n_1$$

8.54.1. Shape of valley curves

In the case of valley curves, a large component of the centrifugal force will act vertically downwards and exert extra pressure on the tyres and springs of the vehicle. If a vehicle is allowed to enter suddenly to the curve, the centrifugal force will change from zero to a finite value at once thus causing greater strain than it would have caused if it grew up gradually. Sudden application of a force causes twice the strain than gradual or static application does. The centrifugal force also sets up oscillations and alternating stresses. In the case of heavily laden goods trucks and passenger buses, this extra strain on the tyres and springs due to centrifugal force and its sudden application would be particularly important. Moreover, the sudden entry of the vehicle from a straight path $(R = \infty)$ to a curved path with finite curvature causes a shock and a feeling of discomfort to the travellers. Centrifugal force varies with the square of the speed of travel and therefore can lead to trouble if the roads are not planned to be safe for high speed traffic.

Thus we should aim at making the centrifugal force virtually static, *i.e.,* it must grow up gradually. Therefore the curve should be such that its radius of curvature decreases gradually as its length increases. Three types of transition curves, namely the spiral, the lemniscate and the cubic parabola can serve this purpose well. Moreover when the polar deflection angles are less than 4°, these curves are practically identical. In the field, it has been observed that the polar deflections seldom exceed this limit. The cubic parabola is recommended for valley curves, since it can be set out easily in the field.

8.54.2. Characteristics of valley curves

Since the polar deflection angles for the valley curves in actual practice are very small, the cubic parabola and the spiral are almost congruent. Therefore the properties of the spiral can be applied to find out the characteristics of the valley curves.

It has been proved in equation 8.20 that for any point on the curve:

$$l = c\sqrt{\phi}$$

where l = Length of the transition curve from the start to that point

ϕ = Change of slope from the start to that point

c = Constant = $\sqrt{2 \times \text{radius of curvature at that point} \times l}$

$= \sqrt{2rl}$

or $$l = \sqrt{2rl\,\phi} \qquad \ldots(8.50)$$

or $\quad l^2 = 2rl\,\phi$

or $\quad r = \dfrac{l}{2\phi}$ ...(8.51)

It has been found (see equation 8.1) that the rate of increase in radial acceleration (C') is given by

$$C' = \frac{v^3}{lR}$$

From point T_1 to O, the length of the curve is $\dfrac{L}{2}$

$\therefore \quad C' = \dfrac{2v^3}{LR}$ ...(8.52)

For comfortable journey, the value of C' should not exceed a certain permissible limit. IRC has recommended a value of 0.61m/sec³ for vertical curves.

Now at the point of the curve T_1T_2, as shown in Fig. 8.66.

$$r = R;\, l = \frac{L}{2} \text{ and } \phi = \frac{N}{2}$$

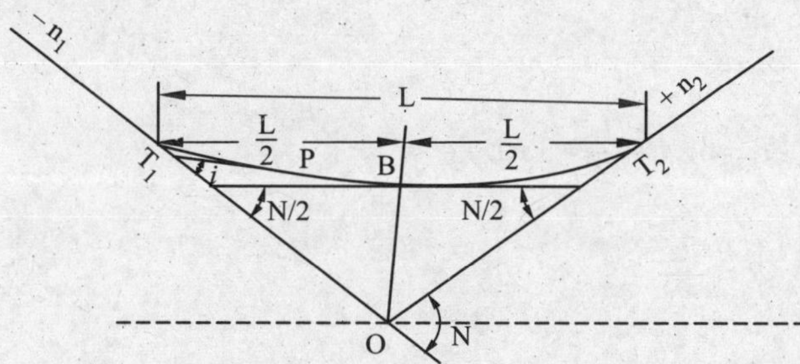

Fig. 8.66 Valley curve.

From equation 8.50, we get

$$\frac{L}{2} = \sqrt{2R \cdot \frac{L}{2} \cdot \frac{N}{2}}$$

$\therefore \quad L = \sqrt{2\,RLN}$

But from equation 8.52, $LR = \dfrac{2v^3}{C'}$

If V is taken in km/hr;

then $v = 0.28\,V$

and also $C' = 0.61$ m/sec^3

$$\therefore \qquad LR = 2\times\frac{(0.28\,V)^3}{0.61}$$

$$= 0.072\,V^3$$

Substituting for LR, we get

$$L = \sqrt{2\times 0.072\,V^3 N}$$

or $$L = 0.38\,(NV^3)^{1/2} \qquad \ldots(8.53)$$

Also the radius of curvature at any point is

$$r = \frac{l}{2\phi}$$

Putting the values for point O i.e, $r = R$, $l = \dfrac{L}{2}$ and $\phi = \dfrac{N}{2}$

$$R = \frac{L}{2N}$$

The cartesian equation of the cubic parabola is

$$y = \frac{x^3}{b} : \text{where } b = 6\,rl$$

At point O

$$b = 6\times R\times\frac{L}{2} = 3\,RL$$

But $$R = \frac{L}{2N}$$

$$\therefore \qquad b = \frac{3}{2}\frac{L^2}{N} \qquad \ldots(8.54)$$

In actual practice, the length of the curve calculated from the above formula is adjusted to the nearest multiple of 10 m. This is done for the purpose of getting equal number of chords (each of 10 m or 5 m length) for setting out.

As per IRC recommendations, the headlight beam distance should equal the stoppping sight distance in case of a valley curve. This requirement is necessary for safe travel at night.

For a square parabolic curve with headlight height equal to or greater than 0.75 m, divergence of the light beam from the longitudinal axis of vehicle equal to unity, the stopping sight distance S and curve length, L are related by the following formulae:

$$L = \frac{NS^2}{(1.5 + 0.035S)} \quad \text{when } L \text{ is greater than } S, \text{ and}$$

$$L = 2S - \frac{(1.5 + 0.035\,S)}{N} \quad \text{when } L \text{ is less than } S.$$

Length of the valley curve for various grade differences is given in Fig. 8.67.

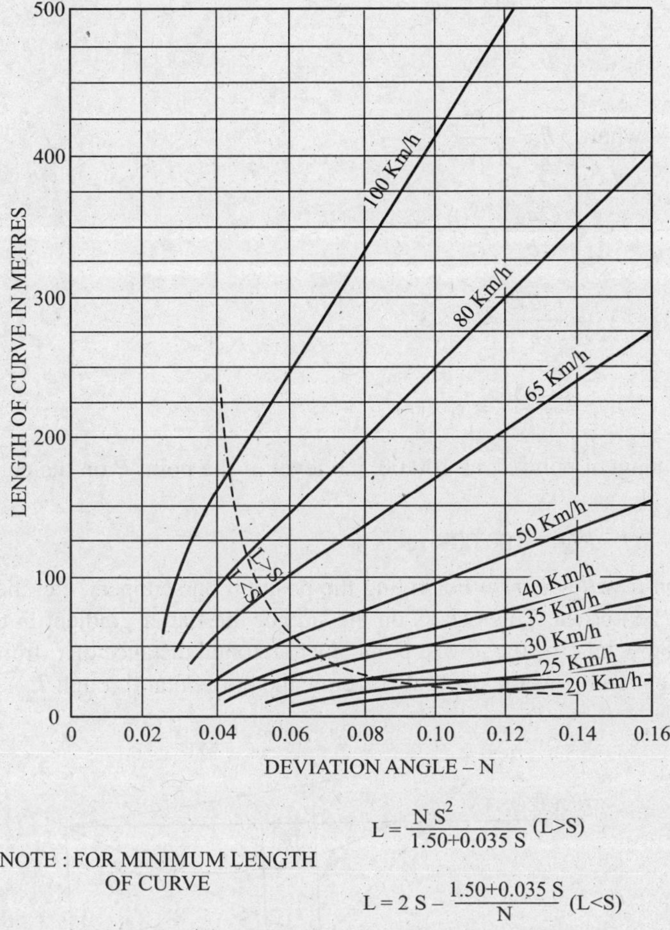

NOTE : FOR MINIMUM LENGTH OF CURVE

$$L = \frac{N S^2}{1.50 + 0.035\, S} \quad (L > S)$$

$$L = 2\,S - \frac{1.50 + 0.035\, S}{N} \quad (L < S)$$

Fig. 8.67 Length of valley curve.

8.54.3. Computing ordinates of valley curves

A valley curve always comprises two transition curves *i.e.,* cubic parabolas, equal and reversed. Let one of these curves T_1C be divided into equal chords, each of u-metre length (Fig. 8.68).

The equation of the curve is

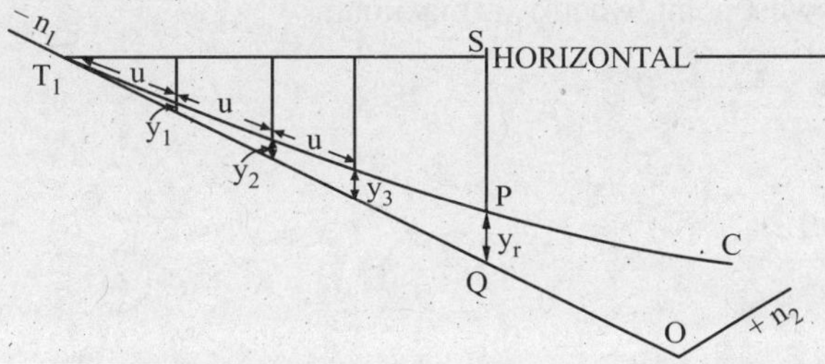

Fig. 8.68 Ordinates of valley curve.

$$y = \frac{x^3}{b}$$

Ordinate $y_1 = \frac{u^3}{b}$ where $b = \frac{3}{2}\frac{L^2}{N}$

$$y_2 = \frac{(2u)^3}{b} = y_1 \times 2^3$$

$$y_3 = \frac{(3u)^3}{b} = y_1 \times 3^3$$

$$y_r = \frac{(ru)^3}{b} = y_1 \times (r)^3$$

Taking R.L. of tangent point T_1 as 100.00, the level of the point P on the curve at the end of the chord is:

$$(100.00 - SQ - PQ) = 100.00 - r.u.n_1 + y_r \qquad ...(8.55)$$

Lowest point on a valley curve. For fixing the position on scuppers, etc. the lowest point on a valley curve must be known. This occurs on the side of the flatter gradient in the case of unequal gradients. In Fig. 8.69, let L be the lowest point at a horizontal distance of x_1 from T_2. Assuming OT_2 as the flatter gradient and y_1 as the depth of L below the horizontal through T_2.

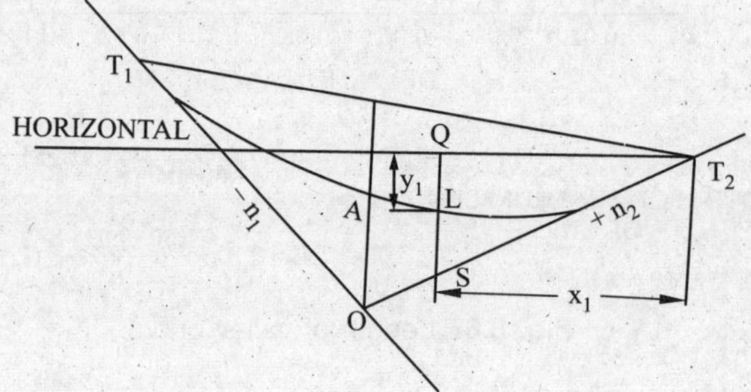

Fig. 8.69 Lowest point on a valley curve.

$$y_1 = QL = QS - LS$$

$$= n_2\, x_1 - \frac{x_1^3}{b}$$

Now L will be the lowest point, when y_1 is maximum

or $$\frac{dy_1}{dx_1} = 0 \text{ i.e., } n_2 - \frac{3x_1^2}{b} = 0$$

or $$x_1 = \sqrt{\frac{bn_2}{3}}$$

but $$b = \frac{3}{2}\frac{L^2}{N}$$

$\therefore$ $$x_1 = \sqrt{\frac{3}{2}\frac{L^2}{N}\frac{n_2}{3}} = L\left(\frac{n_2}{2N}\right)^{\frac{1}{2}}$$

This formula is applicable only when a descending gradient meets an ascending gradient.

8.55. IMPACT FACTOR

This is defined as the ratio of the maximum centrifugal force to the weight of the vehicle and is expressed as a percentage.

The centrifugal force F, is given by

$$F = \frac{W}{g} \times \frac{v^2}{r} \text{ kg}$$

where W = Weight of the vehicle in kg

v = Speed in m/s

r = Radius of curvature in m at that particular instant.

This will be maximum when r is minimum, which occurs at the end of the transition curve, *i.e.,* when $r = R$.

$$\therefore \quad F_{max} = \frac{W}{g} \times \frac{v^2}{R}$$

If I is the impact factor, then

$$I = \frac{F_{max}}{W} \times 100 \text{ per cent}$$

$$= \frac{v^2}{gR} \times 100 \text{ per cent}$$

But $\quad R = \frac{L}{2N}$;

$v = 0.28\,V$, where V is in km/hr

and $\quad g = 9.81 \text{ m/sec}^2$

$$\therefore \quad I = \frac{0.28^2\,V^2}{9.81\,L} \times 2N \times 100 \text{ per cent}$$

$$= 1.59\,\frac{NV^2}{L} \text{ per cent.}$$

Impact causes additional strain on the tyres and body of the vehicle and also on the structure of the road. Moreover it is a source of discomfort to the passengers. Impact should, therefore, be as low as possible and a value of 25 per cent in recommended. This does not take into account undulations of the road structure, badly built-up curves, etc. which add to impact. To allow for all these, IRC has suggested that the impact factor should not exceed 17 per cent.

Problem 8.23 *A falling gradient of 1 in 20 meets a rising gradient of 1 in 40 on a Major District Highway in plain country. Find the length of the valley curve, which should provide safe driving at night.*

Solution. Deviation angle, $N = \frac{1}{20} - \left(-\frac{1}{40}\right)$

$$= 0.05 + 0.025$$

$$= 0.075$$

For a Major District Highway in plain country

Design speed = 80 km/hr; and

Stopping distance = 120 m

(*i*) Length of valley curve $= 0.38\,(NV^3)^{1/2}$

$$= 0.38 \times (0.075 \times 80^3)^{1/2}$$

$$= 0.38 \times 195.96$$

$$= 74.46 \text{ m}$$

say 75 m

(*ii*) Assuming the length of the valley curve to be greater than the stopping distance *i.e.*, $L > S$.

$$L = \frac{NS^2}{1.5 + 0.035\,S}$$

$$= \frac{0.075 \times (120)^2}{1.5 + 0.035 \times 120}$$

$$= \frac{0.075 \times 120^2}{1.5 + 4.2}$$

$$= \frac{0.075 \times 120^2}{5.7}$$

$$= 189.47 \text{ m}$$

say 190 m

Since this value is greater than stopping distance of 120 m, hence the assumption is correct and 190 m length of valley curve is essential for safe driving at night.

From Fig. 8.65 also the length of the valley curve is 190 m for $N = 0.075$ and $V = 80$ km/hr.

Hence adopt $L = 190$ m

(*iii*) Check for impact

$$\text{Impact factor} = \frac{1.59\,NV^2}{L}$$

$$= \frac{1.59 \times 0.075 \times 80^2}{190}$$

$$= 4.02\,\%$$

which is less than the maximum allowable impact factor of 17%.

Hence 190 m length of the valley curve is adequate.

Problem 8.24. *Design valley curve on a National Highway in plain terrain where a falling gradient of* $\frac{1}{25}$ *meets a rising gradient of* $\frac{1}{20}$.

Solution. Deviation angle, $N = \frac{1}{20} - \left(-\frac{1}{25}\right)$

$$= 0.05 + 0.04$$

$$= 0.09$$

For a National Highway in plain country, the design speed is 100 km/hr.

$$\text{Length of the curve} = 0.38\,(NV^3)^{1/2}$$

$$= 0.38\left\{0.09 \times (100)^3\right\}^{1/2}$$

$$= 0.38 \times 300$$

$$= 114 \text{ m}$$

say 120 m

For safe driving at night

For $V = 100$ km/hr

Stopping sight distance, $S = 180$ m

Assuming L greater than S

$$L = \frac{NS^2}{1.5 + 0.035S}$$

$$= \frac{0.09 \times 180^2}{1.5 + 0.035 \times 180}$$

$$= 373.85 \text{ m}$$

$$= \text{Say } 380 \text{ m}$$

Hecne adopt $L = 380$ m.

Check for impact

$$\text{Impact factor} = \frac{1.59\,NV^2}{L}$$

$$= \frac{1.59 \times 0.09 \times 100^2}{380}$$

$$= 3.77\,\%$$

which is less than the maximum permissible value of 17%. Hence O.K.

Layout design

Dividing the length into 38 stations, 10 m apart

$$b = \frac{3L^2}{2N}$$

$$= \frac{3 \times 380^2}{2 \times 0.09} = 2406667$$

The first ordinate y_1 is given by

$$y_1 = \frac{u^3}{b} = \frac{10^3}{2406667} = 0.0004$$

The design chart is worked out in Table 8.26. If the lowest point occurs at a distance x_1 from the side of the flatter gradient, then

$$x_1 = L\left(\frac{n_1}{2N}\right)^{1/2}$$

Table 8.26 Design for valley curve

First half of the curve (1 in 25)					Second half of the curve (1 in 20)				
No. of stations from B.V.C.	*Chainage from B.V.C. (m)*	*R.L. of point on first grade line (m)*	*Ordinates between curve & 1st grade line (m)*	*R.L. of station on curve (m)*	*No. of stations from B.V.C. (m)*	*Chainage from . B.V.C. (m)*	*R.L. of point on second grade !ine (m)*	*Ordinates between curve & 2nd grade line (m)*	*R.L. of station on curve (m)*
Read downwards		Fall 0.40 m per 10 m							
0	0 (B.V.C.)	100.00	0.000	100.000	38	380	101.90	0.000	101.900
1	10	99.60	$y_1 = 0.0004$	99.600	37	370	101.40	0.0004	101.400
4	40	98.40	$4^3y_1 = 0.026$	98.426	34	340	99.00	0.026	99.926
7	70	97.20	$7^3y_1 = 0.137$	97.337	31	310	98.40	0.137	98.537
10	100	96.00	$10^3y_1 = 0.400$	96.400	28	280	96.90	0.400	97.300
13	130	94.80	$13^3y_1 = 0.879$	95.679	25	250	95.40	0.879	96.279
16	160	93.60	$16^3y_1 = 1.638$	95.238	22	220	93.90	1.638	95.538
19	190	92.40	$19^3y_1 = 2.744$	95.144	19	190	92.40	2.744	95.144
				Nearest to third decimal	Read upwards		Rise 0.5 m per 10 m		

$$= 380\left(\frac{0.04}{2 \times 0.09}\right)^{1/2}$$

$$= 179.13 \text{ m from the beginning of the curve.}$$

Intercept between the curve and the tangent

$$= \left(\frac{179.13}{10}\right)^3 \times 0.0004$$

$$= 2.299 \text{ m}$$

Level along the tangent at 179.13 m from the beginning of the curve

$$= 100.00 - 0.04 \times 179.13$$

$$= 92.84 \text{ m}$$

Level of the lowest point

$$= 92.84 + 2.299$$

$$= 95.139 \text{ m}$$

8.56. LATERAL AND VERTICAL CLEARANCES AT UNDERPASSES

Many times a road has to be taken throughout an underpass (short passage beneath a grade separated structure) below another road, railway line or irrigation facility like aqueduct. In order that capacity, speed and safety of travel are not affected, the lateral and vertical clearances at underpasses must be adequate.

8.56.1. Lateral clearance on rural roads

Single carriageway. Desirably the full roadway width at the approaches should be carried through the underpass. This implies that the minimum lateral clearance on either side must equal the shoulder width. This rule should be relaxed only in exceptional circumstances. Normal and exceptional values of lateral clearance for different classes of highways are given in Table 8.27.

Table 8.27 Normal and exceptional lateral clearances

Sr. No.	*Road classification*	*Lateral clearance (m)*	
		Normal	*Exceptional*
1.	National and State Highways	2.5	2.0
2.	Major District Roads and Other District Roads	2.0	1.5
3.	Village Roads	1.5	1.0

If a footpath is needed on a road, lateral clearance in the underpass portion should be the width of the footpath plus one metre. (See Fig. 8.70). Footpath width depends upon the expected pedestrian traffic and may be fixed with the help of the following capacity, subject to a minimum of 1.5 m.

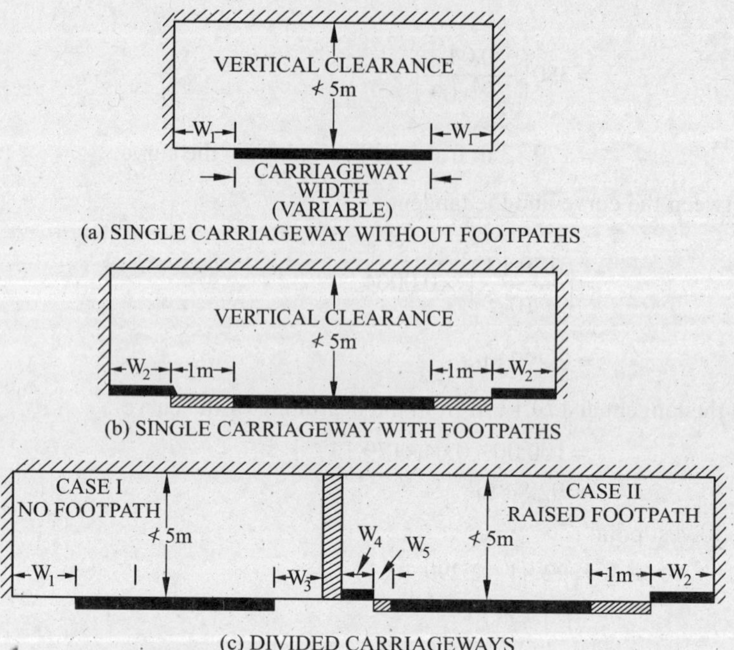

Fig. 8.70 Lateral and vertical clearances for rural roads.

Anticipated capacity (No. of persons/hr)	*In both directions*	*Required footpath width (m)*
1200	800	1.5
2400	1600	2.0
3600	2400	2.5

Divided carriageway. When an underpass is built for a divided highway, left hand side clearance should be in accordance with the above provisions.

Lateral clearance on the right to a pier or column in the central median should be 2 m desirably and 1.5 m minimum. Where the central median is kerbed, the carriageway width should be increased by the side safety margin of 0.5 m, as shown in Fig. 8.70. Lateral clearance in that event could be reduced to 1.5 m (desirable value) or 1 m (exceptional). If the median is not wide enough to permit these clearances, either it should be widened gradually at the approaches or a single span structure provided across the full cross-section, thereby avoiding a central pier.

8.56.2. Vertical clearance

Minimum vertical clearance of 5 m should be ensured over the full width of the roadway at all underpasses, and similarly at over-hanging cliffs and any semi-tunnel sections, etc. However, in urban areas, this should be increased to 5.5 m so that double decker buses could be accommodated. Due allowance for any future raising or strengthening of the pavement should also be made.

INDIAN ROADS CONGRESS RECOMMENDATIONS FOR ALIGNMENT

8.57. HORIZONTAL ALIGNMENT

Uniformity of design standards is one of the essential requirements of a road alignment. In a given section, there must be consistent application of a design element to avoid unexpected situations being created for the drivers. For instance, a short sharp curve in an otherwise good alignment is bound to act as an accident-prone spot if the designer is not vigilant. Similarly, any unnecessary break in horizontal alignment at cross-drainage structures should be avoided.

As a general rule, the horizontal alignment should be fluent and blend well with the surrounding topography. A flowing line which conforms to natural contours is aesthetically preferable to one with long tangents slashing through the terrain. This would not only help in limiting the damage to the environment but also assist in preservation of natural slopes and plant growth. Due consideration should also be given to the conservation of existing features. This aspect is dealt with at length in IRC special publication No. 21-1980 "Guidelines on Landscaping and Tree Plantation".

Long tangent sections exceeding 3 km in length should be avoided as far as possible. A curvilinear alignment with long curves is better from the point of safety and aesthetics.

As a normal rule, sharp curves should not be introduced at the end of long tangents since these can be extremely hazardous.

Short curves give appearance of kinks, particularly for small deflection angles, and should be avoided. The curves should be sufficiently long and have suitable transitions to provide pleasing appearance. Curve length should be at least 150 metres for a deflection angle of 5 degrees, and this should be increased by 30 metres for each one degree decrease in the deflection angle. For deflection angles less than one degree, no curve is required to be designed.

Reverse curves may be needed in difficult terrain. It should be ensured that there is sufficient length between the two curves for introduction of requisite transition curves.

Curves in the same direction separated by short tangents, known as broken-back curves should be avoided as far as possible in the interest of aesthetics and safety and replaced by a single curve. If this is not feasible, a tangent length corresponding to 10 seconds travel time must at least be ensured between the two curves.

Compound curves may be used in difficult topography but only when it is impossible to fit in a single circular curve. To ensure safe and smooth transition from one curve to the other, the radius of the flatter curve should not be disproportional to the radius of the sharper curve. A ratio of 15 : 1 should be considered the limiting value.

To avoid distortions in appearance, the horizontal alignment should be coordinated carefully with the longitudinal profile, keeping in mind that the road is a three-dimensional entity and does not consist simply of a plan and L-section.

The siting of the bridges and the location of the approaches should be properly coordinated keeping in view the overall technical feasibility, economy, fluency of alignment and aesthetics. The following criteria may be followed in general;

(*i*) For major bridges above 300 metres span, proper siting of the bridge should be the principal consideration and the approach alignment matched with the same;

(*ii*) For small bridges less than 60 metres span, fluency of the alignment should govern the choice of the bridge location; and

(*iii*) For spans between 60 and 300 metres, the designer should use his discretion keeping in view the importance of the road, overall economic considerations and aesthetics.

8.58. VERTICAL ALIGNMENT

The vertical alignment should provide for a smooth longitudinal profile consistent with category of the road and lay out of the terrain. Grade changes should not be too frequent as to cause kinks and visual discontinuities in the profile. Desirably, there should be no change in grade within a distance of 150 m.

A short valley curve within an otherwise continuous profile is undesirable since this tends to distort the perspective view and can be hazardous.

Broken-back grade lines, *i.e.*, two vertical curves in the same direction separated by a short tangent, should be avoided due to poor appearance and preferably replaced by a single long curve.

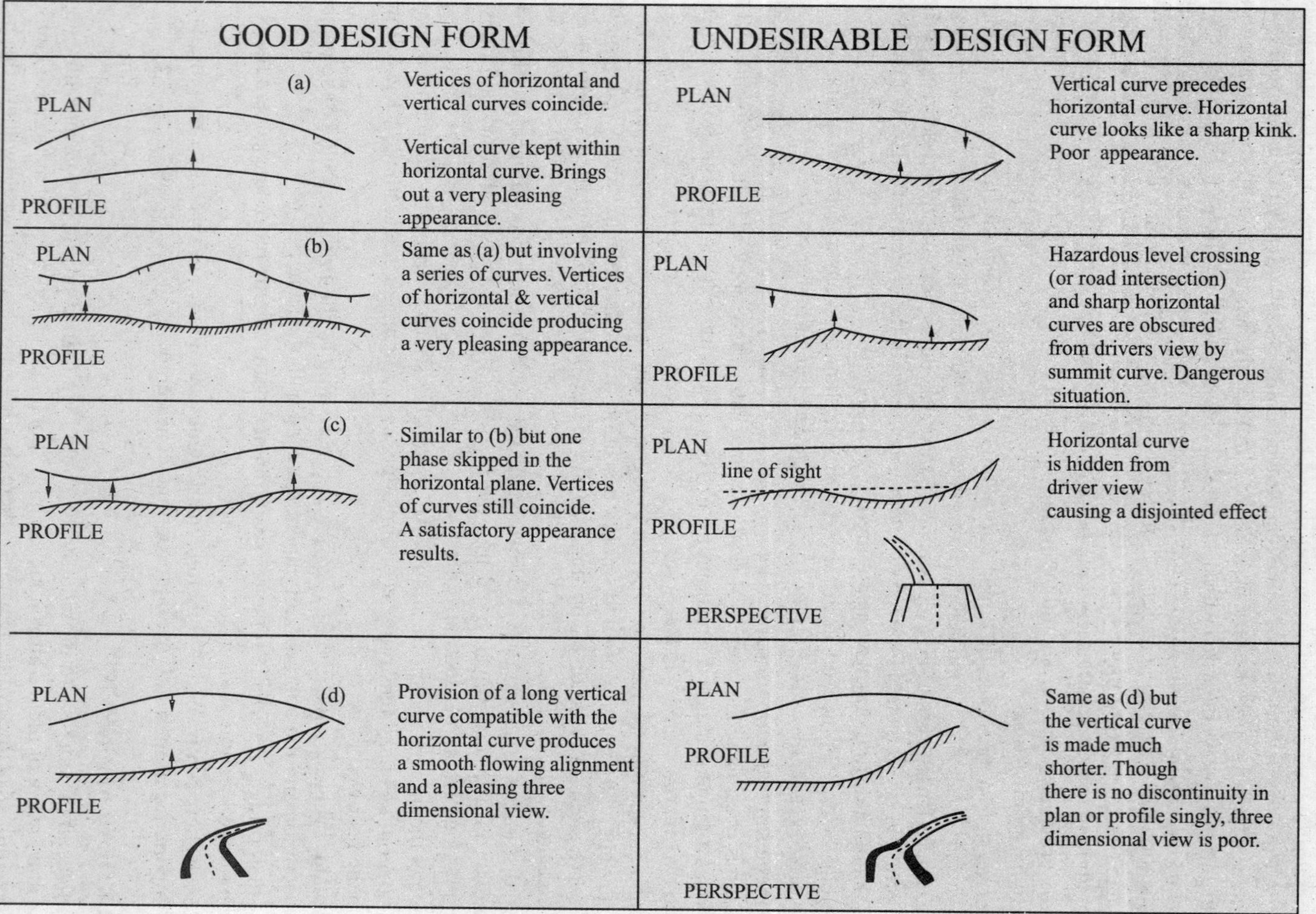

Fig. 8.71 Typical cases of good and bad alignment coordination.

Decks of small cross-drainage structures, (*i.e.*, culverts and minor bridges) should follow the same profile as the flanking road section, without any break in the grade line.

The longitudinal profile should be coordinated suitably with the horizontal alignment.

8.59. COORDINATION OF HORIZONTAL AND VERTICAL ALIGNMENTS

The overall appearance of a highway can be enhanced considerably by judicious combination of the horizontal and vertical alignments. Plan and profile of the road should not be designed independently but in unison so as to produce an appropriate three-dimensional effect. Proper coordination in this respect will ensure safety, improve utility of the highway and contribute to overall aesthetics.

The degree of curvature should be in proper balance with the gradients. Straight alignment or flat horizontal curves at the expense of steep or long grades, or excessive curvature in a road with flat grades, do not constitute balanced designs and should be avoided.

Vertical curvature superimposed upon horizontal curvature gives a pleasing effect. As such the vertical and horizontal curves should coincide as far as possible and their length should be more or less equal. If this is difficult for any reason, the horizontal curve should be somewhat longer than the vertical curve.

Sharp horizontal curves should be avoided at or near the apex of pronounced summit/sag vertical curves from safety considerations.

Fig. 8.71 illustrates some typical cases of good and bad alignment coordination.

GEOMETRIC DESIGN STANDARDS

For Express-ways in plain terrain

1. Design speed (KPH)	:	120
2. Land width (metres)	:	90 – extra to be provided where warranted
3. Building lines (metres)	:	10 metres beyond right of way
4. Road width (metres)		
- For 4-lane divided carriageway	:	27
- For 6-lane divided carriageway	:	34
- On culverts	:	Same as for road sections
5. Carriageway width (metres)		
- 4-lane divided carriageway	:	2 × 7.5
- 6-lane divided carriageway	:	2 × 11
6. Shoulder width (metres)		
- Treated shoulder	:	2.5
- Untreated shoulder	:	1.0
- Total width	:	3.5
7. Median width (metres)	:	6
8. Camber (per cent)		
- Carriageway	:	2.5
- Treated shoulder	:	3.0
- Untreated earth shoulder	:	4.0
9. Sight distance (metres)		
- Safe stopping sight distance (minimum)	:	250
- Desirable	:	500
10. Radius of horizontal curve (m)	:	Min. 700, desirable 2600 (This radius requires no superelevation and normal cross section will suffice)

11. Superelevation : As per formula

$$e = \frac{V^2}{225R}$$

Subject to a maximum of 4%

12. General notes on horizontal alignment

(*i*) Long tangent sections exceeding 6 km should be avoided.

(*ii*) Broken-back curves should be avoided or at least separated by 500 metres straight length.

(*iii*) Minimum curve length should be 150 metres for 5° deflection angle and increased at the rate of 30 m for 1° decrease thereafter.

13. Length of transition curve (metres) : $\frac{0.0215V^3}{CR}$ Where $C = 0.5$

V-the design speed, and

R-the radius of circular curve.

14. Maximum gradient

- Ruling : 1 in 50
- Absolute : 1 in 40

15. Summit and valley curves (metres) : To be designed for sight distance mentioned at S. No. 9 and minimum length = 0.6 V

16. Vertical profile : 1 m clearance between HFL and subgrade

Vertical = 6 m

17. Clearance through road over passes : Horizontal : Same normal section expressway to continue.

18. Design standard for inter-change elements.

(*i*) Speed, sight distance and radius

	Design speed (kph)	Radius (m)	Stopping sight distance (m)
Desirable	80	230	130
Minimum	60	130	80

(The direct ramps/diagonal connection should be designed for the desirable design speed and the design speed of loops may be near the minimum)

(*ii*) Maximum grade (per cent) : Desirable-4

Absolute – 6

(*iii*) Summit and valley curves (metres) : To be designed as per stopping sight distances formulae and minimum length = 0.6 V

(*iv*) Cross-section elements : (*a*) Carriageway

Desirable-2-Lanes

Minimum – Intermediate lane

(*b*) Shoulder: 2 metres each

(*v*) Length of speed change lanes (m)

	Ramp/loop speed	
	80KPH	60 KPH
Acceleration lane	300	400
Declearation lane	130	150

Note: An expressway is a divided arterial highway intended for through traffic with full control of access and generally provided with grade separations at intersections. No slow moving traffic or pedestrians will be permitted on expressways.

❑❑❑

PART – III

Chapter

SOIL CLASSIFICATION

GENERAL

The object of classifying soils is to arrange them into groups on the basis of common properties in respect of their use. Some properties could be anticipated from a knowledge of the soil type. Engineers have resorted to standard classification systems, in order to make it possible to predict the behaviour of the soil.

The existing systems of soil classification can be divided into the following groups:

1. ***Based on the mechanical composition***

Textural classification, according to which soils are divided into groups on the basis of their mechanical analysis. Charts have been prepared by several agencies to facilitate the determination of textural class, when the mechanical analysis of a soil is known. (Charts by Bureau of Public Roads and Department of Agriculture, USA).

2. ***Based on plasticity and engineering characteristics of soil in disturbed condition.***

(*a*) Public roads classification (HRB/AASHO) system.
(*b*) Unified soil classification system which is a modification of Casagrande classification system.
(*c*) Federal Aviation Agency (formerly known as Civil Aeronautics Administration) classification system.
(*d*) Compaction classification system.

3. ***Based on the properties of soil in undisturbed condition.***

(*a*) Descriptive classification based on general description of soil derived from field examination supplemented by detailed testing, if needed.
(*b*) Geological classification based on the geological history of each deposit.
(*c*) Pedological soil classification (devised by agronomists based on the classification of the parent material, scope of weathering, climate, and other environmental factors).

Description of classification systems widely used for classifying soils for highway and airport purposes follows.

VARIOUS CLASSIFICATIONS

9.1. PUBLIC ROADS CLASSIFICATION (HRB OR AASHO) SYSTEM

The Bureau of Public Roads, U.S.A. in 1928 introduced a system of soil classification based upon the engineering characteristics and performance of subgrade materials. In this system all soils were divided into eight major groups, which were designated by the symbols A–1, A–2 *etc*. The soil groups were based on a number of soil properties, and the centrifuge and field moisture equivalents.

After the original system had been in use for about 15 years, a committee of the Highway Research Board suggested a revision and its revision was adopted in 1945. Since 1945, it has become known as the Highway Research Borad (HRB), American Association of State Highway Officials (AASHO), or Modified Bureau of Public Roads system. In this system, the number of the physical properties on which classification is based was reduced to mechanical analysis, liquid limit, and plasticity index.

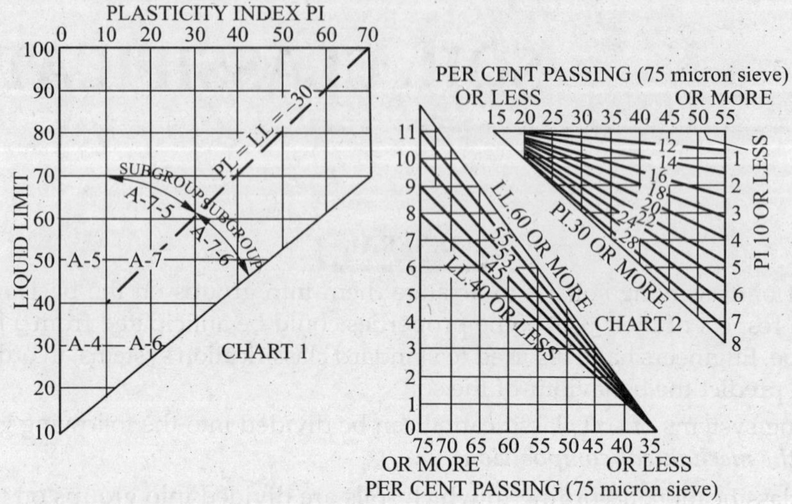

Fig. 9.1 Determination of group-index (group index is the sum of readings on a vertical scale of chart 2).

This system is very widely used for classifying soils for highway purposes. The details of this system are shown in Table 9.1 and in Fig. 9.1.

Soils are divided into two major groups, granular materials containing 35 per cent or less passing IS : 75 micron (ASTM No. 200) mesh sieve, and clay and silt-clay materials, containing more than 35 per cent passing 75 micron sieve.

The system consists of seven groups A–1 to A–7 which can be subdivided into 12 sub-groups as shown in Table 9.1. Also a system of group index numbers was devised for in-group rating of soils. The group index (GI) is an integrated term for the percentage of material passing IS : 75 micron (ASTM No. 200) sieve, liquid limit and plasticity index.

$$GI = 0.2\,a + 0.005\,ac + 0.01\,bd \qquad \ldots(9.1)$$

where a = that portion of the percentage passing IS : 75 micron (ASTM No. 200) sieve greater than 35 and not exceeding 75 expressed as a positive whole number from 0 to 40.

b = that portion of the percentage passing IS : 75 micron (ASTM No. 200) sieve, greater than 15 per cent and not exceeding 55 per cent, expressed as a positive whole number from 0 to 40.

d = that portion of the numerical plasticity index greater than 10 and not exceeding 30, expressed as a positive whole number from 0 to 20.

From Fig. 9.1, the group index is the sum of the readings on the vertical scale of chart 2.

The group index is expressed to the nearest whole number and is written in parentheses after the group or sub-group designation. For example, an A–5 soil with a group index 10 is written A–5 (10). The solution of equation (9.1) for group index is made easy and rapid by charts given in the system.

With appropriate data available, a soil is classified by comparing its properties with the specification limits shown in the table, beginning at the left side of the chart and progressing to the right. The group heading shown at the top of the first column in which the test data fit the specification gives the classification of soil.

Table 9.1

HRB, AASHO, modified PRA classification system (with suggested subgroups)

General classification	*Granular materials (35% or less passing IS 75 micron Sieve)*							*Silt-clay materials (more than 35% passing I.S. 75 micron sieve ASTM. NO. 200)*			
	A-1			*A.2*				*A-4*	*A-5*	*A-6*	*A-7*
Group classificaiton	*A-1-a*	*A-1-b*	*A-3*	*A-2-4*	*A-2-5*	*A-2-6*	*A-2-7*				*A-7.5 - A.7.6*
Sieve analysis, percentage passing											
I.S. Sieve / ASTM Sieve											
2 mm / No. 10	50 max.										
425 mm / No. 40	30 max.	50 max.	51 min								
75 micron / No. 200	15 max.	25 max.	10 max.	25 max.	35 max.	35 max.	35. max.	36 min	36 min	36 min	36 min
Characteristic of fraction passing IS 425 micron (No. 40)											
Liquid limt				40 max.	41 min.	40 max.	41 min.	41 min.	41 min.	40 max.	41 min.
Plasticity index	6 max.		N.P.*	10 max.	10 max.	11 min	11 min.	10 max.	10 max.	11 min.	11 min.
Group index	0		0	0		4 max.		8 max.	12 max.	16 max.	20 max
Usual types of significant constitutent materials	Stone fragments gravel and sand		Fine sand	Silty or clayey gravel and sand				Silty soils		Clayey soils	
General rating as subgrade	Excellent to good							Fair to poor			

Note: Plasticity index of A-7-5 subgroup is equal to or less than LL minus 30. P.I. of A-7-6 subgrade is greater than LL minus 30.

* N.P. means non-plastic.

9.2. UNIFIED CLASSIFICATION SYSTEM

Unified classification system is also widely used. This system is an outgrowth of the Airfield Classification developed by Casagrande. It was adopted jointly by the U.S. Corps of Engineers and the U.S. Bureau of Reclamation in 1952. A similar system primarily used in Britain is known as extended Casagrande classification system.

Soils are classified based on field and laboratory tests and the system recognised three main soil groups-the coarse grained, fine grained and fibrous soils within which there are many subdivisions. All soils are classified into 15 groups, each group being designated by two letters, each letter having a particular meaning as follows.

C—gravel
S—sand
M—silt
C—clay
pt—peat
H—high liquid limit
O—organic
W—well graded
P—poorly graded
U—Uniformly graded
L—low liquid limit

Combinations of above letters are used to identify the soils. For example, SP indicates a sand that is poorly graded, and CL and CH indicate clays with low and high liquid limits respectively.

The essentials of unified classification system are given in Table 9.2 and characteristics pertinent to roads and airfields are shown in Table 9.3.

Soil components used in the unified classification system are as follows.

Cabbles—above 7.5 cm
Gravel—7.5 cm to I.S. : 4.75 mm (ASTM No. 4) sieve
Coarse sand—I.S. : 4.75 mm to I.S. : 2 mm sieve (ASTM No. 4 to No. 10)
Medium sand—I.S. : 2 mm to 425 micron sieve (ASTM No. 10 to No. 40)
Fine sand—I.S. : 425 micron to 75 micron sieve (ASTM No. 40 to No. 200)
Fine—passing I.S. : 75 micron sieve (ASTM No. 200)

Laboratory tests specified for silts and clays are the determination of the liquid limit and the plastic limit. To accomplish this, the plasticity chart shown in Fig. 9.2 is utilized. Values from liquid and plastic-limit tests made on minus 425 micron (No. 40) portion of the soil sample are entered in the plasticity chart, and an appropriate classification symbol is assigned to the soil. A liquid limit value of 50 per cent distinguishes between high and low compressible soils.

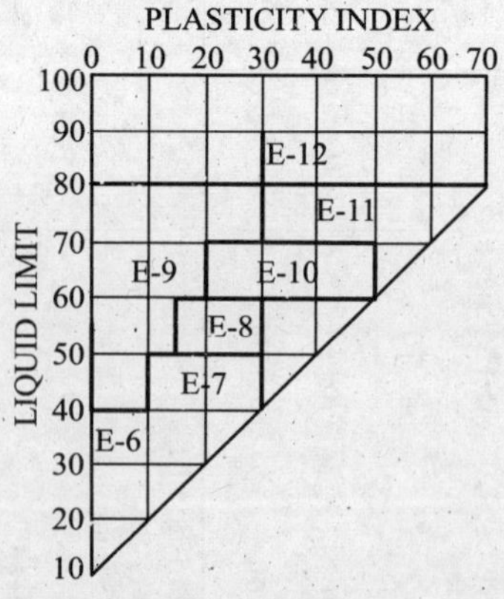

Fig. 9.2 FAA classification chart for fine grained soil.

Table 9.2 Unified classification System

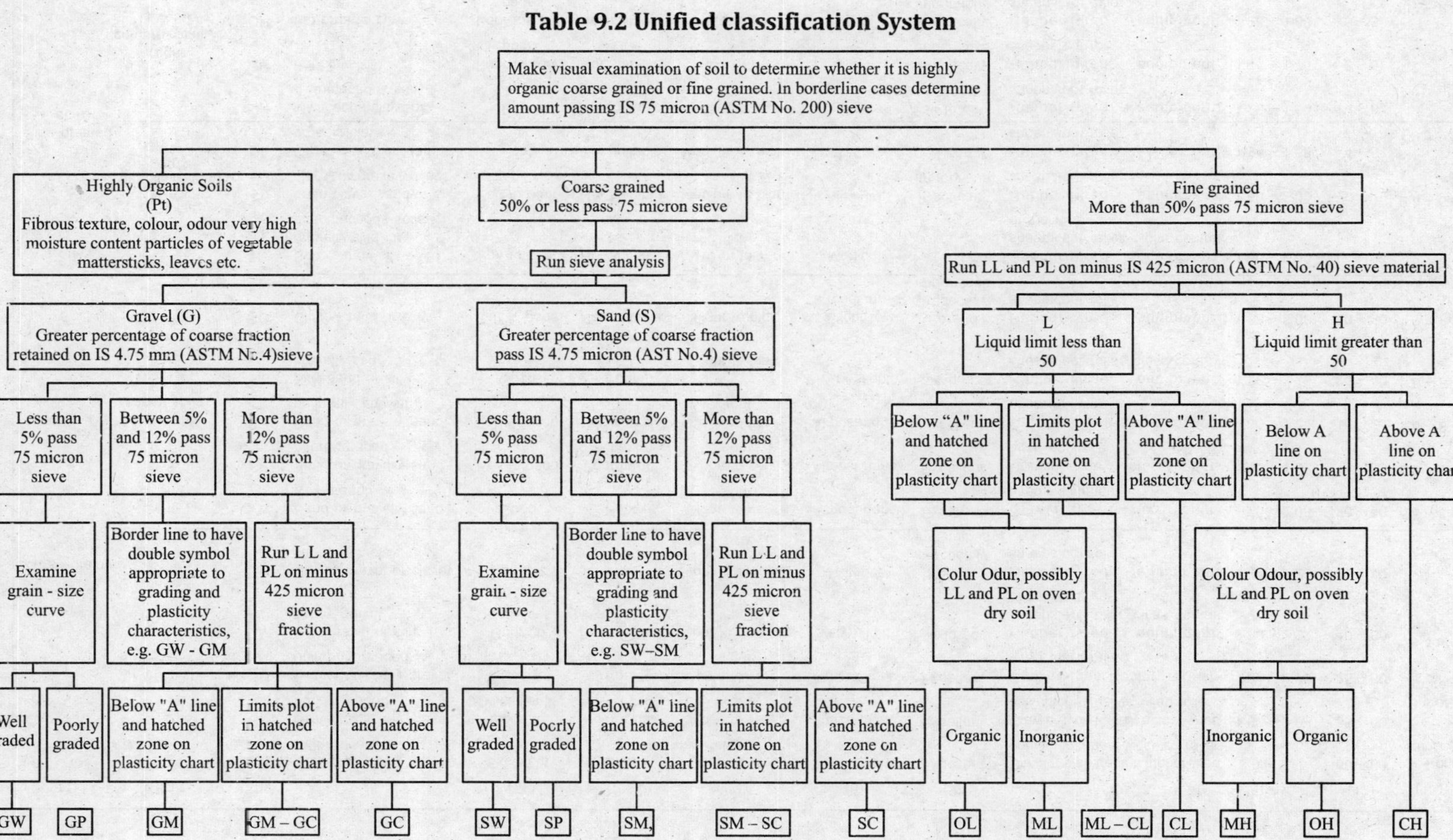

Table 9.3 Characteristics pertinent to road and runway foundation

Major	Divisions	Letter	Name	Value as foundation when not subject to frost action	Value as base direct under wearing surface	Potential frost action	Compressibility and expansion	Drainage characteristics	Compaction equipment	Unit dry weight g/cm³	Field CBR	Subgrade Modulus k kg/cm³
(1)	(2)	(3)	(4)	(5)	(6)	(7)	(8)	(9)	(10)	(11)	(12)	(13)
Coarse grained soils	Gravel and gravelly soils	GW	Gravel or sandy gravel, well graded	Excellent	Good	None to very slight	Almost none	Excellent	Crawler-type tractor, rubber-tired equipment, steel wheeled roller	2.0 – 2.25	60 – 80	8.4 or more
		GP	Gravel or sandy gravel, poorly graded	Good to excellent	Poor to fair	None to very slight	Almost none	Excellent	Crawler-type tractor, rubber-tyred equipment steel-wheeled roller	1.92 – 2.08	35 – 60	8.4 or more
		GU	Gravel or sandy gravel, uniformly graded	Good	Poor	None to very slight	Almost none	Excellent	Crawler-type tractor, rubber-tyred equipment	1.84 – 2.00	25 – 50	
		GM	Silty gravel or silty sandy gravel	Good to excellent	Fair to good	Slight to medium	Very slight	Fair to poor	Rubber-tyred equipment, sheepsfoot roller, close control of moisture	2.08 – 2.32	40 – 80	8.4 or more
		GC	Clayey gravel or clayey sandy gravel	Good	Poor	Slight to medium	Slight	Poor to practically impervious	Rubber-tyred equipment; sheepsfoot roller	1.92 – 2.25	20 – 40	8.4 or more
	Sand and sandy soils	SW	Sand or gravelly sand, well graded	Good	Poor	None to very slight	Almost none	Excellent	Crawler-type tractor, rubber-tyred equipment	1.75 – 2.08	20 – 40	5.6 – 8.4
		SP	Sand or gravelly sand, poorly graded	Fair to good	Poor to not suitable	None to very slight	Almost none	Excellent	Crawler-type tractor, rubber-tyred equipment	1.68 – 1.92	15 – 25	5.6 – 8.4
		SU	Sand or gravelly sand, uniformly graded	Fair to good	Not suitable	None to very slight	Almost none	Excellent	Crawler-type tractor, rubber-tyred equipment	1.60 – 1.84	10 – 20	5.6 – 8.4
		SM	Silty sand or silty gravelly sand	Good	Poor	Slight to high	Very slight	Fair to poor	Rubber-tyred equipment, sheepsfoot roller, close control of moisture	1.92 – 2.16	20 – 40	5.6 – 8.4
		SC	Clayey sand or clayey gravelly sand	Fair to good	Not suitable	Slight to high	Slight to medium	Poor to practically impervious	Rubber-tyred equipment, sheepsfoot roller	1.68 – 2.08	10 – 20	5.6 – 8.4
Fine grained soil	Low compressibility LL < 50	ML	Silts, sandy silt, gravelly silts, or diatomaceous soils	Fair to poor	Not suitable	Medium to very high	Medium	Fair to poor	Rubber-tyred equipment, sheepsfoot roller, close control of moisture	1.60 – 2.00	5 – 15	5.6 – 8.4
		CL	Lean clays, sandy clays, or gravelly clays	Fair to poor	Not suitable	Medium to high	Medium	Practically impervious	Rubber-tyred equipment, sheepsfoot roller	1.60 – 2.00	5 – 15	2.8 – 5.6
		OL	Organic silts or lean organic clays	Poor	Not suitable	Medium to high	High	Poor	Rubber-tyred equipment, sheepsfoot roller	1.44 – 1.68	4 – 8	2.8 – 5.6
	High compressibility LL > 50	MH	Micaceous clays or diatomaceous soil	Poor	Not suitable	Medium to very high	High	Fair to poor	Rubber-tyred equipment, sheepsfoot roller	1.28 – 1.60	4 – 8	2.8 – 5.6
		CH	Fat clays	Poor to very poor	Not suitable	Medium	High	Practically impervious	Rubber-tyred equipment, sheepsfoot roller	1.44 – 1.75	3 – 5	2.8 – 5.6
		OH	Fat organic clays	Poor to very poor	Not suitable	Medium	High	Practically impervious	Rubber-tyred equipment, sheepsfoot roller	1.28 – 1.68	3 – 5	1.4 – 2.8
	Peat and other fibrous organic soils	Pt	Peat, humus, and other	Not suitable	Not suitable	Slight	Very high	Fair to poor	Compaction not practical			1.4 – 2.8

Table 9.4 Indian Standard System of Soil Classification (IS 1498)

Division		Sub-Division		Group letter symbol	Typical names	Value as subgrade when not subject to frost action	Unit Dry weight g/cm^3	CBR Value percent
Coarse-Grained Soils More than half of material is larger than 75 micron IS Sieve Size (The 75 micron IS Sieve size is about the smallest particle visible to the naked eye)	Sands More than half of coarse fraction is smaller than 4.75 mm IS sieve size	(For visual classification the 5 mm size may be used as equivalent to the 4.75 mm IS Sieve size)	Clean gravels (little or no fines)	GW	Well graded gravels, Gravel sand mixtures; little or not fines	Excellent	2.00-2.24	40-80
				GP	Poorly graded gravels or gravel sand mixtures; little or not fines	Good to excellent	1.76-2.24	3-60
			Gravels with fines (appreciable amount of fines)	GM	Silty gravels, poorly graded gravel-sand clay mixtures	d*-good to excellent u*-good	2.00-2.32 1.84-2.16	40-60 20-30
				GC	Clayey gravels, poorly graded gravel-sand clay mixtures	Good	2.08-2.32	20-40
	Gravels More than half of coarse fraction is larger than 4.75 mm IS sieve size		Clean sands (little or no fines)	SW	Well graded sands, gravelly sands; little or not fines	Good	1.76-2.08	20-40
				SP	Poorly graded sands or gravelly sands; little or no fines	Fair to good	1.68-2.16	10-40
			Sands with fines (appreciable amount of lines)	SM	Silty sands, poorly graded sand silt mixtures	d*-fair to good u*-fair	1.92-2.16 1.60-2.08	15-40 10-20
				SC	Clayey sands, poorly graded sand clay mixtures	Poor to fair	1.60-2.16	5-20

(Contd.)

Division	*Sub-Division*	*Group letter symbol*	*Typical names*	*Value as subgrade when not subject to frost action*	*Unit Dry weight g/cm³*	*CBR Value percent*
Fine-Grained Soils More than half of material is smaller than 75 micron IS Sieve Size. (The 75 micron IS Sieve size is about the smallest particle visible to the naked eye)	Silts and clays with low compressibility and liquid limit less than 35	ML	Inorganic silts and very fine sands, rock flour, sility or clayey fine sands or clayey silts with none to low plasticity	Poor to fair	1.44-2.08	15 or less
		CL	Inorganic clays, gravelly clays, sandy clays, silty clays, lean clays of low plasticity	Poor to fair	1.44-2.08	15 or less
		OL	Organic silts and organic silty clays of low plasticity	Poor	1.44-1.68	5 or less
	Silts and clays with medium compressibility and liquid limit greater than 35 and less than 50	MI	Inorganic silts, sility or clayey fine sands or clayey silts of medium plasticity	Poor to fair	1.44-2.08	15 or less
		CI	Inorganic clays, gravelly clays, sandy clays, silty clays, lean clays of medium plasticity	Poor to fair	1.44-2.08	15 or less
	Silts and clays with high	OI	Organic silts and organic silty clays of medium plasticity	Poor	1.44-1.68	5 or less
	compressibility and liquid limit greater than 50	MH	Inorganic silts of high compressibility, micaceous or diatomaceous fine sandy or silty soils, elastic silts	Poor	1.28-1.68	10 or less
		CH	Inorganic clays of high plasticity, fat clays	Poor to fair	1.44-1.84	15 or less
		OH	Organic clays of medium to high plasticity	Poor to very poor	1.28-1.76	5 or less
Highly organic soil		Pt	Peat and other highly organic soils with very high compressibility	Not suitable	-	-

*d-Where liquid limit is ≤ 25% and P.I. ≤ 5. *u – when liquid limit is > 25% and P.I. > 5.

Laboratory tests for coarse grained soils are based on the grain size analysis. Coarse grained materials are those containing 50 per cent or less passing I.S. : 75 micron sieve; find grained are those with more than 50 per cent passing I.S.: 75 micron sieve. Gravels are those large grained materials with the greater percentage of the coarse fraction retained on I.S.: 4.75 mm sieve whereas the sands are those with the greater percentage of coarse fraction passing an I.S.: 4.75 mm sieve.

9.3. BUREAU OF INDIAN STANDARD CLASSIFICATION SYSTEM

This is a modification of the unified classification system. It consists of 18 subgroups, which are based on field identification procedure and laboratory methods as for unified classification system. The various subgroups together with other characteristics such as value as subgrade when not subjected to frost action, unit dry weight and CBR value are shown in Table 9.4.

9.4. FEDERAL AVIATION AGENCY CLASSIFICATION SYSTEM

This system was known as Civil Aeronautics Administration system and comprises 13 groups designated *E*—1 to *E*—13. This classification is shown in Table 21.1 in Chapter 21 and is based on sieve analysis of that portion of the sample passing I.S. : 2 mm (ASTM No. 10). Sieve No. 270 distinguishes fine fraction of the soil from the coarse fraction. Liquid limit and plastic limit are used to distinguish between silty and clayey fractions of the soil. FAA classification chart for fine grained soil is shown in Fig. 9.3. Its use is limited to a subgrade classification for airfield pavement design.

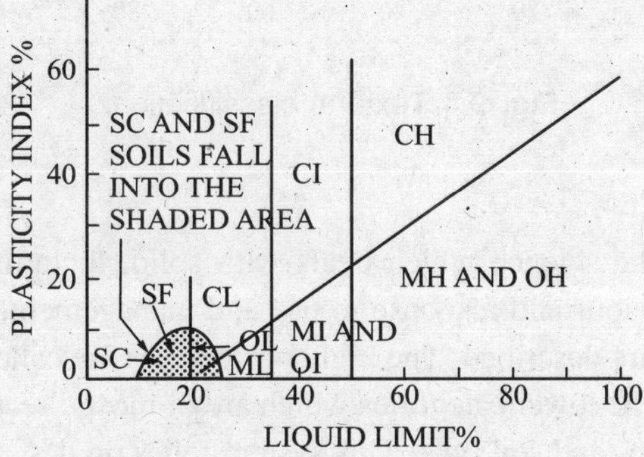

Fig. 9.3 Plasticity chart for the extended system.

9.5. COMPACTION CLASSIFICATION SYSTEM

This system was devised by Prof. K.B. Woods for use in a limited area where the soils are of similar characteristics and where the small differences in the type can be correlated by a single compaction test. This is an example of a very simple system where difference in type can be correlated with the maximum dry density as determined by the standard compaction test. This system simply divides soils into groups as shown in Table 9.5.

Table 9.5 Compaction classification

Max dry density	*General value as a foundation*
Over 2.08	Excellent
1.92–2.08	Good
1.76–1.92	Fair
1.60–1.76	Poor
1.12–1.60	Very poor

9.6. TEXTURAL CLASSIFICATION SYSTEM

A soil is composed of particles of various sizes and shapes which give it a distinctive appearance and feel and the term which describes these features is the texture of the soil. Textural classification system utilizes mainly descriptions of particle size distributions and one such chart is shown in Fig. 9.4.

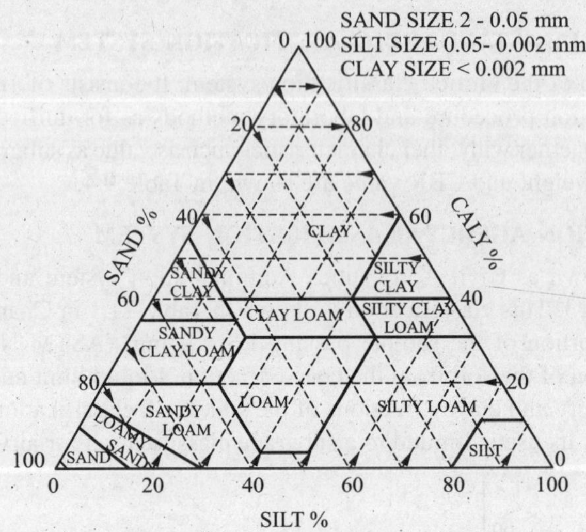

Fig. 9.4 Textural classification.

9.7. PEDOLOGICAL SYSTEM

Pedology is defined as the science which deals with solid, including their nature, properties, formation, functioning behaviour and response to use and management. In the pedological system, six main categories of soils are described. The higher categories are called the order, suborder, great soil group and family while the lower categories which are of most direct interest to the engineer are known by the terms soil series and soil type. This system relies on descriptive names to classify the soils and is based upon the classification of the parent material, scope of weathering, climate and other environmental factors.

Problem 9.1 *Classify the soil, given the following data:*

I.S. : sieve	*Percentage passing*
2.36 mm	*86*
2.00 mm	*83*
480 micron	*58*
425 micron	*58*
100 micron	*28*
75 micron	*26*
50 micron	*26*
8 micron	*8*
2 micron	*6*

$LL = 36$

$PI = 18$

Solution.

(i) Highway research board or AASHO system

The soil is classified by entering table 9.1. Since the soil has less than 35 per cent passing I.S. : 75 micron sieve, it falls into the granular materials category. Starting from left to right, it is seen that the soil cannot be placed in A–1 group as the maximum gradation and plasticity specifications are exceeded. For the same reasons, it cannot be considered in the A–3 group. The gradation meets the requirements of A–2 group, so this has to be further examined. Since the soil has a liquid limit of 36 and plasticity index of 18, it can only meet the requirements of the A–2 (6) subgroup.

The group index of the soil is calculated as follows:

$$GI = 0.2\ a + 0.005\ ac + 0.01\ bd$$

$$= 02\ (0) + 0.005\ (0)\ (0) + 0.01 \times 11 \times 8$$

$$= 1 \text{ approx.}$$

The HRB or AASHO classification for this soil is A–2–6

(ii) Unified classification system

Follow the procedure laid down in Table 9.2. Since 74 per cent (100-26) of the soil is retained on I.S. : 75 micron sieve , it falls into the coarse-grained soil category. Since 60 per cent of the particles pass the 2.36 mm sieve but is retained on 75 micron sieve, the soil falls into the sand group and is given the suffix S.More than 10 per cent passes 75 micron sieve, so the particle size curve should be examined and the plasticity index and liquid limit plotted in plasticity chart. An examination of the particle size curve shows that the soil is well graded and the plot in the plasticity chart does not fall inside the hatched zone and lies above 'A' line.

Thus the soil classification is SC.

(iii) IS : classification system

Since more than half of the total material is larger than I.S. : sieve 8, it falls under the coarse grained soils. Since 58 per cent of the soil is passing I.S. : sieve 480, more than half of the coarse grains are smaller than the I.S.: sieve 480, the soil is sandy.

Further amount of gravel present = 100 – 86 = 14 per cent

Sand content = 83 – 23 = 60 per cent

(2 mm – 50 micron)

Clay content = 6 per cent

(2 micron)

Plasticity Index = 18

From Table 9.4, the soil classification is SC.

(iv) Textural classification

Textural classification chart, shown in Fig. 9.4 can classify soils only on the basis of amount of sand, silt and clay present in the soil. When the soil contains more than 10 per cent of coarse material (gravel or stone), this must be excluded and the percentages of sand, silt and clay adjusted in order to utilize the chart.

In this example, the soil contains 17 per cent of the material retained on 2 mm sieve. In order to use the triangular chart, the percentages of sand, silt and clay are determined as follows:

$$\text{Sand (2 to 50 micron)} = 60 \times \frac{100}{83} = 72\%$$

$$\text{Silt (50 micron to 2 micron)} = 17 \times \frac{100}{83} = 21\%$$

$$\text{Clay (less than 2 micron)} = 6 \times \frac{100}{83} = 7\%$$

From the triangular chart, it is seen that the textural classification of the adjusted soil is sandy loam.

Thus the complete designation of the textural classification of the soil is "*gravelly sandy loam*".

Chapter

EARTHWORK

Earthwork encompasses all construction operations required to convert the road land from its natural condition and configuration to the sections and grades prescribed in the plans. It is also referred to as grading and drainage operations in a road project.

Choice of equipment and scheduling of operations are highly important to successful completion of earthwork.

The purposes of earthwork operation are:

1. To bring the subgrade, together with shoulders, gutters and slopes to the contour called for by the plans.
2. To eliminate unsuitable materials from the subgrade and from other portions of embankments and their foundations or to disperse such materials through the earth mass in such a manner that they will not be harmful.
3. To prevent saturation of soil in service, or if the water cannot be eliminated, to neutralize any bad effects by the use of appropriate materials that are stable even when saturated.

10.1. ESTIMATING EARTHWORK QUANTITIES

Estimates of the quantity of earthwork are essential prior to construction and are invariably calculated in the design stage of the project. Determination of earthwork quantities is usually based on the cross-sectioning method since it is most practical. In addition the cross-sections not only give information about surface configuration but they can also depict graphically subsurface conditions, drainage structures, utility lines, and other features so that their influence can also be estimated.

Cross-sections are taken at regular intervals, usually 15 m, or more depending on the general topography and also, where major surface irregularities occur. Elevations and distances from the centre line are determined at sufficient points along a line at right angles to the centre line to give a reasonably accurate picture of the ground surface. All these points are then plotted to scale along with subsurface data and the final cross-section of the proposed formation of the roadway.

Areas inscribed by the original and final cross-sections can be determined by planimeter, differentiating areas of excavation and areas of embankment. While calculating areas of excavation, distinction is made between solid rock and soil because of their potential difference in cost of excavation and also because of their influence on design.

The volume of earthwork between two consecutive cross-sections is approximated by multiplying the average of two cross-sectional areas by the distance between the cross-sections and expressing their quantity in terms of cubic metres. Correction factors for irregularities are sometimes, applied. If the composition of material is sufficiently variable, volumes of solid rock and soil should be measured and computed separately.

10.2. DISTRIBUTION OF EARTHWORK

After the earthwork quantities are calculated, analysis of relationship between cut and fill may be performed. Planned movement of materials from cuts to fills makes the construction easier and also reduce construction costs. This analysis should indicate how much material is to be moved and where it should be placed for maximum economy and utility.

As a general rule the earthwork should be so designed that all suitable materials encountered in excavation of roadway cuts will be used in the embankment. Occasionally, however, surplus excavated material will have to be wasted and in other cases material will have to be brought in from outside sources to construct part or whole of the embankment.

The simplest form of balancing excavation and embankment quantities consists of merely summing up the quantities of excavation and embankment until points are reached where quantities of excavation, adjusted for shrinkage (for soils) and swell (for solid rock), equal the amount of material required for embankment. These points are known as balance points, between which no borrow or waste should be necessary. This can also be represented by mass diagram which is plotted directly below the profile.

Even though the cut and fill quantities are approximately balanced, it may be desirable for economy reasons to waste roadway excavation and haul in material from outside sources for use in embankment. Economical haul is the distance to which material may be hauled more economically than it can be wasted or borrowed. Economical haul distances vary greatly depending on methods and type of excavation and also depending on local conditions as it may be difficult to dispose off waste in some locations, and likewise borrow may not be readily available. The limit of economical haul may be expressed as:

$$E = B/O + F$$

where O = Limit of economical haul in m

B = Cost of borrow per thousand cubic m

O = Cost of extra lead per thousand cubic m

F = Free haul distance in m.

If it is apparent, when the analysis is made, that more satisfactory material will be excavated from certain cuts that is required for construction of adjacent embankment to the standard section between limit of economical haul, the surplus material should be used for some good purpose, such as uniformly widening and flattening of slopes, approaches, filling low spots in road land, etc.

Suitable material excavated in the construction of nearby ditches, channels, structures foundations and similar features, if not needed for some purpose such as construction of protective works or back-filling structures, should also be used in embankments and should be included in the estimate while making study of economical haul.

10.3. SITE PREPARATION

This denotes clearing and grubbing operations which include removal, from the construction area, of trees, underbrush, stumps, rubbish, abandoned drainage structures and sometimes other obstructions such as buildings, fences, utility lines, etc. In any unusual situation, notes on plans should make clear just what is or what is not included.

It may be desirable to preserve trees, shrubs, etc. growing within road limits. Such trees should be carefully marked and, if necessary, protected against accidental damage during construction by putting tree guards around them. Trees to be left standing within fill slopes should have, tree-guards of dry masonry built around them. Whenever conditions permit, living trees, valuable in finished landscaping of the project, should be rigorously preserved.

The depth to which grubbing of roots and stumps must be carried usually requires that the cutting and grubbing should be carried to a depth where all brush and stumps, roots with diameter larger than

7.5 cm are at least 1 m below final subgrade elevation and should not extend more than 0.3 m above the original ground. Greater heights of stumps above ground would interfere with the operation of equipment in distribution and compacting soil properly. All holes from which stumps, etc. have been removed must be back-filled and compacted unless they lie within the areas to be excavated.

If the salvaging of timber is an important factor, a contract specially for that purpose is desirable, with the work to be completed in advance of clearing and grubbing for the road construction. All waste material, if combustible, should be removed from road land to waste areas, or preferably burnt away from immediate roadway site. Suitable incombustible material, such as old masonry, may be used for fill. In urban areas, much greater care is needed in the removal or artificial obstructions because maintaining utility services, protecting adjacent building, and numerous other considerations have to be given due attention.

After the site is prepared, it should be inspected for satisfactory condition before earthwork is started otherwise organic material or debris intended for removal may get burried.

EXCAVATION

10.4. CLASSIFICATION OF EXCAVATION

All the materials involved in excavation are classified into the following groups:

(a) *Ordinary soil.* This comprises of vegetable or organic soil, turf, sand, silt, loam, clay, mud, peat, black cotton soil, soft shale or loose moorum, a mixture of these and similar material which yields to the ordinary application of pick and shovel, rake or other ordinary digging implements. Removal of gravel or any other nodular material having diameter in any one direction not exceeding 75 mm occurring in such strata are deemed to be covered under this category.

(b) *Hard soil.* This includes:

(*i*) Stiff heavy clay, hard shale, or compact moorum requiring grafting tools or pick or both and shovel, closely applied.

(*ii*) Gravel and cobble stone having maximum diameter in any one direction between 75 and 300 mm.

(*iii*) Soling of roads, paths, etc. and hard core.

(*iv*) Macadam surfaces such as water bound and bitumen/tar bounds;

(*v*) Lime concrete, stone masonry in lime mortar and brick work in lime/cement mortar, below ground level;

(*vi*) Soft conglomerate, where the stones may be detached from the matrix with picks; and

(*vii*) Generally any material which requires the close application of picks, or scarifiers to loosen and not affording resistance to digging greater than the hardest of any soil mentioned in (*i*) to (*vi*) above.

(c) *Ordinary rock* (not requiring blasting). This includes:

(*i*) Limestone, sand stone, laterite, hard conglomerate or other soft or disintegrated rock which may be quarried or split with crowbars;

(*ii*) Unreinforced cement concrete which may be broken up with crowbars or picks and stone masonry in cement mortar below ground level;

(*iii*) Boulders which do not require blasting, having maximum diameter in any direction of more than 300 mm; found lying loose on the surface or embedded in river bed, soil talus, slope wash and terrace material of dissimilar origin; and

(*iv*) Any rock which in dry state may be hard, requiring blasting but which when wet becomes soft and manageable by means other than blasting.

(*d*) ***Hard rock* (requiring blasting).** This comprises of

(*i*) Any rock or cement concrete for the excavation of which the use of mechanical plant or blasting is required.

(*ii*) Reinforced cement concrete (reinforcement cut through but not separated from the concrete) below ground level; and

(*iii*) Boulders requiring blasting.

(*e*) ***Hard rock (blasting prohibited).*** Hard rock requiring blasting, as described under (*d*) but where blasting is prohibited for any reason and excavation has to be carried by chiselling, wedging or any other method.

(*f*) ***Marshy soil.*** This includes soil excavated below the original ground level of marshes and swamps and soils excavated from other areas requiring continuous pumping out of water.

Border line cases frequently arise in which there may be some dispute as to the proper classification of a portion of the work. Many organizations obviate this difficulty by using the term "*unclassified excavation*" to describe the excavation of all materials, regardless of their nature. This avoids controversies over the classification of weakly cemented or highly weathered rock encountered on the project and for this reason it has become customary to discard classification of rock on highway projects, particularly in roadway excavation.

Bridge excavation is an integral part of foundation construction and for purposes of earthwork, the excavation required for foundations for bridges, culverts and other structures is classified separately as "*structure excavation*" or "*excavation in foundation*". Ordinary roadway excavation and trench excavation differ substantially from one another in purposes and problems, and are considered separately.

10.5. CONSERVATION OF SPECIAL MATERIALS

The earthwork operations should be carried out in such a way that the best available materials will be saved for use in topping embankments and in subgrades. In rocky country, such materials will sometimes exist as overburden on the solid rock formations. In other cases also, the materials will be so located that they must be moved before less satisfactory finishing materials. If so, economy may dictate that they may be excavated and stock-piled until they can be placed in the desired position at the top of the embankments or in the cut subgrades, even though this will require double loading and hauling and correspondingly increased cost. Frequently, it will be desirable to select and save stone boulders or other coarse rock for masonry work or for use in constructing or repairing embankment slopes and stream channels to protect them from erosion, or for laying sealing coat, *etc*.

10.6. STRIPPING AND STORING TOP SOIL

In localities where most of the grading materials are not conducive to plant growth, it is usually a good practice to strip top soil existing over the surface of roadway cuts, embankment foundation areas and borrow-pits and store it for covering embankment slopes, cut slopes, and other disturbed areas where revegetation is desirable.

ROCK EXCAVATION

There are almost limitless varieties of rock materials in nature, and the general character in any case determines the most desirable treatment in excavation and sometimes the usefulness of the material elsewhere on the project. As already mentioned, because of variable hardness of rock met with during excavation in projects, it has become customary to discard classification of rock on highway projects, and to combine all rock and soil under the heading of unclassified excavation. However, in order to get reasonable bid prices, a certain amount of exploration in rock is required to indicate qualitative information on rock and rock-soil combinations. As the size and complicity of projects increase, the need for more subsurface data increases.

10.7. BLASTING

Blasting is usually resorted to when rock to be excavated is hard and it becomes necessary to loosen the material and reduce it to sizes feasible for loading and hauling. Blasting necessitates drilling into the rock formation to place the charge. Drilling is generally done by pneumatic equipment.

For deep holes beyond 15 or 20 m and also when large diameter holes 15 to 20 cm are required to be drilled for large quarrying operations, it is usual to use churn drills such as those used by well-drillers which offer the best results at reasonable cost. These large diameter holes are spaced about 6 m apart and the same distance back from the quarry face. The holes should extend from 1.25 to 1.75 m below the floor of the quarry, and each hole should be loaded with 12 to 20 kg of 60 per cent explosive per m of hole for not more than about two-thirds of their total depth. However, the amount of explosive used will depend upon the type best suited to various kinds of work. Blaster's handbook (E.I. dupont de nomous and company) should be consulted which gives the types and estimate best suited to various kinds of work. For ordinary rock excavation 60 or 75 per cent dynamite or blasting gelatine is generally satisfactory. Common practice is to fire 12 to 15 holes together, but sometimes greater number of holes are fired at the same time. The effectiveness of multiple blasts can often be increased and their objectionable shock can be diminished by split second time intervals between the firing of the charges with the help of millisecond delayed detonators.

Unless there is some special reason to the contrary, deep excavations can be worked most efficiently in stages or benches 3 to 9 m in thickness. In this manner each bench can be cleaned in the loading and hauling operation while drilling is progressing on top of the face that is adjacent.

For removing smaller quantities of rock and for depths of holes as great as 15 to 20 m and upto 7.5 cm diameter, wagon drills are used. Such a hole will take about 6 kg of explosive per m of length and should not be loaded more than about two-third of its total depth. For estimating, half kg of 40 per cent dynamite can be considered sufficient to break up and move a m of rock on an average.

Hand drills (jack hammers) are adequate if the holes are less than 5 cm in diameter and not greater than 3 m in depth. These small charges are particularly useful in secondary clean-up work and breaking large blocks of stone that are too heavy to handle.

Explosive should be tamped into holes with a wooden tamping pole; metal rods should never be used. The effectiveness of dynamites is increased by tamping moist soil or other non-combustible material into the hole on top of the charge. This is known as "*Stemming*" and its length should not be less than one-third the depth of the hole and half the length of the charge.

10.8. EQUIPMENT FOR LOADING AND HAULING ROCK EXCAVATION

Equipment for loading and hauling normally involves dipper shovel for loading and necessary number of trucks or rock wagons to keep the shovel occupied almost constantly. Shovel output depends upon the class of material, depth of cut, angle of swing to deposit each dipperful and also on job and management factors. Power shovel bucket capacity ranges from 3/8 cu m to 3 cu m and capacity of bucket should be selected to fit the job.

10.9. SOIL EXCAVATION

It is essential that full consideration be given for the best possible use of the soil materials encountered in the excavation or otherwise available for the construction. Any soil which may cause unstability of the subgrade or embankment or which may have some other detrimental effect should not be used unless adequately treated to make it satisfactory. Some soils which are unsuitable for certain portions of the roadbed such as the subgrade may be used in the bottom or centre of the embankment mass where their detrimental effect will be minimized.

It is not a good practice in writing specifications for earthwork to state that unstable soils shall be removed and be replaced with satisfactory material. Since in this sort of generalization, often no indication is given to the degree of stability required for the purpose in view and no measure for the stability of soils is provided either for that removed or for that by which it is replaced; such a generalization tends to become a dead letter. Instead, the practice should be to show on plans for individual projects what soils are to be removed and with what they are to be replaced, accompanying the plans where necessary with a special provision giving the methods. The scheme, adopted in each instance is based on careful soil surveys and laboratory investigations by persons who specialize in this work. The replaced material to be used may vary from rock and gravel to suitable granular soils. Sources from which such materials are to be obtained should usually be shown on the plans.

Slopes. All slopes, except in solid rock, should be finished accurately in a workman-like manner, free from humps and hollows and should conform to the slopes shown on the plan. However, during excavation, it is necessary at all times, to observe the action of the materials encountered to modify slopes, where safety or economy dictates such modification, and to recommend such changes in grade as seen proper during the process of the work.

Borrow-pits. When sufficient material for the formation of embankments and other elements of the roadway structure is not available from excavation performed within the limits of road land, additional suitable material is generally taken from borrow-pits.

All borrow-pits should generally be dug to one depth. The depth of borrow-pits should be so regulated that these do not cut an imaginary line having a slope of 1 vertical to 4 horizontal projected from the edge of the final section of the bank, as shown in Fig. 10.1. No borrow pit should be dug within 5 m of the toe of the final section taking future extension into consideration. Borrow-pits should not be greater than 0.3 m in depth,when located within 800 m of villages or towns.

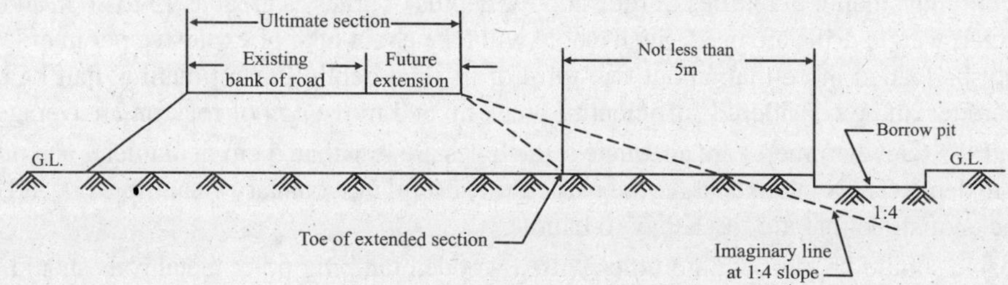

Fig. 10.1 Location of Borrow-pits

Borrow-pits should be rectangular in cross-section and one of its sides should be parallel to the central line of the road. Borrow-pits should be well drained to prevent breeding of mosquitoes. For this, the bed level should slope down progressively towards the nearest cross drain and should not be lower than the bed of the cross drain.

When it becomes essential to borrow earth from a cultivable land, the depth of borrow-pit should be limited to 1 m. The top 15 cm soil is stripped and stacked aside and the soil from further depth upto 85 cm is used for constructing the embankment. The stripped soil is spread back on the land so that the fertility of the land remains preserved.

Borrow-pits may be dredged from under water-sources if available in adequate quantities and soil is of satisfactory quality.

10.10. EQUIPMENT FOR EXCAVATION

The technique and economics of earthwork have been revolutionized during the past fifty years through mechanization. However, in some regions, particularly in India and some other Asian and African countries, hand excavation is still used on a large scale on highway projects. Hand excavation, for longer leads that can be economically hauled manually, is generally used in conjunction with sometimes animal and usually motor truck for hauling.

A good labourer can pick and lead about 3 cu m of loam, 0.2 m of clay or about 0.08 cu m of hardpan (cemented gravel) per hour. He can shovel about 0.4 to 0.5 cu m of well-loosened earth into a wagon or small truck in an hour. If the wagon or truck is high, the output is reduced. Sometimes, loosening by picking or ploughing is necessary as a preparatory step to hand excavation. A man can loosen with pick from 2 to 3.5 m of ordinary loam or from 0.3 to 0.6 cu m of hardpan per hour.

A plough drawn by two bullocks or horses will loosen about 20 to 25 cu m of ordinary load per hour. However, in stiff clay or gravel four horses or mules will be needed, and in hardpan, six.

Power excavators employ crawler tractor units or rubber-tyred tractor units. These units are generally powered by diesel engines and are built in a variety of models which have different weights and power. Light tractors may also be run with petrol engines.

10.10.1. Scraper units

Rubber-tyred self loading scrapers are the backbone of soil excavation, where haul distances are not too long and there is sufficient space for manoeuvre. This equipment will dig, load, carry, dump, and spread earth in an independent, self-sufficient operation. Basic elements of a scraper unit include a cutting edge which may be lowered into the ground to make a shallow cut; the apron in front of the bowl, which opens and closes in order to regulate the flow of earth into and out of the bowl, the bowl itself in which the earth is carried; and, in some scrapers, a tail gate or other ejection mechanism which is shoved from the rear of the bowl forward in order to push the earth out of the bowl; during the unloading operation. These units have capacities generally ranging from 0.25 to 0.6 cu m and they can easily operate at speeds of 30 km per hr. Sometimes, it is desirable or it may be necessary for a pusher tractor to assist in loading operation. One pusher tractor can serve two or more scrapers depending upon the haul distance.

10.10.2. Dozer units

Another group of earth moving tools which utilizes the tractor unit is the "*dozer family*". The bulldozer basically consists of a heavy curved blade which is mounted on the front of the tractor. The bulldozer blade, fastened to the tractor by means of two long beams, one on either side of the tractor unit may be raised or lowered by means of cable control or hydraulic means. Many modifications have been made to the basic bulldozer unit for special job applications. For example, blade may be angled from horizontal position to dig one corner into the ground when it is known as angling blade. Dozer blade may be replaced with a large, heavy circular steel plate for use as pusher tractor for scrapers as explained above. Similarly, "tree dozer" replaces blade by an extra-heavy beam for pushing a tree.

In rough or highly eroded topography it may be necessary to break down or ramp over projecting ground or gullies before scrapers can operate. Bulldozers can push earth for short distances economically less than 75 m when work can be done more economically than by scrapers. Similarly, for earthwork on hill sides which are relatively steep and where the finished section will be in combined cut and fill, probably the dozer will be very advantageous for excavating and levelling.

10.10.3. Shovel and Shovel-crane units

A tractor may be fitted with a front-end shovel and thus serve as a digging, hauling, and loading unit. The standard sizes of shovel, hoe dippers, dragline, clamshell, or orange-peel buckets are $\frac{3}{8}, \frac{3}{4}, 1, 1\frac{1}{2}$, 2 and $2\frac{1}{2}$ cu m, these being the volume of buckets or dippers at struck off level. The cranes vary in maximum capacity from 4 to 60 tonnes. Commercial models of power cranes used as excavating machines are available mounted on crawler, truck, and wheel. Wheel mounting is similar to truck mounting but utilizes the crane motor to drive the vehicle whereas the truck mounting has its own motor which is independent of the one which drives the crane mechanism. As already mentioned in rock excavation, the shovel is preferable to other types of excavators where the material is firm or hard or conditions permit the development of a good face or bank against which the shovel can work. These may also perform very useful functions in miscellaneous earthwork operations such as back filling trenches or around structures, excavation in limited areas, cleaning-up aggregate around piles.

Dargline (scoop or open bucket) is best suited for excavation in soft materials and at or below the existing ground level. The bucket works well under water, and the dragline, is therefore, useful for digging large ditches and drainage canals.

Clamshell (composed of two halves or shells) has its merit in its ability to dig deep holes extending far below the surface upon which the machine rests. Hence, it is very useful for excavating sheet excavations and in dredging wells in bridge foundations. The most common application of clamshell is in stock-piling such materials as broken stone and loading them into bins and hoppers at a central mixing plant.

Hoe (also known as back hoe) combines the digging action of a shovel with the pulling action of a dragline. It is primarily a device for excavating below ground level. It can dig harder material than clamshell but the depth to which it can dig is limited by the length of beam and dipper stick. Its principal use is for digging trenches and small cellars. It opens into three or more sections and closes to form a hemisphere rather than a semicylinder; it is very useful for digging circular excavations and shafts. It can also handle occasional humps and even a fairly large loose rock, more readily than clamshell but clamshell is more effective for digging clean earth.

The shovel-crane units mentioned above are not hauling units and if the excavated earth is to be transported beyond the range of the shovel or crane boom, this must be accomplished by a separate hauling unit which generally consists of dump trucks of various sizes depending on job conditions.

10.10.4. Loaders

These are large tractor-drawn scrapers fitted with belt conveyors that pick up earth. As the tractor moves along, a cutting edge is forced into the ground and the earth thus dug is forced onto a conveyer belt which is then carried up and to one side where it falls off the belt into a truck or wagon. They work best in sand, clay and other materials that are relatively easy to excavate. They are effective only when used in conjunction with large capacity hauling truck and having large flaring bodies affording big targets for the earth delivered by the conveyer belt. A big unit can fill as much as 0.63 cu m truck or trailer in less than a minute.

Difference in properties of soil have limited influence on equipment and methods of soil excavation. Position and land form and length of haul are more important considerations in the selection of equipment.

EMBANKMENTS

10.11. EMBANKMENT FOUNDATIONS

The success of embankment construction depends to a large extent on the preparation of its foundation. Embankment should be placed on sound foundations nearly as level as possible. If it is necessary to excavate material of lower bearing power before placing embankment, such work should be thoroughly done. Where necessary suitable benches should be graded out before embankments are begun. Investigations by means of borings will usually reveal the adverse underground conditions to be met and should be made frequently both during preliminary investigation and during construction. Prior to the beginning of construction of important embankments, the foundation condition under embankments should be given careful study.

Foundation conditions which may cause trouble if proper corrective measures are not adopted:

(*a*) Water in the form of spring and seepage.

(*b*) Foundation materials which are soft, saturated or otherwise unsuitable.

(*c*) Underground inclined slippage planes, usually lubricated with water.

(*d*) Steep slopes, especially if very smooth, wet or covered with vegetation ; and

(*e*) Steep rock slopes.

Periodical corrective measures are set forth as under:

In all, but the last of the above cases, water is nearly always the cause of part of the unsatisfactory condition. Therefore, the most essential step in the preparation of an embankment foundation is taking care of the drainage.

Solution of surface water problems is usually fairly obvious and simple but discovering the presence and source of underground water and making adequate provision to eliminate its harmful effects may be difficult. Nevertheless, these problems must not be neglected, and should be given careful attention.

Sometimes, foundation material will be encountered which is so unsatisfactory that the only effective remedy will be to avoid it or to remove it and substitute it with better material. This will particularly apply to soils such as peat, muck and some types of clay and silt. In general, materials which are unsatisfactory for embankment are like-wise unsatisfactory for foundations although in the latter case they have the advantage of being more adequately confined because of better lateral support, but they have the disadvantage that remedial action is more difficult.

Even though foundation material in its natural undisturbed condition may appear entirely satisfactory it may change considerably when the soil is subjected to the pressure of the dead weight of the embankment and to the weight, impacts and vibrations of traffic. These forces may develop severe shearing stresses, which may cause slippage on underground planes lubricated with water. This may result in sufficient compaction to cause settlements, or may even force some of the water held in films around soil particles to be released as undesirable free water in the soil.

Vertical sand drains are often a satisfactory method for draining off water situated deep under the area to be occupied by the embankment. With such drains spaced 3 to 4.5 m, water has only a short distance to travel from any point in the deposit to sand drain and thus consolidation under the additional fill is completed in a short time.

If the unsuitable material is sufficiently dry and firm, it may be removed by means of ordinary manual or mechanical methods, such as by the use of bulldozers, power shovel and scrapers, etc. Usually, however, the material will be too soft or too wet for such equipment. Dragline may then be effective. Sometimes, removal can be effected by gradually building up the fill which displaces the underlying peat or muck. In some instances, removal can be most easily and economically accomplished by blasting. Blasting will convert the organic soil to an essentially liquid condition, and the displacement of this material by the soil is facilitated. Proper planning for this method is important.

If the natural ground surface in the embankment area is steep (more than about 1 in 3) it should be stripped off sod and humus, then scarified or ploughed to provide a better bond with the embankment and avoid a slippage plane. In particularly bad cases terracing or benching of slope may be necessary to adequately key together the existing ground and the embankment to be built there on. These measures are especially necessary if the ground surface, in addition to being steep, is also hard and smooth.

10.12. SELECTION AND PLACEMENT OF MATERIALS IN EMBANKMENT AND SUBGRADES

Only suitable and satisfactory material obtained from roadway and drainage excavation, excavation from structures, or from borrow-pits, should be placed in embankment. The materials used in embankments are soil, moorum, gravel, reclaimed material from pavement, fly ash, pond ash or a mixture of these. Materials used should be free of logs, stumps, roots, rubbish or any other ingredient likely to deteriorate or affect the stability of the embankment.

If due to lack of sufficient good materials, it is essential to use some materials which are undesirable, they should be placed where their adverse effect will be minimum. These and other highly expansive material exhibiting marked swell and shrinkage properties, should be deposited in the bottom edges of the fills, and in no case be placed in embankment within 50 cm of the finished subgrade. If high capillary soils have to be used, they should be isolated from the source of capillary water by a layer of granular material. The best subgrade material should always be saved for top portion of the embankment. If the material is rocky, but contains some fines, care should be taken to place the finer material in such a way that the interstices between the rocks will be filled so that mass will be dense enough to prevent the loss of the subgrade soil which might otherwise tend to slip downward under the vibrations caused by traffic. If there is insufficient fine material to accomplish the objective throughout the depth of the fill, the fines should be concentrated as much as practicable in the top of the embankment. In some cases, it may be desirable to bring in non-plastic fine material from an outside source to form an insulating layer over the rock to prevent loss of the subgrade material.

Except highly organic materials which should never be used in embankments, soils of any description can be placed and distributed in layers conducive to densification by compaction. Unit dry weights for compacted soil at optimum moisture content for various types of soils are given in Table 9.4 in previous chapter. However, some soils are not suitable for embankment construction and subgrades. Various types of unsuitable materials are:

The following types of material shall be considered unsuitable for embankment construction.

(*a*) Materials from swamps, marshes and bogs.

(*b*) Peat, log, stump and perishable materials; any soil that classifies as OL, OI, OH or Peat in accordance with IS 1498.

(*c*) Materials susceptible to spontaneous combustion.

(*d*) Materials in a frozen condition.

(*e*) Clay having liquid limit exceeding 70 and plasticity index exceeding 45.

(*f*) Materials with salts resulting in leaching of the embankment.

(*g*) Expansive clay having free swelling index exceeding 50 percent. However, in case of non availability of suitable soil in adjoining areas, the available soil shall be suitably modified.

The materials listed above are also not suitable for subgrade construction.

Ordinarily, only the materials satisfying the density requirements given in Table 10.1 shall be employed for embankment construction.

Table 10.1 Density requirements of embankment and subgrade materials

S. No.	*Type of work*	*For NHs/SHs/MDRs Maximum Laboratory Dry Unit Weight when Trested as per IS 2720 (Part 8)*	*For Rural Roads Maximum Laboratory Unit Weight when Trested as per IS 2720 (part 7)*
1.	Embankments upto 3 metre height, not subjected to extensive flooding	Not less than 15.2 kN/cu.m	Not less than 14.4 kN/cu.m
2.	Embankment exceeding 3 metre height or embankment of any height subject to long period of inundation	Not less than 16.0 kN/cu.m	Not less than 15.2 kN/cu.m
3.	Subgrade and earthen shoulders/ verges/backfill	Not less than 17.5 kN/cu.m	Not less than 16.5 kN/cu.m

Note : Table 10.1 is not applicable for light weight fill material, *e.g.* cinder, pond ash/flyash, *etc.*

The thickness of each layer soil for embankment depends upon the character of the soil and the equipment used for soil compaction. With the usual equipment and construction methods, fine-grained and cohesive soils should not be placed in loose depths exceeding 25 cm. Granular and cohesionless materials can be placed satisfactorily upto 30 cm loose depth provided pneumatic rollers or other rollers providing kneading action are to be used for compaction.

If erosion due to wave action or water current is likely to occur, this must be guarded against by rip rapping or slope paving. It the water level fluctuates, the saturation-resistant material must extend to high-water line. If a coarse rock fill is used for embankment, the upper portion of this fill should be protected with granular material suitably graded, so that the normally dry material above, if occasionally saturated, will not flow down into the interstices of the rock. The same treatment should also be provided under stone riprap used for protection of embankment comprising of soil.

COMPACTION OF SOIL

Compaction of soil is the process whereby soil particles are constrained to pack more closely together through a reduction in air voids, generally by mechanical means. The object in compacting soil is to improve its properties, and in particular to increase its strength and bearing capacity, reduce its compressibility and decrease its ability to absorb water due to reduction in volume of voids. Due to above reasons, tendencies for volume change *i.e.,* shrinkage and swell are reduced and the soil mass becomes more uniform and less susceptible to differential settlement. Due to compaction, resistance to frost action is generally increased since heat and moisture transfers are retarded.

Compaction is measured quantitatively in terms of dry density of the soil *i.e.,* weight of soil solids per cubic m of the soil in bulk. The moisture content of the soil is the weight of moisture present expressed as a percentage of the weight of dry soil, and dry density is thus determined from the bulk density of the soil by deducting the weight of the moisture present.

Important factors which influence the increase in the dry density of soil produced by compaction are moisture content and the amount as well as method of application of the compactive effort. With a given amount of compactive effort there exists for each soil, a moisture content at which a maximum dry density is obtained. This moisture content is known as optimum moisture content.

Since the compaction of embankments and subgrade is referred to a given laboratory procedure, it is essential that compactive effort in the laboratory should be related in some way to the compactive effort imparted by equipment operating on the job. In most cases, the standard proctor compaction test (AASHO—T-test) using 0.001 cu m mould with an internal diameter of 10 cm and height of 11.5 cm is applied, unless a high degree of density is desired to be obtained and heavier compaction equipment is available on the job. In the standard test, soil is compacted in the mould with a metal rammer having 5 cm diameter circular face and weighing 2.5 kg and falling 30 cm, the soil is compacted in three equal layers by 25 blows. When heavier compaction is desired, modified AASHO compaction test using 4.54 kg rammer with height of fall 40 cm is used to compact the soil in five equal layers, each layer being compacted by 25 blows. Other methods of laboratory compaction such as kneading compactor, Harvard miniature compaction apparatus and static compression, etc. can also be specified provided they are suitably correlated with compaction obtainable in the field with the compaction equipment available.

10.13. COMPACTION REQUIREMENTS

All embankments and subgrades are compacted to certain specified densities. Table 10.2 and Table 10.3 give specification requirements for soil compaction in embankment and subgrade respectively as developed by IRC : 36-2010 Compaction requirements specified in India are given in Table 10.4.

Table 10.2 Specification requirements for embankment soil compaction

Condition 1 Fills 3 m or less and not subject to extension flooding		*Condition 2* Fills exceeding 3 m in height or fills of any height subject to long periods of flooding	
Max laboratory dry density gm per cu m	*M in field-compaction % laboratory max dry density*	*Max laboratory dry density gm per cu cm*	*M in field compaction, % of laboratory max dry density*
Less than 1.44	Not suitable	Less than 1.52	Not suitable
1.44 to 1.164	100	1.52 to 1.64	102
1.65 to 1.75	98	1.65 to 1.75	100
1.76 to 1.91	96	1.76 to 1.91	98
1.92 and more	95	1.92 or more	96

Table 10.3 Specification requirements for minimum subgrade soil compaction

Max laboratory dry density (gm per cu cm)	*Min subgrade compaction requirement (% of laboratory max dry density)*
Less than 1.65	Not suitable for top 30 cm layer
1.65 to 1.75	102
1.76 to 1.91	100
1.92 or more	98

Table 10.4 Compaction requirements for embankment

Sl. No.	*Type of work/material*	*Relative Compaction as Percentage of Max. Laboratory Dry Density as per IS : 2720 (parts 7 or 8)*
1.	Subgrade and earthen shoulders	Not less than 97 percent
2.	Embankment	Not less than 95 percent
3.	Expansive Clays (Soils having free swelling index exceeding 50% when tested as per IS 2720 part 40)	
	(*a*) Subgrade	Not allowed
	(*b*) Embankment (allowed after suitable treatment)	Not less than 95 percent

In order to achieve the desired densities, most earthwork specifications also specify layer thickness, description of equipment with regard to size, weight, method of operation, number of passes of coverage of the equipment over a given layer, etc. More over the moisture content during compaction must also be specified and carefully controlled during construction to achieve the maximum density by the selected method of compaction.

10.14. COMPACTION EQUIPMENT

In the field the soil is compacted by applying energy in one of three ways (*i*) pressure, *i.e.*, rolling, (*ii*) impact, *i.e.*, ramming or (*iii*) vibration.

Rollers include smooth wheel, pneumatic tyred, and sheep's foot rolles.

Smooth-wheel rollers. The three-wheel roller is the conventional type under this heading and has long been used in pavement construction and compaction of soils. In this type of roller, the load per cm width and the diameter of compaction roll control the pressure in the surface layer of the soil, while the dimensions of the roll affect the rate with which this pressure decreases with depth. Thus, for compaction work, it is important to specify the load per unit width as well as the gross weight of a smooth wheel roller. Present classification of these rollers generally ranges in three weight groups of 5 to 6 tonnes, and 10 to 12 tonnes. Compression on the drive rolls ranges from 25 to 70 kg per linear cm of tyre width of rear wheels. Recommendations of three wheel rollers most adaptable to compaction of various soils are listed in Table 10.5.

Table 10.5 Generalized classification of three-wheel rollers most adaptable to compaction of various soils

Soil	*Weight group and pressure, (Weight per linear cm of width of rear wheels)*
Clean, well graded sands, uniformly graded sands (one size), and some gravelly sands having little or no silt or clay	Cannot be rolled satisfactorily with three -wheel rollers
Friable-silt and clay-sand soils which depend largely on their frictional qualities for developing bearing capacity	5–6 tonnes, 25 to 45 kg
Intermediate group of clayey silts and lean clayey soils of low plasticity (PI < 10)	7–9 tonnes, 40 to 55 kg
Well-graded sand-gravels containing sufficient fines to act as filler and binder	10–12 tonnes, 55 to 60 kg
Medium to heavy clayey soils	10–12 tonnes, 55 to 70 kg

Experience indicates that smooth-wheel rollers are most suitable for compacting gravels, sands, hardcore, crushed rock and any material where a crushing action is needed. They are also used to roll the entire subgrade after it has been properly shaped.

Effectiveness of this type of roller is limited because it compacts by static compression rather than compression supplemented by kneading, tamping, or vibration which are effective in compacting soils. Because of this, modifications of smooth-wheel roller have been devised. Included in these are single drum vibrating roller, the self-propelled tandem roller with vibrating intermediate roll, segmented guide rolls, and similar other types.

Sheep's-foot or tamping rollers. Tamping or sheep's-foot rollers consist of hollow cylindrical steel drum on which projecting feet are mounted. Rollers of this type are now in widespread use for soil compaction. The drums can be ballasted either with water or wet sand to obtain the proper pressure on the feet. They are mounted either singly or in pairs on a welded steel frame. These rollers are not self-propelled, but rather are towed by other construction equipment such as dozer or other form of tractor. The performance of these rollers in compacting soil is principally affected by the feet pressure and coverage of ground obtained per pass.

Table 10.6 gives the contact pressures and sizes of tamping feet for best compacting different soils with sheep's foot rollers.

Table 10.6 Contact pressure of tamping feet best suited for compacting different soils with sheep's foot-rollers

Soil type	*Contact area sq cm*	*Contact pressure kg per sq cm*	*Remarks*
Friable silty and clayey sandy soils, which depend largely on their frictional qualities for developing bearing capacity	45–80	5–9	Based for use in compacting densities of about 95 per cent standard AASHO density.
Intermediate group of clayey silts, clayey sands, and lean clay soils which have low plasticity	40–65	7–14	Much heavier contact pressures with increased contact area are necessary if high field densities are to be produced.
Medium to heavy clays	30–50	10.5–21	

Effectiveness of the roller is dependent upon tamping action, and other things being equal, the greater the contact pressure the fewer the operations required to achieve desired density. However, contact pressure cannot be increased beyond the limits of bearing offered by the soil. That is the reason why contact area of feet is increased for friable soils as shown in Table 10.7. These rollers are most suitable for compacting cohesive soils. As the soil approaches truly cohesionless characteristics, the sheep's-foot roller become less suitable for compaction, until with completely cohesionless soils such as clean sands, disturbance rather than compaction occurs under sheep's-foot roller.

Pneumatic rollers. These rollers consist of pneumatic tyres which are inflated to maintain the air pressure in the various tyres. The actual number of wheels varies according to the make and gross weight of the roller. One type of pneumatic tyred roller known as "Webble-wheel roller', has wheels mounted at a slight angle with respect to the axle which provides greater kneading action giving improved results in certain soils. Pneumatic rollers may be self-propelled or towed by tractor.

The performance of pneumatic tyred roller is affected by contact pressure and the area of contact between the tyre and the ground. Increasing gross weight of the roller does not increase its effectiveness till it is accompanied by increase in contact pressure or an increase in loaded area.

These rollers compact the soil by kneading action from passage of pneumatic tyres as well as due to compression from passing load. Trial runs at the outset of a project may be helpful in establishing the loading of the roller, inflation pressure of tyres, and thickness of each layer most suitable for compaction of soils available on the job. The total weight of the roller can be varied to produce an operating weight to fit the soil conditions. Table 10.7 gives recommendations in respect of contact pressures for compacting different soils.

Table 10.7 Contact pressures of pneumatic tyred rollers best suited for compacting different soils

Soil group	*Contact pressure*
Clean sand and some gravelly sands	1.4 to 2.8 kg/cm^2 inflation pressure, the greater pressure with the large size tyres
Friable-silty and clayey sands which depend largely on their frictional qualities for developing capacity	2.8–4.35 kg/cm^2 inflation pressure
Clayey soils and very gravelly soils	4.36 kg/cm^2 and up inflation pressure

For cohesionless soils, relatively large contact areas are more effective because the soil must be confined in order to densify it. Semi-cohesion soils benefit from kneading action as well as pressure from tyres and for this reason such soils and also more granular materials used in sub-bases and bases are best compacted by Webble-wheel roller mentioned above.

Impact rammers. These are dropping weight, including programmers, piling equipment and pavement breakers which may be internal combustion type or pneumatic type. They compact the soil by impact forces of the dropping weight. Their main application lies in compaction of backfill and bed in trenches for drainage structures.

Vibrators. Vibrators consist of a vibrating unit of either the out of balance weight type or a pulsating hydraulic type mounted on a screed plate or roller. In general, vibration is highly effective on cohesionless soils, and relatively ineffective on fine-grained and cohesive soils. Greater depth of layer can be compacted by this method. The recommended depth of each layer for embankment is 0.5 m. However, backfill behind bridge abutments where the soil is confined at the edges may make it feasible to compact a granular soil to depths of several metres in one operation. Even vibratory smooth-wheel rollers have been used for compaction by vibration. The application of vibrators to soil compaction is still in experimental stage and due to their low output, their use is limited.

As will be clear from the above discussion, the performance of compaction plant is dependent on soil type, particle-size distribution, and also its moisture content. These factors should be taken into account in selecting compaction plant for a particular job. In general, smooth-wheel rollers are most suited to crushed rock, hard-core, mechanically stable gravels and sands; pneumatic rollers to closely graded sands, nonplastic silt and silty soils; and sheep's foot rollers to fine-grained cohesive soils. Regardless of the type of compacting equipment or the degree of cohesion of the soil, the effectiveness of the compaction procedure depends to a large extent on moisture content of the soil which should be adequately controlled during compaction.

Table 10.8 gives the general guidelines for the selection of compaction equipment for different types of soils.

Table 10.8 General guidelines for the selection of compaction equipment for different types of soil

Sl. No.	*Type of material*	*Suitability of compaction equipment for different types of soil*	*Remarks*
1.	Rock fill (except soft material)	Vibratory roller	
2.	Broken concrete, brick (burnt and un-burnt), colliery shale, fly ash etc.	Vibratory roller, Smooth wheeled roller, Pneumatic-tyred roller	Pneumatic-tyred roller to be used for fly ash only
3.	Coarse-Grained soils	Vibratory roller, Pneumatic-tyred roller, Smooth wheeled roller	
4.	Fine-Grained soils	Sheep foot roller, Smooth wheeled roller, Pneumatic-tyred roller Vibratory roller, Vibro rammer, Power rammer, Plate compactor	Sheep foot rollers are most suitable for clayey soil

NOTE : For more details of compaction procedures and equipment, reference may be made to: HRB-Special report No. 3 "State-of-the-Art: Compaction of Earthwork and Sub-grades".

10.15. COMPACTION CONTROL

The usual method of control of compaction during construction is to determine the dry density of the soil in situ in order to make certain that the specified degree of density or unit weight is achieved in the embankment or subgrade.

Main methods of in-place density measurement are (*i*) sand replacement method (*ii*) core cutter method (*iii*) volumenometer method and (*iv*) rubber balloon method. All the methods of determining dry density of the soil in place have limitations and each should be properly evaluated before it is adopted for use as a routine control measure.

Measurement of moisture content in the field can be made by determining the in-place moisture content. In controlling compaction, the rapid determination of the soil moisture content is sometimes made with proctor penetration needle. A calibrating curve between penetration resistance and moisture content for the soil is made during compaction test in the laboratory. To determine the moisture content in the field, a sample of the wet soil is compacted into the standard mould under the same condition as in the laboratory compaction test, and the penetration resistance determined. The moisture content is read off from the calibration curve. This method is particularly accurate for fine-grained cohesive soils and is not applicable to cohesionless material.

In instances where results of a large number of moisture density relations of soils are available, the use of typical moisture density curves is feasible for estimate and control, which requires only one point tests rather than tests involving a series of points giving the moisture density relation.

❑❑❑

PART – IV

11. Low-cost Roads
12. Soil Stabilized Roads
13. Macadam Roads
14. Bituminous Materials
15. Bituminous Surface Treatments
16. Carpet Coat, Road-mix and Intermediate type Bituminous Plant mix Surfaces
17. High-type Bituminous Pavements, Base Courses and their Design
18. Portland Cement Concrete Pavements and Base Courses

Chapter

LOW-COST ROADS

INTRODUCTION

We may classify the various wearing surfaces or pavements into three broad groups as follows:

1. ***Low cost surfaces:***
   (*a*) Natural soil roads.
   (*b*) Natural soil treated roads.
   (*c*) Roads of other local materials including slag.
   (*d*) Well burnt brick and brick aggregate pavements.
   (*e*) Stabilized soil roads.
   (*f*) Water-bound, traffic-bound and wet-mix macadam roads.
2. ***Intermediate-type surfaces:***
   (*a*) Bituminous surface treatment over water-bound macadam, stabilized soil or roads of other local materials.
   (*b*) Carpet coats with adequate base courses.
   (*c*) Low-cost bituminous plant mixes with adequate base courses.
   (*d*) Bituminous plant mixes with adequate base courses.
   (*e*) Cement-bound macadam.
   (*f*) Bituminous macadam.
   (*g*) Rock asphalt.
3. ***High-type surfaces:***
   (*a*) Bituminous concrete.
   (*b*) Mastic asphalt.
   (*c*) Portland cement concrete.

Roads may also be classified as rigid and flexible pavements. Flexible pavements consist of a relatively thin wearing surface built over a base course and subbase course and they rest on the compacted subgrade. On the other hand, rigid pavements are made up of portland cement concrete and may or may not have a base course between the pavement and subgrade.

The essential difference between the two types of pavements is the manner in which they distribute the load over the subgrade. The rigid pavements distribute the load over a relatively wide area because of its rigidity and high modulus of elasticity. Rigid pavements possess considerable resistance to bending and the major factor in the design is the structural strength of the concrete. For this reason, minor variations in subgrade strength have little influence upon the structural capacity of the pavement.

Pavements, other than portland cement concrete listed above which possess very little resistance in bending are known as flexible pavements.

Flexible pavements consist of a series of layers. Hence the strength of a flexible pavements is the result of building up thick layers and distribution of the load over the subgrade, rather than by the bending action of the slab.

Selection of types of bases and surface courses. The best type and combination of base and surface course in a particular case is determined principally by the following considerations:

(*i*) Volume and character of traffic.

(*ii*) The availability of funds.

(*iii*) The local availability of the various road building materials.

(*iv*) Climatic conditions.

(*v*) Nature of subgrade and drainage conditions.

However, fundamental basis is the traffic to be served. Roads carrying up to 400 vehicles per day may be considered low traffic volume roads; from 400 to 1000 vehicles per day, medium traffic volume and over 1000 vehicles per day, high traffic volume roads. Fortunately, there are almost innumerable combinations of different kinds and sizes of aggregates that will give satisfactory results, with various kinds of binders, so that full advantage can usually be taken of locally available materials.

LOW-COST ROADS

A very large kilometreage of roads in every country is not subjected to sufficient traffic to justify large expenditures for improvement, and the proper utilization of low-cost, low-type wearing surface becomes very important in their development.

A low-cost road is a road constructed at a low-cost and capable of being maintained at a low-cost. The construction of low-cost roads consists of the maximum use of material found in the vicinity of the roads. For proper utilization of local materials, the engineer must decide as to what extent the local materials will provide an adequate wearing surface for the traffic under the climatic conditions to be met. Laboratory procedures designed to test the properties of materials will help him to accomplish his goal.

Materials having a high volume change when exposed to water should not be used. The volume change may be determined by procedures used for testing soil.

11.1. NATURAL SOIL ROADS

For the lowest traffic count road, the natural soil from the sides of the road is bladed to the centre to form a crown with a ditch along each side to which water sheds. This is the lowest form of the surface used, and is the first stage in the development of a road which is to be further developed as increasing traffic requires. The surface obtained is generally dusty and ruts are quickly formed, destroying the crown so that water pockets are developed on the surface.

Distribution of traffic over the full width of road is very desirable to preserve the stability of road bed and to distribute the wear. With a moderate crown of 4 to 6 cm per m, the traffic will tend to be distributed over the entire road surface. Minimum formation width of 5 m has been recommended by IRC for village roads and 7 m for other district roads. A minimum width of 7 m is desirable.

11.2. NATURAL SOIL TREATED ROADS

A natural soil bladed road can be improved considerably by treatment with bituminous material or calcium chloride.

Many clay and silt soils are extremely dusty in dry weather and soften readily in wet weather, and the use of oils to prevent dust and to waterproof the surface has been a logical development in many areas, particularly where suitable surfacing materials for light traffic roads are not available locally and where bituminous materials are relatively cheap.

Immediately after the road has been brought to a good grade and cross-section, two litres per sq m of bituminous material such as SC-O, SC-1 or MC-0 is applied by a pressure distributor or other suitable device directly to the surface and allowed to penetrate without any cover. As soon as this application has soaked in, a second application of the same material in the amount 1 to 2 litres per sq m is applied and is also left to be absorbed without cover. This will take more time to absorb than the first application. The surface should be watched and the road may be opened to traffic when this application has been absorbed. Additional applications of 0.5 to 1 litre per square m may be applied on the road from time to time as needed. This is all that is done, except to see that the surface is maintained by patching, as necessary. Patching is done with a mix of the natural road material with the same kind of bituminous material that was used on the road. This mix is made up and stockpiled, to be used as required.

In the case of clay subgrades, good results can be obtained by treatment with bituminous material and a thin blotter of gravel or gravelly sand. With such soils it is only necessary to water-proof the surface of the subgrade and to provide a thin layer of material that will prevent the traffic from immediate contact with the clay soil. Gravel used should all pass IS : 12.5 mm sieve and 15 to 50 per cent should pass 2 mm sieve.

For treatment of highly capillary clay liable to be waterlogged, a bituminous seal with hot application of asphalt cement of penetration 250 at the rate of 2 to 3 litre per sq m is provided 30 cm below the finished surface of the road. After it is cooled, a 30 mm layer of earth is placed and consolidated and further treated with bituminous material and blotter as already explained above.

Calcium chloride treatment. Instead of completing the construction with bituminous material, calcium chloride can be used as a dust palliative. About ½ kg per sq m is applied initially, with the treatment repeated with ¼ kg per sq m as found necessary. Its value lies in its ability to absorb moisture from the air and thus keep the road surface slightly damp. It is not helpful for clayey soils as it tends to make them slippery and easily rutted; nor is it usually effective on sandy soil because the salt does not stay on the surface of open soils. Moreover, the quantities required may not work out to be economical as there is a tendency for the salt to be washed away.

Work carried out in India has shown that calcium chloride can keep the crust wet only, where the relative humidity of the atmosphere is above 31 per cent. Therefore, for areas where the relative humidity is likely to be below 31 per cent, calcium chloride as a dust palliative is not effective.

11.3. ROADS OF OTHER LOCAL MATERIALS

It is quite likely that at places where hard stone required in water-bound macadam construction is not economically available, there are large deposits of naturally occurring soft aggregates which can be used for low-cost roads. Sometimes such material is used alternately with the natural soil surface, as it occurs in the road bed. The natural surface is used where soil and drainage conditions are good. When sections of proper soils are encountered, and where drainage is poor, the better material is hauled in and spread. All these materials are low in wear resistance, may become dusty and rut under traffic. They are sometimes covered with a heavier treatment of a harder crushed stone or gravel to add to their wear resistance. They may also be improved by light oil or chloride treatment as outlined above. For higher intensities of traffic, bituminous surface treatment or carpet coat can be laid over the surface.

Suitable and easily processed local materials for low-cost road are described below:

Kankar is an impure form of limestone which exists in hard or soft form. The hard type occurs in nodular form whereas the soft variety occurs in layers. It is mostly calcium carbonate mixed with some other siliceous material like soil, etc. Good kankar is as strong as some of the stones but weak kankar can be easily broken when allowed to fall from a height of 0.5 to 1 m. This material has been successfully used in low-cost road construction in India.

Kankar, used in base courses, is 8 to 10 cm in size. For wearing courses, which are generally about 10 cm thick, the material generally used is of 63 to 53 mm size. Grading for material used for wearing course is as follows:

Total passing	IS : 80 mm sieve	:	100 per cent
	IS : 25 mm sieve	:	Not more than 10 per cent
	IS : 12.5 mm sieve	:	Nil

Kankar is genrally consolidated during rainy season, as plenty of water is needed for good consolidation. The kankar is compacted with a light roller of 6 to 8 tonnes, with profuse watering after spreading it on the base to the required thickness and cross-section. If the material is very soft, compaction is done with hand tampers. The next day following the compaction , small screenings or kankar dust is spread on the surface and compacted again. No traffic is allowed on the road till the kankar has completely set.

Moorums. Moorums are iron-stone gravels mixed with red clay, resulting from disintegration of rocks by weathering agencies.

Laterites. When recovered it is soft to cut. After exposure to the air for a few months it becomes hard due to formation of hydrated iron oxides. They are perforated and cellular in structure having deep brown red colour. It is not easy to distinguish visually between hard and soft laterites. Laterites can be used in water-bound macadam type of construction.

Slag. In regions where blast furnace slag is readily available, low cost roads may be built of slag using method similar to water-bound macadam. The depth of slag is usually 15 to 20 cm or more. To be satisfactory, slag should have per cent Los Angeles Abrasion value of not more than 50. Typical grading of coarse aggregate is as follows.

Passing:	IS 63 mm sieve	:	100 per cent
	IS : 50 mm sieve	:	90—100 per cent
	IS : 40 mm sieve	:	35—75 per cent
	IS : 25 mm sieve	:	0—15 per cent

Grading for fine aggregate or screening is as follows:

Passing:	IS : 20 mm sieve	:	100 per cent
	IS : 12.5 mm sieve	:	90—100 per cent
	IS : 150 micron sieve	:	0—30 per cent.

However, any grading as specified for crushed stone in water-bound macadam roads is suitable for slag roads.

Volcanic cinders. Volcanic cinders occur widely in many countries e.g. Japan, U.S.A. etc. and for long have been employed in the construction of low-cost roads. This aggregate is pre-coated either with a matrix of clay and water, or a bituminous sand slurry and consolidated by rolling with 3 to 5 tonne roller. Experience has indicated that these can also be used successfully as a base-coat material for a thin surface treatment or carpet coat on more heavily travelled roads. Careful selection is required.

Shells. Deposits of oyster, clam and similar shells, usually obtained by dredging, can be used as a surfacing for low-cost roads. The mud that is dredged with the shells serve as binder. When traffic is heavy, the surface should be protected by a bituminous treatment or stabilized with Portland cement.

Low-grade iron ore. Consisting of hematite or limonite is used in construction similar to water or traffic-bound macadam, and, having high natural cementing value, it compacts and binds together to form a firm and smooth surface.

Other local materials are chert gravels, and coliche which is a calcium formation consisting of sands and gravels cemented together by coats of calcium carbonate.

11.3.1. Evaluation of soft aggregates

Physical characteristics of soft aggregates vary widely. Such aggregates should, therefore, be selected carefully by specifying their physical and mechanical characteristics. Laboratory tests like Los Angeles test, aggregate crushing value test, aggregate impact value test and Deval attrition test can all be used for this purpose. The aggregate impact test is the simplest and also suitable. The apparatus for this test is portable and can be used easily in the field. CBR test may also be used for evaluation of road aggregates.

To test the materials when they are saturated, the standard aggregate impact value test has been modified and used for testing laterites under wet conditions at the Central Road Research Institute, India. In this case, the oven dried, weighed sample in the measure is immersed in water for 66 hours and is tested after surface drying.

The following values of aggregate impact test as recommended by Central Road Research Institute, India, in dry and wet state may be taken as suitable for use of soft aggregates for low-cost roads. These values are tentative and may be modified by observation of local roads in which soft aggregates have been used.

Table 11.1. Tentative limits suggested for the impact value of soft aggregates

Condition of test sample	*Type of construction*	
	Sub-base and base	*Surface course*
	Maximum aggregate impact value (per cent)	
Dry	50	32
Wet	60	39

11.4. WELL BURNT BRICK AND BRICK AGGREGATE PAVEMENTS

In India, wherever good hard road building materials are not available or are costly as they have to be carried from a very long distance, well burnt bricks have been used as base course (soling) and well burnt or over-burnt crushed brick aggregates have been used as wearing course. Well burnt or slightly over-burnt bricks have also been laid as paving material which may be protected with bituminous surface treatment or carpet coat as the increased traffic demands.

Due to necessity of using a cheaper and suitable road material in modernization work over existing kankar surface, over-burnt brick ballast produced from bricks has been used quite extensively in place of stone metal in Uttar Pradesh (India) for the last several years. Hundreds of kilometers of such roads with bituminous surface treatment have been laid and are performing quite satisfactory under different intensities of traffic.

The other alternative for the construction of low-cost roads using bricks is to lay brick pavement either direct on the subgrade where it is feasible or to use such bricks for modernization purposes by laying them on an existing kankar or other soft aggregate surface with sand or earth cushion and then leaving the surface as such under traffic and provide a bituminous surface treatment as the traffic conditions demand.

Experimental brick trackways were laid on Delhi-Mathura road during the year 1941 in order to see if such a pavement could be used to take occasional heavy motor traffic as an emergent measure in places where usual road materials may not be available and water is scarce. Bric on-edge (10 cm thick) was laid on compacted earth subgrade. Observations indicate that brick trackways gave satisfactory performance with and without surface treatment, after the joints between the bricks had been compacted with sand or earth under traffic.

In Holland, due to bad subsoil condition, the road pavement becomes very uneven due to large settlements in the first five or six years of laying a new fill. Brick roads, being cheaptest to repair, Holland still prefers to lay brick pavements on all new roads, even though the cost of brick roads is almost the same as that of cement concrete or high-type bituminous pavement.

Other advantages of brick pavements are (*i*) the surface is dust free, (*ii*) relaying can be carried out any time of the year and can be so programmed as these utilize the village labour during slack season, (*iii*) maintenance cost is low, (*iv*) does not need water for consolidation which is an advantage in areas where water is scarce or during the period of the year when water is scarce.

Selection of brick soil. Not all soils yield good bricks. Suitable soil must, therefore, be selected for making good quality bricks. The soil tests can reveal if it is necessary to blend the soils from different strata or places in order to get the right type of materials.

Table 11.2. Physical properties of soils found suitable for producing good quality bricks are mentioned below.

Properties	*Limiting values*
1. Composition Clay	15 to 30 per cent
Clay and silt	40 to 60 per cent
Sand	35 to 55 per cent
2. Atterberg limits	
Plasticity index	9—15
Liquid limit	23—40
3. Shrinkage factors	
Volumetric shrinkage	25—45
4. Classification of soil,	
P.R.A. classification	A—4 or A—6
Textural classification	Clay loam or sand loam or sandy clay loam
5. General	The soil should be free from coarse aggregates, roots and plants

Size. The size of brick used in standard size is 23 cm × 11 cm × 8 cm $\left(9'' \times 4\frac{1''}{2} \times 3''\right)$ without frog.

Over-burnt brick aggregate. The bricks used for this purpose are over burnt and not merely well burnt. A 50 kg sample should conform to the following requirements.

(*i*) The whole quantity to pass through a screen 12.5 mm bigger than the specified gauge.

(*ii*) Not more than 25 per cent to pass through a screen 12.5 mm smaller than the specified.

(*iii*) No portion to pass through a screen having square meshes 40 mm smaller than the specified gauge.

The usual thickness of base course is 10 cm loose and size of aggregate used is 40 to 50 mm.

Construction method. If brick pavements are laid on old surface foundation, any variations in the old surface are first corrected by levelling up with water-bound macadam or other suitable material. Before laying bricks, locally available soil or sand may be used as a levelling or bedding course over the old surface and also in filling the joints between the bricks.

Brick pavement may consist of flat brick (8 cm) or brick-on-edge (11 cm). Three types of bonds are usually adopted (*i*) Herring bone bond; (*ii*) Header Bond; (*iii*) Stretcher (transverse) Bond.

Transverse bond is simple to lay and also its behaviour has been observed better under traffic than other bonds and this is usually adopted.

The brick pavement may be protected with bituminous surface treatment. Brick joints may some times be filled with bituminous material, but this is however expensive and is often used for high-type brick pavement.

Regarding use of over-burnt brick aggregate, the material is spread on the road to the required thickness, proper camber and grade provided and is compacted like water-bound macadam with 6 to 8-tonnes roller. Necessary clay or other binding soil is added to fill up the voids.

As brick and brick aggregate are absorptive in nature, prime coat, as explained later, has to be used first before applying surface treatment.

Chapter

SOIL STABILIZED ROADS

Soil Stabilization is not new, for man has sought to accomplish it by various means almost since the first roads were built, but it is only in recent years that scientific methods have been applied for soil stabilization.

Definition

For the purpose of highway and runway construction, stabilization may be defined as the process of blending and mixing materials with a soil to improve certain properties of the soil. A stabilized material may be considered as a combination of binder soil and aggregates preferably obtained at or near the site of stabilization manipulated and treated with or without admixtures, and compacted so that it will remain in its compacted state without detrimental change in shape or volume under the force of traffic and exposure to weather. To perform satisfactorily, a stabilized road mixture must not dust or ravel appreciably during dry weather nor become slippery when wet. Proper construction procedure using an optimum moisture content is essential.

Thus stabilization denotes improvement in both strength and durability which are related to performance. Increase in strength may be expressed quantitatively in terms of compressive strength, shearing strength, or some measure of bearing value or load deflection to indicate the load bearing quality; and in terms of absorption, softening, and reduction in strength, or in terms of direct resistance to freezing and thawing, and drying to indicate the durability of the stabilized soil mixture.

Stabilization is a method of processing available materials for the production of low-cost roads. In this type of road design and construction, the emphasis is definitely placed upon the effective utilization of local materials, with a view to decreasing the construction cost. In some areas naturally occurring soils and soil-aggregate combinations exist which require a minimum of processing for successful stabilization, while in other places the natural soils are of an unfavourable character and require modification through the use of suitable mineral constituents such as gravel, crushed stone or soft aggregates or clay binder. In still other areas, admixtures such as bituminous materials, lime or Portland cement must be used for effective stabilization. The type and degree of stabilization required in any given case is largely a function of the availability and cost of the required materials, as well as the use to be made of the stabilized soil mixture.

12.1. NECESSITY OF STABILIZATION

The several reasons for using stabilization are:

1. Poor subgrade conditions.
2. Border line base materials.
3. Construction of superior bases.
4. Moisture control.
5. Dust control.
6. Salvaging old roads.

Stabilization used to improve poor subgrade conditions cuts down the pavement thickness. Sometimes, border line base material are available. For example those with high plasticity, and these can be made suitable by reducing plasticity index on adding lime, cement, etc. to the base. Use of superior bases such as cement stabilized and bituminous stabilized is often justified for use in both bituminous and concrete pavements. These give stiffness to the pavement and added resistance against fatigue failures. Admixtures for moisture control have been used. These retain moisture in a soil and help in getting good compaction in dry seasons. Some times chemicals are used to dry up extremely wet soils, which are otherwise, difficult to compact. Similarly admixtures for dust control can be added.

The strength of an old road can be increased by the use of cementing material. Such stabilization is often required, for example in city streets, where it is necessary to increase the strength of the existing pavement and grade lines are established and cannot be changed.

12.2. EFFECTIVENESS OF STABILIZATION

Pavement design is based on the premise that minimum specified structural strength will be achieved for each layer of material in the pavement system. Each layer must resist shearing, avoid excessive deflections that cause fatigue cracking within the layer or in overlying layers and prevent excessive permanent deformation through densification. As the quality of a soil layer is increased, the ability of that layer to distribute the load over a greater area is generally increased so that a reduction in the required thickness of the pavement layers may be permitted. Some of the attributes of soil modification/ stabilization are indicated below.

1. ***Quality improvement.*** The most common improvements achieved through stabilization include better soil gradation, reduction of plasticity index or swelling potential and increase in durability and strength. In wet weather, stabilization may also be used to provide a working platform for construction operations. These types of soil quality improvement are referred to as soil modification. Stabilization can enhance the properties of road materials and give pavement layers the following qualities:
   - (*i*) A substantial proportion of their strength is retained even after they become saturated with water.
   - (*ii*) Surface deflections are reduced.
   - (*iii*) Resistance to erosion is increased.
   - (*iv*) Materials in the supporting layer cannot contaminate the stabilized layer.
   - (*v*) The elastic moduli of granular layers constructed above stabilized layer are increased.
   - (*vi*) Lime stabilized material is suitable for use as capping layer or working platform when the in-situ material is excessively wet or weak and removal is not economical.
2. ***Thickness reduction.*** The strength and stiffness of a soil layer can be improved through the use of additives to permit a reduction in design thickness of the stabilized material compared with an un-stabilized or unbound material.
3. ***Possible problems.*** The increase in the strength of pavement layers is also associated with the following possible problems:
   - (*i*) Traffic, thermal and shrinkage cracks can cause stabilized layers to crack.
   - (*ii*) Cracks can reflect through the surfacing and allow water to enter the pavement structure.
   - (*iii*) If carbon dioxide has access to the material, the stabilization reactions are reversible and the strength of the layers can decrease.
   - (*iv*) The construction operations require more skills and control than for equivalent un-stabilized materials.

MATERIALS

The soil may be classified into two broad groups, namely, aggregate and fine soil.

12.3. AGGREGATE

Aggregate is that part of a natural deposit or of a stabilized mixture which is retained on IS : 75 micron (ASTM 200) sieve or greater than 0.75 mm diameter. In mixture, it is often referred to as the granular fraction. With the possible exception of mical flakes and peat particle, all material retained on 75 micron can be considered suitable granular materials. Aggregates may thus consist of sand, crushed stones, slag or similar materials of more or less local occurrence. In selecting the source, consideration should be given to the durability of the material.

Important properties of granular materials are (*i*) high unit weights, (*ii*) not affected greatly by moisture, (*iii*) excellent drainage properties, and (*iv*) excellent mechanical stability in wet condition.

In a stable soil mixture, the granular fraction furnishes strength and hardness and otherwise functions much the same as the aggregate in concrete. Controlling characteristics of granular materials are particle size, gradation, and the binding properties of the fine soil fraction. An aggregate well grade from coarse to fine is more stable than a poorly graded or '*one size*' aggregate.

12.4. FINE SOIL

The fine soil includes the silts and clays and consists of all the material passing 75 micron sieve. This is often referred to as binder or fine grained soil. Important properties of fine grained soils are (*i*) low unit weights, (*ii*) good supporting power when dry (*iii*) high capillarity, (*iv*) poor drainage characteristics, and (*v*) very low supporting power when they contain excessive moisture. Its resistance to deformation is dependent almost entirely on its cohesive property. The fine-soil fraction serves as a filler for the aggregate fraction, provides capillary or water-holding properties and acts as a cementing medium to bind all the particles into a strong durable mass.

Clay (finer fraction of fine-soil) due to its preferential adsorption of water in a mixture, is sometimes termed active fraction. It has two objectionable qualities that make it the most trouble-some of the materials to be dealt with. It swells when subjected to wetting and shrinks on drying.

When a clay swells, the individual soil particles are pushed apart and the internal friction between the grains is reduced. In this state the resistance to deformation depends upon the cohesive properties. This cohesion is inversely proportional to the moisture content and, consequently, is small when the clay has absorbed considerable water.

During a dry period, some of the water in the clay evaporates and the resistance of the clay to deformation is greatly improved. However, drying is accompanied by shrinkage which process tends to pull the clay particles away from the more granular material which it is supposed to bind them together. This shrinkage leaves large voids throughout the soil mass which act as passageways for surface water, the entrance of which results in a decrease in the resistance of the road to deformation.

12.5. CLAY MINERALS AND SOIL COLLOIDS

In soil stabilization, clays must always be considered as very important and often determinant soil component; therefore a sound knowledge of their structure, composition and morphology is of great importance.

The surface of every soil particle carries a negative electric charge. The intensity of the charge depends to a large extent on the mineralogical character of the particle. In nature every soil particle is surrounded by water. Since water molecules are polar, the negative charge on the surface of the soil particle attracts the positive (hydrogen) ions of the water molecules. However, in the immediate vicinity of the boundary between solid and water, the water molecules are arranged in a definite pattern. Beyond this zone, to a certain distance from the boundary, the molecular structure of the water is influenced by molecular chain action. The water located within the zone of influence is known as absorbed water. Near the surface the absorbed water has the properties of a solid; at the middle of absorbed layer it resembles a very viscous liquid and at the outer surface of zone of influence, the properties of water become normal.

The properties of soil and consequently the nature of the soil vary tremendously, depending upon the type of ion concentration on the colloids. For instance, the hydrogen colloids is strongly acidic and is easily dispersed into a colloidal suspension. The sodium colloid is strongly alkaline and is the most easily dispersed. The calcium colloid is mildly alkaline and tends to remain in the flocculated state. These facts are of particular importance in the development of soil.

As the hydrogen and sodium ion colloids are readily dispersed, they are quite mobile and under conditions of normal rainfall and normal porosity of the soil, they are entirely transplanted from the surface soil to a lower level.

Another effect of ions is that depending upon the material of which they are composed they absorb water films of varying thicknesses. At one end of the scale is potassium ion which holds only 16 molecules of associated water and at the other end is lithium which holds 120 molecules of associated water. Thus potassium clay becomes the most stable and lithium clay the most unstable. Other ions depending upon number of molecules of associated water occupy relative positive positions between potassium and lithium.

Chemical and mineralogical studies show that the particles of very fine fraction of clays are commonly crystalline and they contain chiefly silicon, aluminum, oxygen and water. The aluminium may be partly replaced by iron or magnesium, and in some instances the silicon may be partly replaced by potassium.

Colloids high in silicon adsorb thick films of water and their volume undergoes great changes upon wetting and drying. Iron and aluminium colloids adsorb thin films of moisture and, therefore undergo less volume change upon wetting and drying. The silica-sesquioxide ratio, that is the ratio of silica to the sesquioxides of aluminum and iron, is a very important chemical characteristics, as soils having colloids with low ratio are less sticky and plastic than soils having colloids with a high ratio. According to the chemical combinations, most of the minerals contained in the very fine soil fraction can be devided into three principal groups, montmorillonite, illite, and kaolinate.

Physical properties such as consolidation, swelling, shrinkage, cohesion, plasticity of very fine soil fractions (size less than 0.002 mm) are all influenced by the thickness of adsorbed layer. For example, cohesion is the shearing strength of the adsorbed layers that separate the grains at these points. When water content of fine-grained soils is reduced, the volume of voids occupied by liquid water decreases whereas the volume occupied by the adsorbed substance remains unchanged and hence the cohesion. The greater the tendency to structure formation in the clay the higher is its shrinkage limit. Also, the presence of ions furthering structure formation increases the shrinkage limit, while the presence of dispersing ions decreases the shrinkage limit. The plastic state may be considered to begin as soon as the active surfaces in the soil system are covered with films of water that are sufficiently thick to be continuous and to have lubricating properties. Therefore, other factors being equal, the larger the amount of internal surface, i.e. the larger the clay content, the higher the plastic limit. However, dispersed clays may require thinner moisture film for lubrication than aggregated clay particles. For this reason, the plastic limit sometimes falls with increasing dispersion of clay fraction.

The liquid limit represents theoretically a moisture volume which permits the independently acting primary or secondary soil particles to rotate freely. Therefore, the larger the proportion of plate-shaped particles and the larger the ratio of the volume of the rotation ellipsoids to the actual volume of particles the higher will be the liquid limits.

The practical aspect of the phenomenon of colloid science explained above is that it is possible to change the physical properties of such soils by chemical treatment which procedure is quite extensively practised if this type of soil is to be used as a construction material. Chemical stabilization is one of the several important methods of soil stabilization in connection with the highway and airport construction. Theoretically, the simplest method of stabilizing cohesive soils is to (*i*) deprive the component particles of their water affinity and (*ii*) to cement these particles together.

DIFFERENT TYPES OF STABILIZATION

Following are the various types of soil stabilization used in highway and airport construction.

1. Mechanical stabilization
2. Chemical stabilization. (*a*) lime and lime with other pozzolanas and additives, such as lime-flyash, lime-cement and lime-bitumen (*b*) calcium chloride; (*c*) sodium chloride; (*d*) resins; (*e*) lignin; (*f*) sodium silicate; (*g*) molasses.
3. Portland cement stabilization.
4. Bituminous stabilization.
5. Thermal stabilization.
6. Electrical stabilization.

Another method of classifying soil stabilization is according to the properties imparted to the soil and is indicated in Table 12.1. Types of admixtures used include cementing agents, modifiers, water proofing agents, water-retaining agents, water-retarding agents, and different chemicals. The behaviour of each of these admixtures very from that of others; each has its particular use and limitations.

12.6. MECHANICAL STABILIZATION

Mechanical stabilization is a process in which soils of two or more gradations or a soil and aggregates are mixed or blended to obtain a material meeting the required specification. The process may be carried out either at the construction site or in a central plant or in a borrow pit area. The blended material is then spread and compacted.

Mechanical stabilization has been successfully applied for sub base and base courses; as well as for surface courses for low cost roads with low traffic volume and village roads where rainfall is less than 135 mm.

The desirable properties of soil aggregates mixtures are strength, incompressibility, fewer changes in volume and stability with variations in moisture content; good drainage; less frost susceptibility and ease of compaction. The stability of a soil aggregate mix can be increased by increasing its dry density. Thus it is aimed to achieve maximum dry density.

Gradation is the most important factor in the design of the mix. The particle size distribution that gives maximum density is the one to be achieved. Each ingredient of a mechanically stablilized soil has a definite function to perform if the mass is to give satisfactory service.

12.6.1. Theories for proportioning of soil mixtures

What constitutes the best mixture of soils to form a stabilized material obviously depends on what soil property is considered to be the most important. For subgrades, strength is the most important property with incompressibility a close second. For large fills incompressibility or optimum dry density is the most important with strength a close second.

Both of these vital properties are influenced by soil density. Field experience and laboratory tests show that the strength of a soil is increased (upto a certain point) and the compressibililty is decreased by an increase in density. As a corollary to this, it is often argued that when two different soils are compacted by the same method, the denser will be the stronger and less compressible. On this basis, therefore the determination of the best soil mix for mechanical stabilization resolves itself into the determination of the mix which yields the greatest compacted density. At first glance this appears to be logical, but a careful consideration shows that the two conditions are not quite compatable. In spite

Table 12.1 Admixture stabilization

	Type	*Admixture*	*Primary mechanics of stabilization*	*Use*	*Situations best suited*	*Approximate quantity by weight*
Increase strength by cementing action		Portland cement	Principally hydration some modification of clay minerals	Base and sub-base	Sandy soils or lean clays	A–7 9 –15% to A–2 5 –9%
		Lime	Change water film, floccultion, chemical	Some base and subbase shoulders	Granular materials or lean clays	2–5%
	Cementing agents	Lime flyash	Pozzolanic action of lime and silica, some modificiation of clay minerals			2–5% lime 10–20% flyash
		Bitumen		Base and sub-base	Granular	2–5%
	Increase strength by cementing action					
Improve plasticity May or may not increase in strength		Cement	Modification of clay, change water film	Improve poorly graded base and subbase	Improve existing road metal, clays	1/2—40%
		Lime	Change water film, modification of clay	Do	Do	1/2—40%
	Modifiers	Bitumen	Retards moisture absorption	Do	Improve existing road metal	1—3%
Little or no increase in strength		Bitumen	Retards moisture absorption by coating soil grains	Primarily sub-base	Sandy soils or poor quality base materials, some clays	4—6%
	Water proofing agents	Membranes	Prevents movement of free water and water vapour	Primarily sub-base and subgrade	Soils that may be improved by compaction	
		Calcium chloride	Deliquescent properties, lowers freezing point, base exchange	Construction expedient, traffic binding	Graded aggregate	$\frac{1}{2}$—$2\frac{1}{2}$%
	Water retaining	Sodium chloride agents	Deliquescent properties lowers freezing point	Do	Do	$\frac{1}{2}$—$2\frac{1}{2}$%
Little or no increase in strength	Water retarding	Organic cationic compounds	Alters clay minerals to act as a hydrophobic agent	Sub-base		Trace

of this inconsistency, most research on mechanical stabilization has been directed towards obtaining the densest possible mixture.

Two different approaches have been followed in obtaining the maximum density. One involves the ideal gradation concept. This is illustrated in the following manner. A quantity of the largest particles are arranged in their most dense state; and the next smaller grains added are those that will just fill in the voids between the largest grains. Each succeeding smaller size is that which will fit into the voids between the next larger grains. If this process is continued to infinity, the result will be a solid mass. For any given shape of particle, the grain size curve of the ideal gradation for maximum density will follows a definite mathematical progression extending to infinity. For practical purposes this may be expressed by Fuller's formula:

$$\text{Percentage passing any sieve, } P = 100\sqrt{\frac{d}{D}}$$

where P = per cent finer than diameter 'd' (mm) in the material

D = diameter of the largest particle, mm

Another approach is the aggregate-binder concept. The soil is considered to be made up of two components; the aggregate, composed of the larger grains; and the binder, consisting of the finer grains and the clay. The aggregate should be compacted to its densest stage. The voids between the grains are filled with compacted binder to produce the maximum density. This approach was first suggested by Macadam's method of pavement construction, but it has been extended by more recent research. In soil work, there are serious objections in that the division of a soil into aggregate and binder must be arbitrary and artificial, and, although it is not too difficult to compact the aggregate alone, there is no way at the present time to introduce compacted binder into the voids of the compacted aggregate.

Particular attention has been paid to the quality of the binder component and various methods for control of the binder quality. Besides ideal dense grading, the clay portion of the mass must not be excessive or of such an abnormally active nature as to destroy the stability of the mass when subjected to fluctuations of water content.

12.6.2. Materials for soil-aggregate in sub-base and surface courses

The quality and particle-size gradations of materials that have been found to produce satisfactory results in the construction of sub-base, base and surface courses are covered by IRC specification.

According to the Indian Roads Congress (IRC), the materials for the sub-bases and shoulders should conform to one of the three grading given in Table 12.2. The mixed material should be free from organic or other deleterious constituents.

Table 12.2. Grading for granular sub-base materials

Sieve designation	*Grading 1*	*Grading 2*	*Grading 3*
80 mm	100	100	100
63 mm	90–100	90–100	90–100
4.75 mm	30–70	40–50	50–100
75 micron	0–20	0–25	0–30
CBR value (Minimum)	30	25	20

The following are the recommended values of the liquid limit and plasticity index for the material passing 425 micron Sieve, to be used for mechanical stabilization:

	Base course	*Surface course for gravel roads*
Liquid limit	25% max.	35% max
Plasticity index	6% max	5 to 10%

The gradations suggested above are intended to provide guidance in the selection or mixing of aggregates for mechanical stabilization, and is not intended to be applied as a rigid specification. The main requirement is that mixtures shall contain a fair proportion of all the particle-size fractions from the largest to the smallest with sufficient of the smallest to provide cohesion, especially in the wearing surface. Successful stabilization has been possible with gradation and consistency values different from those laid down in these specification. The considerations for deciding an economical design mixture are that locally available soil materials should be used to the best advantage and the combination should be so processed that maximum density and maximum bearing power are obtained.

Mechanically stabilized roads using soft aggregates and protected by bituminous treatment of carpet coat have given satisfactory service for a traffic intensity of about 200 tonnes per day. In areas with heavier rainfall, roads built using soft aggregate should invariably be given a bituminous surface treatment for satisfactory service. If on the surface of such roads, stone grafting with 25 mm gauge stone metal using 7 cu m per 100 sq m is rolled and surface treated, it has been found to be strong enough for a traffic intensity of 500 tonnes per day.

12.6.3. Design of Mixtures

In case the natural local material does not fulfil the requirements for a stable mixture, it may be necessary to combine two or more materials from different sources in order to secure a properly proportional mixture for the purpose at hand. There are, broadly speaking, three methods to determine the proportions in which the material should be mixed.

Trial method. This method is to make up series of mixtures, the proportions of which are based on experience with the materials. These mixtures are then tested and the one which best meets the grading and plasticity requirements is selected for use in the field. Experts may have no difficulty in using this method and hitting upon a satisfactory mix on the first or second trial. For the less experienced engineers, the other two methods have the advantage of offering guidance which may shorten the duration of the work.

Combining for sieve analysis. This method consists in proportioning the materials to be combined either graphically or mathematically by formula so that the grading of the mixture will meet the desired specifications and then a sample of the mix is tested to see if the liquid limit, plasticity index, and grading are satisfactory.

An easy and rapid method is that the difference between percentages passing each sieve for the specified grading and each of the two materials are calculated and added neglecting the sign. The ratio of the two summations is the ratio required. For example, Table 12.3 gives the gradings of the two materials *A* and *B*, the specified grading and the desired grading of the mixture. In this case, the desired grading has been taken as the average of the specification limit except that the percentage passing IS : 75 micron sieve is taken as somewhat less than the average in order to hold down the amount of binder. The desired grading is, however, selected within the specified limits keeping in view the grading of local materials in order to arrive at the most economical mix.

Table 12.3

Sieve	*Specified grading % passing*	*Desired grading mixture E % passing*	*Material A % passing*	*Difference E—A*	*Material B % passing*	*Difference E—B*	*Combined mixture % passing*
25 mm	100	100	98	2	100	0	99
10 mm	60–100	80	65	15	98	– 18	79
4.75 mm (No 4)	50–85	68	54	14	91	– 23	69
2 mm (No 10)	40–70	55	35	20	79	– 24	53
425 micron (No 40)	25–45	35	25	10	15	20	21
75 micron (No 200)	10–25	10	10	0	9	1	10
				61		86	

From Table 12.3 the summation of the differences between the desired grading and materials A and B is 61 for A and 86 for B. According to this method, the proportions of these two materials for the final mixture will be one part of B and 86/61=1.41 parts of A. Thus, the mixture will contain 100/ 2.41 = 41 percent of B and 59 per cent of A. Applying these percentages to the sieve analysis of material A and B we can find sieve analysis of combined mixture. Final mix per cent passing = 0.59 A + 0.41 B for each sieve, which is given in the last column of Table 12.3.

Following the determination of the presumptive grading of the mixture of the two materials, sample is tested for compliance with the specifications, by sieve analysis, liquid limit and plastic limit.

Rothfuchs graphical method for proportioning mateials. The method proposed by Rothfuchs is reasonably quick and simple and can be applied to mixtures of any number of components. It comprise of the following steps:

1. The cumulative curve of the required aggregate particle-size distribution is plotted, using the usual linear ordinates for the percentage passing but choosing a scale of sieve size such that the particle size distribution plots as a straight line. This is readily done by drawing an inclined straight line and marking on it the sizes corresponding to the various percentage passing.

2. The particle-size distribution curves of the aggregates to be mixed are plotted on this scale. It will generally be found that they are not straight lines.

3. With the aid of a transparent straight edge, the straight lines that most nearly approximate to the particle size distribution curve of the single aggregate are drawn. This is done by selecting for each curve a straight line such that the areas enclosed between it and the curve are a minimum and are balanced about the straight line.

4. The opposite ends of these straight lines are joined together, and the proportions for mixing can be read off from the points where these joining lines cross the straight line representing the required mixture.

Problem 12.1 *Find the proportions of the materials A, B and C (by Rothfuchs graphical method) so that the mixture may approximate to the desired grading, using the data given below.*

IS sieve Designation	*Specified grading*	*Percent passing*		
		Material A	*Material B*	*Material C*
40 mm	100	95	—	—
20 mm	85—100	70	—	—
10 mm	65—100	21	—	—
4.75 mm	55—85	11	100	—
2.36 mm	40—70	7	85	—
425 micron	25—45	2	35	—
75 micron	10—25	Trace	nil	100

Solution.

IS sieve designation	*Percent passing*		*Percent (from graph)*		
	Specified grading	*Desired grading (average)*	*Material A*	*Material B*	*Material C*
40 mm	100	100			
20 mm	85-100	92.5			
10 mm	65—100	82.5			
4.75 mm	55—85	70			
2.36 mm	40—70	55	37	45	18
425 micron	25—45	35			
75 micron	10—25	17.5			

Draw the Rothfuchs graph, as outlined above, (Fig. 12.1) taking the average of specified grading and plotting it as straight line. Read the proportions of materials *A*, *B* and *C* for mixing from the graph. The proportions are 37 per cent *A*, 45 per cent *B* and 18 per cent *C*.

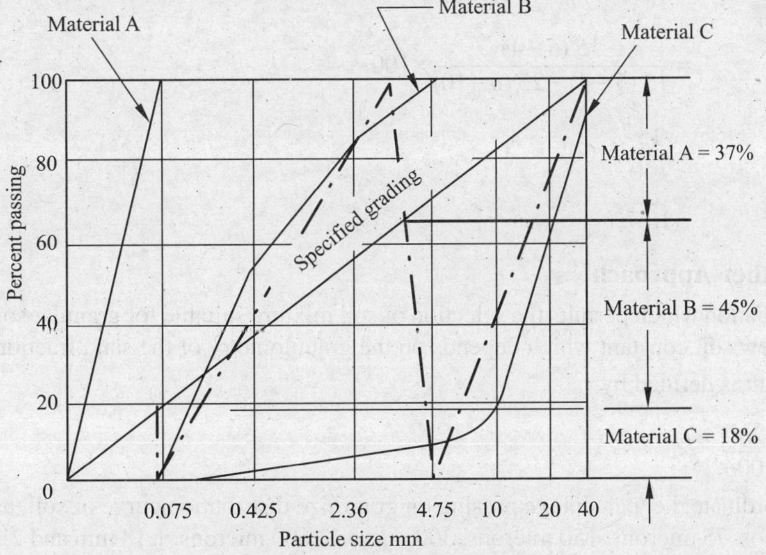

Fig. 12.1. Rothfuchs graph.

Combining for plasticity index. The third method is to proportion the materials by means of formula or with the help of tables so that plasticity index of the mixture may be expected to be within the allowable range. Then a trial mixture is made and tested to determine if the plasticity index is as expected, if the liquid limit is within the allowable limits and if the gradation is reasonably close to the specification.

Let there be two soils A and B which are to be proportioned to get a mixture of required plasticity index P.

Determine the plasticity index of the two soils. Let these be P_A and P_B for the soils A and B.

Determine the percentage passing 425 micron sieve by sieve analysis. Let these be S_A and S_B for the soils A and B respectively.

Then the percentage of soil A to be mixed with soil B to get the required plasticity index P is given by:

$$\text{Soil } A\% = \frac{S_B(P-P_B)}{S_B(P-P_B)-S_A(P-P_A)}$$

Problem 12.2 *Determine the percentage of soil binder which should be added to the road aggregates, from the following data:*

P.I. of binder soil = 10

P.I. of road aggregate = 4

P.I. of desired mixture = 5

Percentage of aggregate pasing 425 micron sieve = 15

Percentage of binder passing 425 miron sieve = 25

Solution.

Here $P = 5, P_A = 10, P_B = 4$

$S_A = 25; S_B = 15$

$\therefore$ Percent of binder A to be mixed with aggregate B is given by

$$\text{Binder } A\% = \frac{S_B(P-P_B)}{S_B(P-P_B)-S_A(P-P_A)}$$

$$= \frac{15\,(5-4)}{15\,(5-4)-25\,(5-10)}\times 100$$

$$= \frac{15}{140}\times 100$$

$$= 10.7\%$$

12.6.4. Another Approach

Another equation which permits the selection of soil mixture suitable for granular soil stabilization introduces a new soil constant which depends on the granulometry of the sand fraction.

This constant is defined by

$$a = \frac{\Sigma y}{100n} \qquad \ldots(12.1)$$

in which y = ordinate, i.e. percentage passing on grain size distribution curve of soil, using IS : sieve Nos. 75 microns, 150 microns, 300 microns, 600 microns, 1.18 mm and 2.36 mm sieves.

n = number of ordinates (six)

The characteristics of a mixture of two or more soils can be determined by the general expression:

$$K = \frac{A_1 a_1 k_1 + A_2 a_2 k_2}{A_1 a_1 + A_2 a_2}$$

in which K = characteristics property of the mixture (liquid limit, plastic limit, shrinkage limit, plasticity index, linear shrinkage, field moisture equivalent)

k_1 and k_2 = corresponding characteristics of the soils to be mixed.

A_1 and A_2 = weight percentage of the soils in the mixtures.

a_1 and a_2 = constants defined by the equation (12.1).

The values of plasticity index and liquid limit of sandy soils which due to its granular nature cannot be determined directly can be derived from the respective values of mixture of equal parts of sand and the reference clay. For example, if the reference clay used has liquid limit = 61, plasticity index = 35 and constant a_1 = 0.98 and constant a_2 for sandy soil = 0.48 and the mixture of 50 per cent clay and 50 per cent sand gave L.L. = 46 and P.I. = 25, then

$$46 = \frac{0.50 \times 0.98 \times 61 + 0.50 \times 0.48 \times \text{L.L.}}{0.50 \times 0.98 + 0.50 \times 0.48}$$

giving liquid limit of sandy soil =15, and

$$25 = \frac{0.50 \times 0.98 \times 35 + 0.50 \times 0.48 \times \text{P.I.}}{0.50 \times 0.98 + 0.50 \times 0.48}$$

giving plasticity index of sandy soil = 4.

12.7. STABILIZATION WITH LIME

The use of lime for strengthening granular soils and sands is one of the oldest methods known to man. Such materials have been used in the construction of the Appian Way in Rome, massive footings for bridge piers in India and China, and plasters throughout the world. But it is only in the past few years that some outstanding advancements have been made in the use of soil lime mixtures.

Lime is a broad term which is used to describe:

(*i*) Quick lime – CaO or CaO . MgO

(*ii*) Slaked or hydrated lime – $Ca(OH)_2$

(*iii*) Carbonate of lime – $CaCO_3$

Commercial lime is manufactured by heating a crushed carbonate rock such as lime-stone, $CaCO_3$ to above 1100°C, causing release of carbon dioxide, CO_2, and leaving quick lime, CaO. A second type of carbonate rock termed dolomite is often used for lime manufacture. Dolomite consists of equal molar parts of $CaCO_3$ and $MgCO_3$. The resulting lime, is a mixture of CaO and MgO.

$$CaCO_3 + \text{heat} = CaO + CO_2$$

$$CaCO_3 + MgCO_3 + \text{heat} = CaO + MgO + 2\,CO_2$$

Quick lime is lime composed of calcium or calcium and magnesium oxides. Quick lime comes from rotary kilns is pebble form (6.3 to 0.63 mm) whereas vertical kilns produce quick lime in lumps (15 to 20 cm). Both types are then ground and sieved prior to slaking or use as powder.

Quick lime reacts readly with water to produce slaked lime or hydrated lime in which all the CaO is converted to $Ca(OH)_2$.

$$CaO + H_2O = Ca(OH)_2$$

Following hydration, limes are usually passed through a hammer mill or ring-roll mill and graded with an air separator. Standard hydrated lime has 99 per cent passing 600 micron sieve and 90 per cent passing 75 micron.

Hydraulic lime is obtained by calcination of impure lime stones containing 15 to 24 per cent clay. The resulting mixture will set under water, but retains some of the plastic properties of lime. It may be regarded as an intermediate between lime and Portland cement. Hydraulic limes are widely used in some countries.

Calcium carbonate is not used for stabilization, although it is used in agriculture as a soil additive to adjust pH value.

12.7.1. Reaction of Lime on Soil

Addition of lime to soil generally results in the following:

(a) Changes the plasticity properties of soil. Calcium ions reduce plasticity of cohesive soils so that they become more friable and more easily workable. Changes in effective grain size distribution are observed almost immediately following the addition of lime to a clayey soil. The major change occurs within the first hour. The new grains produced as a result of lime treatment are mostly silt or sand sized though these are relatively weakly bonded. Aggregation is caused by addition of 1 or 2 per cent lime.

As an example of effectiveness, on one project the addition of 2 to 3% lime produced a material with *PI* averaging 8 from untreated soil having *PI* ranging from 15 to 20. In another instance, *PI* was lowered from 44 to 16. Experiments conducted showed that addition of lime lowered *PI* from about 25 to 17 and in another case from about 10 to 5.

This reaction is due to alteration of the water film surrounding the clay minerals. The strength of the linkage between two clay minerals is dependent on the charge, size, and hydration of attracted ions. The lime (calcium ion) is divalent and serves to bind the soil particles close together. This in turn decreases plasticity and results in a more open and granular structure.

(b) Results in decreased soil density and delay in compaction. The moisture content needed to achieve maximum density for a given compactive effort usually increases sometime rather significantly. Lime in excess of about 5 per cent by weight of soil, generally produces little additional increase in optimum moisture content.

(c) Increases soil strength due to reaction of lime with soil components to form new chemicals. The two principal components of soil which react with lime are alumina and silica. This reaction known as pozzolanic action results in a slower long-term cementation of compacted soil-lime mixtures. This is somewhat similar to hydration of cement. However, soil-lime and soil-cement differ primarily in the rate of strength gain during curing. Soil-cement mixes gain strength rapidly during curing process, but soil-lime mixes show increased strength over relatively long periods of time. This slow setting provides more flexibility in road construction.

12.7.2. Lime Specifications

Pozzolanic reactions are extremely sensitive to chemical class of lime, and the type desired should be specified in some detail. Lime used for stablisation is calcium hydroxide (hydrated lime) or calcium oxide (quick lime). The requirements for lime are given below:

	Quick lime (CaO)	*Hydrated lime* ($Ca(OH)_2$)
Calcium and magnesium oxides	not less than 92%	not less than 95%
Carbon dioxides-at kiln	not more than 3%	not more than 5%
elsewhere		not more than 7%

When lower quality lime produced from kilns is used in place of quality lime, quantities are suitably increased.

The choice of quick lime versus hydrated lime is a matter of convenience and economy. Quick lime is cheaper since it contains no water. However, its caustic nature will require special consideration in handling, since it will attack equipment corrosively and precautions are also needed against the risk of severe skin burns to personnel. Suitable handling methods such as fully mechanized or bottom dump handling equipment, or protective clothing worn by the operators are used. Wind direction should be taken into account to minimise the dust problem and consequent eye or skin irritation to any personnel involved or in the vicinity.

Dust in case of quick lime can also be minimized by the use of coarse quick lime and a light sprinkling of water prior to spreading. The dust and steam hazard during slaking of quick lime can be virtually eliminated by applying water in a fine mist in several applications.

Another method used for quick lime is to hydrate it in a tank and spray as a slurry. The amount of water in slurry is adjusted to give the optimum moisture content in a slurry.

Hydrated lime, on the other hand, can be purchased in bulk and spread with spreaders, purchased in bags and spread by hand, or applied in a slurry.

Because both quick lime and hydrated lime tend to recarbonate readily by combination with carbon dioxide from air, storage should be in tight containers and under dry conditions. Specifications usually cover the maximum carbon dioxide contents in both quick-lime and hydrated lime.

12.7.3. Soil Type and Quantity of Lime

Data available in various countries on proper mix proportions for different kind of soils suggest the following guidelines on the use of lime for stabilisation.

(a) *Fine-grained soils*

Effectiveness of lime to stabilize fine-grained soils, depends by and large, upon the type of clay minerals in the soil. Montmorillonite, for example reacts very quickly with lime, and the clay mineral loses its plasticity at an early stage. The quantity of lime required to stabilize most soils will vary between 4 to 10 per cent by weight. Perhaps the most beneficial effect of lime in fine-grained soils is the reduction in the plasticity index, but pozzolanic action also has its effect. Lean clays and silty clays may or may not be pozzolanic in nature, depending upon the composition of the soil minerals.

(b) *Sandy soils*

As a general rule, sands do not react with lime because they do not contain clay minerals. Sand-clay mixtures can be stabilized satisfactorily depending upon the pozzolanic nature of the clay minerals in the mixture.

(c) *Gravels and gravel mixtures*

As was the case for fine-grained and sandy soils, the adequacy of stabilizing gravel mixtures with lime is dependent to a large extent upon the type of minerals in the mix. Generally with such soils a substantial reduction in plasticity index is achieved and great increase in compressive strength is obtained due to pozzolanic reaction. In addition, the gravel or coarse material creates a satisfactory gradation so that high stability is obtained. For such soils only small quantities of lime, about 1 to 4 per cent by weight are needed and thus with small extra cost, it is possible to convert an otherwise unsatisfactory local base material into a highly satisfactory mix.

The gradation limits and the material characteristics for lime-stabilised soils are given in Table 12.4.

Table 12.4 Material Characteristics for Lime/Modified Soils

Properties	*Specified Value*
Passing 75 mm sieve	100%
Passing 26.5 mm sieve	95-100%
Passing 75 micron sieve	15-100%
Plasticity index	≥10
Organic content	< 2%
Total SO_4 content	< 0.2%
Minimum lime content	2.5%*
Degree of pulverisation	> 60%
Minimum temp. for mixing in field	10°C
Unconfined compressive strength (MPa)	As per Contract Specifications

* In case better mechanical equipment for spreading of lime, for breaking clods and blending is used, the minimum percentage of lime for stabilization could be 0.5 percent. However, extensive lab testing must be done to arrive at this minimum percentage. Sample at site of blended loose soil be collected and remoulded in lab to confirm that the desired CBR can be achieved.

12.7.4. Evaluation and Criteria for Mix Design

Preparation of trial mixes in the laboratory is done simulating field mixing. Plasticity tests can be performed on samples of the material with and without the lime to evaluate the benefits derived from the addition of lime for the purpose of reducing plasticity. These tests can be made immediately after mixing lime, as well as after a waiting period to give lime time to react with the soil.

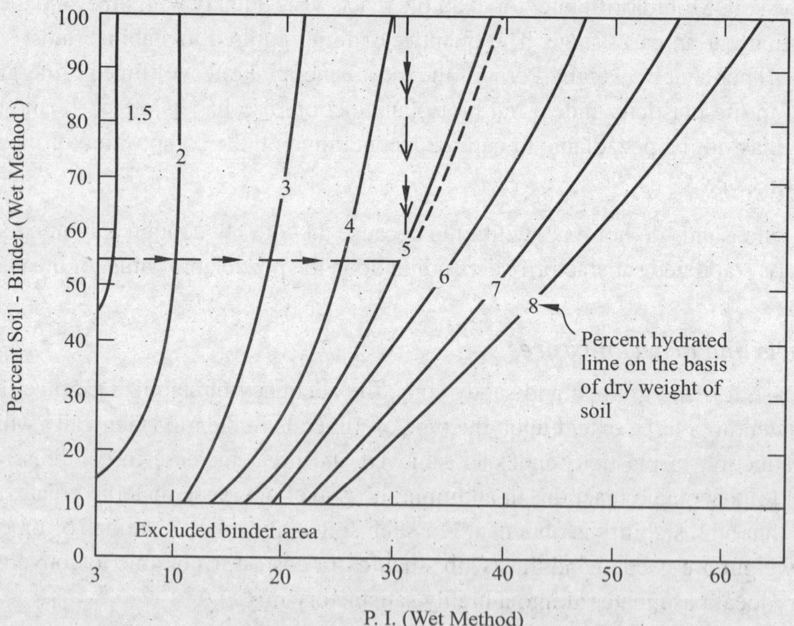

Fig. 12.2 Chart for estimating required lime contents.
(From AASHTO test designation T-220).

Recommended amount of lime for stabilization of subgrades and bases have been given in AASHTO standard T 220. Fig. 12.2 is based on the plasticity index and grain size distribution of the soil. Enter P.I. and percent minus 425 micron sieve of untreated soil on Fig. 12.2 and determine the percentage of lime to be added to the soil.

The percentage of lime for a given project generally are determined by testing lime-soil mixtures using the unconfined compression test. AASHTO T 220 recommends that generally, an unconfined compressive strength of 7 kg/cm^2 is satisfactory for final course of base construction and it is desirable that materials for such courses contain a minimum of 50 plus 425 micron (No. 40) material before treatment. It further recommends that various soil materials may be treated for subbase and, in such cases, the minimum suggested unconfined compressive strength is 3.5 kg/cm^2. Roads and bridges specification require the compressive strength between 7 and 20 kg/cm^2 at 7 days.

Other strength tests used to evaluate soils lime mixtures are triaxial or CBR test. These strengths are correlated with field performance to establish safe minimum requirements.

Often increase in strengths is relatively small by addition of lime beyond a certain percentage, the optimum being in the range of 2 to 10 per cent.

In freezing climates, durability is the major requirement. One method for measuring durability is to measure the decrease in unconfined compressive strength after cycles of freeze-thaw. A durability ratio, defined by British Road Research Laboratory, is the strength after weathering divided by the strength obtained by curing for the same length of time. A ratio of 80 per cent after 74 cycles of freezing and thawing is regarded as satisfactory.

12.7.5. Salient Features of Lime Stabilization

To sum up, some of the salient features of stabilization with lime are:

(1) Lime stabilization is effective for clayey soils including heavy clays, moorum and other soils met in the alluvial plains. For better results, a soil should have a fraction passing 425 micron sieve not less than 15 per cent and Plasticity Index at least 10 per cent.

(2) The percentage retained on 425 micron sieve should be well graded with uniformity coefficient not less than 5.

(3) Organic matter in the soil selected for soil stabilization should not be more than 2 percent and sulphate content should not exceed 0.2 percent.

(4) pH value of 10 or 11 is desired for pozzolanic reaction to take place between clay minerals and lime for the formation of cementitious compounds.

(5) Soils having organic matter and soluble carbonate/sulphate contents in excess of 2.0 percent and 0.2 percent respectively require special studies.

(6) Some materials contain amorphous silica which although has low plasticity but reacts with lime to form the necessary cementation products and should thus be considered for stabilization with lime.

(7) Materials containing high Kaolinite as the basic clay mineral usually have a fairly low PI with a high liquid limit and in such cases lime should be considered for stabilization.

(8) In case of highly plastic soils, two stage stabilization is adopted. In this case soil is first treated with a small quantity of lime. Later on the soil may be treated with remaining quantity of lime or with cement to achieve the desired strength and stability.

12.8. STABILIZATION WITH LIME POZZOLANAS

Pozzolana is defined as a silicious or alumino-silicious material which in itself possesses little or no cementitious value but when added in finely divided form and in the presence of moisture will chemically react with alkali and alkaline earth hydroxides at ordinary temperatures, to form or to assist in forming compounds possessing cementitious properties.

Some natural materials such as pumice or volcanic ash react much more readily with lime than do ordinary soils. These reactive material are named "*pozzolanas*". Pozzolanas characteristically have a glassy or non-crystalline ionic structure. The most abundant natural pozzolans are volcanic materials such as volcanic lava which is sometimes so porous that it will float on water. It is pulverized before use as a pozzolana.

The other major class of natural pozzolanas depends for its activity on opal, or amorphous hydrous silica. Diatomaceous earth is composed of opaline shells or microscopic plants called diatoms. One of the most active pozzolanas is opaline chert, a hard siliceous rock which is ground for use as a pozzolana.

Pozzolanic materials vary in activity, depending mainly on the degree to which they are crystalline since crystallinity reduces activity. With the exception of some volcanic dusts, all natural pozzolanas are ground before use to increase the area of reactive grain surfaces. A surface area of at least 20,000 sq cm per gm (Blaine method) is sometimes recommended for pozzolanas for use in concrete.

Lime-Flyash. One type of pozzolana in particular that has created considerable interest in recent years is flyash which is the grey dust-like ash and is the residue from burnt powdered coal. This is a waste material from coal burning plants.

The cheap availability of flyash in India has increased tremendously as a waste product from Thermal Power Plants. Lime-flyash will provide a useful means of disposal for a huge amount of this waste product. Table 12.5 depicts the physical requirements for Flyash as a Pozzolana.

Table 12.5 Physical requirement for flyash as a pozzolana

Sl. No.	*Characteristics*	*Requirement*
1	Fineness-specific surface in m^2/kg by Blaine's permeability test, Min	250
2.	Particles retained on 45 micron IS sieve, Max	40
3.	Lime reactivity in N/mm^2, Min	3.5
4.	Soundness by autoclave test expansion of specimen in per cent, Max	0.8
5.	Soundness by Lechatelier method-expansion in mm, Max	10

Lot of laboratory and field studies regarding stabilization of fine and coarse-grained soils with lime-flyash admixtures have recently been done and the following conclusions may be drawn from existing literature.

(*i*) The addition of small amount of hydrated lime and flyash develops concrete like composition of high strength at relatively early ages. When compacted with optimum moisture, unconfined compressive strength of the order of 70 kg/cm^2 in 28 days is developed. When cured at higher temperature, which accelerates the pozzolanic reaction of lime and flyash, the compressive strength obtained in 7 days is of the order of 25 to 100 kg/cm^2 depending upon the type of the soil.

(*ii*) The weathering resistance of the composition appears to be exceptionally good. The test of wetting and drying, and freezing and thawing show excellent resistance to deterioration after 12 cycles of treatment.

(*iii*) With addition of lime and flyash, maximum strength is obtained when compacted at a moisture content equal to or slightly less than the optimum.

(*iv*) Strength increases with the degree of pulverization.

(*v*) Strength increases with increase in curing time.

(*vi*) Elevated temperature curing greatly increases strength.

(*vii*) Specimen sealed from contact with atmosphere gives greater strength after six months or more than unsealed specimen.

(*viii*) Small percentage of calcium chloride gives substantial increase in strength of stabilized clayey and sandy soil.

(*ix*) Addition of lime-flyash improves the consistency limits and shrinkage properties of soil.

12.8.1. Lime and Lime-pozzolana Soil Stabilization Mix-design

The choice of lime versus lime-pozzolana stabilization depends on economic factors, kind of soil, climate and use. Selection of proper mix proportions for soil lime pozzolana is complicated by the fact that there are three materials. Fortunately, the mix proportions are seldom critical for a given soil, and many mixes could be used. The problem is then to select proportions which given the needed amount of cementation for the lowest cost. Trial batches are mixed in the laboratory, compacted, cured, and tested, so that the best mix may be chosen.

One way to short-cut the testing programme is to eliminate all uneconomical mixes. First a maximum allowable lime percentage is selected which is economically competitive with other types of construction, such as soil cement, crushed stone, *etc*. As a hypothetical example, this is plotted as point P in Fig. 12.3. Next the cost of handling pozzolan is estimated and expressed as its equivalent in percentage lime. This is subtracted from P to give point Q. Starting at point Q, an equal-cost line is drawn with a negative slope equal to the cost of flyash divided by the cost of lime, both, on a total – dry mix basis. Trail mixes are then selected from the area below this line, since proportions above the line are uneconomical. A second limitation which can be imposed is to require a minimum 3 percent lime, since lower lime contents may lead to lean areas in the field construction from imperfect mixing. This limit is represented by line RS. A lower minimum content may be permissible if lime is applied in a slurry.

The selection of trial mixes within triangle QRS is partly a matter of judgment. If maximum strength is desired, equal-cost points are selected at P and along the line QS, *e.g.* 10 per cent lime, 90 per cent soil (pt P); 7 per cent lime, 8 per cent pozzolana and 85 per cent soil; then 79:5:16 and 73:3:24. Intermediate points can be filled, if desired. Ordinarily one of these mixes will give the highest strength and durability. If the resulting strength is excessive for the proposed use, costs can be cut by using less pozzolana, less lime or less of both. For example, if T and S are found to be equally over designed, the more economical ratio 82:3:15 (pt. U) could be tried. Or, if T is the best mix; trials could be made at point intermediate to T and V, thus maintaining the same lime: pozzolana ratio. If none of the trial mixes gives satisfactory strength and durability, the lime or pozzolana or soil type may be at fault or chemical accelerators such as sodium carbonate or calcium chloride can be tried.

Under certain circumstances stabilization with lime alone will be most satisfactory and economical. For example, if economy dictates a lower equal cost line and thus limits the cost triangle to low percentages of pozzolana, stabilization should be given first consideration. Trail mixes may be prepared with different percentages of lime, often strength are increased relatively little by additions beyond a certain percentage of lime, the optimum being in the range 2 to 10 per cent.

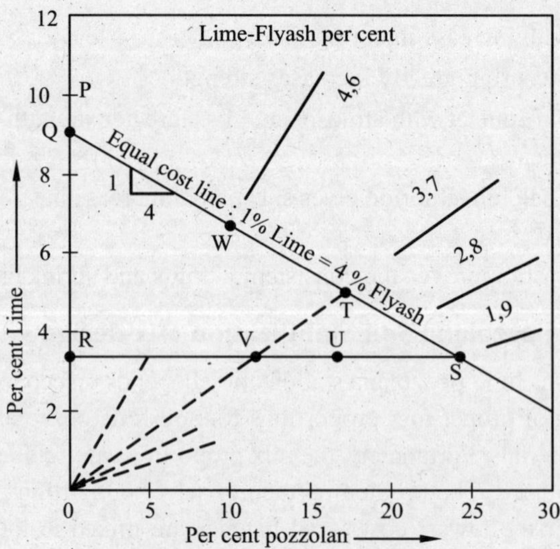

Fig. 12.3 Lime-Flyash ratio.

Lime-Portland cement. Some soils, which require a very high percentage of cement on account of their high colloidal contents, can be economically stabilized, if first treated with lime to materially reduce the plasticity of the binder soil and so facilitate the mixing of cement, and then stabilized with comparatively smaller concentration of cement. This process has been found very satisfactory for stabilization of black cotton soils in India and elsewhere. Hydrated lime has also been used as an admixture to cement stabilized soil mixtures to improve the cement reaction of some organic soils that exhibit retarded setting or which give low strength when mixed with Portland cement alone.

Lime-Bitumen. Both bitumen and tar have certain qualities that are most desirable for stabilization (as discussed later), but they are difficult to mix with fine-grained soils. As little as 1 per cent of hydrated lime will facilitate greatly the manipulation of bitumens into the soil. Thus, by mere thorough mixing a superior stabilization job is possible at reduced labour cost.

Work in India has shown that the addition of 4 per cent wood tar, and 5 per cent lime reduces the water absorption from 35 to 2 per cent and compressive strength of samples treated with wood tar and lime are very much higher than those treated with lime alone, when tested after saturation in contact with water.

Quality of lime. Lime is only partially quality controlled. IS : 1514-1959 suggest methods for testing hydrated lime and quick lime.

Use limitations. Unlike cement, lime sets slowly over a much longer period of time. If severe frost occurs only two or three weeks after a job is completed and is then followed by a long, severe winter, the lime base never fully sets or develops its full strength and serious damage of permanent nature can be inflicted on lime base. This danger of frost damage can be minimized to some extent by using greater amount of lime, by adding pozzolans and accelerators like calcium chloride or sodium carbonate which provide some initial frost protection and accelerate setting of lime. Calcium chloride may benefit the reaction by holding moisture and also by providing more calcium ions for incorporation into the cementitious gel, increasing long term strengths. Sodium carbonate reacts in two ways. It has immediate reaction with lime to produce sodium hydroxide and calcium carbonate, the latter acting as a cement. Sodium hydroxide alters the surface of minerals and pozzolans to increase their pozzolanic reactivity.

However, this slow setting provides more flexibility in road construction.

12.9. STABILIZATION WITH CALCIUM CHLORIDE

The use of calcium chloride as a dust palliative is recommended as a stabilizing agent in sand-clay or aggregate-clay mixtures.

12.9.1. Materials Requirements

The use of calcium chloride in stablilization construction does not eliminate the need for a close control of the gradation of the soil. In the use of this admixture, the resistance of the stabilized mass depends almost entirely upon the soil and moisture; the chemical acts primarily to preserve a more or less constant moisture content.

12.9.2. Properties of Calcium chloride

Deliquescence and hygroscopicity. Deliquescence is the ability of a material to absorb moisture from the air and thus to be dissolved and become liquid. Hygroscopicity is the ability to absorb and retain moisture without necessarily becoming liquid. Clacium chloride possesses these two characteristics to a marked degree. Data in Table 12.6 illustrates these two properties clearly.

Table 12.6 Lowest relative humidity at temperature at which calcium chloride will dissolve

Relative humidity	*Temperature °C*
20	38
30	24
40	7
43	0

Hygroscopicity

Water taken up by one kg of flake calcium chloride at different humidities.

Relative Humidity	*Temperture, °C*	*kg of water taken up by one kg of $CaCl_2$*
30	25	1.0
60	25	1.6
70	25	2.0
80	25	2.8
85	25	3.5
90	25	5.0
95	25	8.4

Decreased vapour pressure of calcium chloride over that of water. Vapour pressure is the tendency shown by a substance to pass from the liquid or solid into the gaseous state. Vapour pressure is a direct measure of the speed of evaporation and varies with temperature and the strength or concentration of the material. This means that water in a solution containing calcium chloride will evaporate at a slower rate than pure water and soil containing calcium chloride in solution will remain moist longer than if the chloride was not present.

Vapour pressure of saturated calcium chloride solution at 25°C = 7 mm of mercury. Vapour pressure of water at 25°C = 23.8 mm of mercury. The rate of evaporation from a free surface is reduced by $23.8 \div 7 = 3.4$ times.

The deliquescence, or relative humidity at which calcium chloride will start absorbing water from air is $\frac{7}{23.8} \times 100 = 29.4\%$.

The hygroscopicity, or ultimate amount of water absorbed, will be enough to reduce the vapour pressure of the solution equal to the vapour pressure of water in the atmosphere.

Increased surface tension of calcium chloride solution over that of water. Surface tension also affects the rate of evaporation in such a manner that a solution possessing high surface tension tends to vaporize less rapidly than one having low surface tension, with other factors being constant. As drying takes place and pore water decreases, the water films can cause further tightening and consolidation of compacted material. Calcium chloride has shown to cause upto 15 per cent increase in density of already compacted soil, increasing strength or stability several times. The slow rate of drying allowed by the presence of calcium chloride produces uniform shrinkage and densification throughout the treated road layer. A 15 per cent increase in density would give a lineal shrinkage of about 5 per cent, but the resulting fine cracks do not appreciably weaken the road. The initial drying after compaction is therefore very important just as the period of curing.

All these properties help in reducing evaporation of moisture from calcium chloride treated soils in comparison to untreated soils. This fact is of great importance in stabilization since one of the fundamental factors of stabilization is not only to supply the correct amount of water during construction in order to obtain maximum density by compaction but also to prevent the thickness of water film from changing.

There is some evidence to indicate that the increased surface tension of the calcium chloride solution may make the resultant moisture films around the clay particles stronger and increase the cohesion of the fine particles, and thus aid in stabilization.

It is also likely that the phenomenon of base exchange plays some part. Base exchange as already explained, means the substitution of a base for another base or hydrogen in a soil. The physical properties of calcium clay are different than those of hydrogen clay and potassium clay. However, the exact determination of this difference in a stabilized road and how it affects the stability of the road is not yet definitely known.

Freezing point. The freezing point of calcium chloride solutions vary according to the strength of the solution. How the addition of small percentages of calcium chloride helps in reducing the freezing point below that of water is shown in Table 12.7.

Though this property is not directly related to the use of calcium chloride in stabilization, it is an aid in preventing detrimental effects due to frost heaving.

The use of calcium chloride facilitates compaction so that the desired density is obtained with less compactive effort. The increase in compacted density varies with the soil and with the same compactive effort it is usually of the order of 1 to 7 per cent. Alternatively the number of roller passes for the same density may be reduced by as much as two-thirds. This increase in density results because calcium chloride solutions are better lubricants, and improve compaction by lubricating between grains.

In order for a calcium chloride solution to have any effect on stabilization, it is necessary for it to move throughout the material being treated and stabilized.

Table 12.7

Chemical percentage of weight of water	*Freezing point °C*
5	–2
10	– 5
15	–6
20	–12
25	– 24
30	– 26
35	– 35
40	– 43
45	– 48
50	– 51

Calcium chloride has also been tried in India for stabilization of soil by adding very small quantities in the optimum moisture for electrolytic effect during compaction. The results of the investigation show that there is an increase in the density and compressive strength of all soils except sand. However, these results have not yet been tried on field scales.

The quantity used in usually 0.1 kg per sq m per cm of thickness.

12.10. STABILIZATION WITH SODIUM CHLORIDE

The use of sodium chloride in stabilization is similar in many respects to the use of calcium chloride. However, the functions of the two chemicals are different in many respects.

General principles. *"The effects of sodium chloride on soil mixtures used for road surfaces arise from properties it imparts to properly proportioned soil mixtures through water retention, crystallization, increase in surface tension, physical and chemical changes in the clay components"*.

Sodium chloride solutions have a lower vapour pressure than water which reduces the rate of evaporation of moisture from the soil mixture. However, with prolonged evaporation of moisture, the sodium chloride solution in the surface becomes concentrated to the point of saturation and fine crystals are deposited within the surface forming a dense hard mat with the stabilized surface. This crystallization in the pores near the surface may enhance stability while inhibiting further penetration of water, as well as retarding evaporation of water already present.

Quantity of sodium chloride varies somewhat but 0.1 kg per sq m per cm of thickness of material represents average usage.

12.11. STABILIZATION WITH RESINS

There are two types of resinous materials. (*i*) resinous water proofing materials and (*ii*) resinous bonding materials. For the purpose of soil stabilization, resinous water-proofing materials are considered to be those natural or synthetic resins whose chief function is to maintain the moisture content of a soil at or below optimum moisture by preventing entry of water into the treated and compacted mixture. Little or no cementing action is obtained from these materials. This type of resin stabilizer is used only for soils of sufficient clay content to possess internal cohesion under dry and moist conditions. The use of this material as a soil stabilizer lies in its ability to act as a water repellent.

The molecules of natural resin contain hydrophilic acid and bulky hydrocarbon groups (the latter acting as strong dipoles). They are insoluble in acid solutions and almost insoluble in pure water, but their solubility increases in an alkaline medium.

In a water medium, partly neutralized, powdered natural resins break down into smaller particles of ultra micron size. These ultra microns attach themselves to or orientate themselves about the surfaces of clay aggregates or particles, so that, in the process of stabilization, the hydrophobic groups are directed outwards from the clay surfaces, thus repelling the entry of water. The presence of a film of moisture around the clay particles appears to enable effective orientation to take place.

Unlike bonding agents, whose effectiveness generally increases with the quantity used, water-proofing resins usually attain maximum effectiveness when applied in small quantities, 1 to 3 per cent by weight of the treated soil. The water-proofing effect decreases and water absorption increases with higher resin contents which suggests that the stabilizer content is critical within one per cent.

The resinous materials available commercially for soil stabilization are all natural products or obtained by processing natural resins. Vinsol resin available commercially is a dark brown powdered resin which is manufactured from the resinous residues obtained after distillation of pine tree stumps for turpentine. Resin is obtained during steam distillation of oleo-resins from pine trees during turpentine extractions. Another resin is known as Resin 321 in which only 25 per cent of the resin is neutralized with sodium hydroxide. A considerable number of natural and partially neutralized resins have been investigated but the most effective water proofing agents have been only a few mentioned above.

Another resinous agent tried is stabinol which is composed of 80 per cent Portland cement and 20 per cent of a complex resinous compound. When used in small quantities, the cement fraction cannot impart appreciable bonding to the treated soil but acts as a carrier for the resinous material and as a possible modifier for the soil to allow more effective use of the resin. A maximum of 2% has been recommended for water repellent purposes.

Most promising are derivatives of ammonia, such as amines and quaternary ammonium salts, having a cation-active nitrogen group and long-chain aliphatic group with 8 to 22 carbon atoms. These are expensive chemicals but only trace amounts, 0.1 to 0.8 per cent are required to water-proof a soil. Following compaction, curing is by air drying. Some field trials have recently been reported.

Another soil conditioner in market is an aniobic polymer of acrylonitrite. This is not a true water-proofing materials, but when present in trace amounts, it acts as a resinous glue linking positively charged ions of adjacent soil-clay particles. Its main value in highway-work is to reduce erosion on embankment slopes until vegetation can take over. This is also expensive. The use of resins reduces the susceptibility of a soil to frost damage.

Requirements for an ideal resin can be summarized as follows:

It should be water-soluble or water dispersible when being mixed with the soil, and it should bond to and link soil mineral particles and become insoluble after the reaction is completed. Moreover, it should be cheap to use, and it should resist biochemical decomposition.

12.12. STABILIZATION WITH LIGNIN

Lignin is the natural cement that binds the fibres of wood together in plants, and lignin derivatives are a major by-product of the paper-making industry. Lignin is extracted from the wood by treatment with sulphite chemical, resulting in calcium lignosulphonic acid. This material, at one time, was wasted. Now the acid is treated with lime to give a calcium lignosulphonate, sometimes called lignin sulphonate or, even lignin. The liquid containing 10 per cent calcium lignosulphonate is sold under the name sulphite road binder. To save transportation charges the liquid is concentrated by evaporation to about 50 per cent solids and known as sulphite road binder concentrate.

Ligno-sulphonates are water-soluble polymers with high and variable molecular weight. When added to soil, they cement down soil particles thus greatly increasing dry strength of the soil. Associated wood sugars, may act as hygroscopic agents and also aid as dust palliative.

Lignin is a very good dispersing agent for clays. In wet weather, dispersed clay in the road swells and plugs the pores thus reducing water penetration. During dry weather, lignin reduces rate of evaporation. This factor of reduced permeability also reduces the frost susceptibility of soils.

Lignin, being water-soluble, is lost from roads by leaching. Thus suitable soil types for treatment with lignin are light, dense-graded soil aggregate mixes. The quantity required for stabilization varies from ½ to 1 per cent of chemical solids by weight of the dry soil. A light surface application is then necessary to touch up dry spots, and surface applications are repeated as and when necessary. When the stabilized layer is protected under a bituminous wearing surface 2 to 3 per cent chemical solids by weight of the dry soil gives maximum strength and also provides extra chemical to prolong the life of the road.

As a dust palliative, as little as 4 litres of 8 per cent solution per sq m has proved to be effective. Treatment are repeated as needed.

As lignin is water-soluble and needs periodical replenishment, an alternative is to chemically polymerize the calcium lignosulphonate to produce an insoluble resin. Chromium ions introduced as sodium or potassium dichromate turn the lignosulphonate into dark, water-absorbent gel like mass which swells and seals soil pores. Field trials of chrome-lignin indicated troubles both from mixing and from development of shrinkage cracks. This led to the idea of making stabilized soil briquettes or moulding under pressure and these briquettes which harden rapidly to give strength of 20 to 70 kg/sq. cm. can be used as aggregate.

Usually 5 to 10 per cent chrome lignin is required to stabilize soil. Although the dichromate forms but a small percentage of the total additive, due to the high cost of chromium salt, n [illegible] make the mix uneconomical. Other cheaper admixtures could also probably be used with lignin.

12.13. STABILIZATION WITH SODIUM SILICATE

Laboratory work carried out in India has shown that in pure sand and non-plastic soils, the addition of small quantity of sodium silicate imparts greater strength to the mixture than cement. The abrasive resistance of the treated mass also increases. On account of the fact that such treatment is usually meant for dry areas, there is no possibility of loss in strength as a result of leaching. Since silicates can be readily manufactured from sand, this material may be economically available for stabilization of sand in desert areas where no other fine soil clay is available for mechanical stabilization.

Regarding plastic soils, research indicates that beneficial results could be expected from treating kaolinitic and illitic clay minerals with sodium silicate but treatment of montmorillonitic clay material will most likely be detrimental to the stability of such a soil.

The use of sodium silicate is also due to its ability to react in aqueous solution with soluble calcium salts, forming insoluble and gelatinous calcium silicates. The chemical is either sprayed on to, or injected into the soil in a fairly concentrated solution, while the calcium may be either derived from the compounds that are already present in the soil, e.g. chalk, or it may be added in aqueous solution. The process can only be used economically in areas where chalky soils or low-grade limestone aggregates are abundant, thus avoiding the necessity of adding a second solution.

The silicate solution employed should be rich in silica, having a ratio of alkali to silica about 1:3:5 and a specific gravity from 1.32 to 1.36. At higher concentrations the silicated limestone tends to stick to the roller and dry out too rapidly. The quantity is usually in the range of 1 to 10 per cent by weight of aggregate which may include stone up to 5 cm in size and certain proportion of limestone filler.

12.14. STABILIZATION WITH MOLASSES

Molasses is a by-product from sugar industry. The addition of molasses increases the internal cohesion of soil and thus helps in stabilization.

To prevent molasses from being washed out during rains, mixing of lime and charcoal with molasses has been tried. Molasses have also been converted into resinified compound with a mixture of coal tar or bitumen in the presence of acids. This compound is insoluble in water. Molasses being hygroscopic also acts as dust palliative.

12.15. PORTLAND CEMENT STABILIZATION

Portland cement has been used with great success to improve existing gravel roads, as well as to stabilize natural soils. It can be adopted for base courses and sub-bases of all types. It can be used for granular soils, but it is not suitable for organic materials which cause delayed setting and reduction in strength. Since addition of cement increases strength of the natural material, it is very often used for base-course construction.

Stabilization of soil with cement consists of adding cement to pulverized soil, and permitting the mixture to harden by hydration of cement. The factors that affect the physical properties of soil-cement include:

(1) Soil type
(2) Quantity of cement
(3) Degree of pulverization and mixing
(4) Time of curing
(5) Dry density of compacted mixture

Brief description of these factors follows.

12.15.1. Soil Type and Quanity of Cement

Granular materials, gravel, sand, lateritic soils, sandy silty material, crushed slag, crushed concrete, brick metal and kankar, etc. can be stabilized with cement. The gradation requirements for cement bound materials for different layers of pavement structure (base, sub-base, capping layer) are summarised in Table 12.8.

Table 12.8 Gradation requirement for cement cound materials for base/sub-bases/capping layer

Sieve size	*Grading I*	*Grading II*	*Grading III*	*Grading IV*
75.0 mm		100		100
53.0 mm	100	80-100	100	-
45.0 mm	95-100	-	-	
37.5 mm	-	-	-	95-100
26.5 mm		55-90	70-100	55-75
22.4 mm	60-80			
11.2 mm	40-60			
9.5 mm		35-65	50-80	-
4.75 mm	25-40	25-55	40-65	10-30
2.36 mm	15-30	20-40	30-50	-
0.600 μ		-	-	-
0.425 μ	8-22	10-35	15-25	-
0.300 μ	-	-	-	-
0.075 μ	0-8	3-10	3-10	0-10
7 days Unconfined Compressive Strength (MPa) for cement bound materials or 28 days strength for lime-fly ash and lime-cement-fly ash bound materials	12*/6**	7*/4.5**	3*/1.5**	1.5*/0.75**

* Average value of a batch of 5 cubes

** Minimum strength of an individual cube within the batch. For Grading IV the unconfined compressive strength and CBR requirement are equally acceptable alternatives.

The quantity of cement added varies from 5 to 12 per cent by weight. However, a minimum of 2 per cent is recommended.

For checking the suitability of soils, the following criterion may be kept in view:

Generally granular soils free of high concentration of organic matter or deleterious salts are suitable for cement stabilization. For checking the suitability of soils, it would be advantageous to keep the following criterion in view:

(*a*) Plasticity Product (PP), expressed as product of PI of soil and percentage fraction passing 75 micron sieve should not exceed 60.

(*b*) Uniformity coefficient of soil should be greater than 5 and preferably greater than 10.

(*c*) Highly micaceous soils are not suitable for cement stabilization.

(*d*) Soils that are having organic content higher than 2 per cent and also those soils having sulphate and carbonate concentration greater than 0.2 per cent are not suitable for cement stabilization.

(*e*) Silty or fine sandy materials may exhibit a high liquid limit because of the high surface area of the particles. This material generally will not react with lime because of lack of clay particles and can be stabilized with cement. However, cement stabilization with high doses of cement may tend to make stabilization uneconomical.

12.15.2. Degree of Pulverization and Mixing

Research results show that the quality of silty and clayey cement-treated soil mixtures is highest when 100 per cent of the soil, exclusive of stone, is pulverized to pass a 4.75 mm (ASTM No. 4) sieve. However, the quality is not seriously affected by the presence of as much as about 20 per cent unpulverised soil provided the lumps are moist (at or slightly above optimum moisture content) at the time of compaction. Generally, specification require that 80 to 100 per cent of the material pass a 4.75 mm sieve. IRC has recommended that the degree of pulverisation should be greater than 60 per cent.

Efficiency or uniformity of mixing influences both the strength and durability of soil-cement. Degree of mixing can be measured either by determining cement content of field samples by method of titration or as ratio of strength of specimens moulded form field mix to the strength of specimens moulded from the field mix after additional laboratory mixing. British experience indicates that 60 per cent efficiency is typical on their construction projects i.e. if the compressive strength of the laboratory remixed material is 28 kg/cm^2, the corresponding value for the field mix is 17 kg/cm^2. Studies show that if a given strength is desired, the cement requirements increase as mixing efficiency decreases. Field experience has shown that values as high as 90 per cent have been achieved.

12.15.3. Timing of Curing

Data on properties of cement-treated soil mixtures, on factors that influence properties and performance of these mixtures are all based on the assumption that adequate moisture is retained in the soil-cement mixtures during the curing period of 7 days or for a longer specified period.

12.15.4. Density

The strength and durability of soil cement mixture are strongly influenced by density. The relationship between strength and density approaches a straight line for some soils and cement contents. A 5 per cent decrease in relative compaction may result in greater strength reduction than a drop of 10 to 15 per cent in cement content (from 10 per cent cement to 9 or 8.5 per cent).

Specifications normally require compaction to continue till minimum field density achieved is 95% of maximum modified AASHTO field density. Compaction is completed before cement hydration takes place. The maximum time allowed for completing compaction is 2 hours and the temperature for mixing should be greater than 10°C.

12.15.5. Criteria for Mix Design

Most agencies have adopted the unconfined compression test for evaluation of soil-cement mixtures. These compression tests are made in the conventional manner with the specimen moulded to the desired moisture and density conditions and with various amounts of admixture added. The specimens are cured in a moist room for a period of 7 days before testing. Figures 12.4 and 12.5 show compressive strength requirements.

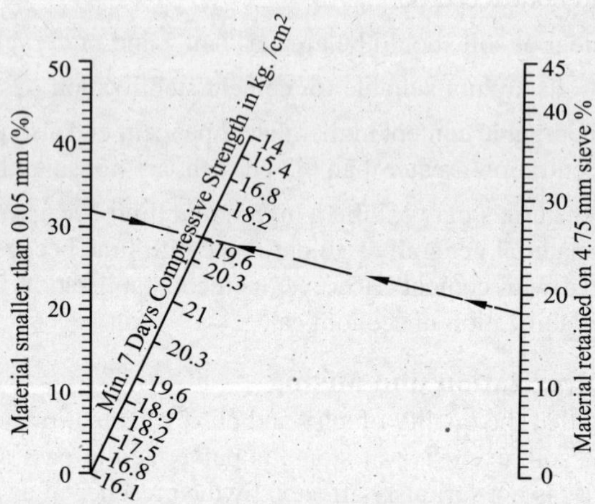

Fig. 12.4 Minimum 7-days compressive strength required for soil-cement mixtures, containing material retained on IS 4.75 mm. (From Portland Cement Association).

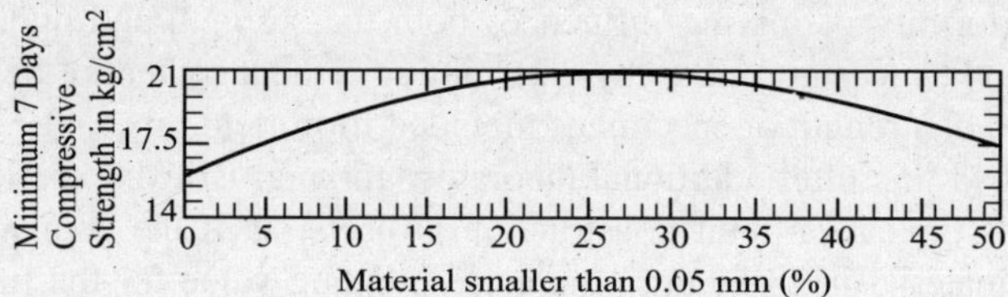

Fig. 12.5 Minimum 7-day compressive strengths required for soil cement mixtures not containing material retained on IS : 4.75 mm (ASTM No. 4) sieve. (From Portland Cement Association).

12.15.6. Cement Modified Soil

Small quantities of cement upto about 4 per cent can sometimes be incorporated into a soil to bring about certain desired modifications in properties. For example, a reduction in the liquid limit, plasticity index and water absorption may be obtained with clay soils. The resulting product is known as cement-modified soil which is a result of increasing the average particle-size of the soil, thus conferring new properties to the soil which are independent of the structure of the soil-cement as a whole. The design criteria for the use of cement modified soil is covered in IRC : 50—1973.

Table 12.9 Summarises the main requirements for cement modification or stabilization of subgrade soil

Table 12.9 Soil characteristics for cement modified soil/improved subgrade/capping layer

Properties	*Specified value*
Liquid Limit (%)	< 45
Plasticity Index	< 20
Organic content (%)	< 2
Total SO_4 content	0.2% Max
Minimum Laboratory CBR at specified density (%)	15
Minimum cement content (%)	2*
Degree of pulverisation (%)	> 60
Temperature for mixing	More than 10°C
Time for completing compaction	2 hrs Max

*In case better mechanical equipment for spreading of cement for breaking clods and blending is used, the minimum percentage of cement for stabilization could be 0.5 percent. However extensive lab testing must be done to arrive at this minimum percentage. Sample at site of blended loose soil be collected and remoulded in lab to confirm that the desired CBR can be achieved.

12.15.7. Improvement of Strength of Soil-cement with Additives

The effectiveness of cement as soil stabilizer can be enhanced with some additives. To reduce the deleterious influence of humus substances, 0.6 to 1.0 per cent of calcium chloride is sometimes used, to accelerate the setting time of cement. The use of lime for this purpose is also reported. Calcium choloride and lime can also be used in combination.

Since a number of additives that are effective are relatively cheap chemicals, they can permit economical improvement of soil cement. The effectiveness may be limited, however, to certain soils.

Another type of combination of additives has been reported in Russia. The chernozem soil having colloidal cement not less than 20 per cent, organic content not less than 5 per cent and pH value not less than 6 has been successfully stabilized with the addition of sand, cement, and certain additives. The approximate composition of the mix of this type of treatment is:

Soil and sand 40 – 50 per cent + lime 0.5 – 0.7 per cent + calcium chloride 0.25 – 0.35 per cent + ferrous chloride 0.10 – 0.15 per cent.

For better preservation of the soil-cement pavement, bituminous surface treatment of carpet coat, etc. is recommended depending on intensity and weight of traffic anticipated on the road.

12.16. BITUMINOUS STABILIZATION

The basic principles of soil-bitumen stabilization, as applied to highway and airfield construction are methods of designing and mixing local soil or aggregate with bituminous material to form a stable and water-proof base course. Properly constructed soil-bitumen base courses resist deformation through the cementing action of bitumen which binds the soil particles together. The thin coating of bitumen around the soil particles also provides a high degree of water-proofing which further aids in providing resistance to deformation.

Thus, bitumen may be used (*i*) for binding the soil particles together for the purpose of supplying cohesion for non-mechanically stabilized granular material or (*ii*) water-proofing, mechanically stabilized granular mixtures or (*iii*) for water-proofing cohesive soils.

The principle controlling the stabilization of granular materials is one of coating soil particles with a film of bitumen, of an optimum thickness, sufficient to supply adequate cohesion, without destroying the natural frictional resistance of the particles.

For cohesive materials, the principle is of adding sufficient bitumen to block the capillaries within the small soil aggregates to hinder moisture changes within these aggregates and partly to sheath these aggregates. Complete water-proofing is considered to be neither necessary nor desirable.

In many areas where suitable flexible base materials are not available or may have to be imported at considerable expense, local soils can be mixed with bituminous materials to form satisfactory base courses at a lower cost. Inferior flexible base course materials can be made satisfactory by soil-bitumen stabilization.

12.16.1. Factors Affecting Bituminous Stabilization

Factors affecting bituminous stabilization include (*i*) soil factor, (*ii*) bitumen type, (*iii*) bitumen content, (*iv*) moisture content and mixing, and (*v*) aeration and compaction.

(i) Soil factor. This requires proper selection of the best available soils. Field tests are made in an organized sequence to record differences in soil types and constructional materials. These data are used in determining thickness of treatment, proportioning of materials and in determining to some extent, the construction procedures.

A pre-requisite to any soil-bitumen stabilization construction project is a thorough laboratory investigation of the soils proposed for use.

A mechanical soil analysis showing the grain sizes usually provides the answer as to whether or not the soil is suitable for bituminous stabilization. Soils having a combined silt and clay content of more than 45 per cent are usually rejected. Pulverization of the soil is a very important factor in soil-bitumen stabilization; therefore, the effectiveness of pulverizing equipment to be used will have some bearing upon the maximum per cent of slit and clay the soil may contain. The higher the silt and clay content, the more difficult pulverization becomes.

The best results are obtained when reasonably well graded soils are used. The following limits are suggested:

Maximum size of material—not greater than approximately one-third of the compacted thickness.

Passing 4.75 mm mesh sieve—greater than 50 per cent
425 micron (No 40 ASTM) sieve—35 to 100 per cent
75 micron (No 200 ASTM) sieve—10—50 per cent.

Liquid limit less than 40, Plasticity index less than 18.

It is very difficult to stabilize cohesive soils containing relatively high moisture contents. With sands, however, there is no such difficulty since excess water may be squeezed out by rolling provided excessive quantities of bitumen are not added. In case of wet sands, wet sand-mix process can be used for stabilization with bitumen in which about 4 to 10 per cent of cut-back bitumen is mixed with wet sand to which one to two per cent of hydrated lime is added to assist coating. For this process the sand should not contain more than 5 per cent of material passing 75 micron sieve otherwise the mixture becomes too spongy and difficult to compact. In warm climate where sand-bitumen roads are constructed, the specifications allow the use of sand containing 12 per cent of material passing 75 sieve and when wind blown or dune sand is used, the limit may be increased to 25 per cent.

The amount of bitumen required depends essentially on the affinity of the bitumen for the soil material. As bitumen predominantly acts as an electro-negatively charged colloid, the amount of bitumen required increases with the number of electronegative charges located on the surface of the material to be stabilized. In the case of clay soil and a given particle grading, the percentage of bitumen required increases with the activity of the clay material as represented by its base-exchange capacity of its silica-sesquioxide ratio. It also increases with decrease in the valiancy of the adsorbed ions and with increases in the amount of organic matter.

Of the possibly adsorbed ions, sodium is usually detrimental. Thus, soils with sodium salts are not suitable with bitumen except with certain additives like calcium chloride. Soils with sulphate content more than 0.5 per cent are also not recommended. The presence of acid organic matter from leached forest soils is also often detrimental though the basic organic matter with the charnozems is not. Free calcium leads to detrimental conditions by promoting bacterial activity leading to the decomposition of bitumen. This particularly applies to heavy clay soils.

The percentage of bitumen required varies with number of other factors such as void ratio, climate and type and source of bitumen used.

(ii) Bitumen type and content. Selection of the type, grade, and amount of bituminous materials to be mixed is the next step in the preliminary laboratory investigations. The soil characteristics govern the type of bitumenous stabilizer to be used for the most satisfactory results.

In the following discussion, more general reference is made to bitumens, tars are equally suitable except when the use of bituminous materials of very low viscosity is prescribed. It may be taken that a tar of equivalent viscosity may be used under similar conditions. With tar, more careful control is required than in the case of bitumens.

Rapid curing cut-back or emulsified bitumens produce best results when mixed with extremely sandy soils or those containing a minimum of silt and clay particles. Soils having a plasticity index of 5 or less are included in this range. Soils having little or no plasticity are cohesionless and depend upon the cementing value of the bituminous material for stability. Paving bitumen penetration grades 80—100 in summer and 120—150 in winter can also be used. However, the use of paving bitumens requires heated sands, hot mixing and hot laying. It provides the highest type of sand asphalt. For higher stability requirement, harder paving bitumen can be used within limits.

Medium-curing cut-back having a kerosene type liquefier usually produces a more homogeneous mass when mixed with soils having a plasticity index range of 6 to 10. As the plasticity of a soil increases, cohesion also increases; therefore the stability of a mixture of this classification of soil depends upon both the cementing value of the bituminous material and the cohesion of clay particles in the soil. Emulsified bitumen is also satisfactory with some soils within this P.I. range.

Slow curing cut-back bitumen having gas oil or heavy residuals as liquefier penetrate clay soils more rapidly than other types of bituminous material. They, therefore perform better when mixed with soils having plasticity index of more than 10 or silt and clay contents of more than 30 per cent. These more highly cohesive soils have greater inherent binding characteristics and depend less upon the cementing value of the bituminous material.

In case a soil considered for use borders on any two of the foregoing classifications, preliminary laboratory investigations should include two types of bituminous materials.

Regardless of the type of bituminous material selected for a project, it is a good practice to use the heaviest grade that can be readily worked into the soil. The type of mixing equipment available is a factor in selecting the grade of bituminous material to be used. For example, if a MC type of cut-back bitumen has been selected for a particular job, it woule be best to use MC-2 grade with conventional blade-mixing equipment so that this cut-back grade containing a greater quantity of solvent would allow more time for mixing. On the other hand, if an efficient, modern travel-mixing plant is available, a much heavier grade of cut-back bitumen, such as MC-4, could be used with equal results.

Climatic conditions also have some bearing on the grade of material that may produce the best results. For warmer climates, a heavier grade can be used than for cooler climates. In warm weather, rapid-curing cut-backs tend to cure too rapidly and should not be used.

(iii) Bitumen content. To determine the grade and quantity of the bituminous material most suitable for a particular soil, a number of mixes are prepared in the laboratory and tested for stability and moisture absorption. Laboratory mixes prepared with bituminous material with 3.5 to 7.0 percent in increments of 0.5 per cent will locate the optimum bitumen content. Experience with a particular soil may reduce laboratory tests to only three or four with different bitumen contents.

After the laboratory mixes are moulded and cured according to the particular design method in use, they are tested for dry and wet stabilities and moisture absorption and graph plotted between per cent bitumen content and stability and per cent moisture absorption. From the laboratory tests, a well-balanced mix design should be selected, one which does not exceed the maximum moisture absorption allowed and does not fall below the minimum stability requirements. Mixes having low bitumen content may produce vary high stabilities, but may be much too susceptible to moisture absorption.

Moisture-absorption tests be made by placing the moulded specimen in a capillary tank for 7 days or by complete immersion for 24 hours.

Several methods of testing stability are in use: most of them are the same as those used for testing high-type bituminous concrete, such as the Hubbard Field, Hveem, Marshall and Triaxial tests which are described in detail in chapter 15. Other modified punch and shear tests are also used in some cases.

Criteria for satisfactory performance with some of the tests are given below:

1. *Hubbard-field test.* With 5 cm diameter and 5 cm high specimens, the swelling should not exceed 5 per cent; the water absorption not more than 7 per cent and the stability not less than 180 kg. If the strength is measured without soaking the specimen, it should not be less than 450 kg but this unsoaked procedure is not recommended except as an emergency measure. Details of testing method are covered by ASTM D 915–47 T for soil bituminous mixture.

2. The apparatus and testing procedure are the same as the original method except, that the specimens are not treated in a bath at 60°C but are tested at room temperature.

For hot mix sand asphalt, original Hubbard-field design method is used and the minimum stability value specified is 360 kg at 60°C. This test should be used with care because it principally measures cohesive strength. Mixtures containing excess of asphalt may have satisfactory cohesion but low frictional resistance.

3. Hveem method and Marshall methods can also be used in design. Usually no voids criteria are used with sand asphalt, and optimum bitumen content is selected on the basis of strength. In the upper 5 cm of sand asphalt base, the Hveem stability should be 20 or more; and below the upper 5 cm a minimum of 17 is allowed.

4. For Marshall method of design for sand asphalt (hot-mix) a minimum of 225 kg Marshall stability is required.

5. For sand stabilized with bitumen cut backs, Hubbard-field test is used except that water bath is not used and specimens are tested in air at 25°C. Minimum stability of 545 kg is specified.

6. *Cone penetrometer test.* Measurement made of resistance offered by compacted soil-bitumen sample to penetration of right angled cone shall not be less than 80 kg per sq. cm.

7. *Florida bearing value test.* This test was devised to determined whether sand is suitable or not for stabilization with bitumen emulsion or tar. 1.75 kg per sq. cm bearing value is required for satisfactory performance for bituminous treatment. The test is briefly as follows: 10.5 cc of water is added to 600 gm of air-dried sand. As much of the wet sand as can be pressed by hand is placed in the mould and compacted by means of a 550 kg load applied to a bearing plate approximately 76 mm diameter resting on the sand. After compaction the test plunger is placed on the sand surface, and loaded by running lead shot into the bucket until failure occurs and plunger sinks in.

8. *Modified Florida bearing value test.* This is used for testing emulsion sand or sand tar mixes. The probable requirement of emulsion as percentage by weight (W) is determined from the formula

$$W = 0.0375\ A + 0.075\ B + 0.375\ C$$

where A = percentage of sand retained on 2 mm sieve.

B = percentage of sand passing 2 mm sieve and retained on 75 micron sieve

C = percentage of sand passing 75 micron sieve after wet sieving.

Mixture compacted in a mould 7.5 cm high and 10 cm in diameter under a static load of 11360 kg (140 kg per sq. cm) is tested at 60°C in the apparatus for Florida bearing value test. The criterion for satisfactory performance is at a bearing value of 10.5 kg per sq. cm.

9. *Unconfined compressive strength test.* This is often carried out when bitumen emulsion and cement are used for stabilization. Criteria to be employed in road construction should have a minimum compressive strength of 10.5 kg per sq. cm after immersion for 7 days, while for foot paths or cycle tracks, a compressive strength of 7 kg per sq. cm is recommended. For soil-bituminous mixtures, the bitumen content used should be that which has the highest dry-wet strength ratio when tested in unconfined compression test as long as the minimum value is 5 kg per sq. cm.

Table 12.10 summarizes material and bitumen requirements for soil stabilization.

Table 12.10 Summary of material and bitumen types for stabilization

Sr. No.	*Material types*	*Maximum values for adequate stabilization*	*Bitumen types and grades*	*Approximate amount of bitumen (%)*
1.	Fine-grained soils	Max. LL = 40% Max. PI = 18%	MC and SC RT 3–6	4 to 8
2.	Sands	Max passing No 75 micron sieve = 25% Max PI = 12%	PB penetration 80-100 and 120–150, RC 1–3 RT 6–10	4 to 10
3.	Gravel and sand gravel	Max. passing No. 75 micron sieve = 15% Max. PI = 12%	RC 1–3 RT 4, 5 PB (Paving bitumen)	2 to 6

(iv) Moisture content and mxing. To produce a homogeneous mass with a minimum of mixing effort, both in the laboratory and on the project, the soil should contain some moisture for cut-back bitumen or tar mixes and 10 per cent or more for emulsified-bituminous mixes, otherwise, emulsions break immediately upon contact with soil. The reason for moisture is that it acts as a carrier for the bitumen and is an aid to mixing.

In case of cut-backs the mixing process has a considerable influence on the performance of bituminous water-proofing agents. Excessive mixing can reduce water-proofing which may be due to bitumen spreading over a larger area in thinner films which are not water-proof.

(v) Aeration and compaction. Upon completion of mixing operation the next step in the construction of soil-bitumen base course are aeration and compaction. Compaction of soil bitumen mixes retaining relatively high percentage of moisture and volatiles results in low densities and too low stabilities. Therefore, the mixture must be aerated to reduce the amount of both moisture and volatiles. Aeration is accomplished by manipulating the mixture in a manner that will permit evaporation of moisture and volatiles. Moisture content should be reduced to approximately three-fourths of optimum and 65 to 75 per cent of volatiles in the cut-back should be removed before compaction. When emulsified bitumen is used as the stabilizing agent, the moisture content of the mix should be reduced to approximately three-fourth of optimum moisture for raw soil, before the mix is compacted.

The moisture content at the time of compaction should be such so as to give optimum fluid content for maximum density.

12.16.2. Additives as Aids to Bitminous Stabilization of Fine-grained Soils

The relatively hydrophobic soils and rocks containing carbonates and oxides, are most easily wetted by bitumen to produce good adhesion and when once protected by bitumen offer a substantial resistance to moisture penetration into the bitumen-soil material interface.

Adhesion between bitumen and the relatively hydrophilic material such as clay and rock silicates is generally poor, and water, if it can reach the bitumen soil material interface, quickly replaces the bitumen as a sheath round such particles. Therefore, the incorporation of a small percentage of wetting agent whose function is a surface-chemical reaction to change the sign of the electro-static force field on the particle surfaces is usually essential with such materials.

Types of wetting agents used include fatty amines, amino salts, lime, cement aniline-furfural, resorcinol formaldehyde, resin, zinc steartate, etc. Bitumens containing a high wax content, cracked oils, and high temperature tars may sometimes be used without wetting agents. The desired percentage of wetting agents has been found to vary with climate and the base-exchange capacity of the clay material.

Studies to investigate the effect of certain chemical additives in low concentration, less than 1 per cent by weight have reported that addition of phosphorous pentoxide, certain epoxy resins, and certain organic isocynates to bitumen cut-back significantly improved the strength characteristics of bitumen stabilized soil as measured by compressive strength after immersion in water. It was concluded that chemical additives hold considerable promise in soil bitumen, offering opportunities for broader and more effective use of bitumen for treatment of fine-grained soil. All these additives improve method of incorporation of bitumen merely enhancing its characteristics water-proofing ability.

12.16.3. Surface Protection

Soil-bitumen bases usually contain only enough bituminous material to make them stable and resistant to moisture absorption. Therefore, they will not withstand loss by abrasion from traffic. So to preserve the soil bitumen base course, a durable wearing course must be provided. After allowing the soil-bitumen base to weather, or cure, for approximately one week, a prime coat and wearing course are applied. The wearing course may be either bituminous surface treatment or the higher type bituminous surface which will be governed by the intensity and weight of traffic anticipated on the road.

12.17. THERMAL STABILIZATION

One of the methods for reducing the plasticity of highly clayey soils is heat treatment. Instances are available where cohesive soil is baked *in situ* or in kilns for soil stabilization. This is one of the oldest methods of soil stabilization which has been used extensively by the Australian agencies in making patheways. For heat treatment of soil *in situ,* a travelling furnace capable of handling large amounts of soil at temperature over 500°C is employed. This burning of the soil *in situ* by mobile furnaces was found very costly and this machine could burn the soil only to a depth 8 to 10 cm of top layer which may not be sufficient. Experience shows that the heating of the soil *in situ* would be successful if some improvement in the design of the machine was made to reduce the cost of operation and to increases the depth of burning.

The heat treatment of pulverized black cotton soil to reduce its plasticity in laboratory was studied in India. These studies show that the soil becomes non-plastic after being heated to about 500°C. It takes four hours of heat treatment to convert the soil clods into aggregates. The modified impact values of burnt black-cotton soil clods after 7 days of immersion in water was found to be of the order of 20 to 22 per cent as compared to values of the order of 40 to 45 per cent for well burnt bricks tested under similar conditions.

These results indicate that the black-cotton soil clods can be used with advantage in place of brick-bats or any other soft aggregates in low-cost road construction. Depending upon the method of burning these will always be a mixture of pulverized soil and clods after heat treatment which gives a good gradation. The CBR values of this material after 4 days soaking was found to be 110 to 140 per cent.

There is also a possibility of further improving the heat treated soil with cement or bitumen. To increase its reactivity with cement, some raw clayey soil may have to be added to heat treated soil. The optimum clay content to be added can be determined in the laboratory based on compressive strength tests. However, such an improvement of heat treated soil by addition of cement or bitumen may prove costly.

Several types of kilns for burning the highly clayey soils in the field have been suggested. Kilns used for calcinations of limestone can also be used. These kilns are of continuous type and can be fired with wood.

12.18. GENERAL RECOMMENDATIONS FOR STABILIZATION OF SILT-CLAY MATERIALS

Constructing roads in areas having highly clayey soils has always presented serious problems because of the extremely low bearing capacity of the subgrade when wet and of its high swelling and shrinkage characteristics during wetting and drying. One such soils is black-cotton soil in India. An analysis of typical black-cotton soil in India shows the values: $LL = 90$ per cent; $P.I = 44$ per cent; $pH = 8.5$; silica sesquioxide ratio = 4:3.

The crucial difficulty involved in the stabilization of cohesive soils that do not contain granular skeleton is due to the fact that the seat of resistance providing cohesion is also the seat of water affinity. The cohesion decreases with the increase in water content.

Theoretically, the simplest method of stabilizing such materials is to deprive the component particles of their water affinity and to cement these particles together. This exactly is achieved by burning clay soils in thermal stabilization. However, the practical difficulties and economics should be well explored before applying this method. The soil type and the effectiveness of thermal stabilization should also be studied when this method is contemplated.

Another method is to replace the water attractive inorganic exchangeable cations by means of water repellent organic cations and to cement the soil constituents, that have now lost their water affinity, together by means of an organic cement. This can be done using aniline-furfural and similar resins already discussed under "stabilization with resins".

Another procedure is to create strongly water resistant secondary soil aggregates which are then cemented together by means of inorganic binders such as Portland cement, lime, lime-flyash combinations. Combination of lime and bituminous material may be quite effective. About 4 to 5 per cent lime and 10 per cent cement has been suggested for stabilizing some black-cotton soils in India. About 4 per cent bitumen and 5 per cent lime were also found to substantially reduce the water absorption and increase the compressive strength.

A less positive but still effective method, if properly used, is to stabilize the moisture content of a soil system by preventing the intake of excess water. This water-proofing may be achieved by various materials such as bituminous and pyrogenous bitumens, natural and artificial resins, fats, waxes, etc.

If economical, classification of soil can be upgraded by addition of granular material and attempt may be made to stabilize such cohesive soils by mechanical method. If more economical, percentage of granular material may be reduced than that required for mechanical stabilization, and further stabilization of modified soil should be tried with admixtures such as lime, cement, bitumen or combinations thereof with or without surface active additives or electrolytes.

In certain cases treatment of soil only with cement, lime and bituminous materials does not ensure necessary effect or requires a high amount of binder to achieve it. For example, humus substances are harmful with cement; and salinized soils, especially of sulphite or soda salinization, treated with bitumens do not achieve water-resistance and usually swell. In many cases it is very difficult to treat clayey soils because it is very troublesome to achieve a finely divided and uniform distribution of binding materials. To overcome these difficulties, use is made of binding materials in combination with surface active additives and electrolytes. The aim of using these additives is to change the conditions of interaction and hardening of binder in the mass of stabilized soil. Optimum quantity and correct selection of suitable surface-active additives and electrolytes should provide for the best use of the cementatious properties of bitumens, cement or lime.

The following combinations are considered suitable for stabilization:

Soil + Portland cement + bituminous emulsion
Soil (acid type) + Portland cement + calcium chloride + lime
Soil + cut-back + hydrated lime
Soil (soda salinization) + cut back + lime or calcium chloride
Soil + Portland cement + naphtha soap

Some of the other trace chemicals (approximately 1 per cent) which improve the strength with cement and bituminous materials have already been discussed under respective methods of stabilization.

12.19. SELECTION OF STABILIZER

The selection of the stabilizer is based on plasticity and particle size distribution of the material to be treated. The appropriate stabilizer can be selected according to the criterion shown in Table 12.11. Some control over the grading can be achieved by limiting the coefficient of uniformity to a minimum value of 5; however, it should preferably be more than 10. If the coefficient of uniformity lies below 5, the cost of stabilization will be high and the maintenance of cracks in the finished road would be expensive. If the plasticity of soil is high-there are usually sufficient clay minerals which can be readily stabilized with lime. Cement is more difficult to mix intimately with plastic material but this problem can be alleviated by pre-treating the soil with approximately 2 per cent lime.

Table 12.11 Guide to the type of stabilization likely to be effective

Type of stabilization	*Soil Properties*					
	More than 25% passing 0.075 mm sieve			*Less than 25% passing 0.075 mm sieve*		
	PI < 10	10 < PI < 20	PI > 20	PI < 6, LL < 60	PI < 10	PI > 10
Cement	Yes	Yes	-	Yes	Yes	Yes
Lime	-	Yes	Yes	No	-	Yes
Lime-Pozzolana	Yes	-	No	Yes	Yes	-

12.19.1. Two Stage Stabilization Using Lime followed by Cement

Cement can be used to stabilize most of the soils. The principal exceptions are those that contain organic matter in a form which retards the hydration of cement and soils which are difficult to mix with cement on account of their high clay content. In comparison with cement, the potential use for lime in soil stabilization is more restricted; used in equivalent amount it generally produces lower strengths than does cement and its main application is for use with clayey soils which are difficult to stabilize with cement. For these reasons the use of a two stage lime/cement stabilization process appears attractive as it offers the possibility of extending the range of soil which can be effectively stabilized. To achieve the maximum effect the lime and cement would not be blended but would be added separately. Lime would first be added to modify the properties of soil and this would be followed by the addition of cement to bring out a long term increase in strength, when lime treated soil is stabilized with cement.

CONSTRUCTION METHODS

Three basic procedures are used in the construction of stabilized soil roads. These methods are:

1. Mix-in-place or road-mix construction.

2. Travelling plant construction

3. Stationary plant or central plant construction.

In all these methods the basic processes are the same. These are: the soil elements must be properly selected and proportioned, the soils pulverized and uniformly blended, water and stabilizer or other additives, if desired, added and uniformly mixed with the soil; the blended materials spread in a thin layer of uniform thickness and properly compacted. The several steps may be performed separately or some of the steps may be combined in one operation.

12.20. MIX-IN-PLACE METHOD

In this method of road-mix construction the proper proportions of soil elements which are to form the base for wearing surface are mixed directly on the surface of the subgrade or sub-base. The sequence of various steps are:

Scarifying and pulverization of soil. The proportion of the soil which is largely available from nearby sources such as road-bcd, shoulders, side slopes, etc. that is to be used for stabilization is first cut and spread on the surface of the road by hand labour or by a blade grader. Any other suitable plant such as a plough or robust tiller with a positive depth control may be used for scarifying the soil. The depth of scarifying must be carefully controlled in order to obtain the desired compacted thickness of the stabilized layer. If the soil is too wet, it is allowed to dry and when this scarified soil attains a satisfactory moisture condition it is thoroughly pulverized manually or by the use of suitable farm equipment such as rotary tillers, disc harrows, ploughs, etc. When pulverization is complete, about 80 per cent or more of the soil, exclusive of stones, should pass 4.75 mm sieve. The moisture content is one of the factors that determines the ease of pulverizing.

IRC has specified soil pulverization requirements for lime and cement stabilization, which are given in Tables 12.12 and 12.13 respectively.

Table 12.12 Soil pulverization requirements for lime stabilisation

Sieve designation	*Minimum per cent by weight passing the sieve*	
	For black cotton soils	*Other soils*
25 mm	100	100
4.75 mm	50	60

Table 12.13 Soil pulverization requirements for cement stabilization

Sieve designation	*Minimum per cent by weight passing the sieve*
25 mm	100
4.75 mm	80

If any imported material from borrow-pits is to be added to the existing soil, the same is hauled onto the road and spread in proper amounts. The previously prepared soil and the imported soil are combined in proper proportions.

In general, pulverizing operations are planned in such a way that no more road material is pulverized ahead of construction that can be processed in two days.

Application of stabilizer. Soil stabilizer, if any, may be added at this stage. With solid stabilizers such as cement or lime, the bags are spotted at required spacing in preparation for opening and spreading in traverse rows. These are split open and emptied by hand. The spreading may be attained by hand using garden rakes or by dragging a spike-teeth harrow or nail drag slowly over the section. On large jobs cement is sometimes spread from bulk cement trucks, containing as much as 25 tonnes of cement. Liquid stabilizers are sprayed on the soil by some suitable spraying devices or from pressure distributors.

Mixing. As soon as stabilizer spreading is done, mixing is begun with field cultivators, and rotary tillers or with special soil mixers. Dry maxing is usually completed in two or three passes of the machine. Where no machinery is available, dry mixing and wet mixing may be done manually. Dry mixing is normally followed by wet mixing with the application of water. It is essential that the materials be blended to a high degree of uniformity if best results are to be obtained. The usual criterion followed in this respect is that the mixing is continued till the mixture throughout the depth of treatment has a uniform colour. There is no other rapid means of assessing the uniformity of mixing.

Frequent checks are made on the depth of treatment to be sure that the equipment is properly adjusted and the depth of treatment controlled.

Addition of water. Raw soil samples are obtained at 100 to 150 m intervals ahead of stabilizer spreading operations to indicate the minimum quantities of water to be added that day. Further, by knowing the moisture content of the raw soil and comparing it with the sample after mixing is completed, it is possible to estimate the evaporation losses during the mixing period. This is very helpful to work out accurately the total quantity of water to be added. Pressure distributors may be used to ensure a uniform spread of water over the entire surface of the roadway. Gravity distributors place much less water below the ends of the spray bars and introduce aggravating problems in bringing the roadway to a uniform moisture content.

The water added should be sufficient to bring the mixture to or slightly above the optimum moisture specified. This will permit some evaporation losses during compaction. The moisture content at compaction should not be less than the optimum moisture content corresponding to IS : 2720 (Part VIII) and not more than two per cent above it.

Compaction. When a uniform mixture of soil, stabilizer if any, and water has been obtained, the entire roadway is loosened and graded to give a loose mass of uniform thickness which can be effectively compacted from bottom.

A typical compacted thickness of layer of this type is about 100 mm when compacted. The maximum thickness may be as high as 200 mm. Bases and wearing surfaces which are much thicker than the above limits are constructed in two or more courses.

Initial compaction may be accomplished by sheep's foot roller except while compacting very sandy soils containing little or no minus 75 micron material in which case compaction can be done by pneumatic tyred equipment. Following initial compaction, the surface is generally shaped to the proper section by the use of a blade grade or manually and final rolling is achieved by the use of pneumatic tyred or 8 to 10 tonnes smooth wheel rollers. It is absolutely essential that adequate compaction be secured in this type of construction until density achieved is at least 100 per cent of the maximum dry density of the material.

The success of final smooth rolling to get proper surface finish depends on the use of a proper weight tandem roller together with a pneumatic-tyre roller and the maintaining of optimum moisture in the surface during the operations.

In soil cement work, it is desirable to complete compaction within 2 hr but not more than 5 hr after the beginning of wet mixing. In soil cement care should also be taken to avoid overstressing the processed soil by the use of a roller that is too heavy, or by rolling for too long. This might reduce the strength by cracking or breaking the initial structure formed by the cement as it hydrates.

Protection and cover. Many stabilizers require a period of curing after mixing before they become fully effective. This particularly is necessary with soil cement which should be protected against rapid drying for a period of 7 days by covering with a layer of moist soil or straw or by frequent spraying of water.

***Advantages and dis-advantages*:** Mix-in-place method has the following advantages: (*i*) the number of machines required can be adjusted to the size of the job; (*ii*) the machines required are simple, cheap, easily transportable and easily repairable; (*iii*) the whole processed section is ready for compaction at the same time; (*iv*) in wet climate, road mix is the only way of getting rid of excess water; and (*v*) a large average output may be maintained.

The disadvantages are : (*i*) it is not easy to obtain a uniform thickness of treatment, because of the difficulty of setting the machines to a given depth; (*ii*) the mixing is not as uniform as with other methods which is the chief drawback of this method.

12.21. TRAVELLING PLANT METHOD

This construction process is not greatly different from the mix-in-place method. However, the operation of pulverization and mixing the various soil elements, the stabilizer and water are accomplished by the travel plant usually in a single pass of a machine. An elevating loader supplies material to a hopper with a measuring gate from where it is discharged to a pugmill where water or a fluid stabilizer may be added through spray nozzles and mixed into the soil. In case of cement, it is spread on top of wind-rowed raw soil. With most travel plants, the materials, which are more thoroughly and uniformly mixed than is usually possible in road mix method, are discharged behind the travel plant and spread by a grader. Final finishing and rolling are done in the same manner as for road mix method. In recent years a number of special machines have been designed which perform virtually all the steps involved in soil stabilization in a single pass. One such machine is P and H single pass stabilizer. The machine advances slowly (at speeds upto 10 m per min) and the soil is cut, pulverized, mixed and water and fluid stabilizers are added and uniformly mixed. Powdered stabilizers are spread on the soil in front of the machine. The finished material is spread behind the machine and the road is ready for compaction.

The advantages of travel plant method are : (*i*) uniform mixing, (*ii*) accurate proportioning of added water; (*iii*) highest output for a given expenditure of plant and labour; (*iv*) short mixing time; and (*v*) depth of treatment can be more accurately controlled.

The disadvantages are: (*i*) high initial cost of plant; (*ii*) work may be stopped for a minor breakdown of one part of plant; and (*iii*) need for the plant to work continuously at full capacity due to its high initial cost.

12.22. STATIONARY PLANT METHOD

Stationary plants are either of the two types: (*i*) continuous mixers or (*ii*) batch mixers.

The principle of continuous mixer is the same as for travelling mixer. The mixed material is, in this case, discharged into trucks. The size of central mixing plant will depend on the output required.

Batch mixers are used on small jobs or for patching work. Ordinary concrete mixers can be employed for mixing soil, stabilizer and water. The best results are obtained by double-paddle mixers, pug-mill or roller pan type mixers, in which soil lumps are easily carried to the site. For patching small areas, hand or pneumatic tampers can be used.

Advantages of this method are:

(*i*) accurate proportioning of the mixture

(*ii*) easy control of the depth of treatment

(*iii*) small evaporation losses during mixing and transport of material

(*iv*) concrete mixers can be utilized where available.

The disadvantages are:

(*i*) expensive if soil at site of work in processed.

(*ii*) material has to be compacted as delivered, and not as complete section.

12.23. FIELD CONTROL

The purpose of field control is to ensure the quality of construction prescribed. Tests essential for proper inspection and control are listed in table 12.14 which also include post-construction tests that are of interest in studies of the behaviour of stabilized mixture and the continued improvement of these mixtures. Post construction tests are not run as routine but can be performed if desired.

Careful field control is needed in stabilized construction and amongst the items that need checking are degree of pulverization of the soil, its moisture content, dry density-moisture content relationship, quality of mixed material and dry density of compacted layer.

Table 12.14 Construction and post construction tests

Test	*Purpose of test*	*Frequency*
A—Construction tests		
1. Screening over 4.75 mm sieve	Pulverization	Every 500 m^2
2. Moisture-density relations of soil-cement mixtures.	To form basis for specification requirements.	Every 5000 m^2
3. In situ moisture content and density.	To determine compliance with specifications	Every 5000 m^2
4. Cement content of mixture.	To check cement application and uniformity of mixing.	Every 5000 m^2
5. Compressive strength test on material taken from construction site before compaction.	To determine effectiveness of mixing by comparison of compressive strength from field mixing and laboratory mixing.	Every 10000 m^2
B—Post-construction tests		
1. Cement content	Same as 4 above	
2. Compressive strength of cores	To check on effectiveness of field proportioning of water and cement and on mixing, compacting and curing procedure.	
3. Wetting and drying test on cores.	—do—	
4. Moisture content	To determine moisture loss during curing.	
5. Measurement of crack interval and opening.	To provide data on cracking characteristics for further studies for improvement in mix design.	

❑❑❑

Chapter

MACADAM ROADS

The oldest type of highway pavement used in modern times is known as Macadam, named after John L. Macadam, the Scottish engineer.

The term Macadam as applied to the road at the present time, has come to mean road surface and bases which are constructed of crushed or broken stone fragments cemented together by the action of rolling or traffic and water. This term is also applied to roads in which the aggregates are bound together by bituminous materials or Portland cement.

Modern broken stone roads were first introduced in France by Tresaguet about 1765, and by Telford in England about 1805. Both these engineers used large pieces of broken stones in the lower course and placed a layer of finer crushed stones as a wearing surface. Telford advocated the use of large flat pieces of stones, which were hand placed to form the base. Telford bases employing flat pieces of ledge stone 7.5 to 20 cm in thickness and from 10 to 40 cm in width and length and placed by hand are still sometimes employed in subgrades having low supporting power. These bases are however expensive particularly where manual labour is not cheap. Macadam however, insisted upon the use of smaller stones, about 40 mm maximum size, for the entire thickness of the pavement. The invention of the stone crusher in 1858 by Blake and also the invention of steam roller in France in about 1860 greatly advanced Macadam construction.

The following design principles employed by Macadam are as basically sound today as they were in his time:

1. Elevate the earth foundation above the surrounding ground to provide good drainage.

2. Crown the subgrade and drain the surface water into the side ditches.

3. Use clean, broken stone for surfacing without admixture of clay or organic materials which will hold water or be affected by frost action.

4. Design the road in accordance with traffic conditions.

5. Drain and compact the subgrade soil to support the loads.

6. Compact the several layers of uniform size stones to form fine and uniform surface.

13.1. TYPES OF MACADAM ROADS

There are five types of macadam roads which are often important in modern highway construction and are defined as follows:

(1) ***Water-bound macadam*** is the layer composed of broken stone aggregate which are bound together by stone dust and water which is applied during construction, in continuation, with consolidation of the layer by a heavy roller. This type of macadam road closely resembles those, so widely used in the early days.

(2) ***Traffic-bound macadam*** is the wearing surfacing composed of broken stones or gravel which is consolidated by the action of traffic, intermittent blading or dragging and rain. This type of surface is generally built up gradually by the successive application of two or more layers, each of which is 2.5 or 5 cm in compacted thickness.

(3) ***Wet Mix macadam.*** It consists of clean, crushed graded aggregates premixed with other granular materials and water. This is rolled to a dense mass on a prepared surface. Wet mix macadam (WMM) construction is an improvement upon the conventional WBM and is an alternative and more durable pavement layer.

(4) ***Bituminous penetration macadam.*** is the layer of construction in which clean crushed stone reasonably uniform in size is compacted and the layer of crushed stone is then heavily sprayed with bituminous materials, much of which penetrate into the voids, binding these stones together. A thin layer of smaller aggregate or choking stone is then uniformly spread to fill the surface voids of the first course which is rolled and is followed by another light application of bituminous material and a thin layer of still smaller aggregate and more rolling. This type is frequently termed as "*penetration macadam*".

(5) ***Cement-bound macadam*** is the type which resembles the bituminous macadam defined above, except that bituminous materials are replaced by Portland cement mortar or grout which is forced into the voids of the compacted stone layer. This type of construction is not very popular at the present time.

WATER-BOUND MACADAM

Water-bound macadam is constructed in thickness ranging from about 8 to 30 cm depending upon the purpose for which they are intended. Their principal use is as bases for flexible pavements, although their use as wearing course in permitted in some cases. Invariably water-bound macadam layers are covered with some sort of bituminous wearing surfacing soon after their construction. One course construction is permitted up to 12.5 to 15 cm compacted thickness, 22.5 cm by two course construction and greater thickness in three lifts. Each compacted layer is generally expected to compress from 75 to 80 per cent of loose thickness. For example, a 15 cm loose layer might become 11.25 to 12.00 cm.

13.2. MATERIALS REQUIREMENTS

Coarse aggregate. The coarse aggregate should be crushed stone of uniform quality, crushed gravel, and crushed air-cooled blast furnaces slag.

Trap rock is one of the best road stones with specific gravity ranging from 2.8 to 3.1. Hard limestone, dolomite and granite are also often satisfactory. The specific gravity of these stones ranges from 2.6 to 2.8. Sandstone, and quartzite whose specific gravity varies from 2.4 to 2.7 are somewhat less satisfactory, especially for wearing course. The quality of different deposits of stones varies widely and they should be carefully tested before acceptance. Crushed boulders, crushed gravel, and blast furnaces slag are also sometimes used in this type of construction, particularly for the bottom course.

Choking materials. Fine products resulting from crushing coarse stone, or naturals and which should have suitable binding quality.

IRC specifications. Standard specifications and code of practice for water bound macadam roads are covered by IRC : 19—2005 (Third revision).

Coarse aggregates may be either crushed or broken stones; crushed slags, or over-burnt brick aggregate. The coarse aggregates should conform to one of the grading given in Table 13.1 provided however, the use of grading No 1 should be restricted to sub-base courses only.

Table 13.1 Grading requirements of coarse aggregates

Grading No.	*Size range and compacted thickness for layer*	*Sieve designation (IS : 460)*	*Per cent by weight passing the sieve*
1	90 mm to 45 mm (100 mm)	125 mm	100
		90 mm	90—100
		63 mm	25—60
		45 mm	0—15
		2.4 mm	0—5
2	63 mm to 45 mm (75 mm)	90 mm	100
		63 mm	90—100
		53 mm	25—75
		45 mm	0—15
		2.4 mm	0—5
3	53 mm to 2.4 mm (75 mm)	63 mm	100
		53 mm	90—100
		45 mm	65—90
		2.4 mm	0—10
		1.2 mm	0—5

The coarse aggregates conform to the physical requirements set forth in Table 13.2.

Table 13.2 Physical requirements of coarse aggregates for water bound macadam

S.No.	*Type of construction*	*Test +*	*Test method*	*Requirements*
1	Sub-base	Los Angeles Abrasion value* or	IS : 2386 (Part 4)	Max. 50%
		Aggregate Impact Value*	IS : 2386 (Part 4) or IS : 5640**	Max. 40%
2.	Base course with bitumnous surfacing	Los Angeles Abrasion value* or	IS : 2386 (Part 4)	Max. 40%
		Aggregate Impact value*	or IS : 2386 (Part 4)	Max. 30%
		Flakiness Index***	IS : 2386 (Part 1)	Max. 20%
3.	Surfacing course	Los Angeles Abrasion value* or	IS 2386 (Part 4)	Max. 40%
		Aggregate Impact value*	IS 2386 (Part 4) or IS 5640**	Max. 30%
		Flakiness Index***	IS 2386 (Part 1)	Max. 15%

Notes:

* Aggregates may satisfy the requirements of either the Los Angeles test or Aggregate Impact Value Test.

** Aggregates like brick metal, kankar, laterite, etc., which get softened in presence of water should invariably be tested for impact value under wet conditions in accordance with IS 5640.

*** The requirement of Flakiness Index shall be enforced only in the case of crushed/broken stone and crushed slag.

\+ Samples for tests shall be representative of the materials to be used and collected in accordance with the procedure set forth in IS 2430.

IRC has further specified that screenings to fill voids in coarse aggregate should generally consist of the same material as the coarse aggregate. However, predominantly non-plastic material such as moorum or graval (other than rounded river borne material), if used for this purpose, should have liquid limt and plasticity index below 20 and 6 respectively and fraction passing 75 micron sieve should not exceed 10 per cent. As far as possible, the screenings should conform to the gradings given in Table 13.3. Screenings of type A should be used with coarse aggregates of grading 1 in Table 13.1. With grading 2, type A or B may be used, but with coarse aggregates of grading 3, only type B screening should be used. The use of screenings may be omitted in case of soft aggregates such as brick metal, kankar and laterite.

Table 13.3 Gradings for screenings for WBM

Grading classification	*Size of screening (IS:460)*	*Sieve designation (IS:460)*	*Per cent by weight passing the sieve*
A	13.2 mm	13.2 mm	100
		11.2 mm	95—100
		5.6 mm	15—35
		180 micron	0—10
B	11.2 mm	11.2 mm	100
		5.6 mm	90—100
		180 micron	15—35

13.3. SUBGRADE FOR MACADAM ROADS

Careful preparation of the subgrade is a necessity for any type of macadam construction. Macadam roads are flexible in nature and the failure and the deformation in the subgrade would certainly be expected to show up in the base and wearing surface in due course of time. For this reason and, in addition, because of the fact that surface irregularities in this type of road are much more difficult to correct than in some other types of construction. The condition of subgrade, before the construction of macadam layer begins, is of vital importance. Weak spots in the subgrade must be corrected before the base or surface course is placed, and the subgrade soil brought to a high degree of uniformity, density and stability. Stabilization may be necessary for some soils. The subgrade is brought to the desired elevation and cross-section before any stone is laid. It is also important that this underlying layer should be well drained. It is a common practice to spread 2.5 cm course of fine stone, or stone and sand, directly over the prepared subgrade before placing of a water-bound macadam base. This layer serves the purpose of preventing the intrusion of fine soil into the lower portion of the macadam under service conditions, and it is especially important in areas where fine-grained soils, such as silt and clay are encountered in the subgrade. This insulation course may be composed on one type of stone of proper gradation or it may be a layer of course stone with finer stone screening or sand placed on the top of the spread layer of coarse stone. This portion of the base is frequently rolled with 10-tonnes, 3-wheeled rollers to a satisfactory density and is carefully shaped to the desired section before the next layer is placed.

13.4. CONSTRUCTION METHOD

Spreading and dry rolling. On the prepared subgrade stone soling is spread uniformly to a proper camber. The stone may be spread directly from trucks by the use of spreader boxes or from trucks of stones placed along the road side. On some of the recent jobs, use of paving machine whose primary function is the laying of bituminous paving mixture has been made in spreading coarse aggregate in this type of construction. The advantage in using this type of equipment is that the hand labour usually associated with this kind of work is largely eliminated. Before rolling begins, all surface irregularities are eliminated and liberal use is made of templates and long straight edges in checking the uniformity of the spread material. The irregularities are much easier to correct in the loose layer than later. Rolling is done with 6 to 10 tonnes 3-wheeled power rollers or tandem or vibratory rollers. Rolling is usually begun at the sides and progresses towards the centre except on superelevated curves where rolling begins on the low side and progresses towards the high side. Passes of roller are made parallel to the centre line and are over-lapped in such a fashion that the entire surface will be covered by the rear wheels as rolling proceeds. Rolling is generally continued until the material does not creep or wave ahead of the roller. Irregularities and weaknesses in the coarse-stone layer are corrected as rolling proceeds, with additional stone being added as required. The objective of the roling is to thoroughly key the coarse material. Slight sprinkling of water may be done during rolling, if necessary. Rolling should not be done if the subgrade is soft or yielding.

Application of screening and rolling. The next step is the application and keying of stone screening of choking material into the coarse layer. The coarse screening may be spread by hand or by the use of mechanical spreader. Spreading, brooming and rolling operations are usually carried out at the same time and in conjunction with one another. Enough screenings are applied to fill the surface voids of the coarse layer and the rolling is conducted until the surface is firm and thoroughly compacted. The brooming may be accomplished by means of fibre or wire brooms attached to the roller units. Additional screenings may be added as needed, and any excess of fine material is removed before the next step begins.

Application of water and wet rolling. Following this application of stone screenings and dry rolling, the surface of the layer is sprinked with water and rolled again. The sprinkling and rolling are continued until all the voids are filled, wave of grout flushes ahead of the roller. This would indicate that all the voids are filled with choking material. Water may be applied manually or through a pressure distribution mounted on pneumatic tyres, and brooms may be used to distribute the wet stone screening evenly over the surface. The amount of water and screening applied is dependent upon a number of factors, including the size and nature of coarse stone, the properties of the screening and the type of surface desired of the completed base course. The amount of each of these items is largely dependent on the judgment of the engineer in the field.

A suitable filler material (Plasticity index not more than 9 in the case of W.B.M. and 6 in the case of surface treated W.B.M.) is then applied at uniform and slow rate in two or more successive thin layers. After each application of filler material, the surface is copiously sprinkled with water, the resulting slurry swept in with hand brooms or mechanical brooms or both to fill the voids properly and surface rolled by a roller, water being applied at the wheels to wash down the binding material that may stick to the wheels. The spreading of filler material, sprinkling of water, sweeping with brooms and rolling is continued until the slurry that is formed will, after filling the voids, form wave ahead of wheel of the moving roller.

Curing and application of second course. In case, this completed layer is to be the bottom or intermediate course of a multiple course water-bound macadam, it is allowed to dry thoroughly before the next course is placed. After this course is cured satisfactorily, the second course is placed and handled in the same manner as the first course. Well compacted shoulders with vertical face to the full depth of the course, thereby serving the same purpose as forms are usually built up before the placing of an additional course to aid in holding the stone in place. The top course may contain more stone screening than the lower course in the interest of placing a more tightly bonded and smoother surface.

Checking surface. At the time the top course is completed, the surface is finally checked for irregularities and deviations of more than 6.3 or 12.5 mm, as measured by 3 m straight edge from the specified cross-section and longitudinal grade, are corrected.

Curing and application of bituminous surface treatment. The completed macadam is frequently allowed to cure under traffic for a period of 14 to 30 days before bituminous surface is placed. During this period additional sprinkling and rolling may be required if needed to satisfactorily bond the surface. In some cases, calcium chloride is applied to the completed surface during this curing period.

TRAFFIC-BOUND MACADAM

This type of surface is produced by applying gravel, crushed stone or slag to the natural soils and allowing it to be rolled into place by motor vehicles. The granular material used in the first course is driven into the subgrade which supplies the internal friction necessary to support the traffic load. Subsequent layers when placed are compacted by traffic and fair surface is, thereby built up. This type of construction is not recommended where the traffic volume of more than few hundred vehicles a day is anticipated. Maintenance costs run high where higher traffic volumes are encountered. Lime stone is generally preferred on account of its cementing properties. Best results are obtained with material having maximum size of 12.5 to 20 mm. Stone is applied in layers not more than 2.5 to 4 cm thick and next layer is added when the first layer gets consolidated under traffic. The usual grading requirements relative to crushed stone as given in Table 13.4.

Table 13.4 Grading requirements relative to crushed stone

Sieve designation	*Per cent passing (by weight)*
25 mm	100
20 mm	90—100
10 mm	20—55
4.75 mm (No. 4)	0—10
2.36 mm (No. 8)	0—5

This type of surface is in reality nothing but a mechanically stabilished road and the modern trend is towards gravel stabilization than this type where the materials and construction are scientifically controlled.

WET MIX MACADAM ROADS (WMM)

Conventional Water Bound Macadam (WBM) construction is time consuming and manual with copius use of water. Another disadvantage of WBM is that segregation of aggregate takes place in the mix and the work results in non-uniformity in the finished surface.

In Wet Mix Macadam (WMM) construction, crushed graded aggregates and granular material like graded coarse sand are mixed with water in a mixing plant and rolled to a dense mass on a prepared surface. It has the following advantages over WBM construction.

(1) Superior gradation of aggregates

(2) Faster rate of construction

(3) Higher standard of densification

(4) Less consumption of water; and

(5) Strict standards of quality achieved.

The work can be done in many layers.

The thickness of an individual layer should not be less than 75 mm and could be upto 200 mm provided suitable type of compacting equipment is available. Suitable specification can be adopted for sub-base and base courses.

The guidelines for Wet Mix Macadam are covered under IRC:109-1997 and are discussed below:

13.5. AGGREGATES

Physical requirements: Coarse aggregates shall be crushed stone/crushed gravel/shingle, not less than 90 per cent by weight of gravel/shingle pieces retained on 4.75 mm sieve and shall have at least two fractured faces. The aggregates shall conform to the physical requirements set forth in Table 13.5.

If the water absorption value of the coarse aggregates is greater than 2 per cent, soundness test shall be carried out on the material as per IS: 2386 (Part V).

Table 13.5 Physical requirements of coarse aggregates for wet: mix

Test	*Test Method*	*Requirements*
1. Los Angeles abrasion value	IS:2386 (Part IV)	40 per cent (Max.)
or		
Aggregate impact value	IS:2386 (Part IV) or IS:5640	30 per cent (Max.)
2. Combined Flakiness and Elongation indices (Total)	IS:2386 (Part I)	30 per cent (Max.)

Grading requirements: The aggregates shall conform to the grading given in Table 13.6.

Table 13.6 Grading requirements of aggregates for wet mix macadam

IS Sieve Designation	*Per cent by Weight Passing Sieve*	
	Grading 1	*Grading 2*
53.00 mm	100	
45.00 mm	95-100	
26.50 mm		100
22.40 mm	60-80	50-100
11.20 mm	40-60	
4.75 mm	25-40	35-55
2.36 mm	15-30	
600 micron	8-22	10-30
75 micron	0-8	2-9

Material finer than 425 micron shall have Plasticity Index (PI) not exceeding 6.

13.6. CONSTRUCTION OPERATIONS

13.6.1. Weather

The work of laying of wet mix macadam is not done during rainy season.

13.6.2. Preparation of Base

The surface of the sub-grade/sub-base/base to receive the WMM course shall be prepared to the specified lines and cross-fall (camber) and made free of dust and other extraneous matter. Any ruts or softy yielding places shall be corrected and rolled until firm surface is obtained, if necessary by sprinkling water.

As far as possible, laying of WMM course over an existing thick bituminous layer may be avoided since it will cause problems of internal drainage of the pavement at the interface of two courses. It is desirable to completely excavate the existing thin bituminous wearing course where WMM is proposed to be laid over it. However, where the intensity of rain is low (less than 1300 mm), and the interface drainage is efficient, WMM can be laid over the existing thin bituminous surfacing by cutting 50 mm × 50 mm furrows at an angle of 45 degrees to the centre line of the pavement at one metre intervals on the existing road. The directions and depth of furrows shall be such that they provide adequate bondage and also serve to drain water to the existing granular base course beneath the existing thin bituminous surface.

13.6.3. Provision of Lateral Confinement of Wet Mix

While constructing WMM, arrangements are to be made for the lateral confinement of wet mix. This shall be done by laying materials adjoining shoulders alongwith that of wet mix layer. The sequence of operations shall be such that the construction of the shoulder is done in layers each matching the thickness of the adjoining pavement layer. Only after a layer of pavement and corresponding layers in shoulder have been laid and compacted, the construction of the next layer of pavement and shoulder is taken up.

13.6.4. Preparation of Mix

WMM shall be prepared in an approved mixing plant of suitable capacity having provision for controlled addition of water and forced/positive mixing arrangement like, pugmill or pan type mixer. For small quantity of wet mix work, mixing may be done in ordinary concrete mixers.

Optimum moisture for mixing shall be determined in accordance with IS:2720 (Part VIII), after replacing the aggregate fraction retained on 19 mm sieve with material of 4.75 to 19 mm size. However, the OMC and required number of passes to achieve the desired density may be determined at site during proof rolling, using the roller selected for compaction. While adding water, due allowance should be made for evaporation losses. However, at the time of compaction, water in the wet mix should not vary by more than ± 1 per cent.

13.6.5. Spreading of Mix

Immediately after mixing, the mixed material shall be transported to site and spread uniformly and evenly upon the prepared subgrade/sub-base/base in required quantities. Hauling of the mix over a freshly completed stretch is not permitted.

The mix may be spread either by a paver finisher or motor grader or a combination of both. However, the use of paver finisher should be preferred to motor grader for spreading. For portions where mechanical means cannot be used, manual method of spreading can be adopted. The equipment used for spreading shall be capable of spreading the material uniformly all over the surface. Its blade shall have hydraulic controls suitable for initial adjustments and maintaining the same so as to achieve the specified slope and grade.

The surface of the layer as spread shall be carefully checked with templates and all high or low spots remedied by removing or adding wet mix material as may be required. The layer thickness may be checked by depth blocks during construction. No segregation of coarse or fine particles should be allowed. The layer as spread should be of uniform gradation and should not have pockets of fine materials.

13.6.6. Compaction

After the mix has been laid to the required thickness, grade and cross-fall/camber, the same is uniformly compacted to the full depth with a suitable roller. If the thickness of the single compacted layer does not exceed 100 mm, a smooth wheel roller of 80 to 100 kN (8 to 10 tonnes) weight may be used. For compacting single layer of higher thickness upto 200 mm, the compaction shall be done with the help of vibratory roller of minimum 80-100 kN static weight or equivalent capacity to achieve the desired density. The speed of roller shall not exceed 5 km/hr.

Along forms, kerbs, walls or other places not accessible to the roller, the mix should be thoroughly compacted with mechanical tampers or a plate compactor.

After completing, the finished surface must present a well-closed appearance, free from movement under compaction equipment or any compaction marks, ridges, cracks and loose material. All loose, segregated or otherwise defective areas shall be made good to the full thickness of the layers and recompacted.

Longitudinal joints and edges are constructed true to the delineating line parallel to the centre line of the road. All longitudinal and transverse joints are cut vertical to the full thickness of the previously laid mix before laying the fresh mix.

13.7. IMPORTANT CONSIDERATIONS IN CONSTRUCTION PROCESS

While due care and attention is required on the whole process of WMM construction, the following are important points needing more attention:

(*i*) Sometimes because of moisture in the fines, these will not flow out from the bin of the three-bin feeder to the belt. In such situation, it would be necessary to have a small vibrator fitted on one of the side walls of the bin to intermittently shake it.

(*ii*) Control on water in the mix is of utmost importance; hence there should not be any variation in the grading, particularly of fines as it will effect the moisture content and uniform mixing. Similarly excessive fluctuations in the moisture content of the fines should be avoided. If necessary, slight increase may be made in the moisture content to account for the moisture loss in transit to the laying site.

(*iii*) Excessive silt or clay in fines should not be permitted, as besides spoiling the quality of mix, it will cause clogging in pugmil and storage silo.

(*iv*) The mixed material should be transported directly to site. Stockpiling of mixed material should be discouraged as excessive handing is the cause of segregation and moisture loss, both of which are detrimental to the quality of the wet mix macadam.

(*v*) There should be minimum joints in laying wet mix macadam. To ensure this, the daily output should at least be 500 linear metres. The width of laying also should be so adjusted to avoid the necessity of laying narrow strips *e.g.* against kerbs.

(*vi*) Single paver of 7 m width or two pavers each of 3.5 m width working in tandem within the short distances should be used for obtaining good results.

13.8. SETTING AND DRYING

After final compaction of the wet mix macadam course, the road is allowed to dry for 24 hours before overlaying with any bituminous layer.

Layout of a wet mix macadam plant with a capacity of 60 tonnes per hour is shown in Fig. 13.1.

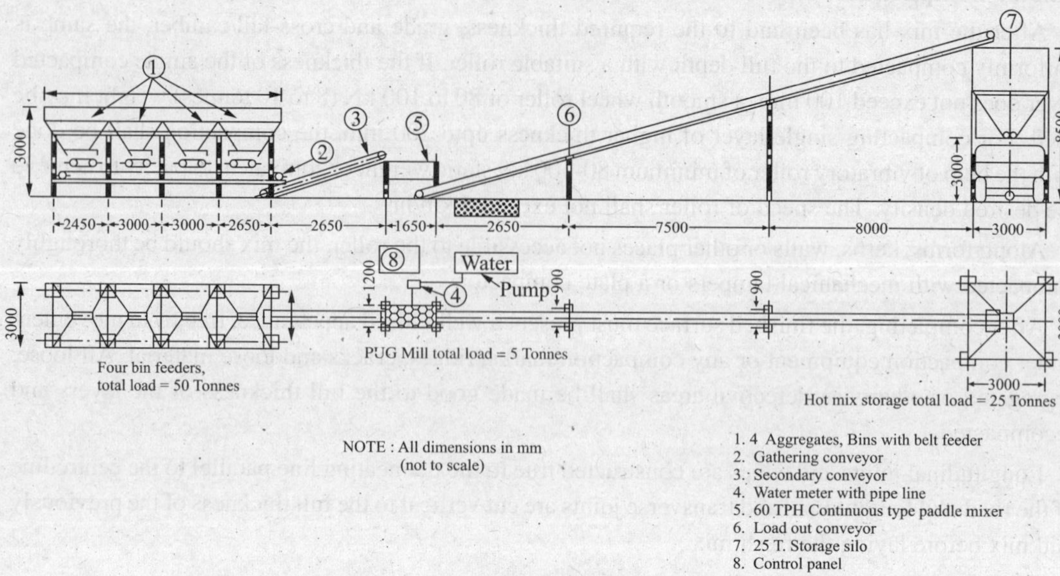

Fig. 13.1 Layout of wet mix Macadam plant (Capacity = 60 TPH)

BITUMINOUS MACADAM

In India, two types of Bituminous Macadam are commonly used:

1. Open graded bituminous macadam, in short known as bituminous macadam (B.M.)

2. Dense graded bituminous macadam, in short known as dense bituminous macadam (D.B.M).

13.9. BITUMINOUS MACADAM (B.M.)

This consists of a graded aggregate coated with bituminous binder. The mixture derives its main source of strength from the mechanical interlocking of the aggregate particles, and the cohesion imparted by the binder. The binder binds the aggregates together and during construction, lubricates the aggregate particles, thus facilitating the compaction of the mixture. Because of the open-graded aggregate matrix, the void content can be as high as 20-25 per cent.

A layer of bituminous macadam serves as a base/binder course, laid immediately after mixing on a previously prepared base.

The layer is generally laid in thicknesses of 50 mm to 100 mm.

13.9.1. Materials

Aggregates: The two commonly used gradings are given in Table 13.7 and the physical requirements of aggregates are specified in Table 13.8.

Binder: The bitumen content varies from 3 to 4 per cent. The binder used is straight run bitumen of penetration grade S.35 (30/40), S.65 (60/70) or S.90 (80/100). The choice of the grade of bitumen depends upon the intensity of traffic and the season of the year. Binders of high viscosity are required in regions of heavy traffic or difficult site conditions and during summer season. A reduction is viscosity may be permitted during winter months.

There is no need for a mix design. The rolling is continued till the density achieved is a minimum of 91 per cent of Percentage Refusal Density (PRD).

Table 13.7 Aggregate gradings for Bituminous Macadam (BM)

Mix Designation	*BM Grading 1*	*BM Grading 2*
Nominal Aggregate size	40 mm	19 mm
Layer Thickness	75-100 mm	50-75 mm
IS Sieve (mm)	Cumulative per cent by weight of total aggregate passing	
45	100	-
37.5	90-100	-
26.5	75-100	100
19	-	90-100
13.2	35-61	56-88
4.75	13-22	16-36
2.36	4-19	4-19
0.3	2-10	2-10
0.075	0-8	0-8
Bitumen Content, Per cent by Weight of Total Mixture[1]	3.1 to 3.4	3.3 to 3.5
Bitumen Grade	35 to 90	35 to 90

These bitumen contents have been found suitable for tropical Indian climates. Appropriate bitumen contents for conditions in cooler areas of India may be upto 0.5 per cent higher.

Table 13.8 Physical requirements for coarse aggregates and key aggregates for Bituminous Macadam

Property	*Test*	*Specification*
Cleanliness	Grain size analysis[1]	Max 5 per cent passing 0.075 mm sieve
Particle shape	Flakiness and Elongation Index (Combined)[2]	Max 30 per cent
Strength*	Los Angeles Abrasion Value[3]	Max 40 per cent
	Aggregate Impact Value[3]	Max 30 per cent
Durability	Soundness[4]	
	Sodium Sulphate	Max 12 per cent
	Magnesium sulphate	Max 18 per cent
Water Absorption	Water absorption[5]	Max 2 per cent
Stripping	Coating and Stripping of Bitumen Aggregate Mixtures[6]	Minimum retained coating 95 per cent
Water Sensitivity	Retained Tensile Strength[7]	Min 80 per cent

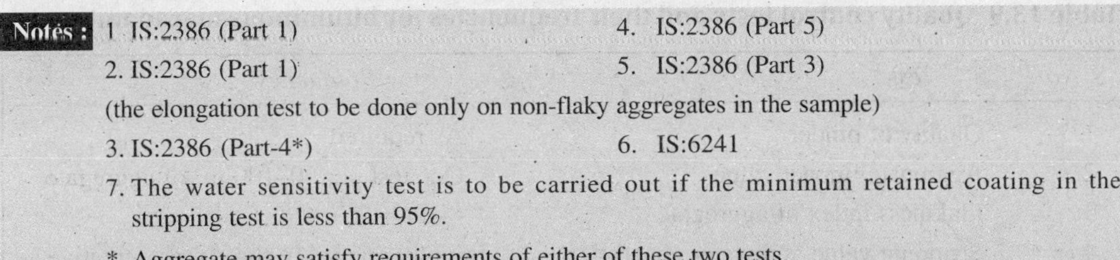
Notes :
1. IS:2386 (Part 1)
2. IS:2386 (Part 1)
(the elongation test to be done only on non-flaky aggregates in the sample)
3. IS:2386 (Part-4*)
4. IS:2386 (Part 5)
5. IS:2386 (Part 3)
6. IS:6241
7. The water sensitivity test is to be carried out if the minimum retained coating in the stripping test is less than 95%.

* Aggregate may satisfy requirements of either of these two tests.

13.9.2. Construction

The base on which bituminous macadam is to be laid is prepared, shaped and conditioned to the specified lines, grade and cross-section. The surface is then thoroughly swept and scrapped clean and free from dust and foreign matter. A tack cost is also applied over the base except when the laying of bituminous macadam is being preceded by bituminous levelling course. Bituminous macadam is not laid during rainy weather or when the base course is damp or wet.

Preparation and transport of mix. Hot mix plant of adequate capacity is used for preparing the mix. The temperature of binder at the time of mixing should be in the range of 150° to 165°C and that of aggregate in the range of 125° to 150°C, provided that the difference in temperature between the binder and the aggregate at no time exceeds 25°C. Mixing must be thorough to ensure that a homogeneous mixture is obtained in which all particles of the aggregate are coated uniformly. The mixture is then transported from the mixing plant to the site in suitable vehicles.

Spreading. The mix is spread immediately by means of a self-propelled mechanical paver with suitable screeds capable of spreading, tamping and finishing the mix true to the specified lines, grades and cross-sections. However, in restricted locations and in narrow widths, where the available plants can not operate, laying of the mix may be done manually. The temperature of the mix at the time of laying should be in the range of 110° to 135°C. In multilayer construction, the longitudinal joint in one layer should offset that in the layer below by about 150 mm. However, the joint in the topmost layer should be at the centre line of the pavement. Longitudinal joints and edges should be constructed true to the delineating lines parallel to the centre line of the road. All joints should be cut vertical to the full thickness of the previously laid mix and surface painted with hot bitumen before placing fresh material.

Rolling. After the spreading of mix, rolling is done by 8 to 10 tonne power rollers. Rolling is started as soon as possible after the material has been spread. Rolling should be done with care to keep free from unduly roughening of the pavement surface. Rolling of the longitudinal joint is done immediately behind the paving operation. After this, the rolling is commenced at the edges and progressed towards the centre longitudinally except that on super-elevated portions where the rolling is done from the lower to the upper edge parallel to the centre line of the pavement.

When the roller has passed over the whole area once, then any high spots or depressions which become apparent should be corrected by removing or adding fresh material. The rolling should then be continued till the entire surface has been rolled to compaction, there is no crushing of aggregates and all roller marks have been eliminated. Each pass of the roller should uniformly overlap not less than one third of the track made in the preceding pass. The roller wheels may be kept wet, in necessary, to avoid the bituminous material from sticking to the wheels and being picked up. In no case should fuel lubricating oil be used for this purpose. Rolling operations should be completed in every respect before the temperature of the mix falls below 80°C. The bituminous macadam should be provided with final surfacing without any delay.

13.9.3. Quality Control Tests

Quality control tests and their frequencies for bituminous macadam roads are given in table 13.9.

Table 13.9 Quality control tests and their frequencies for bituminous macadam roads

S. No.	*Test*	*Frequency*
1.	Quality of binder	As required
2.	Aggregate impact value	One test per 50–100 m^3 of aggregate
3.	Flakiness index of aggregate	—do—
4.	Stripping value and water absorption	Initially one set of three representative

		specimens for each source of supply. Subsequently, when warranted by changes in the quality of aggregate.
5.	Grading of aggregate	Two tests per day per plant on the individual constituents and mixed aggregates from the dryer
6.	Binder content (vide ASTM 2172)	Periodic subject to two tests per day per plant
7	Control of temperature of binder and aggregate for mixing and of the mix at the time of laying and rolling	At regular close intervals
8.	Rate of spread of mixed material	Regular control through checks on layer thickness.

13.10. DENSE BITUMINOUS MACADAM (D.B.M.)

The only difference from the Bituminous Macadam is that the aggregates are more closely graded and the voids are less. As a result, the mixture is dense. It may be laid to thicknesses varying from 50 to 200 mm. When the thickness is greater than 100 mm, laying is carried out in not less than two layers.

Dense Bituminous Macadam is used as a base/binder course for pavements subjected to heavy traffic.

The requirements for dense bituminous macadam are listed below:

13.10.1. Requirements for Dense Bituminous Macadam

Minimum stability (kN at 60°C)	9.0
Minimum flow (mm)	2
Maximum flow (mm)	4
Compaction level (Number of blows)	75 blows on each of the two faces of the specimen
Per cent air voids	3-6
Per cent voids in mineral aggregate (VMA)	See Table 16.8
Per cent voids filled with bitumen (VFB)	65-75

The requirements for minimum per cent voids in mineral aggregate (VMA) are set out in Table 16.8 (see page 520).

BITUMINOUS PENETRATION MACADAM

As already indicated the term bituminous penetration macadam refers to a type of macadam road in which the aggregates are compacted into a layer into which bituminous binder is introduced and bonded together by bituminous material. Use of this method of construction was quite wide spread decades ago and it has been most widely developed especially in areas in which suitable stone is readily available.

Instead of coating the stone with the binder, before they are laid in position on the road as in the case of road mixes and plant mixes, in this case, the binder is added after the stones are laid in position on the road.

The binder penetrates into the layer through the voids and binds the stone aggregates. A layer of smaller aggregates called key aggregates is spread on the surface and rolled so as to fill in the surface voids in the coarse aggregate layer. A seal coat is provided to make the surface more impervious to water.

The penetration macadam construction is not advocated for use in heavily trafficked regions, for which more efficient coated macadam is preferred. In India, it is still used in remote area where it is difficult to transport mixing and laying equipment. It can also be used as a temporary emergency repair to a pavement damaged by rains and floods.

13.11. MATERIALS REQUIREMETNS

Aggregates. The aggregate consists of crushed gravel (shingle) or broken slag of uniform quality which is free from flat or elongated pieces. They should be clean, strong, durable, of fairly cubical shape and free from disintegrated pieces, salt, alkali, vegetable matter, dust or other deleterious matter and adherent coatings. The aggregates should preferably be hydrophobic and of low porosity, satisfying the physical requirements set forth in Table 13.7 for bituminous macadam.

IRC has laid down that the coarse and key aggregates should conform to the grading given in Table 13.10.

Table 13.10 Grading requirements for coarse aggregates and key aggregates for Bituminous penetration Macadam

IS Sieve designation	*Cumultative Penetration Macadam of Total Aggregate Passing*			
	For 50 mm compacted thickness		*For 75 mm compacted thickness*	
	Coarse aggregate	*Key aggregate*	*Coarse aggregate*	*Key aggregate*
63 mm	—	—	100	—
45 mm	100	—	58-82	—
26.5 mm	37-72	—	—	100
22.4 mm	—	100	5-27	50-75
13.2 mm	2-20	50-75	—	—
11.2 mm	—	—	—	5-25
5.6 mm	—	5-25	—	—
2.8 mm	0-5	0-5	0-5	0-5
Approx. Loose Aggregate Quantities Cu.m/m^2	0.06	0.015	0.09	0.018
Binder Quantity (Penetration Grades[1] (Kg/m^2)	5		6.8	

Note: 1. If cutback is used, adjust binder quantity such that the residual bitumen is equal to the values in this Table.

Bituminous materials. The following bituminous materials are used as cementing agents for penetration macadam roads.

1. Paving Bitumen of penetration S-35 to S-90 as per IS:73

Lower penetration paving bitumens are used in warm climates and higher penetration are used in colder climates. In India, in warm climate: bitumen of penetration S-35 (30/40) has been used for surface course.

2. Road tar grades: In India RT—4 or RT—9 are used.

3. Bitumen emulsion RS—1 or RS—2. These are particularly useful when it is difficult to keep aggregate free from moisture as in rainy season.

Quantities of materials. The quantities of material required for this type of construction per 10 sq metre of road surface are given in Table 13.11.

Table 13.11 Quantities of materials required per 10 sq metre of road surface

Compacted thickness	*Binder*		*Coarse aggregate*	*Key aggregate*
	Straight run bitumen	*Road tar RT—4/RT—9*		
50 mm	50 kg	60—65 kg	0.60 m^3	0.15 m^3
75 mm	68 kg	82—88 kg	0.90 m^3	0.18 m^3

In semi-growing specifications as the binder needs to penetrate only half the depth of stone layer, the stones in the lower half-being held in place by a water-bound slurry similar to water-bound macadam, the material requirements in respect of bituminous material are less than for full grouted specifications. For 50 mm consolidated thickness, 30 kg of bitumen is required instead of 50 kg needed for full grout specifications.

The quantities of materials recommended by the Bureau of Public Roads for wearing surfaces 75 mm compacted thickness are given in Table 13.12. The quantities of aggregates are based upon a bulk specific gravity of 2.65. The amount of bituminous material given in the table are approximate and may be varied as desired by the engineer.

Table 13.12

	Aggregate (kg per sq m)	*Bituminous material (litres per sq. m)*
Coarse aggregate	120	—
Bituminous material, 1st application	—	6.75
Key aggregate, 1st spreading	14	—
Bituminous material, 2nd application	—	2.25
Aggregate, 2nd spreading	11	—
Bituminous material, 3rd application	—	1.35
Stone chips	7	—
Stone chips, supplemental stone pile	5	—
	157	10.35

13.12. CONSTRUCTION METHOD

After the base of subgrade is brought to uniform grade and cross-section and swept free of all loose and foreign matter, the coarse aggregate is spread uniformly in a loose layer to the correct thickness by means of suitable spreading devices. The surface of loose aggregate is carefully shaped by hand or by mechanical means and all high and low spots are remedied by removing or adding aggregate as may be required. Side forms are not generally used to retain the aggregate. Instead, well compacted shoulders are constructed with vertical face to the full depth of the course, thereby serving the same purpose as forms.

The coarse aggregate is then thoroughly rolled with power driven 8 to 10 tonnes roller with a minimum weight of 55 kg per cm width of rear wheel. Rolling starts longitudinally at the sides and proceeds towards the centre of the pavement with over-lapping at least half the width of the rear wheel on successive trips. It is a good practice to start rolling on the edge of the pavement with at least half of the rear wheel over-lapping the shoulder and continuing as described. Rolling continues until fragments of the aggregates are firmly inter-locked. Rolling is the most essential operation in macadam pavement construction to obtain the highest degree of interlocking which should never be overlooked or neglected. Although steel-wheeled rollers have for many years produced adequate compaction in macadam base courses, the newly developed vibratory compactors are showing a high degree of efficiency in this type of construction. Irregularities that appear during or after rolling are corrected by the addition or removal of the aggregate as the case may be and is followed by further rolling until compacted to a uniform surface.

After a true-to-grade satisfactory surface has been obtained without any further movement of the stone under the roller, an application of hot paving bitumen, road tar, emulsified bitumen at the rate specified in table 13.11 is made by means of pressure distributor or approved sprayer. Immediately after this application, fine aggregate is applied in sufficient quantity to prevent the roller wheel from sticking and sufficient also to fill the surface voids. The surface is then rolled with a 8 to 10 tonnes smooth steel wheeled roller (along with vibratory roller, if available) until it is firmly bound together and shows no movement under the roller. Key stone may be applied in one or two applications. Seal coat is applied immediately or within few days of completion of grouted surface. Before seal coat is applied, all excess aggregate lying on the surface should be removed. If the excess aggregate is allowed to remain loose on the surface and a seal coat is applied, there would be much larger sized aggregate for the amount of bituminous application which would result in the seal coat or surface treatment soon wiping off under traffic because of lack of binding material to hold it in place.

It is important that the temperature of the paving bitumen for emulsified bitumen, at the time of application, should be as specified. Usually limits are from 135 to 175°C for paving bitumen; between 50 to 65°C for bitumen emulsion and 80 to 120°C for road tar.

When the paving bitumen or road tar is applied at temperature lower than those mentioned above, the application is not uniform and road will not bind the aggregate into a unit mass. Such applications will develop rutting, ravelling and pot-holing. In this method, care must be taken not to over seal the voids in the second application of bitumen, otherwise there may be flushing of bitumen after the road is in use.

13.13. QUALITY CONTROL

In order to control the quality of bituminous penetration macadam roads, the following tests (Table 13.13), along with their frequencies, are prescribed by IRC.

Table 13.13 Quality control tests

S. No.	*Test*	*Frequency*
1.	Quality of binder	As required
2.	Aggregate impact value	One test per 200 m³ of aggregate
3.	Flakiness index	—do—
4.	Stripping value and water absorption of aggregate	Initially one set of three representative specimens for each course of supply. Subsequently when warranted by changes in the quality of aggregates.
5.	Aggregate grading	One test per 100 m³ of aggregate
6.	Temperature of binder at application	At regular close intervals
7.	Rate of spread of binder	One test per 500 m³ of aggregate

Semi grout construction method. This method is more economical than full grouting as less bitumen is used as the binder need penetrate only half the depth of the stone layer, the stones in the lower half being held in place by a water-bound slurry. This specification is recommended only where the traffic is not sufficiently heavy to warrant a full grout but is too much for surface dressing. The method is similar to the full grout specification except that the slurry is produced by laying the stone on a thin bed of sand having a high clay content. The stone is lightly watered during the rolling process and the slurry thus formed is allowed to rise about half-way up to the stone layer. This is controlled by the amount of water applied during rolling. After drying out, the surface is grouted with the bituminous binder and seal coated in the same way as specified for full grouting.

CEMENT-BOUND MACADAM

Cement-bound macadam roads are not popular in modern highway construction. Perusal of latest available specifications of nearly all the state highway departments of India show that none of these organizations includes this type of construction in its standard specifications. Such pavements were at one time covered by patents in the U.S.A. and were known as "Hassam' pavements. The first pavement of this type in the U.S.A. was built in Massachusetts in 1905. At one time it was quite popular in Europe and Australia also. Cement-bound macadam consists of a combination of clean coarse aggregate bound together by a cement-sand grout which is poured on the surface of a uniform layer of the compacted aggregate until all the voids in the mass have been filled. The effect is to produce concrete surface without mixing.

13.14. MATERIAL REQUIREMENTS

Coarse aggregate. Broken stone, broken slag or crushed gravel are used. The maximum size of coarse aggregate generally ranges from 40 to 63 mm and a reasonably uniform size is desirable, so that the void spaces will be open and permit the cement grout to penetrate and completely fill the layer of coarse aggregate. The size of aggregate is as per grading 1 and 2 in Table 13.7 used for bituminous macadam. The course aggregate should be reasonably free from dust and other particles, so that, after placing, the grout will not be prevented from penetrating and filling the entire layer.

Sand for grout. Clean and uniformly graded grout from IS. 2.36 mm sieve to 150 micron sieve size, with not more than 15 per cent passing 300 micron, is recommended.

Grout mixture. The grout mixture consist of 1 part cement and $1\frac{3}{4}$ to 2 parts of sand. The quantity of water should be sufficient for the grout to flow readily into the coarse aggregate but not great enough to cause the sand to separate from the cement and water. Weight measurement of sand is desirable but volume measurement, if carefully done, may be satisfactory.

The grout should preferably be made with the help of special mixture on the job or in truck which is essentially a colloidal mill which helps removal of any film of air or other gas thus ensuring a perfect mix which remains stable and does not separate until it sets.

The quantity of grout required depends on voids; the closer the stones are packed, the less the quantity of grout needed.

13.15. CONSTRUCTION METHOD

Construction process includes preparation of subgrade, setting of side forms, spreading and rolling of coarse aggregates, application of fluid grout, finishing and curing.

The broken stone should be lightly sprinkled before the grout is applied. After application of the grout, the surface is lightly rolled again with 6-tonne roller to squeeze the mortar up, until a uniform and smooth surface is obtained. Additional grout is used and uneven spots are set right with small size (10 – 12.5 mm) broken stone.

After proper finishing with floats similar to those used on concrete pavements, excess grout is swept from the surface by means of brooms or by working wet burlap belts over the surface. Curing is done for 7 to 21 days before the traffic is allowed.

□□□

Chapter

BITUMINOUS MATERIALS

Bituminous materials commonly denote substances in which bitumen is present or from which it can be derived. In India, Bitumen is defined as a viscous liquid or a solid consisting essentially of hydrocarbons and their derivatives, and which is soluble in carbon trichloroethylene and is substantially non-volatile and softens gradually when heated. It possesses water proofing and adhesive properties. American terminology uses the name 'asphalt' known as bitumen in India.

Bituminous materials include both bitumens and tars. Bitumens may be derived from the refining of petroleum or may occur as such in nature, either in a pure state or associated with varying quantities of mineral matter. Tars are bituminous condensates produced by the destructive distillation of organic materials such as coal, wood, etc.

As far as highway engineering is concerned, bituminous materials are principally used as binders for road aggregates and as water-proofing materials. Both bitumens and tars are produced commercially in a number of different types and grades.

14.1. COMPARISON BETWEEN BITUMENS AND TARS

Although bitumens and tars are very much similar in appearance, being black to dark brown in colour, they are produced by vastly dissimilar processes and are quite different materials, both chemically and physically. They differ in certain properties which are of importance in road work. Also, they have many properties in common, and the choice between the two is often a matter of availability.

Tars contain more free carbon which is insoluble in carbon trichloroethylene and they are also readily distinguished by odour.

Chemically speaking, the difference is concerned with the type of molecules present. The molecules in bitumens and tars are very large and complex and the chemistry of these substances is not well developed. In general, tar, especially derived from coal, is very resistant to the disintegrating action of water. For this reason, tar has greater utility of use below ground as water proofing material and for soil stabilization. However, tars are inferior in weather resistance to high quality bituminous materials when used in film form.

With regard to physical properties, the chief difference lies in their temperature susceptibility. Tar is more temperature susceptible than bitumen, becoming softer at summer temperatures and more brittle when cold. However, tars will usually wet and coat mineral aggregates with more facility than bituminous materials, especially when the aggregate is damp or dusty. Moreover, tar coating may be better retained in the presence of water.

The durability or resistance to weathering caused by progressive hardening of bitumen and tar is caused principally by oxidation and evaporation of volatile constituents. Other factors which can contribute towards their deterioration are polymerization and changes in physical structure. In general, road tars contain oil constituents quite volatile in nature and harden more rapidly when exposed to air. However, the weathering depends on a number of external factors such as temperature, surface area exposed, action of light, possible leaching action of water and the advantages, and disadvantages of each of these materials should be considered for selection for a given use.

The production, refining and testing of the broad classes of bitumens and tars are briefly discussed in the following paragraphs.

14.2. NATIVE BITUMENS

Natural deposits of bitumens are found in many widely separated sections of the world. These are usually designated as (*i*) Lake bitumens, (*ii*) Bitumenites, and (*iii*) Rock bitumens. All such natural bituminous materials are considered to have been derived from petroleum.

Lake bitumens. These are found in depressions in earth's surface which have accumulated in lakes from the discharge of springs. Some of these deposits have been of commercial importance such as Bermudez lake bitumen and Trinidad lake bitumen. Trinidad lake, just off the north coast of South America has an area about 45 hectares and is 40 m deep in the centre. Bermudez lake extends over 450 hectares and varies in depth from 0.6 to 2.4 m.

Most lake bitumens contain more or less moisture, hygroscopic or in the form of emulsion and this water is removed by dehydration process by heating in a suitable open container. The refined bitumen has very low penetration (20 to 30 for Bermudez bitumen and 1.5 to 4.0 for Trinidad bitumen at 25°C) and this material is softened by fluxing with a petroleum flux to produce a material of desired consistency for paving purposes. Also, it may be blended with bitumens produced from petroleum to get blended bitumens.

Bitumenites. This is one of the natural bitumen like substances presumably derived from metarmorphosis of petroleum. It is a hard, brittle, practically pure bitumen. Penetration at 25°C is 0 to 3. They are characterized by high fusing points. They are divided into three classes—gilsonite, glance pitch and grahamite. It is mined from seams and crevices in rock formations. Because of its higher cost and limited production, it plays a minor role in road work.They have been used in powdered form in road construction. They are principally used in the manufacture of varnishes, battery boxes, bitumen floor tiles, brake lining, printing inks, *etc*.

Rock bitumens. These are deposits of limestone or sandstone naturally impregnated with bitumen and have been extensively developed in various parts of the world such as Italy, France, Germany, Russia and the U.S.A. The bitumen content of rock may vary from trace to as much as 20 per cent and may be very soft or very hard. These deposits are stripped, drilled and blasted as per ordinary quarrying operations. The excavated pieces are then crushed and pulverized and blended to make a uniform and satisfactory material for paving.

14.3. PETROLEUM BITUMENS

Petroleum bitumens are derived from petroleum crude oil taken from numerous oil wells throughout the world. The crude oil is transported by pipe lines, tank cars, or barges to the refineries where it is separated into its various components by continuous flow refining process.

Essentially the process is one of the fractional distillation wherein heat is used to boil off from the crude oil such distillates as bitumens or gasoline, kerosene, diesel oil, *etc*. and to leave behind road oil or bitumen as a residue. This residual bitumen may be of various grades depending upon the degree to which distillates are removed. These residual bitumens may be further processed by air blowing, solvent extraction, blending, compounding and admixing with other ingredients to produce a great variety of products used in road work and for other purposes such as sealing, roofing, water-proofing, paints and various other industrial applications.

14.4. STRAIGHT RUN BITUMENS

The character of petroleum bitumen is dependent to some degree upon the nature of crude oil from which it is extracted. Crude petroleum may generally be divided into three classes (*i*) bituminous base, (*ii*) semi-bituminous base, and (*iii*) paraffinic base. In bituminous base crude oil, solid paraffins are either absent or are present only in traces. In paraffinic base or non-bituminous petroleums, solid paraffins are usually present. Since waxy materials are not desirable in road work, the most desirable bitumens for the road work are straight run materials produced by refining of bituminous base crudes.

The straight run bitumens are of two types; Type 1 and Type 2. Type 2 relates to bitumen derived from crude petroleum with excessive wax contents. Depending upon the penetration value, the straight run bitumens are futher sub-divided into six grades; as follows

Penetration Test Value	*Type 1*	*Type 2*
30/40	S-35	A-35
40/50	S-45	-
50/60	S-55	A-55
60/70	S-65	A-65
80/100	S-90	A-90
175/225	S-200	

Dehydration of Petroleum. Nearly all crude petroleums carry more or less water, some being entrained mechanically and in other cases held in a state of emulsion. Crude oils containing less than 2 per cent of water are accepted by the pipe lines and refineries. Where it is necessary to remove water and bring it within the prescribed limit, usually the crude oil is held in storage for a long period for the water and other impurities, such as sand and salt to be removed by a simple process of sedimentation. If water is present in emulsified form, it may be necessary to add emulsion-breaking agents to reduce water content. If the salt content of the crude oil reaching the refinery is so high as to cause corrosion problems, it may be necessary to remove it in a washing operation at the refinery before distillation begins.

Distillation. Petroleum is separated into various commercial products by means of distillation by which part of the hydrocarbon materials in petroleum are vaporized by heating them above their boiling points and then are condensed to liquids by cooling back to normal temperature.

Refining of crude oil is a continuous flow operation carried out by first pumping the crude through a tube still where its temperature is elevated and then immediately introducing the crude into a fractional tower. The fractional tower is an upright tall cylindrical vessel containing a series of platforms one above the other. As the hot crude is injected near the middle of the tower, lighter vapour or fractions gather on the uppermost platforms and are carried off to a condenser. Farther down from the top of the tower, heavier grades of crude are taken off for condensation until only the very heavy residue containing bitumen remains to be piped off from near the bottom of the tower. The temperature in such towers varies from top to bottom, decreasing upward.

In order to remove high-boiling point constituents such as heavy lubricating oils without changing them chemically, the pressure in the system must be reduced and/or steam distillation must be used. Action of steam is based on physical law that boiling point of a pair of non-mixible or slightly mixable liquids is lower than that of the pure components. Introduction of steam, therefore, serves material to lower the boiling point of petroleum and produces the maximum yield of heavy lubricating oils. Bitumens produced by such means are said to be vacuum-reduced or steam-refined.

Increasing the temperature of the crude through the tube still and increasing the vacuum in the tower results in an end product of low-penetration-range bitumen cement; conversely, decreasing the temperature of the crude and decreasing the vacuum, produces a higher penetration or softer material.

When the residue is distilled to a definite consistency without further treatment, the distillation is known as "*straight running*" and the residue, is termed "*straight-run bitumen*".

Cut-back bitumens in certain cases, a portion or fraction of the distillate is mixed with the residue at the close of distillation which is termed "*cutting back*", the object of which is to modify the properties of residual product or to dispose of economically a fraction which otherwise has little value commercially. They have the advantage, as compared to straight-run bitumens, that they can be used with cold aggregates and with a minimum of heat.

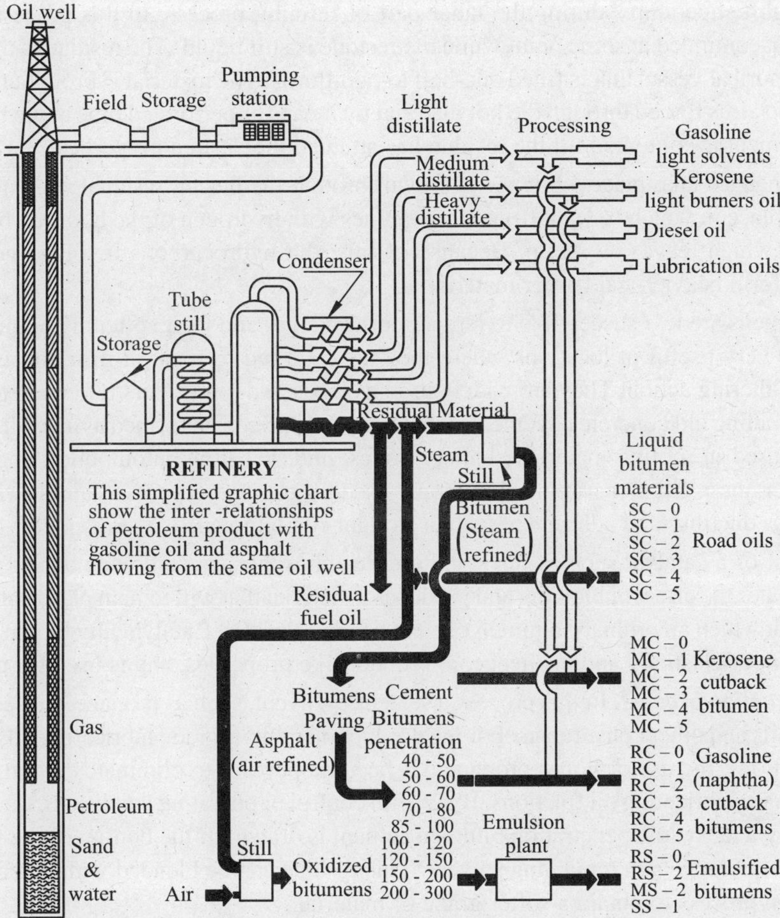

Fig. 14.1 Distillation of crude oil.

When a low-boiling solvent such as naphtha or gasoline is used to temporarily soften or cut-back the residual bitumen, the liquid bitumen produced is called rapid-curing (RC) cut-back bitumen.

When intermediate-boiling point solvent such as kerosene is used, the liquid bitumen is called medium-curing (MC) cut-back bitumen. When this type of liquid bitumen is used, the solvent will not evaporate as rapidly as the gasoline range solvent used in the manufacture of RC cut-backs.

Similarly liquid bitumens may be manufactured by either fluxing bitumen with a high-boiling point less volatile distillate such as gas oil or by controlling the rate of flow and temperature of the crude during distillation process. These are slow-curing (SC) cut-back bitumens. Since the solvent in this case is a semi-volatile material, it requires much longer time for curing.

These three types of liquid bitumens are made in different grades which are distinguished on the basis of viscosity. In standard specifications there are six grades of RC and MC cut-backs, numbered from 0 to 5 (RC-0, RC-1, RC-2, RC-3, RC-4, and RC-5) and seven grades of SC material numbered from 0 to 6. Bitumen content and viscosity increase with the grade number, the 0 grade having the lowest viscosity and the 5 to 6 grade, the highest. For example, to make RC-0, the blend would be approximately 45 per cent bitumen whereas RC-5 grade would require approximately 15 per cent solvent and 85 per cent bitumen cement.

Air-blown bitumens. These are produced by blowing air through residual semi-solid bitumens at moderately high temperatures during the latter part of refining process. In this process the regular distillation is discontinued at some point while the residue is still liquid. The residue is then placed in an upright cylindrical vessel that is filled one-half to two-thirds. The material is brought to the desired temperature and air is forced through the hot material by means of perforated pipe at the bottom of the still. This treatment is continued till the residue has attained the desired properties.

Chemical changes take place while air is blown through the heated residual oil. An exothermic reaction takes place in which oxygen from air combines with hydrogen of the hydrocarbon molecule to form water which leaves as steam. Because of this, the hydrocarbons involved condense and polymerize to form heavier and harder material.

Blown bitumens are less susceptible to temperature changes and their softening points are higher. Thus, they are very useful in locations where they are required to retain a firm consistency when exposed to weathering action. They are widely used as crack and joint fillers for concrete pavements and for under-sealing old concrete pavements where cavities have been formed as a result of pumping. They are also used in roofing materials, battery boxes, under-coating automobiles, and waterproof paints. However, these bitumens are not commonly used for paving purposes as air blowing produces bitumens of less ductility and adhesiveness than straight-run bitumens.

The addition of a catalyst such as ferric chloride or phosphorous pentoxide during the blowing process accelerates the effect of blowing and produces a material that will remain plastic at temperature far below that at which an ordinary bitumen cement becomes brittle. Catalytically blown bitumen has a certain amount of elasticity and resilience, *i.e.* rubber-like properties, and is used for canal lining.

Solvent extraction process. In this process a selective solvent, such as propane, is used to separate the paraffinic oils and other constituents. For the production of high-grade lubricating oils from heavy petroleum fractions, the use of liquid propane has become popular to eliminate certain constituents present in ordinary lubricating oil fractions. By careful control of operating conditions, the end product which is usually a fairly low penetration bitumen cement is drawn off the bottom of the vessel and is steam stripped to remove any remaining solvent. This residue can be blended with other soft residue from distillation process to obtain a softer grade of material.

Cracking process. This process of refining is used to obtain a higher percentage of motor fuel (gasoline). In this process the objective is to cause chemical change by intense heat and high pressure so that a certain portion of the higher boiling point constituents decompose or break down, forming correspondingly larger yields of low-boiling point constituents.

The chemical change in the cracking process produces a bituminous type of materials which are generally regarded less durable or weather-resistant than straight-run materials and as such, are not widely used for road work. These materials are known as "*pressure or petroleum tar*" or "*cracked oils*". These cracked residues can be used for oiling of natural earth roads where they are cheap. They can also be mixed in suitable proportions with straight-run bitumen or air-blown bitumen to produce bitumen of desired grades. There is some evidence that cracked oils are useful as water proofing materials and for stabilization of fine-grained plastic soils.

Emulsified bitumen. Emulsified bitumen is an intimate mixture of water, bitumens, and emulsifying agent. This is another method of rendering an bitumen cement in liquid form. It has advantages over hot bitumen in that it can be used with cold as well as hot aggregate and with aggregate that is dry, damp, or wet. Its use with wet aggregate is an advantage for this material over the other type of liquid bitumens which are cut-backs. In addition, emulsified bitumen does not need any heat for its application.

Bitumen cement cannot dissolve in water, so it is necessary that bitumen and water exist in separate phases, one dispersed in the other in the form of minute droplets. In a properly made emulsion bitumen droplets range in size from 0.001 to 0.01 mm diameter. In order to prevent the bitumen droplets from coalescing, an emulsifying agent is necessary which is a material usually containing polar and nonpolar molecular structure, so that the molecules will concentrate at the interfacial layer between the two liquids and orient themselves to give the required action. Soaps are commonly used for this purpose. In addition to an emulsifying agent, the emulsion may contain a stabilizing agent, usually some form of protein, which will make it still more difficult for the particles to coalesce.

Bitumen emulsions do not depend upon evaporation of solvent for their setting characteristics but when emulsion is spread out thin on the surface of aggregate, it will set by virtue of coagulation of bitumen particles and the squeezing out of water between them. For this reason, emulsified bitumen may set faster and more completely than cut-back bitumens that cure over a long period of time. However, dry weather helps to remove the water quickly for road building.

Emulsified bitumen is usually manufactured by the use of a high speed disintegrator or colloid mill. Great shearing is applied as a mixture of water, bituminous materials and emulsifying agent is passed between a revolving and a stationary plate or two plates rotating in opposite directions which are separated by a very small space (0.05 to 0.075 mm). The bituminous material is reduced in the range of colloidal size and is dispersed in water. It is necessary to heat the bitumen before mixing to render it fluid. As temperature of finished emulsion cannot be high because of rapid loss of water, the hardness of bitumen used is usually limited up to 40 penetration.

Emulsified bitumens are made in three principal types-as per IS : 8887 rapid-setting (RS), medium-setting (MS), and slow-setting (SS). The type is controlled by the type and amount of emulsifying agent used.

Based on the type of emulsifier used, bituminous emulsion can be :

(1) Anionic-bitumen globules being electro-negatively charged and the aqueous phase being alkaline.

(2) Cationic-bitumen globules being electro-positively charged and the aqueous phase acidic.

The classification as per AASHTO/ASTM is more elaborate and has also been adopted in the Ministry's specifications.

This is outlined below:

Anionic	***Cationic***
RS-1	CRS-1
RS-2	CRS-2
MS-1	
MS-2	CMS-2
MS-2h	CMS-2h
HFMS-1	
HFMS-2	
HFMS-2s	

SS-1	CSS-1
SS-1h	CSS-1h

Notes:

(*i*) The numbers 1, 2 relative to viscosity.
(*ii*) C stands for cationic, while absence of C denotes anionic.
(*iii*) *h* stands for harder
(*iv*) *s* stands for softer.

The disadvantage of emulsified bitumens is that freight charges for dispersing water have also to be paid which is an added expense, the usual amount of water contained in emulsion is in the range of 30 to 50 per cent.

14.5. ROAD TARS

As already mentioned, tars generally are refined as a substance obtained by the condensation of distillates, resulting from the destructive distillation of organic and bituminous materials. Many organic substances may be treated to yield tar. However, the following tar products are of importance in highway industry.

Coke-oven coal tar. This is produced as a by-product in the manufacture of coke from bituminous coal for the steel industry and other uses. Tar is produced when bituminous coal is heated in the absence of air so that it decomposes into volatile substances and coke. Distillation is done continuously in series of ovens, each containing 8 tonnes or more of coal which is heated to temperatures of 1200°C or higher. The volatile substances include gases, light oils, tars and chemicals. Properties of tar vary with the coal used, design of oven, temperature used, and pressure in the system as it affects the time in which the volatile gases are exposed to incandescent coke.

The vapour is collected and passed through various forms of condensers and tar is removed by scrubbing with ammonia liquor in scrubbers. Ammonia liquor itself is a by-product of the operation. The tar and ammonia liquor are separated by decantation.

Water gas tar. The other main source of crude tar is from the cracking of petroleum in the carburated-water gas process in which water gas is produced for heating. Water gas is produced by allowing steam to pass through coke which is heated to a point of incandescence. This water gas (carbon monoxide and hydrogen gas) are too low in BTU value for heating purposes and it is enriched by the addition of hydrocarbon gases formed from the cracking of petroleum distillate (gas oil) as they are sprayed into a closed chamber carburator) maintained at a correct temperature where it is vaporized and combined with the gas produced in the heating chamber. The cracking of petroleum oils produces, in addition to hydrocarbon gases, a residue of heavy hydrocarbons which condenses to form tar.

Gas-work coal tar. In the process of eliminating gas from bituminous coal, tar is produced as a by-product. Coal is heated in ovens with water gas obtained by passing air and steam through incandescent coke. Vapours are driven off at temperatures ranging from 700 to 1000°C. These volatile materials are collected in a closed through or hydraulic main which is partially filled with water. The temperature of vapour is reduced to about 65°C in hydraulic main and a large percentage of the tar condenses in water from which it is later removed. The remaining vapours are also converted into tar by passing them through a sequence of operations utilizing condensers, tar extractors, scrubbers and purifiers.

The crude tar, whether from coal or petroleum is then sent to the tar refining plant where it is dehydrated if necessary and then fractionally distilled to produce residual products of various consistencies known as road tars and tar pitches. The crude tar contains materials such as benzole which are valuable as fuel and for the production of chemicals of many different kinds (*e.g.* dyestuffs, plastics, explosives, medicines). Moreover, crude tar being a thin material is quite unsuitable for use as road binder. It is therefore subjected to a distillation process whereby it is separated into a number of different fractions, some of which may later be blended together to form road tar and some of which are used in industry for other purposes. The distillation process of crude tar to produce various grades of road tars is similar to the distillation process for producing petroleum bitumens.

Road tars are prepared in 14 grades designated as RT-1 to RT-12 and RTCB-5 and RTCB-6. Grades RT-1 to RT-5 and RTCB-6 are light tars which can be applied cold or at temperatures below 65°C. RT-7 to RT-12 are heavier grades of road tar which require heating up to 120°C. The grades are progressively more viscous as the number designation increases. Only two cut-back road tars RTCB-5 and RTCB-6 are used where low-temperature application and quick setting are desired. No emulsified tar materials are available in market at present for use as binders for road-building purposes.

14.6. WEATHERING OF PAVING BITUMINOUS MATERIALS

In order to serve satisfactorily as a binder, a paving bituminous material must remain plastic. When bitumen or tar is exposed to the weather in a thin film, it eventually looses much of its plasticity and becomes brittle because of chemical and physical changes. This natural deterioration is known as weathering.

The weathering of bitumen or tar in a pavement caused by progressive hardening of the bitumen eventually results in the formation of cracks. As weathering progresses still further , these cracks become wider. Unless remedial measures are taken, surface water enters into the open cracks to soften the underlying base, or to freeze, causing the bituminous pavement to break up. Deterioration of bituminous binder can also result in abnormal surface abrasion of the pavement.

Weathering of paving bituminous material is principally caused by oxidation and volatilization. Oxidation consists of chemical attack on the bitumen or tar by oxygen in air. Volatilization consist of evaporation of the lighter hydrocarbons, in the bituminous material. The effect of both factors is a progressive permanent hardening of the bitumen, which may be measured by the penetration test. It has been shown that when bitumen in a pavement has been reduced to penetration of 30, it is very likely to become brittle and form cracks and when it has reduced to a penetration of 20 it is virtually sure to crack. Thus, it is evident that an advantage is gained by using the softest (highest-penetration) grade of bitumen or tar. From practical consideration, proper grade depends primarily on temperature range and secondly on traffic. A pavement using 40 penetration bitumen is very close to the border line of brittleness (30 penetration) and will obviously crack much sooner than will a pavement containing 100 penetration bitumen.

The rate of both oxidation and volatilization is accelerated greatly by raising the temperature of the bituminous material. Approximately, the reaction doubles for each 10°C rise in temperature. For instance, the rate of oxidation and volatilization that occurs in a mix agitated at 180°C is eight times as great as that would occur at 150°C. This explains why overheating can seriously injure bitumen or tar even though overheating is only by a few degrees.

Weathering due to oxidation and volatilization also depends upon the exposed surface area of bitumen or tar. Both rate of oxygen absorption and the rate of evaporation loss are directly proportional to the surface area exposed and inversely proportional to the volume. It is due to this effect of surface-area-volume ratio which accounts for the fact that a large mass of bitumen may be held in storage tank at a temperature of 150°C for several weeks or even months without undue hardening whereas this same bitumen when agitated in a film form in a mix at 150°C will loose several points in penetration in less than a minute in a pugmill. This also explains why in the design of high type bituminous pavements, smallest practicable minimum air voids are kept in the compacted mix in order to reduce to minimum the area exposed to oxidation.

Another factor which causes weathering of bituminous materials is the action of light. The active rays of the sun may destroy the bituminous molecules breaking them down into water and water-soluble products. This process is called photo-oxidation because the reaction is essentially one of oxidation, accelerated by the action of light. Fortunately, the destructive rays are able to penetrate only a few molecular layers past the surface of the bitumen. Also, they cannot penetrate through the aggregate particles in the pavement. For these reasons photo-oxidation is of secondary importance in thick pavements but it can contribute considerably to the weathering of light seal coats and very thin bituminous surfaces.

TESTING METHODS AND THEIR SIGNIFICANCE

Large number of different laboratory tests are performed upon bituminous materials for testing their properties for identification, checking the quality, uniformity and their suitability for any given use and also for checking compliance with the specifications. Most of these tests are conducted in accordance with methods of tests established by standards institution such as BIS, ASTM, AASHTO. The usual laboratory tests of bituminous materials used in road work along with their designation are given in Table 14.1.

Brief description of test methods and their significance is given below. The reader is referred to the standard test for details of the test procedure in each case.

14.7. GENERAL TESTS

Specific gravity. This is the ratio of the weight of a given volume of the material to the weight of an equal volume of water at a given temperature, usually 25°C. Specific gravity may be measured directly by means of hydrometer for the fluid grades and by displacement method for solid grades by weighing in air and weighing when completely immersed in distilled water at the standard temperature.

The test is usually performed with a pycnometer bottle. By filling the bottle only partially with bituminous material, the bitumen volume can be determined by the difference between the total volume of the bottle and the volume of water required to complete the filling; and thus specific gravity can be calculated.

The principal use of specific gravity is in converting from volume to weight measurements and vice versa for billing purposes and for mixture calculations.

The usual value of specific gravity for bituminous material varies from about 0.92 for thin liquid materials to about 1.06 for semi-solid materials. For road tars, it varies from 1.08 to 1.24. This test has some value for identification purposes and for checking uniformity.

Flash point. Two methods are in common use. For cut-back bitumens (RC and MC), flash point is generally determined by the use of Tagliabue open cup apparatus in which heating takes place in a glass cup held in a water bath. For other bituminous materials Cleveland open cup is used. In this bitumen is heated in metal container suspended in air.

Table 14.1 Laboratory tests for bituminous materials used in road-work

Name of test	*IS designation*	*AASTHO designation*	*Applicability*
General			
Specific gravity	IS : 2720 (Part 3) - 1980	T 43	Bitumens and tars
Flash point (open cup)	IS:1209-1978	T 48	Bitumens only
Flash point (Tag open cup)		T 79	Bitumens only
Water content	IS:2720 (Part 2) - 1973	T 55	Bitumens and tars
Consistency			
Penetration	IS:1203-1978	T 49	Bitumens only
Viscosity (saybolt-furol)	IS:1206-1978	T 72	Bitumens only

Specific Viscosity (Engler)	—	T	54	Tars only
Float test		T	50	Bitumens and tars
Softening point	IS:1205-1978	T	53	Bitumens and tars
Solubility				
Solubility in carbon disulphide or	IS:1216-1978	T	44	Bitumens and tars
Solubility in carbon tetrachlorate or trichloroethylene		T	45	Bitumens only
Spot test		T	102	Bitumens only
Ductility	IS:1208-1978			
Ductility test		T	51	Bitumens only
Volatility				
Loss on heating and penetration after loss on heating test	IS:1212-1978	T	47	Bitumens only
Distillation				
Cut-back bituminous materials	IS:217-1988	T	78	Bitumens only
Tar products	IS:73-1992	T	52	Tars only
Bitumen residue of Specified penetration		T	56	Bitumens only

The test is performed by heating a sample of the material in an open cup at a specified rate and determining the temperature at which a small flame passed over the surface will cause the vapours to temporarily ignite or flash. The usual value of flash point for bitumen cement is above 175°C. For harder grades, it is above 230°C. Cut-back bitumens flash at low temperature, less than 40°C, RC cut-back may flash at as low as 25°C.

Flash point requirements are included in the specifications for bituminous materials for safety reasons. Since the test is intended to determine the temperature to which the material may be heated safely, it is considered a good practice not to heat bitumen cements higher than 10°C below their flash point. Cut-back bitumens are commonly used at temperatures above their flash point and as such all of these materials present some danger in use and should be handled carefully. The more volatile the solvent in the liquid bitumen, the more hazardous is its use.

Water content. The water content present in bituminous material is determined by mixing a specified weight of material with a water free petroleum distillate in a metal still or glass flask attached to water-cooled condenser with a glass graduated trap. Heating and fluxing is continued till all the water is transferred to the bottom of the glass trap where it is measured and expressed as a percentage by weight of the original material.

Determination of water content in bituminous material is necessary as low water contents are necessary to prevent foaming of the substance heated above the boiling point of water.

14.8. CONSISTENCY TESTS

The consistency of bituminous materials ranges from a very thin liquid for 0 grade of cut-back, to that of a stiff semi-solid resembling sealing wax in the case of blown bitumen cement. Because of this wide range, there is no one instrument which will satisfactorily measure the consistency of all bituminous materials. The following methods of measuring consistencies are now in common use:

1. Penetration
2. Furol viscosity
3. Engler specific viscosity

4. Float test

5. Softening point

Penetration is the consistency test used to designate grades of bitumen cements. As bitumen cement being semi-solid at room temperature, it is not practicable to determine its consistency by Furol viscosity apparatus even when it is heated to decrease the viscosity.

Penetration is the distance in tenths of a millimeter that a standard needle can penetrate the sample under specified conditions of time, temperature, and load on the needle. Standard penetration, which is implied, unless other conditions are stated, is for a load of 100 g applied for 5 sec at a temperature of 25°C. Obviously, softer materials will have higher penetration values. Other test conditions used are (*i*) 0°C, 200 g, 60 sec; and (*ii*) 45°C, 50 g, 5 sec.

Penetration test is an identification test to identify bitumen as to its hardness and grade. Lower the penetration, stiffer is the bitumen cement and more strongly it will bind the stones together in bituminous mix. For these practical considerations consistency is determined at 25°C which is about the average room temperature.

This test is not suitable for tars because of high surface tension exhibited by these materials and the fact that they contain relatively large amounts of carbon.

Furol viscosity (IS:3117). Viscosity is the broad general term used for consistency and is a measure of resistance to flow. Thus higher the viscosity of a liquid, the more nearly it approaches a semi-solid state in consistency. This test is applied to liquid bitumens.

In this test the time in seconds is measured for 60 cc of the material to flow through a standard orifice of 3.8 mm diameter under standard falling head conditions and at a standard temperature. It is not practicable to measure Furol viscosity at the same temperature on all liquid bitumens. The viscosity of bitumen is a function of the temperature, the viscosity decreasing with increased temperature. Because of this, the temperature at which each grade is tested is chosen so as to give a practical time of efflux. Standard temperatures used are 25, 50, 60, 80 and 100°C.

Engler specific viscosity. In this test the time in seconds for 50 cc of road tar to flow through the standard orifice of the Engler viscosity meter at a specified temperature is divided by the time in seconds for 50 cc of water to flow through the same orifice at 25°C. The result is called specific viscosity. The test is similar to Furol viscosity except that the dimensions of the oil tubes are different. In Saybolt apparatus, the column of oil above the orifice is relatively high in relation to its diameter while in Engler apparatus, the height of oil column is relatively low in relation to its diameter.

The viscosity tests are used to control the consistency, which in turn determines the grade of cut-backs and tars.

Viscosity is an important characteristic of a bituminous material because it determines in large measure how the material will function when used. For example, low viscosity materials (less than 150 sec viscosity) are recommended for priming. Similarly, materials with Furol viscosity less than 300 sec. at 60°C are recommended for road mixing with aggregates containing 5 to 15 per cent material passing 75 micron sieve. With coarser aggregates with less dust, more viscous grades can be used.

Float test is used to measure the consistency of materials too soft for the penetration test and too viscous at the desired test temperature for the viscosity test. It is also, a useful consistency test in those cases where the quantity of material tested is small. Usually bitumens which are more viscous than grade 5 liquid bitumens cannot be conveniently tested by the Furol viscosity test. Also, those which have a penetration higher than 300 can not be conveniently tested by the penetration test. For this reason, the Float test is specified to the residue from the distillation of SC materials and heavier grades of road tars.

The test is performed by filling a small brass collar with the material to be tested, cooling collar and contents to 5°C, screwing the collar to an aluminium float, and floating the assembly in water at a specified temperature (50° C for SC residues). The float test time in sec. is from placing the float in the water until the water displaces the plug of material in the collar and breaks through in the float.

Softening point. The test is performed by forming a sample in a small brass ring (1.6 cm inside diameter and 0.62 cm high) and chilling it in a melting ice bath. After placing a 1 cm steel ball on the sample surface, the water bath temperature is raised at the rate of 4°C per minute. The temperature at which the sample sags under the weight of steel ball and touches the bottom of the container or other surface 2.5 cm below the sample is the softening point temperature.

This test may be classed as a consistency test in that it measures the temperature at which the bituminous material reaches a given consistency. It is used in specifications, of bitumens for crack filling, joint sealing of roofing, etc. where materials are used in thick films to ensure that they will not flow during application.

This is also a simple direct method of determining temperature susceptibility characteristics of bitumens. Bitumen having a higher softening point for a given penetration at 25°C is less susceptible to consistency change due to temperature. This is particularly so in case of blown bitumens as air blowing raises the softening point in relation to the hardness. The temperature susceptibility of bitumens also varies with the crude oil from which it was produced, but this variation is minor compared to one which exists between blown and paving bitumens.

Temperature susceptibility may also be measured by comparing consistency measurements such as penetration or viscosity at two or more temperatures.

The softening point for paving bitumens of penetration 40 to 300 ranges from about 55°C down to around 35°C. For blown bitumens, this test is highly significant because it is important that softening point of these materials be well above the temperature that they will reach when exposed to the sun which may be as high as 65°C.

14.9. SOLUBILITY TEST

The test is included in specifications for bituminous materials to determine the total bitumen present and to guard against the inclusion of undesirable amounts of foreign materials (salt, dirt, carbon and minerals). As the definition implies, it is necessary to use carbon disulphide as solvent to determine the bitumen content. However, carbon disulphide is highly inflammable and an unpleasant solvent to use. Carbon tetrachloride, because it is non-inflammable and is practically as good a solvent for bitumen as carbon disulphide is frequently substituted. For tars, carbon disulphide is always used.

In performing the test, a small sample of about 2 g of the bituminous material is dissolved in 100 ml of solvent and the solution filtered through an asbestos mat formed in the bottom of Gooch crucible. Insoluble material retained in the crucible is dried, cooled and the increase in weight of crucible is determined and expressed as a percentage of the weight of original sample.

As already stated, cracked bitumens weather much more rapidly than uncracked bitumens and many paving specifications are written to eliminate cracked bitumens. Cracked bitumens, when subjected to solubility test in carbon tetrachloride, yield 0.5 per cent or more of a black carbonaceous residue (carbons-soluble in carbon disulphide but insoluble in carbon tetrachloride). Bitumens which are badly cracked may be easily identified. Uncracked bitumen has a glossy mirror-like surface. The one which is badly cracked has a dull surface.

Solubility test is used in specifications to ensure due care in refinery operations and to detect contamination in transit.

14.10. SPOT TEST

When the degree of cracking is slight, solubility test does not indicate these characteristics. For detection of overheating or cracking, a more sensitive means of detection is employed by means of spot test. This test measures whether bitumen is hetrogeneous or homogeneous. Overheated or cracked bitumen will be colloidaly unstable.

In the standard test, 2 g of bitumen is dissolved in 10.2 ml of standard naphtha solvent. After one hour and again after 24 hr a drop of the solution is placed on a filter paper. If the stain on the paper is of a uniform colour, the test is negative and the bitumen is considered to be uncracked and is acceptable. If the spot forms a dark brown or black circle in the centre with a surrounding annular ring of lighter colour, the result is positive and the material is considered to be cracked.

This test is controversial as it detects overheating but it does not always detect blend of cracked and normal materials. Moreover, some bitumens due to the nature of crude oil show positive result even though they are not cracked. In such cases naphtha is employed which meets the distillation requirements of the standard solvent but which is made from the parent crude oil. Because of these short comings, there is no standard method for this test.

14.11. DUCTILITY TEST

Ductility is the property which permits bitumens to undergo great deformation (elongation) without breaking.

The standard test is performed by determining the distance in cm that a briquette of material having a cross section of 1 cm sq will stretch without breaking when the rate of pull is 5 cm per min and the test temperature is 25°C.

Most paving bitumens have ductility values of over 100 cm but highly air blown bitumens used for joint and crack fillers have ductility values of 10 cm or less.

High ductility is associated with materials of high temperature susceptibility. Also, materials with high ductility generally are adhesive and have good cementing properties. From the stand point of these two factors, it is felt that some reasonable value of ductility is desirable, but extremely high ductility is not necessarily good. Ductility may be required at two different temperatures to obtain some indication of the tendency of the material to change in temperature. Low temperature value may have more significance to indicate ductility of a material than standard test value at 25°C. Standard test method provides also for performing the test at 4°C and a rate of pull of 1 cm per minute. Some agencies also perform the test at 0°C with a rate of pull of 1 cm per minute.

14.12. VOLATILITY TESTS

Loss on heating test. This test measures the loss in weight when a 50 g sample of bitumen is placed in 85 gm metal can and exposed to a temperature 163°C for 5 hr in a special oven with a revolving shelf. The loss in weight is expressed as a percentage of the original material. Also, the penetration of the bitumen after the test is determined, and expressed as a percentage of the original penetration.

This test is intended as a quality test as it measures the loss and hardening which the bitumen undergoes when heated. This is regarded as an accelerated test, the concept being that losses which occur on heating under the test conditions somewhat resemble the losses which occur in the construction process and during the service life of the pavement.

Distillation test. This test is performed on cut-back bitumens, road oil and tars to determine the type and quantity of both the volatile distillate and the residue from distillation.

For cut-back bituminous materials, distillation is carried out by applying heat to 200 ml of the material in a glass flask. For tars, 100 g sample is distilled and the condenser is air-cooled rather than water-cooled as for bituminous materials. For tars, vapour (rather than liquid) temperature in the flask is measured, and the distillation result expressed as a percentage by weight of the total material rather than as a percentage by volume of the total distillate as applicable for bituminous materials.

Distillation test requirements are used to control the rate of curing of cut-back, road oil or tar.

The test supplies information regarding the kind and amount of volatile material which has been used in fluxing or cutting back liquid bitumens. Moreover, by removal of volatiles, the residue is considered to represent the amount and character of material that will remain in the road after the liquid material has cured. The residue may then be tested as for straight run bitumens or tars to see whether or not it has the qualities which are deemed desirable and necessary.

Because the materials in service are not subjected to high temperatures that are provided in the distillation test, correlation between curing rate on the road and distillation test results is not as good as desired. There are a number of variable factors which affect curing time. For example, properties of material such as volatility of the solvent, amount of solvent and penetration of base bitumen affect the curing time. External factors include temperature, surface area exposed and wind velocity past the surface. Due to these variables it is practically impossible to predict the absolute curing time to be expected in the field. However, a better measure of curing rate than provided by the distillation test is needed.

Bitumen residue of specified penetration. This test is intended to provide a method for determining the amount of 100 penetration bitumen in a SC bitumen or road oil. The test is conducted by heating a 100 g sample at a temperature between 250 and 260°C in a 170 gm metal can until the material reaches penetration at 25°C of 100 ± 15. The amount of residue at this point is weighed and expressed as a percentage of the original material. Heating is continued for indefinite time depending upon the operator's judgment and is stopped only when residue obtained is of desired penetration.

Since it is known that bituminous substances may be created during the heating process from non-bituminous materials, interpretation of test results is somewhat obscure. Moreover, in the test much of hardening is by loss of volatiles, whereas, there is evidence that hardening in service is primarily by oxidation. Further, changes take place in the prolonged heating of the test that do not occur in service. The time required to reach 100 penetration varies greatly with different materials and thus time of heating is an important factor. For these reasons authorities do not agree on the value of this test and the use of this test has decreased in recent years.

Some agencies use a vacuum distillation test.

14.13. TEST FOR SULFONATION INDEX FOR TARS

Sulfonation index of road tar is the number of millimeters of unsulfonated residue per 100 gm of tar when determined in accordance with ASTM designation D 872 or AASHTO test T 108. Sulphuric acid (H_2SO_4) is added to a sample of the distillate that was taken at specified temperature or in a given temperature range. After repeated shaking to mix the liquids intimately, the sample is centrifuged at a speed of approximately 1000 rpm which separates the oils not acted upon by the acid.

Presence of high percentage of parafinic hydrocarbons causes distress in tar roads because either the residual tars become greasy and fail to hold the cover stone or priming materials do not set. This test is based on wide differences in reactivity of different hydrocarbons present in tars and bitumens. These are: (*i*) the parafinic or saturated straight-chain group, (*ii*) the naphthenic or cyclic saturated compounds, (*iii*) the olefinic or unsaturated straight-chain group, and (*iv*) the aromatics, which are the derivatives and homologs of benzene. Parafinic and naphthenic compounds are not attacked appreciably by sulphuric acid whereas the others are oxidized or converted into sulfonic acids.

Although AASHO specifications indicate that the test is applied only to tar grades RT-1 to RT-6, many agencies specify this control on all grades of tar.

14.14. TESTS OF EMULSIFIED BITUMENS

Emulsified bitumens are specified on the basis of the results of a complete series of tests. These are tested for composition, consistency, stability in mixing and in storage, and for character of residue. Composition tests include water content, residue by distillation, and residue by evaporation. Consistency is measured by Furol viscometer at 25°C, or at 50°C in the case of RS-2. Stability in mixing tests include demulsibility, cement mixing, and stone coating. Storage stability tests include sieve test, miscibility with water, modified miscibility with water, settlement and freezing. Tests on residue include penetration, ductility, solubility in carbon-trichloro-ethylene and ash content.

Standard methods of testing emulsified bitumens are designated as IS-8887. Brief description and their significance is given below.

Settlement. 50 ml sample of emulsion is taken in two graduated cylinders and allowed to stand for 5 days. Samples are taken from the top and bottom parts of each cylinder. The samples are heated to allow the water to evaporate; leaving behind residue. The difference in weight of residue between bottom and top sample gives the settlement of bitumen globules. In case the percentage of bitumen residue at various levels in the container is considerably different, serious difficulty will be encountered in controlling the bitumen content of the finished pavement as the emulsion has to be stored for some duration before use.

Demulsibility. A 100 g sample is mixed with a dilute solution of calcium chloride. Two concentrations of calcium chloride are specified: 0.02 N if quick setting emulsion is being tested and 0.1 N solution if medium and slow-setting emulsions are being tested. Calcium chloride solution is added to accelerate breaking down of emulsion into bitumen and water. The bitumen will solidify and adhere to the vessel while the water and calcium chloride solution are poured off through 1.18 mm sieve. Démulsibility, which is stated as a percentage, is the ratio between the weight of bitumen separated from a sample by the demulsibility test and that obtained by distilling a sample of equal weight.

This test determines the relative resistance of emulsions to breaking down on contact with an aggregate surface.

Sieve test. The percentage by weight of a 1000 g sample that fails to pass 850 micron (ASTM No. 20) sieve is determined. This test is used to determine whether or not an emulsion is of satisfactory character for application through pressure distributor and also whether or not the bitumen has properly emulsified. As a small amount of coarse-sized globules will plug spray nozzles, the low limit of 0.10 per cent or less as specified is absolutely necessary.

Miscibility with water. Approximately 50 ml of sample is mixed with 150 ml of water, and the visible coagulation after 2 hr is observed. This test is used to demonstrate the stability of an emulsion when diluted with large quantities of water without the emulsion breaking down.

For slow-setting emulsions, a modified miscibility test is specified. After stirring 50 ml of emulsion diluted with 150 ml of water, the mixture is permitted to stand for 2 hr. The samples taken from three levels in the container are placed in an oven at 163°C for 2 hr to drive off water. Weighing before and after permits calculation of the percentage of bitumen residue at each level. Results are reported as the maximum numerical difference in percentage of bitumen content between any two of the three levels.

Cement mixing. 100 ml sample is mixed for one minute with 50 gm of high early strength cement. After this 150 ml of distilled water is added and mixing continued for 3 more minutes. The percentage by weight of the coagulated material that fails to pass 1.18 mm (No. 14 ASTM) sieve is dried at 163°C and weighed. The weight in grams of dried material retained is reported as the percentage emulsion broken. A maximum of 2 per cent of allowed.

This test is included for slow-setting emulsions to ensure an emulsion which will mix with all types of aggregates and this test has been found to approximate the severed condition of mixing emulsion with fine aggregate in field work.

Coating test. The ability of a 35 gm sample to coat a specified stone when mixed therewith for 3 minutes is determined visually. This test is used for medium-setting emulsions.

Freezing test. A 400 gm sample is frozen three times at – 17.8°C for 12 hr each time, and the coagulation is observed visually. Samples that remain homogeneous or can be made so by stirring and pass the test are reported as "*homogeneous*".

❑❑❑

Chapter 15 BITUMINOUS SURFACE TREATMENTS

About 1915, when motor vehicle traffic became sufficiently heavy to cause serious damage to the gravel and water-bound type of surfacings, the bituminous treatment was adopted as a means of saving these rapidly disintegrating surfacings. With increasing number of automobiles, it was experienced that even the best water-bound macadam roads would not withstand their action and the roads ravelled rapidly. The tractive effort of the rear wheels causing thrust backward on the individual stones and the thrust forward under the front wheels, combined with partial vacuum under the vehicle which removed the binding material from the interstices of stone metal by suction, largely accounted for loosening process which soon completely disintegrated the top surface. Another cause of disintegration was weakening of the foundation by penetration of rain water into the subgrade and thus settlement of the water-bound macadam. Further, in macadam surfaces employing comparatively less wear resistant local materials in low-cost roads, the disintegration of the surface occurred due to abrasion by traffic. Macadam and other low-cost surfaces can, therefore, by improved to cater for intermediate traffic range with bituminous surface treatments.

Bituminous surface treatment includes the following:

(*a*) Prime coats.

(*b*) Tack coats.

The standard specifications and code of practice for Prime and Tack Coats are covered under IRC-16-2008.

(*c*) Single coat and multiple coats surface dressings.

(*d*) Liquid seal coats, colour coats and non-skid coats.

(*e*) Fog spray.

(*f*) Built-up spray grout.

Sometimes carpet coats are also included in the surface treatment because of their thickness upto 15 mm but due to their higher cost these have been included in the next chapter along with bituminous road mixes.

PRIME COATS

15.1. DEFINITION

When a bituminous wearing surface is to be placed on a previously untreated compacted foundation layer, such as earth, gravel, stabilized soil, kanker, laterite, or sometimes water-bound macadam or similar absorptive bases, the base is generally given a single application of liquid bituminous material called "*Prime coat*".

Neither prime coats nor tack coats, which are dealt with in the next article, are properly called wearing surfaces but they have their proper function in the construction of bituminous wearing surfaces.

The introduction of prime coat is the most important development in surface treatments for successful functioning of many types of low-cost bases.

Purposes

(*i*) Serves to promote adhesion or bond between the base and the wearing surface. This is probably its most important function.

(*ii*) Binds together any loose aggregate which may be present on the base. This serves to consolidate the surface on which the new treatment is to be placed.

(*iii*) Serves to plug capillary voids, and functions as a deterent to the rise of capillary moisture where capillary moisture is a factor.

(*iv*) Provides a temporary seal against the infiltration of surface water.

15.2. MATERIALS

The bituminous material used for prime coats should have high penetrating qualities so that it can penetrate deep into the base course to about 10 mm depth. It should be of such a nature that after curing, it will leave a high viscosity residue in the void spaces of the upper portion of the treated surface. Thus, suitable material is bitumen emulsion or road tar of low viscosity. Rapid curing grades are not generally used because of their rapid increase in viscosity after application. Medium curing grades are most widely used which best serve the purpose. The grade and rate of application will be governed by the condition of the existing surfaces.

Cationic bitumen emulsions normally used are SS-1h, CSS-1 or CSS-1h grades conforming to IS 8887/ASTM D2397. These serve as primer. The quantity required for various types of granular surfaces is given in Table 15.1.

Table 15.1 Quantity of cationic bitumen emulsion for different types of surfaces

Sr. No.	*Type of Surface*	*Rate of spray kg/m²*
1.	WMM/WBM	0.7-1.0
2.	Mechanically stabilised soil base, lime/cement stabilised soil and lime cement base	0.9 to 1.2
3.	Gravel Base, Crusher run Macadam and Crushed rock base	1.2 to 1.5

Medium curing cut back bitumen conforming to IS:217 is also used as a primer. The type and quantity of cut-back bitumen to be used as a primer is as given in Table 15.2.

Table 15.2 Quantity of MC cut-back bitumen for different types of surfaces

Sr. No.	*Type of Surface*	*Type of cut-back*	*Rate of spray kg/m²*
1.	WMM/WBM	MC 30	0.6 to 0.9
2.	Mechanically stabilised soil base, lime/ cement stabilised soil and lime cement base	MC 70	0.9 to 1.2
3.	Gravel Base, Crusher run Macadam and Crushed rock base	MC 250	1.2 to 1.5

MC 30 and MC 70 grades are used for low porosity and medium porosity surfaces while MC 250 is used for loosely bonded, open-textured surfaces.

Bitumen emulsion is applied on a just wet surface with light sprinkling of water. Cut-back bitumen primer is applied on a dry surface. Primer should not be applied during dust storm, rainy, foggy or windy weather.

Because of wide difference in density and surface texture of different types of granular bases, a rather wide range in limits of quantity of primer is required. The ideal rate of application is the maximum that will, under favourable weather conditions, be completely absorbed by the base material within 24 hr from the time of application. Some judgment should, therefore, be exercised in specifying a definite rate of application for a given job.

When prime coat is not entirely absorbed within reasonable length of time, usually 24 hr, it is customary to apply a very light sand blindage to blot up the excess primer. Sand used is dry, clean, granular material passing 4.75 mm sieve and grading requirements are not generally included in specifications. Prior to laying the surface course, all loose or excess sand is swept free from the base.

15.3. CONSTRUCTION METHOD

Preparation of road surface. (*a*) If the existing surface contains pot-holes, or ruts, these are repaired by removal of all loose and defective material and replaced with suitable patching material to produce a tightly bonded surface.

(*b*) Bumps and waves on the surface are levelled by scarifying and recompaction of the material.

(*c*) The surface is swept clean of all loose and foreign material prior to the application of the primer.

Application of primer. Upon the prepared surface, the primer is applied uniformly at the rate specified through an approved type of pressure distributor or sprayer. It is applied to a surface which is dry or slightly damp and at a temperature of not lower than 10°C. After the application of the primer; at least 24 hr should elapse before applying sand blindage, if required.

Application of sand blindage, if required. The primer should preferably be entirely absorbed by the base course and, therefore, requires no sand blindage. If, after 24 hr application it has not been completely absorbed a light sand blindage is applied to blot up the excess primer to prevent picking up under traffic. Prior to laying the surface course, all loose or excess sand is swept from the base.

This treatment does not require rolling. The application of a prime coat to a newly constructed flexible base course may sometimes cause fluffing of the upper 6 mm of the base. In this case, the fluffed material immediately after curing may be compacted by rolling. This will then provide a much better seal against the infiltration of surface water. The surface should not be opened to traffic during the period when sufficient fluid material remains on the surface to be picked up by the wheels of the moving traffic. This time is usually 48 hr from the time of application of primer.

TACK COATS

15.4. DEFINITION

Tack coat is a very light application of low viscosity liquid bituminous material on surfaces which have previously been treated or prepared such as existing bituminous, Portland cement concrete, brick or block surfaces. A track coat would, therefore, be needed as a part of the construction of a new wearing surface, upon a new bituminous base course, a previously primed granular type base, or old bituminous wearing surface which is being re-surfaced, a new or old concrete base, or an old concrete, brick or block pavement which is to be covered with bituminous wearing surface.

15.5. PURPOSES

Simply to insure adhesion between the existing surface and the new bituminous surface.

A tack coat would not usually be required if wearing surface is to be of a single coat or multiple coat surface treatment.

15.6. MATERIALS

The binder used for tack coat is either cationic Bitumen Emulsion (RS-1) conforming to IS:8887/ ASTM D2397 or suitable low viscosity paving bitumen of viscosity grading (VG) 10 i.e. 80-100 penetration grade conforming to IS:73. For emergency application, when the atmospheric temperature is below 0°C, cut-back bitumen RC-70, conforming to IS:217 may be used.

The rate of application for tack-coat on various types of surfaces is given in Table 15.3.

Table 15.3 Rate of application of tack-coat.

Sr. No.	*Type of surface*	*Rate of spray (Emulsion) kg/m²*	*Rate of spray (Bitumen VG-10) kg/cm²*
1.	Normal bituminous surfaces	0.20-0.25	0.30-0.35
2.	Dry and hungry bituminous surfaces	0.25-0.30	0.35-0.40
3.	Granular surface treated with primer	0.25-0.30	0.35-0.45
4.	Non-bituminous surfaces		
	(*a*) Granular base (not primed)	0.35-0.40	0.45-0.50
	(*b*) Cement concrete pavement	0.30-0.35	0.40-0.45

Tack coats can do more harm than good when they are applied in too large a quantity as it will unduly enrich the superimposed course and lessen its stability. Care must be taken not to cover so large an area at one time that the tack coat will cool and harden before the mat is placed on it.

15.7. CONSTRUCTION METHOD

As tack coats are very frequently applied to old surfaces of various types, as a first step in many cases, the old surface will require extensive correction and remedial treatment because of rough or inadequate condition of the old surface. The old surface is therefore, first brought to satisfactory condition. Operations regarding patch repairs to the damaged portion of the road and pot holes, removing of inequalities and waviness on the old surface have been described in detail in the next chapter dealing with carpet coat which may be referred to. Major reconstruction measures may be required when the existing pavement may not be of adequate thickness. Substantial changes in grade, profile or crown are called for in the new surface. Space does not permit to cover all such possible situations which may arise when the old surface must be reconstructed in part or whole before resurfacing.

In many cases the preparation of surface may consist simply of the removal of loose material appearing on the surface. Removal of this material may be accomplished by the use of rotary brooms or power blowers. If excessively dusty or unconsolidated areas remain after this treatment, these may have to be primed before construction proceeds.

Application of tack coat. After the base has been brought to the desired condition, the tack coat is applied on dry surface at a uniform specified rate by a suitable device in a single application. In case of emulsion, surface may be slightly damp without harmful effect. The treated surface is protected from traffic and is allowed to dry until it reaches the proper stage of stickiness or tackiness for the application of bituminous wearing surface. Any break which may occur in the tack coat may be corrected by the application of additional material at these points.

SURFACE DRESSING

The term surface dressing, also known as surface treatment, refers to thin surface covering of bituminous material and mineral aggregate which are usually not more than 25 mm in thickness. These are very widely used for surfacing of primary and secondary roads which carry light to moderate amounts of traffic. This thickness usually varies from 12.5 to 20 mm and may be slightly more when multiple treatments are used.

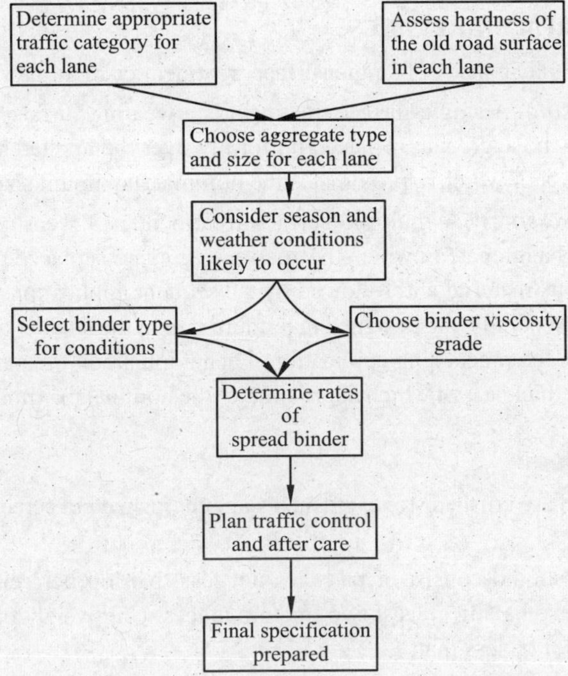

Fig. 15.1 Flow diagram for the surface during design process.

Surface constructed by this method is frequently quite open in character and excellent non-skid and visibility characteristics.

This type of surface is normally used in treatments of granular bases such as water-bound macadam. The surface is so thin that it has little load-supporting value by itself and must depend upon adequate base and subgrade to function satisfactorily. This treatment is relatively inexpensive and can be used to improve many of the low-cost roads to cater for increased traffic.

This type of surface is sometimes designated as 'Armour coats, "*Multiple lift surfaces*" and "*Inverted penetration surfaces*".

15.8. PRINCIPAL FUNCTIONS

(*a*) To protect underlying layer from ravelling and abrasive action of traffic.

(*b*) For water-proofing, *i.e.* to prevent or greatly minimize the penetration of surface water into roadway layer and, thus avoiding the weakening of subgrade..

(*c*) For abating (doing away with) or reducing dust nuisance.

(*d*) To prevent gravel loss.

(*e*) To protect freshly laid WBM or similar unbound bases from the action of pneumatic tyred vehicles, which otherwise tend to draw out the fines from the macadam and thus reduces its stability.

(*f*) To retard the deterioration of an existing bituminous surface which is showing signs of distress.

(*g*) To increase the skid resistance of smooth surfaces.

Surface dressings are not highly resistant to the effects of severe traffic, especially when wheel loads are heavy.

15.9. TYPES OF SURFACE DRESSINGS

This type of surface may consist of single surface treatment or multiple surface treatments.

Single surface dressing. A single surface treatment is a wearing surface of bituminous material and aggregate in which the aggregate is placed uniformly over the applied bituminous material in a single layer, the thickness of which approximates the nominal maximum size of the aggregate used.

Multiple surface dressings. A multiple surface treatment is a wearing surface composed of bituminous material and aggregate in which the coarser aggregate is placed uniformly over an initial application of bituminous material and followed by subsequent applications of bituminous material and smaller aggregate. Generally, the designed maximum size of the smaller aggregate is one half that of the aggregate used in the preceeding application. Each application of aggregate should be placed uniformly in a layer, the thickness of which approximates the nominal maximum size of the aggregate.

15.10. MATERIALS

Stone chippings used are crushed stone, crushed slag, uncrushed but screened gravel, and crushed gravel.

The crushed gravel should consist of particles not less than 60 per cent by weight of portion retained on 4.75 mm sieve having at least two fractured faces. Unit weight of crushed slag per cubic cm compacted should not be less than 1.12 gm.

The chippings should be clean, hard, tough, durable, free from excess amount of flat or elongated pieces, dust, clay film and other objectionable material and are required to be hydrophobic in nature.

Requirements in respect of deleterious substances are shown in Table 15.4.

Table 15.4 Requirements of deleterious substances

Deleterious substances	*Max percentage by weight of total sample*
Clay lumps	0.1
Soft particles	2.5
Chert that will readily disintegrate (soundness test, five cycles)	1.0
Oven-dry material floating on a liquid with a specific gravity of 2.0 (not applicable to slag)	1.0
Flat pieces or elongated pieces	5 to 10

Wear, by Los Angles abrasion machine, should not exceed 40 per cent.

15.11. GRADATION

The size of the aggregate will vary with the type of construction contemplated as to whether it is single surface or multiple surface dressing. The grading requirements recommended by I.R.C. are given in Table 15.5.

Table 15.5 Grading requirements for chips for surface dressing

IS Sieve Designation (mm)	*Cumulative per cent by weight of total aggregate Passing for the following nominal sizes (mm)*			
	19	13	10	6
26.5	100	-	-	-
19.0	85-100	100	-	-
13.2	0-40	85-100	100	-
9.5	0-7	0-40	85-100	100
6.3	-	0-7	0-35	85-100
4.75	-	-	0-10	-
3.35	-	-	-	0-35
2.36	0-2	0-2	0-2	0-10
0.60	-	-	-	0-2
0.075	0-1.5	0-1.5	0-1.5	0-1.5
Minimum 65 per cent by weight of aggregate	Passing 19 mm, retained 13.2 mm	Passing 13.2 mm, retained 9.5 mm	Passing 9.5 mm, retained 6.3 mm	Passing 6.3 mm, retained 3.35 mm

Good results under heavy traffic are most frequently obtained by the use of 13 mm chippings with relatively sharp edges produced by crushing. Chippings which are too large are more easily turned away from the road under traffic. Chipping which are too small become completely embedded and their use may lead to bleeding . Under heavy traffic and on roads with relatively soft surfaces even 13 mm aggregate may prove too small and the 19^{+} mm size would be more suitable, to ensure that they are not completely submerged in the old road surface. On the lightly trafficted roads and those with hard surfaces, the smaller size 10 mm is preferable. On extremely hard surfaces such as concrete it is generally better to use a 6 mm chipping or even 3 mm or coarse sand, under all conditions of traffic.

Table 15.6 gives the recommended nominel sizes of stone chippings in mm for various types of surfaces.

Table 15.6 Recommended nominal sizes of stone chippings (millimetres)

Type of Surface	*Approximate Number of Commercial Vehicles with an Unladen Weight Greater than 1.5 tonnes currently carried per day in the Lane Under Consideration*				
	2000-4000	1000-2000	200-1000	20-200	Less than 20
Very Hard	10	10	6	6	6
Hard	13	13	10	6	6
Normal	19+	13	10	10	6
Soft	*	19^{+}	13	13	10
Very soft	*	*	19	13	10

Note : The size of stone chippings is related to the mid-point of each lane traffic category. Light traffic conditions may make the next smaller size of stone more appropriate.

+ Very particular care should be taken when using 19 mm chipping to ensure that no loose materials remain on the surface when the road is opened to unrestricted traffic as there is a high risk of windscreen breakage.

* Unsuitable for surface dressing.

Single-sized aggregate are used because smaller aggregate may prevent the larger ones from making sufficient contact with the binder film and some of the larger ones are then lost. Moreover, the film of binder may be either too thick for the smaller ones which may become too deeply embedded or too thin for larger aggregate which will cause excessive fly off under traffic.

15.12. BITUMINOUS MATERIALS

The selection of the grade of bituminous material used and its rate of application govern the final performance of surface treatment and should be given careful consideration. The season of the year in which the work is to be performed, the pavement design, whether it be a single or multiple surface treatments and the character of the aggregate are principal factors to be considered. The most important factor is the season of the year or prevailing weather condition when the work is executed. The use of paving bitumens, high viscosity road tars and heavy grades of liquid bitumen is practicable in warm weather.

Bituminous material most suitable for use during the warmer months are 80+penetration bitumens, RC-4, RC-5 and MC-5 grades of liquid asphalts. Tars, RT-5 to RT-12 are also suitable.

It is also preferable to construct bituminous surface treatment during the warmer months but when it is necessary to construct these during winter months, a lighter grade of liquid bitumen or emulsified bitumen should be used. As the general atmosphere temperature comes down, so should the viscosity of the bitumens material. RC-2 and RC-3 are better suited for construction in cooler weather because they remain tacky or soft until the aggregate is spread.

Emulsified bitumens grades RS-1 and RS-2 have broader seasonal range of use and perform equally well in both warm and cool weather. Both grades of emulsified bitumen should be used at temperature above 5°C because neither of them will withstand freezing. Emulsified bitumen provides better results than other bituminous materials when used with wet aggregates.

Quantities. The amount and size of aggregates to be used and the rate of application of the bituminous material will depend on the thickness to be obtained by the treatment and the number of applications of binder and aggregate required to obtain that thickness. No more aggregate should be used in one application than will contact the film or previously applied bitumen. Secondly the amount needed is that which will cover the area to a depth of one particle aggregate. The coarsest aggregate is used in the first application and smaller size in subsequent upper applications.

Rates of application of bituminous material and aggregate as recommended by IRC are given in Table 15.7.

Table 15.7 Rates of spread of binder and chippings

Nominal Size of Chippings (mm)	*Binder (pen. grade) kg/10m²*	*Chippings m³/m²*
19	12.0	0.015
13	10.0	0.010
10	9.0	0.008
6	7.5	0.004

When cutbacks and emulsions are used, the quantity of binder must be increased so that the residual quantity of binder, after dispersment of the cutting agent or the water portion of the emulsion, is as given in Table 15.7.

Bituminous material should be increased slightly (5 to 10 per cent) when it is to be applied with little or no heating.

It is further suggested in the specifications that quantity of bituminous material should be increased by 2 to 10 per cent where the surface treatment is subjected to a large volume of traffic. The quantities of both aggregates and bituminous materials should be increased when the gradation of the aggregate approaches the coarser limits of specification. Likewise, the quantities should be decreased with aggregate approaching the finer limits of specification. This increase or decrease should not be more than 20 per cent of the quantity given in Table 15.7.

General rule is 0.1 litre of bituminous material for each kg of aggregate used.

15.13. CONSTUCTION METHOD

The base is first brought to satisfactory condition and is primed, if necessary, as described for prime coat application. After the prime coat, if applied, has cured, bituminous binder material is applied directly upon the prepared base at a specified rate of application by a suitable device at the proper temperature of application. Manually operated sprayer or pressure distributor may be used for this purpose.

Overlapping of application of bituminous materials at the junction of two applications results in excess of binder which flushes to the surface and causes unstable as well as unsightly condition. Thin applications at junctions result in little or no aggregate retention and immediate patching is required.

To eliminate thin applications or overlapping of bituminous material at the end of one application and the beginning of another, proper precautions such as the use of heavy wrapping paper prior to the beginning of application should be made.

Immediately, after the application of binder material, the aggregate is applied manually or by means of mechanical spreader at the specified rate of application. Aggregate should be placed as quickly as possible after the bituminous material has been applied in order to take advantage of the softness of the binder and to get the maximum amount of aggregate embedded in the bituminous material. Under some conditions, it may be necessary to reduce the length of bituminous applications to that which can be covered with aggregate within a maximum specified time.

As soon as the aggregates have been spread over the freshly applied bituminous material, the entire surface should be rolled with one complete coverage of a steel wheel roller. The surface may then be broomed to obtain more uniform distribution of the aggregate. Steel-wheel or pneumatic-type rolling should then continue until all the aggregate is thoroughly embedded in the bituminous material.

In some instance, this may complete the construction but if double or triple surface treatment is to be constructed, the successive courses of bituminous material and aggregate are applied in the same manner as described in the first coat. Generally speaking very little time need elapse between construction of successive lifts except when emulsified bitumens are used, and then the time between application of bituminous material may be required to be at least 24 hr. Similar waiting period may be enforced by some organizations when other materials are used, particularly, in the top layer of multiple surface treatment. Surface treatment work is most successfully done in warm weather and should not be attempted when air temperature are low, say 10°C or less.

Surface treatment method of construction should preferably be used whenever the intensity of total mixed traffic is not less than 200 tonnes per day, so that there is enough traffic to work up the binder around the aggregate to prevent excessive fly off. When the intensity of traffic is less than about 200 tonnes per day, carpet coat as described in the next chapter may be used.

SEAL COATS, COLOUR COATS AND NON-SKID COATS

15.14. SEAL COATS

Seal coat may be defined as a very thin surface treatment which is either applied as a final step in the construction of certain bituminous surfaces or to existing surface which have cracked or oxidized over a number of years and have begun to ravel.

15.14.1. Purposes

(*a*) To water-proof or seal the surface and, thereby prevent the deterioration of the mix from moisture and air.

(*b*) To increase resistance to skidding.

(*c*) To improve night visibility.

(*d*) To protect the mat from the abrasion caused by heavy traffic.

15.14.2. Types

Seal coats are of two types, namely:

Type A. Liquid seal coat comprising of an application of a layer of bituminous binder followed by a cover of stone chippings.

Type B. Premixed seal coat comprising of a thin application of fine aggregate premixed with bituminous binder.

15.15. COLOUR COATS

Colour coats is an application of bituminous material and an aggregate of a particular colour applied to either old or new surface for the purpose of obtaining a desired colour effect. The most common use of colour coat is on bituminous concrete runway where a light colour surface may be desired.

15.16. NON-SKID COATS

Non-skid coat is a relatively light application of bituminous material and aggregate applied to existing surface where an increase in skid resistance is considered necessary.

These may consist of a single application of bituminous material which is covered by a light spreading of fine aggregate or sand. Specifications for seal coats are not usually given separately but are made a part of the specifications for each particular type of wearing surface.

15.17. MATERIAL REQUIREMENTS

There are many factors involved in determining the grade of bituminous material, its rate of application, and the maximum size and quantity of aggregate for use in seal, colour and non-skid coats and all these deserve proper consideration. The primary factors to be considered are:

(i) Condition of the existing surface and the performance of the sealing coat. The bitumen requirement is variable for old surfaces having different surface characteristics. For open texture and if the surface has oxidized, a portion of the binder will penetrate the existing surface and will partially act as a primer thereby requiring a slightly heavier application than for more non-absorptive surfaces having a smooth texture.

Bitumen requirements for non-skid coats are generally much less than those for the usual seal coats. The reason for this is that a surplus of the bituminous material that was applied originally has flushed to the surface and when the fresh application is made, the old bituminous material is softened, producing the effect of slightly heavier application.

Colour coats and non-skid coats usually require application of 0.80 to 1.4 litres/sq metre for satisfactory results.

(ii) Season of the year in which the work is to be performed. Prevailing climatic conditions and the season of year in which the sealing is to be performed has considerable bearing on the selection of the type and grade of bituminous material. For best results, the work should be performed during the early part of the warm season so that the newly sealed surface may be subjected to a considerable amount of traffic during the warmer months. Generally, soft penetration grades of paving bitumen and high viscosity tar are used when sealing work is to be done during the early part of the summer season. However, if the work needs to be performed during the cooler part of the year or if seal is to be applied to an old surface that is ravelling or has oxidized, the most satisfactory results under these conditions are obtained by using a cut back bitumen.

High-viscosity, rapid-setting emulsified bitumen is satisfactory for all seasons of the year except when the temperature is approaching freezing point.

(iii) Character and size of the aggregate. Character of the aggregate, e.g. whether it is absorptive or non-absorptive, and its maximum effective size are important in governing the rates of application of both bituminous material and the aggregates. The finished thickness results from the maximum effective size of the aggregate. Therefore, large aggregate requires a heavier application of bituminous material.

The ratio of bituminous material to aggregate for seal coats, without allowances for absorption of old surface and aggregate is usually 0.1 litre of binder to 1 kg aggregate per sq.metre.

Quantities used according to specifications followed in the country are: The quantity of bitumen per 100 square metres, shall be 98 kg for type (A) and 68 kg for type (B) seal coat. Where bitumen emulsion is used as a binder, the quantities shall be 150 kg and 105 kg respectively.

Stone chips for type (A) seal coat should consist of angular fragments of clean, hard, tough and durable rock. These should be free of soft or disintegrated stones, organic and other deleterious matter. Stone chips should be of 6.7 mm size defined as 100 per cent passing 11.2 mm sieve and retained on 3.36 mm sieve. The quantity used for spreading is 0.9 cubic metre per 100 square metre area.

Aggregate used for type (B) seal coat is sand or grit consisting of clean, hard, durable, uncoated particles free from dust, soft or flaky/elongated material, organic matter or other deleterious substances. The aggregates shall pass 2.36 mm sieve and retained on 180 micron sieve. The quantity used for premixing is 0.6 m^3 per 100 m^2 area.

The ideal quantity of aggregate is that where there will be one layer of bonded aggregate about 60 per cent embedded in bitumen. Too little cover results in excessive bleeding under heat and traffic and an excess aggregate grinds up under traffic, dislodge some of the embedded particles and is dissipated to the sides.

15.18. CONSTRUCTION METHOD

Construction method is generally the same as described under single surface treatment. Traffic should not be permitted on the new seal coat until rolling is complete and the binder has cured or set and has developed high degree of bond with the embedded aggregate. Increased rolling can substantially increase aggregate retention which tends to minimize bleeding tendencies.

15.19. QUALITY CONTROL

The tests and their frequencies for seal coats and single/double coat surface dressings are given in Table 15.8.

Table 15.8 Control tests and their frequencies

S.No.	Test	Frequency
1.	Quality of binder	As required
2.	Aggregate impact value	One test per 50 m^3 of aggregate
3.	Flakiness Index	—do—
4.	Stripping value of aggregate and water absorption	Initially one set of three representative specimens for each course of supply. Subsequently when warranted by changes in the quality of aggregates.
5.	Grading of aggregates	One test per 25 m^3 of aggregate
6.	Temperature of binder at application	At regular close intervals
7.	Rate of spread of material	One test per 500 m^3 of aggregate

FOG SPRAY

A fog spray is also called fog seal. It is a light application of a slow-setting emulsion diluted with water and applied to an existing surface.

15.20. USES/PURPOSES

(1) To renew an old bituminous surface that has become dry and brittle with age.

(2) To seal small cracks and surface voids.

(3) To control ravelling and whip-off of chippings by traffic on newly finished surface dressing.

It is mainly a maintenance treatment and may not be regarded as a substitute for surface dressing renewal, seal coat or slurry seal. By penetrating into the cracks and voids and coating the loose aggregates on the surface it can prolong the life of pavements and may delay the time when major maintenance or rehabilitation is needed.

Materials: Grades of bituminous emulsion normally used are SS-1, SS1h, CSS-1 or CSS-1h. The grades SS-1h and CSS-1h are, however, to be preferred. The SS grade of emulsion specified in IS:8887-1995 has the same range of viscosity as the above grades and may also be considered.

Quantities: The quantity of emulsion used is usually in the range of 5 to 10 litres per 10 sq.m of diluted materials, prepared by adding emulsion and water in equal proportions.

The precise amounts to be used, will depend on the degree of cracking or ravelling, dryness and texture of the surface over which the fog spray is to be applied. If the applied emulsion is found to be in excess, a light sand dusting may have to be used as in the case of prime coat.

BUILT-UP SPRAY GROUT

It consists of a two-layer composite construction of compacted crushed coarse aggregates with application of bituminous binder after each layer and key aggregates on top for the second layer. It is similar to Penetration Macadam but with less interlock and binder. The thickness of a single course is 75 mm.

Use: It is used only for emergency repair work and other temporary construction.

15.21. MATERIALS

***(i)* Aggregates:** The aggregates should satisfy the various physical requirements laid down in Table 15.8. The grading requirements of coarse and key aggregates are given in Table 15.9. The quantity of coarse aggregates in the first and second layers is 0.5 cum per 10 sq. m while the key aggregate on second application of binder should be 0.13 cum. per 10 sq.m.

Table 15.9 Grading requirements for coarse aggregates and key aggregates for built-up spray grout

IS Sieve Designation	*Cumulative Per cent by Weight of Total Aggregates passing*	
(mm)	*Coarse Aggregate*	*Key Aggregate*
53.0	100	-
26.5	40-75	-
22.4	-	100
13.2	0-20	40-75
5.6	-	0-20
2.8	0-5	0-5

(*ii*) **Binder:** The bitumen should be paving bitumen of penetration grade as per IS:73 and of penetration grade 30/40, 60/70 or 80/100. The actual grade of bitumen should be appropriate to the region, traffic, rainfall and other environmental conditions. The quantity of binder to be used should be 15 kg/10 sq. m (in terms of residual bitumen) for each of the first and second sprays.

❑❑❑

CARPET COAT, ROAD-MIX AND INTERMEDIATE TYPE BITUMINOUS PLANT MIX SURFACES

CARPET COAT

Quite frequently, in order to improve the wearing surface of relatively heavily trafficked roads, a type of wearing surface is applied which is thin enough to be classified as surface treatment but differs from the usual method of construction of surface treatment, in that the aggregate and bituminous material are thoroughly mixed before being rolled. This type of wearing surface is being used for treatment of new bases, reconstructed bases and old wearing surfaces of various types. This is quite a popular type of treatment on portions of the national highways in India and in other countries which have to carry relatively more traffic. This type of surface treatment can rightly be called plant mix surface treatment.

16.1. PRINCIPAL FUNCTIONS

(*a*) Provision of wear-resistant surface.
(*b*) Water-proofing or sealing of the surface.

16.2. MATERIALS REQUIREMENTS

Aggregates. Aggregates may be crushed rock, crushed slag, river shingle or gravel of uniform quality and should be clean and free from excess of dust, flat or elongated flaky pieces, soft and disintegrated material, vegetable or other deleterious material. The limits or requirements in respect of deleterious material have already been given in chapter 15 under the head "Surface dressing" (see page 504).

The grading requirements according to typical specifications followed in India specify aggregates having usually maximum size of 20 mm or 25 mm, the maximum size of aggregate should coincide approximately with the thickness of cover. Typical gradation followed is shown in Table 16.1.

Table 16.1 Aggregate gradation for semi-dense carpet

	Per cent by weight passing the sieve	
Sieve designation	*for 25 mm thickness*	*for 20 mm thickness*
20 mm	100	—
12.5 mm	75—100	100
10 mm	60—85	75—100
4.75 mm	35—55	35—55
2.36 mm	20—35	20—35
600 micron	10—22	10—22
300 micron	6—16	6—16
150 micron	4—12	4—12
75 micron	2—8	2—8

Table 15.9 Grading requirements for coarse aggregates and key aggregates for built-up spray grout

IS Sieve Designation	*Cumulative Per cent by Weight of Total Aggregates passing*	
(mm)	*Coarse Aggregate*	*Key Aggregate*
53.0	100	-
26.5	40-75	-
22.4	-	100
13.2	0-20	40-75
5.6	-	0-20
2.8	0-5	0-5

(*ii*) Binder: The bitumen should be paving bitumen of penetration grade as per IS:73 and of penetration grade 30/40, 60/70 or 80/100. The actual grade of bitumen should be appropriate to the region, traffic, rainfall and other environmental conditions. The quantity of binder to be used should be 15 kg/10 sq. m (in terms of residual bitumen) for each of the first and second sprays.

❑❑❑

Chapter

CARPET COAT, ROAD-MIX AND INTERMEDIATE TYPE BITUMINOUS PLANT MIX SURFACES

CARPET COAT

Quite frequently, in order to improve the wearing surface of relatively heavily trafficked roads, a type of wearing surface is applied which is thin enough to be classified as surface treatment but differs from the usual method of construction of surface treatment, in that the aggregate and bituminous material are thoroughly mixed before being rolled. This type of wearing surface is being used for treatment of new bases, reconstructed bases and old wearing surfaces of various types. This is quite a popular type of treatment on portions of the national highways in India and in other countries which have to carry relatively more traffic. This type of surface treatment can rightly be called plant mix surface treatment.

16.1. PRINCIPAL FUNCTIONS

(*a*) Provision of wear-resistant surface.
(*b*) Water-proofing or sealing of the surface.

16.2. MATERIALS REQUIREMENTS

Aggregates. Aggregates may be crushed rock, crushed slag, river shingle or gravel of uniform quality and should be clean and free from excess of dust, flat or elongated flaky pieces, soft and disintegrated material, vegetable or other deleterious material. The limits or requirements in respect of deleterious material have already been given in chapter 15 under the head "Surface dressing" (see page 504).

The grading requirements according to typical specifications followed in India specify aggregates having usually maximum size of 20 mm or 25 mm, the maximum size of aggregate should coincide approximately with the thickness of cover. Typical gradation followed is shown in Table 16.1.

Table 16.1 Aggregate gradation for semi-dense carpet

Sieve designation	*Per cent by weight passing the sieve*	
	for 25 mm thickness	*for 20 mm thickness*
20 mm	100	—
12.5 mm	75—100	100
10 mm	60—85	75—100
4.75 mm	35—55	35—55
2.36 mm	20—35	20—35
600 micron	10—22	10—22
300 micron	6—16	6—16
150 micron	4—12	4—12
75 micron	2—8	2—8

The aggregates should also satisfy the properties set out in Table 16.2.

Table 16.2 Physical requirements of aggregates for carpet coat

Sr. No.	*Property*	*Value*	*Method of test*
1.	Abrasion value using Los Angeles Machine or Aggregate impact value	max. 35% Max. 30%	IS 2386 (Part IV) -do-
2.	Flakiness index	Max. 30%	IS:2386 (Part I)
3.	Stripping value	Max. 25%	IS: 6241
4.	Water absorption (except in case of slages)	Max. 2%	IS:2386 (Part III)
5.	Soundness: Loss with sodium sulphate 5 cycles (in case of slag only)	Max. 12%	IS:2386 (Part V)
6.	Unit weight or bulk density (in case of slag only)	Min. 1120 kg per m^2	IS:2386 (Part III)

The quantities of aggregate and binder per 10 m^2 of road surface for 20 mm compacted thickness according to typical Indian specifications are as follows:

	Per 10 m^2
1. Stone chipping or gravel	
(*a*) 12.5 mm IRC standard size passing 20 mm square mesh sieve and retained on 10 mm square mesh	0.18 cu m
(*b*) 10 mm IRC standard size passing 12.5 mm square mesh and retained on 6.3 mm square mesh	0.09 cu m
	0.27 cu m
2. Binder for tack coat	
(a) On a water-bound macadam surface	7.3 to 9.8 kg
(*b*) On an existing black top surface	4.9 to 7.3 kg
3. Binder for premixing	
(*a*) For 12.5 mm size @ 52 kg per cu m	9.5 kg
(*b*) For 10 mm size @ 56 kg per cu m	5.1 kg
	14.6 kg

When tar is used as binder, quantity used for priming coat on water-bound macadam is 12.2 to 14.7 kg 10 m^2 and quantity for tack coat on an existing black-top surface is 7.3 to 9.8 kg per 10 m^2. The quantity of tar binder used for coating the stone chippings is 72 kg per 10 m^2. The quantity of tar for premixing would thus be 19.7 kg per 10 m^2.

Bituminous material satisfactory for use include the following:

(*a*) Paving bitumen viscosity grade VG-10 (80/100) or softer penetration complying with IS:73-1961.

These are used if mixing of aggregate and binders are done in a mixer, the softer grades being adaptable to lighter traffic volumes and when the work is to be executed in cold weather.

(*b*) Cut-back bitumen RC-3 and RC-4.

(*c*) Various grades of emulsified bitumen.

(*d*) Road tars grade RT-7 to RT-12.

16.3. CONSTRUCTION METHOD

Typical specifications adopted in this country are described below:

The work involved in laying carpet coat consists of the following operations:

(*a*) Patch repairs to the damaged portion of the road and pot-holes.

(*b*) Removal of inequalities and waviness.

(*c*) Laying 25 mm finished consolidated thickness of bitumen precoated stone chips or bajri carpet.

(*d*) Sand flushing.

(*e*) Placing seal coat.

The above mentioned work is carried out as follows:

(a) Patch repairs to the damaged portion of the road and pot-holes. The work is completed at least a fortnight before the application of precoated bitumen and broken graded stone chips or waterworn gravel carpet and is carried out as below:

(*i*) Each and every pothole and damaged portion of the road is cut to square or rectangular shape and all the loose material from the damaged portion of the sides and the bed of the area is removed.

(*ii*) Sides and bottom of the cut portion are cleaned and painted with 80/100 penetration bitumen heated to 175°C and not exceeding 190°C.

(*iii*) The potholes are filled with precoated bitumen and broken stone 10 mm to 20 mm gauge graded chips or gravel precoated with 50 kg of bitumen per cu m of broken graded stone chips or waterworn gravel (whichever is cheaper) rammed and rolled to finish; the patched surface is kept slightly proud of the road surface to prevent further complication by traffic.

(*iv*) Dry sand or stone dust is then spread and the road opened to traffic.

(b) Removal of inequalities in shape and surface and removing waviness. For this it would be essential to prepare very accurate longitudinal sections and cross sections of the existing road surface at every 7.5 m interval. From the longitudinal section the levels of proposed longitudinal section along the centre line of the road and cross-sections are decided by the engineer-in charge. These sections are called designed longitudinal sections and cross-sections. In arriving at final longitudinal section along centre line not more than 20 mm deep cutting of the existing road crust is allowed anywhere; inequalities are otherwise made up and waviness removed by filling the depressions with precoated asphalt 10 mm to 20 mm gauge graded broken stone chips or gravel in places where filling is not more than 25 mm. In case of depths exceeding 25 mm bigger guage upto 40 mm grade broken stone precoated with 50 kg of 80/100 penetration bitumen heated upto 175°C is used for filling. The surface to be treated is tack coated with 80/100 penetration bitumen heated upto 175°C at 0.75 kg per 100 sq m. Such fillings have the top cover of 20 mm thickness of precoated bitumen 20 mm to 10 mm gauge graded broken stone chips or bajri. On the treated surface thus produced, clean medium coarse sand or stone dust is spread and the road is opened to traffic.

This rectification of inequalities and waviness and imperfections both longitudinally and cross-wise in the above described manner when carried out upto this stage should not have the variation exceeding plus or minus 1 mm in every metre length longitudinally and plus or minus 0.3 mm in the camber of 1 in 60 on either side of the centre line of the road when checked with template. After repairing, patching the road and removal of inequalities and waviness as described above, the road is opened to traffic for a period of three weeks. After the expiry of this period, inequalities, waviness and any other deformities are again set right by the use of 6 mm to 10 mm gauge graded broken stone chips or gravel precoated with bitumen to bring the variation to plus or minus 1 mm from the designed longitudinal section in every metre length and plus or minus 0.5 mm in cross-section on either side of the road.

(c) Laying 25 mm compacted thickness of graded broken stone chips precoated with 80/100 penetration bitumen per 100 sq metre.

(*i*) Surface is prepared by removing caked earth or any or all foreign matter with wire brushes.

(*ii*) Sweeping with brooms and brushes.

(*iii*) Final dusting with old gunny bags.

(*iv*) Tack coat is applied by a spray, at the rate already mentioned. Spraying of paving bitumen 80/100 heated upto 175°C is done for a length of about 15 m over previously prepared surface in which variations in the longitudinal section do not exceed 6 mm from designed longitudinal section and 3 mm from designed cross-section.

For precoating of graded broken stone chips or gravel 1.2 m^3 of aggregate, stone chips or gravel of described grading is placed in a mixer and thoroughly mixed and dried and then required quantity of bitumen added and mixed well until each chip is thoroughly coated with bitumen. Precoated aggregate is emptied on stretchers or wheel-barrows or sheet platform. In no case the precoated aggregate is stored on a dirty surface.

(*v*) *Spreading premix or precoated graded broken stone chips or gravel*. Immediately after applying the tack coat, the above mentioned premixed broken stone chips or gravel is spread with rakes to thickness which will give compacted thickness of 25 mm finished carpet to absolutely correct camber and to correct designed longitudinal section using steel template. As soon as precoated chips or gravel for a length of 15 m is laid on the road, compaction with a three wheel roller of 8 to 12 tonnes weight or with power vibrated screeds is commenced. Rolling is started at edges and progressed towards centre. When rolling had been done twice over the precoated aggregate laid, all high spots or depressions exceeding plus or minus 1 mm per meter length from the designed longitudinal section and 1 mm from the corss-section are brought to proper levels and section and re-compacted. Final rolling is done with 8 to 10 tonnes tandem roller or suitable pneumatic roller.

(d) Seal coat in low rainfall areas where rainfall is under 150 cm per year, a premixed sand-seal coat should be applied immediately after laying the carpet coat. Material required for seal coat are given below:

	Per 10 m^2 of road surface
1. Medium coarse sand or fine grit passing IS : sieve No. 170 and retained on IS : sieve no 180 microns	0.06 cu m
2. Binder	6.8 kg

(e) In high rainfall (over 150 cm per year) areas, a liquid seal with chippings instead of sand is applied immediately after laying the carpet. The binder heated to the required temperature is applied to the cleaned surface, blinded with chipping and rolled. Materials required are given below:

	Per 10 m^2
1. Coarse aggregate 6.3 mm; passing IS 10 mm square mesh and retained on 2.36 mm mesh	0.09 cu m
2. Binder	9.8 kg
For emulsions	
Coarse sand	0.045 cu m
Binder	4.9 kg

When tar carpet is used, the seal coat is applied immediately after laying the carpet coat. Road tar heated to the required temperature is sprayed evenly at 9.8 kg per 10 sq m and then it is blinded evenly with medium coarse dry sand at the rate of 0.06 cu m per 10^2 m.

Traffic is allowed on the road 24 hours after providing the seal coat.

(f) ***However,*** in some countries such as the U.S.A., the mixing of aggregate and binder is accomplished on the surface of the road. After first application of binder with pressure distributor, the requisite amount of coarse aggregate is spread by an approved mechanical spreading device and the remainder of the binder is then applied and the two applications of bituminous material and aggregates are thoroughly mixed usually with the help of multiple drag-line.

(g) ***Quality control tests.*** (IRC recommendations)

The tests, given in Table 16.3, should be carried out to control the quality of carpet coat.

Table 16.3 Quality control tests for carpet coat

Serial No.	*Test*	*Frequency*
1.	Quality of binder	As required
2.	Aggregate impact value	One test per 50–100 m³ of aggregate
3.	Flakiness index of aggregate	—do—
4.	Stripping value	—do—
5.	Mix grading	One set of tests on individual constituents and mixed aggregates from the dryer for each 100 tonnes of mix subject to a minimum of two sets per plant per day.
6.	Control of temperature of binder in boiler, aggregates in the dryer and mix at the time of laying and rolling	At regular close intervals
7.	Control of binder content and gradation in the mix (Binder content test vide ASTM D-2172)	One test for each 100 tonnes of mix subject to a minimum of two tests per day per plant.
8.	Rate of spread of mixed material	Regular control through checks on layer thickness

ROAD MIX AND INTERMEDIATE TYPE BITUMINOUS PLANT-MIX SURFACE

The bituminous wearing surfaces discussed under this heading are generally classified as intermediate-type surfaces and would usually be more than 25 mm thick. They are used in location in which the road is subjected to relatively heavy amount of traffic and may be used on old or new bases and in the re-surfacing of old pavements. They enjoy a preference over bituminous-macadam when a surface course is to be laid over an existing base of adequate strength, in that they can be made as thin as 40 to 50 mm.

The term road-mix or mixed-in-place refers to a type of construction in which the aggregate and bituminous material are thoroughly blended by mixing on top of the existing surface or base.

The term plant-mix refers to a type of surface in which the aggregate and bituminous material are proportioned and mixed at a central mixing plant. The mixture is then transported to the job site and the surface is prepared by spreading, compacting and finishing the prepared mixture. The designation intermediate type bituminous plant mix is introduced to avoid confusion and to distinguish these surfaces from the high-type bituminous pavements which are described in the next chapter.

16.4. ROAD-MIX SURFACES

The road-mix pavement is a surface course of 40 to 80 mm compacted thickness. It is placed on a base course of gravel, slag, dry stone, or it may be newly built. In every case, where the base is not newly constructed, preliminary work on the base may be necessary before the surface is applied. Surface-course mixes of this type are of a lower quality and lower cost than plant-mixes because there is no positive gradation control.

16.4.1. Bituminous road-mixes are of three general types:

(*i*) ***Open graded type.*** This is marked by the use of aggregate of essentially uniform size somewhat similar to that used in surface treatment. Usual thickness varies from 25 to 80 mm, majority of surfaces being 50 to 75 mm thick. This is also sometimes referred to as macadam aggregate type.

(*ii*) ***Dense-graded type.*** This is distinguished by the use of more or less uniformly graded aggregate locally available from such locations as river beds, pits, etc. Usual thickness range is 50 to 100 mm.

(*iii*) ***Mixed-in-place sand asphalt*** in which naturally occurring sand is mixed with liquid bitumen to form the wearing surfaces. Thickness varies from 80 mm to 150 mm. No separate base may be employed in this type of construction; surface may rest directly on the natural or stabilized sand subgrade. Construction methods are similar to other road mixed surfaces. This type of construction is of particular importance where naturally occuring sands are the only aggregates which are economically available for the construction of bituminous wearing surfaces. This is similar to sand bituminous stabilized roads.

16.4.2. Road-mix surfaces: Open-graded type

They are advantageously used on well bonded base or old surface in areas where suitable aggregates are readily available. The Road-mix process is also appropriate when the work to be done is not enough to warrant moving in and setting up a mixing plant or the road is not dense enough to justify the existence of permanent central mixing plant.

Materials Requirements

Aggregates include crushed stone, crushed slag and crushed gravel. They are required to have a per cent of wear of not more than 40 in the Los Angeles abrasion test. Crushed slag, when compacted should not weigh less than 1.12 gm per cu cm.

Only one size of coarse aggregate is used to form the main thickness of wearing surface. The surface is then usually finished by the application of seal coat. Table 16.4 gives an example of aggregate gradings which are typical of this type of construction.

Table 16.4 Gradation-road-mix surfaces, open-graded type

Sieve designation	*50 mm max. size*	*40 mm max. size*	*25 mm max. size*	*Seal coat*
50 mm	100	—	—	—
37.5 mm	90—100	100	—	—
25 mm	—	90—100	100	—
20 mm	30—65	40—75	90—100	—
12.5 mm	—	15—35	—	100
10 mm	0—10	—	20—55	90—100
4.75 mm (No. 4)	0—5	0—5	0—10	10—30
2.36 mm (No. 8)	—	—	0—5	0—10
75 micron (No. 200)	—	—	—	0—2

One of these sizes may be used. Maximum size of aggregate depends upon the thickness of the layer to be constructed.

According to usual specifications followed in this country, a combination of 6.6 cu m of 40 mm IRC standard size crushed stone and 2.4 cu m of 25 mm IRC standard size are used per sq m for constructing a wearing surface 6 cm thick. Quantities for 40 mm thick compacted surface are 4.5 cu m of 25 mm standard size and 1.5 cu m of 20 mm standard size. For seal coat 6.3 mm standard size aggregate is used.

Bituminous material. Bituminous material used may be one of the following cut-back bitumens RC-3, RC-4, RC-5, MC-3, MC-4 or MC-5 emulsified bitumens, medium or slow setting type tars RT-6 to 9.

Heavier grades are used during warm weather and lighter grades during other times of the year. The exact grade of bituminous material selected will probably depend largely upon previous experience gained on similar projects.

Typical specifications in this country suggest bitumen at 56 kg/m^3 for 3.75 cm, 2.5 cm and 2 cm IRC standard sizes stone and 64 kg/m^3 for 6.3 mm IRC standard size.

16.4.3. Road-mix surfaces: Dense-graded type

This differs from open-graded type principally in the use of a wider range of aggregate including naturally occurring gravels which are graded from relatively coarse to fine sizes. This has particular applicability to the reconstruction of old gravel or crushed stone surfaces so that the existing surface may supply all or part of the aggregate for the new wearing surface.

Materials Requirements

Both crushed and uncrushed aggregates are used.

Coarse aggregate. Crushed rock, crushed gravel, or other hard material retained on 2.36 mm sieve. When crushed gravel is used as an aggregate, not less than 90% by weight of the crushed material retained on the 4.75 mm sieve shall have atleast two fractured faces.

Fine aggregage. Screenings from coarse aggregate or sand or combination of the two passing the 2.36 mm sieve and retained on the 75 micron sieve. This fine aggregates shall have a sand equivalent value of not less than 50 as per IS 2720 (Part 37). The plasticity index of the fraction passing the 0.425 mm sieve shall not exceed four.

Filler. Finely divided mineral matter should consist of rock dust, hydrated lime or cement within the grading limits shown in Table 16.5.

Table 16.5 Cumulative per cent of finely divided aggregates passing by weight of total aggregate

IS Sieve (mm)	*Cumulative per cent passing by weight of total aggregate*
0.6	100
0.3	95-100
0.075	85-100

It should have a plasticity index not greater than 4. This requirement does not apply if filler is lime or cement. When the coarse aggregate is non-calcareous material, 2 per cent by mass of total aggregate of Portland Cement or hydrated lime should be added and the percentage of fine aggregate reduced accordingly. Cement or hydrated lime is not required when the coarse aggregate is limestone. The incorporation of 2 per cent of hydrated lime as above is recommended in all cases for mixtures, which fail to meet the water sensitivity requirement given in Table 16.6.

The physical requirements for coarse aggregate and the composition of the dense graded mix are given in Table 16.6 and 16.7.

Table 16.6 Physical requirements for coarse aggregates for Dense Bituminous Macadam

Property	*Test*	*Specification*
Cleanliness (dust)	Grain size analysis[1]	Max 5 per cent passing 0.075 mm sieve
Particle shape	Flakiness and Elongation Index (Combined)[2]	Max 30 per cent
Strength*	Los Angeles Abrasion value[3]	Max 35 per cent
	Aggregate Impact value[4]	Max 27 per cent
Durability	Soundness[5]	
	Sodium Sulphate	Max 12 per cent
	Magnesium Sulphate	Max 18 per cent
Water Absorption	Water absorption[6]	Max 2 per cent
Stripping	Coating and Stripping of Bitumen Aggregate Mixtures[7]	Minimum retained coating 95 per cent
Water Sensitivity	Retained Tensile Strength[8]	Min 80 per cent

Notes:

1. IS:2386 Part I
2. IS:2386 Part I (the elongation test to be done only on nonflanky aggregates in the sample)
3. IS:2386 Part 4*
4. IS:2386 Part 4*
5. IS:2386 Part 5
6. IS:2386 Part 3
7. IS:6241
8. AASHTO T283**

* Aggregate may satisfy requirements of either of these two tests.

** The water sensitivity test is only required if the minimum retained coating in the stripping test is less than 95 per cent.

Table 16.7 Composition of Dense Graded Bituminous Macadam Pavement Layers

Grading	1	2
Nominal aggregate size	40 mm	25 mm
Layer Thickness	80-100 mm	50-75 mm
IS Sieve[1] (mm)	Cumulative % by weight of total aggregate passing	
45	100	
37.5	95-100	100
26.5	63-93	90-100
19	-	71-95
13.2	55.75	56.80
9.5	-	-
4.75	38-54	38-54
2.36	28-42	28-42
1.18	-	-
0.6	-	-
0.3	7-21	7-21
0.15	-	-
0.075	2-8	2-8
Bitumen content per cent by mass of total mix[2]	Min 4.0	Min 4.5
Bitumen grade (pen)	65 to 90	65 to 90

Note: [1]The combined aggregate grading shall not vary from the low limit on one sieve to the high limit on the adjacent sieve.

[2]Determined by the Marshall method.

16.5. MIX DESIGN

Apart from conformity with the grading and quality requirements for individual ingredients, the mixture shall meet the requirements set out in Table 16.7 and the minimum per cent voids in mineral aggregate (VMA) in Table 16.8.

Table 16.8 Minimum Voids in Mineral Aggregates for Dense Bituminous Macadam

Nominal Maximum Particle Size[1] (mm)	Minimum VMA, Per cent Related to Design Air Voids, Per cent[2]		
	3.0	4.0	5.0
9.5	14.0	15.0	16.0
12.5	13.0	14.0	15.0
19.0	12.0	13.0	14.0
25.0	11.0	12.0	13.0
37.5	10.0	11.0	12.0

Notes: [1]The nominal maximum particle size is one size larger than the first sieve to retain more than 10 per cent.

[2]Interpolate minimum voids in the mineral aggregate (VMA) for design air void values between those listed.

Quantity of bituminous material varies from 4 to 7 per cent by weight of the dry aggregate. Both the types and the amount of bituminous material must be adjusted to the aggregate used on the basis of preliminary laboratory and field tests. The greater the proportion of fines, the less viscous should be the bituminous material and greater will be the amount required.

If stability tests are not carried out for determining the exact amount of bituminous material to be used, it may be determined by means of some proportioning formula based on mechanical analysis of aggregate or based on surface area and absorption of aggregate. One such formula is given below:

For cut-back:

$$p = 0.02a + 0.045b + 0.18c$$

For bitumen emulsions:

$$p = 0.05a + 0.1b + 0.50c$$

Where p = per cent of oil required by weight

a = per cent of aggregate above 2 mm sieve by weight.

b = per cent of aggregate passing 2 mm and retained on 75 micron sieve by weight.

c = per cent of aggregate passing 75 micron sieve.

Formulae were also developed for working out bituminous content by surface area method. Total surface area per kg of aggregate is then computed by multiplying the percentage of aggregate in such of several size groupings by its appropriate surface-area constant and taking the sum of the products.

The surface area constants used in U.S.A. are given below:

Sieve No.		*Surface area constant*
Passing	*Retained*	
75 micron (No. 200)	—	250
425 micron (No. 40)	75 micron (No. 200)	80
2 mm (No. 10)	425 micron (No. 40)	18
25 mm	2 mm (No. 10)	4

The formula developed in Colorada state, U.S.A. is given below:

$$Q = \text{Oil ratio} = K \times .0067 \sqrt{\text{surface area aggregate}}$$

Where the constant K is determined in the laboratory and usually taken as unity.

It was observed that some mixtures designed according to these formulae became dry and had tendency to ravel. In other cases the percentage of oil worked out by the formulae gave a mix so rich that under traffic, free bituminous material came to the surface. This was the case when the aggregate contained large amount of blown sand or other hard, fine material. California state now uses "centrifuge kerosene equivalent (CKE)" to determine the amount of bituminous material. Aggregate passing 4.75 mm sieve is saturated with kerosene oil and is centrifuged for 2 minutes under a force 400 times that of gravity. The quantity of kerosene retained by the aggregate determined by the increase in weight after centrifuging and expressed as percentage on dry weight of aggregate is called CKE.

$$\text{Oil ratio} = \frac{0.85CKE + 2.5}{100} \sqrt{\frac{\text{per cent passing 4.75 mm sieve}}{100}}$$

This formula is applicable for cut-back bitumen.

16.6. CONSTRUCTION METHOD

There are two fundamental and somewhat different road-mixing processes which are employed for this type of construction.

Blading and dragging. Blade mixing of aggregate and bituminous material is one of the very old methods of constructing bituminous surface. Construction methods are very simple and utilize the most common road building equipment and some farm tools.

The aggregate is placed in a flattened window of uniform depth and width and is sprayed with heavy application of bituminous material. The required amount of binder is equally divided into two or more applications. Following each application of bituminous material, the aggregate and bituminous material are mixed by blading the materials back and forth across the road or mixing area until the binder has been uniformly dispersed. Application of bituminous material and repeated manipulations are continued until the final bitumen content is well dispersed.

Mixing with motor grader and disc-harrows does not provide the kneading action of pugmill mixers used in plant mixes and therefore, a longer period of time is required to obtain a proper mix. Also for same reason it is more practicable to use a lower viscosity cut-back bitumen.

Travel plant. Travel plant of various types are in general use for mixing the bituminous material and aggregate. The most common type is the mechanical mixer which lifts the aggregate from a window and passes it through a pugmill mixer in a continuous operation. As the aggregate enters the pugmill mixer, the bituminous material is sprayed into the mixing chamber at a pre-determined rate. To obtain proper control of the amount of bituminous material required, it is necessary to have the aggregate in a uniformly sized window so that the rate of discharge of the bituminous pump may be correlated with the speed of the plant and the size of the window.

Some types of travel plants receive the aggregate from trucks which dump the material into a hopper rather than pick it up from a window.

Other types of travel plants are those which take the aggregate from an inplace position or which operate from a flattened window.

Another type of mixing plant is the one which requires that the aggregate be in a uniform loose depth prior to mixing operations. Several passes of the mixer are usually required for a satisfactory mix.

After the material has been satisfactorily spread, it is then rolled by a 3-wheel roller weighing from 6 to 8 tonnes until it is tested for smoothness. Irregularities greater than 6.3 mm as measured by a 3 m straight edge are removed by means of a long-base planner of a blade grader.

Aeration of travel plant and blade mixes before compaction is very essential. The nominal moisture content of most aggregates is an aid to mixing but can be harmful if the mixture is compacted with more than 2 per cent moisture. It is very necessary that these mixtures be given sufficient manipulations after mixing to reduce most of the volatile and moisture content of the mix by evaporation. Spreading and compaction of the mixture should not be performed until the volatile content has been reduced to less than 25 per cent of the original and the moisture content to not more than 2 per cent.

INTERMEDIATE TYPE BITUMINOUS PLANT MIX SURFACES

The plant mix process, however, permits more accurate control both of aggregates and bituminous materials and therefore, normally results in the production of a superior job. The other advantages are:

(*i*) It prolongs the construction season, since both the aggregates and bituminous materials can be heated (*ii*) More viscous grades of bituminous materials can be used; (*iii*) Wet weather, with delays and with danger of inferior results, is a much less serious problem; (*iv*) Traffic is much less inconvenienced. Because of these various considerations there is a trend towards the plant-mix process except where aggregate is to be reused or where some other special condition exists.

16.7. MATERIALS REQUIREMENTS

Aggregates for intermediate type bituminous plant mixtures are practically the same as for open-graded road mixes.

Many different gradings of aggregates give satisfactory results in dense graded plant-mix pavements. Table 16.9 shows typical gradings which are usually adopted for surface courses. For base courses the maximum size may be increased.

The aggregate for all the gradings in Table 16.9, except E, E-1 and E-2, should consist of coarse crushed gravel or stone, with filled of fine crushed stone or sand. For gradings E, E-1 and E-2, aggregate should consist of fine gravel and sand, disintegrated granite, etc. At least 10 per cent of the total should be retained on 2 mm sieve. A per cent of wear not greater than 50 is desirable.

Table 16.9 Requirements for grading of aggregate

Sieve designation	*25 mm maximum size Percentage by weight passing square mesh sieve*				
	Grading A	*Grading B*	*Grading C*	*Grading D*	*Grading E*
25 mm	100	100	100	100	100
20 mm	75—100	75—100	85—100	85—100	85—100
4.75 mm (No. 4)	35—45	40—60	45—65	50—70	60—95
2.0 mm (No. 10)	20—35	20—45	30—50	35—55	45—80
75 micron (No. 200)	2—7	3—8	5—10	5—12	5—15

Sieve designation	*20 mm maximum size Percentage by weight passing square mesh sieve*				
	Grading A-1	*Grading B-1*	*Grading C-1*	*Grading D-1*	*Grading E-1*
20 mm	100	100	100	100	100
4.75 mm	35—50	45—65	50—70	55—75	60—95
2.0 mm	25—45	30—50	35—55	40—60	45—80
75 micron	2—7	3—8	5—10	5—12	5—15

Sieve designation	*12.5 mm maximum size Percentage by weight passing square mesh sieve*				
	Grading A-2	*Grading B-2*	*Grading C-2*	*Grading D-2*	*Grading E-2*
12 mm	100	100	100	100	100
4.75 mm	40—55	50—70	55—75	60—80	65—100
2 mm	30—45	35—55	40—60	45—65	50—85
75 micron	2—7	3—8	5—10	5—12	5—15

Bituminous Materials

For open graded type. Rapid-curing bitumen RC-4, RC-5; Slow curing bitumen SC-6; paving bitumen 80—100 to 150—200; Emulsified bitumen medium setting type; Tars of high viscosity.

For dense graded type. Rapid-curing cut-back bitumen RC-3, RC-4 or RC-5; Medium-curing cut-back bitumen MC-4, MC-5; Slow-curing asphalt SC-6; paving bitumen 80—100, 100—120, 120—150, 150—200; Tars of high viscosity.

16.8. CONSTRUCTION METHOD

The road bed is repaired and is swept clean of all foreign materials. A primer or tack coat, as the base may call for, is used. The mineral aggregate is passed through a rotary mixer driven to reduce the moisture content to 1 per cent or less, and to heat the aggregate to the mixing temperature, which should not exceed the proper temperature for the bituminous material by more than 10°C. The aggregate is then screened into two or more sizes for proportioning and is stored in bins.

It is weighed from the bins and put into the mixer, where it is combined with the pre-determined amount of bituminous material and mixed thoroughly for at least ½ min or more if needed. Either a rotary drum batch type mixer, a pugmill batch type mixer, or a continuous following pugmill-type mixer may be used. The mixed material is hauled to the job in clean trucks, the metal bodies of which have been sprayed with suitable materials to prevent adhesion. The trucks should have insulated bottoms if the length of haul requires, and the mix should be covered for protection from the weather. The material should be spread on the road when the atmospheric and ground temperatures are not less than 5°C. A combination of spreading and finishing machine is desirable; otherwise blade graders are used for spreading and shaping. When the desired cross section has been obtained, compaction is secured by rolling with 6 to 8 tonne weight roller. During rolling the surface is dragged or bladed to secure requisite smoothness.

❑❑❑

Chapter

HIGH-TYPE BITUMINOUS PAVEMENTS, BASE COURSES AND THEIR DESIGN

High-type bituminous pavements are widely used on roads which are subjected to large volumes of traffic and severe service conditions. Properly designed and constructed surface under this category are capable of carrying almost unlimited volumes of passenger, mixed, or truck traffic, provided they are supported by adequate thickness of foundation layers. The economic life of the majority of these surfaces may be expected to be over twenty years.

All the high-type bituminous paving mixtures are prepared in central mixing plants which can achieve the following objectives:

1. Uniform drying of aggregate to desired moisture content
2. Preheating of aggregate to the required temperature
3. Accurate control on the grading and quantity of aggregate
4. Accurate control on the quantity and temperature of bitumen and mix
5. Uniform coating of stone particles with bitumen.

The thickness of these surfaces may very from 20 mm to 100 mm or more, depending on type of surface and its purpose. Wearing surfaces of this type have very slight crown-about one cm per metre. High type bituminous pavements are the combination of appropriate grade of semi-solid bituminous binder with completely graded aggregates proportioned (as in the case of cement concrete) so that fine materials fill the voids in the coarse material and the bitumen fills the voids in the fine material, creating a very dense mixture. Bituminous concrete differs from the dense graded plant mixes already described under the intermediate-type mixes in the care with which it is proportioned, mixed and shaped and consequently in the uniformity, quality and cost of finished product.

CLASSIFICATION OF HIGH-TYPE BITUMINOUS SURFACES

The high-type bituminous surfaces can be broadly categorized as follows and these will be discussed in the same sequence.

A. Hot-mix hot-laid pavements

1. (*a*) Bituminous concrete

(*b*) Mastic asphalt

(*c*) Sand asphalt

2. Tar concrete

B. Hot-mix or cold mix, cold-laid pavements.

(*a*) Bituminous concrete

(*b*) Tar concrete.

C. Miscellaneous bituminous concrete mixtures

Modified binder bituminous Concrete pavements

D. High performance bases.

Before giving details of the various types listed above, factors relating to high-type bituminous paving mixtures will be outlined.

DESIGN OF HIGH-TYPE BITUMINOUS PAVING MIXTURES

The general requirements or objectives for satisfactory bituminous mixes are:

(i) Stability. To be stable, the mix should be able to resist displacement by traffic and have sufficient strength or load carrying capacity to withstand the loads imposed on it. Unstable pavements are marked by rutting, waviness and deformation under the tyres of vehicles.

(ii) Durability. To be durable; the mix should be able to withstand the effect or natural weathering and traffic without excessive deterioration or loss of strength.

(iii) Skid resistance. The surface texture should be such that it will grip the tyre, even when the pavement is wet. Research has indicated that as carbonate content of the fine aggregate decreases and silica content increases, the skid resistance increases. Other factors in case of fine bituminous mixtures are shape and size of aggregates, gradation and ratio of diameters of coarse to fine particles.

(iv) Workability. The mix should have sufficient workability to permit the placement of the mix as an efficient construction operation.

(v) Air voids. The range of air voids in the total compacted mix should be between 2 to 7 per cent to provide space for the expansion of bitumen and for a slight amount of degradation of aggregates under traffic. Complete densification accentuates lubrication action of bitumen with consequent loss of stability. A high volume of voids, however, would be conducive to the hardening of bitumen through weathering which may shorten the life of the pavement. Experience indicates to place the upper limit of voids at 7 per cent to avoid undue weathering of bitumen.

(vi) Economics. The mix should use the least expensive materials which will produce a stable, durable and satisfactory pavement.

Factors affecting design

1. Grading, type and quality of aggregates.
2. Bitumen content, consistency and quality of bituminous material used in the mix.

As regards bituminous material is concerned, the following statements should hold good:

Low bitumen contents invite ravelling. Too much bitumen results in unstable pavement.

Use of very hard bitumen may result in a brittle pavement with resultant ravelling and excess cracking.

Uncracked bituminous materials meeting the standard bitumen cement specifications are commonly accepted for use from the stand-point of quality. Low quality bitumens eventually produce a brittle pavement.

17.1. AGGREGATES

Regarding aggregates, the following fundamental considerations hold good:

17.1.1. Aggregate factors affecting stability

Particle shape. When compacted, irregular or angular particles of aggregate tend to become interlocked, thus possessing a mechanical resistance to displacement. The particles may be in actual contact at only a few points; the mass has, therefore, little frictional resistance to motion and has no cohesion. It may however be very stable if sufficiently angular and well compacted because displacement of one particle may require movement of several others, also against irregular interlocked surfaces.

The stability of open type mixes, with coarse aggregate particles in contact only at a few points is almost entirely due to the effect of mechanical interlocking.

Conversely, aggregates composed of rounded particles develop no mechanical interlock and are as a rule unstable unless mixed with sufficient suitable binder. Vary little stability can be developed when compacted without binder.

Wide-spread recognition of the value of mechanical interlock of angular particles in developing adequate stability is shown by the specifications which require that some minimum percentage of gravel particles be crushed for use in bituminous mixes. ASTM specifications in respect of gravel requires that not less than 40 per cent by weight retained on 4.75 mm sieve should have at least one fractured face.

Frictional resistance. The bituminous material undoubtedly acts as a lubricant in a bituminous mix. In a properly designed mix, however, the lubricating effect which tends to decrease frictional resistance to displacement would tend to be offset by the cohesive strength of the bitumen. In order that any appreciable amount of frictional resistance may be developed in the mix, the amount of bituminous material must be small enough so that the voids are not completely filled. The aggregate particles must be forced by compaction into close contact with one another, with only a thin film of bitumen between them.

The amount of friction that can be developed will depend to a great extent upon the closeness of contact and the area of contact (or the number of contact points) of the aggregate particles. A dense graded aggregate in which the large particles are separated by numerous smaller particles in the voids, has many more points of contact than an open graded material and may be expected to develop greater frictional resistance to displacement. A large portion of the stability of dense graded mixes is due to the friction between aggregate particles. The basic properties of the aggregate, *i.e.* surface roughness and surface attraction between particles will be a limiting factor to the amount of frictional resistance that can be developed under given conditions.

Gradation. One of the most important characteristics of aggregates affecting the stability of a mix is the gradation. As noted above the dense-graded mixes tend to have many more points of contact between particles, resulting in greater frictional resistance. The increased number of contacts also results in a greater area for load transfer from one particle to another, decreasing the possibility of crushing of the individual aggregate particles by point loadings.

Logically, it would seem that the best method of increasing stability would be to use the densest aggregate gradation possible, *i.e.* to approach a solid as closely as possible by packing the maximum amount of aggregates into the given volume, with just enough bituminous material to bind the aggregate particles together. Numerous investigations have proposed aggregate gradations for maximum density. The basic idea of the theory is that the amount of material of a given size is just sufficient to fill the voids between particles of the larger sizes. In actual practice, the smaller particles tend to wedge between the larger ones, increasing the voids that must be filled by the smaller sizes. As a result, maximum densities are actually produced by gradations having a little excess of the small sizes compared to the theoretical values.

The work done by various research workers shows conclusively that aggregate gradation is a very important factor in the stability of bituminous mixes, and that those gradations which approach the theoretical gradation for maximum density are the best.

Degradation of aggregates. The aggregate particles used in bituminous mixes tend to be broken or "*degraded*" by the loads imposed upon them, both during construction and later by the action of traffic. Degradation may take place both by compressive failure from concentrated loads at points of contact between aggregate particles and by abrasive action when the individual particles move with respect to the others.

Degradation of aggregates results in changes in gradation and partial filling of the voids by the small fragments grounded from the larger ones. In a mix containing very few voids, this action may result in over-filled voids and an unstable mix. The amount of degradation is of course, affected by both the magnitude of the loads and the resistance to crushing or abrasion of the individual aggregate particles.

Studies show that the amount of degradation is also dependent upon the original grading. The greatest amount of degradation was found in open gradings, the least in Fuller's gradation. There was a tendency to degrade, towards, Fuller's maximum density curve indicating that it is the most stable gradation. The amount of degradation under road rollers has been correlated with the loss at 100 revolutions in the Los Angeles abrasion machine. The Los Angeles abrasion test, then, can be used to measure the relative amounts of degradation to be expected with various aggregates. This test is usually specified as one of the standard tests for aggregates for bituminous mixes.

In dense graded mixes, if aggregate having reasonable crushing strength is used, little degradation should occur. However, considerable per cent of voids should be included in the mix to preclude the possibility of over filled voids. Open gradations are more subject to degradation, but also have a greater amount of voids thus, making over filling less probable. The greatest danger from degradation in open mixes is the possibility of increasing the surface area so much that inadequate coating results. To prevent excessive change, however, materials having greater crushing resistance would be required for open mixes than for dense graded materials.

17.1.2. Aggregate factors affecting durability

Durability of aggregate. Although aggregate durability is a big factor in the durability of Portland cement concrete, it is believed that the same is not true for bituminous mixes. Failures of concrete due to non-durable aggregates usually takes place by the disruption of rigid mortar due to changes in volume of the aggregate caused by freezing, temperature change, or chemical reaction. Since the bituminous binder is flexible, such changes in volume of the aggregate should occur without any damage to the mix.

Water has a great influence on the effect and severity of freezing, temperature change and chemical reaction on aggregates. Since the bituminous films tend to protect the aggregate from water, any effect of such factors would be decreased. No case is known to the author where lack of aggregate durability has definitely caused failure of bituminous mixes.

Most specifications include the sodium sulphate or magnesium sulphate soundness to evaluate the durability of the aggregate. Although this test, in one form or another, has been used for about 150 years, no investigator has yet been able to correlate its results with any filed performance or service record. The literature, however contains many references to its inadequacy. Apparently, a test of very doubtful value is being used to evaluate the susceptibility of aggregates to a probably non-existent problem so far as bituminous mixes are concerned.

Stripping. Stripping or separation of the bituminous film from aggregate through the action of water is probably the greatest single problem in the durability of bituminous mixes. A large amount of investigation and research has been carried out on this problem.

Some material and aggregates appear to have a greater affinity for water than for bituminous mateials, and bituminous films may be more or less easily displaced from them by water. These materials are called "*hydrophillic*" (water-loving) and those that have a greater affinity for bituminous material are designated "*hydrophobic*" (water-hating). The terms are only relative, however, a material may appear to be hydrophobic under some conditions of use and hydrophillic under others.

Typical examples of hydrophobic aggregates are limestone, basalt, and dolomite which are basically calcareous in nature. A typical example of hydrophillic aggregate which is easily stripped is quartzite which is siliceous in nature.

Many factors affect the resistance to stripping in a bituminous aggregate mix, the major factors affecting adhesion being:

(*a*) Character of aggregate, that is, the chemical, crystalline, and mechanical properties and the extent and type of weathering.

(*b*) Binder type and consistency.

(*c*) Modification of the binder-aggregate interface due to adsorption and/or chemical reaction of materials such as water, gases, binder constituents, and additives or wetting agents.

(*d*) History of the system, *i.e.* extent of interface equilibrium, previous exposure conditions and sequence of conditions.

(*e*) Characteristics of the mix, *i.e.* binder content, aggregate gradation, amount of voids, etc.

Studies regarding the effect of surface area and completeness of coating on adhesion and stripping show that stripping may occur even with "hydrophobic" aggregates if the minus 75 micron sieve size materials has a large amount of very small "clay-size" material in it. Inadequate coating would enable stripping to start more easily.

Tars are usually more resistant to stripping than most bitumens.

In general, binders of high viscosity resist stripping more readily than those of low viscosity under comparable condition. However, low viscosity binders will wet the aggregate easily and rapidly. In order to compromise these two contradicting requirements, advantage is taken of the viscosity/temperature characteristics of bituminous binders by obtaining initial adhesion with a hot binder having low viscosity, the viscosity of which will increase markedly on cooling to road temperature to ensure better protection against stripping.

Addition of filler to a mixture increases viscosity of the binder and is therefore helpful in controlling stripping. Certain fillers such as hydrated lime and portland cement have both actions, physical to increase the viscosity and also chemical to form surface-active calcium salts which ensure better adhesion. Such chemically active fillers may also be used in the mixing of cold and wet aggregates with bituminous binders where they enable coating to be achieved under adverse conditions.

Research has indicated that it is possible to improve adhesion in bituminous materials and to prevent damage due to wet weather by adding surface-active chemicals to the binder.

Absorption. Another characteristic of the aggregate that may affect the durability of the bituminous mix is its absorptive capacity. The absorption of the bituminous material results in leaner bituminous surfaces or films that are no longer near optimum strength and toughness, weather and harden faster and may be more susceptible to trouble from water. Some of the tests indicate that upto 25 per cent of the binder film may be absorbed by some aggregates. Very porous aggregates may continue to absorb a part of the bituminous material over a considerable period. Selective absorption of the less viscous portions of the bituminous film may cause it to be thinner and more brittle and more susceptible to weathering. Some field failures have been attributed to the absorption of bituminous material by highly absorptive cherts and sandstones. Various correction factors for absorption have been suggested based on the centrifuge kerosene equivalent (CKE).

With bitumen cements, the problem of absorption would be less than with liquid bituminous materials. For very absorptive aggregates, however, it would probably be wise to make some corrections regardless of the type of bituminous material used, since selective absorption may take place slowly over a long period of time.

It should be noted that in dense graded mixes, the effect of compaction and degradation by traffic is to decrease the void space, tending toward overfilling the voids with bituminous material. This effect and that of absorption by the aggregate tend to counteract each other. In open gradations, however, degradation may have the effect of greatly increasing the surface area without decreasing the voids sufficiently to make over-filling a problem. In this case, the problem of a great increase in area to be coated by the bituminous materials could be aggravated by the absorption of the aggregate. In making corrections for absorption by the aggregate, the other characteristics of the mix which may affect the result should also be considered.

17.2. CALCULATION OF PER CENT MAXIMUM DENSITY, PER CENT VOIDS, AND VOIDS IN THE MINERAL AGGREGATE

The density of the compacted mixture is generally calculated at room temperature by determining the weight of a test specimen in air and the weight of the same specimen in water. The same method may be used to determine the density of a sample taken from a completed pavement. Thus

$$G = \frac{W_a}{W_a - W_w}$$

where G = density (specific gravity) of compacted mixture.

W_w = weight of test specimen suspended in water (grams)

W_a = weight of test specimen in air (grams)

The specific gravity of the total blended mineral aggregate is calculated after determining the specific gravities of different aggregates used in the mix.

$$G_a = \frac{100}{\frac{W_1}{g_1} + \frac{W_2}{g_2} + \frac{W_3}{g_3} + \frac{W_4}{g_4}}$$

where G_a = specific gravity of combined aggregate.

W_1, W_2 *etc.* = respective per cents by weight of aggregate 1, aggregate 2, *etc.*

g_1, g_2, *etc.* = respective specific gravities of aggregate 1, aggregate 2, *etc.*

While making the design, three types of specific gravity are employed *e.g.* bulk specific gravity which involves several volume of the aggregate particle including its capillaries; apparent specific gravity which involves only the impermeable portion of stone exclusive of the volume of the capillaries which become filled with water upon 24 hr soaking; or effective specific gravity which involves the volume of the impermeable stone and volume of capillaries unfilled with bitumen as it exists in the pavement. Some organizations employ average of bulk and apparent specific gravities. When the effective specific gravity is employed, the determination is made on the total aggregate as it will exist in the mix. For other two specific gravities, determination is made on each individual aggregate used in the mix and specific gravity of combined aggregate is calculated as given below.

Foot note 1.

Specific gravity of combined aggregate, G_a

Average density of combined aggregate, $G_a = \frac{W}{V}$

As individual weights are expressed as percentage of total weight

$\therefore$ $W = 100$

As ρ is cancelled

$$\therefore \quad G_a = \frac{100}{\frac{W_1}{g_1} + \frac{W_2}{g_2} + \frac{W_3}{g_3} + \frac{W_4}{g_4}}$$

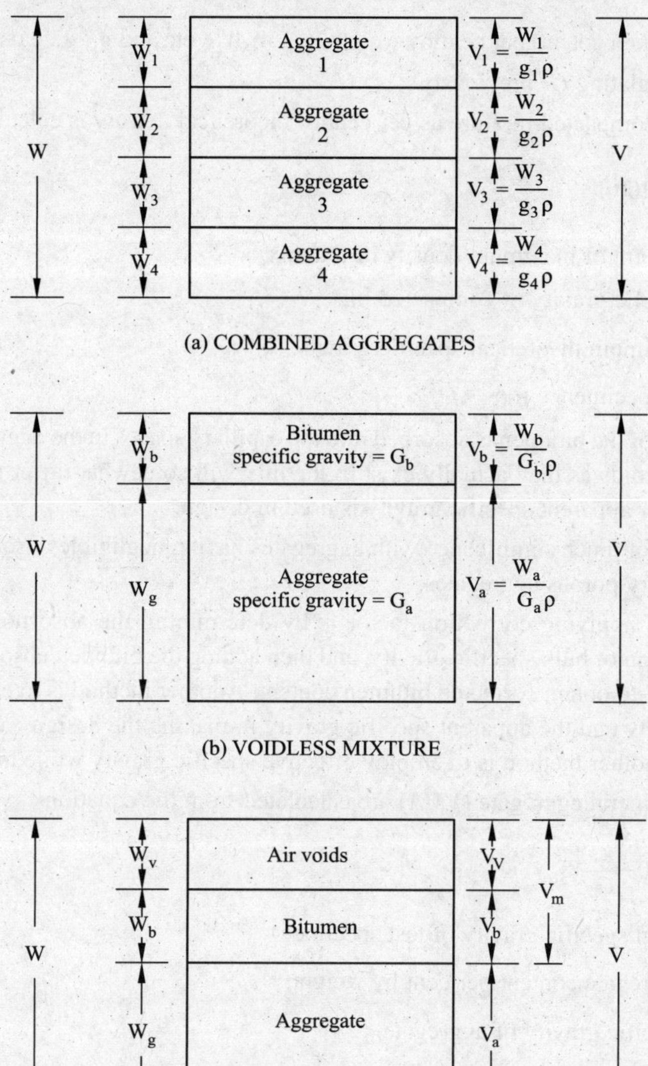

Fig. 17.1 Different types of mixtures.

The theoretical maximum specific gravity which is the theoretical density of a voidless mixture of a bituminous paving mix may be expressed as follows:

$$G_t = \frac{100}{\frac{(100 - W_b)}{G_a} + \frac{W_b}{g_b}}$$ (See foot note 2 on next page)

where G_t = maximum theoretical specific gravity at 25°C.

W_b = bitumen cotent, per cent by weight. $G_a, W_1, W_2,$ etc and g_1, g_2, etc are the same used for calculating G_a previously.

The density of compacted mixture as per cent of theoreitcal density is calculated as follows:

$$M = \frac{G}{G_t} \times 100$$

where M = per cent of maximum density of the mix.

G = specific gravity of compacted mix.

G_t = maximum theoretical specific gravity.

Per cent voids in specimen $= 100 - M$

Actually, some of the bitumen is absorbed into the capillary spaces in the aggregate and as such the percentage of air voids as they actually exist in the mix will somewhat differ than those calculated when either bulk or apparent specific gravity is used in design.

This fact is not of much significance with aggregates having negligible absorption but it becomes significant with very porous aggregates.

One method of applying correction factor is by determining the absorption of the aggregate, predicting the design of bulk specific gravity, and then adding the bitumen absorbed into the stone to the design quantity to obtain corrected bitumen content. Another method is to employ the average of bulk specific gravity and the apparent specific gravity in making the design calculation which is an approximation. Another method is to employ effective specific gravity while making the design.

The voids in mineral aggregate (*VMA*) are calculated from the equation

$$VMA = 100 - \frac{G}{G_a} W_a$$

Where G = actual specific gravity of test specimen.

W_a = aggregate content per cent by weight.

G_a = specific gravity of aggregate.

$$\text{Per cent voids, filled with bitumen} = \frac{VMA - (100 - M)}{VMA}$$

Foot note 2. Theoretical maximum specific gravity G_t

$$\rho G_t = \frac{W}{V_a + V_b} = \frac{W}{\dfrac{W_a}{G_a \rho} + \dfrac{W_b}{G_b \rho}}$$

$$\therefore \quad G_t = \frac{W}{\dfrac{W_a}{G_a} + \dfrac{W_b}{G_b}} \qquad \text{(cancelling } \rho\text{)}$$

When W_a and W_b are expressed as weight percentages, then

$$G_t = \frac{100}{\dfrac{W_a}{G_a} + \dfrac{W_b}{G_b}} = \frac{100}{\dfrac{(100 - W_b)}{G_a} + \dfrac{W_b}{G_b}}$$

Problem 17.1 *Determine the per cent voids*

(*i*) *in mineral aggregate*

(*ii*) *in compacted mix*

(*iii*) *filled with bitumen*

from the following data:

Bitumen content, $W_b = 5.0$ *kg. per 100 kg of aggregate.*

Specific gravity of bitumen, $g_b = 1.002$

Bulk specific graivty of mineral aggregate, $G_a = 2.800$

Bitumen absorption, weight per cent = 0.56

Specific gravity of test specimen, $G = 2.479$

Solution. Assume 100 ml of compacted mix

Wt of mix per 100 ml $= 100 \times 2.479 = 247.9$ gms.

Wt of aggregate, per 100 ml of mix $= \dfrac{274.9 \times 100}{105}$

$= 236.1$ gms

Volume of aggregate including pores $= \dfrac{236.1}{2.800} = 84.3$ ml

Voids in mineral aggregate $= 100 - 84.3$

$= 15.7$ ml

Weight of total bitumen $= 247.9 - 236.1$

$= 11.8$ gms

Weight of absorbed bitumen $= \dfrac{236.1 \times 0.56}{100}$

$= 1.3$ gms

Weight of unabsorbed bitumen $= 11.8 - 1.3$

$= 10.5$ gms

Volume of air voids $= 15.7 - 10.5 = 5.2$ ml

Per cent voids filled with bitumen $= \dfrac{15.7 - 5.2}{15.7} \times 100$

$= 66.9$

Summary of results

Per cent of voids in mineral aggregate $= 15.7$

Foot note 3. Per cent of volume of voids in mineral aggregate, *VMA*

$$VMA = \left(\frac{V - V_a}{V}\right) 100 = \left(1 - \frac{V_a}{V}\right) 100$$

$$= \left(1 - \frac{W_a / G_a \rho}{W / G\rho}\right) 100$$

$$= 100 - W_a \frac{G}{G_a} \qquad [\because W = 100\%]$$

Per cent voids in compacted mix = 5.2

Per cent voids filled with bitumen = 66.9

Problem 17.2 *Determine the unit weight of a bituminous mix containing 70 per cent coarse aggregate, 24 per cent fine aggregate and 6 per cent bitumen by weight of the mixture. The air voids after-compaction are 8 per cent. The specific gravities of the materials are as under:*

Sp. gravity of coarse aggregate = 2.80

Sp. gravity of fine aggregate = 2.66

Sp. gravity of bitumen = 1.00

Solution. $\frac{\text{Vol of voids}}{\text{Total volume}} = \frac{V_v}{V} = 0.08$

$\therefore$ Volume of voids, $V_v = 0.08\ \text{V}$

Let the total weight of mix = W

Volume of coarse aggregate, $V_c = \frac{0.70\ W}{2.8} = 0.25\ W$

Volume of fine aggregates $V_f = \frac{0.24\ W}{2.66} = 0.090\ W$

Volume of bitumen $V_b = \frac{0.06\ W}{1} = 0.06\ W$

$\therefore$ $V = 0.08\ V + 0.25\ W + 0.09\ W + 0.06\ W$

or $V\ (1 - 0.08) = W(0.25 + 0.09 + 0.06)$

or $\frac{W}{V} = \frac{0.92}{0.4} = 2.3$

$\therefore$ Unit weight of mix = 2.3 g/cm^3

Problem 17.3 *Determine the specific gravity of combined aggregates in a bituminous mix having maximum theoretical specific gravity of 2.4. The bitumen content is 8 per cent by weight of the mix and its specific gravity is 1.00.*

Solution. Maximum theoretical specific gravity $G_t = 2.4$

Bitumen content, $W_b = 8$ per cent

Specific gravity of bitumen $g_b = 1$

The maximum theoretical specific gravity is given by

$$G_t = \frac{100}{\frac{(100 - W_b)}{G_a} + \frac{W_b}{G_b}}$$

Where G_a is the specific gravity of the combined aggregate

$$\therefore \quad 2.4 = \frac{100}{\frac{(100-8)}{G_a} + \frac{8}{1}}$$

or $$2.4\left(\frac{92}{G_a} + 8\right) = 100$$

or $$\frac{92}{G_a} = \frac{100}{2.4} - 8 = 33.67$$

$$\therefore \quad G_a = \frac{92}{33.67}$$

$$= 2.733$$

17.3. STEPS IN DESIGN PROCEDURE

The various steps in the rational design of high-type bituminous paving mixtures are as follows:

1. Select grading to be used.
2. Select aggregates to be used in the mix.
3. Determine the specific gravity of the combined aggregate and bituminous material. Specific gravity of bituminous material is rarely determined in the laboratory, as it is usually furnished by the producer.
4. Determine the proportion of each aggregate required to produce the designed grading.
5. Make up trial specimens with varying bitumen contents.
6. Determine specific gravity of each compacted specimen.
7. Make stability tests on the specimens.
8. Calculate per cent voids in each paving specimen and, if the design method requires it, calculate VMA and the per cent voids filled with bitumen.
9. Select the optimum bitumen content from the data obtained.

These are discussed below:

Selection of grading. Specifications, such as ASTM, AASHTO, Indian Roads Congress or other highway departments define the limits of particle sizes within which densely graded mixes, varying in top size of aggregate should lie for satisfactory performance. For new construction a base course or binder course usually employs larger top size mixes than does the surface course. Other factors being equal, the larger top size mixes have greater stability. When the aggregate cost for the large and small top size aggregates is about the same, it is desirable to employ a mix with as high a top size as practical because a lower bituminous content will be required, resulting in a less expensive mix. Base course or binder courses usually range from 25 to 50 mm top size and surface courses most commonly in use are 2 to 12.5 mm top size.

Selection of aggregates. An effort should be made to select the most economical materials that will perform satisfactorily. Not only the materials should be of proper quality but the several aggregates must be of such grading that they may be combined to produce the desired grading. From the stand point of ease of control, it is desirable that as few aggregates as possible be used to make the completed combination. However, from two to five different aggregates may be required.

Quite often, the cost of transporting an aggregate to the plant site is more than the cost of the aggregate itself. For this reason, the least expensive aggregates are those which can be obtained locally. Uncrushed river gravels can be made suitable aggregates by local production because expensive crushing equipment it not required. However, such materials frequently have low stability characteristics. When sufficient quantity of bituminous concrete is included in the project and when adequate deposits of high quality lime-stones or other high-stability aggregates are near to the site, it is feasible to set up crushing equipment and produce the material locally.

Proportioning of aggregate to produce grading. The aggregate proportions selected to produce a given grading may be derived mathematically, graphically or by actually blending them in the laboratory on a trial and error basis until statisfactory grading is obtained.

Preparation and stability test of compacted specimens. The exact method of preparing, compacting and testing the specimen is a function of the test method employed. The four most commonly used methods are : (*i*) Marshall method (*ii*) Hveem method (*iii*) Hubbard-field method and (*iv*) Smith triaxial method.

Brief details of these methods and criteria will be given in the next article. Other popular strength tests used for design of bituminous mixes are unconfined compressive test and immersion compression test.

Calculations of per cent maximum density, per cent voids, and voids in mineral aggregate are made as explained in the previous article.

Selection of optimum bitumen content. This is determined by interpreting the results of laboratory tests in accordance with the particular test method employed. Requirements relative to stability and density are given under each test method. When a well equipped laboratory is not available and the job is vary small, optimum bitumen content may be estimated by CKE test alone.

Following is a general guide line for adjusting the trial mix, but the suggestions outlined may not necessarily apply in all cases:

(*a*) *Voids low, stability low.* Voids may be increased in a number of ways. As a general approach to obtain higher voids in the mineral aggregate (and therefore providing sufficient void space for an adequate amount of bituminous binder and air voids) the aggregate grading should be adjusted to deviate from a maximum density curve (Fuller or Wyoming). More coarse or more fine aggregate may be used to obtain this deviation. When the grading curve plotted on a semi-logarithmic chart has a tendency toward *S* shape, the gradation may be improved by adjusting the sizes necessary to give a straight line. If the dust content is high, a reduction in this fraction usually will increase the aggregate voids.

If the bitumen content is higher than normal and the excess is not required to replace that absorbed by the aggregate, a reduction in the quantity of bitumen may be made to increase the voids. It must be remembered, however, that a reduction in the bitumen content reduces the film thickness, and as a result the durability of the pavement. Too great a reduction in film thickness also may lead to brittleness. If the above adjustments do not produce a stable mix, consideration should be given to a change in the type of aggregate. It is usually possible to improve the stability and increase the aggregate void content of a mix by increasing the amount of crushed materials. With some aggregates however, the freshly fractured faces are as smooth as the water worn faces and an appreciable increase in stability is not possible. This is generally true of quartz or similar rock types.

(*b*) *Voids low, stability satisfactory.* Low void content may result in instability or crushing after the pavement has been exposed to traffic for a period of time because of reorientation of particles and additional compaction. It also may result in sufficient void spaces for the amount of bitumen required for high durability, even though stability is satisfactory. Degradation of the aggregate under the action of traffic may also lead to instability and crushing if the void content of the mix is not sufficient. For these reasons, mixes low in voids should be adjusted by one of the methods given above, even though the stability appears satisfactory.

(*c*) *Voids satisfactory, stability low.* Low stability when voids are satisfactory usually indicates poor quality aggregate. Consideration should be given to improve the quality as discussed above.

(*d*) *Voids high, stability satisfactory.* High voids frequently are, though not always, associated with high permeability. High permeability, by permitting circulation of air and water through the pavement, may lead to premature hardening of the bitumen. Even though stabilities are satisfactory, adjustments should be made to reduce the voids. This usually may be accomplished by increasing the dust mineral content of the mix. In some cases however, it may be necessary to select or combine aggregates closely to the approximate gradation of a maximum density grading curve (Fuller or Wyoming).

(*e*) ***Voids high, stability low.*** Two steps are necessary when the voids are high and the stability is low. Firstly the voids are adjusted by the methods discussed above. If this adjustment does not improve the stability, the second step should be to consider a change in aggregate quality.

17.4. THEORY OF BITUMINOUS MIX DESIGN

There are two fundamental principles underlying the design of bituminous-aggregate mixtures; namely:

(1) Sufficient quantity of binder should be used in the mixture so that the voids are filled, aggregates thoroughly coated and a durable impermeable mix is obtained; and

(2) Excess binder should not be added which may lower the stability of the mixture below some minimum value by the traffic and environmental conditions.

Thus, there is an 'optimum' binder content for each mixture. There are two approaches to determine this optimum value, *viz*.

(1) Surface area concept ; and

(2) Voids concept

17.4.1. Surface area concept

Optimum binder content is considered to be a function of the surface area of the aggregate which is to be covered. The surface area of aggregate is dependent on the size, shape and surface texture of the mineral particles, their absorptive capacities, gradation and specific gravity and the type of binder. Typical formula, relating surface area to binder content, proposed by Nebraskan, is as follows:

$$P = AG\,(0.02a) + 0.06b + 0.10c + Sd$$

where $P =$ per cent by weight of bitumen residue in the bituminous mixture prior to laying.

$A =$ Absorption modifying factor for aggregates retained on $300\ \mu\text{m}$ (micron) sieve

$$G = \frac{2.62}{\text{Apparent specific gravity of aggregate}}$$

$a =$ Per cent by weight of aggregate passing $300\ \mu\text{m}$ sieve and retained on 150 μm sieve.

$c =$ Per cent by weight of the aggretage passing 150 μm sieve and retained on 75 μm sieve.

$d =$ Per cent by weight of the aggregate passing 75 μm sieve.

and $S =$ An experimental factor depending on the fineness and absorptive characteristics of the material passing 75 μm sieve.

The use of the surface area formulae is confined to determine the amount of binder to be used in low cost and intermediate types of bituminous roads and for estimating the quantities of binder to be used in laboratory design procedures based on voids concept.

17.4.2. Voids concept

The amount of binder to be added is controlled by (*i*) the void space in the aggregate which in turn depends upon the aggregate gradation and method of compaction and (*ii*) the desired volume of voids in the final compacted mixture. The compacted mixture should have a residue of air voids to allow for:

(*a*) Expansion of the binder and entrained air under hot weather conditions.

(*b*) To provide the mixture with a certain amount of elasticity; and

(*c*) To provide safety space for further compaction of the surfacing under heavy traffic.

However, the residue of air voids should not be so large as to allow air and moisture to enter the road surface and cause disintegration. Recommended minimum and maximum values of air voids range from 2 to 3 per cent to as high as 7 per cent maximum.

There are two important limitations about the void concept. These are:

(1) It is applicable to only dense bituminous mixtures and not to open graded ones.

As a guide, voids concept is normally applied to mixtures whose gradations lie about those obtained by applying Fuller's gradation requirements for maximum unit weight. If the concept is applied to open graded mixtures, the binder content may be so great as to significantly reduce stability.

(2) It is not used to any great extent as a means to determine the optimum binder content.

MIX DESIGN METHODS

Many empirical and semi-empirical design methods have been evolved which attempt to evaluate various properties of bituminous mixtures and then determine the binder content based on these properties. Some of the more widely used methods of mix design are:

(*i*) Marshall test

(*ii*) Hubbard field extrusion test

(*iii*) Hveem method

(*iv*) Smith triaxial method

(*v*) Wheel tracking simulative test

(*vi*) Splitting test

(*vii*) Lee-rigden test

(*viii*) OTL bearing index test.

The first four methods, which are commonly used, are described below:

17.5. MARSHALL TEST

The Marshall stability test is a type of unconfined compressive strength test. A cylinderical specimen, 101.5 mm diameter and 63.5 mm high, is compressed radially at a constant rate of strain of 50.8 mm per minute. The results are expressed in terms of *Marshall stability value* which is the maximum load in newtons sustained by the specimen and the *Marshall flow value* which is the deformation in mm at failure.

However prior to the stability tests, determination of unit weight and voids are carried out on the test specimens.

The optimum binder content selected for design is a compromise value which meets specific requirements for stability deformation and void content.

The various steps followed in the Marshall test are given below:

(*i*) Prepare a series of test specimens for a range of different binder contents so that the stability test data show a well defined optimum value at some binder content. To estimate the optimum content, either use surface area equation or determine on the basis of experience. Use binder contents in increments of ½ per cent above and below the estimated optimum value.

Each Marshall test specimen requires approximately 1.2 kg of aggregate and three test specimens are usually prepared for each bitumen content used. Thus for six tests in a series would normally require 18 test specimens. The minimum quantity of aggregates required for one series of tests would be 22 kg and five litres of bitumen.

The mixing and compacting temperatures are determined on the basis of viscosity control. It is required that the viscosities of 85 ± 10 seconds Saybolt Furol and 140 ± 15 seconds Saybolt Furol should be taken as mixing and compacting temperatures respectively. The corresponding viscosities for tar have been suggested to be Engler specific viscosity of 25 ± 3 for mixing and 40 ± 5 for compacting.

Compaction is done by a hammer having a flat, circular tamping face 98 mm, equipped with 4.54 kg weight and constructed to obtain a specified drop of 457 mm. Compaction mould has inside diameter of 101.5 mm and a height of approximately 76 mm. Compaction is done on both sides with face of the hammer heated (in water bath). Thirty-five blows on both top and bottom of the specimen are normally applied for compaction of paving mixes designed for light traffic, fifty blows for medium traffic and seventy-five for heavy and very heavy traffic.

(*ii*) The bulk unit weight of each specimen is determined. Depending upon the surface texture, either of the following two methods is used.

(*a*) *For compact smooth surface*, weigh the specimen in air and in water. Then apply the following relation:

$$\gamma_d = \frac{W_A}{V} = \frac{W_A}{W_A - W_W}$$

where γ_d = bulk unit weight g/cm^3

W_A = weight of specimen in air, *g*

V = volume of specimen, cm^3

and W_W = weight of specimen in water, *g*

(*b*) *For open and porous surface*, the specimen must be covered with a paraffin coating before being palced in water.

then $$\gamma_d = \frac{W_A}{V} = \frac{W_A}{W_{PA} - W_{PW} - \left(\dfrac{W_{PA} - W_A}{G_P}\right)}$$

where W_{PA} = weight of specimen plus paraffin coating in air, *g*

W_{PW} = weight of specimen plus paraffin coating in water, *g*

G_P = specific gravity of paraffin.

(*iii*) Calculate the percentage of air voids in each compacted specimen. To do this first calculate the maximum theoretical unit weight. The difference between the maximum theoretical unit weight and the actual unit weight, expressed as a percentage of the total volume of the specimen, will give percentage of air voids.

The maximum theoretical unit weight is that unit weight which would result if the specimen had been compacted so that there were no voids in the aggregate-binder mixture.

Thus

$$\gamma_t = \frac{W_A}{v_b + v_c + v_f + v_{mf}}$$

$$= \frac{W_A}{\dfrac{w_b}{G_b} + \dfrac{w_c}{G_c} + \dfrac{w_f}{G_f} + \dfrac{W_{mf}}{G_{mf}}}$$

where γ_t = maximum theoretical unit weight, g/cm³

W_A = weight of specimen, g

v_b = volume of binder in the specimen, cm³

v_c, v_f and v_{mf} = volume of coarse, fine and mineral filler fractions respectively, of the aggregates in the specimen, cm³.

w_b = weight of binder in the specimen, g

w_c, w_f and w_{mf} = weights of coarse, fine and mineral filler fractions respectively of the aggregates in the specimen, g

G_b = specific gravity of the binder; and

G_c, G_f and G_{mf} = apparent specific gravities of coarse, fine and mineral filler fractions respectively of the aggregates in the specimen.

The percentage of air voids in the compacted specimen is then calculated as follows:

$$\% \text{ VTM} = \frac{\gamma_t - \gamma}{\gamma_t} \times 100$$

where % VTM = voids in total mixture, *i.e.* in the specimen, %

γ_t = theoretical maximum unit weight, g/cm³

γ_d = bulk unit weight, g/cm³.

(*iv*) For each specimen, calculate the percentage of voids in the compacted mineral aggregate framework which is filled with binder. This involves first determination of amount of voids in the aggreagate framework (VMA) and then calculating the percentage filled with binder.

VMA is the volume of voids which in theory is available for filling with binder. This is obtained by subtracting the volume occupied by the aggregate in the compacted specimen from the bulk volume of the compacted specimen.

Thus, $\text{VMA} = V - v_c - v_f - v_{mf}$

$$= \frac{W}{\gamma_d} - \frac{w_c}{G_c} - \frac{w_f}{G_f} - \frac{w_{mf}}{G_{mf}}$$

where VMA = voids in the mineral aggregate frame-work in cm³. The voids in the mineral aggregate framework are often expressed as a percentage of the total volume V of the specimen.

Thus

$$\% \text{ VMA} = \frac{\text{VMA}}{V} \times 100$$

The percentage of voids in the aggregate frame work which is filled with binder is determined from

$$\% \text{ voids filled with binder} = \frac{v_b \times 100}{\text{VMA}}$$

The same percentage can also be calculated as follows:

$$\% \text{ voids filled with binder} = \frac{\% \text{ VMA} - \% \text{ VTM}}{\% \text{ VMA}}$$

(*v*) Determine the Marshall stability value and Marshall flow value for each specimen. The test is carried out in the apparatus shown in Fig. 17.2. Load is applied at a constant rate of strain of 50.8 mm/min. A guage dial is used to measure the vertical deformation of the specimen. The maximum load, in newtons, required to produce failure of the specimen (heated to a prescribed temperature) gives the Marshall stability value. The deformation in mm at the failure load point is the Marshall flow value.

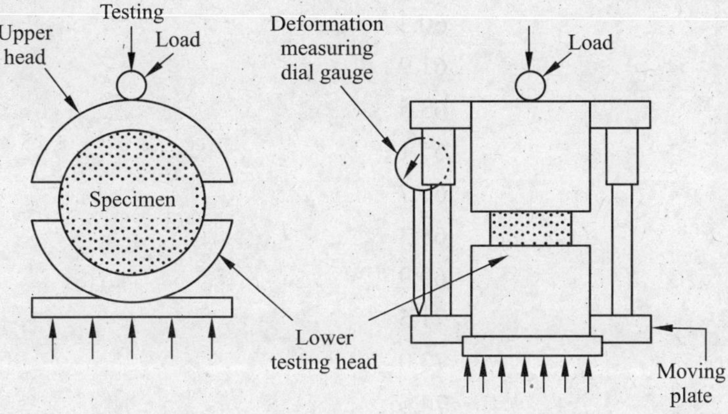

Fig. 17.2 Marshal stability apparatus.

(*vi*) The measured stability values are corrected to those which would have been obtained if the specimen had been exactly 63.5 mm high. This is done by multiplying each measured stability value by a correlation ratio, given in Table 17.1.

Table 17.1 Marshall stability correlation values

Volume of specimen cm^3	*Approximate thickness of specimen, mm*	*Correlation ratio*
200–213	25.4	5.56
214–225	27.0	5.00
226–237	28.6	4.55
238–250	30.2	4.17
251–264	31.8	3.85
265–276	33.3	3.57
277–289	34.9	3.33
290–301	36.5	3.03
302–316	38.1	2.78
317–328	39.7	2.50

329–340	41.3	2.27
341–353	42.9	2.08
354–367	44.5	1.92
368–379	46.0	1.79
380–392	47.6	1.67
393–405	49.2	1.56
406–420	50.8	1.47
421–431	52.4	1.39
432–443	54.0	1.32
444–456	55.6	1.25
457–470	57.2	1.19
471–482	58.8	1.14
483–495	60.3	1.09
496–508	61.9	1.04
509–522	63.5	1.00
523–535	65.1	0.96
536–546	66.7	0.93
547–559	68.3	0.89
560–573	69.9	0.86
574–585	71.5	0.83
586–598	73.0	0.81
599–610	74.6	0.78
611–625	76.2	0.76

(*vii*) Plot the following graphs:

(*a*) Binder content *versus* corrected Marshall stability value.

(*b*) Binder content *versus* Marshall flow value.

(*c*) Binder content *versus* percentage of voids in the total mix (%VTM)

(*d*) Binder content *versus* percentage of voids in the mineral aggregate framework filled with binder.

(*e*) Binder content *versus* unit weight.

Typical Marshall test graphs are shown in Fig. 17.3.

(*viii*) Determine the optimum binder content. This is taken as the average of four binder contents corresponding to maximum stability, maximum unit weight, appropriate percentages of voids in total mix (VTM) and aggregate voids filled with binder. These are determined from the graphs plotted in step (*vii*).

(*ix*) Check that the optimum binder content meets the design specifications. The values of variables are determined by re-entering the curves drawn in step (*vii*) with the optimum binder content determined in step (*viii*). These values should be within the specified ranges, otherwise the mix requires re-designing and the tests performed again.

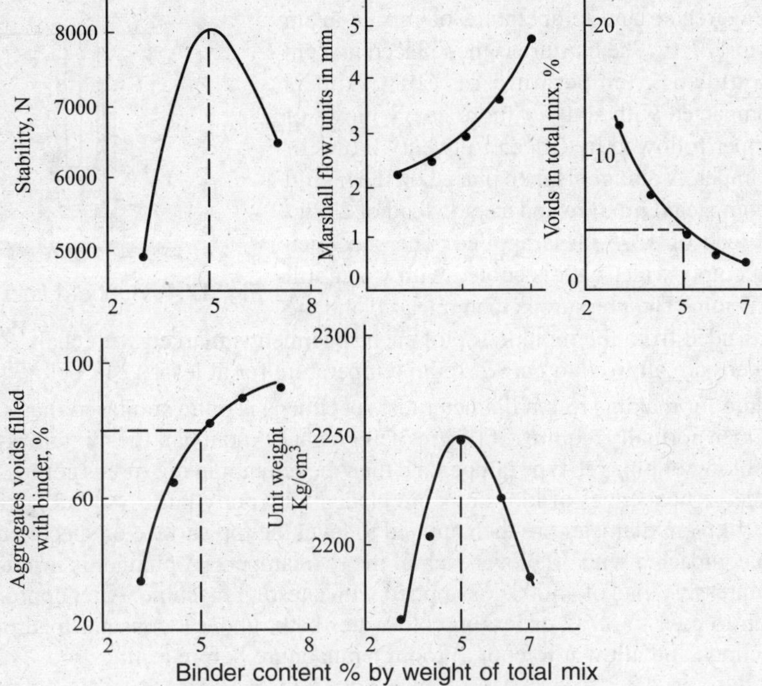

Fig. 17.3 Typical Marshall test data.

Discussion

Plots in step (*vii*) have been found to follow a reasonably consistent pattern for dense graded bituminous paving mixes. Usual trends noted are given below:

(*a*) The stability value increases with increasing bitumen content up to a certain maximum, after which the stability decreases.

(*b*) The curve for unit weight of total mix is similar to the stability curve, except that the maximum unit weight normally occurs at a slightly higher bitumen content than the maximum stability.

(*c*) The flow value increases with increasing bitumen content.

(*d*) The per cent of voids in total mix decreases with increasing bitumen content, ultimately approaching a minimum void content.

(*e*) The per cent of aggregate voids filled with bitumen increases with increasing bitumen content, ultimately approaching a maximum value.

17.6. HUBBARD-FIELD METHOD

The original Hubbard-filed method is primarily applicable to the design of sheet mixes containing bitumen and aggregate gradings having at least 65 per cent passing the 2 mm sieve. The modified Hubbard-field is applicable to paving mixtures using bitumen and containing more than 35 per cent of coarse aggregate with maximum size of 20 mm or less. The original method for sheet type mixes uses standard test specimens, 5 cm in diameter and 2.5 cm height. The modified method for coarse aggregate mixes uses test specimens, 15 cm in diameter and from 7.0 to 7.5 cm height. Difference in size of test specimen is the principal difference between the original and modified methods.

In preparing test specimens the aggregates, with the proper gradation, are thoroughly dried and heated to a temperature of 175°C on hot plate. Each test (specimen 5 cm in diameter) will require about 100 gm of mixture. Two test specimens are generally prepared for each bitumen content. The stipulated amount of bitumen, also heated separately to a temperature of 175°C is thoroughly mixed together with aggregate and temperature of mix is again raised to 120° to 127°C. The mixture is then placed in 5 cm preheated mould to a temperature of 120°C and is thoroughly compacted with sixty or more heavy blows of blade type tamper followed by fifteen to twenty blows of plunger type tamper. A plunger is then placed on the mould on top of the compacted mixture and a static load of 21 kg/cm^3 or a total load of 4275 kg is applied. The specimen thus formed in empty water bath is cooled with water still under load for 5 min. The pressure is then released and the specimen is extruded from the mould. Top of each specimen is marked with chalk for identification and the specimens are allowed to cure at room temperature for at least 12 hr before testing.

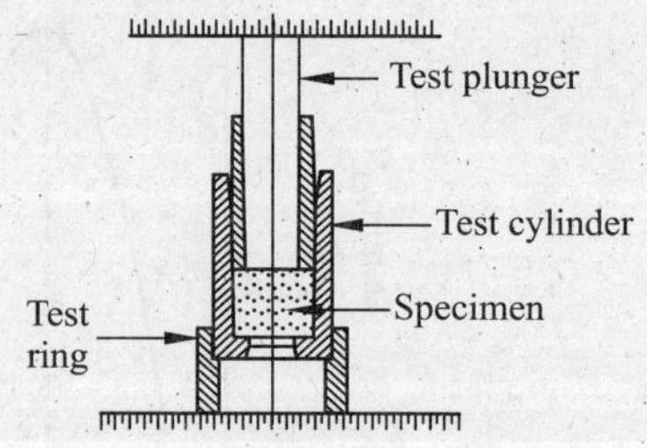

Fig. 17.4 Hubbard field test apparatus.

The procedure for making 15 cm diameter test specimens is quite similar to that described above. Each test specimen normally requires 3000 gm of aggregate. About half the mixture is first compacted with 30 strong blows of plunger-type tamper and then the remaining mixture is compacted in the same manner. Outside vertical edges of loose mix are bladed with putty knife. Additional 30 blows of the second tamper 14.6 cm diameter are then applied to level the top surface of specimen. The other top surface is then compacted with 30 blows each of the two tampers. A plunger is inserted in the mould and a total compressive load of 4500 kg is applied with a testing machine. After approximately 2 min, allow specimen to cool to 38°C or less in cold water bath, under a constant load of 4500 kg. Dry surface of specimen and allow it to cool to room temperature before testing.

Test. The test specimens are placed in a water bath maintained at a temperature of 60°C for at least one hour before testing. 60°C is usually used as this is supposed to be the maximum temperature which will be attained in surface course mixtures under service conditions. The specimen is then placed in assembly, original top end downward and complete assembly (Fig. 17.4) is positioned in water bath on Hubbard-field stability machine. The load is applied to the specimen by forcing the plunger downward at a rate of 2.5 cm in 25 sec. The stability value is measured as the maximum load resistance in kg developed by the specimen when forced through a ring 4.45 cm inside diameter in original method and 14.6 cm inside diameter in modified test procedure. Standard test procedure is covered by ASTM D1138.

After the completion of test, density and voids analysis is made for each series of test specimens. Graphical plots for unit weight versus asphalt content, stability versus bitumen content, per cent voids, total mix versus bitumen content and per cent aggregate voids versus bitumen content are prepared.

Design Criteria	*Extremely heavy traffic*		*Ordinary heavy traffic*	
method	*Min*	*Max*	*Min*	*Max*
Original Stability, kg	900	—	545	900
Voids total mix, per cent	2	5	2	5
Modified Stability, kg	1575	2700	1125	2700
Voids total mix, per cent	2	5	2	5

Determination of optimum bitumen content. Normally bitumen content indicated by 3 or $3\frac{1}{2}$ per cent voids in total mix is read from voids curve and corresponding stability read from stability curve is checked as to whether the stability value falls within the limits of design criteria. In case, this condition is not satisfied, the mix is redesigned and deficiency corrected. The curve for per cent aggregate voids, though not used directly in design is very helpful in comparing different mix designs. If the test values for a particular blend and gradation of aggregates are borderline or fail to meet the requirements of the criteria, the mix should be redesigned using a different blend and gradation of aggregates. Where practical, the mix design should have an optimum bitumen content that will yield maximum stability.

The final selection should take into consideration the economics of the aggregate to be used as well as the suitability of the mix from the standpoint of test values and test requirements. For equal economic considerations, the mix design showing the higher stability should normally be adopted.

17.7. HVEEM METHOD

The Hveem method as developed and used by the California division of highways is applicable to paving mixtures using both penetration grades and liquid grades of bitumen and containing aggregate up to 2.5 cm maximum size.

The Hveem method uses standard test specimens of 102 mm diameter and 63.5 mm height. This test utilizes a special triaxial type testing cell (Fig. 17.5) for measuring the resistance of the compacted mix to lateral displacement under vertical loading. The cohesiometer test measures the cohesive or tensile resistance of the compacted mix: the swell test measures the resistance of the mix to the action of water. The test specimens are maintained at 60°C for both the stability and cohesiometer tests, whereas the swell tests are performed at room temperature.

The first step in Hveem method of mix design is to determine the estimated optimum bitumen content by the centrifuge-kerosene-equivalent method. The amount of kerosene retained as a per cent of dry aggregate weight (Passing 4.75 mm sieve) after centrifuging saturated aggregate for 2 min at a force 400 times gravity is called CKE. With a calculated surface area and factors obtained by the CKE method for a particular aggregate or combination of aggregates, the estimated optimum bitumen is determined by using a series of charts. When well equipped laboratory is not available, the optimum bitumen content may be estimated by the CKE test.

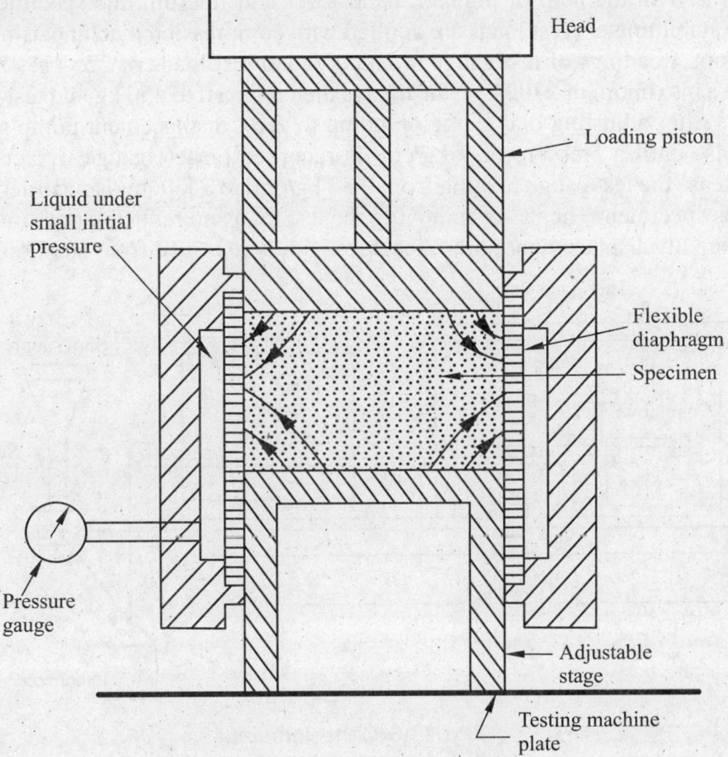

Fig. 17.5 Hveem stabilometer.

A series of test specimens are prepared for a range of bitumen contents, both above and below the estimated optimum bitumen content indicated by CKE procedure. One test specimen for each bitumen content for stability test and one duplicate test specimen for swell test are required. Each test specimen normally requires 1,200 gm of aggregate.

Both aggregates and bitumen are heated to the mixing temperature and are thoroughly mixed and kept in oven for 15 hrs curing period at $60°C \pm 2°C$. After curing the mixture is reheated to 110°C before compaction. The compaction of the specimen is accomplished by means of kneading compactor (Fig. 17.5) which consolidates the material by a series of individual kneading action impressions made by a roving ram having a face shaped as sector of 102 mm diameter circle. At each application of the ram a pressure of 35 kg/cm^2 built up by a compressor is applied subjecting the specimen to a kneading action without impact over an area of approximately 20 sq. cm. Each pressure is maintained for approximately 2 to 5 sec. After rodding the mixture 1000 gm in 2 layers with a bullet-nosed steel rod 10 mm in diameter and 40 cm in length, 20 times in the centre and 20 times around the edge, approximately 20 tamping blows at 10.5 kg/cm^2 are applied in order to consolidate the mix so that it does not get displaced under compaction load of 35 kg/cm^2. Finally the compaction is done by applying 150 tamping blows under a foot pressure of 35 kg/cm^2. Then after inverting mould and pushing specimen to the opposite end, a 70 kg/cm^2 static load is applied with original top surface supported on the lower plate of the testing press and the height of the specimen is recorded.

Test. For swell test, compacted specimen is allowed to stand at room temperature for at least 1 hr to permit rebound after compaction and swell is measured with the help of dial gauge 24 hr after 500 ml of water is introduced at the top of the specimen in the mould.

For stabilometer test, the specimens contained in moulds are kept in oven at 60°C for a minimum period of one hour before testing. Compression machine is adjusted for a head speed of 0.125 cm per min and displacement of stabilometer is checked with dummy metal specimen to adjust it to give 2.00 ± 0.05 turns. With the help of plunger, hand lever and fulcrum, the specimen is forced from mould into the stabilometer. Test loads are applied with compression machine using a head speed of 0.125 cm per min. Readings of test gauge are recorded at test loads of 225 kg, 450 kg and 450 kg thereafter up to a maximum of 2700 kg. The load is then reduced to 450 kg and test gauge is adjusted to 0.35 kg/cm^2. After adjusting dial gauge on pump to zero, displacement pump's handle is turned rapidly clockwise until a pressure of 7 kg/cm^2 is recorded on test gauge. Exact number of turns required to increase the test gauge reading from 0.35 kg/cm^2 to 7 kg/cm^2 is recorded which measures displacement on specimen. Then after removing the test load and reducing pressure on test gauge to zero, and backing off displacement pump, the specimen is removed from stabilometer cell.

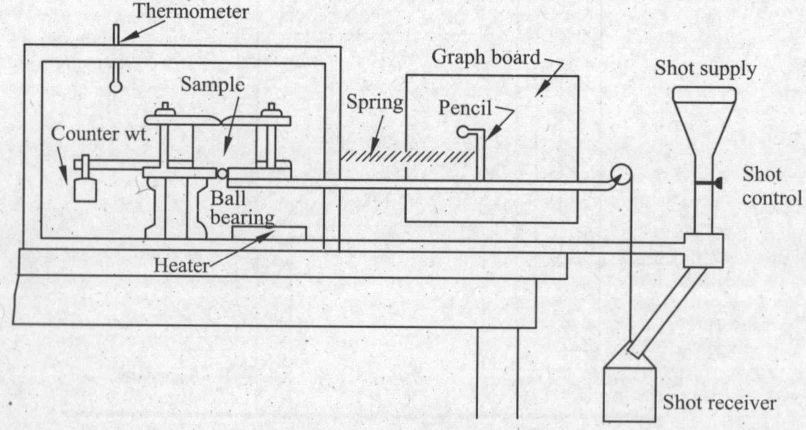

Fig. 17.6 Cohesiometer.

For Cohesiometer test, it is performed on the same specimen previously used in stabilometer and bulk-density tests. Specimen placed in oven at 60°C for approximately 2 hr before test sample is centred and clamped firmly in cohesiometer device (Fig. 17.6) with top plates parallel to the top surface of specimen. The shot are allowed to fall in bucket until the specimen breaks.

Calculations

$$\text{Stabilometer value, } R = \frac{100}{\dfrac{2.5}{D_2}\left(\dfrac{P_v}{P_h}-1\right)+1}$$

where R = relative stability

D_2 = displacement on specimen

P_v = vertical pressure of 28 kg/cm^2.

P_h = horizontal pressure corresponding to P_v = 28 kg/cm^2.

$$\text{Cohesiometer value, } C = \frac{0.4\,L}{0.315\,H + 0.0276\,H^2}$$

where L = shot weight, gm

H = height of specimen, cm

C = cohesiometer value (gm per cm width, corrected to 7.5 cm height)

For more than one layer, the cohesiometer value for the combined layer is determined taking two layers at a time and calculating equivalent value by the following relation:

$$C = C_1 + \left(\frac{t_2}{t_1+t_2}\right)^2 (C_2 - C_1)$$

where C = equivalent cohesiometer value of the two layers

C_1 = cohesiometer value of the top layer

C_2 = cohesiometer value of the next lower layer

t_1 = thickness of the top layer

t_2 = thickness of the next lower layer.

The density and voids analysis is made as for other methods.

Design criteria	*Light traffic*	*All other traffic*
Stabilometer value	30 +	35 +
Cohesiometer value	20 +	20 +
Swell	Less than 0.75 mm	

Although, consideration of per cent air voids is not included as a routine part of this design method, an effort is made to provide a minimum per cent in the total mix of approximately four.

The optimum bitumen content should be the highest percentage of bitumen the mix will accommodate without loss of stability. Its cohesiometer value is relatively low, an increased value can usually be obtained by an adjustment of aggregate grading. For well graded mixes, an increased cohesiometer value can normally be obtained by using increased amounts of mineral dust or a harder penetration grade of bitumen.

17.8. SMITH TRIAXIAL METHOD

The smith triaxial method is applicable to dense paving mixtures using penetration grades of paving bitumen and containing aggregate up to 2.5 cm maximum size. This method uses standard test specimens of 9.7 cm diameter by 20 cm height. For mixes containing aggregates with a maximum size greater than 2.5 cm, the size of test specimen and triaxial cell are increased accordingly. In such cases, the diameter of test specimen should be at least 4 times the maximum dimension of the largest particle used and the height-diameter ratio should be equal to or slightly in excess of 2.0. The method utilizes a closed-system triaxial cell test in which the lateral transmitted pressures are determined for corresponding vertical loadings on the test specimens at ordinary room temperatures of $24°C \pm 1°C$.

Aggregates heated to a temperature of 135 to 150°C and bitumen heated separately to a temperature of 150 to 175°C are thoroughly mixed and the mixing is completed as quickly as possible to prevent excessive cooling of the mixture. After mixing is complete the specimen is compacted immediately in a double plunger split steel mould first by rodding well by any suitable tamper and then by applying a load of 175 kg/cm^2, with compression machine for one minute. The specimen, after a short cooling period is removed from the split mould and is allowed to cool to room temperature before testing.

After the density and voids analysis has been made, the specimen is tested in the triaxial compression cell. The specimen encased in neoprene rubber sleeve along with base plate is kept in the temperature cabinet at $24°C \pm 1°C$ for at least one hour before test. After applying an initial vertical seating load of 0.14 kg/cm^2 on the top of specimen a lateral pressure of 0.14 kg/cm^2 is also applied and deformation gauges are adjusted to zero. Incremental loads are next applied to the specimen and each load increment is held constant until rate of deformation is less than 0.25 mm per min when vertical load and corresponding lateral pressure are recorded. Vertical loads ranging from 0.07 to 10.5 kg/cm^2 with 0.14 kg/cm^2 load increments are usually adopted. A sufficient number of load readings are obtained to plot a representative load curve between lateral transmitted pressure as abscissa and vertical applied load (kg/cm^2) as ordinate. From the upper portion of the curve where the vertical loads are approximately proportional to lateral pressures, a tangent to the curve is drawn and slope of this tangent is calculated.

Angle of internal friction is calculated from the following equation:

$$\text{Slope of tangent} = \tan^2\left(45° + \frac{\phi}{2}\right)$$

Intercept of tangent with vertical axis is read from the graph and cohesion C is determined from the equation:

$$\text{Intercept} = 2C \tan\left(45° + \frac{\phi}{2}\right)$$

Design criteria. The suitability of mix design is determined by the use of a test evaluation chart evolved by Smith and shown in Fig. 17.7. The values of ϕ and C obtained from the test are plotted on the chart. If the plotted point falls in unshaded (satisfactory) area of the chart and the voids content of the mix is between 5 and 10 per cent, the mix is considered satisfactory. If the plotted point falls in the shaded (unsatisfactory) area of the chart, it will be necessary to repeat the test on a new sample with different aggregate gradation or bitumen content or both. The final selection is based on the mix having the highest bitumen content and satisfactory stability commensurate with the density and voids requirements.

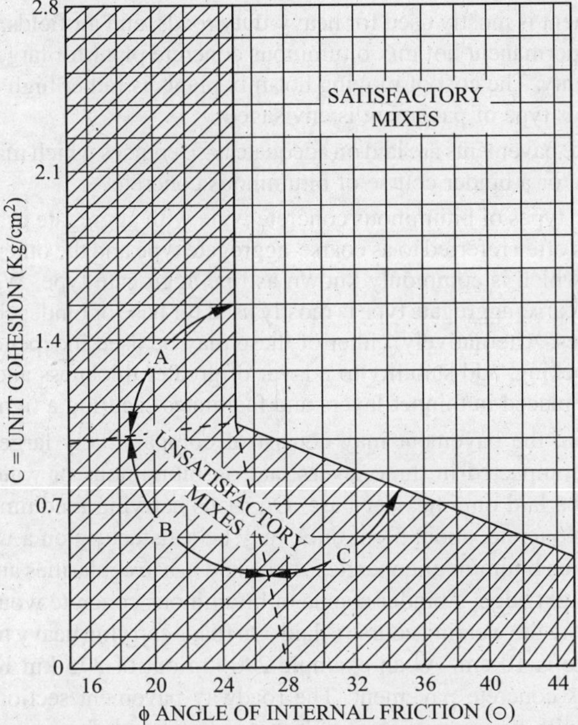

Fig. 17.7 Test evaluation chart for mix design by Smith triaxial test.

Mixes lacking sufficient internal friction and falling within region *A* are usually altered by reducing the bitumen content. Mixes lacking sufficient cohesion and falling within region *C* are usually altered by reducing the bitumen content if the voids are near or below the lower limit. Cohesion can also be increased by the addition of a higher percentage of crushed particles or by increasing the mineral filler or slightly by using a harder grade of bitumen. Mixes lacking both cohesion and internal friction and falling within region *B* may be either due to a large portion of river gravel in the mix or in case of crushed rock mixes due to a high bitumen content. The former type can be corrected by incorporating a sufficient amount of crushed gravel or crushed rock to increase the interlocking and the latter type can be corrected by merely reducing the bitumen content.

HOT-MIX HOT-LAID BITUMINOUS CONCRETE

This is the highest type of bitumen pavement which is composed of an appropriate grade of paving bitumen and completely graded aggregate so that the fine materials fill the voids in the coarse material and bitumen fills the voids in the fine material, thus creating a very dense mixture. As already explained under "stripping" in this chapter, the pupose of the filler by the fine material is to stiffen and strengthen the bitumen so that the voids in the sand are filled with filler/bitumen mixture of higher strength and lower temperature susceptibility than that of bitumen alone. This type of paving mixture differs from the dense-grade plant mixes under the intermediate type mixes in the care with which it is proportioned, mixed and shaped and consequently in the uniformity, quality and cost of finished product.

Hot-mix hot-laid bituminous concrete (British term "rolled asphalt") is manufactured in a central mixing plant where the bitumen and aggregates are heated to a temperature of approximately 150°C, properly proportioned, mixed, and placed while still quite hot. In the plant, after the aggregates are heated and dried they are separated into specific sizes and recombined in accordance with scientific designs. They are then mixed with paving bitumen and the completed mixture is hauled by trucks to the mechanical paver, where it is placed in a smooth layer and compacted by rollers while still quite hot.

This type of pavement is mostly used for heavy-duty roads, and air fields. Many cities and densely populated areas have permanent hot mix bituminous concrete plant for large and small projects with equal degree of economy. The cost of moving hot-mix plants is rather high and should be taken into consideration when this type of pavement is envisaged.

Bituminous concrete pavements are laid on adequate base courses which may consist of old concrete, waterbound macadam, or a binder course of bituminous concrete.

There are two basic types of bituminous concrete : one with aggregate graded from top sizes of 40 mm or 25 mm which is often referred to as coarse-aggregate type and the other with top size aggregates of 20 mm or 12 mm which is commonly known as fine aggregate type. When both these types are used in combination, coarse-aggregate type is mostly used for base or binder courses and fine aggregate type for surface courses. Alternatively, either of these may be used independently. Coarse aggregate type, due to its interlocking, add stability to a layer of greater thickness and on the other hand, fine aggregate type can be placed in thinner layers and facilitates securing a tight and smooth surface.

In new construction, the pavement may consist of comparatively large thickness about 15 cm bituminous concrete compacted in three layers laid on bituminous or water-bound macadam or a lesser thickness may be laid on a heavier base. On roads carrying medium to light traffic, a single course of bituminous concrete, about 5 cm thick may be satisfactory on a well prepared base. If the existing surface on which bituminous concrete is to be laid has irregularities and depressions exceeding 10 mm under 3 m straight edge, a binder course of bituminous concrete would be required to correct these irregularities before the wearing course is laid. On roads carrying heavy traffic, the usual thickness of bituminous concrete is 10 cm. As an example 290 km out of 375 km Kansas Turnpike, U.S.A. consists of bituminous concrete pavement. The roadway pavement section is designed to support 8000 kg wheel loads with tyre pressures of 8.75 kg/cm^2 and with frequency of about 1 in 30. Load repetitions of 25000 per day were expected on this road out of which approximately 28 per cent would be contributed by multiple axle trucks. On this turnpike, the roadway pavement structure consists of a subgrade with CBR of $4\frac{1}{2}$, a 25 cm sub-base of aggregate with minimum CBR of 60; and 20 cm base course of aggregate with minimum CBR of 75; and 10 cm of bituminous concrete composed of 60 mm of binder course and 40 mm of surface course.

17.9. MATERIALS REQUIREMENTS

17.9.1. Aggregate

Coarse aggregate may be crushed stone, crushed slag or gravel. Fine aggregate may be natural sand or crusher run screening or combinations thereof.

Mineral filler may be lime; stone dust, Portland cement, hydrated dolomite dust, fly-ash or other suitable mineral matter.

Other requirements in respect of coarse aggregate are:

(*i*) ***Crushed pieces in gravel.*** Not less than 40 per cent by weight retained on 4.75 mm sieve should have at least one fractured face.

(*ii*) ***Soundness.*** After 5 cycles, loss in soundness test should not be more than 12 per cent when sodium sulphate is used and not more than 18 per cent when magnesium sulphate is used.

(*iii*) ***Abrasion.*** Loss in Los Angeles abrasion test should not exceed 40 per cent for surface courses and 50 per cent for base courses.

Grading requirements. Grading requirements are specified giving limits of percentage passing various sieves for aggregates of different top sizes. ASTM standard D 947—55 covers grading for different aggregates. Grading requirements for composition of bituminous concrete mixture as covered by this standard are given in Table 17.2. Requirements in respect of mineral filler are covered by ASTM and IRC : 29—1968 and these are shown in Table 17.3.

Table 17.2 Composition of bituminous concrete mixture (ASTM)

	Nominal maximum size of aggregate					
Sieve size	*50 mm*	*40 mm*	*25 mm*	*20 mm*	*12.5 mm*	*10 mm*
	Grading of total aggregate (coarse plus fine, plus filler if required). Amounts finer than each laboratory sieve (square opening) per cent by weight					
63 mm	100	—	—	—	—	—
50 mm	95 to 100	100	—	—	—	—
40 mm	—	95 to 100	100	—	—	—
25 mm	60 to 80	—	95 to 100	100	—	—
20 mm	—	60 to 80	—	95 to 100	100	—
12.5 mm	40 to 60	—	60 to 80	—	95 to 100	100
10 mm	—	—	—	60 to 80	—	90 to 100
4.75 mm (No. 4)	20 to 40	25 to 45	30 to 50	40 to 60	50 to 70	60 to 80
2.36 mm (No. 8)	10 to 30	15 to 35	20 to 40	25 to 45	30 to 50	40 to 60
300 micron (No. 50)	2 to 18	3 to 20	5 to 25	5 to 25	5 to 25	8 to 30
75 micron (No. 200)	0 to 4	0 to 5	0 to 10/1–71	2 to 10/2.8	2 to 10/2–9	2 to 12/2–10
	Bitumen, per cent by weight of total mixture					
	3.5 to 7.5	4.0 to 8.0	4.0 to 8.5	4.5 to 9.0	5.0 to 9.5	5.5. to 10.0

Table 17.3 ***Grading requirements of mineral filler***

S. No.	*Sieve designation*	*Percent passing by weight as per*	
		ASTM	*IRC*
1.	600 micron (No. 30)	100	100
2.	300 micron (No . 50)	95–100	—
3.	150 micron (No. 100)	90–100	90
4.	75 micron (No. 200)	70–100	Not less than 70

Composition of the paving mixture can conform to one of the compositions by weight given in Table 17.2. A job-mix-formula for the mixture is designed which comes within the specified limits and that is suitable for the traffic, climatic conditions, and specific gravities of the aggregates used. The design will specify a single definite percentage of aggregate passing each screen sieve, a single definite percentage of bitumen to be added and a single definite temperature at which the mixture is to be delivered. Variations from the job-mix formula should not exceed the limits given in Table 17.4.

Table 17.4 Permissible variations from the job mix formula

Serial No.	*Description of ingredient*	*Permissible variation by weight of total mix as per*	
		ASTM	*IRC*
1.	Aggregate passing 4.75 mm sieve	–	± 5%
2.	Aggregate passing 2.36 (No. 8) sieve	± 5%	± 4%
3.	Aggregate passing 600 micron sieve	–	± 3%
4.	Aggregate passing 75 micron (No. 200 sieve)	± 2%	± 1%
5.	Bitumen content	± 0.3%	± 0.3%
6.	Temperature of mix on delivery	7° C	–

According to IRC : 29—1968 specifications, the mineral aggregates including mineral filler should be so graded or combined as to conform to the gradings given in Table 17.5. Unless otherwise specified, for compacted layer thickness of 24 to 40 mm, any of the two gradings can be used, but for layer thickness of 40 to 50 mm, only grading No. 2 can be used.

Table 17.5 Aggregate gradation for bituminous concrete

Sieve designation	*Per cent by weight passing the sieve*	
	Grading I	*Grading 2*
20 mm	—	100
12.5 mm	100	80–100
10 mm	80–100	70–90
4.75 mm	55–75	50–70
2.36 mm	35–50	35–50
600 micron	18–29	18–29
300 micron	13–23	13–23
150 micron	8–16	8–16
75 micron	4–10	4–10

17.9.2. Bitumen requirements

In earlier practice, the harder grades such as 30–40 or 40–50 penetrations were favoured. More recent studies have shown that there is an advantage in using softer grades, as weathering and oxidizing are slow, which results in much less cracking. It has been shown that when the bitumen in pavement has been reduced to a penetration of 30, it is virtually sure to crack. Thus, an advantage is gained by using the softest (highest penetration) grade of bitumen practicable in the construction of pavement. Proper grade depends primarily on temperature range, and secondly on traffic. Following grades may be used:

	Light or moderate traffic	*Heavy traffic*
Warm climates	80–100	60–70
Moderate climates	70–80, 80–100	80–100
Cold climates	100–120, 120–150	80–100

For airfield construction, for warm, moderate and cold climates, 60–70, 80–100 and 120–150 penetration bitumen may be used respectively. For taxiways respective grades of bitumen recommended are 60–70, 80–100 and 80–100.

Penetration normally used for mastic asphalt in streets is 60–70.

Any advantage gained by the use of softer bitumen may be lost at the mixing plant by lack of control over aggregate temperatures and for this reason aggregate temperature may be kept at the lowest level consistent with proper mixing and placing.

IRC specifies that the binder content should be so fixed as to achieve the requirements given in Table 17.6 and should be in the range of 5 to 7.5 per cent by weight of total mix.

Table 17.6 Requirements of bituminous concrete mix

Serial No.	*Description*	*Requirement*
1.	Marshall stability (ASTM designation D 1559) determined on Marshall specification compacted by 50 compaction blows on each end.	340 kg minimum
2.	Marshall flow (0.25 mm)	8–16
3.	Per cent voids in mix	3–5
4.	Per cent voids in mineral aggregate filled with bitumen	75–85
5.	Binder content per cent by weight of mix	5–7.5

17.10. PLANT REQUIREMENTS

Following are the requirements for plants suitable for producing bituminous paving mixtures.

17.10.1. Paving plant

(a) Equipment for preparation of bitumen, *i.e.* Bitumen storage and heating tanks in which adequate circulating system for bitumen is provided to ensure proper and continuous circulation between storage tank and mixer during the entire operating period. Heating is accomplished by stem coils, electricity or other means such that no flame should come in contact with the heating tank. The oldest method is the use of steam which is passed through a series of pipes or coils located inside and near the bottom of the storage tank. The coils have to be kept free of leaks, otherwise steam leaking through joints will condense and cause foaming of the bitumen. The most modern type of heating equipment is thermostatically controlled hot-oil heater in which hot oil is circulated through a system of pipes or coils inside the storage tank. This type of heating can maintain a uniform temperature at a nominal fuel cost. To minimize the loss in temperature of bitumen during overnight or temporary shutdown, above ground storage tanks are often insulated. Underground installation of storage tanks are more suitable from heat loss point of view and can be adopted for permanently located plants. All pipe lines and fittings are properly insulated. Storage tank capacity should be sufficient for at least one day's run.

(*b*) ***Feeder for drier*** is an accurate mechanical control for adjustment of cold-bin gates so that each cooled aggregate will be fed within a very close range of the amount required for uniform production at a uniform temperature.

(*c*) ***Drier*** can be of any satisfactory design for drying and heating the mineral aggregate to the specified temperature. The drier usually consists primarily of a large rotating steel cylinder lined with firebrick and containing projecting radial fins. The drier is mounted at an angle with the horizontal. Heat is supplied by a jet of flame, fed by steam or air-vaporized oil, coal dust, or gas which is directed into the lower end of the drier. As the cylinder rotates, the aggregate is carried by the fins from where it falls at the outlet and by the time the aggregates have travelled the length of the drier, they are free of moisture and heated to the required temperature. For cold-mixes where cooling of aggregates is required after drying, suitable cooling devices are provided, *e.g.* movement on an open conveyor to the next-step, the use of another drier without heat as cooling unit or longer retention time in the storage bins, *etc*.

(*d*) ***Dust collectors*** are dust collecting systems and devices for feeding the material uniformly to the mixture. The collector serves to eliminate dust nuisance which exists around the plant. In older plants, a smokestack was used to create a draft through the drier which carried with it much of the fine dust introduced with the aggregate. This not only wasted dust but also created a public nuisance in populated areas. Often these lost fines had to be replaced by expensive mineral filler. Modern plants are equipped with a blower to produce a draft and a dust collector which traps the fine particles and returns them to the heated aggregates below the drier.

(*e*) ***Screens*** are usually flat shaking screens for separating the hot aggregates into different sizes having normal capacities slightly in excess of the production capacity of the mixer.

(*f*) ***Bins*** are storage bins for aggregates having sufficient capacity to ensure uniform and continuous operation. Bins are divided into at least three compartments arranged to ensure separate and adequate storage of appropriate fractions of the aggregate. Each compartment is provided with over-flow pipe. Dry storage is also provided for mineral filler, when required.

(*g*) ***Bitumen control units*** are suitable devices which by weighing, metering, or volumetric measurements will obtain the proper amount of bitumen up to one per cent accuracy of the amount required. When bitumen is controlled by metering, provision is made to check the metered delivery by actual weight.

(*h*) ***Thermometric equipment*** are thermometers of suitable range fixed (*i*) in bitumen feed line at suitable location, (*ii*) discharge end of drier, to register automatically temperatures of bitumen and heated aggregate respectively.

(*i*) ***Control of mixing time*** is a positive means to govern the time of mixing and to maintain it.

17.10.2. Mixing plants

These consist of -

(*a*) Batch mix plants, or

(*b*) Continuous mix plants.

Figures 17.8 and 17.9 depicts the various operations involved in these plants.

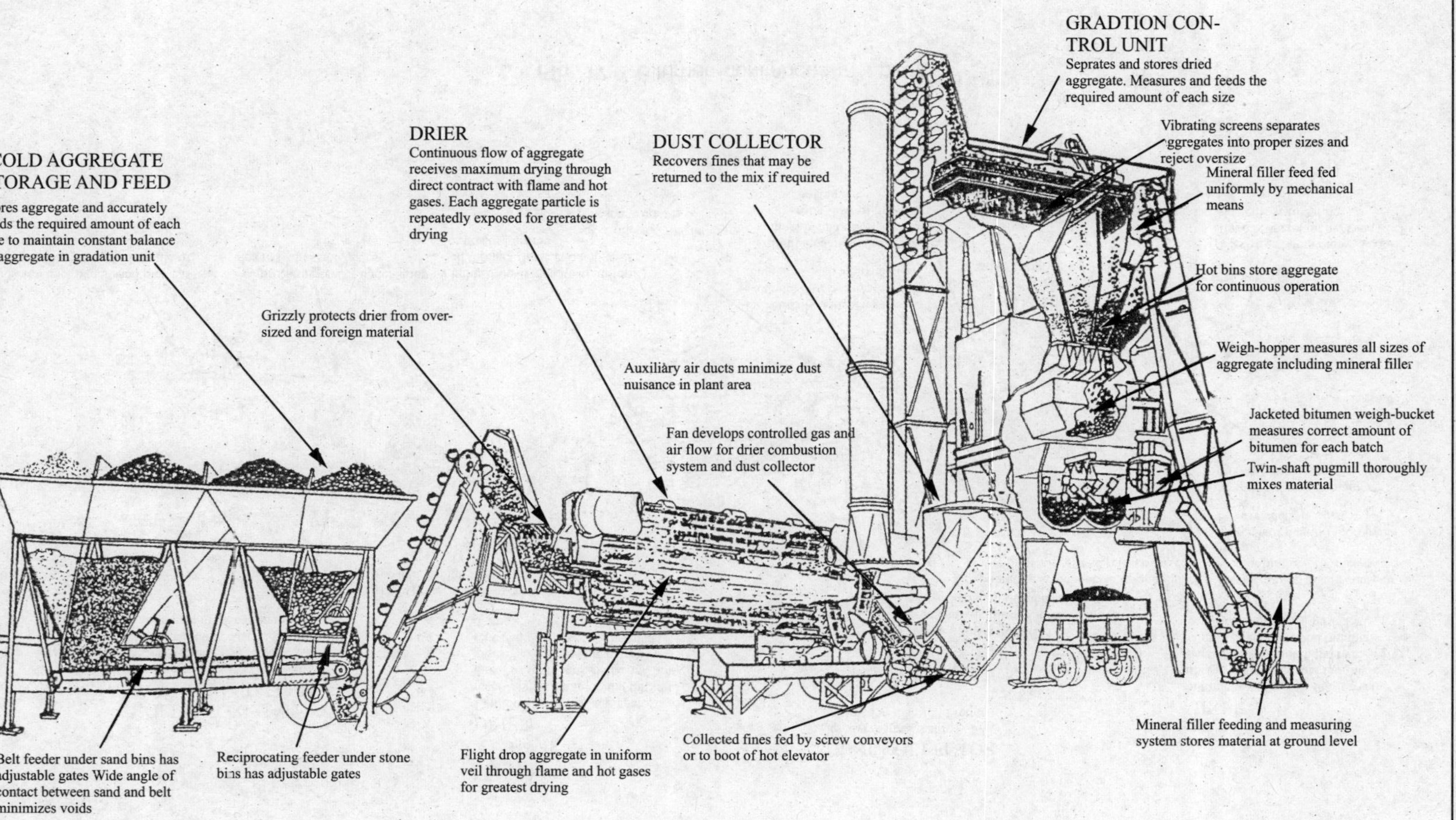

Fig. 17.8 Bitumen batch mix plant.

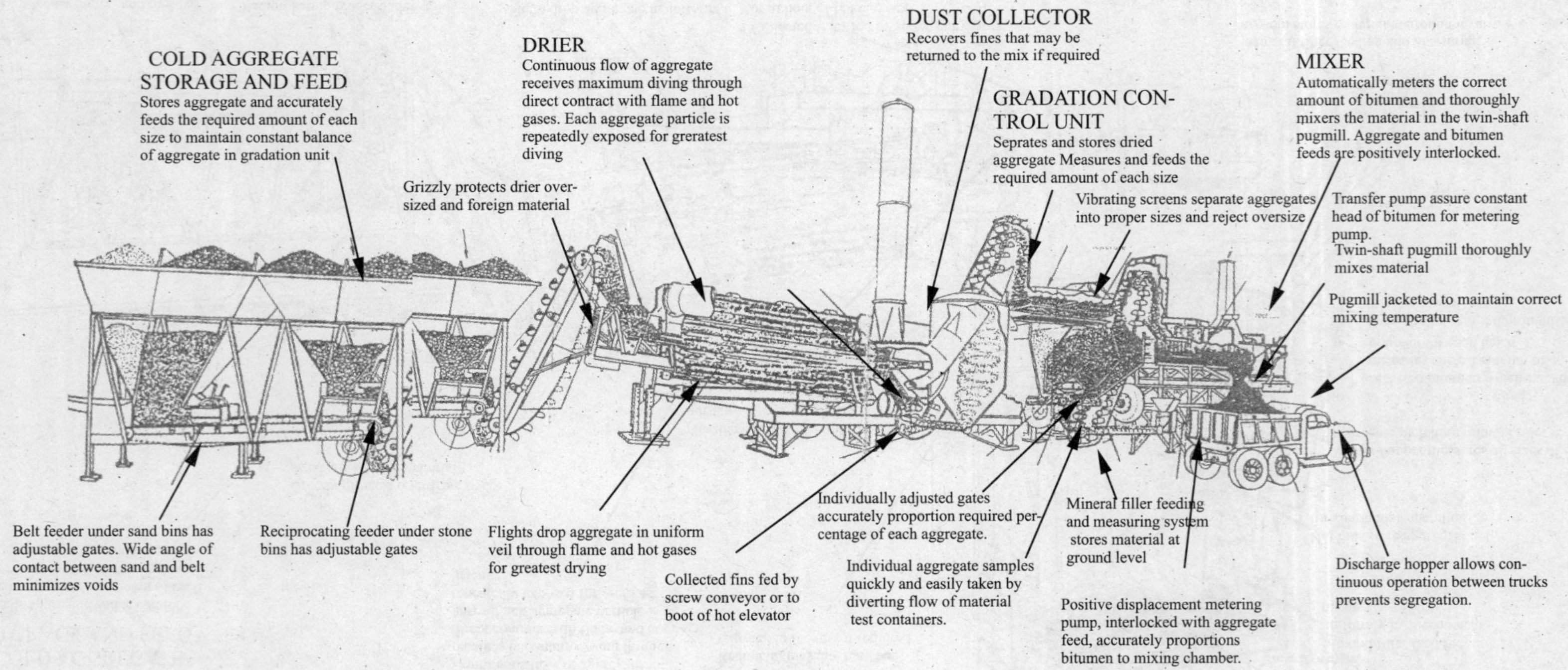

Fig. 17.9 Bitumen continuous mix plant.

Batch mix plants include:

(*i*) ***Weigh box or hopper*** which is a means to accurately weigh each size of aggregate in a hopper suspended on scales. Gates, both on hopper and bins are so constructed as to prevent leakage when they are closed.

(*ii*) ***Aggregate-scales*** are scales for any weigh-box or hopper which are either beam or springless dial-type and accurate up to 0.5 per cent of the indicated load.

(*iii*) ***Bitumen bucket*** is for weighing the bitumen. It should have a capacity of at least 4.0 per cent greater than the maximum percentage of bitumen required per batch of mixture.

(*iv*) ***Mixer unit for batch method*** includes a batch mixer of an approved rotary type or twin pugmill type which is capable of producing a uniform mixture within the permissible job mix tolerances. The mixer has an accurate time lock to control the operation of complete mixing cycle. The control of timing is flexible and capable of being set at intervals of not more than 5 sec throughout cycles upto 3 min. A mechanical batch counter is attached to register completely mixed batches.

Continuous mix plants include:

(*i*) ***Gradation control unit*** which includes means for accurately proportioning each bin size of aggregate either by weighing or volumetric measurement.

(*ii*) ***Synchronization of aggregate and bitumen feed*** is a means to afford positive interlocking control between the flow of aggregate from the bins and the flow of bitumen from the metered or other proportioning source.

(*iii*) ***Weight calibration of bitumen and aggregate feed*** is a means of calibrating gate openings and meters by means of weight test samples.

(*iv*) ***Mixer unit for continuous method*** includes a continuous mixer of an approved twin pugmill type which is capable of producing a uniform mixture within permissible variation in job mix formula.

17.11. CONSTRUCTION METHOD

The fundamental steps in the construction of hot-mix hot-laid bituminous concrete pavements are:

(*a*) ***Preparation of mixture.*** The aggregate and mineral filler are dried, heated to the proper temperature (usually not exceeding 177°C), screened, stored in an adequate number of separate compartments to ensure maintenance of proper gradation. Proper proportions of hot aggregate are then carefully weighed or continuously fed and thoroughly mixed with hot bitumen in a suitable mixer. The temperature of binder at the time of mixing should be in the range of 150° to 177°C and of aggregates in the range 155°C to 163°C, provided that at no time the difference in temperature between the aggregates and binder exceeds 14°C. If an attempt were made to mix the combined material from the drier with bitumen to produce the mix, it would be extremely difficult to control the grading because of segregation. To alleviate this, separate screens are used for separating hot aggregates into different sizes.

There is always some material in each bin which is smaller than the smallest screen represented by the bin. This is because a certain amount of finer material is always trapped and carried over into the next bin. Other things beinge equal, the carry over usually increases as the screen size becomes smaller. For this reason screens smaller than 2 mm (No. 10 ASTM) are never employed in a hot mix plant.

There are two major steps requiring attention for preparing uniform mix. These are:

(*i*) Adjustment of cold-bin gates so that each cooled aggregate will be fed within a very close range of the amount required.

(*ii*) Proportioning material from each of the hot bins and the amount of bitumen to be added.

During production, handling of wet aggregates may create a plant problem. If an aggregate stock becomes saturated from rain, much of the surface moisture can be reduced by turning the stock-piles and aerating the aggregate to the greatest extent. If wet aggregates are to be used throughout a project, two driers may be obtained and also complete removal of moisture from aggregate may be ensured.

Precautions should be taken to ensure proper temperature control. It is usually necessary to carry the temperature of aggregate measured by a pyrometer at the outlet of the drier from 4° to 5° higher than the temperature desired in the finished mix. The bitumen should not be heated above specification limits. As already pointed out, there is a trend towards softer grades of bitumen in order to prevent rapid aging and cracking. Tests have shown that over-heating of bitumen while in contact with oven-heated aggregates reduces the penetration. Over-heating of the binder in storage tanks is not so serious, because the liquid is in bulk rather than in thin films and as such no air is present and oxidation cannot take place.

A special series of tests to determine loss of penetration yielded the following results:

Original penetration	*Penetration after heating at 163°C*				*Loss in weight after*
	(penetration measurement at 40°C)				*5 hr (percent)*
	5hr	16 hr	21 hr	24 hr)	
75	68	58	53	49	0.7

The above results show that any advantage gained by the use of softer bitumen may be lost at the mixing plant by lack of control over aggregate and bitumen temperatures.

Transportation of mixture. Mixture is discharged from the plant into suitable vehicles usually trucks for transportation to the job site. Truck bodies should be tight, well cleaned, wetted with soluble oil solution to prevent sticking and when necessary covered with canvas for weather protection. A lining of 6.3 mm plywood provides cheap and effective lining.

Placing of paving mixture. Mixture is laid on the prepared base. Practically all bituminous concrete and high type bituminous paving mixtures are now laid by machine known as pavers which is a combination of mechanical spreading and finishing machines. The truck backs up the paver and dumps the hot-mix into a hopper from where it is carried back and deposited on the road at a uniform loose depth. One of the machines gives the paving mixture some additional compaction by means of tampers.

The paver can be adjusted to spread a depth of material such that, after proper compaction, the course will have the specified thickness. This fluff usually amounts to about 6.3 mm for a 5 cm compacted course but may range up to as much as 12.5 mm for 5 cm course depending on the characteristics of the mix and the type of finishing machine used. This requisite depth is determined by trial and error or by making density tests at the start of the job. Once a density sample is obtained, the average thickness of compacted material may be controlled to a very close figure by regulating the weight of material added per linear metre.

In areas where mechanical spreading is not possible, hand spreading and finishing are done.

All inequalities and irregularities are corrected before rolling commences. When bituminous concrete containing top size of over 6.3 mm are raked, the larger particles come to the top, leaving an unsightly appearance. For this reason, no more raking should be done than is absolutely necessary. When it is necessary to rake, the appearance of raked portion can be improved by raking the large aggregate to the side and discarding it.

Compaction of mixture. After each course has been laid and screened, it is rolled while still hot until it is thoroughly compacted. No set rule can be given for the temperature at which the rolling should take place. When a mix is rolled at too high a temperature, it will creep out rather than compress downward or will be subjected to sizeable cracks. Also, blisters will sometimes form behind the roller as it passes. When this occurs, the irregularities thus created should be quickly repaired by raking. Rolling if performed, after the pavement has partially chilled, may be responsible for some cracking. High speed rolling, turning the roller too sharply, poor bond between courses and lean mixes also contribute to surface cracking.

Rolling of hot mix is commenced as soon as the spread mixture will sustain the roller without excessive displacement. Some highly stable mixes can be rolled at 120°C or more whereas a very fine mix with low stability may not be able to be rolled until the temperature has reached 95°C or less. In practice rolling operations are usually regulated by trial and error and the only criterion needed is the distance that the initial roller should stay behind the finishing machine from which the mixture is coming out at a fairly uniform temperature.

17.11.1. Spreading and rolling temperatures

Asphalt Institute has recommended the minimum spreading temperatures and rolling to be completed after placing the mixture at site as given in Table 7.7.

Table 17.7 Recommended minimum spreading temperatures °C

Base temp. (on which mix placed °C)	*Minimum spreading temperature °C for*							
	Mat thickness; cm							
	1	*2*	*2.5*	*4*	*5*	*7.5*	*9*	*10 and greater*
7–0	—	—	—	—	—	—	135	127
0–5	—	—	—	—	146	138	132	127
5–10	—	—	—	149	141	135	129	124
10–15	—	—	149	146	138	132	127	124
15–20	—	149	143	141	135	129	124	121
20–27	149	143	141	138	132	129	124	121
27–32	143	138	135	132	129	127	121	121
32+	138	135	132	129	127	124	121	121
Rolling completed after placing time (minutes)	4	6	8	12	15	15	15	15

Compaction is usually accomplished in three stages. However second stage i.e. intermediate rolling by pneumatic roller is sometimes not specified. The three stages are:

(i) Break down rolling. This is compaction by tandem 2-axle minimum 8 tonnes or 2-axle minimum 12 tonnes roller as soon as the roller can safely get on to the mixture. Breakdown rolling compacts the material to obtain practically all the density it will receive. This requires usually 2 to 3 passes of roller. Maximum speed of breakdown roller is 5 km per hour.

(ii) Intermediate rolling. This is done by employing pneumatic roller, minimum 3.15 kg per sq cm tyre pressure immediately behind the breakdown roller. It imparts a very high density, particularly in top 2.5 cm or so by kneading the mixture into a compact position without crushing particles and also provides a sealing action which cannot be achieved by steel wheel rolling. This is particularly useful on airport runways to seal the surface against excessive penetration of oil and fuel spillage from aeroplanes and thus prevent ravelling. Pneumatic roller can run at 25 km/hr without displacement.

(iii) Finish rolling. This is accomplished by tandem minimum 3-axle, 8–10 tonnes roller which adds little more density to the material and removes all irregularities left by breakdown roller. The mix is finish rolled, until the roller marks are eliminated.

Compaction is done by adequate number of rollers which are adjusted according to the mixture laid per hour. Usual specifications require one roller for each 25 tonnes of material spread per hour or each 325 sq metre covered whichever is less.

The degree of compaction obtained by rolling is determined by the field density test. This is made by cutting a sample about 30 cm sq from the finished course and dividing the sample into four smaller samples of approximately equal size. Each sample is coated in the field-laboratory with paraffin, weighed in air, and then weighed in water. The specific gravity is then determined by the equation:

$$G = \frac{A}{B - C - (B - A)\, G_p}$$

where G = specific gravity of specimen

A = weight of specimen in air

B = weight of specimen and paraffin in air

C = weight of coated specimen in water

G_p = specific gravity of paraffin

Compaction is done upto a certain minimum percentage usually 95 per cent of laboratory density and till all roller marks are eliminated.

(iv) Surface finish. Most specifications have straight-edge requirements for the finished surface, specifying a maximum variation of 0.5 mm in any one metre. These measurements should be made while the pavement is still hot so that necessary corrections can be made.

(v) Construction of joints. Longitudinal joints of different courses are never superimposed directly one upon the other. Most specifications require an offset of at least 50 cm to one metre. In order to ensure this, it is desirable to decide, before laying is begun, the width of each lane for both the base and the surface courses in order to provide the specified offset.

When work is temporarily stopped, the roller is driven off thus forming a ramp. This ramp should be cut back to the point where the pavement is of full depth and then lightly primed with a thin coat of cut-back or emulsion before fresh material is placed adjacent to it. Same treatment should also apply for longitudinal joint when side lane is laid. Transverse joints in binder and surface courses and longitudinal binder joints are some times permitted to be placed cold without painting to avoid the danger of excess bitumen normally resulting from primed joints.

(vi) Tack coat. A tack coat is always essential in resurfacing operations and it is a good practice to employ a tack coat between successive courses in new construction. Tack coats are frequently used on the top of the binder courses to offset wind-deposited and traffic-placed dust. This tack coat is very helpful in minimizing roller cracking. Where dust or foreign matter does not separate the binder and the surface courses, generally good bond results without the use of tack coat.

The tack coat should not be placed at a rate more than 50 kg per 100 sq m as this can do more harm than good when they are laid in too large a quantity. An excessive amount of tack coat applied to the existing surface will be absorbed by the freshly laid hot material causing a higher bitumen content and a resultant loss of stability.

17.12. CONSTRUCTION CONTROL

Construction control is governed by performance tests in the field/laboratory.

The tests and their frequencies, as per IRC, for the bituminous concrete pavements are given in Table 17.8.

TABLE 17.8 Control tests and their frequencies for bituminous concrete

Serial No.	*Test*	*Frequency*
1.	Quality of binder	As required
2.	Aggregate impact value	One test per 50–100 m^3 of aggregate
3.	Flakiness index of aggregate	—do—
4.	Stripping value of aggregate	—do—
5.	Mix grading	One set of tests on individual constituents and mixed aggregates from the dryer for each 100 tonnes of mix subject to a minimum of two sets per plant per day.
6.	Control of temperature of binder in boiler, aggregate in the dryer and mix at the time of laying and rolling	At regular close intervals
7.	Stability of mix (vide ASTM D-1559)	For each 100 tonnes of mix produced, a set of three Marshall specimens to be prepared and tested for stability and flow value, density and void content, subjected to a minimum of two sets being tested per plant per day.
8.	Binder content and gradation in the mix (Binder content test vide ASTM D-2172)	One test for each 100 tonnes of mix subject to a minimum of two tests per day per plant.
9.	Rate of spread of mixed material.	Regular control through checks on weight of mixed material and layer thickness
10.	Density of compacted layer	One test per 500 m^2 of aggreagate

MASTIC ASPHALT

Mastic asphalt is a type of paving material consisting essentially of fine aggregate and mineral filler, uniformly coated and mixed with paving bitumen in a suitable plant. It is used almost exclusively as a wearing course laid on a binder course of fine-aggregate type of bituminous concrete or directly upon a prepared base course such as old concrete, brick pavement, bituminous or waterbound macadam base course. Each course is ordinarily laid from 25 to 50 mm thickness. The binder course of bituminous concrete, is an economic measure since it requires about half the bitumen content than that in mastic asphalt surface.

Binder course. The binder course is made, laid, compacted and finished as for hot-mix hot-laid bituminous concrete. The composition of bituminous concrete mixture for binder course conforms to grading requirements as already set forth in Table 17.4 for nominal maximum size of either 25 mm or 20 mm. They generally are 40 mm in thickness.

Mastic asphalt. At the present time, mastic asphalt is used principally in surfacing of city streets and heavily stressed areas i.e., junctions, toll plazas, etc due to their added virtue of being (*i*) extremely smooth surface, (*ii*) noiseless riding surface which are (*iii*) water-proof, and (*iv*) easily cleaned, (*v*) capable of withstanding heavy wheel loads and very heavy volume of traffic without damage and (*vi*) requires very low maintenance costs.

These are somewhat more difficult to design and their construction is slightly more difficult to control than that of bituminous concrete. They tend to become more slippery with age than bituminous concrete pavements. Their principal fault is their susceptibility to contraction cracking in cold weather.

17.13. MATERIALS REQUIREMENTS

Mastic asphalt is an intimate homogeneous mixture of selected well graded aggregates, filler and bitumen in such proportions as to yield a plastic and voidless mass, which when applied hot can be trowelled and floated to form a very dense and impermeable surfacing.

The grade and thickness of mastic asphalt paving and grading of coarse and fine aggregates are as per Table 17.9 and 17.10.

Table 17.9 Grade and thickness of mastic asphalt paving, and grading of coarse aggregate

Application	*Thickness range (mm)*	*Nominal size of Coarse aggregate (mm)*	*Coarse aggregate content % by mass of total mix*
Road and carriageway	25-50	13	40±10
Heavily stressed areas i.e. junctions and toll plazas	40-50	13	45±10
Nominal size of coarse aggregate IS Sieve (mm)		13 mm Cumulative % passing by weight	
19		100	
13.2		88-96	
2.36		0-5	

Table 17.10 Grading of fine aggregate (inclusive of filler)

I.S. Sieve	*Percentage by weight of aggregate*
Passing 2.36 mm but retained on 0.600 mm	0-25
Passing 0.600 mm but retained on 0.212 mm	10-30
Passing 0.212 mm but retained on 0.075 mm	10-30
Passing 0.075 mm	30-55

Construction method. Mastic asphalt is proportioned, mixed, transported, placed, compacted and surface tested as for hot-mix hot-laid bituminous concrete pavements.

SAND ASPHALT

Sand asphalt is a surface or base course formed from a mixture of sand and paving bitumen, with or without mineral filler. This type of mixture is extensively used in areas in which suitable sands are economically available. The main difference between mastic asphalt and sand asphalt lies in the fact that sand-asphalt mixtures are not carefully specified or controlled, and for this reason are usually somewhat less dense and less stable than comparable to mastic asphalts.

They are used on roads which carry moderate amount of traffic.

The maximum thickness of a base course of this type is 80 mm while a surface course may be about 50 mm in compacted thickness.

Bitumen usually used is paving bitumen of penetration grade S65 (60/70) or S90 (80/100). Surface courses usually contain 5 to 10 per cent of bitumen and base courses may contain slightly less. Grading requirements of sand and physical requirements are given in Table 17.11 and the requirements for sand asphalt base course are as per Table 17.12.

Table 17.11 Sand grading and physical requirements

Sieve size (mm)	*Cumulative percentage by weight of total aggregate passing*
9.5	100
4.75	85-100
2.36	80-100
1.18	70-98
0.60	55-95
0.30	30-75
0.15	10-40
0.075	4-10
Plasticity Index (%)	6 max.
Sand equivalent (IS:2720, Part 37)	30 min.
Los Angles Abrasion Value (IS : 2386, Part 4)	40 max.

Note: Maximum thickness for sand asphalt is 80 mm.

Table 17.12 Requirements for sand asphalt base course

Parameter	*Requirement*
Minimum stability (kN at 60°C)	2.0
Minimum flow (mm)	2
Compaction level (Number of blows)	2 × 75
Per cent air voids	3-5
Per cent voids in mineral aggregate (VMA)	> 16
Per cent voids filled with bitumen (VFB)	65-75

Requirement relative to density and stability is also frequently included as a part of specification for sand-asphalt, with requirements in this respect being generally somewhat less severe than for mastic asphalt or bituminous concrete mixtures.

HOT-TAR CONCRETE

This is a type of bituminous concrete which utilizes the heavier grades of road tar as binding material. This type of mixture is prepared in central plant as in the case of hot-mix bituminous concrete with the difference that temperatures used are considerably less than those normally adopted in similar bituminous concretes, with the temperature of the plant mixture generally from 65° to 105°C.

The rolling temperature is usually 60°—80°C and temperature of mixture when delivered on site may be 85° to 100°C. If temperature at mixing rises above 115°C, fuming starts and becomes most objectionable and hardening of tar occurs due to loss of volatile oils resulting in poor durability.

RT-11 and *RT*-12 are most widely used but *RT*-9 and *RT*-10 are also sometimes used.

Two-courses or one-course is used in making this type of pavement. A seal coat frequently applied to this type of mixture is generally the same as for bituminous concrete mixture.

HOT-MIX COLD-LAID BITUMINOUS CONCRETE

This is the nearest approach to hot-mix hot-laid bituminous concrete. Mixtures of this type can be manufactured, shipped, and placed immediately or can be stockpiled for future use over a period of 6 to 8 months. This type of plant mixes are suitable for small jobs where hot-mix plant facilities are not available or where setting up of a plant is not economically justified. This type of mixture also provides a high type maintenance patching material.

17.14. MATERIALS REQUIREMENTS

Aggregates gradation and mix design procedure is the same as already described for hot-mix hot-laid bituminous concrete, except for the use of a higher-penetration bitumen, a lower mixing temperature, an aggregate primer, and the addition of water in the mix. The most suitable penetration range of paving bitumen is within 250—800 penetration. Alternately emulsified bitumen may be used. The uses of bitumen in the cold laid mix is given in Table 17.13. Bituminous primer used is cut-back bitumen MC—0, to which a wetting agent or anti-stripping material is added.

Table 17.13 Uses of bitumen in cold mix

Type of construction	*Emulsified Bitumen*									*Cutback Bitumen*						
	Anionic					*Cationic*				*Medium Curing (MC)*				*Slow Curing (SC)*		
Cold-Laid Plant Mix Pavement base and surface	MS-2, HFMS-2	MS-2h, HFMS-2h	HFMS-2s	SS-1	SS-1h	CMS-2	CMS-2h	CSS-1	CSS-1h	70	250	800	3000	250	800	3000
Open-graded Aggregate	*	*				*	*									
Well-graded Aggregate			*	*	*			*	*		*	*	*	*	*	*
Patching, Immediate Use				*	*			*	*		*	*			*	
Patching, Stockpile											*	*		*	*	

* The corresponding grades in IS:8887 are only broadly classified as RS, MS and SS and further sub-classification is not available at present.

Preparation of mixture. The hot-mix cold-laid bituminous concrete is manufactured in a closely controlled central hot-mix plant as in the case of hot-mix hot-laid bituminous concrete. After the aggregates are dried and before they are introduced into the pugmill mixer, they are cooled to approximately 80°C. This is usually accomplished in a double-drum drier which removes all moisture while the aggregate flows through its inner drum. The material is cooled to the desired mixing temperature while it passes through its outer drum before moving to the hot bins of the plant. The reason for cooling the aggregate to approximately 80°C is to obtain a discharge temperature of $75°C \pm 5°C$ for the completed mix. This enables to evaporate 60 to 65 per cent of original amount of water added during the mixing operation. Consequently, only 3/4 per cent to 1 per cent is retained in the mix which is desirable for mixes designed for immediate placement and stockpiling respectively. These moisture contents are then reduced to negligible amounts during normal placement operations.

The addition of water causes the mix to fluff and therefore to remain workable after cooling to normal temperature. Minimum of 2 per cent water renders a mixture workable for several days and this percentage is therefore used for mixtures which are to be shipped, unloaded and placed immediately. However, 3 per cent water is necessary if the material is to be stockpiled and used over a period of several months.

The combined aggregates after having been weighed separately from each bin are introduced into the mixer and after dry-mixing cycle of about 10 sec, the bituminous primer is added and mixed for another 10 sec. Approximate quantity of primer used is 0.75 per cent which is enough to barely stain the aggregate but it ensures proper bond of paving bitumen in the presence of water.

Immediately after the primer is mixed, water and high-penetration bitumen are added into the mix simultaneously and the mixing is completed.

COLD-PLANT MIX COLD-LAID BITUMINOUS CONCRETE

These differ from the hot-plant mixes in that they use low-viscosity cut-back bitumens and emulsified bitumens and are mixed at normal temperatures. This is a slightly lower type of pavement than the hot-mix hot-laid type and is usually used for maintenance of small jobs where moving in of a hot-mix plant is not economically justified. The manufacture and placement of cold mixes are usually confined to the warmer months. When it is necessary to use this type of mixture during cold weather, some heating of aggregate and bituminous material may be required.

17.15. MATERIAL REQUIREMENT

Aggregate and mix design procedure is the same as for hot-plant mixes. The following bituminous materials are used for dense-graded surface course mixes containing 35 to 45 per cent of aggregate passing 2 mm sieve and required for immediate placement.

Cut-back bitumen, RC—1, RC—2 or MC—2.

Medium setting emulsified bitumen MS—2.

For mixes which are required to be stockpiled, medium curing cut-back bitumen MC-2 is used which affords the necessary workability.

Drying of aggregate for cut-back bitumen is necessary only when the aggregate is saturated or has some surface moisture. Emulsified bitumen mixes may be made with wet aggregate. Actually addition of water, in this case, is necessary for proper mixing especially when the aggregate contains a high percentage of material passing 2 mm sieve.

Cold-plant mixes made with cut-back bitumens have to be thoroughly aerated by blading on the roadway until a large percentage of volatiles has evaporated. When thoroughly aerated, these mixes appear to be very stiff, but they will have sufficient workability to be spread smoothly.

Cold mixes made with MS-2 emulsified bitumen are placed without aeration and compacted. High atmospheric temperature and low humidity during warm weather enable these mixes to cure rapidly.

The amount of material passing 2 mm sieve considerably affects the workability of cold mixes and for this reason, mixes using cut-back bitumen, which are to be stored for future use contain 5 to 10 per cent less of minus 2 mm material than mixes prepared for immediate placement.

MISCELLANEOUS BITUMINOUS CONCRETE MIXTURES

17.16. MODIFIED BINDER BITUMINOUS CONCRETE MIXTURES

Several projects making use of modified binder in bituminous concrete have been undertaken on airfields to develop a bituminous pavement more resistant to jet fuel. Jet engines dump fuel when they are shut down or when they are decelerated. Repeated fuel spillage soon dissolves the bitumen binder and the pavement disintegrates. These areas exposed to traffic and blast of jet engines soon become worse. Loose pavement particles, resulting from these effects, are a hazard to jet engines. This resulted in the use of modifiers in the binder.

For use in high quality bituminous courses, a small amount of modifier is being added to bitumen to improve the quality of bituminous pavement. Rubberized Bitumen/Polymer Modified Bitumen and fibres are the emerging materials. The use of these materials results in:

1. Extended service life of bituminous pavements subjected to heavy traffic loads and in extreme climatic conditions.
2. Lower temperature susceptibility
3. Higher resistance to ageing
4. Higher fatigue life
5. Higher resistance to cracking; and
6. Better adhesion between aggregates and binder.

The bitumen modifiers should :

1. be compatible with bitumen
2. resist degradation of bitumen at mixing temperature.
3. be capable of being processed by conventional mixing and laying machinery.
4. produce coating viscosity at application temperature.
5. maintain good properties during storage, application, in service etc.
6. be cost effective on a life cycle cost basis.

The common types of modifiers and their indicative doses are given in Table 17.14.

Table 17.14 Type of additives for bitumen modification and their indicative doses

Sl. No.	*Type of modifier*	*Examples*	*Indicative dose level, per cent by wt. of bitumen*
1.	Plastics		
	Thermoplastics	Low Density Polyethylene (LDPE), Ethylene Vinyl Acetate (EVA), Ethylene Butyl Acrylate (EBA)	3-6
	Thermosets	Epoxy Resins	3-5
2.	Elastomers		
	Natural Rubber	Latex or Dry Rubber Powder	2-4
	Synthetic Elastomers	Styrene Butadiene Styrene (SBS) Block Copolymer (BC), Styrene Butadiene Rubber (SBR)	3-5
3.	Reclaimed Rubbers	Tyre Crumb Rubber Powder from discarded truck tyres further improved by additives	10-12

The choice of grade of a modified bitumen as per Table 17.14 should be based on the minimum and maximum atmospheric temperatures in the region. For cold climatic areas, properties, *viz*, penetration at 4°C and Fraass breaking point values should be taken into account.

The new construction projects using such material have given variable and questionable results and as such definite conclusions cannot be drawn. For this reason, Portland cement concrete pavements are now coming in vogue on air fields particularly on critical areas such as aprons, warm-up pads, the end 300 metre wash racks, calibration hardstands, etc., which are subjected most to fuel spillage and effects of heat and blast.

HIGH PERFORMANCE BASES

High performance base is an improved bituminous paving mixture for the base course of the pavement, which is the main structural layer. This is used on heavily trafficked roads. Since its design procedure measures performance-based mechanical properties, it permits the pavement thickness design to be optimized for the particular situation of traffic and climate. This leads itself to the optimum use of locally available materials.

17.17. MATERIALS

(A) ***Aggregates:*** The coarse aggregates, fine aggregates and mineral filler should conform to the physical requirements given for dense bituminous macadam. Alternative materials may be considered if they meet the volumetric and mechanical property requirements.

The maximum aggregate size and filler content should be 37.5 mm and 5 to 10 per cent respectively. The target grading may be calculated using the following equation, developed on the basis of field experience:

$$P = \frac{(100 - F)\,(d^n - 0.075^n)}{(D^n - 0.075^n)} + F$$

where P = percentage passing a sieve of aperture size d mm

D = maximum nominal aggregate size in mm

F = filler content (per cent)

n = an exponent, between 0 and 1.

The filler content F should be between 5 and 10 per cent; typically a value of 6 per cent is used.

The value of n should be at least 0.5 to ensure that the target grading is coarser than the maximum density grading so as to avoid a mixture, which could potentially be compacted to too low a void content which may not have adequate resistance to permanent deformation. Values of n greater than 0.7 on the other hand will be prone to segregation. A minimum of three gradings with different values of n should be selected.

(B) ***Binder:*** The binder should be either a penetration grade bitumen meeting the requirements of IS:73 or a base penetration bitumen to which a "*modifier*" has been added to or blended with.

17.18. MIXTURE DESIGN

Flow chart is presented in Fig. 17.10. The procedure involves the following four steps:

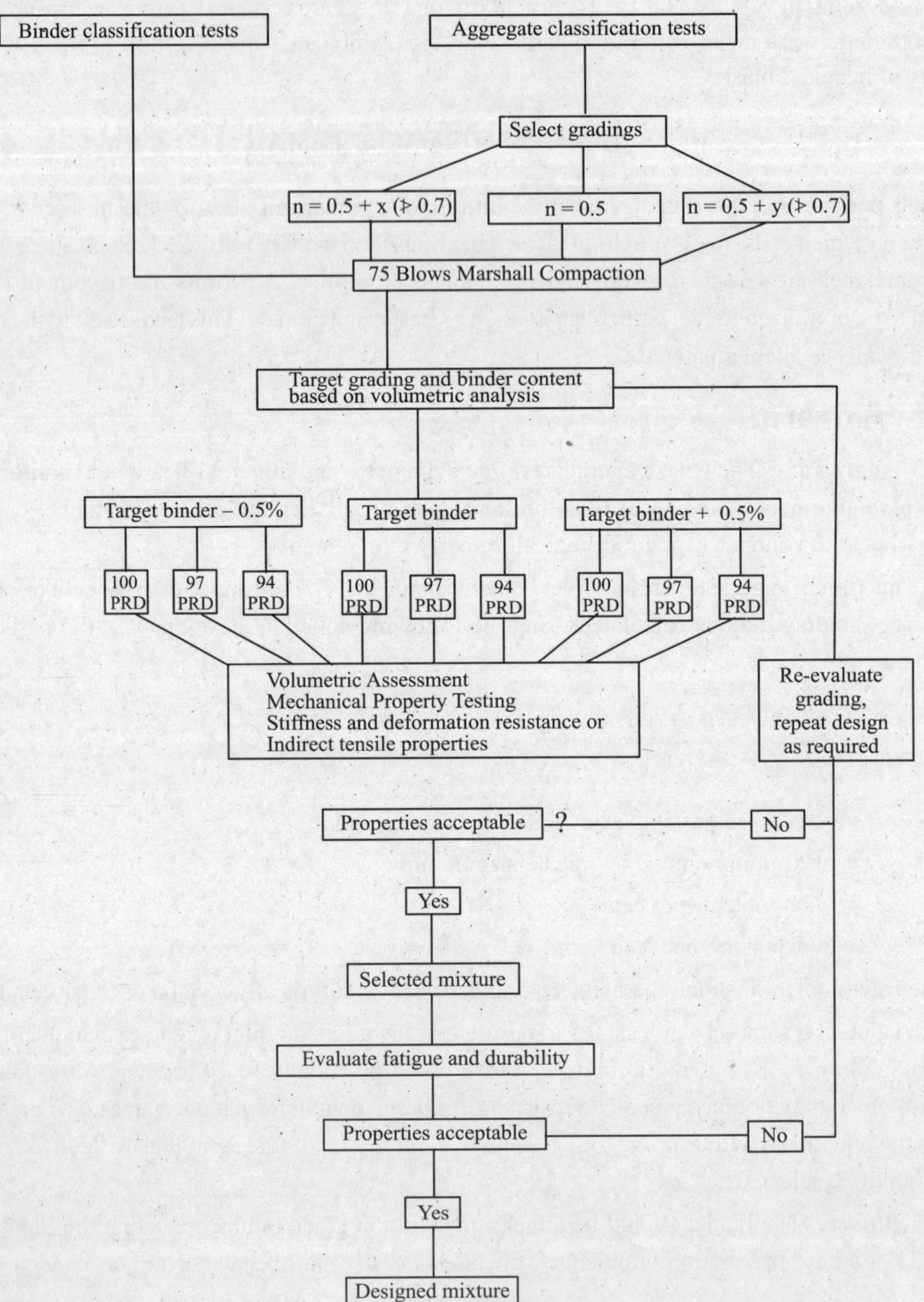

Fig. 17.10 Flow chart for evaluation of high performance base.

(1) ***Selection of aggregate grading and target binder content.*** For each of the three gradings selected specimens should be prepared in accordance with the general requirements of the Marshall method described in Asphalt Institute Manual MS-2, using 75 blows compaction and a range of binder contents to encompass at least those yielding an air void content of 4 ± 2. For selection of the optimum grading and target binder content, the key parameters listed in Table 17.15 should be considered. The Marshall stability and flow testing are not required for this evaluation.

Table 17.15 Key volumetric requirements for high performance base

Parameter	*Requirement*
Per cent air voids (V_A)	3-5
Per cent voids in mineral aggregate (VMA)	See Table 4.12
Per cent voids filled with bitumen (VFB)	65-75

Selection of the preferred grading and target binder content should be based on that which most satisfactorily meets the established criteria. The key points are summarized below:

(*a*) The VMA should be slightly to the left ("dry") side of the VMA/binder content curve and gradings which fall below the minimum values in Table 4.12 at any binder content should not be selected.

(*b*) The design range of air voids (3-5 per cent) is the ultimate level to be achieved in the field after several years of traffic. The ideal per cent air voids just after construction is thus about 8 per cent. Mixtures which ultimately compact to less than 3 per cent air voids in the field are likely to be susceptible to permanent deformation, while mixtures with initial air voids greater than 8 per cent may lead to greater binder oxidation, ageing and/or an increased probability of water damage. High air voids may also be associated with deformation.

(*c*) Use of the VFB criterion helps to prevent the design of mixtures with marginally acceptable VMA and also restricts the allowable air voids for mixtures near the minimum VMA.

(*d*) At the same binder content. air voids and VMA decrease with increasing compactive effort and the binder content at minimum VMA shifts. It is important, therefore, to ensure that the compactive effort, which is selected, simulates that expected from the design traffic.

(*e*) The grading and binder content selected should be those, which best fit, these criteria and are least sensitive to small variations in binder content. The influence of the pavement structure in which the high performance base will be incorporated should also be considered, particularly the specific project conditions of traffic and climate. Thus, mixture with binder contents on the high side of the range are likely to be less susceptible to cracking becuase of the additional flexibility whereas mixtures on the low side of the range are likely to be less susceptible to rutting.

(2) ***Compaction level and binder content.*** Having selected the target grading and binder content (as in the previous paragraph), the effects of variations of binder content above and below the target value must be investigated at three levels of compaction. These must be chosen to encompass the expected production tolerances for binder content and to reflect states at, above and below the level of compaction likely to be achieved in the field.

Different compaction methods may give differences in mechanical properties for the same material, particularly in respect of resistance to permanent deformation. It is known that the mechanical properties of specimens manufactured by Marshall method of compaction, in particular, are not representative of those determined on field compacted specimens, and Marshall compaction is not, therefore, recommended. At the present time, apparatus used for the Percentages Refusal Density (PRD) test is recommended as a suitable means of achieving the required levels of compaction. The procedure to be followed is:

(*i*) Compaction to refusal (100 PRD), *i.e.* 2 minutes compaction on each face of the specimen with the vibrating hammer. Material temperature = 160°C.

(*ii*) Intermediate level of compaction, *i.e.,* 1 minute compaction on each face of the specimen. Material temperature = 100°C (About 97 PRD).

(*iii*) Low level of compaction, *i.e.*, 30 seconds compaction on each face of the specimen. Material temperature = 90°C (about 94 PRD).

The dimensions of the moulded specimens should be 152 mm in diameter by approximately 100 mm in height.

After cooling and demoulding, a 100 mm diameter core should be cut from each specimen, the core then being trimmed to a length of about 70 mm using a diamond tipped masonry saw.

The recommended binder contents for this assessment are the target value determined and values 0.5 per cent above and below. Triplicate specimens should be made at each binder content and each level of compaction, resulting in 27 specimens.

(3) ***Selection of the designed mixture.*** For the determination of the volumetric properties of the samples prepared as in (2) above, the following measurements should be made to enable the air void contents of the samples to be determined in accordance with ASTM D3203:

(*i*) The maximum density of the mixture, obtained using the theoretical maximum specific gravity of the loose mixture, determined in accordance with ASTM D2041 and converted to relative density using the appropriate correction factor.

(*ii*) The bulk density of the specimens, determined in accordance with BS:598, Part 104 Clause 4, as the bulk density required by ASTM D3203, except that the specimens should be coated with self-adhesive aluminium foil rather than the wax specified in the British Standard when weighing the specimens in water.

The determined air voids at refusal density and at 94 and 97 PRD should be evaluated in the light of the requirement for an absolute minimum air voids at refusal of 1 per cent and the considerations enumerated in the preceding para above.

Depending on the specific demands of the site, minimum air voids at refusal may need to be increased to 2 per cent (fast moving heavy traffic) or 3 per cent (slow moving heavy traffic, high axle loads). Acceptable ranges for void content and VMA with a cut-off for binder volume of 8 per cent are shown in for air voids at refusal of 1, 2 and 3 per cent. The volumetric screening described enables the rejection of unsuitable screening formulations before mechanical testing is carried out.

All formulations which pass the volumetric screening are then subjected to mechanical property testing. Two procedures are permitted Procedure A should ordinarily be used, unless otherwise instructed.

Procedure A: The stiffness of each specimen should be determined using the Indirect Tensile Stiffness Modulus (ITSM) test at 20°C in accordance with BSDD 213. The deformation resistance of each specimen should be determined in accordance with the repeated Load Axial Test (RLAT) described in BS DD 226, under the standard conditions of 40°C temperature, 100 kPa axial load and 2 hour test duration. The ITSM test is non-destructive.

The criteria of acceptance will depend upon the specific circumstances of the pavement structure and the traffic and climate to which it will be exposed, subject to the following minimum threshold values.

(*i*) Stiffness (ITSM) not less than 2000 MPa at 20°C.

(*ii*) Deformation resistance (RLAT) at 40°C and greater than 2 per cent cumulative permanent strain at the end of the test.

Procedure B: A simple modification to the Marshall test may be used. This involves application of the load through 10 mm wide strip loading platens as described in ASTM D4867, so that the indirect tensile (splitting) strength can be determined. This procedure should be carried out on the specimens prepared as described in the preceding para above. Two parameters are required for evaluation.

(*i*) The indirect tensile strength in MPa.

(*ii*) The area under the load/time curve in kNs.

The optimum mixture combination will be that which best satisfies the requirements of maximum load and maximum area under the load/time curve. The latter can be thought of as the "*energy dissipated*" during failure and larger values may , thus be, associated with higher crack resistance. The following minimum values should be achieved:

(*i*) Indirect tensile strength, not less than 2.0 MPa at 4°C

(*ii*) Area under load/time curve, not less than 50 kNs at 40°C.

(4) *Verification of the designed mixture.* Having selected the optimum design mixture on the basis of a consideration of mixture volumetrics and stiffness and deformation resistance for indirect tensile strength, the durability (water sensitivity) of the designed mixture should be assessed using the procedure described in AASHTO T283. The minimum retained tensile strength should be 80 per cent.

Where the specialized testing equipment described earlier is available, the fatigue resistance of the designed mixture can be evaluated using the Indirect Tensile Fatigue Test (ITFT). Ten additional specimens should be prepared as described earlier at the 97 PRD level of compaction, and the compacted specimens oven aged (120 hours at 85°C) to simulate in-service conditions. The ITFT should then be carried out on this additional set of samples. The requirements for fatigue resistance will be pavement specific, subject to the following minimum threshold criterion:

Minimum ITFT life at 100 micro strain, 3,50,000 cycles. Provided all the design criteria are met on the selected mixture this becomes the design mixture. As indicated in Fig. 17.10, if at any stage in the procedure the required design criteria are not met, the mixture design must be repeated with appropriate modifications.

❑❑❑

Chapter

PORTLAND CEMENT CONCRETE PAVEMENTS AND BASE COURSES

Portland cement concrete pavement consists of a relatively rich mixture of Portland cement, sand and coarse aggregate laid as a single course. Concrete was not, however, widely used as a paving material until the motor vehicle became an important unit of transportation. Since 1910, vast sums of money have been spent on research of cement concrete roads. It is now possible to construct first class concrete pavements built under rigid inspection to carefully drawn specifications.

Advantages of concrete pavements

1. When properly designed and constructed, concrete roads are capable of carrying almost unlimited amount of any type of traffic with ease, comfort and safety.
2. Wear and tear on properly laid concrete roads are comparatively small.
3. Concrete roads can effectively withstand the effects of repeated fuel spillage and heat and blast of jet aircrafts and are, therefore, best suited for paving airfields for jet aircraft.
4. Concrete roads have low tractive resistance.
5. Concrete roads are not slippery when wet.
6. Concrete roads have low maintenance.
7. Concrete roads are more comfortable due to its smoothness and absence of dust.
8. Old concrete pavement provides excellent base course for some resurfacing material, *e.g.* bituminous concrete, sheet, brick or block pavement.

Disadvantages of concrete pavement

1. With ordinary portland cement, concrete roads when constructed have to be closed to traffic during curing period of minimum 7 days which may cause inconvenience to traffic.
2. As the colour of concrete pavement is light, the glare from reflected sun is uncomfortable for day light driving.
3. Initial cost is comparatively high and thus are only suitable for heavily trafficked roads or runways for jet aircrafts.

MATERIALS

Materials used in concrete for highway construction are coarse aggregate, find aggregate, cement and water. Sometimes certain additives and admixtures are also added to concrete in order to modify its properties.

18.1. COARSE AGGREGATES

This consists of natural, crushed stone, gravel, or aircooled iron blast furnace slag, or a combination thereof. In order to make good concrete, it is important to avoid crushed aggregate of poor shape. Very angular, flaky, elongated or splintery aggregates give a harsh mix of low workability.

Grading requirements. Grading of coarse aggregate is very important in securing an economical and superior concrete. It is generally specified that coarse aggregates be furnished in two separate sizes, single sized *e.g.,* from 63 mm to 20 mm and from 20 mm to 4.75 mm or graded aggregates from 40 mm to 20 mm and from 20 mm to 2.36 mm, etc. These coarse aggregates are combined in such proportions that the resulting mixture has the lowest per cent of voids. This will usually require about twice as much of large size aggregates as of small size. However, the exact proportions to produce the densest mixture should be determined experimentally when the source of aggregate has been decided.

Maximum size of aggregate should not exceed one fourth of the pavement slab thickness. In case of pavements having reinforcement, maximum size of aggregate should also not exceed one-fourth of minimum clear spacing between reinforcing bars.

As far as gradation limits are concerned, IS specifications IS—383 calls for the gradation as given in Table 18.1.

The aggregates should be free from chert, flint or silica in the form that can react with the alkalies in the cement.

The amount of deleterious substances in coarse aggregate should not exceed the following limits.

Item	*Maximum per cent by weight of total sample*	
Clay lumps	0.25	
Soft particles	5.0	
Chert that will readily disintegrate (soundness test 5 cycles)	1.0	
Material finer than 75 micron sieve	1.0	This may be increased to 1.5 provided minus 75 mm material consists of dust of fractured rock.
Coal and lignite	1.0	This does not apply to blast furnaces slag coarse aggregate.

Weight of slag. Weight of slag should have a compact weight of not less than 1.12 gm per cu cm.

Soundness. When subjected to 5 cycles, weighted loss should not be greater than 12 per cent when sodium sulphate is used or 18 per cent when magnesium sulphate is used.

Coarse aggregate, failing to meet the requirement of soundness as given above may be accepted provided, experience under similar weathering conditions warrant or where no service records are available and the concrete when subjected to freezing and thawing tests should give satisfactory results. These aggregates should produce concrete of adequate strength.

Abrasion. When tested for abrasion in Los Angeles machine, the loss should not be greater than 50 per cent for sub-base course and 35 per cent for concrete wearing course. Aggregate having abrasion loss greater than the above values may be used provided the aggregate produce satisfactory strengths in concrete of the proportions selected for the work and that aggregates are stronger than the concrete, of which they form a part, their inherent strength is not likely to influence the strength of the concrete, either in crushing or in flexure.

No aggregate which has water absorption more than 2 per cent shall be used in the concrete mix. Aggregates having more than 2 per cent water content may be allowed subject to conforming to soundness test. However maximum water absorption should not exceed 5 per cent.

The coarse aggregates should not have flakiness index less than 40.

Table 18.1 Coarse aggregates requirements—for single sized and graded aggregates

IS Sieve Designation	*Percentage passing for single size aggregate for nominal size*						*Percentage passing for graded aggregate of nominal size*			
	63 mm	40 mm	20 mm	16 mm	12.5 mm	10 mm	40 mm	20 mm	16 mm	12.5 mm
(1)	(2)	(3)	(4)	(5)	(6)	(7)	(8)	(9)	(10)	(11)
80 mm	100	-	-	-	-	-	100	-	-	-
63 mm	85 to 100	100	-	-	-	-	-	-	-	-
40 mm	0 to 30	85 to 100	100	-	-	-	95 to 100	100	-	-
20 mm	0 to 5	0 to 20	85 to 100	100	-	-	30 to 70	95 to 100	100	100
16 mm	-	-	-	85 to 100	100	-	-	-	90 to 100	-
12.5 mm	-	-	-	-	85 to 100	100	-	-	-	90 to 100
10 mm	0 to 5	0 to 5	0 to 20	0 to 30	0 to 45	85 to 100	10 to 35	25 to 55	30 to 70	40 to 85
4.75 mm	-	-	0 to 5	0 to 5	0 to 10	0 to 20	0 to 5	0 to 10	0 to 10	0 to 10
2.36 mm	-	-	-	-	-	0 to 5	-	-	-	-

18.2. FINE AGGREGATES

This should consist of clean natural sand, crushed stone sands or a combination thereof. Certain crushed stone sands being angular in shape produce harsh mixes and should be investigated before use. In some cases, it may prove advantageous to use a mixture of naturally occurring sand and crushed stone and if the former is not obtainable in adequate supply or where its grading is poor.

Bulking due to the presence of moisture in the fine aggregate should be accounted for when volumetric batching is employed.

Table 18.2 Fine aggregates requirements

IS Sieve Designation	*Percentage Passing for*			
	Grading Zone I	*Grading Zone II*	*Grading Zone III*	*Grading Zone IV*
10 mm	100	100	100	100
4.75 mm	90-100	90-100	90-100	95-100
2.36 mm	60-95	75-100	85-100	95-100
1.18 mm	30-70	55-90	75-100	90-100
600 micron	15-34	35-59	60-79	80-100
300 micron	5-20	8-30	12-40	15-50
150 micron	0-10	0-10	0-10	0-15

Note: 1. Where the grading falls outside the limits of any particular grading zone of sieves other than 600-micron IS: Sieve by a total amount not exceeding 5 per cent, it shall be regarded as falling within that grading zone. This tolerance shall not be applied to percentage passing the 600-micron IS : Sieve or to percentage passing any other sieve size on the coarse limit of grading Zone I or the finer limit of grading Zone IV.

2. For crushed stone sands, the permissible limit on 150-micron IS: Sieve is increased to 20 per cent.

3. Zones here do not depict the location/region. They depict the gradation of fine aggregates in a descending order.

Fine aggregates are further required to be free of injurious amounts of impurities such as soft particles, clay, shale, loam, cemented particles, mica and organic and other foreign matter. When used in concrete that will be subject to wetting, extended exposure to humid atmosphere or contact with moist ground, fine aggregates should not contain any materials that are deleteriously reactive with the alkalies in the cement in an amount sufficient to cause excessive expansion of mortar or concrete. Otherwise, fine aggregates may be used with a cement containing less than 0.6 per cent alkalies calculated as sodium oxide or additive may be used to prevent harmful expansion due to alkali-aggregate reaction.

Limits for deleterious substances as set forth in IS-383 specification are given below:

Item	*Maximum per cent by weight of total sample*
Clay lumps	4.0
Material finer than 75 micron sieve	4.0 per cent in natural sand and 15 per cent in sand produced by crushing rock.
Coal and lignite	1.0

Grading limits. Requirements for grading of fine aggregate by 1S-383 specification are given in Table 18.2.

Loss in soundness test when subjected to 5 cycles should not exceed 10 per cent when sodium sulphate is used or 15 per cent when magnesium sulphate is used except under conditions as already outlined under coarse aggregate.

18.3. CEMENT

Portland cement for use in concrete pavements should comply with standard specifications prevalent in the country. In regions subject to alterations of freezing and thawing weather, and especially where chlorides are used on pavement to facilitate travel under winter conditions which are conducive to surface scaling, materials have been developed that cause the entrainment of air in concrete while it is being mixed so as to control surface and also to add to the durability of concrete. The most widely used of these are sold under the names of Vinsol Resin (hydrocarbon extract of pine wood used in solution with NaOH) and Darex AEA, salt of sulphonated hydrocarbon. There are types of cements which have such materials ground with them during manufacture and are known as air-entraining cements. The air-entraining concrete can, therefore, be made either by using air-entraining cement or by adding air-entraining agent at the mixer. The quantity used is 0.01 to 0.05 per cent by weight of cement.

The increase in concrete volume due to air-entrainment should be between 3 and 6 per cent. A smaller increase than 3 per cent will not provide the desired additional durability; a larger increase than 6 per cent will reduce the strength of the concrete without offsetting increase in durability.

Air-entraining cement generally be specified for concrete that will be used under conditions conducive to surface scaling, otherwise, the cement normally supplied should be used. The cement should whenever feasible to tested in advance so that 7 and 28 days tests can be completed before the cement is used.

High-alumina cement may be used in making speedy repairs to roads. Rapid-hardening cement, because of high early strength obtained is commonly used for repair work where it is essential to open the road to traffic with the minimum delay.

The following types of cements, alongwith their code numbers, are commonly used, for the construction of cement concrete pavements in India.

S. No.	*Type of cement*	*IS code number*
1.	Ordinary Portland Cement 33 grade	IS : 269
2.	Ordinary Portland Cement 43 grade	IS : 8112
3.	Ordinary Portland Cement 53 grade	IS : 12269
4.	Portland blast-furnance slag cement	IS : 455
5.	Portland puzzolana cement (flyash based)	IS : 1489 Part I

18.4. WATER

For mixing concrete, water should be clean and clear, free from salt, oil, acid, injurious alkali and vegetable matter or other substances harmful to the finished concrete. Where water must be taken from a shallow source, an intake should be constructed that will be free from silt, mud, grass, etc. and the intake pipe should be at least 0.6 m above the bottom. It shall meet the requirements stipulated in clause 4.3 of IS : 456–1978.

18.5. ADMIXTURES AND ADDITIVES

Various materials are sometimes added to a concrete mix in order to modify some particular property. Their use should not be regarded as substitute for good material, mix design and workmanship and they should not be used indiscriminately. Some of them, referred to as admixtures are used in small quantities and generally give rise to a chemical or physico-chemical change. Others, used in larger quantities, impart their own characteristics to the concrete and are usually referred to as additives. The various classes of admixtures and additives, the effect they produce on concrete and the quantities in which they are to be used are summarized in Table 18.3.

Table 18.3 Admixtures and additives

Classification of admixtures	*Purpose for which required*	*Composition of admixture*	*Quantity to be used*
1. Workability agents	Increase in workability	1. Finely divided minerals	
		(*a*) bentonite clay; diatomaceous earth	3 to 5 per cent by weight of cement
		(*b*) Fly ash; finely divided silica clay; fine sand; hydrated pulverised stone	Upto 20 per cent by weight of cement
		2. Water-reducing agents-sulphonated organic compounds; carbo-hydrate salts	To be determined by test
		3. Aerating agents	2 to 4 per cent air
2. Puzzolanic materials	1. Reduction in heat of hydration	1. Natural flyash; volcanic ash	10 to 30 per cent by weight of cement
	2. Resistance to sea water and sulphate bearing soils	2. Artificial-heat treated diatomaceous earth; heat-treated raw shales or clay	
	3. Slight improvement of workability	3. Surkhi ground to the same fineness as cement	
	4. Reduction of bleeding and segregation		
3. Retarders	1. Retard the setting time of cement	Carbohydrate derivatives; calcium lignosulphonates	Fractions less than 1 per cent by weight of cement
	2. Remedy premature set of certain cements		
4. Accelerators	1. Increased rate of early strength development	1. Calcium chloride	2 to 3 per cent by weight of cement
	2. Early removal of forms	2. Tri-ethanolamine; soluble carbonates, silicates and fluosilicates	To be determined by test
	3. Reducing curing period	3. Aluminous cement	
	4. Emergency repairs		
	5. Concreting in cold weather		

5. Water repelling agents	1. Decreased capillary rise of moisture	1. Stearates of calcium, ammonium, aluminium and sodium; butyl stearate	0.1 to 0.2 per cent by weight of cement
	2. Increased impermeability	2. Soaps, oils, waxes	Small amounts to be determined by test.
	3. Reduction of efflorescence		0.005 to 0.05 percent by weight of cement
6. Air-entraining agents	1. Protection against frost attack 2. Improved workability 3. Reduced bleeding	Natural wood resins and their soaps; fats and cement oils, their fatty acids and soaps; alkali salts of sulphonated or sulphated organic compounds	
7. Gas forming agents	1. Reduced bleeding 2. Improved bond 3. Improved grout penetration and pumpability	Unpolished aluminium powder	0.005 to 0.02 per cent by weight of cement
8. Natural cementing materials	1. Increased workability 2. Reduced bleeding, segregation and heat of hydration	Natural cements, hydraulic limes; water-quenched blast furnance slag; mixtures of blast furnace slag and lime	Substituted for 10 to 25 per cent by weight of cement

In addition to its various decorative and architectural uses, coloured concrete is sometimes used on roads to distinguish traffic lanes, pedestrian crossings, kerbs, etc. Coloured cements may be used in preference to mixing pigment itself at the mixer as with the latter method, it is difficult to produce uniform colour, and resulting concrete will look patchy. Some of the best pigments are the metallic oxides. As coloured cements are more expensive and also have lower strengths, their use is restricted in road slabs to the top 2.5 or 5.0 cm of concrete. The plain and coloured concretes are placed simultaneously thus ensuring adequate bond between the two.

The following admixtures are recommended in India:

(*i*) Fly ash IS : 3812 (Part I)-2003. Maximum dosage : 20% by mass of cementitious materials.

(*ii*) Granulated blast furnace slag (IS-12089)

(*iii*) Metakaoline

(*iv*) Silica fume (IS : 15388 and IS 456, IRC : SP : 70)

The silica fume (very fine non-crystaline silicon dioxide) is a by-product in the manufacture of silicon, ferrosilicon, etc. from quartz and carbon in electric arc furnaces. It is recommended that 5 to 10 per cent of cementitious material be added. This is more useful for higher grades of concrete i.e. M 50 and above. This is more suitable for high performance concrete e.g. where higher abrasion resistance is required.

18.6. QUALITY CONTROL TESTS ON MATERIALS

The quality control tests on the materials and the frequency of tests are given in IRC : SP. 11 "*Handbook of quality control for construction of roads and runways*". These are reproduced in Table 18.4 for ready reference.

Table 18.4 Quality control tests on materials

Material	*Test*	*Test method*	*Minimum desirable frequencies*
1. Cement	Physical and chemical tasts	IS : 269/445 1489/8112	Once for each source of supply and occasionally when called for in case of long and/or improper storage.
2. Coarse and fine aggregates	(*i*) Gradation	IS : 2386/Pt (I)	One test for 15 m^3 of each fraction of coarse and fine aggregate
	(*ii*) Deleterious constituents	IS : 2386/Pt (II)	—do—
	(*iii*) Moisture content	IS : 2386/Pt (III)	Regularly as required subject to a minimum of one test per day for coarse aggregate and two tests per day for fine aggregate.
	(*iv*) Bulking of fine aggregate (for volume batching)	—do—	Once for each source for deriving the moisture content-bulking relationship.
3. Coarse aggregate	(*i*) Los Angeles abrasion value/aggregate impact test	IS : 2386/ (Pt IV)	Once for each source of supply and subsequently when warnanted by changes in the quality of aggregate.
	(*ii*) Soundness	IS : 2386/ (Pt. V)	As required.
	(*iii*) Alkali aggregate reactivity	IS : 2386/ (Pt VII)	—do—
4. Water	Chemical test	IS : 456	Once for approval of source of supply and subsequently only in case of doubt.

CONCRETE-MIX DESIGN METHODS

18.7. DESIGN CRITERION

Road slabs generally fail in direct tension or in bending, the stresses causing failure being essentially tensile in character. Concrete is therefore usually proportioned on the basis of minimum flexural strength (modulus of rupture). The use of tensile strength as a measure of quality of concrete has certain drawbacks as it is difficult to measure it accurately and further the results obtained are variable. The testing of concrete in flexure yields rather more consistent results than those obtained with the tensile test. Moreover, the flexural test is more easily carried out, and may even be more convenient than the crushing test in the field, since much smaller loads are required. Three methods of loading the specimen in flexure test have been commonly used: two point loading, central loading, and cantilever loading. In this order, they have been observed to give increasingly high results and further the uniformity of results is obtained highest with two point loading and least with cantilever loading. For this reason flexural strength as a quality control of concrete for road slabs is usually specified at two point loading.

For economical design, the design value adopted for flexural strength of paving concrete should not be less than 40 kg/sq cm at 28 days. The mix is so designed as to ensure the minimum structural strength in the field with the desired tolerance level. To achieve the desired minimum strength in the field, the mix in the laboratory is designed for somewhat higher strength, making due allowance for the type and extent of quality control feasible in the field.

The following relation is applied:

$$S = \frac{\overline{S}}{\left(1 - \frac{tv}{100}\right)}$$

where S = average strength of concrete at 28 days for which the mix is designed in the laboratory

$\overline{S}$ = minimum strength of concrete in the field at 28 days.

t = dimensionless factor for the desired tolerance level.

v = Assumed standard deviation (per cent) of field test samples, based on the knowledge of the type of control viz. excellent, very good, good or fair feasible at site.

The values of tolerance level factor t, and coefficient of variation v for normal paving concrete are given in Table 18.5 for different degrees of control.

Table 18.5 Values of tolerance level factor *t*, and coefficient of variation *v*

**Degree of quality level*	*Accepted proportion of low results (tolerance)*	*Tolerance factor, t*	*Standard deviation, ν*
Excellent	1 in 100	2.33	5
Very good	1 in 40	1.96	5
Good	1 in 20	1.65	5
Fair	1 in 15	1.50	4

Where a heavy duty wear-resistant concrete pavements are desired, such as for manoevcring and storage areas of tracked vehicles or for anticipated concentrations of steel tyred traffic, concrete with minimum field compressive strength of 450 kg/sq cm at 28 days should be adopted for wearing course, the 28 days minimum flexural strength requirement in the field remaining at 50 kg/sq cm.

The minimum cement content for the mix corresponding to any grade should not be less than 325 kg/m^3 of concrete inclusive of admixtures and at 28 days should be maximum of 425 kg per m^3 of concrete; exclusive of admixtures.

For the sources of materials that will be used, design of the mix is established to obtain a concrete of the required strength and workability at the lowest cost by a suitable choice of materials and of the proportions in which they are used. It is not the intention to discuss in details the method of designing concrete mixes which is readily available in books pertaining to concrete technology. In this chapter

---

Excellent quality control: Control with weigh batching, use of graded aggregates moisture content of aggregates. Rigid and high level of quality control.

***Very good quality control*:** Control with weigh batching, use of graded aggregates, moisture determination of aggregates, etc. Rigid and constant supervision by the quality control team.

Good quality control: Control with weigh batching, use of graded aggregates, moisture determination of aggregates, etc. Constant supervision by the quality control team.

Fair quality control: Control with volume batching for aggregates, occasional checking of aggregate moisture. Occasional supervision by the quality control team.

discussion will only be limited to problems specially connected with concrete for road slabs. Briefly, procedures for the design of the mix are as follows:

18.8. GRADING CURVES METHOD

(*a*) From the specified minimum strength, the average strength required is estimated, according to the degree of control to be exercised. The amount of this variation depends on several factors, but principally on the accuracy of batching and control operations and on the uniformity of the raw materials used. The usual values of minimum strength as percentages of average strength are : 75 per cent of good control with weigh batching and use of graded aggregate, constant supervision, etc; 60 per cent with fair control with weigh batching, use of two sizes of aggregates only, water content left to the mixer driver's judgment and occasional supervision; and 40 per cent in case of poor control with inaccurate volume batching and practically no supervision.

(*b*) The water/cement ratio needed to give the estimated average strength is found from a series of curves relating strength and water-cement ratio at various stages. The figures given in these curves are average values for fully compacted concrete and cured at normal temperature.

Further, they refer to the ratios of the effective water content of the mix to the weight of the cement. The effective water content includes free water in the aggregate and water added while mixing, but not the absorbed water in the aggregate. When the amount of water required in the mix has been estimated, the amount of water to be added while mixing is found by subtracting the free water in the aggregate. If the aggregate contains no free water, the particles may not be saturated and may therefore absorb water from that added at the mixer. In this case the amount required to saturate the particles should be added to the estimated requirement. If dry aggregate is used the absorption moisture content should also be determined and amount of mixing water increased accordingly.

(*c*) The workability required is chosen according to the conditions of the particular job. The selection of degree of workability will also depend upon the conditions and methods of placing of concrete.

Table 18.6 provides a guide to the degree of workability required for various classes of work.

Table 18.6 Degree of workability for various requirements

Degree of workability	*Slumps in mm/cm*	*Use for which concrete is suitable*
Very low	0 to 10/0 to 1	Vibrated concrete in roads or other large sections
Low	10 to 20/1 to 2	Roads vibrated by hand operated machines or manually compacted. Mass concrete foundation without vibration. Simple reinforced sections with vibration.
Medium	20 to 40/2 to 4	For normal reinforced work without vibration and heavily reinforced sections with vibration
High	40 to 70/4 to 7	For sections with congested reinforcement. Not normally suitables for vibration.

(*d*) The cement/aggregate ratio that will give the required workability and the aggregate grading that will give this workability with lowest cement content, are found with the help of prepared tables and curves and test specimens are checked in the laboratory. These tables and curves are prepared from laboratory test-data.

Leaner mixes may be used with large size aggregates. To obtain economical mix, therefore, the maximum size should be as large as conveniently possible, but it should not normally be greater than one-quarter of the smallest dimension in the structure. Larger aggregates say 63 to 40 mm maximum size can be used in slabs which are of normal thickness and which contain relatively small amount of reinforcement. The maximum size of aggregate may be considerably less when the slab is to be heavily reinforced.

(*e*) By a graphical method or by other suitable means the proportions are estimated in which the available materials should be used to provide a grading approximating to that chosen for getting the most economical mix.

As the job progresses, daily tests are made, both of the accuracy of the mix and the strength of the concrete. A new source of any material component, or failure of the concrete to meet strength requirements requires a revision of the mix formula.

Where adequate facilities for test control are available, the procedure of providing quality control of concrete on specified strength outlined above is desirable.

18.9. FINENESS MODULUS METHOD

(*a*) From the specified minimum strength the average strength required is estimated as already explained in step (*a*) of grading curves method.

The average of the mean strength can also be determined from the equation:

$$f = m - rj$$

where f = Specified minimum strength for a specified confidence level.

m = Average or mean strength for which the concrete mix is to be designed initially.

r = Factor for the desired or specified level of confidence.

j = Standard deviation (known or anticipated) $= \dfrac{mi}{100}$

i = Co-efficient of variation $= \dfrac{j}{m}.100$

The values of co-efficient of variation are taken as 10 per cent for good control, 15 per cent for fair control and 30 per cent for poor control. The American Concrete Institute classify quality control into four standards, *viz*., excellent, good, fair, and poor, corresponding to the values for co-efficient of variation of less than 10, 10—15, 15—20 and more than 20 respectively.

Values of '*r*' for different levels of confidence are taken from Table 18.7.

Table 18.7 Values of *r* for different levels of confidence

Extent of strength values that can be tolerated below the specified minimum strength	*r*
1 in 3.2 (31%)	0.5
1 in 6.25 (16%)	1.0
1 in 15.40 (6.5%)	1.5
1 in 40.00 (2.5%)	2.0
1 in 100.00 (1.0%)	2.33
1 in 666.00 (0.15%)	3.00

Keeping in view the working conditions prevailing in this country, Ghosh, and Dhir in I.RC. Journal, Aug 1966 have suggested a probability corresponding to r equal to about 1.50 for highway and airfield pavements.

Thus, for minimum specified strength of 280 kg/sq. cm, for factor r equal to 1.5 and for good control *i.e.*, co-efficient of variation equal to 10 per cent, the mean strength for which the concrete mix is to be designed works out as follows:

$$280 = m - rj = m - \frac{r \cdot m \cdot i}{100} = m\left(1 - \frac{ri}{100}\right)$$

$$= m\left(1 - \frac{1.5 \times 10}{100}\right) = m \cdot \frac{85}{100}$$

or $$m = 280 \times \frac{100}{85} = 331 \text{ kg/sq.cm}$$

The values of m for fair control and poor control (i, equal to 15 and 30 per cent respectively), for the same level of confidence work out to be 363 kg/sq cm and 511 kg/sq cm respectively.

It can be readily seen from the above results that as co-efficient of variation is reduced through quality control, the strength for which concrete mix is to be designed is reduced. Thus, in other words it would mean that for the same minimum strength, concrete mix has to be designed for higher strengths with the lowering of quality control standards.

(*b*) The water/cement needed to give the estimated average strength is found from curves as for step (*b*) in the previous method. Table 18.8 also gives the relation between the average cube crushing strength in metric units and the water-cement ratio by weight.

Table 18.8 Relation between cube crushing strength and water cement ratio by weight for fully compacted concrete (ordinary Portland cement)

Water-cement ratio by weight	*Cube crushing strength in kg/sq cm*	
	7 days	*28 days*
0.35	400	530
0.40	350	470
0.45	300	420
0.50	250	370
0.55	220	320
0.60	180	280
0.65	150	250
0.70	130	220
0.75	110	200
0.80	105	180

(*c*) The workability required is chosen knowing the requirements of the work from Table 18.6 as for step (*c*) of previous method.

(*d*) Then using Table 18.9, the quantity of water by weight per cubic metre of concrete may be determined.

Table 18.9 Water content per cubic metre of concrete for 75 mm slump

Max. size of coarse aggregate (mm)	*15*	*20*	*25*	*40*	*50*	*75*
Water in kg.						
(*a*) For rounded coarse aggregate	200	184	178	166	158	148
(*b*) For angular coarse aggregate	212	200	193	180	172	164

For each 25 mm increase or decrease in slump, increase or decrease the water content by 3 %.

(*e*) This quantity of water determined in step (*d*) above divided by the water-cement ratio determined in step (*b*) will give the required quantity of cement.

(*f*) The quantity of water and cement per cu. metre of concrete being found, their absolute volumes are calculated by dividing their weights by their absolute specific gravities and unit weight of water. The absolute specific gravity of cement for this purpose may be taken as 3.15.

(*g*) Then the absolute volume of mixed aggregate is one cubic metre minus the absolute volumes of water and cement.

(*h*) The proportions of coarse and fine aggregates to produce optimum workability are obtained from fineness modulus method. From the fineness modulii of the coarse and fine aggregates (F_c and F_f respectively) found in laboratory and from the fineness modulus of mixed aggregates (F_m) obtained from Table 18.10, the percentage of fine aggregate in the mixture of coarse and fine aggregates is obtained, using the formula:

$$\text{Percent fine aggregate in the mixture} = \frac{F_c - F_m}{F_c - F_f} \times 100$$

Table 18.10 Fineness moduli of mixed aggregate for different sizes of aggregates

Max. size of coarse aggregate (mm)	*15*	*20*	*25*	*40*	*50*	*70*
Fineness moduli of mixed aggregate						
Min.	4.5	4.8	5.0	5.4	5.7	5.9
Max.	5.0	5.3	5.5	6.0	6.3	6.5

(*i*) Once the proportions of fine and coarse aggregate are fixed as above, their absolute volumes can be obtained from these proportions. Their weights are determined by multiplying their absolute volumes by their absolute specific gravities and unit weight of water. For this purpose the absolute specific gravities of fine and coarse aggregates may be taken as 2.65 and 2.55 respectively.

(*j*) Then the nominal mix by weight is obtained by dividing the weights of the various components by the weight of cement.

Problem 18.1 *Design concrete mix by fineness modulus method for good, fair and poor quality controls, given the following data:*

Specified minimum strength, $f = 280$ *kg/sq. cm.*

Fineness modulus of coarse aggregate, $F_c = 7.50$

Fineness modulus of fine aggregate $F_f = 2.78$

Specific gravity of coarse aggregate $= 2.60$

Specific gravity of fine aggregate $= 2.62$

It is desired to follow good control with weigh batching for which co-efficient of variation, i is taken as 10 per cent.

Solution. Level of confidence or tolerance level is assumed as 1 in 15 for which the value of '*r*' from Table 18.5 is 1.5.

The mean strength for which the concrete mix is to be designed works out as follows:

$$280 = m - rj = \left(m - r\frac{mi}{100}\right)$$

$$= m\left(1 - \frac{ri}{100}\right)$$

Substituting the values of *r* and *i*

$$280 = m\left(1 - \frac{1.5 \times 10}{100}\right) = m \cdot \frac{85}{100}$$

$$m = 280 \times \frac{100}{85}$$

$$= 331 \text{ kg/sq.cm.}$$

Water/cement ratio needed to give the estimated average strength of 331 kg/sq. cm from Fig. 18.1 or Table 18.8 is 0.54.

Degree of workability for the job condition is kept low and slump assumed is 25 to 50 mm.

Water content required per cubic metre of concrete from Table 18.9 for maximum size of coarse aggregate of 20 mm and for 75 mm slump is 184 kg. Since slump actually to be employed at the job is 25 to 50 mm, water content is decreased by 3 percent for each 25 mm decrease in slump.

Thus, for average slump between 25 to 50 mm, decrease water content by $4\frac{1}{2}$ per cent.

∴ Water content for 20 mm size aggregate for 25 to 50 mm slump

$$= 184 - 184 \times \frac{4\frac{1}{2}}{100} = 175.72 \text{ kg}$$

∴ Weight of cement/cubic metre of concrete $= \dfrac{175.72}{0.54} = 325.4$ kg

$$\text{Volume of cement} = \frac{\text{weight of cement/cubic metre of concrete}}{\text{Specified gravity of cement} \times \text{density of water}}$$

$$= \frac{325.4}{3.15 \times 1000} = 0.103 \text{ cu.m./cu.m.of concrete}$$

$$\text{Volume of water} = \frac{175.72}{1 \times 1000} = 0.176 \text{ cu.m./cu.m of concrete}$$

Therefore, volume of mixed aggregate per cu. m. of concrete

= 1 cu.m. concrete – absolute volumes of water and cement

= 1 – 0.103 – 0.176 = 0.721 cu.m,

$$\text{Ratio of fine to coarse aggregate} = \frac{F_c - F_m}{F_c - F_f} \times 100$$

For calculating the above, the average value of F_m is taken from Table 18.10. For maximum size of coarse aggregate of 20 mm the average value from this table is $\frac{4.8+5.3}{2}=5.05$.

$$\therefore \quad \text{Ratio of fine to coarse aggregate} = \frac{7.50-5.05}{7.50-2.78}\times 100$$

$$= 52 \text{ per cent i.e., } 52:100 \text{ by weight}.$$

The next step is to convert the proportions of fine to coarse aggregates by weight to volume. Ratio, volume of fine aggregate: volume of coarse aggregate.

$$= \frac{52}{2.62\times 1000} : \frac{100}{2.60\times 1000} = 1:2$$

(2.62 and 2.60 are specific gravities of fine and coarse aggregates respectively).

Therefore, volume of fine aggregate per cu.m. of concrete $=\frac{1}{3}\times 0.721 = 0.24$ cu.m. and volume of coarse aggregate per cu.m. of concrete $= 0.721 - 0.24 = 0.481$ cu. m.

Therefore, mix, by volume $= 0.103 : 0.240 : 0.481$ and, mix by weight

$$= 0.103\times 3.15 : 0.240\times 2.62 : 0.481\times 2.60.$$

Then the nominal mix by weight is obtained by dividing the weights of the various components by the weight of cement which works to 1 : 1.94 : 3.86.

The *w/c* ratio for the above mix is 0.54 as already determined.

Similarly, nominal mixes for fair *i.e.* $f = 363$ kg/sq. cm. and for poor control *i.e.* $f = 511$ kg/sq. cm. work out to be 1 : 1.82 : 3.61 and 1 : 1.23 : 2.45 respectively. The *w/c* ratios are 0.51 and 0.37 respectively.

It can be readily seen from the above results that as co-efficient of variation is reduced through quality control, the mix for the same minimum designed strength of concrete becomes cheaper.

18.10. IRC METHOD OF MIX DESIGN

Guidelines for cement concrete mix design for pavements have been outlined in IRC-44-2008 and are reproduced below:

(1) For design of mix, the following tests are to be carried out

(*a*) Cement (*i*) Compressive strength at 3, 7 and 28 days (IS : 4031)

(*ii*) Specific gravity of cement (IS : 4031)

(*b*) Aggregates (*i*) Specific gravity (IS : 2386)

(*ii*) Per cent water absorption (IS : 2386)

(*iii*) Sieve analysis (IS : 2386)

(*iv*) Fineness modulus of fine aggregate

(*c*) Water : IS : 456

(2) Selection of aggregate grading:

The grading of coarse and fine aggregates is selected from Table 18.1 and 18.2 respectively

(3) Consider standard normal tolerance variate (from table 18.5) t = 1.65 for 1 in 20 tolerance

(4) Take assumed Standard Deviation from table 18.5

(5) Determine design strength from the relation:

$$f'_{ck} = f_{ck} + vt$$

where f'_{ck} = target mean compressive strength at 28 days

f_{ck} = characteristic compressive strength at 28 days

v = standard deviation

t = standard normal variate.

(6) Select water cement ratio from Table 18.11 for the grade of concrete.

Table 18.11 Preliminary selection of water/cement ratio for the given grade

Sr. No.	*Grade of concrete*	*Approx water/cement ratio*
1	M 25	0.50
2	M 30	0.45
3	M 35	0.42
4	M 40	0.38
5	M 50	0.34
6	M 60	0.28

(7) Select water content per cubic metre of concrete from Table 18.12.

Table 18.12 Approximate water content per cubic metre of concrete for nominal maximum size of aggregate (without admixture)

Nominal Maximum size of Aggregate (mm)	*Suggestive water content (kg)*
10	208
20	186
40	165

Note: (*i*) Reduce water content by 10% if plasticizing admixture is added

(*ii*) Reduce water content by 20% if super plasticiser is added

(8) Determine cement content by dividing water content by water-cement ratio

(9) Find out coarse and fine aggregate contents from Table 18.13.

Table 18.13 Volume of coarse aggregate per unit volume of total aggregate for different zones of fine aggregate as per IS : 383

Nominal Maximum Size of Aggregate (mm)	*Volume of Coarse Aggregate Per Unit Volume of Total Aggregate for Different Zones of Fine Aggregate*			
	Zone IV	Zone III	Zone II	Zone I
10	0.50	0.48	0.46	0.44
20	0.66	0.64	0.62	0.60
40	0.75	0.73	0.71	0.69

Note: (*i*) Volumes are based on aggregates in saturated surface dry condition

(*ii*) These volumes are for angular aggregate and suitable adjustments may be made for other shapes of aggregate.

(*iii*) The above Table is valid for W/C ratio of 0.50. For every decrease of 0.05 in W/C ratio, the above ratio (volume of coarse aggregates) will be increased approximately by 1 per cent. The above table is applicable for crushed aggregates. For rounded/semi crushed aggregates, the above values of the volumes may be slightly increased.

(10) To determine the contents of coarse and fine aggregates, adopt the following procedure:

(*i*) Determine the absolute volume of cement, water and chemical admixtures by dividing their mass by the specific gravity.

(*ii*) Subtract from 1, the total volumes obtained above to get the volume of aggregates

(*iii*) Find the mass of coarse aggregate by multiplying the result in (*ii*) by its proportion from table 18.13 and multiplying by 1000.

(*iv*) Now determine the mass of fine aggregate as in (*iii*).

Problem 18.2 *Design a concrete mix for a highway pavement as per IRC method, given :*

(*a*) *Grade of Mix: M50*

(*b*) *Type of cement: OPC 43 grade*

(*c*) *Maximum nominal: 40 mm*
size of aggregate

(d) *Minimum cement content = 325 kg/m³*

(*e*) *Maximum cement content = 425 kg/m³*

(*f*) *Maximum water-cement ratio = 0.5*

(*g*) *Workability: 20 ± 5 (slump)*

(*h*) *Degree of supervision: Good*

(*i*) *Type of aggregate: Crushed angular graded aggregate*

(*j*) *Chemical admixture: Super plasticizer*

(*k*) *Specific gravity*

(*i*) *Cement: 3.15*

(*ii*) *Coarse aggregate: 2.75*

(*iii*) *Fine aggregate: 2.60*

(*l*) *Water absorption*

(*i*) *Coarse aggregate: 0.5 per cent*

(*ii*) *Fine aggregate: 1.0 per cent*

(*m*) *Free moisture (surface)*

(*i*) *Coarse aggregate: Nil*

(*ii*) *Fine aggregate: Nil*

(*n*) *Sieve analysis*

(*i*) *Coarse aggregate*

IS Sieve sizes mm	*Analysis of coarse Aggregate fraction % passing*	
	I *40 to 20 mm*	*II* *20 mm down*
40	100	100
20	30	82
10	2.6	46
4.75	Nil	3.5

(*ii*) *Fine aggregate*
Conforming to grading zone III of Table 18.2

(*o*) *Compressive strength of concrete = 45 N/mm²*

Solution.

(1) *Design strength*

Design compressive strength for mix proportioning is given by

$$f'_{ck} = f_{ck} + vt$$

From Table 18.5, for good control $v = 5$ and $t = 1.65$

$$\therefore \quad f'_{ck} = 45 + 5 \times 1.65$$
$$= 45 + 8.25$$
$$= 53.25 \text{ N/mm}^2$$

Design flexural strength at 28 days is given by

$$f_f = 0.7\sqrt{f'_{ck}} = 0.7 \times \sqrt{53.25} = 0.7 \times 7.297$$
$$= 5.1 \text{ N/mm}^2$$

$\therefore$ Flexural strength at 90 days $= 5.1 \times 1.2$

$$= 6.12 \text{ N/mm}^2$$

(2) *Mix the coarse aggregates in the ratio of 60% and 40%*

$\therefore$ Coarse aggregates

IS Sieve sizes mm	*Analysis of coarse aggregate fraction % passing*		*Percentage passing of different fractions*			*Percentage passing for graded aggregate as per Table 18.1*
	I	II	I (60%)	II (40%)	Total	100 % Combined
40	100	100	60	40	100	95-100
20	30	82	18	32.8	50.8	30-70
10	2.6	46	1.5	18.4	19.9	10-35
4.75	Nil	3.5	Nil	1.4	1.4	0-5

Thus the grading of coarse aggregates conforms to the requirements prescribed.

(3) *Water-cement ratio*

From Table 18.11, *W/C* ratio for M50 concrete

= 0.34 which is less than the maximum *W/C* ratio of 0.5.

Hence O.K.

(4) *Water content*

From Table 18.12 water content for 40 mm aggregate

= 165 kg/m^3 at *W/C* ratio = 0.5

Since super plasticizer is being used, water content can be reduced upto a maximum of 30 per cent. Adopting 15 per cent reduction in water content at W/C ratio of 0.34,

Reduced water content = 165 × 0.85

= 140.25 kg/m^3

Say 140 kg/m^3

(5) *Cement content*

W/C ratio = 0.34

Water content = 140 kg/m^3

$\therefore$ Cement content $= \dfrac{140}{0.34} = 411.76$; say 412 kg/m^3

Check for minimum and maximum cement content as per IRC : 15

Minimum cement content as per IRC : 15 = 325 kg/m^3 < 412 kg/m^3

Hence O.K.

Maximum cement content as per IRC : 15 = 425 kg/m^3 > 412 kg/m^3

Hence O.K.

(6) *Proportions of volume of coarse aggregate and fine aggregate*

From Table 18.13 Volume of coarse aggregate corresponding to 40 mm size aggregate and fine aggregate grading Zone III = 0.73 per unit volume of total aggregate.

This is valid for W/C ratio of 0.5. As water-cement ratio is actually 0.34, the ratio is taken as 0.75 to reduce sand content.

$\therefore$ Volume of fine aggregate content $= 1 - 0.75 = 0.25$ /unit volume of total aggregate

(7) *Mix calculations*

(*i*) Volume of concrete $= 1 \text{ m}^3$

(*ii*) Volume of cement $= \dfrac{\text{Mass of cement}}{\text{Specific gravity}} \times \dfrac{1}{1000} = \dfrac{412}{3.15} \times \dfrac{1}{1000}$

$= 0.13 \text{ m}^3$

(*iii*) Volume of water $= \dfrac{140}{1} \times \dfrac{1}{1000}$

$= 0.14$

(*iv*) Volume of admixture (super plasticizer) @ 0.6% by mass of cementitious material $= \dfrac{0.6 \times 412}{100 \times 1.2} \times \dfrac{1}{1000} = 0.002 \text{ m}^3$

(*v*) Volume of total aggregate $= 1 - (0.13 + 0.14 + 0.002)$

$= 0.728 \text{ m}^3$

(*vi*) Mass of coarse aggregate $= 0.728 \times 0.75 \times 2.75 \times 1000$

$= 1501.5$ say 1502 kg/m^3

(*vii*) Mass of fine aggregate $= 0.728 \times 0.25 \times 2.6 \times 1000$

$= 473.2$ say 474 kg/m^3

(8) *Mix proportions based on aggregate with absorbed moisture conditions*

Cement $= 412 \text{ kg/m}^3$

Water $= 140 \text{ kg/m}^3$

Fine aggregate $= 474 \text{ kg/m}^3$

Coarse aggregate $= 1502 \text{ kg/m}^3$

Water-cement ratio = 0.34

(9) *Mix proportions based on aggregates in dry condition*

$$\text{Cement} = 412 \text{ kg/m}^3$$

$$\text{Fine aggregate} = 474 - 1\% \text{ of } 474$$
$$= 474 - 4.74$$
$$= 469.26 \text{ kg/m}^3$$

$$\text{Coarse aggregate} = (1502 - 05\% \text{ of } 1502)$$
$$= 1502 - 7.51$$
$$= 1494.49 \text{ kg/m}^3$$

$$\text{Water} = 140 + 4.74 + 7.51$$
$$= 152.25 \text{ kg/m}^3$$

18.11. CONSTANT CEMENT FACTOR METHOD

Alternatively, proportions are based on a constant cement factor, regardless of the source or characteristics of the aggregate so long as they meet the gradation and other tests required for the aggregate.

This method is much simpler than the above two methods and very large number of highway departments still specify a definite cement factor and slump for concrete for road slabs. Fixed cement factor method is apt to be wasteful, as generally the strengths obtained are considerably above the usual permitted minimals. On the other hand, those who favour it argue that it provides a highly desirable factor of safety at small added cost and further it makes proportioning easier and simplifies payment to contractors.

CONCRETE SLAB DESIGN CONSIDERATIONS

18.12. THICKNESS OF CONCRETE PAVEMENT SLABS

Design of cross-section has undergone several marked changes since concrete roads were first built. Originally, pavements were thicker along the centre line than at the edges, similar to old macadam surfaces. For simplicity, some were later built of uniform thickness. The advent of heavy truck with solid rubber tyres, crowded to outer edge of insufficient pavement width caused a great many corner breaks along outer edges of uniform thickness pavements. In an effort to stop this failure, thickened-edge cross-sections were developed. Over the years, cross-sections of concrete pavement were given a variety of forms involving thickening of the edges of slabs in various ways and degrees. The centre thickness was kept approximately 2/3rd of edge thickness in which case stresses due to load in the centre of the slab are approximately equal to those at the outer edge.

Latest research indicates that when combined effects of load and temperature are considered, the pavement slab of uniform thickness is at least as effective as the thickened edge design. This is particularly true with the present wider pavements when wheels of vehicles no longer travel along the extreme edges of pavements.

For this reason, there has been a decided trend towards a uniform slab thickness and this is now being practically followed every where except in a few departments. The usual thicknesses used for highway pavements are 20 cm, 22.5 cm, or 25 cm. Minimum thickness suggested is 11.25 to 12.5 cm. Thickness of slab is determined from design methods discussed in Chapter 21. However, the use of such design formulae is restricted by the impracticability of varying the thickness of pavement to take into account variations in subgrade conditions, due to forms having been standardized in cm widths and also due to difficulty in control of different thicknesses in the same contract. Besides, the increment in the overall cost of a highway project that results from an added fraction of a cm of pavement thickness is small. Hence, a large measure of standardization has been incorporated into highway practice as far as thickness of concrete pavements is concerned.

18.13. OTHER ELEMENTS OF SLAB DESIGN

Besides thickness, there are four elements of slab design that are closely inter-related. These are :

1. The use and spacing of expansion joints.
2. The use and spacing of contraction joints.
3. The use of dowels or other load transfer devices, at joints.
4. The use of steel reinforcement, or distributed steel (Tie rod).

A typical pavement layout generally adopted in a two-lane road is shown in Fig. 18.1

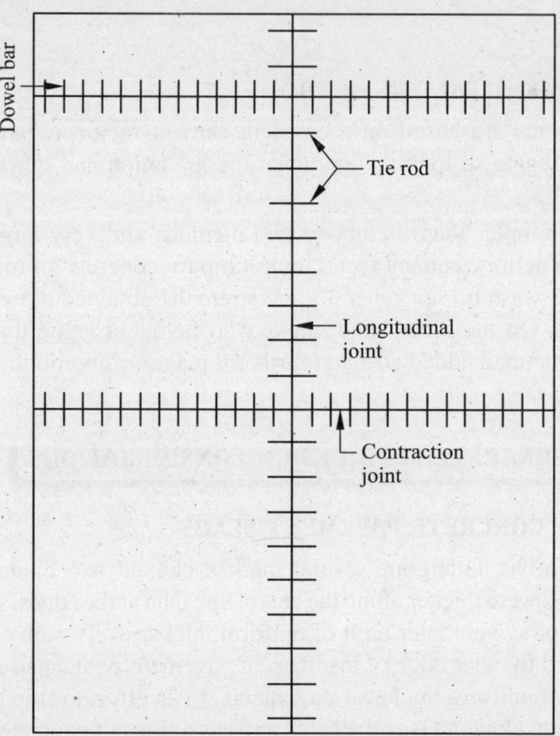

Fig. 18.1 Pavement configuration of two lane road.

8.14. EXPANSION JOINTS

Placed transversely, expansion joints permit increase in the length of a slab as its temperature rises and provide relief to compressive stresses caused by expansion due to certain combinations of temperature and moisture. In order to prevent blow-ups, relief is frequently provided by the installation of transverse expansion joints. However, the blow-ups cannot be attributed solely to expansion of the concrete but it may also be due to infiltration of dirt and other incompressible foreign matter into joints and cracks.

Based on a study by the Central Road Research Institute, New Delhi, from a consideration of daily and annual temperature variations in the pavements in different parts of the country, degree of foundation roughness as well as the season of construction, the maximum recommended spacing of expansion joints, as per IRC : 15-1981 are given in Table 18.14.

Table 18.14 Recommended spacing of joints in rigid pavements for highways: expansion joint spacings (based on CRRI study) (for 20-25 mm wide expansion joint)

Period of construction	*Degree of foundation roughness*	*Maximum expansion joint spacing (m) slab thickness (cm)*		
		15	*20*	*25*
Winter	smooth	50	50	60
(Oct-March)	rough	140	140	140
Summer	Smooth	90	90	120
(April-Sept)	Rough	140	140	140

Design considerations have been discussed in chapter 21 pertaining to rigid pavement design.

The present trend is in the direction of eliminating expansion joints except at comparatively great spacing, with properly designed contraction joints being provided at relatively short intervals. Expansion joints are not as necessary as were once considered to be and they are expensive to install and difficult to maintain.

They are usually from 20 mm to 25 mm in width and extend the full depth of the slab. Details of construction vary somewhat depending on whether the joint is required for alternative-bay construction or for continuous construction and also whether the concrete is tamped by hand-operated compactors or by mechanically propelled compactors. They are made by embedding in the concrete a vertical filler strip of non-extruding compressible material, usually 20 mm or 25 mm thick. The filler is held about 3.75 cm below the top surface of slab, leaving a slot which is later sealed with a plastic material. These are known as *preformed joints*. Alternatively, a removable bulkhead may be used in place of the filler strip. When the pavement has hardened sufficiently, the bulkhead is withdrawn and the joint is poured, usually with paving bitumen. They are referred to as *poured joints*. Poured joints are about the least expensive type but possess certain disadvantages: the removal of the form or bulkhead, unless performed very carefully, may break the upper portion of adjacent concrete, and then it is difficult to retain the bituminous material in the joints.

The filler strip or the bulkhead must be held in exact position as the concrete is placed which can be accomplished in conjunction with load transfer device.

18.14.1. Joint filler and sealing materials

The main functions of joints in concrete roads are to allow expansion, contraction and warping of the concrete to occur without restraint. All joints form breaks in the continuity of concrete, and water and grit are liable to enter. Entry of water into a joint may cause deterioration of subgrade and may lead to failure of the concrete, and if grit and stones become packed into the joint they may prevent free expansion of the concrete and thus cause spalling and in extreme cases blow-ups as already explained. It is a usual-practice, therefore, to fill the top 2 to 3 cm with the sealing compound in order to prevent, or minimize, the entry of water and grit into the joint. In case of preformed joint explained above, the filler strip in addition to forming a gap, also provides a support for the sealing compound. There are thus two types of material used in joints : a sealing compound which fills the top 2 to 3 cm of the joint to resist the entry of grit and water and a joint filler which separates adjacent slabs so that they can expand. IRC : 57-2006 gives recommended practice for sealing of joints in concrete pavements.

Joint filler materials: In order to perform its function, a joint filler must be (*i*) compressible without extruding; (*ii*) elastic, so that after compression it recovers most of its original thickness when the joint opens; (*iii*) durable and rot proof; and (*iv*) sufficiently rigid to facilitate its support during construction. Materials known to have given satisfactory performance are listed below:

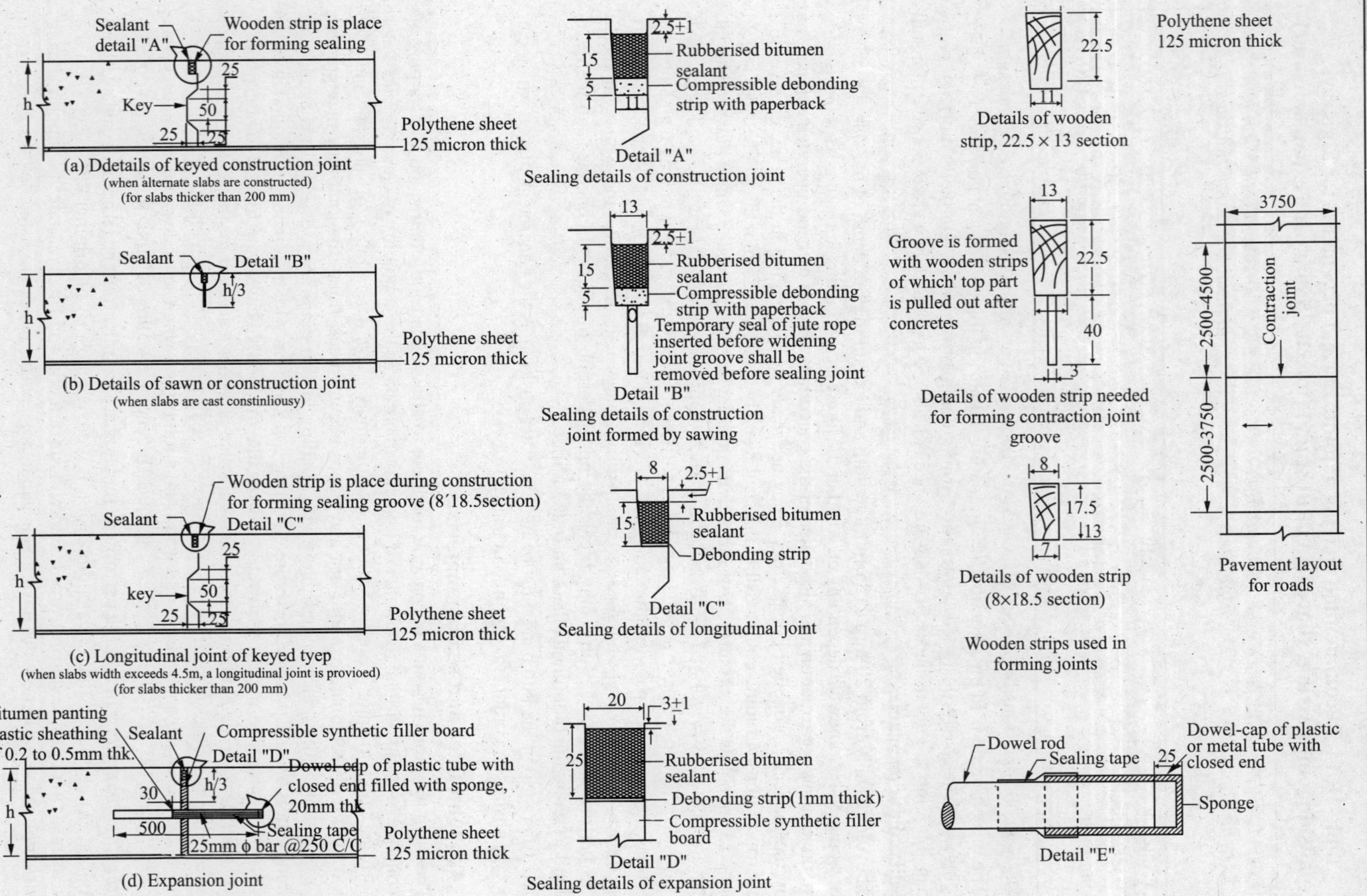

Fig. 18.2 Details of joints in cement concrete pavement.

Soft wood free from knots. Carefully selected boards of soft wood, e.g. long-leaf pine, spruce, cypress, redwood or red cedar have been found to be the most satisfactory joint fillers. They are usually thoroughly soaked in water before using as otherwise swelling of the wood produced by water from the fresh concrete may disrupt the newly finished surface. Because the wood has considerable strength, less elaborate provisions for retaining and supporting it during pouring will suffice. An advantage claimed for wood filler is that they offer considerable resistance to compression which results in smaller openings at joints and cracks. It also shows reasonably good recovery particularly when it is wet. Further, softwood does not extrude. On the other hand, there have been some instances in which certain joint fillers of wood disintegrated completely and permitted the joint to fill with dirt.

Cork. Several varieties of cork joint fillers are available e.g. cork-rubber, strips of natural cork cemented together, cork bound with resin. These are also non-extrusive and recovery after compression is good.

Impregnated fibre board. Cane fibre-board is normally impregnated with creosote to rot proof it. It is readily compressible with negligible extrusion and recovery after compression is good.

Bituminous-type preformed filler. This widely used filler, consists of felt or similar material combined with bitumen or tar and a mineral filler like one of those used in bituminous concrete. Layers of felt at each surface provide strength at high temperatures when the bituminous mixture is soft. The mineral filler reduces brittleness at low temperatures. This type of joint filler has no voids and with the expansion of concrete, the filler is extruded upward into the roadway surface, causing a bump. It shows no recovery after extrusion, and the joints remain partly open which may fill up with foreign matter.

Glass fibre filler. This has been developed recently and consists of glass fibre and bitumen. It is manufactured in strips of various thicknesses and is faced on both sides with a paper. It has a low resistance to compression and is practically free from extrusion, as its bitumen content is limited. Being composed mainly of non-perishable materials; it is a very durable filler.

Other materials used are cellular rubber and cork granules bound with bitumen. Cellular rubber is compressible and elastic but it is pliant (yielding) and requires a rigid support when the concrete is being placed and it is also liable to work up out of the joint due to its extreme pliability. Cork granules ground with bitumen also extrude from the joint when the slabs expand and is not a satisfactory joint filler. CRRI (India) has made trials with coconut pith. Successful joint filler material conforming to ASTMD 549 has been obtained by hot pressing the mixture of coconut pith, rubber latex with certain vulcanizing materials and accelerators.

The tests specified by Indian Standard IS : 1838-1961 or ASTM D 549 are adopted to determine the suitability of joint fillers.

Sealing materials. In order to seal the joint effectively for a long period without maintenance, a joint sealing compound should have : (*i*) good adhesion to concrete; (*ii*) extensibility without fracture; (*iii*) resistance to the ingress of grit; (*iv*) resistance to flow either into or along the joint in hot weather; (*v*) durability; and (*vi*) easy to apply and to resist damage by traffic a short time after laying. Requirements (*i*) and (*ii*) require that the material be fairly soft while (*iii*) and (*iv*) require that it must be a hard material. It is therefore, necessary to strike a balance between these incompatible requirements. For most purposes an intermediate degree of hardness is suitable. On a road over clay it is particularly important to prevent water from passing through the joints which warrants use of a fairly soft compound. On the other hand, roads near gravel pits often have considerable quantities of grit dropped on them, so that grit resistance becomes of prime importance and the use of harder compound is indicated. Types of sealing compounds used are : mainly of two categories; namely:

(1) Hot poured sealants (120–180°C): and

(2) Cold poured sealants.

Table 18.15 Details of various sealants in use

Sealant Type	*Specification*	*Properties*
Hot Poured Joint Sealants		
Rubberized Bitumen Sealant	IS 1834	Self-leveling
Polymeric Asphalt Based	AASHTO M0173	Self-leveling
	ASTM D34005	Self-leveling
	US Federal Highways Administration Specification	Self-leveling
	SS-S-1401 C	Self-leveling
	ASTM D1190	Self-leveling
Polymeric Sealant	ASTM D 3405	Self-leveling
Low Modulus	Modified	Self-leveling
Elastomeric Sealant	US Federal Highways Administration Specification SS-S-1614, ASTM D3406	
Coal Tar, PVC	ASTM D 3406	Self-leveling
Cold Poured Sealants/Single Components		
Silicone Sealant	ASTM 5893-96	Non sag, Toolable, low modulus
Silicone Sealant	ASTM 5893-96	Self-leveling (no tooling), low modulus
Polysulphide Sealant	BS 5212-1990 IS 11433-1986 (Reaffirmed in 1995)	Self-leveling (no tooling), low modulus
Polyurethene Sealant	BS 5212	

It is necessary to apply a primer to the sides of the concrete surface between the concrete surface and sealing compound. Road Research Laboratory, London has recommended a primer with the following approximate composition by weight. 200 penetration bitumen 66%; light creosote oil 14%; solvent naptha 20%. The bitumen and creosote should be blended hot and naphha added, when the mixture is cold.

A joint sealant may fail mechanically in adhesion , cohesion or extrusion. The adhesive failure is a loss of bond between the sealant and the joint wall caused by a tensile load. The cohesive failure is a tearing of the sealant material, also under tension. The extrusion failure occurs under the combined action of compression and traffic. The material under compression is extruded above the roadway surface and then folded and flattened under the action of traffic.

Indian Standard IS : 1834–1961 or ASTM D 1191 is used to determine the suitability of sealing compounds.

18.15. CONTRACTION JOINTS

Contraction joints are usually "*planes of weakness*" provided at intervals to localize and control transverse cracking of the slab during curing or with a fall in temperature. They are most often of "*dummy*" joint type. A temporary strip of wood or metal is embedded in the fresh concrete to form a groove 6.3 mm to 20 mm wide and extending from the top about ¼ to 1/3 the depth of the slab. After the slab has been consolidated and finished the strip is removed and edges of slot are finished and rounded. Before the pavement is put to use, the joint is filled with a suitable sealing compound already described. Sealing compound may be replaced by material such as tarred rope, chopped fibre-board, etc. upto within 2.5 cm of top of joint. When the slab contracts, cracks will normally form at these weakened planes. This dummy type of joint is most suitable where concrete roads are being laid and compacted in a continuous process by mechanical methods.

For the alternate-bay construction, contraction joints may be formed simply by painting the end of the first laid slabs with a bituminous material after the forms have been removed. This will prevent bond with the adjacent slabs, but in order to seal the joint at the surface, a groove for sealing compound should be provided at the surface which can be formed by fixing a strip of wood or other suitable material of the required size to the top of the form before the first slabs are concreted. The strips are left in position until the concreting of the intermediate bays has been completed.

Another type of contraction joint for continuous construction has been formed by fixing a strip of preformed joint filler in position at the bottom of the slab and then concreting straight over without interruption. This method suffers from the serious disadvantage that an irregular and a slight crack may be formed at the surface. The crack is liable to spall and cannot be easily sealed as there is no groove. This type is not popular.

Several organization are now experimenting with sawed joints cut in finished hardened concrete with special saws. For economy such sawn slots are 6.3 mm wide and about the same thickness as that of the slab. This makes it difficult to apply and maintain a satisfactory seal.

Because concrete is very much weaker in tension than in compression, contraction joints normally have to be spaced at closer intervals than expansion joints. Spacing employed for unreinforced and reinforced slabs based on IRC : 15 – 1981 are given in Table 18.16. Design consideration have been discussed in chapter 21 pertaining to rigid pavement design.

Table 18.16 Contraction joint spacings

Slab thickness (cm)	*Maximum contraction joint spacing (m)*	*Weight of reinforcement in welded fabric (for reinforced pavements only) (kg/cm²)*
Unreinforced slabs		
10	4.5	—
15	4.5	–
20	4.5	–
Reinforced slabs		
10	7.5	2.2
15	13.0	2.7
20	14.0	3.8

Note: Where reinforcement is used in the form of mild steel bars, equivalent sectional areas corresponding to the sectional areas of the welded wire fabric should be employed.

18.16. LOAD TRANSFER DEVICES

Load transfer devices of one sort or another are usually used at expansion joints, and sometimes at contraction joints, to prevent the full effect of wheel load from coming onto the end of a slab as the load approaches or leaves the joint, and also to maintain vertical alignment of the adjacent slabs at the joint. These devices most often take the form of dowel-bars 20 mm to 25 mm in diameter, 30 to 35 cm long in each slab and spaced 30 to 37.5 cm centre to centre arranged at least in theory to slip back and forth, longitudinally, in one of the slabs as expansion and contraction occurs. A typical section through a dowel-bar expansion joint in shown in Fig. 18.3.

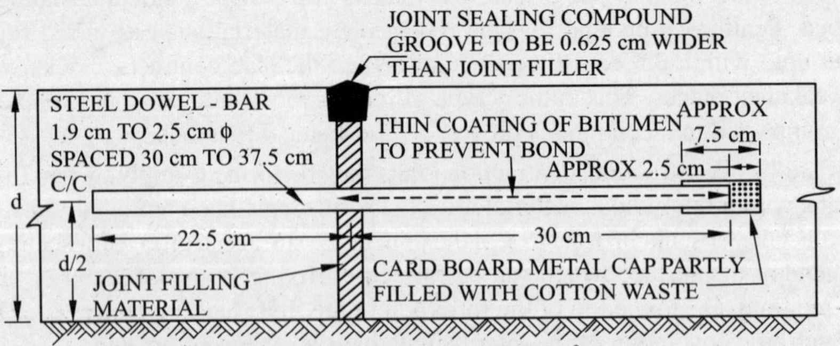

Fig. 18.3 Dowel bar expansion joint-Zimmerman/Megreor.

The bond may be broken by giving the dowel-bar a turn after the concrete has set, or by coating the bar with a thin film of bitumen. Former method can be used only where the slabs are concreted in alternate layers; the latter method is used for continuous construction and is therefore, more generally used. Other coating materials such as stainless steel sleeves, greased water-proof paper and cardboard sleeves have also been used.

When a weakened-plane contraction joint cracks through, the break in the concrete is irregular and rough. Inter-locking of the aggregate in the cracked portion may adequately transfer the load across the joint. The effectiveness of inter-lock will be reduced if the joint opens, and studies show that it becomes ineffective when the width of opening exceeds 1 mm. Where the joint is liable to open by more than this amount, dowel-bars should be provided. Satisfactory size and spacing of dowel-bars for contraction joints are given below.

Reinforced slabs, 15–17.5 cm or	15–22.5 cm, 20 mm dia and 37.5 cm un-reinforced slabs: centre to centre
Reinforced slabs, 20–22.5 cm:	20 mm dia; 30 cm centre to centre
Reinforced slabs, 25 cm and over:	25 mm dia; 30 cm centre to centre

Dowel-bars are greased or painted for one-half of their length to break the bond with the concrete and permit the dowel-bars to slip within one of the abutting slab-ends.

Design considerations for selecting dowel-bar's diameters, spacings and lengths are discussed in Chapter 21.

Accurate alignment of dowel-bars is most important. They should be placed parallel to each other and parallel to the surface and centre line of the road. Failure to do so may result in longitudinal movements being restricted and very high pressures being set up between the dowel-bars and the concrete, either by wedging action if the dowel-bars are not parallel to each other or by a lifting or sideways movement of the slabs if the dowel-bars are not parallel to the surface or centre line of the road. For error in alignment, ranging from 0.5 mm to 12.5 mm per 30 cm length, based on closure of joint from 20 mm to 12 mm width, shear force in a dowel-bar 25 mm diameter may range from 90 kg to 2300 kg respectively per dowel-bar. Similar range for 20 mm diameter dowel-bar is 45 kg to 1170 kg; for 12 mm diameter it is 18 to 500 kg. For 30 mm diameter, the permissible tolerances in dowel-bar alignment both vertical and horizontal directions are ± 1 mm in 100 mm for dowels of 20 mm and smaller diameters and ± 0.5 mm in 100 mm for dowels of diameter greater than 20 mm.

Many different types of dowel-bar assembly are in use. Most of them, in order to obtain accurate alignment, essentially consist of a frame work to hold the dowel-bars in correct positions while one half of the joint is concreted or by using a rigid assembly which can be cast into the concrete. Most of these devices are expensive in their installation or are lacking in durability.

Frequently, the dowel-bar appears to bind in the concrete socket and become immobilized. It has been found that a single dowel-bar which, by corrosion has become immobilized on the sliding end, may resist sliding force upto 150 kg per sq cm which works out to be $4\frac{1}{4}$ tonnes for a dowel-bar 25 mm diameter and 60 cm long. On the other hand, a new and smooth bright mild steel bar, covered with a thin uniform layer of bitumen will offer a resistance of less than 0.014 kg per sq. cm. To ensure free sliding, the dowel-bars should have a smooth surface and the ends must be cleanly sawn. The bond is broken, as already explained, either by giving the dowel-bar a turn after the concrete is set for alternate-bay construction or by coating the bar with a thin film of bitumen or other suitable material.

Space has to be provided at the end of half of the sliding dowel-bar to allow for closure of the joint. With alternative bay construction this may be done by pulling the bar out slightly after the first slab has been concreted. With continous construction either a compressible pad may be fixed to the dowel bar, or, as is more usual, a cap provided. Such caps may be of cardboard or pressed metal and should be a close sliding fit on the bar. The cap is set to allow sufficient space for closure of the joint and it is partly filled with a wad of cotton waste or a piece of joint filling material. Even punchings from the dowel-bar holes are suitable. The edge of the cap should be sealed during the painting of the bar, in order to prevent it from becoming filled with mortar.

For reasons explained above, load transfer devices are an unsatisfactory part of a concrete pavement, and this has stimulated efforts to eliminate their use.

The concrete around the dowel-bars must be thoroughly compacted because the effective operations of joint depends entirely upon the bars being surrounded by concrete of high strength.

AASHO tests have indicated that in slabs upto 12.5 cm thick, the location of dowel-bars several times formed the starting point of longitudinal cracks; this tendency being totally absent in the case of thicker slabs. This finding may indicate that no dowel-bars may be used at joints in slabs upto 12.5 cm thick. British practice also excludes dowel-bars in slabs less than 15 cm thick.

18.17. DISTRIBUTED STEEL REINFORCEMENT

Distributed steel reinforcement is composed of sheets of welded steel-wire fabric or of steel bars. These sheets are placed 5 to 7.5 cm below the top surface of the slab. The principal function of this reinforcement is to localize cracks in the slab, to prevent relative vertical movement of slabs on opposite side of cracks. If the slab is not reinforced, a crack is liable to open particularly under heavy traffic and this is quickly followed by abrasion of the faces of the crack and spalling. Cracking of unreinforced slabs can be reduced to a minimum by increasing the slab thickness so as to reduce loading stresses, or by providing warping or contraction joints at small intervals of 4.5 m so as to reduce temperature stress. Neither alternative is very desirable and control of cracking by reinforcement may be considered preferable. In reinforced slabs, hair cracks often occur at frequent intervals but such cracks show no tendency to open or deteriorate.

The usual amount of steel reinforcement considered satisfactory is 4 kg per sq metre for moderate traffic and up to 8 kg per sq metre under very heavy traffic. About 5.5 kg per sq metre with the greater proportion placed in the longitudinal direction is the minimum used in heavily trafficked roads. Design considerations for calculating distributed steel reinforcements are given in Chapter 21 dealing with design of rigid pavements.

The spacing of joints in reinforced concrete slab will depend upon the amount of reinforcement to be used and the thickness of slabs. Suitable spacings have already been discussed under expansion joints.

General recommendations. Recent practice reveals two distinct schools of thought with regard to joint spacing and the use of load transfer devices and reinforcement. At one extreme, expansion joints are located at comparatively frequent intervals, usually from 10 to 30 metres, besides being used at all structures, manholes, *etc*. Contraction joints are either not used at all, or are used sparingly between expansion joints. Load transfer devices are used both at contraction and expansion joints. Slabs are reinforced quite heavily between joints to prevent the formation of transverse cracks.

At the other extreme, expansion joints are omitted except at bridges and other structures; contraction joints are spaced at 4.5 or 6 metres intervals. Reinforcement is usually limited to areas near bridge approaches, *etc*. and load transfer devices are used only at expansion joints and closely adjacent contraction joints.

The Portland Cement Association U.S.A. lays down four conditions that should be met to justify the omission of expansion joints and steel reinforcement, which are:

1. Pavement be constructed of material with normal expansion characteristics.
2. Pavement be constructed during periods of year when normal construction temperatures prevail.
3. That the pavement be divided into relatively short panels by contraction joints so placed as to prevent the formation of intermediate cracks.
4. That the contraction joints be properly maintained to prevent the infiltration of relatively incompressible materials.

18.18. OTHER TRANSVERSE JOINTS

18.18.1. Longitudinal Joints

To prevent the formation of unsightly, irregular longitudinal cracks and to allow for transverse warping and for unequal settlement of the subgrade, longitudinal joints are constructed on the centre line of two lane pavements and between the lanes of wider pavements. On clay, the longitudinal joint is also required to allow for differential shrinkage or swelling of the soil due to larger and more rapid changes of moisture content under edges than under centre of the road. The longitudinal joint is thus required to act as a hinge to relieve flexural stresses and also to make it possible to construct the road in convenient widths. Where a road is constructed between rigid ends, such as foundation of buildings, the longitudinal joint will be required to allow for expansion and contraction of the concrete.

If for some reason, it is desired to omit longitudinal joints, reinforcement with heavier transverse bars must be provided.

Two-lane pavements are surfaced one lane at a time, or the full width of the road may be laid in one operation. In the first case in order to obtain effective load-transfer, a joint with inter-locking concrete projections together with tie bars are used to hold the faces together in tight contact. The form for the inner edge of the slab is made with a triangular key about 2.5 cm deep. The key is stopped about 15 cm short of each expansion joint. The two slabs are tied together with tie-bars. These may be 12.5 mm diameter deformed bars, 0.75 m long and spaced 0.75 m centre to centre although larger bars, longer and spaced somewhat farther apart are sometimes used.

When the entire width of pavement is laid in one operation, the longitudinal joint is often formed by installing a metal parting strip, to be left in its place with its top 6.3 or 12.5 mm below the surface. A convenient way to make the installation is to cap the strip with an installing shield. This is removed after the pavement is laid. A groove then extends from the top of the strip to the pavement surface, and is filled with a sealing compound.

Many organizations employ "dummy" joint or plane of weakness like those for contraction joints. Principal difference is that deformed reinforcing bars replace the dowel-bars. This is very suitable when full width of road is to be constructed in one operation.

Some organizations use a plain butt joint with tie bars which are sufficiently rigid to provide for load transfer. This is the simplest longitudinal joint and is formed by painting the joint face with bitumen after the forms have been removed from the first line of slabs concreted.

Tie-bars are inconvenient to install and their use involves drilling the road forms. If during construction, the tie bars are left projecting from the joint they will probably be an obstruction to the construction equipment. To avoid this, tie-bars up to 15 mm diameter may be bent aside temporarily and straightened later, or linked or screwed tie bars may be used.

Tie-bars and load-transfer devices are sometimes omitted from longitudinal joints in order to simplify construction, particularly where mechanically propelled machines are used. This is not desirable as openings of joints may occur by creeping of slabs due to expansion and contraction of concrete, and may occur on clay soils due to shrinkage of the clay during periods of dry weather. On many roads, particularly 7 m wide which have central longitudinal joint, there is a large concentration of wheel loadings along the longitudinal joint that may cause transverse cracks originating at the longitudinal joint by high edge stresses due to loading .

18.18.2. Warping Joints

These are simply breaks in the continuity of concrete but opening of joints is prevented by tie-bars or reinforcement. Such joints allow a small amount of angular movement to occur between the slabs and so prevent stresses due to restrained warping. These joints do not allow for contraction of the concrete. The use of dummy warping joints instead of dummy contraction joints might reduce the amount of progressive opening of the joints as a series of bays might tend to expand and contract as a single unit.

18.18.3. Construction Joints

These are formed when work has to be un-expectedly interrupted due to breakdown of plant or sudden onset of bad weather at a point where no joint would otherwise be required or sometimes when repair work has to be carried out. Normally the construction should be planned so that work is broken off at expansion joint, or at a contraction or warping joint. For construction joint, concrete laid before and after should be bonded together as effectively as possible. The joints should be tied together either by continuing the reinforcement across the joint or by installing tie-bars. They should be provided with a groove for sealing against the entry of water or grit.

18.18.4. Pumping of Joints

This is a problem encountered in the design and maintenance of joints in concrete pavements. With a certain combination of factors present, the movement of slab ends under traffic loads causes the extrusion or pumping of a portion of the subgrade material at joints, cracks and along the edges of the pavement. Pumping through joints will only occur under the following circumstances:

1. Frequent occurrence of heavy wheel loads
2. Existence of surplus water in the subgrade soil; and
3. The presence of a subgrade soil which is susceptible to pumping, *i.e.* soils containing 45 per cent or more of silt and clay.

All these three elements must be present in order for pumping to occur. The amount of soil removed by pumping may be sufficient to cause a sizeable reduction in subgrade support for the slab and may result in eventual failure of the pavement. The problem is especially severe in connection with expansion joints in existing pavements located on subgrade soils which contain high percentages (45 per cent or more) of silt and clay with plasticity index more than seven.

Design measures intended to prevent pumping may be placed in the following classifications:

(*i*) Since much difficulty is experienced at expansion joints and obvious solution is to minimize or eliminate them altogether. This consideration has doubtlessly been a contributing factor in the present trend to minimize the use of expansion joints.

(*ii*) Also, since pumping is associated with fine-grained soils, another obvious solution is to replace or improve these so-called "pumping soils". This may be done by the use of nominal thickness of granular sub-base or by stabilization of the existing subgrade soil by the use of cement or bituminous material, etc.

(*iii*) Other measures employed include the use of metal plates, fibre boards and similar devices placed horizontally beneath transverse joints. This method is not satisfactory as it is liable to cause differential settlement between the beam and the subgrade.

Maintenance procedures to prevent pumping include (*i*) providing adequate subgrade drainage, (*ii*) correcting faulty drainage and (*iii*) sealing joints and cracks. Where pumping has progressed to an appreciable degree, this condition is corrected by mud-jacking which consists of drilling holes into the slab and forcing in suitable material or slurry to fill the voids between subgrade and the slab. As the slurry is forced into the drilled hole, pumping of slurry is continued till all the voids are filled and the settled slab is raised to its proper position. The holes are then plugged. A mechanical lifting jack is sometimes used to raise the settled slab into position while slurry is being placed.

Slurry mixtures vary a great deal but mostly consist of loam, cement and water or bituminous material. Widely used mixtures are:

1. Soil 60 to 84 per cent and cement 40 to 16 per cent.

2. Soil 77 per cent, cement 16 per cent and cut-back bitumen (SC-2, MC-1, RC-3) 7 per cent.

3. Paving bitumen (low penetration and high melting point air-blown bitumen).

4. Special coal tar pitch.

CONSTRCUTION METHODS

In the present day construction of Portland cement concrete pavements, machinery plays an increasing part. Some work even now has to be done with hand tools, but it is steadily decreasing. Principal machines employed are batching and mixing plants, spreading equipment and compacting and finishing machines. The exact methods and machines used in the construction process vary somewhat from job to job and plant is available for jobs having a wide range of size and importance.

Sequence of separate steps in construction processes are : (*i*) preparation of the subgrade and sub-base (*ii*) placing of forms ; (*iii*) installation of joints; (*iv*) batching of aggregates and cement; (*v*) mixing and placing concrete; (*vi*) spreading, consolidation and finishing concrete; and (*vii*) curing of concrete.

Presented below is the discussion of each of the above steps together with the equipment which may be used to perform each of these operations.

18.19. PREPARATION OF SUBGRADE AND SUB-BASE

18.19.1. General

The subgrade or sub-base for laying of paving concrete slabs should comply with the following requirements:

1. that no soft spots are present in the subgrade or sub-base;

2. that uniformly compacted subgrade or sub-base extends at least 300 mm on either side of the width to be concreted;

3. that the subgrade is properly drained;

4. that the minimum modulus of subgrade reaction obtained with a plate bearing test shall be 5.5 kg/cm^2.

The manner of achieving these requirements shall be determined depending upon the type of subgrade or sub-base on which concrete is to be laid, and the following requirements in respect of the various types should be satisfactorily met.

18.19.2. Subgrade

Where the type of soil in the formation of the road is of a quality which ensures the above requirements, no intermediate sub-base need be used. The top 150 mm layer of the formation will be compacted at or slightly above the optimum moisture content to the exact profile shown in the drawing. It shall be checked for trueness by means of a scratch template (see IRC : 43—1972 for details) resting on the side forms and set to the exact profile of the base course. The template is drawn along the forms at right angles to the centre line of the road. Uneveness of the surface as indicated by the scratch points should not exceed 12 mm in 3 m. The surface irregularities in excess of this are properly rectified and the surface rolled or tampad until it is smooth and firm. The subgrade is prepared and checked at least two days in advance of concreting.

Where no sub-base is considered necessary and concrete is to be laid directly on the prepared subgrade, the subgrade should be in moist condition at the time the concrete is placed. If necessary, it should be saturated with water not less than 6 hours and not more than 20 hours in advance of placing concrete. If it becomes dry prior to the actual placing of the concrete, it shall be sprinkled with water taking care to see that no pools of water or soft patches are formed on the surface. It is desirable to lay a layer of water-proof paper whenever concrete is laid directly over soil subgrade. Where such a layer of waterproof paper is proposed to be placed between concrete and the subgrade, the moistening of the subgrade prior to placing of the concrete is omitted.

18.19.3. Sub-base

Where the subgrade is of a type not satisfying the requirements, a sub-base layer is provided before laying concrete. The sub-base may be granular material, stabilized soil or semi-rigid material as listed below:

18.19.3.1. Granular material

(*i*) One layer flat brick soling having joints filled with sand under one layer of water bound macadam conforming to IRC : 19—2005.

(*ii*) Two layers of water bound macadam.

(*iii*) Well-graded granular materials like natural gravel, crushed slag, crushed concrete, brick metal, laterite, kankar etc. conforming to IRC : 63—1976.

(*iv*) Well-graded soil aggregate mixtures conforming to IRC : 63—1976.

18.19.3.2. Stabilised soil

Local soil or moorum stabilized with lime or lime-fly ash or cement, as appropriate to give a minimum soaked CBR of 50 after 7 days curing. For guidance as regards design of mixes with lime or cement, reference may be made to IRC : 51 and 50 respectively.

18.19.3.3. Semi-rigid material

(*i*) Lime-burnt clay puzzolana concrete. The lime-puzzolana mixture should conform to IS : 4098—1967. The 28 days compressive strength of the concrete should be in the range of 40—60 kg/cm^2.

(*ii*) Lime-fly ash concrete conforming to IRC : 60—1976.

(*iii*) Lean cement concrete or lean cement-fly ash concrete conforming to IRC : 74—1979.

Thickness of sub-base should be 15 cm when the material used is of any of the types listed in (*a*) and (*b*). This may, however, be reduced to 10 cm for semi-rigid material in (*c*). The sub-base should be constructed in accordance with the respective specification and the surface finished to the required lines, levels and cross-section.

Where the subgrade consists of heavy soils (L.L. > 50) such as black cotton soil, the sub-base should be laid over a 15 cm thick blanket course consisting of non-plastic granular material like local sand, gravel, kankar, etc. or local soil stabilized with lime.

In water-logged areas and where the sub-grade soil is impregnated with deleterious salts such as sodium sulphate etc. in injurious amounts [Sulphate concentration (as sulphur trioxide) more than 0.2% in subgrade soil and more than 0.3% in ground water] a capillary cut-off should be provided before constructing the sub-base, vide details given below.

The sub-base shall be in moist condition at the time the concrete is placed. There shall, however, be no pools of water or soft patches formed on the sub-base surface. In case where a sand layer is placed between the sub-base and pavement concrete, a layer of water-proof paper shall be laid over the sand layer. No moistening of the sub-base shall be done in this case.

18.20. CAPILLARY CUT-OFF

As a result of migration of water by capillarity from the high water table, the soil immediately below the pavement gets more and more wet and this leads to gradual loss in its bearing value besides unequal support. Several measures such as depressing the sub-soil water table by drainage measures, raising of the embankment and provision of a capillary cut-off are available for mitigating this deficiency and should be investigated for arriving at the optimum solution. However, where deleterious salts in excess of the safe limits are present in the subgrade soil, a capillary cut-off should be provided in addition to other measures.

The capillary cut-off may be a layer of coarse or fine sand, graded gravel, bituminized material or an impermeable membrane. Layer thickness recommended for different situations are given in Table 18.17.

Table 18.17 Recommended thickness of sand/graded gravel layer for capillary cut-off

Serial No.	*Situation*	*Thickness of layer (cm)*		
		Coarse sand (mean dia 0.64 mm)	*Fine sand (mean dia 0.18 mm)*	*Graded gravel (40 mm and down without fines)*
1.	Water table at the same level as the subgrade surface	15	45	15
2.	Embankment about 0.6—1.0 m high	12	35	11
3.	Embankment about 0.6—1.0 m high but with the top 15 cm subgrade layer being of sandy soil having PI of 5 or less and sand content not less than 50 per cent	10	30	8

Cut-off bituminized or other materials may be provided in any of the following ways:

18.20.1. Bituminous impregnation using primer treatment

50 per cent straight run bitumen (80—100) with 50 per cent high speed diesel oil or its equivalent in two applications of 1 kg/sq m each, allowing the first application to penetrate before applying the second one. These applications should be given under the roadbed as well as onto the sides.

18.20.2. Heavy-duty tar felt

Enveloping sides and bottom of the roadbed with heavy-duty tar felt.

18.20.3. Polyethylene envelope

Enveloping sides and bottom of the roadbed with polyethylene sheets of at least 400 gauge.

18.20.4. Bituminous stabilized soil

Providing bituminous stabilized soil in a thickness of at least 4 cm.

18.21. SUB-BASE IN FROST AFFECTED AREAS

In frost affected areas, the sub-base may consist of granular material stabilized soil or semi-rigid material. However the compressive strength of the stabilized or semi-rigid material cured in wet condition shall be at least 35 kg/cm^2 at 7 days. For moderate conditions, such as those prevailing in areas at an altitude of 3,000 m and below, the thickness of frost affected depth will be about 45 cm. For protection against frost, the balance between the frost depth (45 cm) and total pavement thickness should be made up with non-frost susceptible material.

For extreme conditions, such as those prevailing in areas above an altitude of 3,000 the foundation may be designed individually for every location after determining the depth of frost.

The suggested criteria for the selection of non-frost susceptible materials are as follows:

(*i*) Graded gravel. Not more than 8 per cent passing 75 micron sieve. Plasticity index not more than 6. Liquid limit not more than 25.

(*ii*) Poorly graded sands, generally 100 per cent passing 4.75 mm sieve.
Max 10 per cent passing 75 micron sieve.
Max 5 per cent passing 50 micron sieve.

(*iii*) Fine uniform sand, generally 100 per cent passing 425 micron sieve.
Max 18 per cent passing 75 micron sieve.
Max 8 per cent passing 50 micron sieve.

18.22. EXISTING MACADAM SUB-BASE

When concrete pavement is laid over existing water bound macadam road, it shall be ensured that the existing macadam road constituting the sub-base extends over the required width and has a minimum thickness of 150 mm. Where the general unevenness of the surface varies by more than 25 mm from the required cross-section, the surface can be reconditioned after scarifying and adding suitable quantities of metal over the entire area including additional metal for correcting the camber and grade as required. Alternatively, depressions in the surface may be levelled by using lean cement concrete or lime-puzzolana concrete or lime-fly ash concrete or lean cement-fly ash concrete, properly compacted prior to laying of the concrete pavement slab, in addition to the depth that may be required for correction of the camber.

Where the width of the existing water bound macadam surface falls short of the width to be concreted by not more than 300 mm on either side and the condition of the surface is sound enough for receiving the paving concrete, the extra width may be made up by placing at least 100 mm depth of lean cement concrete or lime-puzzolana concrete or lime-fly ash concrete or lean cement fly ash concrete in trenches of required width at the sides of the existing metalling after taking care to see that the bottom of such trenches is well compacted by suitable tampers before placing of the new sub-base material. The correction to the unevenness of the surface and for camber shall follow the same guide lines as in the preceding paragraph.

Where the existing water bound macadam surface falls short of the width proposed to be concreted by more than 300 mm on either side, the sub-base shall be reconstructed by removing the entire material of the macadam and forming a fresh macadam surface of the required width by excavation and addition of new metal to the extent necessary over the entire width satisfying the requirements stated earlier.

18.23. EXISTING BLACK-TOPPED SURFACE

Where concrete slabs are to be laid over existing black-topped surfaces, no special treatment is necessary where the surface extends over the required width and is worn uniformly under traffic. Otherwise, steps suggested for macadam sub-base should be followed. Concrete is not laid on black-topped surface having soft spots caused by excessive bitumen or where thick premixed carpets have been rutted under traffic. In such cases the entire bituminous surfacing material is removed upto the top of the compacted macadam surface and the surface shall be prepared as detailed in the case of existing macadam sub-base.

18.24. PLACING OF FORMS

Steel forms are almost used now-a-days. Heavy wooden planks were used previously. With the advent of finishing machines that travel on the side forms there was need for the greater strength and durability, and steel forms were developed. Metal forms are now specified except where sharp curvature or other special conditions make them impracticable. Some areas where timbering is a major industry, permit option between steel and wooden side forms. Steel forms commonly used are straight 3 m sections aligned vertically and horizontally by slip joints and held in position by three or more steel stakes 2.5 cm dia, 45 to 75 cm long driven at intervals at the back of form. Forms are available in heights 15 to 30 cm and with corresponding base widths over a similar range. Forms are made of 4, 5 or 6 mm steel plate. They have a channel cross section and is stiffened at intervals.

Steel forms are carefully set along one side of the first lane of the pavement to be poured. These are set from tacked stakes 15 m apart or closer on curves as needed. The line of forms is checked by measurement and by eye. The second line of forms is set from the first line, giving close attention to proper crown and superelevation.

When set to grade and staked in place, the maximum deviation of the top surface of any section from a straight line should not exceed 3 mm in the vertical plane and 5 mm in the horizontal plane. The method of connection between sections should be such that the joint formed is free from difference in level play or movement in any direction.

The earth under the forms should be thoroughly compacted with mechanical tampers or otherwise, since the pavement can not be smoother than the forms on which the finishing machine operates. It is usually required that forms shall be kept set for at least 150 m ahead of the paver.

Machines are sometimes used for placing of forms. One form grader cuts a trench of exact size in the correct position to receive the forms and a second machine may then be used to clamp the forms securely in place.

18.25. INSTALLATION OF JOINTS

Extreme care must be taken in all operations accompanying the construction of joints if they are to function properly. The face of transverse joints must be straight, at right angle to the centre line of the pavement, and perpendicular to the surface of the finished slab. Load-transfer device *e.g.* dowel-bars where used in expansion joints must be placed and aligned accurately so that they will not inhibit free movement of slab ends in longitudinal direction and that they may perform their function of load transfer properly. Joint assemblies of various types supporting joint filler and dowel-bars are sometimes used to assist in the installation of joints.

18.26. BATCHING OF AGGREGATES AND CEMENT

Materials for paving concrete are carefully proportioned at a batching plant. The type of weigh-batching plant will depend on the size of the job and the capacity of the mixer to be employed. Small weigh-batching plant sufficient to keep 0.30 cu metre mixer fully engaged are available which are of types (*i*) swinging hopper, (*ii*) hopper-rails, (*iii*) fixed-hopper, (*iv*) simple weigh bridge on which the loaded mixer-skip rests. They may be completely mobile or semimobile and the hoppers may tip or may have bottom discharge doors. The capacity of these machines is about 5 cu metre of concrete per hour. This plant is most suitable for small jobs or where mixing site has to be changed frequently. For small jobs, a batching plant, may be improvised by arranging platform scales so that hand barrows or skips running on light railway track can be run on the platform and then filled to the correct weight, or by making a simple lever machine using a spring balance for weighing gauge boxes.

To feed mixers upto 0.75 cu metre size, semi-mobile batching plant is designed. They can be assembled in a short time and may be moved along as job progresses. They are fed by gravity from bins charged by draglines or conveyors.

Fixed batching plants are designed to feed mixers up to 1.5 cu metre and more. They may be set up at the aggregate source, a convenient railway siding, or some other carefully selected location. They are economical only in large scale construction due to the time and labour needed for their erection. They usually consist of a group of elevated aggregate bins equipped with a weigh box and multiple beam scales for weighing the aggregates. Commonly, the plant includes a separate bin and scales for storing and weighing bulk cement.

There should usually be six stockpiles of aggregates three in use after completion of tests, and three being built up. Each set of three will include large stone, and sand piles. Piles should be far enough not to run into one another, or alternatively, should be separated from one another by adequate partitions. Each pile should be of size sufficient to supply material for at least one day's work. In order to prevent segregation in building the piles, the materials should not be heaped into a single large cone-shaped pile but should be spread in layers not more than 1 m thick, commencing with an area large enough for the finished pile. If suitable hard, well drained ground is not available, the ground should be covered with wooden planks, or a concrete slab. It is essential that the conditions be such that stockpiled materials have a uniform moisture content. Washed aggregate should be held in the stock-pile for 24 hr to drain. Sometimes a sprinkling system is used to maintain uniform moisture conditions on both stone and sand.

Careful handling of coarse aggregate at the batching plant is essential if uniform grading is to be maintained. The variations in the grading may be reduced by ordering material in several sizes and recombining in correct proportions when batching, e.g. 20 mm to 10 mm and 10 to 5 mm for aggregate of 20 mm nominal maximum size, and 40 mm to 20 mm, 20 mm to 10 mm and 10 mm to 5 mm for aggregate of 40 mm nominal maximum size. The grading of sand cannot be adjusted easily but uniformity can be ensured by regular sieve tests which will show whether there is any gradual departure from the specified grading.

Aggregate and cement may be batched either by volume or by weight. Because of its greater accuracy, weigh-batching is generally preferred for highway work. All materials other than cement are calculated on the basis of one or more whole bags of cement, taking the weight of cement as 1440 kg/m^3. The use of weigh-batching will eliminate the high variations normally associated with volume-batching.

As already discussed it is not necessary to associate massive and expensive storage hoppers filled with drag lines or bucket elevators, and elaborate devices for weighing for adopting weigh-batching as such plant is only economical for large-scale road schemes or airports. The principles of weigh-batching can be applied, just as easily on small jobs. Once the engineers and contractors begin to appreciate and start using weigh-batching, they will continue to use it in preference to volume batching.

Weighing devices should be regularly maintained since large errors may develop with time and use. Regular checks should be made on their accuracy.

Where volume-batching must be employed the gauge boxes should be deep and of small cross-section to avoid excessive errors arising from filling. Also, allowance should be made for bulking of the fine aggregate.

18.27. MIXING AND PLACING CONCRETE

For large works, pavers which consist of a non-tilting mixer mounted on crawler tracks are now being used. Many of these are of the single drum type, with all mixing accomplished in a single horizontal cylinder. More recently, dual mixers with two drums, end to end, have been developed in which partial mixing in one drum is followed by transfer to and final mixing in the second. In this manner, the one mixer handles two batches at a time. The paver usually travels at the side of the lane being laid so as not to disturb the prepared subgrade and deposits the concrete on the subgrade by means of a boom and a bucket. The cement and aggregate are batched in a central batching plant and transported in lorries to the paver. A mixer is loaded by backing a batch truck into the large skip at its forward end. If the cement is in sacks distributed along the mixer path, labourers dump it into the skip while the truck is unloading. If cement and aggregate are being transported together in batch truck, the cement should be put into the truck on top of the coarse aggregate and the fine aggregate deposited on it, the truck being covered with tarpaulin to prevent loss. After the truck has discharged a batch and pulled away, the skip in raised, and the material slides into the mixer. At the same time, a measured quantity of water flows into the drum. Where a water main is not available at least two water tank lorries must be provided. The amount of water added is very carefully controlled.

Pavers are generally made in 0.75 cu m and 1 cu m sizes which are large enough to provide sufficient concrete for a slab 3.8 m wide being constructed with power propelled vibrating machines.

A separate device may be employed to add an air-entraining agent where required or it may be done by hand.

For smaller jobs, where pavers are not employed, the type, size and number of mixers should be chosen to provide the required output without overloading. Following types of mixers are available:

Non-tilting mixers. These are made in sizes from 0.15 to 3 cu metre.

Tilting-drum mixers. These are made in smaller sizes. However, few have been made up to 2 cu metre capacity.

Pan mixers. These are usually small and are used on very small jobs.

Continuous mixers. They do batching by volume and once they have been set, the batching is done more or less automatically. They can produce from 7 to 10 cu metre of concrete per hour. Continous mixer can, however, also employ batching by weight.

Truck mixers. They may be used to deliver freshly mixed concrete to jobs in congested areas or where it is uneconomical to erect a fixed plant. They are generally used in conjunction with central batching plant. The weighed aggregates and cement are delivered into the mixer drum and water is weighed in a separate tank so that it can be added and mixing begun when the truck is at or near the site. Truck mixers range in capacity from 0.75 to 4.5 cu metre.

18.28. SPREADING, RECONSOLIDATION AND FINISHING CONCRETE

The mixer should deposit the concrete near its final position so as to require as little rehandling as possible. The concrete is then spread with mechanical spreaders or shovels to the required depth and cross-section. For reinforced slabs, it is first placed to a level at which the reinforcement is to be placed. The reinforcement is then laid on the concrete, secured as required and the remaining concrete completed to the full slab thickness. Various types of spreading machinery is available and may be adjusted according to the job.

While spreading, care should be taken not be displace tie rods, dowel-bars, etc. No material should be deposited by the mixer within one metre of an expansion joint where a load transfer device is used. Concrete should be deposited on each side of the joint and shoveled simultaneously against both sides of the filler.

The placing of the concrete should be continuous from transverse joint to transverse joint. Preferably, no construction joint should be laid within 7.5 metre or an expansion joint.

Consolidation and finishing operations were performed originally by hand labour. However, increasing labour costs and demands for uniform high quality have resulted in gradual introduction of finishing machines. Hand-labour methods are now limited to transitions or other sections with variable width, along the forms and at the joints. Hand propelled vibrating machines run on road forms and are propelled by two hand wheels. The vibration is produced by rotating out-of-balance weights driven by a small petrol engine. Vibrating screeds which can be lifted by two men and in which vibration is generally produced by out-of balance weights driven by an electric or petrol engine or by blows from a hammer driven by compressed air, can also be used for consolidation. Small vibrating finishers are available for use in conjunction with some of these vibrating screeds which employ a smaller amplitude of vibration than the screeds used for compaction and when the level of the finished concrete is approximately correct, they can be pushed along the forms by two men. Even, the screeds used for compaction can be employed for finishing if it is fitted with some form of leading edge such as bull nose.

Consolidation and finishing of a concrete pavement is best done with a power driven finishing machine. The machine usually has two strike-off screeds, the leading screed being a vibrating tamper. The machine is driven by a petrol or diesel engine and runs on the road forms or rails. The vibration is produced by rotating out-of-balance weights or an eccentric-camshaft driven from the same engine. The frequency of vibration is generally between 3000 and 4200 vibrations per minute with amplitudes from 0.25 to 0.60 cm giving accelerations of 4 g to 9 g. During compaction, the speed may be varied up to a maxmum of 3 metre per min.

Concrete is placed to a depth a little above the top of the forms. The leading screed is set a little high, which pushes most of the excess material ahead of it while allowing a slight amount to pass beneath it. This is picked up and spread by the second screed operating at a lower frequency and amplitude than the compacting screeds. The second screed shapes the pavement surface to the correct height and crown. If the finishing machine has only one screed, this is raised off the forms slightly for the first pass, and is brought to true elevation in successive passes at a slightly higher forward speed. Intersection, driveways and other irregular areas are screeded by hand.

As soon as possible, after screeding operation is completed, the concrete should be further smoothened and finished with a longitudinal float which may be machine or hand opeated. It is simultaneously moved forward along the pavement and sidewise across the pavement with a shaking motion. It carries ahead of it any thin mortar or laitance on the pavement surface which is pushed to and wasted over the side forms. The surface is then tested for smoothness with 3 metre straight-edge which is held in successive positions parallel to the centre line, from one side of the slab to the other. All variations of more than 3 mm from the true surface contour are corrected by filling or planning and involved areas are re-finished.

When this has been done, and just before the concrete surface becomes nonplastic, any float marks or other surface blemishes are removed by belting or dragging the surface. The belt is a strip of canvas or burlap about 15 cm wide and 60 cm longer than the slab width. It is operated by a man on either side of the slab, with short strokes perpendicular to the centre line, accompanied by advance along the centre line. A drag consists of piece of burlap about 75 cm wide and 60 cm longer than the slab width with one edge fastened to a pole. The drag is rolled on pole and is thoroughly wetted. When dripping has stopped, the drag is held above and transverse to the slab and is gradually unrolled and lowered until 45 to 60 cm burlap are in contact with the concrete. Maintaining this contact area, the drag is then drawn slowly along the surface so as to leave an even texture behind it. This is sometimes followed by brooming the pavement to produce shallow, 1.5 mm transverse grooves on the surface.

The edges of the slab are then finished with an edging tool to round them to a radius of about 6.30 mm. Edges of transverse joints are carefully finished from a bridge which does not touch the concrete.

Recently, travelling or slip-form paver has been developed which constructs the pavement slab without the use of conventional side forms. The use of this equipment considerably reduces the cost of the pavement.

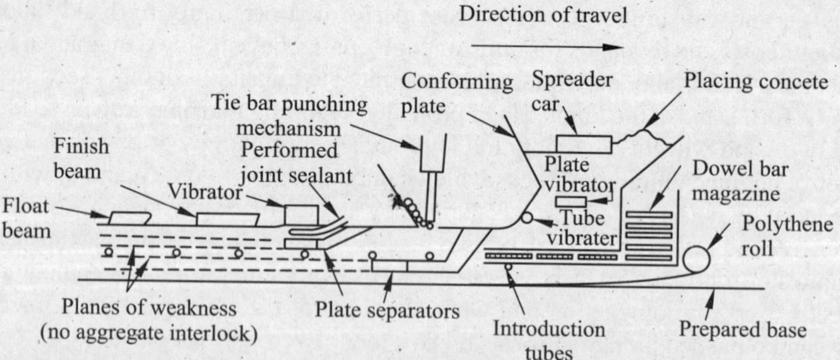

Fig. 18.4 Gunter and Zimmerman/Megregor modified slip form paver.

This slip-form paver is mounted on crawler tracks and rides on either side just outside the travelling forms. Trailing the machine, is approximately 15 metres of sliding form held apart at the proper distance by trusses spanning the slab. Concrete is deposited immediately in front of the paver and all finishing operations are carried out before the end of form is reached and final surface texture is given by a burlap drag at the near end.

If concrete of uniformly low slump is used, the concrete will remain firm after the forms are removed.

As the machine rides on the sub-base, great care must be taken to ensure that sub-base conforms to uniform grade otherwise waviness will be reflected in finished pavement.

To avoid manipulation of the concrete after the machine has passed, all joints, where used, are generally sawed after the concrete has hardened.

18.29. CURING OF CONCRETE

Proper curing of finished concrete is very important. Following are the accepted methods in use (*i*) ponding or earth cover kept wet; (*ii*) hay or straw cover kept wet; (*iii*) cover of felt mats, kept wet; (*iv*) cover of cotton mats, kept wet; (*v*) cover of burlap, usually 2 layers, kept wet; (*vi*) sawdust cover, kept wet; (*vii*) continuous sprinkling ; (*viii*) cover of-water-proof paper; (*ix*) impervious membrane, sprayed on ; (*x*) surface application of calcium chloride ; (*xi*) surface application of sodium silicate.

Many organizations require that preliminary curing with wet burlap should precede final curing by the selected methods listed above. The burlap will not mar the surface even though it is placed immediately after finishing and it offers a means for maintaining surface moisture during the critical early curing period.

Forms are not removed until concrete has set for at least 12 hours. Special devices are usually used to pull the forms so that the pavement will not be damaged in this process. After forms are removed, the ends of all joints are cleaned full width, and all honey-combed areas in the sides of the slab are pointed up. Earth is then shoveled against the sides of the slab to the top of the slab level.

At the end of curing period and before opening the road to traffic, the surface grooves in longitudinal and transverse joints are cleaned and filled with sealing compound.

During placing of the slab, test beams are made and are cured in the same manner as the slab and are tested for flexural strength as soon as it is believed they will show a modulus of rupture 40 kg per sq. cm. Traffic is not ordinarily allowed on the pavement until this strength has been attained.

CONCRETE BASES

Cement concrete is often used as base course, to provide a support for a high type bituminous pavement or for brick or block wearing surface. What has been said about cement concrete pavements applies also, in general, to bases with the following exceptions and additions.

Cement. Ordinary Portland Cement may be used without the addition of air-entraining materials.

Reinforcement. Reinforcement is often dispensed with but sometimes light weight wire mesh is provided, especially where heavy traffic is anticipated. Reinforcement is used to effectively control cracking which will be generally reflected in wearing surface in due course of time.

Joints. Frequently, bases are built without joints. Where joints are used, they are usually dummy joints to control contraction cracks. There is little uniformity in spacing of transverse joints, but they need not be as closely spaced as in pavements. When the full width of base is placed in one operation, a longitudinal dummy joint should be provided at the centre line. Because of less temperature warping stresses, longitudinal joint may not be very necessary.

Thickness. Slabs are generally of uniform thickness from 15 to 20 cm or more depending on the foundation and traffic.

Strength requirement. It is customary to use concrete in the base which has slightly less strength than that required for paving. Usual requirement is 30 kg per sq cm flexural strength against 45 kg per sq cm for wearing surface which is justified as temperature warping stresses are less.

Construction method. Except for omission of expansion joints, construction methods are essentially the same as for concrete pavements. The only difference of importance is the surface texture to be obtained. Instead of light brooming, the surface of base should be heavily scored transversely with a stable broom which will roughen the surface to a degree that will prevent sliding of bituminous surface on the base. For brick and block surface, the concrete base is usually given a smooth finish. The crown of the base should be substantially true to that of the finished road, so that the super-imposed bituminous material will have a uniform thickness.

PRESTRESSED CONCRETE PAVEMENTS

Some prestressed concrete pavements have been built in Europe and U.S.A.

Prestressing involves placing of concrete slab under continual compression. Steel reinforcement is placed and provision is made to prevent bond with the concrete. After the concrete hardens, a tension is placed in the steel by suitable means, and thus compressive stresses are induced in the concrete. This post-tensioning of steel adds to the bending resistance of concrete and probably also provides other load-supporting benefits. Because of this, crack formation is either prevented or, if it does occur, controlled. Obviously, a thinner slab than normal may be used. Due to thinner slab, a smaller temperature differential results and hence smaller warping stress is produced.

In France, base course of high bearing value granular material is provided under the pavement to help prevent local wheel-load distortion of the pavement slab. Also, this sub-base is covered with water-proofing paper to reduce to minimum loss of prestress and subgrade restraint becomes important and causes additional tensile stresses which are maximum at the centre of slab.

The amount of prestressing required depends upon the length of slab and subgrade restraint, etc. and in pavements built so far, prestress between 7 to 35 kg per sq cm has been used. A prestress between 7 to 14 kg compression in concrete is considered desirable. Greater length should require an unduly large prestress in order to overcome subgrade restraint and shorter lengths would be less economical due to larger number of joints and jacking points for a given length of a road. A length of about 125 metres is considered to be the most practical length for a prestressed slab.

Width of slab is kept up to 3.8 m otherwise transverse prestressing becomes necessary. The slabs in different lanes are tied together by tie bars. Some of the difficulties that must be overcome in prestressed pavements are: method of anchoring prestressed concrete slabs due to concentration of prestressing force near the anchorages owing to the thinness of slab; possibility of slab buckling upon release of stress from the anchors; and construction of suitable wide joints between the slabs.

❑❑❑

PART – V

19. General Principles of Pavement Design
20. Flexible Pavement Design Methods
21. Rigid Pavement Design Methods

Chapter

GENERAL PRINCIPLES OF PAVEMENT DESIGN

For the Purpose of design, pavements are divided into two categories: (*i*) flexible pavement, and (*ii*) rigid pavement. The essential difference between these two types of pavements is the manner in which they distribute the load over the subgrade. The design of a flexible pavement is based on the principle that a load of any magnitude may be dissipated by carrying it deep into the ground through successive layers of granular material. The intensity of a load diminishes in geometrical proportion as it is transmitted downward from the surface, by virtue of spreading over an increasingly larger area. Consequently, strength of each layer can be reduced with increased depth, with the highest quality materials at or near the surface. Thus, the strength of the subgrade primarily influences the thickness design of the flexible pavement.

The principle of design of rigid pavement is opposed to that of flexible pavement in that the design of rigid pavement is based on providing sufficient strength in a structural slab composed of Portland cement concrete to resist the destructive action of traffic. The rigid pavement, because of its rigidity and high modulus of elasticity, tends to distribute the load over a relatively wide area of soil. This because, although the downward deflection is greater for a low than for a high modulus of subgrade reaction, the deflections are spread over a much wider area and thus, the amount of bending of slab, upon which the stress in concrete depends, does not differ greatly in two cases. For this reason, minor variations in subgrade strength have little influence on the structural capacity of rigid pavement. Tensile stress as high as 50 kg per sq cm can develop in proper quality concrete pavements which are thus able to bridge small weaknesses and depressions in the subgrade.

COMPONENTS OF A ROAD

Road may be built up in several layers, each layer having a special function. The concrete pavement may rest directly on properly prepared subgrade. On poor subgrades a sub-base is required underneath the concrete slab to distribute the load further and to provide a satisfactory surface on which to construct the slab. This also controls objectionable qualities of subgrade soil such as swelling and shrinkage, drainage, pumping and frost action.

In a flexible pavement, the load is largely distributed by the base, the chief function of the surfacing being to provide a wearing surface. On some subgrades particularly clay subgrades, sub-base courses generally made up of locally available materials, *e.g.*, compacted stone or stabilized soil may be required under the base to distribute the load further and to prevent the clay working up into the base. The base courses are of high quality processed materials.

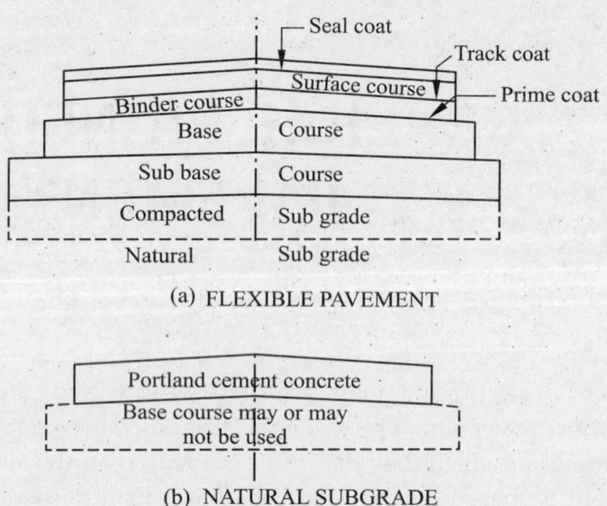

Fig. 19.1 Component parts of flexible and rigid pavements.

19.1. SUBGRADE

Subgrade is the foundation layer, which must eventually support all the loads which come onto the pavement. In some cases this layer will simply be the natural earth surface. In other and more usual instances, it will be the compacted soil existing in the upper layer of an embankment section or a cut section. In the fundamental concept of design of flexible pavements, the combined thickness of sub-base, if used, base and wearing surface must be great enough to reduce the stresses occurring in the subgrade to values which are not sufficiently great to cause deterimental distortion or displacement of the subgrade soil layer. Because of this, the design of a successful base course of a pavement begins with the design of the subgrade upon which it is to rest. No base course or wearing course can last long unless the subgrade is properly prepared.

Soil is a highly variable material and as such subgrade will vary considerably. It is, therefore, necessary to make a thorough study of subgrade soils to determine the design of pavements. This investigation can be accomplished by a proper exploration programme.

Desirable properties which the subgrade should possess are strength, drainage and ease of compaction.

Proper compaction at optimum moisture content pertaining to the compactive effort used is essential for subgrade. Compaction increases density and reduces the extent of subsequent settlement of the subgrade and the extent of water absorption by the soil even in the event of subsequent saturation. Both of these factors result in an increase of strength. However, it is important to keep in view the relationship existing between strength and density. Soil strength is dependent on both moisture content and density. For a clay-like soil, samples compacted at relatively low moisture contents, even though show greater strength when moulded, will swell more on saturation with subsequent losses in strength than those compacted at higher moisture contents. For such soils, it is important to compact them at moisture contents near or slightly in excess of the optimum moisture content. Laboratory tests and observations made in trenches excavated after traffic indicate that in some fine-grained soils, compacted at optimum moisture content, CBR values dropped considerably under further increase in density under traffic with practically very little variation in moisture content. This condition is associated with pore pressure developments because a portion of the applied load is taken up by the pore water. If the laboratory tests indicate that the subgrade will lose strength with increasing densities due to over-compaction, careful consideration should be given to closely controlling the moisture content. It may be necessary to limit the maximum weight of roller to be used.

Compaction is also unlikely to be advantageous for clay in cutting, where the destruction of the natural structure of the undisturbed soil results in a loss of strength.

Design for swelling subgrades may also involve building up a pavement to a sufficient thickness to impose a weight great enough to hold down the subgrade.

Regarding depth of compaction of subgrade, there is a wide variation in specifications followed by various organizations. The required depth of compaction is generally kept between 15 to 25 cm for highways but for airfield it is more stringent. For airfields it may range from 15 cm to as much as 1.35 m for 100 per cent density for heavy load pavements. For 85 per cent compaction the depth of compaction is increased and may be as much as 4.2 m for heavy-load pavements for cohesionless materials and 2.5 m for cohesive materials. However, in the design of layered system, the depth of controlled compaction can also be designed depending on the increase in strength of compacted soil over that of raw soil without compaction. Similarly, degree of compaction for various layers, if necessary, can also be worked out.

19.1.1. Subgrade Drainage

Subgrade drainage is of great importance. Due consideration should be given to both ground water and surface infiltration. Subgrade should be given a proper camber for good drainage. Side ditches or subdrains may be required depending on local conditions. When highway or airport pavements are built in relatively flat areas, formation should be raised above the natural ground level so that it is at least 1.2 to 1.5 m above the water table level depending upon the height of capillary rise for a particular soil. In areas of potential frost problems, special consideration should be given to make certain that sources of free water are removed.

It is important to detect any soft or weak areas in the subgrade and to make necessary corrections before any base course is placed. This is sometimes done by rolling with a heavy rolling equipment such as pneumatic roller, 30-tonnes or more with a minimum of 5 to 15 coverages over the weak subgrade areas.

19.2. SUB-BASE

The purpose of a sub-base is to permit the building of relatively thick pavement at a low cost. Economy is the essential item in the design of sub-base courses. The greatest possible use should be made of locally available materials. These may consist of selected materials, such as natural gravels, which are stable but are not completely suitable as base courses. They may also consist of stabilized soil or merely select borrow. Thus, the quality of sub-bases can vary within wide limits, as long as the thickness design criteria are fulfilled for flexible pavements.

If the soils and drainage conditions are good, the sub-base may be omitted. It is usually on fine-grained soils that sub-bases are needed. The purpose of sub-base is to serve one or more of the following functions:

(*a*) Increase the structural support for the base and surface course.

(*b*) Improve drainage.

(*c*) Eliminate frost heave and salt heave.

(*d*) Prevent the base and surface courses from being affected detrimentally by the poor qualities of the underlying soil.

The sub-base may be continuous or intermittent and its depth may vary considerably on different sections of the road. IRC recommends a minimum thickness of 10 cm for flexible pavements.

Density and moisture requirements are determined from the results of laboratory or field design tests.

19.3. BASE COURSE

Base course is that portion of the roadway super-structure which lies immediately under the wearing course or pavement. Base course usually conforms to more or less rigid specifications. The main considerations are that they should have: (*i*) sufficient thickness to adequately distribute the heavy wheel load pressures to the sub-base or subgrade, (*ii*) sufficient structural stability to resist the vertical pressures and horizontal shear stresses produced by standing or moving wheel loads, (*iii*) sufficient density to provide resistance to consolidation within themselves that would result in distortion of the wearing surface, and (*iv*) sufficient resistance to weathering so that materials whose stability is affected by water should be avoided, where possible, and frost susceptible materials should also be excluded.

Suitable materials for base courses such as suitable local soft aggregates, stabilized soil, macadam, bricks, slag, *etc*. have already been described in great detail.

Base courses are extended to some distance beyond the edge of the wearing surface. This ensures that load applied at the edge of the pavement will be supported by the underlying layers. If the layers are built with an abrupt vertical face, loads applied at the surface are likely to cause failure due to the lack of support at pavement edge. The base course may be extended upto 30 cm or even more beyond the pavement if special situations so warrant.

Purpose. The function of base course varies according to the type of pavement. In a flexible pavement the purpose of base course is to provide a stress-distributing medium which will spread the load applied to the surface, so that shear and consolidation deformation will not take place in the subgrade.

These also increase the load supporting capacity of the pavement by providing added stiffness and resistance to fatigue.

For proper design, the thickness of base and sub-base, if any, should be sufficient to prevent overstressing the subgrade. The minimum thickness of base course kept is 7.5 cm. The base course must be of good quality to prevent failure due to high stress concentrations immediately under the pavement.

For this reason, the minimum CBR that is generally permitted is 80 per cent. In some situations, this CBR may not be possible due to economic consideration in which case inferior materials upto CBR 50 may be used. However, inferior materials must be compensated by increasing the thickness of the wearing surface. The minimum thickness of surfacing required over a base with 50 CBR is 7.5 cm and no surfacing is laid directly on a base with CBR less than 50 per cent, in case of flexible pavement.

Provision of a base under rigid pavement may be made for the following reasons. Unless suitable base material can be obtained, it may be uneconomical to increase the structural strength of road by providing a base course.

(*a*) Prevention of pumping on fine-grained soils.

(*b*) Protection against frost on frost-susceptible soils.

(*c*) Prevention of volume changes of the subgrade on highly active soils.

(*d*) To form a working surface on clays and silts which will enable construction to proceed during wet weather without damage to the subgrade or its protective seal. Thus it helps to expedite construction.

(*e*) To provide a levelling course on roughly shaped formation.

(*f*) For increased structural stability where it is more economical or practicable than a corresponding much smaller extra thickness of slab.

Whenever any base course is used under rigid pavement, it should not constitute an additional source of differential movement which may cause slabs to crack. Where granular materials are used in base, only those which can be thoroughly compacted should be used. Concrete slabs laid on a rigid base have shown somewhat unexpected tendency to crack as they cannot accommodate themselves to the warped shape of the slab which may then crack as a result of non-uniform support. For this reason, it is possible that bases of lean-mix concrete or cement stabilized soil be unsuitable for rigid pavements.

When a base is considered necessary under a rigid pavement, its thickness may range from a minimum of 7.5 cm to prevent pumping to as much as the full depth of frost penetration to prevent frost action. Another major factor which determines the thickness of base courses under rigid pavements is construction techniques. If it is difficult to run heavy equipment over the subgrade, it may be necessary to increase the thickness of the base as a construction expedient. Extra thickness may also be required to backfill the soft spots in the subgrade.

Regarding choice of materials for base course, it depends on the purpose for which the base course is provided. To prevent pumping, a base course should be well graded, free of excessive amount of fines, and should be compacted to a relatively high density. To provide drainage, the base may or may not be well graded material, but it should contain little or no fines. Base course for frost action should be non-frost susceptible and free draining. A base course need not be free draining to provide adequate structural stability, but it should be well graded and should resist deformation due to loading.

19.4. WEARING COURSE

This is the component of a road with which the wheels of vehicles are in actual contact. Its principal purpose is to provide a smooth surface for a comfortable journey and will resist destructive pressures exerted by the traffic. It should also prevent or minimize penetration of surface water into the roadbed, which might materially lower the supporting value of the base and subgrade. In flexible pavement it should be flexible so that it will not fail if consolidation of subgrade or base course takes place. In addition, it should prevent the base from ravelling and disintegrating effects of traffic. Properly designed and constructed wearing course also adds appreciable strength to the entire road structure.

Surface courses are composed of a wide variety of substances ranging from low-cost surfaces to bituminous or Portland cement concrete depending on the intensity and weight of traffic.

Thickness requirements for surface courses are dependent largely upon the applied load and the stability of base course. For base course with a high CBR value, surface courses of lesser thickness can be used than for base courses of low CBR. The choice of thickness for a surface for flexible pavement is usually arbitrary and wearing surface for highways varies considerably from one organization to another. The thickness of surface course is generally assumed according to some typical specifications or it forms part of the design method and the thickness of remaining layers is determined by subtracting the assumed thickness of surface course from the total thickness of pavement required to protect the subgrade.

The selection of the type of wearing surface and thickness of surface depends upon the cost in the light of benefits derived from the paving structure. It is now well recognized that the load-distributing qualities of high-type bituminous paving mixtures are much superior to granular base course and as such by using thicker high-type bituminous surface it can result in overall reduction in the total pavement thickness. WASHO (Western Association of State Highway Officials) road test; USA, indicates that a 10 to 30 cm reduction in the total pavement thickness resulted with a 10 cm surface over that of 5 cm surface. However, this can only be justified when the cost of additional high type bituminous surface is less than the additional base course.

COMPARISON OF HIGHWAY AND AIRPORT PAVEMENTS

There is no fundamental difference between roads and airfields and the general principles of design apply to both of them. However, some distinct differences exit between the two types of pavements, notable among them are: (*i*) magnitude of applied load, (*ii*) tyre pressure, (*iii*) geometric section of the pavement, and (*iv*) number of repetitions of load applied to the pavement during the design life. In addition, the rapid development of jet aircraft in recent years has a profound effect on pavement design concepts.

The total weight of aircraft is usually greater than that of a truck. The design load for a major highway is ordinarily in the vicinity of 8200 kg on dual tyres and for conventional truck tyres, pressure is in the vicinity of 4.2 to 6.3 kg/cm^2. On the other hand, a heavy commercial jet aircraft with gross weight of 310,000 kg (Boeing 747) may have wheel loads in excess of 100,000 kg with tyre pressure as high as 14 kg/cm^2. Heavier bombers have higher gross weight and higher tyre pressure. For example B-52 bomber has tyre pressure of 18.2 kg/cm^2.

The number of repetition of loads is much greater on highways than on airports. On a road the expected repetition may be as much as 1,000 to 2,000 trucks per day on major roads but on airports only 20,000 to 40,000 coverages may be considered for the life of the pavement.

For a given wheel load and a given tyre pressure, highway pavements should be thicker than airfield pavements, because repetition of load on a highway is much higher and also the loads are applied closer to the pavement edge. However, because gross loads on airfields are much higher, in actual practice, airfield pavements are thicker.

Regarding the geometry of the pavement, lateral placement of traffic on roads is such that nearly all truck traffic travel within 1 or 1.20 m of the edge of the pavement. On the other hand, pattern of wheel load application varies across the width of a runway and is concentrated in the central portion.

Analysis of about 9 m of the centre of the pavement with bicycle-type and steerable type landing gears have further resulted in channelized traffic, and studies show that about 75 per cent of this type of traffic will occur on about 2.25 m of pavement. This indicates the possibility of some measure of the saving in cost that could be made by taking into account this variation in wheel load application in determining the thickness of the pavement. As a result, variable thickness section may be designed instead of uniform thickness section for a runway.

Similarly, traffic follows a designated line along the critical areas such as aprons, taxiways, and at runway ends and as such will need to be built of thicker cross-section than the central portion of a runway.

19.5. EFFECT OF JET AIRCRAFT

Probably no other item has been so controversial and has received such wide spread attention during the past 10 years in the design of airfields than the type of pavement most suitable for jet aircraft.

The characteristics of jet aircraft which played a most prominent role in the development of airfield pavement are:

(*a*) Jet fuel spillage and heat and blast of jet engine exhausts.

(*b*) High pressure tyres, with their small contact areas.

(*c*) Heavy wheel loads.

(*d*) Channelized traffic.

Jet engines dump fuel when they are shut off or when they are decelerated. The quantity of fuel spilled is from 0.75 to 1.25 litre for each engine cutoff. This amount of spillage occurs in about 2 min after the engine is cutoff. Repeated spillage at the same spot softens and leaches out bituminous binder leaving aggregate exposed in a loose state. Occasional spillage evaporates and causes no significant damage to dense hot-mix bituminous concrete. Studies show that repeated fuel spillage has no adverse effect on tar or rubberized-tar concrete. Sometimes, scavengers are provided with the aircraft to catch the spilled fuel.

Areas exposed to heat and blast of jet engines are quickly burnt and eroded by the hot high velocity jet blasts. The heat induced in the pavement is dependent to a vary large degree on the angle of tailpipe and its height above the surface. Maximum temperature developed is 200°C or even more. Usual pavement temperatures are less than 150°C. Studies conducted by the U.S. Corps of Engineers, show that the heat and blast effects become critical (bituminous cement melts and the blast erodes the pavement) on tar concrete, bituminous concrete and rubberized-tar concrete pavement at 120, 150 and 157°C respectively.

Loose pavement particles, resulting from the effects mentioned above, are a hazard to jet engines. This indicates the necessity of a pavement at critical areas which is resistant to jet fuel and heat and blast. Jet blast is critical for bituminous pavements at the ends of runway and on aprons where maintenance run-ups are made. No completely satisfactory jet-fuel resistant seal coats for bituminous pavement have been found.

Seal coats and surface treatments are seriously affected by blasts. Seal coats are often made with bitumens that melt at lower temperatures than the bitumens used in bituminous concrete. Blast erodes seal coats quicker than bituminous concretes. Also, thin layers such as surface treatments can be lifted up bodily and blown aside by the blast.

Results of experience and tests indicate that Portland cement concrete pavements are resistant to the effects of repeated fuel spillage, and heat and blast.

Regarding heavy loads, high pressure tyres and channelized traffic, there are certain special problems. Peculiar ground operating characteristics, due to bicycle type landing gear of the B-47 bomber, coupled with channelized traffic, have resulted in porpoising. This phenomenon results from sudden changes in longitudinal grade which cause the plane to undulate, pilot technique in landing, or in applying brakes. Because of peculiar loading, pavement weaknesses are quickly located by the heavy bomber and depressions thus formed cause them to porpoise. Once it starts, porpoising becomes more prominent and the depressions get deeper until complete failure occurs. This phenomenon is particularly apparent on some flexible pavements.

Traffic along centre lines of taxiways and along taxilanes on apron can cause rutting. All this is due to repetitive loading imposed in a narrow band due to channelized traffic.

The above discussion indicates that it is almost impossible to construct a flexible pavement of uniform quality throughout its multiple layered system in airfields catering for heavy jet aircrafts.

Weakness in any one of the layers would be indicated in a short time. For this reason, all primary pavements are now specified to be constructed of Portland cement concrete.

Under ordinary conditions the policy of some departments responsible for design and construction of airports is to construct aircraft parting aprons and runway ends in Portland cement concrete. For the rest of paved areas the choice of pavement surfacing material is made on the basis of economy.

As for the design of channelized traffic is concerned, extra 20 per cent thickness, both in the flexible and rigid pavement, is usually recommended in the central 7.5 m wide portion of primary taxiways, full width of taxiway turns and to the central 30 m width on runway ends for a distance of 150 metres.

There are a few other miscellaneous developments in pavement requirements for jet aircraft. The blast of jet exhausts quickly erodes the unprotected ground areas beyond pavements edges. Soil is blasted out. Grass areas are denuded. Pavements become litered with debris and the debris when sucked into a jet intake damage the engines. To protect these unstable areas, blast pads, i.e., shoulder paving have to be provided on taxiway shoulders, at turns, on the shoulder of warm-up pads, and on 45 m of the over-run area at runway ends.

At bomber bases, in addition to shoulder blast protection mentioned above, there is the problem of providing a stable surface to support the outrigger gear as for example in B-47, and B-52. This may be solved by paving shoulders of taxiways, warm-up pads and aprons.

This paving at the shoulders may consist of 5 cm of bituminous concrete.

Further, due to high landing speeds of jet aircrafts and also due to their long landing rolls and small wheels having high tyre pressures require some degree of strength to be built into the over-run areas 300 m long and as wide as the runway plus shoulders for heavy bombers. This will avoid damage to aircraft and minimize injury to pilot in case he lands short or overshoots. These overruns are required to be stabilized. However, the width of the stabilization is limited to the width of the runway. This stabilized portion is designed for the proper load but for a minimum number of coverages. This may need pavement about half as thick as that of the standard-capacity design. The surface may be protected by double bituminous surface treatment.

The entrance of jet aircrafts into commercial use has presented problems in the design of civil airfields similar to those encountered for military fields. Finally, recognition must be given to the fact that modern aircraft, both military and civil, are rapidly becoming faster, larger, heavier and more numerous. Pavements must be adequately designed and constructed to meet the requirements, not only of the present aircraft but also those which can reasonably be anticipated in the future. Unless due allowance is made now for future increases in loading, pavement maintenance and repair, problems will reach enormous proportions. Today we are faced with such a situation because of the rapid growth of air traffic which was not fully anticipated.

FACTORS AFFECTING PAVEMENT STABILITY

The thickness of pavement depends on a number of variables. Experience indicates that there are five elements which exert controlling influences on the stability of constructed roadbeds. These are:

1. ***Vehicle and traffic factors.*** These are represented by the vehicle type, volume and character of the traffic and mode of operation of the vehicle which will use the proposed road.
2. ***Moisture factors.*** These represent factors resulting from changes in moisture content of the subgrade from any condition of precipitation, capillarity and irrigation, etc.
3. ***Climatic factors.*** These represent factors resulting by extremes of temperature with resultant frost penetration and reaction.
4. ***Soil factors.*** These represent the conditions of the natural foundation soil in cuts, or under shallow embankments, or the soil which is used to construct the embankment immediately underlying the subgrade surfaces. They deal with the measurement of subgrade supporting power.
5. ***Stress distribution factors.*** These represent function of pavement and base in transmitting load to the subgrade.

The thickness of pavement must vary to accommodate the combination of these five elements which it is anticipated will be met in the proposed road. For example, a road utilizing high quality materials with thick bituminous surface (stress distribution factor) built in arid plains with low precipitation (moisture factor) sandy soil (soil factor), low frost penetration (climatic factor) and a small traffic volume with a low percentage of truck traffic (traffic factor) would not require the same thickness which would be required on a road utilizing poor quality aggregate with thin bituminous surface built in an irrigated mountain valley having high precipitation, extreme temperatures and frost penetration and a soil type composed essentially of silt and clay, and carrying a high volume of traffic including an abnormal number of heavy trucks. The same applies to airport pavements. A number of variables influence each of these principal factors and its should be realized that the procedure of evaluating these variables is by no means a simple task. A brief discussion of various variables involved in these factors is given below:

19.6. VEHICLE AND TRAFFIC FACTORS

19.6.1. Design wheel load

In evaluating the magnitude of the wheel load to be selected as a design criterion, consideration should be given to the maximum legal load limits prevailing in the area concerned and the results of traffic survey made on routes similar to the one for which the design is being prepared. Unless two wheels are sufficiently closely spaced for the area of subgrade stressed by each wheel to overlap, it is the maximum wheel load that is important for pavement design and not the gross weight of the vehicle. Plan view of several basic types of wheel configuration is shown in Fig. 19.2.

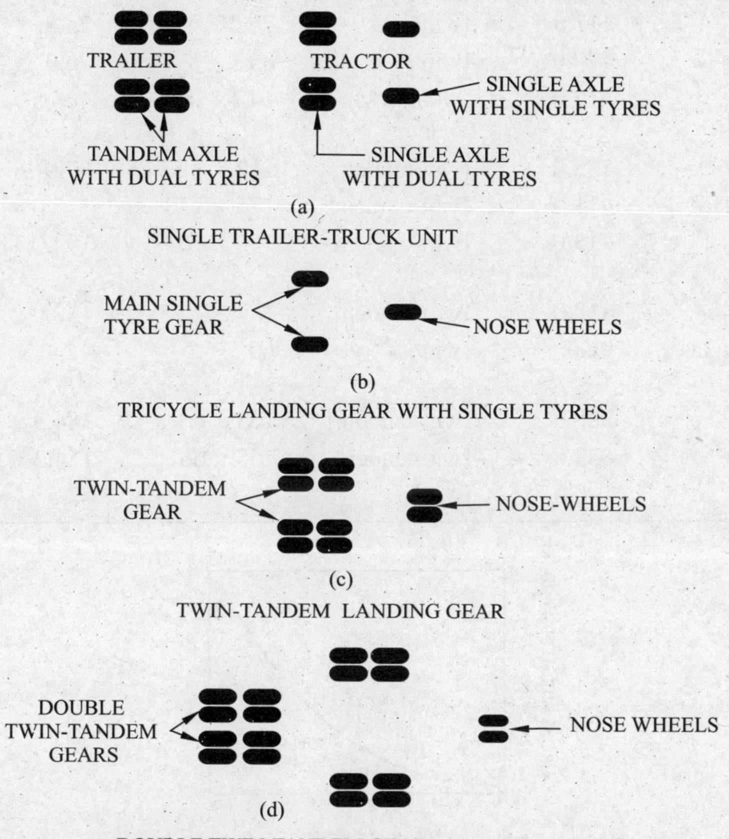

Fig. 19.2 Basic types of wheel configuration.

The design wheel load is the maximum wheel load of the predominant heavy vehicle likely to use the pavement in the normal course. For highways, the legal single axle load for standard vehicle in India is 8200 kg (equivalent to 14500 kg tandem axle load) which implies that the maximum wheel load will be 4100 kg. If greater loads are required, a tandem axle is added. The effects of other vehicles are normally accounted for in the design by the use of equivalent 8200 kg single axle load (EAL).

The design wheel load, in the case of airport pavements, may be that of the largest air-craft using the pavement. Typical data for several air-crafts is given in Table 19.1. There is a marked increase in the gross aircrafts weight during the last 50 years. Fig. 19.3 shows the trend. Extrapolation of the curve shows that future aircraft may approach 10^5 kg gross weight. Modern aircraft utilizes either bicycle or tricycle landing gears. For the purpose of design, the multiple-wheel gear is converted into an equivalent single wheel load (ESWL). In the design of airport pavements, only the critical or design vehicle is considered e.g. the heaviest or most damaging aircraft and the effects of other vehicles is ignored. This method has fixed traffic approach.

Table 19.1 Data for several typical aircraft

Type of plane	*Max gross weight (453.6 kg)*	*Type of gear*	*Main gear dimensions (cm)*	*Max load Each main Assembly (453.6 kg)*	*Tyre pressure (kg/cm^2)*
Boeing 707-320 C	336.0	Twin-tandem	140 × 86.25	157.0	12.6
Boeing 707-120B	258.0	Twin-tandem	140 × 85	120.0	11.9
Boeing 737	111.0	Twin	76.25	25.8	10.36
Boeing 737-100	170.0	Twin	85	76.9	11.62
Boeing 747	713.0	Double-twin-tandem	145 × 110	166.5	14.28
Convair Cv880	185.0	Twin-tandem	112.5 × 53.75	87.0	10.5
Lockheed L 1011-1	411.0	Twin-tandem	175 × 130	115.0	12.25
Mc Donnel-Douglas DC 10-10	413.0	Twin-tandem	135 × 160	194.0	12.25
Douglas DC 8-43	318.0	Twin-tandem	137.5 × 75	148.0	12.39
Mc Donnel Douglas DC 9-15	91.5	Twin	60	42.4	8.89
Concorde	388.0	Twin-tandem	165 × 66	184.3	12.88
Concorde	833.0	Twin-tandem	165 × 66	184.3	12.88
BAC 1-11-500	100.0	Twin	52.5	47.5	12.18

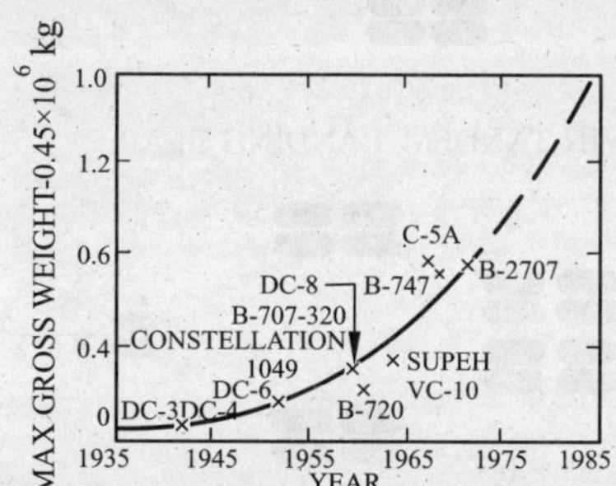

Fig. 19.3 Trends in air-craft gross-weight.

The different types of truck and airplane wheel arrangements can be divided into several basic catogories, including (*a*) single and dual wheels, (*b*) single and tandem axles, and (c) wheel nose tricycle, and bicycle landing gears. The wheel may be arranged in several combinations of those listed above.

19.6.2. Effect of multiple wheels and tandem axles

For the purpose of design the multiple-wheel gear or tandem axles are converted into an equivalent single wheel load (ESWL). ESWL is defined as the load on a single tyre that will cause an equal magnitude of a preselected parameter (stress, strain, deflection or distress) at a given location within a specific pavement system to that resulting from a multiple-wheel load or tandem axle at the same location. Depending upon the procedure selected, either the tyre pressure or contact area of the ESWL may be equal to that of one tyre of the multiple gear assembly.

Table 19.2 gives a summary of various ESWL criterion that are used by major airfield pavement design agencies. For flexible pavements, the two important criterion are equal vertical subgrade stress and equal interface deflection. For rigid pavement design the criterion is based upon maximum tensile stress in the concrete pavement. For highway design, equal axle load (EAL) is mostly employed. The various criteria are briefly discussed below.

Table 19.2 Agency ESWL method

Flexible pavements

Criterion	*Flexible pavements Agency*	*Remarks*
Vertical subgrade stress	Canadian (Mc Leod) Federal Aviation Administration U.S. Navy	One-layer theory, equal Ac. One layer theory, equal Ac based on typical gear spacing One-layer theory, equal Ac, ESWL calculated at h = 75 cm only
Interface deflection	U.S. Corps of Engineers, Asphalt Institute	One-layer theory, equal Ac Two-layer theory, equal Ac restricted to general aviation aircraft only (27 tonnes or less)
	Ragid Pavements	
Maximum tensile	U.S. Corps of Engineers LCN Portland Cement Association Federal Aviation Admn, U.S. Navy	Westergaard free edge stress Ac = 1722.5 cm^2 ESWL generally not calculated, stresses computed directly by PCA Computer program, influence charts assumption of typical gear spacings

Equal stress ESWL. The method of equal maximum vertical subgrade stress is based upon Boussinesq stress distribution concept in homogeneous, one layer system. The depth for a flexible pavement at which stresses in the pavement resulting from dual wheels are equal to those of a single wheel depends upon the spacing of the wheel. Near the surface the wheels act independently and at a depth of approximately half the clear spacing between the wheels i.e. *d*/2, the wheels cease to act independently and stresses in the pavement result from the combined effect of the two wheels. At greater depths the stresses overlap but become smaller as depth increases, until a point is reached where the overlap of stresses is negligible. Analysis shows that at a depth of about twice the *c/c* distance between the wheels *i.e.* equal to $2S_d$, the overlap of stresses becomes negligible, and the equivalent single wheel load is equal to the load on the whole gear. Between these two critical thicknesses it is assumed that the equivalent single wheel load varies as the logarithm of the thickness of the pavement.

To simplify the calculations, use is made of logarithmic paper. The diagram is plotted allowing the equivalent single wheel load to the load on one wheel until pavement thickness becomes greater than *d*/2 *i.e.* half the clear spacing between wheels, then increasing to the value equal to the whole gear load at a pavement thickness of twice *c/c* spacing of wheels. The equivalent single wheel load at any other pavement thickness can then be obtained directly from the plot (See Fig. 19.4). This applies both for dual wheel assembly as well as dual tandem assembly. For tandem gears, after the ESWL for the dual portion of the gear is found at a given depth, the procedure is repeated between the tandem distance.

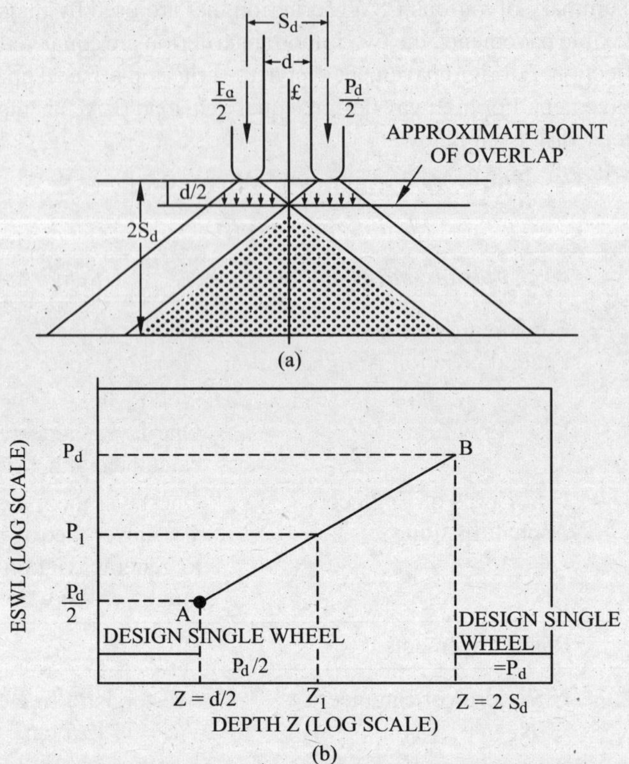

Fig. 19.4 ESWL analysis for equal vertical subgrade stress (*a*) influence of multiple wheels on stresses (*b*) method of determining ESWL for any dual-wheel load.

U.S. Navy department assumes that the stress is inversely proportional to the square of the distance to the point under consideration. The basic concepts of this analysis are given in Fig. 19.5. In this method, the ESWL is always computed at a depth of 75 cm for all aircraft types. The ESWL for dual tyre s given by:

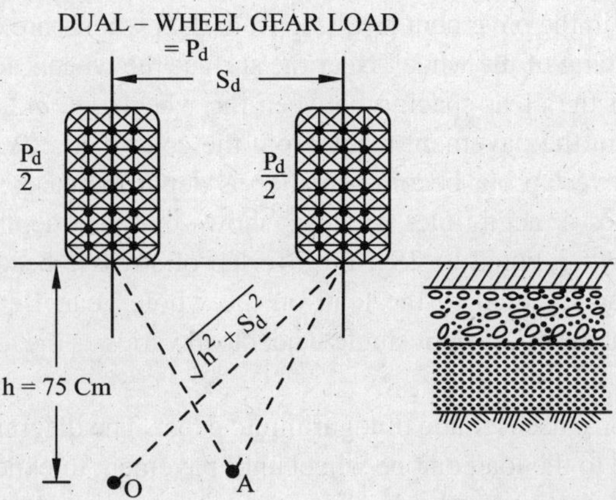

Fig. 19.5 ESWL Analysis using the U.S. Navy procedure.

$$P_s = \frac{P_d}{1 + \dfrac{S_d}{100}} \qquad \text{...(19.1)}$$

For dual tandem configuration ($S_d \times S_t$), the relation used is:

$$P_s = \frac{P_{dt}}{\left(1+\frac{S_d}{100}\right)\left(1+\frac{S_t}{100}\right)} \quad \ldots(19.2)$$

where P_{dt} represents the gross weight on the dual tandem gear.

Equal deflection ESWL. This method is developed by Corps of Engineers, U.S.A. The maximum interface deflection at a given depth or thickness is equated to that produced by the equivalent single wheel. Boussinesq one layer theory is used. The contact areas, Ac, of the ESWL and one tyre of the multiple gear are assumed equal.

Most methods of design are based upon a limiting deflection at the point representing failure. The values for flexible pavements range between 2.5 and 5 mm. For rigid pavements, Burmister assumed 1.25 mm as limiting deflection. When the displacement under wheel loads is indirectly correlated as in CBR test, CBR is calculated at 2.5 or 5 mm penetration of the piston. In some design methods actual deflection measurements under equivalent wheel loads are made as a guide for pavement evaluation. This basis, though quite satisfactory can sometimes give misleading results. For example, deflection caused by a low-pressure tyre with a high load may give the same deflection as a high pressure tyre but stresses acting on pavement will be much higher for the small radius of curvature under a high pressure tyre, than for the high radius of curvature under a low pressure tyre.

The deflection measurement as design criterion, subject to limitations pointed out above, is satisfactory if used with discretion regarding the limiting deflection value in a particular case.

In general, allowable deflection decreases as stiffness of the pavement increases. Hveem has recognized this principle in establishing tentative values of permissible deflection. He has suggested maximum permissible deflection for design purposes for different types of pavements as 1.25 mm for 1.25 cm surface treatment; 0.6 mm far 5 cm plant mix on gravel base; 0.4 mm for 10 cm bituminous concrete ; 0.3 mm for 15 cm cement treated base surfaced with bituminous pavement; and 20 cm for Portland cement concrete pavement.

Two layer ESWL. The Asphalt's Institute, USA developed ESWL concept involving Burmister's two-layer interface deflection theory and equal contact pressure. This is applicable to aircraft having less than 27000 kg gross weight. Simplified charts shown in Fig. 19.6 have been developed for determining the ESWL for dual wheel. The assumptions made in drawing these charts are that the critical modulus of bitumen layer is 7000 kg per sq cm and the modulus of subgrade soil is equal to 1500 times the CBR. Thus the interface deflection charts in Fig. 19.6 can be converted to an equivalent CBR for each modulus ratio *e.g.*

$\frac{E_1}{E_2} = 10$ gives approximately CBR = 7.

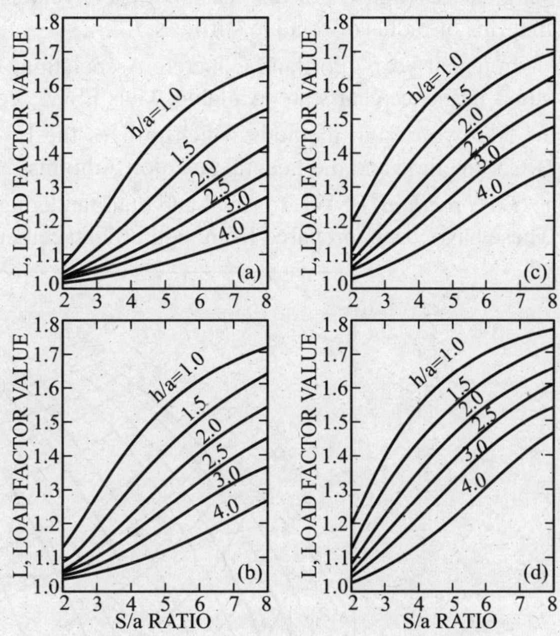

Fig. 19.6 Dual-wheel load factors (full depth bituminous pavement) only (a) CBR = 3% (b) CBR = 7% (c) CBR 15% (d) CBR 30%.

The charts have a direct application. For a given dual tyre assembly operating on a full depth pavement, the load factor value, L can be read directly from the appropriate CBR chart as a function of the dual spacing, S_d and thickness.

Problem 19.1 *Determine the ESWL for a dual gear having a gross weight of 10,000 kg operating on a 45 cm full-depth pavement. The dual wheel spacing is 60 cm and the contact radius = 15 cm. The CBR value of the subgrade is 15%.*

Solution

$$\frac{h}{a} = \frac{45}{15} = 3$$

$$\frac{S_d}{a} = \frac{60}{15} = 4$$

$$\text{CBR} = 15\%$$

From chart (c) of Fig. 19.6.

Load factor value, $L = 1.13$

$$\therefore \quad \text{ESWL, } P_s = \frac{P_d}{L} = \frac{10000}{1.13} = 8850 \text{ kg}$$

19.6.3. Rigid pavement ESWL

All design methods for rigid pavements make use of maximum tensile concrete stress as the salient distress parameter; though there is a difference as to where the stress is computed *i.e.* edge, centre or corner. Stresses in rigid pavements depend in part on the value of relative stiffness which in turn varies with the thickness *h*; Poisson's ratio μ, modulus of elasticity of concrete *E* and modulus of subgrade reaction *k*. For this reason, it is convenient, when determining ESWL, to express the value in terms of radius of relative stiffness, *l*.

In recent years, Portland Concrete Association, USA has developed computer programme. Also stress influence charts are available. Thus ESWL concepts are often not necessary for design.

The two design methods which utilizes the ESWL concept are the load classification number (LCN) method and the Federal Aviation Administration (FAA) method.

LCN method. ESWL for dual and dual-tandem gears can be determined from Fig. 19.7 and 19.8. These have been prepared from the Portland cement association computer programme. The contact

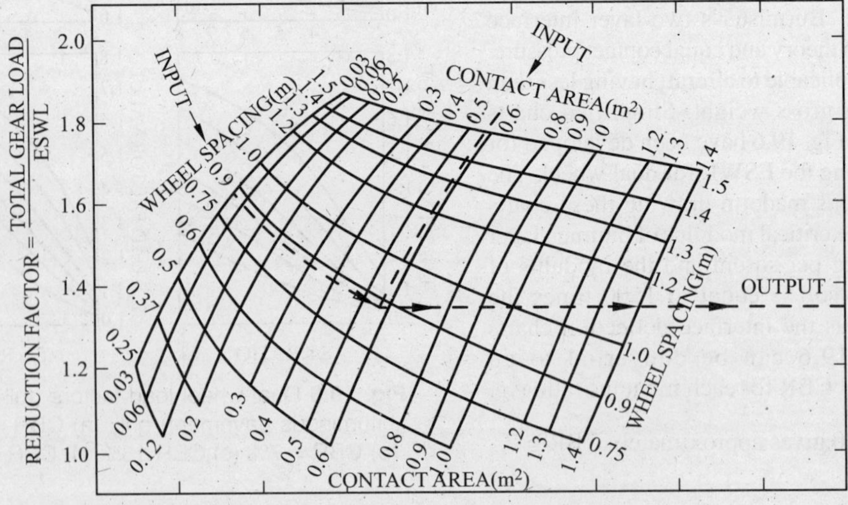

Fig. 19.7 ESWL analysis LCN Method-Dual wheels.

area shown in the figures is the sum of the contact areas of all wheels in the gear assembly. It has been assumed that the tyre pressure is for the ESWL and multiple-gear. Also particular values of E, μ, l and k have been used.

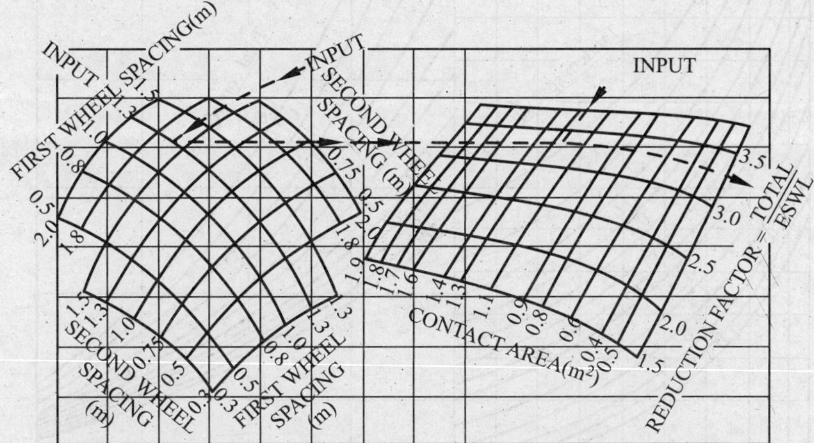

Fig. 19.8 ESWL analysis, LCN Method-Dual tandem wheels.

FAA method. In the LCN method, no dependence of ESWL is made upon pavement thickness or subgrade reaction. These factors are taken into consideration in the FAA method. Fig. 19.9 and 19.10 illustrate the procedure.

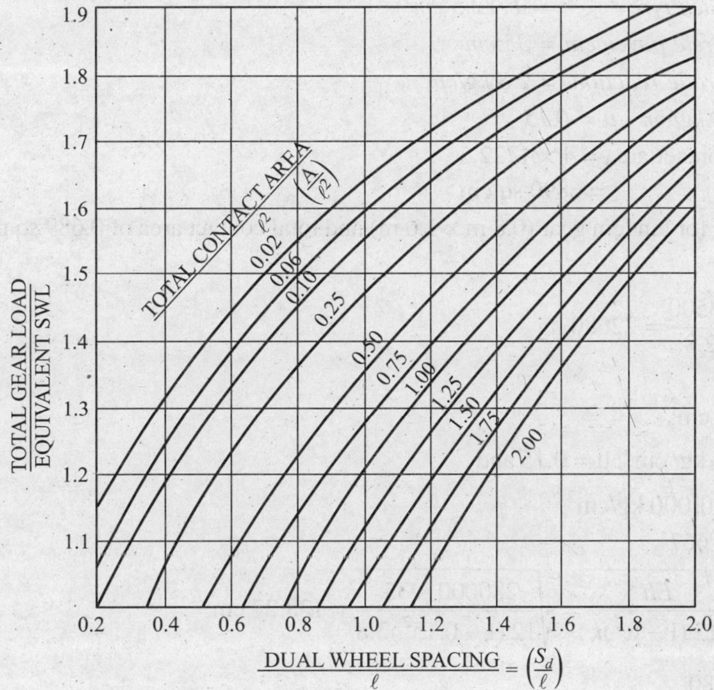

Fig. 19.9 ESWL analysis, dual wheel, concrete pavement FAA method.

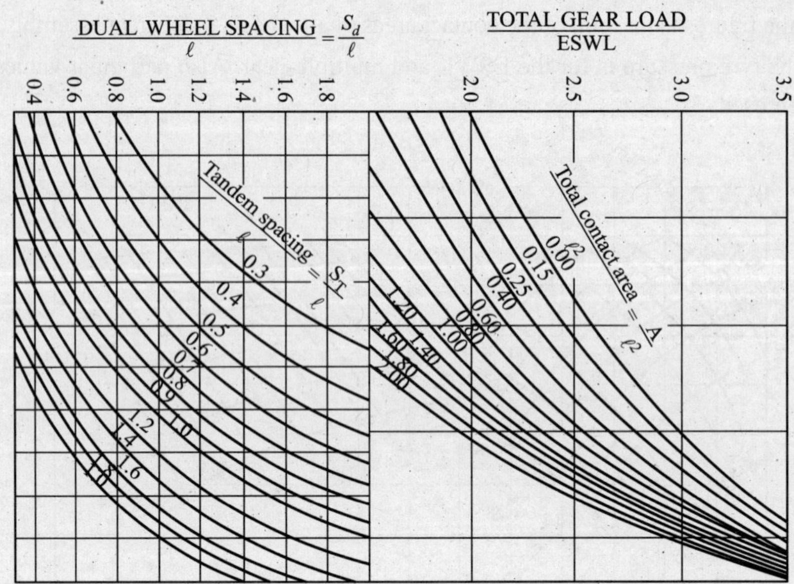

Fig. 19.10 ESWL analysis, dual-tandem wheels, concrete pavement FAA method.

Problem 19.2 *Determine the ESWL by the LCN and FAA methods, from the following data:*

Dual tandem gear (80 cm × 160 cm)

Gross-weight = *63500 kg*

Contact area of each tyre = *1722.5 cm²*

Thickness of concrete pavement = 35 cm

Modulus of sub-grade reaction = *2.8 kg/cm³*

Take E = 280,000 kg/cm², μ = 0.15

Solution Total contact area = 4 × 1722.5

= 6890 sq cm

From Fig. 19.8 for tandem gear (0.8 m × 1.6 m) and total contact area of 0.689 sq m, the reduction factor = 2.8.

$$\therefore \quad \text{ESWL} = \frac{63500}{2.8} = 22680 \text{ kg}$$

FAA Method

$$h = 35 \text{ cm};$$

$$k = 2.8 \text{ kg/cm}^3; \mu = 0.15 \text{ and}$$

$$E = 280{,}000 \text{ kg/cm}^2$$

As given on page 667

$$l = \sqrt[4]{\frac{Eh^3}{12\,(1-\mu^2)k}} = \sqrt[4]{\frac{280000 \times 35^3}{12\,(1-0.15^2)2.8}} = 138.27 \text{ cm}$$

$$\therefore \quad \frac{S_d}{l} = \frac{80}{138.27} = 0.579$$

$$\text{and,} \quad \frac{S_t}{l} = \frac{160}{138.27} = 1.157$$

$$\frac{A}{l^2} = \frac{4\,(1722.5)}{(138.27)^2} = 0.36$$

From Fig. 19.10, reduction factor = 2.2

$$\therefore \quad \text{ESWL} = \frac{63500}{2.2} = 28860 \text{ kg}$$

19.6.4. Impact

Impact effect depends on wheel load, vehicle speed, vehicle characteristics, tyre type and road irregularity. It is not customary to increase the static wheel load used in the design to allow for impact as far as flexible pavements are concerned. This is based on the fact that the wheel load is applied rapidly and in general the strength of a material tested rapidly is higher than the values determined by a static test. This increased resistance to deformation resulting from transient loads cancels the impact effect. Test data observed on the WASHO and AASHTO test roads also indicate that measured stresses and deflection decrease with increasing speeds.

Regarding rigid pavements, loads travelling at transverse joints and cracks may create impact. Tests conducted by the Bureau of Public Roads, U.S.A. indicate that impact due to a load passing over an obstruction may be as much as 150 per cent of the static load. This obviously depends upon the amount a slab is depressed at the joint or crack with respect to adjacent slab. Plain joints without load transfer devices generally depress more than joints with adequate load transfer devices. Similarly, in the case of transverse cracks, the amount of temperature steel, type of subgrade or base course will affect the extent of faulting. For this reason, an impact allowance of 20 per cent is commonly used for rigid pavements.

19.6.5. Repetition of loads

The repeated application of loads to pavements may result in sufficient cumulative permanent deformation to cause failure although deformation for a single application of load may be small. This may be due to accumulated plastic deformation of subgrade or due to fatigue in paving materials or due to both.

If the subgrade is overstressed or poorly compacted, repeated loading may produce a permanent and probably non-uniform deformation of the subgrade. This is particularly significant for rigid pavements as the reduced support given by the subgrade to the slab in these areas may cause the slab to fail. Moreover, if concrete pavements are constructed on pumping soils, high repetitions of heavy wheel loads may result in pavement pumping.

Laboratory and field tests indicate that if a stress less than the peak stress is applied to soil and released a number of times, the stress only remaining for a limited period; there is a gradual increase in the modulus of deformation and the additional permanent strain decreases for the repeated application of the same stress. This may be due to increase in bulk density because of closer packing of the soil particles, and therefore increased rigidity of soil structure. If after a certain number of repeated load applications, the additional permanent strain remains constant, this may be due to plastic flow. Some soils also exhibit susceptibility to complete breakdown after repeated load applications.

Laboratory tests have also indicated that soils under repeated loading will deform much more than identical specimens subjected to sustained static loads of equal magnitude.

Both field plate loading tests and laboratory tests indicate that in general, the amount of deformation under repeated load varies directly with the logarithm of the load applications. This principle can be used with advantage to extend load deformation data from few tests to a large number of repetitions and thus to evaluate the subgrade supporting power for the anticipated load repetitions during the design life of the pavement. This will, however, be true for soils which do not result in breakdown of resistance to deformation after a large number of load applications. Another factor which complicates the analysis is that of mixed traffic.

The above concepts have been used for design purposes in a variety of ways. Firstly, the number of logarithm of designed stress applications can be introduced directly into the design formula or secondly, the repetition concept can be used to vary the design depending on anticipated load repetitions and per cent design thickness required can be correlated with the number of stress repetitions. In the case of airfields, considerable reduction in pavement thickness can be effected in areas where the anticipated traffic is relatively low.

Another concept to take into account repetitions of load is to convert the various loads to some constant equivalent wheel load. Equivalent wheel loads will be those which require identical thickness and quality of pavement, taking into account the repetitions of each load.

19.6.6. Equivalent wheel load factor (EWLF)

An equivalent wheel load factor defines the damage per pass caused to a specific pavement system by the vehicle under consideration relative to the damage per pass of an arbitrarily selected standard vehicle moving on the same pavement system.

Thus by definition

$$F_j = \frac{d_j}{d_s} = \frac{N_{fs}}{N_{fj}} \qquad \ldots(19.3)$$

F_j = equivalency factor or EWLF

d_j = Damage done by any vehicle

d_s = Damage done by standard vehicle

N_{fs} = Number of repetitions to failure of vehicle s

N_{fj} = Number of repetitions to failure of vehicle j

Equation 19.3 shows that the equivalent damage factor F_j is a function of the number of repetitions to failure. This in turn is a direct function of how the design criteria has been developed *i.e.*, theoretically or empirically.

The design criterion can be either on the basis of principal tensile strains at the bottom of a flexible pavement or tensile stress in the case of rigid pavements. The relations are of the general form:

$$F_j = \left(\frac{\epsilon_j}{\epsilon_s}\right)^c \quad \text{Where } \epsilon \text{ refers to strain}$$

$$F_j = \left(\frac{\sigma_j}{\sigma_s}\right)^c \quad \text{Where } \sigma \text{ refers to stress}$$

Both laboratory and field studies have shown that c ranges between 3 and 6, the common value used being 4.

A general form of EWLF or F_j can be expressed in terms of percent gross weight of the vehicle to that of gross weight of standard vehicle. The relation is given by

$$F_j = \left(\frac{X}{100}\right)^4 \qquad \ldots(19.4)$$

Where X is the per cent gross-weight of the vehicle.

AASHO factors. In highway design, equivalency factors developed on the basis of AASHO road test are mostly used. This number of repetitions to failure, for flexible or rigid pavement, are expressed in terms of (*i*) pavement rigidity or stiffness value, (*ii*) load characteristics, and (*iii*) the terminal level of serviceability selected.

For flexible pavements, the rigidity is expressed by the structural number (SN). SN is defined as an index number obtained by the summation of the products of layer thickness and layer coefficient expressed numerically:

$$SN = a_1 D_1 + a_2 D_2 + a_3 D_3 \quad \dots (19.5)$$

where a_1, a_2 and a_3 are coefficients for surface, base and subbase respectively and D_1, D_2, D_3 are the respective thicknesses.

For rigid pavements, the rigidity is expressed in terms of the thickness D.

Vehicle characteristics are denoted by L, the axle load (single or dual) and L_2, axle code (one for single axle and 2 for tandem axle).

Common values of terminal serviceability, p_t, are 2.0 and 2.5.

A summary of the F_j values for both flexible and rigid pavements are given in Tables 19.3 and 19.4. It may be observed that the F_j values are not significantly effected by either the SN, D value or p_t value (factor of safety).

Fatigue of pavement materials by repeated stressing. Repeated loading may produce fatigue in concrete. The relation between the factor of safety and the number of repetitions required to induce failure has been given by Sheets for concrete in flexure. This relationship is reproduced in Fig. 19.11.

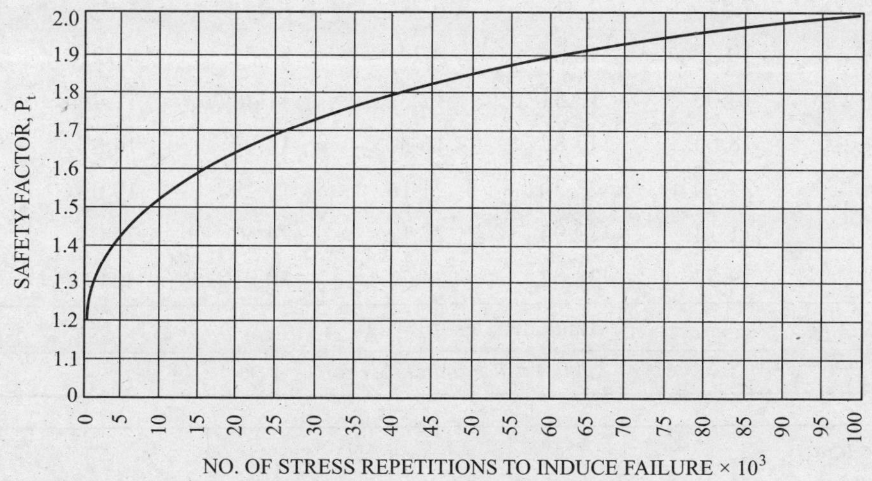

Fig. 19.11 Fatigue of concrete in flexure.

This shows that plain concrete can be subjected to flexural stress almost indefinitely as long as extreme fibre stress is less than about 50 per cent of the modulus of rupture. Thus, all pavement slabs should be designed to carry the predominating wheel load with a factor of safety of at least 2. The predominating wheel load means the heaviest wheel load which will cause at least 100,000 repetitions of stress during the life for which the pavement is being designed.

Table 19.3 AASHO equivalence factors—flexible pavement

Single axle, $p_t = 2.0$

Axle load (Kips)*	Structural number, SN 1	2	3	4	5	6
2	0.0002	0.0002	0.0002	0.0002	0.0002	0.0002
4	0.002	0.003	0.002	0.002	0.002	0.002
6	0.01	0.01	0.01	0.01	0.01	0.01
8	0.03	0.04	0.04	0.03	0.03	0.03
10	0.08	0.08	0.09	0.08	0.08	0.08
12	0.16	0.18	0.19	0.18	0.17	0.17
14	0.32	0.34	0.35	0.35	0.34	0.33
16	0.59	0.60	0.61	0.61	0.60	0.60
18	1.00	1.00	1.00	1.00	1.00	1.00
20	1.61	1.59	1.56	1.55	1.57	1.60
22	2.49	2.44	2.35	2.31	2.35	2.41
24	3.71	3.62	3.43	3.33	3.40	3.51
26	5.36	5.21	4.88	4.68	4.77	4.96
28	7.54	7.31	6.78	6.42	6.52	6.83
30	10.38	10.03	9.24	8.65	8.73	9.17
32	14.00	13.51	12.37	11.46	11.48	12.17
34	18.55	17.87	16.30	14.97	14.87	15.63
36	24.20	23.30	21.16	19.28	19.02	19.93
38	31.14	29.95	27.12	24.55	24.03	25.10
40	39.57	38.02	34.34	30.92	30.04	31.25

Tandem axles, $p_t = 2.0$

Structural number, SN

Axle load (Kips)	1	2	3	4	5	6
10	0.01	0.01	0.01	0.01	0.01	0.01
12	0.01	0.02	0.02	0.01	0.01	0.01
14	0.02	0.03	0.03	0.03	0.02	0.02
16	0.04	0.05	0.05	0.05	0.04	0.04
18	0.07	0.08	0.08	0.08	0.07	0.07
20	0.10	0.12	0.12	0.12	0.11	0.10
22	0.16	0.17	0.18	0.17	0.16	0.16

24	0.23	0.24	0.26	0.25	0.24	0.23
26	0.32	0.34	0.36	0.35	0.34	0.33
28	0.45	0.46	0.49	0.48	0.47	0.46
30	0.61	0.62	0.65	0.64	0.63	0.62
32	0.81	0.82	0.84	0.84	0.83	0.82
34	1.06	1.07	1.08	1.08	1.08	1.07
36	1.38	1.38	1.38	1.38	1.38	1.38
38	1.76	1.75	1.73	1.72	1.73	1.74
40	2.22	2.19	2.15	2.13	2.16	2.18
42	2.77	2.73	2.64	2.62	2.66	2.70
44	3.42	3.36	3.23	3.18	3.24	3.31
46	4.20	4.11	3.92	3.83	3.91	4.02
48	5.10	4.98	4.72	4.58	4.68	4.83

* $1 kip = 453.6$ kg

Table 19.3 (continued)

Single axle, $p_t = 2.5$

Structural number, SN

*Axle load (Kips)**	*1*	*2*	*3*	*4*	*5*	*6*
2	0.0004	0.0004	0.0003	0.0002	0.0002	0.0002
4	0.01	0.004	0.004	0.004	0.003	0.002
6	0.01	0.02	0.02	0.01	0.01	0.01
8	0.03	0.05	0.05	0.04	0.03	0.03
10	0.08	0.10	0.12	0.10	0.09	0.08
12	0.17	0.20	0.23	0.21	0.19	0.18
14	0.33	0.36	0.40	0.39	0.36	0.34
16	0.59	0.61	0.65	0.65	0.62	0.61
18	1.00	1.00	1.00	1.00	1.00	1.00
20	1.61	1.57	1.49	1.47	1.51	1.55
22	2.48	2.38	2.17	2.09	2.18	2.30
24	3.69	3.49	3.09	2.89	3.03	3.27
26	5.33	4.99	4.31	3.91	4.09	4.48
28	7.49	6.98	5.90	5.21	5.39	5.98
30	10.31	9.55	7.94	6.83	6.97	7.79
32	13.90	12.82	10.52	8.85	8.88	9.95

34	18.41	16.94	13.74	11.34	11.18	12.51
36	24.02	22.04	17.73	14.38	13.93	15.50
38	30.90	28.30	22.61	18.06	17.20	18.98
40	39.26	35.89	28.51	22.50	21.08	23.04

Tandem axles, $p_t = 2.5$

Structural Numbers, SN

*Axle load (Kips)**	*1*	*2*	*3*	*4*	*5*	*6*
10	0.01	0.01	0.01	0.01	0.01	0.01
12	0.02	0.02	0.02	0.02	0.02	0.01
14	0.03	0.04	0.04	0.03	0.03	0.02
16	0.04	0.07	0.07	0.06	0.05	0.04
18	0.07	0.10	0.11	0.09	0.08	0.07
20	0.11	0.14	0.16	0.14	0.12	0.11
22	0.16	0.20	0.23	0.21	0.18	0.17
24	0.23	0.27	0.31	0.29	0.26	0.24
26	0.33	0.37	0.42	0.40	0.36	0.34
28	0.45	0.49	0.55	0.53	0.50	0.47
30	0.61	0.65	0.70	0.70	0.66	0.63
32	0.81	0.84	0.89	0.89	0.86	0.83
34	1.06	1.08	1.11	1.11	1.09	1.08
36	1.38	1.38	1.38	1.38	1.38	1.38
38	1.75	1.73	1.69	1.68	1.70	1.73
40	2.21	2.16	2.06	2.03	2.08	2.14
42	2.76	2.67	2.49	2.43	2.51	2.61
44	3.41	3.27	2.99	2.88	3.00	3.16
46	4.18	3.98	3.58	3.40	3.55	3.79
48	5.08	4.80	4.25	3.98	4.17	4.19

* 1 Kip = 453.6 kg

Table 19.4 AASHO equivalence factors-rigid pavement

Single axles, $p_t = 2.0$

h-slab thickness (cm)

*Axle load (kips)**	*15.0*	*17.5*	*20.0*	*22.5*	*25.0*	*27.5*
2	0.0002	0.0002	0.0002	0.0002	0.0002	0.0002
4	0.002	0.002	0.002	0.002	0.002	0.002
6	0.01	0.01	0.01	0.091	0.01	0.01
8	0.03	0.03	0.03	0.03	0.03	0.03

10	0.09	0.08	0.08	0.08	0.08	0.08
12	0.19	0.18	0.18	0.18	0.17	0.17
14	0.35	0.35	0.34	0.34	0.34	0.34
16	0.61	0.61	0.60	0.60	0.60	0.60
18	1.00	1.00	1.00	1.00	1.00	1.00
20	1.55	1.56	1.57	1.57	1.58	1.59
22	2.32	2.32	2.35	2.38	2.40	2.41
24	3.37	3.34	3.40	3.47	3.51	3.53
26	4.47	4.69	4.77	4.88	4.97	5.02
28	6.59	6.44	6.52	6.70	6.85	6.94
30	8.92	8.68	8.74	8.98	9.23	9.39
32	11.87	11.49	11.51	11.82	12.17	12.44
34	15.55	15.00	14.95	15.30	15.78	16.18
36	20.07	19.30	19.16	19.53	20.14	20.71
38	25.56	34.54	24.26	24.63	25.36	26.14
40	32.18	30.85	30.41	30.75	31.58	32.57

Tandem axles, $p_t = 2.0$

h-slab thickness (cm)

*Axle load (kips)**	*15.0*	*17.5*	*20.0*	*22.5*	*25.0*	*27.5*
10	0.01	0.01	0.01	0.01	0.01	0.01
12	0.03	0.03	0.03	0.03	0.03	0.03
14	0.06	0.05	0.05	0.05	0.05	0.05
16	0.10	0.09	0.08	0.08	0.08	0.08
18	0.16	0.14	0.14	0.13	0.13	0.13
20	0.23	0.22	0.21	0.21	0.20	0.20
22	0.34	0.32	0.31	0.31	0.30	0.30
24	0.48	0.46	0.45	0.44	0.44	0.44
26	0.64	0.64	0.63	0.62	0.62	0.62
28	0.85	0.85	0.85	0.85	0.85	0.85
30	1.11	1.12	1.13	1.14	1.14	1.14
32	1.43	1.44	1.47	1.49	1.50	1.51
34	1.82	1.82	1.87	1.92	1.95	1.96
36	2.29	2.27	2.35	2.43	2.48	2.51
38	2.85	2.80	2.91	3.04	3.12	3.16
40	3.52	3.42	3.55	3.74	3.87	3.94
42	4.32	4.16	4.30	4.55	4.74	4.86
44	5.26	5.01	5.16	5.48	5.75	5.92
46	6.36	6.01	6.14	6.53	6.90	7.14
48	7.64	7.16	7.27	7.73	8.21	8.55

* 1 Kip = 453.6 kg

Table 19.4 (continued)

Single axles, p_t = 2.5

	h-slab thickness (cm)					
Axle load (kips)	*15.0*	*17.5*	*20.0*	*22.5*	*25.0*	*27.5*
2	0.0002	0.0002	0.0002	0.0002	0.0002	0.0002
4	0.003	0.002	0.002	0.002	0.002	0.002
6	0.01	0.01	0.01	0.01	0.01	0.01
8	0.04	0.04	0.03	0.03	0.03	0.03
10	0.10	0.09	0.08	0.08	0.08	0.08
12	0.20	0.19	0.18	0.18	0.18	0.17
14	0.38	0.36	0.35	0.34	0.34	0.34
16	0.63	0.62	0.61	0.60	0.60	0.60
18	1.00	1.00	1.00	1.00	1.00	1.00
20	1.51	1.52	1.55	1.57	1.58	1.58
22	2.21	2.20	2.28	2.34	2.38	2.40
24	3.16	3.10	3.23	3.36	3.45	3.50
26	4.41	4.26	4.42	4.67	4.85	4.95
28	6.05	5.76	5.92	6.29	6.61	6.81
30	8.16	7.67	7.79	8.28	8.79	9.14
32	10.81	10.06	10.10	10.70	11.93	11.99
34	14.12	13.04	12.94	13.62	14.59	15.43
36	18.20	16.69	16.41	17.12	18.33	19.52
38	23.15	21.14	20.61	21.31	22.74	24.31
40	29.11	26.49	25.65	26.29	27.91	29.90

Tandem axles, p_t = 2.5

	h-slab thickness (cm)					
*Axle load (kips)**	*15.0*	*17.5*	*20.5*	*22.5*	*25.0*	*27.5*
10	0.01	0.01	0.01	0.01	0.01	0.01
12	0.03	0.03	0.03	0.03	0.03	0.03
14	0.06	0.05	0.05	0.05	0.05	0.05
16	0.10	0.09	0.08	0.08	0.08	0.08
18	0.16	0.14	0.14	0.13	0.13	0.13
20	0.23	0.22	0.21	0.21	0.20	0.20
22	0.34	0.32	0.31	0.31	0.30	0.30
24	0.48	0.46	0.45	0.44	0.44	0.44
26	0.64	0.64	0.63	0.62	0.62	0.62
28	0.85	0.85	0.85	0.85	0.85	0.85
30	1.11	1.12	1.13	1.14	1.14	1.14
32	1.43	1.44	1.47	1.49	1.50	1.51
34	1.82	1.82	1.87	1.92	1.95	1.96

36	2.29	2.27	2.35	2.43	2.48	2.51
38	2.85	2.80	2.91	3.04	3.12	3.16
40	3.52	3.42	3.55	3.74	3.87	3.94
42	4.32	4.16	4.30	4.55	4.74	4.86
44	5.26	5.01	5.16	5.48	5.75	5.92
46	6.36	6.01	6.14	6.53	6.90	7.14
48	7.64	7.16	7.27	7.73	8.21	8.55

* 1 Kip = 453.6 kg

Application of fatigue principle can be used to develop a rational basis for the design of rigid pavements where the traffic includes only a few of the heaviest loads. The safety factors to be used with each of the less frequent heavier loads should be of such a value that the pavement will have adequate structural life when subjected to the anticipated volume of these heavier loads. Such pavements must be designed to carry the heaviest legal wheel loads, but because of their location they will be called upon to carry only a limited number of such loads. From the fatigue curve the required safety factor can be determined for the predicated total number of repetitions of each load. The working stress obtained by dividing the modulus of rupture by the safety factor can then be used in designing the thickness of the pavement. For airfield pavement, a factor of safety of 1.5 will be able to withstand about 10,000 repetitions of load during its life, and design of normal traffic usage are based on this number of load repetitions. If a pavement is required to be such that the number of load repetitions to be carried is increased from 10,000 to 40,000, the flexural strength used in original design will need a factor of safety of about 1.8.

When the factor of safety is less than 2, typical concrete fatigue curves may be used. Fig. 19.12 shows two such curves developed by Bradbury and Portland cement association for airfield design.

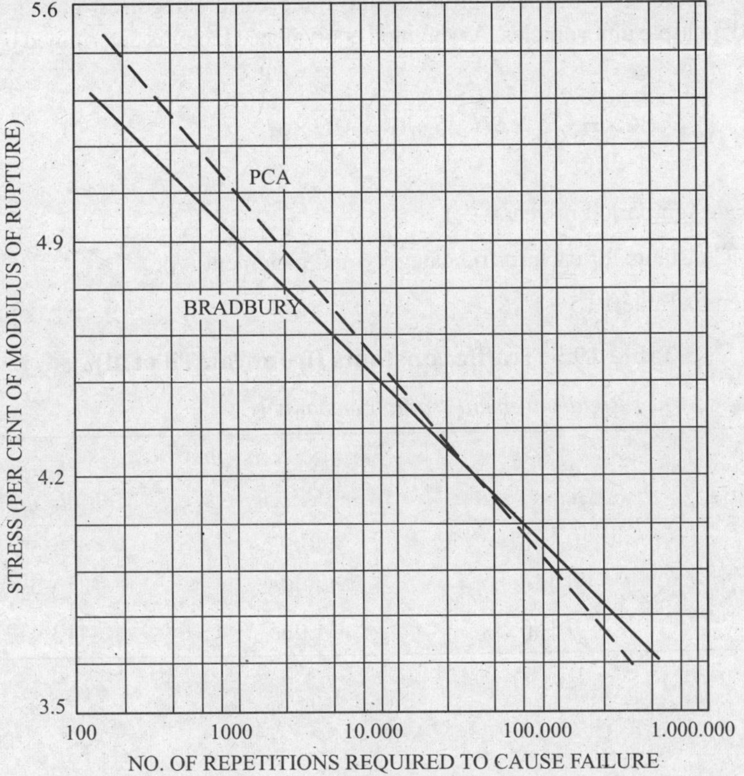

Fig. 19.12 Fatigue curves for plain concrete in flexure.

Regarding flexible pavements, bituminous concrete may also fail by fatigue. Map cracking in bituminous concrete surface may result from repeated load applications. However, weak or improperly compacted base courses can augment this type of failure. Further, base course materials may be affected by repeated load applications in two ways. (*i*) some aggregates may break down, particularly soft aggregate, and (*ii*) subgrade soil may intrude more into the base course particularly in soils containing large percentage of fines and when free water is available in the base.

This concept of effects of repeated load applications can permit economical use of available materials and can also lead to construction by stages. Initially, thin pavement, at less cost, can be laid and later as traffic and life increase, the thickness can be increased.

19.6.7. Equivalent Axis Load (EAL) for Highway Design

The various methods used for determining the equivalent axle loads for highway design are briefly discussed below.

(1) *AASHO test method*

The most exact method of analysis involves the use of a detailed traffic mix analysis, by axle load groupings corresponding to the AASHO equivalency factors given in Tables 19.3 and 19.4.

(2) *Van Til method*

(*a*) In one method, all vehicles are classified by axle type (*i.e.* number of axles) and each axle type may be represented by an average equivalency factor F_j. The present number of average daily trucks in both directions is multiplied by the expansion factor based on growth trends. This will give expanded average daily trucks at the end of 10 years. This is multiplied with the equivalent 8200 kg single axle loads. The sum of these values will yield the total annual value used in the design.

(*b*) In another method, the traffic is grouped into three basic categories; passenger cars, single unit vehicles, and multiple unit vehicles. A weighted equivalence factor is determined by the following relation

$$N_{8200} = DP\frac{C_1\overline{PC}\times P + C_2\times\overline{SU}\times S + C_3\times\overline{MU}\times M}{10^6} \quad \text{...(19.5)}$$

where DP = design period in years

C_1, C_2, C_3 = Constants for each traffic category and pavement type.

These are given in Table 19.5

Table 19.5 Traffic constants (from Van Til et al)

Rigid-pavement traffic constants (C)

Highway class	*Traffic constants*		
	Passenger cars	*Single units*	*Multiple units*
I	0.146	44.995	421.57
II	0.146	44.995	413.91
III	0.146	44.995	413.91

Flexible-pavement traffic constants (C)

Highway class	Passenger cars	Single units	Multiple units
	Traffic constants		
I	0.146	42.705	345.655
II	0.146	39.785	337.260
III	0.146	35.770	289.810
IV	0.146	9.855	78.840

P, S, M = per cent values of ADT in the design lane for each category *i.e.* $\overline{PC}, \overline{SU}, \overline{MU}$, = two directional ADT values for each traffic category.

The classification of highway categories is as follows:

Highway class

I All roads or streets four or more lanes.

II Two or three lanes, design traffic > 1000 ADT.

III Design ADT between 400 and 1000.

IV ADT < 400 vehicles per day.

(3) *Asphalt Institue method*

The method is bsed upon an extensive loadometer studies conducted in the U.S.A. It was established that the equivalent 8200 kg axle load (N_{8200}) was correlated to the single-axle load limit, average gross weight of heavy trucks and number of heavy trucks (daily average in the design lane). A heavy truck is defined as a two axle 6 tyre truck or larger. Fig. 19.13 is a nomograph used for the method. The initial traffic number (ITN) represents the daily number of N_{8200} load repetitions on the design lane. The directional split of traffic is assumed on equal basis. If there are two-lanes in each direction, 80 to 100 per cent of the loads are assumed to occur in the design lane. For three or more lanes in each direction, the above percentage may be taken as 60 to 80.

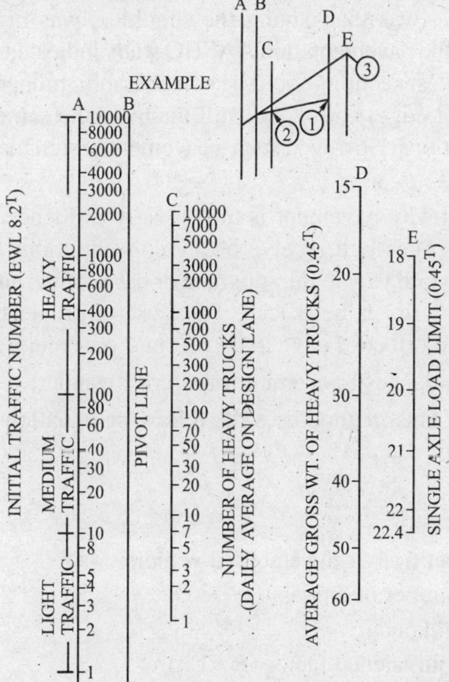

Fig. 19.13 Asphalt Institute nomograph chart for solving EAL_{8200} kg.

19.6.8. Disturibution of wheel loads

Vehicles moving on a highway or airfield do not travel in the exact transverse location because they possess some lateral wander. The degree of wander varies significantly with the pavement type. Two considerations are mainly taken into account (*i*) the probable location of the maximum load repetitions and hence the probable maximum damage location; and (*ii*) the relative degree of damage per pass as the wander effect is increased. The standard deviation for various types of pavements is given in Table 19.6.

Table 19.6 Standard deviation (σ) for different pavements

Pavement type	*Standard deviation*	
	Range	*Normal*
Highways		0.3 m
Taxiways	0.6 to 1 m	1 m
Runway		
Take off	2.25 to 4.5 m	4.8 m
Landing	4 to 6 m	

For highways it is assumed that each movement results in one stress (strain) repetition because of extremely channelized (low σ) effect.

Traffic is highly channelized on highway pavements at a distance of 0.5 to one metre from the edge depending upon the pavement width. Flexible pavements show greater distress near the edges because shear failures and lateral shoving are likely to occur through the shoulder, and secondly the subgrade soil near the edges is subjected to severe fluctuations in moisture conditions. For the same reasons, on multilane road the outer wheel path will show greater distress than the inner wheel path. This effect of edge loads was also confirmed in the WASHO and AASHO road tests. The conclusion drawn in the test project referred to above was that paving the shoulders was highly effective in reducing the amount of distress to the flexible pavements and AASHO study indicated that inner wheel path pavement (closest to the centre line) of test sections could be considerably thinner than pavement in outer wheel path (closest to the pavement edge) and would still maintain the same level of serviceability at any given number of load applications. However, such a pavement design may not prove to be economically justified.

Performance of rigid highway pavement is also greatly influenced by edge loading conditions. This is due to the fact that edges will receive heavy concentrations of rain water runoff from the pavement and this water may find way to the subgrade or base course and secondly, increased stiffness of the pavment slab at the interior is beneficial. Thus, use of proper base course or paved shoulder will minimize the detrimental effect of edge loads on rigid pavements.

There are three basic approaches to account for aircraft wander:

Asphalt Institute method. This method uses the following equation in the design

$$n_e = \sum_{j=1}^{J} p_j \, f_{j(a-b)} F_j \qquad \ldots(19.7)$$

where n_e = Equivalent repetition of the standard vehicles.

p stands for number of passes

f stands for frequency

F stands for equivalence factor (EWLF)

Equivalency diagrams for 22 different commerical jet aircrafts have been drawn.

Portland Cement Association (PCA) method. The load repetition factor (LRF) is introduced to account for the lateral wander effect of aircraft. A LRF of 1 denotes that each pass is a full load repetition. LRF values for important aircrafts for different standard deviations (σ), for taxiways and runways are given in Table 19.7.

TABLE 19.7 Load repetition factors for several aircraft

Aircraft	*Load repetition factor (tentative design values)* Taxiway		Runway	
	$\sigma = 60$ *cm*	$\sigma = 120$ *cm*	$\sigma = 240$ *cm*	$\sigma = 480$ *cm*
DC-3	0.12	0.07	0.05	0.03
B-727	0.41	0.23	0.13	0.09
DC-8 and B-707	0.83	0.46	0.25	0.17
B-747	0.58	0.38	0.33	0.28
C5A	0.74	0.61	0.37	0.25
B-2707	0.52	0.39	0.22	0.15
Concorde	0.83	0.44	0.23	0.15
DC-10-10 and L10-11	0.57	0.40	0.22	0.12
Future No. 4	1.33	0.84	0.44	0.24

Corps of engineers method. The lateral distribution of traffic is assumed to have a uniform distribution in contrast to normal distribution. In this method, the number of passes or movements to achieve one coverage is determined. One coverage indicates that each point of the pavement wihin the design traffic width receives one application of load.

The concept of number of coverages (C) per operation (O) is similar to the LRF developed by Portland Cement Association.

$$\frac{C}{O} = \frac{\alpha n_t w_t}{100\,T} \qquad \text{...(19.8)}$$

where α = per cent traffic in the design lane

n_t = number of tyres in the gear

w_t = width of one tyre in cm

T = width of traffic lane is m.

The value commonly used are as below:

	α	T
Taxiway	75%	2.25 m
Runway	75%	11.25 m

As already discussed, for airfield pavements, the design of pavement can be based on number of repetitions anticipated at various sections of runway and taxiway depending on the channelization of traffic. Data indicate that high stress repetitions occur at the ends of runways and taxiways, at turnoffs from taxiways to the aprons, and at the centre of runway pavement. Because of this, great saving in initial cost can be achieved by permitting less thickness at relatively less trafficked portions.

19.6.9. Traffic Growth

Traffic intensity is a growing phenomenon, the heaviest intensity will occur at the end of the design life of a pavement. It is therefore essential to predict traffic growth. However it is generally considered adequate if the traffic is projected to a period of 20 years after construction.

For traffic prediction on main highways in India, the following correlation is used:

$$T = p\,(1+r)^{n+20}$$

where T = design traffic intensity in terms of number of commercial vehicles (laden weight > 3 tonnes) per day

p = traffic intensity at last traffic count

r = annual rate of increase of traffic intensity, and

n = number of years since last traffic count and commissioning of the new pavement.

The traffic intensity p for assessment of design traffic intensity T should normally be a seven day average based on 24-hours counts, in accordance with IRC : 9—1972 : Traffic census on non-urban roads (First Revision). However, in exceptional cases, where such data are not available, an average of three days counts may be used as an approximation. Based on growth rate of traffic over the past few years, value of 75 per cent is suggested for 'r' for rural roads for the time being wherever actual data are not available.

In case of new highway links, where no traffic count data will be available, data from highway for similar classification and importance may be used to predict the design traffic intensity.

The pavement classification based on design traffic intensity, suggested for adoption for rigid pavement design, is given in Table 19.8.

Table 19.8

Traffic classification for rigid pavement design

Traffic classification	*Design traffic intensity : Vehicles (laden weight > 3 tonnes) per day at the end of design life*
A	0—15
B	15—45
C	45—150
D	150—450
E	450—1500
F	1500—4500
G	Above 4500 and all expressways

A generalized plot of equivalent 8160 kg single axle loads (EAL) versus time is shown in Fig. 19.14. The accumulated EAL over the design period will be the area under the curve. The relation can be expressed as:

$$\sum_{0}^{n} EAL = \frac{EAL_0\,(365)}{\log_e (1+r)}\,(1+r)^n - 1 \qquad \ldots(19.9)$$

where EAL_0 = Initial daily EAL on the day road is opened to traffic.

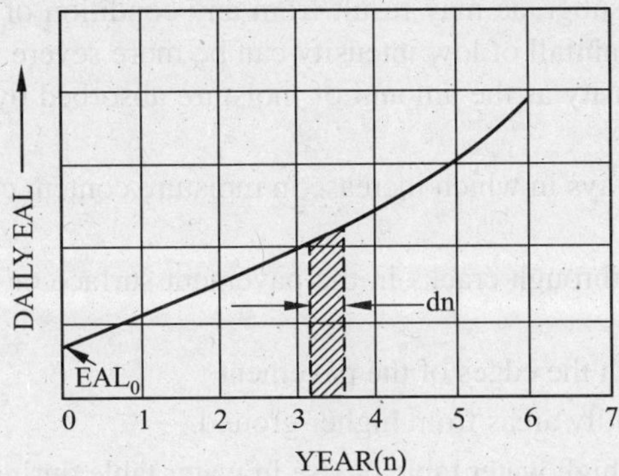

Fig. 19.14. Relationship between EAL and years (n) for a growth rate (*r*) per cent.

19.6.10. Contact Area and Tyre Pressure

Tyre imprint areas and dimensions for specific loadings and tyre pressures generally are available. When such data are not readily available, they may be closely approximated as follows:

Knowing the weight on the gear or axle, the weight on each tyre can be easily obtained. The tyre contact area can be computed by dividing the tyre load by the tyre pressure, making due allowance for the support furnished by the sidewalls of the tyre. The usual value based on investigations for support furnished by sidewalls is about 10 per cent of the total support. Thus, the contact area is equal to 0.9 × wheel load/tyre pressure.

The contact area can also be worked out by taking actual impression of tyre imprints.

Regarding the effect of tyre pressures on the design of pavement it is the total wheel load which influences the thickness requirement for flexible pavements. Tyre pressures vary but little with the thickness of pavement but the requirement for a high degree of stability for high pressure tyres for flexible pavement is most pronounced in the surface and base course of the pavement.

For rigid pavements, the theoretical analysis as well as investigations indicate that for a 0.7 kg/cm^2 change in tyre inflation pressure there is approximately (0.7 to 1.0) kg/cm^2 variation in stress induced in the slab. With usual method of design one kg/cm^2 change in stress represents a difference in thickness of less than 6.3 mm. This means that increasing the tyre pressure by 0.7 kg/cm^2 would necessitate an increase in slab thickness of about 6 mm.

19.7. MOISTURE FACTORS

If the moisture content of the subgrade changes after construction, it may swell or shrink and change the strength. This may cause the surface of the road to deteriorate. For example, a road built on a clay subgrade may fail owing to inadequate strength, following a comparatively small increase in the moisture content of the clay. Nearly all soils decrease in strength with increasing moisture content for a given dry density and this effect is most marked in clayey soils. Likewise rainfall has an immediate effect on pumping of rigid pavements and surface infiltration at crack joints and pavement edges adds to the severity of the problem. For pumping to take place in rigid pavements three factors must be present: (*i*) high repetition of heavy axle loads, (*ii*) pumping susceptible soils, particularly plastic clays, and (*iii*) free water immediately under the pavement.

Decrease of moisture content, on the other hand, may increase the strength of the subgrade but long drying period causes the soil near the edges of the road to dry out more than that of the subgrade with the centre of the road. In clayey soil this may cause considerable differential shrinkage of the soil with the result that longitudinal cracks may result in a flexible pavement.

Increase in moisture in subgrade may result from any condition of precipitation, capillarity and irrigation. Periods of long rainfall of low intensity can be more severe than concentrated rainfall for a short period of high intensity as the amount of moisture absorbed by the soil is greater under the former condition.

There are a number of ways in which increase in moisture content can occur in the subgrade of a road. These are:

(*a*) Percolation of water through cracks in the pavement surface or due to open gradation of the surface.

(*b*) Entry of water through the edges of the pavement.

(*c*) Seepage of water in hilly areas from higher ground.

(*d*) Capillary rise from a high water table or rise in water table during rainy season.

(*e*) Transfer of water vapour through soil due to difference in vapour pressure which may arise from differences in moisture content or from differences in temperature existing between different parts of the soil.

(*f*) Lateral movement from the shoulders.

Many of the causes of increase in moisture content can be controlled by proper design. The construction of side ditches or sub-soil drains, where needed will help to intercept or remove water from the subgrade moving through the soil under the action of gravity. The infiltration of water through the surface and edges of the pavement can be greatly minimized by providing a water-proof surface and by periodical sealing, and reshaping of the shoulders. In areas of high rainfall the use of impervious shoulder materials is also very helpful. However, some of the water is retained in the soil by surface forces and cannot be removed by drains. The distribution of this held water is determined by the equilibrium conditions of suction and vapour pressure in the soil. This held water is known as *equilibrium moisture content*.

For economical design of flexible pavement, as assessment of the strength of subgrade should be made at equilibrium moisture content. The equilibrium moisture content depends on a number of factors and its estimation will depend on the knowledge of seasonal moisture variation at the site concerned. This involves a lengthy procedure and in many cases it is necessary to assess the value from the local knowledge of the worst conditions that arise at the wettest period of the year. In the case of extension to existing roads and airfields, the equilibrium moisture content is obtained by digging trial holes through any existing pavement over similar soils and determining the moisture content of the soil as it exists under the pavement.

When it is anticipated that moisture content under the subgrade may approach 100 per cent saturation during rainy seasons, the strength of subgrade should be assessed on soaked specimens. This is the assumption made in most of the design tests and the total thickness of pavement is designed on the assumption that the subgrade and/or base course will possess a high degree of saturation for most of its life. This is the standard procedure which has been used by most agencies using CBR method of design. However, this procedure of soaking specimens before making the test may sometimes be too severe where equilibrium moisture content is lower than the saturation moisture content.

Another method of taking rainfall into consideration is to use certain modifications of the design curves or formula based on intensity of rainfall or drainage conditions.

19.8. CLIMATIC FACTORS

Another important factor affecting pavement performance is that of climate, which includes temperature differential and frost action, both frost heave and loss of subgrade support during the thaw period.

Temperature differential between top and bottom of concrete pavements will affect the maximum spacing of expansion and contraction joints in the pavement. These are a function of solar radiation received by the pavement surface at the location, losses due to wind velocity, etc. and thermal diffusivity of concrete and are thus affected by the geographical features of the pavement location. As far as possible, values of actual anticipated temperature differentials at the location of the pavement should be adopted for pavement design. However, guidance can be had from Table 19.9 prepared on the basis of actual observations by CRRI, New Delhi.

The term frost heave refers to rise of a portion of the road as a direct result of the formation of ice crystals in a frost-susceptible subgrade or base course. For frost heave to occur, the factors are : (*i*) a frost-susceptible soil, (*ii*) slow temperature decrease below the freezing point, and (*iii*) free supply of water.

Table 19.9 Recommended temperature differentials in concrete roads

Zone	*States*	*Temp. differential in °C in slabs of thickness*				
		10 cm	*15 cm*	*20 cm*	*25 cm*	*30 cm*
I	Punjab, U.P., Uttarakhand, Rajasthan, Gujarat, Haryana and North M.P. exclusing hilly regions and coastal areas	10.2	12.5	13.1	14.3	15.8
II	Bihar, Jharkhand, West Bengal, Assam and Eastern Odisha, excluding hilly regions and coastal areas	14.4	15.6	16.4	16.6	16.8
III	Maharashtra, Karnataka, South M.P., Chattisgarh, Andhra Pradesh, Western Odisha and North Tamil Nadu, excluding hilly regions and coastal areas	14.75	17.3	19.0	20.3	21.0
IV	Kerala and South Tamil Nadu, excluding hilly regions and coastal areas	13.2	15.0	16.4	17.6	18.1
V	Coastal areas bounded by hills	12.8	14.6	15.8	16.2	17.0
VI	Coastal areas unbounded by hills	13.6	15.5	17.0	19.0	19.2

The water in the large capillary openings freezes at or near normal freezing temperature, whereas the water in the small capillary openings get converted into ice crystals and there is an increase in size. This enlargement progresses as long as capillarity continues to supply water, and eventually well defined ice layers are built up and the soil above them heaves. The force exerted by the heaving soil and by the impact on affected areas, damages the wearing surface of pavement. This failure progresses when thawing occurs and when additional water from melting snows and rains enters the subgrade directly through the failed surface.

Studies have indicated that all inorganic soils which contain greater than 3 per cent by weight of particles finer than 0.02 mm are susceptible to frost. However, the degree of susceptibility depends upon soil type and the different categories of soils, in the increasing order of susceptibility, are given in Table 19.10. General difficulties with frost heave occur in fine silts and sand silt mixtures.

Table 19.10 Frost-susceptible soils

Group	*Description*
F1	Gravelly soils containing between 4 and 20 per cent finer than 0.02 mm by weight.
F2	Sands containing between 3 and 15 percent finer than 0.02 mm by weight.
F3	(*a*) Gravelly soils containing more than 20 per cent finer than 0.02 mm by weight, and sands, except fine silty sands, containing more than 15 per cent finer than 0.02 mm by weight (*b*) Clays with plasticity indices of more than 12, except (*c*) Varved clays existing with uniform conditions.
F4	(*a*) All silts including sandy silts, (*b*) Fine silty sands containing more than 15 per cent finer than 0.02 mm by weight, (*c*) Lean clays with plasticity indices of less than 12, (*d*) Varved clays with non-uniform subgrade.

The estimation of maximum depth of frost can be based on field penetration data or temperature data or can be determined by theoretical formulae and charts.

Soil freezing depends to a large extent upon the duration of depressed air temperatures. Usual method to measure time and temperature is by degree days. One degree day represents one day with a mean air temperature one degree below freezing. Thus, twently degree days may result when the air temperature is – 1°C for twenty days or when air temperature is – 20°C for one day. A cumulative plot of degree days versus time is drawn and the difference between the maximum and minimum points on this plot is termed as *freezing index*. The freezing index has been correlated with depth of frost penetration. However, this correlation will also depend on the type of cover i.e., pavement type in regard to both quality and colour. The temperature of soil under dark objects (flexible pavement) is higher than the natural soil, whereas that under white objects (rigid pavements) is lower. The physical characteristics of the soil is also to be considered as it affects its ability to conduct and absorb heat and therefore, its behaviour under depressed air temperature. Rate of heat transfer depends upon soil density, moisture content and soil texture. Similarly, the coefficient of thermal conductivity is greater for frozen soils than for unfrozen soils. The Corps of Engineers, U.S.A. has suggested an empirical curve which relates depth of frost penetration to freezing index for a well-drained non-frost susceptible base course. This curve is shown in Fig. 19.15 and can be used to estimate depth of penetration under pavement kept free of snow and ice. A number of formulae have also been presented for predicting depth of frost penetration.

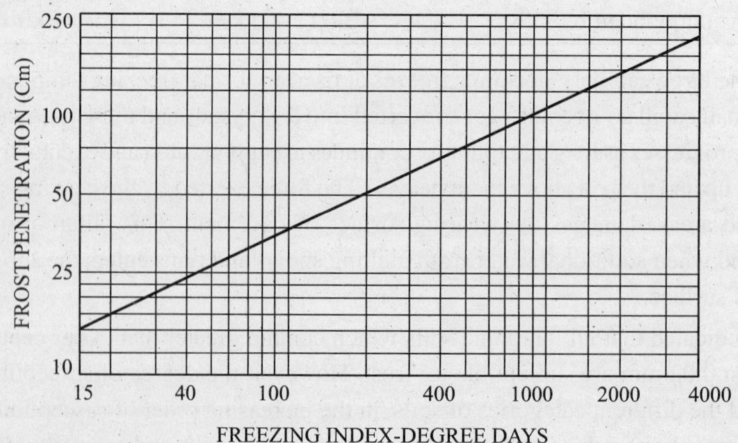

Fig. 19.15 Freezing index and frost penetration in a well drained non-susceptible base course.

Application of the principles explained above can be incorporated in the design of pavement in certain ways. Non-frost susceptible sub-base may be used to estimate depth of penetration sufficient to overcome the loss in stability. It may be possible to build base courses to a depth equal to maximum depth of anticipated frost penetration if an abundant supply of non-frost susceptible materials is available cheaply. Otherwise, pavement built to a great depth would be very costly. For such cases where great depths are uneconomical, pavement of lesser depths designed to withstand loss of subgrade support due to frost action would be justified. Proper drainage should be provided to remove water and for lowering the ground water table. A layer of clean sand or gravel as a capillary cutoff can be utilized for reducing frost action.

Since clean sands and gravels remain relatively stable even during thawing, adverse frost effects are minimum in such cohesionless materials. Every effort should be made to select clean, well drained sand and gravel formations as alignment locations and to use such soils for fill and embankments wherever possible.

Base course materials used should be clean and non-frost susceptible. Normal specification for base course may be adopted except that (*a*) scruplus care should be exercised to avoid inclusion of small lumps, thin layers or other small amounts of potentially troublesome fine-grained frost-susceptible materials and (*b*) gradation of the material should be checked following compaction to ensure that no untoward gradation changes have resulted as a result of placement and compaction breakdown.

For water-bound specification, mixing of or placing of layer of silt and clay on top of base course should be completely avoided.

The bottom 10 cm of base course or sub-base course may be graded so as to provide filter action against intrusion of the fine subgrade soil into the base course under the kneading action of traffic particularly after the frost-melting period.

A potential source of water for frost heaving may be due to surface infiltration. Adequate surface drainage may be considered as pre-requisite to the design against damage due to frost action. In areas of high rainfall, the use of impervious shoulder material is to be recommended.

In permafrost areas where the ground is permanently frozen such as in arctic regions, the design of pavements is based on the principle, that it is necessary to preserve the soil in its frozen state. For this purpose, relatively thick blankets of insulation material are used under the roadway section.

Salt heave. Salt heave is the phenomenon of heaving of the earth in certain areas due to the presence of excessive quantity of soluble salts particularly sodium sulphate and sodium carbonate in the soil. If climatic conditions are such that low temperatures below 20°C and high humidity above 70 per cent prevail, the alternate absorption of moisture by these salts during the night followed by evaporation during the day moves the salts to the road surface. The formation of dehydrate of sodium sulphate results in an increase in volume, about 1.8 times that of anhydrous salt. This causes the superimposed soil or road crust to heave. On dehydration, the salt dissolved in its own water of hydration tends to move towards surface by capillary action. These cycles of hydration and dehydration result in the disintegration of road surface, rendering it loose and fluffy. This phenomenon is noticeable in tropical areas.

If climatic conditions are such that the sodium sulphate in the soil remains either hydrate or dehydrated, the road surface is not liable to damage due to salt heave.

Remedial measures include adding sufficient quantity of calcium chloride to the soil to react chemically with sodium sulphate or carbonate resulting in the formation of relatively harmless calcium sulphate or calcium carbonate or by interposing a layer of sand of suitable thickness to cutoff rise of capillary water and also to accommodate the volume increase due to hydration of salt.

19.9. SOIL FACTORS

To determine the physical properties of a soil and to provide an estimate of its behaviour under various conditions, it is necessary to conduct certain soil tests. In this regard, a number of field and laboratory tests have been developed and standardized. Soil strength not only differs with soil type, but is also affected by bulk density, moisture content, permeability, internal structure of the soil and upon the way in which the stress is applied. Moreover, the soil beneath a foundation is seldom homogenous and large variations may occur in both the vertical and horizontal planes.

In general, the strength of different types of soils has a tendency to decrease with decreasing particle size. For any given soil, strength usually increases with increasing density and decreasing moisture content. However, moisture and density are inter-related, and in some cases it is possible to cause a decrease in soil strength by increasing the moisture content. The permeability of soil determines the degree of drainage attained during a test on a saturated soil; the greater the drainage the greater will be the shearing resistance because of the increase in bulk density and decrease in moisture content due to consolidation. Regarding the structure of soil, the strength of clay remoulded at natural moisture content may be considerably less than the strength of undisturbed soil which is due to the destruction of the natural structure during moulding of the soil particles. This loss in strength depends on the degree of sensitivity of clay. Besides, the soil strata often being anisotropic, the strength of undisturbed specimens may be affected by the direction in which they are taken. The same holds good for specimens cut from larger samples of remoulded and compacted soil, when the compactive effort has been applied in one direction only. Further the stress-strain characteristics of soils are affected by rate of loading. Tests made for design of pavements may be static tests or tests in which the load is applied at a relatively low rate and this method of applying stress may lead to test values that are lower than when soil is tested under transient loads.

Soils have only limited elastic properties. Even when stress less than the ultimate is applied and released, some permanent deformation usually takes place. The permanent deformation may be due to either an increase in bulk density of the soil or due to plastic flow.

As already pointed out under effects of repeated loads, the soils under repeated loading will deform more than identical specimens subjected to sustained static load of equal magnitude and further each repeated application of stress will produce some additional permanent strain. This factor of effect of load repetition makes the prediction of soil behaviour very complex as measured by static tests.

For a given wheel load, the most important single consideration in determining the thickness design of a flexible pavement is the subgrade support.

The common strength criteria, compressive strength and tensile strength, are of limited significance in evaluation of subgrade support, because soils do not function as elastic bodies which follows Hookes' law. These factors have led investigators to develop design methods which employ other strength criteria, most of which are empirical and pavement design procedures are generally based upon a correlation of the test values with field performance. The commonly used methods are described below.

19.10. MEASUREMENT OF SUBGRADE SUPPORTING POWER

Any test performed for this purpose should be made with the soil in a condition which it is anticipated will exist in the field. Requirement relative to density and moisture content are of special importance.

In general, the tests used to determine the strength properties of soil can be divided into three broad groups: (*i*) shear tests, (*ii*) bearing tests, and (*iii*) penetration tests.

19.10.1. Shear tests

These tests are made on comparatively small samples in the laboratory and thus determine only the strength properties at a point in the soil mass. In order to determine the variation in strength of the soil with depth, a number of tests on samples from different points have to be made in order to obtain an overall evaluation of the strength of soil mass. The usual function of shear tests is to determine the apparent cohesion and angle of shearing resistance under test loading and drainage conditions similar to those that will occur in the field. These constants are then used to evaluate the strength of subgrade. Some shear tests can also be used to determine modulus of deformation to be used for calculating stresses in an elastic medium, requiring a value of the modulus of elasticity. Since most of the strain is non-recoverable, the modulus of deformation is generally taken as the slope of the initial straight portion of the stress/strain curve. Where there is no easily discernible straight portion at all, an approximate modulus can be taken as the slope of the straight line which most nearly fits the curve over the range of stress considered. Three tests are commonly carried out.

(i) Unconfined compression test. This test is useful for rapid testing of clayey soils reasonably free from stones. Portable type apparatus can be used in the field. The test is usually made on undisturbed samples. When a definite maximum load is not reached, as is the case for plastic failure, the test is continued untial 20 per cent strain has been reached. By assuming that the volume of the specimens remains constant and that the shape remains a cylinder, it is possible to calculate the cross-sectional area at any deformation. The assumption is made that the shear strength (the apparent cohesion) is equal to half the compressive strength for fully saturated clays.

(ii) Direct shear test. In this test, failure is caused in a predetermined plane in the shear box which is divided horizontally into two halves and which contains a rectangular soil sample. Shear strength and normal stress are measured directly. Plot between normal stress on a number of identical specimens and shearing resistance is drawn as a straight line. The equation of this line is $S = c + \sigma \tan \phi$ where ϕ and c are the angle of shearing resistance and apparent cohesion respectively and σ is the normal stress on shear plane.

(iii) Triaxial compression test. In this test shear strength of a soil is determined under lateral pressure. For pavement design purposes, lateral pressures of 0.7 and 1.4 kg/cm^2 are satisfactory. The axial stress at failure at each of the lateral confining pressure is determined. Values of confining pressure and total vertical pressure at failure are plotted and circles are drawn through these points. The horizontal distance from the origin to the nearest intercept is the lateral pressure employed in the test. The distance from the origin to the outer intercept is the ultimate vertical pressure. Each test yields data for construction of one half-circle. The curve drawn tangent to the circles is shear curve for the material under test, and is called the Mohr rupture envelope. A straight-line repture has the equation $S = c + \sigma \tan \phi$.

The classic Mohr diagram yields, for elastic materials, a straight-line rupture envelope. Subgrade soils rarely produce a straight-line. They are usually in the form of a curve.

When the value of modulus of deformation is desired for the design of flexible pavements, it can be determined by dividing the slope of the initial linear part of the stress/strain curve by the original length. For an approximate value of the modulus of deformation for a specimen having little or no linear part of the stress/strain curve, the best straight line should be drawn through all the points over the required range of stress which will exist in the pavement and modulus of deformation calculated from the slope of this line.

For design of pavements, quick undrained test in which vertical load is applied rapidly (about 1.25 mm/min) with no drainage of the sample permitted during the test is recommended. This is because laods applied on raods are of very short duration and it is doubtful whether any drainage takes place during the loading cycle. In some cases, it may be desirable to run quick drained tests so that samples are allowed to drain and consolidate under lateral pressure equal to load of the pavement before the application of axial loading which is applied in a rapid manner.

Sometimes, Vane test (two blades fixed at right angles) is used in the field to measure the shear strength for clayey soils.

Stabilometer and cohesiometer test. A compacted sample of the subgrade which is 63 mm high and 102 mm in diameter is tested in Hveem stabilometer already described under high-type bituminous pavements for testing bituminous concrete. The specimen is subjected to the confined compression test at the rate of 1.25 mm per min. Vertical displacement and horizontal pressure is read at a unit vertical pressure of 11.25 kg per sq. cm. The stabilometer resistance value is calculated from the following equation:

$$R = 100 - \frac{100}{\frac{2.5}{D_2}\left(\frac{P_v}{P_h} - 1\right) + 1} \quad \text{...(19.10)}$$

where R = resistance value

P_v = Applied vertical pressure, 11.25 kg/cm^2 (1000 kg total force)

P_h = transmitted horizontal pressure

D_2 = Displacement of stabilometer fluid necessary to increase horizontal pressure from 0.35 to 7 kg/cm^2 measured in revolutions of a calibrated pump handle.

The value of R ranges from 0 to 100. Zero indicates a fluid condition, 100 a completely rigid condition. The value of R which is an empirical quantity to measure the resistance value is used for the design of flexible pavements.

Cohesion is measured by means of cohesiometer already described for testing asphaltic concrete. Generally, average cohesion values for paving materials are assumed.

19.10.2. Bearing tests

These are field bearing tests in which several load/ deflection measurements at each site using circular steel plates are made on the surface of the subgrade. The results in this test are affected by variations in soil properties within the volume of soil stressed, and therefore give an overall measurements of the strength of this part of the subgrade. The field plate bearing tests have the following advantages:

(*i*) The equipment is mobile and can be easily transported.

(*ii*) Exact increment load can be applied and exact deflections and residual settlements measured.

(*iii*) Failure loads can easily be applied and ultimate load conditions investigated.

(*iv*) A very wide range of load and contact area can be investigated with the same equipment.

The plate loading test data are applicable for the design of both flexible and rigid pavements. This test can also be used to evaluate the supporting power of bases and completed pavements.

A round metal plate is placed flush on the ground and is subjected to increasing increments of load by means of hydraulic jacks. To ensure an even bearing surface, the bearing plate may be bedded in a thin layer of dry fine sand or plaster of paris. The reaction for the load is supplied by heavy mobile equipment tied to a suitable steel beam. Deflection of plate is measured to the nearest 0.025 mm by means of Ames dials placed usually at one-third points of the plates near its outer edge. These dials are fixed on deflection beams supported on independent support away from the plate. To minimize the bending of plate, a series of decreasing diameter plates are used. For example, for 75 cm diameter plate, 60, 45 and 30 cm plates are usually placed on the large plate and the load is applied to the 30 cm plate.

Plates 75 cm in diameter are most widely used. For the design of flexible pavements, plates equal in area to that of the contact area of the tyre should be used. For this purpose, plates 50, 45 or 30 cm diameter are used. A 45 cm diameter plate is often employed which is thought to reproduce more closely the state of stress produced by the tyre of the normal road vehicle. For rigid pavement design 75 cm plate is used when determining the value of modulus of subgrade reaction (k). However, when smaller plates are used for convenience to avoid large dead load reaction, the value of k obtained with a smaller plate must be corrected by a factor to obtain a value corresponding to a 75 cm plate. Stratton (proc ASCE 1944) has worked out empirical relationship based on experimental data between k and diameter of bearing plate.

The rate of loading should be such that the maximum load increment should not exceed 10 per cent of the maximum wheel load. Each increment should be maintained until the rate of settlement is negligible which will generally be 0.25 mm per hr or about 0.05 mm per minute. When the full load is obtained it should be released in increments, in the same magnitude as those in loading but not at the same rate. To expedite tests, release load at 5 min intervals and record dial readings immediately prior to each load decrement. When the final load is removed, continue to take dial readings until the movement is less than 0.025 mm in 30 cm.

Sometimes, repetitive system of alternating applying and releasing the load is used to evaluate the effect of load repetitions.

***Evaluation of modulus of subgrade reaction* (*k*).** The settlement of the plate is plotted against the total load divided by the area of the plate and $k = \left(\frac{P}{\Delta}\right)$ is the slope or secant modulus of this graph for the first 1.25 mm deflection. The results usually give a curve which is convex upwards. k is therefore the slope of the line passing through the origin and the point on the curve corresponding to 1.25 mm settlement *i.e* $k = P/0.125$ kg per sq cm per cm where P is the pressure in kg per sq cm to cause 1.25 mm settlement.

In case of revised U.S. Corps of Engineers procedure, only the settlement corresponding to a pressure of 0.7 kg/cm^2 (3100 kg for 75 cm plate) is determined. A study of results indicates that the most representative value for k may be obtained using a load intensity of 0.7 kg/cm^2. k, in this case, is simply equal to $10/d$ (kg per sq cm per cm) where d is the settlement in cm corresponding to a load of 0.7 kg/cm^2.

It is sometimes necessary to make an allowance for worst possible moisture conditions of the subgrade to approximate the service conditions to be expected under the road. It is impracticable to do this directly by artificially wetting the test area and so a secondary soil test is used to convert the k value for a soaked subgrade. Two series of consolidation tests or simple compression tests are made on the subgrade soil sample, one on the soil in its natural condition of moisture content and dry density and other after the soil has been soaked under a surcharge of 0.35 kg per sq. cm. On the assumption that the ratio of deflections in the unsaturated and saturated tests will be approximately the same as the ratio of the deflections in the field on an unsaturated and saturated subgrade, modified value of k may be determined by the equation

$$k_s = \frac{d}{d_s} k \quad \text{...(19.11)}$$

where k_s = modulus of subgrade reaction allowing for soaking

d = deformation produced in the laboratory sample by 0.7 kg/cm^2 for unsoaked conditions

d_s = deformation produced in the laboratory sample by 0.7 kg/cm^2 load for soaked conditions.

Similarly for any other pressure, if P is the pressure required to cause a settlement of 1.25 mm and P_s is the pressure in the consolidation test on the soaked specimen to cause deformation equal to that produced by pressure P in the consolidation test on the unsoaked specimen, the value $k_s = P_s/P$.

If the plate bearing test cannot be performed for evaluating k, it may be possible to find approximate value by correlation with CBR values. One such relationship is shown in Fig. 19.16 which is given by Portland Cement Association.

Table 9.2 in chapter 9 dealing with soil classification shows, in its last column, k values typical of soils in the unified classification system. Data such as these can be used for preliminary design purposes for estimating pavement thickness. The effect of k on rigid pavement thickness is such that a high degree of accuracy in estimating k is not required in its determination.

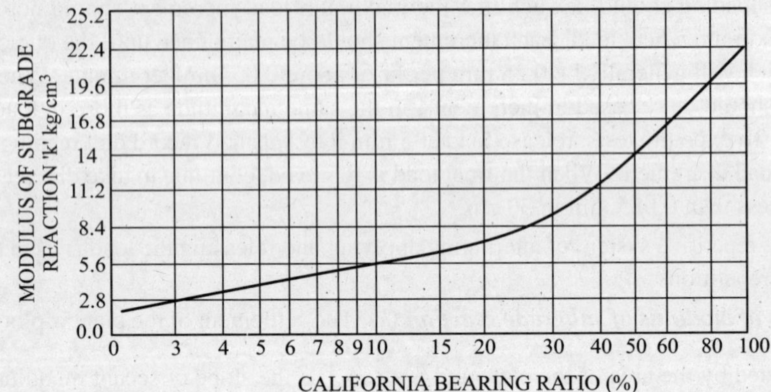

Fig. 19.16 Correlation between modulus of subgrade reaction (k) and CBR.

For the design of flexible pavements bearing value at 2.5, 5 and 12.5 mm have been employed by various investigators as a criterion of the strength of the soil. Sometimes, stress/strain curve as obtained by the plate bearing tests is used to determine modulus of deformation of subgrade in situ which is used in the thickness design of flexible pavements.

19.10.3. Penetration tests

These tests are either made in the field or in the laboratory and may be considered as small scale bearing tests in which the ratio of penetration to size of loaded area is much greater than in bearing test. Like shear tests, they only determine effectively the strength properties at a point in the soil mass.

These tests are arbitrary and may either be correlated with results of shear tests to obtain values of C and ϕ or with past experience of the behaviour of pavements on soil of similar strength as is the approach of empirical methods of designing the thickness of pavements.

California bearing ratio (CBR) test. In this test, the bearing test is made by plunger 19.35 sq cm end area and flush against the surface of the soil and forcing it into the soil at a rate of 1.25 mm per minute (see Fig. 19.17). The ratio of the bearing value at 2.5 mm penetration to that of a standard unit load is the CBR value.

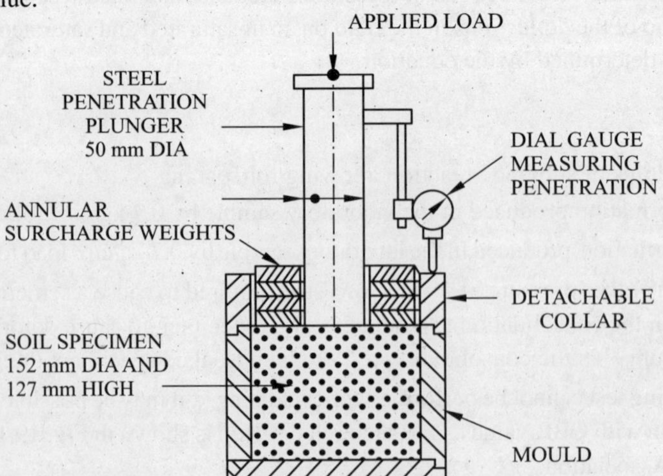

Fig. 19.17 Laboratory California Bearing Ratio test.

The standard values were obtained by testing a high quality crushed stone and are 70, 105, 135, 150 and 180 kg/cm^2 for 2.5, 5, 7.5, 10 and 12.5 mm penetrations. If the CBR value at 5 mm penetration is higher than at the 2.5 mm penetration, the value at 5 mm penetration is used in the design.

The soil sample is 152 mm diameter and about 127 mm high and is prepared in a mould with a bronze displacer disc placed at the bottom of the mould. The sample of soil may be undisturbed specimen which may be taken by fitting to the mould a steel cutting edge of 152 mm internal diameter, and pushing the mould as gently as possible into the ground. Any loose space in the mould where the specimen does not touch the sides of the mould should be filled with paraffin wax so that the soil receives proper support from the sides of the mould during the test.

For remoulded specimens, the sample is compacted at the desired density using proper amount of water. Soil used is passed through 20 mm sieve and all material retained on the sieve is replaced by an equal amount of material between 20 mm and 4.75 mm sieve. The samples may be either statically or dynamically compacted. For highways, standard compaction hammer with a height of fall of 30 cm and weighing 2.5 kg is generally used and the sample is compacted in three equal layers with fifty-five blows per layer. For the design of air field pavements, the modified compaction hammer with a height of fall of 40 cm and weighing 4.5 kg is generally used and the sample is compacted in five equal layers with fifty-five blows per layer.

Sufficient surcharge weights, equal to the actual or estimated weight of pavement are placed on top of the specimen before testing. Each 2.25 kg weight is approximately equivalent to 6 cm construction. The minimum surcharge weight kept if 4.5 kg.

Moisture contents used for preparing remoulded specimens should be equal to equilibrium moisture content. Where it is desired to bring the specimen to the worst moisture condition, which may exist subsequent to construction, the test is carried out after soaking the specimen for 4 days with free water maintained at least 2.5 cm above the top of the specimen. During soaking surcharge weights as explained above are placed on the top before soaking. The test is carried our after draining the specimen for a period of fifteen minutes.

The CBR test can also be made in the field. The basic test procedure is the same as that used in the laboratory. As in the field confining effect of mould is absent, each 2.25 kg surcharge weight represents 2.5 cm of construction and the minimum surcharge weight used is 13.5 kg instead of 4.5 kg as used in the laboratory.

When the load/penetration curve is concave upwards for low values due to surface irregularities, correction is applied by drawing a tangent to the curve at the point of greatest slope and by shifting the origin to the point when the tangent cuts the horizontal axis. (see Fig. 19.18).

North Dakota cone test. This is a cone penetration test which is sometimes used for the design of flexible pavements similar to that associated with CBR. The test is simpler and can easily be used for field tests.

The penetrometer consists of a shaft with a sharp cone, half angle equal to 7 degrees 45 minutes, attached to one end. The relative movement of shaft and supporting frame is measured by means of a Vernier scale. The reading of cone penetration are noted for loads of 5, 10, 20 and 40 kg. The results are expressed as a bearing pressure, taken as the load divided by the cross-sectional area of the cone at the surface level. Correction for zone error due to point of the cone being rounded is added to obtain the true penetration readings. Since, theoretically the penetration for 10 kg load should be one half that for 40 kg load, the value of correction $C = \Delta_{40} - 2\Delta_{10}$ where Δ_{40} and Δ_{10} are penetrations at 40 kg and 10 kg loads respectively.

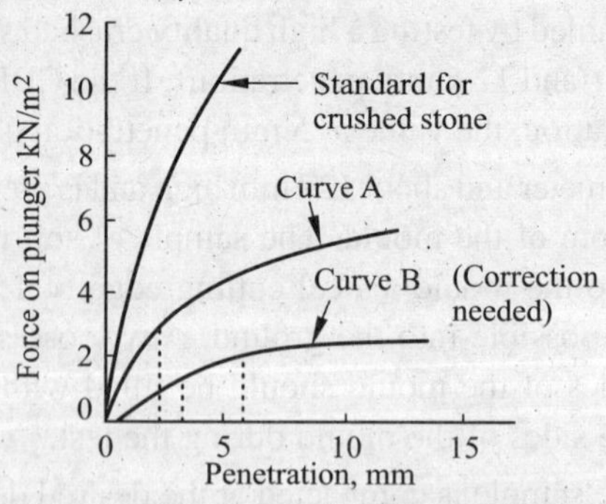

Soil type	P I	CBR % WELL DRAINED*	CBR % POORLY DRAINED
Heavy clay	70	2	1
	60	2.	1.5
	50	2.5	2
	40	3	2
Silty clay	30	5	3
Sand clay	20	6	4
	10	7	5
Silt	–	2	1
Sand (Poorly graded)	Non-plastic	20	10
Sand (Well graded)	"	40	15
Well graded sandy gravel	"	60	20

* Water table at least 0.6 m below formation level

Fig. 19.18 California bearing ratio test data.

The cone bearing value is given by:

$$b = \frac{P}{0.058 p^2} \quad \text{...(19.12)}$$

where b = cone bearing value in kg/cm²

P = load on cone in kg

p = corrected penetration of the cone in cm.

The test is reliable only for fine-grained soils particularly clayey soils. Presence of even small amounts of stone in an otherwise clayey soil can make the results quite unreliable.

19.11. SUMMARY OF CAUSES FOR CHANGE IN STREGNTH OF SUBGRADE

As explained in the preceding paragraphs, the strength of a subgrade may change during the life time of a road. The strength may be higher at times but this change nearly always takes the form of reduction in strength. These changes in the strength of subgrade are known as regression and the following is a summary of its causes. The design of a road pavement should be based, therefore, not on the strength of subgrade when the road is constructed but on the strength at some future time consistent with the estimated life of the pavement.

1. Increase of moisture content.
2. Repeated application of loads due to traffic.
3. Action of frost.
4. Decrease in moisture content due to shrinkage of clay soils.

19.12. STRESS DISTRIBUTION FACTORS

A knowledge of the distribution of stresses and of stress/strain relationship in the component parts of a road structure is of fundamental importance for the design of pavement and for estimating settlement due to consolidation of a soil. In a flexible pavement ideally speaking each layer is built to a depth where stresses on any given layer will not cause undue rutting, shoving (bulging), and other differential movements resulting in an uneven surface and further total thickness of various layers will distribute the wheel load to the subgrade so that allowable soil pressure of subgrade is not exceeded.

The application of a concentrated vertical load to the horizontal surface of any solid body produces a set of vertical stresses on every horizontal plane within the body. Maximum stresses occur on the vertical plane passing through the point of load application. The pressure is maximum at shallow depths and theoretically becomes zero at infinity. The theory of elasticity can be used for estimating stresses within flexible pavement and subgrade due to externally applied loads. In non-elastic soil masses the elastic theory may be applied to cases in which stresses and strains may reasonably be assumed to adhere to constant ratios.

19.12.1. Layered system concepts

In a layered system concept, the road is considered to comprise of different layers such as subgrade, sub-base or base and surface layer. The general concept of a multi-layered system is shown in Fig. 19.19.

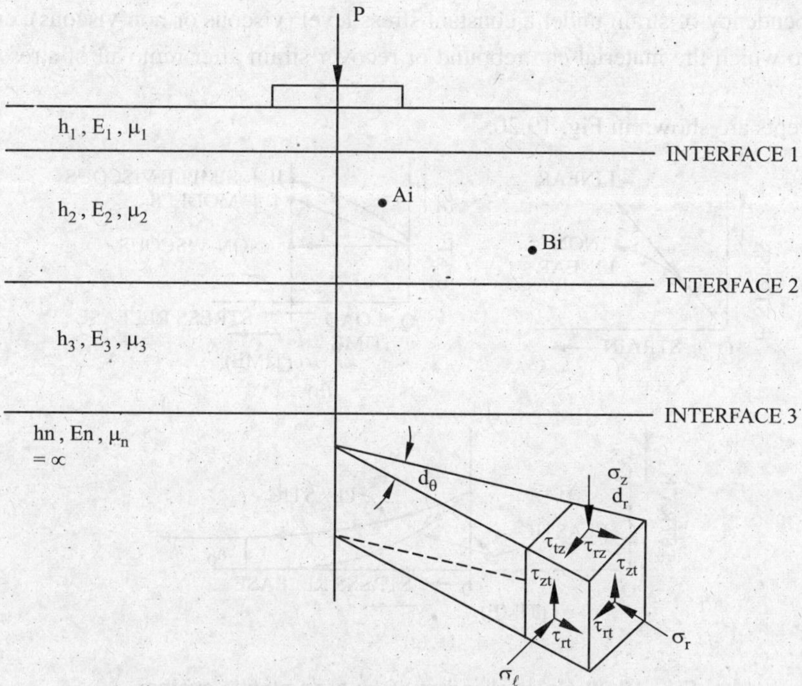

Fig. 19.19 Material characteristics at the interface (*a*) linearity (*b*) viscous effects (*c*) recoverable effects.

The various assumptions made in the analysis are:

(*i*) The properties of the material of each layer are homogeneous.

(*ii*) Each layer, except the bottom layer, has a finite thickness.

(*iii*) Each layer is isotropic *i.e.* the property at every point is the same in all directions.

(*iv*) Full friction is developed between layers at each interface.

(*v*) Surface shearing forces are not present at the surface; and

(*vi*) The stress solutions are characterized by two material properties *i.e.* Poisson's ratio, μ and modulus of elasticity, *E*.

At any given point within each layer, nine stresses exist. These are three normal stresses acting normal to the faces and six shearing stresses acting parallel to the face. The shear stresses acting on intersecting face are equal under static equilibrium conditions. The normal stresses under the condition that the shear stresses acting on each faces are zero, are termed principal stresses and are denoted by σ_z (major), σ_r (intermediate) and σ_t (minor stress). The bulk stress, θ is the sum of the principal stresses at a point. Knowing the stresses, the strains can be computed from the following relations.

$$\epsilon_z = \frac{1}{E}\left[\sigma_z - \mu\left(\sigma_r + \sigma_t\right)\right]$$

$$\epsilon_r = \frac{1}{E}\left[\sigma_r - \mu\left(\sigma_t + \sigma_z\right)\right]$$

$$\epsilon_t = \frac{1}{E}\left[\sigma_t - \mu\left(\sigma_r + \sigma_z\right)\right]$$

The type of theory used refer to the three properties of the material behaviour response. They are:

(*i*) relationship between stress and strain (linear or non-linear)

(*ii*) time dependency of strain under a constant stress level (viscous or non-viscous); and

(*iii*) degree to which the material can rebound or recover strain after removal of stress (plastic or elastic).

These concepts are shown in Fig. 19.20.

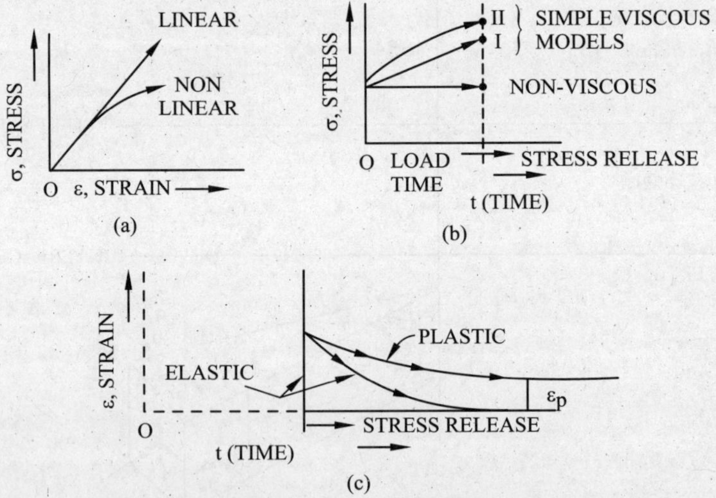

Fig. 19.20 Generalised multi-layered elastic system.

The theory widely used these days is the multilayered linear elastic theory; which is applied in the following cases:

Boussinesq's analysis (one layer system). Distribution of stresses due to the application of vertical point load was first derived by Boussinesq on the assumption that the semi-infinite medium is homogeneous, isotropic and elastic. According to Boussinesq, the vertical stress at any depth below the earth's surface due to a point load at the surface is as follows:

$$\sigma_z = \frac{P}{z^2} \frac{\frac{3}{2\pi}}{\left[1+\left(\frac{r}{z}\right)^2\right]^{5/2}}$$

where σ_z = vertical stress

P = applied load

r = depth

z = distance radially from point load.

In the above equation only two space variables are involved which are the depth z and the ratio r/z. It is seen that the vertical stress is dependent on the depth and the radial distance and is independent of the properties of the transmitting medium.

On each horizontal plane beneath a point load, there is a symmetrical bell shaped distribution of vertical stress and the peak height of the bell decreases as the square of the distance of the plane from the loaded surface increases. Moreover, if any line is drawn from the loaded point at an angle to the axis of symmetry it will cut these horizontal planes at a series of points at which vertical stress also decreases as the square of the distance of the plane from the loaded surface increases. Stress contours can be plotted and the resulting family of equal stress surfaces forms a pressure bulb.

Having obtained the distribution of stress beneath a point load it is possible by integration of Boussinesq's equation, theoretically, to obtain the distribution of stress beneath any loading system. In practice, a complete solution can only be obtained for a relatively few simple shapes of loaded area, *e.g.*, line strip, rectangle, square and circle, under load distributions which are either simple or vary in a uniform manner.

The above equation has been integrated over a circular area and vertical stresses on a vertical plane passing the centre of the plate with 'a' as the radius is given by:

$$\sigma_z = P\left[1 - \frac{z^3}{(a^2+z^2)^{3/2}}\right]$$

where P = unit load on circular plate.

The vertical stress beneath a corner of uniformly loaded rectangle can be expressed as

$$\sigma_z = \frac{q}{4\pi}\left[\frac{2mn\,(m^2+n^2+1)^{1/2}}{m^2+n^2+m^2n^2+1} \cdot \frac{m^2+n^2+2}{m^2+n^2+1} + \tan^{-1}\frac{2mn\,(m^2+n^2+1)^{1/2}}{m^2+n^2+1-m^2n^2}\right]$$

where q = the intensity of the uniform load

$m = \frac{B}{z}; n = \frac{L}{z}$

B = breadth of the rectangle

L = length of the rectangle

z = depth of the point beneath corner of rectangle.

The second term in the above equation is angle in radian and is less than $\pi/2$ when m^2+n^2+1 is larger than m^2n^2. Otherwise, it is between $\pi/2$ and π. L and B or m and n are interchangeable.

When the stress is desired at a point that is not below the corner of the loaded area, it may be determined by considering area to be a combination of four rectangles each with a corner above the point at which the stress is desired. The distribution of vertical stress can be obtained by adding and subtracting the values of σ_z for the constituent rectangles. For a point below the centre of a rectangular area, the actual loaded area must be considered to consist of four equal quadrants, the point desired being below their common corner.

Elastic strain at some depth z below the surface due to applied load under the centre of flexible plate is given by:

$$\Delta = \frac{Pa}{E} \cdot F$$

where Δ = elastic displacement

E = modulus of elasticity

P = applied pressure

a = radius of plate

$F = \dfrac{3}{2} \dfrac{1}{[1+(z/a)^2]^{1/2}}$ and is dimensionless quantity which depends only on depth-radius ratio and termed as ***deflection factor.***

The above deflection results from stress between z and infinity and elastic deformation from the surface down to z are not considered. This assumption neglects deflection in the base and assumes that the only significant deflections are in the subgrade.

For vertical displacement at the surface ($z = 0$), F becomes 1.5 and thus:

$$\Delta = 1.5 \frac{Pa}{E}$$

The above relationship for deflection assumes Poisson's ratio, $\mu = 0.5$.

As most soils do not have a definite elastic limit, nor do they exhibit stress/strain curves which are straight lines, use is made of the term modulus of deformation as already explained.

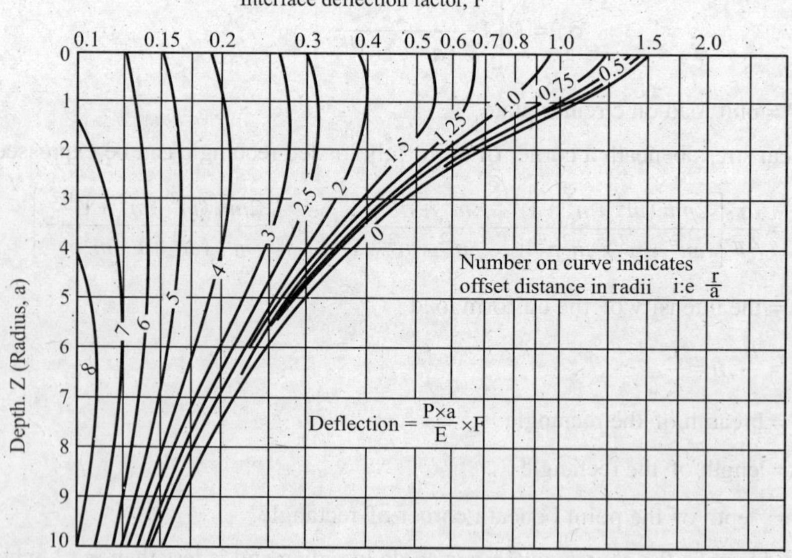

Fig. 19.21 Vertical deflection at the interface of a two layered system.

The above relationship for elastic deformation of the pavement apply to design methods which design for a limitng deformation of the pavement.

For rigid circular plate, as in the case of plate loading test, the loading is not uniformly distributed and the theoretical displacement of the plate assuming Poisson's ratio $\mu = \frac{1}{2}$ is given by:

$$\Delta = 1.18 \frac{Pa}{E}$$

Determination of deflection factor, F is facilitated with the help of a chart. This chart has been prepared to calculate deflection factor for semi-infinite elastic masses and is given in Fig. 19.21. This chart is applicable at any depth and at any point on the horizontal as well as under the centre of the plate. The depth and off set distance in radii are z/a and r/a where z is depth and r is offset distance.

Problem 19.3 *Using Boussinesq's analysis, calculate the vertical stress beneath a circular tyre imprint at a depth of 37.5 cm for the following conditions.*

Gross load on tyre = 20,000 kg

Tyre pressure = 7 kg/cm²

Also determine the elastic deformation if the subgrade has a modulus of deformation of 56 *kg/cm²* *and the thickness of pavement is* 40 *cm.*

Solution. Neglecting effect of side walls of tyre,

$$a\text{, radius of tyre imprint} = \sqrt{\frac{20{,}000}{\pi \times 7}} = 30.16 \text{ cm}$$

$$\sigma_z\text{, at depth of 37.5 cm} = P\left[1 - \frac{z^3}{(a^2 + z^2)^{3/2}}\right]$$

$$= 7\left[1 - \frac{37.5^3}{\{(30.16)^2 + (37.5)^2\}^{3/2}}\right]$$

$$= 7\,(1 - 0.473)$$

$$= 3.69 \text{ kg/cm}^2$$

Further $\qquad \Delta = \frac{Pa}{E} F$

where $\qquad F = \frac{3}{2} \frac{1}{\left[1 + \left(\frac{z}{a}\right)^2\right]^{1/2}}$

$$= \frac{3}{2} \frac{1}{\left[1 + \left(\frac{40}{30.16}\right)^2\right]^{1/2}} = 0.90$$

$$\therefore \qquad \Delta = \frac{7 \times 30.16}{56} \times 0.9 = 3.39 \text{ cm}$$

Stresses in layered systems. It would be observed from Boussinesq's analysis that the vertical stress is independent of the properties of the transmitting medium. This would not be so if the material had varying elastic properties. The equations are only correct provided the modulus of elasticity is constant throughout. The modulus varies through the layers of sub-base, base and surfacing. Even for any given soil the approximate modulus of elasticity varies with the moisture content, dry density and stress conditions. The approximate modulus for soil can be anything from under 70 to 100 kg per sq. cm, that for flexible bases from 700 to 7000 kg per sq cm and for concrete from 140000 to 420000 kg per sq cm. Also, soils are neither homogeneous nor isotropic.

As a result of these deviations from assumptions made in Boussinesq analysis, the true stresses and deflection depart from those given by the theoretical analysis.

Studies conducted by various investigators show that the stress and deflection patterns follows the ideal, but quantitative values do not. The comparison of calculated and measured stresses under 60 cm of aggregate base as reported by Horner are indicated in Fig. 19.22. These stresses are at the bottom of base when tyre pressure is applied at the top of base. These measured stresses were obtained in large-scale model study where subgrade was made of series of springs.

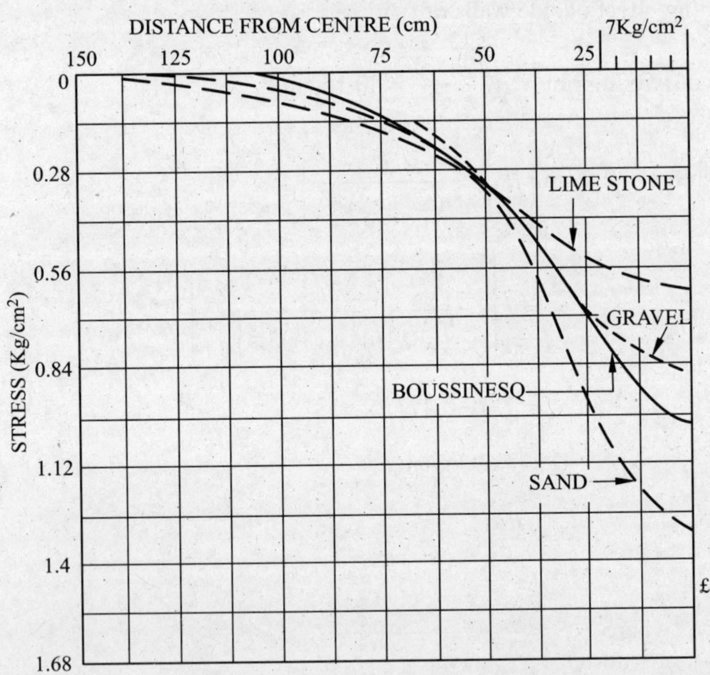

Fig. 19.22 Comparison of calculated and measured stresses at the bottom of an aggregate base 60 cm thick.

The data presented in Fig. 19.22 indicate that the stress-distributing properties of different base course materials are not similar. Crushed limestone distributed the load to a greater extent than gravel or sand.

Burmister's analysis (two layered system). Burmister analysed the stresses and strains in a two layered system, consisting of an elastic slab, infinite in the horizontal plane only, but of finite depth, placed on a semi-infinite solid of lower modulus of elasticity, the system being subjected to a uniformly distributed load acting over a circular area and applied to the upper surface of the slab. This approaches actual conditions as typical flexible pavements are composed of layers so that the modulii of elasticity decrease with depth.

The stress and deflection values as obtained by Burmister are dependent upon the strength ratio of the layers, E_1/E_2, where E_1 and E_2 are the modulii of the base and subgrade layers respectively. Fig. 19.23 shows the vertical stress values under the centre of a circular plate for the two-layer system as worked out by Burmister. Fig. 19.24 gives the comparison of vertical stress distribution as calculated from Boussinesq one layer and two layers theories.

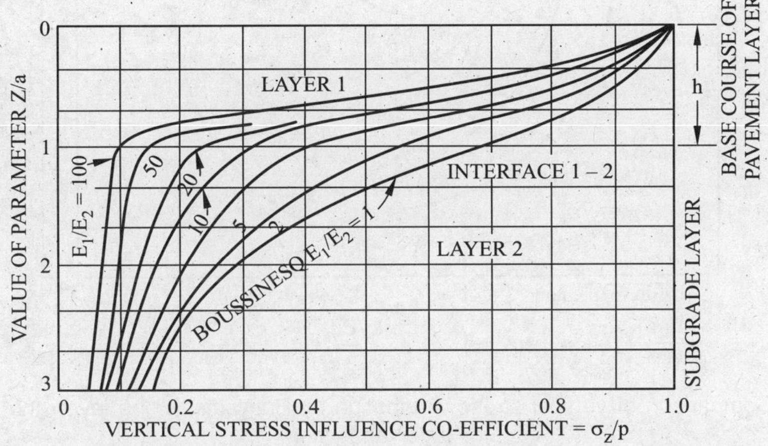

Fig. 19.23 Burmister two-layer stress influence curves.

The data given in Fig. 19.24 show that the effect of stiff top layer is pronounced. For $E_1/E_2 = 1$, vertical stress at the base-subgrade interface is about 70 per cent of the applied pressure, whereas for $E_1/E_2 = 100$, the stresses are about 10 per cent only. The effect of higher quality base is to reduce stress and deflections in the subgrade from those obtained from the ideal homogeneous case.

Another significant factor brought out by Burmister's analysis is the stress gradient obtained by two-layer theory. Although at great depths ($z/a = 3$), the two analysis approach a common level, they are vastly different near the base-subgrade contact. This shows that in addition to depth, the quality of base and surface is also important in stress distribution. As already pointed out, on the basis of WASHO findings, 10 to 30 cm reduction in the total pavement thickness resulted with a 10 cm bituminous concrete surface over that of a 5 cm surface.

AASHO road tests also indicate that flexible pavement bases treated with bituminous materials or cement were far superior to untreated crushed limestone or gravel bases in resisting loss of serviceability under load. An indication of this superiority can be seen in Table 19.11 extracted from AASHO road test report No 7.

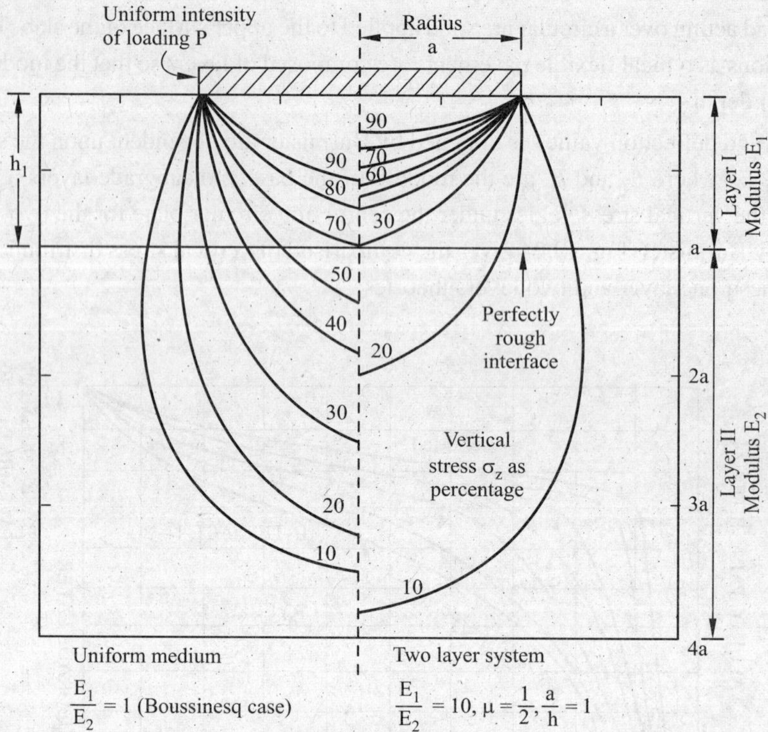

Fig. 19.24 Comparison of vertical stress distribution by Boussinesq one layer and Burmister two layer analysis.

Table 19.11

Thickness of base in cm required to maintain serviceability at 2.5 at 1,14,000 load repetitions, from graphical analysis

Load (kg)	*Cement treated uncrushed sand/gravel*	*Crushed limestone*	*Bitumen treated uncrushed sand/gravel*
5440 S	17.5	25.0	7.5
10890 T	27.5	30.0	10.0
8200 S	20.0	32.5	12.5
14520 T	20.0	32.5	17.5
11070 S	25.0	37.5	17.5
18140 T	25.0	37.5	17.5
13610 S	30.0	50.0	27.5
21770 T	27.5	45.0	25.0

Note: All these bases were used with 7.5 cm bituminous concrete wearing surface and 10 cm thick sub-base (*ii*) S stands for single axle; and T for tandem axle.

It has been demonstrated by actual tests in the field that the load-carrying capacity for granular materials per cm of thickness decrease quite rapidly after reaching a maximum at a thickness approximately equal to diameter of the loaded area, which is about 30 cm for heavy highway vehicles. For example, in one case it was found that the 12.5 cm increment in pavement thickness from 75 to 87.5 cm added load-supporting value at an average rate of about only 22 kg/cm thickness as against an average increase in load support of about 62 kg/cm thickness for the first 26.25 cm of pavement, which was the optimum thickness for this particular case. Consequently, the use of greater thickness of granular bases is an inefficient way to employ granular material to increased wheel load supporting capacity.

Burmister computed the vertical displacement at the surface under the centre of applied load, when the interface is perfectly rough, for various ratios of E_1/E_2 and for various ratios of the depth of the top layer to the radius of circular area of the applied load. Deflection for the two-layer system can be obtained by the following equations

$$\Delta = 1.5 \frac{Pa}{E_2} F_w \text{ – for flexible plate}$$

$$= 1.18 \frac{Pa}{E_2} F_w \text{ – for rigid plate}$$

where P = unit load on circular plate

a = radius of plate

E_2 = modulus of elasticity of lower layer

F_w = displacement factor which is dimensionaless and is dependent on the ratio E_1/E_2 as well as the depth to radius ratio.

Curves of F_w plotted against the thickness, h in multiples of radius are shown in Fig. 19.25.

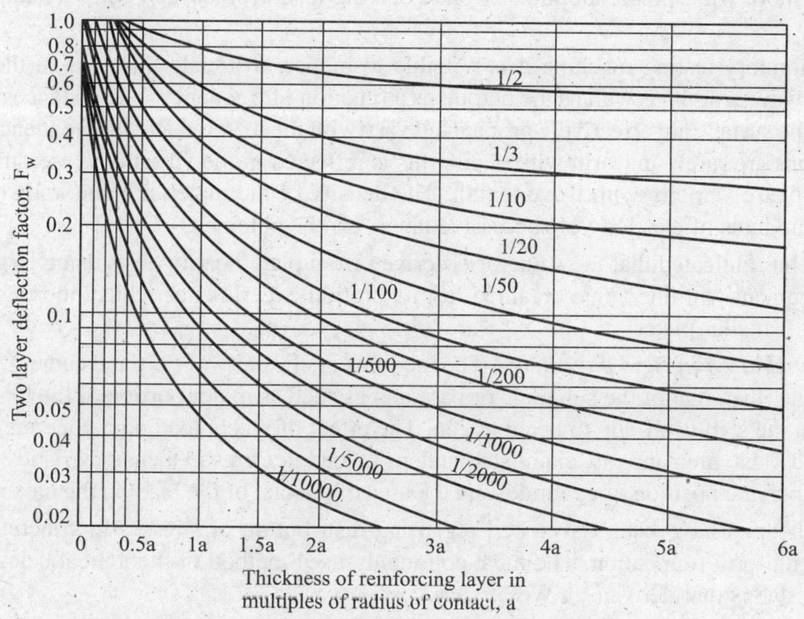

Fig. 19.25 Influence values-two layer Burmister theory.

Burmister has suggested a design method using the results of his analysis by assuming the value of limiting displacement under the wheel. The modulus for the base course and for the subgrade can be determined by plate-bearing tests or triaxial tests.

A source of error in using the Burmister equations for deflections results from a popular misconception that the elastic properties of the soil below a certain depth have only a negligible effect on the stresses and displacements in the pavement and subgrade. The error is small as regards stresses, but is not small in regard to displacements. Calculations show that this may be in considerable error if thickness of soil layer of modulus E_2 is not much greater than radius of loaded area.

Another common error, in using the Burmister equation is to assume Poisson's ratio to be 0.5 for all layers. The assumption in regard to Poisson's ratio is not important in the Boussinesq's case, but it is of considerable importance for layered systems. Poisson's ratio is not 0.5 for all soils. It may be as much as 0.4 for clay and as small as 0.25 for sand or crushed rock.

19.13. DETERMINATION OF MODULUS OF ELASTICITY OF VARIOUS LAYERS

The above relationship would indicate that the calculation of pavement deflections by application of elastic theory to a layered system requires knowledge of the modulii of the individual layers. This determination of an appropriate modulus value for the subgrade is rather difficult as it is affected by many factors such as number of repetitions of stress; magnitude and rate of repeated stress; thixotropy; method of compaction; density and moisture content at the time of compaction; and moisture content and density changes after compaction. All these factors should be taken into consideration when selecting a subgrade modulus for design of a flexible pavement. As the subgrade is not a perfectly elastic material, no definite modulus, can be obtained; the subgrade behaves as a material of varying modulus. But, as this variation cannot be introduced into the theory of elasticity, it becomes necessary to determine a relative value of the modulus of elasticity. This value may be obtained from the slope of the linear portion of the stress-strain curve determined from a triaxial compression test. The modulus determined in this manner depends largely upon the judgement of the individual making the test. This has led to a more widespread adoption of plate-bearing tests from which better results are usually obtained.

With a vibratory testing machine it is possible to induce artificial vibrations in the soil and to measure the wave length, as well as the output, as a function of frequency. The distances to which the vibration at the surface are effectively propagated vary with the material and the frequency. As natural soil formations are rarely uniform with depth, this is reflected in the vibrational measurements by a variation in phase velocity with wave-length. The density of the material being known at different depths, the modulus of elasticity of successive layers can be obtained.

Research has indicated that the value of E derived from plate-bearing tests using large plates was in good agreement with the value obtained from vibration experiments, whilst poor agreement was obtained with smaller plate.

***Stress distribution in rigid pavements*.** The modulus of elasticity of Portland cement concrete slab is much greater than that of the subgrade or base and as such as major portion of the load-carrying is derived from the slab itself due to beam action. The effect of wheel load on a concrete slab extends over a considerable area and pavement slab under the load deforms in a characteristics saucer shape depending upon the position, magnitude, and the area of contact of the load on the pavement surface.

Several theories have been developed regarding distribution of stresses in concrete pavements resting on a uniform foundation. The most commonly used method of theoretical calculation of the stresses in a slab is that derived by Westergaard.

Westergaard considered that a concrete pavement slab functions essentially as a flat plate resting upon a continuous but a yielding support. The tendency of the slab to deflect downward when a wheel load is applied on the pavement slab; is restrained to a certain extent by an upward induced reaction exerted by the subgrade. The degree of resistance to slab deflection offered by the subgrade is dependent upon the pressure/deformation properties of the subgrade material and the tendency of the slab itself to deflect will depend upon the properties of flexural stiffness. The resultant deflection of the slab which is also the deformation of the subgrade becomes a function of the relative stiffness of the slab to the subgrade.

Westergaard defined the relative stiffness by the following equation:

$$l = \sqrt[4]{\frac{Eh^3}{12\,(1-\mu^2)\,k}}$$

where l = radius of relative stiffness, cm

E = modulus of elasticity of concrete, kg/cm^2

h = thickness of slab, cm

μ = poisson's ratio of concrete

k = modulus of subgrade reaction (kg per sq cm per cm)

The value of l depends upon properties of both slab and subgrade. For $\mu = 0.15$ and $E = 280000$ kg/cm^2. The value of l for 25 cm thick slab varies from 159.33 cm for k = 1 kg/cm^3 to 134.0 cm for k = 2 kg/cm^3.

Westergaard made the following assumptions in the elastic subgrade theory:

(*i*) The pavement slab consists of a homogeneous, isotropic and elastic material.

(*ii*) The slab is uniform in thickness.

(*iii*) The thickness dimension of the slab in comparison to its length and breadth is so small that the ordinary theory of thin plates can be applied.

(*iv*) The subgrade reaction is vertical only and proportional to the deflection. This means that the support given to the slab is similar to that given by a dense fluid and hence the subgrade has no shear strength.

Deflection produced due to load on a slab extends over a considerable area and in the case of concentrated load P at the interior of slab given by the equation:

$$\text{Deflection} = \text{coefficient} \times \frac{P}{kl^2}$$

The deflections produced by concentrated load P at interior of slab as worked out by Westergaard are shown in Fig. 19.26.

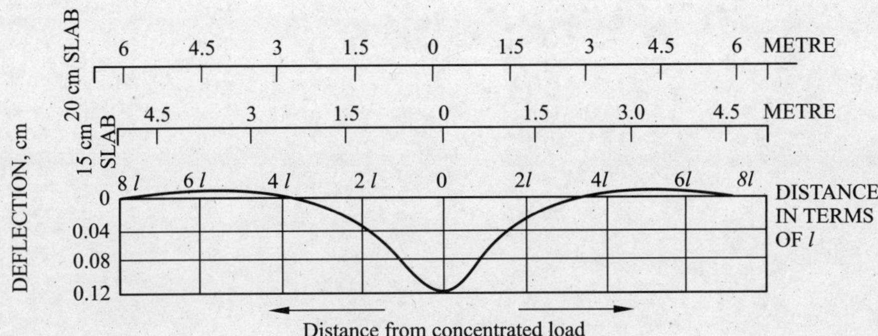

Fig. 19.26 Deflection produced by concentrated load (*P*) at interior of concrete slab.

The analysis made by Westergaad clearly shows that although the downward deflection of the slab is greater for a low than for a high modulus of subgrade reaction, the deflection are spread over a much wider area and thus the amount of bending of the slab, upon which the stress in concrete depends, does not differ greatly in the two cases. On the basis of these calculations, therefore, the soil has little effect on the stresses in the concrete when the slabs are uniformly supported.

Further, calculations show that k has more effect on the required thickness for high loads than it does for light loads. For this reason, thickness requirements for highways are practically independent and are affected to a minor extent by values of modulus of subgrade reaction.

Another point worth consideration is that when two beams deflect to an equal amount but with different radii of curvature, the beams with the sharper bending surface has the greater stress. Hence, for equal deflections, stiff subgrades result in higher stresses in concrete slab than in the case of lesser stiffness because of different radii of curvatures.

From the above discussion, it is seen that the distribution of stress in rigid pavements due to wheel loads depend in part upon values of relative stiffness which in turn depends upon the thickness, Poisson's ratio, modulus of elasticity of the concrete and the modulus of subgrade reaction.

❑❑❑

FLEXIBLE PAVEMENT DESIGN METHODS

Chapter

The general principles for the design of airfield and highway pavements are the same. But several distinct differences exist between the two types of pavement. Most important of these are the magnitude and the type of load, tyre pressure, geometric section of the pavement and the number of repetitions of load applied during the life time of the pavement.

In recent years, there has been a rapid development in the gross weight of the jet aircraft. Highway engineers are mostly concerned with the effect of vehicular traffic on the road whereas the design of airfield have to account for both *i.e.* the effect of the air-craft on the pavement, as well as effect of pavement on the aircraft. Jet engines get damaged by debris sucked into the air intake. The repeated fuel spillage and heat from the air engine may cause softening of bituminous binder.

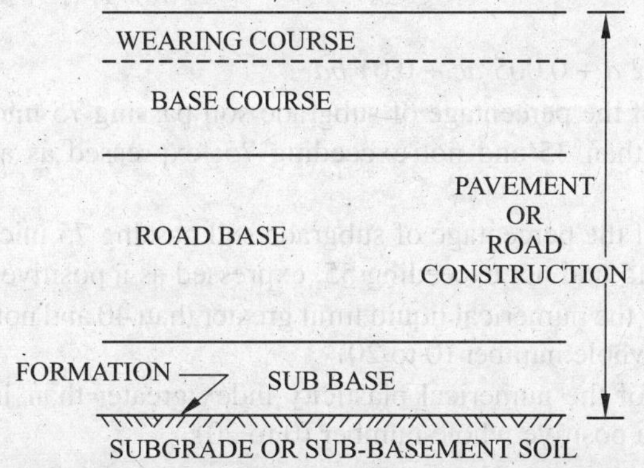

Fig. 20.1 Structural elements of a flexible pavement.

Type of load is another important factor to the considered in the case of airfields. Tyre pressurs as high as 28 kg/cm^2 have been noticed. The punching effect of such tyres necessitates the use of high quality pavement. The bicycle loading gears results in channelised traffic.

There has been a considerable shift in the design concepts of flexible pavements in recent years. More reliance is being placed on the performance serviceability concept *i.e.* roughness as the criterion of design in place of the methods based strictly upon structural analysis. Functional failure is, therefore, given precedence over structural failure.

Broadly speaking, flexible pavement design procedures can be classified into four distinct groups:

(*i*) Empirical methods based on soil classification and other factors such as climate and moisture.

(*ii*) Empirical methods based upon some arbitrary soil strength test which has been correlated with pavement performance.

(*iii*) Methods based on performance-serviceability concept.

(*iv*) Theoretical and partly theoretical methods based on mathematical analysis of stress/strain characteristics of the various materials and subgrade.

Each of the methods mentioned above will be considered in detail. Only those methods which have survived the test of time and used by one or more organizations will be described.

EMPIRICAL METHODS USING SOIL CLASSIFICATION TESTS

The approach in these empirical methods using no soil strength test is that if moisture content and density of the subgrade soil are controlled by adequate subgrade drainage and compaction, the strength of the subgrade and hence the thickness of construction depends largely on the composition and structure of the subgrade soil. The various methods in this group differ only in the classification used to define composition and in a way the results of classification tests are combined to evaluate the strength of the subgrade or thickness of construction required.

20.1. GROUP INDEX METHOD

The method is based on revised Public Roads Administration (PRA) classification which uses an empirical quantity called group index which is claimed to be an inverse measure of the thickness of sub-base required. The higher the group index, the lower is its strength and greater is the thickness of sub-base required. The thickness of base and surfacing, on the other hand, is varied according to volume of commercial traffic expected.

Revised PRA classification system comprises seven groups A-1 to A-7 (Table 9.1 in chapter 9) which can further be subdivided into twelve sub-groups. Classification is based only on three soil properties: (*i*) particle size distribution, (*ii*) liquid limit, and (*iii*) plasticity index. The formula for group index is:

$$\text{Group Index} = 0.2\,a + 0.005\,ac + 0.01\,bd \qquad \ldots(20.1)$$

where a = that portion of the percentage of subgrade soil passing 75 micron or No. 200 (ASTM) sieve greater than 35 and not exceeding 75; expressed as a positive whole number (0 to 40)

b = that portion of the percentage of subgrade soil passing 75 micron or No. 200 sieve greater than 15 and not exceeding 55, expressed as a positive whole number (0 to 40)

c = that portion of the numerical liquid limit greater than 40 and not exceeding 60, expressed as a positive whole number (0 to 20)

d = that portion of the numerical plasticity index greater than 10 and not exceeding 30, expressed as a positive whole number (0 to 20).

Charts are also given to calculate more easily the value of group index.

According to this method, subgrades have been divided into five groups, excellent (group index 0); good (group index 0 to 1) fair (group index 2-4); poor (group index 5-9); very poor (group index 10-20). Traffic is divided into three groups; light (less than 50 commercial vehicles per day); medium (50-300 commerical vehicles per day) and heavy (greater than 300 commercial vehicles per day). It is further assumed that the percentage of heavy wheel loads (4100 kg or more) does not exceed 15.

Fig. 20.2 contains correlation data is respect of tentative total pavement thickness.

The design chart shows that thickness of 152, 229 and 305 mm of combined surface and base would be required on excellent or good (*P.I.* < 6, stone fragments, gravel and sand) subgrades for light, medium and heavy traffic volumes respectively. No sub-base is required when subgrade material falls into either of these two classifications. Similarly, under all types of traffic, base and surface would be strengthened by the construction of 103, 203 and 305 mm of sub-base for fair, poor and very poor subgrades respectively. Thus, under heavy traffic for roads and with subgrade soil having group index 10 to 20, 305 mm of sub-base would be required under 305 mm of surface and base.

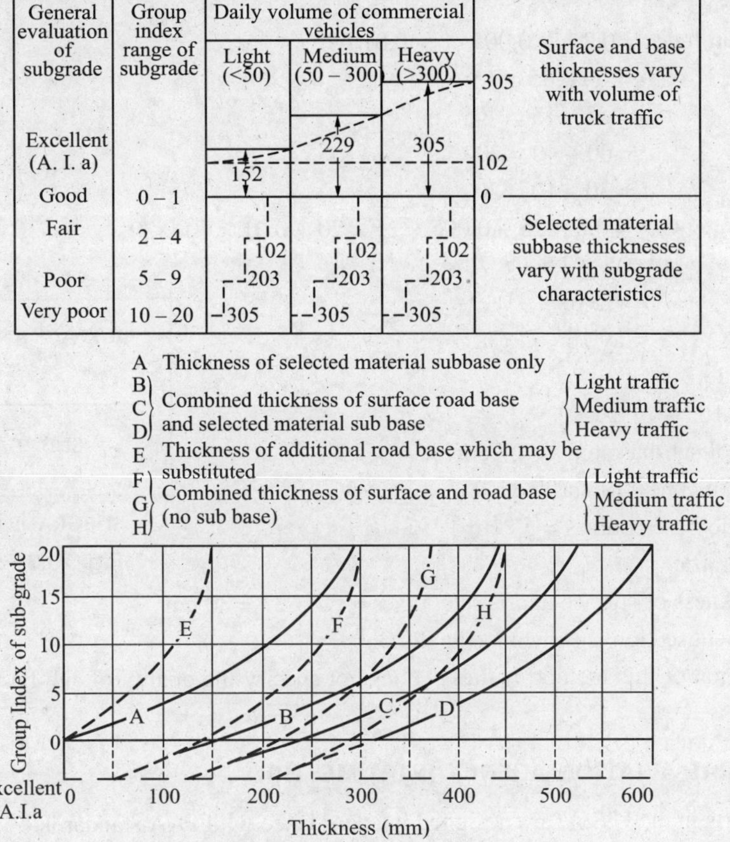

Fig. 20.2 Group index design curves.

These curves are only approximate and are subjected to modification by the engineer to suit local conditions as dictated by experience.

With regard to climate, compaction and drainage conditions, the following assumptions were made in preparing the design chart:

(*i*) The overall thickness shown should be adequate for most climatic conditions. However, in areas of deep frost penetration thickness and material requirement should be based on local experience.

(*ii*) Subgrade will be compacted to not less than 95 per cent standard AASHO maximum density and sub-base and base materials will be compacted to not less than 100 per cent standard AASHO maximum density.

(*iii*) Drainage conditions are assumed to be such as to permit proper compaction of subgrade prior to placing the base or sub-base and soil drainage or sufficient embankment height to be provided to permit the water table to be maintained at least 1 to 1.25 m below the surface.

Problem 20.1 *A sub-grade soil sample has the following properties:*

Soil passing 75 micron sieve = 60 per cent
Liquid limit = 65 per cent
Plastic limit = 45 per cent

Design the pavement section by group index method for heavy traffic with over 400 commercial vehicles per day. What are the limitations of this method?

Solution.

Group Index $= 0.2\,a + 0.005\,ac + 0.01\,bd$

Here $a = 60 - 35 = 25$

$b = 55 - 15 = 40$

$c = 60 - 40 = 20$

$d = 30 - 10 = 20$

$$\text{G.I.} = 0.2 \times 25 + 0.005 \times 25 \times 20 + 0.01 \times 40 \times 20$$
$$= 5 + 2.5 + 8$$
$$= 15.5$$

Say 16

From Fig. 20.2

For G.I. of 16,

Thickness of sub-base only = 27 cm

Combined thickness of surface, base and sub-base courses = 57 cm for heavy traffic

∴ Thickness of surface and base courses = 57 – 27 = 30 cm

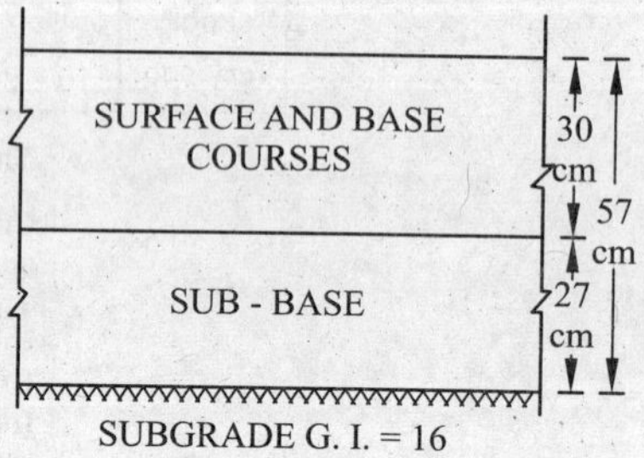

The pavement section is shown in Fig. 20.3.

Fig. 20.3 For Prob. 20.1

The limitation of this method is that is does not specify the quality of sub-base and base course materials.

20.2. FEDERAL AVIATION AGENCY (FAA) METHOD

Civil airports in the U.S.A. are designed in accordance with FAA standards.

The subgrade soil is located in a classification table based on FAA classification system and shown in Table 20.1. The table allots the soil a strength symbol varying according to the likelihood of severe frosts and the site drainage conditions.

Poor drainage in this method is defined as a condition where the subgrade may be rendered unstable due to (*i*) inadequate internal drainage caused by the character of the soil profile, (*ii*) capillary rise from a high water table, (*iii*) topographic features such as flat terrain at elevations only slightly above water level, or (*iv*) any other cause that may result in instability or produce saturation of the subgrade.

Good drainage is defined as a condition where (*i*) the internal drainage characteristics are such that there will be no accumulation of water that would envelop spongy areas in the subgrade, (*ii*) the water table is at such an elevation that the soil will not become water-logged either by percolation from above or capillarity from below, and (*iii*) topography is such that surface water will be removed rapidly.

Only granular subgrades (E_1 to E_4) are rated in good drainage condition. Even these materials may have to be assigned 'poor drainage' category if they are found in conjunction with a high water table, relatively flat terrain or near impervious strata.

It is emphasized in the method that classification of drainage must be established by a detailed study of surface and sub-surface characteristics of the site. In addition, if the assumption, that good drainage will prevail, is made, it will be necessary to provide adequate surface and sub-surface drainage installations where necessary. If, however, possibility exists that surface drainage installations will provide ponding of water between runways and taxiways, provision should be made for this by increasing the pavement thickness for poor drainage conditions.

Table 20.1 FAA Subgrade classification for flexible airport pavement

Soil group	Retained on 2 mm sieve (per cent)	Mechanical analysis			Liquid limit frost	Plasticity index frost	Subgrade class			
		Coarse sand pass 2 mm sieve ret. No. 60 (per cent)	Fine sand pass No. 60 ret. No. 270 (per cent)	Combined silt and clay pass No. 270 (per cent)			Good drainage		Poor drainage	
							No frost	Severe frost	No	Severe
E-1	0.45	40+	60–	15–	25–	–6	Fa	Fa	Fa	Fa
E-2	0.45	15+	85–	25–	25–	–6	Fa	Fa	F1	F2
E-3	0.45	—	—	25–	25–	–6	F1	F1	F2	F3
E-4	0.45	—	—	35–	35–	10–	F1	F1	F2	F4
E-5	0.45	—	—	45–	40–	15–	F1	F2	F3	F5
E-6	0.55	—	—	45+	40–	10–	F2	F3	F4	F6
E-7	0.55	—	—	45+	50–	10–30	F3	F4	F4	F7
E-8	0.55	—	—	45+	60–	15–40	F4	F5	F5	F8
E-9	0.55	—	—	45+	40+	30–	F5	F6	F7	F9
E-10	0.55	—	—	45+	70–	20–50	F5	F6	F8	F10
E-11	0.55	—	—	45+	80–	30+	F6	F7	F9	F10
E-12	0.55	—	—	45+	80+	–	F7	F8	F10	F10
E-13	MUCK AND PEAT (FIELD EXAMINATION)				NOT SUITABLE FOR SUBGRADE					

Note : Classification is based on sieve analysis of the portion of the sample passing 2 mm sieve. When a sample contains material coarser than the 2 mm sieve in amounts equal to or greater than the maximum limit shown in the table, a raise in classification may be allowed provided the coarse material is reasonably sound and fairly well graded.

With respect to frost action, a "*severe frost*" condition exists if the depth of frost penetration for the particular site is greater than the anticipated thickness of surfacing, base and sub-base as determined for "*no frost*" and the drainage condition as defined above. Otherwise, the condition of "*no frost*" prevails.

For each soil group there are corresponding subgrade classes based on the performance of the particular soil as a subgrade for flexible pavements. The subgrade class is determined from the results of soil tests and the information obtained by means of soil survey and a study of climatological data. For example, Fa subgrades class furnishes adequate subgrade support without the addition of sub-base material. A soil of particular group may fall in several subgrade classes depending on drainage and frost conditions. For example, E-4 is classed as F1 subgrade for good drainage and both for severe and no frost conditions, F2 for poor drainage and no frost and F4 for poor drainage and severe frost condition.

If considered advantageous, CBR tests may be conducted on the subgrade soils and the CBR value converted to the FAA subgrade class with the help of Table 20.2.

Table 20.2 CBR-FAA subgrade class correlation

FAA Class	F10	F9	F8	F7	F6	F5	F4	F3	F2	F1	Fa
CBR	3.5	4.5	5.5	6.5	7.5	8.5	10.0	12.0	14.5	18.0	20.0

After sugbrade class has been determined, the required thickness of sub-base, base and surface are determined from the appropriate design charts, shown in Fig. 20.4 to 20.6.

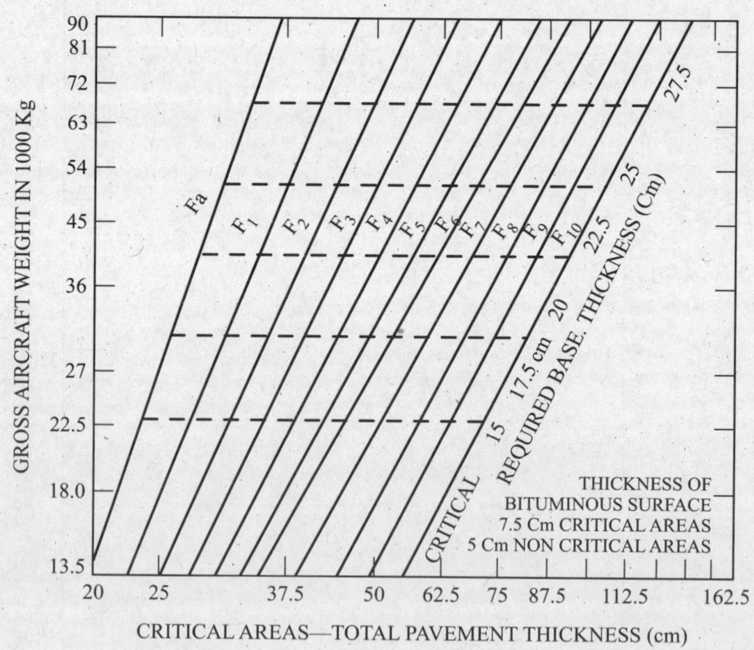

Fig. 20.4 FAA design curves single wheel gears.

The curves for single, dual and dual-tandem gears aircraft have been prepared on the assumption that 95 per cent of the gross weight is carried by the main gears. The design charts are applicable to aircrafts having gross weight greater that 13600 kg. These curves are not applicable to wide bodied jet aircrafts (e.g. B-747, DC-10, etc.) For such aircraft, curves shown in Fig. 20.7 and 20.8 are used. These have been derived on the basis of revised Corps of Engineers design procedure and are applicable to 100,000 passes.

The effect of traffic volume is considered in the following manner. Fig. 20.4 to 20.6 are based on capacity operation, which is equivalent to 5000 design coverage to failure. However FAA makes direct use of departures. 5000 coverages are approximately equal to 1200 annual departures, assuming service life of 20 years. The thickness of the pavement is determined as follows:

(*a*) If the anticipated traffic mix contains no wide bodied aircraft, calculate the equivalent departures of the critical aircraft. If this value is less than 1200, use curves in Fig. 20.4 to 20.6.

(*b*) If the equivalent departures in the above case are greater than 1200, then increase the thickness found in (a) by the following amounts:

(*i*) 2.5 cm for surface thickness.

(*ii*) Allow percentage increase as per Table 20.3 for base and sub-base courses.

Table 20.3 Base and sub-base thickness percentage increase

Equivalent annual departures	(1200-3000)	(3000-600)	(> 60000)
Per cent increase in thickness	10	20	30

(*c*) When the traffic mix includes wide bodied aircraft, first determine the equivalent critical departures as described above. In addition, substitute the wide-bodied jet aircraft repetitions on a 1 : 1 basis for a DC-8-60 series aircraft. The annual equivalent critical departures are then used to calculate the pavement thickness, applying correction as per table 20.3.

Alternately, use Figs. 20.7 and 20.8 directly.

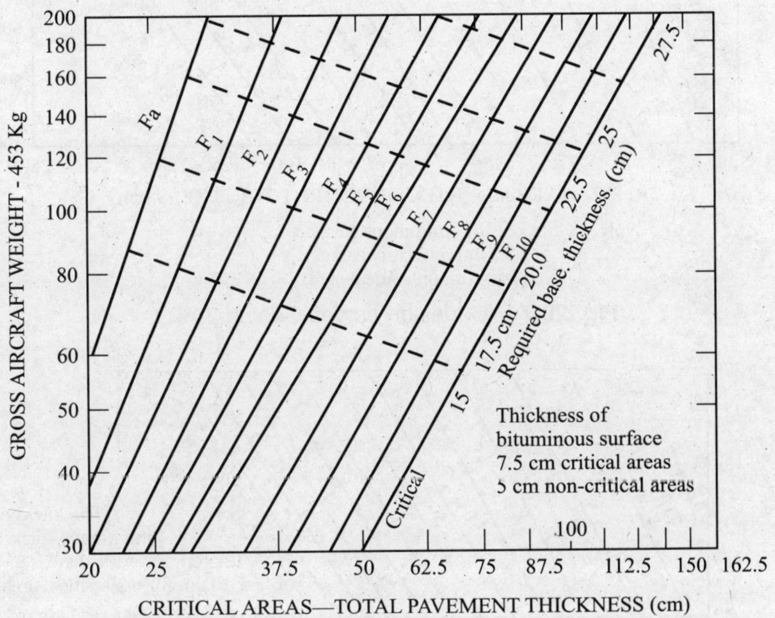

Fig. 20.5 FAA design curves-dual wheel gear.

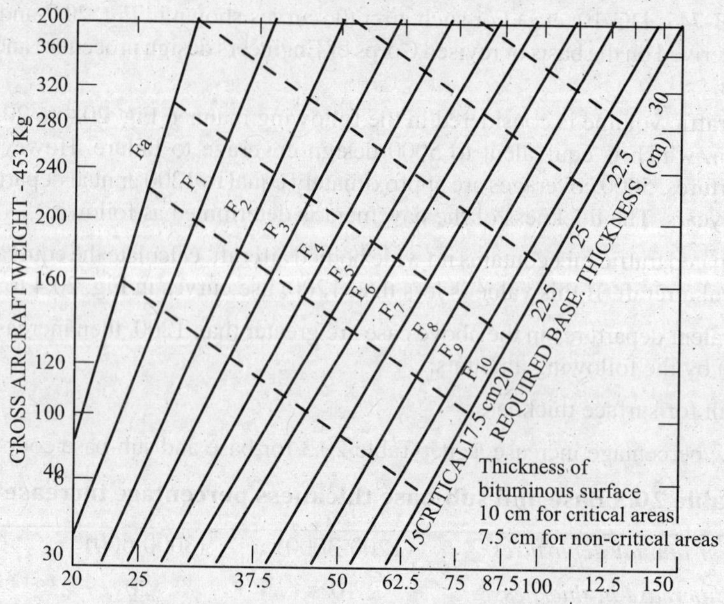

Fig. 20.6 FAA design curves-dual tandem gear.

Alternately use Fig. 20.7 and 20.8 directly.

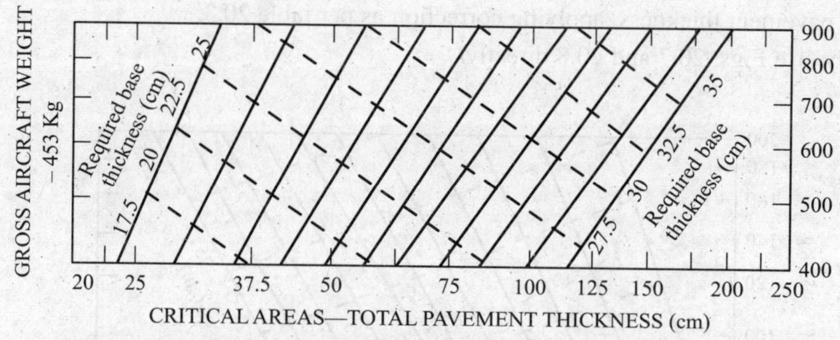

Fig. 20.7 FAA design curves Boeing 747.

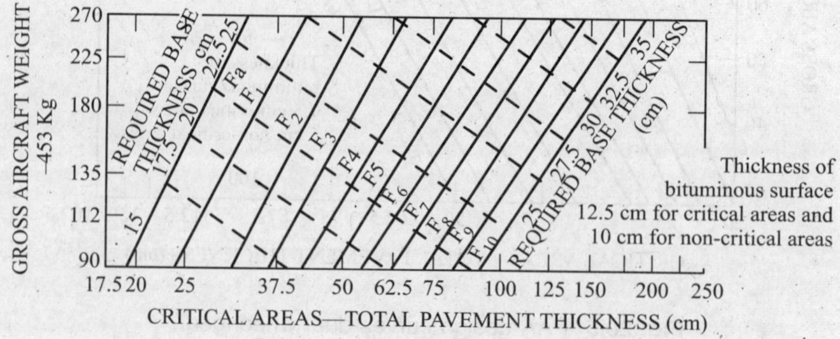

Fig. 20.8 FAA design curves-DC-10.

The final design thickness will be the maximum of the two designs.

The equivalent departures is based upon the following relation,

$$\log p_{es} = \left(\frac{w_i}{w_s}\right)^{0.5} \log (CP_i) \qquad \ldots(20.2)$$

where p_{es} = equivalent passes of the critical (standard) air-craft.

p_i = number of anticipated passes of the ith aircraft.

w_i = load on one wheel of ith aircraft

w_s = load on one wheel of standard aircraft

C = constant to account for gear type difference.

The values of C are given in Table 20.4.

Table 20.4 C values used in equivalent damage equation

To convert from	*to*	*C value*
Single wheel	Dual wheel	0.80
Single wheel	Dual tandem	0.50
Dual wheel	Dual tandem	0.60

The method of developing the design curves is entirely empirical and is based on service records of pavements on different types of soils under variables drainage and climatic conditions. For the heavier wheel loads, service records have been supplemented by the results of experimental work on varying thickness of especially constructed surfaces and bases. The various materials used in the subbase, base and surface course are described in Table 20.5.

Table 20.5 FAA description of pavement materials

Surface	*Base course*	*Subbase course*
Bituminous concrete	Bituminous base	Granular subbase
	Crushed aggregate	Dry-bound macadam
	Caliche	Water-bound macadam
	Lime rock	Aggregate base
	Shell	Sand-clay
	Penetration macadam	Mixed-in-place
	Emulsified bitumen aggregate	Soil cement
	Cement-treated	

In making adjustments by substituting one material for another the following equivalents are used:

$1\frac{1}{2}$ cm of sub-base = 1 cm of stabilized base or subbase

$1\frac{1}{2}$ cm of sub-base = 1 cm of base (when subbase works out to be less than 7.5 cm)

Minimum thickness of stabilized layers, in any pavement area, are 10 cm for bituminous and 15 cm for cement treated bases.

The above method of designing thickness is based solely on soil classification.

Since uniform subgrade soil conditions rarely exist, a method proposed by FAA of utilizing subgrade profile is given by the equation:

$$z = y - \left[\frac{t\,(y-x)}{(y+x)}\right] \qquad \ldots(20.3)$$

where z = required thickness of sub-base
x = sub-base thickness for first layer of subgrade soil
y = sub-base thickness for second layer of subgrade soil
t = thickness of first layer of soil.

For secondary airports, consideration should be given towards using stabilized subgrades for sub-bases and bases. Surface courses can be of road mix type, since wheel loads are relatively light. FAA design curve used is shown in Fig. 20.9.

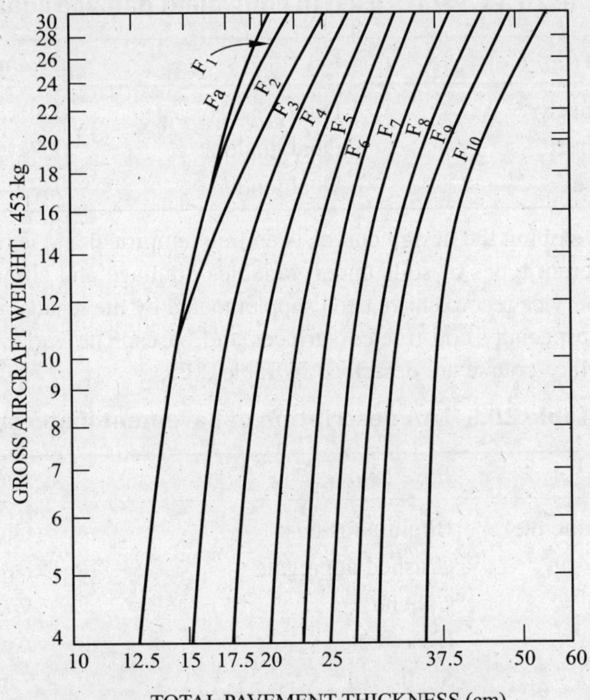

Fig. 20.9 FAA design curves-light aircraft.

For light aircraft, aggregate turf strips may be built. The thickness requirement for stabilized turf runways as recommended by FAA are given in Table 20.6.

Table 20.6 Aggregate turf thickness in (cm)

(Gross load 4500 kg or less)

Soil group	*Good drainage*		*Poor drainage*	
	No frost	*Severe frost*	*No frost*	*Severe frost*
E-1 to E-5	10-15	10-15	10-15	10-15
E-6, E-7	15 (min)	15 (min)	15 (min)	15 (min)
E-8 to E-12	15 (min)	20 (min)	20 (min)	25 (min)
E-13	NOT SUITABLE			

20.2.1. General observations

The empirical methods described above are adopted primarily because of their simplicity.

There are certain limitations in following these design procedures. Factors such as moisture and density are not taken into account and it is assumed that soil with identical group index number or subgrade class will give identical strengths after compaction in the field. However, strength with identical classification can vary within wide limits. Moreover, it becomes essential that compaction and construction specifications are followed rigidity at all times. In addition, the effect of repetition of loads is not considered rationally.

Besides, these methods do not furnish a definite procedure for designing the thickness of various courses nor for the blending of the various materials which might be economically available for the base and sub-base courses.

Atterberg limits, which are used as classification tests, are greatly affected by testing procedure and for this reason standard techniques should be used at all times.

Thus, with these design methods in some cases, over designed pavement may result, while in other cases, an under designed pavement will result.

EMPIRICAL METHODS USING ARBITRARY SOIL STRENGTH TESTS

In these methods an arbitrary soil strength is selected which is considered to stress the subgrade in a manner similar to an actual wheel load. The factors of load repetitions and distribution of stress and strain through the pavement layers and subgrade is taken into account indirectly by determining the thickness of pavement by relating the results of this arbitrary strength tests to the performance of roads already in use on subgrades of similar strength.

20.3. METHODS BASED ON CBR TEST

The most popular of these methods is the well-known California bearing ratio (CBR) method. This method was originally devised by Porter in California State Highway Department, but it has since been developed and modified by various organizations for different conditions and wheel loads. The details of test procedure have already been described in the previous chapter. The load required to cause a cylindrical plunger, 19.35 sq cm end cross-section, to penetrate a soil sample at the rate of 1.25 mm per minute to cause a penetration of 2.5 or 5.0 mm expressed as percentage of standard load is the CBR value.

The strength of the subgrade soil and other construction materials is determined from an adhoc penetration test and the required thickness of pavement obtained from design curves evolved from experience. A number of different CBR charts are available for design depending upon the local conditions. Some of the important methods are described in the following paragraphs:

20.4. DESIGN OF FLEXIBLE PAVEMENTS IN INDIA

In India the CBR method of design adopted in Britain for traffic classified according to the number of heavy vehicles is followed. This is described in detail in IRC : 37-1970. Basic data for the formation of our own CBR charts taking into consideration the unique pattern of traffic on our roads, the varying climatic and environmental conditions, and the country-wide practice of generally using thin bituminous surfacing are not yet available. Hence for the present the CBR design curves evolved by the Road Research Laboratory, U.K., which owing to their simplicity and experience with their use have been found reasonably well suited for the Indian conditions, are recommended for design. These are reproduced in Fig. 20.10. Thickness of different layers of sub-base, base and surfacing are determined by repeated use of these curves.

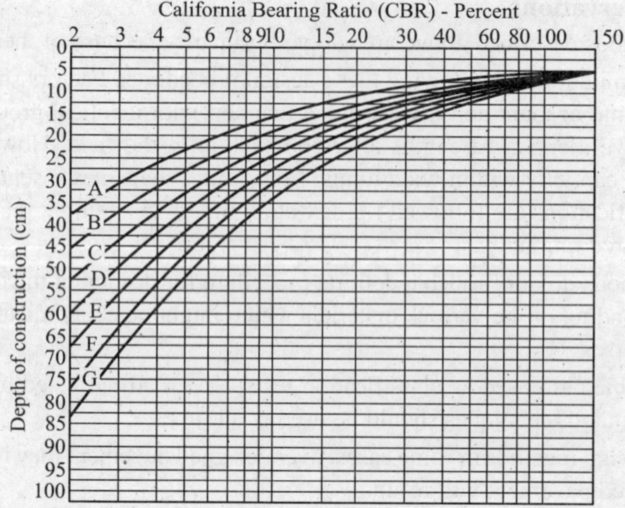

Fig. 20.10 CBR curves for flexible pavement design.

Since CBR test is an ad hoc penetration test, it is carried out strictly in accordance with the method described in IS : 2720 (part XVI)-1965 on remoulded samples of soils. The test specimens should be prepared by static compaction but if possible dynamic method may be used as an alternative.

For a given soil, the CBR value and consequently the design, will depend largely on the density and moisture content of the test sample.

For new roads the sample of soil or the sub-base material, should be compacted to a dry-density corresponding to the minimum state of compaction likely to be achieved in practice paying due regard to the compaction equipment used and the compaction limits specified. By and large proctor density, conforming to IS : 2720 (Part VII) is made use of. In the case of existing roads, the moulding density should correspond to the actual density of the subgrade, determined by sand replacement method, laid down in IS : 2720 (Part XXVIII).

The choice of moisture content of the test specimens is not so simple. As a general practice, the design of new construction is based on the strength of the samples prepared at optimum moisture content (OMC) corresponding to proctor compaction and soaked in water for a period of four days prior to testing. In case of existing roads requiring strengthening, the soil is moulded at the field moisture content and soaked similarly for four days prior to testing. The field moisture content used for moulding is determined preferably immediately after the rainy season.

The measurement of field density and moisture content, in the case of design for strengthening of existing roads, should as far as possible be carried out at a distance of 0.6 to 1 m from the pavement edge below the pavement.

However, it should be realized that soaking for four days may be an unrealistically severe moisture condition in certain cases. Cases in this category would be:

(*i*) Subgrades of roads where a comparatively thick bituminous surfacing of impermeable nature is provided on top such as a well laid and sealed dense carpet, and where simultaneously (*a*) water table is too deep to affect the subgrade adversely (*i.e.* greater than 1 m in sands to 3 m in sandy clays and 6 m in heavy clays) and (*b*) well-shaped verges exist facilitating quick drainage for the surface water to the side drains and or pavement base layer is continued across part/full width of the verge with the same objective.

(*ii*) Subgrade in areas where the climate is arid throughout the year, *i.e.* the annual rainfall is of the order of 50 mm or less and the water table is too deep to affect the subgrade adversely.

In the above mentioned situations, it is anticipated that the most severe moisture condition in the field will be far behind that of the sample at the end of four days soaking, resulting in unduly conservative design based on soaking procedure adopted. Hence the procedure of soaking for four day could be discarded in such cases and the specimens tested immediately after compaction at moisture contents indicated below.

For the determination of CBR the soil specimens, of category (*i*) roads, should be compacted at OMC in the Proctor compaction test. The soil specimens of category (*ii*) roads should be prepared at natural moisture content of the soil at subgrade depth for finding the CBR value.

The design curves give the total thickness of the pavement in terms of the volume of traffic carried by the road. For purposes of this design method, the traffic is considered in units of heavy vehicles per day in both directions (irrespective of whether the design is for a two-lane or a dual carriageway) and divided into seven categories as indicated in table given below. The terms heavy vehicle is meant to apply to any vehicle with a laden weight of 3 tonnes or more.

Classification of traffic for design

Traffic (heavy vehicles per day of laden weight exceeding 3 tonnes)	*CBR design curve applicable*
0–15	A
15–45	B
45–100	C
150–450	D
450–1500	E
1500–4500	F
Exceeding 4500 and all expressways	G

In the above classification, due allowance exists for presence of a nominal number of excessively heavy axle loads in the traffic stream. But there can be situations, such as on approach roads to docks or in the neighbourhoods of heavy industrial complexes, where the number of such loads may be inordinately high. It is recommended that in these cases the designer should use his discretion in selecting a curve of higher traffic category than otherwise applicable.

Since the design method is empirical, it is not essentially related to any particular value of axle load or wheel load repetitions. But for the purposes of design, curves should be considered operative upto single axle loads of 8160 kg and tandem axle loads of 14500 kg. Beyond axle loads of this magnitude, an increase over the thickness obtained from these curves will be necessary.

The design curves are not meant to be used on the basis of traffic immediately carried by the road (in the case of existing roads) or that anticipated (in the case of new constructions). In every case, before design, an estimate should be made of the traffic likely to be carried by the road at the end of its expected life. In preparing these estimates due consideration should be given to the existing traffic data, possible changes in road network and land use of the area served, and the probable growth of traffic during the period of projection under consideration. It is considered appropriate that major through routes should be designed for at least 20 years of life. Less important routes such as residential streets or rural roads, where the growth of traffic may be small, may however be designed for a different duration based on judgment and depending on the rate of growth of the traffic expected.

Very often it may not be possible to provide the full thickness of pavement needed ultimately right at the time of initial construction. Stage construction techniques should be resorted to in such cases and roads be strengthened as traffic increases.

The following formula may be used for traffic predictions on the main through routes:

$$A = P(1 + r)^{n+15}$$

where A = number of heavy vehicles per day for design;

p = Number of heavy vehicles per day at last count;

r = Annual rate of increase in the number of heavy vehicles; and

n = Number of years between the last count and the year of completion of construction.

The census figure p used in the formula should normally be a seven day average of heavy vehicles derived from 24 hour counts. However, in exceptional cases where this information is not available, an average of three day counts could also be used. Based on growth data of heavy vehicles over the past few years it is recommended that the value of parameter 'r' in the above formula may be assumed at 6 per cent for the time being unless more authentic data be available. However, this value will require a review periodically. The value of 6 per cent is meant primarily for rural routes; in the case of urban roads it is necessary that it should be applied judiciously.

To account for the concentration of wheel load repetitions on a single lane, carriageways should be designed for twice the traffic for which two-lane roads may be designed at the same location.

Problem 20.2 *Design a flexible pavement, using CBR curves, given the following data:*

(i)	*Subgrade soil (soaked) CBR*	= 5%
(ii)	*Laterite sub-base (soaked) CBR*	= 15%
(iii)	*Water bound macadam base CBR*	= 95%
(iv)	*Number of heavy vehicle per day in September 2008*	= 150
(v)	*Design life*	= 15 years
(vi)	*Annual rate of increase in the heavy vehicles*	= 5%

The road is proposed to be completed in September, 2013.

Solution. Average daily design traffic A is given by

$$A = p\,(1+r)^{n+15}$$

where p = Number of heavy vehicles per day = 150

r = Annual rate of increase in heavy vehicles = 5%

n = Number of years between the year of completion and the year of last count
= 2012 – 2007 = 5 years

$$\therefore \quad A = 150\left(1+\frac{5}{100}\right)^{5+15}$$

$$= 150 \times 1.05^{20}$$

$$= 150 \times 2.65$$

$$= 39.8$$

∴ Use curve D of Fig. 20.10 to design the various components of the flexible pavement.

Total depth of construction for CBR of 5 = 39 cm

Depth of construction for CBR of 15 = 22 cm

Depth of cosntruction for CBR of 95 = 7 cm

∴ Depth of sub-base material = 39 – 22 = 17 cm

Depth of WBM base = 22 – 7 = 15 cm

Provide a wearing course consisting of 5 cm of bituminous macadam and 2 cm of semi-dense bituminous carpet.

The section of the road is shown in Fig. 20.11.

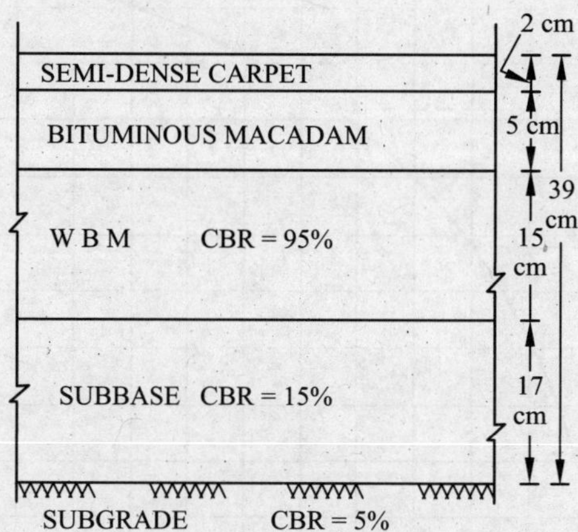

Fig. 20.11 For Problem 20.2.

Problem 20.3 *The CBR test carried out on a subgrade soil gave the following readings:*

Penetration (mm)	*Load (kg)*
0.0	*0.0*
0.5	*4.0*
1.0	*14.0*
1.5	*30.0*
2.0	*41.0*
2.5	*50.0*
3.0	*58.0*
4.0	*70.0*
5.0	*77.5*
7.5	*93.2*
10.0	*102.5*
12.0	*110.8*

The different pavement materials available near the construction site are as follows:

(*i*) *Sandy soil with CBR value = 10%*

(*ii*) *Soil-kankar mix with CBR value = 25%*

(*iii*) *Broken stone and gravel with CBR value = 90%*

(*iv*) *Bituminous concrete for surfacing = Minimum 5 cm thick.*

Design the pavement structure for commercial vehicles of 2000 per day.

Solution. For determining the CBR value of the subgrade soil, a graph (See Fig. 20.12) is plotted between the load in kg and the penetration in mm. After applying corrections due to concavity, the loads at 2.5 and 5.0 mm penetrations are 58 and 82 kg respectively.

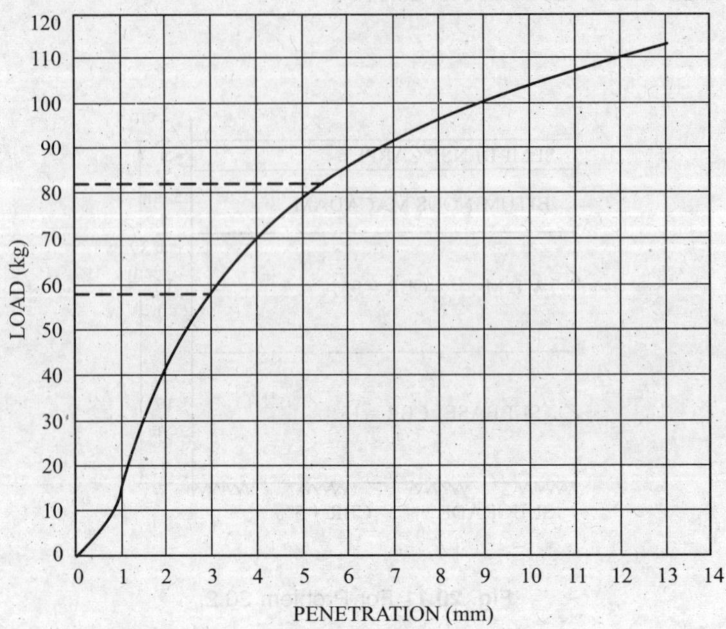

Fig. 20.12 Graph between load and penetration

$$\therefore \quad \text{Bearing value at 2.5 mm penetration} = \frac{\text{Load}}{\text{Area of plunger}}$$

$$= \frac{58}{19.35} \text{ kg/cm}^2$$

$$\text{CBR value at 2.5 mm penetration} = \frac{\text{Bearing value at 2.5 mm penetration}}{\text{Standard unit load}} \times 100$$

$$= \frac{58}{19.35} \times \frac{100}{70}$$

$$= 4.21$$

$$\text{CBR value at 5.0 mm penetration} = \frac{82}{19.35} \times \frac{100}{105}$$

$$= 4.04$$

Adopt CBR value of 4.21 say 4

For 2000 commercial vehicles per day, use curve F in Fig. 20.10.

Depth of construction for CBR of 4 = 55.0 cm

Depth of construction for CBR of 10 = 32.5 cm

Depth of construction for CBR of 25 = 20.0 cm

Depth of construction for CBR of 90 = 7.5 cm

Depth of sandy soil = 55 – 32.5

= 22.5 cm

Depth of soil kankar mix = 32.5 – 20.0

= 12.5 cm

Depth of broken stone and gravel = 20 – 7.5

= 12.5 cm

The section is shown in Fig. 20.13.

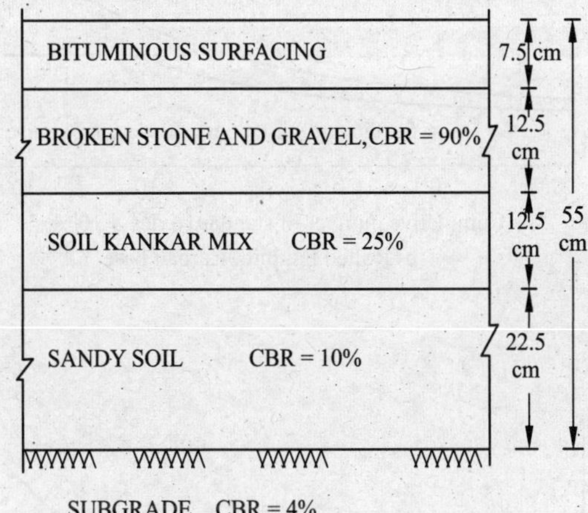

Fig. 20.13 For Problem 20.3.

20.5. BRITISH REVISED CBR METHOD

The design curves shown in Fig. 20.10 have been replaced by a new set of design charts given in Figs. 20.14 to 20.18. In these charts, only one set of curves utilizes the CBR procedure *i.e.*, the sub-base curves. The other curves specify independently the thicknesses of base course and surfacing to be used with given materials (bituminous concrete, dense macadam, lean concrete, soil cement and wet mix and dry bound macadam) for varying conditions of traffic flow *i.e.*, for different design lives. This is a major development as it enables the engineer to plan stage construction.

The British design procedure is explained briefly in the following steps:

***Step* 1.** Determine the number of commercial vehicles expected to use the roadway on the day it is opened *i.e.* present or construction traffic. This is obtained from traffic survey data. In the absence of detailed traffic survey, the guidelines given in Table 20.7 are used.

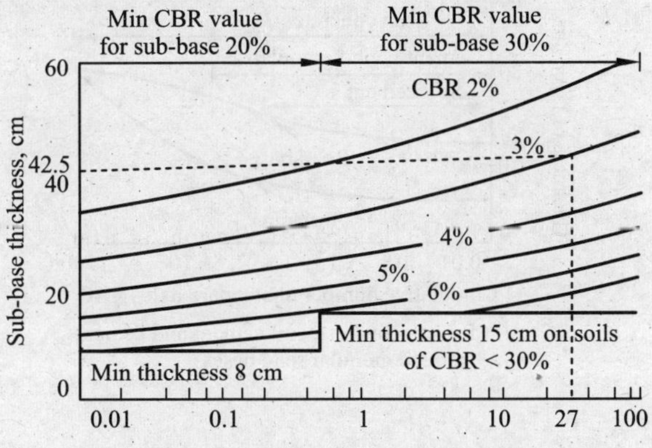

(a) Sub-base

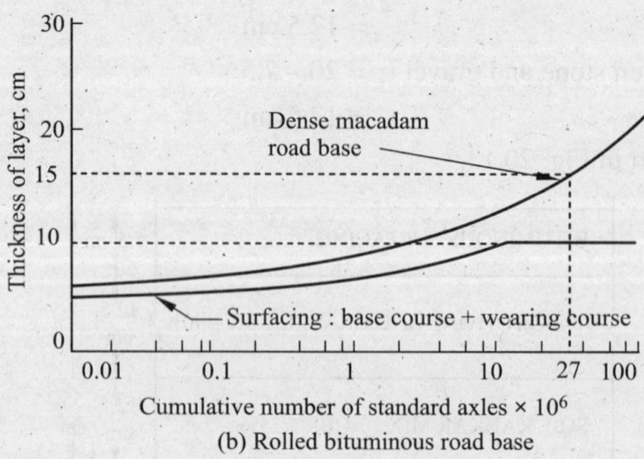

(b) Rolled bituminous road base

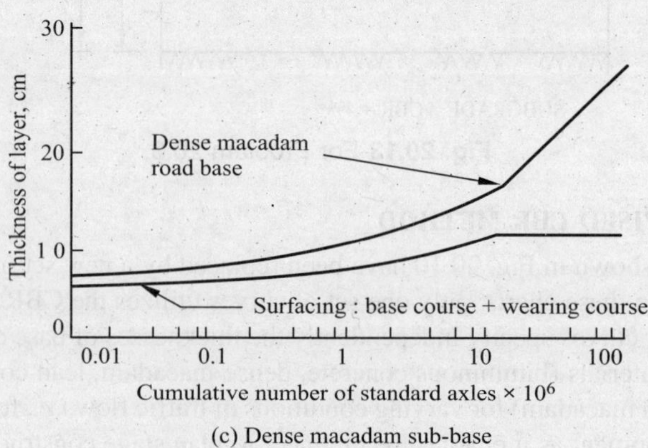

(c) Dense macadam sub-base

Fig. 20.14 to 20.16. Revised British design method

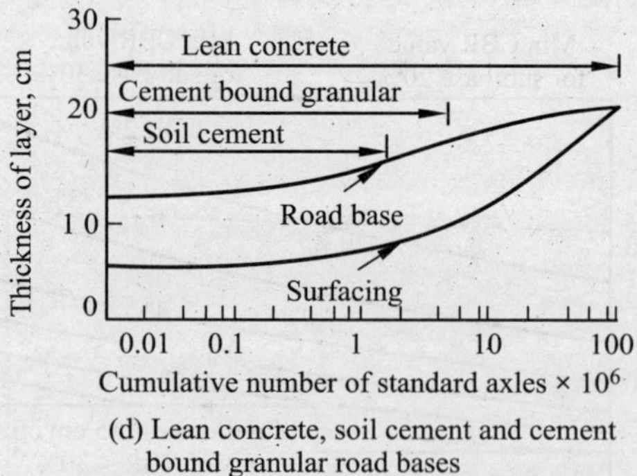

(d) Lean concrete, soil cement and cement bound granular road bases

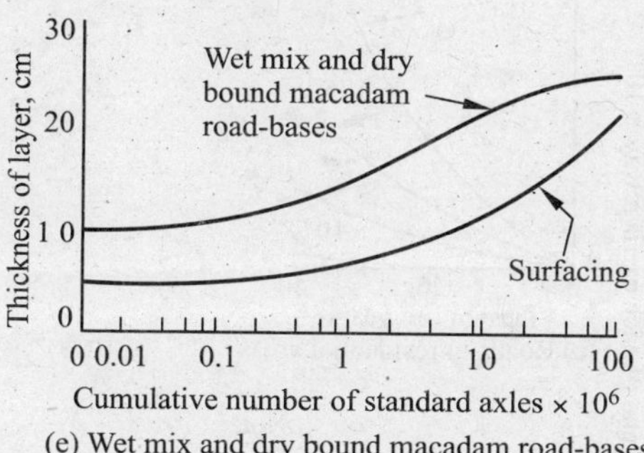

(e) Wet mix and dry bound macadam road-bases

Fig. 20.17 to 20.18 Revised British design method.

Table 20.7 Estimated traffic flow

Type of road	*Estimated traffic flow in each direction comm. Veh/day*
1. Minor residential roads	10
2. Through roads and roads carrying regular bus routes involving up to 25 *psv*/day in each direction	75
3. Major through roads carrying regular bus routes involving 25-50 *psv*/day in each direction	175
4. Main shopping centre of a large development carrying goods deliveries and main through roads carrying more than 50 psv/day in each direction	350

***Step* 2.** Project the traffic over the life time of the roadway. This is done with the help of Fig. 20.19. The curves are drawn for both rural and urban roads, assuming varying growth rates. Normally a life period of 20 years is considered adequate and a growth rate of 4 per cent is assumed.

***Step* 3.** Convert the number of commercial vehicles found in step 2 into the equivalent single axle load to be used for design factor given in tables 19.3 and 19.4 and determined from AASHO road test. Lacking similar British data, use of Table 20.8 is made to obtain equivalent number of standard axles from the number of commercial vehicles.

***Step* 4.** Determine the sub-base thickness. This is done by entering Fig. 20.14 with the appropriate cumulative number of standard axles and the subgrade CBR value. It may be noted that if the CBR of the subgrade is greater than the minimum requirement for the sub-base, then obviously no sub-base need be provided.

***Step* 5.** Determine the road base and surfacing thicknesses from Figs. 20.14 to 20.18 for the selected material. For very high quality road base materials e.g. rolled bituminous and dense coated macadam, the thickness of surfacing increases upto 100 mm with increasing traffic; above 11 × 10^6 standard axles any further increase in pavement strength is obtained only by increasing the road base.

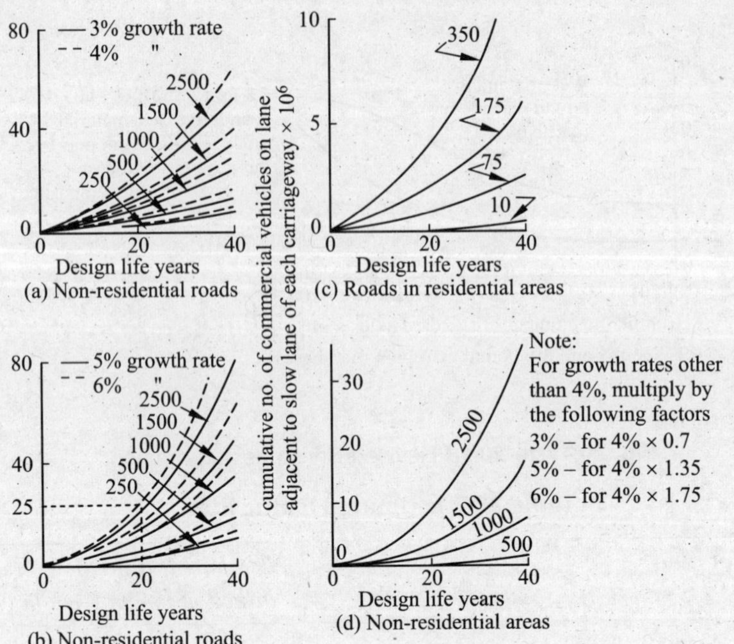

Fig. 20.19 Relationship between cumulative number of commercial vehicles in each lane and design life (in years).

Table 20.8 Equivalence factors and the damaging power of different axle loads

Axle load kg	*Equivalence factor*	*Axle load kg*	*Equivalence factor*
910	0.0002	9980	2.3
1810	0.0025	10890	3.2
1720	0.01	11790	4.4
3630	0.03	12700	5.8
4540	0.09	13610	7.6
5440	0.19	14520	9.7
6350	0.35	15420	12.1
7260	0.61	16320	15.0
8160	1.00	17230	18.6
9070	1.50	18140	22.8

Problem 20.4 *Design a flexible pavement for a two-lane coarriageway in rural area to carry 1500 commercial vehicles per day (in each direction) at the time of construction. The traffic growth rate may be assumed as 5 per cent. The design life is 20 years. The CBR value of subgrade is 3 per cent.*

Solution.

Step 1. The present traffic is 1500 commercial vehicles per day per carriageway.

Step 2. From Fig. 20.19 (b) the cumulative number of commercial vehicles in each lane, at the end of 20 years, will be 25×10^6, for 5% growth rate.

Step 3. From table 20.8 the number of standard axles to be used for design purposes is $1.08 \times 25 \times 10^6 = 27 \times 10^6$.

Step 4. From Fig. 20.14 the required sub-base thickness is 42.5 cm of material with a CBR of 3 per cent.

Step 5. Assuming a rolled bituminous road base (Fig. 20.15) the thickness of roadbase required is 15 cm for dense macadam and the surfacing thickness is 10 cm.

20.6. WYOMING CBR METHOD

The design curves are shown in Fig. 20.20. The factors of rainfall, frost action, ground water, existing conditions of subgrade and traffic are all taken into account and the subgrade is evaluated by the CBR test. Each of these factors is rated as indicated in Table 20.9, and the sum of the appropriate rating number is used to choose the design curve.

Traffic is determined by converting all wheel loads to equivalent single wheel loads by multiplying with equivalent factors as outlined in the previous chapter under repeated loads.

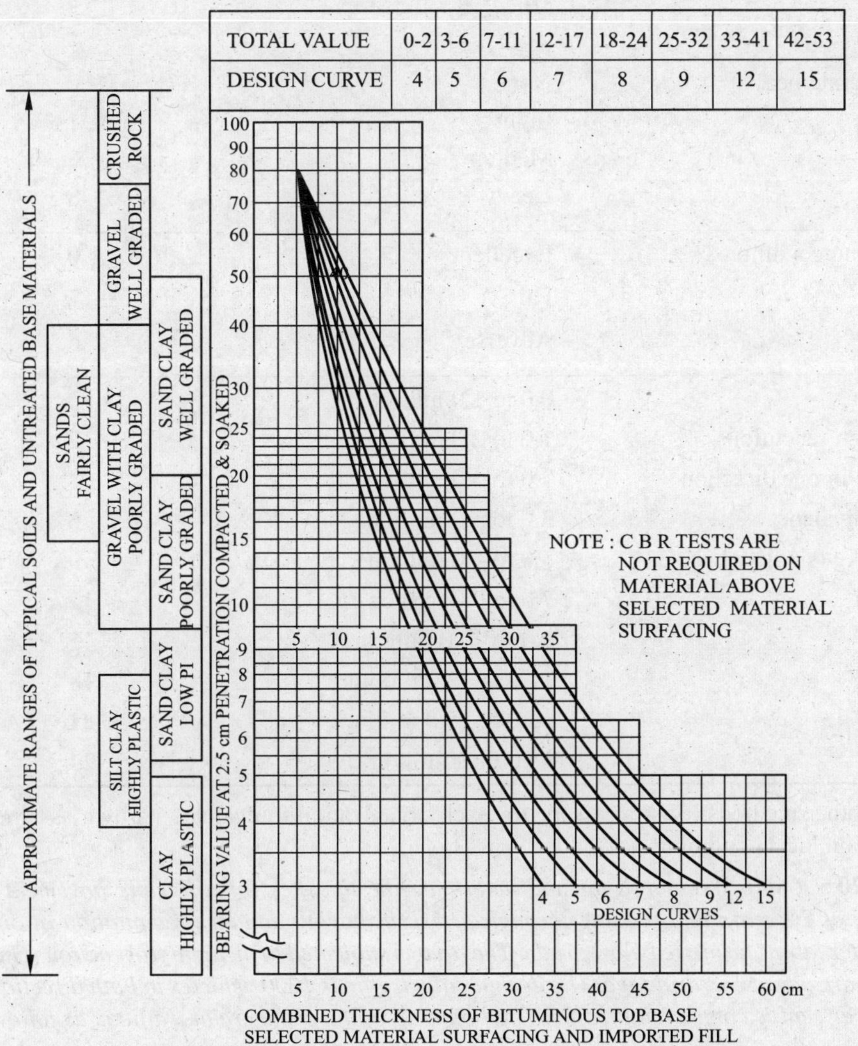

Fig. 20.20 Wyoming CBR design chart.

Table 20.9 Assigned values

Item	*Limits*	*Value*
Annual precipitation	12.5 to 25 cm	0
	25 to 37.5 cm	1
	37.5 to 50 cm	3
	50 to 62.5 cm or light irrigation	6
	62.5 to 125 cm or heavy irrigation	10
Water table	None evident	0
	3 to 1.75 m below grade	1
	1.75 to 1.25 m below grade	3
	1.25 to 0.5 m below grade	5
Frost action (Inducting heave)	None	0
	Light	1
	Medium	3
	Heavy	8
Existing conditions	Excellent	0
	Fair	2
	Adverse	6
Traffic (Design repetitions traffic in one direction in equivalent 2268 kg (single wheel loads)	0.0 to 1.0 million	0
	1.0 to 2.0 million	2
	2.0 to 3.0 million	4
	3.0 to 5.0 million	6
	5.0 to 7.0 million	9
	7.0 to 9.0 million	12
	9.0 to 11.0 million	15
	11.0 to 13.0 million	18
	13.0 to 15.0 million	21
	15.0 plus million	24

As for other methods, it is necessary for each organization to evaluate its own assigned values depending on local conditions.

Problem 20.5 *Using Wyoming's method determine the thickness of a flexible pavement with life expectancy of 20 years and passing through a terrain having annual precipitation of 50 cm and ground water table one metre below grade. The area is subjected to medium frost action. The existing conditions are adverse to drainage. The design daily traffic is 1000 vehicles in both directions, out of which 40 per cent is commercial traffic. The break-up of the commercial traffic is as follows:*

Wheel load (kg)	*Per cent of total*
2268	40
2722	20
3176	15
3630	10
4083	6
4537	5
4991	4

CBR value of subgrade for 2.5 mm penetration = 4

Solution.

Traffic in both directions = 1000 vehicles/day

Traffic in one direction $= \frac{1000}{2} = 500$ vehicle/day

Commercial vehicles in one direction $= 500 \times \frac{40}{100} = 200$ vehicles/day

The equivalent wheel load is determined below:

Serial No	*Wheel load kg*	*Average daily traffic*	*Equivalent factor (from table 19.2)*	*Equivalent Wheel load*
1	2268	200 × 0.4 = 80	1	80
2	2722	200 × 0.2 = 40	2	80
3	3176	200 × 0.15 = 30	4	120
4	3630	200 × 0.1 = 20	8	160
5	4083	200 × 0.06 = 12	16	192
6	4537	200 × 0.05 = 10	32	320
7	4991	200 × 0.04 = 8	64	512
				Total: 1464

Annual traffic $= 1464 \times 365 = 534.36 \times 10^3$

Traffic for 20 years $= 534.36 \times 10^3 \times 20$

$= 10.69 \times 10^6$

The assigned values for different items are as follows:

Item	*Assigned value*
Precipitation (50 cm)	6
Water table (1 m)	5
Frost action (medium)	3
Existing condition (adverse)	6
Traffic (10.69 million)	15
	Total: 35

For a total value of 35, use curve No. 12 in Fig. 20.20.

Against CBR value of 4, the design thickness of pavement from curve No. 12 in Fig. 20.20 is 46 cm.

20.7. CBR DESIGN METHOD FOR AIRFIELD PAVEMENTS

The U.S. Corps of Engineers developed design curves for various wheel loads, tyre pressures, gear configurations and traffic areas. These are contained in the design manuals published by the U.S. Corps of Engineers. These curves were based upon the following relationships between flexible pavement thickness to wheel loads, tyre pressures and repetition of loads and are applicable for CBR values less than 10 to 12.

$$t = \sqrt{P\left[\frac{1.75}{\text{CBR}} - \frac{1}{p\pi}\right]} \qquad \ldots(20.4)$$

$$t = \left(\frac{23.1 \log C + 14.4}{100}\right)\sqrt{P\left(\frac{1.75}{\text{CBR}} - \frac{1}{p\pi}\right)} \qquad \ldots(20.5)$$

where t = thickness of pavement in cm

P = wheel load in kg (ESWL)

p = tyre pressure in kg/cm^2

C = number of coverages. 100 per cent design is equivalent to 5000 coverages.

The design has now been simplified by categorizing airfields into three load conditions, i.e. light load, medium load and heavy load airfields. Each category has a set of critical aircraft load and configuration parameters. The design gear loads are 11350, 45350, 120000 kg for light, medium and heavy airfield respectively. The design curves are shown in Figs. 20.21 to 20.23.

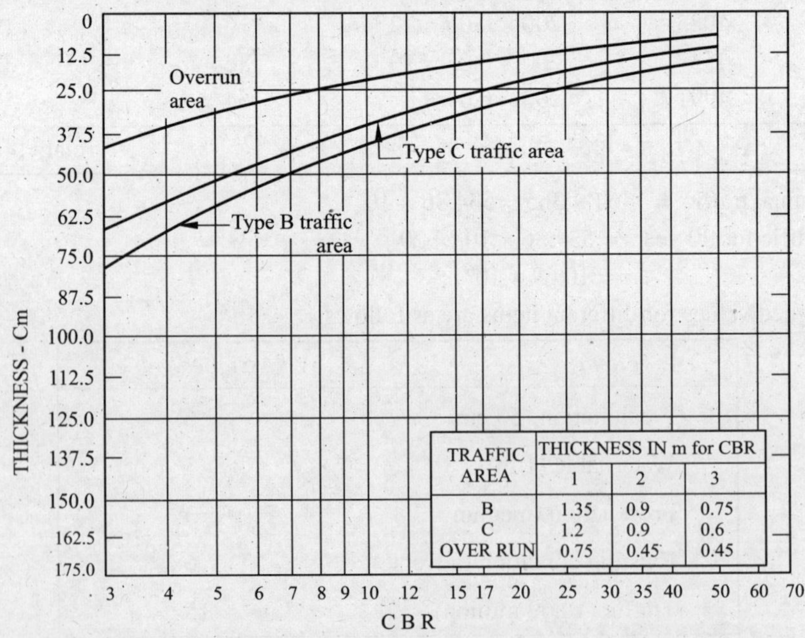

TRAFFIC AREA	THICKNESS IN m for CBR		
	1	2	3
B	1.35	0.9	0.75
C	1.2	0.9	0.6
OVER RUN	0.75	0.45	0.45

Fig. 20.21 Flexible pavement design curves for light load pavements single wheel.

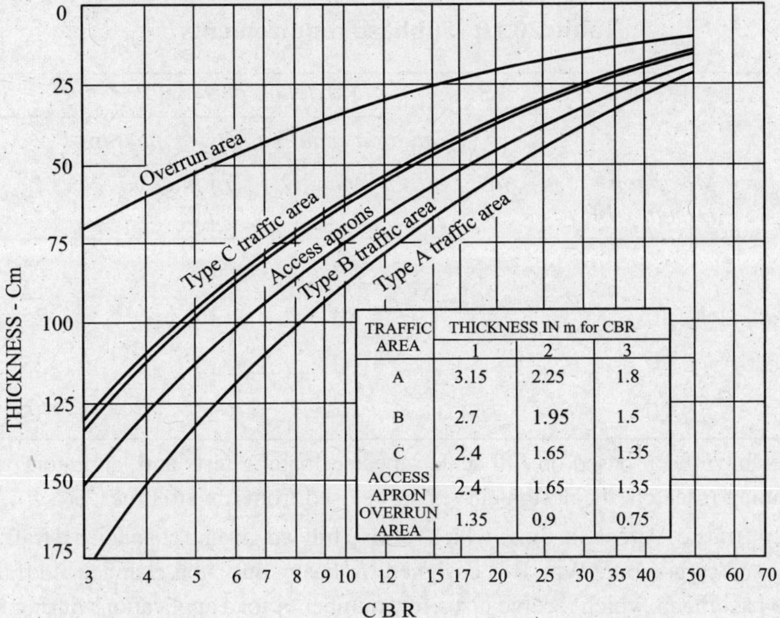

TRAFFIC AREA	THICKNESS IN m for CBR		
	1	2	3
A	3.15	2.25	1.8
B	2.7	1.95	1.5
C	2.4	1.65	1.35
ACCESS APRON	2.4	1.65	1.35
OVERRUN AREA	1.35	0.9	0.75

Fig. 20.22 Flexible pavement design curves for medium load pavement twin wheels.

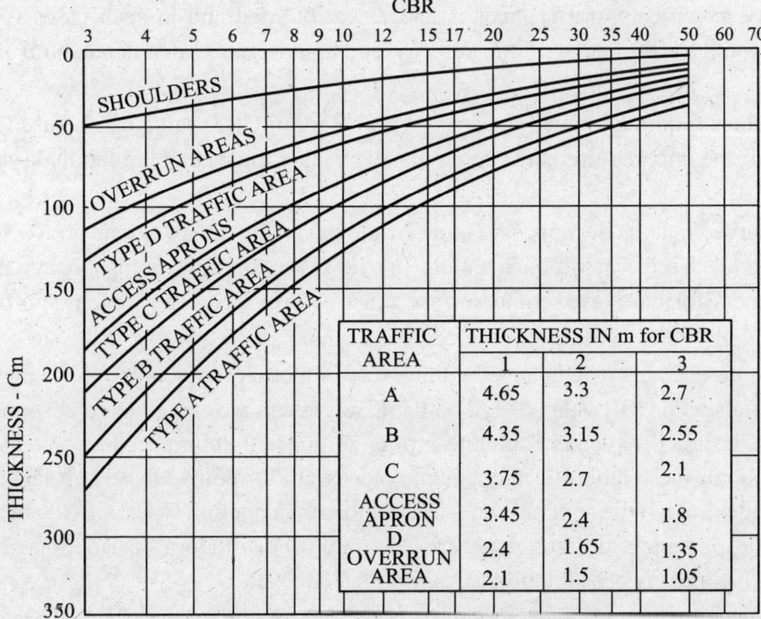

TRAFFIC AREA	THICKNESS IN m for CBR		
	1	2	3
A	4.65	3.3	2.7
B	4.35	3.15	2.55
C	3.75	2.7	2.1
ACCESS APRON	3.45	2.4	1.8
D	2.4	1.65	1.35
OVERRUN AREA	2.1	1.5	1.05

Fig. 20.23 Flexible pavement design curves for heavy load pavements-twin-twin wheel.

Table 20.10 indicates the maximum design values of CBR for sub-base courses. The gradation requirements, liquid limit and plasticity index are also specified. In addition to these requirements, laboratory CBR value of the sub-base course material must be greater than the design values selected from table 20.10.

Table 20.10 Subbase requirements

Material	*Maximum Design CBR*	*Maximum permissible value*				
		Gradation requirements, % passing				
		Size cm	*2 mm*	*75 micron Sieve*	*LL*	*PI*
Subbase	50	7.5	50	15	25	5
Subbase	40	7.5	80	15	25	5
Subbase Select	30	7.5	100	15	25	5
Material	20	7.5	—	25	35	12

These curves have been based on full scale repeated loading tests and laboratory tests and as a result of continuing research, the design curves are revised from time to time.

In these charts, traffic A areas are those which receive full weight and channelized traffic of multiple wheel bombers and cargo craft. Type B is designed for heavy duty non-channelized traffic, whereas type *C* refers to pavements which receive just a few number of load applications during the pavement life. Type D pavements are those where traffic volumes are extremely low and weight of aircraft is much lower than the design weight. Thickness requirements decreases for traffic class *A* through *D*.

For light-duty pavements, traffic areas *A* and *D* are omitted and in such cases type *B* denotes taxiways aprons and 3000 metres of the runway ends; and type *C* denotes central portion of the runway.

Thickness values required above a material with a certain CBR value can be used for design of successive layers, the only requirements being that each layer must be of higher quality than the layer below it.

It will be observed that the design curves for airfield pavement do not extend above 50 CBR which means that materials used for sub-base should not be given strength rating greater than 50 CBR. Minimum thickness for base course and surface must be provided even though sub-base materials with CBR excess of 50 are used.

A small table on each chart indicates the thickness in metre of pavement plus subgrade required above soft materials with CBR value of 1, 2 and 3. If soft layers are encountered at some depth below surface and the designed pavement thickness plus compacted subbrade does not meet thickness requirement set up in these tables, it may become necessary to remove these soft layers to preclude consolidation under load. If the subgrade consists of more than one soil types, CBR values are selected for each area and pavement designs made for each area. For random distribution of soil areas, a uniform design thickness is based upon the minimum CBR value.

Sub-base courses may consists of variety of materials. In some cases the subgrade itself might meet the requirements for a sub-base, or it may be stabilized to function as a sub-base, provided the liquid limit and plasticity index requirements for subbases are met by the natural subgrade soil.

Design thickness requirements for pavements and bases are dominated by considerations of stability and durability. Tyre contact pressure is more important than loading in base and surface determinations, though both must be considered. Design CBR values for base course materials as recommended by U.S. Corps of Engineers, is shown in Table 20.11.

Table 20.11 Design CBR for base course

Type	*Design CBR*
Graded crushed aggregate	100
Water-bound macadam	100
Dry-bound macadam	100
Bituminous base course, central plant hot-mix	100
Lime rock	80
Stabilized aggregate	80

Stabilized aggregate is permitted to be used only in areas subjected up to 7 kg/cm^2 tyre pressure and for type *D* traffic areas.

Compaction requirements of minimum of 100 per cent of modified AASHO density for sub bases and base course is required to be met at all times in order to control consolidation resulting from repeated wheel loads.

The minimum thickness of surface and base course for the wheel loads in Fig. 20.21 to 20.23 as recommended by U.S. Corps of Engineers, is given in Table 20.13. As will be observed, this requirement contemplates the use of a high type pavement and base material which have CBR rating of 80 or higher.

Procedure is also given for design for protection against frost action. A summary of the frost design methods for airfield pavements is given in Table 20.12.

Table 20.13 Pavement and base thickness design criteria

HEAVY-LOAD DESIGN

Twin-twin assembly, bicycle; spacing 94-155-94 cm centre to centre; contact area 1722.5 cm^2 each wheel

*Minimum thickness (cm)**

	100 CBR base			*80 CBR base*		
Traffic area	*Pavement*	*Base*	*Total*	*Pavement*	*Base*	*Total*
A	12.5	25	37.5	15.0	22.5	37.5
B	10.0	22.5	32.5	12.5	20.0	32.5
C	10.0	22.5	32.5	12.5	20.0	32.5
D	7.5	15.0	22.5	7.5	15.0	22.5
Access aprons	7.5	15.0	22.5	7.5	15.0	22.5
Shoulders	5.0	15.0	20.0	5.0	15.0	20.0

MEDIUM-LOAD DESIGN

Twin assembly, tricycle; spacing 94 cm centre-to-centre; contact area 1722.5 cm^2 each wheel

*Minimum thickness (cm)**

	100 CBR base			*80 CBR base*		
Traffic area	*Pavement*	*Base*	*Total*	*Pavement*	*Base*	*Total*
A	10.0	15.0	25.0	12.5	15.0	27.5
B	7.5	15.0	22.5	10.0	15.0	25.0
C	7.5	15.0	22.5	10.0	15.0	25.0
Access aprons	7.5	15.0	22.5	7.5	15.0	22.5

Table 20.12 Summary of methods for design of airfield pavement for frost conditions

	Horizontal variability of subgrade soil and moisture conditions			
	Uniform	*Slightly variable*	*Variable*	*Extremely variable*
Design Method	Variations affecting heave potential virtually undetectable by ordinary methods of investigation. Negligible differential frost heave and thaw settlement may be anticipated under reduced subgrade strength design	Small variations of subgrade conditions apparent by ordinary methods of investigation.	Subgrade conditions moderately variable. Widespread cracking of rigid pavements and appreciable surface deformation would be expected if reduced subgrade strength design method were used.	*Very large, frequent, and abrupt change in subgrade frost heave potential not permitting use of transition sections.*
(1)	(2)	(3)	(4)	(5)
Complete Protection	*Required for flexible and rigid pavements:*			Applicable only under exceptionally adverse conditions for F3 and F4 subgrades.
Limited Subgrade frost penetration	1. Over F4 subgrade soil 2. Over other frost-susceptible subgrade soils when: *a.* Cracking of rigid pavements or unacceptable pavement roughness caused by nonuniform frost heave may be expected with lesser design thickness, or *b.* Limited subgrade frost penetration design required less combined thickness or is otherwise more economical than reduced subgrade strength design.			
Reduced subgrade stregnth	Applicable for flexible and rigid pavements over F1 through F3 subgrades when objectionable differential heave or cracking will not occur.	Application for flexible pavements over F1 through F3 subgrades when objectionable diffe- rential heave or cracking will not occur.	Applicable for flexible pavements F1 through F4 subgrades when pavements are minor, slow speed, and noneritical and heave can be tolerated, except not to be used for F4 subgrade under adverse moisture conditions.	

LIGHT-LOAD DESIGN

Single wheel, tricycle; contact area 645 cm^2

*Minimum thickness (cm)**

	100 CBR base			*80 CBR base*		
Traffic Area	*Pavement*	*Base*	*Total*	*Pavement*	*Base*	*Total*
B	7.5	15.0	22.5	10.0	15.0	25.0
C	7.5	15.0	22.5	7.5	15.0	22.5
Access aprons	7.5	15.0	22.5	10.0	15.0	25.0

* These minimum thicknesses apply when layer directly under the base course has a design CBR of 50. When the underlying has a design CBR of 80, the minimum thickness of base shall be 15 cm.

Problem 20.6 *The CBR value of subgrade soil is 5 per cent. Calculate the total thickness of pavement using design formula developed by U.S. Corps of Engineers. Assume wheel load = 4082 kg. Tyre pressure = 7 kg/cm^2.*

Solution. The thickness of pavement is given by equation 20.4.

$$t = \sqrt{P}\sqrt{\frac{1.75}{\text{CBR}} - \frac{1}{p\pi}}$$

Here $P = 4082$ kg

$\text{CBR} = 5$

$p = 7 \text{ kg/cm}^2$

$$t = \sqrt{4082}\sqrt{\frac{1.75}{5} - \frac{1}{7\times\pi}}$$

$$= 63.89\sqrt{0.35 - 0.046}$$

$$= 63.89 \times 0.55$$

$$= 35.14 \text{ cm}$$

say 36 cm

Problem 20.7 *A flexible pavement is to be designed for a heavy load airfield to cater for Type A traffic area; neglecting frost problem. The material properties are given in Table 20.14 and the depth of compaction requirements for the plastic clay in Table 20.15.*

Table 20.14 Material characteristics

Material	*Test CBR %*	*Percent passing*		*Liquid limit %*	*Plasticity index %*
		2 mm	*No. 75 micron*		
Subgrade (Natural)	5			35	15
Subgrade (compacted)	10			35	15
Subbase No, 1	25	85	15	20	7
Subbase No. 2	55	45	10	15	5
Base (meets rquirements for graded crushed aggregates)					
Surface (meets requirements for hot-mix asphaltic concrete)					

Table 20.15 Depth of compaction

Per cent compaction	*100*	*95*	*90*	*85*	*80*
Depth of compaction (cm)	*60*	*90*	*125*	*200*	*250*

Solution. A summary of the design data is givne in Table 20.16, also showing the source of information.

Table 20.16 Design data summary

Material	*Design CBR %*	*Thickness above layer, cm*	*Source of information*
Subgrade (natural)	5	195	Fig. 20.23
Subgrade (compacted)	10	125	Fig. 20.23
Subbase No. 1	25	65	Fig. 20.23
Subbase No. 2	50	35	Fig. 20.23
Base	100	15 (min)	Table 20.13
Surface	NA	10 (min)	Table 20.13

The design CBR values for the two sub-base materials are selected from the following considerations.

For sub-base No. 1, the material characteristics would allow a maximum design CBR = 40 per cent from table 20.10 since the P.I. value is higher than the maximum permissible value. However the actual CBR value of 25 per cent is less than that from table 20.10. Therefore, the tested CBR value is used as the design CBR.

For sub-base No. 2, Table 20.10 gives maximum CBR = 50 per cent.

Since this value is less than the actual CBR of 55 per cent, adopt a design value of 50 per cent.

From the results shown in Table 20.16, the design thickness of the various layers are:

Surface (Bituminous concrete layer)	: 10 cm
Base course (gramular crushed ston)	: 15 cm
Sub base No. 2	: 35 cm
Sub base No. 1	: 65 cm
Compacted subgrade	: 70 cm with at least 90% AASHO density
Natural sub-grade	: Must have atleast 85% of modified AASHO density between a depth 125 to 200 cm and at least 80 per cent compaction below this depth.

If the above compaction requirements are not met with by the natural subgrade, either of the following modifications may have to be resorted to:

(*i*) compacted in place

(*ii*) excavated and recompacted as a fill material; and

(*iii*) the thickness of the pavement increased to provide the required in place depth-compaction criteria.

20.8. GENERAL OBSERVATIONS OF CBR METHOD

The CBR design procedure is relatively simple and was most widely used for the design of flexible pavements.

The principal advantage of this method lies in the test that can be performed in the laboratory on comparatively small samples.

The design procedure utilizing CBR test is entirely empirical and does not give the designer a measure of shearing resistance of soil and bears no discernible relationship to the basic elements of the design of flexible pavements. However, such a procedure can prove perfectly satisfactory if the variables affecting the stability of the various layers can be properly evaluated and if CBR tests are correlated with pavement performance. Vast amount of such correlation exists in this method.

The chief difficulty in CBR method is to decide under what conditions of moisture content and dry density to test the subgrade in order to allow for changes subsequent to construction, particularly in moisture content. The procedure often adopted of soaking specimens before making penetration test may be too severe in certain cases. This is evident from the investigations of airfields conducted in Canada and reported by Mcleod. A further objection to soaking procedure is that it produces a specimen with more sudden change in density and moisture content within a small volume, owing to swelling and softening of the top 12.5 mm of soil, than are likely to occur in practice. However, some of these objections can be minimized by determining CBR value in situ.

The main draw-back of this method is non-uniformity of test results with certain soils. With clayey gravels and clean sands, it is difficult to obtain reliable test results-results which would correlate with the observed performance of these materials in the roadway.

It appears, however that much more research is needed to perfect the testing procedure and to assure uniform correlation with the fundamental soil properties. For this, it is necessary to determine proper moisture content, probably between the plastic and liquid limits of plastic soil at which CBR should be tested to establish relationship between CBR value and other fundamental properties of soils. An other step toward the improvement of the usefulness of the CBR method would be perfection of a reliable method for determining CBR values for granular materials with coarse aggregate and variously coarse graded base mixtures. Until these steps are taken, the CBR method should be used with caution and judgment.

Another limitation of this method is that even though thickness of various layers of road can be designed, it is assumed that stress distribution is independent of the quality of the various layers which assumption for high quality materials is incorrect.

Another criticism of this method, as for any other empirical method, is that safety factor actually being used cannot be correctly determined.

20.9. NORTH DAKOTA CONE METHOD

A method similar to the CBR method has been developed by the North Dakota State Highway Department, U.S.A. Boyd, who developed the North Dakota cone test proposed the following equation for determining pavement thickness which was obtained by a curve separating failure from non-failures on a graph between subgrade cone bearing value and thickness of construction.

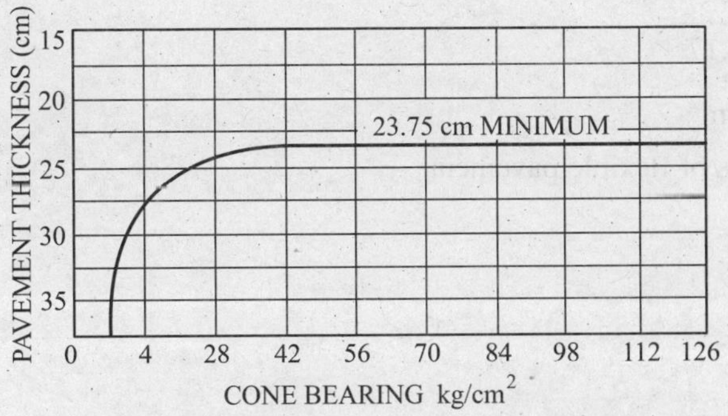

Fig. 20.24 North Dakota cone design curve

$$h = \frac{72.45}{b^{0.388}} \qquad \ldots(20.6)$$

where h = thickness of pavement (cm)
b = cone bearing value (kg/cm²)

Equation 19.12 (page 656) gives a method for the determination of b.

The above equation was developed on the basis of cone bearing value data from many highways. The curve shown in Fig. 20.24 is used in arriving at the design.

When the cone bearing value is 28 kg/cm² or more, a standard minimum section 24 cm thickness is used. This standard section consists of 12.50 cm of sub-base, 5 cm of stabilized aggregate base and 6.5 cm of bituminous concrete wearing surface. Additional sub-bases are required whenever the cone value is less than 28 kg/cm².

The test method in this case is simpler than CBR test and this design method has scope for future development in India.

Problem 20.8 *Calculate the total thickness of a flexible pavement by North Dakota cone method, given the following data for the subgrade:*

Load on cone (kg)	*Penetration reading (cm)*
5	*2.60*
10	*3.78*
20	*5.45*
40	*7.95*

Solution. Correction for zero due to blunt end

$$C = \Delta_{40} - 2\Delta_{10}$$
$$= 7.95 - 2 \times 3.78$$
$$= 0.39 \text{ cm}$$

∴ Corrected reading p for 10 kg load, P

$$= 3.78 + 0.39$$
$$= 4.17 \text{ cm}$$

Subgrade cone bearing value

$$b = \frac{P}{0.058\, p^2}$$
$$= \frac{10}{0.058 \times 4.17^2}$$
$$= 9.92 \text{ kg/cm}^2$$

∴ Total thickness of flexible pavement

$$h = \frac{72.45}{b^{0.3888}}$$
$$= \frac{72.45}{9.92^{0.388}}$$
$$= 29.75 \text{ cm}$$

Say 30 cm.

PLATE-BEARING TEST METHODS

These methods recommend plate-bearing tests to evaluate the strength of the subgrade and/or base course materials.

20.10. U.S. NAVY METHOD FOR AIRFIELD PAVEMENTS

This method of design consists of three distinct steps, namely:

1. Burmister's elastic theory is used for obtaining first approximation of the required flexible pavement thickness.
2. Trial sections are constructed in which the base is made equal to the theoretical thickness determined in step 1 above, $\frac{2}{3}$ of the theoretical thickness, and $1\frac{1}{2}$ times the theoretical thickness.
3. Plate-bearing tests are run on these trial-sections and the final thickness is chosen on the basis of these tests which produces a deflection of 5 mm.

In this method, the compacted densities and moisture contents of the sub-base and base course materials of the trial sections must be the same in all test sections. The subgrade must be the same with respect to type, depth and degree of compaction and moisture content throughout. The only variable kept is pavement thickness. Moreover, the construction of the trial sections must conform to the specification requirements that are to govern and apply in the construction of the airfield pavement. The usual compaction requirements are 95 per cent of standard AASHO density for light planes with wheel loads less than 6800 kg and 95 per cent of the modified AASHO value for greater loads.

The design method is based on the assumption that if the downward deflection of the pavement does not exceed 5 mm under the maximum anticipated load the structural strength of the pavement and base is satisfactory.

Step 1. A plate-bearing test is run by a circular plate on the subgrade and sufficient load-deflection data are obtained to plot a curve of load versus deflection upto at least 6.25 mm. From these data, the value of modulus of elasticity of subgrade, E_2 is obtained by means of Burmister's theory of elasticity for two-layer system for a rigid plate by the equation:

$$E_2 = \frac{1.18P.a}{\Delta} \quad \text{...(20.11)}$$

where E_2 = modulues of elasticity of the subgrade, kg/cm^2
P = unit load on plate, kg/cm^2
a = radius of plate, cm
Δ = deflection, or settlement of plate, cm

After E_2 is known, a base course about 15 to 30 cm thick is built over the same area. Base material used is the same which is to be employed in actual construction. The plate-bearing test is now repeated in the centre of this area. The test section should be at least 4.5 m × 4.5 m or larger upto 6 m × 6 m. Using the load-deflection data obtained from this test, the value of settlement factor F_w is calculated from the Burmister's deflection equation for the two layer system for rigid plate. This equation as already given in the previous chapter is:

$$\Delta = 1.18\frac{P.a}{E_2}F_w$$

$$\therefore \quad F_w = \frac{E_2\,\Delta}{1.18\,P.a} \quad \text{...(20.12)}$$

Knowing F_w, h and a, the ratio E_1/E_2 is found from Fig. 19.25 given in the previous chapter. Knowing E_2 from the plate-bearing test and E_1/E_2, the value of E_1, the modulus of elasticity of base course is calculated.

The procedure so far used was adopted to establish the value of E_1, and it does not matter as to what subgrade was used in making this test so as to get the value of E_2. In any case, the value of E_2 for actual subgrade is established by plate-bearing test. Knowing E_1 and E_2, the theoretical pavement thickness is determined from Burmister's theory relating to flexible plate.

The radius of tyre imprint, a, is calculated by obtaining contact area by dividing the total anticipated load, W by the anticipated contact pressure, P. This is on the assumption that supports furnished by sidewalls of tyre are ignored. The U.S. navy assumes support by side wall as 10 per cent and thus contact pressure is greater than tyre pressure by a factor of 1.1 and contact area is equal to wheel load divided by 1.1 × tyre pressure or 0.9 × wheel load divided by tyre pressure.

To obtain the value of F_w, in flexible plates, substitute in equation

$$\Delta = \frac{1.5\, P.a}{E_2} F_w$$

the value of Δ which is assumed as 0.5 cm

$$\therefore \quad F_w = \frac{0.5 \times E_2}{1.5\, P.a} = \frac{E_2}{3\, P.a} \qquad \ldots(20.13)$$

Knowing the values of F_w and E_1/E_2, the values of the thickness h, as multiple of 'a' can be read directly from the Fig. 19.25. This gives the value of thickness as theoretically calculated from Burmister's elastic theory for any condition of loading.

Steps 2 and 3. Trial sections are now built with a base equal to the theoretical thickness, 2/3 of theoretical thickness and $1\frac{1}{2}$ times the theoretical thickness. The design procedure recommends that trial sections be constructed for three different soil conditions, one on typical fill section, another on typical cut section and another on natural grade making a total of nine test sections. Plate-bearing tests are made on each of these sections.

In making these tests, a plate is employed which has a radius corresponding to the effective tyre radius, a.

Correction for saturation of subgrade is made on the assumption that the subgrade will never get wetter than optimum plus 2 per cent. Un-confined compressive tests are made on subgrade samples at optimum moisture content and at moisture content 2 per cent greater than the optimum. The samples used are 5 × 10 cm cylinders. The ratio of strength at optimum plus 2 per cent, is multiplied by the actual deflection to give a corrected deflection to account for subsequent increase in moisture.

The corrected settlement for each test section for the required loading condition is now plotted against the thickness of the respective sections. More accurate thickness is determined from this plot which produces a deflection of 5 mm.

The above is the recommended design thickness for base course. The added bituminous concrete surfacing is taken as an additional safety factor and is not considered in arriving at the thickness of the base. The recommended thickness of bituminous concrete pavement for various types of wheel loads is given in Table 20.16.

Table 20.16 Pavement surface thickness

Wheel Load (kg)	*Thickness (cm)*
6800 or less on soil cement or macadam bases	3.75
6800 or less, other bases	5.00
6800 to 11375	6.25
11375 to 22750	7.50
22750 and greater	10.00

Base courses can be made of several types and they should satisfy certain minimum strength requirements. The values range from 60 CBR to 80 CBR (or greater) for tyre pressures upto 10.5 kg/cm^2, or greater than 10.5 kg/cm^2 respectively. Sub-bases may also be included for very heavy traffic.

Problem 20.9 *A plate bearing test was carried out on subgrade soil and for 0.125 cm deflection, the unit pressure on 75 cm diameter plate was 0.8 kg/cm^2. On a test section of base course 15 cm thick, the pressure on the plate was found to be 2.1 kg/cm^2 for 0.125 cm deflection. Using U.S. Navy method, determine the thickness of flexible pavement required to sustain a single wheel load of 22500 kg with a tyre inflation pressure of 14 kg/cm^2.*

Solution (*i*) Modulus of elasticity of the sub-grade

$$E_2 = \frac{1.18\, P.\, a}{\Delta}$$

$$= \frac{1.18 \times 0.8}{0.125} \times \frac{75}{2}$$

$$= 283.2 \text{ kg/cm}^2$$

(*ii*) For plate bearing test on base course, settlement factor,

$$F_w = \frac{E_2 \Delta}{1.18\, P.a}$$

$$= \frac{283.2 \times 0.125 \times 2}{1.18 \times 2.1 \times 75}$$

$$= 0.38$$

$$\frac{h}{a} = \frac{15}{75/2} = 0.4$$

From Fig. 19.25, for $F_w = 0.38$ and $h = 0.4\, a$

$$\frac{E_1}{E_2} = 200$$

∴ Modulus of elasticity of base course

$$E_1 = 200\, E_2 = 200 \times 283.2$$

$$= 56640 \text{ kg/cm}^2$$

(*iii*) For flexible pavement

Radius of contact,

$$a = \sqrt{\frac{W}{P\pi}} = \sqrt{\frac{22500}{14\pi}}$$

$= 22.62$ cm

Deflection equation for flexible circular plate

$$\Delta = \frac{1.5\,Pa}{E_2} F_w$$

$\therefore$ Settlement factor for flexible pavement

$$F_w = \frac{\Delta\, E_2}{1.5\,Pa}$$

$$= \frac{0.125 \times 283.2}{1.5 \times 14 \times 22.62}$$

$$= 0.075$$

From Fig. 19.25, for $F_w = 0.075$ and $\frac{E_1}{E_2} = 200$

$$\frac{h}{a} = 2.7$$

$\therefore$ Thickness of pavement

$$h = 2.7a = 2.7 \times 22.62$$

$$= 61 \text{ cm}.$$

METHOD BASED ON PERFORMANCE SERVICEABILITY CONCEPT

20.11. THE ASPHALT INSTITUTE (TAI) DESIGN METHOD

The method is unique among all other flexible pavement design methods in that the designed pavement thickness comprises of entirely bituminous mixtures. TAI define the pavement having bituminous mixtures for all courses above the subgrade or improved subgrade as '*depthfull pavement*'. The analysis of WASHO, AASHO and British road tests showed that the unbound granular bases and subbases could be replaced by bituminous concrete by using suitable substitution ratios.

The basis for the design is primarily the analysis of bituminous pavement performance at the AASHO road test. Thus indirectly the use of performance serviceability concept is made by adopting a terminal serviceability (p_t) value of 2.5 to indicate failure. The use of equivalent 8200 kg single-axle load repetitions is manifested in the method by design traffic number (DTN). The DTN denotes the average daily number of equivalent 8200 kg single axle load repetitions for the design lane during the design period.

The methods for determining the DTN have been explained in chapter 19. If use of nomograph given in Fig. 19.13 is made, the initial traffic number (ITN) is obtained. This represents the DTN for the first year. An adjustment is then made to account for projected traffic growth as well as for design periods less than or greater than 20 years. The adjustment factors for different design periods and traffic growth rates are given in Table 20.17. This has been worked out from the relation :

$$\text{Adjustment factor} = \frac{(1+r)^n - 1}{20\,r}$$

Table 20.17 Initial traffic number (ITN) adjustment factors

Design period years (n)	Annual growth rate, percent (r) 0	2	4	6	8	10
1	0.05	0.05	0.05	0.05	0.05	0.05
2	0.10	0.10	0.10	0.10	0.10	0.10
4	0.20	0.21	0.21	0.22	0.22	0.23
6	0.30	0.32	0.33	0.35	0.37	0.39
8	0.40	0.43	0.46	0.50	0.53	0.57
10	0.50	0.55	0.60	0.66	0.72	0.80
12	0.60	0.67	0.75	0.84	0.95	1.07
14	0.70	0.80	0.92	1.05	1.21	1.40
16	0.80	0.93	1.09	1.28	1.52	1.80
18	0.90	1.07	1.28	1.55	1.87	2.28
20	1.00	1.21	1.49	1.84	2.29	2.86
25	1.25	1.60	2.08	2.74	3.66	4.92
30	1.50	2.03	2.80	3.95	5.66	8.22
35	1.75	2.50	3.68	5.57	8.62	13.55

Full depth pavement thickness (T_A) is a function of both DTN and the subgrade strength evaluated in terms of either CBR or cohesiometer value R, plate bearing test results (30 cm dia, 5 mm deflection and 10 repetitions). The correlation between CBR value and plate bearing test are given in Table 23.4 (page 849) or Fig. 19.16 (page 654). The design subgrade strength value on a given project is based on 90th percentile value obtained from random sampling techniques.

For ITN less than 10, use of Fig. 20.25 is made for adjustment.

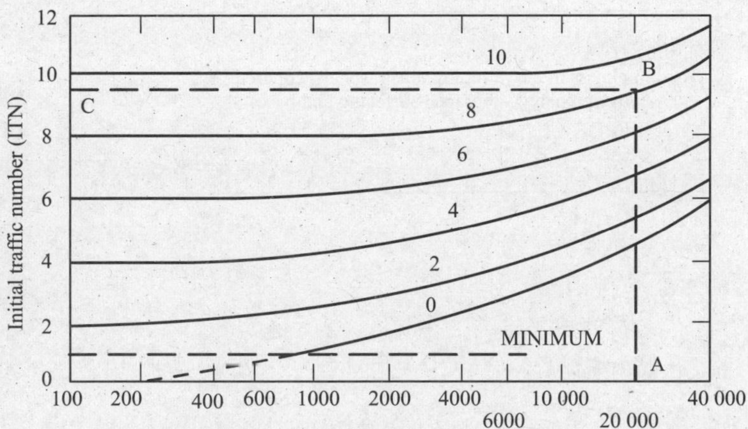

Fig. 20.25 Adjustment of ITN for light-traffic conditions.

The total depth thickness (T_A) can be determined by applying the following realations:

$$T_{A(\text{cm})} = \frac{23 + 10 \log DTN}{(CBR)^{0.4}} \qquad \ldots(20.17)$$

or $$T_{A(\text{cm})} = 2.5\left[6.37 + 2.75 \log DTN - 0.0893\, DTN^{0.119}(R-20)\right] \quad \ldots(20.18)$$

for $DTN \leq 20$; and

$$T_{A(\text{cm})} = 2.5\left[6.37 + 2.75 \log DTN - 0.117\, DTN^{0.0279}(R-12)\right] \quad \ldots(20.19)$$

for $DTN > 20$.

The solution of these equations is given in nomograph shown in Fig. 20.26. By entering the design traffic number (DTN) and the design subgrade strength (CBR or R or bearing plate value), the total thickness (T_A) in cm of bituminous concrete above prepared subgrade can be determined.

Alternate structural designs can be evolved by using hot-mix sand asphalt bases, liquid and emulsified bituminous bases, or untreated granular base or subbase with the help of Fig. 20.26 and using the substitution ratios, given in Table 20.18.

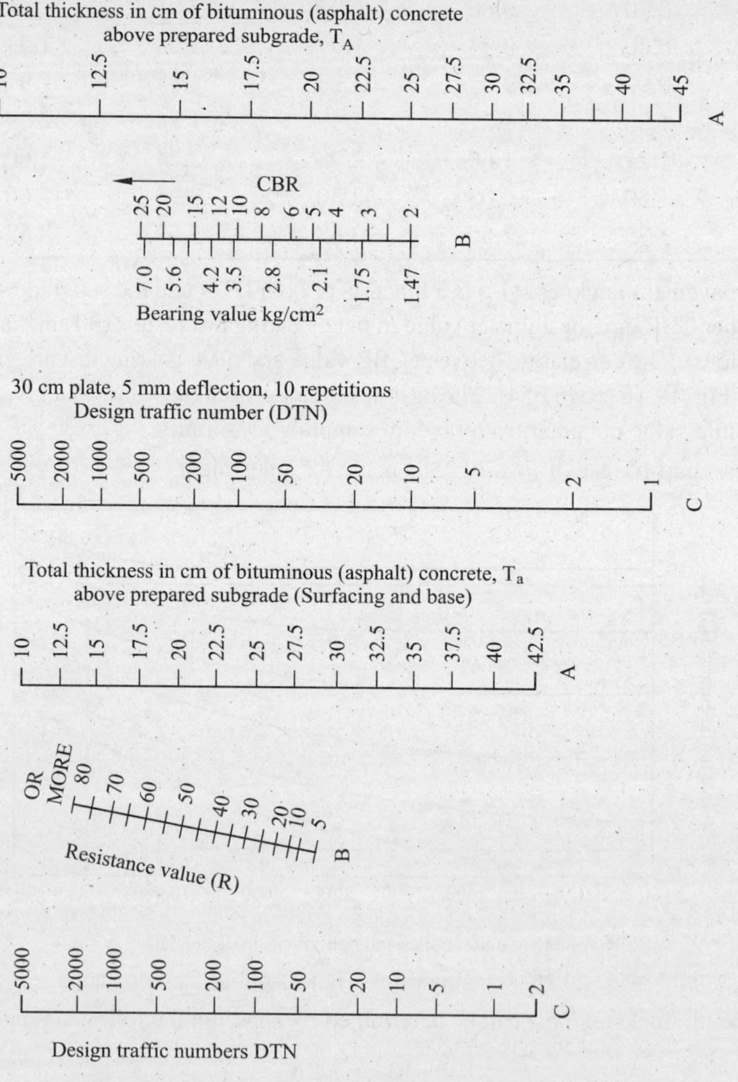

Fig. 20.26 The Asphalt Institute (TAI) design curves.

Table 20.18 Substitution ratios

S. No.	*Type of base/subbase*	*Substitution ratio*
(1)	Hot-mix sand asphalt bases	1.3 : 1
(2)	Liquid emulsified bituminous bases	1.4 : 1
(3)	Untreated granular bases	2 : 1
(4)	Untreated granular sub-bases	2.7 : 1

The minimum thickness recommended for bituminous concrete surfacing is shown in Fig. 20.27 for untreated granular bases. Similar requirement for bituminous stabilized bases is given in Table 20.19.

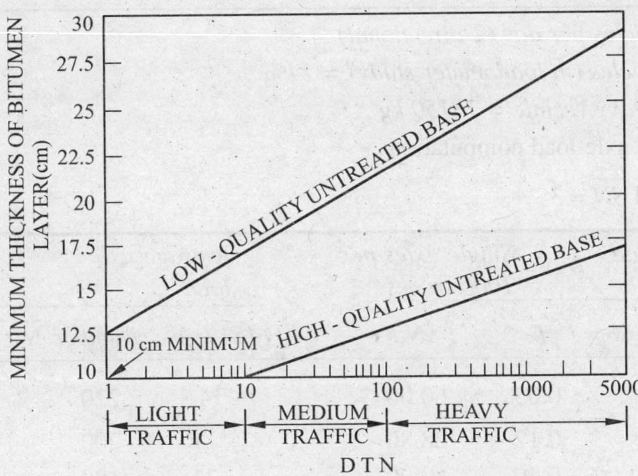

Fig. 20.27 Recommended surface thickness for untreated base pavements.

Table 20.19 Surface thickness requirements

		Liquid/emulsified bitumens	
Design DTN	*Hot mix-sand asphalt (cm)*	*A (cm)*	*B (cm)*
< 10	5.0	5.0	7.5
> 10 and < 100	7.5	7.5	10.5
≥100	10.0	10.0	12.5

Problem 20.10. *A four lane bituminous pavement is to be designed for the subgrade and traffic conditions given below. Determine:*

(a) full depth designs using bituminous concrete surface, hot mix sand asphalt, and emulsified bituminous bases.

(b) Pavement structure with bituminous surface and untreated granular base and subbase; and

(c) Pavement structure with a granular subbase material along with a 10 cm and 12.5 cm hot-mix sand asphalt base.

Traffic analysis period may be taken as 20 years assuming growth rate of 4 per cent. $\overline{SN}$ *value is 3.*

CBR data

9, 6, 12, 7, 8, 7, 10, 9, 10, 11

Traffic data

Single axle		*Tandem axle*	
Axle load (in 454 kg)	*Axles/day/1000 trucks*	*Axle load (in 454 kg)*	*Axle/day/1000 trucks*
≤8	1200	≤14	220
8-12	240	14.20	200
12-16	200	20.26	180
18-20	160	26-30	200
20-22	15	30-32	5
22-24	8		

ADT : 12,000 vehicles per day (2-directional)

Percent heavy vehicles (in loadometer study) = 14%

Gross weight of heavy vechile = 18150 kg

Solution. Equivalent axle load computations

$p_t = 2.5$ and $\overline{SN} = 3$

Average axle load	*Single axles per 1000 trucks*			*Average axle load*	*Tandem axles per 1000 trucks*		
(454 kg)	*Number, N*	*F*	*N.F.*	*(454 kg)*	*Number, N*	*F*	*N.F*
≤8	1200	0.05	60.00	≤14	220	0.04	8.8
10	240	0.12	28.80	17	200	0.09	18.0
14	200	0.40	80.00	23	180	0.27	48.6
19	160	0.83	132.80	28	200	0.55	110.0
21	15	1.98	29.70	31	5	0.80	4.0
23	8	2.63	21.04				189.4
			352.34				

Total equivalent 8200 kg single axle per 1000 trucks $= 352.34 + 189.40$

$$= 541.74$$

Assuming that 90 per cent of the trucks are in the design lane, the initial daily equivalent 8200 kg single axle loads is

$$= \frac{12000}{2} \times (0.90)\,(0.14) \left(\frac{541.74}{1000} \right) = 409.14 \text{ say } 410.$$

From Fig. 19.13, for 410 heavy trucks in the design lane, with average gross laden weight of 18150 kg and legal axle load of 8200 kg, the initial traffic number (ITN) = 275.

From table 20.17, the adjustment factor to account for the 4 per cent growth rate and 20 years design period is 1.49.

The design DTN $= 275 \times 1.49 \simeq 410.$

The 90th percentile design CBR can be obtained by the use of normal distribution characteristics or by simply calculating a cumulative frequency distribution of the values. Adopting the latter method, design CBR = 7.

From Fig. 20.26, design $T_A = 22.5$ cm.

The various design alternative are:

(1) For the full depth design, thickness of asphalt concrete = 22.5 cm.

If a hot-mix sand asphalt base is used, the minimum surface thickness of bituminous concrete is 10 cm from Table 20.19. This leaves 12.5 cm of the bituminous concrete to be replaced by hot-mix sand asphalt base. The substitution ratio of this material is 1.3. Thus 12.5 × 1.3 = 16.25 cm of hot-mix sand asphalt base is adequate. Similarly for bituminous emulsion base, thickness = 12.5 × 1.4 = 17.5 cm. Thus the two design alternatives are 10 cm of bituminous concrete surfacing over either 16.25 cm of hot-mix sand asphalt base or 17.5 cm of bituminous emulsion base.

(2) The minimum thickness of surfacing over an untreated granular material is shown in Fig. 20.27. For the DTN = 410, the minimum bituminous concrete is 14 cm for high-quality base material. The substitution ratio for granular base is 2.0. Thus, the required thickness is (22.5 – 14) × 2.0 = 17 cm of granular base. This is the only design using high quality granular base material.

(3) The minimum bituminous concrete requirement using hot-mix sand asphalt base is 10 cm, leaving 12.5 cm of equivalent bituminous concrete material to provide support for traffic. A 10 cm thick layer of hot-mix sand asphalt is equivalent to 7.75 cm of bituminous concrete $\left(\frac{10}{1.3} = 7.75 \text{ cm}\right)$. Thus the granular subbase thickness must provide an equivalent (22.5 – 10 – 7.75) = 4.75 cm of bituminous concrete. The required thickness for granular subbase would be 4.75 × 2.7 = 12.83 cm.

The required design for a 12.5 cm of hot mix sand asphalt base would be 10 cm bituminous concrete surface, 12.5 cm of hot mix sand asphalt base and 8 cm of granular subbase. The equivalent T_A of the pavement structure would be:

$$10 + \frac{12.5}{1.3} + \frac{8.0}{2.7} = 10.0 + 9.6 + 2.96 = 22.56 \text{ cm}$$

THEORETICAL AND SEMI-THEORETICAL METHODS

Most of the semi-theoretical methods make use of the Boussinesq's displacement equation $\left(\Delta = \frac{1.5P.a}{E_2} F_w\right)$ and design to a limiting deformation and make some allowance for the stiffness of the pavement relative to the subgrade. Use is made of the triaxial test to determine the value of modulii of pavement and subgrade. This is the approach followed in Kansas triaxial design method.

Some methods compare the shear strength of the soil when saturated as determined by triaxial test or unconfined compressive strength test for clayey soils with a shear stress curve derived from Boussinesq's analysis. Hicks of North Corolina has prepared curves for definite dual assemblies and for a number of wheel loads, ranging from 1815 kg to 4540 kg which have been developed from the Boussinesq's theory with the help of Newmark's chart.

Some methods assume a certain angle of spread of wheel load through the pavement and the pressure on the subgrade obtained is equated to the bearing capacity derived from Terzaghi's or Prandtl's formula divided by a factor of safety. This method has, however, not been widely used due to the reason that a relation for calculating the ultimate bearing capacity under a footing for a single application of load is not suitable without modification for designing pavement for which repeated applications of load have to be considered.

Burmister has suggested a theoretical design method based on analysis of stresses and strains in a two-layered system by choosing an arbitrary value for limiting displacement.

20.12. IRC DESIGN METHOD

Indian Roads Congress in their publication *"Guidelines for the Design of Flexible Pavements"* published in 2001 has developed a model considering flexible pavement as a three layer structure. Stresses and strains have been computed using the linear elastic model FPAVE developed under the Ministry of Road Transport and Highwasy research scheme 'Analytical Design of Flexible Pavements'. Refer to page 716 for details.

20.13. KANSAS TRIAXIAL METHOD

Method dealing with theoretical stresses in ideal masses using the Boussinesq's equation was first proposed by Palmer and Barber and was later on expanded by Barber. A modification of a formula developed by these investigators is the basis for Kansas method of design. The limiting deflection for flexible pavement is assumed as 2.5 mm. The equation for design thickness is as follows:

$$T = \left[\sqrt{\frac{3\,Pmn}{2\pi CS}} - a^2\right]^{*} \sqrt[3]{\left[\frac{C}{C_p}\right]} \qquad \text{... 20.20}$$

Where T = thickness required

C_p = modulus of deformation of pavement or surface course

C = modulus of deformation of subgrade or sub-base

P = base wheel load

m = traffic coefficient based on volume of traffic

n = saturation coefficient based on rainfall

a = radius of area of tyre contact corresponding to P

S = permitted deflection of surface (assumed 2.5 mm).

Both C_p and C are obtained from a triaxial test on the material in question. As the design procedure calls for all samples to be tested in a saturated condition, saturation coefficient n was introduced in the original formula proposed by Palmer and Barber to account for the possibility of the soil not being saturated after construction which gives a decreased thickness in localities of less rainfall. Similarly, traffic coefficient m was introduced to account for variation of traffic. The stiffness factor $\sqrt{C/C_p}$ was proposed on the basis of stiffness factor for slabs. The factor takes into account the stress distributing qualities of the paving surface as related to those of the subgrade soil. The values of m and n as proposed by Kansas Highway Department are given in Table 20.20 and 20.21 respectively.

TABLE 20.20 Values of *m*

Traffic coefficient m	*Total traffic (vehicles per day)*
1/2	50—400
2/3	401—800
5/6	801—1200
1	1,201—1800
7/6	1,801—2700
8/6	2,701—4000
9/6	4,001—6000
10/6	6,001—9,000
11/6	9,001—13,500
12/6	13,501—20,000

* See the derivation on next page.

Table 20.21 Values of n

Saturation coefficient n	*Average annual rainfall (cm)*
0.5	37.5—49.9
0.6	50.0—62.4
0.7	62.5—74.9
0.8	75.0—87.4
0.9	87.5—99.9
1.0	100—125

Derivation of formula: (Equation 20.20).

$$\Delta = \frac{p \cdot a}{E_2} \cdot F_w \text{ (Boussinesq's displacement equation)}$$

$$= \frac{p \cdot a}{E_2} \cdot \frac{3}{2} \cdot \frac{1}{\left[1+\left(\frac{h}{a}\right)^2\right]^{1/2}}$$

$$= \frac{3p \cdot a}{2E_2} \cdot \frac{a}{[a^2+h^2]^{1/2}}$$

$$= \frac{3P}{2\pi E_2} \cdot \frac{1}{[a^2+h^2]^{1/2}} \qquad \left[\text{Since } P = \pi a^2 p\right]$$

$$\therefore \quad h^2 = \left[\frac{3P}{2\pi E \Delta}\right]^2 - a^2$$

Using notations followed in the design formula

$$T^2 = \left[\frac{3P}{2\pi CS}\right]^2 - a^2$$

$$\therefore \quad T = \sqrt{\left(\frac{3P}{2\pi CS}\right)^2 - a^2}$$

Introducing the stiffness factor $\sqrt{\frac{C}{C_p}}$ and factors *m* and *n*, the formula is easily derived as given in equation 20.20.

Combination thickness. With the stiffness factor, the equation gives the thickness of a bituminous mat placed directly upon the subgrade soil for the assigned loading conditions. After obtaining the total thickness of surface course by the above formula the equivalent combination of two types of flexible layers, *i.e.* base course and surface pavement are obtained by the relation:

$$t_B = (T - t_p)\sqrt[3]{\frac{C_p}{C_B}}$$

where t_B = thickness of base course

T = thickness of surface course as previously calculated

t_p = assumed thickness of wearing surface

C_p = modulus of elasticity of wearing surface

C_B = modulus of elasticity of base course.

The above formula can also be used to calculate equivalent thickness of sub-base. In this case, suitable thickness of base course will be assumed.

For determining the values of C_p and C from the stress/strain curve in a triaxial test, a lateral pressure of 1.4 kg/cm^2 is commonly used which is considered comparable under usual field conditions. As the stress/strain curve for a soil is normally curved for the entire portion, values of stress differences are chosen which represent the calculated stress conditions.

The above equation applies to a single tyre load, whereas the design may be required for dual tyres. To avoid detailed calculations in each case, Kansas Department has prepared curves for different values of m and n, assuming C_p = 1050 kg/cm^2. Tests have established a minimum value of C_p for a well designed bituminous pavements as 1050 kg/cm^2 for a dense graded surface course. However, well designed bituminous concrete may have value of C_p as high as 1750 kg/cm^2 or more.

This method, however, useful has shortcomings when viewed from theoretical stand point. Firstly the design is based on displacement equation which gives subgrade deflection and it is assumed that the pavement deflection is negligible. For ratio E_1/E_2 or C_p/C greater than 100, the pavement deflection is much smaller than the subgrade deflection and can be safely neglected. But if the modular ratio is smaller than 100, this assumption no longer holds good. In actual practice, the modular ratio of flexible pavement is seldom greater than 100 and the pavement deflection should be taken into consideration. Actual test, such as AASHO road test show that pavement deflection often amounts to 30 per cent or more of subgrade deflection.

Another draw-back of this method is the assumption of a fixed and arbitrary value for the allowable deflection regardless of the pavement type. The usual value of 2.5 mm is rather large. Performance studies, including the AASHO road test, indicate that 1 mm would be a more realistic value to prevent the early deterioration of the pavement. However, if this is introduced in the Kansas formula, unrealistically high thickness will result.

In spite of its theoretical inconsistencies the Kansas method of design has been reportedly used with success.

Problem 20.11 *Calculate the thickness of sub-base, base and wearing course, using Kansas triaxial method, given the following data:*

Wheel load : 4080 kg

E value of subgrade soil = *90 kg/cm^2*

E value of sub-base = *270 kg/cm^2*

E value of base = *540 kg/cm^2*

E value of wearing course = *9000 kg/cm^2*

Tyre pressure = *6 kg/cm^2*

Total traffic (vehicles per day) = *5000*

Average annual rainfall = *80 cm*

Permitted deflection = *2.5 mm*

Solution. From Table 20.20, for traffic volume of 5000 vehicles per day, traffic coefficient, m = 1.5.

From Table 20.21, for average annual rainfall of 80 cm, saturation coefficient, n = 0.8.

According to Kansas triaxial method, design thickness

$$T = \left[\sqrt{\frac{3Pmn}{2\pi CS} - a^2}\right]\left[\left(\frac{C}{C_p}\right)^{1/3}\right]$$

Here $P = 4080$ kg

$m = 1.5$

$n = 0.8$

$C = 90$ kg/cm^2

$S = 0.25$ cm

$$a = \sqrt{\frac{P}{p\pi}} = \sqrt{\frac{4080}{6\times\pi}} = 14.71 \text{ cm}$$

$C_p = 9000$ Kg/cm^2

$$\therefore \quad T = \sqrt{\left(\frac{3\times4080\times1.5\times0.8}{2\pi\times90\times0.25}\right)^2 - (14.71)^2 \times \left[\left(\frac{90}{9000}\right)^{1/3}\right]}$$

$= 22.19$ cm

Assume thickness of wearing course $= 5$ cm

$\therefore$ Thickness of remaining pavement $= 22.19 - 5$

$= 17.19$ cm

Assume thickness of base course as 20 cm

This 17.19 cm pavement thickness can be converted into an equivalent thickness of base course, t_B by the formula

$$t_B = (T - t_p)\sqrt[3]{\frac{C_p}{C_B}}$$

$$= (22.19 - 5)\sqrt[3]{\frac{9000}{540}}$$

$= 43.87$ cm

$\therefore$ Thickness of sub-base

$$t_s = (t_B - 20)\times\sqrt[3]{\frac{C_B}{C_s}}$$

$$= (43.87 - 20)\sqrt[3]{\frac{540}{270}}$$

$= 30$ cm

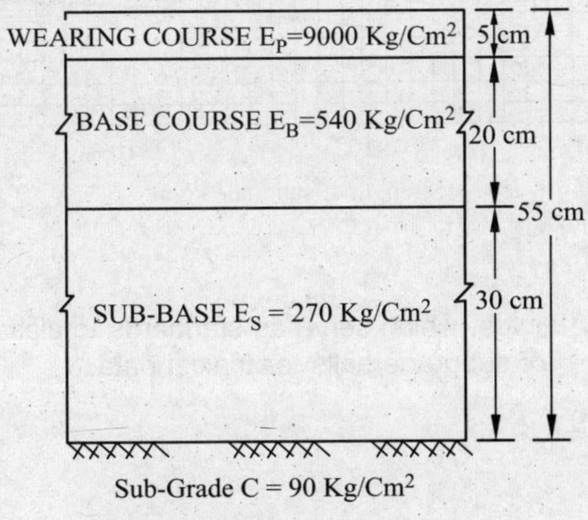

WEARING COURSE E_P=9000 Kg/Cm^2 5 cm

BASE COURSE E_B=540 Kg/Cm^2 20 cm

55 cm

SUB-BASE E_S = 270 Kg/Cm^2 30 cm

Sub-Grade C = 90 Kg/Cm^2

Fig. 20.28 Fig. for Prob. 20.11

20.14. TEXAS TRIAXIAL METHOD

In this method the samples are tested in triaxial test at a rate of strain of 3.75 mm per minute after the samples are subjected to capillary water until equilibrium is reached. Six samples are tested and from the data, Mohr's circles are plotted and rupture envelope is drawn. By plotting the rupture envelope of subgrade shown in Fig. 20.29, a mat classification number is derived. Curves are arbitrary curves established by Texas engineers which serve as standards to classify the strength of subgrade soils from triaxial test data.

Design curves for thickness are shown in Fig. 20.30. The upper horizontal scale is used to represent the wheel load for the heavily travelled long life highways and the lower scale is based on the concept of "*per cent design*" which takes into account traffic data. Percent design is defined as:

$$\text{Per cent design} = \frac{\text{depth used}}{\text{depth required for long life design} - 3.75 \text{ cm}}$$

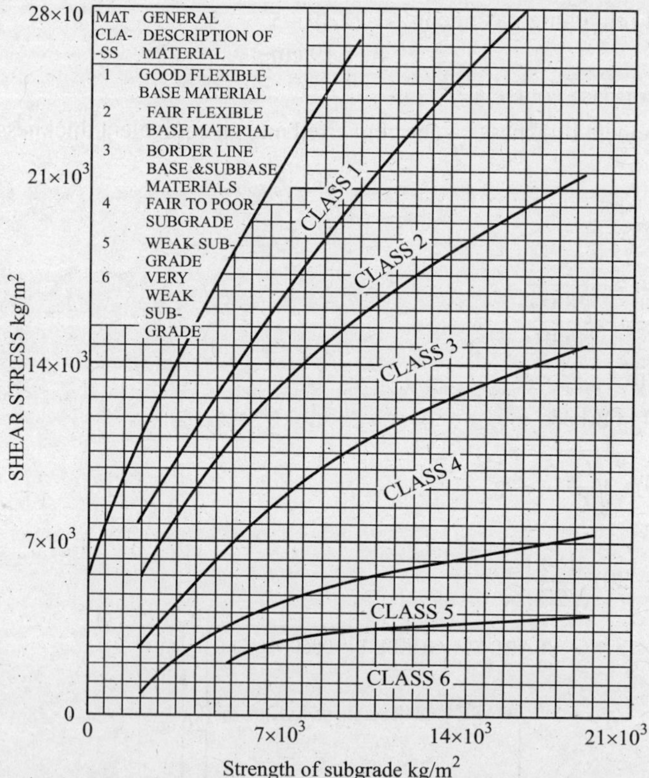

Fig. 20.29 Arbitrary curves, which serve as standards to classify the strength of subgrade soils for triaxial data.

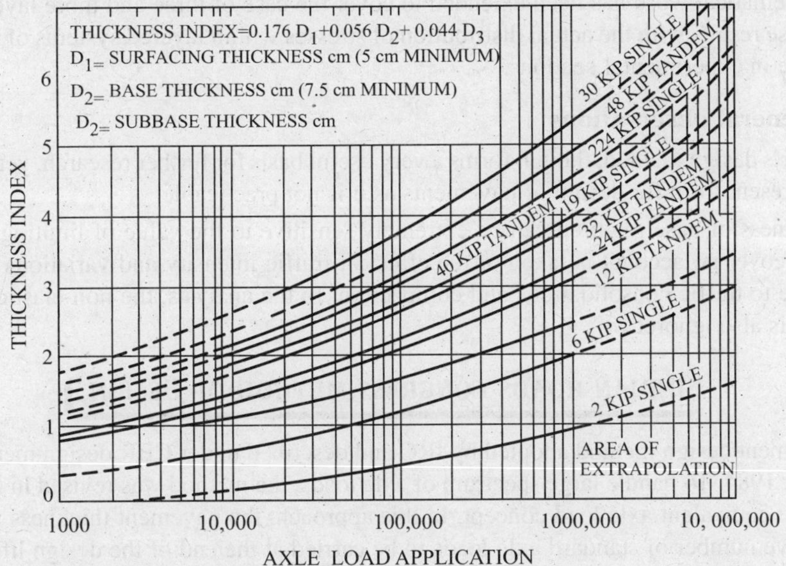

Fig. 20.30 Flexible pavement design-load relationship ($p_t = 2.5$)

According to McDowell, good long life pavements may be approximately 3.75 cm thinner than required. For this reason, it is the practice in Texas to reserve 2.5 cm or 5 cm of surfacing for future application. As already mentioned the equation relating length of life to percent design is number of years of life = $0.96 \times 10^{0.01465r}$ where r is per cent design.

20.14.1. General observations

Methods employing triaxial tests are advantageous in that the theory can be used provided appropriate correlation coefficients can be determined. The method can be extended to include a variety of conditions. The test method takes into account true stress/strain relationship of subgrade and various materials used in construction which are compacted at appropriate densities.

The method has the disadvantage that the test is quite complicated and costly. Also, the rate of loading affects the test results and for soils the values obtained under slower rates of strain may have lower value of modulus than those tested under transient conditions.

20.15. BURMISTER'S DESIGN METHOD

Based on analysis of stresses and strains in a two-layered system, as already explained, the vertical elastic displacement at the surface was obtained from the equation:

$$\Delta = \frac{1.5\,P.a}{E_2} F_w$$

Displacement factor F_w was plotted against the thickness of the top layer for various ratios of E_1/E_2, which plot is given in Fig .19.25 (page 665).

Analysis was based considering flexible pavement, base course and sub-base as top layer and the subgrade as the bottom layer. The thickness of the top layer was selected so that the displacement under the wheel is limited to an arbitrary quantity usually taken as 5 mm as suggested by Burmister but may be changed in the light of experience.

Tentative design curves for flexible runway pavements, using 5 mm as limiting deformation have been drawn assuming approximate values of modulus of elasticity for various types of bases.

The mathematical work has been extended to cover the case of three and more layers and also to correlate these results with the actual distribution of stresses within layered systems of road materials and subgrade in experimental sections.

20.15.1. General observations

Burmister's design method, though forms a very useful basis for further research, yet it is not being used in its present form for design of pavements as it is not practicable.

The thickness of base required can be extremely sensitive to the value of limiting displacement chosen. Moreover no account is taken of the effect of traffic intensity and variations in strength of subgrade due to traffic consolidations and compaction. In the analysis, the non-elastic behaviour of the material is also ignored.

INDIAN ROADS CONGRESS METHOD OF DESIGN

The pavement design method adopted by IRC and described under CBR design method was used upto the year 1984. To handle large spectrum of axle loads, the method was revised in the year 1984, following the equivalent axle load concept. In this approach, the pavement thickness was related to the cumulative number of standard axle loads to be carried at the end of the design life for different subgrade strengths. These were based on semi-emperical approach depending upon the past experience and judgement of the designer. Design curves developed were applicable to cater traffic upto 30 million standard axle (msa).

As has already been pointed out, the volume of commercial traffic and the axle load have increased tremendously in the last one or two decades. Moreover there may be situations such as approach roads to docks and in the neighbourhood of heavy industrial complexes, where the cumulative standard axle loads may be much higher than 30 msa. The emperical methods have their limitations regarding their applicability and extrapolation. Thus with the rapid growth of heavy traffic, the pavements are required to be designed analytically for heavy volumes, of the order of 150 msa.

The Indian Roads Congress developed a mathematical model of the pavement structure, using multiple layer elastic theory. Flexible pavement is considered to be composed of three layers namely bituminous surface, granular base and subbase course conforming to IRC standards. Stresses and strains are computed at critical locations using the linear elastic model FPAVE. To account for performance, three types of pavement distresses resulting from repeated application of traffic loads are considered:

(*i*) Vertical compressive strain at the top of the subgrade.

(*ii*) Horizontal tensile strain at the bottom of the bituminous layer.

(*iii*) Pavement deformation within the bituminous layer.

The thickness of the granular and bituminous layers are selected using the analytical design approach so that strains at the critical points are within the allowable limits. Permanent deflection within the bituminous layer can be controlled by the mix design requirements. To calculate tensile strain at the bottom of the bituminous layer, the Elastic Modulus of Dense Bituminous Macadam (DBM) layer with 60/70 bitumen has been used in the analysis. The average annual temperature has been taken as 35°C.

20.16. PAVEMENT THICKNESS AND COMPOSITION

For easy application by the field engineer, the results have been provided in the form of simple design curves and a catalogue of pavement designs. Pavement thickness design curves are shown in Fig. 20.31 and 20.32. These relate the pavement thickness to the cumulative number of standard axles to be carried over the design life for a CBR values of subgrade varying from 2 to 10 percent. Fig.

20.31 is for carrying traffic in the range of 1 to 10 msa whereas Fig. 20.32 give the design thickness for traffic in the range of 10–150 msa.

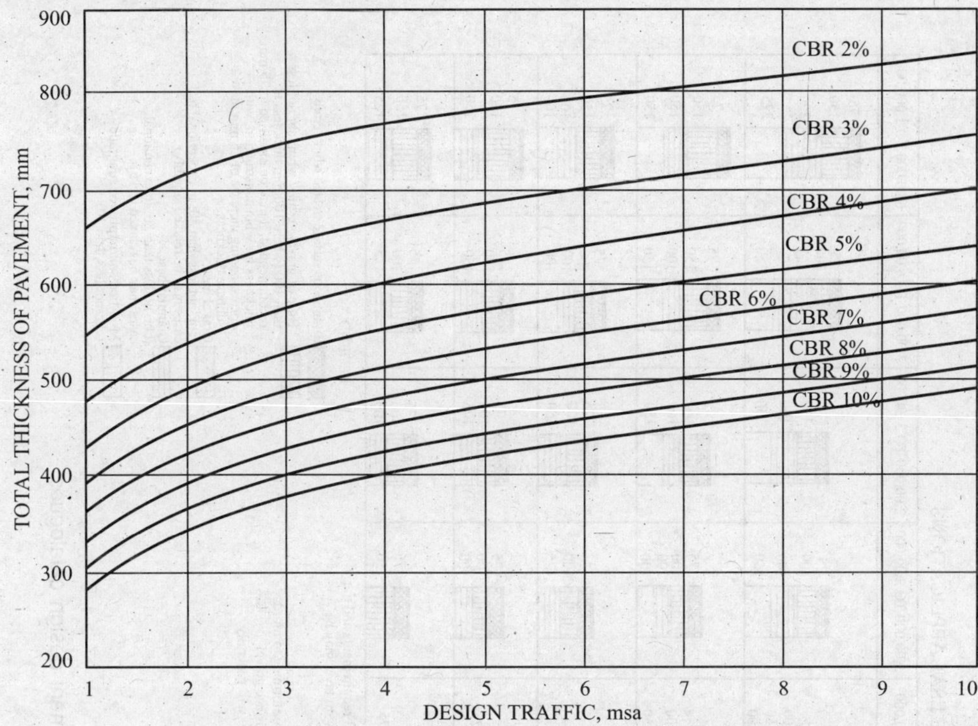

Fig. 20.31 Pavement thickness design chart for traffic 1–10 msa

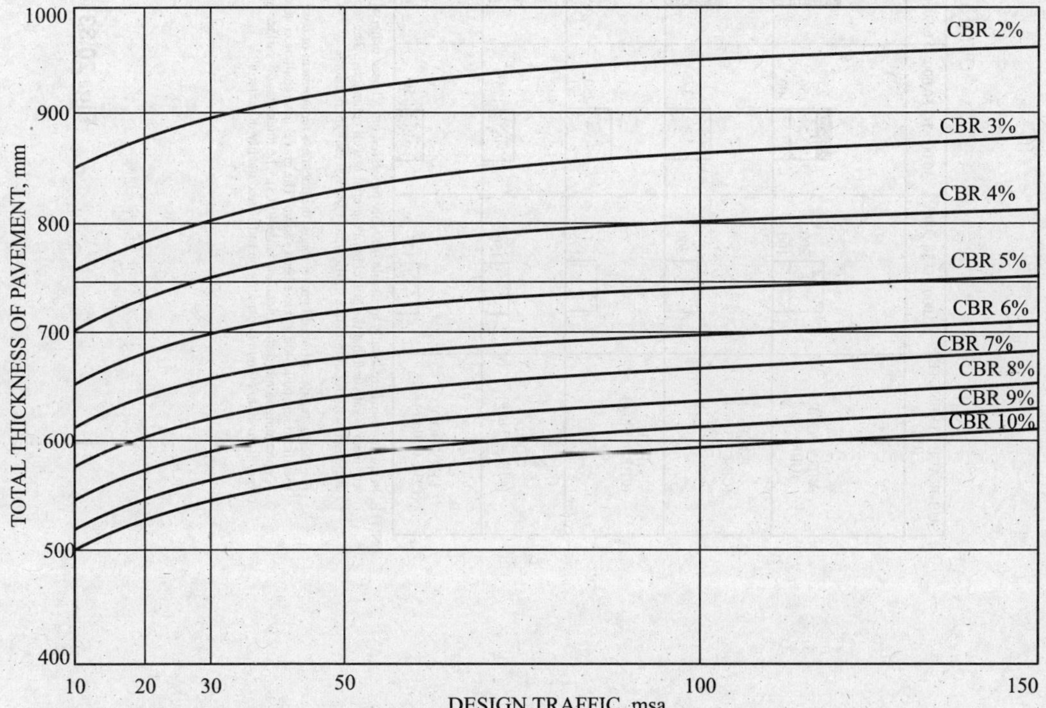

Fig. 20.32 Pavement thickness design chart for traffic 10–150 msa.

CUMULATIVE ESAL APPLICATIONS

SUBGRADE STRENGTH (CBR)	10,000 TO 30,000	30,000 TO 60,000	60,000 TO 1,00,000	1,00,000 TO 2,00,000	2,00,000 TO 3,00,000	3,00,000 TO 6,00,000	6,00,000 TO 1,000,000
VERY POOR (CBR = 2)	200 100	75 150 100	75 100 150 100	75 100 100 150	75 100 150 150	75 100 225 150	75 150 200 225
POOR (CBR = 3 to 4)	200	275	75 100 150	75 100 100 100	75 100 100 150	75 150 100 150	75 150 150 150
FAIR (CBR = 5 to 6)	175	250	275	75 100 125	75 100 150	75 100 100 100	75 150 100 100
GOOD (CBR = 7 to 9)	150	175	225	75 100 100	75 100 125	75 100 150	75 150 150
VERY GOOD (CBR = 10 to 15)	125	150	175	75 150	75 100 100	75 100 125	75 150 125

NOTE: In situations where locally available /suitably processed gravel base material fulfills all requirements and the engineer is satisfied that the gravel base material is well compacted, The top 75 mm wbm layer may be dispensed with for cumulative traffic upto 1,00,000 esal applications.

For the aggregate. surfaced / gravel roads, the thickness requirement of base gravel (Conforming to mord specifications, table 400.2) have been shown. The gravel base shall be covered with surface gravel (Conforming to mord specifications table 400.3). The thickness of surface gravel layer will generally vary from 40 to 50 mm depending on traffic and quality of material

LEGEND

- Bituminous surface treated wbm / crmb
- Base of gravel / crmb / wbm; cbr not less than 100 (where 100 mm thickness is recommended, it may be modified to 75 mm thickness for wbm construction, with corresponding increase of 25 mm in subbase thickness
- Gravel base (cbr. not less than 80: in exceptional cases may be relaxed suitably)
- Granular base (cbr. notless than 20: in exceptional cases may be relaxed to 15)
- Modified soil/improved subgrade (cbr. not less than 10)

Fig. 20.33 Pavement design catalogues.

These curves give the total pavement thickness to be provided, comprising of granular sub-base, granular base and bituminous surfacing. The requirements for the minimum thickness and composition of the various layers are depicted in the pavement Design Catalogue as given in Fig. 20.33.

20.16.1. Design Procedure

(1) Determine the design CBR value in accordance with IS : 2720 (Part 16) code.

(2) Convert the given axle load into standard axle load (8160 kg) by multiplying with equivalance factor from Table 20.8 (page 688) or calculate from the following relations :

$$F = \left(\frac{W}{W_s}\right)^4$$

where W = axle load in kg or kN of the commercial vehicle in question

and W_s = Standard axle load

= 80 kN or 8160 kg for single axle

and = 146.75 kN or 14968 kg for tandem axle

(3) For estimating design traffic, the following information is required:

(*i*) Initial traffic after construction in terms of commercial vehicles per day (CVPD) based on at least 7 days, 24 hours classified traffic count.

For new roads, if is based on potential land use and traffic on similar existing roads in the area.

(*ii*) Traffic growth rate during the design life in percentage; in accordance with IRC-108 'Guidelines for traffic prediction on Rural Highways.

In the absence of adequate data, adopt an average annual growth rate of 7.5%.

(*iii*) Design life in number of years

(*a*) As per IRC recommendations, adopt design life as follows:

N.H. and S.H. : 15 years

Expressways and Urban roads : 20 years

Other categories : 10 to 15 years

(*b*) Determine the cumulative number of standard axles (8160 kg) from the following formula

$$N = \frac{365\{(l+r)^n - 1\}}{r} \times A \times D \times F$$

where N = Cumulative number of standard axles in msa

r = Annual growth of commercial vehicles expressed in decimals.

n = Design life in years

A = Initial traffic in the year of completion in terms of CVPD

$= P\,(1+r)^x$

where P = Commercial vehicles on last count

x = Number of years between the last count and the year of completion

D = Lane Distribution factor (from table 20.22)

Table 20.22 Load distribution factor

Sr. No.	*Type of Road*	*D as % of commercial vehicles*
1	Single-lane	100% in both directions
2	Two lane single carriage-way	75% in both directions
3	Four-lane single carriage-way	40% in both directions
4	Dual carriage way lanes	
	(*a*) 2 lanes	75% in each directions
	(*b*) 3-lane	60% in each directions
	(*c*) 4-lane	45% in each directions

F = Vehicle damage factor; from Table 20.23

Table 20.23 Vehicle damage factor, F

Initial traffic volume in CVPD	*F for terrain*	
	Rolling/plain	*Hilly*
0-150	1.5	0.5
150-1500	3.5	1.5
> 1500	4.5	2.5

(4) Find the total pavement thickness for design CBR value and N from Fig. 20.31 or 20.32

(5) Then interpolate pavement composition from Fig. 20.33, for design CBR value.

Problem 20.12 *Design the pavement for a new road in plain terrain with a two-lane single carriageway; given the following data:*

(*i*) *Initial traffic volume in the year of completion of construction in both directions = 500 commercial vehicles per day (CVPD)*

(*ii*) *Traffic growth rate per annum = 7.5%*

(*iii*) *Design life = 12 years*

(*iv*) *Design CBR of subgrade soil = 4%*

Solution.

(*i*) From table 20.22, the load distribution factor = 75%

∴ Factor D, for a two-lane single carriageway = 0.75

(*ii*) From table 20.23, the vehicle damage factor, F, in plain terrain for traffic of 500 C.V.P.D. = 3.5

(*iii*) Cumulative number of standard axles to be catered in the design

$$N = \frac{365\left\{(1+r)^n - 1\right\}}{r} \text{A.D.F}$$

Here $r = 7.5\% = 0.075$

$n = 12$ years

$A = 500$ CVPD

$D = 0.75$

$F = 3.5$

Substituting the values

$$N = \frac{365\left[(1+0.075)^{12} - 1\right]}{0.075} \times 500 \times 0.75 \times 3.5$$

$$= \frac{365\,(2.382-1)}{0.075} \times 500 \times 0.75 \times 3.5$$

$$= \frac{365 \times 1.382 \times 500 \times 0.75 \times 3.5}{0.075} = 8827525$$

$$\simeq 8.83 \text{ msa}$$

(*iv*) From Fig. 20.31, total pavement thickness for CBR = 4% and traffic 8.83 msa = 680 mm say 700 mm

(*v*) Pavement composition from Fig. 20.33 for CBR = 4%.

(*a*) Bituminous surfacing = 40 mm Bituminous concrete + 80 mm Dense bituminous mix

(*b*) Granular Base = 250 mm W.B.M

(*c*) Sub base = 330 mm granular material of CBR > 30 %

Problem 20.13 *An existing two lane single carriageway National Highway is proposed to be widened to 4-lane divided highway. Design the new pavement from the following data:*

(*i*) *4-lane divided carriageway*

(*ii*) *Initial traffic in each direction in the year of completion of construction = 5640 CVPD*

(*iii*) *Design life = 10 years*

(*iv*) *Design CBR value = 6%*

(*v*) *Traffic growth rate = 8%*

(*vi*) *Axle load using the road = 11800 kg*

Solution.

(*i*) From table 20.22 load distribution factor $D = 0.75$

(*ii*) F = Vehicle damage factor or equivalency factor $= \left(\frac{W}{W_s}\right)^4 = \left(\frac{11800}{8160}\right)^4 = (1.446)^4$

(*iii*) $A = 5640$ CVPD $= 4.37$ say 4.5

(*iv*) Design life $n = 10$ years

(*v*) $r = 0.08$

(*vi*) Cumulative number of standard axles to be carried

$$N = \frac{365\{(1+0.08)^{10}-1\}}{0.08} \times 5640 \times 0.75 \times 4.5$$

$$= \frac{365\,(2.1589-1)}{0.08} \times 5640 \times 0.75 \times 4.5$$

$$= 100647206$$

$$= 100.6 \text{ msa}$$

say 100 msa

(*vii*) From Fig. 20.32 for CBR = 6% and N =100 msa

Total pavement thickness = 700 mm

(*viii*) Pavement composition from Fig. 20.33 for CBR = 6%

(*a*) Bituminous surfacing = 50 mm Bituminous concrete + 140 mm Dense Bituminous mix

(*b*) Base = 250 mm WBM

(*c*) Sub-base = 260 mm granular base with CBR > 30 %

❑❑❑

RIGID PAVEMENT DESIGN METHODS

Rigid pavements are designed primarily on the basis of their resistance to bending. Road slabs generally fail in direct tension or bending, the stresses causing failure being essentially tensile in character. Direct tension tests of concrete are difficult to make and the results obtained are variable. The testing of concrete in flexure yields rather more consistent results than those obtained with the tensile test. For this reason, strength of concrete for design of rigid pavements is usually measured as modulus of rupture which is the maximum tensile bending stress at the instant of failure. This modulus of rupture $\overline{MR}$ is defined by the equation $\overline{MR} = M/Z$, where M is the bending moment in the beam at failure, and Z is the section modulus of the beam calculated in the manner as for truly elastic materials. Three methods of loading the beam may be used: (*i*) third point loading, (*ii*) central loading, and (*iii*) cantilever loading. Third point loading method gives more uniform results than other methods and is mostly specified for determining flexural strength of concrete.

Modulus of rupture is influenced by many variables, such as quality of aggregate, surface characteristics of aggregate, curing of concrete, strength of cement, handling of specimens, moisture condition of specimen at the time of testing, and also by the temperature of the specimen, when tested. In order to get uniform results and to avoid erroneous results, close adherence to standard testing methods is necessary.

Design of rigid pavements for the average highway is based largely on experimental, standardized practices, and performance. Theoretical formulae have been modified and semi-empirical equations have been developed on the basis of field performance as well as tests.

For rigid airfield pavements, several design methods are used including those of Portland Cement Association, U.S. Corps of Engineers, U.S. Navy and Federal Aviation Agency. Before actual design methods are considered, it is necessary to discuss the various forces, both external and internal which might combine unfavourably to overstress the slab.

STRESSES IN CONCRETE PAVEMENTS

Factors stressing the slab are: (*i*) wheel loads, (*ii*) cyclic changes in temperature, (*iii*) changes in moisture content, and (*iv*) volumetric changes in subgrade or base course.

21.1. STRESSES DUE TO WHEEL LOADS

General considerations regarding distribution of stresses due to wheel loads have already been discussed in Chapter 19.

Westergaard Theory. Westergaard in his theory of stress analysis of concrete pavements considered three conditions of loading (*a*) interior, (*b*) edge, and (*c*) corner. Corners are formed by the intersection of transverse joints or cracks with slab edges or longitudinal joints. These are shown in Fig. 21.1 and 21.2.

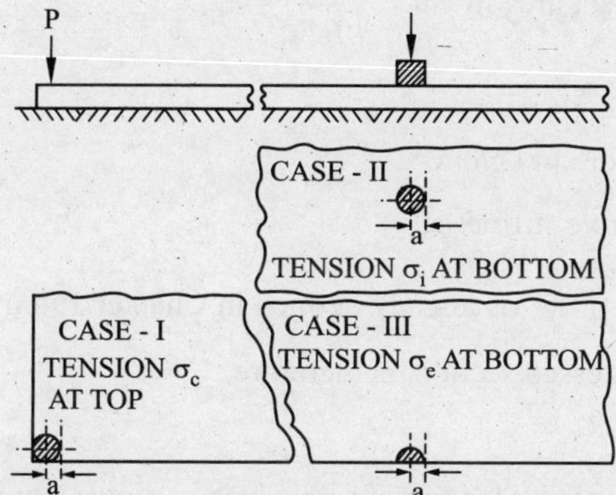

Fig. 21.1 Three cases of loading considered by Westergaard in his original analysis.

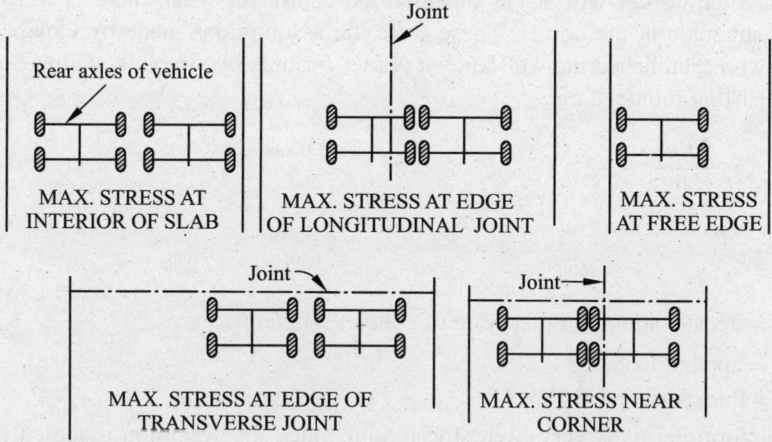

Fig. 21.2 Loading positions for critical stresses in concrete road slabs.

In the case of interior and edge loading the maximum bending moment is positive and causes tension at the bottom and compression at the top of the slab. For both these conditions stress intensity is highly concentrated in the immediate vicinity of load, being maximum at the centre of load application and decreasing rapidly as the distance from the load increases. For interior loading the stress acts radially from the load and is uniform in all directions. For edge loading the maximum stress is unidirectional and parallel to the edge of the slab.

Corner loading causes the loaded portion of the slab to act as a cantilever which develops negative bending moment at the top causing maximum tensile stresses at the top of the slab parallel to the bisector of the corner angle. In the case of corner loading, the stress intensity is not highly concentrated as in other two cases and the maximum stress occurs not at the load but at some distance from it and is distributed over a full diagonal section across the corner of the slab. Westergaard assumed it to be straight line at 45 degrees extending across the corner of the slab and located at perpendicular distance from the apex of the corner given by the formula:

$$X = 2.38\sqrt{al}$$

where X = distance in cm from the apex of slab corner to the section of maximum stress measured along corner bisector

a = radius of contact area in cm = $\sqrt{\dfrac{P}{p\pi}}$

P = applied load in kg

p = the tyre pressure in kg/cm^2

l = radius of relative stiffness

$= \sqrt[4]{Eh^3/12\,(1-\mu^2)k}$ as already defined in Chapter 19 on page 667.

The length of the most stressed section, is therefore,

$$2X = 4.76\sqrt{al}$$

The simplest solution for corner loading is one which assumes that pavement acts as a cantilever with the load concentrated at the corner produced either by crack or transverse joint through which no dowel passes. Such corner will act as unsupported cantilever when there is warping of slab or settlement of subgrade at the corner. These were the assumptions made by Goldbeck (1919) and Older (1920) who established the well-known corner formula by simply equating external moment and internal resisting moment, *i.e.*,

$$PX = S_c\,\frac{2Xh^2}{6}$$

or $$S_c = \frac{3P}{h^2}$$

where S_c = tensile stress in outer fibre of concrete (kg/cm^2)

P = applied load, kg

h = thickness of slab, cm.

The corner formula gives very high stress value since the condition assumed rarely exists in pavements.

Westergaard improved upon his method by using Ritz's method of successive approximations which is based on the principle of minimum energy used to determine the deflection along the bisection axis. The minimum energy principle states that the potential energy of the load should be equal to the sum of the energies stored in the slab and foundation. The moment was found to be maximum at a distance $d = 2\sqrt{a_1 l}$ or $2.38\sqrt{al}$ form the corner where a_1 is the distance from the corner to the centre of the load and equals $a\sqrt{2}$. The tensile stress at the top was found by the equation:

$$S_c = \frac{3P}{h^2}\left[1 - \left(\frac{a_1}{l}\right)^{0.6}\right] \qquad \text{...(21.1)}$$

For the interior and edge loading, Westergaard utilized the theory of thin plates resting on an elastic foundation. Two theories were used by Westergaard called the "Ordinary theory" and the "*Special theory*". In the ordinary theory, it is assumed that straight lines through the slab remain straight and perpendicular to the neutral axis under load. In the special theory, this assumption is abandoned in the immediate vicinity of the load.

Quoting from Westergaard "*with slabs of proportion as found in pavements, the theory based on these assumptions (the ordinary theory) leads to a satisfactory determination of deflections at all points. At the point of application of a concentrated force, this ordinary theory leads to peak in the diagrams of the bending moments, with infinite values at the point of load itself. When the force is applied at the top of the slab, the tensile stresses at the bottom are not, in fact, infinite. One may say then that the effect of thickness of the slab is equivalent to a rounding off the peak in the diagrams of moments. In order to find out as to what extent the diagrams are rounded off, it is necessary to abandon the assumption of the straight lines drawn through the slab remaining straight as applying to the immediate neighbourhood of the load, a special theory is required*". Westergaard found that the special theory had to be used only if the radius of the circular area 'a' was less than 1.724 h, where h is the thickness of the slab. Consequently with large contact area, such as in airports, the ordinary theory suffices; however, for highways, both theories have to be used.

The Westergaard equation for the interior load for $\mu = 0.15$ is:

$$S_i = \frac{0.315P}{h^2}\left[4\log_{10}\left(\frac{l}{b}\right) + 1.069\right] \qquad \text{...(21.2)}$$

For edge loading

$$S_e = \frac{0.572P}{h^2}\left[4\log_{10}\left(\frac{l}{b}\right) + 0.359\right] \qquad \text{...(21.3)}$$

For large values of 'a' the ordinary theory is used and $b = a$. For values of 'a' less than 1.724, the special theory applies and $b = \sqrt{1.6a^2 + h^2} - 0.675h.$

It will be observed that elastic constants E and μ do not appear in the equations. In original representation the equations were limited to the case where E = 210,000 kg/cm^2 and $\mu = 0.15$. Subsequently, the equations were generalized and were restated as below. Westergaard modified his original formula for the interior loading to allow for the reactions of the subgrade under and immediately around the load, being greater than the reaction when it is assumed to be proportional to the deflection.

$$S_c = \frac{3P}{h^2}\left[1 - \left\{\frac{12\,(1-\mu^2)\,k}{Eh^3}\right\}^{0.25} \left\{a\sqrt{2}\right\}^{0.6}\right] \qquad \text{...(21.4)}$$

$$S_i = \frac{0.275\,P}{h^2}(1+\mu)\left[\log_{10}\left\{\frac{Eh^3}{kb^4}\right\} - 44.54Z\left\{\frac{1}{L}\right\}^2\right] \qquad \text{...(21.5)}$$

$$S_e = \frac{0.529\,P}{h^2}(1+0.54\mu)\left[\log_{10}\left\{\frac{Eh^3}{kb^4}\right\} - 0.71\right] \qquad \text{...(21.6)}$$

where S_c, S_i and S_e are maximum tensile stresses at corner, interior and unbroken edge of the slab. Other notations carry the usual meaning as used in concrete slab design. The equation for corner case is the same as already stated in the form

$$S_c = \frac{3P}{h^2}\left[1-\left(\frac{a_1}{L}\right)^{0.6}\right]$$

In the case of interior loading, L is the maximum radial distance in cm from the centre of the load within which redistribution of subgrade reaction is made; Z is a ratio of reductions of maximum deflections. Teller and Sutherland suggested the use of $Z = 0.2$ and $L = 5l$ though both quantities vary with pavement and subgrade stiffness.

For airfield pavements, Westergaard modified the equation taking into account the large tyres of airplanes. For aircraft wheels the tyre contact area was assumed an ellipse with the equation in the form $x^2/a^2 + y^2/b^2 = 1$. The equations for the interior and edge loading cases are as follows:

$$\left.\begin{matrix} S_x \\ S_y \end{matrix}\right\} = \frac{P}{h^2}\left[0.275(1+\mu)\log_{10}\frac{Eh^3}{k\left(\frac{a+b}{2}\right)^4} \pm 0.239\,(1-\mu)\,\frac{a-b}{a+b}\right]$$

In the above expression, the minus sign corresponds to S_y and plus sign to S_x. S_y and S_x are tensile stresses in the interior in the y and x directions.

$$S_c = \frac{2.2\,(1+\mu)\,P}{(3+\mu)\,h^2}\log_{10}\frac{Eh^3}{100\,k\left(\frac{a+b}{2}\right)} + \frac{3\,(1+\mu)\,P}{(3+\mu)\,h^2}$$

$$\left[1.84-\frac{4}{3}\mu+(1+\mu)\,\frac{a-b}{a+b}+2\,(1-\mu)\,\frac{a-b}{(a+b)^2}+1.18\,(1+2\mu)\,\frac{b}{l}\right]$$

21.1.1. Conclusions

The following general conclusions can be drawn from Westergaard's analysis and equations for stresses due to loading:

1. Maximum tensile stress is proportional to applied load.
2. For usual loading conditions greatest stresses occur when load is located at the edge and corner.
3. There is a marked reduction in stress with increased slab thickness.
4. There is a considerable reduction of maximum stress if load is spread over as large an area as possible.
5. Effect of change in modulus of subgrade reaction is small.

21.1.2. Short comings

Westergaard's original analysis was presented in 1925. Since then wheel loads have increased in magnitude and frequency. Because of this and especially near joints, where mud pumping removed a portion of the subgrade, extremely non-uniform subgrade support developed and one of the assumptions made by Westergaard regarding uniform subgrade support was no longer valid.

Similarly, no account was taken of secondary stresses caused by temperature gradients or other causes which are appreciable. Because of this warping upward of the corners, subgrade support at the corners and exterior edges, were not uniform and under this condition the corner stresses were higher than would exist, had the support remained constant.

21.1.3. Modifications of Westergaard theory

Several modifications of the original Westergaard formulae have been made to take into account curling action and other shortcomings.

In 1938, *Bradbury,* in order to account for the partial subgrade support of the slab at the edges rather than full support as assumed by Westergaard, introduced an empirical formula based on the lines of Westergaard's theory. Bradbury assumed that modulus of subgrade reaction k under the corner region and for some distance away from the corner is only one fourth of the normal subgrade modulus for the slab. By reducing k to 1/4th of its normal value, l for the corner region becomes $\sqrt{2}$ times the normal l for the slab.

Kelley introduced an empirical formula of the Westergaard type in 1939. This formula was based on Arlington test carried by Teller and Sutherland. The empirical formula was developed by modifying the exponent in Westergaard equation for corner in such a manner as to cause the computed values to coincide more nearly with the observed data for the case of incompletely supported slab corners. According to Teller and Sutherland the assumed direction of corner warping was based upon the relative temperatures of the upper and lower surfaces of the slab. When the top and bottom temperatures were the same, the slab was assumed to be not warped. Over the range of a_1/l represented by the conditions in the test sections reported by Teller and Sutherland, the equation proposed by Kelley gives about 40 to 50 per cent higher values than the equation proposed by Westergaard which is in good agreement with the observed values, as reported.

Spangler, in 1942, proposed an empirical equation which is a simplification of Kelley's formula which yields substantially the same results over a considerable range of a_1/l values. He eliminated the fractional component of the quantity a_1/l and raised the numerical coefficient to 3.2.

The semi-empirical formulae described above for corner stresses in concrete pavement slabs are:

$$S_c = \frac{3P}{h^2}\left[1-\left(\frac{a_1}{l}\right)^{0.6}\right] - \text{Westergaard} \quad \ldots(21.7)$$

$$S_c = \frac{3P}{h^2}\left[1-\left(\frac{a_1}{\sqrt{2}l}\right)^{0.6}\right] - \text{Bradbury} \quad \ldots(21.8)$$

$$S_c = \frac{3P}{h^2}\left[1-\left(\frac{a_1}{l}\right)^{1.2}\right] - \text{Kelley} \quad \ldots(21.9)$$

$$S_c = \frac{3.2P}{h^2}\left[1-\frac{a_1}{l}\right] \quad \text{Spangler} \quad \ldots(21.10)$$

$$S_c = \frac{3P}{h^2} \quad -\text{Older} \quad \ldots(21.11)$$

where S_c = maximum tensile stress in kg/cm²

P = design wheel load in kg

h = thickness of concrete slab, cm

a_1 = distance from the corner of the slab to the centre of the area of load application and is equal to $a\sqrt{2}$ where a is the radius of circle equal in area to the loaded area

l = radius of relative stiffness, cm

E, k are the usual notations already explained. All these equations are for unprotected corners.

Pickett's modifications. In 1951, Gerald Pickett proposed semi-empirical formulae: one for "*protected*" corners and the other for "unprotected" corners.

A protected corner is one in which provision has been made to transfer at least 20 per cent of the load from one slab corner to the other by some mechanical device or by aggregates interlock. An unprotected corner is one in which no provision has been made for transferring any of the load and one corner must carry more than 80 per cent of the load.

The formulae are:

$$S_c = \frac{3.36P}{h^2}\left[1 - \frac{\sqrt{a/l}}{0.925 + 0.22\,a/l}\right] \quad \text{Protected corners} \qquad \ldots(21.12)$$

$$S_c = \frac{4.2P}{h^2}\left[1 - \frac{\sqrt{a/l}}{0.925 + 0.22\,a/l}\right] \quad \text{Unprotected corners} \qquad \ldots(21.13)$$

P is the static wheel load increased by a factor to provide adequate allowance for the impact of moving loads. For protected corners the stress is 8/10 that of stress in unprotected corners under the same load.

This modification is applicable for a wider range of values of a_1/l. The previous modifications, except that by Bradbury, all give zero stress for a_1/l equal to unity and very low and rapidly decreasing stress when a_1/l is greater than about 0.6 or 0.7. Pickett, while choosing semi-empirical equation allowed 40 per cent increase in stress due to non-uniform distribution of moment along sections perpendicular to the bisector of the corner angle. Further allowance was made for the lack of subgrade support. The formula was selected to give stress within 3 per cent of those given by either Kelley or Spangler formula in the range of a_1/l between 0.1 to 0.4 so that the stresses may be in good agreement with the Arlington experimental data.

The Westergaard formula for edge loading was also modified by *Teller* and *Sutherland* to allow for the variation in support due to warping of slab. The modified correlation is as follows and is also adopted by IRC.

$$S_c = 0.529\frac{P}{h^2}(1 + 0.54\mu)\left(4\log_{10}\frac{l}{b} + \log_{10} b - 0.4048\right) \qquad \ldots(21.14)$$

where l = radius of equivalent stiffness, cm

$$= \frac{\sqrt{Eh^3}}{4\sqrt{12\,(1-\mu^2)\,k}}$$

b = radius of equivalent distribution of pressure

$= a$ for $\frac{a}{h} \geq 1.724$; and

$$= \sqrt{1.6a^2 + h^2} - 0.675h \text{ for } \frac{a}{h} \le 1.724$$

and a = radius of load contact areas, cm, assumed to be circular. The values of l and b can be ascertained directly from Table 21.1 and 21.2.

Table 21.1 Radius of relative stiffness, *l*, for different values of pavement slab thickness, *h*, and foundation reaction modulus, *k* for concrete $E = 3.0 \times 10^5$ (kg/cm²)

Radius of relative stiffness l (cm) for different values of k (kg/cm³)					
h (cm)	*k = 6*	*k = 8*	*k = 10*	*k = 15*	*k = 30*
15	61.44	57.18	54.08	48.86	41.09
16	64.49	60.02	56.76	51.29	43.31
17	67.49	62.81	59.40	53.67	45.15
18	70.44	65.56	62.01	56.03	47.04
19	73.36	68.28	64.57	58.39	49.05
20	76.24	70.95	67.10	60.63	50.99
21	79.08	73.59	69.60	63.89	52.89
22	81.89	76.20	72.08	65.13	54.77
23	84.66	78.80	74.52	67.33	56.62
24	87.41	81.35	76.94	69.31	58.45
25	90.13	83.88	79.32	71.68	60.28

Table 21.2 Radius of equivalent distribution of pressure section, *b* in terms of radius of contact, *a* and slab thickness, *h*

a/h	*b/h*	*a/h*	*b/h*
0.0	0.325	1.0	0.937
0.1	0.333	1.1	1.039
0.2	0.357	1.2	1.143
0.3	0.387	1.3	1.250
0.4	0.446	1.4	1.358
0.5	0.508	1.5	1.470
0.6	0.580	1.6	1.582
0.7	0.661	1.7	1.695
0.8	0.747	1.724	1.724
0.9	0.840	> 1.724	*a/h*

The Teller and Sutherland's modified formula for corner loading is the same as given by Kelley.

The above formulae are suitable for highway loading and are widely used for rigid highway pavements but for airports, because of very large contact areas, some modification may be desirable.

21.2. STRESSES DUE TO CYCLIC CHANGES IN TEMPERATURE

Cyclic changes in air temperature result in: (*i*) changes in temperature gradient through the slab which cause warping stresses, and (*ii*) change in the mean temperature of the slab causing it to expand and contract which is resisted by friction of the subgrade thus causing stresses in concrete slab.

21.2.1. Warping Stresses

The surface of the slab is subjected to a wide range of temperature during the daily cycle, whereas the temperature of the bottom of the slab in contact with subgrade or base remains relatively more constant. This temperature gradient through the slab causes differential expansion or contraction between top and bottom of the slab.

During the day when the top surface of the slab becomes hotter than the bottom, the top surface expands more than the bottom and hence the slab tends to assume a shape which is convex upward. If the slab was weight-less and unrestrained, it would warp into a shape of part of a spherical surface and there would be small internal stress only if temperature gradient varied with depth. The weight of the slab and load transfer devices or friction at joints will restrain free warping and will tend to bend it back into its original shape, and hence tensile bending stresses are created in the bottom and compressive stresses would occur at the top of the slab. At night, reverse effects are created because the temperature at top becomes lower than the bottom of the slab; the sides and corners tend to warp upward and might actually leave the subgrade. In this position, the weight of raised portions of slab tends to bend them down; thus creating tension in the top of the slab and compression in the bottom. Fig. 21.3 shows crack formation in slab due to warping.

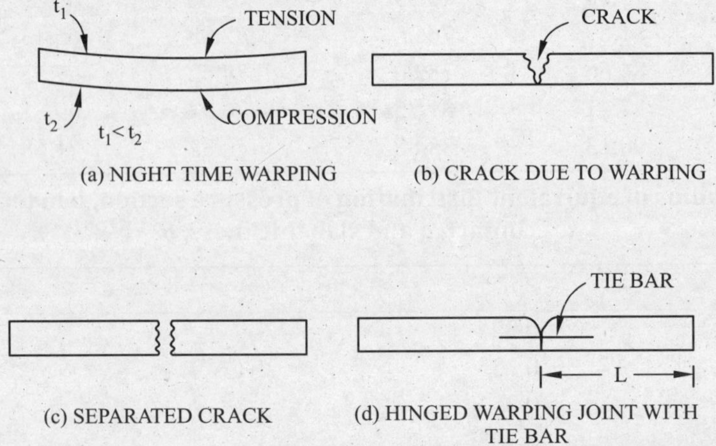

Fig. 21.3 (a-c) Crack formation in slab due to warping (d) Load transfer, using tie-bar.

Air temperatures are largely governed by the local climate and also by the latitude, which in turn controls the angle of incidence of sun's rays. The insulating effect of the concrete delays the transmission of heat through the slab. Because of this as the daily cyclic changes take place rather rapidly, the differential in temperature between the top and bottom is not as big as might be expected from the extreme daily range of air temperature. Field observations indicate that the maximum temperature differentials in terms of slab thickness may be assumed as 0.3 to 0.5°C per cm of slab thickness during day time when the top of slab is warmer, and 0.2°C per cm of slab thickness when the top is cooler than the bottom. These figures are usually assumed for design purposes. Thus for a slab 25 cm thick, the top surface can be $25 \times 0.5 = 12.5$°C warmer than the bottom during the day and can be 5°C cooler during the night. Data collected at Kharagpur, India indicates that for the weather conditions prevailing in the test section, the maximum temperature differential in terms of slab thickness can be 2°C per 2.5 cm thickness when the top of the slab is warmer, and 1°C per 2.5 cm thickness when top is cooler than the bottom.

Analysis of warping stresses was presented by Westergaard (Proc HRB 1926) on the assumption that temperature gradient was constant throughout its thickness. This subject was also treated by Kelley (Public Roads, July, 1939), who slightly altered the general equations for calculating warping stresses in the edge and interior of slab developed by Bradbury (Wire Reinforced Institute, 1938). Thomlinson (Concr. Const. Engg. 1940) proposed a method of calculating temperature gradients and the resulting stresses. Thomlinson assumed that temperature gradient varies with depth which may be calculated if variation in temperature at top surface and the thermal properties of concrete are known. For detailed discussion of warping stresses, it is suggested that the reference given above be consulted.

For practical purposes and neglecting the biaxial effect of Poisson's ratio, the maximum intensity of warping stress may be determined by the following formula proposed by Bradbury:

$$S_t = \frac{CE\alpha t}{2} \qquad \text{...(21.15)}$$

where S_t = warping stress, kg/cm^2

E = modulus of elasticity of concrete, kg/cm^2

α = thermal coefficient of concrete

t = maximum temperature differential, °C during day between top and bottom of slab

C = a coefficient which depends on slab length and radius of relative stiffness.

This is known as Bradbury's coefficient, which can be ascertained from Bradbury's chart against values of L/l and W/l; where

L = slab length or spacing between consecutive contraction joints

W = slab width; and

l = radius of relative stiffness.

Values of the coefficient 'C' based on the curves given in Bradbury's chart, are given in Table 21.3.

Table 21.3 Values of coefficient 'C' based on Bradbury's chart

L/l or W/l	*C*
1	0.000
2	0.040
3	0.175
4	0.440
5	0.720
6	0.920
7	1.030
8	1.075
9	1.080
10	1.075
11	1.050
12 and above	1.000

The above formula applies to a narrow strip along the edge of the slab, and this gives a somewhat smaller warping stress than would exist in the centre of the slab. Considering the effect of Poisson's ratio, μ, the warping stress in either direction in the interior is:

$$S_t = \frac{E\alpha t}{2}\left(\frac{C_1 + \mu C_2}{1-\mu^2}\right) \qquad \text{...(21.16)}$$

where C_1 = coefficient for slab length in desired direction

C_2 = coefficient for slab length in the perpendicular direction.

The values of C_1 and C_2 are determined from Fig. 21.4.

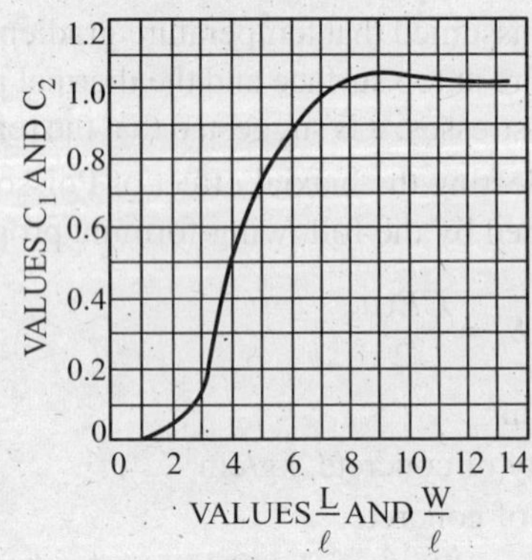

Fig. 21.4 Warping stress coefficients.

When $C_2 = 0$ and $\mu = 0$, the above formula takes the same form as the formula given by equation 21.15. For most practical purposes, warping stresses may be calculated assuming

$$C = 1.10,\ E = 3\times10^5 \text{ kg/cm}^2,$$

$$\alpha = 10\times10^{-6}$$

and $t = 0.5° \times$ slab thickness.

Warping stresses were more accurately calculated by Kelley and Thomlinson. Reference to these results indicates that warping stresses alone can be very large and may range from 21 to 28 kg/cm^2 and exceed half the modulus of rupture. Thus, they alone can cause cracking of pavement slabs having sufficient length to develop such high warping stresses. They are particularly important in the design of plain, unreinforced slabs. On the other hand, if warping stresses produce transverse cracks in reinforced slabs, these cracks are held tightly together by the steel reinforcement and no particular harm results. Because of this, warping stresses may be neglected in the design of reinforced concrete slabs.

Bradbury has also proposed a relation for approximate warping stresses at a slab corner $\left(S_t = \dfrac{E\alpha t}{3\,(1-\mu)}\sqrt{\dfrac{a}{l}}\right)$ but the calculated warping stresses at a corner are so small as to be practically negligible and are usually neglected in the slab design.

As already stated Thomlinson assumed that temperature gradient varies with depth which may be calculated if variation in temperature at top surface and the thermal properties of concrete are known. To simplify the calculation of stresses, Thomlinson determined the coefficients of stress in terms of $\dfrac{E\alpha t}{1-\mu}$ for various thicknesses of slab, expressed in terms of $\gamma = \dfrac{h}{\phi}\sqrt{\dfrac{\pi}{T}}$ and for various restraints,

where

ϕ^2 = the diffusivity of the material, that is, thermal conductivity divided by heat capacity per unit volume.

= 0.009 CGS units as assumed by Thomlinson.

Table 21.4 Stresses due to temperature gradients for various conditions of restraint of the slab expressed in terms of $\frac{E\alpha t}{1-\mu}$

*Slab thickness**	*Internal stress*			*Restrained warping stress*		*Internal stress +restrained warping stress*			*Internal stress +end restraint stress*		*Stress due to complete restraint*	
	Top	*Bottom*	*Mid-depth*	*True*	*Approx*	*End restraint stress*	*Top*	*Bottom*	*Top*	*Bottom*	*Top*	*Bottom*
0.0	0.000	0.000	0.000	0.000	0.000	1.000	0.000	0.000	1.000	1.000	1.0	1.000
0.2	0.005	0.005	0.003	0.128	0.128	0.905	0.128	0.124	0.908	0.905	1.0	0.819
0.4	0.023	0.021	0.011	0.232	0.232	0.818	0.245	0.217	0.820	0.822	1.0	0.670
0.6	0.047	0.042	0.022	0.314	0.314	0.741	0.347	0.284	0.741	0.745	1.0	0.549
0.8	0.077	0.066	0.036	0.379	0.379	0.667	0.434	0.332	0.667	0.667	1.0	0.449
1.0	0.112	0.091	0.051	0.429	0.429	0.607	0.508	0.362	0.611	0.627	1.0	0.368
1.2	0.148	0.116	0.066	0.466	0.466	0.551	0.571	0.380	0.561	0.584	1.0	0.301
1.4	0.187	0.141	0.081	0.492	0.494	0.500	0.625	0.387	0.524	0.548	1.0	0.247
1.6	0.225	0.163	0.096	0.509	0.512	0.454	0.670	0.386	0.496	0.521	1.0	0.202
1.8	0.265	0.183	0.110	0.519	0.524	0.413	0.710	0.380	0.482	0.501	1.0	0.165
2.0	0.302	0.200	0.123	0.523	0.531	0.376	0.743	0.368	0.477	0.485	1.0	0.135
2.2	0.338	0.215	0.135	0.524	0.534	0.344	0.772	0.355	0.481	0.474	1.0	0.111
2.4	0.373	0.227	0.145	0.517	0.535	0.315	0.793	0.336	0.492	0.464	1.0	0.091
2.6	0.406	0.236	0.153	0.509	0.532	0.290	0.813	0.318	0.507	0.456	1.0	0.074
2.8	0.437	0.243	0.161	0.499	0.529	0.268	0.828	0.299	0.526	0.450	1.0	0.061
3.0	0.466	0.247	0.168	0.487	0.525	0.248	0.842	0.280	0.546	0.442	1.0	0.050
3.2	0.493	0.249	0.173	0.476	0.521	0.231	0.852	0.266	0.567	0.435	1.0	0.041
3.4	0.518	0.250	0.176	0.461	0.516	0.215	0.863	0.246	0.588	0.427	1.0	0.033
3.6	0.541	0.249	0.179	0.446	0.512	0.201	0.871	0.227	0.608	0.419	1.0	0.027
3.8	0.563	0.247	0.180	0.431	0.508	0.189	0.878	0.212	0.628	0.410	1.0	0.022
4.0	0.582	0.244	0.181	0.416	0.506	0.179	0.883	0.196	0.646	0.401	1.0	0.018

* Slab thickness is expressed in term of γ.

T = period of temperature cycle.
= 24 hours in the case of daily variation.

Coefficients of stress as worked out by Thomlinson are given in Table 21.4.

21.2.2. Stress due to Subgrade Friction

Changes in mean temperature of a slab will cause the slab to expand or contract. If the slab is free to move and there is no friction between the slab and the subgrade, no stress will result. However, due to friction between the slab and subgrade there is restraint which results in stresses. On the underside of the slab this stress will be compressive during expansion and tensile during contraction. In long slabs, the stresses due to subgrade restraint may be sufficient to cause overstressing of the concrete which may result in cracking. Excessive expansion may cause blow-ups to occur.

During contraction, for equilibrium conditions, the sum of frictional forces from the centre of slab to the free end must be equal to the total tension in the concrete. A contracting slab will move more at its free end than in the centre, with the result that frictional resistance varies along the slab from the centre to the free edge. Minimum amount of displacement required for friction to be fully mobilized is 1.5 mm. Kelley (Public Roads, Vol. 20, 1939) has suggested the distribution of frictional resistance which is fully mobilized for a distance 1000/temperature drop in °F, from the free end to be linear after which distribution is parabolic in shape upto the centre of slab. The subgrade frictional resistance is commonly measured by the average coefficient of subgrade resistance.

Tests by Bureau of Public Roads show that the highest value of coefficient of friction was 2.3 on firm subgrades and 1.15 on the soft subgrade. The composition of subgrade also affected the coefficient of friction. The resistance offered by subgrade is composed of two elements; (i) a resistance caused by simple sliding friction, and (ii) shearing resistance in the subgrade due to deformation within the soil. Evidently, the frictional resistance will vary from a value equal to twice the weight of the slab where large movement takes place to nearly zero when the slab moves very little over the subgrade. In a long slab, the greatest frictional resistance would be at the ends of the slab and the least would be at the centre, where practically no movement takes place.

The tensile stress due to subgrade restraint may be roughly calculated by equating the total force of subgrade friction from the free end and centre of slab (or any other section in question) to the total tension produced in the slab. For one metre width of slab:

$$\text{Total frictional resistance upto the centre of slab} = w \times \frac{L}{2} \times f$$

$$\text{Total tension in concrete} = S \times 100h$$

Thus, $$100\,Sh = \frac{wLf}{2}$$

or $$S = \frac{wLf}{200h} \qquad \ldots(21.17)$$

where S = unit stress in tension in concrete (kg/cm^2)

w = weight of slab per square metre

L = length of slab in metre

f = average coefficient of subgrade resistance. An average value of 1.5 is usually taken for design calculations

h = depth of slab in cm.

More exact calculation could be made by taking into account the variation in friction between the free end and the centre of slab.

The above results are also affected due to warping of the slab due to temperature differential as under those conditions the entire slab is not in intimate contact with the subgrade.

The value of direct tensile stress produced will depend on the length of slab. For example, for a slab 15 cm thick and 30 m long, the value of *S* works out to be 5.40 kg/cm^2, taking unit weight of concrete as 2400 kg/m^3.

The tensile stress due to subgrade restraint for usual slab dimensions is relatively small in amount and in general would be of importance only when added to the tension produced by other forces. However, there is a possibility of transverse tension cracks developing because of relatively large decrease in temperature immediately after the concrete has set. At that period its tensile strength is very low and could be exceeded by the direct stress produced by subgrade friction.

21.3. STRESSES DUE TO CHANGES IN MOISTURE CONTENT

Concrete shrinks when it dries and expands when it is kept wet. Each wetting and drying cycle causes respectively expansion and shrinkage. Studies show that concrete specimens 24 hr after casting and kept wet expand approximately 0.0001 cm per cm and when allowed to dry contract upto approximately 0.0005 cm per cm. Recently, admixtures have been found which claim to reduce concrete shrinkage beneficially.

Stresses produced either by shrinkage or swelling due to temperature changes are in the same manner as due to moisture gradient between top and bottom of slab. However, stresses produced are small and much less than those due to changes in temperature. Moreover, the effects of moisture often tend to oppose those of temperature. For example, when slab is very hot, the moisture content will be low and when slab is wet, the temperature will be moderate. The stress under worst condition of temperature is, therefore, unlikely to be increased by the effect of moisture and in fact, it may be decreased. Due to lack of precise data about changes in moisture content and also due to small magnitude of stress, the effect of volumetric change due to changes in moisture content is unimportant.

21.4. STRESSES DUE TO VOLUMETRIC CHANGES OF SUBGRADE

Due to changes in moisture content, subgrade soils are subjected to volume changes. Clay soils are much more affected than other soils. There is a greater change in moisture near the shoulders than under the middle of pavement because in dry season moisture evaporates from the shoulders and in wet season shoulders take on moisture. The subgrade under the pavement, except near the sides, changes very little. Due to this non-uniform distribution of moisture, there is non-uniform swelling and shrinkage with the result that non-uniform support results under the slab. This non-uniform subgrade support will result in high stress concentrations at certain portions of the slab.

This non-uniform support also results when subgrade material varies.

In certain types of granular subgrade, at intermediate values of pressure, shearing force due to horizontal displacement of a slab results in rearrangement of the particles in the upper layers of the subgrade and causes an increase in void ratio of dense sand. This expansion of soil due to shear at a constant value of pressure is called dilatency. During expansion and contraction of a slab, the horizontal movement is greatest at the ends of the slab. Movements in both directions produce an upward deflections of the slab. Sparkes (Struct. Engr. 1939) has given formulae for long and short slabs to calculate maximum tensile stresses due to dilatency. Stresses due to dilatency are small and for usual lengths of slabs do not exceed 0.7 kg per sq. cm.

21.5. CRITICAL STRESS CONDITIONS

As explained in previous paragraphs stress conditions result from a combination of loading and volume-change stresses. Stresses due to (i) loading, (ii) temperature gradient, and (iii) subgrade restraint in case of long slabs attain very high values than all the other conditions. Let us now consider the conditions under which these stresses will combine to produce the greatest resultant stress.

The worst combination depends mainly upon the time at which the temperature variations produce the greatest stress since loading may occur at any time. The following are the three conditions likely to give rise to particularly high stresses:

(i) Loading at corner of slab warping upward by a temperature gradient. During a clear, cool night following a hot, sunny day, the edges and corners of a slab will warp upwards and the corners may then be only partially supported or even unsupported. This condition will give the maximum stresses due to loading. The stresses due to subgrade restraint, restrained warping and the effects of moisture will be negligible.

(ii) Loading at interior or edge of slab warping downwards by a temperature gradient. Large temperature gradient during the day will warp a slab to the convex shape, causing maximum tensile stresses at the bottom near the ends. The rising temperature will also cause the slab to expand and thus produce compressive stresses in the slab which will be greatest at the mid-point and diminishing towards the ends. The resultant stress due to these two opposing factors will vary along the length of the slab and its maximum value will be at a point nearest to the end. In this condition the stress due to moisture gradient will also be compressive at the bottom. Thus, the resultant maximum tensile stress for the above condition will be that due to loading and temperature gradient minus that due to subgrade restraint and moisture gradient.

(iii) Loading at interior or edge of slab contracting due to falling temperature. An appreciable fall in mean temperature of a slab at night or in cold weather will produce appreciable tensile stresses at or near the mid-point of the length of the slab if the slab is long. The maximum tensile stress will occur when a load is applied at the interior or edge of the slab, close to its mid-point when the stress due to subgrade restraints is at a maximum. The temperature gradient in such conditions will counteract the stress due to subgrade restraint and the maximum of these two may occur when the fall in the air temperature is slow in which condition the accompanying temperature gradient will be small so that it will have a negligible effect in reducing stress due to subgrade restraint. Thus the maximum tensile stress for the above condition will be stress due to temperature gradient and moisture gradient.

Under the action of load application, maximum stress is induced in the corner region, as the corner is discontinuous in two directions. The edge being discontinuous in one direction only, has lower stress, while the least stress is induced in the interior where the slab is continuous in all directions. Further more, the corner tends to bend like a cantilever, giving tension at the top, interior like a beam giving tension at bottom. At edge, main bending is along the edge like a beam giving maximum tension at bottom.

The maximum combined tensile stresses in the three regions of the slab will thus be caused when effects of temperature differentials are such as to be additive to the load effects. This would occur during the day in case of interior and edge regions, at the time of maximum temperature differential in the slab. In the corner region the temperature stress is negligible, but the load stress is maximum at night when the slab corners have tendency to lift up due to warping and lose partly the foundation support. Considering the total combined stress for the three regions, viz. corner edge and interior , for which the load stress decreases in that order while the temperature stress increases, the critical stress condition is reached in the edge region where neither of the load and temperature stresses are the minimum. It is, therefore, felt that both the corner and the edge regions should be checked for total stresses and design of slab thickness based on the more critical condition of the two.

Problem 21.1 *Determine the warping stresses for a 25 cm concrete pavement with 15 metre transverse joints; width of lane is 3.6 metres. The modulus of sub-grade reaction (k) is 2.8 kg/cm². Assume temperature differential for day conditions to be 1°C per cm for computing warping stress at the corner. Take the following values of various factors:*

Thermal coefficient of concrete 'α' = 8×10^{-6} per °C
Modulus of elasticity of concrete 'E' = 2.8×10^5 kg/cm²
Poisson's ratio, μ = 0.15
Radius of contact area = 16 cm

Solution. The radius of relative stiffness

$$l = \sqrt[4]{\frac{Eh^3}{12\,(1-\mu^2)\,k}}$$

$$= \sqrt[4]{\frac{2.8\times10^5\times25^3}{12\,(1-0.15^2)\times2.8}}$$

$$= 107.43 \text{ cm}$$

The ratio of length of slab to radius of relative stiffness:

$$\frac{L}{l} = \frac{15\times100}{107.43} = 13.95$$

The ratio of lane width (which is perpendicular to slab length) to radius of relative stiffness:

$$\frac{W}{l} = \frac{3.6\times100}{107.43} = 3.35$$

From Fig. 20.4, for

$$\frac{L}{l} = 13.95;\ C_1 = 1.03; \text{ and for}$$

$$\frac{W}{l} = 3.35;\ C_2 = 0.35$$

According to Bradbury, warping stress in the interior is given by

$$St_i = \frac{E\alpha t}{2}\left(\frac{C_1+\mu C_2}{1-\mu^2}\right)$$

$$= \frac{2.8\times10^5\times8\times10^{-6}\times25}{2}\left(\frac{1.03+0.15\times0.35}{1-0.15^2}\right) \qquad [t = 1 \times 25 = 25]$$

$$= 31.01 \text{ kg/cm}^2$$

Stress at the longitudinal edge region

$$St_e = \frac{C_1\,E\alpha t}{2}$$

$$= \frac{1.03\times2.8\times10^5\times8\times10^{-6}\times25}{2}$$

$$= 28.84 \text{ kg/cm}^2$$

Stress at the transverse edge region

$$St_t = \frac{C_2 E\alpha t}{2}$$

$$= \frac{0.35 \times 2.8 \times 10^5 \times 8 \times 10^{-6} \times 25}{2}$$

$$= 9.8 \text{ kg/cm}^2$$

Warping stress at the corner region is given by

$$St_c = \frac{E\alpha t}{3(1-\mu)} \sqrt{\frac{a}{l}}$$

$$= \frac{2.8 \times 10^5 \times 8 \times 10^{-6} \times 25}{3(1-0.15)} \sqrt{\frac{16}{107.43}}$$

$$= 8.48 \text{ kg/cm}^2$$

Problem 21.2 *Determine the stress for a 25 cm thick concrete pavement using Thomlinson method. Given:*

$$E = 3 \times 10^5 \, kg/cm^2$$

$$\alpha = 10 \times 10^{-6} \, cm \text{ per } °C$$

$$\mu = 0.15$$

Temperature differential $= 1°C$ *per cm.*

Solution. Factor $\frac{E\alpha t}{1-\mu} = \frac{3 \times 10^5 \times 10 \times 10^{-6} \times 1 \times 25}{1-0.15}$

$$= 88.24$$

$$h = 25 \text{ cm}$$

$$\phi^2 = 0.009 \qquad \text{(See page 732)}$$

$$T = 24 \times 60 \times 60 \text{ sec}$$

$$\gamma = \frac{h}{\phi}\sqrt{\frac{\pi}{T}} = \frac{25}{(0.009)^{1/2}} \sqrt{\frac{\pi}{24 \times 60 \times 60}} = 1.59$$

From Table 21.4, for slab thickness $\gamma = 1.59$, restrained warping stress coefficient = 0.509.

∴ Restrained warping stress

$$= 0.509 \times 88.24$$

$$= 44.91 \text{ kg/cm}^2.$$

Problem 21.3 *Compare the tensile stresses by Westergaard, Bradbury, Kelley, Spangler and Older at corner of a concrete slab, given the following data:*

Wheel load = *4200 kg*

Modulus of elasticity of concrete = *3 × 10*^{5} *kg/cm*^{2}

Pavement thickness = *20 cm*
Poisson's ratio = *0.15*
Modulus of subgrade reaction = *3 kg/cm³*
Radius of contact area = *20 cm*

Solution. Radius of relative stiffness

$$l = \sqrt[4]{\frac{Eh^3}{12\,(1-\mu^2)\,k}}$$

$$= \sqrt[4]{\frac{3\times10^5\times20^3}{12\,(1-0.15^2)\times3}}$$

$$= 90.88\,\text{cm}$$

Distance from the corner of the slab to the centre of the area of load application:

$$a_1 = \sqrt{2}a = \sqrt{2}\times20 = 28.28\text{ cm}$$

S. No.	*Theory*	*Formula*	*Tensile stress*	*Result*
1.	Westergaard	$\frac{3P}{h^2}\left[1-\left(\frac{a_1}{l}\right)^{0.6}\right]$	$\frac{3\times4200}{20\times20}\left[1-\left(\frac{28.28}{90.88}\right)^{0.6}\right]$	15.86 kg/cm²
2.	Bradbury	$\frac{3P}{h^2}\left[1-\left(\frac{a_1}{\sqrt{2}l}\right)^{0.6}\right]$	$\frac{3\times4200}{20\times20}\left[1-\left(\frac{28.28}{\sqrt{2\times90.88}}\right)^{0.6}\right]$	18.80 kg/cm²
3.	Kelley	$\frac{3P}{h^2}\left[1-\left(\frac{a_1}{l}\right)^{1.2}\right]$	$\frac{3\times4200}{20\times20}\left[1-\left(\frac{28.28}{90.88}\right)^{1.2}\right]$	23.74 kg/cm²
4.	Spangler	$\frac{3.2P}{h^2}\left[1-\frac{a_1}{l}\right]$	$\frac{3.2\times4200}{20\times20}\left[1-\frac{28.28}{90.88}\right]$	23.14 kg/cm²
5.	Older	$\frac{3P}{h^2}$	$\frac{3\times4200}{20\times20}$	31.5 kg/cm²

Problem 21.4 *Determine the thickness of a concrete pavement using Westergaard's corner load formula to support a maximum wheel load of 4100 kg. Allow 10 per cent for impact. The tyre pressure may be taken as 5.5 kg/cm². The modulus of subgrade reaction is 5.5 kg/cm³. The flexural strength of concrete may be taken as 40 kg/cm².*

Use a factor of safety of two.

Also determine the distance from the corner at which the maximum stress occurs.

Solution. Allowable flexural stress $= \frac{\text{Flexural strength}}{\text{Factor of safety}}$

$$= \frac{40}{2} = 20\text{ kg/cm}^2$$

Maximum wheel load = 4100 kg

Impact factor = 10 per cent

Design wheel load $= 4100 + \frac{10}{100} \times 4100$

$= 4510$ kg

Assume slab thickness of 20 cm

Radius of contact area $a = \sqrt{\frac{P}{p\pi}} = \sqrt{\frac{4510}{5.5\times\pi}} = 16.16$ cm

Radius of relative stiffness $l = \sqrt[4]{\frac{Eh^3}{12\,(1-\mu^2)\,k}}$

$$= \sqrt[4]{\frac{3\times10^6\times20^3}{12\,(1-0.15\times0.15)\,5.5}}$$

$= 78.10$ cm

The tensile stress is given by $S_c = \frac{3P}{h^2}\left[1-\frac{a_1}{l}\right]^{0.6}$

$$= \frac{3\times4510}{20\times20}\left[1-\left(\frac{16.16\sqrt{2}}{78.10}\right)\right]^{0.6}$$

$$= \frac{3\times4510}{20\times20}\,(1-0.478)$$

$= 17.66$ kg/cm^2

This is less than the allowable stress. Hence the slab thickness of 20 cm is alright. In case the calculated stress is more than the allowable stress, the second trial may be done by increasing the slab thickness and if it is less, the thickness may be reduced.

The maximum stress produced by a wheel load at corner occurs at a distance $d = 2.38\sqrt{al}$ from the corner. In this example:

$d = 2.38\sqrt{al}$

$= 2.38\sqrt{16.16\times78.10}$

$= 84.55$ cm

Problem 21.5 *Show that the design thickness of a concrete pavement slab is safe for combined load and temperature stresses for longitudinal end conditions, given the following data:*

Design thickness	*= 20 cm*
Maximum wheel load	*= 4080 kg*
Impact factor	*= 10%*
Modulus of elasticity of concrete	*= 3 × 10^5 kg/cm^2*
Modulus of subgrade reaction	*= 6 kg/cm^3*
Poisson's ratio of concrete	*= 0.2*
Tyre pressure	*= 7 kg/cm^2*
Slab dimensions	*= 4.5 m × 3.8 m*

Thermal coefficient of concrete = *8 × 10⁻⁶ per °C*

Temperature difference during the day = *0.5 °C/cm*

Allowable flexural strength of concrete = *35 kg/cm²*

Solution. Radius of relative stiffness

$$l = \sqrt[4]{\frac{Eh^3}{12\,(l-\mu^2)\,k}}$$

$$= \sqrt[4]{\frac{3\times10^5\times20^3}{12\,(1-0.2^2)\,6}}$$

$$= 86.33 \text{ cm}$$

$$\text{Design load } = 4080+\frac{10}{100}\times4080$$

$$= 4488 \text{ kg}$$

$$\text{Radius of contact area } a = \sqrt{\frac{P}{p\pi}}$$

$$= \sqrt{\frac{4488}{7\pi}}$$

$$= 14.29 \text{ cm}$$

Since the radius of the circular area 'a' = 14.29 cm is less than 1.724 h = 1.724 × 20 = 34.48 cm; the special theory has to be applied.

Equivalent radius of resisting section

$$b = \sqrt{1.6a^2+h^2}-0.675\,h$$

$$= \sqrt{1.6\times14.29^2+20^2}-0.675\times20$$

$$= 26.96-13.50$$

$$= 13.46 \text{ cm}$$

According to IRC (Westergaard formula modified by Teller and Sutherland) the tensile stress at the unbroken edge of the slab

$$S_e = \frac{0.529P}{h^2}[1+0.54\,\mu]\left[4\log_{10}\frac{l}{b}-\log_{10}b-0.4048\right]$$

$$= \frac{0.529\times4488}{20^2}[1+0.54\times0.2]\left[4\log_{10}\frac{86.33}{13.46}-\log_{10}13.46-0.4048\right]$$

$$= \frac{0.529\times4488}{400}\times1.108\times3.9527$$

$$= 26.00 \text{ kg/cm}^2$$

According to Bradbury, warping stress along longitudinal edge

$$S_t = \frac{CE\alpha t}{2}$$

$$\frac{L}{l} = \frac{4.5 \times 100}{86.33} = 5.21$$

From Fig. 21.4 or Table 21.3 for $\frac{L}{l} = 5.21, \quad C = 0.75$

$$\therefore \qquad S_t = \frac{0.75 \times 3 \times 10^5 \times 8 \times 10^{-6} \times 0.5 \times 20}{2}$$

$$= 9 \text{ kg/cm}^2$$

Combined stress = Load stress + warping stress

$$= 26.00 + 9.00$$

$$= 35.00 \text{ kg/cm}^2$$

This is equal to the allowable flexural strength of 35 kg/cm^2.

Hence the slab thickness of 20 cm is safe.

Problem 21.6 *Estimate the thickness of cement concrete pavement using Pickett's corner load stress equation (for unprotected corner) from the following data:*

Modulus of elasticity of concrete	*= 4 × 10^5 kg/cm^2*
Modulus of rupture of concrete	*= 50 kg/cm^2*
Design factor of safety	*= 2*
Poisson's ratio of concrete	*= 0.15*
Modulus of sub-grade reaction	*= 3 kg/cm^3*
Wheel load	*= 4200 kg*
Type pressure	*= 7 kg/cm^2*

Solution. Modulus of rupture of concrete = 50 kg/cm^2

Factor of safety = 2

$$\therefore \qquad \text{Allowable stress in concrete} = \frac{50}{2} = 25 \text{ kg/cm}^2$$

Assume slab thickness = 17.5 cm

$$\text{Radius of contact area } a = \sqrt{\frac{P}{p\pi}} = \sqrt{\frac{4200}{7\pi}} = 13.82 \text{ cm}$$

$$\text{Radius of relative stiffness } l = \sqrt[4]{\frac{Eh^3}{12\,(1-\mu^2)\,k}}$$

$$= \sqrt[4]{\frac{3 \times 10^5 \times 17.5^3}{12\,(1-0.15^2)\,3}}$$

$$= 82.22 \text{ cm}$$

According to Pickett, corner load stress (for unprotected corner) is given by:

$$S_c = \frac{4.2P}{h^2}\left[1 - \frac{\sqrt{\frac{a}{l}}}{0.925 + 0.22\frac{a}{l}}\right]$$

$$= \frac{4.2 \times 4200}{17.5^2}\left[1 - \frac{\sqrt{\frac{13.82}{82.22}}}{0.925 + \frac{0.22 \times 13.82}{82.22}}\right]$$

$$= 57.6\left[1 - \frac{0.41}{0.925 + 0.037}\right]$$

$$= 57.6\,(1 - 0.426)$$

$$= 33.05\ \text{kg/cm}^2$$

which is unsafe, since the allowable stress is 25 kg/cm^2.

Try slab thickness of 20 cm

$$l = \sqrt[4]{\frac{3 \times 10^5 \times 20}{12\,(1 - 0.15^2)\,3}}$$

$$= 90.88\ \text{cm}$$

$$\therefore \quad S_c = \frac{4.2 \times 4200}{20 \times 20}\left[1 - \frac{\sqrt{\frac{13.82}{98.88}}}{0.925 + \frac{0.22 \times 13.82}{90.88}}\right]$$

$$= 44.1\left[1 - \frac{0.39}{0.925 + 0.033}\right]$$

$$= 44.1\,(1 - 0.41)$$

$$= 26.02\ \text{kg/cm}^2, \text{ which is still unsafe}$$

Try slab thickness of 21 cm

$$l = \sqrt[4]{\frac{3 \times 10^5 \times 21^3}{12\,(1 - 0.15^2)\,3}}$$

$$= 94.26\ \text{cm}$$

$$\therefore \quad S_c = \frac{4.2 \times 4200}{21 \times 21}\left[1 - \frac{\sqrt{\frac{13.82}{94.26}}}{0.925 + \frac{0.22 \times 13.82}{94.26}}\right]$$

$$= 40\left(1 - \frac{0.383}{0.957}\right)$$

$= 40\,(1 - 0.4)$

$= 24\ \text{kg/cm}^2$, which is within the allowable stress limit.

Hence, slab thickness = 21 cm.

TYPES OF RIGID PAVEMENTS

There are three basic types of concrete pavements. These are:

(*i*) Plain concrete pavements

(*ii*) Simply reinforced concrete pavements; and

(*iii*) Continuously reinforced concrete pavements.

These are shown in Fig. 21.5.

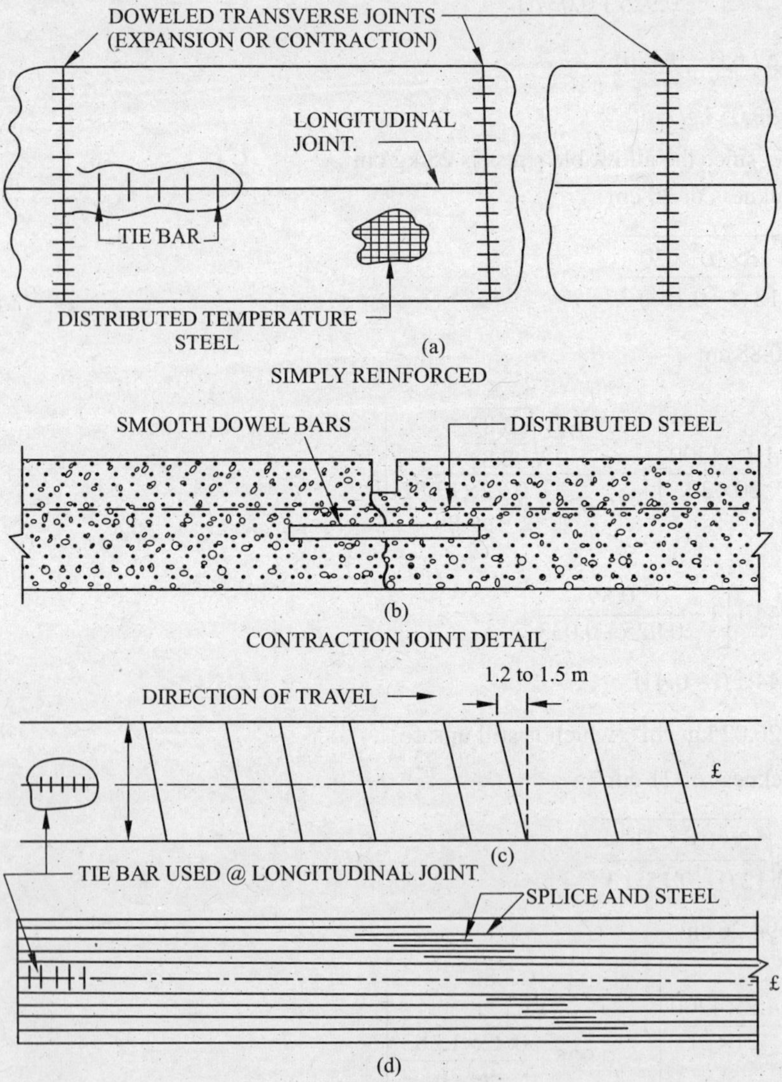

Fig. 21.5 Three basic rigid pavement types (a) simply reinforced (b) contraction joint detail (c) longitudinal joint (d) continuously reinforced concrete pavement.

Plain concrete pavement consist of relatively short slabs 3 to 6 m in length. These pavements contain no steel for crack control or load transfer, except at the longitudinal joints. These pavements must have close contraction joint spacings. It has been a common practice to use diagonal or skewed joints to provide a smooth ride, since only one wheel of the vehicle will traverse a joint at a time. Although plain concrete pavements, by definition, contain no steel, steel tie bars are generally used at the longitudinal joints to prevent the joints from opening and acting as a hinge joint. It is also a common practice to use dowel bars on many short slab pavements.

Simply reinforced concrete pavements have contraction joints at relatively long intervals, 10 to 30 m. Distributed temperature steel is provided throughout the slab. Further, tie bars are used at all longitudinal joints; these joints thus behaving as hinge or warping joints. The use of tie bars may be independent of the use of dowel bars and is primarily controlled by the distance between the joints.

In the case of simply reinforced slabs, it is assumed that the crack will form, generally at the centre of the slab. Temperature steel is provided to keep this crack intact so that it will not open. Since the slabs are relatively long, the joints will open and it is necessary to use load transfer devices at these joints so that they can act as contraction joint.

Continuously reinforced concrete pavements have no transverse joint, except where it is necessary to provide construction joint or expansion joint at bridge, etc. These pavements contain relatively high percentages of steel, 0.6 per cent and higher. Steel splices are made in a staggered manner to prevent an excessive amount of steel occurring at one transverse location. These are provided with tie bars across the longitudinal joints. Thus all types of highway concrete pavements are provided with tie bars across the longitudinal joints.

THICKNESS DESIGN FOR AIRPORT PAVEMENTS

The approach in most rigid-airport pavement design techniques is based on the theoretical stress analysis in elastic slabs, modified by field experience and appropriate factor of safety, which depends largely upon the type of feature i.e., taxiways, runways, etc.

The magnitude of distress of rigid airfield pavements depends upon repetition of load; gear configuration, gross load, tyre pressures and type of feature. Also, the magnitude of stresses in pavements depends upon the modulus of subgrade reaction, the modulus of elasticity of concrete and Poisson's ratio.

The factor of safety which should be applied to the modulus of rupture to give allowable working stress varies with the type of feature. Packard has recommended the following values for the factor of safety:

Apron, taxiway, hard standings, runway ends, hanger floors: 1.7 to 2

Runways (central portion), high speed exit taxiways: 1.4 to 1.7.

It is a common practice to use 110 per cent of the 90 day modulus in the thickness design. However, Packard has suggested the following relation for the design modulus of rupture.

$$\overline{MR}_d = \overline{MR}_{90\text{ days}} \left(1 - \frac{CV}{100}\right) k \quad \text{...(21.18)}$$

Where $\overline{MR}_d$ = design modulus of rupture (kg/cm^2).

$\overline{MR}_{90\text{ days}}$ = average modulus of rupture at 90 days (kg/cm^2)

C.V. = coefficient of variation of modulus of rupture (percent). This is the ratio of the standard deviation to the average value. The values are:

Concrete with excellent construction control : Below 10 per cent

Concrete with good control : 10 to 15 per cent

Concrete with fair control : 15 to 20 per cent

Concrete with poor control : above 20 per cent.

k = factor for the average modulus of rupture during design life (because concrete strength increases with age) = 1.10.

Rigid airport pavements differ from rigid highway pavements in many ways. Some of the important points of difference are:

(*i*) The gross load on the airfield pavement is greater than on a highway.

(*ii*) The number of repetitions of load is much less.

(*iii*) Loads are primarily applied in the centre of the airfield slab; whereas they act near the edges of the highway pavement.

(*iv*) Pumping is not a major problem for rigid airfield pavement but it is very important on highways.

(*v*) Roughness is induced at joints and structural breaks of rigid airfield pavements. Fast moving aircrafts, particularly jet aircraft that use bicycle landing gears, are subject to porpoising.

The various methods of rigid-airfield-pavement thickness design are outlined in the following paragraphs:

21.6. PORTLAND CEMENT ASSOCIATION (PCA) METHOD

Portland cement association use design charts for airport pavements developed from theory (Westergaard equations). These design charts are for specific aircraft. Portland cement association base their design data for the interior load case. The modulus of elasticity of concrete is assumed to be 280,000 kg/cm^2 and Poisson's ratio for concrete as 0.15. One such chart giving rigid pavement design curves for Boeing 747, dual tandem gear, 1.1 m × 1.45 m is given in Fig. 21.6.

The left hand vertical scale is the allowable flexural stress which is determined by dividing the modulus of rupture by the appropriate factor of safety. The modulus of rupture is the beam breaking load. Against the allowable stress in concrete, move horizontally to the design wheel load, then vertically to the modulus of subgrade reaction, k and then horizontally to determine the thickness of rigid concrete pavement on the right hand side vertical scale.

The chart can also be worked backward to determine the stress existing in a pavement for the known characteristics under various wheel loads.

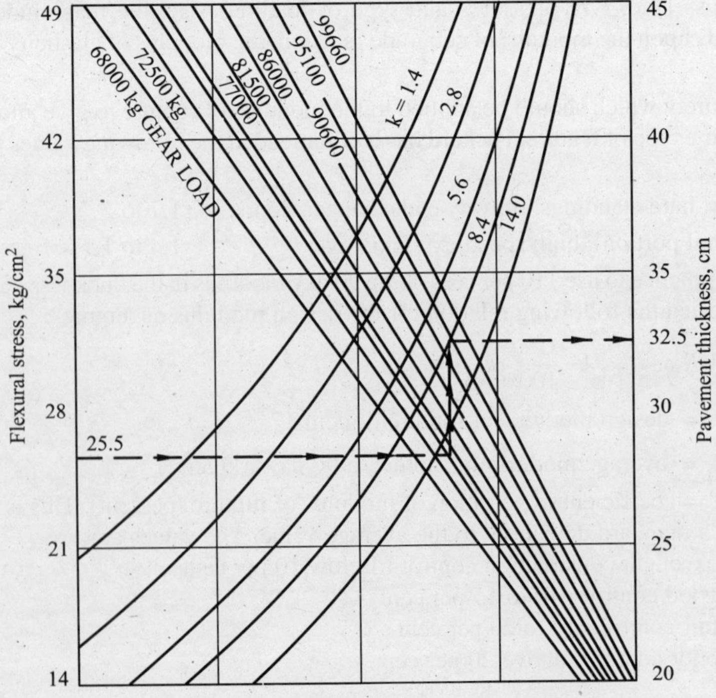

Fig. 21.6 PCA method for boeing 747.

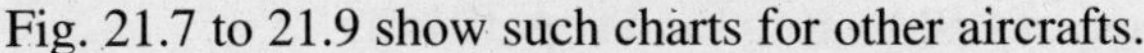

Fig. 21.7 to 21.9 show such charts for other aircrafts.

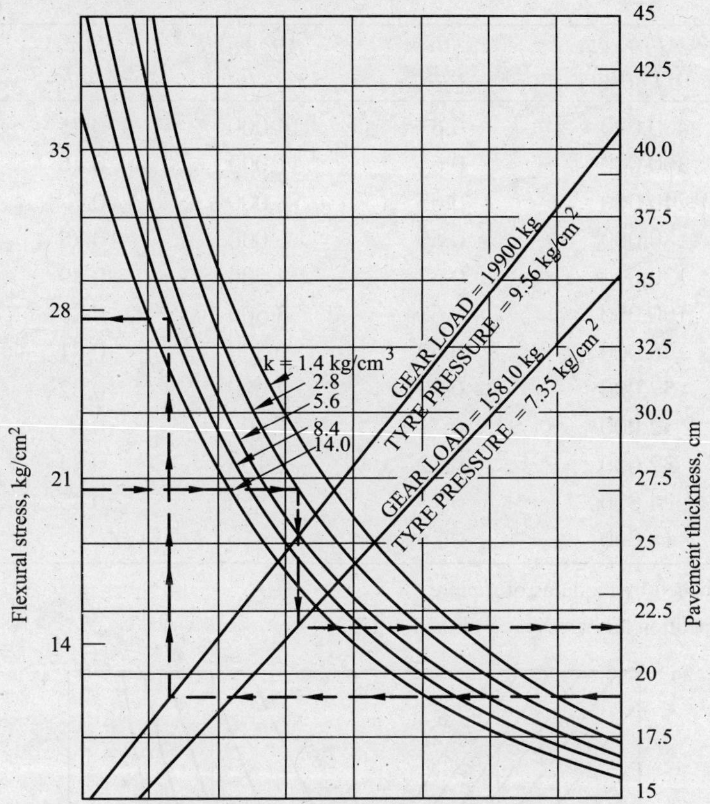

Fig. 21.7 PCA method for DC-9.

Design based upon fatigue: The method outlined above has application to the design of a rigid airfield pavement where the critical design aircraft is known. Fatigue concept is used, if mixed air traffic is to use the pavement. Table 21.5 shows stress ratios (actual stress to allowable stress) for allowable load repetitions as proposed by Portland cement association. The concrete pavement can sustain an indefinite number of load repetitions if the stress ratio is less than 0.51. However with an increase in stress ratio, the allowable number of repetitions decrease.

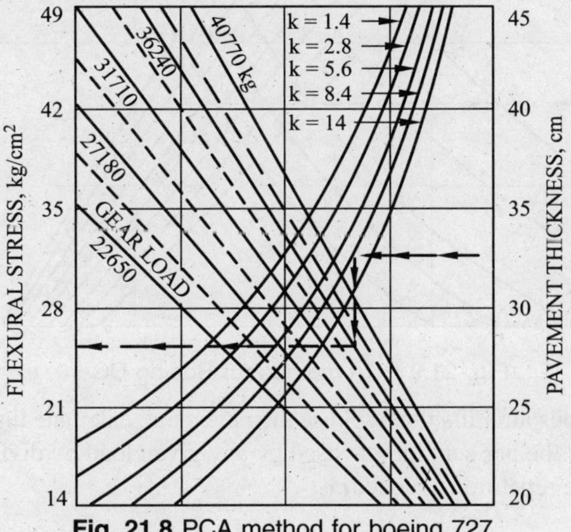

Fig. 21.8 PCA method for boeing 727.

Table 21.5 Stress ratios and allowable load repetitions

Stress ratio*	*Allowable repetition*	*Stress ratio*	*Allowable repetitions*	*Stress ratio*	*Allowable repetition*
0.51	400,000	0.63	14,000	0.75	490
0.52	300,000	0.64	11,000	0.76	360
0.53	240,000	0.65	8,000	0.77	270
0.54	180,000	0.66	6,000	0.78	210
0.55	130,000	0.67	4,500	0.79	160
0.56	100,000	0.68	3,500	0.80	120
0.57	75,000	0.69	2,500	0.81	90
0.58	57,000	0.70	2,000	0.82	70
0.59	42,000	0.71	1,500	0.83	50
0.60	32,000	0.72	1,100	0.84	40
0.61	24,000	0.73	850	0.85	30
0.62	18,000	0.74	650		

*Load stress divided by modulus of rupture.

Unlimited repetition has stress ratio of 0.50 or less.

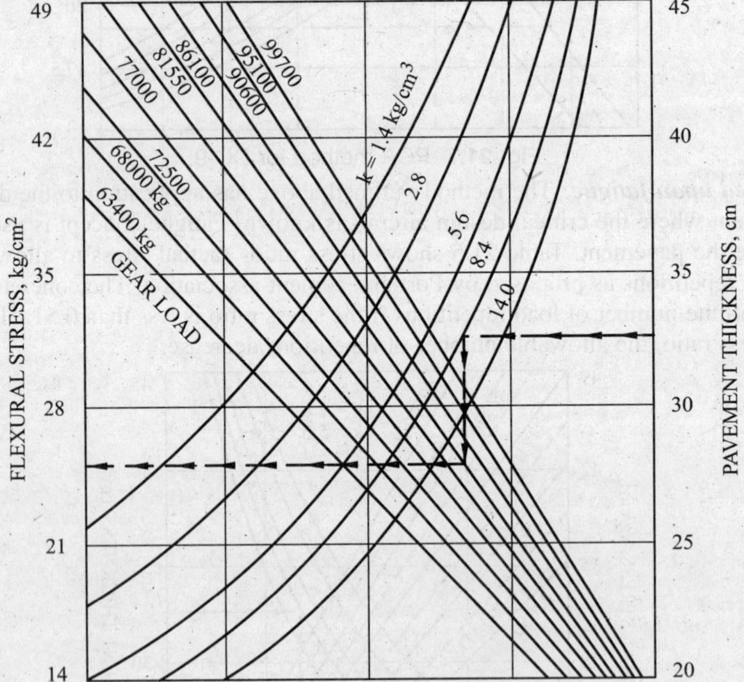

Fig. 21.9 PCA method for Boeing DC-10.

So knowing the various aircrafts that will use the pavement, calculate the stress under each wheel load and then determine the per cent fatigue used by any given load by dividing the actual repetition of load by the allowable repetition of load, *i.e.*

$$\frac{n_1}{n_1'}+\frac{n_2}{n_2'}+\frac{n_3}{n_3'}+\ldots\frac{n_n}{n_n'}\leq 1.0 \text{ to } 1.10$$

where n stands for actual repetitions; and
n' stands for allowable repetitions.

The pavement will not fail if the sum of the above ratio works out to be equal to or less than one. In the limiting case, the sum may be allowed to go as high as 1.10.

Repetitions, coverages and operations. The wander of an aircraft must be taken into account to determine the number of repetitions of load. This is computed from the relation:

$$C = D\times\frac{0.75\,w}{12.0T} \quad \ldots(21.19)$$

where C = Coverages
D = Number of operations at full load
w = Width of contact area of one tyre (cm)
T = Traffic width (m).

Packard has found the relationship of coverages to operations of air-craft and has tabulated values of load repetition factor (LRF). This is shown in Table 21.6 for various types of commonly used aircrafts for both taxiways and runways. The actual number of repetitions for a given aircraft can be found out by multiplying LRF with the expected number of departures.

Table 21.6 *Load repetition factors for different aircraft*

Aircraft	*Load repetition factor (Tentative design value)*			
	Taxi way		*Runway*	
	$\sigma = 60$ *cm*	$\sigma = 120$ *cm*	$\sigma = 240$ *cm*	$\sigma = 450$ *cm*
DC-3	0.12	0.07	0.05	0.03
B-727	0.41	0.23	0.13	0.09
DC-8 and B-707	0.83	0.46	0.25	0.17
B-747	0.58	0.38	0.33	0.28
C5A	0.74	0.61	0.37	0.25
B-2707	0.52	0.39	0.22	0.16
Concorde	0.83	0.44	0.23	0.15
DC-10-10 and L 10-11	0.57	0.40	0.22	0.12
Future No. 4*	1.33	0.84	0.44	0.24

*Projected 453600 kg aircraft, tandem gear 110 × 140 cm, 4 post (2 tracking)

21.7. CORPS OF ENGINEERS METHOD

Design method proposed by Corps of Engineers, U.S.A. is for edge loading case, on the assumption that 25 per cent load is transferred to the adjacent slab. Another design factor is introduced which compensates for temperature stresses, fatigue of concrete and other contingencies that may arise from applying a theoretical analysis to actual design problems. The design charts and discussion on the method is contained in U.S. Corps of Engineers EM 1110-45-303, 1958. The thickness of pavement is determined in a manner similar to the one described in PCA method and is given in Figs. 21.10 to 20.12.

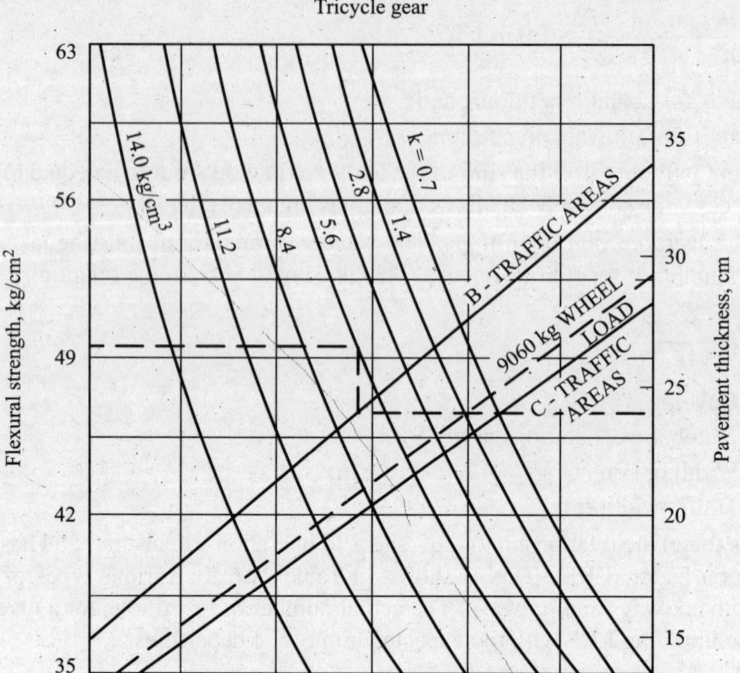

Fig. 21.10 Rigid pavement design curve for light loads.

In these charts, the modulus of rupture used on the left hand vertical scale includes the factor of safety and is determined by test. The design curves are on the basis of light, medium and heavy load pavements. The curves give thickness values for various traffic areas A, B and C, described in chapter 20.

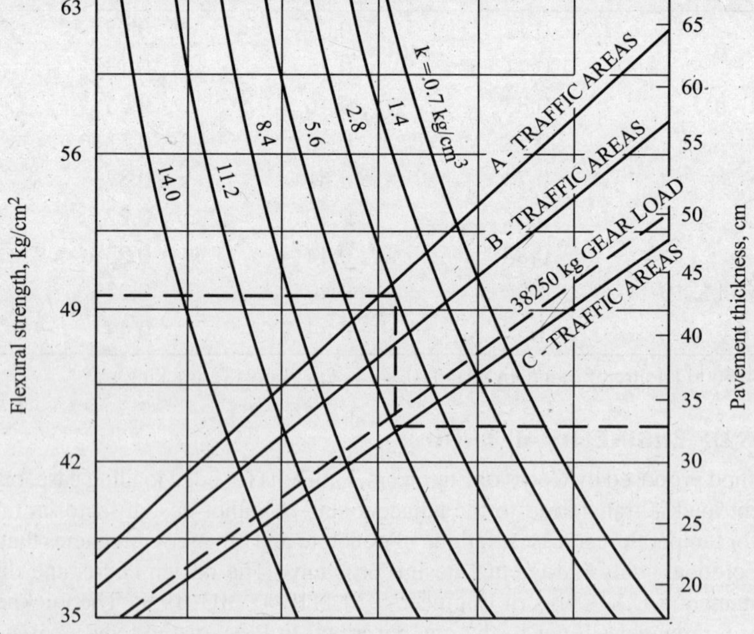

Fig. 21.11 Rigid pavement design curves, for medium loads.

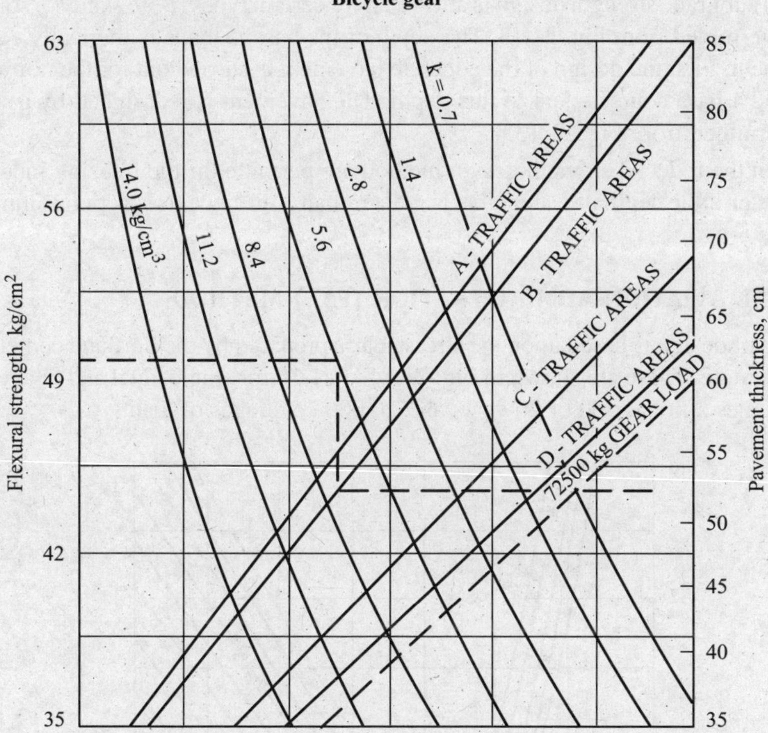

Fig. 21.12 Rigid pavement design curves, for heavy loads.

The Corps of engineers method takes into account the frost action. In such cases, one of the following two cases may control design:

(*a*) Design for limited frost penetration, and

(*b*) Design for loss of strength during the frost melt period.

Where the design is for limited frost penetration, the bottom 10 cm of the base is designed as a filter to prevent intrusion of the subgrade into the base.

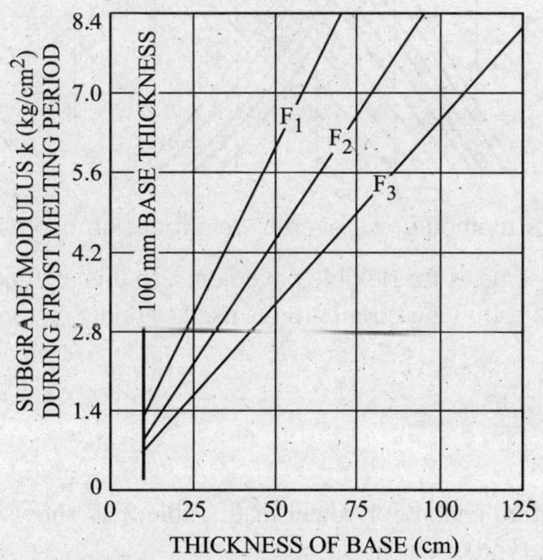

Fig. 21.13 Reduced strength subgrade modulus curves (Corps of engineers)

Reduction in subgrade strength design is allowed over certain types (F_1, F_2 and F_3) soils. A reduced value of k is determined from Fig. 21.13. The depth of the base is made numerically equal to that of concrete pavement. First the design of the concrete pavement is carried out for the normal conditions and then assume a base course equal to this depth. The pavement is redesigned by using a reduced value of k, determined from Fig. 21.13.

Exceptions in the reduced strength design method are permitted if the freezing index is less than 100 degree-days or if the depth of water table is greater than 3 m, since frost action is minimal in these cases.

21.8. FEDERAL AVIATION ADMINISTRATION (FAA) METHOD

FAA method of design is based upon the stress charts prepared by the Portland cement association computer programme. These are shown in Fig. 21.14 to 21.16 for single, dual and dual tandem gears. The design assumes a life period of 20 years and 20,000 coverages of load.

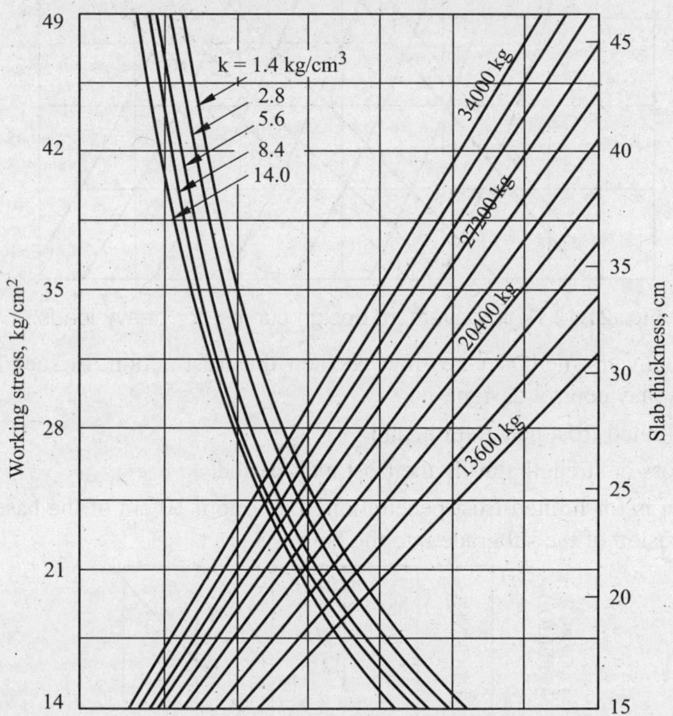

Fig. 21.14 FAA method-gross aircraft weight design curves-single gear.

The traffic analysis is the same as for flexible pavements. In this method, equivalent departures of a critical aircraft is used. The following correlation is used to relate equivalent repetitions of several loads.

$$\log R_1 = \log R_2 \left(\frac{W_2}{W_1}\right)^{1/2}$$

Where R stands for departure and W for wheel load. Table 21.7 shows factors that are applied to the repetitions to account for type of gear.

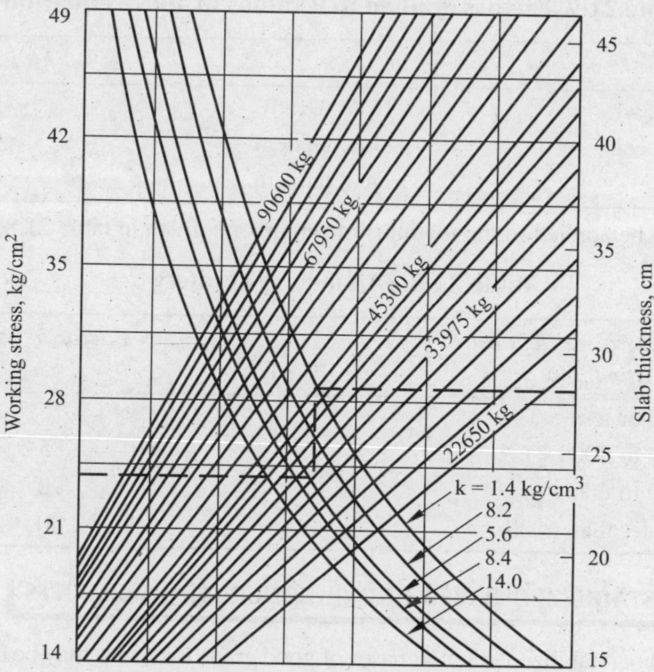

Fig. 21.15 FAA method-gross aircraft weight design curves-dual gear.

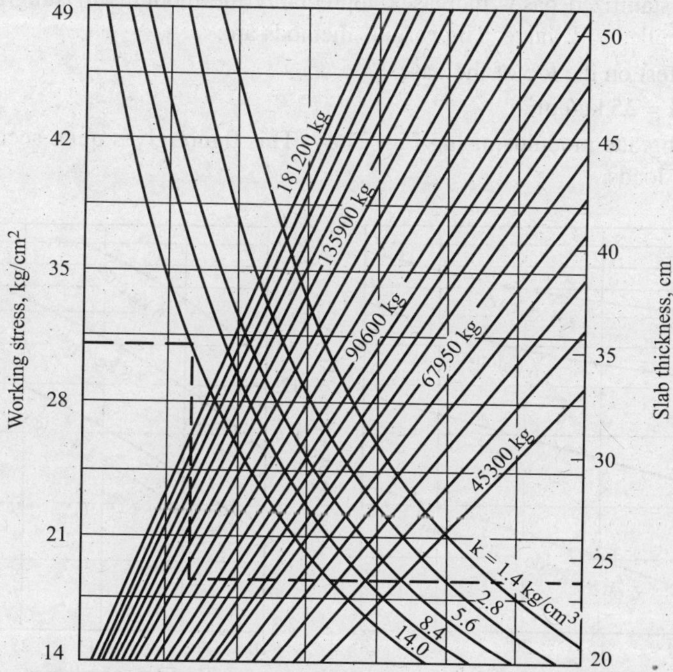

Fig. 21.16 FAA method-gross aircraft weight design curves-dual tandem gear.

Table 21.7 Factors applied to R values in above equation

To convert from	*To*	*Multiply R by*
Single wheel	Dual wheel	0.80
Single wheel	Dual tandem	0.50
Dual wheel	Dual tandem	0.60

Factor of safety to be applied to the modulus of rupture are given in table 21.8.

Table 21.8 FAA factors of safety

Annual equivalent departures of critical aircraft	*Factor of safety*
1200 or less	1.75
1200 to 3000	1.85
3000 to 6000	1.90
Greater than 6000	2.00

STABILIZED BASES UNDER CONCRETE PAVEMENTS

For lighter loads, non-stabilized granular bases of good grade materials may be used. FAA requires that stabilized bases should be used for heavy loads *i.e.*, gross load greater than 90,000 kg and when the critical aircraft is dual tandem.

Since the use of stabilized bases increases appreciably the modulus of subgrade reaction (k), correction may be applied to k value. The various methods are;

(1) Perform plate test on the top of the base,

Maximum $k = 35$ kg/cm^2.

(2) Correct the subgrade modulus, using Fig. 21.17. This figure gives quite accurate data except for very heavy loads.

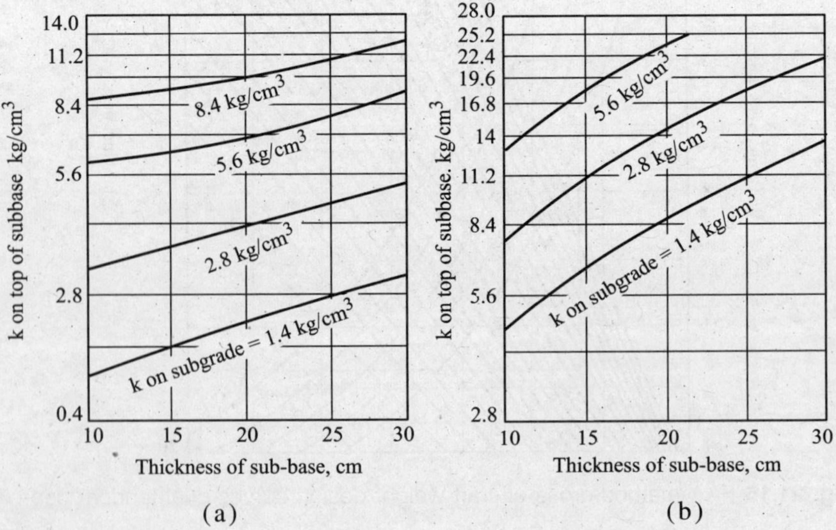

Fig. 21.17 Effect of sub-base thickness on modulus of sub-grade reaction (a) non-treated (b) cement treated.

(3) Use the trial and error method in which the relative stiffness of the slab is corrected applying Packard method, detailed below:

Packard method. Fig. 21.18 gives equivalent thickness of slab and base in terms of the radius of relative stiffness, l. Stabilized bases reduce the radius 'l' since they increase the subgrade modulus, k.

The relative rigidity of two slabs is given by $\sqrt[4]{\frac{D_1}{D_2}}$ and for the elastic solid cases $\sqrt[3]{\frac{D_1}{D_2}}$.

In each case, the rigidity D is computed from the equation

$$D = \frac{Eh^3}{12\,(1-\mu^2)}$$

From above, the ratio $\frac{l_1}{l_2}$ for the elastic solid case equals, $r^{4/3}$ and values of r are given in Fig. 21.18.

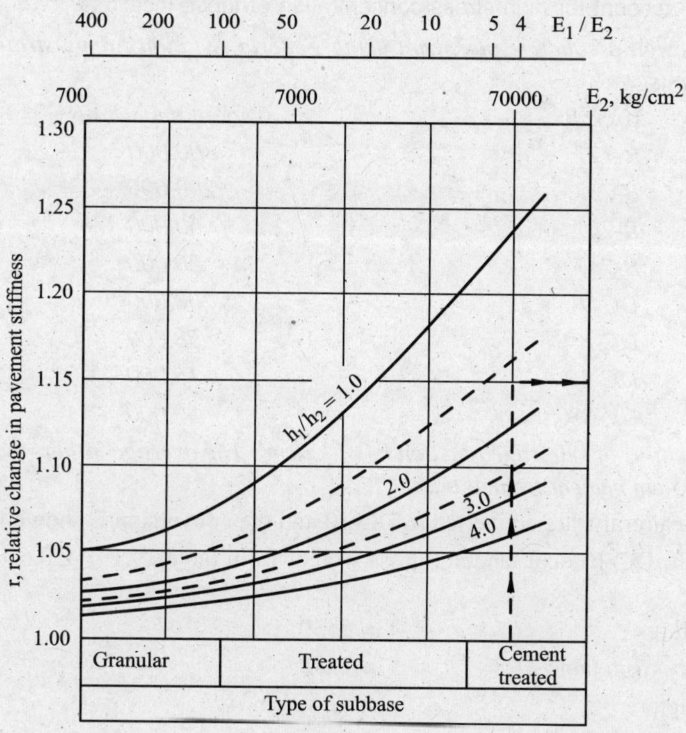

Fig. 21.18 Changes in pavement stiffness due to base-course.

Thickness for various regions. The value of thickness 'T' from the charts is for critical areas. FAA states that the thickness of non-critical areas should be 0.91 T. This reduction applies only to the concrete pavement and not to the base. The edges of the keel section are reduced to 0.71 T, but the total thickness *i.e.*, pavement plus base should be the same as for critical areas. Critical and non-critical areas for runways are shown in Fig. 21.19.

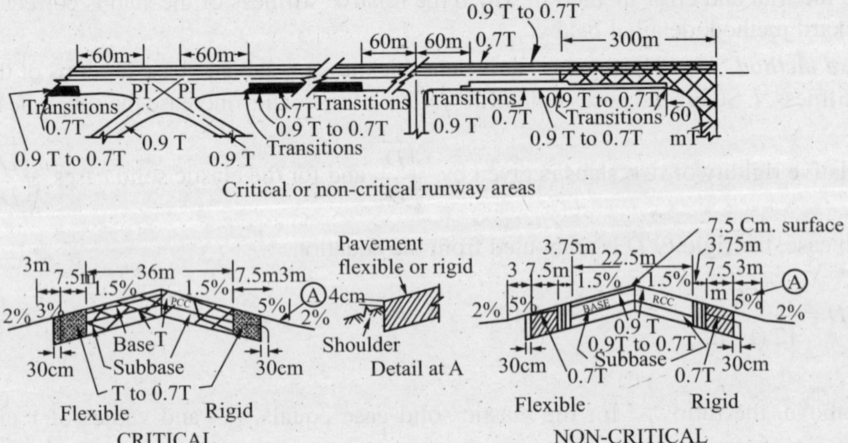

Fig. 21.19 Typical sections and critical areas for jet aircraft showing the 'keel' section. The 'keel' section refers to thickened centres of runways and or taxiways to account for high stress concentration at these locations.

Problme 21.7 *Design a concrete pavement using Federal Aviation Administration method for the following conditions:*

Aircraft	*No. of departures*
B-727	*300,000*
B-737	*200,000*
B-707	*60,000*
B-747	*20,000*
DC-9	*200,000*
DC-8	*50,000*
DC-10	*15,000*
CV-880	*50,000*

The design modulus of subgrade reaction k is 7 kg/cm³. The average 90-day modulus of rupture is 42 kg/cm². Use 15 cm cement treated base.

Solution. All the aircrafts are converted to DC-10 and the conversion is shown in Table 20.9.

Critical aircraft, DC-10 dual tandem, gross weight from table 19.1 = 453.6×413 = 187000 kg (Approx.)

Annual departures = 5530

Factor of safety from table 21.8 = 1.90

Modulus of rupture = 42 kg/cm²

$$\text{Design modulus of rupture} = \frac{42}{1.90} = 22 \text{ kg/cm}^2$$

Subgrade modulus = 7 kg/cm³

Base course is 15 cm of cement treated soil.

The solution to this problem is done by trial and error method wherein the thickness of equivalent pavement is the same as that determined using a subgrade modulus of 7 kg/cm³.

From Fig. 21.16 and using a k value of 7 kg/cm³, the thickness works out to be 45 cm. This would be the design thickness using a modulus of subgrade reaction equal to that of subgrade alone.

TABLE 21.9 Conversion of all aircraft to DC-10

Aircraft	*Type of gear*	*Factor for R*	*Departures R (20 years)*	*Load per wheel (kg)*	*Log R_1*	*R_1*
B-727	Twin	0.6	300,000	175000	$\log 0.6\times300000\left(\frac{17500}{22000}\right)^{1/2}$	48650
B-737	Twin	0.6	200,000	5800	$\log 0.6\times200000\left(\frac{5800}{22000}\right)^{1/2}$	405
B-707	Twin tandem	1.0	60,000	17900	$\log 60,000\left(\frac{17900}{22000}\right)^{1/2}$	20416
B-747	Double twin tandem	1.0	20,000	18900	$\log 20,000\left(\frac{18900}{22000}\right)^{1/2}$	9694
DC-9	Twin	0.6	200,000	9600	$\log 0.6\times200000\left(\frac{9600}{22000}\right)^{1/2}$	2266
DC-8	Twin tandem	1.0	50,000	16800	$\log 50,000\left(\frac{16800}{22000}\right)^{1/2}$	12772
DC-10	Twin tandem	1.0	15,000	22000	log 15000	15000
CV-880	Twin tandem	1.0	50,000	9900	$\log 50,000\left(\frac{9900}{22000}\right)^{1/2}$	1420

Total : 110623 (20 years)
or 5530 per year (approx.)

However, from Fig. 21.18 for E_2 value of 7×10^4 kg/cm², r value for $\frac{h_1}{h_2} = \frac{45}{15} = 3$ is 1.08 and the equivalent thickness is 45 $(1.08)^{1.33}$ = 49.85 cm. This represents an over design.

Assuming a 40 cm pavement and 15 cm cement treated base, then from Fig. 21.18, using $E_2 = 7 \times 10^4$ kg/cm², r is found to be 1.09 and the new equivalent thickness is 40 $(1.09)^{1.33}$ = 45.85 cm. Hence the design will consist of 40 cm of concrete and 15 cm of CTB (Cement Treated Base)

To check for the wide bodied aircraft (B-747), design data are as follows:

Critical aircraft boeing (747), gross load= 32300 kg

Gear load = 75525 kg

Annual departures $= \left(\frac{20,000}{20}\right)$ = 1000

Factor of safety = 1.75

Design modulus of rupture = 22 kg/cm²

Subgrade modulus = 7 kg/cm³

From Fig. 21.6, using k = 7 kg/cm³ and MR_d = 22 kg/cm², the thickness is equal to 38 cm. Since this is less than that for conventional craft using the same subgrade modulus (45 cm), the design of 40 cm of concrete and 15 cm CTB will be used. The final design for the keel section will be as below:

	Critical areas (cm)	*0.9 T (cm)*	*0.7T (edges (cm)*
Concrete pavement	40	36	28
Cement treated base	15	13.5	10.5
	Total 55	49.5	38.5

THICKNESS DESIGN FOR HIGHWAYS

The stress values given by theory are for static loads. These have to be modified to account for repeated load applications. Use is made of fatigue values for concrete in repeated flexure. Plain concrete can sustain an indefinite number of repetition of loads, if the stress ratio is less than 0.50 *i.e.,* the extreme fibre stress does not exceed 50 per cent of the static modulus of rupture. For higher stress ratios, the concrete will sustain a fewer number of load repetitions, given in Table 21.5. These values are used to modify the thickness determined by using static loads.

The various methods used to determine the thickness of rigid highway pavements are discussed below:

21.9. INDIAN ROADS CONGRESS DESIGN METHOD

The Indian Roads Congress (IRC : SP-62-2004) has recommended the following design procedure for concrete pavements for Rural Roads.

Step 1: Select values for the various parameters i.e. design wheel load, concrete flexural strength, modulus of subgrade reaction, modulus of elasticity of concrete, Poisson's ratio, coefficient of thermal expansion of concrete.

Step 2: Decide joint spacing and lane widths (Table 18.12 based on IRC : 15–1981 and Table 8.5).

Step 3: Select tentative design thickness of pavement slab based on design load, k value / CBR and flexural strength of concrete. Select the temperature differential from Table 21.10 for the region and selected thickness.

Table 21.10 Recommended temperature differentials for concrete slabs

Zone	*States*	*Temperature Differential, °C is Slabs of Thickness*		
		150 mm	200 mm	250 mm
I	Punjab, U.P., Uttarakhand, Rajasthan, Haryana, Gujarat and North M.P., excluding hilly regions	12.5	13.1	14.3
II	Bihar, Jharkhand, West Bengal, Assam and Eastern Odisha excluding hilly regions and coastal areas.	15.6	16.4	16.6
III	Maharashtra, Karnataka, South M.P., Chattisgarh, Andhra Pradesh, Western Odisha and North Tamil Nadu excluding hilly regions and coastal areas.	17.3	19.0	20.3
IV	Kerala and South Tamil Nadu excluding hilly regions and coastal areas.	15.0	16.4	17.6
V	Coastal areas bounded by hills.	14.6	15.8	16.2
VI	Coastal areas unbounded by hills.	15.5	17.0	19.0

***Step 4*:** Ascertain maximum temperature stress for the critical edge region from equation 21.15 or Fig. 21.20.

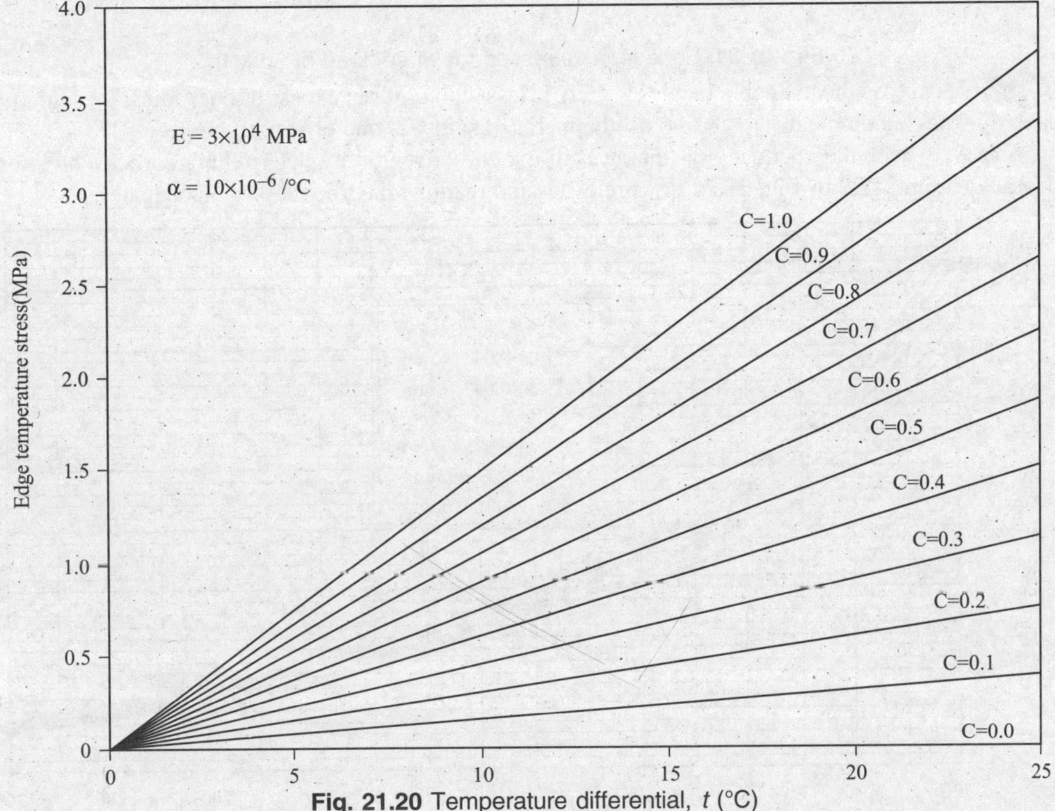

Fig. 21.20 Temperature differential, *t* (°C)

***Step 5*:** Calculate the residual available strength of concrete for supporting traffic loads.

Step 6: Ascertain edge load stress from equation 21.14 or from Fig. 21.21 or 21.23 for the relative wheel load. Calculate factor of safety.

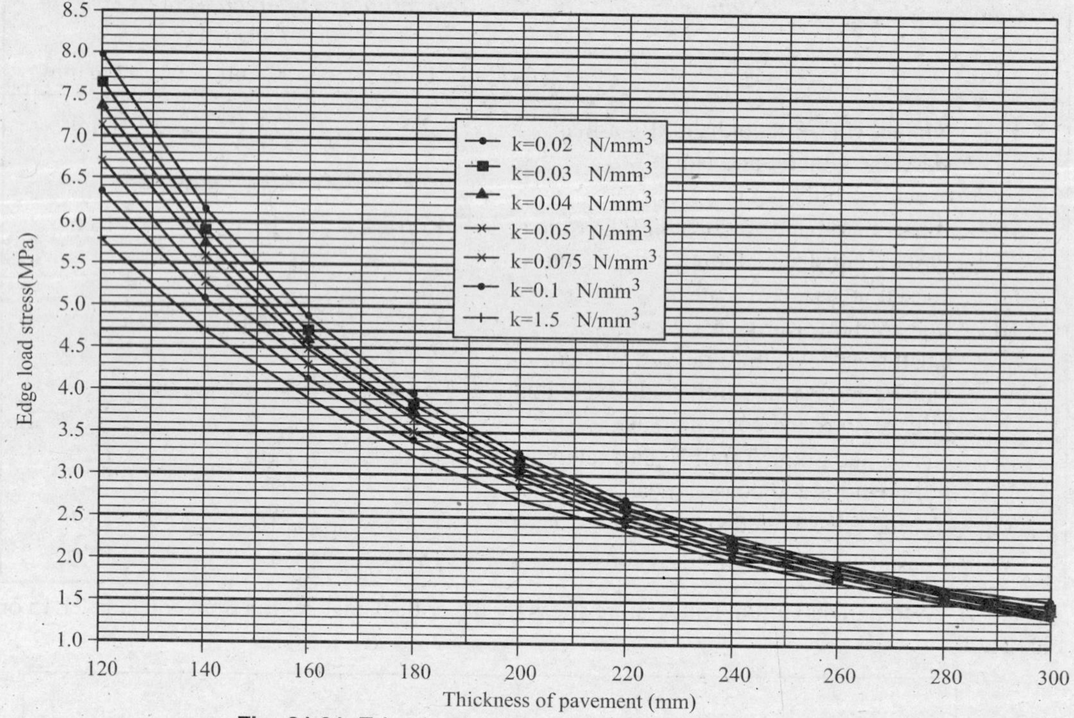

Fig. 21.21 Edge loads stresses for wheel load of 30 kN.

Step 7: In case the available factor of safety is less than or far in excess of one, adjust the tentative slab thickness and repeat steps 3 to 6 till the factor of safety is one or slightly more.

Step 8: Check the adequacy of thickness in the corner region by determining corner load stress from equation 21.22 or Fig. 21.24 or as relevant and readjust the thickness if inadequate.

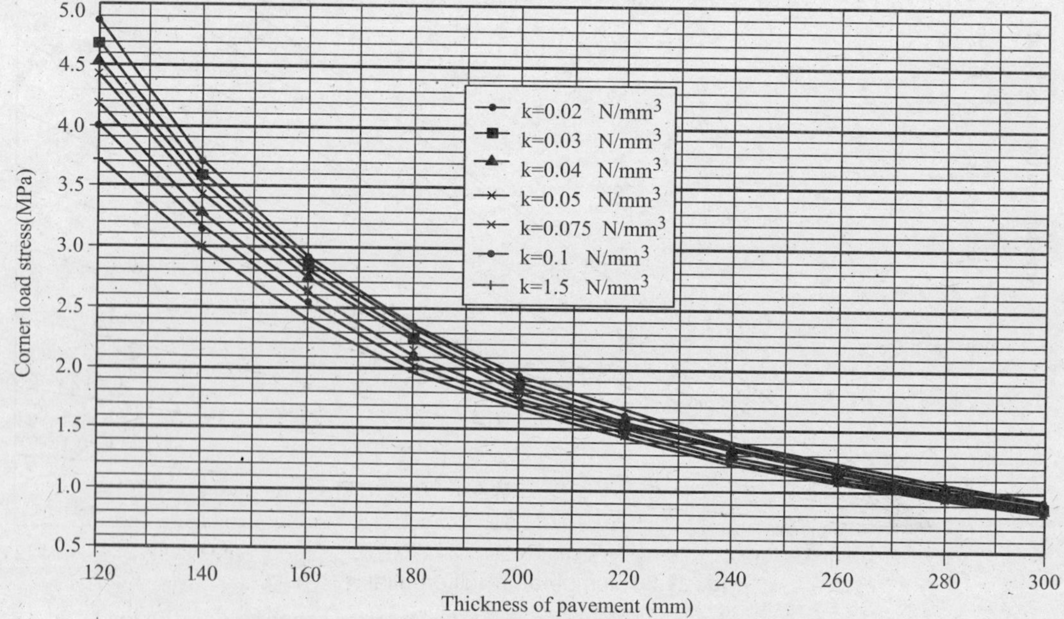

Fig. 21.22 Corner load stresses for wheel load of 30 kN.

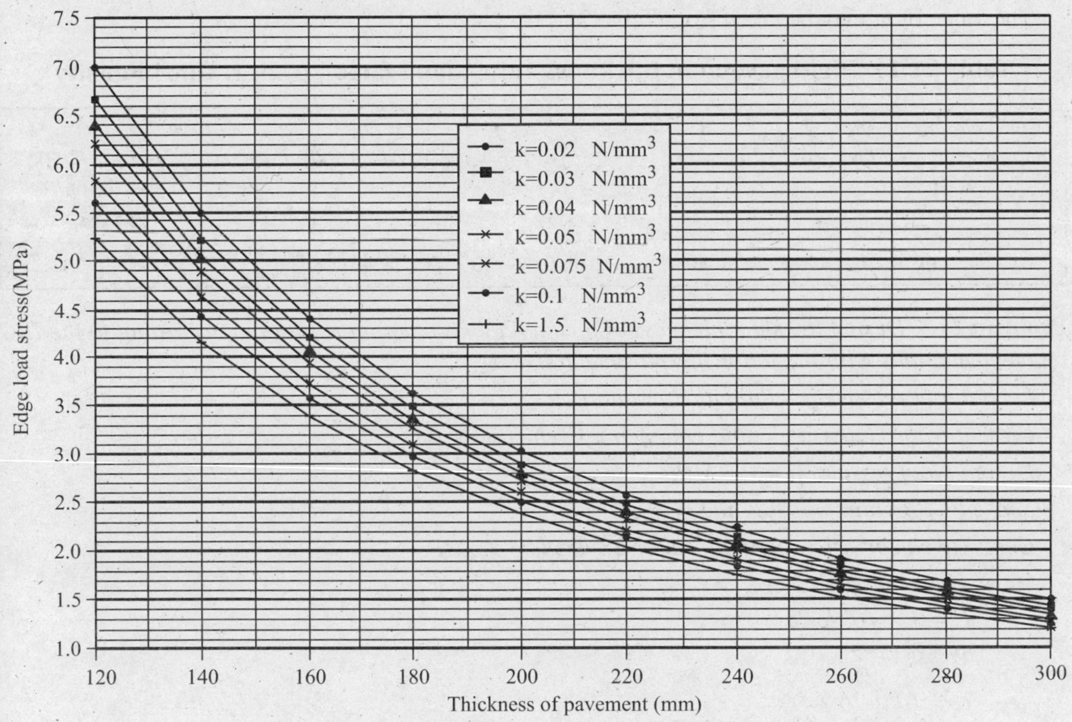

Fig. 21.23 Edge load stresses for wheel load of 51 kN.

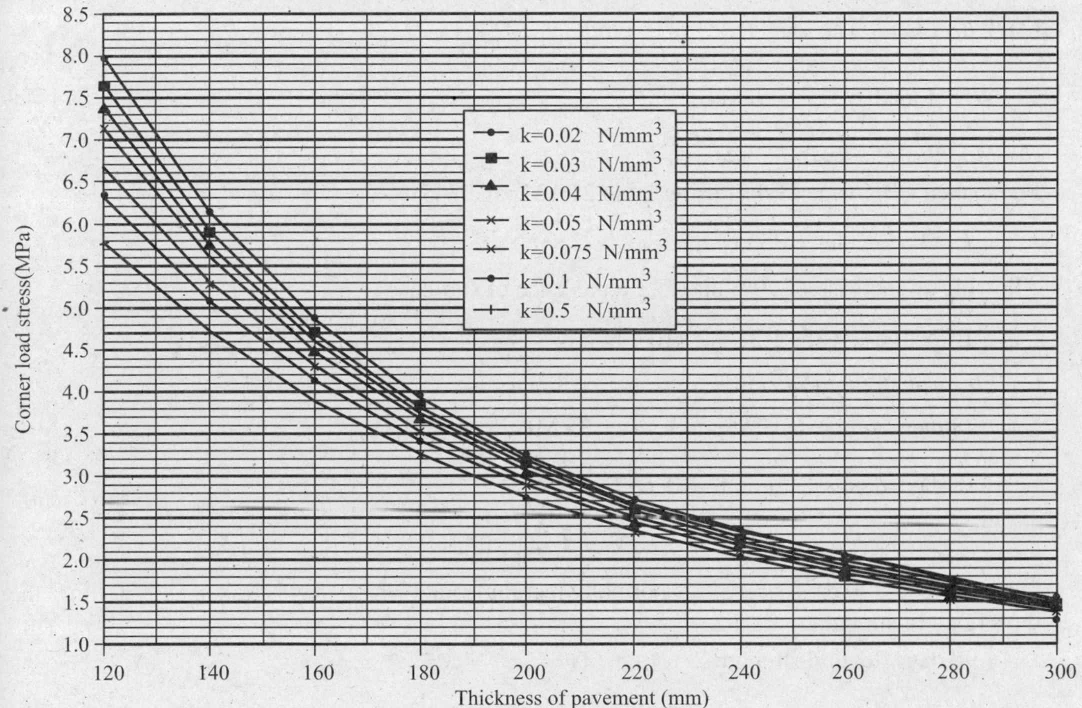

Fig. 21.24 Corner load stresses for wheel load of 51 kN.

The value of h_t may be taken from Table 21.11.

Table 21.11 Rigid pavement thickness adjustment factor, h_t for traffic intensity

Traffic classification/ intensity vehicles/day)	*A (0-15)*	*B (15-45)*	*C (45-150)*	*D (150-450)*	*E (450-1500)*	*F (1500-4500)*	*G Above 4500 and all express ways*
h_t (cm)	– 50	– 50	– 20	– 20	+ 0	+ 0	+ 20

Problem 21.8 *Design the slab thickness of a concrete pavement in rural area according to the IRC recommendations with the following parameters:*

Location of pavement : Punjab/Haryana/U.P.

Present traffic intensity = 300 commercial vehicles/day

Design tyre pressure, p = 0.7 MPa

Soaked CBR value of subgrade = 4%

Concrete compressive strength, (after 28 days) = 35 MPa

Other concrete parameters

$E = 3 \times 10^4$ *MPa*

$\mu = 0.15$

$\alpha = 10 \times 10^{-6}$ /°C

Solution.

Step 1 *Design parameters*

(*a*) *Wheel load:* The legal axle load in India is 102 kN.

$$\therefore \quad \text{Wheel load} = \frac{102}{2} = 51 \text{ kN}$$

(*b*) *Modulus of subgrade reaction, k*

For soaked CBR vaue of 4%, using table 21.12 (page 765)

$$k = 3.5 \text{ kg/ cm}^3 = 35 \text{ kN/ mm}^3$$

Allowing an increase of 20% due to the presence of sub-base

$$\text{Effective } k = 35 \times 1.2 = 42 \times 10^{-3} \text{ N/mm}^3$$

(*c*) *Flexural strength of concrete*

$$\text{28 days compressive strength, } f_c = 35 \text{ MPa}$$

$$\therefore \quad \text{28 days flexural strength} = 0.7\sqrt{f_c}$$

$$= 0.7\sqrt{35} = 4.141 \text{ MPa}$$

For rural roads, 90 days strength is normally recommended which may be taken as 1.20 times the 28 days flexural strength

$$\therefore \quad \text{90 days flexural strength} = 1.2 \times 4.141$$

$$= 4.969 \text{ MPa}$$

(*d*) *Other variables as given are*

(*i*) Modulus of elasticity, $E = 3\times10^4$ MPa

(*ii*) Poisson's ratio, $\mu = 0.15$

(*iii*) Coefficient of thermal expansion $\alpha = 10\times10^{-6}$ / °C

Step 2 From the tables 18.14 and 8.5
adopt joint spacing, L = 45 m = 4500 mm; and
Lane width, W = 3.5 m = 3500 mm

Step 3 Adopt tentative design thickness of pavement = 200 mm
From table 21.10 or Fig. 21.20
The temperature differential in the slab for the region = 13.1

Step 4 Temperature stress

Radius of relative stiffness, $l = \sqrt[4]{\dfrac{Eh^3}{12\,(1-\mu^2)\,k}}$

$$= \sqrt[4]{\frac{3\times10^4\times200^3\times10^3}{12\,(1-0.15^2)\times42}}$$

$$= 469.8 \text{ mm}$$

$$\frac{L}{l} = \frac{4500}{469.8} = 9.58$$

$$\frac{W}{l} = \frac{3500}{469.8} = 7.45$$

For $\dfrac{L}{l} = 9.58$; Bradbury's coeff . C from table 21.3 (page 731) = 1.077 (By interpolation)

$$\sigma_{t_e} = \frac{E\alpha t}{2}C$$

$$= \frac{3\times10^4\times10\times10^{-6}\times13.1}{2}\times1.077$$

$$= 2.116 \text{ MPa}$$

Step 5 Residual available strength of concrete for supporting traffic load
= 4.969 − 2.116 = 2.853 MPa

Step 6 *Edge load stress*

Radius of load contact $a = \sqrt{\dfrac{P}{p\pi}} = \sqrt{\dfrac{51000\times7}{0.7\times22}} = 152.26$ mm

$$\frac{a}{h} = \frac{152.26}{200} = 0.7613 < 1.724$$

$$\therefore \quad b = \sqrt{1.6a^2 + h^2} - 0.675\,h = \sqrt{1.6\times152.26^2 + 200^2} - 0.675\times200 = 147.65 \text{ mm}$$

$$\sigma_{l_e} = 0.529 \frac{P}{h^2}(1+0.54\,\mu)\left(4\log_{10}\frac{l}{b}+\log_{10} b - 04048\right)$$

$$= \frac{0.529\times 51000}{200\times 200}(1+0.54\times 0.15)\left(4\log_{10}\frac{469.8}{147.65}+\log 147.65 - 0.4048\right)$$

$$= \frac{0.529\times 51}{40}(1.081)(4\log 3.18185+\log 147.65-0.4048)$$

$$= \frac{0.529\times 51\times 1.081\times 4\times 0.5027+2.1692-0.4048}{40}$$

$$= \frac{0.529\times 51\times 1.081\times 3.7752}{40}$$

$$= 2.75 \text{ MPa}$$

$\therefore$ Factor of safety $= \frac{2.853}{2.75} = 1.037 > 1$

Hence safe

Step 7 Corner stress

$$\sigma_{l_c} = \frac{3P}{h^2}\left[1-\left(\frac{a\sqrt{2}}{l}\right)^{1.2}\right]$$

$$= \frac{3\times 51000}{200\times 200}\left[1-\left(\frac{152.26\times\sqrt{2}}{469.8}\right)^{1.2}\right] = \frac{153}{40}\left[1-\left(\frac{152.26\times 1.414}{469.8}\right)^{1.2}\right]$$

$$= \frac{153}{40}\left[1-(0.4583)^{1.2}\right]$$

$$= \frac{153}{40}[1-0.392] = \frac{153}{40}\times 0.608$$

$$= 2.3256 \text{ MPa}$$

The corner stress is less than 4.969 MPa and hence the thickness of 200 mm assumed is safe.

21.10. IRC DESIGN PROCEDURE FOR EXPRESSWAYS AND NATIONAL HIGHWAYS

The method outlined above has been revised by IRC for concrete pavement design for expressways and National Highways carrying heavy traffic and the recommended design procedure is outlined in IRC-58-2002 "Guidelines for the Design of Plain Jointed Rigid Pavements for Highways, and is as follows:

Step 1 **:** Stipulate design values for the various parameters.

Step 2 **:** Decide types and spacing between joints.

Step 3 **:** Select a trial design thickness of pavement slab.

Step 4 **:** Compute the repetitions of axle loads of different magnitude during the design period.

Step 5 **:** Calculate the stresses due to single and tandem axle loads and determine the cumulative fatigue damage (CFD).

Step 6 **:** If the CFD is more than 1.0, select a higher thickness and repeat the steps 1 to 5.

Step 7 **:** Compute the temperature stress at the edge and if the sum of the temperature stress and the flexural stress due to the highest wheel load is greater than the modulus of rupture, select a higher thickness and repeat the steps 1 to 6.

Step 8**:** Design the pavement thickness on the basis of corner stress if no dowel bars are provided and there is no load transfer due to lack of aggregate inter-lock.

21.11. EFFECTIVE MODULUS OF SUBGRADE REACTION *k*

An approximate idea of k-value of a homogeneous soil subgrade may be obtained from its soaked CBR value using Table 21.12. It is advisable to have a filter layer above the subgrade for drainage of water to prevent (*i*) excessive softening of subgrade and (*ii*) erosion of the subgrade particularly under adverse moisture condition. Annexure-4 of IRC:37-2001 may be referred for further details.

Table 21.12 Approximate *k*-value corresponding to CBR values for homogeneous soil subgrade

Soaked CBR Value %	2	3	4	5	7	10	15	20	50	100
k-value ($kg/cm^2/cm$)	2.1	2.8	3.5	4.2	4.8	5.5	6.2	6.9	14.0	22.2

The recommendations of IRC:15-2002 shall be followed and if the k-value tested on wet condition of the subgrade is less than 6.0 $kg/cm^2/cm$, cement concrete pavement should not be laid directly over the subgrade. A Dry Lean Concrete (DLC) sub-base is generally recommended for modern concrete pavements, particularly those with high intensity of traffic. The sub-base of DLC should conform to "*Guidelines for the use of Dry Lean Concrete as Sub-base for Rigid Pavement, IRC:SP:49-1998*". In the case of problematic subgrade, such as, clayey and expansive soils, etc. appropriate provisions shall be made for blanket course in addition to the sub-base as per the relevant stipulations of IRC:15-2002.

The approximate increase in k-values of subgrade due to different thickness of sub-bases made up of untreated granular, cement treated granular and dry lean concrete (DLC) layers may be taken from Tables 21.13 and 21.14. Seven days unconfined compressive strength of cement treated granular soil should be a minimum of 2.1 MPa. Dry Lean Concrete should have a minimum compressive strength of 7 MPa at 7 days.

Table 21.13 k-values over granular and cement treated sub-bases

k-value of subgrade ($kg/cm^2/cm$)	*Effective k ($kg/cm^2/cm$) over untreated granular layer sub-base of thickness in cm*			*Effective k ($kg/cm^2/cm$) over cement treated sub-base of of thickness in cm*		
	15	22.5	30	10	15	20
2.8	3.9	4.4	5.3	7.6	10.8	14.1
5.6	6.3	7.5	8.8	12.7	17.3	22.5
8.4	9.2	10.2	11.9	-	-	-

Table 21.14 k-values over dry lean concrete sub-base

k-value of Subgrade $kg/cm^2/cm$	2.1	2.8	4.2	4.8	5.5	6.2
Effective k over 100 mm DLC $kg/cm^2/cm$	5.6	9.7	16.6	20.8	27.8	38.9
Effective k over 150 mm DLC, $kg/cm^2/cm$	9.7	13.8	20.8	27.7	41.7	-

The maximum value of effective k shall be 38.9 kg/cm^2/cm for 100 mm of DLC and 41.7 kg/cm^2/cm for 150 mm of DLC.

21.12. EQUIVALENCE STANDARD AXLE LOAD (ESAL)

The axle load equivalency factors recommended in the AASHTO guide are given in table 21.16. They are used for converting different axle load repetitions into equivalent standard axle load repetitions.

21.13. VEHICLE DAMAGE FACTOR (VDF)

For designing a new pavement, the VDF should be arrived at carefully by carrying out specific axle load surveys on the existing roads. Some surveys have been carried out in the country on National Highways, State Highways and Major District Roads which reveal excessive overloading of commercial vehicles. Therefore, it is recommended that the designer should take the realistic values of VDF after conducting the axle load survey, particularly in the case of major projects. On some sections, there may be significant difference in axle loading in two directions of traffic. In such situations, the VDF should be evaluated direction wise to determine the lanes which are heavily loaded for the purpose of design.

Where sufficient information on axle loads is not available and the project size does not warrant conducting an axle load survey, the indicative values of vehicle damage factor as given in Table 21.15 may be used.

Table 21.15 Indicative VDF Values

Initial traffic volume in terms of number of commercial vehicles per day	*Terrain*	
	Rolling/Plain	*Hilly*
0-150	1.5	0.5
150-1500	3.5	1.5
More than 1500	4.5	2.5

21.14. DISTRIBUTION OF COMMERCIAL TRAFFIC OVER THE CARRIAGEWAY

A realistic assessment of distribution of traffic by direction and by lane is necessary as it affects the total equivalent standard axle load used in the design. In the absence of adequate and reliable data for Indian conditions, it is recommended that for the time being the following distribution may be assumed until more reliable data on placement of commercial vehicles on the carriageway lanes are available.

21.14.1. Single-lane roads

Traffic tends to be more channelised on single-lane roads than two-lane roads and to allow for this concentration of wheel load repetitions the design should be based on total number of commercial vehicles in both directions.

21.14.2. Two-lane single carriageway roads

The design should be based on 75 per cent of the total number of commercial vehicles in both direction.

21.14.3. Four-lane single carriageway roads

The design should be based on 40 per cent of the total number of commercial vehicles in both directions.

21.14.4. Dual carriageway roads

The design of dual two-lane carriageway roads should be based on 75 per cent of the number of commercial vehicles in each direction. For dual three-lane carriageway and dual four-lane carriageway, the distribution factor will be 60 per cent and 45 per cent respectively.

The traffic in each direction may be assumed to be half of the sum in both directions when the latter only is known. However the design should be based on the traffic in the most heavily trafficked lane and the same design will normally be applied for the whole carriageway width.

Table 21.16 Equivalence factors and damaging power of different axle loads

Gross Axle weight kg	*Load Equivalency Factors*	
	Single Axle	*Tandem Axle*
900	0.0002	0.0000
1810	0.002	0.0002
2720	0.009	0.001
3630	0.031	0.003
4540	0.08	0.006
5440	0.176	0.013
6350	0.35	0.024
7260	0.61	0.043
8160	1.00	0.070
9070	1.55	0.110
9980	2.30	0.166
10890	3.27	0.242
11790	4.48	0.342
12700	5.98	0.470
13610	7.8	0.633
14520	10.0	0.834
15420	12.5	1.08
16320	15.5	1.38
17230	19.0	1.73
18140	23.0	2.14
19051	27.7	2.61
19958	33.0	3.16
20865	39.3	3.79
21772	46.5	4.49
22680	55.0	5.28
23587	-	6.17
24494	-	7.15
25401	-	8.20
26308	-	9.4
27216	-	10.7
28123	-	12.1
29030	-	13.7
29937	-	15.4
30844	-	17.2
31752	-	19.2
32660	-	21.3
33566	-	23.6
34473	-	26.1
35380	-	28.8
36288	-	31.7

Note: In case the class mark of the axle load survey does not match with the above axle loads, 4th power law may be used for converting axle loads into equivalent standard axle loads using the following formulae:

Single axle load

Equivalency factor = (axle load in kg/8160)^4

Tandem axle load

Equivalency factor = (axle load in kg/14968)^4

The above equations also give reasonably correct results for practical values of axle loads.

21.15. DESIGN FLEXURAL STRENGTH OF CONCRETE

Though the 28-day flexural strength of concrete is taken for design, concrete gains strength with age. 90 days strength is taken for desigs purposes; which equals 1.2 times the 28 days strength. Further warping caused by temperature gradient is nullified to some degree by the moisture gradient. Thus the above design procedure is likely to result in a much higher life of the pavement than design life.

The critical stress condition are:

(*i*) Flexural stress at the edge due to wheel load + warping stress due to temperature variation between the top and bottom of the slab during the summer period.

(*ii*) Corner stress, if dowel bars are not provided at the transverse joints and there is no possibility of load transfer by aggregate inter lock.

Based on the Westergaard theory modified by Picket and Ray, a computer programme IITRIGID has been developed at I.I.T. Kharagpur to determine the edge stresses for the critical conditions shown in Figs. 21.17 and 21.18 for different slab thicknesses.

However, to facilitate the designer, ready to use charts for the calculation of load stresses in the edge region of rigid pavement slabs for single and tandem axle loads of different magnitude for sub-bases with *k* values in the range of 6, 8, 10, 15 and 30 kg/cm^3 have been prepared; for which reference may be made to IRC-58-2002 publication. These charts are reproduced in Figs. 21.25 to 21.44.

21.15.1. Stress Ratio and Fatigue Analysis

Stress ratio is obtained by dividing the estimated edge stress by the design flexural strength of concrete. If the stress ratio is less than 0.45, the allowable number of repetitions of the axle load are infinite. Cumulative fatigue damage is determined for different axle loads. This value should be less than one for the slab thickness to be safe. Table 21.17 gives the allowable repetitions of load for different stress ratios.

Table 21.17 Stress ratio and allowable repetitions in cement concrete

Stress Ratio	*Allowable*	*Stress Ratio*	*Allowable Repetitions*
0.45	6.279×10^7	0.66	5.83×10^3
0.46	1.4335×10^7	0.67	4.41×10^3
0.47	5.2×10^6	0.68	3.34×10^3
0.48	2.4×10^6	0.69	2531
0.49	1.287×10^6	0.70	1970
0.50	7.62×10^5	0.71	1451
0.51	4.85×10^5	0.72	1099
0.52	3.26×10^5	0.73	832
0.53	2.29×10^5	0.74	630
0.54	1.66×10^5	0.75	477

0.55	1.24×10^5	0.76	361
0.56	9.41×10^4	0.77	274
0.57	7.12×10^4	0.78	207
0.58	5.4×10^4	0.79	157
0.59	4.08×10^4	0.80	119
0.60	3.09×10^4	0.81	90
0.61	2.34×10^4	0.82	68
0.62	1.77×10^4	0.83	52
0.63	1.34×10^4	0.84	39
0.64	1.02×10^4	0.85	30
0.65	7.7×10^3		

Problem 21.9 *Design a cement concrete pavement for a two lane two-way National Highway in Maharashtra; given;*

(*i*) *Traffic in both directions at the end of the construction period = 2000 commercial vehicles per day (CVPD)*

(*ii*) *Flexural strength of cement concrete = 45 kg/cm²*

(*iii*) *Effective modulus of subgrade reaction of the Dry Lean Concrete (DLC) subbase = 8 kg/cm³*

(*iv*) *Elastic modulus of concrete, E = 3 × 10^5 kg/cm²*

(*v*) *Poisson's ratio = 0.15*

(*vi*) *Coefficient of thermal coefficient of concrete = 10 × 10^{-6}/°C*

(*vii*) *Tyre pressure = 8 kg/cm²*

(*viii*) *Rate of traffic increase = 0.075*

(*ix*) *Spacing of contraction joint = 4.5 m*

(*x*) *Width of slab = 3.75 m*

(*xi*) *Design life = 20 years*

The results of the axle load survey are given below:

Single axle load		*Tandem axle load*	
Axle load class tonnes	*Percentage of axle loads*	*Axle load class tonnes*	*Percentage of axle loads*
19-21	0.7	34-38	0.3
17-19	1.8	30-34	0.3
15-17	4.6	26-30	0.5
13-15	10.8	22-26	1.8
11-13	21.1	18-22	1.5
9-11	23.0	14-18	0.8
Less than 9	30.0	Less than 14	2.8
Total	92.0	Total	8.0

Solution.

Present traffic = 2000 cvpd; Design life = 20 years; r = 0.075

∴ Cumulative repetition in 20 years $= 2000 \times 365 \left[\dfrac{(1.075)^{20} - 1}{0.075}\right]$

$= 31,612,417$ commercial vehicles

Design traffic = 25% of cumulative repetition of Commercial Vehicles
= 0.25 × 31612417 = 7,903,104

Since front axles of the commercial vehicles carry much lower loads and cause small flexural stress in the concrete pavements, they are not considered in the pavement design. Only the rear axles, both single and tandem, are considered in the design. The total number of rear axles is 7,903,104. Assuming that mid point of the axle load class represents the group, the total repetitions of the single axle and tandem axle are as follows:

Single axle		*Tandem axle*	
Load in Tonnes	*Expected repetitions*	*Load in Tonnes*	*Expected repetitions*
20	55322	36	23709
18	142256	32	23709
16	363543	28	39516
14	853535	24	142256
12	1667555	20	118547
10	1817714	16	63224
Less than 10	2370931	Less than 16	221287

Trial thickness = 32 cm; Subgrade modulus = 8 kg/cm^3

Design period = 20 years; Load safety factor = 1.2

Axle load AL Tonnes	*AL x 1.2*	*Stress kg/cm²*	*Stress ratio*	*Expected repetitions*	*Fatigue life*	*Fatigue consumed*
(1)	(2)	(3)	(4)	(5)	(6)	(5/6)
Single Axle						
20	24	25.19	0.56	55322	9.41×10^4	0.588
18	21.6	22.98	0.51	142256	4.85×10^5	0.293
16	19.2	20.73	0.46	363543	14.33×10^6	0.025
14	16.8	18.45	0.41	853535	Infinite	0.000
Tandem axle						
36	43.2	20.07	0.45	23709	62.8×10^6	0.0004
32	38.4	18.40	0.40	23709	Infinite	0.000
						Total = 0.9064

Therefore the design is safe since the cumulative fatigue life consumed is less than one.

Check for Temperature Stresses

Radius of relative stiffness

$$l = \sqrt[4]{\frac{Eh^3}{12\,(1-\mu^2)k}} = \sqrt[4]{\frac{3\times10^5\times32^3}{12\,(1-0.15^2)\times8}} = \sqrt[4]{\frac{3\times10^5\times32^3}{12\times0.9775\times8}}$$

$$= 101.168 \text{ cm}$$

$$\frac{L}{l} = \frac{4.5}{1.01168} = 4.448$$

∴ From table 21.3; For $\frac{L}{l} = 4.448$, the value of $C = 0.565$

Edge warping stress $= \frac{CE\alpha t}{2}$

The temperature differential for Maharashtra region may be taken as 21°C (see table 19.9 page 647).

$$\therefore \quad \sigma_{t_e} = \frac{0.565 \times 3 \times 10^5 \times 10 \times 10^{-6} \times 21}{2}$$

$$= 17.8 \text{ kg/cm}^3$$

Total of temperature warping stress and the highest axle load stress $= 17.8 + 25.19 = 42.99$ kg/cm^2

This is less than the flexural strength of concrete, 45 kg/cm^2.

So the pavement thickness of 32 cm is safe under the combined action of load and temperature.

Check for Corner Stress

This is not critical in a dowelled pavement. Corner stress is given by:

$$\sigma_{l_c} = \frac{3P}{h^2}\left[1 - \left(\frac{a\sqrt{2}}{l}\right)^{1.2}\right]$$

The 98 percentile axle load is 16 tonnes.

∴ Wheel load, P = 8 Tonnes

For a single axle dual wheel, the radius of area of contact of wheel is given by

$$a = \left[0.8521 \times \frac{P}{p\pi} + \frac{S}{\pi}\left(\frac{P}{0.5227\,p}\right)^{0.5}\right]^{0.5}$$

where P = Load = 8 tonnes

p = Tyre pressure = 8 kg/cm^2

S = C/C distance between two tyres = 31 cm

$$a = \left[0.8521 \times \frac{8000}{8\pi} + \frac{31}{\pi}\left(\frac{8000}{0.5227} \times \frac{1}{8}\right)^{0.5}\right]^{0.5}$$

$$= [271.23 + 431.60]^{0.5}$$

$$= 26.51 \text{ cm}$$

$$\therefore \quad \text{Corner stress} = \frac{3 \times 8000}{32^2}\left[1 - \left(\frac{26.51 \times \sqrt{2}}{101.1}\right)^{1.2}\right]$$

$$= \frac{3 \times 8000}{32 \times 32}[1 - 0.304]$$

$$= 16.31 \text{ kg/cm}^2$$

The corner stress is less than the flexural strength of concrete i.e. 45 kg/cm^2.

Hence the assumed thickness of slab of 32 cm is safe.

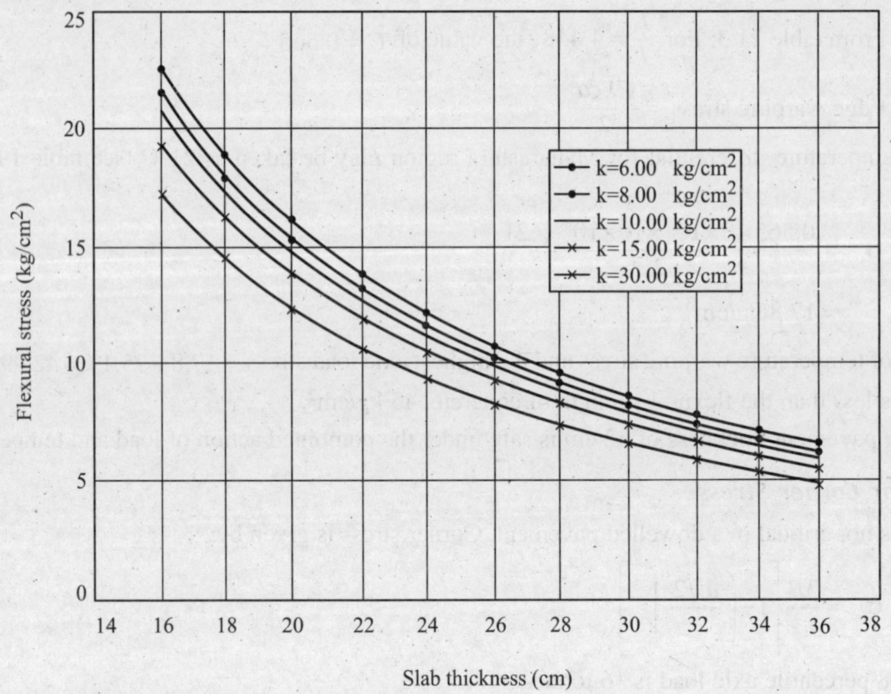

Fig. 21.25 Stresses in rigid pavement (single axle load = 6 tonnes).

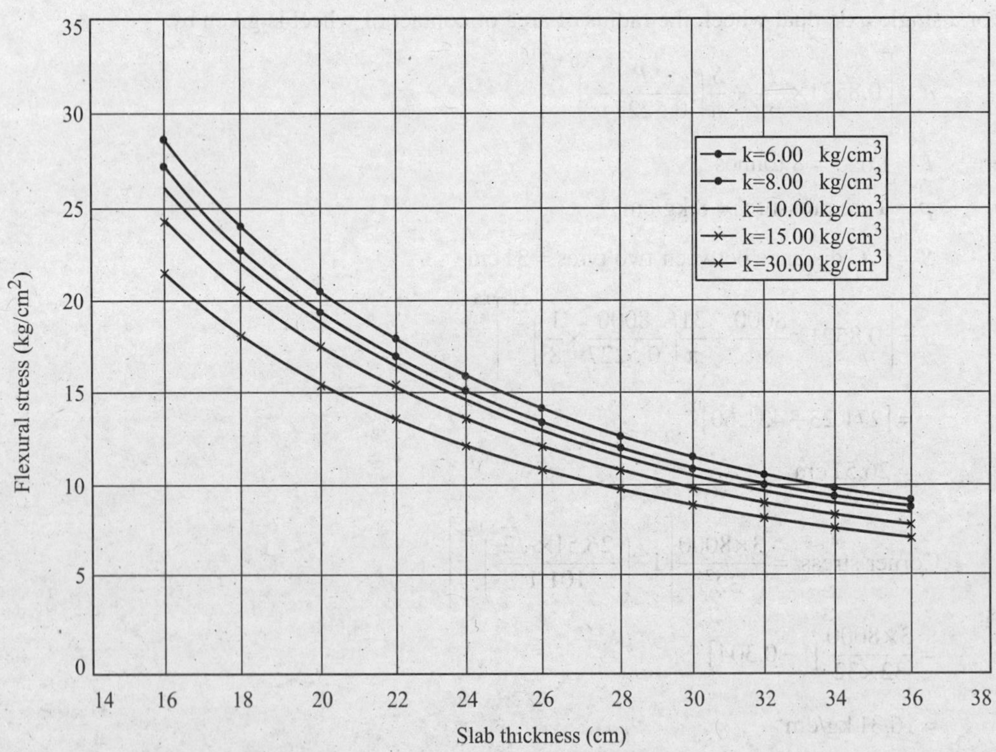

Fig. 21.26 Stresses in rigid pavement (single axle load = 8 tonnes).

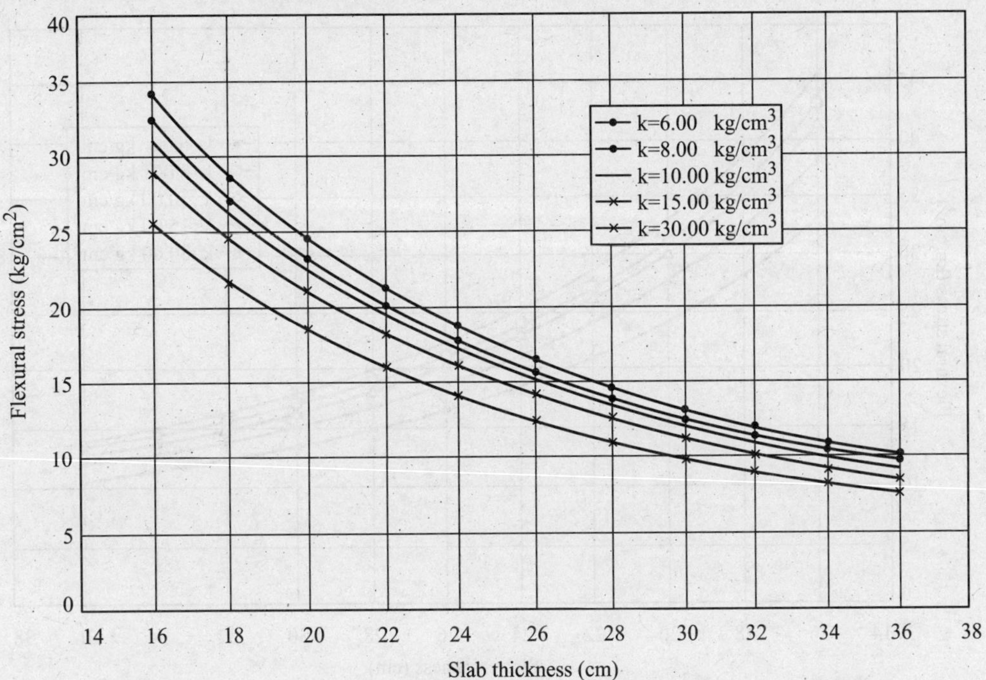

Fig. 21.27 Stresses in rigid pavement (single axle load = 10 tonnes).

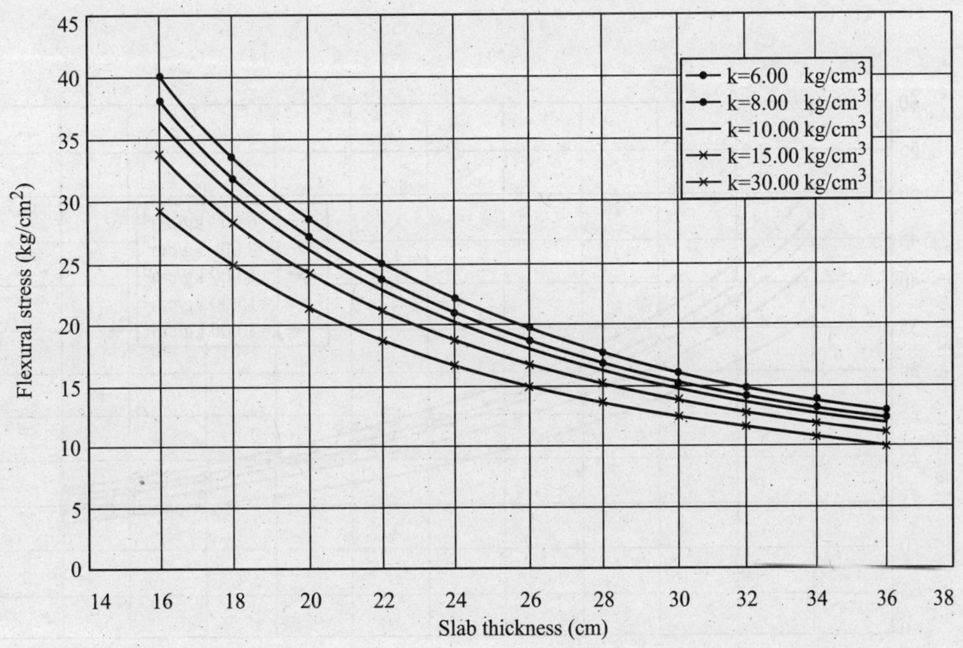

Fig. 21.28 Stresses in rigid pavement (single axle load = 12 tonnes).

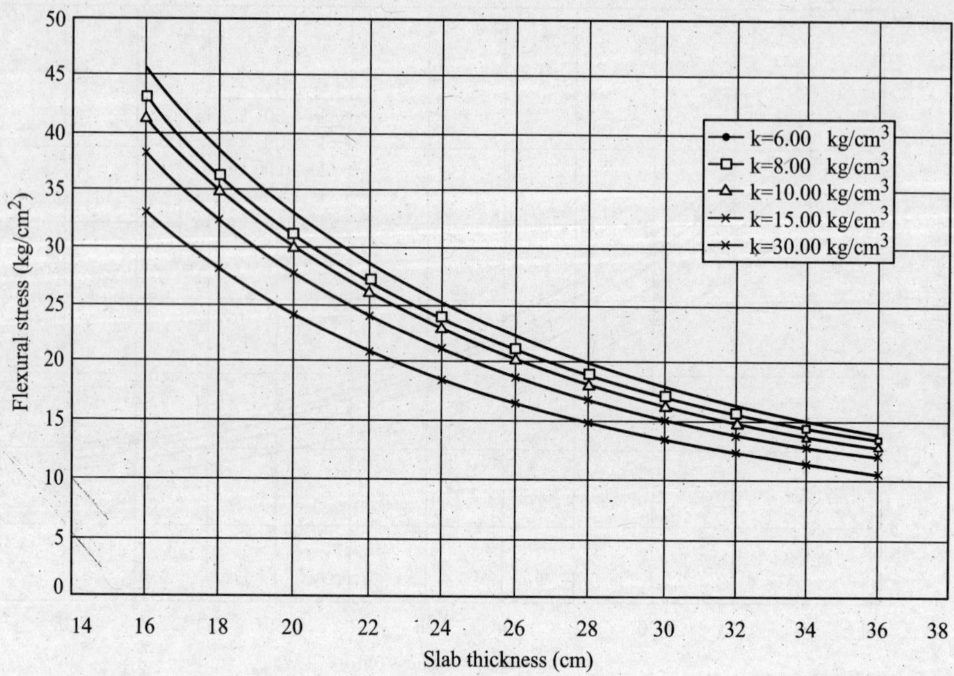

Fig. 21.29 Stresses in rigid pavement (single axle load = 14 tonnes).

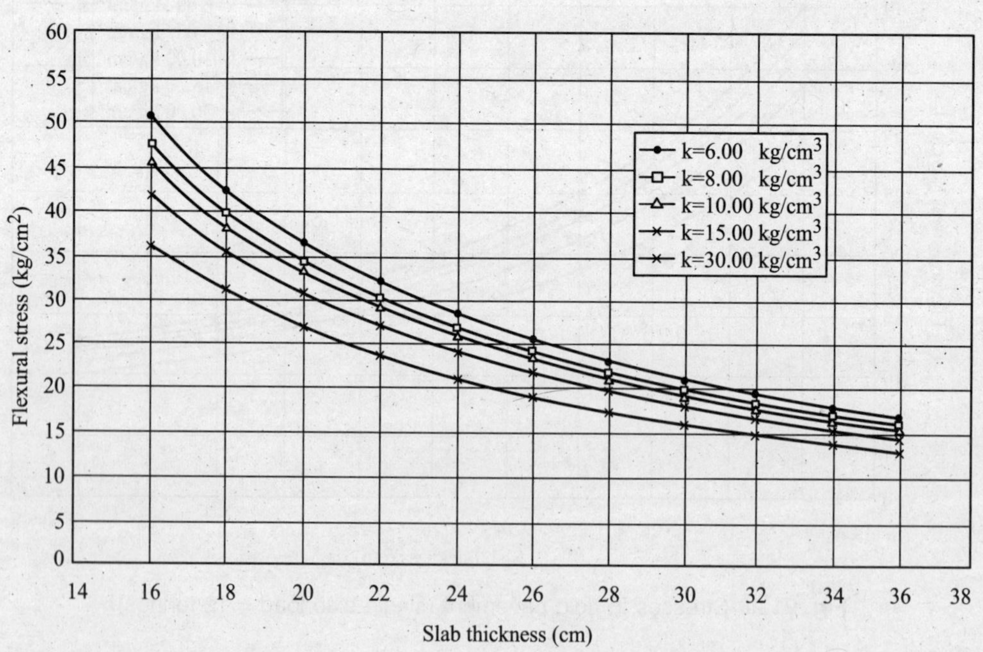

Fig. 21.30 Stresses in rigid pavement (single axle load = 16 tonnes).

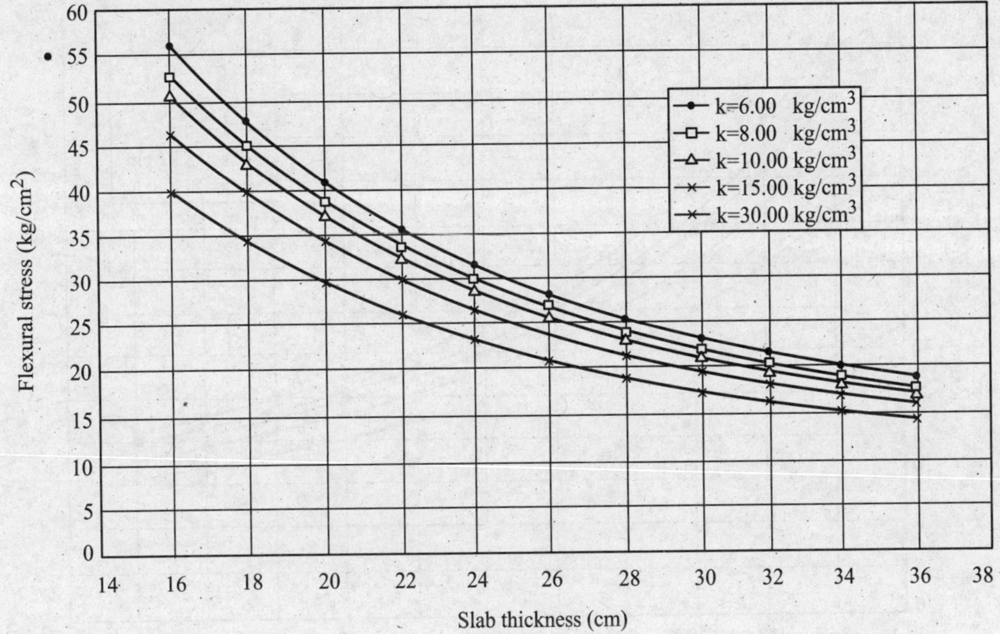

Fig. 21.31 Stresses in rigid pavement (single axle load = 18 tonnes).

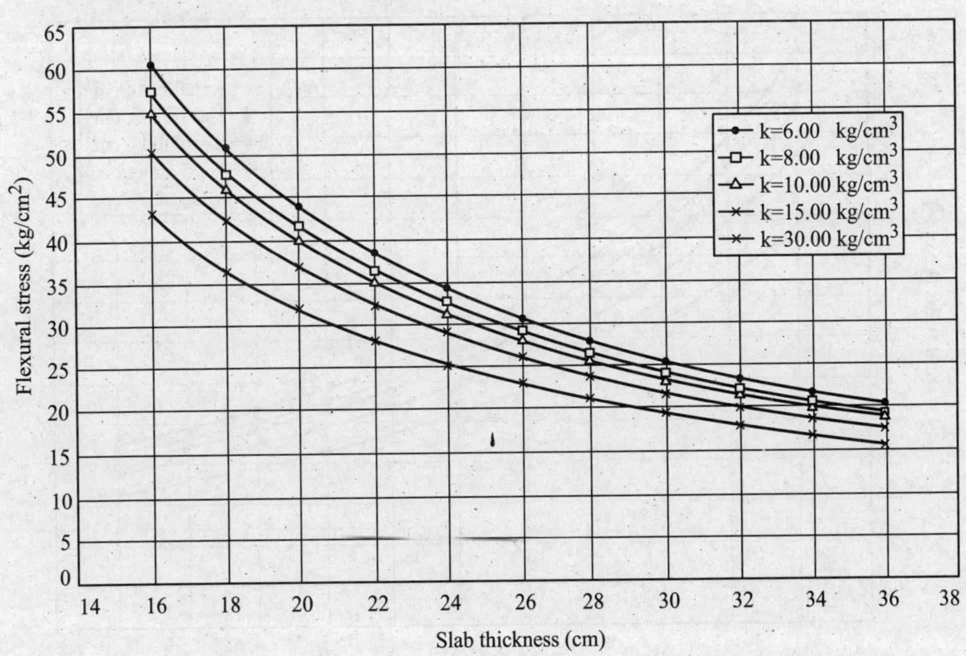

Fig. 21.32 Stresses in rigid pavement (single axle load = 20 tons).

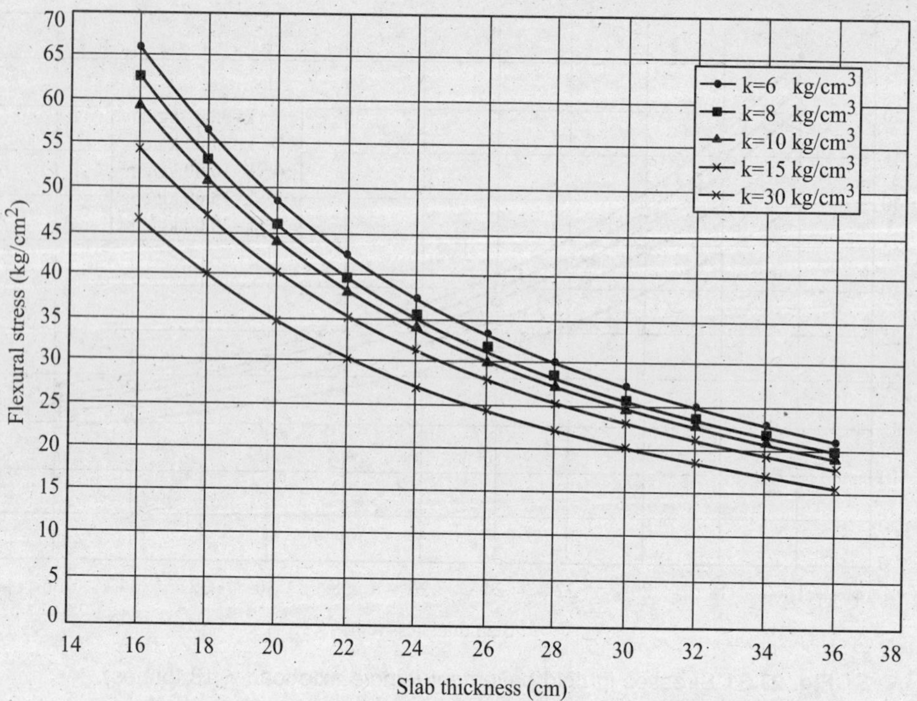

Fig. 21.33 Stresses in rigid pavement (Single axle load = 22 tonnes).

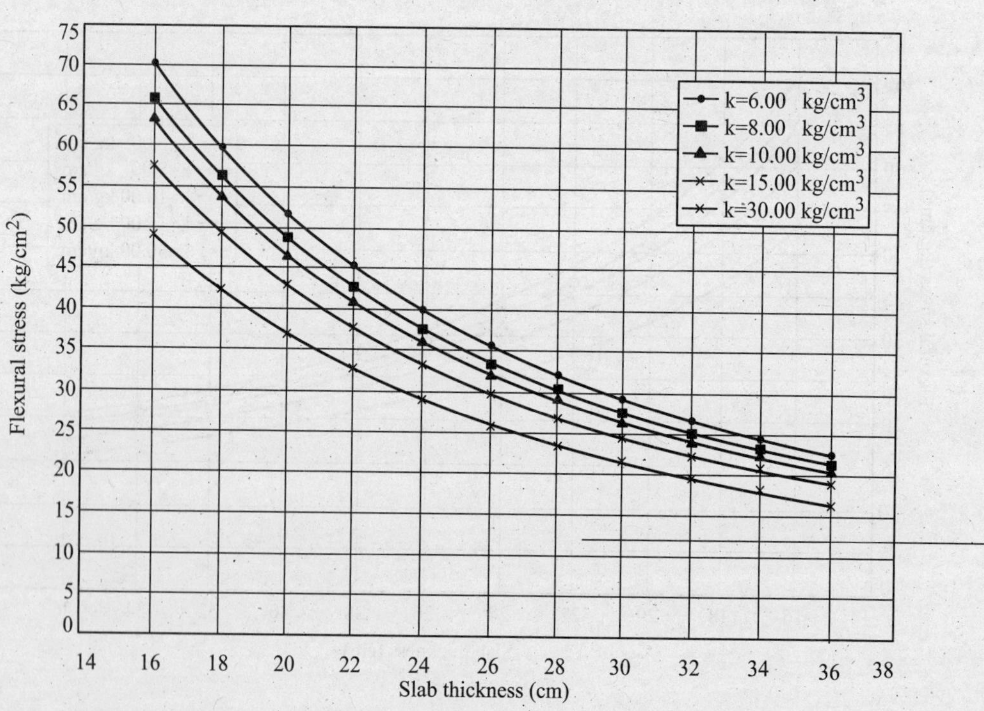

Fig. 21.34 Stresses in rigid pavement (Single axle load = 24 tonnes).

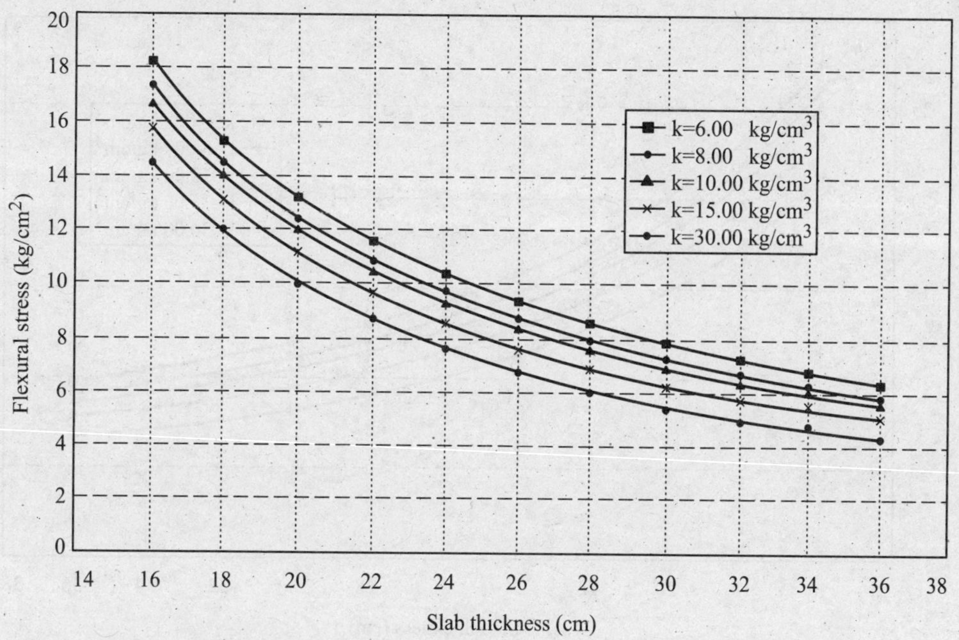

Fig. 21.35 Stresses in rigid pavement (Tandem axle load = 12 tonnes).

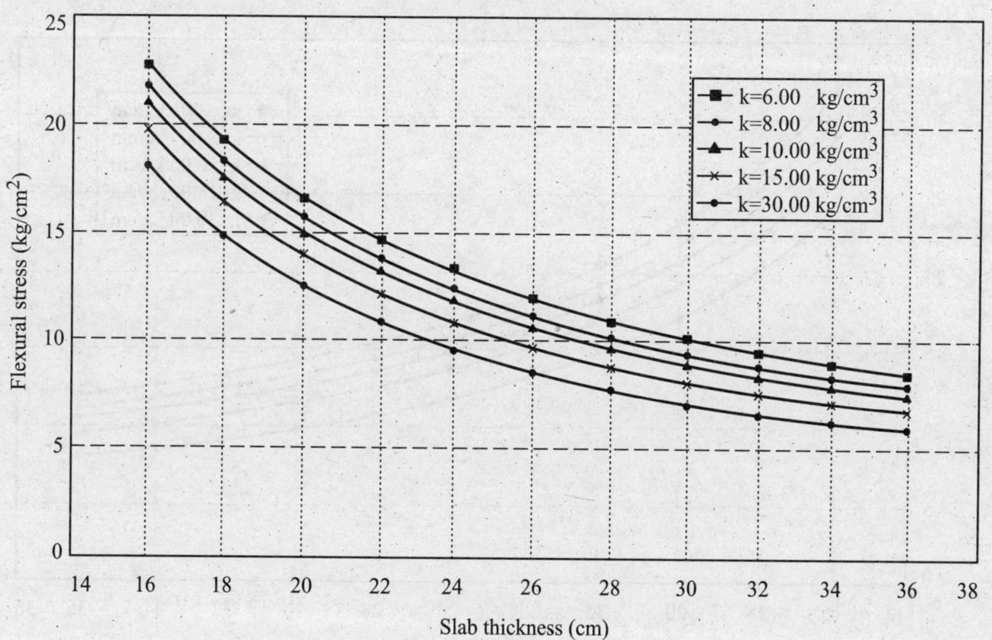

Fig. 21.36 Stresses in rigid pavement (Tandem axle load = 16 tonnes).

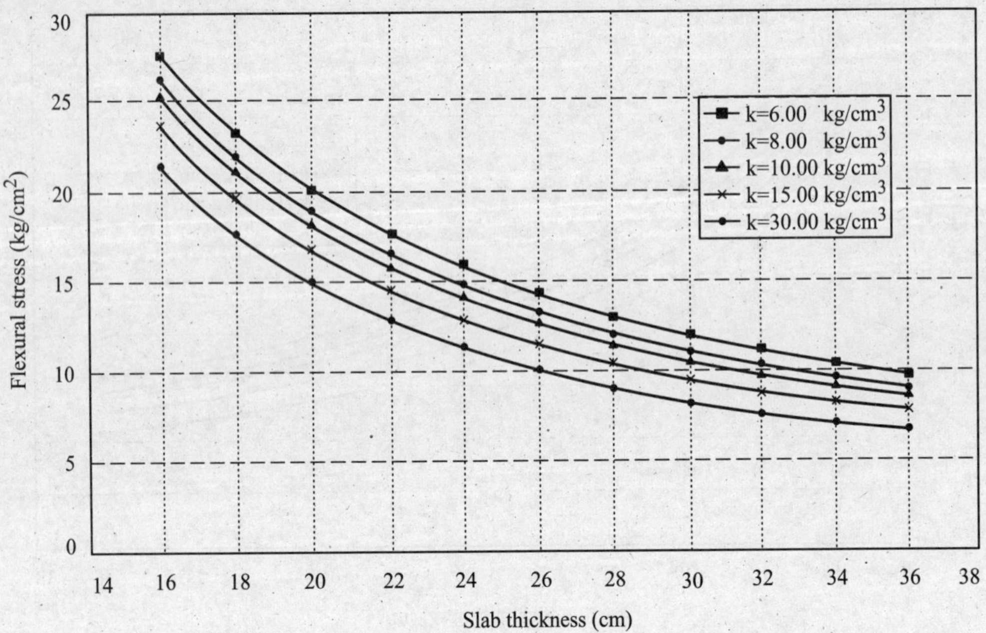

Fig. 21.37 Stresses in rigid pavement (Tandem axle load = 20 tonnes).

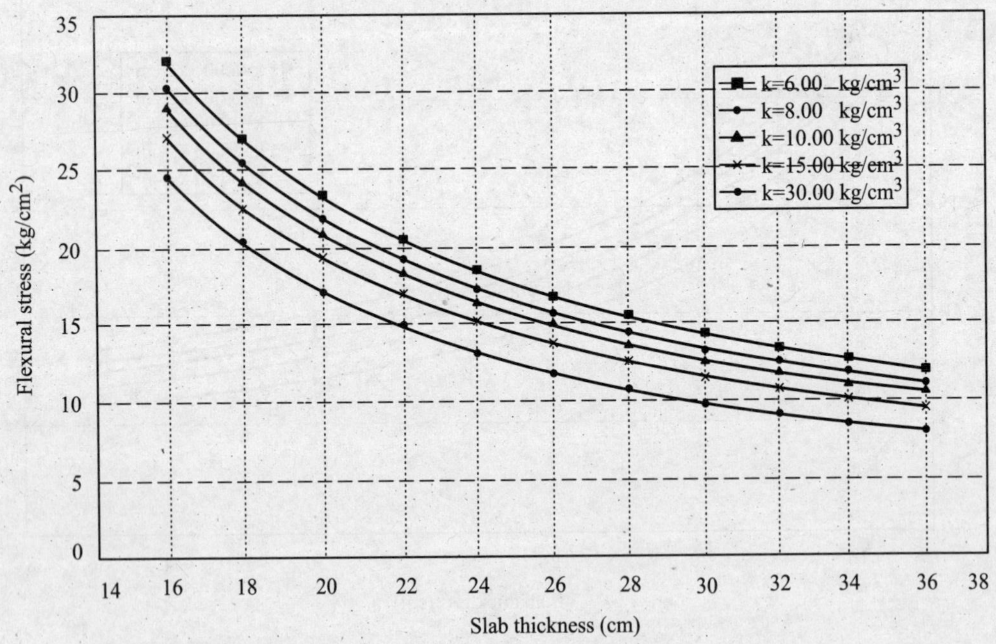

Fig. 21.38 Stresses in rigid pavement (Tandem axle load = 24 tonnes).

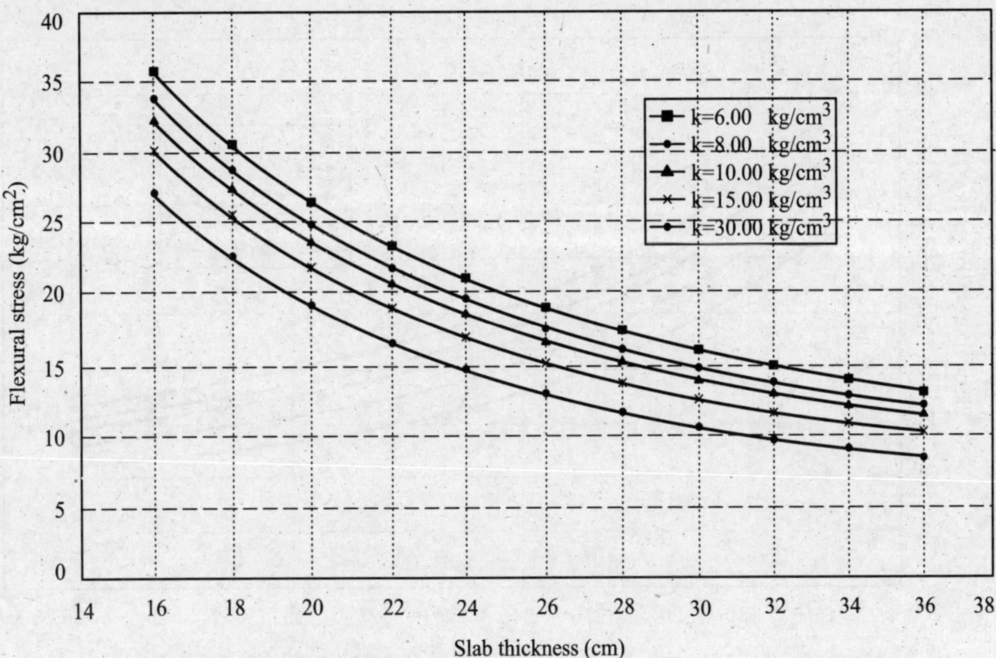

Fig. 21.39 Stresses in rigid pavement (Tandem axle load = 28 tonnes).

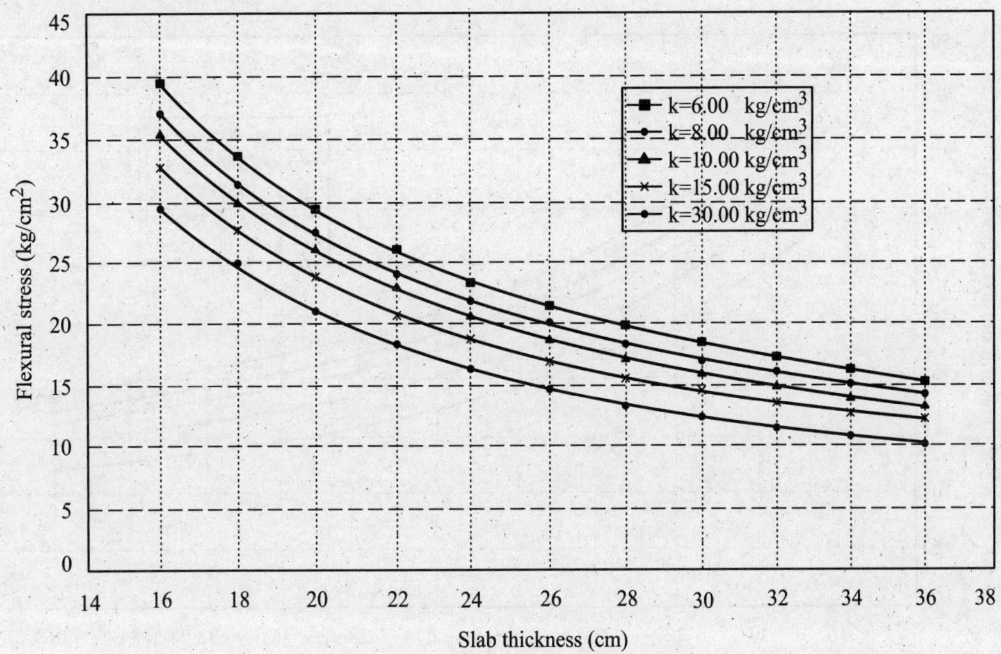

Fig. 21.40 Stresses in rigid pavement (Tandem axle load = 32 tonnes).

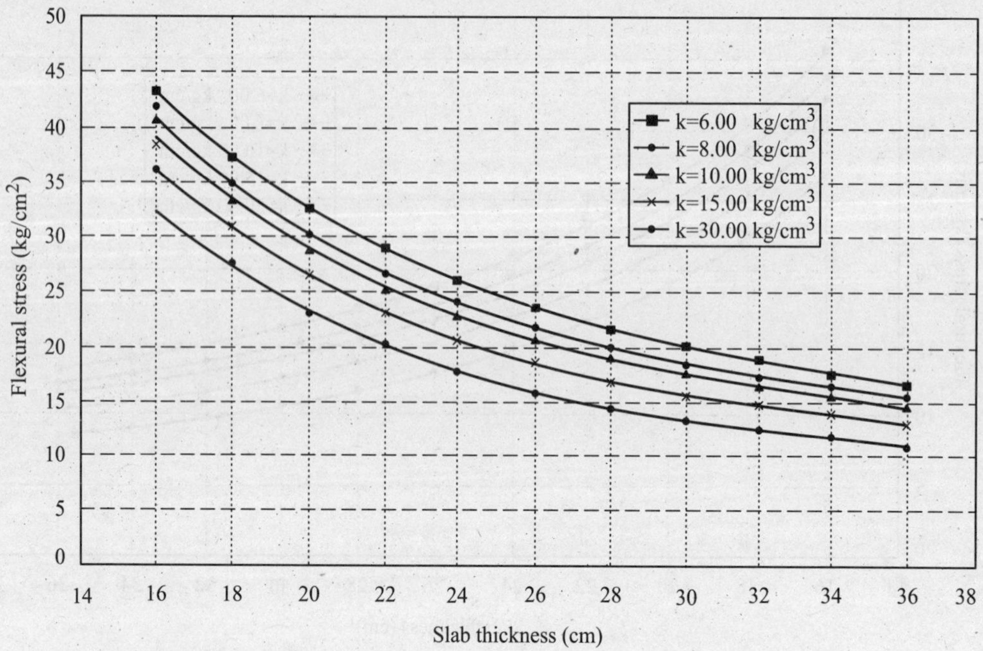

Fig. 21.41 Stresses in rigid pavement (Tandem axle load = 36 tonnes).

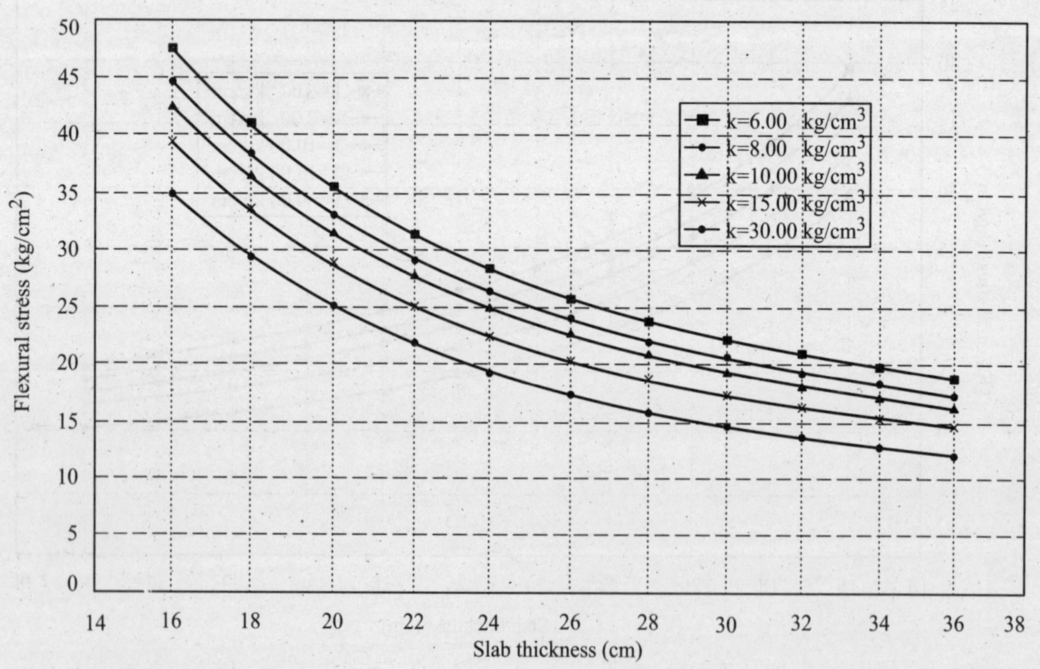

Fig. 21.42 Stresses in rigid pavement (Tandem axle load = 40 tonnes).

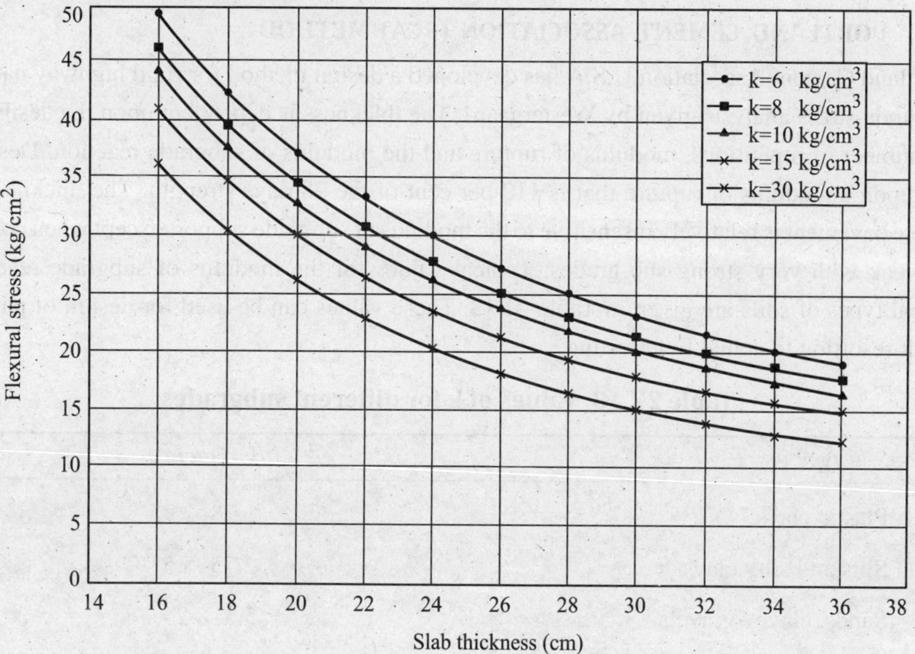

Fig. 21.43 Stresses in rigid pavement (Tandem axle load = 42 tonnes).

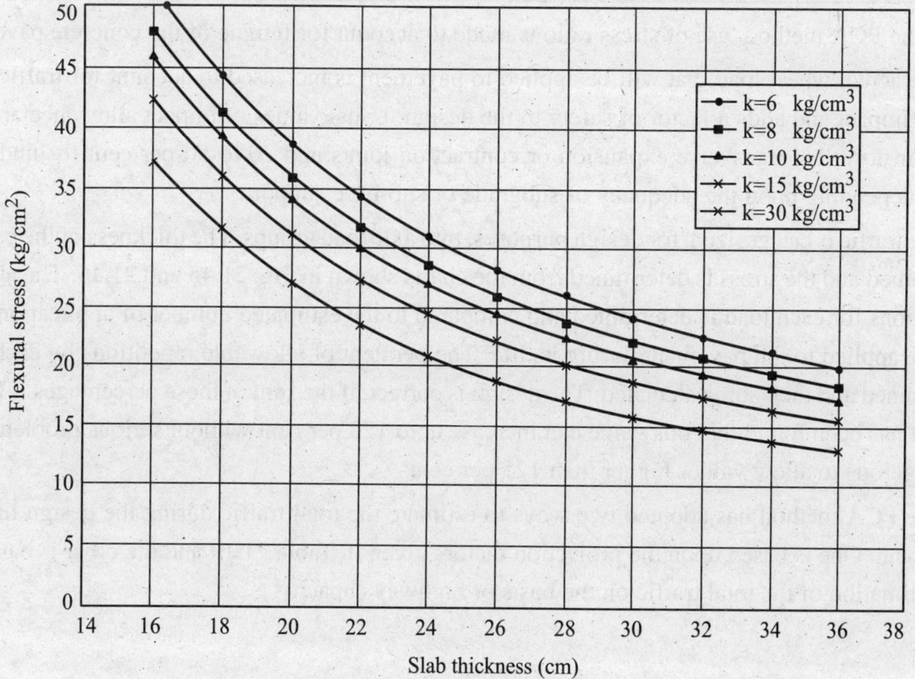

Fig. 21.44 Stresses in rigid pavement (Tandem axle load = 36 tonnes).

21.16. PORTLAND CEMENT ASSOCIATION (PCA) METHOD

Portland Cement Association U.S.A has developed a design method for rigid highway pavements based upon stress analysis given by Westergaard. The thickness is dependent upon the design wheel load, number of repetitions, modulus of rupture and the modulus of subgrade reaction. Designs are based upon a modulus of rupture that is 110 per cent of the 28 days strength. The thickness of the concrete pavement is relatively insensitive to the modulus of subgrade support except when comparing very weak with very strong sub-grades. Typical values for the modulus of subgrade reaction for different types of soils are given in Table 21.18. These values can be used for design of pavements without resorting to actual field testing.

Table 22.18 Values of k for different subgrades

Soil type	*k (kg/cm³)*
Plastic clays	1.38–2.75
Silts and silty clays	2.75–5.5
Sands, clayey gravels	5.5–8.25
Gravels	≥ 8.25
Cement treated bases or asphalt treated bases	≥ 11.00

In the PCA method, use of stress ratio is made to account for fatigue of the concrete pavement.

The actual wheel load that will be applied to pavement is increased to account for traffic growth due to impact and adds a factor of safety to the design. Usual values of impact allowance are 20 per cent for dowelled transverse expansion or contraction joints and 20 to 50 per cent for undowelled joints depending upon the adequacy of subgrade or sub-base support.

The traffic is categorized, for design purposes, into axle load groups. The thickness of the pavement is assumed and the stress is determined from the charts shown in Fig. 21.45 and 21.46. The allowable repetitions for each load is determined and compared to the estimated number of applications which will be applied to the pavement during its life. The per cent of allowable repetitions by each load is determined and their sum calculated. The design is correct, if the sum of these percentages is less than 100. It has been found that this value can increase upto 125 per cent without serious problems, but it is dangerous to allow values higher than 125 per cent.

The PCA method has adopted two ways to estimate the total traffic during the design life of the pavement. One is based upon the projection factors given in Table 21.19 and the other is based upon the estimation of the total traffic on the basis of highway capacity.

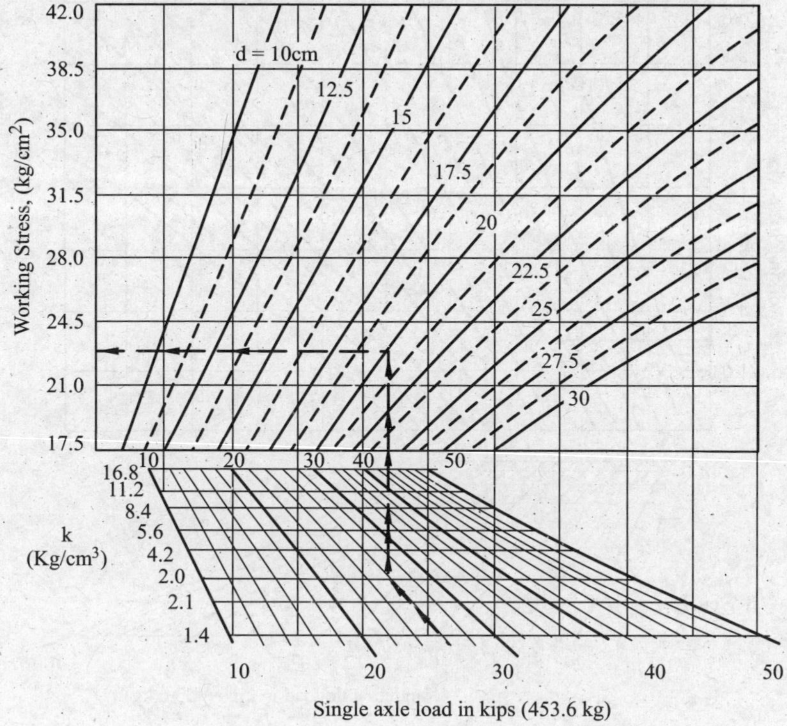

Fig. 21.45 Design chart for single axle truck loads.

Table 21.19 Projection factors for different rates of traffic growth

Yearly rate of traffic growth (per cent)	*Projection factor for 20 years*	*Weighted average projection factor for 40 years*
1.0	1.2	1.2
1.5	1.3	1.3
2.0	1.5	1.5
2.5	1.6	1.7
3.0	1.8	1.9
3.5	2.0	2.2
4.0	2.2	2.5
4.5	2.4	2.8
5.0	2.7	3.2
5.5	2.9	3.6
6.0	3.2	4.1

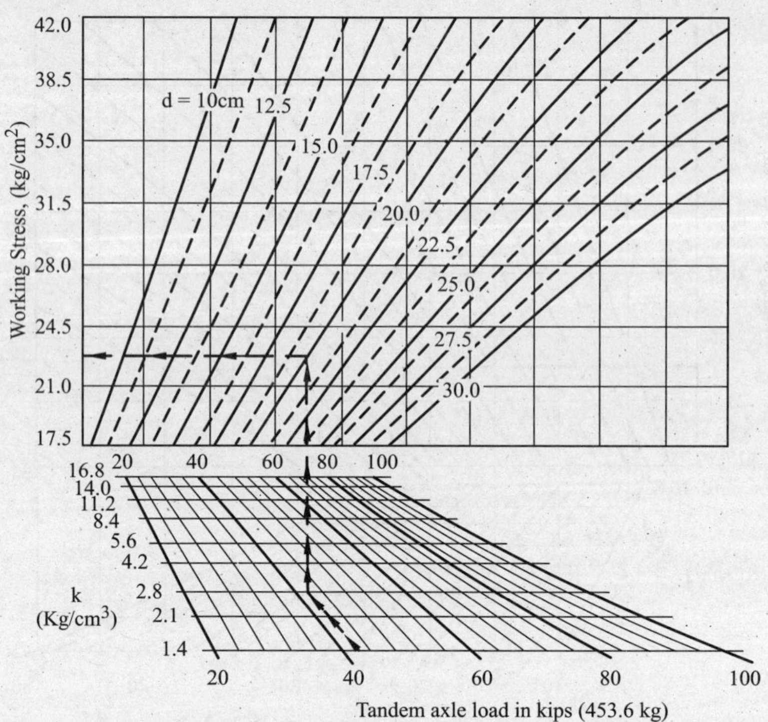

Fig. 21.46 Design chart for tandem axle truck loads.

Problem 21.10 *Design a concrete pavement for the following conditions:*

Traffic	:	*ADT two directions 300 vehicles/day*
Trucks	:	*20 per cent of ADT*
Annual growth	:	*3.5 per cent*

Trucks axle weight distribution is given below:

Axle load (10^3 kg)	*No. axles per 1000 trucks*
7–8 S*	120.9
8–9 S	110.8
9–10 S	60.3
10–11 S	15.6
11–12 T†	1.3
13–14 T	0.9
14–15 T	70.3
15–16 T	34.4
16–17 T	24.0
17–18 T	17.2
18–19 T	16.8
19–20 T	7.5

*S stands for single axle

†T stands for tandem axle

Plate bearing tests have indicated k to be 4.2 kg/cm^3, corrected for saturation. The modulus of rupture of concrete is 45.5 kg/cm^2.

Solution. The PCA method uses two approaches to estimate the total traffic during the design life of the pavement viz:

(1) Utilizing projection factors given in Table 21.19.

(2) Estimation of the total traffic on the basis of highway capacity.

The first method is adopted in this problem.

Thus the total trucks to be considered for a design life of 20 years for annual growth of 3.5 per cent is:

$$\frac{(20)\,(300)\,(2)}{100 \times 2} = 60 \text{ trucks/day in one direction}$$

The design is carried out in the table below:

Modulus of rupture = 45.5 kg/cm^2

Axle load (tonnes) (Average)	*Approximate repetition in 20 years (No. axles per 1000 trucks × 0.06 × 365 × 20)*	*Axle load × 1.2 (tonnes)*	*t = 17.5 cm (assumed)*		*R = 4.2 kg/cm^3*	
			Stress (kg/cm^2)	*Stress ratio*	*Allowable repetition*	*Percent used*
19.5	3285	23.4. T	28.5	0.63	14,000	23.46
18.5	7358	22.2 T	26.5	0.58	57,000	12.90
17.5	7534	21.0 T	26.25	0.58	57,000	13.22
16.5	10512	19.8 T	26.0	0.57	75,000	14.02
15.5	15067	18.6 T	24.5	0.54	180,000	8.37
14.5	30791	17.4 T	22.75	0.50	Unlimited	—
13.5	394	16.2 T	21.75	0.48	Unlimited	—
11.5	569	13.8 T	18.5	0.41	Unlimited	—
10.5	6833	12.6 S	26.5	0.58	57,000	11.99
9.5	26411	11.4 S	24.0	0.53	240,000	11.00
8.5	48530	10.2 S	21.5	0.47	Unlimited	—
7.5	52954	9.0 S	19.5	0.43	Unlimited	—
			Total fatigue used		=	94.96

Since the total fatigue used is 94.96 percent, (< 100), hence the design is O.K. Thus a slab thickness of 17.5 cm is adequate. This is because the traffic is low *i.e.,* 300 vehicles/day in both directions. For heavy traffic, a thickness of 20 to 25 cm may be required.

21.17. AASHO INTERIM GUIDE METHOD

This method was developed subsequent to the AASHO road test. Based on studies of the AASHO test, an equation (similar to that developed by Westergaard) was developed to determine stresses for a corner load. Poisson's ratio was taken as 0.2 and the distance from the corner to the centre of the load was assumed to be 25 cm.

The solution of the AASHO equations are shown in Fig. 21.47 and 21.48. These are for terminal serviceabilities of 2.0 (secondary roads) and 2.5 (primary roads). Acceptable levels for the terminal or final serviceability for important roads may be taken between 2.2 to 2.5. For design of less important roads, a terminal serviceability of 1.5 is considered to be satisfactory.

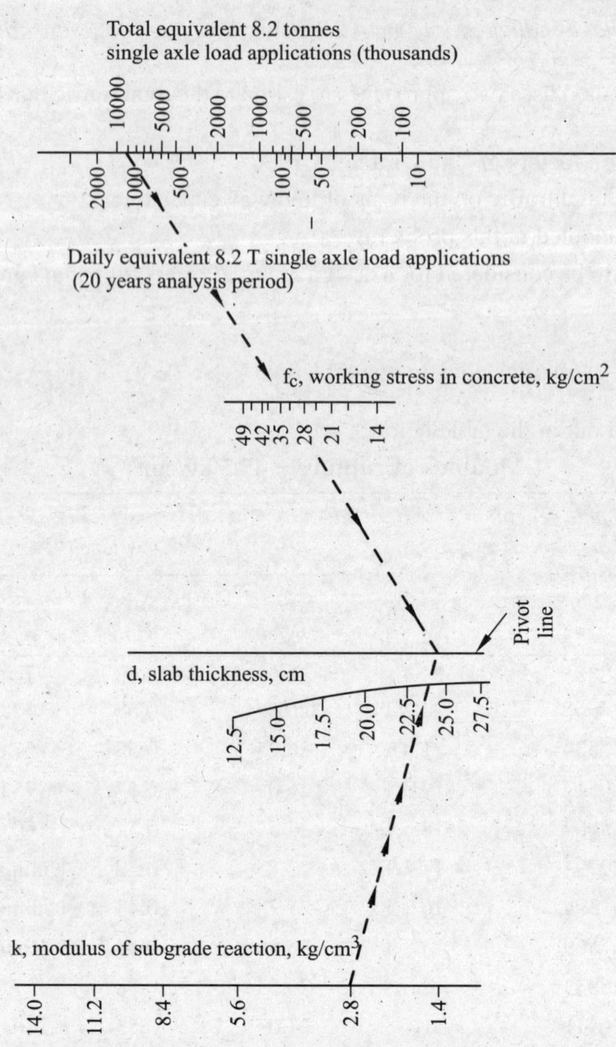

Fig. 21.47 Design chart for rigid pavements (p_t = 2.0).

Allowable stress in concrete is taken as 0.75 time the modulus of rupture *i.e.*, 0.75 MR.

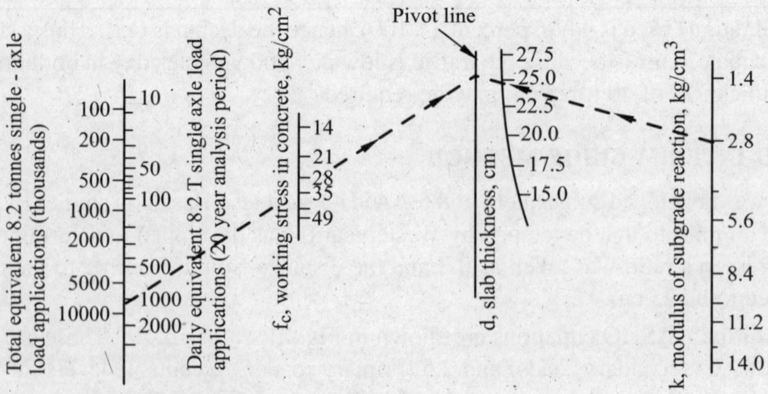

Fig. 21.48 Design chart for road pavements (P_t = 2.5).

It was further assumed that the traffic mix of a given highway can be converted to an equivalent single wheel load. The single wheel load taken in this case was 4100 kg; which is the legal single wheel load limit.

To evaluate the effects of mixed traffic axle loadings, use is made of "Equivalency factors" that were derived from the road test performance equations. The equivalency factor for any given axle load expresses the number of applications of a 8200 kg single-axle load equivalent to one application of the given angle load. 8200 kg single axle equivalency factors are given in Table 21.20.

Mixed-traffic axle loadings can be reduced to the common denominator or basic loading, by grouping the individual axles in the traffic stream into various weight and configuration categories. The sum of the products of the equivalency factors times the corresponding numbers of axle in the various categories gives the total number of equivalent 18-kip (8200 kg) single-axle load applications in the traffic stream.

Table 21.20 Single axle equivalency factors

Single-axle load (kips) or 453.6 kg	*8200 kg single-axle equivalency factor*		*Tandem-axle load (kips) or 453.6 kg*	*8200 kg single-axle equivalency factor*	
	Pt = 2.0	*Pt = 2.5*		*Pt = 2.0*	*Pt = 2.5*
2	0.0002	0.0002	4	0.0005	—
4	0.002	0.002	8	0.005	—
6	0.010	0.010	12	0.030	0.030
8	0.030	0.030	16	0.082	0.085
10	0.082	0.085	20	0.207	0.212
12	0.178	0.183	24	0.443	0.452
14	0.343	0.352	28	0.850	0.850
16	0.603	0.610	32	1.490	1.473
18	1.000	1.000	36	2.467	2.388
20	1.572	1.552	40	3.858	3.673
22	2.363	2.302	44	5.797	5.430
24	3.437	3.300	48	8.412	7.760

1 kip = 453.6 kg.

The above design method takes into consideration most of the design parameters. Powerful computational techniques along with the wealth of experimental data that is being accumulated should advance pavement performance knowledge.

CONTINUOUSLY REINFORCED CONCRETE PAVEMENTS (CRCP)

Continuously reinforced concrete pavements have been used with success on both highways and airports. In this type of pavement no transverse joint is used except at intersections and as construction joint.

Continuously reinforced concrete pavements have the following advantages over other types of pavements:

(1) These have smooth riding, since there are practically no joints.

(2) It has low maintenance cost, if steel reinforcement is properly designed.

The various features of such pavements are discussed below:

21.18. THICKNESS DESIGN

ACI committee on design of concrete pavements have given monograph shown in Fig. 21.49 for the design of continuously reinforced concrete pavements. The monograph is similar to the AASHO Interim guide design charts, with the exception that two additional scales are introduced: scale (B) for the load transfer factor function (J) and scale (E) for the modulus of elasticity of the concrete.

J value of 2.2 is recommended for use on scale B. However this should be evaluated for the given local conditions.

It has been observed that using a reduced thickness to account for the continuous reinforcement has caused some distress to pavements under heavy loads irrespective of the fact whether the pavement is a plain, simply reinforced or continuously reinforced concrete pavement. The steel is not counted for added load transfer, but its primary function is to reduce the number of joints and thereby enhance the riding quality of the pavement.

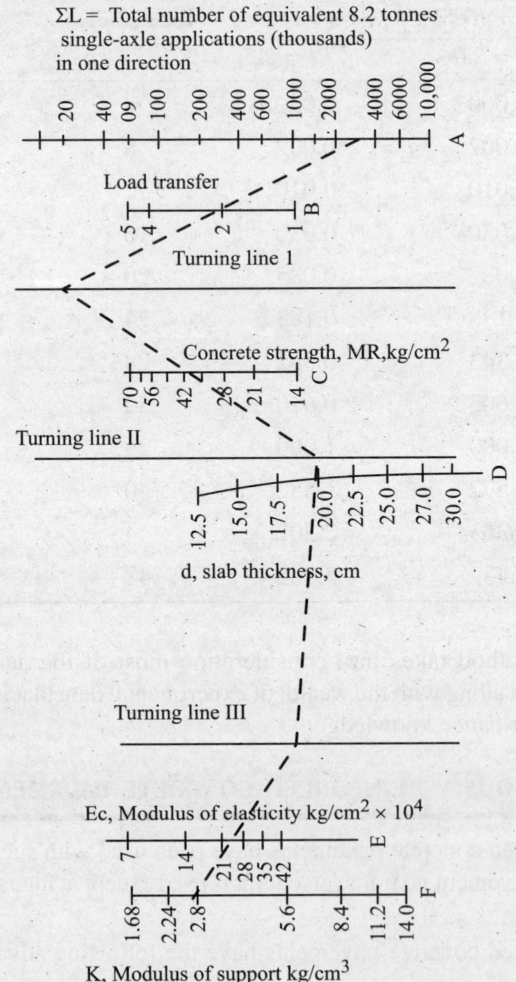

Fig. 21.49 Thickness design method; Continously Reinforced Concrete Pavement (CRCP).

21.19. DESIGN OF STEEL

Taking the above criterion that the thickness of the concrete pavement is the same for all types of pavements, the only factor that remains to be determined is the quantity of steel required. The percentage of steel needed is given by the following relations:

Steel (Per cent)

For shrinkage

$$p = \frac{S'_c}{S_s} \times 100 \qquad \text{...(21.20)}$$

$$p = \frac{S'_c}{S_s + ZE_s - nS'_c} \times 100 \qquad \text{...(21.21)}$$

For temperature drop

$$p = \frac{S'_c}{S_s - nS'_c} \times 100 \qquad \text{...(21.22)}$$

$$p = \left[\frac{S'_c}{2\,(S_s - \Delta t \in E_s)} \right] \times 100 \qquad \text{...(21.23)}$$

where S'_c = ultimate tensile strength of the concrete

S_s = yield strength of steel

Z = coefficient of shrinkage

Δt = coefficient of thermal volume change

$n = \frac{E_s}{E_c}$

$\in$ = coefficient of thermal change

Equations 21.20 and 21.22 give minimum amount of steel required for shrinkage and temperature drop respectively.

Determine the steel required by the above relations and use the largest amount in the design.

Table 21.21 gives typical values of properties to be used in the design of steel for continuous reinforced concrete pavements.

Table 21.21 Typical values of properties for use in design of steel for continuously reinforced concrete pavements

Sr. No.	*Property*	*Value*
1.	Tensile strength of concrete, S'_c	0.4 to 0.5 MR
2.	Coefficient of shrinkage, Z	0.0005 to 0.0006
3.	Coefficient of thermal change, $\in$	0.000005 cm/cm per °F
4.	Yield strength of steel, S_s	4200 kg/cm^2
5.	Safety factor applied to p	1.2 to 1.5
6.	(Larger value used in cold climates)	

Further the following requirements should be met for satisfactory performance of CRCP.

Minimum spacing of longitudinal steel: 10 cm

Maximum spacing of longitudinal steel: 22.5 cm

Minimum cover for steel : 6 cm

Minimum bond area : 0.01 cm²/cm³ of concrete, minimum 48 cm or 30 diameter

All splices (lap) at least

All splices should be staggered.

21.20. TERMINAL AND EXPANSION PROVISIONS

As continuous reinforced concrete pavements have no joints, other than construction, it is necessary to provide expansion space at the end of long slabs and at bridge approaches or wherever one pavement abuts into another. Either the pavement is anchored into the ground (Fig. 21.50) or by providing a space for concrete to expand, as shown in Figs. 21.50 and 21.51. In the first case, an anchor is provided at the end of the concrete pavement and a series of expansion joints are placed next to the bridge or conventional pavement. Though this has worked satisfactorily, the latter method is preferred.

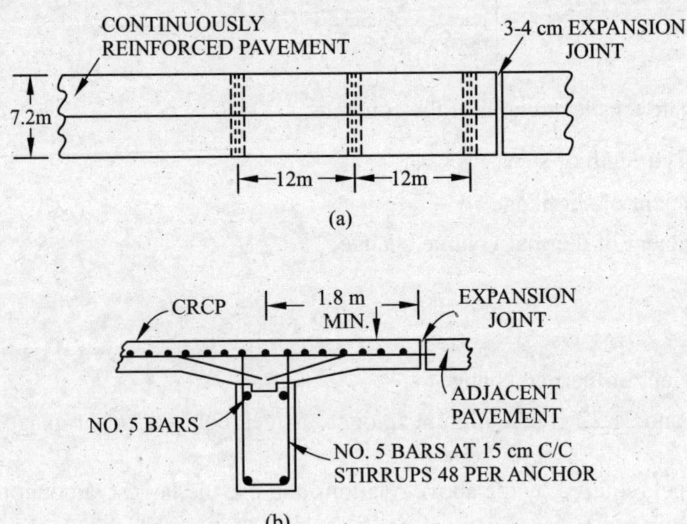

Fig. 21.50 Method of joining to adjacent slab, using lugs.

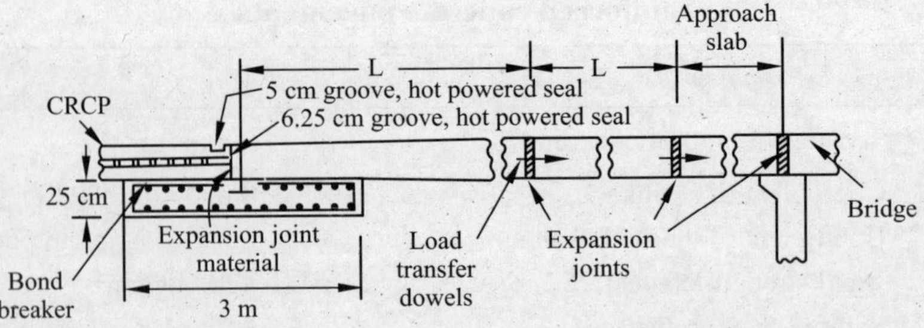

Fig. 21.51 Flanged joint at bridge approach.

A sleeper slab is constructed at the end of CRCP (Fig. 21.51) and a wide flange beam is provided for the slab to slide into and load transfer is then provided by means of the sleeper slab. In Fig. 21.52, a sleeper slab is provided and the space between the CRCP and the conventional pavement is filled with a good grade bituminous material.

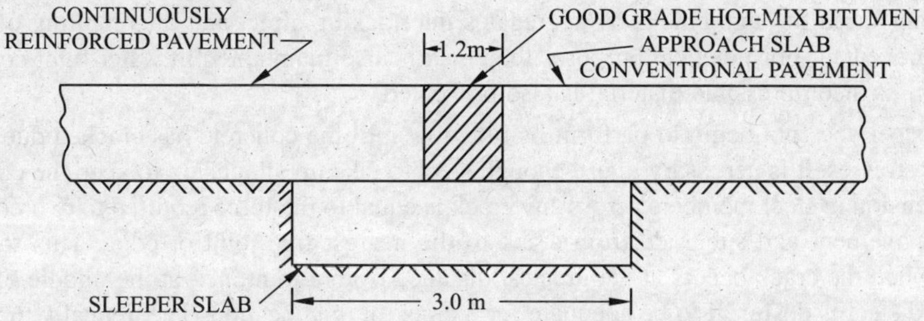

Fig. 21.52 Schematic diagram of use of sleeper slab, without wide flange.

21.21. CONSTRUCTION JOINT

A typical construction joint used in continuous reinforced concrete pavement is shown in Fig. 21.53. Normally a header board is placed at the end of the day's work and the steel is extended through this header board so that it can be continued into the next day's pour. To provide for true continuity from one slab to the other, added steel is provided at the construction joint. Due care must be exercised to thoroughly compact the concrete at the joint, at the beginning of next days work, as poor vibration of concrete at this location will cause distress in certain situations.

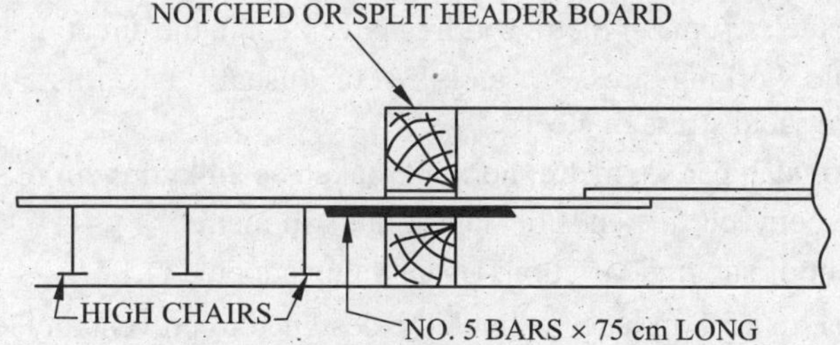

Fig. 21.53 Construction joint for CRCP.

21.22. MINIMUM STEEL

Under normal conditions, a minimum steel of 0.5 per cent is adequate, but under severe cold climates, the minimum percentage of steel should be 0.7 per cent.

21.23. GENERAL REMARKS

The performance of CRCP is greatly influenced by the following factors:

(*i*) Type of sub-base

(*ii*) Drainage of the sub-base

(*iii*) Degree of compaction of the sub-grade.

Therefore, to insure good performance of this type of pavement, use of high quality construction is essential.

REINFORCEMENT AND JOINTS

21.24. DESIGN OF STEEL REINFORCEMENT

Reinforcement is assumed, in most cases, to add nothing to the structural strength of the pavement but may be used to hold together fractured faces of slabs after cracks have formed so as to aid in load transmission at these points. The steel reinforcement will not allow the crack to increase in width and load is transferred through cracks by aggregates interlocking effect and thus faulting of the slab at crack is prevented. For practical purposes the crack is also maintained in water-tight condition and infiltration of incompressible material is also prevented.

Since steel does not begin to perform its function until the concrete has cracked due to warping and other stresses, it is necessary to use enough steel to take up all tension to keep the cracks tightly closed. Tension in steel members across any crack is equal to the force required to overcome friction between pavement and subgrade from crack to the nearest free joint or edge. This force will be greatest when the crack occurs at the greatest distance from a joint, *i.e.*, at the middle of a slab and distributed steel is designed to be adequate for a crack in this location. Theoretically the amount of steel may be reduced as the ends are approached but due to practical difficulties of varying the steel mat and due to extra supervision and labour required to ensure varying weights of steel in correct position, usually uniform weight of steel is used. For longitudinal steel:

Total tension in steel = Total frictional resistance

$$A_s\, f_t = f_w \frac{L}{2}$$

$$\therefore \quad A_s = \frac{Lf_w}{2f_t} \qquad \text{...(21.24)}$$

where A_s = area of steel, sq cm per metre width of pavement in the direction in which L is measured.

f_t = allowable working stress in steel, kg/cm² (usually taken as 50 to 60 per cent of the minimum yield stress of steel)

w = weight of slab per sq. metre; normally taken as 24 kg/m²/cm of slab thickness.

L = distance between free ends of slab (joints), in metre

f = coefficient of subgrade friction. Usual value adopted is 1.5.

For transverse steel, in a 3.8 m lane, it should be designed to prevent any longitudinal crack from pulling apart. As the slab is anchored against lateral movement at the longitudinal joint 3.8 m, from the side, the transverse steel should be capable of sliding one lane width (c/c of two lanes) of concrete during contraction. Design principles are the same as those for determining longitudinal reinforcement:

$$A_s f_t = wLf$$

or

$$A_s = \frac{wLf}{f_t} \qquad \text{...(21.25)}$$

where L = lane width.

Longitudinal cracks seldom develop when longitudinal hinge joints are used and the above value of transverse steel is somewhat conservative. Some organizations reduce this steel considerably, as much as 35 per cent.

Experience indicates that small bars or mesh is more effective than the same area of larger diameter bars because they can be distributed more uniformly in the slab.

As already pointed out, since distributed steel is not intended to act in flexure its position within the slab is not important, except that it should be far enough from either surface to be adequately protected against corrosion. As already explained, common practice is to place reinforcement 5 cm below the finished surface.

Problem 21.11 *Design the reinforcement for a* 12 *cm thick simply supported RCC pavement. The contraction joints are spaced at 12 m intervals. The pavement width is 7.5 m, comprising of two lanes. Allowable stress for steel is* 3000 *kg/cm². Take f* = 1.5.

Solution. The required steel for a 12 cm pavement is determined by applying equation 21.24 for longitudinal steel and equation 21.25 for transverse steel.

$$\text{Longitudinal steel} = \frac{L\,fw}{2f_t}$$

Here $L = 12$ m

$f = 1.5$

$w = 24$ kg/m²/cm of slab = 24 × 12 = 288 kg/m²

$f_t = 3000\ \text{kg/cm}^2$

$$\therefore \quad A_s = \frac{12\times1.5\times288}{2\times3000}$$

$$= 0.864\ \text{cm}^2$$

Provide 25 mm dia bars.

$$\text{Spacing} = \frac{100\times4}{0.864\times\pi\times(2.5)^2}$$

$$= 23.58\ \text{cm}$$

Say 22.5 cm.

$$\text{Transverse steel As} = \frac{wLf}{f_t}$$

Here $L = \frac{7.5}{2} = 3.75$ m

$$\therefore \quad A_s = \frac{288\times3.75\times1.5}{3000}$$

$$= 0.54\ \text{cm}^2$$

Provide 20 mm dia bars,

$$\text{Spacing} = \frac{100\times4}{0.54\times\pi\times(2)^2} = 58.95\ \text{cm say } 57.5\ \text{cm}.$$

Problem 21.12 *Find the spacing between contraction joints for a 3.80 m slab width having a thickness of 15 cm for both plain concrete slab and RCC slab. The ultimate tensile stress values in concrete and steel are 1.62 and 1200 kg/cm². The unit weight of concrete and steel are 2400 and 7500 kg/m³. Coefficient of friction is 1.5 and desired factor of safety is two. Total reinforcement of 3 kg/m² is provided and is equally distributed in both directions.*

Solution. (1) *For plain concrete slabs.*

The contraction of the slab is resisted by the friction between the bottom of the slab and the subgrade.

Total tension in cement concrete = frictional resistance of subgrade upto the centre

or $$S\times100h = \frac{wLf}{2}$$

$$\therefore \quad L = 200\,\frac{Sh}{wf}$$

Here S = allowable stress in tension in concrete

$$= \frac{\text{ultimate tensile stress in concrete}}{\text{factor of safety}}$$

$$= \frac{1.62}{2} = 0.81 \text{ kg/cm}^2$$

$w = 24 \times 15 = 360$ kg/sq metre

$f = 1.5$

$h = 15$ cm

$$\therefore \quad L = \frac{200 \times 0.81 \times 15}{360 \times 1.5}$$

$= 4.5$ m

(2) *For RCC slab*

In the case of R.C.C. slab, the contraction in the slab is resisted by the reinforcement, and the spacing between the contracting joints is given by

$$L = \frac{2 f_t \times A_s}{fw}$$

Taking one metre length of the slab, the cross sectional area of the steel, A_s, in one direction per metre of slab width is given by

$$\frac{\text{Total reinforcement}}{2} = \frac{A_s}{10^4} \times \text{Unit weight of steel}$$

or $$\frac{3}{2} = \frac{A_s}{10^4} \times 7500$$

or $$A_s = \frac{3 \times 10^4}{2 \times 7500}$$

$= 2$ cm^2/sq metre of slab

$$f_t = \frac{\text{Ultimate tensile stress in steel}}{\text{factor of safety}} = \frac{1200}{2} = 600 \text{ kg/cm}^2$$

$f = 1.5$

$w = 24 \times 15$

$= 360$ kg/m^2

The spacing of contraction joint,

$$L = \frac{2 \times 600 \times 2}{360 \times 1.5}$$

$$= 4.44 \text{ m}$$

Say 4.5 m

21.24.1. Design of tie bars

In case opening of longitudinal joints is anticipated in case of heavy traffic, side long ground, expansive subgrade, *etc.* tie bars are provided. These are designed in accordance with the recommendations of IRC: 15-2002 and are given herein.

Tie bars across longitudinal joints are designed on the same principle and using the same equation as given above for transverse steel assuming that slab faces at longitudinal joints will be firmly held together to ensure adequate load transference.

The area of steel required per metre length of joint is given by

$$A_s = \frac{b\,f\,W}{S}$$

where A_s = Area of steel in cm^2; required per metre length of joint

b = lane width in metres

f = coefficient of friction between pavement and the sub-base/base (usually taken as 1.5)

W = Weight of slab in kg/m^2/cm of slab thickness

S = Allowable working stress of steel in kg/cm^2

The length of the tie bar should be at least twice that required to develop a bond strength equal to the working stress of the steel.

$$\therefore \quad L = \frac{2SA}{B' \times P}$$

where L = Legnth of tie-bar in cm

A = Cross-sectional area of one tie bar in cm^2

$= \frac{\pi d^2}{4}$; in which d is the diameter of the tie-bar in cm

P = Perimeter of tie bar $= \pi d$

B' = Permissible bond stress of concrete, usually taken as 17.5 kg/cm^2 for plain tie-bars and 24.6 kg/cm^2 for deformed tie-bars.

To permit warping at the joint, the maximum diameter in case of tie bars may be limited to 20 mm and to avoid concentration of tensile forces, they should not be spaced more than 75 cm apart. The calculated length, L, may be increased by 5 to 8 cm to account for any inaccuracy in placement during construction.

Typical tie-bar details as per IRC : 58-2002 (second revision) for use at central longitudinal joint in double lane rigid pavements with a lane width of 3.5 m and spacing at 5 m joint are given in Table 21.22. The following values of different design parameters are assumed.

f_s = 1250 kg/cm^2 for plain bars and 2000 kg/m^2 for deformed bars; $\mu = 17.5$ kg/cm^2 for plain bars and = 24.6 kg/cm^2 for deformed bars; f = 1.5 ;

W = 24 kg/m^2/cm of slab thickness.

Table 21.22 Details of tie bars for central longitudinal joint of two-lane rigid highway pavement

Slab thickness (cm)	Diameter (mm)	Maximum spacing (cm) Plain bars	Maximum spacing (cm) Deformed bars	Minimum length (cm) Plain bars	Minimum length (cm) Deformed bar
15	8	33	53	41	48
	10	52	83	51	56
20	10	39	62	51	56
	12	56	90	58	64
25	12	45	72	58	64
	14	61	98	65	72
30	12	37	60	58	64
	16	66	106	72	80
35	12	32	51	58	64
	16	57	91	72	80

Length of tie bar shown in the table above is increased by 15–20 cm to compensate for:

(*a*) Loss of bond due to painted length of 10 cm in the middle of tie-bar; and

(*b*) 5 cm to compensate for placement error laterally.

Problem 21.13 *A cement concrete pavement has a thickness of 32 cm and has two lanes of 7.0 metres width with a longitudinal joint. Design the dimensions and spacing of the tie bar. Use the following data:*

Allowable working stress in tension = 1800 kg/cm²

Unit weight of concrete = 2400 kg/m³

Aliowable tensile stress in plain bar = 1250 kg/cm²

(As per IRC : 21-2000)

Allowable tensile stress in deformed bars = 2000 kg/cm²

(As per IRC : 21-2000)

Allowable bond stress in plain bar = 17.5 kg/cm²

Allowable bond stress in deformed bar = 24.6 kg/cm²

Diameter of tie bar = 12 mm

Solution. (1) *Spacing and Length of plain bar*

Consider one metre width of the joint.

For tie bars across longitudinal joint, steel required:

$$A_s = \frac{b\,f\,W}{S}$$

Here $b = \frac{7.0}{2} = 3.5$ m

$f = 1.5$ (usual value adopted)

$$W = \frac{2400 \times 32}{100}$$

$= 768 \text{ kg/cm}^2$

$\therefore \quad A_s = \dfrac{3.5 \times 1.5 \times 768}{1250}$

$= 3.2256 \text{ cm}^2/\text{m}$

The spacing of the tie-bars is given by

$$\frac{100}{A_s \times \frac{\pi d^2}{4}}$$

$$= \frac{100 \times 4}{3.2256 \times \pi \times (1.2)^2}$$

$= 27.4$ cm

Adopt a spacing of 27 cm c/c

The length of the tie-bar is determined from the relation

$$L = \frac{2S \times A}{B' \times P}$$

Here $S = 1250 \text{ kg/m}^2$

$A = \frac{\pi}{4} \times (1.2)^2 = 1.131 \text{ cm}^2$

$P = \pi d = \frac{22}{7} \times 1.2 = 3.771$ cm

$B' = 17.5 \text{ kg/ cm}^2$

$\therefore \quad L = \dfrac{2 \times 1250 \times 1.131}{17.5 \times 3.771}$

$= 42.85$ cm

Say 43 cm

Increasing length by 10 cm for loss of bond due to painting and 5 cm for tolerance in placement

$L = 43 + 10 + 5 = 58 \text{cm}$

Thus the dimensions of the plain tie bar are:

Diameter = 12 mm

Length = 58 cm

Spacing = 27 cm

Spacing and length of the deformed tie-bar

$A_s = \dfrac{b f W}{S}$

$= \dfrac{3.5 \times 1.5 \times 0.32 \times 2400}{2000}$

$= 2.016 \text{ cm}^2/\text{m}$

$\therefore$ Spacing of tie-bar $= \dfrac{100 \times 1.131}{2.016}$

$= 56.1$ cm c/c

Say $= 56$ cm c/c

Length of tie-bar $= \dfrac{2 \times 2000 \times 1.131}{24.6 \times 3.771}$

$= 48.77$ cm

Say 49 cm

Increasing the length by 10 cm for loss of bond due to painting and 5 cm for tolerance in placement

$$L = 49 + 10 + 5$$

$$= 64 \text{ cm}$$

$\therefore$ The dimensions of the deformed tie-bar are:

diameter = 12 mm

length = 64 cm

spacing = 56 cm

21.24.2. Design of dowel bars

As already explained, dowel-bars are load-transfer devices and, thus, they must be fairly heavy and spaced at closer intervals than tie bars to provide resistance to bending, shearing, and bearing on the concrete.

Dowel bar stresses resulting from shear, bending and bearing can be analysed analytically by making certain simplifying assumptions. It is assumed that dowel-bars are perfectly aligned and free to move. The stress analysis of dowel is based upon work presented by Timoshenko. According to him, a dowel-bar encased in concrete will deflect downward at the end when the load is applied there. This will exert pressure at the lower end of the dowel for a certain distance after which a point of counterflexure exists with resulting bearing on top of the dowel, and then, at some distance beyond this, bearing is on the bottom of the dowel-bar.

The relative stiffness of a bar embedded in concrete is given by the equation:

$$\beta = \sqrt[4]{\frac{Kd}{4EI}}$$

where β = relative stiffness

d = diameter of dowel

E = modulus of elasticity of dowel

I = moment of inertia of dowel

K = modulus of dowel support kg/cm^3

The deflection of the bar y resulting from load P is :

$$y = \frac{e^{-\beta x}}{2\beta^3 EI}\left[P \cos \beta x - \beta M_0 (\cos \beta x - \sin \beta x)\right]$$

where x = distance along dowel from face of concrete

M_0 = bending moment on dowel at face of concrete.

Friberg (Trans ASCE, 1940) used this equation for dowel in concrete pavements. If joint-width opening is z and since the concrete is very stiff compared to steel bar, the moment at the face of concrete:

$$M_0 = -\frac{Pz}{2}$$

Substituting this value of M_0 in Timoshenko equation given above where $x = 0$, the deflection, y_0 at the joint is given by:

$$y_0 = \frac{P}{4\beta^3 EI}(2+\beta z)$$

The bearing pressure, P_b, on the concrete at the joint face is:

$$P_b = Ky_0 = \frac{KP}{4\beta^3 EI}(2+\beta z)$$

Values of K, suggested by Grinter and Friberg range between 8300 and 41500 kg/cm³. The American Institute Committee recommends, 41500 kg/cm³. Since β varies as $\sqrt[4]{K}$, large changes in the value of K do not effect the stress calculation greatly.

Bradbury (Proc, HRB 1932) assuming straight line stress distribution from the face of the concrete to the point of counter-flexure worked out the load transfer capacity of one dowel bar by the following formulae; which has also been adopted by IRC.

$$\text{Capacity in shear} = P_s = 0.785\, d^2 f_s' \qquad \text{...(21.27)}$$

$$\text{Capacity in bending} = P_b = \frac{2d^3 f_s}{r+8.8z} \qquad \text{...(21.28)}$$

$$\text{Capacity in bearing on concrete} = P_c = \frac{f_c r^2 d}{12.5\,(r+1.5z)} \qquad \text{...(21.29)}$$

in which f_s' = shearing stress in steel (kg/cm²)

f_s = tensile or bending stress in steel (kg/cm²)

f_c = bearing stress in concrete (kg/cm²)

r = length of dowel embedded in concrete (cm)

z = width of joint opening (cm)

The design of dowels can be based either on Bradbury analysis or Friberg analysis.

The formulae worked out by Bradbury show that the capacity of a dowel in bearing increases with its length, while its capacity in bending decreases with its length. For equal capacity in bending and bearing the length found by equating equations 21.28 and 21.29 is:

$$r = 5d\sqrt{\frac{f_s}{f_c}\cdot\frac{r+1.5z}{r+8.8z}} \qquad \text{...(21.30)}$$

The above formula is not adapted for direct solution of r. The value of load transfer capacity for single dowel can be read easily with the help of graphs and the total load transfer capacity of dowel group can be worked out as explained below by Friberg analysis.

If Friberg analysis is used for design, decision is first made of the total load carried by each dowel. This can be done by calculating the number of effective dowels under the load assuming that the dowel bar immediately under the applied load carries full capacity, decreasing to zero at a distance of 1.8 l from this dowel. Friberg observed that, according to the practical analysis presented by Westergaard, maximum negative moment occurs at a distance 1.8 l from the load. Thus, dowels directly under the point of loading are most effective, and effectiveness of dowels decreases to zero at a distance of 1.8 l from the point of load. The load carrying capacity of the group of dowels is equal to the sum of loads carried by each of these effective dowels. The total load transferred (50 per cent of the applied load for 100 per cent efficiency, usually taken 45 per cent in design due to looseness) divided by number of effective dowels gives the load P carried by the dowel to be used in Friberg equation.

For example, for wheel load of 4000 kg placed at the first dowel from the edge, 50 per cent, *i.e.*, 2000 kg will be transferred through the dowel. If radius of relative stiffness, l is 75 cm then dowel-bar at the edge will carry full capacity, decreasing to zero at a distance 1.8 × 75 = 135 cm. If spacing of dowels is 30 cm centre to centre, the total number of dowel-bars helping to transfer the load is 5. The second dowel from edge will carry $\left(\dfrac{135-30}{135}\right)$ *i.e.*, 105/135 or 0.78 of full capacity. Similarly, 3rd, 4th and 5th dowel will carry 0.56, 0.34, and 0.11 of full capacity. Thus the total number of effective dowels is 1.00 + 0.78 + 0.56 + 0.34 + 0.11 = 2.79. Thus P, load carried by the dowel at the edge is 2,000/2.79 = 717 kg. Similarly, if some small load due to the second wheel on the axle is also carried by this dowel at the edge, this is also added to 717 kg to get the value of P. For example, under the second wheel at the interior of the pavement, there will be 4.58 effective dowels, and if the dowel at the edge happens to be the last dowel, then 2,000/4.58 × 0.11, *i.e.*, 79 kg will be carried by the dowel at the edge. Thus P in this case would be 717 + 79 = 796 kg.

Assuming the diameter of dowel bar and width of joint opening, the value of β and y_0 and bearing stress on the concrete at the dowel face can be calculated. If this stress is less and close to the allowable bearing stress, the assumed diameter is correct. The permissible bearing stress recommended by ACI can be determined by equation:

$$f_b = \frac{(10.16-d)}{9.525}\, f_c'$$

where f_b = permissible bearing stress (kg/cm^2)

d = dowel diameter (cm)

f_c' = characteristic compressive strength of concrete cube (kg/cm^2) after 28 days curing

According to Friberg, cutting a dowel at the second point of contraflexure does not affect the concrete bearing stress. Tests show that this length is about 8 diameters for 2 cm round bars and decreases as dowel size increases. American Concrete Institute recommends dowels 45 cm long for highways and 45, 50 and 60 cm long for airports for diameters of 2.5, 3.1 and 3.75 cm respectively.

Typical dowel bar designs for use in 20 mm wide expansion joints for highway pavements, for 40 per cent load transfer, as recommended by IRC: 58-2002, are given in Table 21.23. The recommended details are based on the following values of different design parameters:

f_s = 1400 kg/cm^2 ; f_c = 100 kg/cm^2 ; E_c = 3.0 × 10^5 kg/cm^2 ;

μ = 0.15; k = 8.3 kg/cm^3; maximum joint width = 20 mm; design load transfer = 40 per cent.

Table 21.23 Design details of dowel bars for rigid highway pavements

Design wheel loading (kg)	*Slab thickness (cm)*	*Dowel bar details*		
		Diameter (mm)	*Length (mm)*	*Spacing (mm)*
4080	15	25	500	200
	20	25	500	250
	25	25	500	300
	30	32	500	300
	35	32	500	300

U.S. Federal Aviation Agency recommends the dowel-bar requirements for airport pavements as given in Table 21.24.

Table 21.24 FAA recommendations for dowel-bars

Slab thickness (cm)	*Dowel diameter (cm)*	*Length (cm)*	*Spacing (cm)*
15–17.5	2.0	40	30
20–25	2.5	40	30
27.5–37.5	3.1	50	37.5
40–50	3.1	60	37.5

U.S. Corps of Engineers recommendations for airport pavements are given in Table 21.25.

Table 21.25 Corps of Engineers recommendations for dowel bars

Slab thickness (cm)	*Dowel diameter (cm)*	*Length (cm)*	*Spacing (cm)*	
Less than 20	2.0	40	30	
20–28	2.5	40	30	
29–38	3.1	50	37.5	2.5 cm ex-strong pipe may be used
35–50	3.79	50	45	3.75 cm ex-strong pipe may be used
51–63	5.0	60	45	5.0 cm ex-strong pipe may be used
over 63	7.5	75	45	7.5 cm ex-strong pipe may be used

AASHO test indicated that in slabs upto 12.5 cm thick, the locations of dowels several times formed the starting point of longitudinal cracks, this tendency being totally absent in the case of thicker slabs. This finding concluded that no dowel bars be used at joints in slabs upto 12.5 cm thick. According to British practice, dowel bars are excluded in slabs less than 15 cm thick.

Problem 21.14 *Design dowel bar length, spacing and copacity factor required of dowel system from the following data; as per IRC recommendations*

Design wheel load, P = *4100 kg*
Design load transfer = *40 per cent*
Slab thickness, h = *20 cm*
Joint width, z = *2 cm*
Permissible flexural stress in dowel bar = *1400 kg/cm²*

Permissible shear stress in dowel bar = *1000 kg/cm²*
k-value on subbase = *8 kg/cm²/cm*
Other concrete parameters:

$$E = 3\times10^5\, kg/cm^2$$

$$\mu = 0.15$$

$$f_c = 100\, kg/cm^2$$

Solution. ***Step 1.*** Dowel length

Assume dowel diameter, $d = 2.5$ cm

Then, for equal capacity in bending and bearing, from equation 21.30, length of dowel embedded in concrete, is given by

$$r = 5d\left[\frac{f_s}{f_c}\frac{r+1.5z}{r+8.8z}\right]^{1/2}$$

Substituting the values

$$r = 5\times2.5\left[\frac{1400}{100}\cdot\frac{r+3}{r+17.6}\right]^{1/2}$$

which on solution gives $r = 40.5$ cm.

$\therefore$ Dowel length, $L = r + z = 40.5 + 2.0 = 42.5$

say 45 cm

Step 2. Load transfer capacity of single dowel

From equations 21.27 to 21.29, the load transfer capacity of a single dowel is obtained in shear, bending and bearing.

$$P_s = 0.785\, d^2 f_s'$$

$$= 0.785\times(2.5)^2\times1000$$

$$= 4900 \text{ kg}$$

$$P_b = \frac{2d^3 f_s}{r+8.8z}$$

$$= \frac{2\times(2.5)^3\times1400}{40.5+8.8\times2}$$

$$= 753 \text{ kg}$$

$$P_c = \frac{f_c r^2 d}{12.5\,(r+1.5z)}$$

$$= \frac{100\times(40.5)^2\times2.5}{12.5\,(40.5+1.5\times2)}$$

$$= 754 \text{ kg}$$

Taking the least of these values for design purposes

$P = 753$ kg

Step 3. Capacity factor required of dowel system

$$\text{Load transfer capacity of the dowel system} = 4100 \times \frac{40}{100}$$

$$= 1640 \text{ kg.}$$

$\therefore$ Required capacity factor, n

$$= \frac{\text{load transfer capacity of the dowel system}}{\text{load transfer capacity of a single dowel bar}}$$

$$= \frac{1640}{753}$$

$$= 2.18$$

Step 4. Spacing of dowel bars

$$\text{Radius of relative stiffness, } l = \sqrt[4]{\frac{Eh^3}{12\,(1-\mu^2)\,k}}$$

$$= \sqrt[4]{\frac{3 \times 10^5 \times (20)^2}{12\,(1-0.15^2) \times 8}}$$

$$= 70.95 \text{ cm}$$

This can also be directly read from Table 21.1. Considering the joint corner, the distance over which the dowel bars are effective in load transfer = 1.8 l

$$= 1.8 \times 70.95 \simeq 128 \text{ cm}$$

Assuming a dowel spacing of 35 cm

$$\text{Available capacity factor } = 1 + \frac{128-35}{128} + \frac{128-70}{128} + \frac{128-105}{128}$$

$$= 2.36$$

which is slightly greater than the required capacity factor of 2.18. Hecne adopt 35 cm as dowel spacing.

21.25. DESIGN OF SPACING OF JOINT

The determination of spacing of transverse joints in concrete pavements depends on several factors. Contraction joints are used at spacing which will control cracking. As for the design of distributed steel, the total frictional resistance of slab and subgrade can be equated to the total stress in the concrete at the centre of slab and theoretical spacing in the slab can be worked out. The equation, $S = WLf/200h$ as already worked out for stresses due to subgrade friction can be used for design of spacing of contraction joints. For this the value of allowable tensile stress of concrete is assumed very low, say equal to 2.1 kg/cm^2 due to the fact that the slab may be subjected to contraction at an early age before much strength is developed.

However, the problem of transverse cracking is so complex that simple theoretical relationships alone are not reliable to determine the best spacing of joints for unreinforced pavements. Pavements subjected to different conditions and made with different materials display cracking tendencies, which cannot be explained by any simple theory.

Crack intervals in a pavement are primarily governed by volume-change characteristics and tensile strength of concrete. High-volume change occurs due to; (*a*) use of expansive aggregates; (*b*) high ratio of the fine to coarse aggregate; (*c*) high cement content; and (*d*) to minor extent, due to high water content. Low flexural strength of concrete may be due to (a) sufficient percentage of low strength coarse aggregate; (*b*) smooth coarse dense aggregate resulting in poor bond between mortar and coarse aggregate; and (*c*) a weak mortar due to certain reasons.

For average conditions in plain concrete pavements, it is a common practice to have joints at spacing of 4.5 metre. Spacing of 6 m is commonly used and it may be decreased to 4.5 m or increased to 7.5 m depending on local conditions as explained above.

For reinforced concrete pavements, the amount of steel required to keep cracks tightly closed increases with the slab length. On the other hand, as slab length decreases the number and cost of transverse joints increases and also cost of joint maintenance increases. Another point to be considered is that if slabs are longer than 7.5 m, the amount of joint opening will be sufficient to reduce aggregate interlock and for such conditions it becomes necessary to provide adequate load-transfer devices.

Thus, within the limits of permissible opening and closure of joints, the spacing in reinforced concrete slabs should be fixed so that the cost of distributed steel, cost of installing joints and cost of joint maintenance is minimum. This spacing will vary with changing costs of steel and labour. The spacing vary from 12 to 21 metres.

Experience has shown that it is not possible to maintain an adequate seal at a joint with present sealing materials if seasonal fluctuations in joints width exceed about 50 per cent of the original width of the sealing material. This represents fluctuation in joint opening of about 5 mm. Calculations using average values of coefficient of thermal expansion and contraction indicate that reinforced joint spacings up to 12 m maintain joint width opening within the maximum permissible values. However, temperature is not the only factor. Moisture content also causes change in volume and usually oppose those due to variation in temperature.

From the above considerations, it is evident that the usual spacings for reinforced concrete slab up to 12 m is quite desirable and may be increased upto 18 m considering the initial cost of steel and cost of joints. However, it becomes more difficult and costly to maintain a joint with spacing more than 15 m due to larger openings.

The present trend is towards decreasing the expansion joints because of their high cost of construction and maintenance. Expansion joints prevent the development of excessive compressive stresses in pavement as a result of expansion caused by increase in temperature and moisture. They are always used between pavements and fixed structures. The expansion space needed is governed by pavement details, aggregates and local climatic conditions. To provide more than 2.5 cm expansion space at any special locations, such as bridges, etc two or three expansion joints of standard widths at spacing about 6 m may be provided. The common width of expansion joint opening is 2 or 2.5 cm.

21.25.1. Joints for airport pavements

Airport pavements will depend on type of concrete, thickness of slab, and width of pouring lane. Usually, concrete in wide area is poured in alternate lanes. Longitudinal construction joints are either keyed or of butt-type. Transverse joints are generally of dummy groove type, either preformed or sawed. Common practice is to have transverse joint spacing greater than longitudinal spacing, though in some cases square slabs are used. Generally, expansion joints are provided at fixed structures and at intersections of runways and taxi-ways. FAA recommends that expansion joints be not over 450 m apart. Expansion joints may be with or without load transfer devices.

Table 21.26 gives the joint spacing recommended by different agencies for airport pavements.

Table 21.26 Joint spacing for airport pavements

Agency	*Slab thickness (cm)*	*Plain concrete*		*Reinforced concrete*	
		Longitudinal (m)	*Transverse (m)*	*Longitudinal (m)*	*Transverse (m)*
Federal Aviation Administration (FAA)	22.5 or less	3.75	4.5		
	22.5–30	6.0	6.0		
	Greater than 30	7.5	7.5		
Portland Cement Association (PCA)	30 or less	3.75 max	4.5–6.0	3.75 max	9.0–12.0
	30 to 37.5	3.75 max	7.5–9.0	3.75 max	15.0
	For channelised traffic more than 37.5 and 30 to 37.5 for non-channelised traffic	varies	7.5–9.0	varies	15.0
Corps of Engineers (CE)	22.5 or less	3.75	4.5 max		
	22.5–30	6.0 max	6.0 max		
	More than 30	7.5 max	7.5 max		

PART – VI

22. Pavement Distress
23. Pavement Evaluation and Strengthening
24. Highway Maintenance

Chapter

PAVEMENT DISTRESS

It is necessary to have a complete knowledge of types of distress or failure in pavements to evaluate fully the factors that affect the design of pavements. Certain types of pavement failure are progressive, leading to eventual failure of the road, whereas some others are non-progressive. Progressive distress is more dangerous.

The cause of pavement distress or failure may be three-fold. Firstly structural failure which indicates a breakdown of one or more of the pavement components, may result from overload including excessive gross loads, repetition of loads, *i.e.*, surface fatigues and high tyre pressure. Secondly climatic conditions as well as environmental factors may cause surface irregularities and structural failure to develop, for example frost heaving, volume change due to wetting and drying, break up resulting from freezing and thawing or improper drainage may be the major cause of pavement distress. A third cause may be disintegration of the paving materials due to weathering *i.e.*, wetting and drying and/or freezing and thawing.

FLEXIBLE PAVEMENT DISTRESS

The types of defects in flexible pavements can be grouped under four distinct categories, namely:

(1) ***Surface defects.*** Which include fatty surfaces, smooth surfaces, streaking and hungry surfaces.

(2) ***Cracks.*** Under which hair-line cracks, alligator cracks, longitudinal cracks, edge cracks, shrinkage cracks and reflection cracks are dealt with.

(3) ***Deformation.*** Includes slippage, rutting, corrugations, shoving, shallow depressions, settlements and upheavals; and

(4) ***Disintegration.*** Covering stripping, loss of aggregate, ravelling, pot-holes, and edge breaking.

Fig. 22.1 depicts the failures in flexible pavement. The arrows indicate the direction of upheaval due to movement of material from the layer. Fig. 22.2 shows the soil deformation under the load. Failure due to improper compaction of subgrade and pavement layer are shown in Fig. 22.3.

Table 22.1 describes the symptoms and causes of these defects and indicates the possible type of treatment. The specifications and construction methods for each type of treatment are given in chapter 24.

In each case of pavement distress, the cause or causes of the distress should be clearly determined before applying corrective measures. Suitable maintenance measures will not only correct the damage but also prevent or delay its recurrence. In many situations, lack of proper drainage is the principal cause for stripping, loss of material from the pavement and shoulder, weakening of the pavement layers. In such situations, the cause should be completely eliminated before taking any maintenance measure.

Table 22.1 Symptoms, causes and treatment of defects in bituminous surfacings

Type of distress	*Symptoms*	*Causes*	*Treatment*
A. SURFACE DEFECTS			
1. Fatty surface	Collection of binder as a film on the surface. This becomes soft in hot weather and may be picked up and spread by traffic. In cold or wet weather, the surface is likely to be slippery which may cause accidents.	1. Excessive binder in a premix surfacing overfilling the voids. 2. Loss of cover aggregates in surface dressing. 3. Non-uniform spreading of cover aggregates in surface dressing. 4. Poor quality of aggregates leading to their fracture, breakdown and eventual loss. 5. Excessive application of binder in surface dressing. 6. Graded cover aggregates with particles so small that they are covered by the binder. 7. Too heavy a prime or a tack coat. 8. Excessively heavy axle loads causing compression of a surfacing, forcing the bitumen to the top.	1. If the bleeding is fairly uniform and the surface is free from irregularities, application of cover aggregates or sand (sand-blotting or sand-blinding) would be successful. The aggregate or sand used shall be of small size, clean and angular, and may be heated, if necessary. 2. An open-graded premix surfacing with a low bitumen content can absorb the excess binder. 3. A liquid seal coat, with special care taken to select the rate of application of the binder and the quantity and size of cover aggregates, can also be effective. 4. Special methods such as the burning of the excess binder. 5. In case of large areas of fatty surface having irregularities, removal of the affected layer in the area and replacing it with a layer having a properly designed mix, may be necessary.
2. Smooth surface	Slippery when wet and has very low skid resistance value. Such a condition is hazardous especially on gradients, bends and intersections.	1. Polishing of aggregates under traffic. 2. Excessive binder.	1. Resurfacing with a surface dressing course or a premix carpet. Selecting hard and angular aggregates. Carpet can be open graded mix. 2. A slurry seal can be used to impart antiskidding texture.

3. Streaking	Presence of alternate lean and heavy lines of bitumen either in longitudinal or transverse direction.	1. Non-uniform application of bitumen across the surface. 2. Mechanical faults, improper or poor adjustment and careless operation of bitumen distributors.	1. Remove the streaked surface and apply a new surface treatment. 2. Manufacturers recommendation of the bitumen distributor should be carefully adhered to.
4. Hungry surface	1. Loss of aggregates from the surface. 2. Appearance of fine cracks.	1. Use of less bitumen in the surfacing. 2 Use of absorptive aggregates in the surfacing.	1. Apply slurry seal in thickness of 2 to 5 mm. 2. As an emergency repair, a fog seal may be used.
B. CRACKS			
1. Hair-line cracks (Fig. 22.4)	Short and fine cracks at close intervals on the surface	1. Insufficient bitumen content. 2. Excessive filler at the surface. 3. Improper compaction; over-compaction, compaction when the supporting layer was unstable, or compaction of too hot a mixture.	See next page
2. Alligator cracks (Fig. 22.5)	Inter-connected cracks forming a series of small blocks resembling the skin of an alligator.	1. Excessive deflection of the surface over unstable subgrade, subbase or base of the pavement, particularly in the wheel tracks. The unstable conditions in the subgrade or lower layers of the pavement might have arisen from saturation. 2. Excessive overloads by heavy vehicles or inadequate pavement thickness, or both. 3. Brittleness of the binder either due to ageing of binder or initial over-heating might cause fine cracks of the alligator pattern, but there will be no deflection of the surface. These cracks are sometimes called 'crazing'.	

3. Longitudinal crack (Fig. 22.6)	Cracks on a straight line, along the road. Appear either at the joint between the pavement and the shoulder or at the joint between two paving lanes.	1. The cracking at the pavement-shoulder joint may be due to alternate wetting and drying beneath the shoulder surface owing to poor drainage or due to depressions in the pavement edge which allow water to stand and seep through the joint. Shoulder settlement or trucks passing over the joint, may also cause these cracks. 2. The lane joint cracks is caused by a weak joint between adjoining spreads in the layers of the pavement. Differential frost heave along the centre line may also be one of the causes.	The treatment will depend on whether the pavement is structurally sound, or unsound. Where the pavement is structurally sound the cracks should be filled with a low viscosity (cutback or emulsion) binder or a slurry seal or fog seal depending on the width of cracks. Slurry seal is used for wide cracks and fog seal for fine cracks. Unsound cracked pavements will need strengthening or rehabilitation treatment.
4. Edge crack	1. Crack parallel to the outer edge of the pavement, usually 0.3 to 0.5 m from the edge. 2. At times some transverse cracks branch out from the edge cracks towards the shoulder.	1. Lack of lateral support from the shoulder. 2. Settlement of yielding of the underlying material. 3. Inadequate surface drainage, especially during flooding conditions. 4. Shrinkage due to drying out of the surrounding earth, generally caused by roots of trees or bushes close to the pavement edge. Highly expansive soils are particularly prone to shrinkage when moisture dries out. 5. Frost heave. 6. Inadequate pavement width forcing traffic too close to the edge of the pavement. 7. Non-provision of extra width of pavement on curves.	
5. Shrinkage cracks (Fig. 22.7)	Cracks in transverse direction or interconnected cracks forming a series of large blocks. Top surfacing appear to	Shrinkage of bituminous layer with age. The binder loses its ductility as it ages and becomes brittle.	

	become old or cracked though the pavement suffers no deterioration or deformation.	Due to joints and cracks in the pavement layer underneath.	
6. Reflection crack (Fig. 22.8)	Sympathetic cracks in bituminous surfacing over joints and cracks in the pavement underneath. The pattern may be longitudinal, transverse, diagonal or block. Most frequently occurs in overlays on cement concrete pavement or on cement-soil bases. Also occurs in overlays or surfacings on flexible pavements where the cracks in the old pavement have not been properly repaired. Also occurs over the junction between the old pavement and the widened strip.		
C. DEFORMATION 1. Slippage	Formation of crescent shaped cracks pointing in the direction of the thrust of wheels.	1. Unusual thrust of wheels in a particular direction. 2. Omission or inadequacy of tack coat or prime coat. 3. Lack of bond between the surface and lower course caused by a layer of fine dust, moisture or both. 4. Failure of bond between two layers due to excessive deflection of the pavement.	Removal of the surface layer around the area affected upto the point where good bond between the surfacing and the layer underneath exists and patching the area with premix material after a tack coat.
2. Rutting (Fig. 22.9)	Longitudinal depression or groove in the wheel tracks. Accumulation of water in the ruts can cause skidding. If accompanied by adjacent bulging, it may be a sign of	1. Heavy channelised traffic. 2. Inadequate compaction of the mix at the surface or in the underlying courses during construction. 3. Improper mix design, lacking in stability of the mix to support the traffic and leading to	1. Filling with premix open graded or dense graded patching material after a tack coat and compacting to the desired level. 2. Situation-indicative of shear failure or subgrade movement require

	subgrade movement or weak pavement.	plastic movement laterally under traffic. 4. Weak pavement. 5. Incidence of high stress caused by heavy bullock-cart traffic. 6. Intrusion of subgrade clay into base course. 7. Aggregates of surface dressing being pressed into the lower supporting bituminous layer.	excavation; and then applying the above treatment.
3. Corrugation	Fairly regular undulation (ripples) across the bituminous surface. These are shallow (about 25 mm) and spaced around 3 m.	1. Lack of stability in the mix (excessive binder, high proportion of fines, too round or too smooth textured coarse or fine aggregate, too soft a binder). 2. Oscillations set up by the vehicle springs can cause alternative valleys and ridges. 3. Faulty laying of surface course.	1. Scarification and relaying of a new surface. 2. Cutting of high spots with a blade and filling of low spots. 3. Spreading of sand bituminous premix with a drag spreader with its blade adjusted to just clear the high spots. The area is then rolled.
4. Shoving	A form of plastic movement within the layer resulting in localised bulging of the pavement surface. Occurs at points where traffic starts and stops (intersections, busy bus stops), on hills where vehicles accelerate or brake, on grades or on sharp curves. The first indication of shoving is the formation of slippage cracks which are crescent shaped cracks with the apex in the direction of the shove.	1. Lack of stability in the mix (excessive binder, high proportion of fines, too soft a binder) in the surface or base course. 2. Lack of bond between bituminous surface and underlying layer. 3. Heavy traffic movement of a start and stop type or involving negotiation of curves and gradients. 4. Use of non-volatile oil on roller wheel.	Removal of the material in the affected area down to a firm base and laying a stable premix patch.
5. Shallow depression	Localised shallow depressions, dipping about 25 mm or more below the desired profile, where water will normally collect.	Settlement of lower pavement layers due to a pocket of inadequately compacted subgrade or pavement layers.	Filled with premix materials, open graded or dense graded and compacted to the desired profile as the surrounding pavement.

6. Settlement and upheaval (Fig. 22.10)	Large deformations of the pavement, extremely uncomfortable to traffic and cause reduction in speed. They are generally followed by extensive cracks in the pavement surface in the affected region.	1. Inadequate compaction of the fill at locations behind bridge abutments, over utility cuts, etc.	1. If indicate inherent weakness in the fill, excavate the defective fill and lay the embankment afresh under properly controlled conditions, using materials having good drainage qualities.
		2. Excessive moisture in subgrade and permeable layer of sub-base and base caused by capillary action or poor drainage.	2. Construct under-drains, where cause of failure is lack of drainage.
		3. Inadequate pavement thickness.	3. Provide properly designed pavement where the cause of deformation is inadequate pavement thickness.
		4. Frost heave conditions.	4. Frost-affected regions may need complete reconstruction of the pavement.
D. DISINTEGRATION			
1. Stripping (Fig. 22.11)	Separation of bitumen binder from the aggregate in the presence of water. This may lead to loss of bond and subsequently to loss of strength and materials from the surface.	1. Use of hydrophillic aggregates. 2. Inadequate mix composition. 3. Continuous contact of water with the coated aggregate. 4. Initial over-heating of the binder or the aggregate or both. 5. Presence of dust or moisture on aggregate when it comes in contact with the bitumen. 6. In the case of surface dressing, poor bond with the surface existing below, delay in spreading the cover aggregate over the sprayed bitumen, or insufficient compaction. 7. Occurrence of rain or dust storm immediately after the construction. 8. Opening the road to fast traffic before the binder has set.	1. Spreading and compacting hot coarse sand, heated to at least 150°C, over the affected area in the case of surface dressing. 2. Replacement with fresh bituminous mix with added antistripping agent in other cases.

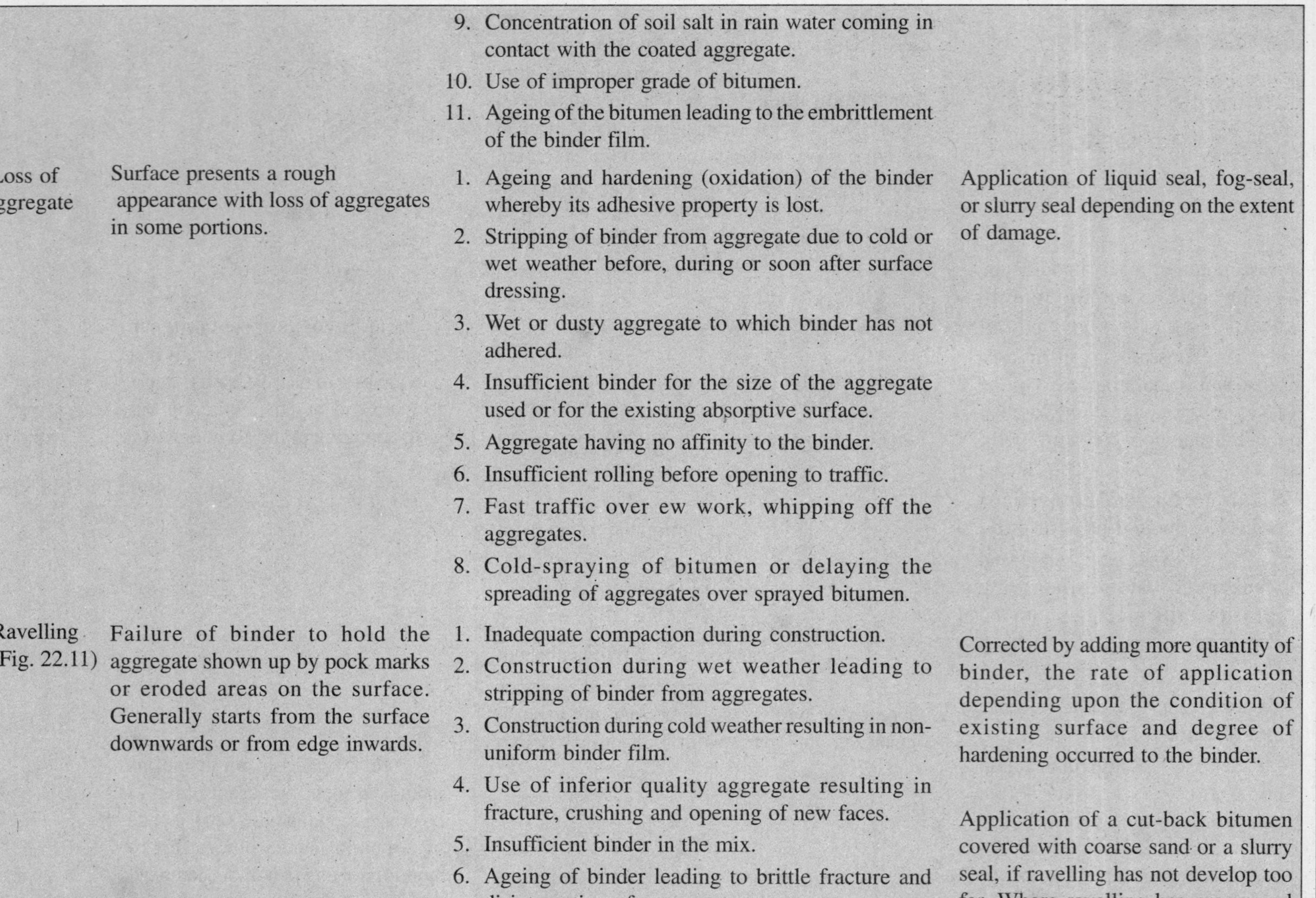

		9. Concentration of soil salt in rain water coming in contact with the coated aggregate. 10. Use of improper grade of bitumen. 11. Ageing of the bitumen leading to the embrittlement of the binder film.	
2. Loss of aggregate	Surface presents a rough appearance with loss of aggregates in some portions.	1. Ageing and hardening (oxidation) of the binder whereby its adhesive property is lost. 2. Stripping of binder from aggregate due to cold or wet weather before, during or soon after surface dressing. 3. Wet or dusty aggregate to which binder has not adhered. 4. Insufficient binder for the size of the aggregate used or for the existing absorptive surface. 5. Aggregate having no affinity to the binder. 6. Insufficient rolling before opening to traffic. 7. Fast traffic over ew work, whipping off the aggregates. 8. Cold-spraying of bitumen or delaying the spreading of aggregates over sprayed bitumen.	Application of liquid seal, fog-seal, or slurry seal depending on the extent of damage.
3. Ravelling (Fig. 22.11)	Failure of binder to hold the aggregate shown up by pock marks or eroded areas on the surface. Generally starts from the surface downwards or from edge inwards.	1. Inadequate compaction during construction. 2. Construction during wet weather leading to stripping of binder from aggregates. 3. Construction during cold weather resulting in non-uniform binder film. 4. Use of inferior quality aggregate resulting in fracture, crushing and opening of new faces. 5. Insufficient binder in the mix. 6. Ageing of binder leading to brittle fracture and disintegration of pavement.	Corrected by adding more quantity of binder, the rate of application depending upon the condition of existing surface and degree of hardening occurred to the binder. Application of a cut-back bitumen covered with coarse sand or a slurry seal, if ravelling has not develop too far. Where ravelling has progressed

		7. Excessively open graded mix. 8. Poor compatibility of binder and aggregate. 9. Over-heating of mix or the binder. 10. Improper coating of aggregates by binder.	far, apply a renewal coat with premix material.
4. Pot-hole (Fig. 22.12)	Bowl shaped holes of varying sizes in the surface layer or extending into the base course caused by localised disintegration of material.	1. The most common cuase of pot-hole formation is the ingress of water into the pavement through the surfacing course. This can happen if the surfacing is open-textured and lacks proper camber. Water can enter the pavement also through the cracks in the bituminous surface. The pavement gets softened as a result, and under the action of traffic a depression soon gets formed. This is aggravated by the use of plastic filler in WBM. If not atteneded to properly, the aggregates in the surface get progressively loosened and a regular pot-hole forms. 2. Lack of proper bond between the bituminous surfacing and the underlaying water bound macadam base can also cause pot-holes. The bond is usually supplied by a tack coat, and any localised inadequacies in these applications can cause pot-holes. 3. Insufficient bitumen content in localised areas of the surfacing layer can cause pot-holes. 4. Too thin a bituminous surface which is unable to withstand the heavy traffic can also cause pot-holes, when associated with improper or inadequate camber. 5. In dense-graded mixtures, pot-holes can be caused by too much fines or too few fines.	Pot-holes are filled with premix open-graded or dense-graded patching, or penetration patching.

5. Edge breaking	Irregular breakage of pavement edge, leading to peeling off the surfacing in large chunks at the edges.	1. Infiltration of water which softens the foundation layers causing the pavement edges to break. 2. Worn out shoulders resulting in insufficient side support to the pavement. 3. Inadequate strength at the edge of the pavement due to inadequate compaction. 4. Lower layer of pavement not being wider than upper layer.	1. Cutting the affected area to regular sections and rebuilding with simultaneous attention paid to proper construction of shoulders. 2. Surface and sub-surface drainage, wherever deficient, should be improved.

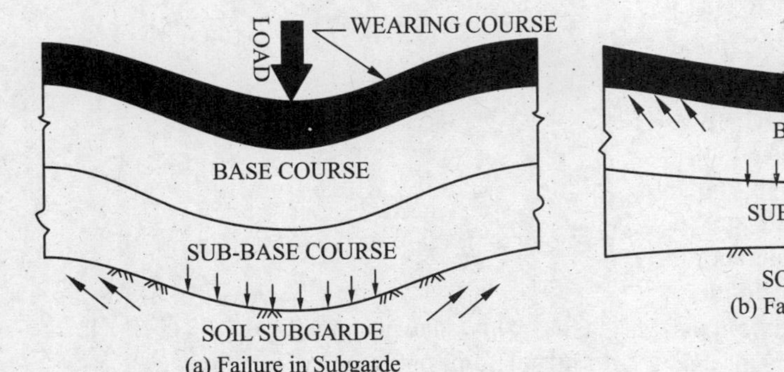

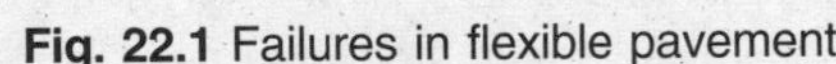
Fig. 22.1 Failures in flexible pavements.

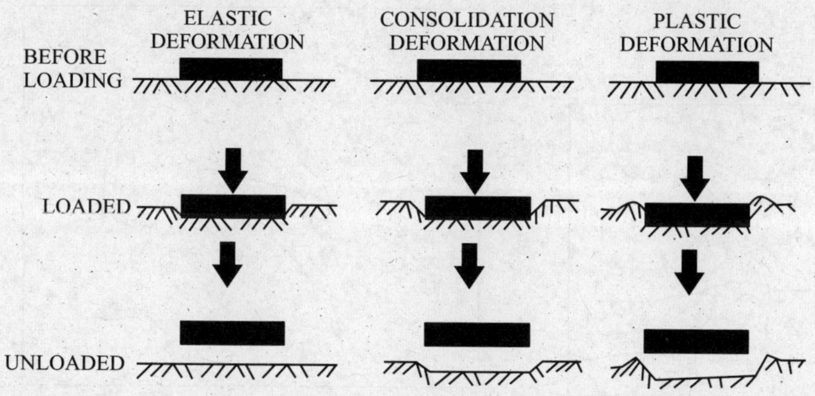

Fig. 22.2 Soil deformation under load.

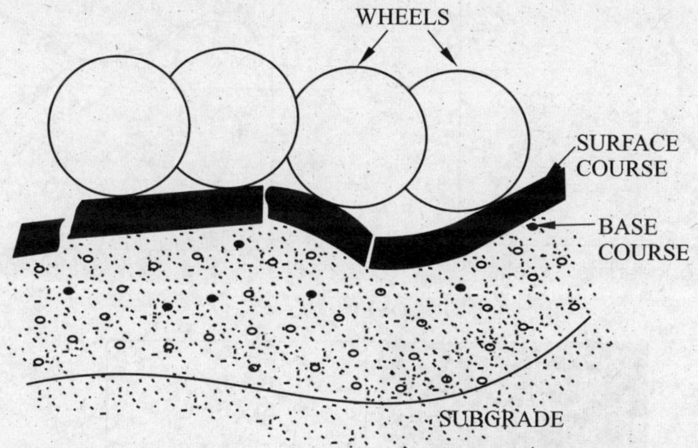

Fig. 22.3 Failure due to improper compaction of subgrade and pavement layer.

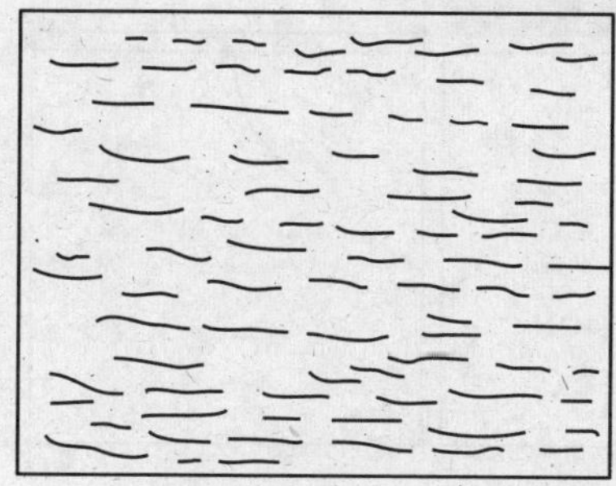

Fig. 22.4 Hair-line cracks.

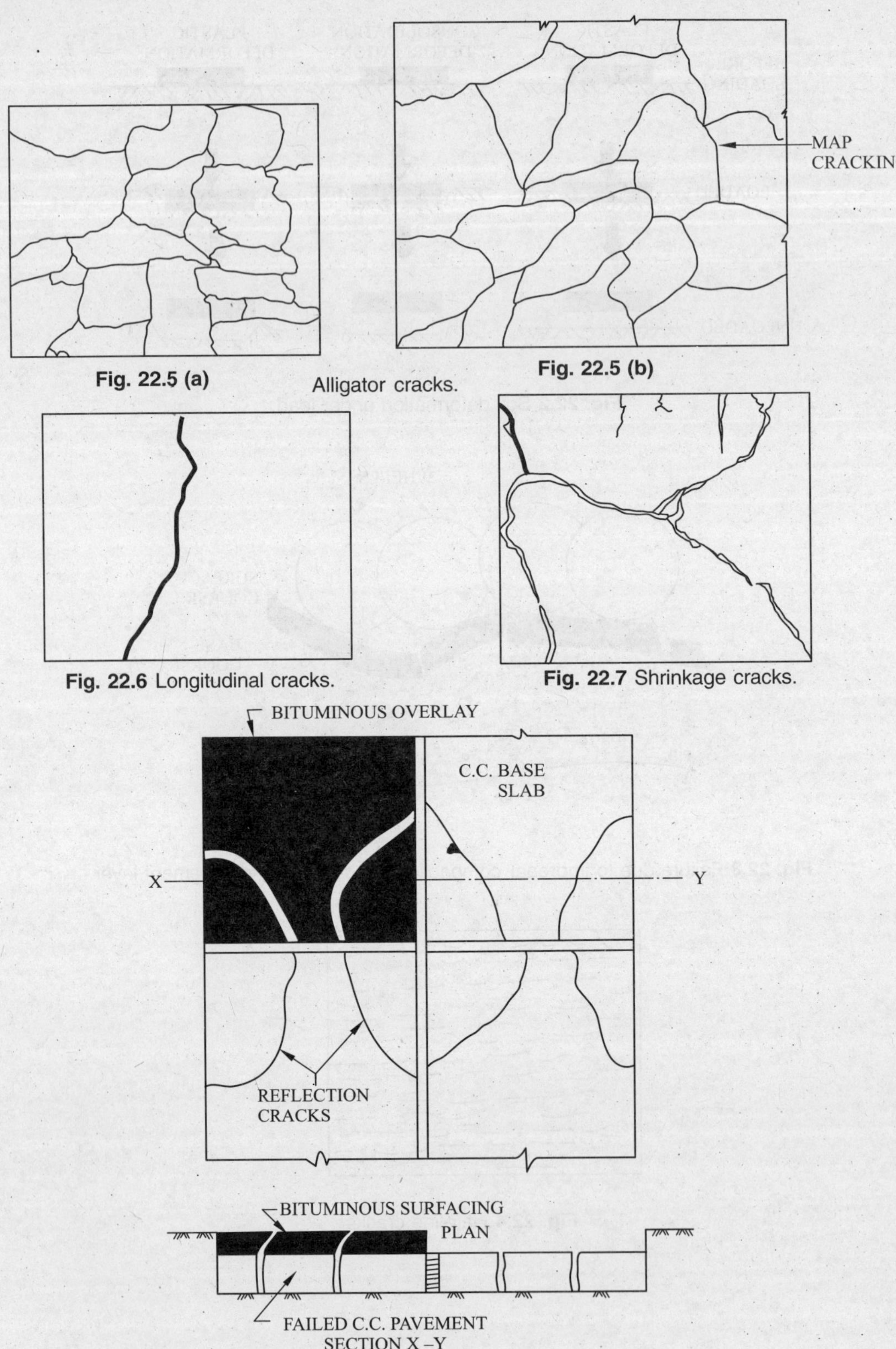

Fig. 22.5 (a) **Fig. 22.5 (b)** Alligator cracks.

Fig. 22.6 Longitudinal cracks.

Fig. 22.7 Shrinkage cracks.

Fig. 22.8 Reflection cracks.

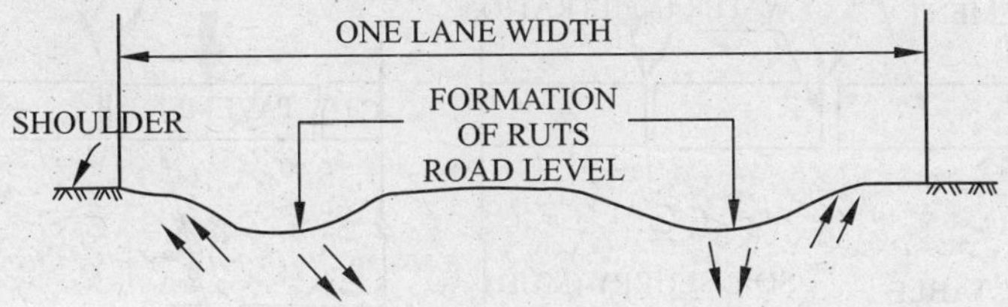

Fig. 22.9 Ruts

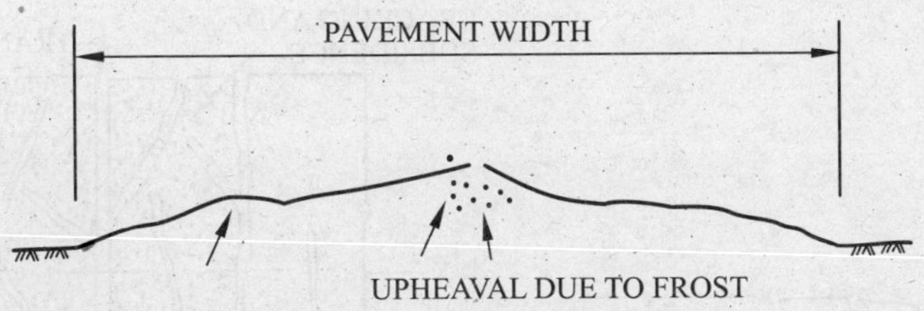

Fig. 22.10 Frost heave failure.

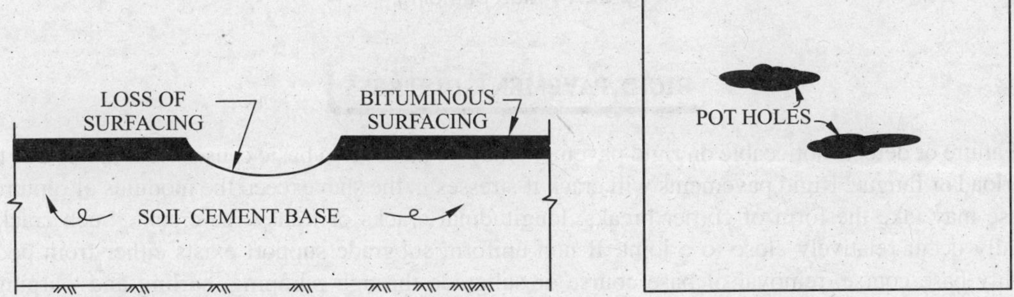

Fig. 22.11 Failure due to lack of binding.

Fig. 22.12 Pot holes.

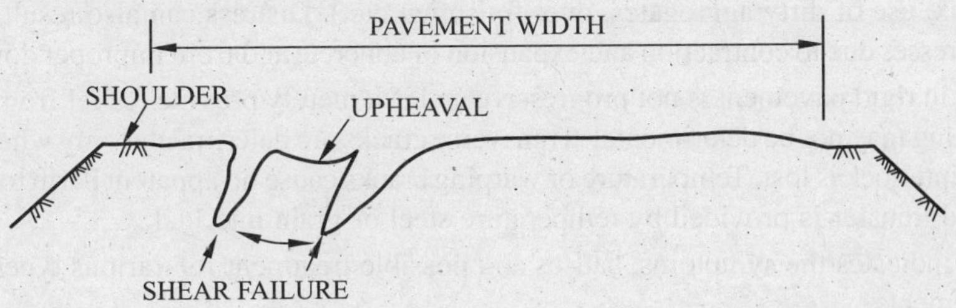

Fig. 22.13 Shear failure cracking in rigid pavement.

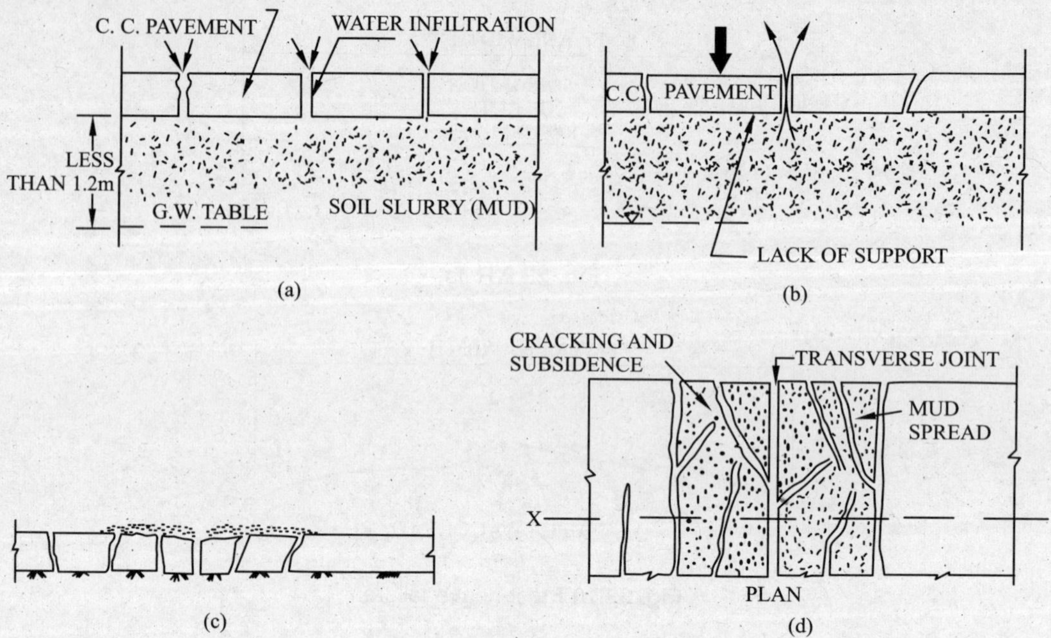

Fig. 22.14 Mud pumping.

RIGID PAVEMENT DISTRESS

Failure or defects noticeable on rigid pavement may be due to two basic causes. The first is due to overload or fatigue. Rigid pavements will crack if stresses in the slab exceed the modulus of rupture. These may take the form of corner breaks, longitudinal cracks or transverse cracks. Such cracks usually occur relatively close to a joint. If non-uniform subgrade support exists either from poor quality base course, removal of base course or subgrade through pumping, curling and warping, high stress concentrations will result at certain points of the slab which may result in cracks in the slab.

The other cause is due to deterioration or the deficiency of the pavement itself. Deterioration of concrete may result from alkali-aggregate reaction, use of non-durable materials, freezing and thawing, scaling resulting from use of salts for ice removal and some other causes (use of high water-cement ratio in the mix, use of dirty aggregates, over finishing, etc.). Distress can also result from warping and curling, stresses due to contraction and expansion of concrete and from improper dowel alignment.

If cracking in rigid pavement is not progressive and if it merely provides relief from high restraint stresses, cracking may not be deterimental. Transverse cracks are deterimental only when load transfer through grain interlock is lost. Temperature or warping cracks cause no apparent harm to the pavement, as long as load transfer is provided by temperature steel or grain interlock.

Table 22.2 indicates the symptoms, causes and possible treatment for various types of distress in concrete pavements.

Table 22.2 Symptoms, causes and treatment of defects in rigid pavements

Type of distress	*Symptoms*	*Causes*	*Treatment*
1. Edge failure	Semi-circular hair-line cracks closely spaced and typically increasing from the joints and pavement edges. This is a progressive distress until nearly complete deterioration may result.	Low durability of concrete due to use of non-durable materials and/or climatic conditions.	1. Provide adequate drainage of the pavement. 2. Maintenance of the joints.
2. Scaling	General deterioration of concrete.	1. Use of mixes which are too wet. 2. Use of dirty aggregate which causes silt and clay to flow to the surface during the finishing process. 3. Use of salts for ice control. 4. Over finishing at the edges. 5. Abrasion action of traffic.	Provide a skid-resistance sand-seal coat
3. Shrinkage cracks	Short cracks of random spacing both in transverse and longitudinal directions.	Shrinkage of concrete itself during the curing period.	
4. Warping cracks	1. Random cracks giving unsightly appearance. 2. Differential movement at the crack. 3 Spalling of longitudinal crack is noted.	1. Absence of longitudinal hinge joints. 2. Crack due to warping which produces high stresses at the centre of the slab.	1. Seal the cracks immediately to prevent infiltration of water and other foreign material.

5. Contraction cracks	Wide cracks, generally 6.3 to 12.5 mm wide.	1. Long length of slab. 2. Transverse cracks from warping. 3. Excessive contraction of the slab. 4. Non-existance of temperature steel.	2. Provide temperature steel at the joints.
6. Frozen dowel bar	1. Spalling on one side of slab near the joints. 2. Cracking of slab.	1. Improper alignment of the dowel bar. 2. Improper lubrication of dowel bars so that freedom of expansion and contraction of the slab is not permitted. 3. Cracks due to shear stresses resulting during the expansion cycle. 4. Deterioration of concrete. 5. Infiltration of incompressible material into the joint.	Maintenance of joints including scaling, cleaning incompressible material from the joint and repair of the joints when defects occur.
7. Blow-ups or tenting	Bumps in the surface which are nuisance to the driving public and may constitute a functional break-down of the highway.	1. Compressive stresses at the pavement joints. 2. Clogging and filling of joints with sand and other incompressible material.	1. Plaining of bitumen from the joint. 2. Removal of joint. 3. Cleaning of joints before resurfacing. 4. Use of pressure relief joints.
8. Structural breaks	Corner breaks, longitudinal cracks or transverse cracks.	1. Over loading. 2. Fatigue due to repetition of loads. 3. Consolidation of the fine grained subgrade.	1. Fill the cracks. 2. Provide an overlay over the surface.
9. Faulted or depression joint	1. Imperfect riding quality. 2. Pumping is apparent. 3. Unequal settlement at the joints.	1. Absence of load transfer devices at the joints. 2. Impact effects of load at the joints.	1. Under-sealing or mud-jacking in conjuction. 2. In some cases use of a granular levelling course between the concrete and surface is required.

10. Structural breaks due to channelised aircraft traffic	Localised structural failure.	Over load with respect to gross load and repetition of load.	Provide a levelling course to restore the riding quality of the pavement before the resurfacing operation.
11. Consolidation and shear failure due to deep foundation movements. (Fig. 22.13)	1. Relative great length of the defected area. 2. Gentle slope	Deep foundation movements.	The levelling course should be kept flat rather than follow the general contour of depression that is being filled.

Chapter 23 PAVEMENT EVALUATION AND STRENGTHENING

Road roughness can be broadly classified into two categories, namely longitudinal distortion and transverse distortion, shown in Fig. 23.1. Longitudinal disturbances primarily affect the rating of pavements.

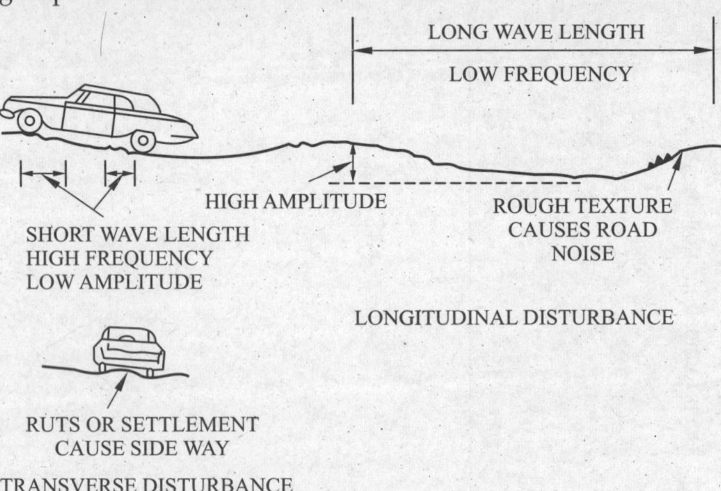

Fig. 23.1 Types of roughness.

Oscillations in the vehicle are set up due to the various frequencies and amplitudes. Long wave-length disturbances can be caused by consolidation of deep foundation layers and normally have high amplitude but low frequency. Another type of longitudinal disturbance, has short wave length, high frequency and low amplitude. Both these disturbances and the rough surface texture of the pavement cause road noise. Similarly transverse disturbances due to ruts will cause discomfort to the passengers.

Measurement of road roughness is important from two considerations:

(*i*) to determine whether pavement is acceptably smooth to carry the intended traffic: or not and

(*ii*) as a correlation factor, which indicates failure of one or more layers within the pavement structure.

23.1. PRESENT SERVICEABILITY INDEX (PSI)

Carey and Irick gave the concept of present serviceability index which correlates users opinions with measurements of road roughness, cracking, patching and rutting.

A group of individuals (raters) drive over the pavement and rate the pavement on a scale shown in Fig. 23.2. The raters mark on the scale, which varies from 0 to 5. A rating of zero indicates an exceedingly poor pavement and a rating of 5, a perfect smooth pavement. The raters are also asked to give their opinion about the objective feature of the pavement that influenced their rating and to indicate whether the road is acceptable for the intended traffic or not. The average of the rating numbers is designated as present serviceability rating (PSR).

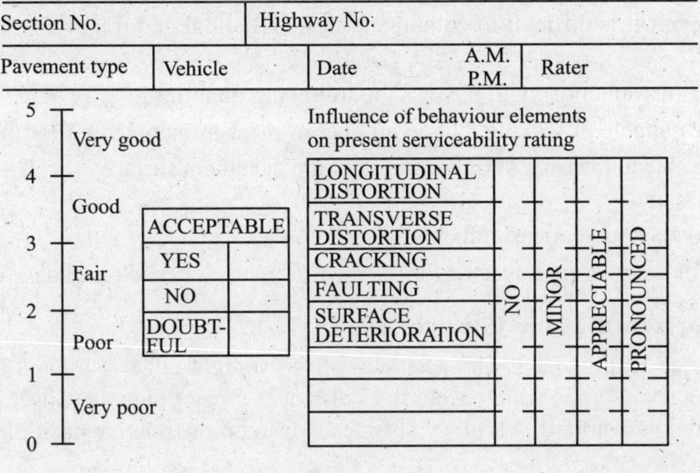

Fig. 23.2 Rating card.

PSR was correlated on the AASHO road test with measurements of roughness, patching and cracking. The regression analysis equations, took the general form as given below:

$$PSI = A_0 + A_1(R) + A_2(F_1) + A_3(F_2) \quad \ldots(23.1)$$

where PSI = Present serviceability index

A = Regression analysis constants

R = Measure of roughness

F = Physical measurements of patching, cracking, etc.

It may be emphasized that PSI only indicates the condition of the pavement at an instant of time, without being influenced by factors such as pavement width, shoulder width and condition, grade, alignment, structural adequacy, traffic and climate.

A serviceability rating of 2.5 is considered an acceptable pavement for the primary system and a rating of 2.0 for the secondary system.

23.1.1. Uses of PSI

The uses of present serviceability index are:

(*i*) It permits rating of pavements on a common scale.

(*ii*) It permits making priority and maintenance programmes in a logical manner.

(*iii*) It helps to establish relationship between objective pavement measurements and subjective rating of the road user. From the PSI value, an engineer can get an idea of the degree of rutting, cracking, patching, etc.

(*iv*) It permits obtaining measurement at various times and the establishment of a parameter that defines pavement condition in design equations.

(*v*) Histories of pavement performance can be related to changes in serviceability with time.

(*vi*) The greatest use is that it permits to estimate the life trends of an in-service pavement.

23.2. CONDITION SURVEY INSTRUMENTS

Roughness is the primary factor which influences road users opinion of a road. It can be longitudinal or transverse. The distortion of a road pavement may consist of both long and short wave length disturbance of varying frequencies. Thus the instrument used must be capable of measuring all these distortions. However, it is difficult to combine both longitudinal and transverse measurements in a single instrument.

An all-purpose instrument should possess the following qualities:

(*i*) It should be capable of making a large number of measurements in a short period of time.

(*ii*) It should be able to measure abrupt changes of the pavement surface as well as long wavelength disturbances; and

(*iii*) It should give accurate measurements.

A description of the instruments used for the measurements of road roughness is given below:

23.2.1. Fixed horizontal plane (Straight Edge)

This is illustrated in Fig. 23.3. The accuracy of the straight-edge is limited by its wheel base length. Small changes in grade which occur over distances greater than the wheel base length of the instrument, are greatly distorted. Accuracy also depends upon the relative position of the measuring wheel.

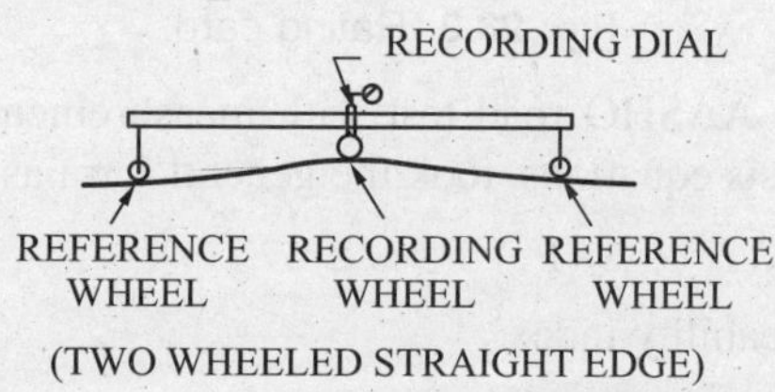

Fig. 23.3 Common straight edge.

The straight-edge has been used for a long period for construction control. However, its use for measuring pavement roughness during its life time, is quite limited.

23.2.2. Multiple-wheel device

Fig. 23.4 illustrates the principle of the multiple-wheel device. This instrument has an advantage over the straight-edge that the hinged wheels at either end tend to iron out the small irregularities as the profilometer (straight-edge) passes over them. Continuous readings can be taken by means of a feeler wheel at the centre of the instrument. The profilometer has an advantage that it can be used for both construction control and pavement evaluation. The main disadvantage is the slow rate of making measurements over large distances. The instrument is portable and one person can operate it. This has a great utility in both highway and airport pavements.

Sometimes the profilometers are mounted on a truck. The truck's frame acts as a straight-edge and a feeler wheel is provided at the centre of the truck, which make continuous readings of pavement profile as the truck moves, over the pavement.

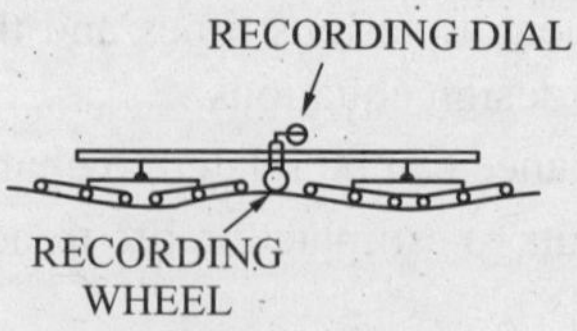

Fig. 23.4 Multiple wheels profilometer.

23.2.3. Slope profilometer

The slope profilometer (Fig. 23.5) works on the principle of an imaginary fixed horizontal or vertical plane which is established by means of a pendulum or gyroscope. To increase the accuracy of the straight-edge, the main frame of the profilometer is provided with a relatively long wheel base and the pendulum or gyroscope maintains the reference of the frame with a fixed horizontal plane. At the end of the profilometer two small wheels, with a relatively small wheel base, are fixed in such a manner that they pick up and measure small disturbances which can be fed electronically into the main measuring system. The data are integrated into that of the fixed horizontal plane. The instrument will then measure the true profile of the pavement.

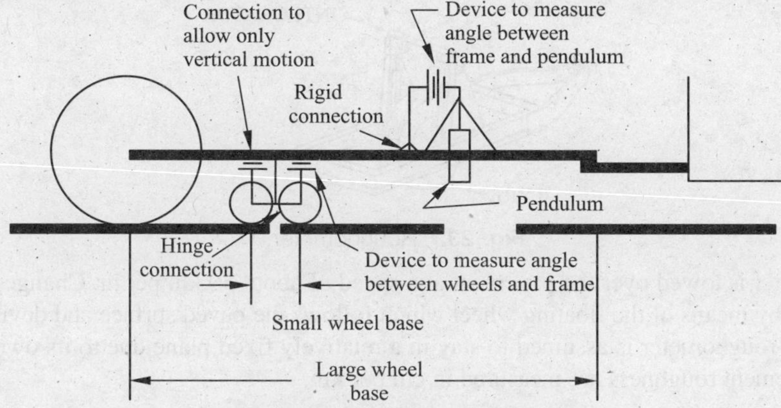

Fig. 23.5 Slope profilometer.

This type of instrument was used in the evaluation of the AASHO road test. The statistic obtained from the device is known as the slope variance (SV) and it is measured in both wheel paths of the pavement. The slope variance is related to the difference in elevation measured by the device by the following relation. Measurements are made at a speed of about 6 to 8 km per hr.

$$SV = \frac{\Sigma y^2 - \frac{1}{n}(\Sigma y)^2}{n-1} \qquad \ldots(23.2)$$

where y = Difference in elevation of two points 0.3 m apart;

and n = number of readings.

23.2.4. Chloe profilometer

This is essentially the same as the slope profilometer, with the exception that the fixed horizontal reference is omitted. Equation 23.2 also holds good in this case. This instrument is widely used for the pavement evaluation.

23.2.5. GMR profilometer

This was developed by the General Motors Corporation U.S.A. (See Fig. 23.6) and works on the principle of a series of accelerometers).

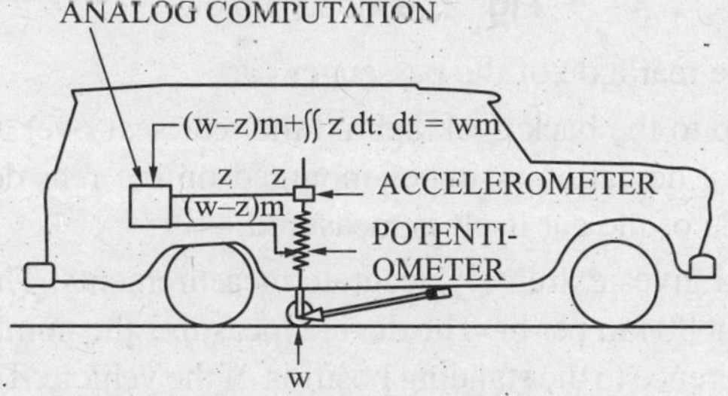

Fig. 23.6 High speed GMR profilometer.

It runs at a high speed, 65 to 80 km per hr. The pavement response is measured by means of a small feeler wheel at the centre of the profilometer.

23.2.6. Roughometer

It consists of a rectangular frame. A single wheel having a pneumatic tyre is provided inside the frame on an axle which is attached to the centre of two single lease springs, one on each side of the wheel. The integrator unit is mounted on its cross-frame over the wheel. The pistons of two damping units are also attached to the cross-frame. The components of a roughometer are shown in Fig. 23.7.

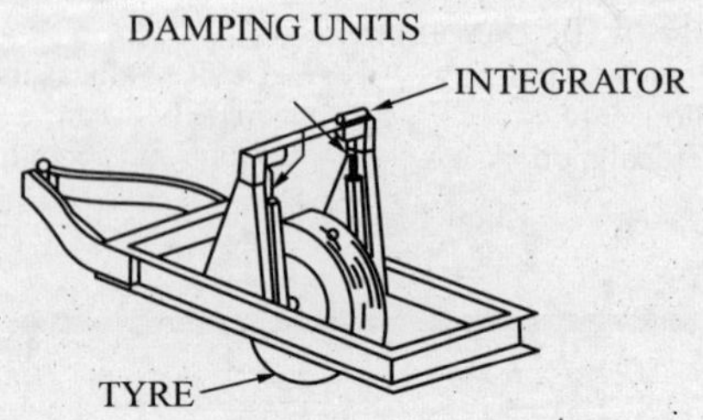

Fig. 23.7 Roughometer.

Roughometer is towed over the pavement at a speed of about 50 km per hr. Changes in elevation are measured by means of the floating wheel which follows the paved surface and deviates from the machine. The roughometer is assumed to stay in a relatively fixed plane due to its own inertia. The values of pavement roughness are measured in cm per km.

23.2.7. Road meter

M.P. Brokaw was the first person to develop a road meter, shown in Fig. 23.8. The vehicle itself is used as the measuring instrument. This is commonly known as PCA road meter.

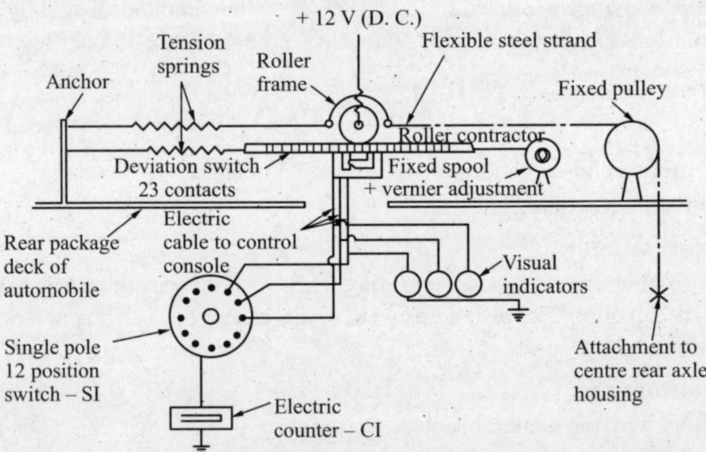

Fig. 23.8 Road meter.

A cable is attached to the rear axle of the passenger car.

The cable then comes up to the back deck behind the rear seat over a pulley and is attached to a wheel that rolls over several deviation switches mounted on the rear deck. Deviation between the body frame and the rear axle of the car itself is measured.

The device is simple and gives extremely accurate measurements. The vehicle is driven over the pavement at a speed of about 80 km per hr. The device measures the number of road-car deviations in 3.2 mm increments with reference to the standing position of the vehicle. The numbers are accumulated on electric counters. The sum of the squares of the deviations are correlated with slope variance and are used as statistic derived from the instrument.

$$\Sigma D^2 = \frac{a + 4b + 9c + 16d + 25e + \ldots}{64}$$

Where a = number of deviation reading equal to 3.2 mm.

b = number of deviation readings equal to 2 × 3.2 mm, and so on.

The road meter has been correlated with the chloe profilometer from which slope variance is calculated.

The relation is $SV = 0.68\,\Sigma\,(D^2) + 0.8$ ...(23.3)

23.2.8. Comparison of methods

It has been found out, on the basis of a research study, that one device offers little advantage over another so far as accuracy is concerned. Therefore, the choice of the type of instrument should be made on the basis of economy, utility and speed of operation in the field.

23.3. SKID RESISTANCE

The discussion so far relates to the pavement condition as determined by riding quality. Another important factor, from the view point of maintenance, is the resistance of the pavement surface to sliding or skidding of the vehicle. Skidding refers to a sliding motion with locked wheels and is evaluated as a skid number, called the *coefficient of friction*. Skidding may result from polishing action of the traffic or by flushing or bleeding of the surface in the case of bituminous surfaces.

The phenomenon of hydroplaning is of equal concern to skidding. The tyre of the vehicle loses contact with the pavement surface because of the presence of a film of water between the tyre and the road surface. Hydroplaning may be caused due to the vehicle moving at a high speed on a road surface having a film of water or due to rutting of the pavement.

The remedial measures to avoid skidding or hydroplaning is to provide a sand seal or by grooving the pavement.

STRUCTURAL EVALUATION

Evaluation of the structural properties of the pavement components is essential to determine the structural adequacy of the pavement and to redesign the same for strengthening or laying overlays. The structural evaluation can be done by the following methods.

23.4. CBR METHOD

Estimate of the strength of the subgrade can be obtained from California bearing ratio test carried out prior to pavement design and construction. However, the test often becomes unreliable due to the densification of the road under traffic and environmental factors.

In-placc ficld CBR tests are sometimes performed, but these tests are time consuming and expensive as large pits are required to be dug. In lieu of in-place CBR tests, undisturbed sample of soil from test pits are tested in the laboratory for CBR values under unsoaked and soaked conditions. These are possible only in the case of fine grained soils which possess cohesion.

23.4.1. Plate bearing tests

Modulus of elasticity of the different components of the pavement and the modulus of subgrade reaction can be determined by performing plate bearing tests. These tests require large sized test pits to be dug, are time consuming and expensive.

23.4.2. Non-destructive field tests

Instruments which measure deflection are used to measure stress-strain relationship of the pavement. These are described in the next paragraph. In addition, instruments are used which induce vibratory forces to the pavement. The response of the pavement is measured by means of velocity transducers. One such device has been developed by the U.S. Corps of Engineers for evaluation of airports. The instrument is mounted on a heavy truck, exerts a static load of 7250 kg and can produce vibratory loads from 0 to 6800 kg at a frequency of 15 Hz.

23.4.3. Laboratory tests

These tests include grain-size distribution, density and moisture content. They determine the quality of the pavement components. Properties of the pavement can also be determined by taking a core of pavement and carrying out compression tests, split tensile tests, etc.

In place density and moisture content tests

Moisture and density data can be obtained using standard sand core, water balloon or nuclear techniques. The density and moisture content of the subgrade will vary with season.

EVALUATION BY DEFLECTION MEASUREMENTS

The stress-strain properties of the pavement will determine its structural adequacy. Structural evaluation of pavements can be done by means of the following methods:

1. Benkelman beam deflection technique.
2. Direct load test method: and
3. Indirect reverse design method.

The first method is normally adopted for flexible pavements and the other two methods for rigid pavements. These are described in the following paragraphs.

23.5. BENKELMAN BEAM DEFLECTION TECHNIQUE

23.5.1. Basic principles of deflection method

The deflection method is based on the concept that pavement sections which have been conditioned by traffic, deform elastically under a load. The deformation or elastic deflection under a given load depends upon subgrade soil type, its moisture content and compaction, the thickness and quality of the pavement course, drainage conditions, pavement surface temperature, etc. Extensive studies have also shown that performance of flexible pavements are closely related to the elastic deflection of pavement under the wheel loads.

Pavement deflection is measured by the Benkelman beam (see Fig. 23.9) which consists of a slender beam 3.66 m long pivoted at a distance of 2.44 m from of tip. By suitably placing the probe between the dual wheels of a loaded truck, it is possible to measure the rebound and residual deflection of the pavement structure. While the rebound deflection is the one related to pavement performance, the residual deflection may be due to non-recoverable deflection of the pavement due to the influence of the deflection bowl on the front legs of the beam. Rebound deflection is used for overlay design.

23.5.2. Procedure for deflection

General

The deflection survey essentially consists of two operations; (*i*) condition survey for collecting the basic information about the road structure and based on this demarcation of the road into sections of more or less equal performance; and (*ii*) actual deflection measurements.

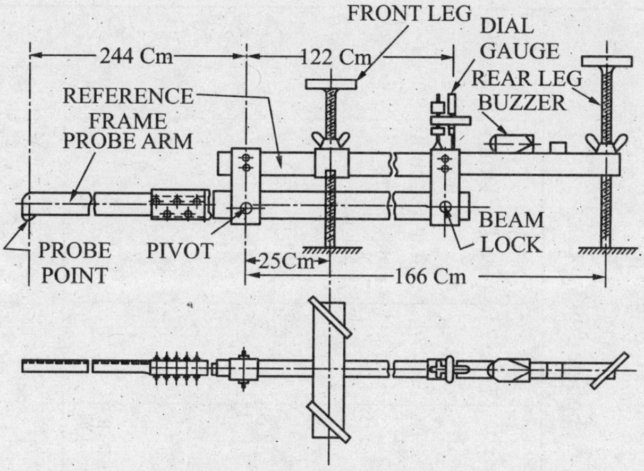

Fig. 23.9 Benkelman beam.

23.5.3. Pavement condition survey

This phase of operation, which should precede the actual deflection measurements, consists primarily of visual observations supplemented by simple measurements for rut depth using a 3 meter straight edge. Based on these, the road should be classified into sections of equal performance in accordance with the criteria given in Table 23.1.

Table 23.1 Criteria for classification of pavement sections

Classification	*Pavement condition*
Good	No cracking, rutting less than 10 mm
Fair	No cracking or cracking confined to single crack in the wheel track with rutting between 10 mm and 20 mm
Poor	Extensive cracking and/or rutting greater than 20 mm

As it is inexpedient to modify the overlay design at frequent intervals, it will be preferable if the length of each section is kept minimum say 500 m.

During condition survey, information should also be collected about drainage characteristics, depth of water table, whether the road is in cut or fill, changes in soil profile, topography, climatic conditions and other relevant features. Test pits should be dug approximately every 250-500 m depending on the uniformity in performance of pavement structure to determine the thickness and composition of the pavement layers. Where it is intended to compare the results with the CBR design method, CBR and other characteristics of the subgrade soil should also be determined.

23.5.4. Actual deflection measurements

In each road section of uniform performance minimum of ten points should be marked at equal intervals in each lane of traffic for making the deflection observations in the outer wheel path. The intervals betwexen the points may be as low as 50 m depending on the length of section under investigation. On roads having more than one lane, the points marked on adjacent lanes should be staggered. In the transverse direction, the measurement points should be 60 cm from the pavement edge if the lane width is less than 3.5 m and 90 cm when the lane width is more.

Variability of deflections in a given section should be considered for detecting spots where extra deflection measurements have to be made. For this purpose highest and lowest values in a group of ten should be compared with mean value. If the highest or lowest values differ from the mean by more than one-third of mean then extra deflection measurements should be made at 25 m on either side of point where high or low values are observed.

Table 23.2 Proforma for recording field information collected during the pavement condition survey

Name of road :
Section :
No. of traffic lanes :

Date of survey :
Traffic intensity :
Annual rainfall (mm) :
Subgrade soil type :

S. No.	Sub-section from ... to	Height of embankment depth of cutting	Pavement condition **	Pavement details						Type of shoulder	Depth of water table	Drainage condition	Remarks ***
				Surface		Base		Sub-base					
				Type	Thickness	Type	Thickness	Type	Thickness				
1	2	3	4	5	6	7	8	9	10	11	12	13	14

Note: 1. **Classify as 'Good, Fair, or Poor' based on the criteria given in Table 23.1.

2. ***Record any special or abnormal conditions such as flooding, submergence, previous failure history if any, etc.

For measuring pavements deflection, several procedures are available and fall under two main categories: (i) testing under static load, and (ii) testing under creep speed. C.G.R.A. procedure determining the rebound deflection of a pavement under static load, or the WASHO procedure based on test under creep speed may be adopted. In both methods, a standard truck having a rear axle weighing 8160 kg fitted with dual tyre inflated to a pressure of 5.60 kg/cm^2 is used for loading the pavement. During actual tests the total load and the tyre pressure are maintained within a tolerance of ± 1 per cent and ± 5 per cent respectively.

Before starting the deflection measurements, the Benkelman beam should be calibrated to ensure that the dial gauge and beam are working correctly. This can be done by using the simple procedure described below. The beam is placed and levelled on a hard level ground. A number of metallic blocks of different thickness (measured accurately with a precision micrometer) with perfectly plane faces are placed under the probe and the dial gauge reading recorded each time. If the beam is in order, the dial gauge on the beam should read one half the thickness of the metallic block on which the probe was placed. Otherwise, the dial gauge should be checked and replaced if necessary. If the dial gauge is functioning correctly, the beam pivot should be checked for free and smooth operation, secondly the striking plate beneath the dial gauge spindle should be checked to ensure that it is tightly secured and has not become grooved by the dial gauge stylus.

23.6. STATIC LOAD DEFLECTION TEST PROCEDURE—C.G.R.A. METHOD

23.6.1. Equipment

The equipment includes:

1. Benkelman beam
   (*a*) Length of probe arm from pivot to probe point : 244 cm
   (*b*) Length of measurement arm from pivot to dial : 122 cm
   (*c*) Distance from pivot to front legs : 25 cm
   (*d*) Distance from pivot to rear legs : 166 cm
   (*e*) Lateral spacing of front support legs : 33 cm
2. A 5 tonne truck is recommended as the reaction. The vehicle should have 8160 kg rear axle load equally distributed over the two wheels, equipped with dual tyres. Spacing between the tyre walls should be 30 to 40 mm. The tyres shall be 10 × 20, 12 ply inflated to a pressure of 5.60 kg/cm^2. The use of tyres with tubes and rib treads is recommended.
3. Tyre pressure measuring gauge.
4. Thermometer (0–100°C) with 1° division.
5. A mandrel for making 4.5 cm deep hole in the pavement for temperature measurement. The diameter of the hole at the surface should be 1.25 cm and at bottom 1 cm.

23.6.2. Procedure

1. The point on the pavement to be tested is selected and marked. For highways, the point should be located 60 cm from the pavement edge if the lane width is less than 3.5 m and 90 cm from the pavement edge for wider lanes. For divided four lane highway, the measurement point should be 1.5 m from the pavement edge.
2. The dual wheels of the truck are centered above the selected point.
3. The probe of the Benkelman beam is inserted between the duals and placed on the selected point.
4. The locking pin is removed from the beam and the legs are adjusted so that the plunger of the beam is in contact with the stem of the dial gauge. The beam pivot arms are checked for free movement.

5. The dial gauge is set at approximately 1 cm. The initial reading is recorded when the rate of deformation of the pavement is equal to or less than 0.025 mm per minute.
6. The truck is slowly driven a distance of 270 cm and stopped.
7. An intermediate reading is recorded when the rate of recovery of the pavement is equal to or less than 0.025 mm per minute.
8. The truck is driven forward a further 9 m.
9. The final reading is recorded when the rate of recovery of pavement is equal to or less than 0.025 mm per minute.
10. Pavement temperature is recorded at lest once every hour inserting thermometer in the standard hole and filling up the hole with glycerol.
11. The tyre pressure is checked at two to three hour intervals during the day and adjusted to the standard, if necessary.

23.6.3. Calculations

1. Subtract the final dial reading from the initial dial reading. Also subtract the intermediate reading from the initial reading.
2. If the differential readings obtained compare within 0.025 mm the actual pavement deflection is twice the final differential reading.
3. If the differential reading obtained do not compare to 0.025 mm, twice the final differential dial reading represents apparent pavement deflection.
4. Apparent deflections are corrected by means of the following formula:

   $X_t = X_A + 2.91$ Y, in which

   X_t = True pavement deflection

   X_A = Apparent pavement deflection

   Y = Vertical movement of the front legs *i.e.*, twice the difference between the final and intermediate dial readings.

23.7. CREEP LOAD DEFLECTION TEST PROCEDURE

23.7.1. Equipment

Same as above

23.7.2. Procedure

1. The point of the pavement where the deflection is to be measured is selected and marked.
2. Move the truck so that its rear wheel is about 1.2 m behind the selected point.
3. Insert probe arm between dual tyres of the vehicle to a distance of about 1.2 m lining up arm by eye in such a position that rubbing of probe arm and tyre walls does not occur.
4. While truck is standing, record initial reading of dial. Turn on vibrator buzzer before taking first reading.
5. Vehicle should be moved slowly (2 km/h) and smoothly forward to at least 3 m past the tip of the beam. The beam operator should watch to see that the probe arm does not rub. The maximum dial reading will occur when wheels are in line with the contact point. Record this value. After a reasonable length of time or when dial needle has come to rest the final reading should be recorded.

23.7.3. Calculations

1. The maximum deflection is twice the difference between the initial and maximum readings.
2. The rebound deflection is twice the difference between the second reading and the final reading.

3. The residual deflection is twice the difference between the initial and final readings.

The deflection measurements and other information collected during the deflection survey should be recorded in the proforma suggested in Table 23.3.

Table 23.3 Proforma for recording pavement deflection data

Name of Road :		Data and time of observation	:
Section :		Climatic conditions	:
No. of traffic lanes :		Ambient temperature	:
		Method of deflection measurement	:
		Whether temperature correction is to be applied	Yes/No
		Whether correction for seasonal variation is to be applied	Yes/No

S. No.	Location of test point and identification of lane	Pavement temperature ° C	Dial gauge reading			
			Initial	Intermediate	Final	Deflection
1	2	3	4	5	6	7

23.8. CORRECTION FOR TEMPERATURE VARIATIONS

Deflections measured by the Benkelman beam are influenced by the pavement temperature. For design purposes, therefore, all deflection values should be related to a common temperature. Measurements made when the pavement temperature is different than standard temperature would need to be corrected. The standard temperature and the procedure for correction are discussed below.

The stiffness of bituminous layers changes with temperature of the binder and consequently the surface deflections of a given pavement will vary depending on the temperature of the constituent bituminous layers. For purposes of design, therefore, it is necessary that the measured deflection be corrected to a common standard temperature. For areas in the country having a tropical climate, the standard temperature recommended is 35°C. Correction for temperature is not applicable in the case of roads with thin bituminous surfacing (such as premix carpet or surface dressing over a non-bituminous base) since these are usually unaffected by changes in temperature. But temperature correction will be required for pavements having a substantial thickness of bituminous construction (*i.e.*, minimum 40 mm). Correction need not however be applied in the latter case if the road is subject to severe cracking or the bituminous layer is substantially stripped.

Correction for temperature variation on deflection values measured at pavement temperature greater than 35°C should be 0.01 mm for each degree centrigrade change from the standard temperature of 35°C. The correction will be positive for pavement temperature lower than 35°C and negative for temperatures higher than 35°C. For example, if the deflection is measured at a pavement temperature of 37°C, the correction factor will be 0.02 mm (= 2 × 0.01) which should be subtracted from the measured deflection to obtain the corrected value corresponding to standard pavement temperature of 35°C.

In colder areas, and areas of altitude greater than 1000 m where the average day temperature is less than 20°C for more than 4 months in a year, the standard temperature of 35°C will not apply. In the absence of adequate data about deflection-performance relationship, it is recommended that the deflection measurements in such areas be made when the ambient temperature is greater than 20°C and that no correction for temperature need be applied.

In cases where temperature correction is to be applied, the pavement temperature should be measured during the deflection survey. The measurement should be made at a depth of 40 mm using a suitable short-stem mercury thermometer. A hole of about 45 mm deep and about 10 mm diameter should be drilled in the pavement and filled with enough glycerol and temperature recorded after about 5 minutes.

23.9. CORRECTION FOR SEASONAL VARIATIONS

Pavement deflections are also affected by seasonal variations in climate. For the purpose of applying correction, it is intended that the pavement deflections should pertain to the period when the pavement is in its weakest condition. In India, this period occurs soon after monsoon, it is, therefore, desirable to conduct deflection measurements during this period. When deflections are measured during the dry months, they will require a correction factor which is defined as the ratio of the maximum deflections immediately after monsoon to that of minimum deflection in the dry months.

Correction for seasonal variation shall depend on type of subgrade soil, its field moisture content (at the time of deflection survey) and average annual rainfall in the area. For this purpose, subgrade soils have been divided into three broad categories, namely sandy/gravelly, clayey with low plasticity ($PI \leq 15$) and clayey with high plasticity (PI > 15). Similarly, rainfall has been divided into two categories, namely low rainfall (annual rainfall $\leq$ 1300 mm) and high rainfall (annual rainfall > 1300 mm). Moisture correction factors (or seasonal correction factors) shall be obtained from Figs. 23.10 to 23.15 for given field moisture content, type of subgrade soil and annual rainfall.

The soil sample for determination of subgrade type and its field moisture content shall be scooped from below the pavement as shown in Fig. 23.16. For this purpose a test pit at the shoulder (adjacent to pavement edge) shall be dug to a depth upto 15 cm below the subgrade level in every kilometer depending on the uniformity of subgrade soil, topography of the area and road profile. A soil sample of weight not less than 100 gm should be collected using an auger from the subgrade underneath the deflection observation points i.e. 0.6 m and 0.9 m from the pavement edge for single and two lane pavements respectively at a depth of 50 mm to 100 mm below the subgrade level as shown in Fig. 23.16. The subgrade soil shall be tested as per IS-2720 for type of subgrade soil, plasticity index and field moisture content. The test pit shall be made good immediately after taking soils sample and study of pavement composition.

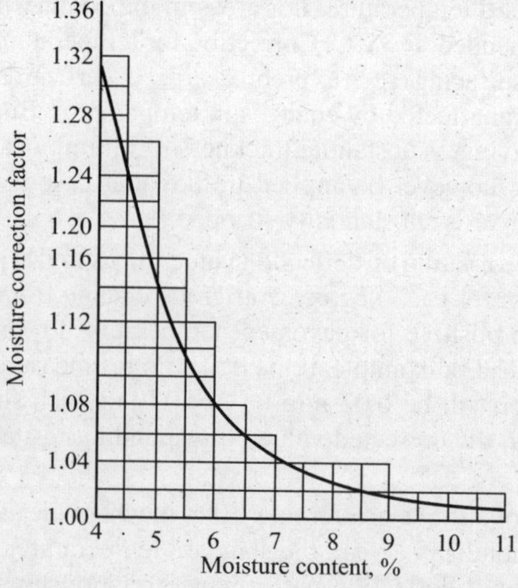

Fig. 23.10 Moisture correction factor for Sandy/Gravelly soil subgrade for low rainfall areas (Annual rainfall $\leq$ 1300 mm)

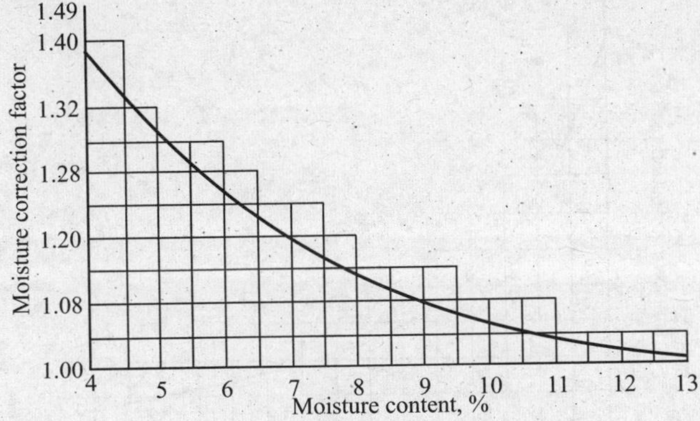

Fig. 23.11 Moisture correction factor for Sandy/Gravelly subgrade for high rainfall areas (Annual rainfall > 1300 mm).

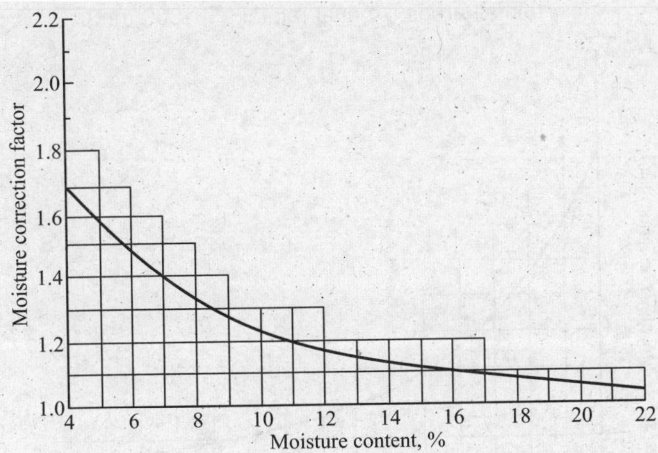

Fig. 23.12 Moisture correction factor for clayey subgrade with low plasticity (PI < 15) for low rainfall areas (Annual rainfall ≤ 1300 mm).

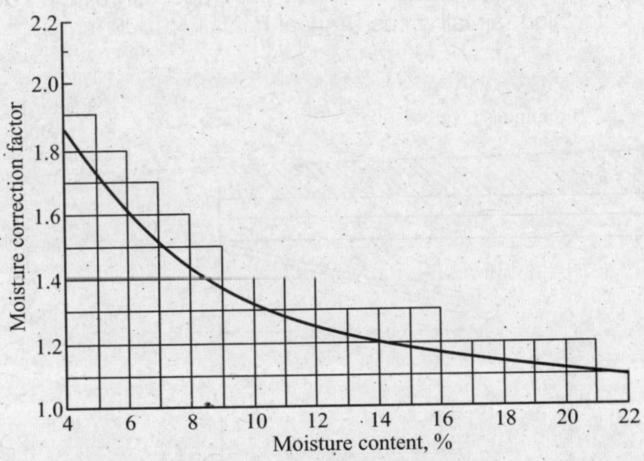

Fig. 23.13 Moisture correction factor for clayey subgrade with low plasticity (PI < 15) for high rainfall areas (Annual rainfall > 1300 mm).

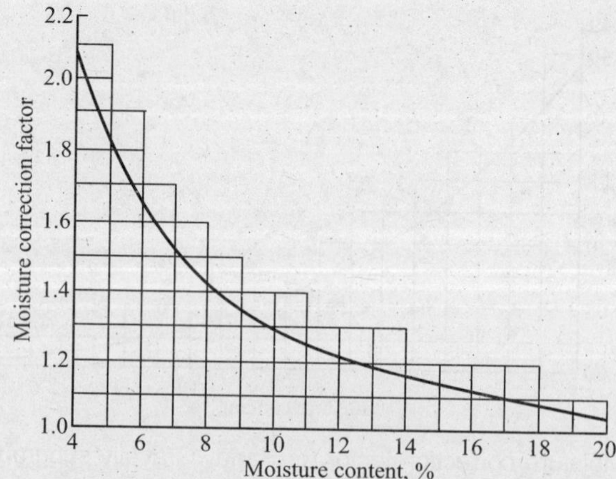

Fig. 23.14 Moisture correction factor for clayey subgrade with high plasticity (PI > 15) low rainfall areas (Annual rainfall ≤ 1300 mm).

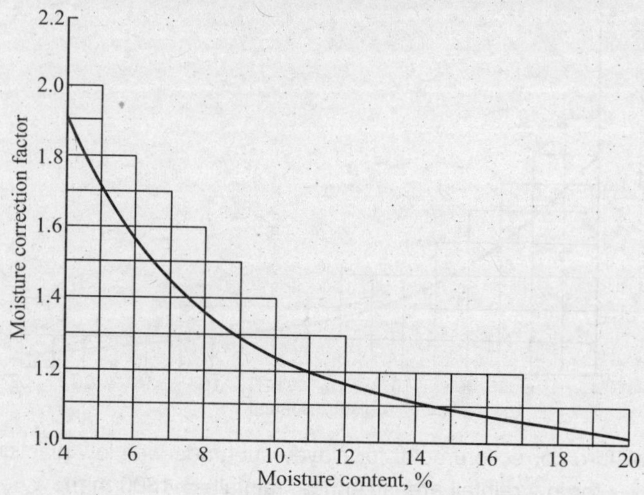

Fig. 23.15 Moisture correction factor for clayey subgrade with high plasticity (PI > 15) for high rainfall areas (Annual rainfall > 1300 mm).

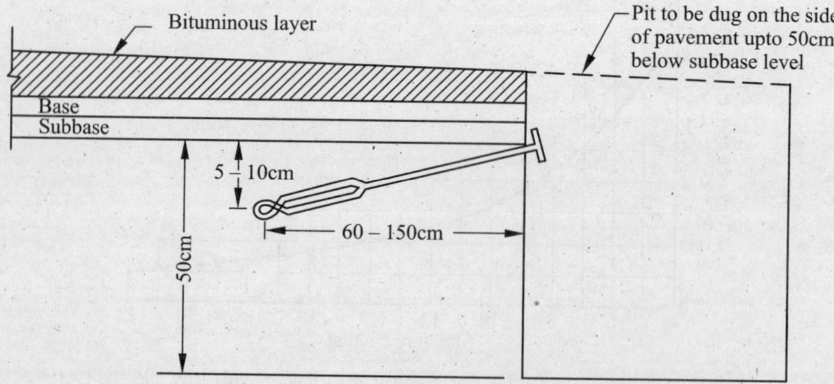

Fig. 23.16 Determination of field moisture content using auger.

23.10. STRUCTURAL EVALUATION OF RIGID PAVEMENTS

The structural evaluation of rigid pavements can be done by two alternative methods, namely the '*direct load test method*' and the "*indirect reverse design method*" which are commonly adopted in the country.

The "direct load test method" provides actual load carrying capacity of the pavement taking into account the interaction between various pavement layers, extent of load transfer at joints *etc*., as actually obtained at site. The "*reverse design method*" on the other hand, involves indirect computation of the pavement strength based on evaluation of the individual design parameters. Since the interaction between different parts of the in-service pavement cannot be taken into account in the case of indirect method, and as the inherent approximations in the method also affect the evaluated strength value, from the point of accuracy of evaluation the direct load test method has a definite edge over the indirect method and as such should be preferred.

The actual choice of the method in individual cases will, however, be directed by other considerations as well, for instance the availability of suitable load testing reaction frame, testing facilities, time available for the study, the period for which airfield can be closed to traffic, *etc*. For important works, every effort should be made to arrange direct load tests. However, when facilities for this are not available, a reasonably approximate assessment of the pavement structural strength can be had from the indirect method. Even if the reverse design method has to be adopted, it would be worthwhile conducting a few direct load tests for comparison. These tests could be done at locations where indirect tests are proposed to be carried out, so that the results could be tied together to assess actual interaction of the pavement components obtaining at site.

Both in the case of direct and indirect methods, after field data have been collected, the structural capacity of rigid pavements can be noted either in terms of LCN (Load Classification Number) or LCG (Load Classification Group). While the former is based on corner loading, the latter is for internal loading conditions. Joints in rigid pavements in the country are in most cases not provided with dowel bars. For these situations, the corner loading condition is more appropriate. As such the LCN rating system is considered more suitable for conditions existing in the country.

The procedures recommended are intended for rigid pavements in sound structural condition. It the pavement has any localized defective spots, such locations should not be selected for overall structural evaluation, as this might result in gross under rating of the pavement. Such defective spots should be grouped together and separately investigated to evaluate the reasons for the defects, as well as the remedy measures like mud jacking, improvement of drainage, removal and replacement, *etc*., and not in superimposition of a thick overlay which may provide only temporary relief.

23.11. DIRECT LOAD TEST METHOD

23.11.1. General

The direct load test method essentially involves the application of static loads through a rigid plate on the existing pavement and noting the response in the form of deflections/strains. Based on the data collected, the structural capacity of the pavement is rated in terms of LCN through the use of appropriate charts. This method has the advantage that it gives the actual load carrying capacity of the pavement at the time of test, taking the different interactions into account. Since the results obtained by this method are dependent on test conditions, such as the time of testing, location of test points etc, it is necessary to follow a common procedure to determine the critical load carrying capacity of the pavement. Guidelines with regard to these conditions are given below:

(1) *Period of testing*: The period of testing should be such that it is critical for foundation strength (*i.e.* when the foundation saturation is at maximum or near maximum) as well as for load transfer especially in the case of dummy joints (*i.e.* when the mean pavement temperature is minimum during the year). Early winter, following the rainy season, can be a good working compromise from these considerations. If it becomes necessary from other considerations to conduct the tests at any other period, appropriate adjustments, based on actual tests or engineering judgement, should be applied in respect of foundation strength and degree of load transfer.

(2) *Locations of test*: In the case of undowelled pavements, 50 per cent tests should be carried out at the junction of transverse expansion joint and longitudinal construction joint, and the remaining at free slab corner at the junction of transverse and longitudinal expansion joint. For pavements having dowelled transverse expansion joints, however, the junction of longitudinal expansion or construction joint and the transverse dummy joint may be chosen as the test locations.

(3) *Number of test locations and selection of test sites*: For airfield evaluation one test is recommended for every 60 m length in case of runways and 60-90 m length in case of taxi tracks and aprons. In general a total of 15-20 determinations may be required for proper statistical evaluation of the test data. These recommendations may be treated as general broad guidelines only, depending on the site conditions the test locations and frequency may be left to the discretion of the engineer.

(4) *Time of testing*: Since tests are conducted on slab corners which are the most vulnerable portions of the slab, as far as possible the tests should be done at the most critical time of the day when the load carrying capacity of the corners will be the minimum from considerations of pavement warping, *viz.* early hours in the morning.

23.11.2. Test procedure

***General*:** The test is performed on selected pavement corner by loading it through 45 cm diameter plate and measuring deflections at the top of the loaded slab corner and the three adjoining slab corners. The loading frame for use in the tests should be of a capacity 50 per cent higher than the equivalent single wheel load corresponding to the original design LCN value of the pavement.

A thin layer of fine sand or plaster of paris may be used below the test plate, where required, to ensure full contact with the pavement. Also a seating load of 3000 kg may be applied initially for about 10 seconds and released before commencement of the test proper.

Basically two procedures (I and II) are available for the test within the scope of LCN method *viz.* load test well beyond cracking and load test upto imminent cracking. A third alternative, that is testing upto standardized value of working deflection to directly obtain the working load, is also suggested as a modified and simpler procedure (III) requiring much less test loads as compared to the first two alternatives. While any of the first two procedures may be adopted independently for the test, the third procedure based on working load deflection may be adopted in conjunction with either of the first two procedures. This procedure will be specially useful where either the testing is to be done expeditiously or where cracking of slabs due to testing is to be avoided.

Procedure I. Load test beyond cracking: In this procedure, in addition to the deflection gauges on slab corners, four additional deflection gauges are located along the bisector of the loaded corner angle as shown in Fig. 23.17. Load is applied in equal increments of 3000 kg and deflection readings taken after each load increment. After occurrence of cracks the gauges on the main slab beyond the corner crack will register a lower or no increase in deflection with increase in load, while in case of other gauges including gauge 1 (Fig. 23.17) there will be a sudden increase. As soon as this change in deflection is noticed, the test is stopped, and the load corresponding to the point of change taken as the failure load. As per the standard LCN procedure, safe working load is obtained by applying a factor of safety of 1.5 for non-channelised traffic areas and 1.8 for channelised traffic areas.

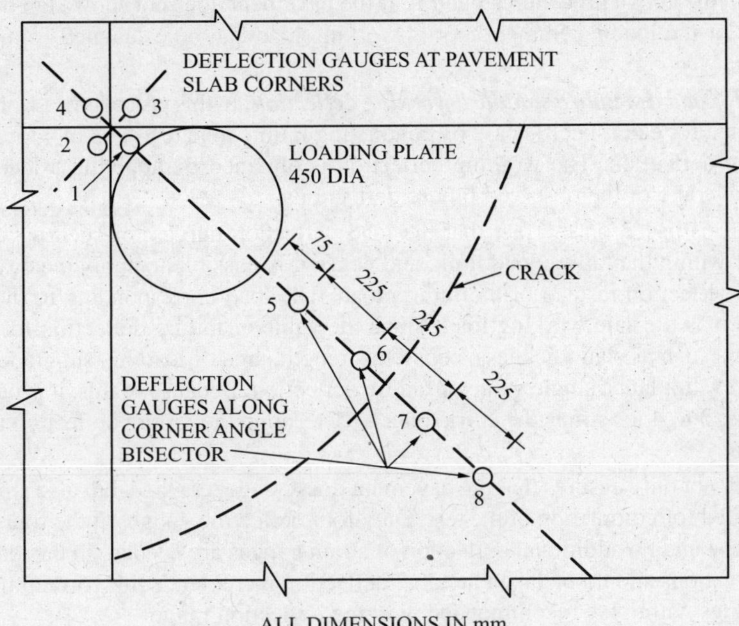

Fig. 23.17 Arrangement of deflection gauges for corner load test on rigid pavements.

***Procedure II. Load test upto imminent cracking*:** In this procedure, instead of deflection gauges four mechanical strain gauges (Fig. 23.18) are used along the bisector of the loaded corner to detect the imminence of cracking, so that the test load may be taken as close as possible to the point of failure without actually causing a crack. As soon as the reading on any one strain gauge starts increasing rapidly relative to the adjacent gauges, the corresponding load is taken as the failure load and the safe working load obtained by applying a factor of safety of 1.5 or 1.8 as in procedure I.

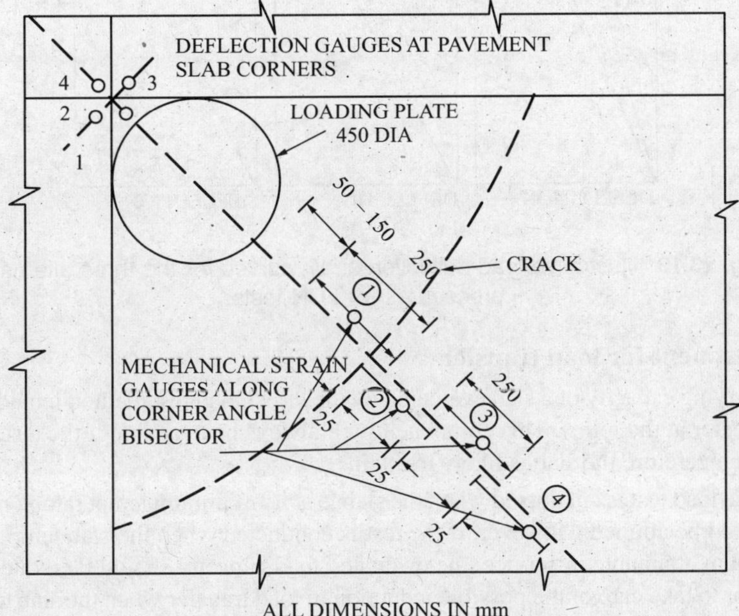

Fig. 23.18 Arrangement of strain gauges for deflection of crack incidence for corner load test.

In case of both the above procedures I and II, if the pavement does not show sign of cracking upto the full capacity of the loading frame, the safe working load can be calculated from the maximum applied load.

Procedure III. Load test upto a standard working deflection. In this procedure, a working deflection value is determined for each specific case by conducting failure load tests at 3 to 4 locations only and noting the failure deflection. The working deflection is obtained by applying a factor of safety of 1.5 or1.8 to the average of the observed failure deflections.

This procedure is based on the fact that the load deflection curves for such pavement tests are practically linear within the failure load limit, and hence it is possible to apply factor of safety of 1.5 or 1.8 to failure deflection instead of the failure load. The load corresponding to the safe working deflection is taken as the safe working load. Since the value of failure deflection for a pavement is affected by factors such as slab thickness, concrete strength, area of loading, subgrade stiffness, etc., it is not possible to stipulate a single value of failure deflection. Because of this, it is necessary that in every case at least 3 to 4 tests must be carried out to determine the failure deflection as the basis for further evaluation.

This procedure not only ensures that the pavement does not get cracked, but also smaller loads and less time are needed for completion of the test. Only four deflection gauges at the four corner tips are required for the test and no additional deflection or strain gauges are needed. In this deflection-based procedure, observations should be taken at equal deflection increments of 0.15-0.25 mm to obtain at least 5 to 6 readings within the recommended working deflection range.

The possibility of an occasional erratic result cannot however be ruled out due to factors such as subgrade pumping, poor local drainage, usually weak spots in foundation or the slab, etc. In such cases, when the slab cracks during the test, the working load should be obtained as per procedure I by applying a factor of safety of 1.5 or 1.8 to the failure load.

Typical load deflection/strain curves for the three procedures are shown in Fig. 23.19.

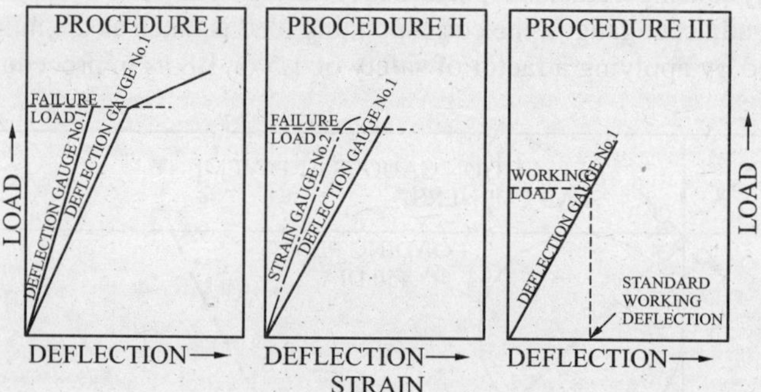

Fig. 23.19 Illustrative load-deflection strain curves for the three alternative procedures for LCN tests.

23.11.3. Adjustment for load transfer

The load carrying capacity of a slab evaluated by the direct load test method includes a component due to load transfer at the joints. This component will always be available if the test is carried out at the minimum temperature the slab is likely to attain.

As such if the load test is conducted when the slab is at its minimum temperature, no correction for load transfer need be applied. However, if the test is conducted when the slab temperature is higher than the minimum attainable in service, the evaluated load capacity should be reduced by a suitable correction factor to take care of the possible reduction in load transfer when the slab temperature falls down to the minimum value.

According to the current state of knowledge, adjustment for load transfer is an arbitrary process based on experience. It will therefore be desirable to conduct the load tests during the coldest part of the year as far as feasible so that the need for load transfer adjustment is obviated. Where this is not possible, the correction factor to be applied may be evaluated on the following lines. The load transfer actually available during the time of test is first determined. The load transfer which will always be available even when the slab is at its minimum temperature is then assessed. The correction factor is taken to be the difference between the two.

The load transfer actually available during time of the test can be evaluated by assuming that its magnitude is directly proportional to the measured deflection of the respective corners. If S_1, S_2, S_3 and S_4 are the recorded deflections of gauges 1, 2, 3 and 4 respectively (see Fig. 23.17 for position of the gauges), the load transfer will work out to:

$$\left(\frac{S_1}{S_1 + S_2 + S_3 + S_4}\right) \times 100 \text{ per cent}$$

The minimum load transfer which will always be available is assumed as the average of the lowest quartile of the observed load transfers subject to a maximum value of 20 per cent.

A frequency distribution table is prepared for the measured load transfers. The lower values below the first quartile (25 per cent frequency) are considered and their average value obtained subject to a maximum of 20 per cent. If the measured load transfer is x per cent and the minimum load transfer y per cent, the corrected load capacity will be $[100 - (x - y)]$ per cent of the measured load capacity.

The above applies to cases where the joints are not provided with load transfer devices. If the pavement contains load transfer devices like dowels or continuous reinforcement, the component of load transfer will be higher and most of it will be available even at the minimum temperature. Some reduction should however be anticipated. It is recommended that a reduction of 10 per cent in the measured load transfer of slab may be applied.

23.11.4. Determination of safe LCN rating

Knowing the safe working load and the contact area of the test plate, the LCN rating for pavement of any test location may be obtained from the standard LCN chart shown in Fig. 23.20.

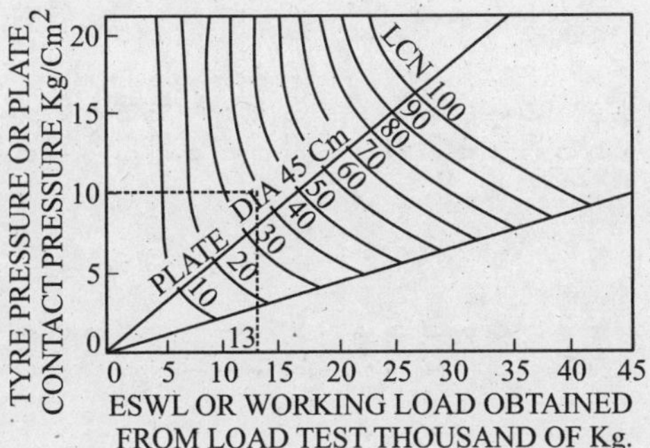

Fig. 23.20 LCN in terms of load, contact pressure/dia of contact area for rigid pavements.

For assessing the safe LCN rating of the pavement as a whole, the safe working load should be calculated statistically for a confidence level of 1 in 15, substracting 1.5 times the standard deviation from the average of the working loads calculated with respect to individual tests.

This value should then be examined in relation to individual LCN test values to see whether any test locations have exceptionally low values, so that the same may be considered separately for appropriate remedial measures.

Illustrative example for calculation of safe LCN for an airfield runway from failure test loads and measured load transfer values.

Location	*Failure load/ max. applied load (tonnes)*	*Safe loads/ Col. (2)/1.5 (tonnes)*	*Measured load transfer (Percent)*	*Adjusted load transfer (Percent)*	*Deductable load transfer (Percent)* Col. 4-Col. 5	*Corrected safe load, x (tonnes) Col. (3) ×* $\left(1-\frac{Col.6}{100}\right)$	*LCN from Fig. 23.13*	$x-\bar{x}$	$(x-\bar{x})^2$	*Calculations for LCN*
1	2	3	4	5	6	7	8	9	10	11
1	34.5	23.0	0	0	0	23.0	80	2.6	6.76	
2	34.5	23.0	24.32	10	14.32	19.7	65	0.7	0.49	Standard deviation
3	34.5	23.0	16.66	10	6.66	21.4	70	1.0	1.00	for the set of safe
4	34.5	23.0	18.03	10	8.03	21.1	70	0.7	0.49	loads given in Col.
5	24.6	16.4	0	0	0	15.7	50*	4.0	16.00	(7):
6	29.6	19.7	0	0	0	19.7	65	0.7	0.49	
7	34.5	23.0	25.66	10	15.66	19.4	65	1.0	1.00	
8	34.5	23.0	8.45	8.45	0	23.0	80	2.6	6.76	$\sigma=\sqrt{\frac{(x-\bar{x})^2}{n-1}}$
9	29.6	19.7	0	0	0	19.7	65	0.7	0.49	
10	27.1	18.0	22.65	10	12.65	15.7	46*	4.7	22.09	
11	34.5	23.0	25.45	10	15.45	19.5	65	0.9	0.81	$=\sqrt{\frac{95.23}{29}}=1.81$
12	32.0	21.3	0	0	0	21.3	70	0.9	0.81	
13	29.6	19.7	0	0	0	19.7	65	0.7	0.49	
14	34.5	23.0	22.05	10	12.05	20.2	67	0.2	0.04	L.C.L. (lower control
15	27.3	18.2	0	0	0	18.2	56	2.2	4.85	limit) of overall
16	29.6	19.7	0	0	0	19.7	65	0.7	0.49	corrected safe load
17	29.6	19.7	0	0	0	19.7	65	0.7	0.49	for pavement for a
18	34.5	23.0	20.80	10	10.80	20.5	67	0.1	0.01	tolerance level of 1
19	34.5	23.0	22.60	10	12.60	20.1	67	0.3	0.09	in 15.
20	32.0	21.3	1.80	1.80	0	21.3	70	0.9	0.81	20.7-1.5 × 1.81
21	32.0	21.3	14.63	10	4.63	20.3	60	0.1	0.01	
22	32.0	21.3	0	0	0	21.3	70	0.9	0.81	
23	34.5	23.0	10.1	10	0.1	23.0	80	2.6	6.76	$=\bar{x}-1.5\sigma$
24	29.6	19.7	10.1	10	0.1	19.7	65	0.7	0.49	$=20.4-1.5\times1.81$
25	32.0	21.3	8.9	8.9	0	21.3	70	0.9	0.81	= 17.7 tonnes.
26	34.5	23.0	2.7	2.7	0	23.0	80	2.6	6.76	The corresponding
27	34.5	23.0	8.7	8.7	0	23.0	80	2.6	6.76	overall safe LCN
28	32.0	21.3	0	0	0	21.3	70	0.9	0.81	for the pavement
29	27.1	18.0	0	0	0	18.0	55	2.4	5.76	(from Fig. 23.13).
30	32.0	21.3	0	0	0	21.3	70	0.9	0.81	= 54*
					Average	$\Sigma\bar{x}=611.5$ $\bar{x}=20.4$		$\Sigma=95.23$		

Note: Examining the individual LCN values vis-a-vis LCN (lower control limit), it is seen that only 2 values marked with astericks fall below LCN out of total of 30 values (corresponding to a tolerance level of 1 in 15). However, in view of relatively low value *vis-a-vis* overall safe LCN rating of 54, the location with LCN 46 may be investigated

23.12. INDIRECT REVERSE DESIGN METHOD

23.12.1. General

Where direct evaluation by load tests is not possible due to any reason, indirect evaluation may be done by reverse design method.

This method requires the same basic information for evaluating pavement structural strength, as in the case of design for a new pavement (with the obvious exception of design wheel load, actual pavement slab thickness), viz:

(*i*) Flexural strength of concrete in the pavement.

(*ii*) Foundation strength in terms of k-value.

(*iii*) Maximum temperature differential over pavement depth.

It is, however, necessary to determine the actual strength of concrete and pavement foundation by appropriate tests. Selection of test locations for this purpose should be done on the same lines as suggested in direct load test method.

However, if effects of fatigue are also to be evaluated, a few additional test locations may be selected along the outer edges which are subject to much less traffic intensity.

23.12.2. Test procedure

Concrete strength. Either beam or core samples may be recovered from the test locations for determination of concrete strength. While it is preferable to have beam samples for direct determination of flexural strength, core samples may have to be resorted to in many cases from considerations of expediency, available equipment, time available for investigation, or the need to keep minimum damage to the pavement from sample recovery.

Concrete samples, whether beam or core, should be carefully examined for quality of compaction, and their dimensions should be accurately measured. Beam samples may be tested in flexure using third point loading and concrete flexural strength determined. For cores, crushing strength results should be corrected for *h/d* ratio before determination of corresponding cube compressive strength. The crushing strength of cylinders with *h/d* ratio between 1 and 2 may be corrected to correspond to standard *h/d* ratio of 2 by multiplying with the correction factor obtained from the following equation:

$$f = 0.11\,n + 0.78$$

Where f = correction factor

n = *h/d* ratio.

The diameter "*d*" of cores recovered should not be less than 10 cm for concrete with maximum aggregate size of 10 mm and not less than 15 cm for concrete with maximum aggregate size of 40 mm.

Cube compressive strength may be taken as 1.25 times the corrected cylinder crushing strength. Conversion from cube compressive to flexural strength may be done using the chart given in Fig. 23.21.

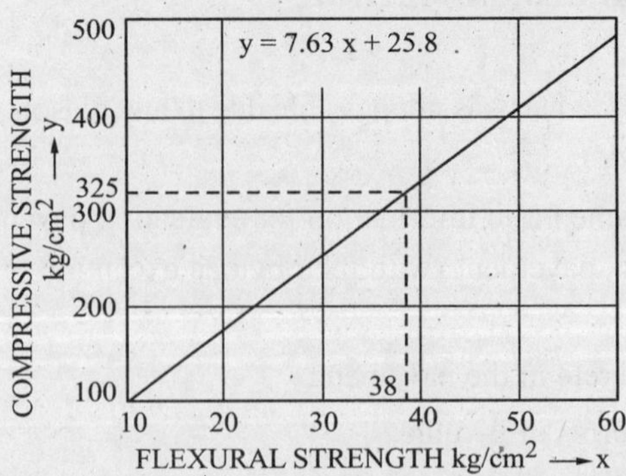

Fig. 23.21 Statistical correlation between compressive and flexural strength of concrete

The strength tests should be done in accordance with the relevant I.S. specifications.

Foundation strength. Foundation strength in terms of its k-value is normally determined directly by conducting plate bearing tests on the foundation on which the slab rests (which could be earth subgrade, or a sub-base if provided) after removing sections of pavement slab. The test could be conducted on a 30 cm dia plate and converted subsequently to the standard k-value for 75 cm diameter plate by the approximate correlation:

$$k_{75} = 0.5\, k_{30}$$

where k_{75} and k_{30} are the k-values for 75 cm and 30 cm diameter plates respectively. It should, however be noted that this correlation is based on homogeneous foundation conditions and in the case of layered construction *i.e.* when the test is conducted on a sub-base, the smaller plate will give a greater weightage to the stronger top layer. In such cases, direct conversion to 75 cm plate value by the above correlation somewhat over-estimates the foundation strength and should be regarded as very approximate.

After conducting the test, the sub-base may be removed upto the subgrade level, noting its type and thickness and a plate bearing test conducted to determine subgrade k-value, if considered necessary.

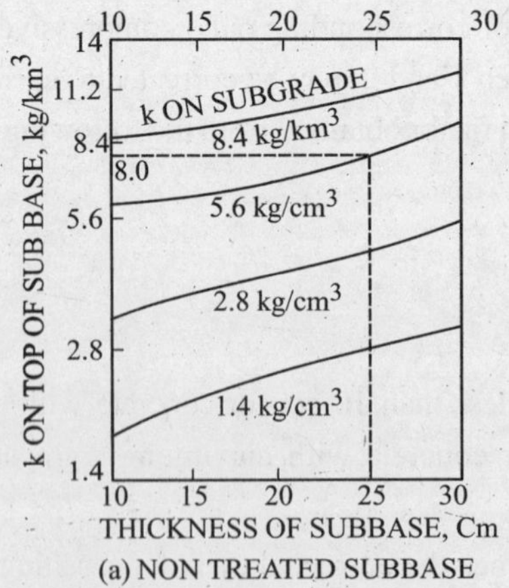

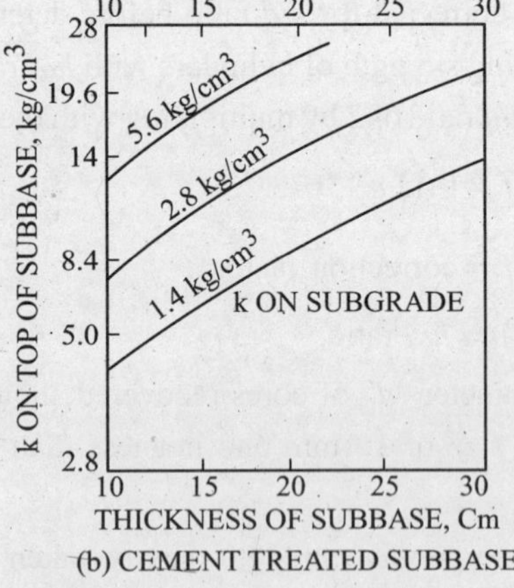

Fig. 23.22 Allowance for increase in (*k*) value due to providing sub-base over subgrade.

If direct determination of foundation *k*-value is not possible, an approximate indirect assessment can be made by conducting insitu CBR test on the subgrade. From this, an approximate idea of the subgrade *k*-value may be obtained using the CBR. *k*-value correlation given in Table 23.4.

If any sub-base is present over the subgrade, due allowance should be made for increase in the foundation *k*-value. For this purpose, the charts given in Fig. 23.22 may be made use of.

Table 23.4 Approximate k-values corresponding to CBR values for homogeneous soil subgrades

CBR value (%)	2	3	4	5	7	10	20	50	100
k-value (kg/cm^3)	2.08	2.77	3.46	4.16	4.86	5.54	6.92	13.85	22.16

Supplementary soil tests. Soil plasticity and grain-size analysis tests should also be carried out for determining the soil classification of the subgrade.

Slab thickness. Slab thickness may be determined by direct measurement of height of core or beam samples recovered from the pavement slab for concrete strength determination.

Number of test locations and selection of test sites.

Same considerations will apply in this case as for the direct load test method.

23.12.3. Analysis of concrete/foundation strength test data

The distribution of test data should be studied carefully both for concrete and foundation strength to see if the pavement section under investigation could be divided into zones of distinctly different concrete and foundation strengths, in which case "typical" strength values should be separately worked out for each zone. In case such sub-division is not possible, the strength data may be examined on an overall basis.

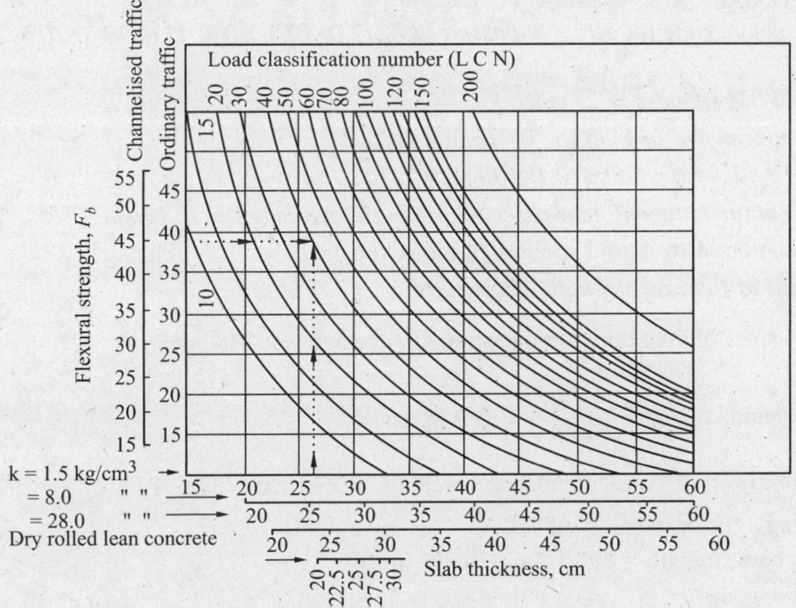

Fig. 23.23 LCN design chart for rigid pavements.

It is suggested that the "typical" strength values should be arrived at from actual test data for ensuring a confidence level of 1 in 15. This can be worked out by reducing the average of all the strength test values by 1.5 times their standard deviation. The design strength values may there after be obtained by applying a factor of safety of 1.1 to the typical values.

23.12.4. Determination of Pavement Structural Strength

Pavement design procedure to be adopted. The LCN method for rigid pavement design may be used. Knowing pavement thickness, concrete flexural strength and foundation k-value, the rated LCN for the pavement can be read directly from the chart shown in Fig. 23.23.

Allowance for load transfer. As in the case of evaluation by direct load test, it is necessary to assess the minimum load transfer that will be available at any location in the pavement. The critical locations from this consideration would be the same as those recommended in direct load test.

Contribution to load capacity due to load transfer may be assessed by conducting actual load tests, noting deflections of adjacent slabs up to anticipated design equivalent single wheel load for the pavement. Only a limited number of tests on typical joints would be adequate. (Similar tests can also be conducted to assess the load transfer capacity of typical cracks in the pavement). The tests will however, have to be conducted in the coldest part of the year when load transfer is at its critical minimum; otherwise a correction factor will have to be applied as explained in direct load test.

The total assessed load carrying capacity of the pavement at any location should be obtained by combining the individual slab load carrying capacity, as calculated from Fig. 23.16 with the load transfer capacity at joints.

Problem 23.1 *Determine the load carrying capacity of the pavement and the corresponding LCN value from the following data.*

Slab thickness, h = 25 cm

Compressive strength (kg/cm^2) of pavement concrete cores, 25 cm high, 15 cm dia (at 10 locations):

385, 355, 320, 360, 340, 295, 325, 360, 330, 310.

CBR values of subgrade (at 10 locations) = 13.0, 12.0, 10.5, 10.0, 11.0, 12.5, 10.5, 10.5, 10.5, 11.5.

Thickness of WBM subbase = 25 cm

Measured load transfer, percent (at 10 locations)

12.66, 13.03, 8.40, 14.63, 6.80, 10.1, 10.1, 8.9, 6.7, 8.7

Tyre pressure of predominant gear assembly using the pavement = 10 kg/cm^2

Solution. Calculation of structural strength of the pavement

(1) *Assessment of flexural strength of concrete*

From the 10 values of core compressive strength, average $\bar{x} = 338$ kg/cm^2

Standard deviation, $\sigma = \sqrt{\sum \frac{(x-\bar{x})^2}{n-1}} = 27.3$ kg/cm^2

$\therefore$ For a tolerance level of 1 in 15, min. core strength $= \bar{x} - 1.5\,\sigma = 338 - 1.5 \times 27.3 = 297$ kg/cm^2

Applying a factor of safety of 1.1,

Design core strength = 297/1.1 = 270 kg/cm^2

Height/diameter ratio of cores, $n = \frac{25}{15} = \frac{5}{3}$

Correction factor for height/diameter ratio:

$f = 0.11\,n + 0.78$

$= 0.1833 + 0.78 = 0.9633$

Corrected core compressive strength

$= 270 \times 0.9633 = 260$ kg/cm^2

Cube compressive strength

$= 1.25 \times \text{core compressive strength}$

$= 1.25 \times 260 = 325 \text{ kg/cm}^2$

From Fig. 23.21 for compressive strength of 325 kg/cm^2, flexural strength of pavement concrete $F_b = 38 \text{ kg/cm}^2$.

(2) *Assessment of k-values of sub-base*

From the 10 values of subgrade CBR,

average, $\bar{x} = 11.5$ per cent

Standard deviation, $\sigma = 1.0$ per cent

For a tolerance level of 1 in 15,

min. CBR value $= \bar{x} - 1.5\,\sigma = 11.5 - 1.5 = 10.0$ per cent

From Table 23.4, *k*-value of subgrade,

(corresponding to CBR value of 10%) = 5.54 kg/cm^2/cm

From Fig. 23.22 (a), for subgrade *k*-value of 5.54 kg/cm^2/cm and subbase thickness of 25 cm, *k*-value on top of subbase = 8 kg/cm^3.

(3) *Assessment of pavement slab LCN*

From Fig. 23.23, for unchannelised traffic (for middle of runway), for $h = 25$ cm, $k = 8$ kg/cm^3 and $F_b = 38$ kg/cm^2,

Pavement LCN = 40

(4) *Correction for load transfer*

(*i*) *Calculation of minimum load transfer*

Since all the values obtained are less than 20 per cent, and none of them is zero, taking the average of all these values, the level to which the higher values of load transfer will get reduced = 10.0 per cent.

∴ Adjusted load transfer values (per cent) = 10.0, 10.0, 8.40, 10.0, 6.8, 10.0, 10.0, 8.9, 6.7, 8.7.

From these values, $\bar{x} = 8.95\ \%,\ \sigma = 1.32\%$

Most probable min. available load transfer

$= \bar{x} - 1.5\,\sigma = 8.95 - 1.5 \times 1.32$

$= 8.95 - 1.98 = 6.97\%$

(*ii*) *Modified pavement slab LCN taking into account load transfer.*

From Fig. 23.20 for tyre pressure of 10 kg/cm^2, equivalent single wheel load (ESWL) corresponding to LCN of 40 (without load transfer) = 13000 kg. Accounting for load transfer,

Actual load carrying capacity of the pavement

$$= 13{,}000 \times \frac{100}{100 - 6.97} = 13980 \text{ kg}.$$

Corresponding LCN (From Fig. 23.20) = 42.

STRENGTHENING OF PAVEMENTS OR OVERLAYS

23.13. TYPES OF OVERLAYS

The various types of overlays are given in Table 23.5 (next page)

The classification is based upon the type of overlay, the type of existing pavement and behaviour of the pavement system. The table indicates the overlay equation to be used and the design criteria.

The overlay may be either flexible or rigid. Flexible overlays are generally bituminous surfaces and can be constructed over existing flexible pavements or existing concrete pavements. A bond breaker comprising of a granular base is sometimes used between a concrete base pavement and overlay. This also reduces deflections to within tolerable limits, due to the building up of a relatively thick layer of overlay.

Rigid overlays may consist of plain, simply reinforced or continuously reinforced concrete pavements. These can be placed on existing flexible or rigid type pavements. Bond breaker between an existing flexible pavement and a rigid overlay may or may not be provided, based upon the factors stated above.

23.14. FLEXIBLE OVERLAYS OVER FLEXIBLE PAVEMENTS, BASED ON DEFLECTION

23.14.1. Benkelman beam method

Overlay thicknesses over flexible pavements are generally determined by means of deflection measurements. Measurements are made using Benkelman beam. The analysis of data for overlay design and the determination of overlay thickness as given in IRC : 81-1997(First Revision) are detailed below:

23.14.1.1. Analysis of data for overlay design

Characteristic deflection. Overlay design for a given section is based not on individual deflection values but on a statistical analysis of all the ten measurements in the section corrected for temperature and seasonal variations. This involves calculation of mean deflection, standard deviation and characteristic deflection. The characteristic deflection for all design purposes should be taken as the mean deflection plus one standard deviation. The formulae to be used in the calculation are as follows:

$$\text{Mean deflection } \bar{x} = \frac{\Sigma x}{n}$$

$$\text{Standard deviation } \sigma = \sqrt{\frac{\Sigma(x-\bar{x})^2}{n-1}}$$

$$\text{Characteristic deflection } \Delta_0 = \bar{x} + \sigma$$

where x = Individual deflection

$\bar{x}$ = Mean deflection

n = Number of deflection measurements

σ = Standard deviation

Δ_0 = Characteristics deflection

These should be recorded in the proforma suggested in Table 23.7.

Traffic. For purposes of this design method, the traffic is considered in units of heavy vehicles (vehicles with a laden weight of 3 tonnes or more) per day, in both the directions in the case of two lane roads and in the direction of heavier traffic in the case of multilane divided highways. Based on

Table 23.5 Design approaches by overlay category

Overlay	*Existing pavement*	*System behaviour*	*Overlay equation*	*Remarks*
Flexible	Flexible		$t_0 = t_n - t_e$	Deflection criteria generally used for highways
Flexible	Composite		$t_0 = t_n - t_e$	Overlay thickness generally based on experience; for highway
Flexible	Rigid	Rigid	$t_0 = 2.5\,(Fh_n - h_e)$	Overlay thickness generally based on experience; for highways
Flexible	Rigid	Flexible	$t_0 = t_n - t_e$	
Rigid	Flexible		$h_0 = h_n$	
Rigid	Rigid	Rigid (bond)	$h_0 = h_n - h_e$	
Rigid	Rigid	Rigid (partial bond)	$h_0 = 1.4\sqrt{h_n^{1.4} - Ch_e^{1.4}}$	Used mostly for airports
Rigid	Rigid	Rigid (unbonded)	$h_0 = \sqrt[2]{h_n^{2} - Ch_e^{2}}$	Used mostly for airports
Rigid	Composite	Flexible	$h_0 = h_n$	

Where h = rigid-pavement thickness

t = flexible-pavement thickness

o = overlay-pavement thickness

n = new pavement

e = existing pavement

F = Factor which is a function of subgrade or sub-base

C = Coefficient depending upon existing pavement (See page 866 under rigid overlays on rigid pavements)

the present traffic, the design traffic, *i.e.* the traffic expected at the end of the design year should be estimated. In preparing these estimates due consideration should be given to the existing traffic, possible changes in road network and land use of the area served, and probable growth of traffic during the period of projection under consideration. It is considered appropriate that major through routes should be designed for at least 20 years of life. Less important roads may, however, be designed for a somewhat lower period based on growth prospects and judgement.

The following formula may be used for traffic projection on the main through routes:

$$T = p\,(1+r)^{n+20} \qquad \ldots(23.4)$$

For details please refer to page 644.

Allowable deflection. Based on the limited experience available in the country tentative values for allowable deflection recommended for different traffic conditions are set out in Table 23.6.

Table 23.6 Recommended values for allowable deflection

S.No.	*Design traffic intensity (Commercial vehicle/day)*	*Allowable deflection when measured by*	
		CGRA method	*WASHO method*
1.	150-450	1.50 mm	1.40 mm
2.	450-1500	1.25 mm	1.10 mm
3.	1500-4500	1.00 mm	0.80 mm

Examination of data. The deflection method is based on the premise that if the characteristic deflection of a section is more than the allowable deflection, further strengthening of the pavement is called for. Before this conclusion is reached for individual sections it will be desirable to cross-check the deflection values with other information collected during the pavement condition survey. This may involve the following steps:

(*i*) For sections of a road carrying the same traffic, compare the characteristic deflection values in the light of performance. If a section rated good shows higher deflection values than the ones with fair or poor performance, a possible error, or influence of an extraneous factor might be present. Such cases should be identified and the cause of discrepancy ascertained through detailed examination at site including fresh deflection measurement, if necessary.

(*ii*) Sections showing inordinately high deflection values compared to values at a adjacent point should be carefully examined at site to see if the situation is on account of poor drainage or other causes. Such sections should first be treated by suitable remedial measures after which fresh deflection measurements should be made for ascertaining the strengthening required.

After identification of problematic sections and ironing out discrepancies in deflection values in the aforesaid manner, the required overlay thickness for different sections should be determined as follows:

23.14.2. Determination of overlay thickness

Having obtained the characteristics deflection and knowing the permissible deflection for a given section (See Table 23.7), following relationship may be used for obtaining overlay thickness:

Table 23.7

Analysis of test data

Name of road :						Date and time of observation :					
Section :						Climatic conditions :					
Nc. of traffic lanes :						Ambient temperature :					
						Pavement temperature :					
						Method of deflection measurement :					
Location of test point	*Measured deflection (from column 7 of table 23.4)*	*Correction for temperature*	*Correction for season*	*Corrected deflection*	*Mean deflection*	*Standard deviation*	*Characteristic deflection mm*	*Design traffic com.veh./day*	*Permissible deflection*	*Overlay required*	*Remarks*
1	2	3	4	5	6	7	8	9	10	11	12

$$h = R \log_{10} \frac{\Delta_0}{\Delta} \qquad \ldots(23.5)$$

where h = Thickness of granular overlay in mm (WBM)

Δ_0 = Characteristic deflection

Δ = Allowable deflection

R = Constant which may be taken as 550.

The above equation is plotted in Fig. 23.24 from which the overlay thickness can be directly taken.

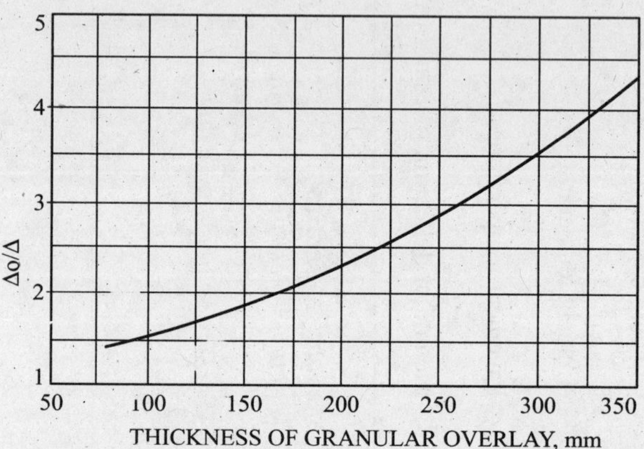

Fig. 23.24 Thickness of granular overlay.

When a long st ch of road is to be strengthened involving a lot of expenditure, the overlay thickness obtained from the above relationship may be treated as indicative only. In such cases, a few sections lower than this thickness and a few greater than this thickness besides those corresponding to design thickness m..y be laid and deflection data collected after allowing the traffic for about a month or so.

The relationship in equation 23.5 gives the thickness of overlay in terms of granular materials such as W.B.M Where bituminous concrete and/or bituminous macadam are to be laid, equivalency factor of these ..ay be taken as 2.

The type of material to be used in overlay construction will depend on several factors such as the importance of the road and the traffic carried by it, the depth of existing bituminous construction, construction convenience and relative economics. For heavily trafficked roads, it will however be advisable to provide bitumen-bound overlays.

23.14.2. California method

The California highways department, U.S.A. has developed a method of overlay design pavements based upon deflection measurements. Figs. 23.25 to 23.27 show the curves used in this method. The highest 80th percentile deflection value is used. Fig. 23.25 depicts the relationship between tolerable deflection and repetition of load as a function of surface thickness. Traffic is evaluated in terms of traffic index (TI), based upon equivalent 2250 kg wheel loads, as described in chapter 20. The first approximation of the thickness of overlay required is carried out with the help of Fig. 23.26.

Fig. 23.27 gives the per cent reduction for various thicknesses of overlay expressed as gravel equivalent. The equivalency factors assumed for the design are shown in the figure.

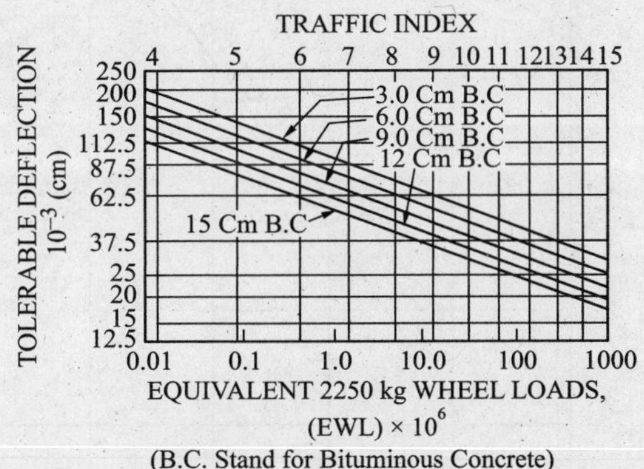

Fig. 23.25 Tolerable deflection chart for varying thickness of bituminous concrete.

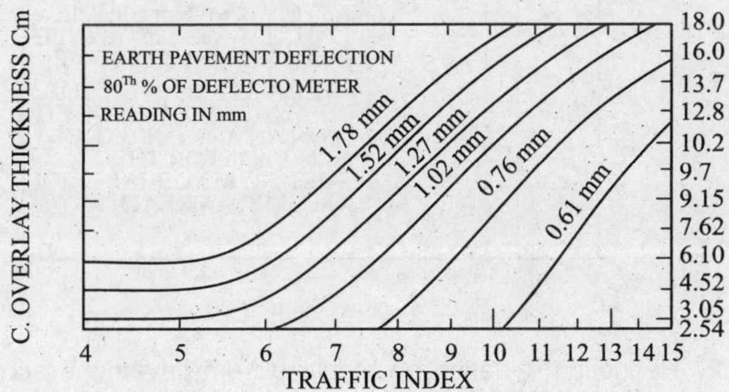

Fig. 23.26 Bituminous concrete overlay design guide.

Problem 23.2 *A flexible pavement consists of 1.25 cm of surface treatment, 15 cm of crushed-stone base and 15 cm of sandy subbase. The 80th percentile deflection is 1 mm. The traffic index for a design life of 20 years is 8. The pavement shows continuous alligator cracking and minor rutting. Design an overlay of bituminous concrete.*

Solution. Assume overlay thickness of bituminous concrete (B.C.) = 6 cm

From Fig. 23.25, tolerable deflection for overlay of 6 cm B.C. and traffic index 8 = 0.65 mm

$$\text{Percentage reduction of deflection required} = \left(\frac{1-0.65}{1}\right)\times 100$$

$$= 35\%$$

From Fig. 23.27, 35% reduction in-deflection requires 10.8 cm of gravel. This is less than the design value of 6 × 1.9 = 11.4 cm, 1.9 being the conversion factor from bituminous concrete to gravel.

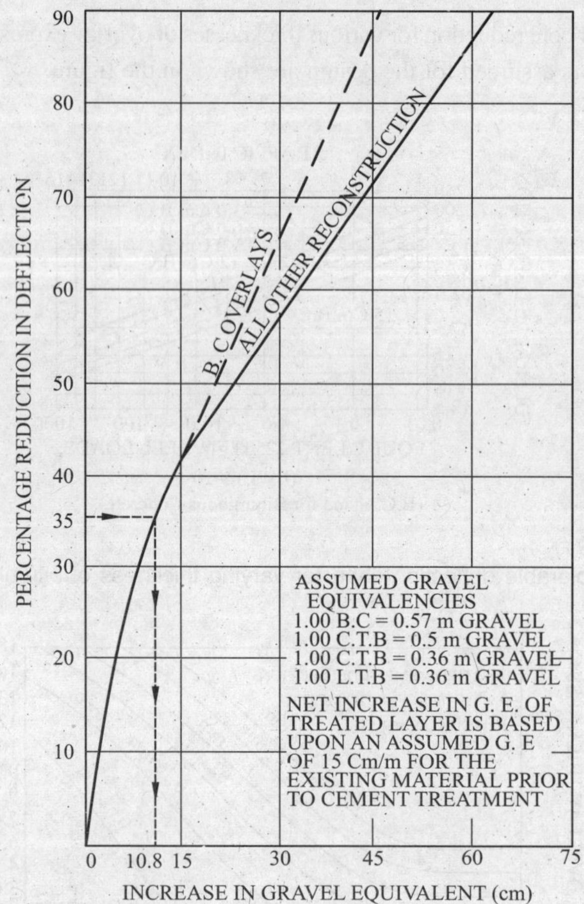

Fig. 23.27 Reduction in deflection resulting from pavement reconstruction.

Therefore, assumed overlay of 6 cm B.C. is considered to be satisfactory.

23.14.3. Asphalt institute method

The method has been developed by the Asphalt Institute U.S.A. and is also based on deflection measurements. Rebound deflection are determined using 8200 kg single axle. The evaluated deflections is the mean deflection plus two standard deviations. Corrections are applied for temperature and the seasonal variation (See Fig. 23.28). The relation used for evaluated deflection is

$$\Delta = (\bar{x} + 2\sigma)\, f\, C \qquad \ldots(23.6)$$

Where Δ = Evaluated deflection

$\bar{x}$ = Arithmetic mean of deflection measurements

σ = Standard deviation

f = Temperature correction

C = Critical period correction

An estimate of the standard deviation can be made by using the factors *d* and *m* given in Table 23.8 and using equations:

$$\sigma = \frac{R}{d} \quad \text{or} \quad \sigma = R_m$$

Where R = Difference between maximum and minimum value ; and

$$m = \frac{1}{d}$$

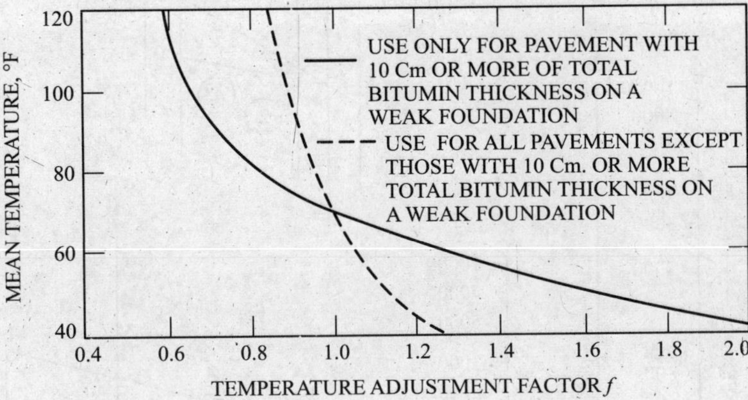

Fig. 23.28 Temperature adjustment factors to apply to Benkelman beam measurements.

Table 23.8 Factors for estimation of standard deviation

Number of values	*Factor*	
(n)	*d*	*m*
2	1.1284	0.8862
3	1.6926	0.5908
4	2.0588	0.4857
5	2.3259	0.4299
6	2.5344	0.3946
7	2.7044	0.3698
8	2.8472	0.3512
9	2.9700	0.3369
10	3.0775	0.3249

The design charts are shown in Fig. 23.28 to Fig. 23.29. The traffic value is determined on the basis of Initial Traffic Number (ITN) and adjusted for traffic growth. (Table 23.9)

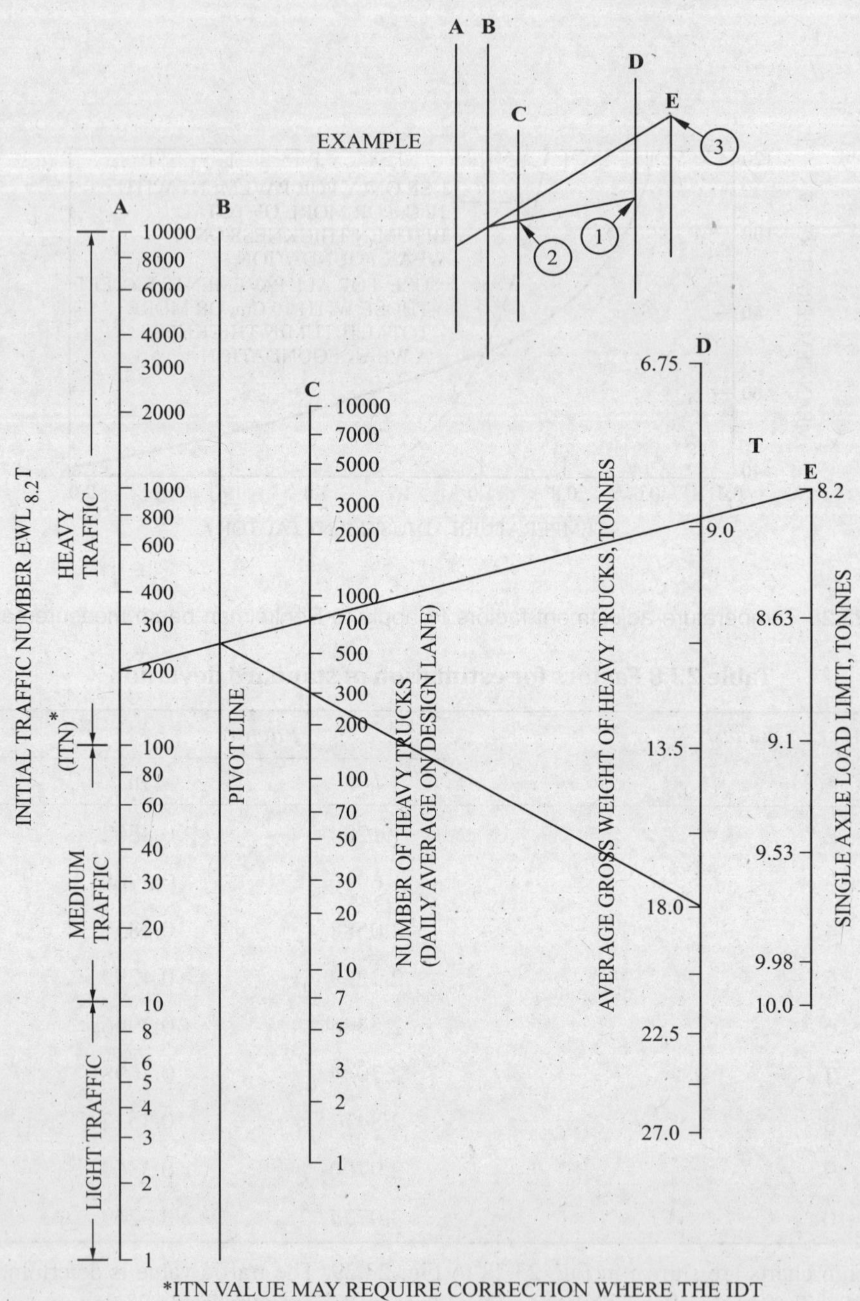

Fig. 23.29 Nomographic solution of EAL (for The Asphalt Institute method)

Table 23.9 Initial traffic number adjustment factors

$$\textbf{Factor} = \frac{(1+r)^n - 1}{20r}$$

where r **= annual growth rate**

n **= design period (yr)**

Design period, (yr) (n)	*ITN adjustment factor for annual growth rate (Per cent) (r)*					
	0	2	4	6	8	10
1	0.05	0.05	0.05	0.05	0.05	0.05
2	0.10	0.10	0.10	0.10	0.10	0.10
4	0.20	0.21	0.21	0.22	0.22	0.23
6	0.30	0.32	0.33	0.35	0.37	0.39
8	0.40	0.43	0.46	0.50	0.53	0.57
10	0.50	0.55	0.60	0.66	0.72	0.80
12	0.60	0.67	0.75	0.84	0.95	1.07
14	0.70	0.80	0.92	1.05	1.21	1.40
16	0.80	0.93	1.09	1.28	1.52	1.80
18	0.90	1.07	1.28	1.55	1.87	2.28
20	1.00	1.21	1.49	1.84	2.29	2.86
25	1.25	1.60	2.08	2.74	3.66	4.92
30	1.50	2.03	2.80	3.95	5.66	8.22
35	1.75	2.50	3.68	5.57	8.62	13.55

and for the volume of trucks in the design lane. The following example makes the method clear.

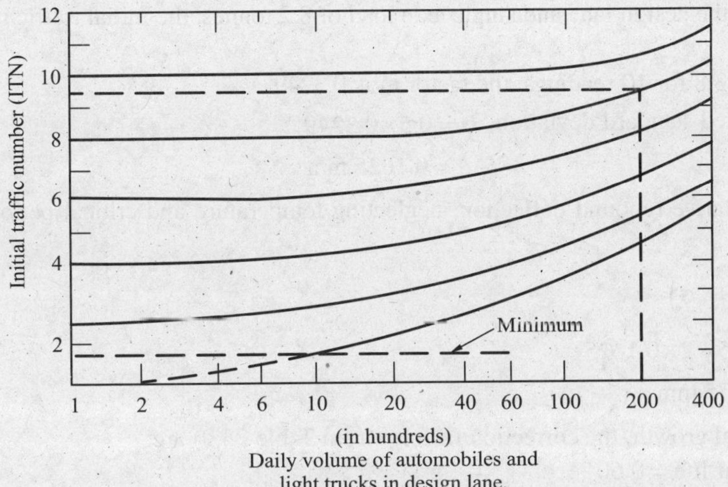

Fig. 23.30 Adjustment of ITN for daily volume of automobiles and light trucks.

Problem 23.3 *Determine the thickness of flexible overlay over an existing flexible pavement comprising of 15 cm base over 15 cm subbase and surface treatment over them from the following data:*

Average gross weight of heavy trucks	*= 18 tonnes*
Number of trucks in design lane (daily)	*= 300*
Traffic growth	*= 4%*
Mean deflection (10 readings)	*= 1.25 mm*
Range in deflection (R)	*= 0.5 mm*

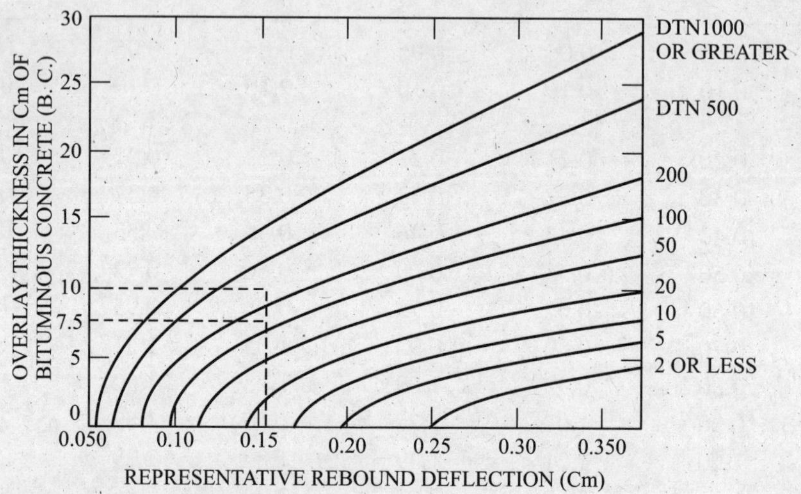

Fig. 23.31 Representative rebound deflection (cm) asphalt concrete overlay thickness required to reduce pavement deflection.

Solution. From Fig. 23.29, against average gross weight of heavy trucks of 18 tonnes, daily heavy trucks of 300 in the design lane and single axle load of 8.2 tonnes, the initial traffic number (ITN) is found to be 200.

From Table 23.8 for 10 readings, the factor m is 0.3249.

∴ Estimated standard deviation, $\sigma = 0.5 \times 0.3249$

$$= 0.1625 \text{ mm}$$

The representative rebound deflection, neglecting temperature and critical period correction, is given by

$$\Delta = \left(\bar{x} + 2\sigma\right)$$

$$= 1.25 + 2 \times 0.1625$$

$$= 1.575 \text{ mm}$$

For 4% annual growth, the correction factors (from Table 23.9) are:

Ten year life = 0.60

Twenty year life = 1.49

Thus the DTN numbers and the required overlay thickness are:

Age (years)	*DTN = ITN × correction factor*	*Overlay thickness (cm) (from Fig. 23.31) against rebound deflection of 1.575 mm*
10	200 (0.60) = 120	7.5 cm
20	200 (1.49) = 298	10.0 cm

23.14.4. Component layer analysis

This method of analysis consists of working backwards the design methods explained in the text. The pavement is designed as if a pavement does not exist at the site. The difference between the required total thickness and the existing thickness gives the thickness of overlay to be added.

It is necessary to make adjustments in the thickness of the existing pavement to take into account the condition of the pavement. This is done by multiplying the actual thickness of the existing pavement with the conversion factors given in Table 23.10 and Table 23.11.

Table23.10 The Asphalt Institute conversion factors

Material type	*Equivalency*
Bituminous Materials	
1 cm B.C. (uncracked, little deformation)	0.9–1.0 cm B.C.
1 cm Bituminous-treated base (other than B.C.)	0.7–0.9 cm B.C.
1 cm Liquid bituminous mix (stable, uncracked, no defect)	0.7–0.9 cm B.C.
1 cm B.C. (fine cracking, slight deformation, stable)	0.7–0.9 cm B.C.
1 cm B.C. (appreciable cracking, little or no spalling, some deformation, but essentially stable)	0.5–0.7 cm B.C.
1 cm B.C. (large cracks, spalls, appreciable deformation)	0.3–0.5 cm B.C.
Portland Cement Concrete	
1 cm P.C.C. (stable, uncracked)	0.9–1.0 cm B.C.
1cm P.C.C. base under A.C.C (stable, little cracking)	0.9–1.0 cm B.C.
1 cm P.C.C. (stable, slight cracking)	0.7–0.9 cm B.C.
1 cm P.C.C. (appreciably cracked, faulted, fragments 1-4 m)	0.5–0.7 cm B.C.
1 cm P.C.C. (broken, pieces 0.6 m or less in max. dimension-with subbase)	0.3–0.5 cm B.C.
1 cm P.C.C. (broken, pieces 0.6 m or less in max. dimension-on subgrade)	0.3–0.5 cm B.C.
Cement-Stabilized	
1 cm Soil cement base (little cracking, stable surface)	0.5–0.7 cm B.C.
1 cm Soil cement base (extensive pattern cracking, unstable)	0.3–0.5 cm B.C.
1 cm Cement-modified base/subbase (soil $PI \leq 10$)	0.2–0.3 cm B.C.
Unbound Granular	
*1 cm Granular base (nonplastic, high quality)	0.3–0.5 cm B.C.
1 cm Granular base or subbase (CBR 20, $PI \leq 6$)	0.2–0.3 cm B.C.
1 cm Granular base or subbase (CBR 20, $PI > 6$)	0.2–0.3 cm B.C.

The assignment of a suitable conversion factor is a matter of judgement and should be made after judicious consideration of each particular problem.

Table 23.11 Conversion factors by different agencies

Material type	*FAA Equivalency*	*U.S. Navy Equivalency*	*U.S. Air Force Equivalency*	*Portland cement Association (U.S.) Equivalency*
Flexible overlays				
1 cm B.C. surface good condition, bituminous overlay)	1.5 cm G. B.	1.0 cm G. B.	1.0 cm G. B.	
1 cm B.C. surface (poor condition)	1.0 cm G. B.			
1 cm B.C. base (good condition, bituminous overlay)	1.5 cm G. B.			
1 cm C.T.B. (good condition)	1.5 cm G. B.			
Rigid overlays				
1 cm P.C.C. (good condition)	1.0 cm PCC	1.0 cm PCC	1.0 cm PCC	1.0 cm PCC
1 cm P.C.C. (initial corner cracking, no progressive cracking)	0.75 cm PCC	0.75 cm PCC	0.75 cm PCC	0.75 cm PCC
1 cm P.C.C. (badly cracked or crushed)	0.35 cm PCC	0.35 cm PCC	0.35 cm PCC	0.35 cm PCC

23.15. BITUMINOUS OR FLEXIBLE OVERLAYS ON FLEXIBLE PAVEMENTS

The required thickness of bituminous or flexible overlay over existing flexible pavements can be easily determined by assigning the appropriate design value to the subgrade. The CBR or other design data will determine the total required thickness of pavement. The difference between the required total thickness and existing thickness is the amount of overlay to be added.

For example, if CBR design procedure is being used, the CBR of the subgrade underlying the existing pavement is determined on the basis of soil tests, drainage and climate. From the basic CBR design curves used in the design of flexible pavements, the thickness of flexible pavement required for the particular wheel loading under consideration, assuming it to be placed directly on the subgrade, is determined. The difference between the total thickness as determined from the design curves and the thickness of existing pavement represents the required thickness of overlay pavement. Certain adjustments to this required thickness may be made, depending on the condition of the existing pavement and character of materials in the overlay pavement as per Table 23.11. FAA recommends that 1 cm of bituminous surfacing in good condition is equivalent to 1.5 cm of aggregate base course. Thus, the required pavement thickness can be reduced if the thickness of bituminous surface of existing and overlay pavement is taken into consideration. The U.S. Corps of Engineers method recommends that the base course thickness in a flexible overlay should not be less than 10 cm. If the resulting overlay results in a base course thickness less than 10 cm it is required that the entire overlay for airfield pavements be of the bituminous type.

Whether the overlay is to be of the flexible type or bituminous type depends entirely on the economics involved in the particular project. For a flexible overlay, as already stated, it is recommended that the thickness of base course be not less than 10 cm. Sub-base materials are not recommended in flexible overlays.

23.16. BITUMINOUS AND FLEXIBLE OVERLAYS ON RIGID PAVEMENTS

The required thickness of flexible overlays on rigid pavements can be determined from the following equation:

$$t_0 = 2.5\,(Fh_n - h_e) \qquad \ldots(23.7)$$

where t_0 = required thickness of flexible overlay

h_n = required thickness of equivalent single slab placed directly on subgrade or sub-base (calculated or read from appropriate design diagram)

h_e = thickness of existing slab

F = factor which is a function of the subgrade or sub-base.

Both the U.S. corps of engineers and FAA use the same formula with minor variations to fit in their nomenclature and subgrade soil classification.

Values of factor F recommended by the U.S. corps of engineers are shown in Fig. 23.32. Curve 1 is used for type A traffic for all multiple-wheel gear designs except twin-twin wheel gear. Curve 2 is used for type A traffic area for twin-twin wheels. Curve 3 is used for type B and C traffic areas for all gear designs. Curve 4 is used for type D traffic area for twin-twin wheel. It is necessary to measure subgrade modulus k to get the factor F before thickness of flexible overlays can be designed. The relationship between k and F have been developed from the results of traffic tests on plain concrete pavements. The factor 2.5 given in the equation was also developed from the results of full scale traffic tests.

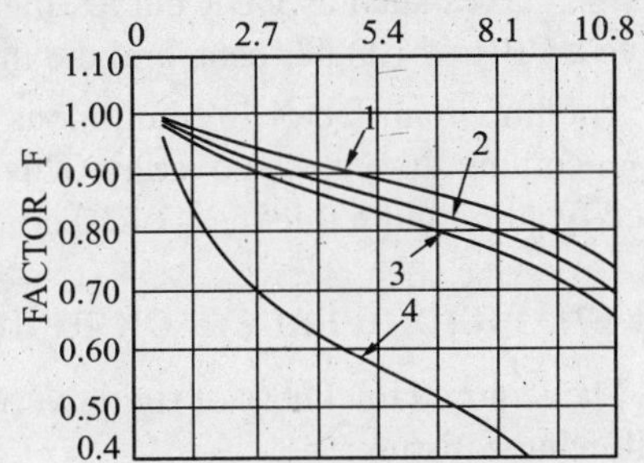

Fig. 23.32 Non-rigid overlay design.

Values of factor F recommended by FAA are given in Table 23.12.

Table 23.12 shows that if adequate sub-base thickness has been provided under the existing concrete pavement in accordance with the requirements of FAA design curves as given in Chapter 21 the value of 0.80 for the factor F can be used in all cases. This condition would be equivalent to R_a subgrade class. If sub-base under older pavements do not conform to these requirements, higher value of F is selected. For example, if design must be based on R_d subgrade and sub-base thickness provided under existing pavement conforms to the requirements for R_b or R_c subgrade, then the value of 0.94 and 0.90, respectively, will be used for F.

The FAA requires a minimum thickness of non-bituminous base course of 15 cm as well as minimum thickness of bituminous surfacing determined from appropriate design curve (see chapter 20) when flexible overlays are constructed on rigid pavements. If all-bituminous overlay is to be applied, the thickness required would be $t_0/1.5$, considering 1 cm of bituminous surfacing to be equivalent to 1.5 cm of aggregate base course.

According to U.S. corps of engineers method, all bituminous overlays are used only when the base course as given by the design formula is less than 10 cm. In addition , the base course materials are required to have a minimum CBR of 100 per cent, and the overlay must consist of hot-mix binder and surface course.

Table 23.12 Value of "F" recommended by FAA when thickness of sub-base under existing pavement conforms to requirements for subgrade class indicated below

Subgrade class	R_a*	R_b	R_c	R_d	R_e
R_a	0.80	-	-	-	
R_b	0.90	0.80	-	-	-
R_c	0.94	0.90	0.80	-	-
R_d	0.98	0.94	0.90	0.80	-
R_e	1.00	0.98	0.94	0.90	0.80

*Figures in this column apply when no sub-base has been provided.

An alternate evaluation procedure to reduce design thickness of flexible overlays can sometimes be employed where the original concrete slab rests on a soil which is classified as a flexible pavement subgrade class R_a. In this case, the concrete slab may be considered as base course and evaluation made on that basis using the appropriate flexible pavement (FAA) design curves. Similarly, other design curves such as CBR curves may be used. For these cases the rigid pavement is assumed to have a CBR of 100 per cent, and the thickness of base and surface can be determined on this basis.

The important factor in bituminous overlays is the prevention of cracks in the underlaying slab from coming through the overlay. This is the reason that a number of airport and highway agencies specify a minimum thickness of 10 cm for bituminous overlays.

23.17. RIGID OVERLAYS ON RIGID PAVEMENTS

The required thickness of rigid overlays over existing rigid pavement can be determined from the following equation:

$$h_n = (h_0^{\,x} - C h_e^{\,y})^z \qquad \ldots(23.8)$$

where h_0 = thickness of concrete overlay slab in cm

h_e = thickness of existing slab in cm

h_n = thickness of equivalent single slab placed directly on the subgrade with a working stress equal to that of the overlay slab

C = coefficient, depending on condition of existing pavement;

= 1, existing pavement in good condition.

= 0.75, existing pavement with initial corner cracks due to loading but no progressive crack.

= 0.35 existing pavement badly cracked or crushed.

Values recommended by U.S. corps of engineers, FAA and PCA for x, y and z are given in Table 23.13.

Table 23.13 Values of x, y and z in equation 23.8

Agency	*Condition*	x	y	z
U.S. corps of engineers	Overlay placed directly on existing pavement	1.4	1.4	1/1.4
	Levelling course used	2.0	2.0	0.5
PCA, FAA	Overlay placed directly on existing pavement	1.87	2.0	0.5
	Levelling course used	2.0	2.0	0.5

Wherever possible, joints on overlay pavements should coincide with those of existing pavement otherwise reflection cracks may result.

Problem 23.4 *Determine the thickness of rigid overlay for a 10 cm thick slab in good condition. The design thickness required, due to increased loading, is 20 cm. Use U.S. corps of engineers method.*

Solution. The required thickness of rigid overlay over existing rigid pavement is given by equation 23.8:

$$h_0 = (h_n{}^x - Ch_e{}^y)^z$$

Here

$h_n = 20$ cm

$C = 1$, since the existing pavement is in good condition

$h_e = 10$ cm.

Overlay can be placed directly over the existing pavement as it is in good condition. Therefore, from Table 23.13,

$$x = 1.4$$

$$y = 1.4$$

$$z = \frac{1}{1.4} = 0.714$$

$$h_0 = (20^{1.4} - 1 \times 10^{1.4})^{0.714}$$

$$= (66.29 - 25.12)^{0.714}$$

$$= (41.17)^{0.714} = 14.22 \text{ cm say 15 cm.}$$

Problem 23.5 *Design a rigid overlay for a badly crushed concrete pavement of 20 cm thickness. The design thickness of equivalent single slab placed directly on the sub-grade with a working stress equal to that of the overlay slab is 30 cm.*

Solution. The overlay thickness is given by

$$h_0 = \left(h_n{}^x - Ch_e{}^y\right)^x$$

Here $h_n = 30$ cm

$C = 0.35$, for badly crushed pavement (page 866)

$h_e = 20$ cm

Since the existing pavement is badly crushed, a levelling course will be required. Taking the values of x, y and z from Table 23.13,

$$x = 2$$

$$y = 2$$

and $z = 0.5$

$$\therefore \quad h_0 = (30^2 - 0.35 \times 20^2)^{0.5}$$

$$= (900 - 140)^{0.5}$$

$$= (760)^{0.5}$$

$$= 27.57 \text{ cm} \qquad \text{Say 28 cm}$$

23.18. BONDED CONCRETE OVERLAY ON RIGID PAVEMENT

Based on the laboratory work done on overlays at Central Road Research Institute, India an experimental stretch, 200 m, in length, on the existing 7.5 cm thick cement concrete road between Hapur and Garhmukteshwar has been provided with a bonded concrete overlay.

The required thickness of monolithic concrete pavement designed for *k*-value obtained by plate-bearing tests on base course; using Westergaard formula as modified by Teller and Sutherland for load stresses, and Thomlinson analysis for temperature stresses, was found to be 17.5 cm.

Assuming that it will be possible to create perfect bond between the overlay and the existing slab, it was decided to provide a 10 cm, thick overlay slab so as to make up the 7.5 cm thickness of the existing pavement to the required thickness of 17.5 cm.

For obtaining bond between the existing pavement and the overlay thereon, the following procedure was adopted. The existing surface was thoroughly cleaned of all dust, traces of bitumen, etc. and was then treated with commercial hydrochloric acid diluted with equal volume of water, at the rate of 0.35 kg/m^2. After the effervescence stopped the surface was washed with water to remove all traces of acid. On the dried surface, a thin slurry coat of 1 : 1 cement—sand mortar was given just in advance of concreting operations for the overlay.

Studies conducted in U.S.A. also indicate that to insure proper bond, the base concrete must be brushed and flushed after acid etching to get a clean surface free of residue and loose particles. The base concrete must be damp but free of standing water. The grout should have a thick paint-like consistency so that it can be easily broomed into the surface. The presence of free water on the pavement surface, or a thin grout with excess water from the resurfacing concrete, could lead to a weak mortar layer at the bond interface.

As the joints in the existing pavement were 10 m apart in the overlay, expansion joints were provided 30 m apart and intermediate dummy contraction joints at 5 m apart.

On a section with cracked old pavement, to asses the efficacy of reinforcement in prevention of reflection cracking in the overlay, M.S. reinforcement was provided at the rate of 4.5 kg/sq m.

It was proposed to cut out cores from the test-track for conducting bond strength test between the existing pavement and the overlay.

The above method is based on the assumption that if perfect bond is attained between the existing pavement and the overlay, the two together will behave as a monolithic slab. In this case when perfect bond is ensured between the two layers, and the existing pavement is in good condition, then the thickness of overlay need just be sufficient to make up the thickness of the existing pavement to the required design thickness of the monolithic slab.

This method, has however, not been given much field trial to justify its unreserved use for overlay design. Experimental data from overlay test track of U.S. Corps of Engineers have established that lesser overlay thickness is needed in case of partially bonded overlays as compared to unbonded overlays. Bond between the two is complicated by the tendency of the new concrete to shrink with respect to the older concrete.

23.19. RIGID OVERLAYS OVER EXISTING FLEXIBLE PAVEMENTS

In this case, the original flexible pavement is considered as sub-base course and the rigid overlay thickness can be determined by using design criteria for rigid pavements. The value of subgrade reaction k of existing flexible pavement can be determined and this value can be used for the design of overlay.

If FAA design procedure is used, overlay pavement will consist of the required thickness of concrete slab plus any deficiency in sub-base that may exist. If the thickness of existing pavement is greater than the requirements for sub-base thickness determined from the FAA design charts, no adjustment in thickness of new concrete slab is permitted.

No reduction in thickness of Portland cement concrete may be made even though the existing flexible pavement may be close to the full thickness required. The reason for this is that the strength of a Portland cement concrete slab is very critical with respect to the thickness. The normal deflection to be expected of a flexible pavement is about 6.25 mm., whereas the deflection of a rigid pavement sufficient to cause the pavement to break is about 2.5 mm. Unless this deflection is limited, the rigid pavement will crack and lose its effectiveness.

23.20. GENERAL REMARKS

A review of the methods, contained in this chapter indicates that they are primarily empirical, in the sense that when a formula is used the values assigned to the terms in the formula are chosen so as to ensure that the answer shall conform with observations of pavements in service or in full scale test tracks. There is nothing wrong with this method of approach. In fact it is very doubtful that an ideal solution will ever be evolved for determining thickness of overlay pavements. However, it is felt that efforts toward the development of an adequate theory should be pursued with vigour, since theory will be helpful in better understanding the performance of overlay pavements.

When considering an overlay over an existing concrete slab, it may work out that the thickness of a bituminous overlay obtained from the charts is often less than a Portland cement concrete overlay. This is rather difficult to explain quantitatively. One reason is of course the fact that a Portland cement concrete pavement is seldom placed less than 15 cm thick under existing construction practices. Another possible reason concerns stresses. A concrete overlay is designed to avoid overstressing in both old and new slabs whereas when a bituminous overlay is used a limited amount of cracking is permitted in the base slab. In this connection, the amount of deflection that a Portland cement concrete slab can stand without causing cracks is small compared to the deflection that can be withstood by the bituminous overlay. The important factor in bituminous overlay is the prevention of cracks in the underlaying slab from coming through the overlay. This is the reason that a number of airport and highway agencies specify a minimum thickness of 10 cm for bituminous overlays.

❑❑❑

Chapter

HIGHWAY MAINTENANCE

Maintenance of a road network involves a variety of operations from planning, programming and scheduling to actual implementation in the field and monitoring. Whatever be the approach or system adopted, the essential objective should be to keep the road surface and appurtenances in good order, and to extend the life of the road assets to the maximum extent possible. Broadly, the activities include identification of defects and the possible causes thereof, determination of appropriate remedial measures, implementation of these in the field and monitoring of the results. This will involve several sub-systems of identification, evaluation, planning, scheduling, management of men, materials and equipment, reporting and performance evaluation.

Some of the important aspects involved in road maintenance are however highlighted in the following paragraphs:

24.1. BASIC MAINTENANCE OBJECTIVES

The basic objective of maintenance functions are to maintain and operate the highway system in a manner such that:

(*a*) Comfort, convenience and safety are afforded to the public;

(*b*) The investment in roads, bridges and appurtenances is preserved;

(*c*) The aesthetics and compatibility of highway system with the environment is preserved; and

(*d*) The necessary expenditure of resources is accomplished with continuing emphasis on economy.

24.2. CLASSIFICATION OF MAINTENANCE ACTIVITIES

The maintenance activities can be grouped under the following categories:

24.2.1. Ordinary repairs

These include routine activities which are required irrespective of traffic or engineering characteristics of the road such as grass cutting, repairs to road furniture, culvert/drain maintenance, etc; and recurrent activities such as repairing potholes and edges, sealing of cracks, etc. which depend on traffic and environmental conditions.

24.2.2. Periodical renewal

This consists of application of a renewal coat of surfacing at periodical intervals. Based on an assessment made either through visual inspections, or objective measurements for riding quality, or a combination thereof, a programme of renewal is prepared. Frequency and type of renewal will depend on the type of original surfacing, traffic, rainfall and other climatic conditions, etc. The renewal cycle should be judicious so that the needy section are not starved of funds.

24.2.3. Special repairs

Need for these may arise on account of damage by floods or other natural calamities, or small geometric improvements which can bring forth positive improvement in traffic operation, etc.

24.2.4. Urgent repairs

Works which cannot be anticipated but have to be invariably carried out on immediate basis fall under this category. Examples are flood damage repairs, making up road breaches and restoring traffic, removal of road blockages, etc.

Assistant Engineers and Junior/Sub-Engineer while going on inspection on roads in their jurisdiction should examine the points mentioned below:

(1) Safety aspects

(*i*) Safety precautions for blockade and breaches taken;
(*ii*) Deep cuts on roads;
(*iii*) Damaged culvert/bridge;
(*iv*) Branches of trees at less height;
(*v*) Power line crossings provided with guard cradles as per IRC:32;
(*vi*) Vertical clearances for power lines should be as per IRC:32;
(*vii*) Horizontal clearances for poles carrying power and telecommunication lines as per IRC:32;
(*viii*) Berms not lower than 25 mm for carriageway; and
(*ix*) For new plantation only, the horizontal clearances to be kept as minimum 5 m wherever possible.

(2) Carriageway and crust conditions

(*i*) Location, magnitude of potholes and patches;
(*ii*) Condition of edges;
(*iii*) Magnitude and location of undulations; and
(*iv*) Location of crust failure, along with their causes.

(3) Berms (shoulders)

(*i*) Width of berms is adequate or not as per PWD specifications;
(*ii*) Cross slope kept as 3 to 5 per cent;
(*iii*) Side slopes;
(*iv*) Berms properly dressed; and
(*v*) If turning exists, whether it is properly cut or not.

(4) Road drainage

(*i*) Cross-sectional area of drains adequate or not;
(*ii*) To check if the drains are blocked or damaged; and
(*iii*) Whether proper disposal is provided to the drains.

(5) Road fixtures

(*i*) Km stone, 5th km stone, 200 m stones and boundary stones exist in proper conditions;
(*ii*) Traffic signs correctly located and maintained;
(*iii*) Location and condition of berms on curves and high embankments;
(*iv*) Painting and numbering of culverts required; and
(*v*) History of the road mentioned on km stones.

(6) Road protection works

(*i*) Retaining walls and pitchings on slopes properly maintained or not;
(*ii*) Condition of drains, spouts and weep holes in retaining walls and in pitching on slopes; and
(*iii*) Condition of parapet walls on culverts, etc.

(7) Roadside trees

(*i*) Check in numbering of trees done or not
(*ii*) Disposal of dead trees; and
(*iii*) Register of trees maintained.

(8) Road Geometrics

(*i*) Horizontal curves laid out properly or not;
(*ii*) Extra width on curves conforms to IRC:38;
(*iii*) Sight distances conform to IRC standards;
(*iv*) Vertical curves properly laid out or not; and
(*v*) Ruling gradients conform to IRC standards.

MAINTENANCE MANAGEMENT

The main objectives of maintenance management are the allocation of available maintenance resources according to actual needs and priorities, increasing the productivity, the development of uniform ways for doing the work, and the monitoring and evaluation of the outputs for the continuous improvement of the system. These can be achieved through a step by step procedure outlined in the following paragraphs:

24.3. DEFECTS, THEIR SYMPTOMS, CAUSES AND TREATMENT

Standardisation of defects, and a clear understanding as to why they occur and most effective treatment that are necessary so that every one involved in the maintenance effort is clear on the objectives and possible solutions, and the problem of differing opinions is minimised. Tables 24.1 provides guidance in this regard in respect of bituminous surfacings. Broad specifications for sealing and patching materials for repairing defects are given in table 24.2.

Table 24.1 Maintenance criteria

Feature	*Criteria*	*Action*	*Priority*
1. Features concerned with safety of traffic			
(*i*) Major Breaches in the roadway	Any type of breach which endangers safety of traffic and causes obstruction to flow of traffic	Steps to be taken as repair of major breaches immediately to restore traffic	Urgent
(*ii*) Minor cuts, ruts or blockades	Cuts or blockades which do not completely obstruct the traffic but endanger safety of traffic	Get blockades removed and get the cuts repaired	Urgent
(*iii*) Branches of trees at height less than 4.5 m over the roadway	Any kind	Get them cut in order of lower ones first	Special attention

2. Carriageway and crust conditions			
(*i*) Cracking not accompanied by rutting	(*a*) Cracking in local areas equal to or less than 25 per cent of the total area	(*a*) Local sealing or filling of the cracks preferably with slurry seal or fog seal as per Ministry's Specifications (*b*) Binder @ 1.5 kg/m^2 of bitumen emulsion or 1 kg/m^2 of cut-back or local sealing	Routine
	(*b*) Cracking in large areas exceeding 25 per cent of the total area	(*c*) Surface Dripping as per Ministry's specifications	Special attention
(*ii*) Stripping	(*a*) In local areas not exceeding 25 per cent of the total area	Apply local sealing	Routine
	(*b*) In long areas exceeding 25 per cent of the total area	Apply surface dressing. Use antistripping compounds	Special attention
(*iii*) Bleeding	(*a*) In local areas not exceeding 25 per cent of the total area	Spread and roll over 6 mm size aggregate, heated to 60°C	Routine
	(*b*) In local areas exceeding 25 cent of total area	Apply surface dressing	Special attention
(*iv*) Rutting	(*a*) Less than 50 mm accompanied by cracking	Apply tack coat @0.5 kg/m^2 and fill bituminous mix using a rake and leaving an excess thickness of about one-third the depth of rut. Compact till surface is levelled and local sealing of cracks.	Routine
	(*b*) More than 50 mm accompanied by cracking	With surface dressing over cracks, overlay required	Work of original nature
(*v*) Potholes	Potholes, as soon as they occur	Local restoration by patching preferable	Special attention
(*vi*) Reflection Cracks	(*a*) Widely spaced cracks	Slurry for fog seal	Recurrent
	(*b*) Closely spaced	Apply surface dressing use of geotextiles	Special attention
(*vii*) Edge subsidence and rutting	Any extent	Patch road edge and repair shoulder	Recurrent

(*viii*) Defective camber	Any extent	Check and correct by reconstructing to proper camber profile	Special attention
(*ix*) Undulations	Any extent	Investigate the cause and rectify	Special attention
(*x*) Loss of material from unpaved road	Any extent	Do regravelling	Special attention
3. Shoulders, side-drains			
(*i*) Deformation or scour of shoulders	Any extent	Fill and compact and bring its surface to desired camber	Routine
(*ii*) Silting of side-drains	Any extent	Clean out the drains	Routine
(*iii*) Damage or scouring of drain	Any extent	Reconstruct to adequate shape and size	Special attention
4. Cross-drainage works			
(*i*) Causeways			
(*a*) Potholes in paved surface	Any extent	Repair by filling	Special attention
(*b*) Erosion at inlet/outlet	Any extent	Repair	Special attention
(*c*) Guide posts/ flood guage missing	Any extent	Repairs/ Replace	Special attention
(*ii*) Culverts			
(*a*) Silting	Any extent	Desilting	Special attention
(*b*) Erosion at inlet/outlet	Any extent	Repairs	Special attention
(*c*) Settlement cracks	Any extent	Reparis	Special attention
5. Other works			
(*i*) Road furniture and warning signs dirty or corroded or damaged	Any extent	Clean and repair/replace	Routine
(*ii*) Missing road signs	Any extent	Fix new ones	Special attention

24.4. EVALUATION AND REPORTING

There should be a continuous evaluation of the quantum, standard and cost of work turned out by the gangs to detect deviations from the planned norms so that corrective action could be taken. Corrective action would be in the form of re-direction of work emphasis, removal of bottle necks, special direction or training to gangs etc. Cumulative summary of cost and effort would also highlight road section needing high maintenance effort so that such sections could be taken up for detailed investigations for appropriate corrective measures.

24.4.1. Maintenance inspection

Frequent inspections of the road can help in the correct identification of defects and the reasons there of, determination of priorities and prompt and appropriate remedial measures. A road should be inspected by a responsible officer at least twice a year, once during the wet season and the other during the dry season.

A basic inventory of each road section must be available for reference during inspections as also for planning and execution of maintenance operations. This should be in the form of Road Register. These should be updated regularly.

A check list of important items to be seen during inspection is indicated in Table 24.1. It will be preferable if the different items to be inspected are listed out on an inspection card with columns for each 200 m of the road so that the inspecting officer would just tick or cross each column for further evaluation and decision on the appropriate action to be taken.

24.4.2. Prioritisation of maintenance operations

It will be necessary to allocate the limited resources in a way that most nearly satisfies the objectives of maintaining the road system to the desired standard. From this angle, the different operations should be graded in the following order of priority:

(*i*) ***Urgent.*** Emergency repairs to road breaches and other flood damages, removal of debris, dead animals and other road blockages, collection of accident information in the prescribed format, etc.

(*ii*) ***Routine drainage work.*** Cleaning and deepening of drains, and checking and cleaning of bridge/ culvert openings.

(*iii*) ***Recurrent drainage work.*** Filling scoured areas, building check dams and scour control devices, repair of structures, etc.

(*iv*) ***Recurrent work on pavement.*** Patching and filling potholes and ruts.

(*v*) ***Other routine works.*** Grading shoulders, slopes, and repairing/replacing road furniture.

(*vi*) ***Periodic work.*** Surface renewal.

(*vii*) ***Special repairs.*** These are as a result of maintenance inspection but are in the nature of original or improvement works.

Guidance regarding assignment of priorities is given in Table 24.1.

24.4.3. Highway maintenance

It has been defined as the preserving and keeping of each roadway, structure and facility as nearly as possible in its original condition or as subsequently improved, and such additional work as is necessary to keep the traffic moving safely. This falls into maintenance of : (*i*) road surface, (*ii*) shoulders, (*iii*) roadway drainage, (*iv*) bridges and other structures, and (*v*) road sides. This also includes items for safe operation of the highway such as erecting signs and traffic controls, painting traffic strips and facilitation of traffic through removal of snow and spreading sand, *etc.*

Highway location, design and construction have a bearing on maintenance costs and conditions. At the time of alignment studies, proper considerations should be given to drainage problems, stream crossings, the avoidance of possible landslide conditions and unsuitable soils, and directness of route. Shorter route generally means less maintenance cost and less maintenance problems. Insufficient pavement or base thickness or improper construction methods soon results in expensive patching or surface repair.

Shoulder maintenance becomes a serious problem where narrow lanes or small roadway width force heavy traffic to travel with one set of wheels off the pavement. Proper design of gradients through-out will result in least amount of erosion problem and at the same time will have enough fall to provide good drainage. Similarly design of waterways and structures should be adequate to handle the runoff drainage. In regions of snow fall the proper location and grade line is extremely important otherwise very difficult snow-removal problems would result. The type of surface should be selected keeping in view predicted volume and character of traffic for ease of maintenance and reduction in maintenance cost.

MAINTENANCE OF ROAD SURFACES

Whatever is the type of road surface, maintenance, for economy, should be preventive. The higher the road type, greater the importance of preventive maintenance. Defects should be corrected before they become serious, keeping the surface in good repair rather than allowing it to fall into disrepair and then correcting it. When a rut, pothole or other defect appears, it progresses very rapidly and timely maintenance is more economical and safe than allowing the road surface to disintegrate.

About 50 per cent of the maintenance budget is used for maintenance of road surfaces. Most of this work is performed by the various highway agencies themselves which is in direct contrast to construction, where most of the work is done by contract. The principal reason for this is that maintenance work is so diverse, so subject to variation from the expected, and on occasions so hurried that it does not lend itself to competitive bidding. A brief description of maintenance operations relating to various surfaces will now be given.

24.5. BITUMINOUS SURFACES

Maintenance of bituminous surfaces consists primarily of patching, base repairs, surface treatments, and resurfacing. Regular and frequent inspections are essential for repairing failures in early stages. Rainy weather inspections are very helpful in that surface defects can be detected readily and marked for future correction.

24.5.1. Patching

Patch repairs may be needed for the damaged portion of the road and potholes or for removing of inequalities in shape and surface and removing waviness in order to smoothen a rough or uneven riding surface. Pothole is usually due to a localized poor gradation of surface materials provided there is no foundation trouble. Corrugations on the road surface are often created by incorrect gradation of aggregates, excessive bitumen in the mix, traffic overloads, or excessive moisture in bituminous mix. These potholes and corrugations or any type of unevenness which might create discomfort or hazard to traffic should be repaired immediately by patching.

The areas of bituminous patching vary in size from small potholes to as much as full width of the roadway.

Potholes in the road surface should be cut to square or rectangular shape, and should be excavated and extended until sound materials are encountered. Holes should be cleaned of all loose aggregate, dust, foreign matter and water before the patch is applied. The sides and bottom of the cut portion, after cleaning are lightly primed and patching mixture should be placed as soon as possible after application of prime coat. Patching mixture usually consists of premixed material made by proper mixing of graded mineral aggregate and a suitable bituminous material such as cut-backs, emulsified bitumen or paving bitumen or tar.

Table 24.2 gives the broad specifications for patching materials used for repairing defects.

Table 24.2 Specification for patching material

Treatment	*Binder*	*Aggregate*	*Specification in brief*
1. Sand premix	Penetration grade, or cut-back RC or MC	Fine grit 1.7mm to 180 micron	Apply tack coat at 7.5 kg per 10 sq m. Spread mix consisting of 0.06 cu m of grit and 6.8 kg of binder per 10 sq m and roll.
2. Open-graded premix	Paving grade, RC or MC cut-back, MS emulsion	12.5 mm and 10 mm size aggregate	Apply tack coat, prepare the mix as per IRC : 14-2004, spread and roll. Where cutback is the binder, the premix should be prepared at least 3 days in advance of use. The final surface should be provided with seal coat.
3. Dense graded premix	Paving grade	Well graded as per IRC : 16	Apply tack coat and prepare, spread and compact mix as per IRC : 16-2008.
4. Penetration patching	Penetration grade RC or MC cut-back	Coarse and key aggregates as per IRC : 20-1996	Apply tack coat, spread coarse aggregate and dry roll; apply binder and key aggregate and roll as per IRC : 20. For patching 50-75 mm thickness, use built up spray grout as per IRC : 47-1972.

It is necessary to compact all patching material so that subsequent consolidation and ravelling will not take place. If the hole is more than 7.5 cm deep, the patch should be made in two or more layers and each layer tamped or rolled. Small holes can be hand tamped and the larger patches should be rolled. The surface of the finished patch may be kept slightly proud of the existing surface to allow for further compaction by traffic but care must be taken not to cause a bump.

Where the road-width is narrow or the road shoulders are inadequate, a frequent location of ravelled areas is along the edges of the road metal. These are patched as mentioned above, care being taken to cut-back the damaged surface sufficiently to establish a soled foundation for the patch, and to compact the new material solidly against the side of the hole. Along the free edge, a stepped back construction should be used, the lowest layer of the patch extending into the shoulder area with each succeeding layer drawn in so that the top layer will line with the established edge.

Corrugations, both longitudinal and crosswise are rectified up to designed cross sections by cutting off the high ridges. The inequalities and waviness is removed by filling the depressions with premixed material after applying tack coat over the cleaned surface. Fig. 24.1 shows the correct and incorrect methods for providing corrective courses for short sags and depressions.

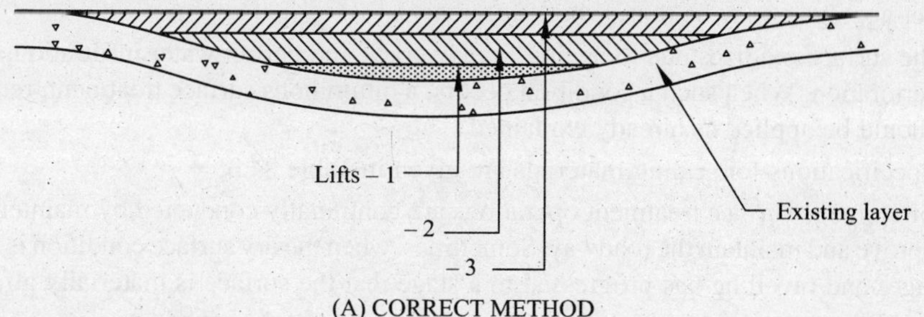

(A) CORRECT METHOD

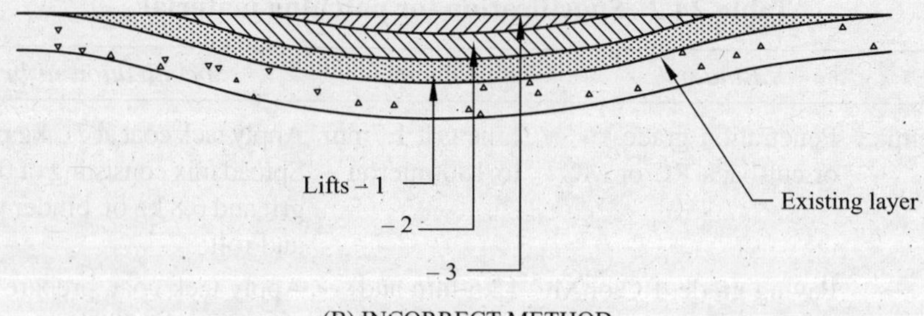

(B) INCORRECT METHOD

Fig. 24.1 Methods for providing corrective course for short sags and depressions.

Note: Profile corrective course material to be in accordance with the lift thickness.

24.5.2. Base course repairs

Structural failures of base course in flexible pavements may result from inadequate thickness of base, inadequate subgrade support, excess moisture in the base or subgrade, or combination of these causes. The failure may be due to consolidation or shear developing in the subgrade, sub-base, base course, or surface.

In case the failure is due to consolidation of one or more of the road layers, rut is not accompanied by any upheaval of the surface; whereas subgrade shear failures exhibit surface upheaval at some distance from the depressed rut and shear failure in surface course will result in upheavals relatively close to the tyre track. Investigation by trenching the pavement and visual inspection or measurements should determine the cause of base failure and necessary corrections should be decided before any repair work is carried out.

If the failure is due to improper removal of surface water, proper drainage should be provided. Small local failures due to slight inadequate base thickness can be economically corrected by adding additional surface thickness.

However, when areas of local failure become general and it is decided to increase the base thickness, the old surface and base is loosened by scarifying to the full depth. The subgrade is removed to the depth necessary to give the needed additional thickness of new base. New suitable material is added as necessary and after mixing with the old material, the road is brought to a proper cross-section and is recompacted. Sometimes, satisfactory materials salvaged from the old base can be used in the bottom of the new repair. The new base is thoroughly compacted in layers not more than 7.5 cm thick. The new base should be covered with suitable surface treatment or a bituminous mat.

24.5.3. Surface treatment

If bituminous surface contains too much bitumen, it will bleed and become slippery when wet and corrugations or rutting or shoving may develop. Bleeding usually occurs soon after construction of the surface. It should be corrected soon after occurrence by application of blotting material such as aggregate chips (maximum size 12 mm or 15 mm), or coarse sand, reasonably free from dust. Rolling of this cover aggregate may be done where necessary.

When the surface oxidizes due to ageing, it may crack and permit water infiltration, thus causing further deterioration. When such a condition occurs, a bituminous surface treatment, renewal coat or seal coat should be applied as already explained.

Broad specifications for sealing materials are given in Table 24.3.

These bituminous surface treatment operations are continually conducted by maintenance staff in order to improve and maintain the roadway. Sometimes, when the dry surface condition is not corrected in early stages and ravelling has progressed to a stage that the surface is materially affected, it may become necessary to apply double surface treatment as already described.

Table 24.3 Specification for sealing materials

Treatment	*Binder*	*Aggregate*	*Brief specification*
1. Liquid seal	Penetration grade, cut-back or emulsion	6.3 mm size	Spray binder uniformly at 9.8 kg/ 10 sq m. spread aggregate at 0.09 cu m/ 10 sq m and roll.
2. Fog seal	Slow setting emulsion RS diluted with equal amount of water		Spray at 0.5 to 1 litre/sq m. Allow traffic after the seal sets.
3. Slurry seal	Slow setting emulsion	Well graded material between 4.75 mm and 75 micron (grading-specified)	Apply tack coat with diluted emulsion at 2.5 to 3.5 kg/10 sq m. Apply slurry mix consisting of 18 to 20% emulsion and 10 to 12% water by weight of aggregate at the rate of 200 sq m per tonne.

24.5.4. Resurfacing

When the road base is structurally sound but the bituminous mat may become distorted resulting in poor riding qualities, it may be more economical to provide an additional mat on the top of existing surface rather than scarifying the existing surface and strengthening the base. After the existing surface is repaired by suitable patching a light tack coat is applied and the new surface is laid by road mix or plant mix method as already described.

24.5.5. Summary of maintenance of bituminous pavements

A summary of major types of defects their symptoms, probable causes and treatment are given in Table 24.4.

Table 24.4 Symptoms, causes and treatment of defects in bituminous surfacing

Type of distress	*Symptoms*	*Probable causes*	*Possible types of treatment*
1. Surface defects			
(*i*) Fatty surface	Collection of binder on the surface	Excessive binder in premix, spray or tack coat, loss of cover to aggregates; excessively heavy axle loads	Sand blinding; open-graded premix; liquid seal coat; burning of excess binder; removal of affected area
(*ii*) Smooth surface	Slippery	Polishing of aggregates under traffic. Excessive binder	Resurfacing with surface dressing or premix carpet
(*iii*) Streaking	Presence of alternate lean and heavy lines of bitumen	Non-uniform application of bitumen or at a low temperature	Application of a new surface
(*iv*) Hungry surface	Loss of aggregates or presence of fine cracks	Use of less bitumen or absorptive aggregates	Slurry seal or fog seal

2. Cracks			
(*i*) Hairline cracks	Short and fine cracks at close intervals on the surface	Insufficient bitumen, excessive filler or improper compaction	The treatment will depend on whether pavement is structurally sound or unsound
(*ii*) Alligator cracks	Interconnected cracks forming a series of small blocks	Weak pavement, unstable conditions of subgrade or lower layers, excessive over loads or brittleness of binder	Where the pavement is structurally sound, the cracks should be filled with a low viscosity binder or a slurry seal or fog seal depending upon the width of cracks.
(*iii*) Logitudinal cracks	Crack on a straight line along the road	Poor drainage, shoulder settlement, weak joint between adjoining spreads of pavement layers or differential frost heave	Unsound cracked pavements will need strengthening or rehabilitation treatment
(*iv*) Edge cracks	Crack near and parallel to pavement edge	Lack of support from shoulder, poor drainage, frost heave, or inadequate pavement width	
(*v*) Shrinkage cracks	Cracks in transverse direction or interconnected cracks forming a series of large blocks	Shrinkage of bituminous layer with age	
(*vi*) Reflection cracks	Sympethetic cracks over joints and cracks in the pavement underneath	Due to joints and cracks in the pavement layer underneath	
3. Deformation			
(*i*) Slippage	Formation of crescent-shaped cracks pointing in the direction of the thrust of wheels	Usual thrust of wheel in a direction, lack of failure of bond between surface & lower pavement courses	Removal of the surface layer in the affected area and replacement with fresh material
(*ii*) Rutting	Logitudinal depressions in the wheel tracks	Heavy channelised traffic, inadequate compaction of pavement layers, poor stability of pavement material, or heavy bullock cart traffic	Filling the depressions with premix material
(*iii*) Corrugations	Formation of regular undulations	Lack of stability in mix, oscillations set up by vehicles, springs, or faulty laying of surface course	Scarification and relaying of surfacings, or cutting of high spots and filling of low spots
(*iv*) Shoving	Localised bulging of pavement surface along the crescent-shaped cracks	Unstable mix, lack of bond between layers, or stop type movements and those involving negotiation of curves and gradients	Removing the material to firm base and relaying a stable mix

(*v*) Shallow depress-ions	Localised shallow depressions	Presence of inadequately compacted pockets	Filling with premix materials
(*vi*) Settle-ment and upheaval	Large deformation of pavement	Poor compaction of fills, poor drainage, inadequate pavement or frost heave	Where fill is weak, the defective fill should be excavated and re-done. Where inadequate pavement is the cause, the pavement should be strengthened
4. Disintegration			
(*i*) Stripping	Separation of bitumen from aggregate in the presence of moisture	Use of hydrophilic agg-regate, inadequate mix composition, continuous contact with water, poor bond between binder and aggregate, poor compac-tion, *etc*.	Spreading and compacting hea-ted sand over the affected area in the case of surface dressing; replacement with fresh bituminous mix with added anti-stripping agent in other cases
(*ii*) Loss of aggregate	Rough surface with loss of aggregate in some portions	Ageing and hardening of binder, stripping poor bond between binder and aggregate, insufficient binder, brittleness of binder, *etc*.	Application of liquid seal, fog seal, or slurry seal depending on the extent of damage
(*iii*) Ravelling	Failure of binder to hold the aggregate shown up by pack marks or eroded areas on the surface	Poor compaction, poor bond between binder and or aggregate insufficient binder, brittleness of binder etc.	Application of cutback covered with coarse sand, or slurry seal, or a renewal coat.
(*iv*) Pothole	Appearance of bowl-shaped holes, usually after rain	Ingress of water into the pavement, lack of bond between the surfacing and WBM base, insuffcient bitumen content. *etc*.	Filling potholes with premix material or penetration patching
(*v*) Edge-breaking	Irregular breaker of pavement edges	Water infiltration, poor lateral support from shoulders, inadequate strength of pavement edge, *etc*.	Cutting the affected area to regular sections and re-building with simultaneous attention paid to the proper construction of shoulders.

24.6. LOW-COST SURFACES

Low-cost or low traffic volume surfaces include many types such as natural soil, natural soil treated, soil stabilized, soft aggregates, brick and macadam. These types of surfaces are subjected to different types of deterioration and their proper maintenance procedures, are widely varied.

For proper maintenance of this type of road, it is necessary that the surface be kept smooth, firm and free from excess loose material and that a proper crown be maintained for adequate drainage. Maintenance of this nature usually consist of : (*i*) Patching and repairing local breaks, (*ii*) dragging and blading, (*iii*) scarifying and resurfacing, (*iv*) stabilization and application of dust palliatives, and (*v*) sometimes the application of bituminous surface treatment.

Patching is done in the same manner as explained for bituminous surfaces except that patching material in this case consists of carefully graded material of the sorts used in the original road surface.

Blading and dragging are general maintenance operations carried out on all soils and stabilized surfaces. Since surfaces lacking stability typically fail by rutting under traffic, the purpose of blading and dragging is to fill in the ruts and smoothen out any irregularities in the surface.

A road drag consists of a series of blades attached to a rigid frame. It is hauled behind a truck or tractor, and is weighed as needed to enable it to cut deeper into the surface or plane down the material of the road surface. Blading is usually accomplished with an adjustable steel blade, carried on a frame which may be self-propelled, or which may be dragged after a truck or tractor. Much use is made today of the power driven motor grader because of its flexibility in doing other maintenance operations such as shaping shoulders, cleaning and shaping ditches, and similar work.

Blading and dragging are best done after a rain when the surface materials are sufficiently moist to compact well. In blading material is generally first drawn from edges to centre to establish a crown. The loosened material is then drawn back again and deposited in depressed areas. Maintenance of a crown is necessary on this type of road surfaces so that runoff will occur. Relatively pervious mixtures require less crown than more densely graded materials.

When the local failures are numerous, either through neglect, or as an indication of inadequate subgrade support or improperly constructed surface, complete scarification and reconstruction of road after adding new selected material is necessary. If failure is due to inadequate drainage or due to soft foundation areas, these must be corrected before the surface is reconstructed.

Low-type road surfaces are subject to very rapid deterioration due to blowing away of surface materials and sometimes dust palliatives such a calcium chloride or light penetrating bituminous road oils are used as already described in Chapter 11 relating to low-cost roads.

The action of the dust palliatives is effective only in the top thin layer. For example, when natural soil road is treated with bituminous binder, it gives enough cohesiveness to resist being blown from the road surface or drawn from the road surface by tyre suction. In case of more granular material such as gravel and macadam, a somewhat heavier grade of bituminous material can be used along with sand or other light mineral aggregate to form a protective surface treatment as already described.

24.7. PORTLAND CEMENT CONCRETE SURFACES

Maintenance work on rigid pavements depends on the type of deterioration, each one of which calls for its own remedial measure. The usual type of maintenance operations are : (*i*) filling and sealing joints and cracks in the pavement surface, (*ii*) patching areas where failure has occurred, (*iii*) repairing blow ups, (*iv*) repairing areas due to settlement or pumping, and (*v*) repairing scaling and other surface defects. These are discussed briefly in the following paragraphs:

24.7.1. Sealing cracks and joints

Prevention of infiltration of surface water to reach subgrade under pavement and collection of water in cracks and joints will contribute greatly to life, serviceability and low maintenance cost. Because of this, it is desirable to seal all cracks and joints as completely as possible. Suitable joint filler and sealing materials have already been described in detail in the chapter dealing with Portland cement concrete pavements.

All foreign matter should be removed from the joint or crack before application of seal material. Compressed air can be used to blow the loose material from the cracks and joints. Routine cleaning can be done with small tools such as spades, hand brooms, etc. If the old joint filler has extruded above the surface of the pavement to any great extent, it should be removed by blading. Where rubber compounds are to be used, extreme care must be taken to clean the sides of the crack or joint to promote adherence of the compound. A crack cleaning machine is now available which can satisfactorily clean the sides of crack or joint and can also remove the old filler to the desired depth in one operation. This machine consists of power driven cutting wheel 20 cm diameter with cutting knobs extending about 2.5 cm from the rim. The wheel is made of very hard material and revolves at high speed. Concrete power saws can also be used for this purpose. Compressed air is used with these machines for blowing off the loose particles.

Sealing is generally accomplished by hand methods by pouring heated sealing material with the help of conical-shaped pots with 6 mm openings in the bottom. Machine methods are also used in some cases in which material is applied through hand nozzle from a tank. Care should be taken to fill the crack or joint to the proper level and to avoid the use of excess material. However, surplus material is removed by a scraper.

24.7.2. Patching

Any hole, depression or sharp break in the smooth concrete surface should be repaired without delay. Such irregularities or depressions or local failures can be successfully and economically patched with bituminous materials. Procedures are similar as already described for patching bituminous surfaces.

However, when disintegration or scaling has progressed both in severity and area to the extent that bituminous patching is considered inadequate or when the impaired surface is very much limited in area and it is desired to use concrete patches to retain a uniform surface appearance, it becomes advisable to patch with concrete. Also, concrete patches should be used to replace cuts in the pavement for the purpose of installing public utilities.

The type of pavement break or failure determines the shape of patch. Good practice indicates that the dimension of any side of a patch should not be less than 1.2 m; that a side of a patch parallel to the flow of traffic should not be placed under a wheel track, and that the inside angle of a diamond or triangular shaped patch should not be less than 60° with the points flattened.

Expansion and contraction joints in concrete patches are unnecessary even though there may be a joint in the old pavement at the location of the new patch. However, when patch is placed at bridge ends, suitable joint as already existing should be provided.

The amount of patching to be done determines the methods for removing the old concrete. If only few small patches are to be made, the old concrete pavement can be removed by hand methods, using such tools as sledge hammers, hand drills with chisel points, crowbars, and other hand tools. Otherwise, for large patching work, it is more economical and efficient to remove the old concrete by machine methods, using air hammers and air compressors. Incidentally, the old concrete can be used with advantage for erosion control purposes. Concrete saws can also be used for making the original cut along the outside lines of the proposed patch. This will make it possible to have a more even edge for placing and finishing the new patch and also will ensure a neater joint between the old and new pavement than is provided by other methods.

The sides of the old pavement at the patch should be trimmed as nearly vertical as possible (not more than 3.75 cm out of perpendicular). The top 2.5 cm of vertical edge of old pavement should be cut as straight and as nearly vertical as possible in order to eliminate small projections of either old or new concrete extending across the junction which are likely to cause spalling. The edge or faces of old pavement should be cleaned of all dust, dirt, and loose particles.

All subgrade defects should be corrected and subgrade should be uniformly levelled and compacted to the proper elevation before new concrete is placed in the patch.

The thickness of patch should never be less than that of the existing slab. It is a good practice to make the patch thicker by 2.5 cm or slightly more, depending on subgrade conditions and volume of heavy traffic.

Patches should be finished in such a way as to blend in with surrounding old pavement.

24.7.3. Repairing blow-ups

As already explained in Chapter 18, buckling or blow-ups occur at transverse cracks or joints due to longitudinal expansion so as to cause the pavement to be raised off the subgrade. This raise may be quite large and as much as 60 cm. The pavement will often remain in raised position until broken down during maintenance. In certain cases, the damage from blow-ups results is crumbling of concrete pavement, 0.3 to 0.6 m each side of the crack or joint.

Blow-ups are a hazard to the traffic and should be corrected immediately. Damaged portion or about 15 cm of pavement is removed and replaced with a patch of concrete or bituminous material. If necessary, very temporary repairs can be made with roadside soil which has the advantage of allowing the pavement to settle back to its original position under traffic before permanent patch is applied.

24.7.4. Repairing settlement due to pumping or subgrade failure

One of the major maintenance problems of rigid pavements is correcting settlement of the slab caused by subgrade failure or by pumping. As already explained while discussing concrete pavements, if a rigid pavement has been subjected to pumping action for an extended period, it may be necessary to fill the resulting voids under the pavement by application of a mud jacking or bituminous under-sealing and raising the sections of concrete pavement to a desired elevation.

As already explained, mud jacking consists of pumping slurry of materials such as soil, cement and water through drilled holes in the pavement with a mud jacking machine.

Slurry has the consistency of thick cream, which is the proper consistency to fill the majority of the voids under the pavement. The most widely used mixtures are: (*i*) soil 60 to 84 per cent and cement 16 to 40 per cent (*ii*) soil 77 per cent, cement 16 per cent and cut-back asphalt (SC-2, MC-1, RC-3) 7 per cent. Consistency can be varied so that a fairly stiff mixture can be used to fill the large voids under the slab or to raise it, and a thinner mixture can be used to fill the smaller voids.

Mudjacking is continued until some of the slurry is pumped out of joints or along pavement edges.

Settlement of concrete slab may occur due to subgrade failure or settlement at bridge ends, over culverts, or at other places. As mentioned previously, it is possible to completely lift slabs (if broken not too badly) by mudjacking process using slurry of stiffer consistency. This will restore the riding qualities of the pavement which would otherwise cause unsafe conditions for traffic.

Bituminous undersealing is carried out much in the manner of mudjacking and consists of pumping highly viscous bitumen heated to a temperature of about 205°C through a drilled hole in the concrete pavement. This will cover the affected area of subgrade with a sheet of bitumen, thus supporting and under sealing the pavement by filling the cracks and joints from below. This will control slab pumping and it is a good practice to underseal with bitumen during the earliest detectable stages of pumping and thus keep pavement damage from this cause to a minimum.

Raising pavement slabs by undersealing is not desirable since there is no control possible.

Sometimes, mudjacked pavements will again develop voids under heavy traffic and will require continued mudjack slurry treatment. Instead of repeating this treatment, it has been found by experience that bitumen underseal of the pavement some 2 or 3 weeks after the mudjack treatement gives better and more lasting results.

24.7.5. Repairing scaling and other surface defects

These defects include surface scaling, spalling at joints, disintegration, and wear. Scaling is shelling off thin pieces from the surface of the pavement which may be caused by too much surface application of salts for curing and ice removal and by impurities in the original concrete mix, too much water in the mix, too much finishing and unsuitable aggregate. Scaling once started is aggravated by traffic. The usual corrective practice is to retard further scaling by bituminous patching or surface treatment.

Spalling is chipping or splintering of sound pavement due to frozen dowel-bars and incompressible material in the joints and cracks. This usually occurs along the joints or cracks in the pavement. This is also corrected by bituminous patching.

Disintegration of concrete pavement is breaking of the pavement into small pieces which may be due to unsound aggregate, freezing and thawing, and bad drainage. Disintegration is first reflected by surface cracking which usually starts at the intersection of longitudinal joint with a transverse crack or joint and then extends in all directions until full width is affected. It then extends longitudinally. This phenomenon is sometimes referred to as map cracking, or *D* cracking. If this is not controlled, it will reduce concrete pavement into completely unbound pieces of aggregate which will ravel and rut under traffic. All such areas are cleaned of all loose and foreign material and sealed with one or more application of surface treatment. Poor sub-drainage is corrected where it is one of the contributing causes.

Wear due to abrasion of traffic is a minor problem and requires maintenance only at isolated spots such as stop areas when brakes are sharply applied or on sharp curves where vehicles tend to skid or areas subject to heavy cross traffic from all directions. Abrasion at such spots can be retarded by bituminous surface treatement.

From the above discussion, it is seen that the surface defects can be retarded by bituminous patching and bituminous surface treatment of affected areas. An extensively repaired surface often results in an undesirable mottled appearance which is sometimes corrected by application of bituminous seal coat over the entire area.

MAINTENANCE OF SHOULDERS

The importance of properly maintaining the shoulders or berms becomes apparent when their important functions are understood. The most important functions of shoulders are: (*i*) to provide space for vehicles leaving the travelled surface of the roadway, in emergency, by accident, or in order to park, (*ii*) to provide adequate runoff of surface water from the pavement to the ditches (*iii*) to support the edge of pavement or road surface and also to provide lateral support to some types of pavement. It is desirable also that they have a colour contrast to that of the pavement. The purpose of paved shoulders adjacent to taxiways is to minimize dust and other debris that may result from blast of propellers or jet engines. This point is particularly important for large jet aircraft, where the out board engine extends beyond the paved area.

Maintenance should preserve these functions, regardless of the type of shoulders. Hence shoulders should be kept stable, smooth, and with proper elevation and cross slope. The most common defect is depressions in the shoulders along the edge of the pavement which defeats the most important functions of the shoulders. It creates a traffic hazard; it collects and holds water; and it deprives the edge of the road surface of any support. Other defects such as deep ruts, holes or slippery surface on the shoulders present a traffic hazard to fast moving vehicles. Shoulders must be sloped to provide adequate runoff and also maintained periodically to ensure its proper function. Shoulder slopes generally range between 60 mm and 75 mm per m of width. The amount of work necessary to maintain the shoulder varies directly with the type of shoulder construction, volume and type of traffic and width of the pavement.

A large percentage of shoulders on the existing highways have been constructed with available natural soil and it is the responsibility of the maintenance organization to maintain the shoulder in its natural condition or to improve it as the volume of traffic demands or when the funds permit.

Natural soil shoulder is the most difficult to maintain as it becomes slippery during wet weather, dusty during dry weather, and subject to erosion during heavy rains. Further, it is subjected to rutting during wet periods.

It is essential that the shoulder is maintained to remain in full contact with the pavement edge at all times. Other maintenance operations consist of blading and dragging at frequent intervals to eliminate ruts and to maintain proper shape.

24.8. TURFING

Possibly the first step in shoulder improvement is the development of turf. Increased stability offered by grass during the rainy season is an important consideration. A good growth of turf will also cut down on the erosion of side slopes. The most desired type of turf is usually a combination of grasses and legumes which will develop a dense growth with a minimum of height in shortest time and which will be self perpetuating from either roots or seeds or both. Root perpetuation of grasses is preferred because such plants tend to hold the soil there by reducing erosion and increasing load carrying capacity of the shoulder. On the other hand the plants whose roots die every year tend to loosen the soil thereby increasing erosion and decreasing load carrying capacity.

Turf shoulders need maintenance of firm, tough mat of sod and require that the grass be mowed often enough to keep it at attractive height and also to make the solid character of the shoulder evident to the driver. Light cuttings may be allowed to remain on the shoulder to supply humus to the soil. Holes and ruts in the turfed shoulder due to traffic should be immediately repaired by filling and grass re-established as early as possible in the growing season.

Turf shoulders have a tendency to build up in elevation, gradually through natural growth processes aided by the deposit of foreign matter and by frost action. This build up can be reduced to some extent by rolling the shoulder following wet periods. Otherwise at regular intervals, the shoulder must be planed down and proper elevation and slope early in growing season because the turf will usually re-establish itself from the roots without the expense or re-seeding. A drop-off of approximately 12.5 mm from the pavement edge to the turf shoulder is permissible, in order to increase the time before the shoulder may have to be planed down again.

Turf shoulder offer a good colour contrast between pavement and shoulder and provide inexpensive maintenance. Turf shoulders are considered adequate if the traffic is not too heavy so as not to prevent or offer too much resistance to the growth of turf and maintenance of the shoulder. However, when the pavement is so narrow that there is considerable wheel traffic on the shoulder, it becomes practically impossible to maintain the turf adjacent to the pavement. Under these circumstances, this type of shoulder is unsatisfactory, at least next to pavement. Further, turfed shoulders may be detrimental where the turf is placed adjacent to most types of bituminous surfaces. Turf retards the lateral flow of surface water and tends to increase the saturation of top soil. This shoulder saturation spreads to the subgrade under the bituminous surface resulting in road deterioration, especially where the bases are inadequate.

24.9. SOIL STABILIZED SHOULDERS

Soil stabilized shoulders, practically without any turf, have proved to be satisfactory adjacent to bituminous surfaces as they are well drained, more stable and can be readily shaped. Performance surveys have indicated that increased pavement performance will result whenever stabilized shoulders are used. The usual thickness of shoulders is about 10 to 15 cm. However, thickness of stabilized shoulders largely depends upon economics. Materials for stabilized shoulder may consist of soil gravel, soil sand, crushed stone, bituminous treatments, soil cement or any number of locally available materials. Principles of design are the same as already explained for soil stabilised roads. Proper maintenance of the shoulder by periodically reshaping and retamping is needed. Sometimes, it may be necessary to attempt to plant grass even on granular stabilized shoulders.

Sometimes, the shoulders are paved with Portland cement concrete or high type bituminous plant mixes which require maintenance similar to that for surfaces made of these materials.

It is sometimes, found economical and satisfactory to construct shoulder of two types, the higher type adjacent to the pavement.

MAINTENANCE OF ROADWAY DRAINAGE

Proper drainage maintenance requires that all structures and appurtenances pertaining to removal of water from the highway have to be kept in good working condition so as to provide free and unobstructed flow. Generally, maintenance operations consist of those relating to : (*i*) surface drainage, (*ii*) ditches, and (*iii*) subsurface drainage, storm sewers, culverts, etc.

Good surface drainage is maintained by having the pavement surface as water-proof as possible and by having such crown and surface smoothness that the water will flow freely to the road edge and on to the shoulder.

Similarly, shoulders are maintained flush with the pavement and smooth enough to provide for free flow of water away from the edge of the road and across the shoulder to the side ditch in cut sections or down slope of fill sections.

However, proper highway design can contribute to eliminate maintenance difficulties due to inadequate drainage. For example, if pavement and bases or sub-bases have to be constructed in a trench, a proper subgrade drainage either by constructing shoulders of porous materials or by placing pipe drains through the shoulders can contribute to ease of maintenance.

The side ditch which runs longitudinally and outside the shoulder should be maintained below the grade of subgrade in order that water from the subgrade can be drained to the side ditch. The ditch should be kept about 30 to 45 cm minimum below the top of the subgrade and should be kept sufficiently large to carry the runoff of the area draining into it. The side ditch should be kept clean and free from water pockets at all times. Water standing inside ditches will seep to the subgrade resulting in loss in supporting value. Too small side ditch or steep gradient will cause erosion. If ditch design is inadequate to prevent erosion, it is important that this be checked as soon as discovered. To prevent erosion, sodding the ditch may be effective. Where sod will not control erosion, it is desirable to construct checks in the side ditch to retard the flow or line the ditch with concrete or masonry. Ditch checks can be constructed with timber, stones, concrete, woven wire, etc.

All other elements of a drainage system such as catch basins, manholes, culverts, storm sewers, subsurface drains should be inspected frequently to assure their proper functioning. If inspection reveals that outlets of perforated subsurface drains do not discharge water in the normal manner, the likelihood is that the filter material in the trench has become clogged with silt. The trench will have to be excavated and refilled with clean material.

MAINTENANCE OF BRIDGES AND OTHER STRUCTURES

Bridge maintenance. The fundamental basis of good bridge maintenance is regular field inspections by qualified staff who is able to judge the magnitude of any damage and devise proper method of repair. The usual procedure is that at least once in a year, the bridges and culverts are inspected by a qualified engineer and regular record of these inspections is kept on a proper printed form bound in a register. This register provides a convenient method of noting the defects and showing the condition of the bridge giving proper identification and location of the bridge. In addition to regular annual inspection, the bridge should be inspected following any unusual occurrences affecting the bridge, such as a heavy flood, fire, collision, overloading, or damage of any sort.

During inspection, thorough examination should be made of both the superstructure and the substructure. In the inspection of superstructure, careful examination should be made of the floor; wearing surface; any indication of cracking or disintegration of reinforced concrete structure; drainage performance of the floor; any sign of failure in rails and kerbs; any sign of corrosion or rust on painted surfaces of steel bridges; correct functioning of expansion devices; any unnatural movement of the superstructure, *etc*.

In the examination of substructure, careful examination should be made for evidence of scour, undermining, settling, sliding, tilting, or any movement whatever, of the pier or abutment; any sign of cracking, scaling or disintegration of pier or abutment if it is of concrete; any sign of corrosion and condition of rivets and bolts in case piers are made of metal. The waterway on each side of the bridge should be examined for adequacy. The need for bank protection, if any, should be examined. The conditions of training or protective works should be inspected for soundness.

If any timber happens to form part of super-structure or substructure close inspection should be made to detect any sign of decay or loss of strength due to other causes.

Floors. Most highway bridge floors are of concrete. After a number of years, some of the bridge floors on heavily travelled roads will need replacing due to wear, or deterioration from some other causes.

Sometimes the surface of the floor is rough which may be due to faulty construction and which may increase the impact far beyond what was provided in the original design. The condition of the bridge floor which has deteriorated can be restored to some extent and improved by placing a surface course of bituminous concrete. The thickness may range from 20 mm to 30 mm. On lightly travelled roads, suitable intermediate type of bituminous mixture can be used as a wearing surface.

When it becomes necessary to renew the effective design depth of concrete floor and to level up the riding surface, satisfactory result may be obtained with a thin concrete surfacing which should be at least 3.75 cm thick and preferably 5 to 7.5 cm, provided the added load is not detrimental. The existing floor should be roughened and cleaned of all grease, oil or dirt and thoroughly saturated for 12 hr before the surfacing concrete is placed. As vibration caused by traffic will tend to break the bond between old and new concrete, traffic should not be allowed on the bridge when the new concrete is being placed.

Damage to super-structure because of traffic accidents or other causes should be repaired promptly and the damaged member replaced.

Painting. Another type of routine maintenance activity on steel bridges is painting. In order to protect exposed structural steel, it is necessary to paint it soon after erection and to renew the paint often enough to keep a sufficient covering over the steel to prevent rusting.

Before new paint is applied, the surface of the metal to be painted is thoroughly cleaned to remove all rust, dirt, old paint, oil, grease or other foreign matter. This cleaning is usually done by sand blasting but sometimes, cleaning is done by hand with sand papers or with gas flame. Painting is done in three coats, first the prime coat (red-lead paint) and two remaining coats. Each coat should be thoroughly dry before the succeeding coat is applied.

The usual life of original paint is 8 to 12 years. All exposed surfaces are corrected by spot paint jobs by painting exposed areas and its average life is 7 to 9 years. If for any reason, the spot painting is delayed or neglected until about 50 per cent of the steel has to be cleaned, it may be more economical to clean all the steel by sand blasting and to completely repaint it.

Substructure. Difficult maintenance problems are encountered as a result of undermining of substructures by erosion and scour of the stream bed to a greater depth and width than was provided in the original design, resulting in loss of support. The extent of loss of support determines the magnitude and type of repair. The scouring condition when detected during inspection can be stopped by placing riprap or other type of protection. Scouring at a pier or abutment is very difficult to detect when it is occurring during high floods, but if it is suspected that scour is progressing the action can be retarded and often controlled by dumping large size boulders or rock from the bridge floor on the upstream side of the pier. If scouring is detected when support is partially lost, correction will normally involve lowering the footing with concrete to a point well below the stream bed and to firm foundation.

Surface damage to concrete substructures can be corrected with sand-cement grout under air pressure. If deep repairs are needed, these can be made by removing all loose pieces of the weakened concrete, squaring up the ends, and pouring in new concrete, with proper use of forms. For such repairs, concrete having minimum shrinkage should be used. Such a mix may be obtained by adding suitable metallic admixtures to the concrete. Iron filings can be used for this purpose. One part metallic compound, 2 parts cement, and 3 parts coarse sand are quite satisfactory. Shoring of the affected portions of the superstructure may be needed during replacement. Similarly, cracks in concrete can be repaired by enlarging them into V-shape and filling with non-shrinking mixture mentioned above.

When timber piling has been used in the substructure, the damaged or unsound portion can be repaired by cutting off the piling about 60 cm below the ground or to the point where the piling is in sound condition, and then pouring a concrete block 1 m square and 60 cm thick around the top of the piling. The concrete block should encase 30 cm of the piling already in the ground. After curing, new section of piling is placed on top of the concrete block and held in place at the bottom by means of steel straps embedded in concrete.

Steel piling when used in substructure and requiring repairs due to damage may be strengthened or reinforced by placing a concrete collar around it or by welding additional metal along the section.

Masonry. The use of masonry (brick or stone) in superstructure (arches, *etc.*) is decreasing. The usual maintenance for masonry culverts is filling of joints as and when needed. Stone or brick which has fallen out due to some cause can be readily replaced if discovered in time.

Channel control. In some areas, control of stream channel is the most costly item as far as bridge maintenance is concerned. Some channels continually meander and in such cases, it becomes a continuous maintenance operation to keep the channel straight under the bridge. While constructing the new bridge, the design for such unstable channels should provide for suitable protection works such as guide banks, etc. Such protection works are fairly successful on smaller streams. All riprap on earth slopes needs to be maintained properly when any damage occurs.

On larger streams, it may become necessary to control the location of the current on the upstream side in order to hold the channel in its proper place under the bridge. For this purpose maintenance measures include cutting a new straight channel; construction of suitably spaced spurs (deflectors) which are of several different types; straightening the channel and stabilizing the banks by adequate sloping and paving with stone boulders. Rows of wooden piles and trees starting well back of the bank line and extending into the channel known as tree-retard method can also be used for channel control. Piles are driven at about 3 metre intervals and are angled downstream depending upon the curvature of the bank line and the amount of current deflection required. The top of piles are kept about 0.60 to 1 m above the low water level at the stream end and at ground elevation near the bank. The pile tops are tied together by two lines of 20 mm cables, the ends of which are suitably anchored in the bank and in the stream bed. Trees ranging from 15 to 30 cm butt diameter are tied to the cables, with tops of tree pointing downstream. The trees are spaced so as to form a heavy mat. The spacing of rows of piles should be such that the stream end of one tree-retard will protect the bank end of the retard below it. Due to this arrangement, the current is slowed down sufficiently by the trees to cause deposition of silt. This would result in a fill next to the bank line immediately below each retard. This method would be suitable only on streams that carry a large amount of silt or sediment.

Retaining walls. Retaining walls are of various types such as cement concrete, plain or reinforced, dry or grouted stone, brick masonry, untreated or treated timber. Maintenance requires regular inspection to detect defects. Drain holes (weep holes) placed in most retaining walls must he kept open otherwise the wall can collapse due to water pressure behind the wall. Erosion at the base of the wall should be prevented.

Disintegration of concrete walls can be repaired by pressure grouting methods. Masonry walls need to have the mortar joints repointed at regular intervals. Any stone or brick that may disintegrate or drop should be carefully replaced. Dry stone walls need careful and regular inspections as they require more maintenance. Individual stones as well as portions of the wall should be replaced when damage occurs. Timber is rarely used in walls and where used, the maintenance is confined entirely to replacement of timber pieces.

Guard rails. Maintenance of guard rail consists of painting or treating the timber posts, painting metal posts, realigning posts, and replacing all posts when they have been broken by collision or destroyed by other causes. Wire cable and wire fabric should be kept in proper tension to prevent sagging and reduced protection. Vegetation along guardrail should be removed to prevent possible damage by fire and to improve visibility.

MAINTENANCE OF ROAD SIDES

Roadside maintenance in general may include the care of the area between the travelled surface and the limits of road land. The principal roadside maintenance activities may be catalogued as follows:

1. Mowing and cleaning of roadsides.
2. Weed control and eradication.
3. Control of erosion by seeding, sodding, or planting of vegetation.
4. Care of trees and shrubs.
5. Maintenance of roadside parks, picnic spots, recreational and rest areas.

Much of this work is more or less continuous in its requirements; some is seasonal and some may be done in any season, enabling it to be fitted into a continuing programme, as the pressure of more urgent work permits. Hence, for economy, the work should be carefully scheduled.

Maintenance of cut-and-fill slopes is best done by stabilizing with vegetable growth, where this is feasible. Seeding may be done on relatively flat areas while sodding is necessary on steeper slopes. Erosion should be closely watched, on both cut and fill slopes and checked, if possible, before it causes serious damage. This may require the construction of the intercepting ditches on top of slopes and downspots on fills. Rock slopes should be watched for loose rocks and scaled, if necessary.

SNOW AND ICE CONTROL

Snow is a major winter maintenance problem in affected areas. This is particularly true in areas in high altitudes and regions where snowfall occurs every year during winter and where uninterrupted use of highways throughout the year has to be maintained. To solve it properly, careful organization and advance planning is required.

Design measures to alleviate snow problems should provide for high grades, wide road land, flat slopes, careful location (wind-ward) of borrow-pits, and proper selection and placing of roadside plants.

Measures to minimize the formation of snow-drifts across the roadway should be taken before the coming of winter. Wood-slate snow fence is most commonly used for snow-drift control. This fence is usually 1.2 m high and vertically spaced, approximately 5 cm centre to centre or wider, woven together with galvanized wire. Sections can be rolled up for handling and storage. Support is furnished by posts or angle irons driven into the ground. Snow fence is placed on the prevailing wind side of the highway about 30 to 45 m from centre line of the road. The purpose of snow fence is to reduce wind velocity so that snow is deposited in the space between the fence and the road. The wind will be comparatively free from snow when it passes over the highway. After the snow fence is filled with snow, additional fence may be piaced as an extension above the old fence or may be constructed as a new fence 20 to 30 m back from the first fence.

Removal of snow from road should start soon after snow begins to fall. In rural areas snow on the roadway is bladed or thrown to the roadside. In city areas snow from heavy falls is generally loaded into trucks and hauled away. Where snow fall is less, regular maintenance trucks of $1\frac{1}{2}$ to 2 tonnes.0 size equipped with blade plows are effective. Large trucks, usually four-wheel drive or heavy-duty, tandem-driven motor graders equipped with wing plows or hydraulically operated blade or V-plows, are widely employed in areas of heavier snowfall. Drifts which cannot be penetrated by these large V-plows can be opened by rotary plows. Great mounds of snow left piled on both shoulders of the road after plowing should be removed or pushed beyond the shoulder before the next snow storm or severe blow or thaw occurs.

Ice is formed on the road by freezing rain, drizzle or sleet. This condition of the road reduces coefficient of friction between tyres and surface to 0.05 or less and results in slippery surface that is very dangerous. Curves, hills and sections adjacent to railway crossing are treated by applying abrasives. If traffic is sufficiently heavy to warrant it, the entire road may be treated with abrasives. Suitable abrasive materials are clean, sharp sand (max. size 10 mm), cinders, washed stone, stone screenings (max size 1.0 mm). Normally the abrasives are treated with calcium or sodium chloride in solid or brine form before storage to prevent freezing before application and as an aid to anchoring the material in ice or packed snow so that it will not be blown off the road.

Before treating a road surface with abrasives the snow or loosened ice should be removed as completely as possible. It is preferable to score the ice to roughen the surface with a serrated blade to control skidding. An application of 0.3 to 0.6 kg of treated sand or cinder per sq metre will give an effective cover and will make the road non-skid if it is uniformly distributed. When wind or traffic has whipped off the material, it may be necessary to again apply abrasives in some sections especially on grades, curves and intersections.

Calcium chloride or sodium chloride should be used as sparingly as possible on portland cement concrete pavements because repeated freezing and thawing of concrete in contact with these salts is conducive to scaling. However, pavements protected with bituminous surface treatment or those made from air-entrained concrete are generally not susceptible to damage from ice-control chemicals.

There are a few instances where short stretches of pavement or side walks in cities have been heated, by use of coils placed under the pavement, to melt the snow as it falls.

MAINTENANCE OF EQUIPMENT

Adequate and proper equipment is one of the essentials along with men and material for highway maintenance. A fairly sizeable percentage of all highway maintenance expenditure, is for the purpose, operation, and repair of equipment. The effective use of mechanical equipment is essential for improving and maintaining the highway within reasonable cost. Equipment should be selected for its flexibility and use.

The amount and type of equipment needed for maintenance will depend upon the number of km and type of road to be maintained and the ability to purchase the necessary equipment. Automobiles and trucks are needed for many phases of maintenance work. Other essential pieces of equipment are rollers, tractors, mixers, power shovels, graders and loaders. Paving equipment for bituminous and concrete paving mixtures is also desirable. Special equipment includes snow-plows, pavement marking machine, air compressors, welding machine, sanding machine, and pumps. Some maintenance organizations find it economical to maintain their own crushing plants.

Proper care, lubrication, and servicing of the mechanical equipment is essential. Maintenance equipment should be used in the right manner. Except in emergencies a machine should never be used on work for which the unit was not designed for, and it should never be operated which is not in proper condition.

Rental rates of equipment which are charged to the maintenance of roads should include both operation and depreciation costs without either loss or profit. In order to work out depreciation costs, estimate is first made of the probable life of the machine in km or hours of operations. The original cost including accessories, divided by the probable life will provide the km or hour depreciation rate to fully liquidate the cost of the unit. The depreciation rate plus the operation rate should form the basis for the rental rate. Periodic checks are necessary to establish the correct rate.

❑❑❑

BIBLIOGRAPHY

TEXT BOOKS

1. Agg, T.R., *The construction of roads and pavements* McGraw Hill Book Co, Inc., New York.
2. Balemen, J.H., *Introduction to highway engineering* John Wiley and Sons, Inc., New York.
3. *Bituminous materials in road construction* H.M.S.O., London.
4. Chandola S.P., *A Text book of Transportation Engineering, S.Chand & Co. Ltd. New Delhi.*
5. Chakrobarty P. and Animesh Das: *Principles of Transportation Engineering Prentice Hall of India, New Delhi.*
6. Bruce A.G. and Clarkson J. *Highway design and construction*, International Tex-book Co., Seranton U.S.A.
7. Collins H.J. and Hart C.A. *Principles of road engineering* Edward Arnold Publishes Ltd, London.
8. *Concrete pavement design* Portland Cement Association, U.S.A.
9. *Concrete roads, design and construction.,* H.M.S.O. London.
10. Evans H.K. *Traffic engineering handbook* Institute of Traffic Engineers, Connecticut, U.S.A.
11. Flaherty C.A.O. *Highways* Vol. I and II. Edward Arnold Publishers Ltd., London.
12. Hay, William W. *An introduction to transport engineering* Toppan Company Ltd., Tokyo.
13. Hewes L.I. *American highway practice,* Vol. I and II, John Wiley and Sons Inc., New York.
14. Hickerson. Thomas F. *Route surveys and design* McGraw Hill Book Co. Inc. New York.
15. Hobbs F.D. *Traffic planning and engineering*, Pergamon Press, Oxford.
16. Kadiyali L.R. *Traffic engineering and traffic planning,* Khanna Publishers. Delhi.
17. Khanna S.K. and Justo C.E.G. *Highway Engineering*, Nem Chand and Bros, Roorkee, India.
18. Kadiyali L.R. and N.B. Lal. Principles and *Practice of Highway Engineering* Khanna Publishers Delhi.
19. Khanna S.K. and Arora M.G. *Airport planning and design,* Nem Chand and Bros., Roorkee.
20. Krebs R.D. and Walker R.D. *Highway materials*, McGraw Hill Co. Inc. New York.
21. Luthin J.N. *Drainage engineering* Wiley Eastern Publishing Ltd. New Delhi.
22. Matson T.M., Smith W.S. and Hurd H.W. *Traffic engineering* McGraw Hill Book Co., Inc., New York.
23. Peurify R.L. *Construction, planning, equipment and methods* McGraw Hill Book Co., New York.
24. Rangwala S.C. and others *Bridge Engineering* Charotar Publishing House, Anand
25. Sehgal S.B. and Bhanot K.L. *A text-book of highway engineering and airports,* S. Chand and Co. Ltd., New Delhi.
26. *Soil mechanics for road engineers,* H.M.S.O., London.

27. Vazirani V.N. and Chandola S.P. *Transportation Engineering* Khanna Publishers, Delhi.
28. Vector D.J. *Essentials of bridge engineering,* Oxford and IBH Publishing Co., New Delhi.
29. Wallace H.A. and Martin J.R. *Asphalts pavement engineering* McGraw Hill Book Co. Inc., New York.
30. Westergaard H.M. *Stress in concrete pavements computed by theoretical analysis.*
31. Wohl M. and Martin B.V. *Traffic system analysis for engineers and planners,* McGraw Hill Book Co. Inc. New York.
32. Woods K.B., Berry D.S. and Goetz W.H. *Highway engineering handbook* McGraw Hill Book Co. Inc., New York.
33. Young N.C. *Design of functional pavements,* McGraw Hill Book Co. Inc., New York.
34. Yoder E.J. and Witczak M.W. *Principles of pavement design* John Wiley and Sons, Inc., New York.

INDIAN STANDARD CODES PUBLISHED BY BUREAU OF INDIAN STANDARDS, NEW DELHI

1. IS : 73-1992 Paving bitumen specification (Second Revision)
2. IS : 217-1988 Cut-back bitumen specification (Second Revision)
3. IS : 269-1989 33 Grade ordinary Portland cement (Fourth Revision)
4. IS : 383-1970 Coarse and fine aggregates from natural sources for concrete (Second Revision)
5. IS : 456-2000 Code of practice for plain and reinforced concrete (Fourth Revision)
6. IS : 460-1985 Test Sieves
7. IS : 702-1988 Industrial bitumen (Second Revision)
8. IS : 702-1959 Method of sampling and analysis of concrete
9. IS : 1201 to 1220 – 1958 Method's for testing tars and bitumens
10. IS : 1203-1978 Determination of penetration (First Revision)
11. IS : 1205-1978 Determination of saftening point (First Revision)
12. IS : 1212-1978 Determination of loss of heating (First Revision)
13. IS : 1216-1978 Determination of solubility in carbon disulphide or carbon tetrachlorate or Trichloroethylene (First Revision)
14. IS : 1489 – 1991 Part I Portland pozzolana cement – Flyash based (Third Revision)
15. IS : 1498-1970 Classification and identification of soils for general engineering purposes (First Revision)
16. IS : 1915-1961 Steel bridges
17. IS : 2720 Method of tests for soils
18. Part 2-1973 Determination of water content (Second Revision)
19. Part 4-1985 Grain size analysis (Second Revision)
20. Part 5-1985 Determination of liquid and plastic limits (Second Revision)
21. Part 7-1980 Determination of moisture content/dry density relation using light compaction (Second Revision)
22. Part 8-1983 Determination of water content/dry density relation using heavy compaction
23. Part 11-1993 Determination of the shear strength parameters of a specimen tested in unconsolidated undrained triaxial compression without the measurement of pore-water pressure (First Revision)

24. Part 12-1993 Determination of shear strength parameters of soil from consolidated undrained triaxial test with measurement of pore-water pressure (First Revision)
25. Part 13-1986 Direct shear test (Second Revision)
26. Part 15-1986 Determination of consolidation properties (First Revision)
27. Part 16-1987 Laboratory determination of CBR (Second Revision)
28. Part 27-1977 Determination of total soluble sulphate (First Revision)
29. Part 31-1990 Field determination of CBR (First Revision)
30. Part 37-1976 Determination of sand equivalent values of soils and fine aggregates
31. Part 38-1976 Compaction control test (Hilp method)
32. IS : 3066-1965 Hot asphalt mixing plants
33. IS : 3117-1965 Bitumen emulsion for roads
34. IS : 4332-1967 Methods of tests for stabilized soils
35. IS : 5317-1969 Bitumen mastic for roads and bridge decking
36. IS : 2149-1970 Luminaires for street lighting
37. IS : 5640-1970 Aggregate impact value of soft coarse aggregates
38. IS : 5817-1970 Lime puzzolana concrete in building and roads
39. IS : 6241-1971 Stripping value of road aggregates
40. IS : 6509-1972 Joints in concrete pavements
41. IS : 6579-1972 Coarse aggregates for water bound macadam
42. IS : 7242-1974 Concrete spreaders
43. IS : 7245-1974 Concrete pavers
44. IS : 7537-1974 Road traffic signals
45. IS : 7740-1975 Road gullies
46. IS : 7774-1975 Transport tractors and trailers—Basic terms
47. IS : 8095-1976 Accident prevention tags
48. IS : 9211-1979 Denominations and definitions of weights of road vehicles
49. IS : 9214-1979 Method of determination of modulus of subgrade reaction (k value) of soils in field
50. IS : 10262-1982 Guidelines for concrete mix design
51. IS : SP : 23-1982 Handbook on concrete mixes (Based on Indian Standards)

PUBLICATIONS OF INDIAN ROADS CONGRESS (IRC), NEW DELHI

1. IRC : 2-1968 Route marker signs for national highways
2. IRC : 3-1983 Dimensions and weights of Road Design Vehicles (First Revision)
3. IRC : 5-1998 Standard specifications and code of practice for road bridges, section I—General features of design (Seventh Revision)
4. IRC : 6-2000 Standard specifications and code of practice for road bridges, section II—Loads and stresses (Fifth Revision)
5. IRC : 9-1972 Traffic census on non-urban roads (First Revision)
6. IRC : 10-1961 Recommended practice for borrowpits for road embankments constructed by manual operation

7. IRC : 14-2004 Recommended practice for open graded premix carpets (Third Revision)

8. IRC : 15-2011 Standard specifications and code of practice for construction of concrete roads (Fourth Revision)

9. IRC : 16-12008 Standard specifications and Code of practice for Prime and Tack Coat (Second Revision)

10. IRC : 17-1965 Tentative specification for single coat bituminous surface dressing

11. IRC : 18-2000 Design criteria for prestressed concrete road bridges (Post tensioned concrete) (Third Revision)

12. IRC : 19-2005 Standard specifications and code of practice for water bound macadam. (Third Revision)

13. IRC : 20-1966 Recommended practice for bituminous penetration macadam (full grout)

14. IRC : 21-1972 Standard specifications and code of practice for road bridges, section III—Cement concrete (plain and reinforced)

15. IRC : 22-2008 Standard specifications and code of practice for road bridges, section VI—Composite construction (Limit State Design) (Second Revision)

16. IRC : 23-1966 Tentative specification for two coat bituminous surface dressing

17. IRC : 24-1967 Standard specifications and code practice for road bridges, section V—Steel road bridges

18. IRC : 27-2009 Specifications for bituminous macadam (First Revision)

19. IRC : 28-1967 Tentative specification for the construction of stabilized soil roads with soft aggregate in areas of moderate and high rainfall

20. IRC : 29-1968 Tentative specification for 4 cm asphaltic concrete surface course

21. IRC : 30-1968 Standard letters and numerals of different heights for use on highway signs

22. IRC : 31-1969 Route marker signs for state routes

23. IRC : 32-1969 Standard for vertical and horizontal clearances of overhead electric power and telecommunication lines as related to roads

24. IRC : 33-1969 Standard procedure for evaluation and condition surveys of stabilized soil roads

25. IRC : 34-2011 Recommendations for road construction in areas affected by water logging, flooding and/or salts infestation (First revision)

26. IRC : 35-1997 Code of practice for road markings (First Revision)

27. IRC : 37-2001 Guidelines for the design of flexible pavements (Second Revision)

28. IRC : 38-1988 Guidelines for Design of horizontal curves for highways and design tables (First Revision)

29. IRC : 40-2002 Standard specifications and code of practice for road bridges, section IV—brick, stone, and block masonry (Second Revision)

30. IRC : 43-1972 Recommended practice for tools, equipment and appliances for concrete pavement construction

31. IRC : 44-2008 Guidelines for cement concrete mix design for pavements (Second Revision)

32. IRC : 47-1972 Tentative specifications for built up spray grout

33. IRC : 48-1972 Tentative specification for bituminous surface dressing using precoated aggregates

34. IRC : 49-1973 Recommended practice for the pulverization of black cotton soils for lime stabilization
35. IRC : 50-1973 Recommended design criteria for the use of cement modified soil in road construction
36. IRC : 51-1992 Recommended design criteria for the use of soil lime mixes in road construction (First Revion)
37. IRC : 52-2001 Recommendations about the alignment survey and geometric design of hill roads (Second revision)
38. IRC : 54-1974 Lateral and vertical clearance at underpasses for vehicular traffic
39. IRC : 55-1974 Recommended practice for sand-bitumen base courses
40. IRC : 57-2006 Recommended practice for sealing of joints in concrete pavements (First Revision)
41. IRC : 58-2002 Guidelines for the design of plain jointed rigid pavements for highways (Second Revision)
42. IRC : 59-1976 Tentative guidelines for design of gap graded cement concrete mixes for road pavements
43. IRC : 60-1976 Tentative guidelines for the use of lime-flyash concrete as pavement base or sub-base
44. IRC : 61-1976 Tentative guidelines for the construction of cement concrete pavements in hot-weather
45. IRC : 62-1976 Guidelines for control of access on highways
46. IRC : 63-1976 Tentative guidelines for the use of low grade aggregates and soil aggregate mixtures in road pavement construction
47. IRC : 64-1990 Guidelines for capacity of roads in rural areas (First Revision)
48. IRC : 65-1976 Recommended practice for traffic rotaries
49. IRC : 66-1976 Recommended practice for sight distances on rural highways
50. IRC : 67-1977 Code of practice for road signs
51. IRC : 68-1976 Tentative guidelines on cement flyash concrete for rigid pavement construction
52. IRC : 69-1977 Space standards for roads in urban areas
53. IRC : 72-1978 Recommended practice for use and upkeep of equipment, tools and appliances for bituminous pavement construction
54. IRC : 73-1980 Geometric design standards for rural (non-urban) highways
55. IRC : 74-1979 Tentative guidelines for lean-cement concrete and lean-cement-flyash concrete as a pavement base or sub-base
56. IRC : 75-1979 Guidelines for the design of high embankments
57. IRC : 76-1979 Tentative guidelines for structural strength evaluation of rigid airfield pavements
58. IRC : 77-1979 Tentative guidelines for repair of concrete pavements using synthetic resin
59. IRC : 78-1979 Standard specifications and code of practice for road bridges-section VII-part 1 : Foundations and substructure general features of design
60. IRC : 79-1981 Recommended practice for road delineators
61. IRC : 81-1981 Guidelines for strengthening of flexible road pavements using Benkelman beam deflection technique (First Revision)

62. IRC : 82-1982 Code of practice of maintenance of bituminous surfaces of highways

63. IRC : 84-1983 Code of practice for curing of cement concrete pavement

64. IRC : 86-1983 Geometric design standards for urban roads in plains

65. IRC : 88-1984 Recommended practice for lime flyash stabilized soil base/sub-base in pavement construction

66. IRC : 91-1985 Tentative guidelines for construction of cement concrete pavements in cold weather

67. IRC : 92-1985 Guidelines for the design of interchanges in urban areas

68. IRC : 93-1985 Guidelines on design and installation of road traffic signals

69. IRC : 101-1988 Guidelines for design of continuously reinforced concrete pavement with elastic joints

70. IRC: 104-1988 Guidelines for environmental impact assessment of highway projects

71. IRC : 105-1988 Tentative specification for bituminous concrete (asphaltic concrete) for airfield pavement

72. IRC : 106-1990 Guidelines for capacity of urban roads in plain areas

73. IRC : 107-1992 Tentative specifications for bitumen mastic wearing courses

74. IRC : 108-1996 Guidelines for traffic prediction or rural highways

75. IRC : 109-1997 Guidelines for wet mix macadam

76. IRC : 110-2005 Standard specification and code of practice for design and construction of surface dressing

77. IRC : 111-2009 Specification for dense graded bituminous mixes.

SPECIAL PUBLICATIONS

78. Special publication—5-1967 Road drainage practices around the world

79. Special publication—8-1972 Methodology of flexible pavement design around the world

80. Special publication—11-1984 Handbook of quality control for construction of roads and runways (Second Revision)

81. Special publication—12-1973 Tentative recommendation on the provision of parking spaces for urban areas

82. Special publication—13-2004 Guidelines for the design of small bridges and culverts (First Revision)

83. Special publication—14-1973 A manual for the application of the critical path method of highway projects in India.

84. Special publication—15-1996 Ribbon development along highways and its prevention

85. Special publication—16-2004 Guidelines for Surface evenness of highway pavements (First Revision)

86. Special publication—17-1977 Recommendations about overlays on cement concrete pavements

87. Special publication—19-2001 Manual for survey, investigation and preparation of road projects (Second Revision)

88. Special publication—20-2002 Rural roads (Manual)

89. Special publication—21-1980 Guidelines on landscaping and tree plantation (First Revision)

90. Special publication—22-1980 Recommendations for the sizes for each type of road making machinery to cater to the general demand of road works

91.	Special publication 23-1983	Vertical curves of highways
92.	Special publication 30-1993	Manual on economic evaluation of highway projects in India (First Revision)
93.	Special publication 41-1994	Guidelines on design of at-grade intersections in rural and urban areas
94.	Special publication 42-1994	Guidelines on road drainage
95.	Special publication 48-1994	Hill road manual
96.	Special publication49-1998	Guidelines for the use of dry lean concrete as sub-base for rigid pavements
97.	Special publication 50-1999	Guidelines on urban drainage
98.	Special publication	Design, construction and maintenance of hill roads Seminar Shimla-1995
99.	Special publication 53-2002	Guidelines on use of modified bitumen in road construction (Second Revision)
100.	Special publication 62-2004	Guidelines for the design and construction of cement concrete pavements for rural roads
101.	Special publication 69-2005	Guidelines and specification for expansion joints
102.	Special publication 72-2007	Guidelines for the design of flexible pavements for low volume rural roads
103.	Special publication 79-2008	Tentative specification for stone matrix asphalt
104.	Special publication 85-2010	Guidelines for variable message signs
105.	Special publication 89-2010	Guidelines for soil and granular material stabilization using cement, lime and flyash
106.	Special publication 90-2010	Manual for grade separators and elevated structures
107.	Special Publication 91-2010	Guidelines for road tunnels
108.	MORT & H:01-1983	Mannual for maintenance of roads
109.	MORT & H : 33	Guidelines for expressways Part I and Part II
110.	MORT & H : 2001	Specifications for road and bridge works (Fourth Revision)
111.	MORT & H : 2001	Mannual for construction and supervision of bituminous works
112.	MORT & H : 2002	Pocket Book for highway engineers

❑❑❑

INDEX

A

B

C

D

E

S

T

U

V

W

Y

❑❑❑